Advances in Noncommutative Geometry

Ali Chamseddine • Caterina Consani
Nigel Higson • Masoud Khalkhali
Henri Moscovici • Guoliang Yu
Editors

Advances in Noncommutative Geometry

On the Occasion of Alain Connes' 70th Birthday

Springer

Editors
Ali Chamseddine
Department of Physics
Institute for Mathematics
Astrophysics and Particle Physics
Nijmegen, The Netherlands

American University of Beirut
Beirut, Lebanon

Radboud University
Nijmegen, The Netherlands

Caterina Consani
Department of Mathematics
Johns Hopkins University
Baltimore, MD, USA

Masoud Khalkhali
Department of Mathematics
The University of Western Ontario
London, ON, Canada

Guoliang Yu
Department of Mathematics
Texas A&M University
College Station, TX, USA

Nigel Higson
Department of Mathematics
Pennsylvania State University
University Park, PA, USA

Henri Moscovici
Department of Mathematics
Ohio State University
Columbus, OH, USA

ISBN 978-3-030-29596-7 ISBN 978-3-030-29597-4 (eBook)
https://doi.org/10.1007/978-3-030-29597-4

Mathematics Subject Classification: 58B34

This Springer imprint is published by the registered company Springer Nature Switzerland AG.
The registered company address is: Gewerbestrasse 11, 6330 Cham, Switzerland

Foreword

From March 23 to April 7, 2017, the Shanghai Center for Mathematical Sciences hosted a conference on noncommutative geometry dedicated to its founder and foremost architect, Alain Connes, on the occasion of his 70th birthday. This volume contains a collection of invited surveys and research articles stemming from talks given at the conference and reflecting the scope and the state of the art of this field.

Deeply rooted in the modern theory of operator algebras and inspired by two of the most influential mathematical discoveries of the twentieth century, the foundations of quantum mechanics and index theory, Connes' vision of noncommutative geometry echoes the astonishing anticipation of Riemann that "it is quite conceivable that the metric relations of space in the infinitely small do not conform to the hypotheses of geometry" and accordingly "we must seek the foundation of its metric relations outside it, in binding forces which act upon it."

The radically new paradigm of space proposed by Connes in order to achieve such a desideratum is that of a *spectral triple*, encoding the (generally noncommuting) spatial coordinates in an algebra of operators in Hilbert space, and its metric structure in an analogue of the Fermion propagator viewed as "line-element." For the analytic treatment of such spaces, Connes devised the *quantized calculus*, whose infinitesimals are the compact operators, and where the role of the integral is assumed by the Dixmier trace. On the differential-topological side, Connes has invented a far-reaching generalization of de Rham's theory, *cyclic cohomology* which, in conjunction with KK-theory, provides the key tool for a vast extension of index theory to the realm of noncommutative spaces.

Besides the wealth of examples of noncommutative spaces coming from physics (including space-time itself with its fine structure), from discrete groups, Lie groups (and smooth groupoids), with their rich K-theory, a whole class of new spaces can be handled by the methods of noncommutative geometry and in turn lead to the continual enrichment of its toolkit. They arise from a general principle, which first emerged in the case of foliations. It states that difficult quotients such as spaces of leaves are best understood using, instead of the usual commutative function algebra, the noncommutative convolution algebra of the associated equivalence relation.

An important such new space is the space of adele classes of a global field that Connes has introduced to give a geometric interpretation of the Riemann–Weil explicit formulas as a trace formula. The set of points of the simplest Grothendieck toposes are typically noncommutative spaces in the above sense and the adele class space itself, for the field of rationals, turns out to be the set of points of the Scaling Site, a Grothendieck topos which provides the missing algebro-geometric structure as a structure sheaf of tropical nature.

The pertinence and potency of the new concepts and methods are concretely illustrated in the contributions which make up this volume. They cover a broad spectrum of topics and applications, shedding light on the fruitful interactions between noncommutative geometry and a multitude of areas of contemporary research, such as operator algebras, K-theory, cyclic homology, arithmetic geometry, index theory, spectral theory, geometry of groupoids, and in particular of foliations.

Some of these contributions stand out as groundbreaking forays into more seemingly remote areas, namely high energy physics, algebraic geometry, and number theory.

The Shanghai Center of Mathematical Sciences is a recently founded mathematical research center jointly funded by the Central Government of China, the city of Shanghai, and Fudan University. A major goal of the center is to foster research collaborations among mathematicians from all over the world. The event celebrating Alain Connes' 70th birthday was part of this effort.

We thank the faculty and staff of the Shanghai Center for their exemplary hospitality. We also acknowledge with thanks the support granted by the National Science Foundation, through award no. 1701934, for the participation of US-based researchers. Finally, we would like to thank Springer Verlag and in particular Elizabeth Loew for her care and support during the production of this volume.

Beirut, Lebanon — Ali Chamseddine
Baltimore, MD, USA — Caterina Consani
University Park, PA, USA — Nigel Higson
London, ON, Canada — Masoud Khalkhali
Columbus, OH, USA — Henri Moscovici
College Station, TX, USA — Guoliang Yu

Contents

A survey of spectral models of gravity coupled to matter

Ali Chamseddine and Walter D. van Suijlekom

Dedicated to Alain Connes

Abstract This is a survey of the historical development of the spectral Standard Model and beyond, starting with the ground breaking paper of Alain Connes in 1988 where he observed that there is a link between Higgs fields and finite noncommutative spaces. We present the important contributions that helped in the search and identification of the noncommutative space that characterizes the fine structure of space-time. The nature and properties of the noncommutative space are arrived at by independent routes and show the uniqueness of the spectral Standard Model at low energies and the Pati–Salam unification model at high energies.

1 Introduction

In 1988, at the height of the string revolution, there appeared an alternative way to think about the structure of space-time, based on the breathtaking progress in the new field of noncommutative geometry. Despite the success of string theory in incorporating gravity, consistency of the theory depended on the existence of supersymmetry as well as six or seven extra dimensions. Enormous amount of research was carried out to obtain the Standard Model from string compactification, which even up to day did not materialize. Most compactifications start in ten

---

A. Chamseddine
Department of Physics, Institute for Mathematics, Astrophysics and Particle Physics, Nijmegen, The Netherlands

American University of Beirut, Beirut, Lebanon

Radboud University, Nijmegen, The Netherlands
e-mail: chams@aub.edu.lb

W. D. van Suijlekom (✉)
Institute for Mathematics, Astrophysics and Particle Physics, Radboud University Nijmegen, Nijmegen, The Netherlands
e-mail: waltervs@math.ru.nl

A. Chamseddine et al. (eds.), *Advances in Noncommutative Geometry*,
https://doi.org/10.1007/978-3-030-29597-4_1

dimensions with the Yang–Mills gauge group $E_8 \times E_8$ requiring a very large number of fields to become massive at high energies. In a remarkable paper, Alain Connes laid down the blueprint of a new innovative approach to uncover the origin of the Standard Model and its symmetries [28]. The foundation of this approach was based on noncommutative geometry, a field he founded few years before [27] (see also [29]). Alain realized that by making space slightly noncommutative by tensoring the four-dimensional space with a space of two points, one gets a parallel universe where the distance between the two sheets is of the order of 10^{-16} cm, with the unexpected bonus of having the Higgs scalar field mediating between them. Although this looked similar to the idea of Kaluza–Klein, there were essential differences, mainly in avoiding the huge number of the massive tower of states as well as obtaining the Higgs field in a representation which is not the adjoint. Soon after this work inspired similar approaches also based on extending the four-dimensional space to become noncommutative [43, 44, 45, 46, 23].

In this survey, we will review the key developments that allowed noncommutative geometry to deepen our understanding of the structure of space-time and explain from first principles why and how nature dictates the existence of the elementary particles and their fundamental interactions. In Section 2, we will start by reviewing the pioneering work of Alain Connes [28] introducing the basic mathematical definitions and structures needed to define a noncommutative space. We summarize the characteristic ingredients in the construction of the Connes–Lott model and later generalizations by others. We then consider how to develop the analogue of Riemannian geometry for noncommutative spaces, and to incorporate the gravitational field in the Connes–Lott model. In Section 3, we present a breakthrough in the development of noncommutative geometry with the introduction of the reality operator which led to the definition of KO dimension of a noncommutative space. With this it became possible to present the reconstruction theorem of Riemannian geometry from noncommutative geometry. Section 4 covers the formulation and applications of the spectral action principle where the spectrum of the Dirac operator plays a dominant role in the study of noncommutative spaces. This key development allowed to obtain the dynamics of the Standard Model coupled to gravity in a non-ambiguous way, and to study geometric invariants of noncommutative spaces. We then show that incorporating right-handed neutrinos with the fundamental fermions forces a change in the algebra of the noncommutative space and the use of real structures to impose simultaneously the reality and chirality conditions on fundamental states, singling out the KO dimension to be 6. We show in detail how the few requirements about KO dimension, Majorana masses for right-handed neutrinos and the first order condition on the Dirac operator, single out the geometry of the Standard Model. In Section 5, we present a classification of finite noncommutative spaces of KO dimension 6 showing the almost uniqueness of the spectral Standard model. In Section 6, we give a prescription of constructing spectral models from first principles and show that the spectral Standard Model agrees with the available experimental limits, provided that the scale giving mass to the right-handed neutrinos is promoted to a singlet scalar field. We then show that there exists a more general case where the first order condition on the Dirac operator is removed, the singlet scalar fields become part of a larger representation of the Pati–Salam model. The Standard Model becomes a special point in the spontaneous breaking of the Pati–Salam symmetries. In Section 6, we show that a different starting point where a Heisenberg like quantization condition in terms of the Dirac operator considered as momenta and two possible Clifford-algebra valued maps from the four-dimensional manifold to two four-spheres S^4 result in noncommutative spaces with quantized volumes. The Pati–Salam model and its various truncations are uniquely determined as the symmetries of the spaces solving the constraint. Section 7 contains the conclusions and a discussion of possible directions of future research.

2 Early days of the spectral Standard Model

The first serious attempt to utilize the ideas of noncommutative geometry in particle physic was made by Alain Connes in 1988 in his paper "Essay on physics and noncommutative geometry" [28]. He observed that it is possible to change the structure of the (Euclidean) space-time so that the action functional gives the Weinberg–Salam model. The main emphasis was on the conceptual understanding of the Higgs field, which he calls, the black box of the Standard Model. The qualitative picture was taken to be of a two-sheeted Euclidean space-time separated by a distance of the order of 10^{-16} cm. In order to simplify the presentation, and to easily follow the historical development, we will use a uniform notation, representing old results in a new format. It is therefore more efficient to start with the basic definitions.

2.1 Noncommutative spaces and differential calculus

A noncommutative space is determined from the spectral data $(\mathcal{A}, \mathcal{H}, D, \gamma, J)$ where $\mathcal{A}$ is an associative algebra with unit element 1 and involution $*$, $\mathcal{H}$ a Hilbert space carrying a faithful representation π of the algebra, D is a self-adjoint operator on $\mathcal{H}$ with $\left(D^2+1\right)^{-1}$ compact, γ is the unitary chirality operator, and J an anti-unitary operator on $\mathcal{H}$, the reality structure. The operator J was introduced later in 1994 [30].

In the model proposed in 1988, there were ambiguities in defining the algebra and the action on the Hilbert space. These were rectified in the 1990 paper [33] with John Lott, in what became known as the Connes–Lott model. In order to appreciate the enormous progress made over the years, we will summarize this model in a simplified presentation. A more detailed account can be found in the early reviews [77, 66, 54, 55, 56, 57]. Note that at around the same time, a derivation-based differential calculus was introduced by others in [43, 44, 45, 46] with many similarities to the model proposed by Connes in 1988.

We need to first introduce new ingredients. Given a unital involutive algebra $\mathcal{A}$, the universal differential algebra over $\mathcal{A}$ is defined as

$$\Omega^* (\mathcal{A}) = \bigoplus_{n=0}^{\infty} \Omega^n (\mathcal{A})$$

where we set $\Omega^0 (\mathcal{A}) = \mathcal{A}$, and take

$$\Omega^n (\mathcal{A}) = \left\{ \sum_i a_0^i da_1^i da_2^i \cdots da_n^i, \qquad a_j^i \in \mathcal{A}, \quad \forall i, j \right\}, \qquad n = 1, 2, \ldots$$

Here da denotes an equivalence class of $\mathcal{A}$, modulo the following relations:

$$d (a \cdot b) = da \cdot b + a \cdot db, \qquad d1 = 0, \qquad d^2 = 0$$

An element of $\Omega^n (\mathcal{A})$ is called a form of degree n. One form can be considered as connections on a line bundle whose space of sections is given by the algebra $\mathcal{A}$. A one-form $\rho \in \Omega^1 (\mathcal{A})$ is expressed in the form

$$\rho = \sum_i a^i db^i, \qquad a^i, b^i \in \mathcal{A}$$

and since $d1 = 0$, we may impose the condition $\sum_i a^i b^i = 1$, without any loss in generality. We say that $(\mathcal{H}, D)$ is a Dirac K-cycle for $\mathcal{A}$ if and only if there exists an involutive representation π of $\mathcal{A}$ on $\mathcal{H}$ satisfying $\pi (a)^* = \pi (a^*)$ with the properties that $\pi (a)$ and $[D, \pi (a)]$ are bounded operators on $\mathcal{H}$ for all $a \in \mathcal{A}$. The K-cycle is called even if there exists a chirality operator γ such that $\gamma D = -D\gamma$, $\gamma = \gamma^{-1} = \gamma^*$ and $[\gamma, \pi (a)] = 0$, otherwise it is odd. The action of π on $\Omega^* (\mathcal{A})$ is defined as

$$\pi \left(\sum_i a_0^i da_1^i \cdots da_n^i \right) = \sum_i \pi(a_0^i)[D, \pi(a_1^i)] \cdots [D, \pi(a_n^i)]$$

The space of auxiliary fields is defined by

$$\mathrm{Aux} = \ker \pi + d \ker \pi$$

where

$$\ker \pi = \bigoplus_{n=0}^{\infty} \left\{ \sum_i a_0^i da_1^i \cdots da_n^i \, : \pi \left(\sum_i a_0^i da_1^i \cdots da_n^i \right) = 0 \right\}$$

and

$$d\ker\pi = \bigoplus_{n=0}^{\infty}\left\{\sum_i da_0^i da_1^i \cdots da_n^i \; : \pi\left(\sum_i a_0^i da_1^i \cdots da_n^i\right) = 0\right\}$$

The integral of a form $\alpha \in \Omega^*(\mathcal{A})$ over a noncommutative space of metric dimension d is defined by setting

$$\int \alpha = \mathrm{Tr}_w\left(\pi(\alpha)\, D^{-d}\right)$$

where Tr_w is the Dixmier trace.

2.2 *Two-sheeted space-time*

A simple extension of space-time is taken as a product of continuous four-dimensional manifold times a discrete set of two points. The algebra is $\mathcal{A} = \mathcal{A}_1 \otimes \mathcal{A}_2$ acting on the Hilbert space $\mathcal{H} = \mathcal{H}_1 \otimes \mathcal{H}_2$ where $\mathcal{A}_1 = C^\infty(M)$ and $\mathcal{A}_2 = M_2(\mathbb{C}) \oplus M_1(\mathbb{C})$, the algebra of 2×2 and 1×1 matrices. The Hilbert space is that of spinors of the form

$$L = \begin{pmatrix} l \\ e \end{pmatrix}$$

where l is a doublet and e is a singlet. The spinor L satisfies the chirality condition $\gamma_5 \otimes \Gamma_1 L = L$ where $\Gamma_1 = \mathrm{diag}(1_2, -1)$ is a grading operator. From this we deduce that l is a left-handed spinor and e is right handed, and we thus write $l = \begin{pmatrix} \nu_L \\ e_L \end{pmatrix}$ and $e = e_R$. The Dirac operator is given by $D = D_1 \otimes 1 + \gamma_5 \otimes D_2$ where $D_1 = \gamma^\mu \partial_\mu$ and D_2 is the Dirac operator on $\mathcal{A}_2$ such that

$$D_l = \begin{pmatrix} \gamma^\mu \partial_\mu \otimes 1_2 \otimes 1_3 & \gamma_5 \otimes M_{12} \otimes k \\ \gamma_5 \otimes M_{21} \otimes k^* & \gamma^\mu \partial_\mu \otimes 1 \otimes 1_3 \end{pmatrix}$$

where $M_{21} = M_{12}^*$ and k is a 3×3 family mixing matrix representing Yukawa couplings for the leptons. The 1×2 matrix M_{12} turns out to be the vev of the Higgs field and is taken as $M_{12} = \mu \begin{pmatrix} 0 \\ 1 \end{pmatrix} = H_0$. The elements $a \in \mathcal{A}$ have the representation $a = \begin{pmatrix} a_1 & 0 \\ 0 & a_2 \end{pmatrix}$ where a_1, a_2 are 2×2 and 1×1 unitary valued functions. A quick calculation shows that the self-adjoint one-form ρ has the representation

$$\pi_1(\rho) = \begin{pmatrix} A_1 \otimes 1_3 & \gamma_5 \otimes H \otimes k \\ \gamma_5 \otimes H^* \otimes k^* & A_2 \otimes 1_3 \end{pmatrix}$$

where

$$A_1 = \gamma^\mu \sum_i a_1^i \partial_\mu b_1^i, \qquad A_2 = \gamma^\mu \sum_i a_2^i \partial_\mu b_2^i,$$

$$H = H_0 + \sum_i a_1^i H_0 b_2^i.$$

The quarks are introduced by taking for the finite space a bimodule structure relating two algebras $\mathcal{A}$ and $\mathcal{B}$ where the algebra $\mathcal{B}$ is taken to be $M_1(\mathbb{C}) \oplus M_3(\mathbb{C})$ commuting with the action of $\mathcal{A}$. In addition, the mass matrices in the Dirac operator are taken to be zero when acting on elements of $\mathcal{B}$. The one-form $\eta \in \Omega^1(\mathcal{B})$ has the simple form $B_1 \mathrm{diag}(1_2, 1)$ where B_1 is a gauge field associated with $M_1(\mathbb{C})$. The Hilbert space for the quarks is

$$Q = \begin{pmatrix} q_L \\ u_R \\ d_R \end{pmatrix}, \qquad q_L = \begin{pmatrix} u_L \\ d_L \end{pmatrix}$$

The representation of $a \in \mathcal{A}$ is $a \to (a_1, a_2, \overline{a}_2)$ where a_1 and a_2 are a 2×2 and 1×1 complex valued functions. The Dirac operator acting on the quark Hilbert space is

$$D_q = \begin{pmatrix} \gamma^\mu(\partial_\mu + \cdots) \otimes 1_2 \otimes 1_3 & \gamma_5 \otimes M_{12} \otimes k' & \gamma_5 \otimes \widetilde{M}_{12} \otimes k'' \\ \gamma_5 \otimes M_{12}^* \otimes k'^* & \gamma^\mu(\partial_\mu + \cdots) \otimes 1_3 & 0 \\ \gamma_5 \otimes \widetilde{M}_{12}^* \otimes k''^* & 0 & \gamma^\mu(\partial_\mu + \cdots) \otimes 1_3 \end{pmatrix}$$

where k' and k'' are 3×3 family mixing matrices and $\widetilde{M}_{12} = \mu \begin{pmatrix} 1 \\ 0 \end{pmatrix}$. The one form in $\Omega^1(\mathcal{A})$ has then the representation

$$\pi_q(\rho) = \begin{pmatrix} A_1 \otimes 1_3 & \gamma_5 \otimes H \otimes k' & \gamma_5 \otimes \widetilde{H} \otimes k'' \\ \gamma_5 \otimes H^* \otimes k'^* & A_2 \otimes 1_3 & 0 \\ \gamma_5 \otimes \widetilde{H}^* \otimes k''^* & 0 & \overline{A}_2 \otimes 1_3 \end{pmatrix}$$

where $\widetilde{H}_a = \varepsilon_{ab} H^b$. When acting on the algebra $\mathcal{B}$, the Dirac operator has zero mass matrices and the one-form η in $\Omega^1(\mathcal{B})$ has the representation $\pi_q(\eta) = B_2 \mathrm{diag}(1_2, 1)$ where B_2 is the gauge field associated with $M_3(\mathbb{C})$. Imposing the unimodularity condition on the algebras $\mathcal{A}$ and $\mathcal{B}$ would then relate the $U(1)$ factors in both algebras so that $\mathrm{tr}(A_1) = 0$, $A_2 = B_1 = -\mathrm{tr}(B_2) \equiv \frac{i}{2} g_1 B$. With these we can then write

$$A_1 = -\frac{i}{2} g_2 \sigma^a A_a$$

$$B_2 = -\frac{i}{6} g_1 B - \frac{i}{2} g_3 V^i \lambda_i$$

where g_1, g_2, and g_3 are the $U(1)$, $SU(2)$, and $SU(3)$ gauge coupling constants, σ^a and λ^i are the Pauli and Gell-Mann matrices, respectively. The fermionic actions for the leptons and quarks are then given by

$$\langle L, (D+\rho+\eta) L\rangle = \int d^4x \sqrt{g} \left(\overline{L} \left(D_l + \pi_l(\rho) + \pi_l(\eta)\right) L\right)$$

$$\langle Q, (D+\rho+\eta) Q\rangle = \int d^4x \sqrt{g} \left(\overline{Q} \left(D_q + \pi_q(\rho) + \pi_q(\eta)\right) Q\right)$$

These terms can be easily checked to reproduce all the fermionic terms of the Standard Model.

The bosonic action is the sum of the square of curvatures in both the lepton and quark sectors. These are given by

$$I_l = \mathrm{Tr}\left(C_l \left(\theta_\rho + \theta_\eta\right)^2 D_l^{-4}\right)$$

$$I_q = \mathrm{Tr}\left(C_q \left(\theta_\rho + \theta_\eta\right)^2 D_q^{-4}\right)$$

where

$$\theta_\rho \equiv d\rho + \rho^2$$

is the curvature of ρ, and C_l and C_q are constant elements of the algebra. Since the representation π has a kernel, the auxiliary fields must be projected out. This step mainly affects the potential. After some algebra, one can show that the bosonic action given above reproduces all the bosonic interactions of the Standard Model with the same number of parameters. If one assumes that C_l and C_q belong to the center of the algebra, then one can get fixed values for the top quark mass and Higgs mass. The main advantage of the noncommutative construction of the Standard Model is that one gets a geometrical understanding of the origin of the Higgs field and a unification of the gauge and Higgs sectors. One sees that the Higgs fields are the components of the one form along discrete directions.

2.3 Constructions beyond the Standard Model

The early constructions of the Standard Model provided encouragements to look further into noncommutative spaces. The construction was also complicated with some ambiguities such as the independence of the lepton and quark sectors, the construction of the Higgs potential, and projecting out the auxiliary fields. It was then natural to ask whether it is possible to go beyond the Standard Model. In particle physics, the route taken was to consider larger groups such as $SU(5)$ or $SO(10)$ which contains $U(1) \times SU(2) \times SU(3)$ as a subgroup. The main advantage of GUT is that the fermionic fields are unified in one or two representations, the most attractive possibility being $SO(10)$ where the spinor representation 16_s contains all the known fermions in addition to the right-handed neutrino. The simplicity in the fermionic sector did not make the theory more predictive because of the arbitrariness of the Higgs sector. There are many possible Higgs representations that can break the symmetry spontaneously from $SO(10)$ to $SU(3) \times U(1)$. In the noncommutative construction, the Higgs sector is more constrained which was taken as an encouragement to explore the possibility of considering larger matrix algebras. As an example, if one arranges the leptons in the form $L = \begin{pmatrix} l_L \\ l_R \end{pmatrix}$ where $l = \begin{pmatrix} \nu \\ e \end{pmatrix}$, then the corresponding algebra will be $M_2(\mathbb{C}) \oplus M_2(\mathbb{C})$. A natural possibility is then to consider a discrete space of four points and where the fermions are arranged in the format $\psi = \begin{pmatrix} l_L \\ l_R \\ l_L^c \\ l_R^c \end{pmatrix}$ and the representation π acting on $\mathcal{A}$ is given by $\pi(a) = \operatorname{diag}(a_1, a_2, \overline{a}_1, \overline{a}_2)$ where a_1, a_2 are 2×2 complex matrices. The resulting model has $SU(2)_L \times SU(2)_R \times U(1)_{B-L}$ with the Higgs fields in the representations $(2, 2)$, $(3, 1) + (1, 3)$ of $SU(2)_L \times SU(2)_R$. We can summarize the steps needed to construct noncommutative particle physics models. First we specify the fermion representations then we choose the number of discrete points and the symmetry between them. From this we deduce the appropriate algebra and the map π acting on the Hilbert space of spinors. Finally, we write down the Dirac operator acting on elements of the algebra and choose the mass matrices to correspond to the desired vacuum of the Higgs fields.

To illustrate these steps, consider the chiral space-time spinors $P_+\psi$ to be in the 16_s representation of $SO(10)$, where P_+ is the $SO(10)$ chirality operator, and the number of discrete points to be four. The Hilbert space is taken to be $\Psi = \begin{pmatrix} P_+\psi \\ P_+\psi \\ P_-\psi^c \\ P_-\psi^c \end{pmatrix}$ where $\psi^c = BC\overline{\psi}^T$, C being the charge conjugation matrix while B is the $SO(10)$ conjugation matrix. The finite algebra is taken to be $\mathcal{A}_2 = P_+ \left(\text{Cliff SO}(10)\right) P_+$, and the finite Hilbert space $\mathcal{H}_2 = \mathbb{C}^{32}$. Let π_0 denote the representation of the algebra $\mathcal{A}$ on the Hilbert space $\mathcal{H}$ and let $\overline{\pi}_0$ denote the antirepresentation defined by $\overline{\pi}_0(a) = B\overline{\pi_0(a)}B^{-1}$. We then define $\pi(a) = \pi_0(a) \oplus \pi_0(a) \oplus \overline{\pi}_0(a) \oplus \overline{\pi}_0(a)$. The Dirac operator is taken to be

$$\begin{pmatrix} \gamma^\mu \partial_\mu \otimes 1_{32} \otimes 1_3 & \gamma_5 \otimes M_{12} \otimes K_{12} & \gamma_5 \otimes M_{13} \otimes K_{13} & \gamma_5 \otimes M_{14} \otimes K_{14} \\ \gamma_5 \otimes M_{12}^* \otimes K_{12}^* & \gamma^\mu \partial_\mu \otimes 1_{32} \otimes 1_3 & \gamma_5 \otimes M_{23} \otimes K_{23} & \gamma_5 \otimes M_{24} \otimes K_{24} \\ \gamma_5 \otimes M_{13}^* \otimes K_{13}^* & \gamma_5 \otimes M_{23}^* \otimes K_{23}^* & \gamma^\mu \partial_\mu \otimes 1_{32} \otimes 1_3 & \gamma_5 \otimes M_{34} \otimes K_{34} \\ \gamma_5 \otimes M_{14}^* \otimes K_{14}^* & \gamma_5 \otimes M_{24}^* \otimes K_{24}^* & \gamma_5 \otimes M_{34}^* \otimes K_{34}^* & \gamma^\mu \partial_\mu \otimes 1_{32} \otimes 1_3 \end{pmatrix}$$

where the K_{mn} are 3×3 family mixing matrices commuting with $\pi(a)$. We may impose the exchange symmetries $1 \leftrightarrow 2$ and $3 \leftrightarrow 4$ so that $M_{12} = M_{12}^* = \mathcal{M}_0$, $M_{13} = M_{14} = M_{23} = M_{24} = \mathcal{N}_0$, $M_{34} = M_{34}^* = B\overline{\mathcal{M}}_0 B^{-1}$. Computing $\pi(\rho)$ we get

$$\pi(\rho) = \begin{pmatrix} A & \gamma_5 \mathcal{M} K_{12} & \gamma_5 \mathcal{N} K_{13} & \gamma_5 \mathcal{N} K_{14} \\ \gamma_5 \mathcal{M} K_{12}^* & A & \gamma_5 \mathcal{N} K_{23} & \gamma_5 \mathcal{N} K_{24} \\ \gamma_5 \mathcal{N}^* K_{13}^* & \gamma_5 \mathcal{N}^* K_{23}^* & B\overline{A}B^{-1} & \gamma_5 B\overline{\mathcal{M}}B^{-1} K_{34} \\ \gamma_5 \mathcal{N}^* K_{14}^* & \gamma_5 \mathcal{N}^* K_{24}^* & \gamma_5 B\overline{\mathcal{M}}B^{-1} K_{34}^* & B\overline{A}B^{-1} \end{pmatrix}$$

where

$$A = P_+ \sum_i a^i \gamma^\mu \partial_\mu b^i P_+$$

$$\mathcal{M} + \mathcal{M}_0 = P_+ \sum_i a^i \mathcal{M}_0 b^i P_+$$

$$\mathcal{N} + \mathcal{N}_0 = P_+ \sum_i a^i \mathcal{N}_0 B \overline{b}^i B^{-1} P_-$$

One sees immediately that the Higgs fields $\mathcal{M}$ and $\mathcal{N}$ are in the $16_s \times 16_s$ and $16_s \times \overline{16}_s$ representations. Equating the action of A on ψ and ψ^c will reduce it to an $SO(10)$ gauge field. Specifying $\mathcal{M}_0$ and $\mathcal{N}_0$ determines the breaking pattern of $SO(10)$. One can then proceed to construct the bosonic sector and project out the auxiliary fields to determine the potential. There are very limited number of models one can construct. These models, however, will suffer the same problems encountered in the GUT construction, mainly that of low unification scale of 10^{14} GeV implying fast rate of proton decay which is ruled out experimentally.

2.4 Coupling matter to gravity

The dynamics of the gravitational force is based on Riemannian geometry. It is therefore natural to study the nature of the gravitational field in noncommutative geometry. The original attempt [24, 25] was based on generalizing the basic notions of Riemannian geometry, notably the theory of linear connections on differential forms. (Note that an alternative route that takes vector fields as a starting point ends with a derivation-based differential calculus as in [43] (cf. [65]). In line with the Connes—Lott model, we will instead take differential forms as our starting point. For more details, we also refer to the exposition in [60, Sect. 10.3].)

First one defines the metric as an inner product on a cotangent space. Then one shows that every cycle over $\mathcal{A}$ yields a notion of cotangent bundle associated with $\mathcal{A}$ and a Riemannian metric on the cotangent bundle $\Omega_D^1(\mathcal{A})$. With the connection ∇ the Riemann curvature of ∇ on $\Omega_D^1(\mathcal{A})$ is defined by $R(\nabla) := -\nabla^2$ and the torsion by $T = d - m \circ \nabla$ where m is the tensor product. Requiring ∇ to be unitary and the torsion to vanish we obtain the Levi–Civita connection. If $\Omega_D^1(\mathcal{A})$ is a finitely generated module, then it admits a basis e^A, $A = 1, 2, \ldots, N$, and the connection $\omega_B^A \in \Omega_D^1(\mathcal{A})$ is defined by $\nabla e^A = -\omega_B^A \otimes e^B$. The components of the torsion $T(\nabla)$ are defined by $T^A = T(\nabla) e^A$ then $T^A \in \Omega_D^2(\mathcal{A})$ is given by

$$T^A = de^A + \omega_B^A e^B$$

Similarly, components of the curvature $R_B^A \in \Omega_D^2(\mathcal{A})$ satisfy the defining property that $R(\nabla) e^A = R_B^A \otimes e^B$ so that

$$R_B^A = d\omega_B^A + \omega_C^A \omega_B^C.$$

The analogue of the Einstein–Hilbert action is then

$$I(\nabla) := \kappa^{-2} \left\langle R_B^A e^B, e_A \right\rangle$$

where κ^{-1} is the Planck scale. Computing this action for the product space $M_4 \times Z_2$ one finds that

$$I(\nabla) = 2 \int_M d^4x \sqrt{g} \left(\kappa^{-2} r - 2 \partial_\mu \sigma \partial^\mu \sigma \right)$$

where r is the scalar curvature of the Levi–Civita connection of the Riemannian manifold M_4 coupled to a scalar field σ. Applying this construction to the Connes–Lott model is rather involved because the two sheets are not treated symmetrically, being associated with two different algebras. The complication arise because the projective module is not free and the basis e^A is constrained. The Einstein–Hilbert action in this case is given by

$$I(\nabla) = 2 \int_M d^4x \sqrt{g} \left(\kappa^{-2} \frac{3}{2} r - 2(3 + \lambda) \partial_\mu \sigma \partial^\mu \sigma + c(\lambda) e^{-2\sigma} \right)$$

where $\lambda = \mathrm{Tr}(kk^*)^2 - 1$. To understand the significance of the field σ, we note that by examining the Dirac operator one finds that the field $\phi = e^{-\kappa\sigma}$ now replaces the weak scale. Thus quantum corrections to the classical potential will depend on σ, thus the vev of σ could be determined from the minimization equations.

3 The spectral action principle

Despite the success of the Connes–Lott model and the generalizations that followed in giving a geometrical meaning to the Higgs field and unifying it with the gauge fields, it was felt that the construction is not satisfactory. The first unpleasant feature was the use of the bimodule structure to introduce the $SU(3)$ symmetry and the second is the use of unimodularity condition to get the correct hypercharge assignments to the particles. Another major problem was the existence of mirror fermions as a consequence of the fact that the conjugation operator on fermions gives independent fields. In addition, there was arbitrariness in the construction of the potential in the bosonic sector associated with the step of eliminating the auxiliary fields.

3.1 Real structures on spectral triples

The first breakthrough came in 1995 with the publication of Alain Connes' paper "Noncommutative geometry and reality" [30]. In this paper, the notion of real structure is introduced, motivated by Atiyah's KR theory and Tomita's involution operator J. A hint for the necessity of the reality operator can be taken from physics. We have seen that space-time spinors, which are elements of the Hilbert space, satisfy a chirality condition. The charge conjugation operator, when acting on these spinors, produces a conjugate element, which in general is independent. It is possible to replace the chirality condition, with a reality one, known as the Majorana condition which equates the two. Imposing both conditions, chirality and reality, simultaneously can only occur in certain dimensions. The action of the antilinear isometry J on the algebra $\mathcal{A}$ satisfies the commutation relation $[a, b^o] = 0$, $\forall a, b \in \mathcal{A}$ where

$$b^o = Jb^* J^{-1}, \qquad \forall b \in \mathcal{A} \tag{1}$$

so that $b^o \in \mathcal{A}^o$. This gives a bimodule, using the representation of $\mathcal{A} \otimes \mathcal{A}^o$, given by

$$a \otimes b^o \to aJb^* J^{-1}, \qquad \forall a, b \in \mathcal{A} \tag{2}$$

We define the fundamental class μ of the noncommutative space as a class in the KR-homology of the algebra $\mathcal{A} \otimes \mathcal{A}^o$ having the involution

$$\tau \left(a \otimes b^o\right) = b^* \otimes \left(a^*\right)^o, \qquad \forall a, b \in \mathcal{A} \tag{3}$$

The KR-homology cycle implements the involution τ given by

$$\tau(w) = JwJ^{-1}, \qquad \forall w \in \mathcal{A} \otimes \mathcal{A}^{o} \tag{4}$$

These imply that the KR-homology is periodic with period 8 and the dimension n modulo 8 is determined from the commutation rules

$$J^2 = \varepsilon, \qquad JD = \varepsilon' DJ, \qquad J\gamma = \varepsilon'' \gamma J \tag{5}$$

where $\varepsilon, \varepsilon', \varepsilon'' \in \{-1, 1\}$ are given as function of n modulo 8 according to the table

n	0	1	2	3	4	5	6	7
ε	1	1	−1	−1	−1	−1	1	1
ε'	1	−1	1	1	1	−1	1	1
ε''	1		−1		1		−1	

(6)

It is not surprising that this table agrees with the one obtained by classifying in which dimensions a spinor obey the Majorana and Weyl conditions. The intersection form $K_*(\mathcal{A}) \times K_*(\mathcal{A}) \to \mathbb{Z}$ is obtained from the Fredholm index of D in $K_*(\mathcal{A} \otimes \mathcal{A}^o)$. Using the Kasparov intersection product, Poincare duality is formulated in terms of the invertibility of μ and that D is an operator of order one implies the condition

$$\left[[D, a], b^o\right] = 0, \qquad \forall a, b \in \mathcal{A} \tag{7}$$

Next we consider automorphisms of the algebra $\mathcal{A}$ denoted by $\mathrm{Aut}(\mathcal{A})$. This comprises both of inner and outer automorphisms. Inner automorphisms $\mathrm{Int}(\mathcal{A})$ is a normal subgroup of $\mathrm{Aut}(\mathcal{A})$ defined by

$$\alpha(f) = ufu^*, \qquad \forall f \in \mathcal{A}, \qquad u\,u^* = u^*u = 1 \tag{8}$$

The group $\mathrm{Aut}^+(\mathcal{A})$ of automorphisms of the involutive algebra $\mathcal{A}$ are implemented by a unitary operator U in $\mathcal{H}$ commuting with J satisfying

$$\alpha(x) = UxU^{-1} \qquad \forall x \in \mathcal{A} \tag{9}$$

For Riemannian manifolds M, this plays the role of the group of diffeomorphisms $\mathrm{Diff}^+(M)$, which preserves the K-homology fundamental class of M. Let $\mathcal{E}$ be a finite projective, Hermitian right $\mathcal{A}$-module, and define the algebra $\mathcal{B} = \mathrm{End}(\mathcal{A})$ as the Morita equivalence of the algebra $\mathcal{A}$ with a Hermitian connection ∇ on $\mathcal{E}$ defined as the linear map $\nabla : \mathcal{E} \to \mathcal{E} \otimes_{\mathcal{A}} \Omega^1_D$ satisfying

$$\nabla(\zeta a) = (\nabla\zeta) a + \zeta \otimes da, \qquad \forall \zeta \in \mathcal{E},\ a \in \mathcal{A}$$

$$d(\zeta, \eta) = (\zeta, \nabla\eta) - (\nabla\zeta, \eta), \qquad \forall \zeta, \eta \in \mathcal{E}$$

where $da = [D, a]$ and Ω^1_D is the bimodule of operators of the form

$$A = \sum_i a_i \left[D, b_i\right], \qquad a_i, b_i \in \mathcal{A} \tag{10}$$

Since any algebra is Morita equivalent to itself with $\mathcal{E} = \mathcal{A}$, applying the construction given above yields the inner deformation of the spectral geometry. The unitary equivalence is implemented by the representation $u \to \widetilde{U} = u\left(Ju J^{-1}\right) = u\left(u^{o}\right)^*$ so that the Dirac operator that includes inner fluctuations

$$D_A = D + A + JAJ^{-1} \tag{11}$$

where $A = A^*$ transforms as $D_A \to \widetilde{U} D_A \widetilde{U}^{-1}$ provided that

$$A \to u\,Au^* + u\left[D, u^*\right] \tag{12}$$

This will ensure that the inner product

$$\left(\Psi, D_A \Psi\right) \tag{13}$$

is invariant under the transformation $\Psi \to \widetilde{U}\Psi$. This expression will then take care of all fermionic interactions which, as will be seen in the next section, removes the arbitrariness in specifying the action of the connection on the Hilbert space.

3.2 *The spectral action principle*

The next breakthrough came a year later in 1996 in the work of Chamseddine and Connes entitled "The spectral action principle" [12]. The basic observation is that for a noncommutative space defined by spectral data, the emphasis is shifted from the coordinates x of a geometric space to the spectrum $\Sigma \subset \mathbb{R}$ of the operator D. We postulate the following hypothesis

$$\text{The physical action depends only on } \Sigma \tag{14}$$

The existence of Riemannian manifolds which are isospectral but not isometric shows that the spectral action principle is stronger than the usual diffeomorphism invariance. In the usual Riemannian case the group $\mathrm{Diff}\,(M)$ of diffeomorphisms of M is canonically isomorphic to the group $\mathrm{Aut}\,(\mathcal{A})$ of automorphisms of the algebra $\mathcal{A} = C^\infty(M)$. To each $\varphi \in \mathrm{Diff}\,(M)$ one associates the algebra preserving map $\alpha_\varphi : \mathcal{A} \to \mathcal{A}$ given by

$$\alpha_\varphi(f) = f \circ \varphi^{-1} \qquad \forall f \in C^\infty(M) = \mathcal{A} \tag{15}$$

The prescription to determine the bosonic action with some cutoff energy scale Λ is to first replace the Hilbert space $\mathcal{H}$ by the subspace $\mathcal{H}_\Lambda$ defined by

$$\mathcal{H}_\Lambda = \text{range}\, \chi \left(\frac{D}{\Lambda} \right) \tag{16}$$

where χ is a suitable smooth positive function, restricting both D and $\mathcal{A}$ to this subspace maintaining the commutation relations for the algebra. This procedure is superior to the lattice approximation because it does respect the geometric symmetry group. The *spectral action functional* is then given by the

$$\operatorname{Tr} \chi \left(\frac{D}{\Lambda} \right).$$

For a noncommutative space which is a tensor product of a continuous manifold times a discrete space, the functional $\operatorname{Tr} \chi \left(\frac{D}{\Lambda} \right)$ can be expanded in an asymptotic series in Λ, rendering the computation amenable to a heat kernel expansion. This procedure will be illustrated in the next section. More general methods to analyze the spectral action have also been developed, see [50] for an early result and also the recent book [48]. An interpretation of the spectral action as the von Neumann entropy of a second-quantized spectral triple has been found recently in [20] (cf. [42]).

To summarize, the breakthroughs carried out in the short period 1995–1996, defining the reality operator J and developing the spectral action principle, will allow to remove the ambiguities encountered before in the construction of the noncommutative spectral Standard Model.

4 The spectral Standard Model

At the time that the spectral action was formulated, it was clear that this principle forms a unifying framework for gravity and particle physics of the Standard Model. As said, this led to much activity (cf. [69]) in the years that followed. Also shortcomings of the approach were pointed out quite quickly, such as the notorious fermion-doubling problem [63, 52]. This doubling—or actually, quadrupling—was due to the incorporation of left-right, particle–anti-particle degrees of freedom both in the continuum spinor space and in the finite noncommutative space. At the technical level this was a crucial starting point, allowing for a product geometry to describe gravity coupled to the Standard Model.

Nevertheless, it was a somewhat disturbing feature which, together with the apparent arbitrariness of the choice of a finite geometry and the absence of neutrino mixing in the model, led Connes to eventually resolve these problems in [31]. At the same time John Barrett [4] arrived at the same conclusion (see also the recent uniqueness result [6]), even though his motivation came from the desire to have noncommutative geometry with a Lorentzian signature.

The crucial insight in both of these works is that one should allow for a KO-dimension for the finite space F which is different from the metric dimension (which is zero). More specifically, the KO-dimension of the finite space should be 6 (modulo 8), so that the product of the continuum M with F is 10 modulo 8. The precise structure of the spectral Standard Model (see Section 4.2) is then best understood using the classification of all irreducible finite noncommutative geometries of KO-dimension 6 which we now briefly recall.

4.1 Classification of irreducible geometries

In [14] Chamseddine and Connes classified *irreducible* finite real spectral triples of KO-dimension 6. This lead to a remarkably concise list of spectral triples, based on the matrix algebras $M_N(\mathbb{C}) \oplus M_N(\mathbb{C})$ for some N. We remark that earlier classification results were obtained [58, 68] which were also exploited in a search Beyond the Standard Model (see Remark 5 below).

Definition 1 *A finite real spectral triple* $(A, H, D; J, \gamma)$ *is called* irreducible *if the triple* (A, H, J) *is irreducible. More precisely, we demand that:*

1. *The representations of* A *and* J *in* H *are irreducible;*
2. *The action of* A *on* H *has a separating vector.*

We will prove the main result of [14] using an alternative approach which is based on [75, Sect. 3.4].

Theorem 2 *Let* $(A, H, D; J, \gamma)$ *be an irreducible finite real spectral triple of KO-dimension 6. Then there exists a positive integer* N *such that* $A \simeq M_N(\mathbb{C}) \oplus M_N(\mathbb{C})$.

Proof Let $(A, H, D; J, \gamma)$ be an arbitrary finite real spectral triple. We may then decompose

$$A = \bigoplus_{i=1}^{N} M_{n_i}(\mathbb{C}), \qquad H = \bigoplus_{i,j=1}^{N} \mathbb{C}^{n_i} \otimes (\mathbb{C}^{n_j})^\circ \otimes V_{ij},$$

with V_{ij} corresponding to the multiplicities as before. Now each $\mathbb{C}^{n_i} \otimes \mathbb{C}^{n_j}$ is an irreducible representation of A, but in order for H to support a real structure $J : H \to H$ we need both $\mathbb{C}^{n_i} \otimes (\mathbb{C}^{n_j})^\circ$ and $\mathbb{C}^{n_j} \otimes (\mathbb{C}^{n_i})^\circ$ to be present in H. Moreover, an old result of Wigner [78] for an anti-unitary operator with $J^2 = 1$ assures that already with multiplicities $\dim V_{ij} = 1$ there exists such a real structure. Hence, the irreducibility condition (1) above yields

$$H = \mathbb{C}^{n_i} \otimes (\mathbb{C}^{n_j})^\circ \oplus \mathbb{C}^{n_j} \otimes (\mathbb{C}^{n_i})^\circ,$$

for some $i, j \in \{1, \ldots, N\}$. Then, let us consider condition (2) on the existence of a separating vector. Note first that the representation of A in H is faithful only if $A = M_{n_i}(\mathbb{C}) \oplus M_{n_j}(\mathbb{C})$. Second, the stronger condition of a separating vector ξ then implies $n_i = n_j$, as it is equivalent to $A'\xi = H$ for the commutant A' of A in H. Namely, since $A' = M_{n_j}(\mathbb{C}) \oplus M_{n_i}(\mathbb{C})$ with $\dim A' = n_i^2 + n_j^2$, and $\dim H = 2n_i n_j$ we find the desired equality $n_i = n_j$. □

With the complex finite-dimensional algebras A given as a direct sum $M_N(\mathbb{C}) \oplus M_N(\mathbb{C})$,[1] the additional demand that H carries a symplectic structure $I^2 = -1$ yields real algebras of which A is the complexification. We see that this requires $N = 2k$ so that one naturally considers triples (A, H, J) for which

$$A = M_k(\mathbb{H}) \oplus M_{2k}(\mathbb{C}); \qquad H = \mathbb{C}^{2(2k)^2}. \tag{17}$$

4.2 Noncommutative geometry of the Standard Model

The above classification of irreducible finite geometries of KO-dimension 6 forms the starting point for the derivation of the Standard Model from a noncommutative manifold [17]. Hence, it is based on the matrix algebra $M_N(\mathbb{C}) \oplus M_N(\mathbb{C})$ for $N \geq 1$. Let us make the following two additional requirements on the irreducible finite geometry $(A, H_F, D_F; J_F, \gamma_F)$:

1. The finite-dimensional Hilbert space H_F carries a symplectic structure $I^2 = -1$;
2. the grading γ_F induces a non-trivial grading on A, by mapping

$$a \mapsto \gamma_F a \gamma_F,$$

   and selects an even subalgebra $A^{\mathrm{ev}} \subset A$ consisting of elements that commute with γ_F.

But the first demand sets $A = M_k(\mathbb{H}) \oplus M_{2k}(\mathbb{C})$, represented on the Hilbert space $\mathbb{C}^{2(2k)^2}$. The second requirement sets $k \geq 2$; we will take the simplest $k = 2$ so that $H_F = \mathbb{C}^{32}$.[2] Indeed, this allows for a γ_F such that

$$A^{\mathrm{ev}} = \mathbb{H}_R \oplus \mathbb{H}_L \oplus M_4(\mathbb{C}),$$

[1] The case $N = 1$ was exploited successfully in [47] for a noncommutative description of Abelian gauge theories.

[2] Also other algebras that appear in the classification of irreducible geometries of KO-dimension have been considered in the literature: besides the case $N = 4$ that we consider here the simplest case $N = 1$ is relevant for the noncommutative geometric description of quantum electrodynamics [47] and the case $N = 8$ leads to the "grand algebra" of [40, 38].

where $\mathbb{H}_R$ and $\mathbb{H}_L$ are two copies (referred to as *right* and *left*) of the quaternions; they are the diagonal of $M_2(\mathbb{H}) \subset A$. The Hilbert space can then be decomposed according to the defining representations of A^{ev},

$$H_F = (\mathbb{C}^2_R \oplus \mathbb{C}^2_L) \otimes (\mathbb{C}^4)^\circ \oplus \mathbb{C}^4 \otimes ((\mathbb{C}^2_R)^\circ \oplus (\mathbb{C}^2_L)^\circ). \tag{18}$$

According to this direct sum decomposition, we write

$$D_F = \begin{pmatrix} S & T^* \\ T & \overline{S} \end{pmatrix} \tag{19}$$

Moreover, J_F is the anti-unitary operator that flips the two 16-dimensional components in Equation (18).

The key result is that if we assume that T is non-trivial, then the first-order condition selects the maximal subalgebra of the Standard Model, that is to say, $A_F = \mathbb{C} \oplus \mathbb{H} \oplus M_3(\mathbb{C})$.

Proposition 3 ([17, Prop. 2.11]) *Up to $*$-automorphisms of A^{ev}, there is a unique $*$-subalgebra $A_F \subset A^{\mathrm{ev}}$ of maximal dimension that allows $T \neq 0$ in (19). It is given by*

$$A_F = \left\{ \left(q_\lambda, q, \begin{pmatrix} q & 0 \\ 0 & m \end{pmatrix} \right) : \lambda \in \mathbb{C}, q \in \mathbb{H}_L, m \in M_3(\mathbb{C}) \right\} \subset \mathbb{H}_R \oplus \mathbb{H}_L \oplus M_4(\mathbb{C}),$$

where $\lambda \mapsto q_\lambda$ is the embedding of $\mathbb{C}$ into $\mathbb{H}$, with

$$q_\lambda = \begin{pmatrix} \lambda & 0 \\ 0 & \overline{\lambda} \end{pmatrix}.$$

Consequently, $A_F \simeq \mathbb{C} \oplus \mathbb{H} \oplus M_3(\mathbb{C})$.

The restriction of the representation of A on H_F to the subalgebra A_F gives a decomposition of H_F into irreducible (left and right) representations of $\mathbb{C}$, $\mathcal{H}_L$, and $M_3(\mathbb{C})$. For instance,

$$(\mathbb{C}^2_R \oplus \mathbb{C}^2_L) \otimes (\mathbb{C}^4)^\circ \rightsquigarrow (\mathbb{C} \oplus \overline{\mathbb{C}} \oplus \mathbb{C}^2_L) \otimes \Big((\mathbb{C})^\circ \oplus (\mathbb{C}^3)^\circ \Big). \tag{20}$$

and similarly for $\mathbb{C}^4 \otimes ((\mathbb{C}^2_R)^\circ \oplus (\mathbb{C}^2_L)^\circ)$. In order to connect to the physics of the Standard Model, let us introduce an orthonormal basis for H_F that can be recognized as the fermionic particle content of the Standard Model, and subsequently write the representation of A_F in terms of this basis.

We let the subspace of H_F displayed in Equation (20) be represented by basis vectors $\{\nu_R, e_R, (\nu_L, e_L)\}$ of the so-called *lepton space* H_l and basis vectors $\{u_R, d_R, (u_L, d_L)\}$ of the *quark space* H_q. Their reflections with respect to J_F are the *anti-lepton space* $H_{\overline{l}}$ and the *anti-quark space* $H_{\overline{q}}$, spanned by $\{\overline{\nu_R}, \overline{e_R}, (\overline{\nu_L}, \overline{e_L})\}$

and $\{\overline{u_R}, \overline{d_R}, (\overline{u_L}, \overline{d_L})\}$, respectively. The three colors of the quarks are given by a tensor factor $\mathbb{C}^3$ and when we take into account *three generations* of fermions and anti-fermions by tripling the above finite-dimensional Hilbert space we obtain

$$H_F := \left(H_l \oplus H_{\overline{l}} \oplus H_q \oplus H_{\overline{q}}\right)^{\oplus 3}.$$

Note that $H_l = \mathbb{C}^4$, $H_q = \mathbb{C}^4 \otimes \mathbb{C}^3$, $H_{\overline{l}} = \mathbb{C}^4$, and $H_{\overline{q}} = \mathbb{C}^4 \otimes \mathbb{C}^3$.

An element $a = (\lambda, q, m) \in A_F$ acts on the space of leptons H_l as $q_\lambda \oplus q$, and acts on the space of quarks H_q as $(q_\lambda \oplus q) \otimes 1_3$. For the action of a on an anti-lepton $\overline{l} \in H_{\overline{l}}$ we have $a\overline{l} = \lambda 1_4 \overline{l}$, and on an anti-quark $\overline{q} \in H_{\overline{q}}$ we have $a\overline{q} = (1_4 \otimes m)\overline{q}$.

The $\mathbb{Z}_2$-grading γ_F is such that left-handed particles have eigenvalue $+1$ and right-handed particles have eigenvalue -1. The antilinear operator J_F interchanges particles with their anti-particles, so $J_F f = \overline{f}$ and $J_F \overline{f} = f$, with f a lepton or quark.

The first indication that the subalgebra A_F is relevant for the Standard Model—to say the least—comes from the fact that the Standard Model gauge group can be derived from the unitaries in A_F. We restrict to the *unimodular gauge group*,

$$\mathrm{SU}(A_F) = \left\{u \in A_F : u^*u = uu^* = 1, \det(u) = 1\right\}$$

where det is the determinant of the action of u in H_F. It then follows that, up to a finite Abelian group we have

$$\mathrm{SU}(A_F) \sim U(1) \times SU(2) \times SU(3)$$

and the hypercharges are derived from the unimodularity condition to be the usual ones:

Particle	ν_R	e_R	ν_L	e_L	u_R	d_R	u_L	d_L
Hypercharge	0	-2	-1	-1	$\frac{4}{3}$	$-\frac{2}{3}$	$\frac{1}{3}$	$\frac{1}{3}$

Let us now turn to the form of the finite Dirac operator, and see what we can say about the components of the matrix D_F as displayed in (19). Recall that we are looking for a self-adjoint operator D_F in H_F that commutes with J_F, anti-commutes with γ_F, and fulfills the first-order conditions with respect to A_F:

$$[[D, a], JbJ^{-1}] = 0; \qquad (a, b \in A_F).$$

We also require that D_F commutes with the subalgebra $\mathbb{C}_F = \{(\lambda, \lambda, 0)\} \subset A_F$ which physically speaking corresponds to the fact that the photon remains massless. Then it turns out [31, Theorem 1] (see also [17, Theorem 2.21]) that any D_F that satisfies these assumptions is of the following form: in terms of the decomposition of H_F in particle $(H_l^{\oplus 3} \oplus H_q^{\oplus 3})$ and anti-particles $(H_{\overline{l}}^{\oplus 3} \oplus H_{\overline{q}}^{\oplus 3})$ the operator S is

$$S_l := S|_{H_l^{\oplus 3}} = \begin{pmatrix} 0 & 0 & Y_\nu^* & 0 \\ 0 & 0 & 0 & Y_e^* \\ Y_\nu & 0 & 0 & 0 \\ 0 & Y_e & 0 & 0 \end{pmatrix}, \tag{21}$$

$$S_q \otimes 1_3 := S|_{H_q^{\oplus 3}} = \begin{pmatrix} 0 & 0 & Y_u^* & 0 \\ 0 & 0 & 0 & Y_d^* \\ Y_u & 0 & 0 & 0 \\ 0 & Y_d & 0 & 0 \end{pmatrix} \otimes 1_3, \tag{22}$$

where Y_ν, Y_e, Y_u, and Y_d are some 3×3 matrices acting on the three generations, and 1_3 acting on the three colors of the quarks. The symmetric operator T only acts on the right-handed (anti)neutrinos, so it is given by $T\nu_R = Y_R\overline{\nu_R}$, for a certain 3×3 symmetric matrix Y_R, and $Tf = 0$ for all other fermions $f \neq \nu_R$. Note that ν_R here stands for a vector with three components for the number of generations.

The above classification result shows that the Dirac operators D_F give all the required features, such as mixing matrices for quarks and leptons, unbroken color, and the seesaw mechanism for right-handed neutrinos. Let us illustrate the latter in some more detail. The mass matrix restricted to the subspace of H_F with basis $\{\nu_L, \nu_R, \overline{\nu_L}, \overline{\nu_R}\}$ is given by

$$\begin{pmatrix} 0 & Y_\nu^* & Y_R^* & 0 \\ Y_\nu & 0 & 0 & 0 \\ Y_R & 0 & 0 & \overline{Y}_\nu^* \\ 0 & 0 & \overline{Y}_\nu & 0 \end{pmatrix}.$$

Suppose we consider only one generation, so that $Y_\mu = m_\nu$ and $Y_R = m_R$ are just scalars. The eigenvalues of the above mass matrix are then given by

$$\pm\frac{1}{2}m_R \pm \frac{1}{2}\sqrt{m_R{}^2 + 4m_\nu{}^2}.$$

If we assume that $m_\nu \ll m_R$, then these eigenvalues are approximated by $\pm m_R$ and $\pm\frac{m_\nu{}^2}{m_R}$. This means that there is a heavy neutrino, for which the Dirac mass m_ν may be neglected, so that its mass is given by the Majorana mass m_R. However, there is also a light neutrino, for which the Dirac and Majorana terms conspire to yield a mass $\frac{m_\nu{}^2}{m_R}$, which is in fact much smaller than the Dirac mass m_ν. This is called the *seesaw mechanism*. Thus, even though the observed masses for these neutrinos may be very small, they might still have large Dirac masses (or Yukawa couplings).

Remark 4 Of course, in the physical applications one chooses Y_ν, Y_e to be the *Yukawa mass matrices* and Y_R is the *Majorana mass matrix*. There have been searches for additional conditions to be satisfied by the spectral triple (A_F, H_F, D_F) to further constrain the form of D_F, see, for instance, [11, 8, 59, 36, 37].

4.3 The gauge and scalar fields as inner fluctuations

We here derive the precise form of internal fluctuations A_μ for the above spectral triple of the Standard Model (following [17, Sect. 3.5] or [75, Sect. 11.5]).

Take two elements $a = (\lambda, q, m)$ and $b = (\lambda', q', m')$ of the algebra $\mathcal{A} = C^\infty(\mathbb{C} \oplus \mathbb{H} \oplus M_3(\mathbb{C}))$. According to the representation of A_F on H_F, the inner fluctuations $A_\mu = -ia\partial_\mu b$ decompose as

$$\Lambda_\mu := -i\lambda\partial_\mu\lambda'; \qquad \Lambda'_\mu := -i\overline{\lambda}\partial_\mu\overline{\lambda}'$$

on ν_R and e_R, respectively, and as

$$Q_\mu := -iq\partial_\mu q'; \qquad V'_\mu := -im\partial_\mu m'$$

acting on (ν_l, e_L) and $H_{\overline{q}}$, respectively. On all other components of H_F the gauge field A_μ acts as zero. Imposing the hermiticity $\Lambda_\mu = \Lambda_\mu^*$ implies $\Lambda_\mu \in \mathbb{R}$, and also automatically yields $\Lambda'_\mu = -\Lambda_\mu$. Furthermore, $Q_\mu = Q_\mu^*$ implies that Q_μ is a real-linear combination of the Pauli matrices, which span $i\, su(2)$. Finally, the condition that V'_μ be Hermitian yields $V'_\mu \in i\, u(3)$, so V'_μ is a $U(3)$ gauge field. As mentioned above, we need to impose the unimodularity condition to obtain an $SU(3)$ gauge field. Hence, we require that the trace of the gauge field A_μ over H_F vanishes, and we obtain

$$\mathrm{Tr}|_{H_{\overline{l}}}\left(\Lambda_\mu 1_4\right) + \mathrm{Tr}|_{H_{\overline{q}}}\left(1_4 \otimes V'_\mu\right) = 0 \quad \Longrightarrow \quad \mathrm{Tr}(V'_\mu) = -\Lambda_\mu.$$

Therefore, we can define a traceless $SU(3)$ gauge field V_μ by $\overline{V}_\mu := -V'_\mu - \frac{1}{3}\Lambda_\mu$. The action of the gauge field $B_\mu = A_\mu - J_F A_\mu J_F^{-1}$ on the fermions is then given by

$$B_\mu\big|_{H_l} = \begin{pmatrix} 0 & 0 & \\ 0 & -2\Lambda_\mu & \\ & & Q_\mu - \Lambda_\mu 1_2 \end{pmatrix},$$

$$B_\mu\big|_{H_q} = \begin{pmatrix} \frac{4}{3}\Lambda_\mu 1_3 + V_\mu & 0 & \\ 0 & -\frac{2}{3}\Lambda_\mu 1_3 + V_\mu & \\ & & (Q_\mu + \frac{1}{3}\Lambda_\mu 1_2) \otimes 1_3 + 1_2 \otimes V_\mu \end{pmatrix}. \tag{23}$$

for some $U(1)$ gauge field Λ_μ, an $SU(2)$ gauge field Q_μ, and an $SU(3)$ gauge field V_μ.

Note that the coefficients in front of Λ_μ in the above formulas are precisely the aforementioned (and correct!) hypercharges of the corresponding particles.

Next, let us turn to the scalar field ϕ, which is given by

$$\phi|_{H_l} = \begin{pmatrix} 0 & Y^* \\ Y & 0 \end{pmatrix}, \quad \phi|_{H_q} = \begin{pmatrix} 0 & X^* \\ X & 0 \end{pmatrix} \otimes 1_3, \quad \phi|_{H_{\bar{l}}} = 0, \quad \phi|_{H_{\bar{q}}} = 0, \tag{24}$$

where we now have, for complex fields ϕ_1, ϕ_2,

$$Y = \begin{pmatrix} Y_\nu \phi_1 & -Y_e \overline{\phi}_2 \\ Y_\nu \phi_2 & Y_e \overline{\phi}_1 \end{pmatrix}, \qquad X = \begin{pmatrix} Y_u \phi_1 & -Y_d \overline{\phi}_2 \\ Y_u \phi_2 & Y_d \overline{\phi}_1 \end{pmatrix}.$$

The scalar field Φ is then given by

$$\Phi = D_F + \begin{pmatrix} \phi & 0 \\ 0 & 0 \end{pmatrix} + J_F \begin{pmatrix} \phi & 0 \\ 0 & 0 \end{pmatrix} J_F^* = \begin{pmatrix} S + \phi & T^* \\ T & \overline{(S + \phi)} \end{pmatrix}. \tag{25}$$

Finally, one can compute that the action of the gauge group $\mathrm{SU}(A_F)$ by conjugation on the fluctuated Dirac operator

$$D_\omega = D \otimes 1 + \gamma^\mu \otimes B_\mu + \gamma_M \otimes \Phi$$

is implemented by

$$\Lambda_\mu \mapsto \Lambda_\mu - i\lambda \partial_\mu \overline{\lambda}, \quad Q_\mu \mapsto q Q_\mu q^* - iq\partial_\mu q^*, \quad \overline{V}_\mu \mapsto m\overline{V}_\mu m^* - im\partial_\mu m^*,$$
$$H \mapsto \overline{\lambda} q H,$$

for $\lambda \in C^\infty(M, U(1))$, $q \in C^\infty(M, SU(2))$ and $m \in C^\infty(M, SU(3))$ and we have written the *Higgs doublet* as

$$H := \begin{pmatrix} \phi_1 + 1 \\ \phi_2 \end{pmatrix}$$

For the detailed computation we refer to [17, Sect. 3.5] or [75, Prop. 11.5].

Summarizing, the gauge fields derived take values in the Lie algebra $u(1) \oplus su(2) \oplus su(3)$ and transform according to the usual Standard Model gauge transformations. The scalar field ϕ transforms as the Standard Model Higgs field in the defining representation of $SU(2)$, with hypercharge -1.

4.4 Spectral action

The spectral action for the above spectral Standard Model has been computed in full detail in [17, Section 4.2] and confirmed in, e.g., [75, Theorem 11.10]. Since it would lie beyond the scope of the present review, we refrain from repeating this computation. Instead, we summarize the main result, which is that the Lagrangian derived from the spectral action is

$$
\begin{aligned}
S_B = \int \Bigg(& \frac{48\chi_4\Lambda^4}{\pi^2} - \frac{c\chi_2\Lambda^2}{\pi^2} + \frac{d\chi(0)}{4\pi^2} + \left(\frac{c\chi(0)}{24\pi^2} - \frac{4\chi_2\Lambda^2}{\pi^2}\right) s - \frac{3\chi(0)}{10\pi^2}(C_{\mu\nu\rho\sigma})^2 \\
& + \frac{1}{4} Y_{\mu\nu}Y^{\mu\nu} + \frac{1}{4} W^a_{\mu\nu}W^{\mu\nu,a} + \frac{1}{4} G^i_{\mu\nu}G^{\mu\nu,i} + \frac{b\pi^2}{2a^2\chi(0)}|H|^4 \\
& - \frac{2a\chi_2\Lambda^2 - e\chi(0)}{a\chi(0)}|H|^2 + \frac{1}{12}s|H|^2 + \frac{1}{2}|D_\mu H|^2 \Bigg)\sqrt{g}d^4x,
\end{aligned}
$$

where $\chi_j = \int_0^\infty \chi(v)v^{j-1}dv$ are the moments of the function χ, $j > 0$, $s = -R$ is the scalar curvature, $Y_{\mu\nu}$, $W_{\mu\nu}$, and $G_{\mu\nu}$ are the field strengths of Y_μ, Q_μ, and V_μ, respectively, and the covariant derivative $D_\mu H$ is given by

$$
D_\mu H = \partial_\mu H + \frac{1}{2} i g_2 W^a_\mu \sigma^a H - \frac{1}{2} i g_1 Y_\mu H. \tag{26}
$$

Moreover, we have defined the following constants:

$$
\begin{aligned}
a &= \mathrm{Tr}\left(Y_\nu^* Y_\nu + Y_e^* Y_e + 3Y_u^* Y_u + 3Y_d^* Y_d\right), \\
b &= \mathrm{Tr}\left((Y_\nu^* Y_\nu)^2 + (Y_e^* Y_e)^2 + 3(Y_u^* Y_u)^2 + 3(Y_d^* Y_d)^2\right), \\
c &= \mathrm{Tr}\left(Y_R^* Y_R\right), \\
d &= \mathrm{Tr}\left((Y_R^* Y_R)^2\right), \\
e &= \mathrm{Tr}\left(Y_R^* Y_R Y_\nu^* Y_\nu\right).
\end{aligned} \tag{27}
$$

The normalization of the kinetic terms imposes a relation between the coupling constants g_1, g_2, g_3 and the coefficients χ_0 of the form

$$
\frac{\chi(0)}{2\pi^2} g_3^2 = \frac{\chi(0)}{2\pi^2} g_2^2 = \frac{5\chi(0)}{6\pi^2} g_1^2 = \frac{1}{4}. \tag{28}
$$

The coupling constants are then related by

$$
g_3^2 = g_2^2 = \frac{5}{3} g_1^2,
$$

which is precisely the relation between the coupling constants at unification, common to grand unified theories (GUT). We shall further discuss this in Section 4.6.

4.5 Fermionic action in KO-dimension 6

As already announced above, the shift to KO-dimension 6 for the finite space solved the fermion-doubling problem of [63]. Let us briefly explain how this works, following [31].

The crucial observation is that in KO-dimension $2 \equiv 4+6 \mod 8$ the following pairing

$$(\psi, \psi') \mapsto (J\psi, D_\omega \psi')$$

is a skew-symmetric form on the $+1$-eigenspace of γ in $\mathcal{H}$. This skew-symmetry is in concordance with the Grassmann nature of fermionic fields ψ, guaranteeing that the following action functional is in fact non-zero:

$$S_F = \frac{1}{2}\langle J\xi, D_A \xi\rangle$$

for ξ a Grassmann variable in the $+1$-eigenspace of γ.

This then solves the fermion doubling, or actually quadrupling as follows. First, the restriction to the chiral subspace of γ takes care of a factor of two. Then, the functional integral involving anti-commuting Grassmann variables delivers a Pfaffian, which takes care of a square root. That this indeed work has been worked out in full detail for the case of the Standard Model in [17, Section 4.4.1] or [75, Section 11.4].

4.6 Phenomenological consequences

The first phenomenological consequence one can derive from the spectral Standard Model is an upper bound on the mass of the top quark. In fact, the appearance of the constant a in both the fermionic and the bosonic action allows to derive

$$\mathrm{Tr}\left(m_\nu^* m_\nu + m_e^* m_e + 3m_u^* m_u + 3m_d^* m_d\right) = 2{g_2}^2 v^2 = 8{M_W}^2. \tag{29}$$

It is natural to assume that the mass m_{top} of the top quark is much larger than all other fermion masses, except possibly a Dirac mass that arises from the seesaw mechanism as was described above. If we write $m_\nu = \rho m_{\mathrm{top}}$, then the above relation would yield the constraint

$$m_{\mathrm{top}} \lesssim \sqrt{\frac{8}{3+\rho^2}} M_W. \tag{30}$$

The relations (28) between the coupling constants and $\chi(0)$ suggest that we have grand unification of the coupling constants. Moreover, from the action functional we see that the quartic Higgs coupling constant λ is related to $\chi(0)$ as well via

$$\lambda = 24\frac{b}{a^2}g_2^2.$$

Thus, the spectral Standard Model imposes relations between the coupling constants and bounds on the fermion masses. These relations were used in [17] as input at (or around) grand unification scale Λ_{GUT}, and then run down using one-loop renormalization group equations to "low energies" where falsifiable predictions were obtained.

In fact, the mass of the top quark can indeed be found to get an acceptable value, however, for the Higgs mass it was found that

$$167\,\mathrm{GeV} \le m_h \le 176\,\mathrm{GeV}\,.$$

Given that there were not much models in particle physics around that could produce falsifiable predictions, it is somewhat ironical that the first exclusion results on the mass of the Higgs that appeared in 2009 from Fermilab hit exactly this region. See Figure 1. And, of course, with the discovery of the Higgs at $m_h \approx 125.5\,\mathrm{GeV}$ in [1, 26] one could say that the spectral Standard Model was not in a particularly good shape at that time.

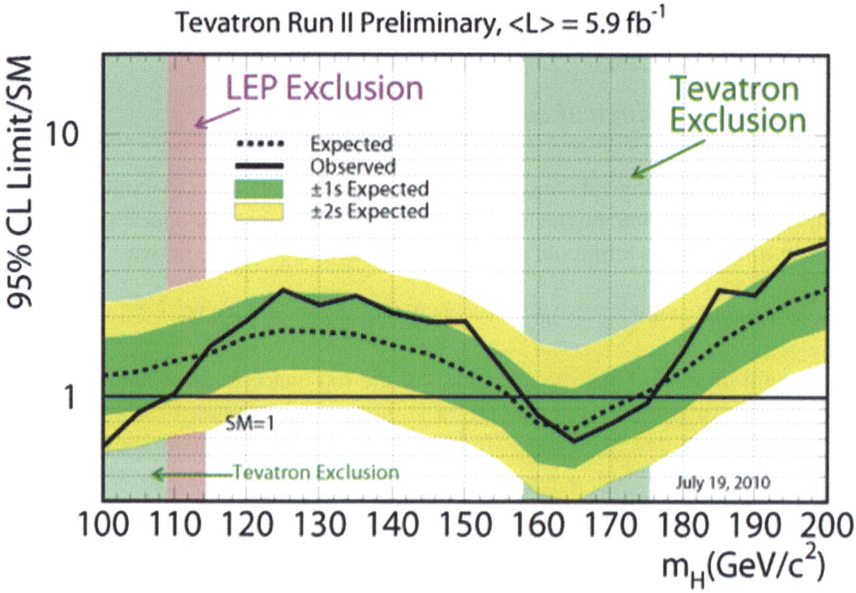

Fig. 1 Observed and expected exclusion limits for a Standard Model Higgs boson at the 95% confidence level for the combined CDF and DZero analyses (Fermilab)

5 Beyond the Standard Model with noncommutative geometry

Even though the incompatibility between the spectral Standard Model and the experimental discovery of the Higgs with a relatively low mass was not an easy stroke at the time, it also led to a period of reflection and reconsideration of the premises of the noncommutative geometric approach. In fact, it was the beginning of yet another exciting chapter in our story on the spectral model of gravity coupled with matter. As we will see in this and the next chapter, once again the input from experiment is taken as a guiding principle in our search for the spectral model that goes beyond the Standard Model.

Remark 5 We do not pretend to give a complete overview of the literature here, but only indicate some of the highlights and actively ongoing research areas.

Other searches beyond the Standard Model with noncommutative geometry include [53, 70, 71, 73, 72, 74], adopting a slightly different approach to almost-commutative manifolds as we do.

There is another aspect that was studied is the connection between supersymmetry and almost-commutative manifolds. It turned out to be very hard—if not impossible—to combine the two. A first approach is [13] and more recently the intersection was studied in [9, 10, 5].

5.1 Resilience of the spectral Standard Model

In 2012 it was realized how a small correction of the spectral Standard Model gives an intriguing possibility to go beyond the Standard Model, solving at the same time a problem with the stability of the Higgs vacuum given the measured low mass m_h. This is based on [16], but for which some of the crucial ingredients surprisingly enough were already present in the 2010 paper [15].

Namely, in the definition of the finite Dirac operator D_F of Equation (19), we can replace Y_R by $Y_R\sigma$, where σ is a real scalar field on M. Strictly speaking, this brings us out of the class of almost-commutative manifolds $M \times F$, since part of D_F now varies over M and this was the main reason why it was disregarded before. However, since from a physical viewpoint there was no reason to assume Y_R to be constant, it was treated as a scalar field already in [15]. This was only fully justified in subsequent papers (as we will see in the next subsections) where the scalar field σ arises as the relic of a spontaneous symmetry breaking mechanism, similar to the Higgs field h in the electroweak sector of the Standard Model. We will discuss a few of the existing approaches in the literature in the next few sections. For now, let us simply focus on the phenomenological consequences of this extra scalar field.

Thus we replace Y_R by $Y_R\sigma$ and analyze the additional terms in the spectral action. The scalar sector becomes

$$S'_H := \int_M \left(\frac{bf(0)}{2\pi^2}|H|^4 - \frac{2af_2\Lambda^2}{\pi^2}|H|^2 + \frac{ef(0)}{\pi^2}\sigma^2|H|^2 \right.$$
$$\left. - \frac{cf_2\Lambda^2}{\pi^2}\sigma^2 + \frac{df(0)}{4\pi^2}\sigma^4 + \frac{af(0)}{2\pi^2}|D_\mu H|^2 + \frac{1}{4\pi^2}f(0)c(\partial_\mu\sigma)^2 \right)\sqrt{g}dx,$$

where we ignored the coupling to the scalar curvature.

We exploit the approximation that m_{top}, m_ν, and m_R are the dominant mass terms. Moreover, as before we write $m_\nu = \rho m_{top}$. That is, the expressions for a, b, c, d, and e in (27) now become

$$a \approx m_{top}^2(\rho^2 + 3),$$
$$b \approx m_{top}^4(\rho^4 + 3),$$
$$c \approx m_R^2,$$
$$d \approx m_R^4,$$
$$e \approx \rho^2 m_R^2 m_{top}^2.$$

In a unitary gauge, where $H = \begin{pmatrix} h \\ 0 \end{pmatrix}$, we arrive at the following potential:

$$\mathcal{L}_{pot}(h, \sigma) = \frac{1}{24}\lambda_h h^4 + \frac{1}{2}\lambda_{h\sigma} h^2\sigma^2 + \frac{1}{4}\lambda_\sigma\sigma^4 - \frac{4g_2^2}{\pi^2} f_2\Lambda^2(h^2 + \sigma^2),$$

where we have defined coupling constants

$$\lambda_h = 24\frac{\rho^4 + 3}{(\rho^2 + 3)^2} g_2^2, \qquad \lambda_{h\sigma} = \frac{8\rho^2}{\rho^2 + 3} g_2^2, \qquad \lambda_\sigma = 8g_2^2. \tag{31}$$

This potential can be minimized, and if we replace h by $v + h$ and σ by $w + \sigma$, respectively, expanding around a minimum for the terms quadratic in the fields, we obtain:

$$\mathcal{L}_{pot}(v + h, w + \sigma)|_{\text{quadratic}} = \frac{1}{6}v^2\lambda_h v^2 + 2vw\lambda_{h\sigma}\sigma h + w^2\lambda_\sigma\sigma^2$$
$$= \frac{1}{2}\begin{pmatrix} h & \sigma \end{pmatrix} M^2 \begin{pmatrix} h \\ \sigma \end{pmatrix},$$

where we have defined the mass matrix M by

$$M^2 = 2\begin{pmatrix} \frac{1}{6}\lambda_h v^2 & \lambda_{h\sigma} vw \\ \lambda_{h\sigma} vw & \lambda_\sigma w^2 \end{pmatrix}.$$

This mass matrix can be easily diagonalized, and if we make the natural assumption that w is of the order of m_R, while v is of the order of M_W, so that $v \ll w$, we find that the two eigenvalues are

$$m_+^2 \sim 2\lambda_\sigma w^2 + 2\frac{\lambda_{h\sigma}^2}{\lambda_\sigma} v^2,$$

$$m_-^2 \sim 2\lambda_h v^2 \left(\frac{1}{6} - \frac{\lambda_{h\sigma}^2}{\lambda_h \lambda_\sigma}\right).$$

We can now determine the value of these two masses by running the scalar coupling constants λ_h, $\lambda_{h\sigma}$, and λ_σ down to ordinary energy scalar using the renormalization group equations for these couplings that were derived in [51], referring to [16, 75] for full details. The result varies with the chosen value for Λ_{GUT} and the parameter ρ. The mass of σ is essentially given by the largest eigenvalue m_+ which is of the order 10^{12} GeV for all values of Λ_{GUT} and the parameter ρ. The allowed mass range for the Higgs, i.e. for m_-, is depicted in Figure 2. The expected value $m_h = 125.5$ GeV is therefore compatible with the above noncommutative model. Moreover, without the σ the λ_h turns negative at energies around 10^{12} GeV. Furthermore, this calculation implies that there is a relation (given by the red line in the figure) between the ratio m_ν / m_{top} and the unification scale Λ_{GUT}.

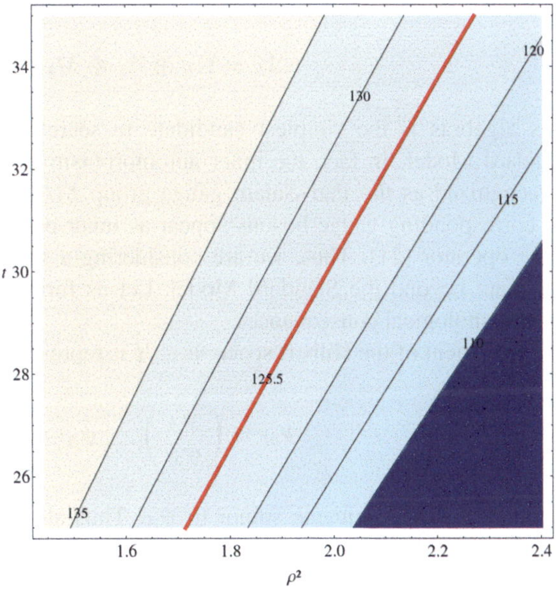

Fig. 2 A contour plot of the Higgs mass m_h as a function of ρ^2 and $t = \log(\Lambda_{GUT}/M_Z)$. The red line corresponds to $m_h = 125.5$ GeV

5.2 Pati–Salam unification and first-order condition

In order to see how we one can use the noncommutative geometric approach to go beyond the Standard Model it is important to trace our steps that led to the spectral Standard Model in the previous section. The route started with the classification of the algebras of the finite space (cf. Equation (17)). The results show that the only algebras which solve the fermion doubling problem are of the form $M_{2a}(\mathbb{C}) \oplus M_{2a}(\mathbb{C})$ where a is an even integer. An arbitrary symplectic constraint is imposed on the first algebra restricting it from $M_{2a}(\mathbb{C})$ to $M_a(\mathbb{H})$. The first non-trivial algebra one can consider is for $a = 2$ with the algebra

$$M_2(\mathbb{H}) \oplus M_4(\mathbb{C}). \tag{32}$$

Coincidentally, and as explained in the introduction, the above algebra comes out as a solution of the two-sided Heisenberg quantization relation between the Dirac operator D and the two maps from the four spin-manifold and the two four spheres $S^4 \times S^4$ [18, 19]. This removes the arbitrary symplectic constraint and replaces it with a relation that quantize the four-volume in terms of two quanta of geometry and have far reaching consequences on the structure of space-time. We will come back to this in the last section.

The existence of the chirality operator γ that commutes with the algebra breaks the quaternionic matrices $M_2(\mathbb{H})$ to the diagonal subalgebra and leads us to consider the finite algebra

$$\mathcal{A}_F = \mathbb{H}_R \oplus \mathbb{H}_L \oplus M_4(\mathbb{C}). \tag{33}$$

This algebras is the simplest candidate to search for new physics beyond the Standard Model. In fact, the inner automorphism group of $\mathcal{A} = C^\infty(M) \otimes \mathcal{A}_F$ is recognized as the Pati–Salam gauge group $SU(2)_R \times SU(2)_L \times SU(4)$, and the corresponding gauge bosons appear as inner perturbations of the (space-time) Dirac operator [21]. Thus, we are considering a spectral Pati–Salam model as a candidate beyond the Standard Model. Let us further analyze this model and its phenomenological consequences.

An element of the Hilbert space $\Psi \in \mathcal{H}$ is represented by

$$\Psi_M = \begin{pmatrix} \psi_A \\ \psi_{A'} \end{pmatrix}, \quad \psi_{A'} = \psi_A^c \tag{34}$$

where ψ_A^c is the conjugate spinor to ψ_A. Thus all primed indices A' correspond to the Hilbert space of conjugate spinors. It is acted on by both the left algebra $M_2(\mathbb{H})$ and the right algebra $M_4(\mathbb{C})$. Therefore, the index A can take 16 values and is represented by

$$A = \alpha I \tag{35}$$

where the index α is acted on by quaternionic matrices and the index I by $M_4(\mathbb{C})$ matrices. Moreover, when the grading breaks $M_2(\mathbb{H})$ into $\mathbb{H}_R \oplus \mathbb{H}_L$ the index α is decomposed to $\alpha = \dot{a}, a$ where $\dot{a} = \dot{1}, \dot{2}$ (dotted index) is acted on by the first quaternionic algebra $\mathbb{H}_R$ and $a = 1, 2$ is acted on by the second quaternionic algebra $\mathbb{H}_L$. When $M_4(\mathbb{C})$ breaks into $\mathbb{C} \oplus M_3(\mathbb{C})$ (due to symmetry breaking or through the use of the order one condition as in [14]) the index I is decomposed into $I = 1, i$ and thus distinguishing leptons and quarks, where the 1 is acted on by the $\mathbb{C}$ and the i by $M_3(\mathbb{C})$. Therefore, the various components of the spinor ψ_A are

$$\begin{aligned}\psi_{\alpha I} &= \begin{pmatrix} \nu_R & u_{iR} & \nu_L & u_{iL} \\ e_R & d_{iR} & e_L & d_{iL} \end{pmatrix}, \qquad i = 1, 2, 3 \\ &= \left(\psi_{\dot{a}1}, \psi_{\dot{a}i}, \psi_{a1}, \psi_{ai}\right), \qquad a = 1, 2, \quad \dot{a} = \dot{1}, \dot{2}\end{aligned} \tag{36}$$

This is a general prediction of the spectral construction that there is 16 fundamental Weyl fermions per family, 4 leptons and 12 quarks.

The (finite) Dirac operator can be written in matrix form

$$D_F = \begin{pmatrix} D_A^B & D_A^{B'} \\ D_{A'}^B & D_{A'}^{B'} \end{pmatrix}, \tag{37}$$

and must satisfy the properties

$$\gamma_F D_F = -D_F \gamma_F \qquad J_F D_F = D_F J_F \tag{38}$$

where $J_F^2 = 1$. A matrix realization of γ_F and J_F are given by

$$\gamma_F = \begin{pmatrix} G_F & 0 \\ 0 & -\overline{G}_F \end{pmatrix}, \qquad G_F = \begin{pmatrix} 1_2 & 0 \\ 0 & -1_2 \end{pmatrix}, \qquad J_F = \begin{pmatrix} 0_4 & 1_4 \\ 1_4 & 0_4 \end{pmatrix} \circ \mathrm{cc} \tag{39}$$

where cc stands for complex conjugation. These relations, together with the hermiticity of D, imply the relations

$$(D_F)_{A'}^{B'} = \left(\overline{D}_F\right)_A^B \qquad (D_F)_{A'}^{B} = \left(\overline{D}_F\right)_B^{A'} \tag{40}$$

and have the following zero components [15]:

$$(D_F)_{aI}^{bJ} = 0 = (D_F)_{\dot{a}I}^{\dot{b}J} \tag{41}$$

$$(D_F)_{aI}^{\dot{b}'J'} = 0 = (D_F)_{\dot{a}I}^{b'J'} \tag{42}$$

leaving the components $(D_F)_{aI}^{\dot{b}J}$, $(D_F)_{aI}^{b'J'}$, and $(D_F)_{\dot{a}I}^{\dot{b}'J'}$ arbitrary. These restrictions lead to important constraints on the structure of the connection that appears in the inner fluctuations of the Dirac operator. In particular, the operator D of the full noncommutative space given by

$$D = D_M \otimes 1 + \gamma_5 \otimes D_F \tag{43}$$

gets modified to

$$D_A = D + A_{(1)} + J A_{(1)} J^{-1} + A_{(2)} \tag{44}$$

where

$$A_{(1)} = \sum a \left[D, b\right], \qquad A_2 = \sum \widehat{a} \left[A_{(1)}, \widehat{b}\right], \qquad \widehat{a} = J a J^{-1} \tag{45}$$

We have shown in [21] that components of the connection A which are tensored with the Clifford gamma matrices γ^{μ} are the gauge fields of the Pati–Salam model with the symmetry of $SU(2)_R \times SU(2)_L \times SU(4)$. On the other hand, the non-vanishing components of the connection which are tensored with the gamma matrix γ_5 are given by

$$(A)_{aI}^{\dot{b}J} \equiv \gamma_5 \Sigma_{aI}^{\dot{b}J}, \qquad (A)_{aI}^{b'J'} = \gamma_5 H_{aIbJ}, \qquad (A)_{\dot{a}I}^{\dot{b}'J'} \equiv \gamma_5 H_{\dot{a}I\dot{b}J} \tag{46}$$

where $H_{aIbJ} = H_{bJaI}$ and $H_{\dot{a}I\dot{b}J} = H_{\dot{b}J\dot{a}I}$, which is the most general Higgs structure possible. These correspond to the representations with respect to $SU(2)_R \times SU(2)_L \times SU(4)$:

$$\Sigma_{aI}^{\dot{b}J} = (2_R, 2_L, 1) + (2_R, 2_L, 15) \tag{47}$$

$$H_{aIbJ} = (1_R, 1_L, 6) + (1_R, 3_L, 10) \tag{48}$$

$$H_{\dot{a}I\dot{b}J} = (1_R, 1_L, 6) + (3_R, 1_L, 10) \tag{49}$$

We note, however, that the inner fluctuations form a semi-group and if a component $(D_F)_{aI}^{\dot{b}J}$ or $(D_F)_{aI}^{b'J'}$ or $(D_F)_{\dot{a}I}^{\dot{b}'J'}$ vanish, then the corresponding A field will also vanish. We can distinguish three cases: (1) Left-right symmetric Pati–Salam model with fundamental Higgs fields $\Sigma_{aI}^{\dot{b}J}$, H_{aIbJ}, and $H_{\dot{a}I\dot{b}J}$. In this model the field H_{aIbJ} should have a zero vev. (2) A Pati–Salam model where the Higgs field H_{aIbJ} that couples to the left sector is set to zero which is desirable because there is no symmetry between the left and right sectors at low energies. (3) If one starts with $(D_F)_{aI}^{\dot{b}J}$ or $(D_F)_{aI}^{b'J'}$ or $(D_F)_{aI}^{\dot{b}'J'}$ whose values are given by those that were derived for the Standard Model, then the Higgs fields $\Sigma_{aI}^{\dot{b}J}$, H_{aIbJ}, and $H_{\dot{a}I\dot{b}J}$ will become composite and expressible in terms of more fundamental fields Σ_I^J, $\Delta_{\dot{a}J}$,

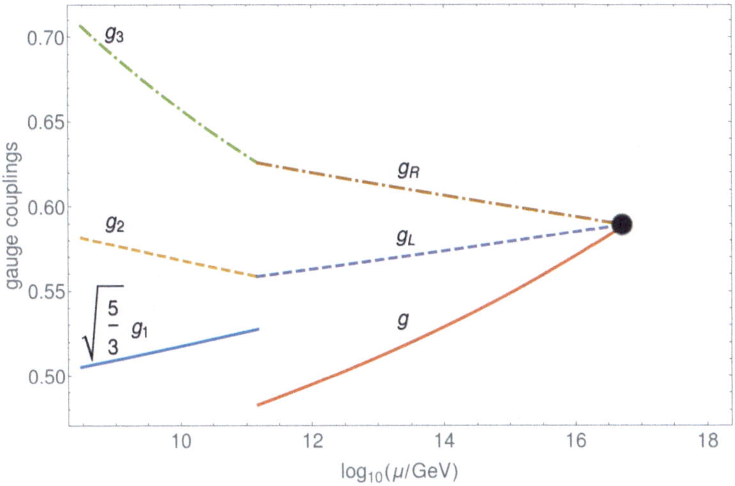

Fig. 3 Running of the gauge couplings of the Standard Model gauge couplings (below scale $m_R \approx 10^{11}$ GeV) and the Pati–Salam gauge coupling (above scale m_R) in case 2

and $\phi^b_{\dot{a}}$. We refer to this as the composite model. It has the scalar field σ discussed in the previous section as a remnant after spontaneous symmetry breaking [21]. In fact, contrary to some claims in the literature it is possible to perform the potential analysis in this case in unitarity gauge and arrive at the conclusion that the field content contains the scalar field σ (cf. Appendix).

Depending on the precise particle content we may determine the renormalization group equations of the Pati–Salam gauge couplings g_R, g_L, g. In [22] we have run them to look for unification of the coupling $g_R = g_L = g$. The boundary conditions are taken at the intermediate mass scale $\mu = m_R$ to be the usual (e.g., [67, Eq. (5.8.3)])

$$\frac{1}{g_1^2} = \frac{2}{3}\frac{1}{g^2} + \frac{1}{g_R^2}, \qquad \frac{1}{g_2^2} = \frac{1}{g_L^2}, \qquad \frac{1}{g_3^2} = \frac{1}{g^2}, \tag{50}$$

in terms of the Standard Model gauge couplings g_1, g_2, g_3. At the mass scale m_R the Pati–Salam symmetry is broken to that of the Standard Model, and we take it to be the same scale that is present in the seesaw mechanism. It should thus be of the order 10^{11}–10^{13} GeV. What we have found in [22] (and this was confirmed by others in [3]) is that in all three cases it is possible to achieve grand unification of the couplings, while connecting to Standard Model physics in the broken, low-energy phase. An example of a running of the gauge coupling is illustrated in Figure 3.

5.3 Grand symmetry and twisted spectral triples

In [40] the next-to-next case[3] in the list of irreducible geometries in Equation (17) was considered: $k = 4$. Thus, one considers

$$A_G = M_4(\mathbb{H}) \oplus M_8(\mathbb{C}); \qquad H_F := \mathbb{C}^{128}. \tag{51}$$

where 128 is exactly the number of spinor and internal degrees of freedom combined (including the aforementioned fermion quadruplication). The geometry is then

$$(C^\infty(M, A_G), L^2(M) \otimes H_F, D_M + \gamma_M D_F)$$

where one has to assume that the spinor bundle on M has been trivialized to gather the spinor and internal fermionic degrees of freedom in a single Hilbert space H_F.

Note that the above geometry is not a direct product of the continuum with a discrete space. In fact, both the algebra and the Dirac operator D_M contain spinor indices. As a consequence the commutator $[D_M, a]$ can become unbounded, thus challenging one of the basic axioms of spectral triples. Instead, it is possible to guarantee that *twisted* commutators are bounded so that this example fits in the general framework of twisted spectral triples developed in [34]. In [41] the authors identify an inner automorphism $\rho = R(\cdot)R$ of A_G such that

$$[D, a]_\rho = Da - \rho(a)D$$

is bounded.

An interesting question that arises at this point is how to generate inner fluctuations of twisted spectral triples. This was analyzed in full detail from a mathematical viewpoint in [61, 62]. One of the intriguing aspects is the self-adjointness of the Dirac operator under fluctuations (even gauge transformations): for this to be respected one has to impose a compatibility between the twist and the fluctuation.

An alternative route was suggested in [39]. Namely, one may drop the above condition of self-adjointness and instead look for operators that are Krein-self-adjoint, using the Krein structure on the Hilbert space that is induced by the operator R (defining the twist ρ). This will have an intriguing appearance of the Lorentzian structure (given by the Krein inner product) from a purely algebraic and Euclidean starting point. Here we also refer to the nice overview given in [64].

[3] The case $k = 3$ was ruled out by physical considerations [40].

5.4 Algebraic constraints on the finite geometry

An interesting question to consider—in particular in light of theories that go beyond the Standard Model—is whether one can *derive* the restricted form of the Dirac operator D_F in (19). We highlight a few approaches to this question that are present in the literature.

First of all, as mentioned already on page 18, the form of the D_F in terms of the matrices Y_ν, Y_e, Y_u, Y_d, and Y_R as in Equations (21) and (22) appears naturally in the study of moduli of finite Dirac operators. The only constraint (in addition to the usual conditions laid out in Section 3.1) there was that the photon remained massless.

An attempt was made to make the latter condition less ad hoc is [49, 7, 8]. They proposed to generalize noncommutative geometry to non-associative noncommutative geometry, thus allowing for non-associative algebras. The crucial idea—which goes back to Eilenberg—is to combine the (differential) algebra and (Hilbert space) bimodule into a single algebra, and understand the conditions such as commutant property and first-order conditions as consequences of associativity of the pertinent algebra B. However, this associativity is a strong constraint and accordingly further restrict the geometry described by D_F. Note that non-associative algebras have also been used in the context of noncommutative geometry and particle physics to predict the number of families (to be three) [76].

Another approach to analyzing the form of the Dirac operator D_F by imposing algebraic conditions is taken by Dabrowski et al. [35, 36]. Here the authors adopt the principle that, similar to differential forms in the continuum, the finite Hilbert space should be a Morita equivalence between A and the Clifford algebra generated by A_F and D_F. One finds that the aforementioned form of D_F does not satisfy this condition but additional entries in D_F should be non-zero. This gives rise to a model beyond the Standard Model: an analysis of the phenomenological consequences is performed in [59, 37]. In [2] it was then found that this model does not exhibit grand unification of the Standard Model couplings.

6 Volume quantization and uniqueness of SM

In the classification of finite noncommutative spaces we arrived at the result that the algebra $\mathcal{A}_F = (\mathbb{H}_R \oplus \mathbb{H}_L) \oplus M_4(\mathbb{C})$ was the first possibility out of many of the form $\mathcal{A}_F = (M_n(\mathbb{H})_R \oplus M_n(\mathbb{H})_L) \oplus M_{4n}(\mathbb{C})$. In addition we made an assumption, that seemed arbitrary, of the existence of antilinear isometry that reduced the algebra $M_{4n}(\mathbb{C})$ to $(M_n(\mathbb{H})_R \oplus M_n(\mathbb{H})_L)$. It is necessary to have a stronger evidence of the uniqueness of our conclusions that helps us to avoid making the abovementioned assumptions. Surprisingly, the new evidence came in the process of solving a seemingly completely independent problem, encoding low dimensional geometries, and in particular dimension four.

6.1 Higher form of Heisenberg's commutation relations

Starting with the simple example of one-dimensional geometries, consider the equation

$$U^* [D, U] = 1, \qquad U^* U = 1$$

where D is self-adjoint operator. Assuming that the one-dimensional space is a closed curve parameterized by coordinate x and the Dirac operator to be $D = -i\frac{d}{dx} + \alpha$ the above equation simplifies to

$$-iU^* dU = dx$$

Writing $U = e^{in\theta}$ we obtain $dx = nd\theta$. Integrating both sides implies that the length of the one-dimensional curve is an integer multiple of 2π, the length of S^1

$$\oint_C dx = n\,(2\pi)$$

To adopt this construction to higher dimensions, we note that we can characterize the circle S^1 by the equation $Y^A Y^A = 1$, $A = 1, 2$, $Y^{A*} = Y^A$. Assembling the two coordinates Y^1, Y^2 in one matrix, define $Y = Y^A \Gamma_A$, where Γ_A, $A = 1, 2$ are taken to be 2×2. In addition we identify $\Gamma_1 = \sigma_1$, $\Gamma_2 = \sigma_2$, the Pauli matrices, and define $\Gamma = -i\Gamma_1\Gamma_2 = \sigma_3$ so that $\Gamma_+ = \frac{1}{2}(1 + \Gamma)$ is a projection operator. We notice that we can write

$$Y = \begin{pmatrix} 0 & Y^1 - iY^2 \\ Y^1 + iY^2 & 0 \end{pmatrix} = \begin{pmatrix} 0 & U^* \\ U & 0 \end{pmatrix}$$

where $U = Y^1 - iY^2$ and $U^*U = 1$. The expression

$$\langle \Gamma_+ Y [D, Y] \rangle = 1$$

where $\langle\rangle$ is defined to be the trace over the Clifford algebra defined by Γ_A, gives back the equation $U^* [D, U] = 1$. For higher dimensional geometries we consider a Riemannian manifold with dimension n and where the algebra $\mathcal{A}$ is taken to be $C^\infty(M)$, the algebra of continuously differentiable functions, while the operator D is identified with the Dirac operator given by

$$D_M = \gamma^\mu \left(\frac{\partial}{\partial x^\mu} + \omega_\mu \right),$$

where $\gamma^\mu = e_a^\mu \gamma^a$ and $\omega_\mu = \frac{1}{4}\omega_{\mu bc}\gamma^{bc}$ is the $SO(n)$ Lie-algebra valued spin-connection with the (inverse) vielbein e_a^μ being the square root of the (inverse) metric $g^{\mu\nu} = e_a^\mu \delta^{ab} e_b^\nu$. The gamma matrices γ^a are anti-Hermitian $(\gamma^a)^* = -\gamma^a$ that define the Clifford algebra $\left\{\gamma^a, \gamma^b\right\} = -2\delta^{ab}$. The Hilbert space $\mathcal{H}$ is the space of square integrable spinors $L^2\,(M, S)$. The chirality operator γ in even dimensions is then given by

$$\gamma = (i)^{\frac{n}{2}}\, \gamma^1 \gamma^2 \cdots \gamma^n$$

Starting with manifolds of dimension 2 we first define the two sphere by the equation $Y^A Y^A = 1$, $A = 1, 2, 3$, $Y^{A*} = Y^A$. Assembling the three coordinates Y^1, Y^2, Y^3 in one matrix, defining $Y = Y^A \Gamma_A$, where Γ_A, $A = 1, 2, 3$ are taken to be 2×2 Pauli matrices. Notice that in this case $\Gamma \equiv -i\Gamma_1\Gamma_2\Gamma_3 = 1$ and to generalize Equation (6.1) to two dimensions the factor Γ can be dropped, and we write instead

$$\frac{1}{2!}\left\langle Y\,[D, Y]^2\right\rangle = \gamma$$

The reason we have to include the chirality operator γ on the two- dimensional manifold M is that the Dirac operator D appears twice yielding a product of the form $\gamma_1\gamma_2 = -i\gamma$. A simple calculation shows that the above equation in component form is given by

$$\frac{1}{2!}\varepsilon^{\mu\nu}\varepsilon_{ABC} Y^A \partial_\mu Y^B \partial_\nu Y^C = \det\left(e_\mu^a\right)$$

which is a constraint on the volume form of M_2. This implies that the volume of M_2 will be an integer multiple of the area of the unit 2-sphere

$$\int_{M_2} d^2x\sqrt{g} = \int \varepsilon_{ABC} Y^A dY^B dY^C$$
$$= n(4\pi)$$

where n is the winding number. An example of a map Y with winding number n is

$$Y \equiv Y^1 + iY^2 = \frac{2z^n}{|z|^{2n} + 1}, \qquad Y^3 = \frac{|z|^{2n} - 1}{|z|^{2n} + 1}, \qquad z = x^1 + ix^2$$

From this we deduce that the pullback $Y^*\,(w_n)$ is a differential form that does not vanish anywhere. This in turn implies that the Jacobian of the map Y does not vanish anywhere, and that Y is a covering of the sphere. The sphere is simply connected, and on each connected component $M_j \subset M_n$, the restriction of the map Y to M_j is a diffeomorphism, implying that the manifold must be disconnected, with each piece having the topology of a sphere. To allow for two-dimensional manifolds with arbitrary topology, our first observation is that condition (6.1) involves the commutator of the Dirac operator D and the coordinates Y. In momentum space D is the Feynman-slashed $\gamma^\mu p_\mu$ momentum and Y are the Feynman-slashed coordinates. This suggests that the quantization condition is a higher form of Heisenberg commutation relation quantizing the phase space formed by coordinates and momenta. We first notice that although the quantization condition is given in terms of the noncommutative data, the operator J is the only one missing. We therefore modify the condition to take J into account. The operator J transforms Y into its commutant $Y' = iJYJ^{-1}$ so that $\left[Y, Y'\right] = 0$. Thus let $Y = Y^A\Gamma_A$ and $Y' = iJYJ^{-1}$ and $\Gamma'_A = iJ\Gamma_A J^{-1}$ so that we can write

$$Y = Y^A\Gamma_A, \qquad Y' = Y'^A\Gamma'_A,$$

satisfying $Y^2 = 1$ and $Y'^2 = 1$ with the Clifford algebras $C_\pm$

$$\{\Gamma_A, \Gamma_B\} = 2\,\delta_{AB}, \qquad (\Gamma_A)^* = \Gamma_A \tag{52}$$

$$\left\{\Gamma'_A, \Gamma'_B\right\} = -2\,\delta_{AB}, \qquad (\Gamma'_A)^* = -\Gamma'_A \tag{53}$$

We immediately see that the Clifford algebra $C_+ = M_2(\mathbb{C})$ and $C_- = \mathbb{H}$. We then define the projection operator $e = \frac{1}{2}(1+Y)$ satisfying $e^2 = e$ and similarly $e' = \frac{1}{2}\left(1+Y'\right)$ satisfying $e'^2 = e'$. From the tensor product of $E = ee'$ satisfying $E^2 = E$, we construct $Z = 2E - 1$ satisfying $Z^2 = 1$ and allowing us to write

$$\frac{1}{2}\left\langle Z\,[D,Z]^2\right\rangle = \gamma$$

A straightforward calculation reveals that this relation splits as the sum of two non-interfering parts

$$\frac{1}{2}\left\langle Y\,[D,Y]^2\right\rangle + \frac{1}{2}\left\langle Y'\left[D,Y'\right]^2\right\rangle = \gamma$$

which in component form reads

$$\frac{1}{2!}\varepsilon^{\mu\nu}\varepsilon_{ABC}\left(Y^A\partial_\mu Y^B\partial_\nu Y^C + Y'^A\partial_\mu Y'^B\partial_\nu Y'^C\right) = \det\left(e^a_\mu\right)$$

We will show later, when considering the four-dimensional case, that this modification allows to reconstruct two-dimensional manifolds of arbitrary topology from the pullbacks of the maps Y, Y'.

For three-dimensional manifolds $\gamma = 1$ and in analogy with the one-dimensional case we write

$$\frac{1}{3!}\left\langle \Gamma_+ Y\,[D,Y]^3\right\rangle = 1$$

where $Y = Y^A\Gamma_A$, $A = 1, \dots 4$, $Y^2 = 1$, $Y = Y^*$, Γ_A are 4×4 Clifford algebra matrices C_+ where $\{\Gamma_A, \Gamma_B\} = 2\,\delta_{AB}$. In this representation of the Γ matrices we have $\Gamma = \Gamma_5 = \Gamma_1\Gamma_2\Gamma_3\Gamma_4 = \begin{pmatrix} 1_2 & 0 \\ 0 & -1_2 \end{pmatrix}$ so that $\Gamma_+ = \frac{1}{2}(1+\Gamma)$ is a projection operator. In $d = 3$, we can write

$$Y = Y^A\Gamma_A = \begin{pmatrix} 0 & U^* \\ U & 0 \end{pmatrix}$$

where U is a unitary 2×2 matrix such that it could be written in the form $U = \exp\left(i\left(\alpha_0 1 + \alpha_a \sigma^a\right)\right)$ so that $U^*U = 1$. It is easy to check that $\left\langle Y\,[D, Y]^3\right\rangle = 0$ and that the component form of the above relation is

$$\det\left(e_\mu^a\right) = \frac{1}{3!}\varepsilon^{\mu\nu\rho}\operatorname{Tr}\left(U^*\partial_\mu U U^*\partial_\nu U U^*\partial_\rho U\right)$$

whose integral is the winding number of the $SU(2)$ group manifold. Again, using the reality operator J we act on the Clifford algebra $Y' = iJYJ^{-1}$ so that $\left[Y, Y'\right] = 0$, then $\Gamma'_A = iJ\Gamma_A J^{-1}$ satisfies $\left\{\Gamma'_A, \Gamma'_B\right\} = -2\,\delta_{AB}$, $(\Gamma'_A)^* = -\Gamma'_A$. Forming the projection operators $e = \frac{1}{2}(1+Y)$, $e' = \frac{1}{2}\left(1+Y'\right)$, we form the tensor product $E = ee'$ we define the field $Z = 2E - 1$, and thus the two sided relation becomes

$$\frac{1}{3!}\left\langle \Gamma_+\Gamma'_+ Z\,[D, Z]^3\right\rangle = 1$$

A lengthy calculation shows that the component form of this relation separates into two parts without interference terms

$$\det\left(e_\mu^a\right) = \frac{1}{3!}\varepsilon^{\mu\nu\rho}\Big(\operatorname{Tr}\left(U^*\partial_\mu U U^*\partial_\nu U U^*\partial_\rho U\right) + \operatorname{Tr}\left(U'^*\partial_\mu U' U'^*\partial_\nu U' U'^*\partial_\rho U'\right)\Big)$$

Finally, for four-dimensional manifolds the Clifford algebras C_+ and C_- defined as in (52) (53) with Γ_A, Γ'_A, $A = 1, \dots, 5$ are known to be given by $C_+ = M_2(\mathbb{H})$ and $C_- = M_4(\mathbb{C})$. The quantization condition takes the same form as the two-dimensional case

$$\frac{1}{4!}\left\langle Z\,[D, Z]^4\right\rangle = \gamma \tag{54}$$

This relation separates into two non-interfering terms

$$\frac{1}{4!}\left\langle Y\,[D, Y]^4\right\rangle + \frac{1}{4!}\left\langle Y'\left[D, Y'\right]^4\right\rangle = \gamma$$

the component form of which is given by

$$\det\left(e_{\mu}^{a}\right) = \frac{1}{4!}\varepsilon^{\mu\nu\kappa\lambda}\varepsilon_{ABCDE}\Big(Y^{A}\partial_{\mu}Y^{B}\partial_{\nu}Y^{C}\partial_{\kappa}Y^{D}\partial_{\lambda}Y^{E} + Y'^{A}\partial_{\mu}Y'^{B}\partial_{\nu}Y'^{C}\partial_{\kappa}Y'^{D}\partial_{\lambda}Y'^{E}\Big)$$

One can verify that similar considerations fail when the dimension of the manifold $n > 4$ as there are interference terms between the Y and Y'. Integrating both sides imply

$$\int_{M_4} d^4x\sqrt{g} = \frac{8}{3}\pi^2\left(N + N'\right)$$

where N, N' are the winding numbers of the two maps Y, Y'. An example of a map Y with winding number n is given by

$$Y \equiv Y^4 1 + Y^i e_i = \frac{2x^n}{x^n\overline{x}^n + 1},$$

$$Y^5 = \frac{x^n\overline{x}^n - 1}{x^n\overline{x}^n + 1},$$

where $x = x^4 1 + x^i e_i$ and e_i, $i = 1, 2, 3$ are the quaternionic complex structures $e_i^2 = -1$, $e_i e_j = \varepsilon_{ijk} e_k$, $i \neq j$.

6.2 Volume quantization

Consider the smooth maps $\phi_\pm : M_n \to S^n$ then their pullbacks $\phi_\pm^*$ would satisfy

$$\phi_+^*(\alpha) + \phi_-^*(\alpha) = \omega, \tag{55}$$

where α is the volume form on the unit sphere S^n and $\omega(x)$ is an n-form that does not vanish anywhere on M_n. We have shown that for a compact connected smooth oriented manifold with $n < 4$ one can find two maps $\phi_+^*(\alpha)$ and $\phi_-^*(\alpha)$ whose sum does not vanish anywhere, satisfying equation (55) such that $\int \omega \in \mathbb{Z}$. The proof for $n = 4$ is more difficult and there is an obstruction unless the second Stiefel–Whitney class w_2 vanishes, which is satisfied if M is required to be a spin-manifold and the volume to be larger than or equal to five units. The key idea in the proof is to note that the kernel of the Jacobian of the map Y is a hypersurface Σ of co-dimension 2 and therefore

$$\dim \Sigma = n - 2.$$

We can then construct a map $Y' = Y \circ \psi$ where ψ is a diffeomorphism on M such that the sum of the pullbacks of Y and Y' does not vanish anywhere. The coordinates Y are defined over a Clifford algebra C_+ spanned by $\{\Gamma_A, \Gamma_B\} = 2\delta_{AB}$. For $n = 2$, $C_+ = M_2(\mathbb{C})$ while for $n = 4$, $C_+ = M_2(\mathbb{H}) \oplus M_2(\mathbb{H})$ where $\mathbb{H}$ is the field of quaternions. However, for $n = 4$, since we will be dealing with irreducible representations we take $C_+ = M_2(\mathbb{H})$. Similarly the coordinates Y' are defined over the Clifford algebra C_- spanned by $\left\{\Gamma'_A, \Gamma'_B\right\} = -2\delta_{AB}$ and for $n = 2$, $C_- = \mathbb{H} \oplus \mathbb{H}$ and for $n = 4$, $C_- = M_4(\mathbb{C})$. The operator J acts on the two algebras $C_+ \oplus C_-$ in the form $J(x, y) = (y^*, x^*)$ (i.e., it exchanges the two algebras and takes the Hermitian conjugate). The coordinates $Z = \frac{1}{2}(Y + 1)\left(Y' + 1\right) - 1$ then define the matrix algebras [18]

$$\mathcal{A}_F = M_2(\mathbb{C}) \oplus \mathbb{H}, \qquad n = 2$$

$$\mathcal{A}_F = M_2(\mathbb{H}) \oplus M_4(\mathbb{C}), \qquad n = 4.$$

One, however, must remember that the maps Y and Y' are functions of the coordinates of the manifold M and therefore the algebra associated with this space must be

$$\begin{aligned} \mathcal{A} &= C^\infty(M, \mathcal{A}_F) \\ &= C^\infty(M) \otimes \mathcal{A}_F. \end{aligned}$$

To see this consider, for simplicity, the $n = 2$ case with only the map Y. The Clifford algebra $C_- = \mathbb{H}$ is spanned by the set $\left\{1, \Gamma^A\right\}$, $A = 1, 2, 3$, where $\left\{\Gamma^A, \Gamma^B\right\} = -2\delta^{AB}$. We then consider functions which are made out of words of the variable Y formed with the use of constant elements of the algebra [32]

$$\sum_{i=1}^{\infty} a_1 Y a_2 Y \cdots a_i Y, \qquad a_i \in \mathbb{H},$$

which will generate arbitrary functions over the manifold which is the most general form since $Y^2 = 1$. One can easily see that these combinations generate all the spherical harmonics. This result could be easily generalized by considering functions of the fields

$$Z = \frac{1}{2}(Y + 1)\left(Y' + 1\right) - 1, \qquad Y \in \mathbb{H}, \quad Y' \in M_2(\mathbb{C}),$$

showing that the noncommutative algebra generated by the constant matrices and the Feynman slash coordinates Z is given by [32]

$$\mathcal{A} = C^\infty(M_2) \otimes (\mathbb{H} + M_2(\mathbb{C})).$$

We now restrict ourselves to the physical case of $n = 4$. Here the algebra is given by

$$\mathcal{A} = C^{\infty}(M_4) \otimes (M_2(\mathbb{H}) + M_4(\mathbb{C})).$$

The associated Hilbert space is

$$\mathcal{H} = L^2(M_4, S) \otimes \mathcal{H}_F.$$

The Dirac operator mixes the finite space and the continuous manifold non-trivially

$$D = D_M \otimes 1 + \gamma_5 \otimes D_F,$$

where D_F is a self-adjoint operator in the finite space. The chirality operator is

$$\gamma = \gamma_5 \otimes \gamma_F,$$

and the anti-unitary operator J is given by

$$J = J_M \gamma_5 \otimes J_F,$$

where J_M is the charge-conjugation operator C on M and J_F the anti-unitary operator for the finite space. Thus an element $\Psi \in \mathcal{H}$ is of the form $\Psi = \begin{pmatrix} \psi_A \\ \psi_{A'} \end{pmatrix}$ where ψ_A is a 16 component $L^2(M, S)$ spinor in the fundamental representation of $\mathcal{A}_F$ of the form $\psi_A = \psi_{\alpha I}$ where $\alpha = 1, \ldots, 4$ with respect to $M_2(\mathbb{H})$ and $I = 1, \ldots, 4$ with respect to $M_4(\mathbb{C})$ and where $\psi_{A'} = C\psi_A^*$ is the charge conjugate spinor to ψ_A [15]. The chirality operator γ must commute with elements of $\mathcal{A}$ which implies that γ_F must commute with elements in $\mathcal{A}_F$. Commutativity of the chirality operator γ_F with the algebra $\mathcal{A}_F$ and that this $\mathbb{Z}/2$ grading acts non-trivially reduces the algebra $M_2(\mathbb{H})$ to $\mathbb{H}_R \oplus \mathbb{H}_L$ [18]. Thus the γ_F is identified with $\gamma_F = \Gamma^5 = \Gamma^1\Gamma^2\Gamma^3\Gamma^4$ and the finite space algebra reduces to

$$\mathcal{A}_F = \mathbb{H}_R \oplus \mathbb{H}_L \oplus M_4(\mathbb{C}).$$

This can be easily seen by noting that an element of $M_2(\mathbb{H})$ takes the form $\begin{pmatrix} q_1 & q_2 \\ q_3 & q_4 \end{pmatrix}$ where each q_i, $i = 1, \ldots, 4$, is a 2×2 matrix representing a quaternion. Taking the representation of $\Gamma^5 = \begin{pmatrix} 1_2 & 0 \\ 0 & -1_2 \end{pmatrix}$ to commute with $M_2(\mathbb{H})$ implies that $q_2 = 0 = q_3$, thus reducing the algebra to $\mathbb{H}_R \oplus \mathbb{H}_L$. Therefore the index $\alpha = 1, \ldots, 4$ splits into two parts, $\dot{a} = \dot{1}, \dot{2}$ which is a doublet under $\mathbb{H}_R$ and $a = 1, 2$ which is a doublet under $\mathbb{H}_L$. The spinor Ψ further satisfies the chirality condition $\gamma\Psi = \Psi$ which implies that the spinors $\psi_{\dot{a}I}$ are in the $(2_R, 1_L, 4)$ with respect to the algebra

$\mathbb{H}_R \oplus \mathbb{H}_L \oplus M_4(\mathbb{C})$ while ψ_{aI} are in the $(1_R, 2_L, 4)$ representation. The finite space Dirac operator D_F is then a 32×32 Hermitian matrix acting on the 32 component spinors Ψ. In addition we take three copies of each spinor to account for the three families, but will omit writing an index for the families. At present we have no explanation for why the number of generations should be three. The Dirac operator for the finite space is then a 96×96 Hermitian matrix. The Dirac action is then given by [17]

$$(J\Psi, D\Psi).$$

We note that we are considering compact spaces with Euclidean signature and thus the condition $J\Psi = \Psi$ could not be imposed. It could, however, be imposed if the four-dimensional space is Lorentzian [4]. The reason is that the KO dimension of the finite space is 6 because the operators D_F, γ_F, and J_F satisfy

$$J_F^2 = 1, \qquad J_F D_F = D_F J_F, \qquad J_F \gamma_F = -\gamma_F J_F.$$

The operators D_M, $\gamma_M = \gamma_5$, and $J_M = C$ for a compact manifold of dimension 4 satisfy

$$J_M^2 = -1, \qquad J_M D_M = D_M J_M, \qquad J_M \gamma_5 = \gamma_5 J_M.$$

Thus the KO dimension of the full noncommutative space $(\mathcal{A}, \mathcal{H}, D)$ with the decorations J and γ included is 10 and satisfies

$$J^2 = -1, \qquad JD = DJ, \qquad J\gamma = -\gamma J.$$

We have shown in [17] that the path integral of the Dirac action, thanks to the relations $J^2 = -1$ and $J\gamma = -\gamma J$, yields a Pfaffian of the operator D instead of its determinant and thus eliminates half the degrees of freedom of Ψ and have the same effect as imposing the condition $J\Psi = \Psi$.

We have also seen that the operator J sends the algebra $\mathcal{A}$ to its commutant, and thus the full algebra acting on the Hilbert space $\mathcal{H}$ is $\mathcal{A} \otimes \mathcal{A}^o$. Under automorphisms of the algebra

$$\Psi \rightarrow U\Psi,$$

where $U = u\widehat{u}$ with $u \in \mathcal{A}$, $\widehat{u} \in \mathcal{A}^o$ with $[u, \widehat{u}] = 0$, it is clear that Dirac action is not invariant.

At this point it is clear that we have retrieved all our conclusions we have before arriving at a unique possibility, which is to have a noncommutative space corresponding to the Pati–Salam Model we considered before, and in the special case where the Dirac operator and algebra satisfy the order one condition, the result is the noncommutative space of the Standard Model. We have thus succeeded in obtaining the Pati–Salam Model and Standard Model as unique possibilities starting with the two sided Heisenberg like Equation (54) thus eliminating all other possibilities obtained in classifying finite noncommutative spaces of KO dimension 6. There is no need to assume the existence of an isometry that reduces the first algebra from $M_4(\mathbb{C})$ to $M_2(\mathbb{H})$, and no need to assume that the KO dimension of the finite space to be 6. These results are very satisfactory and serve to enhance our confidence of the fine structure of space-time as given by the above derived noncommutative space.

7 Outlook: towards quantization

Starting with the simple observation that the Higgs field could be interpreted as the link between two parallel sheets separated by a distance of the order of 10^{-16} cm it took enormous effort to identify a noncommutative space where the spectrum of the Standard Model could fit. Small deviations from the model, such as the need for a real structure and a KO dimension 6, were taken as input to fine tune and determine precisely the noncommutative space. The spectral action principle proved to be very efficient way in evaluating the bosonic sector of the theory. Having identified the noncommutative space, the next target was to understand why nature would chose the Standard Model and not any other possibility. A classification of finite spaces revealed the special nature of the finite part of the noncommutative space identified. Work on encoding manifolds with dimensions equal to four satisfying a higher form of Heisenberg type equation showed that the most general solution of this equation is that of a noncommutative space which is a product of a four-dimensional Riemannian spin-manifold times the finite space corresponding to a Pati–Salam unification model. The Standard Model is a special case of this space where a first order differential condition is satisfied. After a long journey the reason why nature chose the Standard Model is now reduced to determining solutions of a higher form of Heisenberg equation. With such little input, it is quite satisfying to learn that it is possible to answer many of the questions which puzzled theorists for a long time. We now know why there are 16 fermions per generation, why the gauge group is $SU(3) \times SU(2) \times U(1)$, an explanation of the Higgs field, and origin of spontaneous symmetry breaking. The spectral model also predicts a Majorana mass for the right-handed neutrinos and explains the seesaw mechanism. We thus understand unification of all fundamental forces as a geometrical theory based on the spectral action principle of a noncommutative space.

Naturally, there are many questions that are still unanswered, and this motivates the need for further research to address these problems using noncommutative geometry considerations. To conclude, we mention few of the possible directions of future research. One important aspect to consider is the renormalizability properties of the spectral model. Another problem is to study the quantum properties of the Dirac operator and whether it could be related to the pullbacks of the maps used in determining the quanta of geometry. The future of noncommutative geometry in the program of unification of all fundamental interactions looks now to be very promising.

Acknowledgements The work of A. H. C. is supported in part by the National Science Foundation Grant No. Phys-1518371. He also thanks the Radboud Excellence Initiative for hosting him at Radboud University where this research was carried out. We would like to thank Alain Connes for sharing with us his insights and ideas.

Appendix: Pati–Salam model: potential analysis

We here include the scalar potential analysis for the composite Pati–Salam model, as described in Section 5.2 above.

If there is unification of lepton and quark couplings, then $\rho = 1$ so that the Σ^I_J-field decouples. In that case we have

$$\begin{aligned}\mathcal{L}_{pot}(\phi^b_{\dot a}, \Delta_{\dot a I}) &= -\mu^2 \phi^{\dot c}_{a} \phi^{a}_{\dot c} - \nu^2 \left(\Delta_{\dot a K} \overline{\Delta}^{\dot a K}\right)^2 + \lambda_\Sigma \phi^{\dot c}_{a} \phi^{b}_{\dot c} \phi^{\dot d}_{b} \phi^{a}_{\dot d} \\ &\quad + \lambda_H \left(\Delta_{\dot a K} \overline{\Delta}^{\dot a L} \Delta_{\dot b L} \overline{\Delta}^{\dot b K}\right)^2 + \lambda_{H\Sigma} \left(\Delta_{\dot a J} \overline{\Delta}^{\dot a J} \Delta_{\dot c I} \overline{\Delta}^{\dot d I}\right) \phi^{\dot c}_{b} \phi^{b}_{\dot d}\end{aligned}$$

where we have absorbed some constant factors by redefining the couplings $\lambda_H, \lambda_{H\Sigma}$, and λ_Σ.

We choose unitarity gauge for the Δ and ϕ-fields, in the following precise sense.

Lemma 6 *For each value of the fields* $\{\phi^b_{\dot a}, \Delta_{\dot a I}\}$ *there is an element* $(u_R, u_L, u) \in SU(2)_R \times SU(2)_L \times SU(4)$ *such that*

$$u_R \begin{pmatrix} \phi^1_{\dot 1} & \phi^2_{\dot 1} \\ \phi^1_{\dot 2} & \phi^2_{\dot 2} \end{pmatrix} u_L^* = \begin{pmatrix} h & 0 \\ 0 & \chi \end{pmatrix}$$

and

$$u_R \begin{pmatrix} \Delta_{\dot 1 1} & \Delta_{\dot 1 2} & \Delta_{\dot 1 3} & \Delta_{\dot 1 4} \\ \Delta_{\dot 2 1} & \Delta_{\dot 2 2} & \Delta_{\dot 2 3} & \Delta_{\dot 2 4} \end{pmatrix} u^t = \begin{pmatrix} 1+\delta_0 & 0 & 0 & 0 \\ \delta_1 & \eta_1 & 0 & 0 \end{pmatrix}$$

where $h, \delta_0, \delta_1, \eta_1$ *are real fields and* χ *is a complex field.*

Proof Consider the singular value decomposition of the 2×2 matrix $(\phi^b_{\dot a})$:

$$\begin{pmatrix} \phi^1_{\dot 1} & \phi^2_{\dot 1} \\ \phi^1_{\dot 2} & \phi^2_{\dot 2} \end{pmatrix} = U \begin{pmatrix} h & 0 \\ 0 & k \end{pmatrix} V^*$$

for unitary 2×2 matrices U, V and real coefficients h, k. If we define

$$u_R = \begin{pmatrix} 1 & 0 \\ 0 & \det U \end{pmatrix} U^* \in SU(2)_R$$

$$u_L = \begin{pmatrix} 1 & 0 \\ 0 & \det V \end{pmatrix} V^* \in SU(2)_L$$

it follows that

$$u_R \begin{pmatrix} \phi_{\dot 1}^1 & \phi_{\dot 1}^2 \\ \phi_{\dot 2}^1 & \phi_{\dot 2}^2 \end{pmatrix} u_L^* = \begin{pmatrix} h & 0 \\ 0 & k \det U V^* \end{pmatrix} =: \begin{pmatrix} h & 0 \\ 0 & \chi \end{pmatrix}.$$

Next, we consider $\Delta_{\dot a I}$ and write

$$(\Delta_{\dot a I}) = \begin{pmatrix} u_1^* \\ u_2^* \end{pmatrix}, \qquad \text{with } u_a^* = \begin{pmatrix} \Delta_{\dot a 1} & \Delta_{\dot a 2} & \Delta_{\dot a 3} & \Delta_{\dot a 4} \end{pmatrix}$$

for $a = 1, 2$. We may suppose that the vectors u_1, u_2 are such that their inner product $u_1^* u_2$ is a real number. Indeed, if this is not the case, then multiply $\Delta_{\dot a I}$ by a matrix in $SU(2)_R$ as follows:

$$\begin{pmatrix} u_1^* \\ u_2^* \end{pmatrix} \mapsto \begin{pmatrix} \alpha & 0 \\ 0 & \alpha^* \end{pmatrix} \begin{pmatrix} u_1^* \\ u_2^* \end{pmatrix} = \begin{pmatrix} \alpha u_1^* \\ \alpha^* u_2^* \end{pmatrix}.$$

Now the inner product is $(\alpha^* u_1)^* \alpha u_2 = (\alpha)^2 u_1^* u_2$ and we may choose α so as to cancel the phase of $u_1^* u_2$. Moreover, this transformation respects the above form of $\phi_{\dot a}^b$ after a $SU(2)_L$-transformation of exactly the same form:

$$\begin{pmatrix} h & 0 \\ 0 & \chi \end{pmatrix} \mapsto \begin{pmatrix} \alpha & 0 \\ 0 & \alpha^* \end{pmatrix} \begin{pmatrix} h & 0 \\ 0 & \chi \end{pmatrix} \begin{pmatrix} \alpha & 0 \\ 0 & \alpha^* \end{pmatrix}^* = \begin{pmatrix} h & 0 \\ 0 & \chi \end{pmatrix}.$$

Thus let us continue with the vectors u_1, u_2 satisfying $u_1^* u_2 \in \mathbb{R}$. We apply Gramm–Schmidt orthonormalization to u_1 and u_2, to arrive at the following orthonormal set of vectors $\{e_1, e_2\}$ in $\mathbb{C}^4$:

$$e_1 = \frac{u_1}{\|u_1\|}; \qquad e_2 = \frac{u_2 - \frac{u_1^* u_2}{\|u_1\|} u_1}{\|u_2 - \frac{u_1^* u_2}{\|u_1\|} u_1\|}.$$

We complete this set by choosing two additional orthonormal vectors e_3 and e_4 and write a unitary 4×4 matrix:

$$U = \begin{pmatrix} e_1 & e_2 & e_3 & e_4 \end{pmatrix}$$

The sought-for matrix $u \in SU(4)$ is determined by

$$u^t = U \begin{pmatrix} 1_3 & 0 \\ 0 & \det U^* \end{pmatrix}$$

so as to give

$$(\Delta_{\dot{a}I})\, u^t = \begin{pmatrix} u_1^* e_1 & 0 & 0\,0 \\ u_2^* e_1 & u_2^* e_2 & 0\,0 \end{pmatrix} =: \begin{pmatrix} 1+\delta_0 & 0 & 0\,0 \\ \delta_1 & \eta_1 & 0\,0 \end{pmatrix}$$

□

Remark 7 Note that this is compatible with the dimension of the quotient of the space of field values by the group. Indeed, the fields $\phi_{\dot{a}}^b$ and $\Delta_{\dot{a}I}$ span a real 24-dimensional space (at each manifold point). The dimension of the orbit space is then $24 - \dim P$ with P a principal orbit of the action of $SU(2)_R \times SU(2)_L \times SU(4)$ on the space of field values. This dimension $\dim P$ is determined by the dimension of the group and of a principal isotropy group.

First, we see that up to conjugation there is always a $SU(2)$-subgroup of $SU(4)$ leaving $\Delta_{\dot{a}I}$ invariant: it corresponds to $SU(2)$-transformations in the space orthogonal to the vectors $\Delta_{\dot{1}I}$ and $\Delta_{\dot{2}I}$ in $\mathbb{C}^4$. Moreover, one can compute that the isotropy subgroup of the field values

$$(\phi_{\dot{a}}^b) = \begin{pmatrix} 1\,0 \\ 0\,0 \end{pmatrix}; \qquad (\Delta_{\dot{a}I}) = \begin{pmatrix} 1\,0\,0\,0 \\ 1\,1\,0\,0 \end{pmatrix}$$

is given by $\mathbb{Z}_2 \times SU(2)$. Hence, the dimension of the principal orbit is $21 - 3 = 18$ so that the orbit space is six-dimensional. This corresponds to the 4 real fields $h, \delta_0, \delta_1, \eta_1$ and the complex field χ.

We allow for the color $SU(3)$-symmetry not to be broken spontaneously, hence we only choose unitarity gauge in the $SU(2)_R \times SU(2)_L \times U(1)$-representations. That is, we retain the row vector $\Delta_{\dot{2}I}$ for $I = 1, \ldots, 4$ as a variable and write

$$(\Delta_{\dot{a}I}) = \begin{pmatrix} \sqrt{w} + \delta_0/\sqrt{w} & 0 & 0 & 0 \\ \delta_1/\sqrt{w} & \eta_1/\sqrt{w} & \eta_2/\sqrt{w} & \eta_3/\sqrt{w} \end{pmatrix}$$

so that (η_i) forms a scalar $SU(3)$-triplet field (so-called *scalar leptoquarks*). The reason for the rescaling with $\sqrt{w}$ is that it yields the right kinetic terms for δ_0, δ_1, and η. Indeed, from the spectral action we then have

$$\begin{aligned} \frac{1}{2}\partial_\mu H_{\dot{a}I\dot{b}J}\partial^\mu H^{\dot{a}I\dot{b}J} &= \frac{1}{2}\partial_\mu \left(\Delta_{\dot{a}J}\Delta_{\dot{b}I}\right)\partial^\mu\left(\Delta^{\dot{a}J}\Delta^{\dot{b}J}\right) \\ &\sim \sum_{a=0}^{1} \partial_\mu\delta_a\partial^\mu\delta^a + \partial_\mu\eta\partial^\mu\eta^* + \text{ higher order} \end{aligned}$$

The scalar potential becomes in terms of the fields $h, \chi, \delta_0, \delta_1, \eta_i$:

$$\mathcal{L}_{pot}(h,\chi,\delta_0,\delta_1,\eta) = -\mu^2(h^2+|\chi|^2) - \nu^2\left((w+\delta_0)^2+\delta_1^2+|\eta|^2\right)^2/w^2$$
$$+\lambda_{H\Sigma}\left((w+\delta_0)^2h^2+(\delta_1^2+|\eta|^2)|\chi|^2\right)\left((w+\delta_0)^2+\delta_1^2+|\eta|^2\right)/w^2$$
$$+\lambda_H\left((w+\delta_0)^4+2(w+\delta_0)^2\delta_1^2+(\delta_1^2+|\eta|^2)^2\right)^2/w^4+\lambda_\Sigma(h^4+|\chi|^4)$$

As we are interested in the truncation to the Standard Model, we look for extrema with $\langle\delta_1\rangle = \langle\eta_i\rangle = 0$, while setting $\langle h\rangle = v$, $\langle\delta_0\rangle = 0$, $\langle\chi\rangle = x$. Note that the symmetry of these vevs is

$$\left\{\left(\begin{pmatrix}\lambda & 0\\ 0 & \lambda^*\end{pmatrix},\begin{pmatrix}\lambda^* & 0\\ 0 & \lambda\end{pmatrix},\begin{pmatrix}\lambda^* & 0\\ 0 & m\end{pmatrix}\right) : \lambda\in U(1), m\in SU(3)\right\}$$
$$\subset SU(2)_R\times SU(2)_L\times SU(4)$$

In other words, $SU(2)_R\times SU(2)_L\times SU(4)$ is broken by the above vevs to $U(1)\times SU(3)$.

The first derivative of V vanishes for these vevs precisely if

$$2v(w^2\lambda_{H\Sigma}+2v^2\lambda_\Sigma-\mu^2)=0,$$
$$4x^3\lambda_\Sigma-2x\mu^2=0,$$
$$4w(2w^2\lambda_H+v^2\lambda_{H\Sigma}-\nu^2)=0.$$

This gives rise to the fine-tuning of v, w as in [16]:

$$w^2\lambda_{H\Sigma}+2v^2\lambda_\Sigma-\mu^2,\qquad 2w^2\lambda_H+v^2\lambda_{H\Sigma}-\nu^2$$

choosing μ and ν such that the solutions v, w are of the desired orders. Moreover, we find that the vev for χ either vanishes or is equal to $x=\sqrt{\mu^2/2\lambda_\Sigma}$. Note that this latter vev appears precisely at the entry $k^d h$ (or $k^e h$) of the finite Dirac operator, which we have disregarded by setting $\rho = 1$.

If $\langle\chi\rangle = x = 0$, then the Hessian is (derivatives with respect to $h,\chi,\delta_0,\delta_1,\eta$):

$$\begin{pmatrix} 8v^2\lambda_\Sigma & 0 & 8vw\lambda_{H\Sigma} & 0 & 0\\ 0 & -2w^2\lambda_{H\Sigma}-4v^2\lambda_\Sigma & 0 & 0 & 0\\ 8vw\lambda_{H\Sigma} & 0 & 32w^2\lambda_H & 0 & 0\\ 0 & 0 & 0 & -2v^2\lambda_{H\Sigma} & 0\\ 0 & 0 & 0 & 0 & -8\lambda_H w^2-2v^2\lambda_{H\Sigma}w^2\mathbf{1}_3 \end{pmatrix}$$

where the $\mathbf{1}_3$ is the identity matrix in color space, corresponding to the η-field. This Hessian is not positive definite so we disregard the possibility that $\langle\chi\rangle = 0$.

If $x=\sqrt{\mu^2/2\lambda_\Sigma}$, then the Hessian is

$$\begin{pmatrix} 8v^2\lambda_\Sigma & 0 & 8vw\lambda_{H\Sigma} & 0 & 0 \\ 0 & 4w^2\lambda_{H\Sigma}+8v^2\lambda_\Sigma & 0 & 0 & 0 \\ 8vw\lambda_{H\Sigma} & 0 & 32w^2\lambda_H & 0 & 0 \\ 0 & 0 & 0 & w^2\frac{\lambda_{H\Sigma}^2}{\lambda_\Sigma} & 0 \\ 0 & 0 & 0 & 0 & w^2\frac{\lambda_{H\Sigma}^2-8\lambda_H\lambda_\Sigma}{\lambda_\Sigma}\mathbf{1}_3 \end{pmatrix}$$

which is positive-definite if

$$\lambda_{H\Sigma}^2 \geq 8\lambda_H\lambda_\Sigma. \tag{56}$$

Note that this relation may hold only at high-energies. The masses for χ, δ_1, and η are then readily found to be:

$$m_\chi^2 = 4w^2\lambda_{H\Sigma} + 8v^2\lambda_\Sigma,$$

$$m_{\delta_1}^2 = w^2\frac{\lambda_{H\Sigma}^2}{\lambda_\Sigma},$$

$$m_\eta^2 = w^2\frac{\lambda_{H\Sigma}^2 - 8\lambda_H\lambda_\Sigma}{\lambda_\Sigma}.$$

Under the assumption that $v^2 \approx 10^2$ GeV, $w^2 \approx 10^{11}$ GeV we have $m_\chi^2 \approx 10^{11}$ GeV and $m_{\delta_1}^2, m_\eta \approx 10^{11}$ GeV.

The (non-diagonal) h and δ_0 sector has mass eigenstates as in [16]:

$$m_\pm^2 = 16w^2\lambda_H + 4v^2\lambda_\Sigma \pm 4\sqrt{16w^4\lambda_H^2 + v^4\lambda_\Sigma^2 + 4v^2w^2\left(\lambda_{H\Sigma}^2 - 2\lambda_H\lambda_\Sigma\right)}$$

Under the assumption that $v^2 \ll w^2$ we can expand the square root:

$$4\sqrt{16\lambda_H^2 w^4\left(1 + \frac{\lambda_\Sigma^2}{\lambda_H^2}\frac{v^4}{w^4} + \frac{\lambda_{H\Sigma}^2 - 2\lambda_H\lambda_\Sigma}{4\lambda_H^2}\frac{v^2}{w^2}\right)}$$
$$\approx 16\lambda_H w^2\left(1 + \frac{\lambda_{H\Sigma}^2 - 2\lambda_H\lambda_\Sigma}{8\lambda_H^2}\frac{v^2}{w^2}\right)$$
$$= 16\lambda_H w^2 + \frac{2\lambda_{H\Sigma}^2}{\lambda_H}v^2 - 4\lambda_\Sigma v^2.$$

Consequently,

$$m_+ \approx 32\lambda_H w^2 + 2\frac{2\lambda_{H\Sigma}^2}{\lambda_H}v^2,$$

$$m_- \approx 8\lambda_\Sigma v^2 \left(1 - \frac{\lambda_{H\Sigma}^2}{4\lambda_H \lambda_\Sigma}\right).$$

which are of the order of 10^{11} and 10^2 GeV, respectively. This requires that we have at low energies

$$4\lambda_H \lambda_\Sigma \geq \lambda_{H\Sigma}^2, \tag{57}$$

which fully agrees with [16] when we identify $\delta_0 \equiv \sigma$ and with the couplings related via

$$\lambda_H = \frac{1}{4}\lambda_\sigma, \qquad \lambda_{H\Sigma} = \frac{1}{2}\lambda_{h\sigma}, \qquad \lambda_\Sigma = \frac{1}{4}\lambda_h$$

Note the tension between Equations (57) and (56), calling for a careful study of the running of the couplings in order to guarantee positive mass eigenstates at their respective energies.

We have summarized the scalar particle content of the above model in Table 1. In terms of the original scalar fields $\phi_{\dot a}^b$ and $\Delta_{\dot a I}$ the vevs are of the following form:

$$(\phi_{\dot a}^b) = \begin{pmatrix} v & 0 \\ 0 & \sqrt{\mu^2/2\Lambda_\Sigma} \end{pmatrix}$$

$$(\Delta_{\dot a I}) = \begin{pmatrix} w \, 0 \, 0 \, 0 \\ 0 \, 0 \, 0 \, 0 \end{pmatrix}.$$

This shows that there are two scales of spontaneous symmetry breaking: at 10^{11}–10^{12} GeV we have

$$SU(2)_R \times SU(2)_L \times SU(4) \to U(1)_Y \times SU(2)_L \times SU(3)$$

and then at electroweak scale (both v and μ) we have

$$U(1)_Y \times SU(2)_L \times SU(3) \to U(1)_Q \times SU(3)$$

Table 1 Scalar particle content with SM-representations

	$U(1)_Y$	$SU(2)_L$	$SU(3)$
$\begin{pmatrix} \phi_1^0 \\ \phi_1^+ \end{pmatrix} = \begin{pmatrix} \phi_{\dot 1}^1 \\ \phi_{\dot 1}^2 \end{pmatrix}$	1	2	1
$\begin{pmatrix} \phi_2^- \\ \phi_2^0 \end{pmatrix} = \begin{pmatrix} \phi_{\dot 2}^1 \\ \phi_{\dot 2}^2 \end{pmatrix}$	-1	2	1
δ_0	0	1	1
δ_1	-2	1	1
η	$-\frac{2}{3}$	1	3

References

1. G. Aad et al. Observation of a new particle in the search for the Standard Model Higgs boson with the ATLAS detector at the LHC. *Phys. Lett.* B716 (2012) 1–29.
2. U. Aydemir. Clifford-based spectral action and renormalization group analysis of the gauge couplings. arXiv:1902.08090.
3. U. Aydemir, D. Minic, C. Sun, and T. Takeuchi. Pati–Salam unification from noncommutative geometry and the TeV-scale W_R boson. *Int. J. Mod. Phys.* A31 (2016) 1550223.
4. J. W. Barrett. A Lorentzian version of the non-commutative geometry of the standard model of particle physics. *J. Math. Phys.* 48 (2007) 012303.
5. W. Beenakker, T. van den Broek, and W. D. van Suijlekom. *Supersymmetry and noncommutative geometry*, volume 9 of *SpringerBriefs in Mathematical Physics*. Springer, Cham, 2016.
6. F. Besnard. On the uniqueness of Barrett's solution to the fermion doubling problem in noncommutative geometry. arXiv:1903.04769.
7. L. Boyle and S. Farnsworth. Non-Commutative Geometry, Non-Associative Geometry and the Standard Model of Particle Physics, 1401.5083.
8. L. Boyle and S. Farnsworth. A new algebraic structure in the standard model of particle physics. *JHEP* 06 (2018) 071.
9. T. van den Broek and W. D. van Suijlekom. Supersymmetric QCD and noncommutative geometry. *Commun. Math. Phys.* 303 (2011) 149–173.
10. T. van den Broek and W. D. van Suijlekom. Supersymmetric QCD from noncommutative geometry. *Phys. Lett.* B699 (2011) 119–122.
11. C. Brouder, N. Bizi, and F. Besnard. The Standard Model as an extension of the noncommutative algebra of forms. arXiv:1504.03890.
12. A. H. Chamseddine and A. Connes. Universal formula for noncommutative geometry actions: Unifications of gravity and the Standard Model. *Phys. Rev. Lett.* 77 (1996) 4868–4871.
13. A. H. Chamseddine. Connection between space-time supersymmetry and noncommutative geometry. *Phys. Lett.* B332 (1994) 349–357.
14. A. H. Chamseddine and A. Connes. Why the Standard Model. *J. Geom. Phys.* 58 (2008) 38–47.
15. A. H. Chamseddine and A. Connes. Noncommutative geometry as a framework for unification of all fundamental interactions including gravity. Part I. *Fortsch. Phys.* 58 (2010) 553–600.
16. A. H. Chamseddine and A. Connes. Resilience of the Spectral Standard Model. *JHEP* 1209 (2012) 104.
17. A. H. Chamseddine, A. Connes, and M. Marcolli. Gravity and the Standard Model with neutrino mixing. *Adv. Theor. Math. Phys.* 11 (2007) 991–1089.
18. A. H. Chamseddine, A. Connes, and V. Mukhanov. Geometry and the quantum: Basics. *JHEP* 1412 (2014) 098.
19. A. H. Chamseddine, A. Connes, and V. Mukhanov. Quanta of geometry: Noncommutative aspects. *Phys. Rev. Lett.* 114 (2015) 091302.
20. A. H. Chamseddine, A. Connes, and W. D. van Suijlekom. Entropy and the spectral action. *Commun. Math. Phys.* (online first) [arXiv:1809.02944].
21. A. H. Chamseddine, A. Connes, and W. D. van Suijlekom. Beyond the spectral Standard Model: Emergence of Pati-Salam unification. *JHEP* 1311 (2013) 132.
22. A. H. Chamseddine, A. Connes, and W. D. van Suijlekom. Grand unification in the spectral Pati-Salam model. *JHEP* 11 (2015) 011.
23. A. H. Chamseddine, G. Felder, and J. Fröhlich. Unified gauge theories in noncommutative geometry. *Phys. Lett. B.* 296 (1992) 109.
24. A. H. Chamseddine, G. Felder, and J. Fröhlich. Gravity in noncommutative geometry. *Comm. Math. Phys.* 155 (1993) 205–217.
25. A. H. Chamseddine, J. Fröhlich, and O. Grandjean. The gravitational sector in the Connes-Lott formulation of the standard model. *J. Math. Phys.* 36 (1995) 6255–6275.

26. S. Chatrchyan et al. Observation of a new boson at a mass of 125 GeV with the CMS experiment at the LHC. *Phys. Lett.* B716 (2012) 30–61.
27. A. Connes. Noncommutative differential geometry. *Publ. Math. IHES* 39 (1985) 257–360.
28. A. Connes. Essay on physics and noncommutative geometry. In *The interface of mathematics and particle physics (Oxford, 1988)*, volume 24 of *Inst. Math. Appl. Conf. Ser. New Ser.*, pages 9–48. Oxford Univ. Press, New York, 1990.
29. A. Connes. *Noncommutative Geometry*. Academic Press, San Diego, 1994.
30. A. Connes. Noncommutative geometry and reality. *J. Math. Phys.* 36(11) (1995) 6194–6231.
31. A. Connes. On the foundations of noncommutative geometry. In *The unity of mathematics*, volume 244 of *Progr. Math.*, pages 173–204. Birkhäuser Boston, Boston, MA, 2006.
32. A. Connes. Noncommutative geometry year 2000 [MR1826266 (2003g:58010)]. In *Highlights of mathematical physics (London, 2000)*, pages 49–110. Amer. Math. Soc., Providence, RI, 2002.
33. A. Connes and J. Lott. Particle models and noncommutative geometry. *Nucl. Phys. Proc. Suppl.* 18B (1991) 29–47.
34. A. Connes and H. Moscovici. Type III and spectral triples. In *Traces in number theory, geometry and quantum fields*, Aspects Math., E38, pages 57–71. Friedr. Vieweg, Wiesbaden, 2008.
35. L. Dabrowski. On noncommutative geometry of the Standard Model: fermion multiplet as internal forms. In *36th Workshop on Geometric Methods in Physics (XXXVI WGMP) Bialowieza, Poland, July 2–8, 2017*, 2017.
36. L. Dabrowski, F. D'Andrea, and A. Sitarz. The Standard Model in noncommutative geometry: fundamental fermions as internal forms. *Lett. Math. Phys.* 108 (2018) 1323–1340.
37. L. Dabrowski and A. Sitarz. Fermion masses, mass-mixing and the almost commutative geometry of the Standard Model, 1806.07282.
38. A. Devastato, F. Lizzi, and P. Martinetti. Higgs mass in noncommutative geometry. *Fortschr. Phys.* 62 (2014) 863–868.
39. A. Devastato, S. Farnsworth, F. Lizzi, and P. Martinetti. Lorentz signature and twisted spectral triples. *JHEP* 03 (2018) 089.
40. A. Devastato, F. Lizzi, and P. Martinetti. Grand symmetry, spectral action, and the Higgs mass. *JHEP* 1401 (2014) 042.
41. A. Devastato and P. Martinetti. Twisted spectral triple for the Standard Model and spontaneous breaking of the Grand Symmetry. *Math. Phys. Anal. Geom.* 20 (2017) 2.
42. R. Dong and M. Khalkhali. Second quantization and the spectral action. arXiv:1903.09624.
43. M. Dubois-Violette. Dérivations et calcul différentiel non commutatif. *C. R. Acad. Sci. Paris Sér. I Math.* 307 (1988) 403–408.
44. M. Dubois-Violette, R. Kerner, and J. Madore. Classical bosons in a noncommutative geometry. *Class. Quant. Grav.* 6 (1989) 1709.
45. M. Dubois-Violette, R. Kerner, and J. Madore. Gauge bosons in a noncommutative geometry. *Phys. Lett.* B217 (1989) 485–488.
46. M. Dubois-Violette, J. Madore, and R. Kerner. Noncommutative differential geometry and new models of gauge theory. *J. Math. Phys.* 31 (1990) 323.
47. K. van den Dungen and W. D. van Suijlekom. Electrodynamics from noncommutative geometry. *J. Noncommut. Geom. 7,* 433–456 (2013).
48. M. Eckstein and B. Iochum. *Spectral Action in Noncommutative Geometry*. Mathematical Physics Studies. Springer, 2018.
49. S. Farnsworth and L. Boyle. Non-associative geometry and the spectral action principle. *JHEP* (2015) 023, front matter+25.
50. H. Figueroa, J. M. Gracia-Bondía, F. Lizzi, and J. C. Várilly. A nonperturbative form of the spectral action principle in noncommutative geometry. *J. Geom. Phys.* 26 (1998) 329–339.
51. M. Gonderinger, Y. Li, H. Patel, and M. J. Ramsey-Musolf. Vacuum Stability, Perturbativity, and Scalar Singlet Dark Matter. *JHEP* 1001 (2010) 053.
52. J. M. Gracia-Bondía, B. Iochum, and T. Schücker. The standard model in noncommutative geometry and fermion doubling. *Phys. Lett. B* 416 (1998) 123–128.

53. B. Iochum, T. Schucker, and C. Stephan. On a classification of irreducible almost commutative geometries. *J. Math. Phys.* 45 (2004) 5003–5041.
54. D. Kastler. A detailed account of Alain Connes' version of the standard model in noncommutative geometry. I, II. *Rev. Math. Phys.* 5 (1993) 477–532.
55. D. Kastler. A detailed account of Alain Connes' version of the standard model in non-commutative differential geometry. III. *Rev. Math. Phys.* 8 (1996) 103–165.
56. D. Kastler and T. Schücker. A detailed account of Alain Connes' version of the standard model. IV. *Rev. Math. Phys.* 8 (1996) 205–228.
57. D. Kastler and T. Schücker. The standard model à la Connes-Lott. *J. Geom. Phys.* 24 (1997) 1–19.
58. T. Krajewski. Classification of finite spectral triples. *J. Geom. Phys.* 28 (1998) 1–30.
59. M.A. Kurkov and F. Lizzi. Clifford Structures in noncommutative geometry and the extended scalar sector. *Phys.Rev.* D97 (2018) 085024.
60. G. Landi. *An Introduction to Noncommutative Spaces and their Geometry*. Springer-Verlag, 1997.
61. G. Landi and P. Martinetti. On twisting real spectral triples by algebra automorphisms. *Lett. Math. Phys.* 106 (2016) 1499–1530.
62. G. Landi and P. Martinetti. Gauge transformations for twisted spectral triples. *Lett. Math. Phys.* 108 (2018) 2589–2626.
63. F. Lizzi, G. Mangano, G. Miele, and G. Sparano. Fermion Hilbert space and fermion doubling in the noncommutative geometry approach to gauge theories. *Phys. Rev.* D55 (1997) 6357–6366.
64. F. Lizzi. Noncommutative Geometry and Particle Physics. *PoS* CORFU2017 (2018) 133.
65. J. Madore. *An Introduction to Noncommutative Differential Geometry and its Physical Applications*. Cambridge University Press, 1995.
66. C. P. Martín, J. M. Gracia-Bondía, and J. C. Várilly. The standard model as a noncommutative geometry: the low-energy regime. *Phys. Rep.* 294 (1998) 363–406.
67. R. Mohapatra. *Unification and supersymmetry. The frontiers of quark-lepton physics*. Springer, New York, third edition, 2003.
68. M. Paschke and A. Sitarz. Discrete spectral triples and their symmetries. *J. Math. Phys.* 39 (1998) 6191–6205.
69. F. Scheck, H. Upmeier, and W. Werner, editors. *Noncommutative geometry and the standard model of elementary particle physics*, volume 596 of *Lecture Notes in Physics*. Springer-Verlag, Berlin, 2002. Papers from the conference held in Hesselberg, March 14–19, 1999.
70. C. A. Stephan. Almost-commutative geometries beyond the Standard Model. *J. Phys.* A39 (2006) 9657.
71. C. A. Stephan. Almost-commutative geometries beyond the Standard Model. ii. new colours. *J. Phys.* A40 (2007) 9941.
72. C. A. Stephan. Beyond the Standard Model: A Noncommutative Approach, 0905.0997.
73. C. A. Stephan. New scalar fields in noncommutative geometry. *Phys. Rev.* D79 (2009) 065013.
74. C. A. Stephan. A Dark Sector Extension of the Almost-Commutative Standard Model. *Int. J. Mod. Phys.* A29 (2014) 1450005.
75. W. D. van Suijlekom. *Noncommutative Geometry and Particle Physics*. Springer, 2015.
76. I. Todorov and M. Dubois-Violette. Deducing the symmetry of the standard model from the automorphism and structure groups of the exceptional Jordan algebra. *Int. J. Mod. Phys.* A33 (2018) 1850118.
77. J. C. Várilly and J. M. Gracia-Bondía. Connes' noncommutative differential geometry and the standard model. *J. Geom. Phys.* 12 (1993) 223–301.
78. E. P. Wigner. Normal form of antiunitary operators. *J. Mathematical Phys.* 1 (1960) 409–413.

The Riemann–Roch strategy

Complex lift of the Scaling Site

Alain Connes and Caterina Consani

Abstract We describe the Riemann–Roch strategy which consists of adapting in characteristic zero Weil's proof, of RH in positive characteristic, following the ideas of Mattuck–Tate and Grothendieck. As a new step in this strategy we implement the technique of tropical descent that allows one to deduce existence results in characteristic one from the Riemann–Roch result over $\mathbb{C}$. In order to deal with arbitrary distribution functions this technique involves the results of Bohr, Jessen, and Tornehave on almost periodic functions.

Our main result is the construction, at the adelic level, of a complex lift of the adèle class space of the rationals. We interpret this lift as a moduli space of elliptic curves endowed with a *triangular* structure. The equivalence relation yielding the noncommutative structure is generated by isogenies. We describe the tight relation of this complex lift with the GL(2)-system. We construct the lift of the Frobenius correspondences using the Witt construction in characteristic 1.

1 Introduction

This paper presents our latest attempts in the quest of an appropriate geometry to localize the zeros of the Riemann zeta function. The constructions described in this article define a complex geometry that is a "lift" in characteristic zero, of the (tropical) Scaling Site. This project has undergone in the past years several developments that we list below in order to frame and justify this latest work.

A. Connes
IHES, Bures sur Yvette, France

Collège de France, Ohio State University, Columbus, OH, USA
e-mail: alain@connes.org

C. Consani (✉)
Johns Hopkins University, Baltimore, MD, USA
e-mail: kc@math.jhu.edu

A. Chamseddine et al. (eds.), *Advances in Noncommutative Geometry*,
https://doi.org/10.1007/978-3-030-29597-4_2

- The interpretation of the explicit formulas of Riemann-Weil as a trace formula for the scaling action on the adèle class space of a global field [11, 36].
- The interpretation of the Riemann zeta function as a Hasse–Weil counting function [20, 21].
- The discovery of the Arithmetic Site, and the identity between the Galois action on its points over the tropical semifield $\mathbb{R}_+^{\max}$ with the scaling action on the adèle class space of the rational numbers [22, 23].
- The discovery of the Scaling Site, the identity between its points and the adèle class space of the rationals [24].
- In [25] we unveiled the tropical structure of the Scaling Site, proved the Riemann–Roch theorem on its periodic orbits, and developed the theory of theta functions on these orbits.

At this stage, the geometric framework that we built in characteristic one is well understood. The theory of theta functions and the Riemann–Roch formula with *real valued indices* on the periodic orbits of the Scaling Site, provide a convincing reason in support of the strategy of adapting Weil's proof (of the Riemann Hypothesis in positive characteristic) by following the ideas of Mattuck and Tate, and Grothendieck [30, 27]. However, in the process to formulate a Riemann–Roch theorem on the square of the Scaling Site one faces a substantial difficulty. The problem, which is still open at this time, has to do with an appropriate definition of the sheaf cohomology (as idempotent monoid) H^1 (the definition of H^0 is straightforward and that of H^2 can be given by turning Serre duality into a definition). In [26], we have developed the beginning of a general homological algebra machine in characteristic one (i.e. for tropical structures) exactly to aim for a definition of the above H^1. In particular, we proved the existence of non-trivial Ext-functors and we were also able to input the resolution of the diagonal to obtain the tropical analogue of the Čech complex. However, when applied to Čech covers, the presence of the null elements creates unwanted contributions to the cohomology which so far we are unable to handle. The root of this problem had been already unearthed in the Example 6.5 of [44]. This example provides pairs (C, D), (C', D') of tropical curves and divisors on them, for which the tropical invariants $r(D)$ and $r(D')$ entering in the Riemann–Roch formulas [2, 28] as a substitute for the dimension of the modules $H^0(D)$ and $H^0(D')$ are different, while the modules themselves are isomorphic.

It is well-known that the hard part of the Riemann–Roch results of [2, 28] concerns the existence of non-trivial solutions i.e. the proof of a Riemann–Roch inequality. This fact leads us now to concentrate, in our setup, exactly on the *existence theme* and to develop a technique of "tropical descent," with the goal to deduce existence results in characteristic one from available Riemann–Roch theorems in complex geometry.

Already in the appendix of [25], we pointed out the relevance of the tropicalization map in the non-Archimedean resp. Archimedean cases. In both cases the tropicalization associates to an analytic function f in a corona a piecewise affine convex function $\tau(f)$, (on a real interval I), whose tropical zeros are the valuations $v(z_j)$ (resp. $-\log|z_j|$) of the zeros z_j of f. The ensuing technique of "tropical descent" is reported in Section 3.2. In the complex (Archimedean) case the tropicalization of an analytic function f in the corona $R_1 < |z| < R_2$ is a convex function in the interval $-\log R_2 < \lambda < -\log R_1$ and one obtains the real half-line involved in the definition of the Scaling Site by taking $R_2 = 1$ and $R_1 = 0$. Namely, one works with the punctured unit disk $\underline{\mathbb{D}}^* := \{q \in \mathbb{C} \mid 0 < |q| \leq 1\}$ in $\mathbb{C}$.

Moreover, the action by multiplication of $\mathbb{N}^\times$ on the real half-line, which is the key structure in the definition of the Scaling Site, lifts naturally to the operation $f(z) \to f(z^n)$ on analytic functions. This observation provides, as a starting point, the definition of the ringed topos obtained by endowing the topos $\underline{\mathbb{D}}^* \rtimes \mathbb{N}^\times$ (for the natural action of $\mathbb{N}^\times$ on $\underline{\mathbb{D}}^*$ given by $q \mapsto q^n$) with the structure sheaf $\mathcal{O}$ of complex analytic functions. Given a pair of open sets Ω, Ω' in $\mathbb{D}^*$ and an integer $n \in \mathbb{N}^\times$, with $q^n \in \Omega'$ for any $q \in \Omega$, one has a natural restriction map

$$\Gamma(\Omega', \mathcal{O}) \to \Gamma(\Omega, \mathcal{O}), \quad f(q) \mapsto f(q^n).$$

The map $u : \underline{\mathbb{D}}^* \to [0, \infty)$ given by $u(q) = -\log|q|$ extends to a geometric morphism $u : \underline{\mathbb{D}}^* \rtimes \mathbb{N}^\times \to [0, \infty) \rtimes \mathbb{N}^\times$ of toposes.

This development provides a first glimpse of a complex lift of the Scaling Site. The piecewise affine functions obtained as tropicalization using the Jensen formula all have integral slopes, and the zeros have integral multiplicities. To reach more general types of convex functions requires generalizing the original Jensen framework. This is achieved by the extension of the work of Jensen as developed by Jessen, Tornehave, and Bohr [6, 31, 32], to the case of analytic *almost periodic functions*. In this work the Jensen formula, which counts a finite number of zeros, is extended to measure, by the second derivative of a convex function φ, the *density* of the zeros of an analytic function $f(z)$

$$\lim_{T\to\infty} \frac{1}{2T}\{\#z \mid f(z) = 0,\ Re(z) = \alpha,\ |\Im z| < T\} = \varphi''(\alpha),$$

where $f(z)$ is analytic and almost periodic on the lines $Re(z) = \alpha$. In particular, any convex function φ (there are minor restrictions on φ on intervals in which the second derivative of φ is identical to 0) can be obtained as the "tropicalization" of an analytic almost periodic function. This construction resolves the problem of realizing arbitrary functions as tropicalizations and shows (see Section 4) how to reach continuous divisors of the form $\int n(\lambda)\delta_\lambda d^*\lambda$ as "tropical shadows" of discrete almost periodic divisors. This part is a first step, in our project, in order to handle the continuous integrals $\int f(\lambda)\Psi_\lambda d^*\lambda$ of the Frobenius correspondences Ψ_λ involved in the implementation of the Riemann-Roch strategy to a proof of the Riemann Hypothesis (RH).

This analytic construction supplies the useful hint that in order to construct a complex lift of the Scaling Site one needs to implement an almost periodic imaginary direction. This amounts to use the covering of the pointed disk $\underline{\mathbb{D}}^*$ by the closed Poincaré half plane $\bar{\mathbb{H}} := \{z \in \mathbb{C} \mid \Im(z) \geq 0\}$ defined by the map $q(z) := \exp(2\pi i z)$, and to compactify the real direction in $\bar{\mathbb{H}}$ to a compact group G. In fact, the only requirement sought for the group compactification $\mathbb{R} \subset G$ is to be $\mathbb{Q}^\times$-invariant. We take for G the smallest available choice which is the compact dual of the discrete additive group $\mathbb{Q}$. The compactification of the real direction in $\bar{\mathbb{H}}$ then yields the pro-étale covering $\tilde{\underline{\mathbb{D}}^*}$ of the punctured disk $\underline{\mathbb{D}}^*$, described as the projective limit $\tilde{\underline{\mathbb{D}}^*} := \varprojlim(E_n, p_{(n,m)})$

$$E_n := \underline{\mathbb{D}}^*, \quad p_{(n,m)} : E_m \to E_n, \quad p_{(n,m)}(z) := z^a\ , \ \forall m = na, \ z \in E_m = \underline{\mathbb{D}}^*.$$

Here, the indexing set $\mathbb{N}^\times$ is ordered by divisibility. At the topos level one would then consider the semidirect product $\tilde{\underline{\mathbb{D}}^*} \rtimes \mathbb{N}^\times$.

In this paper we prefer to proceed directly at the adelic level and consider the quotient, by the action of $\mathbb{Q}^\times$, of the product of the adèles $\mathbb{A}_\mathbb{Q}$ by G. The noncommutative space which we reach is thus the quotient

$$\mathscr{C}_\mathbb{Q} := \mathbb{Q}^\times \backslash \left(G \times \mathbb{A}_\mathbb{Q}\right)$$

The first key observation of this paper (see Section 5) is that with the above choice of G the quotient space $\mathscr{C}_\mathbb{Q}$ is *identical* to the quotient

$$\mathscr{C}_\mathbb{Q} = P(\mathbb{Q})\backslash \bar{P}(\mathbb{A}_\mathbb{Q}), \quad \bar{P}(\mathbb{A}_\mathbb{Q}) := \left\{\begin{pmatrix} a & b \\ 0 & 1 \end{pmatrix} \mid a, b \in \mathbb{A}_\mathbb{Q}\right\}$$

of $\bar{P}(\mathbb{A}_\mathbb{Q})$ by the action by left multiplication of the affine "$aX + b$" group $P(\mathbb{Q})$ of the rationals. Then, the right action of the affine group $P(\mathbb{R})$ determines a natural foliation on $\mathscr{C}_\mathbb{Q}$ whose leaves are one dimensional complex curves generically isomorphic to the Poincaré half plane $\mathbb{H}$. The sector of the Riemann zeta function is obtained after division by the right action of $\hat{\mathbb{Z}}^\times$ on $\mathscr{C}_\mathbb{Q}$ (which naturally extends its action on the adèle class space).

In Section 5.3, we consider (after division by $\hat{\mathbb{Z}}^\times$) the periodic orbit $\Gamma(p)$ associated with a prime p. We find that $\Gamma(p)$ is the mapping torus of the multiplication by p in the compact group G. This mapping torus is an ordinary compact space and we analyze the restriction of the above foliation by one dimensional complex leaves. We show that this foliation is of type III_λ where $\lambda = 1/p$ and that the discrete decomposition of the associated factor has natural geometric interpretation. We determine the de Rham cohomology in Proposition 5.2.

In Section 5.4 we analyze the restriction to the classical orbit of the above foliation by one dimensional complex leaves. We show that it is of type II_∞ and we give an explicit construction, based on the results of Section 4, of the leafwise discrete lift of continuous divisors.

The second key observation of this paper (see Section 6) is the tight relation of the noncommutative space $\mathscr{C}_\mathbb{Q} = P(\mathbb{Q})\backslash\overline{P(\mathbb{A}_\mathbb{Q})}$ to the GL(2)-system ([16]). The GL(2)-system was conceived as a higher dimensional generalization of the BC-system and its main feature is its arithmetic subalgebra constructed using modular functions. After recalling in Section 6.1 the standard notations for the Shimura variety $Sh(\mathrm{GL}_2, \mathbb{H}^\pm) = \mathrm{GL}_2(\mathbb{Q})\backslash\mathrm{GL}_2(\mathbb{A}_\mathbb{Q})/\mathbb{C}^\times$, we consider in Section 6.2 the natural map

$$\mathscr{C}_{\mathbb{Q}} = P(\mathbb{Q})\backslash\overline{P(\mathbb{A}_{\mathbb{Q}})} \xrightarrow{\theta} \mathrm{GL}_2(\mathbb{Q})\backslash M_2(\mathbb{A}_{\mathbb{Q}})^{\bullet}/\mathbb{C}^{\times} = \overline{Sh^{\mathrm{nc}}(\mathrm{GL}_2, \mathbb{H}^{\pm})} \tag{1}$$

from $\mathscr{C}_{\mathbb{Q}}$ to the noncommutative space $\overline{Sh^{\mathrm{nc}}(\mathrm{GL}_2, \mathbb{H}^{\pm})}$ underlying the GL(2)-system. At the Archimedean place, the corresponding inclusion $\overline{P(\mathbb{R})} \subset (M_2(\mathbb{R}) \setminus \{0\})$ induces a bijection of $\overline{P(\mathbb{R})}$ with the complement in $(M_2(\mathbb{R}) \setminus \{0\})/\mathbb{C}^{\times}$ of the point ∞ given by the class of matrices with vanishing second line. This result shows that the nuance between $\mathscr{C}_{\mathbb{Q}}$ and $\overline{Sh^{\mathrm{nc}}(\mathrm{GL}_2, \mathbb{H}^{\pm})}$ is mainly due to the non-Archimedean components. In Section 6.3 we use the description of the GL(2)-system in terms of adelic $\mathbb{Q}$-lattices and interpret $\mathscr{C}_{\mathbb{Q}}$ in terms of *parabolic* $\mathbb{Q}$-lattices. The key result is Theorem 6.1 which states that the natural inclusion of parabolic $\mathbb{Q}$-lattices among $\mathbb{Q}$-lattices remains injective, except in degenerate cases, into the space of two-dimensional $\mathbb{Q}$-lattices up to scale. The relevance of this fact originates from the richness of the function theory on the space of two-dimensional $\mathbb{Q}$-lattices up to scale which involves in particular modular forms of arbitrary level.

By using the geometric interpretation of $\mathbb{Q}$-lattices up to scale in terms of elliptic curves endowed with pairs of elements of the Tate module, we provide in Section 6.4 the geometric interpretation of the points of $\mathscr{C}_{\mathbb{Q}}$ in terms of elliptic curves endowed with a *triangular structure* (reflecting the parabolic structure of the $\mathbb{Q}$-lattice) and modulo the equivalence relation generated by isogenies (Section 6.5). In Section 6.6 we prove that the natural complex structure of the moduli space of triangular elliptic curves is the same as the complex structure on $\mathscr{C}_{\mathbb{Q}}$ defined in Section 5 using the right action of $P(\mathbb{R})$. The right action of $P(\hat{\mathbb{Z}})$ has a simple geometric interpretation (Section 6.7) and allows one to pass from $\mathscr{C}_{\mathbb{Q}}$ to $\Gamma_{\mathbb{Q}}$. Finally we give in the last section (Section 6.8) the geometric interpretation of the degenerate cases.

The beginning of this paper explains a central philosophy of our strategy which is not to focus on the zeta function itself (as a function "per se") but to find from the start, a geometric interpretation of the zero-cycle of its zeros. We record that the original motivation for H. Bohr was the Riemann zeta function and in particular, we recall that Borchsenius and Jessen proved in [9] (Theorems 14 and 15) the "frightening" result that for any value $x \neq 0$, the zeros of $(\zeta(z) - x)$ have real parts which admit 1/2 as a limit point. More precisely, their results show that for any fixed $\sigma_1 > \frac{1}{2}$, the density of zeros of $(\zeta(z) - x)$, in the strip $\sigma < \Re(z) < \sigma_1$, $\frac{1}{2} < \sigma < \sigma_1$, tends to infinity when $\sigma \to \frac{1}{2}$.

In Section 2, we explain why the adèle class space of the rationals is a natural geometric space underlying the zeros of the Riemann zeta function. First, notice that what is really central in studying these zeros is the ideal that the Riemann zeta function generates among holomorphic functions in a suitable domain. After applying Fourier transform, the key operation on functions of a positive real variable which generates this ideal is the summation

$$f \mapsto E(f), \quad E(f)(v) := \sum_{\mathbb{N}^{\times}} f(nv), \quad \mathbb{N}^{\times} := \{n \in \mathbb{N} \mid n > 0\}. \tag{2}$$

In Section 2 we explain the two geometric approaches suggested by this formula. The first one is of adelic nature and derived from Tate's thesis. It consists in replacing the sum over the monoid $\mathbb{N}^\times$ by a sum over the associated group $\mathbb{Q}_+^\times$, at the expense of crossing the half-line involved at the geometric level in (2) by a non-Archimedean component. This process leads directly to the adèle class space. The second approach is topos theoretic and consists in considering the topos (called the Scaling Site) which is the semidirect product of the half-line by the monoid $\mathbb{N}^\times$. The key fact recalled in Section 2.2 is that the points of the Scaling Site coincide with the points of the (sector of the) adèle class space. Thus while one could be tempted to dismiss at first the adèle class space, the topos theoretic interpretation of its points endows it with a clear geometric status. We explain the unavoidable noncommutative nature of this space in Section 2.1.

The Riemann–Roch strategy, and in particular the technique of tropical descent which allows one to deduce existence results in characteristic one from the Riemann–Roch result over $\mathbb{C}$, are explained in Section 3.

The framework in characteristic 1 is perfectly adapted to the geometric role of the Frobenius. For instance, in the interpretation of the adèle class space as the points of the Arithmetic Site defined over $\mathbb{R}_+^{\max}$, the action by scaling becomes the natural action of $\mathrm{Aut}(\mathbb{R}_+^{\max})$ on these points.

In the lift from characteristic 1 to characteristic 0 one loses the automorphisms $\mathrm{Aut}(\mathbb{R}_+^{\max}) = \mathbb{R}_+^*$. We explain in Section 7 the difficulty created by this loss and show in Section 7.1 how it is resolved by the Witt construction in characteristic 1 achieved in our previous work [17, 18, 19]. Finally in Section 7.2 we discuss the link between our construction of the complex lift and quantization.

Figure 1 gives a visually intuitive global picture at the present time. In particular, the counterpart of $\Gamma_{\mathbb{Q}}$ on the left column is the semidirect product of the pro-étale cover $\tilde{\underline{\mathbb{D}}}^*$ of the punctured unit disk $\underline{\mathbb{D}}^*$ in the complex domain, by the natural action of $\mathbb{N}^\times$.

2 The geometry behind the zeros of ζ

It is important to clarify from the start why the Riemann Hypothesis (= RH), namely the problem of locating the zeros of the Riemann zeta function $\zeta(s)$, is tightly related to the geometry of the adèle class space of the rationals $\mathbb{Q}$. First of all we remark that what characterizes the zeros locus of $\zeta(s)$ is not the zeta function itself rather the *ideal* it generates among complex holomorphic functions in a suitable class. A key role in the description of this ideal is played by the map E on functions $f(v)$ of a real positive variable v that is defined by the assignment

$$f \mapsto E(f), \qquad E(f)(v) := \sum_{n\in\mathbb{N}^\times} f(nv), \quad \mathbb{N}^\times := \{n \in \mathbb{N} \mid n > 0\}. \tag{3}$$

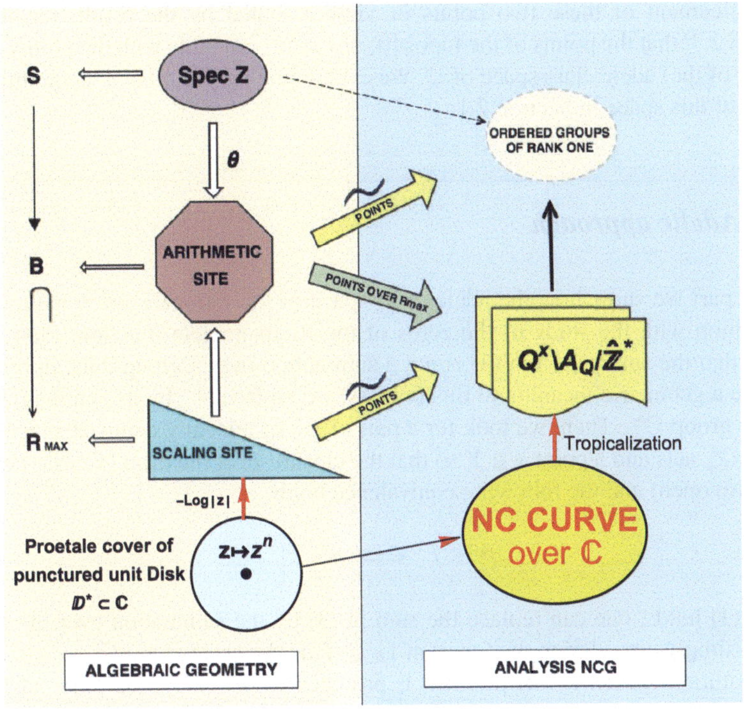

Fig. 1 Global picture

Notice that the map E becomes, in the variable $\log v$, a sum of translations by $\log n$: $\log v \mapsto \log v + \log n$ (i.e. a convolution by a sum of delta functions). Thus, after a suitable Fourier transform, $E(f)$ is a product by the Fourier transform of the sum of the Dirac masses $\delta_{\log n}$, i.e. by the function $\sum e^{-is\log n} = \zeta(is)$. Hence it should not come as a surprise that the cokernel of E determines a spectral realization of the zeros of the Riemann zeta function.

At this point, there are two ways of unveiling the geometric meaning of the map E

1. One may replace in (3) the sum over $\mathbb{N}^\times$ by a summation over the multiplicative group $\mathbb{Q}_+^*$ of positive rational numbers (obtained from the multiplicative monoid $\mathbb{N}^\times$ by symmetrization) with the final goal to interpret E as a projection onto a quotient (of the adèles of $\mathbb{Q}$) by the group $\mathbb{Q}_+^*$. This approach leads naturally to the adèle class space of $\mathbb{Q}$, and more precisely to the sector associated with the trivial character. This construction is described in Section 2.1.
2. Alternatively, one may keep the monoid $\mathbb{N}^\times$ and have it acting on the real half-line $[0,\infty)$. In this way one sees the space $[0,\infty) \rtimes \mathbb{N}^\times$ as a Grothendieck topos. This process yields the Scaling Site of [25] that is reviewed in Section 2.2.

The agreement of these two points of view is stated by the result, recalled in Section 2.3, that the points of the topos $[0, \infty) \rtimes \mathbb{N}^{\times}$ coincide with the points of the (sector of the) adèle class space of $\mathbb{Q}$. We explain the *unavoidable* noncommutative nature of this space in Section 2.1.

2.1 Adelic approach

In this part we show how the adèle class space of the rationals arises naturally in connection with the study of the zeros of the Riemann zeta function. First of all notice that the summation in (3) is not a summation over a group thus, in order to provide a geometric meaning to this process, we replace $\mathbb{N}^{\times}$ by its symmetrization i.e. the group $\mathbb{Q}_{+}^{*}$. Then, we look for a pair (Y, y) of a locally compact space Y on which $\mathbb{Q}_{+}^{*}$ acts and a point $y \in Y$ so that the closure F of the orbit $\mathbb{N}^{\times} y$ is *compact* (and also open) and the following equivalence holds

$$qy \in F \iff q \in \mathbb{N}^{\times}. \tag{4}$$

When (4) holds, one can replace the sum in (3) by the summation over the group $\mathbb{Q}_{+}^{*}$ by simply considering the function $1_F \otimes f$ on the product $Y \times \mathbb{R}$.

A natural solution to this problem is provided by $Y = \mathbb{A}_f$, the finite adèles of $\mathbb{Q}$, and by the principal adèle $y = 1$. One eventually achieves the minimal solution after dividing by $\hat{\mathbb{Z}}^{\times}$. Here, we also note that the action of $\mathbb{Q}_{+}^{*}$ on the idèles cannot be used because it is a proper action.

A basic difficulty that one faces at this point is that the quotient of $Y \times \mathbb{R} = \mathbb{A}_{\mathbb{Q}}$ by the action of $\mathbb{Q}_{+}^{*}$ is noncommutative in the sense that classical techniques to analyze this space are here inoperative. A distinctive feature of a noncommutative space is present already at the level of the underlying "set" since a noncommutative space has the cardinality of the continuum and at the same time it is not possible to put this space constructively in bijection with the continuum. More precisely, any explicitly constructed map from such a set to the real line fails to be injective! From these considerations one perceives immediately a major obstacle if one seeks to understand such spaces using a commutative algebra of functions. The reason why these spaces are named "noncommutative" is that if one accepts to use noncommuting coordinates to encode them, and one extends the traditional tools of commutative algebra to this larger noncommutative framework, everything falls correctly in place. The basic principle that one adopts is to take advantage of the *presentation* of the space as a quotient of an ordinary space (here the adèles) by an equivalence relation (given here by the action of $\mathbb{Q}_{+}^{*}$) but then, instead of effecting the quotient in one stroke, one replaces the equivalence relation by its convolution algebra over the complex numbers.

A distinctive feature of noncommutative spaces can be seen at the level of the Borel structure allowing all sorts of countable operations on Borel functions. In the noncommutative case, the Borel structure is *no longer countably separated*, in the sense that any countable family of Borel functions fails to separate points, i.e. fails to be injective.

Our goal in this section is to show that for whatever choice of the pair (Y, y) fulfilling (4) the resulting quotient space $\mathbb{Q}_+^* \backslash (Y \times \mathbb{R})$ is noncommutative.

In Section 2.1.1 we shall consider the easier case obtained by replacing the pair $(\mathbb{N}^\times, \mathbb{Q}_+^*)$ with $(\mathbb{Z}_{\geq 0}, \mathbb{Z})$. Then we show that any solution (Y, y) for $(\mathbb{Z}_{\geq 0}, \mathbb{Z})$ involves a compactification of the discrete set $\mathbb{Z}_{\geq 0}$. In Section 2.1.2 we prove, for the pair $(\mathbb{Z}_{\geq 0}, \mathbb{Z})$, the stability of "noncommutativity." This means that given a locally compact space R and a homeomorphism $S : R \to R$ whose orbit space is not countably separated, the product action of $T \times S$ on $Y \times R$ is never countably separated, for any auxiliary action (Y, T) as in Section 2.1.1. The strategy we follow in the subsequent Section 2.1.3, in the process of extending this result to the case of the action of $(\mathbb{N}^\times, \mathbb{Q}_+^*)$ on $[0, \infty)$, is reviewed by the following steps

1. In the presence of a fixed point $p \in Y$ for the action of $\mathbb{Q}_+^*$, the quotient of $\{p\} \times \mathbb{R}_+^*$ by $\mathbb{Q}_+^*$ would be $\mathbb{R}_+^*/\mathbb{Q}_+^*$ which is not countably separated and this entails that $(Y \times \mathbb{R}_+^*)/\mathbb{Q}_+^*$ is not countably separated.
2. If instead of a fixed point $p \in Y$ for the action of $\mathbb{Q}_+^*$ one has a fixed probability measure, then the same reasoning applies using Lemma 2.3.
3. Using (4), we construct a $\mathbb{Q}_+^*$-invariant probability measure on Y.

Notice that condition (4) is essential for a meaningful development of the full strategy. Indeed, if one takes the action of $\mathbb{Q}_+^*$ on $Y = \mathbb{Q}_+^*$ by translation, the quotient $(Y \times \mathbb{R}_+^*)/\mathbb{Q}_+^*$ is the standard Borel space $\mathbb{R}_+^*$.

2.1.1 Forward compactification

To understand how to choose the pair (Y, y) as in Section 2.1, one first considers the simpler case of the semigroup $\mathbb{N}^\times$ replaced by the additive semigroup $\mathbb{Z}_{\geq 0}$ of nonnegative integers. Then the symmetrized group is $\mathbb{Z}$ and one looks for a space Y on which $\mathbb{Z}$ acts by a transformation T, and a point $x \in Y$ so that the closure K of $T^{\mathbb{N}}x$ is compact in Y and the following equivalence is fullfilled

$$T^n x \in K \iff n \in \mathbb{Z}_{\geq 0}.$$

Next lemma states that any solution of this problem involves a compactification of the discrete set $\mathbb{Z}_{\geq 0}$

Lemma 2.1 *Let Y be a locally compact space and $T \in \mathrm{Aut}(Y)$ an automorphism. Let $x \in Y$ be such that the closure K of the forward orbit $T^{\mathbb{N}}x$ in Y is compact and the following equivalence holds*

$$T^n x \in K \iff n \geq 0. \tag{5}$$

Then the map $\mathbb{N} \ni n \mapsto T^n x \in K$ *turns* K *into a compactification of the discrete set* $\mathbb{Z}_{\geq 0} = \{n \in \mathbb{Z} \mid n \geq 0\}$.

Proof It is enough to prove that for $n \in \mathbb{N}$ the subset $\{T^n x\} \subset K$ is open (i.e. that $T^n x$ is isolated in K). Note that the complement $V = K^c$ of K in Y is open as well as $T^n V$ for any $n \in \mathbb{Z}$, and that the intersection $T^n V \cap K$ is contained in the closure (in Y) of $T^n V \cap T^{\mathbb{N}} x$. Next, note that (5) is equivalent to $T^u x \in V \iff u < 0$ and this equivalence implies $T^j x \in T^n V \iff j < n$. Thus for $n > 0$, one gets

$$T^n V \cap K = \{T^j x \mid 0 \leq j < n\}.$$

Since a point of K is closed, it follows that each $\{T^j x\}$ is open in K. □

The simplest compactification K of the discrete set $\mathbb{Z}_{\geq 0}$ is the Alexandrov compactification $K = \mathbb{Z}_{\geq 0} \cup \{\infty\}$ obtained by adding a limit point ∞. The open subsets of K containing ∞ are the complements of finite subsets of $\mathbb{Z}_{\geq 0}$. More generally, the Alexandrov compactification of a locally compact space X is obtained by adding a point at infinity and the obtained pointed space $X \cup \{\infty\}$ admits as open sets the open subsets of X and the complements of compact subsets of X. It is described by the following universal property. For every pointed compact Hausdorff space $(Y, *)$ and every continuous map $f : X \to Y$ such that $f^{-1}(K)$ is compact for all compact sets $K \subset Y$ not containing the base point $*$, there is a unique basepoint-preserving continuous map that extends f. When passing to the associated C^*-algebra, the one-point compactification just means adjoining a unit. At the C^*-level, this is the smallest compactification, but since the functor $X \mapsto C_0(X)$ is contravariant, one needs to express this fact dually. From a categorical point of view, it means that the one-point compactification is a *final* object among the compactifications of a given locally compact space X, where morphisms of compactifications are continuous maps $g : X_1 \to X_2$ which restrict to the identity on $X \subset X_j$.

Taking $K = \mathbb{Z}_{\geq 0} \cup \{\infty\}$ yields the following minimal solution (Y, T) of (5).

Lemma 2.2 *Let* $\mathbb{Z}(+\infty)$ *be the union* $\mathbb{Z} \cup \{\infty\}$ *endowed with the topology whose restriction to* $\mathbb{Z}$ *is discrete and where the intervals* $[m, \infty]$ *form a basis of neighborhoods of* ∞. *Then* $\mathbb{Z}(+\infty)$ *is locally compact, the translation* $T(m) := m + 1$, $T(\infty) = \infty$ *defines a homeomorphism of* $\mathbb{Z}(+\infty)$ *and any* $x \in \mathbb{Z}$ *fulfills* (5).

Proof By construction $\mathbb{Z}(+\infty)$ is the disjoint union of the discrete space of negative integers with the Alexandrov compactification $K = \mathbb{Z}_{\geq 0} \cup \{\infty\}$. □

Remark 2.1 *As a topological space the quotient* $Y = \mathbb{A}_f/\hat{\mathbb{Z}}^\times$ *is the restricted product of the spaces* $\mathbb{Q}_p/\mathbb{Z}_p^*$ *each of which is isomorphic to* $\mathbb{Z}(+\infty)$ *using the p-adic valuation. Thus Lemma 2.1 shows that for each rational prime* $\mathbb{Q}_p/\mathbb{Z}_p^*$ *is the minimal solution of* (5) *for the multiplication by p. It is in this sense that* $Y = \mathbb{A}_f/\hat{\mathbb{Z}}^\times$ *is the minimal solution of* (4).

2.1.2 Stability of noncommutative nature of quotients

Let now R be a locally compact space endowed with an action of $\mathbb{Z}$ given by a homeomorphism $S : R \to R$, such that the space of the orbits is *not* countably separated. In this section we show that for any auxiliary action (Y, T) as in Lemma 2.1 the product action of $T \times S$ on $Y \times R$ is *never* countably separated. In order to prove this result (Proposition 2.1) we first state the following standard fact

Lemma 2.3

(i) *Let X be a compact metrizable space. Then the set of compact subsets of X is countably separated.*
(ii) *The quotient of a compact metrizable space by an equivalence relation whose orbits are closed is always countably separated.*
(iii) *The space of probability measures on a standard Borel space is countably separated.*

Proof

(i) For any $\epsilon > 0$ there exists a finite subset F of X such that the union of open balls of radius ϵ centered at points of F cover X. For $n \in \mathbb{N}$, let $\epsilon_n = 2^{-n}$ and F_n an associated finite set. Let $V \subset X$ be an open set. For each n, let $y_n(V) = \{t \in F_n \mid B(t, \epsilon_n) \subset V\}$. The map $V \mapsto (y_n(V))_n$ from open subsets of X to the product $\prod_n 2^{F_n}$ is injective since $V = \cup_n \cup_{y_n(V)} B(t, \epsilon_n)$ and a product of finite sets is countably separated by construction.
(ii) A subset of a countably separated set is also countably separated, and since the orbits are closed they are compact so that they form a subset of the set of compact subsets of X which is countably separated by *(i)*.
(iii) The space of probability measures on a standard Borel space is the state space of the separable C^*-algebra of continuous functions on a compact metrizable space. Using a countable dense set of functions one gets the assertion. □

With the notations of Lemma 2.1 one obtains

Lemma 2.4 *The complement F of the forward orbit $T^{\mathbb{N}}x$ in K is a compact subset of (Y, T) invariant under the action of $\mathbb{Z}$ on Y.*

Proof By Lemma 2.1, for $n \in \mathbb{N}$ the subset $\{T^n x\} \subset K$ is open, thus $F \subset K$ is closed and hence compact. One has $TT^{\mathbb{N}}x \subset T^{\mathbb{N}}x$, $TK \subset K$, and if $y \in F$ and $Ty \notin F$ one has $Ty = T^m x \in T^{\mathbb{N}}x$ for some $m \geq 0$. For $m > 0$ this contradicts $y \notin T^{\mathbb{N}}x$. For $m = 0$ this gives $T^{-1}x \in K$ which contradicts (5). Thus $TF \subset F$. Let then $y \in Y$ with $Ty \in F$. Then $y \in T^{-1}TK = K$ and $y \notin T^{\mathbb{N}}x$ since $TT^{\mathbb{N}}x \subset T^{\mathbb{N}}x$. Thus $y \in F$ and one has $TF = F$. □

One concentrates on the product action of $T \times S$ on $F \times R$. Note that it is enough to show that this action is not countably separated to obtain the same result for the action of $T \times S$ on $Y \times R$. Since F is compact and $\mathbb{Z}$ is an amenable group, one can find a probability measure μ on F invariant for the action of T. Then one considers the quotient Z of $F \times R$ by the product action of $T \times S$, i.e. the space of orbits of this action. Let $\pi : F \times R \to Z$ be the quotient map and denote by $M_1(Z)$ the space of probability measures on Z.

Proposition 2.1

(i) *The map $\rho : R \to M_1(Z)$, $\rho(x) = \pi(\mu \times \delta_x)$ given by the image in Z of the probability measure $\mu \times \delta_x$ is S-invariant and induces an injection in $M_1(Z)$ of the orbit space of S in E.*
(ii) *If Z is a standard Borel space then the orbit space of S in R is countably separated.*

Proof

(i) Since $T\mu = \mu$, and $\pi \circ (T \times S) = \pi$ one has

$$\rho(Sx) = \pi(\mu \times \delta_{Sx}) = \pi((T \times S)(\mu \times \delta_x)) = \pi(\mu \times \delta_x) = \rho(x).$$

We show that ρ is an injection in $M_1(Z)$ of the orbit space of S in R. Let $x, y \in E$ belong to distinct orbits of S. Then the characteristic function h_x of the Borel subset of $F \times R$ given by $F \times S^{\mathbb{Z}}(x)$ is $(T \times S)$-invariant and one has

$$\rho(x)(h_x) = (\mu\times\delta_x)(F\times S^{\mathbb{Z}}(x)) = 1, \quad \rho(y)(h_x) = (\mu\times\delta_y)(F\times S^{\mathbb{Z}}(x)) = 0.$$

Thus one concludes that $\rho(x) \neq \rho(y)$.
(ii) It follows from *(i)* that the map ρ is an injection in $M_1(Z)$ of the orbit space of S in R. By Lemma 2.3 *(iii)*, the space $M_1(Z)$ is countably separated if Z is a standard Borel space. □

2.1.3 The need for the NCG point of view

The quotient of the real half-line $[0, \infty)$ by the action of the multiplicative group $\mathbb{Q}_+^*$ is not countably separated. Indeed, this action is ergodic for the Haar measure on the multiplicative group $\mathbb{R}_+^* \subset [0, \infty)$. Thus any Borel function invariant for the action of $\mathbb{Q}_+^*$ is almost everywhere constant. In this section we show (Theorem 2.1) that for any auxiliary action of $\mathbb{Q}_+^*$ on a locally compact space X such that the forward orbit $\mathbb{N}^\times x$ of some point $x \in X$ has a compact closure in X, the quotient of $X \times [0, \infty)$ by the product action of $\mathbb{Q}_+^*$ is *never* countably separated. The multiplicative group $\mathbb{Q}_+^*$ is the product of an infinite number of copies of $\mathbb{Z}$ parametrized by the set of primes. Its action is denoted simply as multiplication: $(q, x) \mapsto qx$. Let K be the compact closure of $\mathbb{N}^\times x$ in X. We use the compactness property to construct a probability measure μ on X invariant under the action of $\mathbb{Q}_+^*$. To achieve this result we define an increasing sequence of finite subsets $F_k \subset \mathbb{N}^\times$, $k \in \mathbb{N}$, which fulfill the following properties

1. For any integer n, all elements of F_k are divisible by n for k large enough.
2. For any prime p, one has

$$\#(F_k \Delta(pF_k))/\#(F_k) \to 0 \quad \text{for } k \to \infty$$

   where for two subsets A, B of a set C, we denote by $A \Delta B$ their symmetric difference, i.e. the complement of $A \cap B$ in $A \cup B$.

A way to define the set F_k is, for p_j the j-th prime,

$$F_k := \left\{ \prod_{j=1}^{k} p_j^{\alpha_j} \mid k < \alpha_j \leq 2k\,,\ \forall j \right\}.$$

By construction all elements in F_k are divisible by any integer n whose prime factorization only involves the first k primes taken with powers less than k. This fact holds for k large enough and for any given n, thus condition 1 is fullfilled. Moreover, for a given prime $p = p_u$ and any $k \geq u$, one has

$$F_k \Delta(pF_k) = \left\{ \prod_{j=1}^{k} p_j^{\alpha_j} \mid k < \alpha_j \leq 2k\,,\ \forall j \neq u,\ \alpha_u \in \{k+1, 2k+1\} \right\}$$

and from this one derives $\#(F_k \Delta(pF_k))/\#(F_k) = \frac{2}{k}$. Thus also 2 is achieved.

Theorem 2.1 *Let X be a locally compact metrizable space on which $\mathbb{Q}_+^*$ acts by homeomorphisms and assume that for some $x \in X$ the closure of $\mathbb{N}^\times x$ is compact. Then the quotient of the product $X \times \mathbb{R}_+^*$ by the product action of $\mathbb{Q}_+^*$ is not a standard Borel space.*

Proof We define a probability measure μ on the compact space K closure of $\mathbb{N}^\times x$ in X, by taking a limit point μ in the compact space $M_1(K)$ of the sequence of measures

$$C(K) \ni f \mapsto \frac{1}{\#(F_k)} \sum_{n \in F_k} f(nx) = \mu_k(f).$$

For $f \in C(K)$ and any prime p, one has $\mu(f_p) = \mu(f)$, where $f_p(y) := f(py)$. The same property $\mu(f_n) = \mu(f)$ thus holds for any integer n. This proves that, when viewed as a probability measure on X, the measure μ is invariant under the action of $\mathbb{Q}_+^*$. Assume now that the quotient Z of the product $X \times \mathbb{R}_+^*$ by the product action of $\mathbb{Q}_+^*$ is a standard Borel space and let π be the quotient map. One proceeds as in Lemma 2.1 to show that the map $\rho : \mathbb{R}_+^* \to M_1(Z)$ which associates with $\lambda \in \mathbb{R}_+^*$ the image $\pi(\mu \times \delta_\lambda)$ in Z of the probability measure $\mu \times \delta_\lambda$ is $\mathbb{Q}_+^*$-invariant and defines an injection in $M_1(Z)$ of the orbit space of $\mathbb{Q}_+^*$ in $\mathbb{R}_+^*$. Indeed, for any $q \in \mathbb{Q}_+^*$ and $\lambda \in \mathbb{R}_+^*$ one has

$$\rho(q\lambda) = \pi(\mu \times \delta_{q\lambda}) = \pi(q(\mu \times \delta_\lambda)) = \pi(\mu \times \delta_\lambda) = \rho(\lambda).$$

Moreover, the evaluation on the characteristic function h_λ of the Borel subset of $X \times \mathbb{R}_+^*$ given by $X \times \mathbb{Q}_+^*\lambda$ shows that ρ is an injection in $M_1(Z)$ of the orbit space of $\mathbb{Q}_+^*$ in $\mathbb{R}_+^*$. The conclusion follows since the orbit space of $\mathbb{Q}_+^*$ in $\mathbb{R}_+^*$ is not countably separated. □

2.1.4 Classical orbit and cohomological meaning of the map E

The role of the crossed product in encoding noncommutative spaces enters to give a conceptual meaning of the map E as the cyclic homology counterpart of the map between noncommutative spaces connecting the adèle class space to its "classical orbit" which in turn can be understood as a special case of the "cooling procedure" described in [16]. The cooling procedure is nothing but a testifier of the thermodynamical nature of noncommutative spaces. When applied to the BC system the cooling amounts to replace the additive Haar measure on the adèles, for which the multiplicative action of $\mathbb{Q}^*$ is ergodic, by the product of the Haar measure of the idele class group by a power of the module. The adèle class space $\mathbb{Q}^\times\backslash\mathbb{A}_\mathbb{Q}$ contains the idele class group and the cooling process provides a conceptual meaning of the restriction map. It turns out that once re-interpreted on cyclic homology HC_0, the restriction map coincides with the map E.

2.2 *The Scaling Site*

The map E defined in (3) implements the action of $\mathbb{N}^\times$ by multiplication on the real half-line $[0, \infty)$. The notion of Grothendieck topos allows one to interpret this construction geometrically, namely as the Grothendieck topos $[0, \infty) \rtimes \mathbb{N}^\times$ of $\mathbb{N}^\times$-equivariant sheaves (of sets) on the real half-line.

The combinatorial skeleton of this topos is the Arithmetic Site $\mathscr{A} = (\widehat{\mathbb{N}^\times}, \mathbb{Z}_{\max})$ [22, 23]. This is a semiringed topos where $\widehat{\mathbb{N}^\times}$ denotes the topos of sets equipped with an action of $\mathbb{N}^\times$. The structure sheaf of the Arithmetic Site is given by the semiring $\mathbb{Z}_{\max}$ of "max-plus" integers that plays a key role in tropical geometry and idempotent analysis. It is a semiring of characteristic 1, i.e. $1 \in \mathbb{Z}_{\max}$ fulfills the rule $1 + 1 := \max(1, 1) = 1$. Moreover $\mathbb{Z}_{\max}$ is the only semifield whose multiplicative group is infinite cyclic ([25] Appendix B2, Proposition B3). The action of $\mathbb{N}^\times$ on $\mathbb{Z}_{\max}$ (which turns $\mathbb{Z}_{\max}$ into the structure sheaf of $\mathscr{A}$) is an instance of a general result [29] stating that in a semifield of characteristic 1, for any $n \in \mathbb{N}$, the power maps $x \mapsto x^n$ are injective endomorphisms. These maps provide the right generalization of the Frobenius endomorphisms in finite characteristic. By construction, $\mathscr{A}$ is a topos defined over $\mathbb{B} = (\{0, 1\}, \max, +)$, the only finite semifield which is not a field. Even though $\mathscr{A}$ is a combinatorial object of countable nature, it is nonetheless endowed with a 1-parameter semigroup of correspondences on its square [22, 23]. Two further key properties of the Arithmetic Site are now recalled: (1) The points of $\mathscr{A}$ defined over $\mathbb{R}_+^{\max}$ (the multiplicative version of the tropical semifield $\mathbb{R}_{\max}$) form the basic sector $\mathbb{Q}^\times\backslash\mathbb{A}_\mathbb{Q}/\hat{\mathbb{Z}}^*$ of the adèle class space of $\mathbb{Q}$; (2) the canonical action of $\mathrm{Aut}(\mathbb{R}_+^{\max})$ on these points corresponds to the action of the idele class group on $\mathbb{Q}^\times\backslash\mathbb{A}_\mathbb{Q}/\hat{\mathbb{Z}}^*$. These facts lead us to investigate the semiringed topos obtained from the Arithmetic Site by extension of scalars from $\mathbb{B}$ to $\mathbb{R}_+^{\max}$. This space admits $[0,\infty)\rtimes\mathbb{N}^\times$ as the underlying topos, and moreover it inherits, from its construction by extension of scalars, a natural sheaf $\mathscr{O}$ of regular functions. We call *Scaling Site* the semiringed topos

$$\mathscr{S} := \big([0,\infty)\rtimes\mathbb{N}^\times, \mathscr{O}\big)$$

so obtained [24, 25]. The sections of the sheaf $\mathscr{O}$ are convex, piecewise affine functions with integral slopes.

2.3 Geometry of the adèle class space

The relation between $\mathscr{S}$ and the adèle class space of $\mathbb{Q}$ is provided by the following result which states that the isomorphism classes of points of the topos $[0,\infty)\rtimes\mathbb{N}^\times$ form the basic sector of the adèle class space of $\mathbb{Q}$ [25].

Theorem 2.2 *The space of points of the topos* $[0,\infty)\rtimes\mathbb{N}^\times$ *is canonically isomorphic to* $\mathbb{Q}^\times\backslash\mathbb{A}_\mathbb{Q}/\hat{\mathbb{Z}}^*$.

This theorem provides an algebraic-geometric structure on the adèle class space, namely that of a tropical curve in an extended sense. In [25] this structure was examined by considering its restriction onto the periodic orbit of the scaling flow associated with each rational prime p. The output is that of a tropical structure which describes this orbit as a real variant $C_p = \mathbb{R}_+^*/p^\mathbb{Z}$ of the classical Jacobi description $\mathbb{C}^\times/q^\mathbb{Z}$ of a complex elliptic curve. On C_p, a theory of Cartier divisors is available; moreover the structure of the quotient of the abelian group of divisors by the subgroup of principal divisors has been also completely described in *op.cit.* The same paper also contains a description of the theory of theta functions on C_p and finally a proof of the Riemann–Roch formula stated in terms of real valued dimensions, as in the type-II index theory.

The main contribution of the adèle class space to this geometric picture is to provide, through the implementation of the Riemann-Weil explicit formulas as a trace formula, the understanding of the Riemann zeta function as a Hasse–Weil generating function.

In the function field case, the Hasse–Weil formula writes the zeta function as a generating function (the Hasse–Weil zeta function)

$$\zeta_C(s) := Z(C, q^{-s}), \qquad Z(C,T) := \exp\left(\sum_{r\geq 1} N(q^r)\frac{T^r}{r}\right). \tag{6}$$

For function fields, q is the number of elements of the finite field $\mathbb{F}_q$ on which the associated curve C is defined.

In the case of the Riemann zeta function, the analogue of (6) was obtained in [20, 21] by considering the limit of the right hand side of (6) when $q \to 1$. This process was originally suggested by C. Soulé, who introduced the zeta function of a variety X over $\mathbb{F}_1$ using the *polynomial* counting function $N(x) \in \mathbb{Z}[x]$ associated with X. The definition of the zeta function is as follows

$$\zeta_X(s) := \lim_{q\to 1} Z(X, q^{-s})(q-1)^{N(1)}, \qquad s \in \mathbb{R}. \tag{7}$$

When one seeks to apply (7) to get the Riemann zeta function (completed by the gamma factor at the Archimedean place), one meets the obvious obstruction that the exponent $N(1)$ is equal to $-\infty$ due to the infinite number of its zeros. In [20, 21] a simple way to bypass this difficulty is described i.e. one considers the logarithmic derivatives of both terms in (7) and observes that the Riemann sums of an integral appear from the right hand side. Then, instead of dealing with (7) one works with the equation

$$\frac{\partial_s \zeta_N(s)}{\zeta_N(s)} = -\int_1^\infty N(u)\, u^{-s} d^*u \tag{8}$$

which points out to a precise equation for the counting function $N_C(q) = N(q)$ associated with C namely

$$\frac{\partial_s \zeta_{\mathbb{Q}}(s)}{\zeta_{\mathbb{Q}}(s)} = -\int_1^\infty N(u)\, u^{-s} d^*u. \tag{9}$$

In fact, one finds that this equation admits a *distribution* as a solution which is given explicitly as

$$N(u) = \frac{d}{du}\varphi(u) + \kappa(u) \tag{10}$$

where $\varphi(u) := \sum_{n<u} n\, \Lambda(n)$, and $\kappa(u)$ is the distribution that appears in the explicit formula

$$\int_1^\infty \kappa(u) f(u) d^*u = \int_1^\infty \frac{u^2 f(u) - f(1)}{u^2-1} d^*u + cf(1), \qquad c = \frac{1}{2}(\log\pi + \gamma).$$

The conclusion is that the distribution $N(u)$ is positive on $(1, \infty)$ and is given by

$$N(u) = u - \frac{d}{du}\left(\sum_{\rho \in Z} \operatorname{order}(\rho) \frac{u^{\rho+1}}{\rho+1}\right) + 1 \tag{11}$$

where the derivative is taken in the sense of distributions, and the value at $u = 1$ of the term $\omega(u) = \sum_{\rho \in Z} \operatorname{order}(\rho) \frac{u^{\rho+1}}{\rho+1}$ is given by $\frac{1}{2} + \frac{\gamma}{2} + \frac{\log 4\pi}{2} - \frac{\zeta'(-1)}{\zeta(-1)}$.

As explained in [21] the adèle class space provides the geometric meaning of the counting distribution $N(u)$ and thus shows the coherence of our geometric approach.

3 The Riemann–Roch strategy

In relation to the study of the zeros of the Riemann zeta function, the Riemann–Roch strategy consists in trading the question of the location of the zeros for the problem of proving the *non-positivity* of a certain quadratic form $\mathfrak{s}(f, f)$ (see (12)). In the function field case, this inequality derives from an argument of algebraic geometry in finite characteristic and the most conceptual proof was obtained by applying the Riemann–Roch formula on the square of the curve defining the function field [30]. In that case, the function f defines a divisor D on the surface, as a linear combination of Frobenius correspondences. Then, if one assumes the positivity of $\mathfrak{s}(f, f) > 0$ for some f, it is the existence part of the Riemann–Roch theorem which yields a contradiction. More precisely, the assumed positivity $\mathfrak{s}(f, f) > 0$ together with the appearance of $\mathfrak{s}(f, f)$ as the leading term in the topological side of the Riemann–Roch formula show that one can turn the divisor nD for a suitable $n \in \mathbb{Z}$ into an effective divisor and obtain a contradiction. This argument will be reconsidered in more detail in Section 3.1. For function fields, the Riemann–Roch formula relies on algebraic geometry in the same finite characteristic. In the case of the Riemann zeta function, the structure sheaf of the Scaling Site $\mathscr{S}$ is in characteristic 1, thus it seems reasonable trying to develop a Riemann–Roch formalism in that context. Some very encouraging results are obtained in [25], inclusive of a type-II Riemann–Roch formula for the periodic orbits. In this case, the cohomology H^0 is defined using global sections while H^1 is introduced by turning Serre duality into a definition. In order to attack the two-dimensional case of the square of the Scaling Site, one needs to define the intermediate H^1 and a first direct attempt, based on homological algebra in characteristic 1, is developed in [26]. It is striking that the existence results for the Riemann–Roch problem in tropical geometry [2, 28, 37] are deeply related to potential theory and game theory [3, 5] thus pointing to the relevance of these tools in a direct attack to the Riemann–Roch formula needed for RH. Here we develop yet another approach which is based on the construction of a complex lift from a geometry in characteristic 1 to the complex world and the use of the tropicalization map. In Section 3.2, we explain how this *tropical descent* allows one, in the context of the Riemann–Roch problem, to prove the existence results in characteristic 1 from existence results in characteristic 0. Section 3.3 recalls the classical link, in characteristic zero, between the Hirzebruch–Riemann–Roch theorem and the Index theorem. Finally, Section 3.4 lays down our actual strategy which is based on the complex lift of the Scaling Site.

3.1 The role of the existence part of the Riemann–Roch formula in characteristic one

It is known [8] that the RH problem is equivalent to an inequality for real valued functions f on $\mathbb{R}_+^*$ of the form

$$\text{RH} \iff \mathfrak{s}(f,f) \le 0\,,\ \ \forall f \mid \int f(u)d^*u = \int f(u)du = 0. \tag{12}$$

Here, for real compactly supported functions on $\mathbb{R}_+^*$, one lets $\mathfrak{s}(f,g) := N(f \star \tilde{g})$, where $\star$ is the convolution product on $\mathbb{R}_+^*$, $\tilde{g}(u) := u^{-1}g(u^{-1})$, and

$$N(h) := \sum_{n=1}^{\infty} \Lambda(n)h(n) + \int_1^{\infty} \frac{u^2h(u) - h(1)}{u^2 - 1} d^*u + c\,h(1)\,,\ c = \frac{1}{2}(\log \pi + \gamma). \tag{13}$$

It follows from the geometric interpretation of the explicit formulas as in [21] that the quadratic form $\mathfrak{s}(f,f)$ can be expressed as the self-intersection of the divisor on the square of the Scaling Site by the formula involving the Frobenius correspondences Ψ_λ

$$\mathfrak{s}(f,f) = D \bullet D,\ \ D := \int f(\lambda)\Psi_\lambda\, d^*\lambda. \tag{14}$$

The intersection number of divisors is provided by the formula

$$D \bullet D' := < D \star \tilde{D}', \Delta >$$

where $\tilde{D}'$ is the transposed of D' and the composition $D \star \tilde{D}'$ is computed by bilinearity, while the intersection $< D \star \tilde{D}', \Delta >$ is obtained using the distribution $N(u)$ and the fact that Ψ_λ is of degree λ.

The Riemann–Roch strategy seeks to obtain a contradiction by assuming that, contrary to (12), one has $\mathfrak{s}(f,f) > 0$, for some function f. The key missing step is provided by the implementation of a Riemann–Roch formula whose topological side is $\frac{1}{2}D \bullet D$ and to conclude from it that one can make the divisor $D := \int f(\lambda)\Psi_\lambda\, d^*\lambda$ (or its opposite $-D$) effective.

The positivity of the divisor $D + (k)$ would then contradict the fact that the degree and codegree of $D = \int f(\lambda)\Psi_\lambda\, d^*\lambda$ is equal to 0 in view of the hypothesis $\int f(u)d^*u = \int f(u)du = 0$.

3.2 *Tropical descent*

The new step in our strategy is to obtain the existence part of the Riemann–Roch theorem in the tropical shadow from the results on the analytic geometric version of the space. Obviously, the advantage of working in characteristic zero is that to have already available all the algebraic and analytical tools needed to test such formula.

We first explain how the Scaling Site appears naturally from the well-known results on the localization of zeros of analytic functions by means of Newton polygons in the non-Archimedean case and Jensen's formula in the complex case. These results in fact combine to show that the tropical half-line $(0, \infty)$, endowed with the structure sheaf of convex, piecewise affine functions with integral slopes, gives a common framework for the localization of zeros of analytic functions in the punctured unit disk. The additional structure involved in the Scaling Site, namely the action of $\mathbb{N}^\times$ by multiplication on the tropical half-line, corresponds, as shown in (16) and (17), to the transformation on functions given by the composition with the n-th power of the variable. The tropical notion of "zeros" of a convex piecewise affine function f with integral slope is that a zero of order k occurs at a point of discontinuity of the derivative f', with the order k equal to the sum of the outgoing slopes. The conceptual meaning of this notion is understood by using Cartier divisors.

3.2.1 Tropicalization in the p-adic case, Newton polygons

Let K be a complete and algebraically closed extension of $\mathbb{Q}_p$ and $v(x) = -\log|x|$ be the valuation. The tropicalization of a series with coefficients in K is obtained by applying the transformation $a \mapsto \log|a| = -v(a)$ to the coefficients and by implementing the change of operations: $+ \to \vee = \sup$, $\times \to +$, so that $X^n \to -nx$. In this way a sum of monomials such as $\sum a_n X^n$ is replaced by $\vee(-nx - v(a_n))$.

Definition 3.1 Let $f(X) = \sum a_n X^n$ be a Laurent series with coefficients in K and convergent in an annulus $A(r_1, r_2) = \{z \in K \mid r_1 < |z| < r_2\}$. The tropicalization of f is the real valued function of a real parameter

$$\tau(f)(x) := \max_n\{-nx - v(a_n)\} \qquad \forall x \in (-\log r_2, -\log r_1). \tag{15}$$

Up-to a trivial change of variables, this notion is well-known in p-adic analysis, where the function $-\tau(-x)$, or rather its graph, is called the *valuation polygon* of the series [40]. This polygon is dual to the Newton polygon of the series which is, by definition, the lower part of the convex hull of the points of the real plane with coordinates $(j, v(a_j))$. By construction, $\tau(f)(x)$ is finite since, using the convergence hypothesis, the terms $-nx - v(a_n)$ tend to $-\infty$ when $|n| \to \infty$. Thus one obtains a convex and piecewise affine function. Moreover, the multiplicativity property also holds $\tau(fg)(x) = \tau(f)(x) + \tau(g)(x), \forall x \in (0, \infty)$ as well as the following classical result [40].

Theorem 3.1 *Let* $f(X) = \sum a_n X^n$ *be a Laurent series with coefficients in* K*, convergent in an annulus* $A(r_1, r_2) = \{z \in K \mid r_1 < |z| < r_2\}$*. Then the valuations* $v(z_i)$ *of its zeros* $z_i \in A(r_1, r_2)$ *(counted with multiplicities) are the zeros (in the tropical sense and counted with multiplicities) of the tropicalization* $\tau(f)$ *in* $(-\log r_2, -\log r_1)$.

In particular, one can take $r_1 = 0, r_2 = 1$ so that $A(r_1, r_2)$ is the punctured open unit disk $D(0, 1)\backslash\{0\}$. In this case, $\tau(f)$ are convex piecewise affine functions on $(0, \infty)$ and one derives the following compatibility with the action of $\mathbb{N}^\times$ on functions by $f(X) \mapsto f(X^n)$

$$\tau(f(X^n))(x) = \tau(f)(nx)\,, \quad \forall x \in (0, \infty), \; n \in \mathbb{N}^\times. \tag{16}$$

3.2.2 Tropicalization in the Archimedean case, Jensen's formula

Over the complex numbers, unlike the non-Archimedean case, it is not true that for a generic radius r, the modulus $|f(z)|$ (of a complex function $f(z)$) is constant on the sphere of radius r. One replaces (15) with the following

Definition 3.2 Let $f(z)$ be a holomorphic function in an annulus $A(r_1, r_2) = \{z \in \mathbb{C} \mid r_1 < |z| < r_2\}$. Its tropicalization is the function on the interval $(-\log r_2, -\log r_1)$

$$\tau(f)(x) := \frac{1}{2\pi} \int_0^{2\pi} \log|f(e^{-x+i\theta})| d\theta.$$

By construction, the multiplicativity property still holds: $\tau(fg)(x) = \tau(f)(x) + \tau(g)(x), \forall x \in (0, \infty)$.

For $x \in (-\log r_2, -\log r_1)$ such that f has no zero on the circle of radius e^{-x}, the derivative of $\tau(f)(x)$ is the opposite of the winding number $n(x)$ of the loop $\theta \mapsto f(e^{-x+i\theta}) \in \mathbb{C}^\times$. Thus the function $\tau(f)(x)$ is piecewise affine with integral slopes. When the radius e^{-x} of the circle increases, the winding number of the associated loop increases by the number of zeros of f in the intermediate annulus and this shows that the function $\tau(f)(x)$ is convex and fulfills Jensen's formula (cf. [41] Theorem 15.15). Thus we derive the analogue of Theorem 3.1

Theorem 3.2 *Let* $f(z)$ *be a holomorphic function in an annulus* $A(r_1, r_2) = \{z \in \mathbb{C} \mid r_1 < |z| < r_2\}$ *and* $z_i \in A(r_1, r_2)$ *its zeros counted with their multiplicities.*

Then the values $-\log|z_i|$ *are the zeros (in the tropical sense and counted with multiplicities) of the tropicalization* $\tau(f)$ *in* $(-\log r_2, -\log r_1)$.

In particular, one can take $r_1 = 0, r_2 = 1$ so that $A(r_1, r_2)$ is the open punctured unit disk $D(0, 1) \setminus \{0\}$. In that case the $\tau(f)$ are convex piecewise affine functions on $(0, \infty)$ and one has the following compatibility with the action of $\mathbb{N}^\times$ on functions by $f(z) \mapsto f(z^n)$

$$\tau(f(z^n))(x) = \tau(f)(nx)\,, \quad \forall x \in (0, \infty),\ n \in \mathbb{N}^\times. \tag{17}$$

This fact follows from the equality for periodic functions $h(\theta)$

$$\frac{1}{2\pi}\int_0^{2\pi} h(n\theta)d\theta = \frac{1}{2n\pi}\int_0^{2n\pi} h(\alpha)d\alpha = \frac{1}{2\pi}\int_0^{2\pi} h(u)du.$$

3.2.3 Descent from characteristic zero to characteristic one

To explain the general technique that allows one to deduce the existence results in characteristic one from a Riemann–Roch formula in characteristic zero, we first develop the following simple example. Consider an open interval I of the real half-line and an integral (finite) divisor $D = \sum n_j \delta_{\lambda_j}$, with $n_j \in \mathbb{Z}$ and $\lambda_j \in I$. The Riemann–Roch problem in characteristic one asks for the construction of a piecewise affine continuous function f with integral slopes, whose divisor (f) fulfills $D + (f) \geq 0$. Here, (f) is best understood as the second derivative $\Delta(f)$, taken in the sense of distributions. Thus the Riemann–Roch problem in characteristic one corresponds to the solutions f, among piecewise affine continuous function f with integral slopes, of the inequality

$$D + (f) := \sum n_j \delta_{\lambda_j} + \Delta(f) \geq 0. \tag{18}$$

The technique we follow is to lift geometrically the divisor D to a divisor $\tilde{D}$ (in the ordinary complex analytic sense) in the corona

$$\mathscr{C}(I) := \{z \in \mathbb{C} \mid -\log|z| \in I\}.$$

This involves a choice, for each λ_j, of points $z \in \mathscr{C}(I)$ such that $-\log|z| = \lambda_j$, and of multiplicities for these points which add up to n_j. Now, assume that one has a solution as a meromorphic function g in $\mathscr{C}(I)$ such that $\tilde{D} + (g) \geq 0$. We then consider, using Definition 3.2, the tropicalization $f = \tau(g)$. This formula is in fact extended to meromorphic functions by the multiplicativity rule, i.e. using $\tau(h/k) := \tau(h) - \tau(k)$ for $g = h/k$. Then, Theorem 3.2 shows that the divisor $\Delta(\tau(g))$ is the image by the map $u(z) := -\log|z|$ of the divisor of g. This proves that the tropicalization $f = \tau(g)$ fulfills the inequality $D+(f) \geq 0$ of the Riemann–Roch problem.

3.3 *The Hirzebruch–Riemann–Roch formula and the Index theorem*

Here we recall the Hirzebruch–Riemann–Roch theorem. Let E be a holomorphic complex vector bundle of rank r over a compact complex manifold X of dimension n. The Euler characteristic $\chi(E)$ of E is defined by

$$\chi(E) := \sum_{j\geq 0}(-1)^j \dim(H^j(X, E)). \tag{19}$$

The cohomology $H^j(X, E)$ used in the formula is sheaf cohomology and one uses the equivalence between holomorphic vector bundles and locally free sheaves. It is known that the cohomology $H^j(X, E)$ vanishes for $j > n$. The relation with the analytic index is given, with the above notations, by the formula

$$\chi(E) = \mathrm{Ind}_a(\overline{\partial}_E). \tag{20}$$

The analytic index $\mathrm{Ind}_a(T)$ of an operator is defined as

$$\mathrm{Ind}_a(T) := \dim(\mathrm{Ker}(T)) - \dim(\mathrm{Ker}(T^*))$$

and $\overline{\partial}_E$ denotes the "dbar" operator with coefficients in E. The Hirzebruch–Riemann–Roch formula, which is a special case of the Atiyah–Singer Index theorem, is the equality

$$\chi(E) = \langle \mathrm{Ch}(E)\mathrm{Td}(X), [X]\rangle \tag{21}$$

of the Euler characteristic of E with the topological index which is the evaluation on the fundamental class $[X]$ of X of the cohomology class $\mathrm{Ch}(E)\mathrm{Td}(X)$ product of the Chern character $\mathrm{Ch}(E)$ of the vector bundle E and the Todd genus $\mathrm{Td}(X)$ of X.

3.4 *Potential role of the complex lift of the Scaling Site*

In the case of the complex lift of the (square of the) Scaling Site, we expect E to be a line bundle, the Todd genus be equal to 1 and that the relevant term in the topological index comes from the term $\frac{1}{2}c_1(E)^2$ in the Chern character of E.

In this setup, one difficulty is that the self-intersection of the divisor D appears as a trace taken in a *relative* situation. This means that one works with the difference between the adèle class space (divided by $\hat{\mathbb{Z}}^*$), say X, and the ideles (also divided by $\hat{\mathbb{Z}}^*$), which form a subset $Y \subset X$. The explicit formulas are obtained in the form (after a cut-off)

$$(\mathrm{Tr}_X - \mathrm{Tr}_Y)(\pi(f))$$

and this corresponds to the spectral realization as a cokernel of $E : \mathscr{F}(X) \to \mathscr{F}(Y)$. Thus, the trace on this cokernel corresponds to the *opposite* of $(\mathrm{Tr}_X - \mathrm{Tr}_Y)(\pi(f))$ as required by the minus sign in the Explicit Formulas. In fact, a first task should be to understand how to express this difference of traces as an intersection number and then develop an appropriate intersection theory. The advantage of working in a complex framework is that one could replace the naive real intersections by the intersection of complex manifolds and also that everything is compatible with the use of the Fourier transform. In fact, we also speculate that the divergent term in $\log \Lambda$ which enters as coefficient of $f(1)$ for the test function (see [16] Theorem 2.36) is due to the lack of good definition of self-intersection of the diagonal. While one obtains an infinite result when working naively, the implementation of a suitable intersection theory should provide the correct Euler characteristic. Thus adapting the Riemann–Roch strategy comprises the following five steps

1. Construct the complex lift Γ of the Scaling Site.
2. Develop intersection theory in such a way that the divergent term in $\log \Lambda$ (see [16] Theorem 2.36) is eliminated.
3. Formulate and prove a Hirzebruch–Riemann–Roch formula on Γ^2, whose topological side part $\frac{1}{2}c_1(E)^2$ is $\frac{1}{2}\mathfrak{s}(f, f)$ as in (14). This step involves the lifting of the divisor $D(f) = \int f(\lambda)\Psi_\lambda \, d^*\lambda$ in characteristic 1 to a divisor $\tilde{D}(f)$ in the complex setup and the use of correspondences.
4. Use the assumed positivity of $\mathfrak{s}(f, f)$ to get an existence result for $H^0(\tilde{D}(f))$ or $H^0(-\tilde{D}(f))$.
5. Use tropical descent to get the effectivity of a divisor equivalent to $D(f)$ and finally get a contradiction.

The development of step 3 is the most problematic since in the lift from characteristic 1 to characteristic 0 one loses the automorphisms $\mathrm{Aut}(\mathbb{R}_+^{\max}) = \mathbb{R}_+^*$ which are at the origin of the Frobenius correspondences Ψ_λ. We settle this problem in Section 7 using the Witt construction in characteristic 1.

4 Tropical descent and almost periodic functions

In order to lift a continuous divisor $D(f) = \int f(\lambda)\delta_\lambda \, d^*\lambda$ on the Scaling Site (in characteristic 1, Section 3.1) to a *discrete* divisor $\tilde{D}(f)$ on a complex geometric space, one first needs to understand how to generalize Jensen's formula to a case where the Jensen function is no longer a piecewise linear affine convex function with integral slopes but an arbitrary convex function.

In this part we explain how H. Bohr's theory of almost periodic functions, and the theory developed by B. Jessen on the density of zeros of almost periodic analytic functions gives a satisfactory answer to this question. This technique plays a crucial role in the process to extend the tropical descent procedure of Section 3.2 to control the continuous divisors, in characteristic 1, following the Riemann–Roch lifting strategy. This procedure will also suggest a further important information on the need of a suitable compactification G of the *imaginary direction* required for a correct complex lift of the Scaling Site. This part will be developed in Section 5.

4.1 Almost periodic functions

We recall the definition of almost periodic functions (see [6] and [4] for more details). Let H be the locally compact abelian group $\mathbb{R}$ or $\mathbb{Z}$.

Definition 4.1 Let $f : H \longrightarrow \mathbb{C}$ be a bounded continuous function and $\varepsilon > 0$ a real number. An ε-almost period for f is a number $\tau \in H$ such that

$$\|f(.+\tau) - f(.)\|_\infty := \sup_{x\in H} |f(x+\tau) - f(x)| < \varepsilon.$$

The function f is said to be almost periodic if for any $\varepsilon > 0$ the set of ε-almost periods of f is relatively dense, i.e., there is a real number $l = l(\varepsilon) > 0$ such that any interval with length l contains at least one ε-almost period.

The space of almost periodic functions on H is denoted by $\mathrm{AP}(H)$. In the following part we are mostly interested in the case $H = \mathbb{R}$ but will use the case $H = \mathbb{Z}$ when considering sequences.

By construction, $\mathrm{AP}(H)$ is a C^*-subalgebra of the C^*-algebra $C_b(H)$ of bounded continuous functions on H. An important characterization of almost periodic functions was given by Bochner [7].

Theorem 4.1 *A bounded continuous function $f \in C_b(H)$ is an almost periodic function if and only if the family of translates $\{f(.+t)\}_{t\in H}$ is relatively compact in $C_b(H)$, i.e. its closure is compact.*

Bochner's characterization lead von Neumann in [42] to extend the notion of almost periodic function to arbitrary groups by requiring the relative compactness for the uniform norm of the set of translates of f. This definition does not make use of the topology of the group and von Neumann constructed the mean value of a function f using the translation invariant element in the closed convex hull of the translates of the function.

4.2 From Jensen to Jessen and the tropical descent

Jensen's formula in the annular case allows one to define the tropicalization of a holomorphic function. In [31], Jessen extended Jensen's formula to analytic almost periodic functions.

Recall that an analytic function $f(z)$ in the strip $\Re(z) \in [\alpha, \beta]$ is called almost periodic when the function $\mathbb{R} \ni t \mapsto f(\sigma + it)$ is uniformly almost periodic for $\sigma \in [\alpha, \beta]$.

Jessen showed that, for such a function, the following limit exists

$$\varphi(\sigma) := \lim_{T\to\infty} \frac{1}{2T} \int_{-T}^{T} \log |f(\sigma + it)| dt \tag{22}$$

and determines a real convex continuous function $\varphi(\sigma)$ of $\sigma \in [\alpha, \beta]$. The function $\varphi(\sigma)$ is called the Jensen function of f. By convexity, the derivative $\varphi'(\sigma)$ exists at all points of the interval except for a denumerable set E. For σ_j outside E, Jessen proved that the relative frequency of zeros of f in the strip $\Re(z) \in [\sigma_1, \sigma_2]$ exists and is given by the variation of the derivative φ'. More precisely, if $N(T)$ denotes the number of zeros of f with $\Re(z) \in [\sigma_1, \sigma_2]$, and $-T < \Im(z) < T$ one has

$$\lim_{T\to\infty} \frac{N(T)}{2T} = \frac{\varphi'(\sigma_2) - \varphi'(\sigma_1)}{2\pi}. \tag{23}$$

4.3 Discrete lift of a continuous divisor

In this part we describe, following [32], the procedure of lifting a tropical continuous divisor (i.e. a formal integral of delta functions $\int f(\lambda)\delta_\lambda \, d\lambda$) to a discrete, *integer valued* divisor, using the technique of almost periodic lifting.

The formal expression $\int f(\lambda)\delta_\lambda \, d\lambda$ replaces the finite discrete sum as in (18) of Section 3.2.3. We recall that the basic relation defining the divisor $\mathrm{div}(\phi)$, in characteristic 1, of a piecewise affine function $\phi(\sigma)$ is $\mathrm{div}(\phi) = \Delta(\phi)$, where Δ is the Laplacian taken in the sense of distributions. Here, we extend this definition to convex functions in terms of the equation (taken in the sense of distributions)

$$\mathrm{div}(\phi) := \Delta(\phi). \tag{24}$$

Then, the almost periodic lifting of a convex function is the *choice* of an almost periodic analytic function f whose tropicalization gives back the function $\phi(\sigma)$. More precisely (following [32], Theorem 25) one has the next characterization of a Jensen function of an almost periodic analytic function

Theorem 4.2 ([32], Theorem 25) *A real function $\phi(\sigma)$, in the interval $\alpha < \sigma < \beta$, is the Jensen function of an almost periodic analytic function in the strip $\Re(z) \in [\alpha, \beta]$ if and only if $\phi(\sigma)$ is convex and for every compact interval $I \subset (\alpha, \beta)$ there exist a finite set F of $\mathbb{Q}$-linearly independent real numbers and a real number $C < \infty$ such that the positive difference $\phi'(\sigma_2) - \phi'(\sigma_1)$ of slopes of $\phi(\sigma)$ in intervals where it is affine is a rational combination of elements of F with*

$$\phi'(\sigma_2) - \phi'(\sigma_1) = \sum_{\mu \in F} r(\mu)\mu, \qquad \sum_{\mu \in F} r(\mu)^2 \leq C\,|\phi'(\sigma_2) - \phi'(\sigma_1)|^2.$$

These results are based on the characterization of the asymptotic distribution function of almost periodic sequences $U(k) \in \mathbb{R}$, $k \in \mathbb{Z}$. Here, to be almost periodic for sequences means, with $H = \mathbb{Z}$ in Definition 4.1, that for any $\epsilon > 0$ the set of ϵ-periods

$$\chi(\epsilon) := \{j \mid |U(k+j) - U(k)| < \epsilon\,,\ \forall k \in \mathbb{Z}\}$$

is relatively dense in $\mathbb{Z}$ (see Definition 4.1). A distribution function is a non-decreasing function $\mu(\sigma)$ of a real variable $\sigma \in \mathbb{R}$ whose limit, when $\sigma \to -\infty$ is 0 and whose limit, when $\sigma \to \infty$ is 1. One defines $\mu(\sigma \pm 0)$ respectively as the limits on the left and on the right and one disregards the choice of a precise value in the interval $[\mu(\sigma - 0), \mu(\sigma + 0)]$ when the two values are different. This situation only occurs on a denumerable set of values of σ. The asymptotic distribution function of an almost periodic sequence $U(k)$ of real numbers is defined using the densities of the subsets $E_-(\sigma) := \{k \mid U(k) < \sigma\}$ and $E_+(\sigma) := \{k \mid U(k) \leq \sigma\}$. For an arbitrary subset $E \subset \mathbb{Z}$, one defines first the lower and upper densities by the formulas

$$\underline{\rho}(E) := \liminf_I \frac{\#\{E \cap I\}}{\#I}, \qquad \overline{\rho}(E) := \limsup_I \frac{\#\{E \cap I\}}{\#I} \tag{25}$$

where the limits are taken over all intervals $I = [a, b] \subset \mathbb{Z}$ whose length $b - a$ tends to ∞. Finally, the asymptotic distribution of an almost periodic sequence $U(k)$, when it exists, is uniquely determined (as a distribution function in the above sense) as the non-decreasing function $\mu(\sigma)$ such that

$$\mu(\sigma - 0) \leq \underline{\rho}(E_-(\sigma)) \leq \overline{\rho}(E_+(\sigma)) \leq \mu(\sigma + 0). \tag{26}$$

A. Wintner showed that such a distribution function exists for all almost periodic sequences $U(k)$ of real numbers (see [32] Theorem 10 for a simple proof). The almost periodic sequences are the continuous functions on the almost periodic compactification of $\mathbb{Z}$ which is the dual of the additive group $(\mathbb{R}/2\pi\mathbb{Z})_{\rm dis}$ endowed with the discrete topology. This abelian group is uncountable but the Fourier transform of an almost periodic sequence $U(k)$

$$\hat{U}(s) := \lim_{T \to \infty} \frac{1}{2T} \sum_{-T}^{T} U(k) e^{isk}, \quad s \in (\mathbb{R}/2\pi\mathbb{Z})_{\rm dis}$$

vanishes except on a countable subset, called the set of exponents of U in *op.cit.* The subgroup $M \subset \mathbb{R}/2\pi\mathbb{Z}$ generated by the exponents of U is called the "modul" M_U of U. Its intersection with $2\pi\mathbb{Q}$ plays a role in particular in the following result

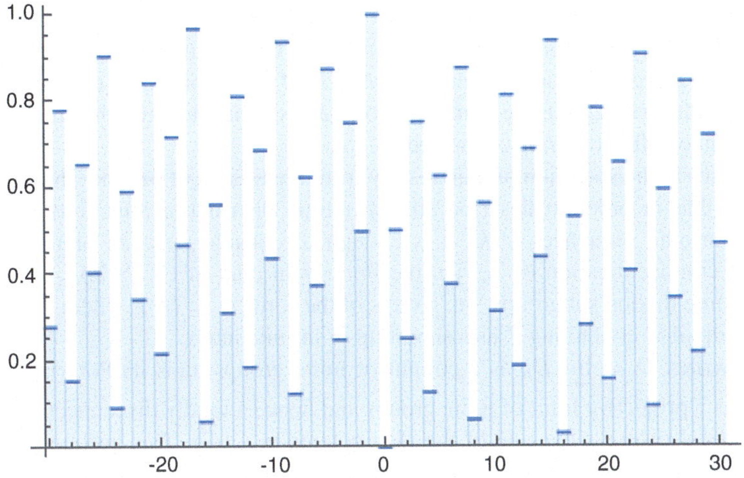

Fig. 2 Jessen sequence $U(k)$

Theorem 4.3 ([32], Theorem 11) *The asymptotic distribution function $\mu(\sigma)$ of an almost periodic sequence $U(k)$ of real numbers is constant in an open interval if and only if the sequence does not take any value in this interval. In this case the value of $\mu(\sigma)$ in this interval is a rational number which belongs to $\frac{1}{2\pi}M_U$.*

The interesting outcome of this result is the rationality of the value $\mu(\sigma) = r$ in the interval where the spectrum is empty. The proof uses the fact that an ϵ-period, for ϵ smaller than the size of the gap, is a true period for the subset where $U(k) < \sigma$ and that the density of a periodic set is a rational number (Figure 2).

The following result (see [32], Theorem 12) characterizes the distribution functions of the almost periodic sequences whose exponents belong to a fixed subgroup $M \subset \mathbb{R}/2\pi\mathbb{Z}$ which is assumed to be everywhere dense for the usual topology.

Theorem 4.4 *Let M be a given dense subgroup of $\mathbb{R}/2\pi\mathbb{Z}$. A distribution function is the asymptotic distribution function $\mu(\sigma)$ of an almost periodic sequence $U(k)$ with exponents in M if and only if it has compact support and the values $\mu(\sigma)$ in constancy intervals belong to $\frac{1}{2\pi}M$.*

To phrase this result in modern terms, note that the asymptotic distribution functions $\mu(\sigma)$ of almost periodic sequences $U(k)$ with exponents in M are the same as the functions of the form

$$\mu(\sigma) = \nu(\{u \mid h(u) \leq \sigma\}$$

where $h \in C(\hat{M})$ is an arbitrary continuous function on the compactification of $\mathbb{Z}$ given by the Pontrjagin dual $\hat{M}$ of M, and where ν is the normalized Haar measure on $\hat{M}$. The constancy intervals of $\mu(\sigma)$ are gaps in the spectrum of h and the corresponding spectral projection given by the Cauchy integral of the resolvent through the gap gives an idempotent in $C(\hat{M})$. If M is torsion free (i.e. M intersects trivially with $2\pi\mathbb{Q}/2\pi\mathbb{Z}$), the compact group $\hat{M}$ is connected and the spectrum of any continuous function h is connected.

The gaps in the spectrum of h arise only from the torsion of M, and Theorem 4.4 suggests that in order to describe all asymptotic distribution functions $\mu(\sigma)$ of almost periodic sequences $U(k)$ it is enough to consider, instead of the almost periodic compactification of $\mathbb{Z}$, (i.e. the Pontrjagin dual of $(\mathbb{R}/2\pi\mathbb{Z})_{\text{dis}}$) the dual G of infinite torsion subgroups, i.e. groups of the form $D = H/\mathbb{Z}$ where $H \subset \mathbb{Q}$ is a subgroup of $\mathbb{Q}$ containing $\mathbb{Z}$ as a subgroup of infinite index.

One writes $D = \varinjlim D_n$ as a colimit of finite groups, thus its Pontrjagin dual $G = \hat{D}$ is a projective limit of finite groups and hence a totally disconnected space. Theorem 4.4 displays the role of the idempotents in $C(G)$. This accounts for the existence of the many idempotents in $C(G)$ associated with constancy intervals of asymptotic distribution functions $\mu(\sigma)$.

To understand in explicit terms how to obtain general distribution functions, we first work out the description of the distribution function $\mu(\sigma) := \sigma$ for $\sigma \in [0, 1]$, for the group H_p of rational numbers with denominator a power of p, with p a fixed prime number.

In this case, the group $D = H/\mathbb{Z}$ is the colimit of the finite groups $D_n := \mathbb{Z}/p^n\mathbb{Z}$, viewed as groups of roots of unity of order dividing p^n, so that the inclusions $D_n \subset D_{n+1}$ turn D into a dense subgroup of $U(1)$. Dually, one thus gets a projective system of the finite cyclic groups $\hat{D}_n$ and an homomorphism $\mathbb{Z} \to \varprojlim \hat{D}_n$. The projective limit $K := \varprojlim \hat{D}_n$ is topologically a Cantor set

Lemma 4.1 *Let p be a prime.*

(i) There exists a unique sequence $U(x)$, $x \in \mathbb{Z}$ such that

$$U\left(\sum_{j\geq 0} a_j p^j\right) = \sum_j a_j p^{-j-1}, \quad \forall a_j \in \{0, \ldots, p-1\},$$

$$U(x) := \lim_{n\to\infty} U(x + p^n). \tag{27}$$

(ii) One has $U(x) \in H_p$, $\forall x \in \mathbb{Z}$, and

$$|U(x + np^m) - U(x)| \leq p^{-m}, \quad \forall m \in \mathbb{N}, \ x, n \in \mathbb{Z} \tag{28}$$

(iii) The sequence $U(x)$ is almost periodic with modul $M = 2\pi H_p$ and has as distribution function $\mu(\sigma) := \sigma$, for $\sigma \in [0, 1]$

Proof

(i) Let $x \in \mathbb{N}$ be a positive integer with expansion in base p given by $x = \sum_{j=0}^{k} a_j p^j$. One has for $m > k$, $U(x + p^m) = U(x) + p^{-m-1}$ thus $U(x)$ fulfills the continuity condition as in (27). For $x \in \mathbb{Z}$, $x < 0$ let $k > 0$ be such that $y = x + p^k > 0$ and let $y = \sum_{0 \le j < k} b_j p^j$ be its expansion in base p. One has

$$\lim_{n\to\infty} U(y + p^n - p^k) = \sum_{0\le j<k} b_j p^{-j-1} + \sum_{k}^{\infty} (p-1) p^{-\ell-1}$$
$$= \sum_{0\le j<k} b_j p^{-j-1} + p^{-k}.$$

Replacing k by $k' \ge k$ replaces y by $y' = \sum_{0\le j<k} b_j p^j + \sum_{k}^{k'-1} (p-1) p^m$ and gives the same result for $U(x)$ since $\sum_{k}^{k'-1} (p-1) p^{-m-1} = p^{-k} - p^{-k'}$. Thus for any $x \in \mathbb{Z}$ the limit $\lim_{n\to\infty} U(x + p^n)$ exists and this shows the existence and uniqueness of the sequence fulfilling (27).

(ii) The proof of (i) shows that $U(x) \in H_p$, $\forall x \in \mathbb{Z}$. Let us prove (28). One can assume $n > 0$ by symmetry and $x > 0$ using (27). Replacing x by $x + np^m$ does not alter the digits a_j of x in base p for $j < m$. Thus one has

$$|U(x + np^m) - U(x)| \le \sum_{j\ge m} (p-1) p^{-j-1} = p^{-m}.$$

(iii) By (28) the sequence $U(x)$ is almost periodic with modul $M = 2\pi H_p$. Let $k > 0$ and $a_j \in \{0, \ldots, p-1\}$ for $0 \le j \le k-1$. Let $I \subset [0, 1]$ be the interval

$$I = \left(\sum_{j<k} a_j p^{-j-1}, \sum_{j<k} a_j p^{-j-1} + p^{-k} \right).$$

One has using the almost periodicity

$$\lim_{T\to\infty} \frac{1}{2T} \#\{x \in \mathbb{Z} \mid |x| \le T,\ U(x) \in I\} = \lim_{T\to\infty} \frac{1}{T} \#\{x \in \mathbb{N} \mid x \le T,$$
$$U(x) \in I\}.$$

Moreover

$$\lim_{T\to\infty} \frac{1}{T} \#\{x \in \mathbb{N} \mid x \le T,\ U(x) \in I\} = p^{-k}$$

since the condition $U(x) \in I$ for $x \in \mathbb{N}$ means that the first k digits of x in base p are equal to the a_j. Thus the density of the subset $\{x \in \mathbb{Z} \mid U(x) \in I\}$ is p^{-k} and coincides with the length (and hence the Lebesgue measure) of the interval I in the range of U. This shows that U has distribution function $\mu(\sigma) = \sigma$ for $\sigma \in [0, 1]$. □

Remark 4.1 *In the above example we have chosen $D = H_p/\mathbb{Z} = \mathbb{Q}_p/\mathbb{Z}_p$ and its dual is the group $\mathbb{Z}_p$ of p-adic integers (using the self-duality of the p-adic numbers $\mathbb{Q}_p$). The sequence U is thus obtained by mapping $\mathbb{Z}_p$ to real numbers by means of (up to an overall factor p) $\phi(\sum_{j\geq 0} a_j p^j) = \sum a_j p^{-j}$. This map is continuous but not additive. It fulfills however the restricted additivity $\phi(x+y) = \phi(x) + \phi(y)$ when no carry over is involved in computing $x + y$.*

Corollary 4.1

(i) Let $h \in C[0, 1]$ be a real valued function and U as in (27). Then the sequence $h(U)(n) := h(U(n))$ is almost periodic and its distribution function μ is the primitive of the image by h of the Lebesgue measure m on $[0, 1]$, i.e. one has $d\mu = h(m)$.

(ii) Let μ be a strictly increasing continuous function in a real interval $[\alpha, \beta]$ with $\mu(\alpha) = 0$ and $\mu(\beta) = 1$. Then with $h \in C[0, 1]$ as in (i), its inverse function, μ is the distribution function of $h(U)$.

Proof

(i) The almost periodicity follows from the continuity of h. One has

$$h(U(n)) \in (a, b) \iff U(n) \in h^{-1}((a, b))$$

and the density of this set of integers is the Lebesgue measure of $h^{-1}((a, b))$. This shows that the measure $d\mu$ is the image by h of the Lebesgue measure on $[0, 1]$.

(ii) follows from (i) since for any pair of real numbers $a < b$ one has

$$h(u) \in (a, b) \iff u \in (\mu(a), \mu(b)).$$

□

In the construction provided in [32] of an almost periodic analytic function f whose associated Jensen function is a given convex function φ, the zeros of f can be taken of the form $z_k = V(k) + ik$ for $k \in \mathbb{Z}$, where $V(k)$ is an almost periodic sequence. The Jensen function φ is related to the asymptotic distribution function μ of V by the equation

$$\varphi'(\sigma) = \mu(\sigma). \tag{29}$$

Next, we apply Corollary 4.1 to construct a lifted divisor associated with the formal expression $\int f(\lambda)\delta_\lambda\, d\lambda$. We write $f = f_+ - f_-$, where $f_\pm$ is positive, and we assume that the support of $f_\pm$ is a compact real interval $I_\pm = [\alpha_\pm, \beta_\pm]$. We also assume for simplicity that $\int_{I_\pm} f_\pm(\lambda)\, d\lambda = 1$. Let us define

$$h_\pm : [0, 1] \to [\alpha_\pm, \beta_\pm], \qquad h_\pm(u) = a \iff \int_{\alpha_\pm}^{a} f_\pm(v)dv = u. \tag{30}$$

Then the construction of [32] provides the following lift

Lemma 4.2 *The following discrete integral divisor lifts the continuous divisor* $\int f(\lambda)\delta_\lambda\, d\lambda$

$$D := \sum_{k\in\mathbb{Z}} \delta_{h_+(U(k))+ik} - \delta_{h_-(U(k))+ik}. \tag{31}$$

Proof For a complex test function ψ defined on the real half-line $(0, \infty)$ one has by definition

$$< \int f(\lambda)\delta_\lambda\, d\lambda, \psi >= \int f(\lambda)\psi(\lambda)\, d\lambda.$$

It is enough to show that when evaluated on the function $z \mapsto \psi \circ \Re(z)$, the discrete divisor D gives the same result after averaging. One has

$$\lim_{T\to\infty} \frac{1}{2T} \sum_{k\in\mathbb{Z}, |k|\le T} < \delta_{h_+(U(k))+ik} - \delta_{h_-(U(k))+ik}, \psi \circ \Re >=$$

$$= \lim_{T\to\infty} \frac{1}{2T} \sum_{k\in\mathbb{Z}, |k|\le T} (\psi(h_+(U(k))) - \psi(h_-(U(k))) =$$

$$= \int_0^1 (\psi \circ h_+)(u)du - \int_0^1 (\psi \circ h_-)(u)du =$$

$$= \int_{\alpha_+}^{\beta_+} \psi(a)d\left(\int_{\alpha_+}^{a} f_+(v)dv\right) - \int_{\alpha_-}^{\beta_-} \psi(a)d\left(\int_{\alpha_-}^{a} f_-(v)dv\right) =$$

$$= \int f(\lambda)\psi(\lambda)\, d\lambda.$$

□

Remark 4.2 *The construction as in* (27) *generalizes when* H_p *is replaced by any infinite subgroup of* $\mathbb{Q}$ *not isomorphic to* $\mathbb{Z}$. *This is also explained in [32] and the connection with the Scaling Site should be explored further. In fact, note that for the periodic orbits of the Scaling Site the restriction on the slopes of the convex functions of the structure sheaf was stated in terms of these slopes belonging to* H_p, *and this condition corresponds to* (29).

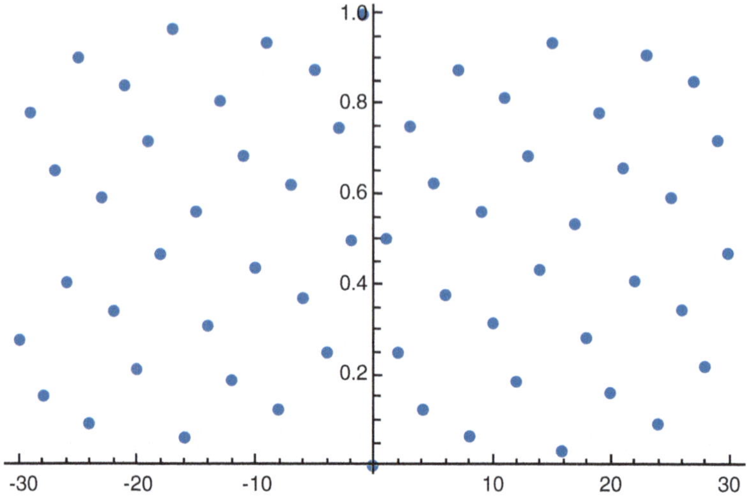

Fig. 3 Jessen sequence $k + iU(k) \in \mathbb{H}$

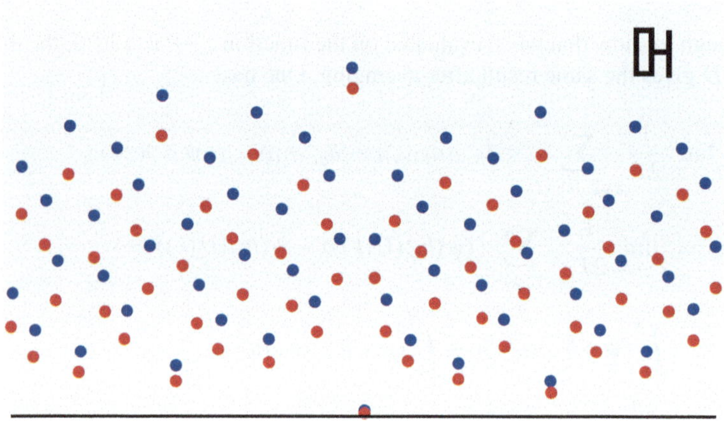

Fig. 4 Divisor for $f(x) = -2x^3 + 3x^2 - x$ written as $f = f_+ - f_-$, with $f_+(x) = x(1-x)$ and $f_-(x) = 2x(1-x)^2$

In the above development (following [32]) we have worked with the complex right-half plane; however in relation with the action of $\mathrm{SL}(2, \mathbb{R})$, the use of the upper-half plane $\mathbb{H} := \{z \in \mathbb{C} \mid \mathrm{Im}(z) > 0\}$ is more convenient.

Figure 3 shows the discrete almost periodic (in the horizontal direction) distribution of points in the upper-half plane $\mathbb{H}$ associated with $k + iU(k) \in \mathbb{H}$.

Next Figure 4 represents the divisor (zeros in blue, poles in red) associated with $f(x) = -2x^3 + 3x^2 - x$, written as $f = f_+ - f_-$ with $f_+(x) = x(1-x)$ and $f_-(x) = 2x(1-x)^2$.

5 The complex lift of the Scaling Site

In this section we refine the framework of Jensen's formula to construct the complex lift of the Scaling Site. The original setup of Jensen's theory and the related tropicalization map, explained in Section 3.2.2, provide us with the following starting point.

We let $\underline{\mathbb{D}}^* := \{q \in \mathbb{C} \mid 0 < |q| \leq 1\}$ be the punctured unit disk in $\mathbb{C}$. The monoid $\mathbb{N}^\times$ acts naturally on $\underline{\mathbb{D}}^*$ by means of the map $q \mapsto q^n$. In this way, one defines a ringed topos by endowing the topos $\underline{\mathbb{D}}^* \rtimes \mathbb{N}^\times$ with the structure sheaf $\mathcal{O}$ of complex analytic functions.

Given a pair of open sets Ω, Ω' in $\mathbb{D}^*$ and an integer $n \in \mathbb{N}^\times$ with $q^n \in \Omega'$ for $q \in \Omega$, there is a natural restriction map

$$\Gamma(\Omega', \mathcal{O}) \to \Gamma(\Omega, \mathcal{O}), \qquad f(q) \mapsto f(q^n).$$

Lemma 5.1 *The map* $u : \underline{\mathbb{D}}^* \to [0, \infty)$, $u(q) := -\log |q|$, *extends to a geometric morphism of toposes* $u : \underline{\mathbb{D}}^* \rtimes \mathbb{N}^\times \to [0, \infty) \rtimes \mathbb{N}^\times$.

Proof This follows from the continuity and $\mathbb{N}^\times$-equivariance of the map $u(q) := -\log |q|$. □

In order to work with continuous divisors as explained in Section 3.2, we consider the covering of $\underline{\mathbb{D}}^*$ defined by the closed upper-half plane $\bar{\mathbb{H}}$

$$q : \bar{\mathbb{H}} \to \underline{\mathbb{D}}^*, \qquad q(z) = e^{2\pi i z}, \qquad \forall z \in \bar{\mathbb{H}} = \{z \in \mathbb{C} \mid \Im(z) \geq 0\},$$

and make a compactification of the real direction in $\bar{\mathbb{H}}$ using a group compactification G of $\mathbb{R}$ motivated by the results of Section 4. The compact group $G \supset \mathbb{R}$ used in this construction is the smallest compactification of $\mathbb{R}$ on which $\mathbb{Q}^*$ acts (by unique divisibility of the dual discrete group) by extending its natural action on $\mathbb{R}$. The dual of G is the additive discrete group $\mathbb{Q}$.

At this point one could proceed in terms of ringed toposes and this would amount, in the construction of $\underline{\mathbb{D}}^* \rtimes \mathbb{N}^\times$, to replace the punctured unit disk $\underline{\mathbb{D}}^*$ by its pro-étale cover given by the projective limit

$$\tilde{\underline{\mathbb{D}}}^* := \varprojlim_{\mathbb{N}^\times} (\underline{\mathbb{D}}^*, z \mapsto z^n).$$

on which $\mathbb{N}^\times$ is acting by lifting the above action (see Proposition 5.4).

In this paper we prefer to proceed at the adelic level and the complex lift that we are going to describe in detail is obtained as the fibered product of the adèle class space of $\mathbb{Q}$ and G. It is thus the quotient of $\mathbb{A}_\mathbb{Q} \times G$ by the diagonal action of $\mathbb{Q}^*$.

The construction of the compactification G is developed, in adelic terms, in Section 5.1. It leads, in Section 5.2, to the adelic definition of the complex lift. In Sections 5.3 and 5.4, we analyze the restriction of the so obtained complex lift to the periodic orbits and to the classical orbit of the adèle class space. In particular, the restriction to the classical orbit turns out to be the projective limit $\tilde{\mathbb{D}}^*$ (Proposition 5.4) of the open unit disks (whose closed version was mentioned above in the topos theoretic context).

The complex lift is naturally endowed with a one dimensional complex foliation that we describe and analyze both on the periodic and classical orbits. It is this foliation that provides the geometric meaning of the discrete lift of continuous divisors of Lemma 4.2. This part is explained in Section 5.4. This foliation retains a meaning on the full complex lift, it is still one dimensional (complex) with leaves being the orbits of the right action of the $aX + b$ group.

By construction of the group G, the scaling action on $\mathbb{A}_{\mathbb{Q}} \times G$ exists for rational values of λ and it extends the action by scaling on $\mathbb{R} \subset G$. At the Archimedean place, this action is simply given by (i.e. induces a) multiplication of the complex variable by $\lambda \in \mathbb{Q}^*$ and thus preserves the complex structure. At the geometric level what really matters in this construction is to have a compact abelian group G which compactifies the additive locally compact group $\mathbb{R}$ and is such that the action of $\mathbb{Q}^*$ on $\mathbb{R}$ by multiplication extends to G. After implementing such data one forms the double quotient

$$C(G) := \mathbb{Q}^* \backslash (\mathbb{A}_{\mathbb{Q}} \times G)/(\hat{\mathbb{Z}}^\times \times \mathrm{Id}). \tag{32}$$

By construction, the space $C(G)$ maps onto the principal factor $\mathbb{Q}^* \backslash \mathbb{A}_{\mathbb{Q}}/\hat{\mathbb{Z}}^\times$ of the adèle class space and the fibers of this projection only involve the compact freedom in G. Writing $\mathbb{A}_{\mathbb{Q}} = \mathbb{A}_f \times \mathbb{R}$, one sees that the element $-1 \in \mathbb{Q}^*$ acts as identity on $\mathbb{A}_f/\hat{\mathbb{Z}}^\times$ but *non-trivially* on $\mathbb{R} \times G$. Moreover, $[0, \infty) \times G \subset \mathbb{R} \times G$ is a fundamental domain for the action of ± 1 and in the quotient the boundary $\{0\} \times G$ is divided by the symmetry $u \mapsto -u$. This fact accounts for the use of the complex half plane $\mathbb{H}$ in the above discussion with this nuance on the boundary. The gain, before the division by $\hat{\mathbb{Z}}^\times$, is that one retains the additive group structure that we expect to play a key role in the definition of the de Rham complex, since the H^2 should be generated by the Haar measure.

5.1 *Adelic almost periodic compactification of* $\mathbb{R}$

The main requirement on an almost periodic compactification G of $\mathbb{R}$ is that the action of $\mathbb{Q}^*$ on $\mathbb{R}$ by multiplication extends to G. Then, by turning to the Pontrjagin duals one derives a morphism $\rho : \hat{G} \to \hat{\mathbb{R}}$ with dense range. The fact that the scaling action of $\mathbb{Q}^*$ on $\mathbb{R}$ extends to G means here that the subgroup $\rho(\hat{G})$ is stable under multiplication by $\mathbb{Q}^*$ and hence is a $\mathbb{Q}$-vector subspace of $\mathbb{R}$.

The simplest case is when this vector space is one dimensional. We shall now describe this special case in detail. Thus, and up to an overall scaling, we assume $\rho(\hat{G}) = \mathbb{Q} \subset \hat{\mathbb{R}}$.

We denote by $G := \mathbb{A}_{\mathbb{Q}}/(\mathbb{Q}, +)$ the compact group quotient of the additive group of adèles by the discrete subgroup $\mathbb{Q} \subset \mathbb{A}_{\mathbb{Q}}$. We first recall how one can interpret the group G as the projective limit of the compact groups $G_n := \mathbb{R}/n\mathbb{Z}$ under the natural morphisms

$$\gamma_{n,m} : G_m \to G_n, \qquad \gamma_{n,m}(x + m\mathbb{Z}) = x + n\mathbb{Z}, \qquad \forall n|m. \tag{33}$$

First, notice that there is a natural group isomorphism

$$(\hat{\mathbb{Z}} \times \mathbb{R})/\mathbb{Z} \simeq \mathbb{A}_{\mathbb{Q}}/(\mathbb{Q}, +)$$

that is deduced by using the inclusion of additive groups $\hat{\mathbb{Z}} \times \mathbb{R} \subset \mathbb{A}_{\mathbb{Q}}$ together with the two equalities

$$\mathbb{Q} \cap (\hat{\mathbb{Z}} \times \mathbb{R}) = \mathbb{Z}, \qquad \mathbb{Q} + (\hat{\mathbb{Z}} \times \mathbb{R}) = \mathbb{A}_{\mathbb{Q}}$$

where the latter derives from the density of $\mathbb{Q}$ in finite adèles $\mathbb{A}_f$.

Next, recall that by construction the group $\hat{\mathbb{Z}}$ is the projective limit of the finite groups $\mathbb{Z}/n\mathbb{Z}$, and hence one obtains

$$(\hat{\mathbb{Z}} \times \mathbb{R})/\mathbb{Z} = \varprojlim(\mathbb{Z}/n\mathbb{Z} \times \mathbb{R})/\mathbb{Z}.$$

Moreover one has a group isomorphism

$$\rho_n : (\mathbb{Z}/n\mathbb{Z} \times \mathbb{R})/\mathbb{Z} \to G_n = \mathbb{R}/n\mathbb{Z}, \qquad \rho_n(j, x) = x - j + n\mathbb{Z}$$

and when $n|m$ one also has

$$\gamma_{n,m} \circ \rho_m(j, x) = x - j + n\mathbb{Z} = \rho_n(j, x).$$

The outcome is that the projective system $(\hat{\mathbb{Z}} \times \mathbb{R})/\mathbb{Z} = \varprojlim(\mathbb{Z}/n\mathbb{Z} \times \mathbb{R})/\mathbb{Z}$ is isomorphic to the projective system $\{\gamma_{n,m} : G_m \to G_n\}$ and one derives in this way a natural isomorphism of the corresponding projective limits.

Let us now understand the action of $\mathbb{Q}^*$ on G by checking that G is a uniquely divisible group. We show that the multiplication by an integer $n > 0$ defines a bijection of G on itself. First, notice that its range contains the subgroup of $(\hat{\mathbb{Z}} \times \mathbb{R})/\mathbb{Z}$ of the classes of the elements $(0, t) \in \{0\} \times \mathbb{R} \subset \hat{\mathbb{Z}} \times \mathbb{R}$ and is therefore dense in G since $\mathbb{Z}$ is dense in $\hat{\mathbb{Z}}$. Since G is compact, the image of the multiplication by n is closed and thus equal to G. Let us now show that the kernel of the multiplication by n is trivial. The equality $(na, ns) = (m, m)$ with $a \in \hat{Z}$ and $m \in \mathbb{Z}$ implies that m is divisible by n since one has $m/n \in \hat{\mathbb{Z}}$. Hence one obtains $(a, s) \sim (0, 0)$ and the multiplication by n is therefore proven to be bijective.

Next lemma provides several relevant details on the chosen almost periodic compactification G of $\mathbb{R}$.

Lemma 5.2 *Let $G := \mathbb{A}_{\mathbb{Q}}/(\mathbb{Q}, +)$ be the compact group described above.*

(i) The homomorphism $\mathbb{R} \ni t \mapsto a(t) = (0, t) \in \mathbb{A}_{\mathbb{Q}}/(\mathbb{Q}, +)$ determines an almost periodic compactification of $\mathbb{R}$.
(ii) Let $\alpha \in \hat{\mathbb{A}}_{\mathbb{Q}} \cap \mathbb{Q}^{\perp}$ be a non-trivial additive character of the adèles which restricts to the identity on $\mathbb{Q} \subset \mathbb{A}_{\mathbb{Q}}$. The map $\mathbb{Q} \ni q \mapsto \alpha(q\,\cdot\,)$ identifies the additive group $\mathbb{Q}$ with the Pontrjagin dual $\hat{G}$.
(iii) Let $\alpha = \prod \alpha_v$ be the standard choice of the additive character of the adèles, with $\alpha_\infty(s) = e^{2\pi i s}$. The Pontrjagin dual ρ of the map a as in (i) identifies $r \in \mathbb{Q}$ with the character (element of $\hat{\mathbb{R}}$) given by $\mathbb{R} \ni s \mapsto e^{2\pi i r s}$.

Proof

(i) By construction of the adèles, the subgroup $\mathbb{Q} \subset \mathbb{A}_{\mathbb{Q}}$ is discrete and cocompact thus G is a compact abelian group. The kernel of the homomorphism $a : \mathbb{R} \to \mathbb{A}_{\mathbb{Q}}/(\mathbb{Q}, +)$ is trivial since a non-zero rational has non-zero components at every local place. The homomorphism a has also dense range since the subgroup $a(\mathbb{R}) + \mathbb{Q}$ is dense in the adèles as a consequence of the density of $\mathbb{Q}$ in the finite adèles.
(ii) The pairing of $\mathbb{A}_{\mathbb{Q}}$ with itself given by $\alpha(xy)$ identifies $\mathbb{A}_{\mathbb{Q}}$ with its Pontrjagin dual and the quotient of $\mathbb{A}_{\mathbb{Q}}$ by $\mathbb{Q}^{\perp} = \mathbb{Q}$ with the Pontrjagin dual of $\mathbb{Q}$ (see [43]).
(iii) One has $\alpha(r\,a(s)) = \alpha_\infty(rs) = e^{2\pi i r s}$. □

5.2 The adelic complex lift

Next, we assemble together the adèle class space and the adelic almost periodic compactification of $\mathbb{R}$. Our primary goal is to describe the complex structure that arises from the pair (x, y) of variables at the Archimedean place, and to verify that these variables are rescaled by the same rational number under the action of $\mathbb{Q}_+^*$.

In the following part we shall work with the full adèle class space and postpone the division by $\hat{\mathbb{Z}}^\times$ after this development.

Lemma 5.3 *Let $P(\mathbb{Q})$ be the $ax + b$ group over $\mathbb{Q}$. The left action of $P(\mathbb{Q})$ on the adelic affine plane $\mathbb{A}_{\mathbb{Q}}^2$ defined by*

$$\ell\begin{pmatrix} a & b \\ 0 & 1 \end{pmatrix}(x, y) := (ax + b, ay) \tag{34}$$

preserves the complex structure at the Archimedean place given by $\bar{\partial} = \partial_x + i\partial_y$.

Proof The statement holds because the translation by b commutes with the operator $\bar{\partial}$ and the multiplication by a, being the same on both entries, just rescales the operator. □

Definition 5.1 The adelic complex lift is the adelic quotient

$$\mathscr{C}_{\mathbb{Q}} := P(\mathbb{Q})\backslash\mathbb{A}^2_{\mathbb{Q}}. \tag{35}$$

We denote by $\Gamma_{\mathbb{Q}}$ the further quotient obtained by implementing the action of $\hat{\mathbb{Z}}^\times$ on $\mathbb{A}^2_{\mathbb{Q}}$ given by multiplication on the second adelic variable y.

Recall that for any commutative ring R the algebraic group P is defined as

$$P(R) := \left\{\begin{pmatrix} a & b \\ 0 & 1 \end{pmatrix} \mid a \in R^{-1},\ b \in R\right\}. \tag{36}$$

One has a canonical inclusion of groups $\mathrm{GL}_1(R) \subset P(R)$ given by

$$\mathrm{GL}_1(R) \ni a \mapsto \begin{pmatrix} a & 0 \\ 0 & 1 \end{pmatrix}. \tag{37}$$

For any commutative ring R, we introduce the notation

$$\overline{P(R)} := \left\{\begin{pmatrix} a & b \\ 0 & 1 \end{pmatrix} \mid a \in R,\ b \in R\right\}. \tag{38}$$

By construction $\overline{P(R)} \subset M_2(R)$, moreover there is a canonical set-theoretic identification $\overline{P(R)} \simeq R^2$ given by the first line of the matrix. In the following part we prefer to work with the identification $\iota : R^2 \to \overline{P(R)}$ defined by

$$\iota(x, y) := \begin{pmatrix} y & x \\ 0 & 1 \end{pmatrix}. \tag{39}$$

Lemma 5.4

(i) Let $R = \mathbb{A}_{\mathbb{Q}}$, then the bijection ι of (39) is equivariant for the left action of $P(\mathbb{Q})$ and induces a bijection

$$j : \mathscr{C}_{\mathbb{Q}} = P(\mathbb{Q})\backslash\mathbb{A}^2_{\mathbb{Q}} \xrightarrow{\sim} P(\mathbb{Q})\backslash\overline{P(\mathbb{A}_{\mathbb{Q}})}. \tag{40}$$

(ii) Let K be the compact subgroup $\hat{\mathbb{Z}}^\times \subset \mathrm{GL}_1(\mathbb{A}_{\mathbb{Q}})$, then j induces a canonical bijection

$$j : \Gamma_{\mathbb{Q}} \xrightarrow{\sim} P(\mathbb{Q})\backslash\overline{P(\mathbb{A}_{\mathbb{Q}})}/K. \tag{41}$$

(iii) The action of $P(\mathbb{R})$ by right multiplication on $P(\mathbb{Q})\backslash\overline{P(\mathbb{A}_{\mathbb{Q}})}$ is free on the open subset V determined by the conditions $y_f \neq 0$ and $y_\infty \neq 0$, where $y_\infty \in \mathbb{R}$

(resp. $y_f \in \mathbb{A}_f$) is the Archimedean (resp. non-Archimedean) component of y for $\begin{pmatrix} y & x \\ 0 & 1 \end{pmatrix} \in \overline{P(\mathbb{A}_\mathbb{Q})}$.

Proof

(i) The multiplication rule

$$\begin{pmatrix} a & b \\ 0 & 1 \end{pmatrix}\begin{pmatrix} y & x \\ 0 & 1 \end{pmatrix} = \begin{pmatrix} ay & ax+b \\ 0 & 1 \end{pmatrix}$$

shows the equivariance of the map ι with respect to the action (34).

(ii) The right action of $\mathrm{GL}_1(\mathbb{A}_\mathbb{Q})$ on $\overline{P(\mathbb{A}_\mathbb{Q})}$ is given by the formula

$$\begin{pmatrix} y & x \\ 0 & 1 \end{pmatrix}\begin{pmatrix} u & 0 \\ 0 & 1 \end{pmatrix} = \begin{pmatrix} uy & x \\ 0 & 1 \end{pmatrix}.$$

Thus, under the isomorphism j of (40) the right action of $\hat{\mathbb{Z}}^\times$ corresponds to the action of $\hat{\mathbb{Z}}^\times$ by multiplication on the y-component and hence it defines the required isomorphism.

(iii) The conditions $y_f \neq 0$ and $y_\infty \neq 0$ are invariant under left multiplication by $P(\mathbb{Q})$ since this action replaces y by ay for a non-zero rational number a. Thus these conditions define an open subset V of the quotient $P(\mathbb{Q})\backslash\overline{P(\mathbb{A}_\mathbb{Q})}$. Let $\pi : \overline{P(\mathbb{A}_\mathbb{Q})} \to P(\mathbb{Q})\backslash\overline{P(\mathbb{A}_\mathbb{Q})}$ be the canonical quotient map. The right action of $P(\mathbb{R})$

$$\begin{pmatrix} y & x \\ 0 & 1 \end{pmatrix}\begin{pmatrix} u & v \\ 0 & 1 \end{pmatrix} = \begin{pmatrix} uy & vy+x \\ 0 & 1 \end{pmatrix}$$

leaves y_f and x_f unchanged. The open set $V \subset P(\mathbb{Q})\backslash\overline{P(\mathbb{A}_\mathbb{Q})}$ is invariant under this action. Let $z \in V$ and $g \in P(\mathbb{R})$ be such that $zg = z$, with

$$z = \pi\begin{pmatrix} y & x \\ 0 & 1 \end{pmatrix}, \qquad g = \begin{pmatrix} u & v \\ 0 & 1 \end{pmatrix}.$$

The equality $zg = z$ means that there exists $h \in P(\mathbb{Q})$ with $zg = hz$, thus one derives

$$h = \begin{pmatrix} a & b \\ 0 & 1 \end{pmatrix}, \qquad \begin{pmatrix} y & x \\ 0 & 1 \end{pmatrix}\begin{pmatrix} u & v \\ 0 & 1 \end{pmatrix} = \begin{pmatrix} a & b \\ 0 & 1 \end{pmatrix}\begin{pmatrix} y & x \\ 0 & 1 \end{pmatrix}.$$

The equality $y_f = ay_f$ shows that $a = 1$ since $y_f \neq 0$. Then, the equality $x_f = ax_f + b$ forces $b = 0$. Hence one gets $h = 1$. In turn, the equality $uy_\infty = y_\infty$ proves that $u = 1$ since $y_\infty \neq 0$. Finally, $vy_\infty + x_\infty = x_\infty$ implies $v = 0$. □

Let $P_+(\mathbb{R}) \subset P(\mathbb{R})$ be the connected component of the identity, in formulas

$$P_+(\mathbb{R}) := \left\{ \begin{pmatrix} a & b \\ 0 & 1 \end{pmatrix} \mid a > 0 \right\}.$$

With the notations of Lemma 5.4, we shall use the right action of $P_+(\mathbb{R}) \subset P(\mathbb{R})$ to obtain a foliation of V by one dimensional complex leaves endowed with a natural metric. We let

$$X = \begin{pmatrix} 0 & 1 \\ 0 & 0 \end{pmatrix}, \qquad Y = \begin{pmatrix} 1 & 0 \\ 0 & 0 \end{pmatrix}$$

be the generators of the Lie algebra of $P_+(\mathbb{R})$: one has $[Y, X] = X$.

Proposition 5.1

(i) The free right action of $P_+(\mathbb{R})$ on the (open) space V endows the orbits with a unique Riemannian metric so that the vector fields X, Y form an orthonormal basis (at each point) and a unique complex structure such that

$$\bar\partial(f) = 0 \iff (X + iY)f = 0.$$

(ii) Each orbit as in (i) is isomorphic to the complex upper-half plane $\mathbb{H} = \{x + iy \mid y > 0\}$ with the Poincaré metric.
(iii) The Laplacian $\Delta = -\bar\partial^*\bar\partial$, for the Riemannian metric as in (ii), is equal to $X^2 + (Y - \frac{1}{2})^2 - \frac{1}{4}$.

Proof Recall that the Poincaré complex half plane $\mathbb{H}$ is a one dimensional complex manifold endowed with the Riemannian metric

$$ds^2 = \frac{dx^2 + dy^2}{y^2}.$$

The group $\mathrm{GL}(2, \mathbb{R})^+$ acts by automorphisms of $\mathbb{H}$ as follows

$$\begin{pmatrix} a & b \\ c & d \end{pmatrix} \cdot z := \frac{az + b}{cz + d}.$$

Using the inclusion $P_+(\mathbb{R}) \subset \mathrm{GL}(2, \mathbb{R})^+$ obtained by setting $c = 0$ and $d = 1$, and selecting the point $z = i$, one obtains the left invariant Riemannian metric $ds^2 = a^{-2}(da^2 + db^2)$ on $P_+(\mathbb{R})$, and the complex structure such that $2\bar\partial f = (\partial_a f - i\partial_b f)(da + idb)$. The vector fields which provide the right action of P on itself are $Y = a\partial_a$ and $X = a\partial_b$. Using these fields, the Laplacian $\Delta = a^2(\partial_a^2 + \partial_b^2)$ is given by

$$\Delta = X^2 + \left(Y - \frac{1}{2}\right)^2 - \frac{1}{4}. \tag{42}$$

Indeed, one has $Y^2 - Y = (a\partial_a)(a\partial_a) - a\partial_a = a^2(\partial_a^2)$ and $X^2 = a^2(\partial_b^2)$.

For the orbit L though $x \in V$ one has a bijection defined by $\phi_x : P_+(\mathbb{R}) \xrightarrow{\sim} L$, $\phi_x(g) := xg$, while for another point $y = xg_0$ of the same orbit one has $\phi_y(g) = \phi_x(g_0 g)$. Thus, the full geometric structure of $P_+(\mathbb{R})$, invariant under left translations, carries over unambiguously to the orbit and the three statements follow from their validity on $P_+(\mathbb{R})$. □

5.3 The periodic orbits

The complex foliation of the open invariant set $V \subset P(\mathbb{Q})\backslash\overline{P(\mathbb{A}_\mathbb{Q})}$ as in Proposition 5.1 is, by construction, invariant under the right action of $K = \hat{\mathbb{Z}}^\times$. To describe the geometric structure induced on the quotient $\Gamma_\mathbb{Q} = P(\mathbb{Q})\backslash\overline{P(\mathbb{A}_\mathbb{Q})}/\hat{\mathbb{Z}}^\times$, we start by investigating the induced structure on the periodic orbit associated with a prime p in the adèle class space.

More precisely, we consider the subset $\prod(p)$ of $\mathbb{A}_\mathbb{Q}/\hat{\mathbb{Z}}^\times$ of classes modulo $\hat{\mathbb{Z}}^\times$ of adèles

$$a = (a_v), \ \ a_v \in \mathbb{Z}_v^*, \quad \forall v \notin \{p, \infty\}, \ \ a_p = 0, \ \ a_\infty = \lambda > 0.$$

Any such class is uniquely determined by λ and will be denoted by $\pi(\lambda) \in \prod(p)$.

Lemma 5.5 *The image in $\Gamma_\mathbb{Q} = P(\mathbb{Q})\backslash\overline{P(\mathbb{A}_\mathbb{Q})}/\hat{\mathbb{Z}}^\times$ of $G \times \prod(p) \subset \overline{P(\mathbb{A}_\mathbb{Q})}/\hat{\mathbb{Z}}^\times$ is the compact space*

$$\Gamma(p) := p^\mathbb{Z}\backslash\left((\mathbb{A}_\mathbb{Q}/(\mathbb{Q},+)) \times \prod(p)\right) \tag{43}$$

which is described by the mapping torus of the homeomorphism $\psi : G \to G$ given by multiplication by p.

Proof We recall that the left action of $P(\mathbb{Q})$ on $\overline{P(\mathbb{A}_\mathbb{Q})}$ is given by

$$\begin{pmatrix} a & b \\ 0 & 1 \end{pmatrix}\begin{pmatrix} y & x \\ 0 & 1 \end{pmatrix} = \begin{pmatrix} ay & ax+b \\ 0 & 1 \end{pmatrix}.$$

Two elements $y = \pi(\lambda)$ and $y' = \pi(\lambda')$ in $\prod(p)$ are equivalent under the action of $a \in \mathbb{Q}^\times$ if and only if $\lambda/\lambda' \in p^\mathbb{Z}$, i.e. $a \in p^\mathbb{Z}$. Thus the orbits of the left action of $P(\mathbb{Q})$ are the same as the orbits of $p^\mathbb{Z}$ in $(\mathbb{A}_\mathbb{Q}/(\mathbb{Q},+)) \times \prod(p)$. The group G is compact and the multiplication by p defines an automorphism ψ of G as can be seen on the Pontrjagin dual $\mathbb{Q}$ which is a uniquely divisible group. Thus, as a topological space $\Gamma(p) = p^{\mathbb{Z}}\backslash\left(G \times \prod(p)\right)$ is the mapping torus of the homeomorphism ψ and is a compact space. □

The geometric structure of the space $\Gamma(p)$ as in (43) is described as follows

Theorem 5.1

(i) The foliation of V as in Proposition 5.1 induces on $\Gamma(p)$ the foliation of G by the cosets of the subgroup $a(\mathbb{R})$ combined with the action of $\mathbb{R}_+^*$ on $\prod(p)$.
(ii) The foliation F on $\Gamma(p)$ as in (i) is by one dimensional complex leaves which are Riemann surfaces of curvature -1. All leaves of F, except one, are isomorphic to $\mathbb{H}$. The exceptional leaf is the quotient $p^{\mathbb{Z}}\backslash\mathbb{H}$.
(iii) The foliated space $(\Gamma(p), F)$ is, at the measure theory level, a factor of type III_λ, for $\lambda = \frac{1}{p}$.

Proof

(i) By Lemma 5.5, $\Gamma(p)$ is the quotient of $G \times \mathbb{R}_+^*$ by the action of powers of the map θ, given by $G \times \mathbb{R}_+^* \ni (x, y) \mapsto \theta(x, y) := (\psi(x), py)$, where ψ is as in Lemma 5.5. The right action of $P_+(\mathbb{R})$ on $\Gamma_{\mathbb{Q}}$ induces on $p^{\mathbb{Z}}\backslash\left((\mathbb{A}_{\mathbb{Q}}/(\mathbb{Q}, +)) \times \prod(p)\right)$ the following right action

$$\phi_{u,v}(x, y) := (x + vy, uy), \quad \forall (x, y) \in G \times \mathbb{R}_+^*, \qquad \begin{pmatrix} u & v \\ 0 & 1 \end{pmatrix} \in P_+(\mathbb{R}).$$

This is a translation $x \mapsto x + a(vy)$ in the variable $x \in G$, where $vy \in \mathbb{R}$ by construction, and is the scaling $y \mapsto uy$ by $u > 0$ in the variable $y \in \mathbb{R}_+^*$. The compact group $G = \mathbb{A}_{\mathbb{Q}}/(\mathbb{Q}, +)$ is foliated by the cosets of the subgroup $a(\mathbb{R})$ of Lemma 5.2. This foliation is globally invariant under the action of ψ because the subgroup $a(\mathbb{R})$ is globally invariant under this action. More precisely the foliation of G by the cosets of $a(\mathbb{R})$ derives from the flow $\phi_t(x) := x + a(t)$, $t \in \mathbb{R}$, $x \in G$ and one has

$$\psi(\phi_t(x)) = p\,(x + a(t)) = px + pat = \psi(x) + a(pt) = \phi_{pt}(\psi(x)).$$

The right action of $P_+(\mathbb{R})$ on $G \times \mathbb{R}_+^*$ commutes with θ and thus drops down to the quotient $\Gamma(p)$

$$\theta(\phi_{u,v}(x, y)) = (\psi(\phi_{vy}(x)), p(yu)) = (\phi_{vpy}(\psi(x)), pyu) = \phi_{u,v}(\theta(x, y)).$$

The orbits of the right action of $P_+(\mathbb{R})$ on $\Gamma(p)$ coincide with the leaves of the foliation of the mapping torus of ψ induced by the foliation of G by the cosets of the subgroup $a(\mathbb{R})$, as in Figure 5.

(ii) For $(x, y) \in G \times \mathbb{R}_+^*$, $\begin{pmatrix} u & v \\ 0 & 1 \end{pmatrix} \in P_+(\mathbb{R})$ and $n \in \mathbb{Z}$ one has

$$\phi_{u,v}(x, y) = \theta^n(x, y) \iff u = p^n, \qquad x + a(vy) = p^n x.$$

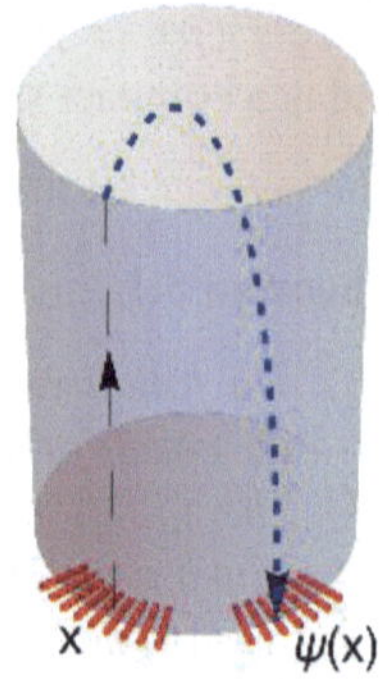

Fig. 5 The foliation of G by the cosets of the subgroup $a(\mathbb{R})$ is preserved by the map ψ and thus extends to a two-dimensional foliation of the mapping torus of ψ

For $n = 0$ this gives $u = 1$ and $v = 0$. Assume now $n \neq 0$. One has $G = \mathbb{A}_\mathbb{Q}/\mathbb{Q}$, $p^n x \in x + \mathbb{Q} + a(\mathbb{R})$. Let $x = (x_f, x_\infty)$ correspond to the decomposition $\mathbb{A}_\mathbb{Q} = \mathbb{A}_f \times \mathbb{R}$. The above condition means that $p^n x_f \in x_f + \mathbb{Q}$, i.e. (since $n \neq 0$) that $x_f \in \mathbb{Q}$. This shows that the right action of $P_+(\mathbb{R})$ on $\Gamma(p)$ is free on the orbit of (x, y) provided $x_f \notin \mathbb{Q}$, i.e. equivalently $x \notin a(\mathbb{R})$. Thus, as in Proposition 5.1, these orbits of the right action of $P_+(\mathbb{R})$ inherit a canonical structure of Riemann surface isomorphic to $\mathbb{H}$. The right action of $P_+(\mathbb{R})$ gives the two vector fields

$$X(f)(x, y) = y\partial_t f(x + a(t), y), \qquad Y(f)(x, y) = y\partial_y f(x, y). \tag{44}$$

The vector fields X and Y verify the Lie algebra of the affine $ax + b$ group:

$$[Y, X] = X. \tag{45}$$

Assume now that $x \in a(\mathbb{R})$. Then the orbit of $(x, y) \in \Gamma(p)$ under the right action of $P_+(\mathbb{R})$ is

$$p^\mathbb{Z}\backslash \left(a(\mathbb{R}) \times \prod(p)\right) \simeq p^\mathbb{Z}\backslash\mathbb{H}$$

and does not depend upon the choice of the base point (x, y). The complex structure makes sense and as a complex space one gets an open subset of the elliptic curve $E = p^\mathbb{Z}\backslash\mathbb{C}^\times$. One has $E \simeq \mathbb{C}/\Gamma$ by the isomorphism $e : \mathbb{C}/\Gamma \to E$, $e(z) := e^{2\pi i z}$ and $\Gamma = \mathbb{Z} + \frac{\log p}{2\pi i}\mathbb{Z}$.

(iii) At the measure theory level, the space of leaves of the foliation of G by the cosets of the subgroup $a(\mathbb{R})$ is the same as the quotient of the finite adèles $\mathbb{A}_f$ by the additive subgroup $\mathbb{Q}$. The action of $\mathbb{Q}$ by addition on the finite adèles $\mathbb{A}_f$ is ergodic and measure preserving. In fact since $\hat{\mathbb{Z}}$ is open in $\mathbb{A}_f$ and $\mathbb{Q}$ is dense, every orbit meets $\hat{\mathbb{Z}}$. Moreover, if $b \in a + \mathbb{Q}$ with $a, b \in \hat{\mathbb{Z}}$ one has $b - a \in \hat{\mathbb{Z}} \cap \mathbb{Q} = \mathbb{Z}$. Also, the action of $\mathbb{Z}$ on $\hat{\mathbb{Z}}$ by translation is ergodic by uniqueness of Haar measure on a compact group and density of $\mathbb{Z}$ in $\hat{\mathbb{Z}}$.

In $\Gamma(p)$ the leaves of the two-dimensional foliation all meet the fiber of the projection $\Gamma(p) \to p^{\mathbb{Z}}\backslash \prod(p)$ over the point $\pi(1)$ and in the leaf space a leaf of the foliation of G by the cosets of the subgroup $a(\mathbb{R})$ gets identified with its image by the map ψ (see Figure 5). Thus the leaf space is the quotient of $\mathbb{A}_f/\mathbb{Q}$ (the quotient of the finite adèles $\mathbb{A}_f$ by the additive subgroup $\mathbb{Q}$) by the further action by multiplication by powers of p. This latter action rescales the invariant measure by a factor of p and thus one obtains a factor of type III_λ, where $\lambda = \frac{1}{p}$. □

Remark 5.1

1. *The Haar measure $dn(x)$ on G gives an invariant transverse measure Λ for the flow ϕ_t, moreover $dn(x)$ is invariant under the automorphism of multiplication by p. But the above transverse measure Λ is not invariant under multiplication by p because it is obtained as the contraction of $dn(x)$ by the flow ϕ_t and this flow is rescaled by multiplication by p.*
2. *Both $dn(x)$ and the measure on $\prod(p)$ given by dy/y, are invariant under multiplication by p and thus the product measure descends to a measure on $\Gamma(p)$ given by*

$$\int f(x, y)dm(x, y) := \int_1^p \int_G f(x, y)dn(x)\frac{dy}{y}. \tag{46}$$

Next, we use the basis of differential forms along the leaves which is dual to the vector fields (45). It is given by $\alpha = y^{-1}dx$, $\beta = y^{-1}dy$ in the cotangent space to the leaves.

Next statement computes the de Rham cohomology of $\Gamma(p)$.

Proposition 5.2 *The canonical projection $\Gamma(p) \to p^{\mathbb{Z}}\backslash \prod(p) = \mathbb{R}_+^*/p^{\mathbb{Z}}$ is an isomorphism in the de Rham cohomology.*

Proof The de Rham complex on $\Gamma(p)$ is described as follows using the Lie algebra L of the affine group and its dual L^*. We take the basis (X, Y) for L and the dual basis (α, β) for L^*. One lets $\Omega^j := \mathscr{A} \otimes \wedge^j L^*$ where $\mathscr{A}$ is an algebra of functions on $\Gamma(p)$ stable under the derivations X, Y. The differential is given by

$$df = X(f)\alpha + Y(f)\beta\,,\ \forall f \in \Omega^0,\ d(f\alpha + g\beta) = df\wedge\alpha + f d\alpha + dg\wedge\beta,\ d\alpha = \alpha\wedge\beta.$$

We first describe the algebra $\mathscr{A}$ of functions on $\Gamma(p)$ stable under the derivations X, Y. Let $\mathscr{B}$ be the algebra of functions on G linearly generated by the characters e_q for $q \in \mathbb{Q}$. Thus the multiplication rule is $e_q e_{q'} = e_{q+q'}$ for all $q, q' \in \mathbb{Q}$. Let $f(y, q)$ be a function on $\mathbb{R}_+^* \times \mathbb{Q}$ and define

$$\hat{f}(x, y) := \sum_{\mathbb{Q}} f(y, q)e_q(x).$$

This definition of $\hat{f}$ is meaningful if one assumes that for each $y \in \mathbb{R}_+^*$ the function $q \mapsto f(y, q)$ has finite support. The condition that $\hat{f}$ defines a function on $\Gamma(p)$ is $\hat{f}(px, py) = \hat{f}(x, y)$ and since $e_q(px) = e_{pq}(x)$, the condition means that

$$f(py, p^{-1}q) = f(y, q), \quad \forall y \in \mathbb{R}_+^*, \; q \in \mathbb{Q}. \tag{47}$$

In terms of the function $f(y, q)$ the derivations[1] X and Y become

$$X(f)(y, q) = 2\pi i y q f(y, q), \qquad Y(f)(y, q) = y\partial_y f(y, q). \tag{48}$$

The group $p^{\mathbb{Z}}$ acts on $\mathbb{Q}$ by multiplication and (47) and (48) show that the de Rham complex is a direct sum over the orbits $\mathcal{O}$ of this action

$$(\Omega, d) = \bigoplus_{\mathcal{O} \in p^{\mathbb{Z}} \backslash \mathbb{Q}} (\Omega(\mathcal{O}), d).$$

The trivial orbit of $0 \in \mathbb{Q}$ corresponds to the pull back by the projection $\Gamma(p) \to p^{\mathbb{Z}} \backslash \prod(p) = \mathbb{R}_+^*/p^{\mathbb{Z}}$ and the vector field X gives 0, thus the contribution of this orbit reduces to the following complex of functions on $\mathbb{R}_+^*/p^{\mathbb{Z}}$

$$df = Y(f)\beta, \; \forall f \in \Omega^0, \;\; d(f\alpha + g\beta) = (-Y(f) + f)\alpha \wedge \beta, \;\; d(f\alpha \wedge \beta) = 0.$$

The map $\mathrm{Id} - Y$ is diagonalized in the basis of characters of $\mathbb{R}_+^*/p^{\mathbb{Z}} \simeq U(1)$ and the eigenvalues are the complex values $1 - 2\pi i n/\log p$, $n \in \mathbb{Z}$. Indeed, with $u = \log y$ the condition $f(py) = f(y)$ becomes periodicity of period $\log p$ and Y becomes ∂_u. It follows that $\mathrm{Id} - Y$ is an isomorphism on smooth functions, since it does not affect the rapid decay of the Fourier coefficients. This shows that the extra part due to the presence of the sub-complex of the $f\alpha$ and $f\alpha \wedge \beta$ does not contribute to the cohomology.

Next, we consider the contribution of a non-trivial orbit $\mathcal{O} = p^{\mathbb{Z}} q_0$ with $q_0 \neq 0$. A function $f(y, q)$ restricted to this orbit can be seen as a function on $\mathbb{R}_+^* \times \mathbb{Z}$ given by $h(y, n) = f(y, p^n q_0)$. Then, condition (47) becomes $h(py, n - 1) = h(y, n)$. This shows that the restriction of f to the orbit is entirely specified by the function on $\mathbb{R}_+^*$ given by $\phi(y) = f(y, q_0)$. Moreover this function is smooth and its support intersects finitely each orbit $p^{\mathbb{Z}} y$. We thus deal with the space $C_c^\infty(\mathbb{R}_+^*)$ of smooth compactly supported functions on $\mathbb{R}_+^*$. Let us compute the operators X, Y in terms of the functions $\phi(y)$. Using (48) we get

$$(X\phi)(y) = (2\pi i q_0) y \phi(y), \quad (Y\phi)(y) = y\partial_y \phi(y).$$

[1] The product is the convolution in the variable q and the ordinary product in the variable y.

Thus the operator X is invertible, and using its inverse X^{-1} one defines a homotopy $s : (\Omega(\mathcal{O}), d) \to (\Omega(\mathcal{O}), d)$, by

$$s(f\alpha + g\beta) := X^{-1}(f) \in \Omega^0(\mathcal{O}), \quad s(f\alpha \wedge \beta) = X^{-1}(f)\beta.$$

Next, we check that $ds + sd = \mathrm{Id}$. This is clear on Ω^0 since $sdf = X^{-1}X(f) = f$. On Ω^2 is also clear since $ds(f\alpha \wedge \beta) = dX^{-1}(f)\beta = XX^{-1}(f)\alpha \wedge \beta = f\alpha \wedge \beta$. On Ω^1 one has

$$\begin{aligned}(ds + sd)(f\alpha + g\beta) &= dX^{-1}(f) + s\left((-Y(f) + f + X(g))\alpha \wedge \beta\right)\\ &= f\alpha + YX^{-1}(f)\beta + X^{-1}(-Y(f) + f + X(g))\beta\\ &= f\alpha + g\beta.\end{aligned}$$

This is because $YX^{-1}(f) + X^{-1}(-Y(f) + f) = 0$ which follows from the commutation relation (45) by multiplying on both sides by X^{-1}. □

5.4 The classical orbit

Consider the subset $J \subset \mathbb{A}_{\mathbb{Q}}/\hat{\mathbb{Z}}^\times$ of classes of adèles modulo $\hat{\mathbb{Z}}^\times$

$$a = (a_v), \quad a_v \in \mathbb{Z}_v^*, \quad \forall v \neq \infty, \qquad a_\infty = \lambda > 0. \tag{49}$$

A class as in (49) is uniquely determined by λ and will be denoted $j(\lambda) \in J$. Two such classes are in the same (classical) orbit for the left action of $\mathbb{Q}^*$ if and only if they are equal. Thus the structure of $\Gamma_{\mathbb{Q}}$ over a classical orbit is simply that of the product

$$\Gamma_{\mathbb{Q},\mathrm{cl}} \simeq G \times \mathbb{R}_+^*.$$

Thus, in order to exploit measure theory and de Rham theory on this plain product it is enough to supply this description for G foliated by the cosets of the subgroup $a(\mathbb{R})$. The right action of $P_+(\mathbb{R})$ on $\Gamma_{\mathbb{Q}}$ induces on $\Gamma_{\mathbb{Q},\mathrm{cl}}$ the right action

$$\phi_{u,v}(x, y) := (x + a(vy), uy), \quad \forall (x, y) \in G \times \mathbb{R}_+^*, \qquad \begin{pmatrix} u & v \\ 0 & 1 \end{pmatrix} \in P_+(\mathbb{R}).$$

Proposition 5.3

(i) The space $\Gamma_{\mathbb{Q},\mathrm{cl}}$ is locally compact.
(ii) The right action of $P_+(\mathbb{R})$ on $\Gamma_{\mathbb{Q},\mathrm{cl}}$ is free and defines a foliation F by Riemann surfaces isomorphic to $\mathbb{H}$.

(iii) The foliated space $(\Gamma_{\mathbb{Q},\mathrm{cl}}, F)$ is, at the measure theory level, a factor of type II_∞.
(iv) The de Rham complex of $(\Gamma_{\mathbb{Q},\mathrm{cl}}, F)$ is the tensor product of the de Rham complex of $\mathbb{R}_+^*$ by the de Rham complex of the foliation (G, W) of G by the cosets of the subgroup $a(\mathbb{R})$.
(v) The de Rham cohomology of (G, W) is one dimensional in degree 0 and 1 and vanishes in higher degrees.

Proof

(i) Follows from the isomorphism $\Gamma_{\mathbb{Q},\mathrm{cl}} \simeq G \times \mathbb{R}_+^*$.
(ii) The freeness of the right action is clear. As in Proposition 5.1 the orbits of the right action of $P_+(\mathbb{R})$ inherit a canonical structure of Riemann surface isomorphic to $\mathbb{H}$.
(iii) The space of leaves of the foliated space $(\Gamma_{\mathbb{Q},\mathrm{cl}}, F)$ is the same as for the foliation of G by the cosets of the subgroup $a(\mathbb{R})$ and is thus the quotient of the finite adèles $\mathbb{A}_f$ by the additive subgroup $\mathbb{Q}$. It is thus ergodic of type II_∞.
(iv) Follows because the foliation $(\Gamma_{\mathbb{Q},\mathrm{cl}}, F)$ is the product of (G, W) by the trivial foliation of $\mathbb{R}_+^*$.
(v) Let $\mathscr{B}$ be the algebra of functions on G linearly generated by the characters e_q for $q \in \mathbb{Q}$. The operator d_W of differentiation along the flow lines fulfills $d_W(e_q) = 2\pi i q\, e_q$. Thus its kernel is one dimensional and spanned by e_0. Its cokernel is given by the linear form L associated with the Haar measure of G, i.e. $L(e_q) = 0$ for $q \neq 0$ and $L(e_0) = 1$. Thus de Rham cohomology of (G, W) is one dimensional in degree 0 and 1 and vanishes in higher degrees. □

Theorem 5.2 *Let $C = \int f(\lambda)\delta_\lambda\, d\lambda$ be a continuous divisor on $\mathbb{R}_+^*$ with compact support. There exists a finite union of graphs $G_j^\pm$ of maps $g_j^\pm$*

$$g_j^\pm : D_j^\pm \to \mathbb{R}_+^*, \quad D_j^\pm \subset \hat{\mathbb{Z}} \subset G \tag{50}$$

such that the leafwise discrete divisor $D := \sum \pm G_j^\pm$ is a lift of C.

Proof This follows from Lemma 4.2. □

We next give a canonical isomorphism of the classical orbit $\Gamma_{\mathbb{Q},\mathrm{cl}}$ with the pro-étale cover $\tilde{\mathbb{D}}^*$ of the punctured open unit disk $\mathbb{D}^*$ constructed from the projective system defined as follows

$$E_n := \mathbb{D}^*, \quad p_{(n,m)} : E_m \to E_n, \quad p_{(n,m)}(z) := z^a\ , \quad \forall m = na, \ z \in E_m = \mathbb{D}^*$$

where the indexing set $\mathbb{N}^\times$ is ordered by divisibility. By construction, $\Gamma_{\mathbb{Q},\mathrm{cl}} \simeq G \times \mathbb{R}_+^*$ is the projective limit

$$G \times \mathbb{R}_+^* = \varprojlim (G_n, \gamma_{n,m}) \times \mathbb{R}_+^* = \varprojlim \mathbb{H}/n\mathbb{Z} \tag{51}$$

where the projective limit in the right hand side uses the canonical projections $\mathbb{H}/m\mathbb{Z} \to \mathbb{H}/n\mathbb{Z}$ for $m = na$ corresponding to the $\gamma_{n,m}$,

$$\gamma_{n,m} : G_m \to G_n, \quad \gamma_{n,m}(x + m\mathbb{Z}) = x + n\mathbb{Z}, \quad \forall n|m$$

Proposition 5.4 *Let* $\tilde{\mathbb{D}}^* := \varprojlim(E_n, p_{(n,m)})$. *The maps* $e_n : \mathbb{H}/n\mathbb{Z} \to \mathbb{D}^*$, $e_n(z) = \exp(2\pi i \frac{z}{n})$ *assemble into an isomorphism* $\exp : \Gamma_{\mathbb{Q},\mathrm{cl}} \to \tilde{\mathbb{D}}^*$.

Proof For each integer n the map $e_n : \mathbb{H}/n\mathbb{Z} \to \mathbb{D}^*$ is an isomorphism. One has the compatibility for $m = na$

$$p_{(n,m)}(e_m(z)) = e_m(z)^a = \exp\left(2\pi i \frac{az}{m}\right) = \exp\left(2\pi i \frac{z}{n}\right) = e_n(z).$$

Thus this gives an isomorphism of the projective systems. □

Remark 5.2 *The formulation of Proposition 5.3 does not reflect the additive structure at the Archimedean place. Instead of* J *as in* (49), *one can consider the subset* $\tilde{J} \subset \mathbb{A}_{\mathbb{Q}}/\hat{\mathbb{Z}}^\times$ *formed of classes of adèles modulo* $\hat{\mathbb{Z}}^\times$

$$a = (a_v), \quad a_v \in \mathbb{Z}_v^*, \quad \forall v \neq \infty.$$

Then, the value of a_∞ *gives an isomorphism* $\tilde{J} \simeq \mathbb{R}$ *and the equivalence for the multiplicative action of* $\mathbb{Q}^*$ *is reduced to the orbit of* ± 1. *In this way one obtains the following refinement of* $\Gamma_{\mathbb{Q},\mathrm{cl}}$

$$\bar{\Gamma}_{\mathbb{Q},\mathrm{cl}} \simeq (G \times \mathbb{R})/\pm 1 = \Gamma_{\mathbb{Q},\mathrm{cl}} \cup (G/\pm 1).$$

Thus the only additional piece is $G/\pm 1$.

6 The moduli space interpretation

In this section we relate the noncommutative space $\mathscr{C}_{\mathbb{Q}} = P(\mathbb{Q})\backslash\overline{P(\mathbb{A}_{\mathbb{Q}})}$ described in Section 5 to the GL(2)-system (see [16]). This system was conceived as a higher dimensional generalization of the BC-system [10] and its main new feature is provided by its arithmetic subalgebra of modular functions. The classical Shimura scheme $Sh(\mathrm{GL}_2, \mathbb{H}^\pm) := \mathrm{GL}_2(\mathbb{Q})\backslash\mathrm{GL}_2(\mathbb{A}_{\mathbb{Q}})/\mathbb{C}^\times$ recalled in Section 6.1 appears as the set of classical points of the noncommutative space $\overline{Sh^{\mathrm{nc}}(\mathrm{GL}_2, \mathbb{H}^\pm)}$ underlying the GL(2)-system. This noncommutative space admits a simple description as the double quotient

$$\overline{Sh^{\mathrm{nc}}(\mathrm{GL}_2, \mathbb{H}^\pm)} = \mathrm{GL}_2(\mathbb{Q})\backslash M_2(\mathbb{A}_{\mathbb{Q}})^\bullet/\mathbb{C}^\times$$

obtained by replacing in the construction of $Sh(\mathrm{GL}_2, \mathbb{H}^\pm)$ the middle term $\mathrm{GL}_2(\mathbb{A}_\mathbb{Q})$ by $M_2(\mathbb{A}_\mathbb{Q})^\bullet := M_2(\mathbb{A}_{\mathbb{Q},f}) \times (M_2(\mathbb{R}) \smallsetminus \{0\})$ i.e. the space of matrices (of adèles) with non-zero Archimedean component.

In Section 6.2 we construct a map $\theta : \mathscr{C}_\mathbb{Q} \to \overline{Sh^{\mathrm{nc}}(\mathrm{GL}_2, \mathbb{H}^\pm)}$ using the natural inclusion $\overline{P(\mathbb{A}_\mathbb{Q})} \subset M_2(\mathbb{A}_\mathbb{Q})^\bullet$ (Lemma 6.1). The important feature of this inclusion is that, at the Archimedean place, the inclusion $\overline{P(\mathbb{R})} \subset M_2(\mathbb{R}) \smallsetminus \{0\})$ induces a bijection of $\overline{P(\mathbb{R})}$ with the complement of a single point in $(M_2(\mathbb{R}) \smallsetminus \{0\})/\mathbb{C}^\times$.

In Section 6.3 we use the description of the GL(2)-system in terms of $\mathbb{Q}$-lattices to give a geometric interpretation of a generic element of $\mathscr{C}_\mathbb{Q}$ in terms of commensurability classes of *parabolic* $\mathbb{Q}$-lattices. More precisely, we show in Theorem 6.1 that the space of parabolic $\mathbb{Q}$-lattices, up to commensurability, is canonically isomorphic to the quotient $\mathscr{C}_\mathbb{Q}^o := P(\mathbb{Q})\backslash(\overline{P(\mathbb{A}_{\mathbb{Q},f})} \times P(\mathbb{R}))$.

In Section 6.4 we interpret these results in terms of elliptic curves E endowed with a *triangular* structure, i.e. a pair of elements of the Tate module $T(E)$ fulfilling an orthogonality relation (Definition 6.3). In Theorem 6.2 we prove that the triangular condition characterizes the range of the map θ. The equivalence relation of commensurability of $\mathbb{Q}$-lattices is then interpreted in terms of isogenies of triangular elliptic curves in Section 6.5.

In Section 6.6 we show that the complex structure on $\mathscr{C}_\mathbb{Q}$ inherited from the right action of $P^+(\mathbb{R})$ coincides with the natural complex structure as a moduli space of elliptic curves. In Section 6.7 we briefly describe the right action of $P(\hat{\mathbb{Z}})$, while the boundary cases are described in Section 6.8.

6.1 Notations

In this part we fix the notations for the Shimura scheme $Sh(\mathrm{GL}_2, \mathbb{H}^\pm)$. The group $\mathrm{GL}_2^+(\mathbb{R})$ acts on the complex upper-half plane $\mathbb{H}$ by fractional linear transformations

$$\alpha(z) = \frac{az+b}{cz+d}, \quad \forall \alpha = \begin{pmatrix} a & b \\ c & d \end{pmatrix} \in \mathrm{GL}_2^+(\mathbb{R}). \tag{52}$$

We identify the multiplicative group $\mathbb{C}^\times$ as the subgroup $\mathrm{SO}_2(\mathbb{R}) \times \mathbb{R}_+^* \subset \mathrm{GL}_2^+(\mathbb{R})$ by the map

$$a + ib \in \mathbb{C}^\times \mapsto \begin{pmatrix} a & b \\ -b & a \end{pmatrix} \in \mathrm{GL}_2^+(\mathbb{R}). \tag{53}$$

The quotient $\mathrm{GL}_2^+(\mathbb{R})/\mathbb{C}^\times$ gets thus identified with the upper-half plane $\mathbb{H}$ by the map

$$\alpha \in \mathrm{GL}_2^+(\mathbb{R}) \mapsto z = \alpha(i) \in \mathbb{H}. \tag{54}$$

In fact, the same map identifies the quotient $\mathrm{GL}_2(\mathbb{R})/\mathbb{C}^\times$ with the disjoint union $\mathbb{H}^\pm$ of the upper and lower half planes. By definition $Sh(\mathrm{GL}_2, \mathbb{H}^\pm)$ is the quotient

$$Sh(\mathrm{GL}_2, \mathbb{H}^\pm) := \mathrm{GL}_2(\mathbb{Q})\backslash \mathrm{GL}_2(\mathbb{A}_\mathbb{Q})/\mathbb{C}^\times = \mathrm{GL}_2(\mathbb{Q})\backslash(\mathrm{GL}_2(\mathbb{A}_{\mathbb{Q},f}) \times \mathbb{H}^\pm), \tag{55}$$

where the left action of $\mathrm{GL}_2(\mathbb{Q})$ in $(\mathrm{GL}_2(\mathbb{A}_{\mathbb{Q},f}) \times \mathbb{H}^\pm)$ is via the diagonal embedding in the product $\mathrm{GL}_2(\mathbb{A}_{\mathbb{Q},f}) \times \mathrm{GL}_2(\mathbb{R})$. $Sh(\mathrm{GL}_2, \mathbb{H}^\pm)$ is a scheme over $\mathbb{C}$ (see [38], Remark 2.10) which is the inverse limit of the Shimura varieties obtained as quotients by compact open subgroups $K \subset \mathrm{GL}_2(\mathbb{A}_{\mathbb{Q},f})$. The space $Sh(\mathrm{GL}_2, \mathbb{H}^\pm)$ has infinitely many connected components. They are the fibers of the map

$$\det \times \mathrm{sign} : Sh(\mathrm{GL}_2, \mathbb{H}^\pm) \to Sh(\mathrm{GL}_1, \{\pm 1\}), \tag{56}$$

where the determinant $\det : \mathrm{GL}_2(\mathbb{A}_{\mathbb{Q},f}) \to \mathrm{GL}_1(\mathbb{A}_{\mathbb{Q},f})$ gives a map to the group of finite ideles. Passing to the quotient gives a map to the idele class group modulo its Archimedean component, i.e. here the group $\hat{\mathbb{Z}}^\times$. The fiber of the map (56) over the point $(1, 1) \in Sh(\mathrm{GL}_1, \{\pm 1\})$ is the connected quotient

$$Sh_0(\mathrm{GL}_2, \mathbb{H}^\pm) := \mathrm{SL}_2(\mathbb{Q})\backslash(\mathrm{SL}_2(\mathbb{A}_{\mathbb{Q},f}) \times \mathbb{H}). \tag{57}$$

By strong approximation (see *op.cit.* Theorem 1.12) $\mathrm{SL}_2(\mathbb{Q})$ is dense in $\mathrm{SL}_2(\mathbb{A}_{\mathbb{Q},f})$, thus one derives

$$Sh_0(\mathrm{GL}_2, \mathbb{H}^\pm) = \mathrm{SL}_2(\mathbb{Z})\backslash(\mathrm{SL}_2(\hat{\mathbb{Z}}) \times \mathbb{H}). \tag{58}$$

Using the identification $\mathrm{SL}_2(\hat{\mathbb{Z}}) = \varprojlim_N \mathrm{SL}_2(\mathbb{Z}/N\mathbb{Z})$, the above quotient is associated with the *modular tower*, that is the tower of modular curves. More precisely, for $N \in \mathbb{N}$, let $Y(N) = \Gamma(N)\backslash\mathbb{H}$ be the modular curve of level N, where $\Gamma(N)$ is the principal congruence subgroup of $\Gamma = \mathrm{SL}_2(\mathbb{Z})$. One has

$$Sh_0(\mathrm{GL}_2, \mathbb{H}^\pm) = \varprojlim_N \Gamma(N)\backslash\mathbb{H} = \varprojlim_N Y(N). \tag{59}$$

6.2 *The relation with the* GL(2)*-system*

In the following part we explain the relation between the arithmetic construction of $\mathscr{C}_\mathbb{Q} = P(\mathbb{Q})\backslash\overline{P(\mathbb{A}_\mathbb{Q})}$ and the GL(2)-system. The noncommutative space underlying the GL(2)-system contains the quotient

$$Sh^{\mathrm{nc}}(\mathrm{GL}_2, \mathbb{H}^\pm) := \mathrm{GL}_2(\mathbb{Q})\backslash(M_2(\mathbb{A}_{\mathbb{Q},f}) \times \mathbb{H}^\pm), \tag{60}$$

and enlarges it by taking cusps into account. It is defined as the double quotient

$$\overline{Sh^{\rm nc}(\mathrm{GL}_2,\mathbb{H}^\pm)} := \mathrm{GL}_2(\mathbb{Q})\backslash M_2(\mathbb{A}_\mathbb{Q})^\bullet/\mathbb{C}^\times, \tag{61}$$

where one sets

$$M_2(\mathbb{A}_\mathbb{Q})^\bullet := M_2(\mathbb{A}_{\mathbb{Q},f}) \times (M_2(\mathbb{R}) \smallsetminus \{0\}).$$

Next lemma defines a canonical map $\mathscr{C}_\mathbb{Q} \stackrel{\theta}{\to} \overline{Sh^{\rm nc}(\mathrm{GL}_2,\mathbb{H}^\pm)}$

Lemma 6.1

(i) The inclusion $\overline{P(\mathbb{R})} \subset (M_2(\mathbb{R}) \smallsetminus \{0\})$ induces a bijection of $\overline{P(\mathbb{R})}$ with the complement in $(M_2(\mathbb{R})\smallsetminus\{0\})/\mathbb{C}^\times$ of the point ∞ given by the class of matrices with vanishing second line.

(ii) The inclusion $\overline{P(\mathbb{A}_\mathbb{Q})} \subset M_2(\mathbb{A}_\mathbb{Q})^\bullet$ induces a morphism of noncommutative spaces

$$\mathscr{C}_\mathbb{Q} = P(\mathbb{Q})\backslash\overline{P(\mathbb{A}_\mathbb{Q})} \stackrel{\theta}{\longrightarrow} \mathrm{GL}_2(\mathbb{Q})\backslash M_2(\mathbb{A}_\mathbb{Q})^\bullet/\mathbb{C}^\times = \overline{Sh^{\rm nc}(\mathrm{GL}_2,\mathbb{H}^\pm)}. \tag{62}$$

Proof

(i) Note that for real matrices the following implication holds

$$\begin{pmatrix} a & b \\ 0 & 1 \end{pmatrix} = \begin{pmatrix} a' & b' \\ 0 & 1 \end{pmatrix}\begin{pmatrix} x & y \\ -y & x \end{pmatrix} \Longrightarrow y = 0 \;\&\; x = 1 \Longrightarrow a = a' \;\&\; b = b'.$$

Thus the induced map $\overline{P(\mathbb{R})} \to (M_2(\mathbb{R}) \smallsetminus \{0\})/\mathbb{C}^\times$ is injective.

Also, and again for real matrices one has, provided c or d is non-zero

$$\begin{pmatrix} a & b \\ c & d \end{pmatrix} = \begin{pmatrix} \frac{ad-bc}{c^2+d^2} & \frac{ac+bd}{c^2+d^2} \\ 0 & 1 \end{pmatrix} \times \begin{pmatrix} d & -c \\ c & d \end{pmatrix}. \tag{63}$$

Thus all matrices in $M_2(\mathbb{R})$ whose second line is non-zero belong to $\overline{P(\mathbb{R})}/\mathbb{C}^\times$. When both c and d are zero, one derives

$$\begin{pmatrix} a & b \\ 0 & 0 \end{pmatrix} = \begin{pmatrix} 1 & 0 \\ 0 & 0 \end{pmatrix} \times \begin{pmatrix} a & b \\ -b & a \end{pmatrix}.$$

This means that when $c = d = 0$ the right action of $\mathbb{C}^\times$ determines a single orbit $\{\infty\}$ provided one stays away from the matrix 0. Thus one obtains a canonical bijection

$$(M_2(\mathbb{R}) \smallsetminus \{0\})/\mathbb{C}^\times = \overline{P(\mathbb{R})} \cup \{\infty\}.$$

(ii) By construction one has the inclusion $\overline{P(\mathbb{A}_{\mathbb{Q}})} \subset M_2(\mathbb{A}_{\mathbb{Q}})^{\bullet}$ and moreover the groups involved on both sides of the double quotient $\overline{Sh^{\rm nc}(\mathrm{GL}_2, \mathbb{H}^{\pm})}$ are larger than those involved on the left hand side, thus one gets the required map θ. □

The proof of Lemma 6.1 shows that one has the identification

$$M_2(\mathbb{A}_{\mathbb{Q}})^{\bullet}/\mathbb{C}^{\times} = M_2(\mathbb{A}_{\mathbb{Q},f}) \times (\overline{P(\mathbb{R})} \cup \{\infty\}). \tag{64}$$

By construction, one has the factorization

$$\overline{P(\mathbb{A}_{\mathbb{Q}})} = \overline{P(\mathbb{A}_{\mathbb{Q},f})} \times \overline{P(\mathbb{R})}.$$

Thus the map θ as in (62), when considered at the Archimedean place, only misses the point at infinity of $\overline{P(\mathbb{R})} \cup \{\infty\}$.

6.3 *Commensurability classes of parabolic* $\mathbb{Q}$*-lattices*

In this section we give a geometric interpretation of the subspace

$$\mathscr{C}^{o}_{\mathbb{Q}} := P(\mathbb{Q})\backslash(\overline{P(\mathbb{A}_{\mathbb{Q},f})} \times P(\mathbb{R})) \subset P(\mathbb{Q})\backslash\overline{P(\mathbb{A}_{\mathbb{Q}})} =: \mathscr{C}_{\mathbb{Q}}.$$

To this end, we introduce the notion of *parabolic* $\mathbb{Q}$-lattice in Definition 6.1. Then, by implementing the commensurability equivalence relation we provide, in Proposition 6.1, the geometric description of $\mathscr{C}^{o}_{\mathbb{Q}}$ in terms of commensurability classes of parabolic $\mathbb{Q}$-lattices. The condition $\det(g_\infty) \neq 0$ which defines the subspace $\mathscr{C}^{o}_{\mathbb{Q}} \subset \mathscr{C}_{\mathbb{Q}}$ is invariant under the left action of $P(\mathbb{Q})$ and defines a dense open set in the naive quotient topology. One obtains the canonical identification

$$\mathscr{C}^{o}_{\mathbb{Q}} = P^{+}(\mathbb{Q})\backslash(\overline{P(\mathbb{A}_{\mathbb{Q},f})} \times P^{+}(\mathbb{R})). \tag{65}$$

We recall (see [16], III Definition 3.17) that a two-dimensional $\mathbb{Q}$-lattice is a pair (Λ, ϕ) where $\Lambda \subset \mathbb{C}$ is a lattice and $\phi : \mathbb{Q}^2/\mathbb{Z}^2 \to \mathbb{Q}\Lambda/\Lambda$ is an arbitrary morphism of abelian groups. The morphism ϕ encodes the non-Archimedean components of the lattice. The action of $\mathbb{C}^{\times}$ by scaling on $\mathbb{Q}$-lattices is given by

$$\lambda(\Lambda, \phi) = (\lambda\Lambda, \lambda\phi), \quad \forall\lambda \in \mathbb{C}^{\times}. \tag{66}$$

The set of two-dimensional $\mathbb{Q}$-lattices is (see [16], III Proposition 3.37) the quotient space

$$\Gamma\backslash(M_2(\hat{\mathbb{Z}}) \times \mathrm{GL}_2^{+}(\mathbb{R})), \tag{67}$$

where $\Gamma = \mathrm{SL}_2(\mathbb{Z})$. The set of two-dimensional $\mathbb{Q}$-lattices up to scaling is therefore identified with

$$\Gamma\backslash(M_2(\hat{\mathbb{Z}}) \times \mathrm{GL}_2^+(\mathbb{R}))/\mathbb{C}^\times = \Gamma\backslash(M_2(\hat{\mathbb{Z}}) \times \mathbb{H}). \tag{68}$$

In this part, we provide the details of this identification. We use, as in *op.cit.* the basis $\{e_1 = 1, e_2 = -i\}$ of $\mathbb{C}$ as a two-dimensional $\mathbb{R}$-vector space to let $\mathrm{GL}_2(\mathbb{R})$ act on $\mathbb{C}$ as $\mathbb{R}$-linear transformations. More precisely,

$$\alpha(xe_1 + ye_2) = (ax + by)e_1 + (cx + dy)e_2, \qquad \alpha = \begin{pmatrix} a & b \\ c & d \end{pmatrix} \in \mathrm{GL}_2(\mathbb{R}). \tag{69}$$

Every two-dimensional $\mathbb{Q}$-lattice (Λ, ϕ) can then be described by the data

$$(\Lambda, \phi) = (\alpha^{-1}\Lambda_0, \alpha^{-1}\rho), \qquad \Lambda_0 := \mathbb{Z}e_1 + \mathbb{Z}e_2 = \mathbb{Z} + i\mathbb{Z} \tag{70}$$

for some $\alpha \in \mathrm{GL}_2^+(\mathbb{R})$ and some $\rho \in M_2(\hat{\mathbb{Z}})$ unique up to the left diagonal action of $\Gamma = \mathrm{SL}_2(\mathbb{Z})$. Let us explain the notation $\alpha^{-1}\rho$ used in (70). The action of $\mathbb{Z}$ by multiplication on the abelian group $\mathbb{Q}/\mathbb{Z}$ extends to an isomorphism of rings $\hat{\mathbb{Z}} = \mathrm{Hom}(\mathbb{Q}/\mathbb{Z}, \mathbb{Q}/\mathbb{Z})$ and $M_2(\hat{\mathbb{Z}}) = \mathrm{Hom}(\mathbb{Q}^2/\mathbb{Z}^2, \mathbb{Q}^2/\mathbb{Z}^2)$. This gives meaning to the notation $ax \in \mathbb{Q}/\mathbb{Z}$ for $a \in \hat{\mathbb{Z}}$ and $x \in \mathbb{Q}/\mathbb{Z}$. We associate with $\rho \in M_2(\hat{\mathbb{Z}})$ the map

$$\rho : \mathbb{Q}^2/\mathbb{Z}^2 \to \mathbb{Q}\Lambda_0/\Lambda_0, \qquad \rho(u) = \rho_1(u)e_1 + \rho_2(u)e_2, \tag{71}$$

where $\rho_1(u) = ax + by$ and $\rho_2(u) = cx + dy$ for $u = (x, y) \in \mathbb{Q}^2/\mathbb{Z}^2$. The action of ρ is similar to the action of α as in (69)

$$\rho((x, y)) = (ax + by)e_1 + (cx + dy)e_2\,, \ \forall (x, y) \in (\mathbb{Q}/\mathbb{Z})^2,$$
$$\rho = \begin{pmatrix} a & b \\ c & d \end{pmatrix} \in M_2(\hat{\mathbb{Z}}). \tag{72}$$

To understand the extra structure on $\mathbb{Q}$-lattices which reduces the group $\mathrm{GL}(2)$ down to the parabolic subgroup P, we first consider the Archimedean component. The natural characterization of the subgroup $P^+(\mathbb{R}) \subset \mathrm{GL}_2^+(\mathbb{R})$ is that its elements g fulfill $\tau \circ g = \tau$, where τ is the projection on the imaginary axis

$$\tau : xe_1 + ye_2 \mapsto ye_2, \qquad \tau = \begin{pmatrix} 0 & 0 \\ 0 & 1 \end{pmatrix}.$$

For $z = x + iy \in \mathbb{C}$, we let $\Im(z) := y$ denote the imaginary part of z, thus with our choice of basis one has $\Im(xe_1 + ye_2) = -y$: we shall keep track of this minus sign here below. This projection defines (Lemma 6.2 (ii)) a character of the elliptic curve $E = \mathbb{C}/\Lambda$ where the lattice Λ is of the form

$$\Lambda = \alpha^{-1}\Lambda_0, \qquad \Lambda_0 := \mathbb{Z}e_1 + \mathbb{Z}e_2 = \mathbb{Z} + i\mathbb{Z}. \tag{73}$$

We define the orthogonal of a lattice Λ by the formula

$$\Lambda^{\perp} = \{z \in \mathbb{C} \mid < z, z' > \in \mathbb{Z}, \forall z' \in \Lambda\}.$$

Here we use the standard non-degenerate pairing defining the duality, given by

$$< z, z' > := \Re(z\bar{z}') = xx' + yy', \qquad \forall z = x + iy, \quad z' = x' + iy'.$$

Lemma 6.2 *Let $\Lambda = \alpha^{-1}\Lambda_0$ be a $\mathbb{Q}$-lattice, with $\alpha \in P(\mathbb{R})$. Then*

(i) $\Im(\Lambda) = \mathbb{Z}$.
(ii) The linear map $\Im$ induces a group homomorphism $\Im : E \to \mathbb{R}/\mathbb{Z}$ from the elliptic curve $E = \mathbb{C}/\Lambda$ to the abelian group $U(1) := \mathbb{R}/\mathbb{Z}$, i.e. a character of the abelian group E.
(iii) The orthogonal lattice $\Lambda^{\perp}$ contains the vector e_2.

Proof

(i) One has the implications

$$\alpha = \begin{pmatrix} a & b \\ 0 & 1 \end{pmatrix} \Rightarrow \alpha^t = \begin{pmatrix} a & 0 \\ b & 1 \end{pmatrix} \Rightarrow \alpha^t e_2 = e_2,$$

thus α^t, the transpose of the matrix α, fulfills $\alpha^t e_2 = e_2$ and

$$\Im(\Lambda) =< \Lambda, e_2 >=< \alpha^{-1}\Lambda_0, \alpha^t e_2 >=< \Lambda_0, e_2 >= \mathbb{Z}.$$

(ii) For $\xi \in E = \mathbb{C}/\Lambda$ the value $\Im\xi$ is meaningful modulo $\Im(\Lambda) = \mathbb{Z}$, thus the group homomorphism $\Im : E \to \mathbb{R}/\mathbb{Z}$ is well defined.
(iii) For Λ as in (73), $\Lambda^{\perp} = \alpha^t \Lambda_0$. This follows from $\Lambda_0 = \Lambda_0^{\perp}$ and

$$< \alpha^{-1}\xi, \eta >=< \xi, (\alpha^{-1})^t\eta >, \ (\alpha^{-1})^t\eta \in \Lambda_0 \iff \eta \in \alpha^t\Lambda_0.$$

Then the orthogonal lattice always contains the vector e_2, since one has $\alpha^t e_2 = e_2$. □

Next, we restrict the homomorphisms ϕ for $\mathbb{Q}$-lattices (Λ, ϕ) in the same way as we restricted the lattices in Lemma 6.2. From (72) and the definition of $\overline{P(R)}$ in (38), one has

$$\rho \in \overline{P(\hat{\mathbb{Z}})} \iff \rho_2(u) = y, \ \forall u = (x, y) \in \mathbb{Q}^2/\mathbb{Z}^2. \tag{74}$$

To write this condition in terms of $\phi : \mathbb{Q}^2/\mathbb{Z}^2 \to \mathbb{Q}\Lambda/\Lambda$, with $\phi = \alpha^{-1}\rho$ and for $\Lambda = \alpha^{-1}\Lambda_0$, $\alpha \in P(\mathbb{R})$, we use the character $\chi = -\Im : E \to \mathbb{R}/\mathbb{Z}$ (sending torsion points to torsion points). One has $\chi \circ \alpha^{-1} = \chi$, since $\alpha^{-1} \in P(\mathbb{R})$ and

$$\chi \circ \phi : \mathbb{Q}^2/\mathbb{Z}^2 \to \mathbb{Q}/\mathbb{Z}, \qquad \chi \circ \phi = \chi \circ \alpha^{-1} \circ \rho = \chi \circ \rho = \rho_2.$$

One thus obtains

$$\chi \circ \phi = \rho_2. \tag{75}$$

Lemma 6.3 *Let (Λ, ϕ) be a two-dimensional $\mathbb{Q}$-lattice described by data $(\Lambda, \phi) = (\alpha^{-1}\Lambda_0, \alpha^{-1}\rho)$, for some $\alpha \in \mathrm{GL}_2^+(\mathbb{R})$ and some $\rho \in M_2(\hat{\mathbb{Z}})$. Then,*

$$(\rho, \alpha) \in \Gamma \backslash \left(\overline{P(\hat{\mathbb{Z}})} \times P^+(\mathbb{R})\right) \iff \Im(\Lambda) = \mathbb{Z} \,\&\, \chi \circ \phi(u) = y,$$
$$\forall u = (x, y) \in \mathbb{Q}^2/\mathbb{Z}^2 \tag{76}$$

where $\chi : \mathbb{C}/\Lambda \to \mathbb{R}/\mathbb{Z}$ is given by $\chi = -\Im$ and $\Gamma = \mathrm{SL}_2(\mathbb{Z})$ acts diagonally.

Proof By Lemma 6.2, one has $\alpha \in \underline{P(\mathbb{R})} \Rightarrow \Im(\Lambda) = \mathbb{Z}$. Moreover, it follows from the above discussion that $\rho \in \overline{P(\hat{\mathbb{Z}})}$ is equivalent to $\rho_2(u) = y$, thus one gets $\chi \circ \phi(u) = y$ for any $u = (x, y) \in \mathbb{Q}^2/\mathbb{Z}^2$. This shows the implication $\Rightarrow$ in (76), since the $\mathbb{Q}$-lattice (Λ, ϕ) associated with (ρ, α) only depends upon the orbit of this pair under the left diagonal action of $\Gamma = \mathrm{SL}_2(\mathbb{Z})$. The character $\chi : \mathbb{C}/\Lambda \to \mathbb{R}/\mathbb{Z}$ is induced by $-\Im : \mathbb{C} \to \mathbb{R}$ and only depends upon the lattice Λ so that the conditions on the right hand side of (76) only depend on the $\mathbb{Q}$-lattice (Λ, ϕ).

Conversely, let us now assume that these conditions hold. The condition $\Im(\Lambda) = \mathbb{Z}$ means that $e_2 \in \Lambda^\perp$ is not divisible, i.e. e_2 does not belong to any multiple of $\Lambda^\perp$. It follows (using Bezout's theorem) that there exists $\xi \in \Lambda^\perp$ such that $\mathbb{Z}\xi + \mathbb{Z}e_2 = \Lambda^\perp$. Define $\beta \in \mathrm{GL}_2^+(\mathbb{R})$ as $\beta(e_1) = \pm\xi$ and $\beta(e_2) = e_2$. One has $\Lambda^\perp = \beta(\Lambda_0)$ by construction and thus one derives

$$\Lambda = (\Lambda^\perp)^\perp = (\beta(\Lambda_0))^\perp \quad = (\beta^t)^{-1}\Lambda_0.$$

For $\beta = \begin{pmatrix} a & b \\ c & d \end{pmatrix} \in \mathrm{GL}_2^+(\mathbb{R})$, one derives from (69), $\beta(e_2) = be_1 + de_2$ and thus $\beta(e_2) = e_2$ is equivalent to $b = 0$ and $d = 1$. In turns these conditions mean that $\alpha' = \beta^t \in P^+(\mathbb{R})$. In this way we have proven that, using the condition $\Im(\Lambda) = \mathbb{Z}$, we can find $\alpha' \in P^+(\mathbb{R})$ such that $\Lambda = \alpha'^{-1}\Lambda_0$. The equality $\alpha'^{-1}\Lambda_0 = \alpha^{-1}\Lambda_0$ shows that $\gamma = \alpha'\alpha^{-1} \in \Gamma = \mathrm{SL}_2(\mathbb{Z})$ and $\gamma\alpha \in P^+(\mathbb{R})$. Thus by replacing (ρ, α) with $(\gamma\rho, \gamma\alpha)$ we can assume that $(\Lambda, \phi) = (\alpha^{-1}\Lambda_0, \alpha^{-1}\rho)$, with $\alpha \in P^+(\mathbb{R})$. The second hypothesis $\chi \circ \phi(u) = y$, $\forall u = (x, y) \in \mathbb{Q}^2/\mathbb{Z}^2$ implies, using (75), that $\rho_2(u) = y$, $\forall u = (x, y) \in \mathbb{Q}^2/\mathbb{Z}^2$ and thus, by (74), that $\rho \in \overline{P(\hat{\mathbb{Z}})}$. □

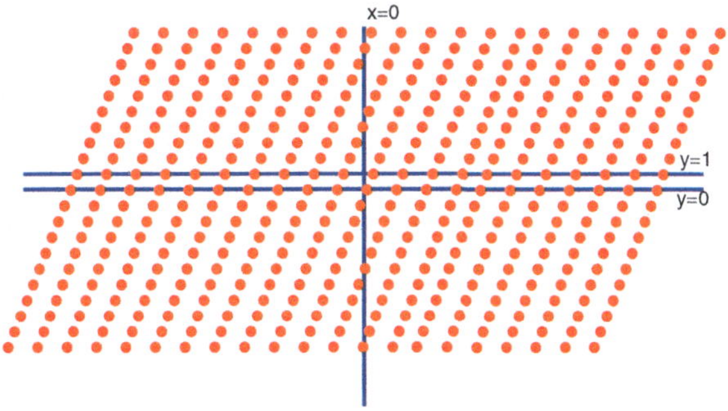

Fig. 6 A parabolic lattice

Definition 6.1 A *parabolic* two-dimensional $\mathbb{Q}$-lattice is a $\mathbb{Q}$-lattice of the form $(\Lambda, \phi) = (\alpha^{-1}\Lambda_0, \alpha^{-1}\rho)$, where $\rho \in \overline{P(\hat{\mathbb{Z}})}$ and $\alpha \in P^+(\mathbb{R})$.

We say that a parabolic $\mathbb{Q}$-lattice (Λ, ϕ) is degenerate when $\rho_1 = 0$.

Notice that for a parabolic two-dimensional $\mathbb{Q}$-lattice (Figure 6), the pair (ρ, α) such that $(\Lambda, \phi) = (\alpha^{-1}\Lambda_0, \alpha^{-1}\rho)$ is unique up to the diagonal action of $\Gamma \cap P^+(\mathbb{R}) = P^+(\mathbb{Z}) \simeq \mathbb{Z}$. Thus the space of parabolic two-dimensional $\mathbb{Q}$-lattices is described by the quotient

$$\Pi := P^+(\mathbb{Z})\backslash(\overline{P(\hat{\mathbb{Z}})} \times P^+(\mathbb{R})). \tag{77}$$

Remark 6.1

1. *When the parabolic two-dimensional $\mathbb{Q}$-lattice (Λ, ϕ) is degenerate, there exists a unique $\alpha \in P^+(\mathbb{R})$ such that $(\Lambda, \phi) = (\alpha^{-1}\Lambda_0, \alpha^{-1}p)$, where $p = \begin{pmatrix} 0 & 0 \\ 0 & 1 \end{pmatrix} \in \overline{P(\hat{\mathbb{Z}})}$.*
2. *Let $\Lambda \subset \mathbb{C}$ be a $\mathbb{Q}$-lattice such that $\Im(\Lambda) = \mathbb{Z}$, then Λ is characterized by the following arithmetic progression in $\mathbb{R}$ with associated lattice $L = \Lambda \cap \mathbb{R}$*

$$A = \mathrm{prog}(\Lambda) := \{u \in \mathbb{R} \mid i + u \in \Lambda\}. \tag{78}$$

Let $(\Lambda, \phi) = (\alpha^{-1}\Lambda_0, \alpha^{-1}\rho)$ be a parabolic $\mathbb{Q}$-lattice, with $\alpha = \begin{pmatrix} y & x \\ 0 & 1 \end{pmatrix} \in P^+(\mathbb{R})$. Then one has $\mathrm{prog}(\Lambda) = y^{-1}(\mathbb{Z} + x)$, with $L = y^{-1}\mathbb{Z}$. The pair (L, ξ), with $\xi : \mathbb{Q}/\mathbb{Z} \to \mathbb{R}/L$, $\xi(u) := \phi(u, 0)$ determines a one dimensional $\mathbb{Q}$-lattice (L, ξ).

The next step we undertake is to describe the meaning of commensurability for parabolic $\mathbb{Q}$-lattices. We recall from [16] the following (see Definition 3.17)

Definition 6.2 Two $\mathbb{Q}$-lattices are said to be commensurable ($(\Lambda_1, \phi_1) \sim (\Lambda_2, \phi_2)$) when

$$\mathbb{Q}\Lambda_1 = \mathbb{Q}\Lambda_2 \quad \text{and} \quad \phi_1 = \phi_2 \text{ mod. } \Lambda_1 + \Lambda_2. \tag{79}$$

Commensurability is an equivalence relation ([16] Lemma 3.18). By applying Proposition 3.39 of *op.cit.*, the space of commensurability classes of two-dimensional $\mathbb{Q}$-lattices up to scaling is given by the quotient space

$$\mathrm{GL}_2^+(\mathbb{Q})\backslash(M_2(\mathbb{A}_{\mathbb{Q},f}) \times \mathbb{H}) \tag{80}$$

here $\mathrm{GL}_2^+(\mathbb{Q})$ acts diagonally by $(\rho, z) \mapsto (g\rho, g(z))$. We then continue by providing the description of the orbits for this action. Let us first consider the orbit of $\Lambda_0 := \mathbb{Z}e_1 + \mathbb{Z}e_2 = \mathbb{Z} + i\mathbb{Z}$.

Lemma 6.4 *Let $g \in \mathrm{GL}_2(\mathbb{Q})$ and assume that $\Lambda = g\Lambda_0$ fulfills $\Im\Lambda = \mathbb{Z}$. Then there exist $h \in P(\mathbb{Q})$ and $k \in \Gamma = \mathrm{SL}_2(\mathbb{Z})$ such that $g = hk$.*

Proof For $g = \begin{pmatrix} a & b \\ c & d \end{pmatrix} \in \mathrm{GL}_2(\mathbb{Q})$, one then obtains

$$g(xe_1 + ye_2) = (ax + by)e_1 + (cx + dy)e_2, \quad \Im(g(xe_1 + ye_2)) = -(cx + dy).$$

Thus $\Im\Lambda = \mathbb{Z}$ means that $c, d \in \mathbb{Z}$ and they are relatively prime. Let then $u, v \in \mathbb{Z}$ be such that $cu + dv = 1$. Set $w = \begin{pmatrix} d & u \\ -c & v \end{pmatrix}$, then one has $w \in \Gamma$ and

$$gw = \begin{pmatrix} ad - bc & au + bv \\ 0 & cu + dv \end{pmatrix} = \begin{pmatrix} ad - bc & au + bv \\ 0 & 1 \end{pmatrix} \in P(\mathbb{Q}).$$

Thus by taking $h = gw$ and $k = w^{-1}$ one obtains the required factorization. $\square$

We recall ([16], Sect. III.5) that the equivalence relation of commensurability on the space of two-dimensional $\mathbb{Q}$-lattices is induced by the partially defined action of $\mathrm{GL}_2^+(\mathbb{Q})$. Indeed, for $g \in \mathrm{GL}_2^+(\mathbb{Q})$ and $(\Lambda, \phi) = (\alpha^{-1}\Lambda_0, \alpha^{-1}\rho)$ such that $g\rho \in M_2(\hat{\mathbb{Z}})$, the $\mathbb{Q}$-lattice $(\alpha^{-1}g^{-1}\Lambda_0, \alpha^{-1}\rho)$ is commensurable to (Λ, ϕ). Moreover all $\mathbb{Q}$-lattices commensurable to a given (Λ, ϕ) are of this form. Here we used, as done above, the description of two-dimensional $\mathbb{Q}$-lattices as

$$\Gamma\backslash(M_2(\hat{\mathbb{Z}}) \times \mathrm{GL}_2^+(\mathbb{R})), \quad (\rho, \alpha) \mapsto (\Lambda, \phi) = (\alpha^{-1}\Lambda_0, \alpha^{-1}\rho).$$

We can now state the following key result on commensurability

Theorem 6.1

(i) Two parabolic two-dimensional $\mathbb{Q}$-lattices $(\Lambda_j, \phi_j) = (\alpha_j^{-1}\Lambda_0, \alpha_j^{-1}\rho_j)$, $j = 1, 2$, with $\rho_j \in \overline{P(\hat{\mathbb{Z}})}$ and $\alpha_j \in P^+(\mathbb{R})$ are commensurable (as $\mathbb{Q}$-lattices) if and only if there exists $g \in P^+(\mathbb{Q})$ such that $\rho_2 = g\rho_1$ and $\alpha_2 = g\alpha_1$.

(ii) The space of parabolic $\mathbb{Q}$-lattices up to commensurability is canonically isomorphic to the quotient $\mathscr{C}^o_{\mathbb{Q}}$ as in (65).

Proof

(i) Since $P^+(\mathbb{Q}) \subset \mathrm{GL}_2^+(\mathbb{Q})$, the existence of $g \in P^+(\mathbb{Q})$ with $\rho_2 = g\rho_1$ and $\alpha_2 = g\alpha_1$ implies the commensurability. Conversely, if two parabolic $\mathbb{Q}$-lattices are commensurable, there exists $g \in \mathrm{GL}_2^+(\mathbb{Q})$ such that $\rho_2 = g\rho_1$ and $\alpha_2 = g\alpha_1$. Because $P^+(\mathbb{R})$ is a group and $\alpha_j \in P^+(\mathbb{R})$, one gets $g \in P^+(\mathbb{R})$ and thus $g \in \mathrm{GL}_2^+(\mathbb{Q}) \cap P^+(\mathbb{R}) = P^+(\mathbb{Q})$.

(ii) Let $h = \begin{pmatrix} a & b \\ 0 & 1 \end{pmatrix} \in \overline{P(\mathbb{A}_{\mathbb{Q},f})}$. We show that there exists $g \in P^+(\mathbb{Q})$ such that $gh \in \overline{P(\hat{\mathbb{Z}})}$. Let $\alpha \in \mathbb{Q}_+^\times$ such that $\alpha a \in \hat{\mathbb{Z}}$. Since $\hat{\mathbb{Z}}$ is open and $\mathbb{Q}$ is dense in $\mathbb{A}_{\mathbb{Q},f}$ with $\hat{\mathbb{Z}} + \mathbb{Q} = \mathbb{A}_{\mathbb{Q},f}$, there exists $\beta \in \mathbb{Q}$ such that $\alpha b + \beta \in \hat{\mathbb{Z}}$. One then derives

$$\begin{pmatrix} \alpha & \beta \\ 0 & 1 \end{pmatrix} \begin{pmatrix} a & b \\ 0 & 1 \end{pmatrix} = \begin{pmatrix} \alpha a & \alpha b + \beta \\ 0 & 1 \end{pmatrix} \in \overline{P(\hat{\mathbb{Z}})}.$$

It follows that all the left $P^+(\mathbb{Q})$ orbits of elements of $\overline{P(\mathbb{A}_{\mathbb{Q},f})} \times P^+(\mathbb{R})$ intersect the open subset $\overline{P(\hat{\mathbb{Z}})} \times P^+(\mathbb{R})$ whose elements yield parabolic $\mathbb{Q}$-lattices. Thus by applying (i) one obtains the required isomorphism. □

Remark 6.2 *From the point of view of noncommutative geometry, the quotient space derived by applying the commensurability relation on the space* (77) *of parabolic $\mathbb{Q}$-lattices is best described by considering the crossed product, in the sense of [33],[39], by the Hecke algebra of double classes of the subgroup $P^+(\mathbb{Z}) \subset P^+(\mathbb{Q})$. We find quite remarkable (and encouraging) that this Hecke algebra is precisely the one on which the BC-system is based.*

6.4 $\mathscr{C}_{\mathbb{Q}}$ and $\Gamma_{\mathbb{Q}}$ as moduli spaces of elliptic curves

In this section we use the description of $Sh^{\mathrm{nc}}(\mathrm{GL}_2, \mathbb{H}^\pm)$ as a moduli space of elliptic curves to obtain a similar interpretation for the spaces $\mathscr{C}_{\mathbb{Q}} = P(\mathbb{Q})\backslash\overline{P(\mathbb{A}_{\mathbb{Q}})}$ and $\Gamma_{\mathbb{Q}} = P(\mathbb{Q})\backslash\overline{P(\mathbb{A}_{\mathbb{Q}})}/\hat{\mathbb{Z}}^\times$. We first formulate Lemma 6.3 in terms of the global Tate module[2] of the elliptic curve $E = \mathbb{C}/\Lambda$. For the theory developed in this paper, we think of the global Tate module as the abelian group

$$TE = \mathrm{Hom}(\mathbb{Q}/\mathbb{Z}, E_{\mathrm{tor}})\,. \tag{81}$$

[2]The global Tate module TE is best described at the conceptual level as the pro-etale fundamental group $\pi_1^{\mathrm{alg}}(E, 0)$, where E is viewed as a curve over $\mathbb{C}$. Given $\rho \in \mathrm{Hom}(\mathbb{Q}/\mathbb{Z}, E_{\mathrm{tor}})$ the corresponding element of $\pi_1^{\mathrm{alg}}(E, 0)$ is given by the $(\rho(\frac{1}{n}))_{n\in\mathbb{N}}$.

We denote by $E_{\rm tor} = \mathbb{Q}\Lambda/\Lambda$ the torsion subgroup of the elliptic curve E. In this way, the morphism ϕ in the definition of a two dimensional $\mathbb{Q}$-lattice is now seen as the map $\phi : \mathbb{Q}^2/\mathbb{Z}^2 \to E_{\rm tor}$. By applying the covariant functor $T := \mathrm{Hom}(\mathbb{Q}/\mathbb{Z}, -)$, we rewrite ϕ as a $\hat{\mathbb{Z}}$-linear map

$$T(\phi) : \hat{\mathbb{Z}} \oplus \hat{\mathbb{Z}} \to TE, \tag{82}$$

that is given by a pair of elements (ξ, η) of TE. The character $\chi : \mathbb{C}/\Lambda \to U(1)$ induces a homomorphism of torsion subgroups $\chi : E_{\rm tor} \to \mathbb{Q}/\mathbb{Z}$, and by applying $\mathrm{Hom}(\mathbb{Q}/\mathbb{Z}, -)$, we obtain a morphism

$$T(\chi) : TE \to \hat{\mathbb{Z}}.$$

The following result characterizes the parabolic $\mathbb{Q}$-lattices among two dimensional $\mathbb{Q}$-lattices.

Proposition 6.1 *Let (Λ, ϕ) be a two-dimensional $\mathbb{Q}$-lattice, $E = \mathbb{C}/\Lambda$ the associated elliptic curve, and (ξ, η) the related pair of points in the total Tate module TE. Then (Λ, ϕ) is a parabolic $\mathbb{Q}$-lattice if and only if*

$$\Im(\Lambda) = \mathbb{Z} \ \& \ T(\chi)(\xi) = 0, \ \ T(\chi)(\eta) = Id, \ for \ \chi = -\Im. \tag{83}$$

Proof The result follows from Lemma 6.3, using the faithfulness of the functor T in the form

$$h \in \mathrm{Hom}(\mathbb{Q}/\mathbb{Z}, \mathbb{Q}/\mathbb{Z}) \ \& \ T(h) = 0 \Rightarrow h = 0.$$

This is clear, by using $T(h)(\mathrm{Id}) = h$. □

Given two elliptic curves $E = \mathbb{C}/\Lambda$ and $E' = \mathbb{C}/\Lambda'$, an isomorphism $j : E \to E'$ is given by the multiplication map by $\lambda \in \mathbb{C}^\times$, such that $\Lambda' = \lambda\Lambda$. It follows that the elliptic curve E endowed with the pair $(\xi, \eta) \in T(E)$ associated with a two-dimensional $\mathbb{Q}$-lattice (Λ, ϕ) determines the latter up to scale. In particular, passing from a parabolic $\mathbb{Q}$-lattice to the associated triple $(E; \xi, \eta)$ is equivalent to assigning the map θ from parabolic $\mathbb{Q}$-lattices to $\mathbb{Q}$-lattices up to scale

$$P^+(\mathbb{Z})\backslash(\overline{P(\hat{\mathbb{Z}})} \times P^+(\mathbb{R})) \xrightarrow{\theta} \Gamma\backslash(M_2(\hat{\mathbb{Z}}) \times \mathrm{GL}_2^+(\mathbb{R}))/\mathbb{C}^\times. \tag{84}$$

Remark 6.3 *If one ignores the non-Archimedean components, the map θ restricts to the map $\theta_\infty : P^+(\mathbb{Z})\backslash P^+(\mathbb{R}) \to \Gamma\backslash\mathrm{GL}_2^+(\mathbb{R})/\mathbb{C}^\times$ induced by the inclusion $P^+(\mathbb{R}) \subset \mathrm{GL}_2^+(\mathbb{R})/\mathbb{C}^\times$. Notice that this restriction is far from being injective. Indeed, let $\alpha \in P^+(\mathbb{R})$ and $\gamma \in \Gamma$. Let $\gamma\alpha = \alpha'\lambda$ be the $P\mathbb{C}^\times$ decomposition of $\gamma\alpha$ as in (63). Then $\alpha' \in P^+(\mathbb{R})$ and $\theta_\infty(\alpha') = \theta_\infty(\alpha)$, while $\alpha' \notin P^+(\mathbb{Z})\alpha$, unless $\gamma \in P^+(\mathbb{Z})$.*

Next proposition shows that the implementation of the non-Archimedean components makes θ injective except in a well understood case, corresponding to the vanishing of the non-Archimedean components.

Proposition 6.2 *The natural map θ as in (84) is injective except when the parabolic $\mathbb{Q}$-lattices are degenerate (Definition 6.1). Furthermore, the following formula defines a free action of $\mathbb{Z}$ on the degenerate parabolic $\mathbb{Q}$-lattices, whose orbits are the fibers of the map θ*

$$\tau(c)(p,\alpha) = (p, t_c(\alpha)), \quad t_c(\alpha) = \begin{pmatrix} 1 & 0 \\ c & 1 \end{pmatrix} \alpha \, (1 + c\bar{z})^{-1}, \quad \forall c \in \mathbb{Z}. \tag{85}$$

Here, $p = \begin{pmatrix} 0 & 0 \\ 0 & 1 \end{pmatrix} \in \overline{P(\hat{\mathbb{Z}})}$, $\alpha = \begin{pmatrix} y & x \\ 0 & 1 \end{pmatrix} \in P^+(\mathbb{R})$, and $z = x + iy \in \mathbb{C}$.

Proof We first test the injectivity of θ. Let (ρ, α) and (ρ', α') be elements of $\overline{P(\hat{\mathbb{Z}})} \times P^+(\mathbb{R})$ and assume that one has an equality of the form

$$(\rho', \alpha') = \gamma(\rho, \alpha\lambda), \quad \gamma \in \Gamma, \ \lambda \in \mathbb{C}^\times.$$

Let $\gamma = \begin{pmatrix} a & b \\ c & d \end{pmatrix}$ and $\rho = \begin{pmatrix} u & v \\ 0 & 1 \end{pmatrix}$. One has $\gamma\rho = \rho' \in \overline{P(\hat{\mathbb{Z}})}$ and thus $cu = 0$. This implies that either $c = 0$ or $u = 0$, since $c \in \mathbb{Z}$ and $u \in \hat{\mathbb{Z}}$.

Assume first that $c = 0$. Then $\gamma\rho = \rho'$ implies $d = 1$ and it follows that $a = 1$ so that $\gamma \in P^+(\mathbb{Z})$. Then the equality $\gamma\alpha\lambda = \alpha'$ implies $\lambda \in P(\mathbb{R})$ and since $\lambda \in \mathbb{C}^\times$, one concludes that $\lambda = 1$ and $\gamma\alpha = \alpha'$. This shows that (ρ, α) and (ρ', α') are equal in $P^+(\mathbb{Z})\backslash(\overline{P(\hat{\mathbb{Z}})} \times P^+(\mathbb{R}))$.

Assume now that $c \neq 0$. Then one has $u = 0$. Moreover since $\gamma\rho = \rho' \in \overline{P(\hat{\mathbb{Z}})}$ one gets $cv + d = 1$. This gives $v = (1 - d)/c$ and since $v \in \hat{\mathbb{Z}}$ and $c, d \in \mathbb{Z}$ one gets that $v \in \hat{\mathbb{Z}} \cap \mathbb{Q} = \mathbb{Z}$. Then, by replacing (ρ, α) with $\delta(\rho, \alpha)$, where $\delta = \begin{pmatrix} 1 & -v \\ 0 & 1 \end{pmatrix}$ one obtains the equality, in $P^+(\mathbb{Z})\backslash(\overline{P(\hat{\mathbb{Z}})} \times P^+(\mathbb{R}))$ of (ρ, α), with an element of the form (p, α''). It remains to see when two such elements are equal in $\Gamma\backslash(M_2(\hat{\mathbb{Z}}) \times \mathrm{GL}_2^+(\mathbb{R}))/\mathbb{C}^\times$. Thus we now assume that $\rho = \rho' = p$. The equality $\gamma\rho = \rho'$ now means that $b = 0$ and $d = 1$. But since $\gamma \in \Gamma$ one gets also that $a = 1$ and thus $\gamma = \begin{pmatrix} 1 & 0 \\ c & 1 \end{pmatrix}$. Now by (63) there exists uniquely $\alpha'' \in P(\mathbb{R})$ and $\lambda' \in \mathbb{C}^\times$ such that $\gamma\alpha = \alpha''\lambda'$ and the equality $(\rho', \alpha') = \gamma(\rho, \alpha\lambda)$ shows that $\alpha' = \alpha''$ and $\lambda' = \lambda^{-1}$. One has

$$\gamma\alpha = \begin{pmatrix} 1 & 0 \\ c & 1 \end{pmatrix} \alpha = \begin{pmatrix} 1 & 0 \\ c & 1 \end{pmatrix} \begin{pmatrix} y & x \\ 0 & 1 \end{pmatrix} = \begin{pmatrix} y & x \\ yc & xc + 1 \end{pmatrix} = \alpha' \begin{pmatrix} xc + 1 & -yc \\ yc & xc + 1 \end{pmatrix}$$
$$= \alpha'(1 + c\bar{z}).$$

Thus $\alpha' = t_c(\alpha)$. The first line of $t_c(\alpha)$ is

$$\left(\frac{y}{y^2c^2+(xc+1)^2}, \frac{(x^2+y^2)c+x}{y^2c^2+(xc+1)^2}\right)$$

and one has $t_{c'+c}(\alpha) = t_{c'}(t_c(\alpha))$ for all $c, c' \in \mathbb{Z}$ since

$$\begin{pmatrix} 1 & 0 \\ c'+c & 1 \end{pmatrix} \alpha = \begin{pmatrix} 1 & 0 \\ c' & 1 \end{pmatrix} \gamma\alpha = \begin{pmatrix} 1 & 0 \\ c' & 1 \end{pmatrix} \alpha'(1+c\bar{z}) \in t_{c'}(t_c(\alpha))\mathbb{C}^\times.$$

By construction, the map θ is invariant under left multiplication by Γ and right multiplication by $\mathbb{C}^\times$, thus one has, with $\gamma = \begin{pmatrix} 1 & 0 \\ c & 1 \end{pmatrix} \in \Gamma$, and $z = x+iy$

$$\theta((p,\alpha)) = \theta((\gamma p, \gamma\alpha)) = \theta((p, \alpha'(1+c\bar{z}))) = \theta((p,\alpha')) = \theta(\tau(c)(p,\alpha)).$$

Finally, we claim that the pairs $(p, t_c(\alpha))$ for $c \in \mathbb{Z}$ are all distinct elements of $P^+(\mathbb{Z})\backslash(P(\hat{\mathbb{Z}}) \times P^+(\mathbb{R}))$. Indeed, if $(p, t_c(\alpha)) = u(p, t_{c'}(\alpha))$ for some $u \in P^+(\mathbb{Z})$ the equality $up = p$ implies $u = 1$, thus it is enough to show that the $t_c(\alpha)$ are all distinct. But since $y \neq 0$, the equality $t_c(\alpha) = t_{c'}(\alpha)$ implies in particular $(x^2 + y^2)c = (x^2+y^2)c'$ and hence $c = c'$. □

Next, we associate a character $\chi \in \mathrm{Hom}(E, \mathbb{R}/\mathbb{Z})$, unique up to sign, with certain elements of $T(E)$.

Lemma 6.5 *Let ξ be an element of the total Tate module TE of the elliptic curve $E = \mathbb{C}/\Lambda$. Let*

$$\xi^\perp := \{\chi \in \mathrm{Hom}(E, \mathbb{R}/\mathbb{Z}) \mid T(\chi)(\xi) = 0\} \subset \mathrm{Hom}(E, \mathbb{R}/\mathbb{Z}) \simeq \Lambda^\perp.$$

Then if $\xi \neq 0$ and $\xi^\perp \neq \{0\}$ one has $\xi^\perp = \mathbb{Z}\alpha$, for a primitive character α unique up to sign.

Proof By fixing a basis of $\mathrm{Hom}(E, \mathbb{R}/\mathbb{Z}) \simeq \Lambda^\perp$ we may identify

$$TE = \hat{\mathbb{Z}} \times \hat{\mathbb{Z}}, \;\; \xi = (u, v), \;\; \xi^\perp = \{(n,m) \in \mathbb{Z}^2 \mid nu + mv = 0\}.$$

Since $\xi \neq 0$, let ℓ be a prime such that $(u_\ell, v_\ell) \neq (0,0)$. Then since $\xi^\perp \neq \{0\}$ there exists relatively prime integers $(n,m) \neq (0,0)$ such that $nu_\ell + mv_\ell = 0$ and any solution of $n'u_\ell + m'v_\ell = 0$ is a multiple of (n,m). If $nu + mv = 0$ in $\hat{\mathbb{Z}}$ one has $\xi^\perp = \{k(n,m) \mid k \in \mathbb{Z}\}$. Otherwise, there exists ℓ' such that $nu_{\ell'} + mv_{\ell'} \neq 0$. Then $k(nu_{\ell'} + mv_{\ell'}) \neq 0$ for any $k \neq 0$ and this contradicts $\xi^\perp \neq \{0\}$. This shows that $\xi^\perp = \mathbb{Z}\alpha$, where $\alpha = (n,m)$. Uniqueness up to sign is clear. □

Next result gives an intrinsic description of the range of the map θ in terms of the geometric interpretation of $\mathbb{Q}$-lattices up to scale given in terms of an elliptic curve E, and a pair of points $\xi = (\xi_1, \xi_2)$ in the total Tate module TE (see also Proposition 3.38 of [16]).

Theorem 6.2 *Let E be an elliptic curve together with a pair of elements (ξ, η) of the total Tate module TE. Assume $\xi \neq 0$. Then the corresponding $\mathbb{Q}$-lattice belongs to the range of the map θ if and only if one has $< \xi^\perp, \eta >= \mathbb{Z}$, where $< \xi^\perp, \eta >:= \{T(\chi)(\eta) \mid \chi \in \xi^\perp\} \subset \hat{\mathbb{Z}}$.*

Proof Assume first that the datum $(E; \xi, \eta)$ arises from a $\mathbb{Q}$-lattice $(\Lambda, \phi) = (\alpha^{-1}\Lambda_0, \alpha^{-1}\rho)$, where $\rho \in \overline{P(\hat{\mathbb{Z}})}$ and $\alpha \in P^+(\mathbb{R})$. The action of ρ is given in (72). This determines

$$\xi \in \mathrm{Hom}(\mathbb{Q}/\mathbb{Z}, E_{\mathrm{tor}}),\ \xi(x) = \alpha^{-1}(axe_1 + cxe_2) \in \mathbb{Q}\Lambda/\Lambda = E_{\mathrm{tor}}$$

and similarly

$$\eta \in \mathrm{Hom}(\mathbb{Q}/\mathbb{Z}, E_{\mathrm{tor}}),\ \eta(y) = \alpha^{-1}(bye_1 + dye_2) \in \mathbb{Q}\Lambda/\Lambda = E_{\mathrm{tor}}.$$

Since $\rho \in \overline{P(\hat{\mathbb{Z}})}$, one has $c = 0$ and thus $T(\chi)(\xi) = 0$, where the character χ of E is associated with the element $\alpha^t(e_2) = e_2$ of the dual lattice $\Lambda^\perp$. Since χ is primitive and $\xi \neq 0$ one has $\xi^\perp = \mathbb{Z}\chi$. Since $d = 1$ one gets that $\chi \circ \eta(y) = y$ for all $y \in \mathbb{Q}/\mathbb{Z}$ and thus one obtains $< \xi^\perp, \eta >= \mathbb{Z}$ as required. Conversely, let us assume that $(E; \xi, \eta)$ fulfill $\xi \neq 0$ and $< \xi^\perp, \eta >= \mathbb{Z}$. By Lemma 6.5 one has $\xi^\perp = \mathbb{Z}\chi$, for a primitive character χ unique up to sign. Moreover since $< \xi^\perp, \eta >= \mathbb{Z}$ one can choose the sign in such a way that $T(\chi)(\eta) = 1$. Consider the pair (E, χ) of the elliptic curve E and the primitive character χ. Let $E = \mathbb{C}/\Lambda$, then $\chi \in \Lambda^\perp$, and using the scaling action of $\mathbb{C}^\times$, we can assume that $\chi = e_2$. Since χ is primitive one has $\Im(\Lambda) = \mathbb{Z}$ and the linear map $-\Im$ induces the group morphism $\chi : \mathbb{C}/\Lambda \to \mathbb{R}/\mathbb{Z}$. We also have $T(\chi)(\xi) = 0$ and $T(\chi)(\eta) = 1$ so that Proposition 6.1 applies showing that the $\mathbb{Q}$-lattice (Λ, ϕ), with $\phi = (\xi, \eta)$ is parabolic. Thus $(E; \xi, \eta)$ is in the range of the map θ. □

Definition 6.3 A triangular structure on an elliptic curve E is a pair (ξ, η) of elements of the Tate module $T(E)$, such that $\xi \neq 0$ and $< \xi^\perp, \eta >= \mathbb{Z}$.

In the following, we shall abbreviate "elliptic curve with triangular structure" by "triangular elliptic curve."

By Proposition 3.38 of [16] a triangular elliptic curve corresponds to a $\mathbb{Q}$-lattice (Λ, ϕ) unique up to scale, and by Proposition 6.2 this datum corresponds to a unique parabolic $\mathbb{Q}$-lattice which we call the associated $\mathbb{Q}$-lattice.

6.5 Commensurability and isogenies

We recall that an isogeny from an abelian variety A to another B is a surjective morphism with finite kernel. In this section we describe how a triangular structure behaves under isogenies. At the geometric level, the commensurability relation is obtained from the following notion of isogeny between triangular elliptic curves

Definition 6.4 An *isogeny* $f : (E, \xi, \eta) \to (E', \xi', \eta')$ of triangular elliptic curves is an isogeny $f : E \to E'$ such that $T(f)(\xi) = \xi'$ and $T(f)(\eta) = \eta'$.

For ordinary isogenies, one can use the dual isogeny to show that the existence of an isogeny $E \to E'$ is a symmetric relation. This result uses the fact that multiplication by a positive integer n is an isogeny. In our setup the multiplication by n gives $\xi' = n\xi$ and $\eta' = n\eta$. This modification does not alter the orthogonal, i.e. one has $\xi'^{\perp} = \xi^{\perp}$. But one has $< \xi'^{\perp}, \eta' >= n\mathbb{Z}$, thus the triangular condition is not fulfilled unless $n = \pm 1$.

The following result determines the equivalence relation generated by isogenies

Proposition 6.3 *Let $(E; \xi, \eta)$ and $(E'; \xi', \eta')$ be two triangular elliptic curves and (Λ, ϕ) and (Λ', ϕ') the associated parabolic $\mathbb{Q}$-lattices.*

(i) Let $f : (E; \xi, \eta) \to (E'; \xi', \eta')$ be an isogeny of triangular elliptic curves. Then $\Lambda \subset \Lambda'$, $f : \mathbb{C}/\Lambda \to \mathbb{C}/\Lambda'$ is the map induced by the identity and $\phi' = f \circ \phi$.

(ii) The parabolic $\mathbb{Q}$-lattices (Λ, ϕ) and (Λ', ϕ') are commensurable if and only if there exist two isogenies $f : (E, \xi, \eta) \to (E'', \xi'', \eta'')$ and $f' : (E', \xi', \eta') \to (E'', \xi'', \eta'')$ to the same triangular elliptic curve.

Proof

(i) By definition, an isogeny $f : E \to E'$ is a holomorphic group morphism $f : \mathbb{C}/\Lambda \to \mathbb{C}/\Lambda'$, $f(z) = \lambda z$, $\forall z \in \mathbb{C}$, where the complex number λ is such that $\lambda\Lambda \subset \Lambda'$. The characters χ and χ' uniquely determined by the triangular structure are given in both cases by minus the imaginary part, and one has $\chi' \circ f = \chi$. This shows that, modulo $\mathbb{Z}$, one has $\Im(\lambda z) = \Im(z)$ for all $z \in \mathbb{C}$, i.e. $\Im((\lambda - 1)\mathbb{C}) \subset \mathbb{Z}$. Thus $\lambda = 1$, $\Lambda \subset \Lambda'$, and f is the map induced by the identity.

(ii) By applying Definition 6.2, when the two $\mathbb{Q}$-lattices $\Lambda_1 = \Lambda$, $\Lambda_2 = \Lambda'$ are parabolic, one derives

$$\Im(\Lambda_j) = \mathbb{Z} \,\&\, \chi \circ \phi_j(u) = y\,, \quad \forall u = (x, y) \in \mathbb{Q}^2/\mathbb{Z}^2$$

for the character $\chi = -\Im$. Then the lattice $\Lambda'' = \Lambda_1 + \Lambda_2$ fulfills $\Im(\Lambda'') = \mathbb{Z}$ and the quotient maps $f_j : \mathbb{C}/\Lambda_j \to \mathbb{C}/\Lambda''$ fulfill $\chi \circ f_j = \chi$, since for $z \in \mathbb{C}$ one has $\Im(z + \Lambda_j) = \Im(z) + \mathbb{Z} = \Im(z + \Lambda'')$. It follows that the two equal maps $\phi := f_j \circ \phi_j$ fulfill the condition

$$\chi \circ \phi(u) = y\,, \quad \forall u = (x, y) \in \mathbb{Q}^2/\mathbb{Z}^2,$$

so that the pair (Λ'', ϕ) is a parabolic $\mathbb{Q}$-lattice. Thus the inclusions $\Lambda_j \subset \Lambda''$ induce isogenies to the same triangular elliptic curve as required. To prove the converse it is enough, using the transitivity of the commensurability relation for $\mathbb{Q}$-lattices, to show that if $f : (E, \xi, \eta) \to (E', \xi', \eta')$ is an isogeny of triangular elliptic curves the associated $\mathbb{Q}$-lattices are commensurable. This fact follows from (i). □

6.6 *The complex structure*

Proposition 6.2 states that the natural map θ as in (84) from parabolic $\mathbb{Q}$-lattices to $\mathbb{Q}$-lattices up to scale is injective except in the degenerate case. Thus θ provides, by pull back, a large class of functions, by implementing the arithmetic subalgebra of the GL_2-system ([16] Chapter 3, §7). The functions in this algebra are holomorphic for the natural complex structure on the moduli space of elliptic curves and in this section we compare this complex structure with the one on the space $\Pi = P^+(\mathbb{Z})\backslash(\overline{P(\hat{\mathbb{Z}})} \times P^+(\mathbb{R}))$ defined using the right action of $P^+(\mathbb{R})$ (Proposition 5.1).

We recall that the complex structure on the moduli space of elliptic curves is obtained by comparing two descriptions of the quotient space $\mathrm{GL}_2^+(\mathbb{R})/\mathbb{C}^\times$. The first one identifies $\mathrm{GL}_2^+(\mathbb{R})/\mathbb{C}^\times$ with the complex upper-half plane $\mathbb{H}$ via the map

$$C : \alpha \in \mathrm{GL}_2^+(\mathbb{R}) \mapsto z = \alpha(i) = \frac{ai + b}{ci + d} \in \mathbb{H}. \tag{86}$$

The second description derives from the space $\mathscr{B}/\mathbb{C}^\times$ of pairs (ξ_1, ξ_2) of $\mathbb{R}$-independent elements of $\mathbb{C}$ up to scale. The maps

$$r : \mathscr{B}/\mathbb{C}^\times \to \mathbb{H}^\pm = \mathbb{H} \cup -\mathbb{H}, \quad r(\xi_1, \xi_2) = -\xi_2/\xi_1 \in \mathbb{C} \setminus \mathbb{R} = \mathbb{H}^\pm \tag{87}$$

and

$$B : \mathrm{GL}_2^+(\mathbb{R}) \to \mathscr{B}/\mathbb{C}^\times, \quad B(\alpha) = (\alpha^{-1} e_1, \alpha^{-1} e_2), \tag{88}$$

fulfill $r \circ B = C$. Indeed, both maps only depend on the right coset in $\mathrm{GL}_2^+(\mathbb{R})/\mathbb{C}^\times$. The right $\mathbb{C}^\times$-coset associated with $z = x + iy \in \mathbb{H}$ contains, in view of (86), the matrix

$$\alpha = \begin{pmatrix} y & x \\ 0 & 1 \end{pmatrix} \in \mathrm{GL}_2^+(\mathbb{R}).$$

One can replace $\alpha^{-1} = y^{-1}\begin{pmatrix}1 & -x\\ 0 & y\end{pmatrix}$, up to scale, by $g = \begin{pmatrix}1 & -x\\ 0 & y\end{pmatrix}$ and obtain

$$g(e_1) = \begin{pmatrix}1 & -x\\ 0 & y\end{pmatrix}\begin{pmatrix}1\\ 0\end{pmatrix} = e_1 = 1,\; g(e_2) = \begin{pmatrix}1 & -x\\ 0 & y\end{pmatrix}\begin{pmatrix}0\\ 1\end{pmatrix} = -xe_1 + ye_2 = -z.$$

This shows that $r \circ B = C$. Notice that it is meaningless to use the action of $\mathrm{GL}_2^+(\mathbb{R})$ on the elements (ξ_1, ξ_2) of a basis because this action does *not* commute with scaling.

Next, we consider the complex structure on $\Pi = P^+(\mathbb{Z})\backslash(\overline{P(\hat{\mathbb{Z}})} \times P^+(\mathbb{R}))$ as defined in Lemma 5.3, namely by means of the "dbar" operator $\partial_x + i\,\partial_y$ and in terms of the map ι of (39), i.e. of the matrix $\begin{pmatrix}y & x\\ 0 & 1\end{pmatrix} \in P^+(\mathbb{R})$. The above calculation then shows that this complex structure is identical to the canonical complex structure on the moduli space of elliptic curves. We now verify that this complex structure can also be described, as in Proposition 5.1, using the right action of $P^+(\mathbb{R})$ on Π.

The "dbar" operator for the latter structure is defined by $X + iY$, where the two vector fields X, Y on Π are defined as $X = y\partial_x$ and $Y = y\partial_y$, corresponding to the Lie algebra elements of the one parameter subgroups $u(\epsilon)$ for X

$$\epsilon \mapsto u(\epsilon) := \begin{pmatrix}0 & \epsilon\\ 0 & 1\end{pmatrix},\quad \begin{pmatrix}y & x\\ 0 & 1\end{pmatrix} u(\epsilon) = \begin{pmatrix}y & x + y\epsilon\\ 0 & 1\end{pmatrix}$$

and $v(\epsilon)$ for Y

$$\epsilon \mapsto v(\epsilon) := \begin{pmatrix}e^\epsilon & 0\\ 0 & 1\end{pmatrix},\quad \begin{pmatrix}y & x\\ 0 & 1\end{pmatrix} v(\epsilon) = \begin{pmatrix}ye^\epsilon & x\\ 0 & 1\end{pmatrix}.$$

The comparison of the complex structures is summarized in the following statement

Proposition 6.4 *The natural map from parabolic $\mathbb{Q}$-lattices to $\mathbb{Q}$-lattices up to scale*

$$\theta : \Pi \to \Gamma\backslash(M_2(\hat{\mathbb{Z}}) \times \mathrm{GL}_2^+(\mathbb{R}))/\mathbb{C}^\times \tag{89}$$

is holomorphic for the canonical complex structure on the moduli space of elliptic curves and the complex structure on Π associated with the right action of $P^+(\mathbb{R})$.

Proof It suffices to check that $(X+iY)(f) = 0$, where the function f is the pullback by θ of the local parameter $z \in \mathbb{H} = \mathrm{GL}_2^+(\mathbb{R}))/\mathbb{C}^\times$. This fact follows from the direct computation

$$f\begin{pmatrix}y & x\\ 0 & 1\end{pmatrix} = x + iy,\quad (X + iY)(f) = y\partial_x(x + iy) + iy\partial_y(x + iy) = 0.$$

□

6.7 The right action of $P(\hat{\mathbb{Z}})$

In order to pass from $\mathscr{C}_{\mathbb{Q}} = P(\mathbb{Q})\backslash\overline{P(\mathbb{A}_{\mathbb{Q}})}$ to $\Gamma_{\mathbb{Q}} = P(\mathbb{Q})\backslash\overline{P(\mathbb{A}_{\mathbb{Q}})}/\hat{\mathbb{Z}}^{\times}$ one needs to divide the component in the adèle class space by the action of $\hat{\mathbb{Z}}^{\times}$ given by multiplication. This action is induced by the right action of $\hat{\mathbb{Z}}^{\times}$ on $\Pi = P^{+}(\mathbb{Z})\backslash(\overline{P(\hat{\mathbb{Z}})} \times P^{+}(\mathbb{R}))$, and it is meaningful even before passing to commensurability classes. In this section we provide its geometric meaning in terms of parabolic $\mathbb{Q}$-lattices. It turns out that this is the special case (obtained by restricting to the subgroup $\hat{\mathbb{Z}}^{\times} \subset P(\hat{\mathbb{Z}})$ of diagonal matrices) of the right action of $P(\hat{\mathbb{Z}})$ on Π, whose geometric meaning is given in the following Proposition 6.5.

Proposition 6.5 *Let $(E; \xi, \eta)$ be a triangular elliptic curve and (Λ, ϕ) the associated $\mathbb{Q}$-lattice. Its image, under the right action of $w = \begin{pmatrix} u & v \\ 0 & 1 \end{pmatrix} \in P(\hat{\mathbb{Z}})$, is the triangular elliptic curve $(E; \xi', \eta')$ where $\xi' = \xi \circ u$ and $\eta' = \eta + \xi \circ v$.*

Proof Note that the condition $< \xi^{\perp}, \eta >= \mathbb{Z}$ of Theorem 6.2 still holds for the transformed pair $(E; \xi', \eta')$, since $(\xi \circ u)^{\perp} = \xi^{\perp}$ and $< \xi^{\perp}, \xi \circ v >= 0$. For $\rho = \begin{pmatrix} a & b \\ 0 & 1 \end{pmatrix} \in \overline{P(\hat{\mathbb{Z}})}$, one has

$$\xi(x) = \alpha^{-1}(axe_1) \in \mathbb{Q}\Lambda/\Lambda, \quad \eta(y) = \alpha^{-1}(bye_1 + ye_2) \in \mathbb{Q}\Lambda/\Lambda$$

and for $w = \begin{pmatrix} u & v \\ 0 & 1 \end{pmatrix} \in P(\hat{\mathbb{Z}})$, one obtains

$$\begin{pmatrix} a & b \\ 0 & 1 \end{pmatrix} \begin{pmatrix} u & v \\ 0 & 1 \end{pmatrix} = \begin{pmatrix} au & av + b \\ 0 & 1 \end{pmatrix} = \begin{pmatrix} a' & b' \\ 0 & 1 \end{pmatrix}.$$

Thus the right action of w determines the new pair (ξ', η')

$$\xi'(x) = \alpha^{-1}(auxe_1) \in \mathbb{Q}\Lambda/\Lambda, \quad \eta'(y) = \alpha^{-1}(bye_1 + ye_2) + \alpha^{-1}(avxe_1) \in \mathbb{Q}\Lambda/\Lambda,$$

and one concludes $\xi' = \xi \circ u$ and $\eta' = \eta + \xi \circ v$. □

6.8 Boundary cases

Theorem 6.2 and Proposition 6.2 show that triangular elliptic curves are classified by the subspace

$$\Pi' := P^+(\mathbb{Z})\backslash \left\{(\rho,\alpha) \in \overline{P(\hat{\mathbb{Z}})} \times P^+(\mathbb{R}) \mid \rho = \begin{pmatrix} u & v \\ 0 & 1 \end{pmatrix},\ u \neq 0 \right\} \subset \Pi.$$

The condition $u \neq 0$ in this definition is meaningful in the quotient since the left action of $P^+(\mathbb{Z})$ leaves u unaltered. Assume now that $u = 0$ and $v \notin \mathbb{Z}$. Then, with the notations of Proposition 6.2, $\rho \notin P^+(\mathbb{Z})p$. This result thus shows that the corresponding triple $(E; \xi, \eta)$ still characterizes the element of Π. One has $\xi = 0$ since $\rho(x, 0) = uxe_1 = 0$, moreover one also gets $\rho(0, y) = vye_1 + ye_2$. Let χ be a character of E, then $\chi = \alpha^t(ne_1 + me_2)$, with $n, m \in \mathbb{Z}$. Thus $T(\chi)(\eta) = nv + m$. Since $v \notin \mathbb{Z}$, while $v \in \hat{\mathbb{Z}}$, one derives $v \notin \mathbb{Q}$. Thus $\chi = \alpha^t(e_2)$ is the only character which takes the value 1 on η, i.e.

$$\exists! \chi \in \xi^{\perp} \text{ such that } T(\chi)(\eta) = 1. \tag{90}$$

Note that if (90) holds, then, when $\xi \neq 0$, Lemma 6.5 shows that there exists a primitive character χ_0 of E with $\xi^{\perp} = \mathbb{Z}\chi_0$. Since $\chi \in \xi^{\perp}$, one thus gets $\chi = n\chi_0$ for some $n \in \mathbb{Z}$. Thus $T(\chi)(\eta) = nT(\chi_0)(\eta)$ and $nT(\chi_0)(\eta) = 1$. But $T(\chi_0)(\eta) \in \hat{\mathbb{Z}}$ and one then gets $n = \pm 1$. This shows that one can refine the definition of a triangular structure using (90) in place of the condition

$$\xi \neq 0 \ \& \ < \xi^{\perp}, \eta >= \mathbb{Z}.$$

Thus, the case $u = 0$ (i.e. $\xi = 0$) and $v \notin \mathbb{Z}$ is covered by the following

Definition 6.5 A degenerate triangular structure on an elliptic curve E is a pair (χ, η) of a character $\chi : E \to \mathbb{R}/\mathbb{Z}$ and an element $\eta \in T(E)$ with $T(\chi)(\eta) = 1$.

This notion also covers the case $u = v = 0$, i.e. of degenerate parabolic $\mathbb{Q}$-lattices. Indeed, in this case Proposition 6.2 shows that one needs to choose the character $\chi \in \xi^{\perp}$ so that $< \chi, \eta >= 1$. More precisely, let $\tilde{\theta}$ be the map which associates with a degenerate parabolic $\mathbb{Q}$-lattice (Λ, ϕ) the degenerate triangular structure on $E = \mathbb{C}/\Lambda$ given by the pair (χ, η), where $\chi = -\Im$ and $\eta(y) = \phi((0, y))$ for all $y \in \mathbb{Q}/\mathbb{Z}$. Then we have the following

Proposition 6.6 *Two degenerate parabolic $\mathbb{Q}$-lattices are the same if and only if the degenerate triangular elliptic curves, associated via the map $\tilde{\theta}$, are isomorphic.*

Proof Let (Λ, ϕ) and (Λ', ϕ') be degenerate parabolic $\mathbb{Q}$-lattices, and $E, (\chi, \eta)$, $E', (\chi', \eta')$ their images under $\tilde{\theta}$. By the degeneracy hypothesis there exists uniquely $\alpha, \alpha' \in P^+(\mathbb{R})$ such that

$$\Lambda = \alpha^{-1}\Lambda_0,\ \phi(x, y) = \alpha^{-1}(ye_2),\ \ \forall x, y \in \mathbb{Q}/\mathbb{Z}$$

and similarly for (Λ', ϕ'). An isomorphism $j : E \to E'$ is implemented by the multiplication by a complex number λ such that $\lambda\Lambda = \Lambda'$. If j preserves the degenerate triangular structure, one has $\chi' \circ j = \chi$, i.e. $\Im(\lambda z) = \Im(z)$ modulo

$\mathbb{Z}$ for all $z \in E = \mathbb{C}/\Lambda$ and thus $\lambda = 1$. Hence one derives $\Lambda = \Lambda'$ and j is the identity. Let us show that $\phi' = \phi$. One has $\phi(x, y) = \eta(y)$, $\forall x, y \in \mathbb{Q}/\mathbb{Z}$, and since by hypothesis $j \circ \eta = \eta'$ one concludes that $\phi' = \phi$. □

Next we consider the degeneracy occurring when in the matrix $\alpha = \begin{pmatrix} a & b \\ 0 & 1 \end{pmatrix} \in P^+(\mathbb{R})$, a tends to 0. We follow the lattice $\Lambda = \alpha^{-1}\Lambda_0$ up to scale. One has

$$a\alpha^{-1} = a\begin{pmatrix} \frac{1}{a} & -\frac{b}{a} \\ 0 & 1 \end{pmatrix} = \begin{pmatrix} 1 & -b \\ 0 & a \end{pmatrix}, \quad a\alpha^{-1}\begin{pmatrix} x \\ y \end{pmatrix} = \begin{pmatrix} 1 & -b \\ 0 & a \end{pmatrix}\begin{pmatrix} x \\ y \end{pmatrix} = \begin{pmatrix} x - by \\ ay \end{pmatrix}.$$

Thus when $a \to 0$ the lattice $\Lambda = \alpha^{-1}\Lambda_0$ up to scale converges pointwise (i.e. for each fixed pair (x, y)) towards the subgroup of $\mathbb{R} \subset \mathbb{C}$ given by

$$\Lambda(b) := \mathbb{Z} + b\mathbb{Z} \subset \mathbb{R} \subset \mathbb{C}. \tag{91}$$

The subgroup $\Lambda(b)$ only depends upon $b \in \mathbb{R}/\mathbb{Z}$ and the quotient $\mathbb{R}/\Lambda(b)$ corresponds to the noncommutative torus $\mathbb{T}^2_b$. In fact, the composition with $\rho = \begin{pmatrix} u & v \\ 0 & 1 \end{pmatrix} \in \overline{P(\hat{\mathbb{Z}})}$ gives the following $\mathbb{Q}$-pseudolattice in the sense of Definition 3.106 of [16]

$$\phi : \mathbb{Q}^2/\mathbb{Z}^2 \to \mathbb{Q}\Lambda(b)/\Lambda(b), \quad \phi((x, y)) = ux + vy - by.$$

When $b \notin \mathbb{Q}$, the subspace $\mathbb{Q}\Lambda(b) \subset \mathbb{R}$ is two-dimensional over $\mathbb{Q}$ and one defines a character on the $\mathbb{Q}$-rational points by setting $\chi(x - by) := y \in \mathbb{Q}/\mathbb{Z}$, for $x - by \in \mathbb{Q}\Lambda(b)/\Lambda(b)$. Using this character one gets $\chi \circ \phi((x, y)) = y$ and this condition characterizes the relevant pseudolattices.

7 Lift of the Frobenius correspondences

In this paper we have constructed the simplest complex lift of the Scaling Site, using an almost periodic compactification of the added imaginary direction. We have illustrated the role of the tropicalization map and found a surprising relation between the obtained complex lift and the GL_2 system of [16].

In order to complete the Riemann–Roch strategy in this lifted framework one meets a fundamental difficulty tied up to the loss of the one parameter group of automorphisms of $\mathbb{R}_{max}$ in moving from characteristic 1 to characteristic zero. The difficulty arises in the construction of the proper lift of the correspondences Ψ_λ which are canonical in characteristic 1. The natural candidates in characteristic zero come from the right action of $P_+(\mathbb{R})$. This choice is justified using the tropicalization map which is given on all terms by the determinant (on 2 by 2 matrices of adèles) and makes the following diagram commutative

$$\begin{array}{ccccc} P(\mathbb{Q}) & \longrightarrow & \overline{P(\mathbb{A}_{\mathbb{Q}})} & \longleftarrow & P_+(\mathbb{R}) \\ \Big\downarrow \mathrm{det} & & \mathrm{det} \Big\downarrow & & \mathrm{det} \Big\downarrow \\ \mathbb{Q}^\times & \longrightarrow & \mathbb{A}_{\mathbb{Q}} & \longleftarrow & \mathbb{R}_+^* \end{array} \tag{92}$$

The problem arises because the right action of $\mathbb{R}_+^* \subset P_+(\mathbb{R})$ does not preserve the complex structure defined in Proposition 5.1. More precisely this action preserves the foliation (since the leaves are precisely the orbits of the right action of $P_+(\mathbb{R})$) but it does not preserve the complex structure of the leaves. Indeed, the right action of $\mathbb{R}_+^* \subset P_+(\mathbb{R})$ is of the form

$$\begin{pmatrix} y & x \\ 0 & 1 \end{pmatrix} \begin{pmatrix} \lambda & 0 \\ 0 & 1 \end{pmatrix} = \begin{pmatrix} \lambda y & x \\ 0 & 1 \end{pmatrix}$$

and since it replaces $x + iy \in \mathbb{H}$ by $x + i\lambda y$ it does not respect the complex structure.

7.1 Witt construction in characteristic 1 *and lift of the* Ψ_λ

In [19] we already addressed the problem of the loss of the one parameter group of automorphisms of $\mathbb{R}_{\max}$ in moving from characteristic 1 to characteristic zero. In that paper we also developed the Archimedean analogue of the basic steps of the construction of the rings of p-adic periods and we defined the universal thickening of the real numbers. In fact we showed that when one applies the analogue of the Witt construction to the real tropical hyperfield $\mathbb{R}^\flat$, the universal W-model of $\mathbb{R}^\flat$ exists and it coincides with the triple which was constructed in [17, 18], by working with the tropical semifield $\mathbb{R}_+^{\max}$ of characteristic one and implementing concrete formulas, involving entropy, which extend the Teichmüller formula for sums of Teichmüller lifts to the case of characteristic one. We use the notation $[x] = \tau(x)$ for the Teichmüller lift of $x \in \mathbb{R}_+^*$. These elements generate linearly the ring W, they fulfill the multiplication rule $[xy] = [x][y]$ and the automorphisms of $\mathbb{R}_+^{\max}$ lift to automorphisms θ_λ of W such that

$$\theta_\lambda([x]) = [x^\lambda], \quad \forall x \in \mathbb{R}_+^{\max}, \ \lambda \in \mathbb{R}_+^*.$$

We shall use complex coefficients so that in first approximation W is the complex group ring of the multiplicative group $\mathbb{R}_+^*$. We disregard here the nuances obtained from various completions of W explored in [19] and concentrate on the algebraic question of showing why the use of W as coefficients resolves the problem of the lack of invariance of the complex structure under the right action of $\mathbb{R}_+^*$. For each

$\lambda \in \mathbb{R}_+^*$ one has a ring homomorphism $\chi_\lambda : W \to \mathbb{C}$ which is $\mathbb{C}$-linear and such that $\chi_\lambda([x]) = x^\lambda$ for any $x \in \mathbb{R}_+^*$. By construction one has $\chi_\lambda \circ \theta_\mu = \chi_{\lambda\mu}$ and the character χ_1 plays a key role in [19] where it is denoted θ. Now given a function with values in W on a leaf given by an orbit of the right action of $P_+(\mathbb{R})$ we say that f is holomorphic when

$$(\lambda X + iY)\,\chi_\lambda(f) = 0\,, \quad \forall \lambda \in \mathbb{R}_+^*, \tag{93}$$

where X, Y correspond to the generators of the Lie algebra of $P_+(\mathbb{R})$ as in Proposition 5.1. We then restore the invariance under the right action of $\mathbb{R}_+^* \subset P_+(\mathbb{R})$ by combining it with the automorphisms $\theta_\lambda \in \mathrm{Aut}(W)$ as we did in the construction of the arithmetic Frobenius in [25]. By construction the right action $R(\mu)$ of $\mathbb{R}_+^* \subset P_+(\mathbb{R})$ extends to W valued functions so that the following equation holds

$$\chi_\lambda(R(\mu)(f)) = \mu^Y\,\chi_\lambda(f)\,, \quad \forall \lambda, \mu \in \mathbb{R}_+^* \tag{94}$$

Proposition 7.1 *For $\mu \in \mathbb{R}_+^*$, the operation Fr_μ^a,*

$$f \mapsto \mathrm{Fr}_\mu^a(f),\ \mathrm{Fr}_\mu^a(f) := \theta_\mu(R(\mu^{-1})(f)) \tag{95}$$

preserves the holomorphy condition (93).

Proof One has

$$\chi_\lambda(\mathrm{Fr}_\mu^a(f)) = \chi_\lambda \circ \theta_\mu(R(\mu^{-1})(f)) = \chi_{\lambda\mu}(R(\mu^{-1})(f)) = \mu^{-Y}\,\chi_{\lambda\mu}(f)$$

using (94) for the last equality. By construction Y commutes with μ^Y, and since $[Y, X] = X$, one has $\mu^Y X \mu^{-Y} = \mu X$ so that

$$(\lambda X + iY)\,\mu^{-Y} = \mu^{-Y}\,(\lambda\mu X + iY)$$

and one gets

$$(\lambda X + iY)\,\chi_\lambda(\mathrm{Fr}_\mu^a(f)) = (\lambda X + iY)\,\mu^{-Y}\chi_{\lambda\mu}(f) = \mu^{-Y}\,(\lambda\mu X + iY)\,\chi_{\lambda\mu}(f) = 0.$$

Thus $\mathrm{Fr}_\mu^a(f)$ fulfills (93) if f does. □

To give a non-trivial example of a W-valued function which is holomorphic in the sense of (93), we take the classical orbit $\Gamma_{\mathbb{Q},\mathrm{cl}}$ which is the pro-étale cover $\tilde{\mathbb{D}}^*$. One can represent the elements of $\Gamma_{\mathbb{Q},\mathrm{cl}}$ as pairs $(x, y) \in G \times \mathbb{R}_+^*$ and the following equality defines a function

$$q : \Gamma_{\mathbb{Q},\mathrm{cl}} \to W,\ \ q(x + iy) := [e^{-2\pi y}]e^{2\pi i x} \tag{96}$$

where, by Lemma 5.2 (iii), $e^{2\pi i x}$ makes sense as a complex number for any $x \in G$.

Proposition 7.2 *The function* $q : \Gamma_{\mathbb{Q},\mathrm{cl}} \to W$ *of* (96) *is holomorphic in the sense of condition* (93). *Moreover it is invariant under the transformations* Fr^a_μ *for any* $\mu \in \mathbb{R}^*_+$. *The same holds for the rational powers* q^r *of* q, $\forall r \in \mathbb{Q}$.

Proof One has by construction

$$\chi_\lambda(q)(x+iy) = e^{-2\pi\lambda y} e^{2\pi i x} = e^{2\pi i(x+i\lambda y)}.$$

Moreover

$$(\lambda y \partial_x + i y \partial_y)(x + i\lambda y) = 0 \Rightarrow (\lambda X + iY)\, \chi_\lambda(q) = 0$$

which gives condition (93). Finally the equality

$$\theta_\mu\left(q\left(x + i\mu^{-1}y\right) = \left[\left(e^{-2\pi\frac{y}{\mu}}\right)^{\mu}\right] e^{2\pi i x} = \left[e^{-2\pi y}\right] e^{2\pi i x} = q(x+iy)\right)$$

shows that q is invariant under the transformations Fr^a_μ. The same proof applies to the rational powers of q which make sense because of Lemma 5.2 (iii). □

Remark 7.1 Proposition 7.2 suggests that for the topos counterpart $\underline{\tilde{\mathbb{D}}}^* \rtimes \mathbb{N}^\times$ of the above adelic description of the complex lift (see the left column of Figure 1) the structure sheaf involves the ring $W[q^r]$ generated by rational powers q^r of q over W.

7.2 *Quantization*

It is still unclear in which precise sense the complex lift constructed in this paper can be used to "quantize" the Scaling Site. One of the origins of the world of characteristic 1 is the inverse process of quantization, it is called "dequantization." It was developed under the name of idempotent analysis by the school of Maslov, Kolokolstov, and Litvinov [34, 35]. One of their key discoveries is that the Legendre transform which plays a fundamental role in all of physics and in particular in thermodynamics in the nineteenth century, is simply the Fourier transform in the framework of idempotent analysis. There is a whole circle of ideas which compares tropicalization, dequantization, and deformation of complex structures (see e.g. [1] and the references there) and these ideas should be carefully identified for the complex lift constructed in our paper. In particular the deformation of complex structures used in Section 7.1 and the interpretation of its limit as a real polarization should be clarified. In the process of quantizing a classical dynamical system the expected outcome is a self-adjoint operator in Hilbert space. We expect here that the operator $X+iY$ will play a role as well as the "transverse elliptic theory" developed in [12]. Indeed, when viewing the adèle class space $\mathbb{Q}^\times\backslash\mathbb{A}_\mathbb{Q}$ as a noncommutative space and the complex lift $\mathscr{C}_\mathbb{Q}$ over it, the complex structure takes place in the transverse direction. In fact, as we have seen in Section 5, the space of points of $\mathscr{C}_{\mathbb{Q}}$ fibers over $\mathbb{Q}^\times\backslash\mathbb{A}_{\mathbb{Q}}$ with fiber the almost periodic compactification G of $\mathbb{R}$. The effect of the almost periodic compactification is occurring purely in the transversal direction and it thus suggests that the $\bar\partial$ operator associated with the complex structure should be viewed as a K-homology class in the relative type II setup. The results of [13, 14, 15] on the transverse structure of modular Hecke algebras should then be brought into play.

Acknowledgements The second named author would like to thank Alain Connes for introducing her to the noncommutative geometric vision on the Riemann Hypothesis and for sharing with her many mathematical ideas and insights.

Caterina Consani is partially supported by the Simons Foundation collaboration Grant no. 353677. She would like to thank Collège de France for some financial support.

References

1. T. Baier, C. Florentino, J. Mourao, J. Nunes, *Toric Kähler metrics seen from infinity, quantization and compact tropical amoebas*. J. Differential Geom. 89 (2011), no. 3, 411–454.
2. M. Baker, S. Norine, *Riemann-Roch and Abel-Jacobi theory on a finite graph*, Advances in Mathematics 215 (2007), 766–788.
3. M. Baker, F. Shokrieh, *Chip-firing games, potential theory on graphs, and spanning trees*. J. Combin. Theory Ser. A 120 (2013), no. 1, 164–182.
4. A.S. Besicovitch, *Almost Periodic Functions,* Cambridge University Press, 1932.
5. A. Bjorner, L. Lovasz, P. W. Shor, *Chip-firing games on graphs*, European J. Combin., 12(4), (1991), 283–291.
6. H. Bohr, *Almost Periodic Functions,* Chelsea Pub. Co., New York, 1947.
7. S. Bochner, *Beitrage zur Theorie der fastperiodischen Funktionen*, Math. Annalen, 96: 119–147.
8. E. Bombieri, J. Lagarias *Complements to Li's criterion for the Riemann hypothesis*. J. Number Theory 77 (1999), no. 2, 274–287.
9. V. Borchsenius, B. Jessen, *Mean motion and values of the Riemann zeta function*. Acta Math. 80 (1948), 97–166.
10. J.B. Bost, A. Connes, *Hecke algebras, Type III factors and phase transitions with spontaneous symmetry breaking in number theory*, Selecta Math. (New Series) Vol.1 (1995) N.3, 411–457.
11. A. Connes, *Trace formula in noncommutative geometry and the zeros of the Riemann zeta function*. Selecta Math. (N.S.) 5 (1999), no. 1, 29–106.
12. A. Connes, H. Moscovici, *The local index formula in noncommutative geometry*, GAFA, Vol. 5 (1995), 174–243.
13. A. Connes, H. Moscovici, *Modular Hecke Algebras and their Hopf Symmetry*, Mosc. Math. J. 4 (2004), no. 1, 67–109, 310.
14. A. Connes, H. Moscovici, *Rankin-Cohen Brackets and the Hopf Algebra of Transverse Geometry*, Mosc. Math. J. 4 (2004), no. 1, 111–130, 311.
15. A. Connes, H. Moscovici, *Transgressions of the Godbillon-Vey class and Rademacher functions*, in "Noncommutative Geometry and Number Theory", pp.79–107. Vieweg Verlag, 2006.

16. A. Connes, M. Marcolli, *Noncommutative Geometry, Quantum Fields and Motives* American Mathematical Society Colloquium Publications, 55. American Mathematical Society, Providence, RI; Hindustan Book Agency, New Delhi, 2008.
17. A. Connes, C. Consani, *Characteristic one, entropy and the absolute point*, " Noncommutative Geometry, Arithmetic, and Related Topics", the Twenty-First Meeting of the Japan-U.S. Mathematics Institute, Baltimore 2009, JHUP (2012), 75–139.
18. A. Connes, *The Witt construction in characteristic one and Quantization*. Noncommutative geometry and global analysis, 83–113, Contemp. Math., 546, Amer. Math. Soc., Providence, RI, 2011.
19. A. Connes, C. Consani, *Universal thickening of the field of real numbers*. Advances in the theory of numbers, 11–74, Fields Inst. Commun., 77, Fields Inst. Res. Math. Sci., Toronto, ON, 2015.
20. A. Connes, C. Consani, *Schemes over $\mathbb{F}_1$ and zeta functions*, Compositio Mathematica 146 (6), (2010) 1383–1415.
21. A. Connes, C. Consani, *From monoids to hyperstructures: in search of an absolute arith- metic*, in Casimir Force, Casimir Operators and the Riemann Hypothesis, de Gruyter (2010), 147–198.
22. A. Connes, C. Consani, *The Arithmetic Site*, Comptes Rendus Mathematique Ser. I 352 (2014), 971–975.
23. A. Connes, C. Consani, *Geometry of the Arithmetic Site*. Advances in Mathematics 291 (2016) 274–329.
24. A. Connes, C. Consani, *The Scaling Site*, C.R. Mathematique, Ser. I 354 (2016) 1–6.
25. A. Connes, C. Consani, *Geometry of the Scaling Site*, Selecta Math. New Ser. **23** no. 3 (2017), 1803–1850.
26. A. Connes, C. Consani, *Homological algebra in characteristic one*. Higher Structures Journal **3** (2019), no. 1, 155–247.
27. A. Connes, *An essay on the Riemann Hypothesis*. In "Open problems in mathematics", Springer (2016), volume edited by Michael Rassias and John Nash.
28. A. Gathmann and M. Kerber. *A Riemann-Roch theorem in tropical geometry*. Math. Z., 259(1):217–230, 2008.
29. J. Golan, *Semi-rings and their applications*, Updated and expanded version of The theory of semi-rings, with applications to mathematics and theoretical computer science [Longman Sci. Tech., Harlow, 1992. Kluwer Academic Publishers, Dordrecht, 1999.
30. A. Grothendieck *Sur une note de Mattuck-Tate* J. reine angew. Math. 200, 208–215 (1958).
31. B. Jessen, Börge; Über die Nullstellen einer analytischen fastperiodischen Funktion. Eine Verallgemeinerung der Jensenschen Formel. (German) Math. Ann. 108 (1933), no. 1, 485–516.
32. Jessen, Tornehave *Mean motions and zeros of almost periodic functions*. Acta math. 77, 137–279 (1945).
33. M. Laca, N. Larsen and S. Neshveyev, *Phase transition in the Connes-Marcolli GL_2 system*, J. Noncommut. Geom. 1 (2007), no. 4, 397–430.
34. V. Kolokoltsov, V. P. Maslov, *Idempotent analysis and its applications*. Mathematics and its Applications, 401. Kluwer Academic Publishers Group, Dordrecht, 1997.
35. G. Litvinov, *Tropical Mathematics, Idempotent Analysis, Classical Mechanics and Geometry*. Spectral theory and geometric analysis, 159–186, Contemp. Math., 535, Amer. Math. Soc., Providence, RI, 2011.
36. R. Meyer, *On a representation of the idele class group related to primes and zeros of L-functions*. Duke Math. J. Vol.127 (2005), N.3, 519–595.
37. G. Mikhalkin and I. Zharkov, *Tropical curves, their Jacobians and theta functions*. In Curves and abelian varieties, volume 465 of Contemp. Math., p 203–230. Amer. Math. Soc., Providence, RI, 2008.
38. J.S. Milne, *Canonical models of Shimura curves*, manuscript, 2003 (www.jmilne.org).
39. M. Laca, S. Neshveyev, M. Trifkovic *Bost-Connes systems, Hecke algebras, and induction*. J. Noncommut. Geom. 7 (2013), no. 2, 525–546.

40. A. Robert, *A course in p-adic analysis*. Graduate Texts in Mathematics, 198. Springer-Verlag, New York, 2000.
41. W. Rudin, *Real and complex analysis*. McGraw-Hill, New York, 1966.
42. J. von Neumann, *Almost Periodic Functions in a Group I*, Trans. Amer. Math. Soc., 36 no. 3 (1934) pp. 445–492
43. A. Weil, *Basic Number Theory*, Reprint of the second (1973) edition. Classics in Mathematics. Springer-Verlag, 1995.
44. S. Yoshitomi, *Generators of modules in tropical geometry*. ArXiv Math.AG, 10010448.

The Baum–Connes conjecture: an extended survey

Maria Paula Gomez Aparicio, Pierre Julg, and Alain Valette

To Alain Connes, for providing lifelong inspiration

Abstract We present a history of the Baum–Connes conjecture, the methods involved, the current status, and the mathematics it generated.

1 Introduction

1.1 Building bridges

Noncommutative Geometry is a field of Mathematics which builds bridges between many different subjects. Operator algebras, index theory, K-theory, geometry of foliations, group representation theory are, among others, ingredients of the impressive achievements of Alain Connes and of the many mathematicians that he has inspired in the past 40 years.

At the end of the 1970s the work of Alain Connes on von Neumann theory naturally led him to explore foliations and groups. His generalizations of Atiyah's L^2 index theorem were the starting point of his ambitious project of Noncommutative Geometry. A crucial role has been played by the pioneering conference in Kingston

M. P. Gomez Aparicio
Laboratoire de Mathématiques d'Orsay, Univ. Paris-Sud, CNRS,
Université Paris-Saclay, Orsay, France
e-mail: maria.gomez@math.u-psud.fr

P. Julg
Institut Denis Poisson, Université d'Orléans, Collegium Sciences et Techniques,
Bâtiment de mathématiques, Orléans, France
e-mail: pierre.julg@univ-orleans.fr

A. Valette (✉)
Institut de Mathématiques, Université de Neuchâtel, Neuchâtel, Switzerland
e-mail: alain.valette@unine.ch

A. Chamseddine et al. (eds.), *Advances in Noncommutative Geometry*,
https://doi.org/10.1007/978-3-030-29597-4_3

in July 1980, where he met the topologist Paul Baum. The picture of what was soon going to be known as the Baum–Connes conjecture quickly emerged. The catalytic effect of IHES should not be underestimated; indeed the paper [BC00] was for a long time available only as an IHES 1982 preprint. It is only in 1994 that the general and precise statement was given in the proceedings paper [BCH94] with Nigel Higson.

1.2 In a nutshell: without coefficients…

The Baum–Connes conjecture also builds a bridge between commutative geometry and noncommutative geometry. Although it may be interesting to formulate the conjecture for locally compact groupoids,[1] we stick to the well-accepted tradition of formulating the conjecture for locally compact, second countable groups.

For every locally compact group G there is a Baum–Connes conjecture! We start by associating to G four abelian groups $K_*^{top}(G)$ and $K_*(C_r^*(G))$ (with $* = 0, 1$), then we construct a group homomorphism, the *assembly map*:

$$\mu_r : K_*^{top}(G) \to K_*(C_r^*(G)) \quad (* = 0, 1).$$

We say that **the Baum–Connes conjecture holds for G if μ_r is an isomorphism for $* = 0, 1$.** Let us give a rough idea of the objects.

- The RHS of the conjecture, $K_*(C_r^*(G))$, is called the *analytical side*: it belongs to noncommutative geometry. Here $C_r^*(G)$, the *reduced C^*-algebra of G*, is the closure in the operator norm of $L^1(G)$ acting by left convolution on $L^2(G)$, and $K_*(C_r^*(G))$ is its *topological K-theory*.

  Topological K-theory is a homology theory for Banach algebras A, enjoying the special feature of Bott periodicity ($K_i(A)$ is naturally isomorphic to $K_{i+2}(A)$), so that there are just two groups to consider: K_0 and K_1. K-theory conquered C^*-algebra theory around 1980, as a powerful invariant to distinguish C^*-algebras up to isomorphism. The first success was, in the case of the free group $\mathbf{F}_n$ of rank n, the computation of $K_*(C_r^*(\mathbf{F}_n))$ by Pimsner and Voiculescu [PV82]: they obtained

  $$K_0\left(C_r^*(\mathbf{F}_n)\right) = \mathbf{Z}, \quad K_1\left(C_r^*(\mathbf{F}_n)\right) = \mathbf{Z}^n,$$

  so that K_1 distinguishes reduced C^*-algebras of free groups of various ranks.

  For many connected Lie groups (e.g., semisimple), $C_r^*(G)$ is type I, which points to using *dévissage* techniques: representation theory allows to define ideals and quotients of $C_r^*(G)$ that are less complicated, so $K_*(C_r^*(G))$ can be computed by means of the 6-term exact sequence associated with a short exact

[1] This is important, e.g., for applications to foliations, see Chapter 7.

sequence of Banach algebras. By way of contrast, if G is discrete, $C_r^*(G)$ is very often *simple* (see [BKKO17] for recent progress on that question); in that case, dévissage must be replaced by brain power (see [Pim86] for a sample), and the Baum–Connes conjecture at least provides a conjectural description of what $K_*(C_r^*(G))$ should be (see, e.g., [SG08]).

- The LHS of the conjecture, $K_*^{top}(G)$, is called the *geometric*, or *topological* side. This is actually misleading, as its definition is awfully analytic, involving Kasparov's bivariant theory (see in this chapter). A better terminology would be the *commutative side*, as indeed it involves a space $\underline{EG}$, the *classifying space for proper actions of* G (see Chapter 4) and $K_*^{top}(G)$ is the G-equivariant K-homology of $\underline{EG}$.

  When G is discrete and torsion-free, then $\underline{EG} = EG = \widetilde{BG}$, the universal cover of the classifying space BG. As G acts freely on EG, the G-equivariant K-homology of EG is $K_*(BG)$, the ordinary K-homology of BG, where K-homology for spaces can be defined as the homology theory dual to topological K-theory for spaces.
- The assembly map μ_r will be defined in Chapter 4 using Kasparov's equivariant KK-theory. Let us only give here a flavor of the meaning of this map. It was discovered in the late 1970s and early 1980s that the K-theory group $K_*(C_r^*(G))$ is a receptacle for indices, see Section 2.3. More precisely, if M is a smooth manifold with a proper action of G and compact quotient, and D an elliptic G-invariant differential operator on M, then D has an index $ind_G(D)$ living in $K_*(C_r^*(G))$. Therefore, the geometric group $K_*^{top}(G)$ should be thought of as the set of homotopy classes of such pairs (M, D), and the assembly map μ_r maps the class $[(M, D)]$ to $ind_G(D) \in K_*(C_r^*(G))$.

1.3 ...and with coefficients

There is also a more general conjecture, called *the Baum–Connes conjecture with coefficients*, where we allow G to act by *-automorphisms on an auxiliary C^*-algebra A (which becomes a $G - C^*$-algebra), and where the aim is to compute the K-theory of the reduced crossed product $C_r^*(G, A)$. One defines then the assembly map

$$\mu_{A,r} : K_*^{top}(G, A) \to K_*\left(C_r^*(G, A)\right) \quad (* = 0, 1)$$

and we say that **the Baum–Connes conjecture with coefficients holds for G if $\mu_{A,r}$ is an isomorphism for $* = 0, 1$ and every $G - C^*$-algebra A**. The advantage of the conjecture with coefficients is that it is inherited by closed subgroups; its disadvantage is that it is false in general, see Chapter 9.

1.4 Structure of these notes

Using the acronym BC for "*Baum–Connes conjecture*," here is what the reader will find in this piece.

- Where does BC come from? Chapter 2, on the history of the conjecture.
- What are the technical tools and techniques? Chapter 3, on Kasparov theory (and the Dirac–dual-Dirac method).
- What is BC, what does it entail, what is the state of the art? Chapter 4.
- Why is BC difficult? Chapter 5, discussing BC with coefficients for semisimple Lie groups and their closed (e.g., discrete) subgroups.
- How can we hope to overcome those difficulties? Chapter 6, on Banach algebraic methods.
- Is BC true or false? For BC without coefficients we don't know, but we know that the natural extension of BC from groups to groupoids is false (see Chapter 7), and we know that BC with coefficients is false (see Chapter 9).

We could have stopped there. But it seemed unfortunate not to mention an important avatar of BC, namely the *coarse Baum–Connes conjecture* (CBC) due to the late John Roe: roughly speaking, groups are replaced by metric spaces, see Chapter 8. An important link with the usual BC is that for a finitely generated group, which can be viewed as a metric space via some Cayley graph, CBC implies the injectivity part of BC.

Finally, it was crucial to mention the amount of beautiful mathematics generated by BC, and this is done in Chapter 9.

1.5 What do we know in 2019?

In Chapter 3 we explain the "*Dirac–dual-Dirac*" method used by Kasparov [Kas95] to prove the injectivity of $\mu_{A,r}$ for all semisimple Lie group G and all $G - C^*$-algebras A; this also proves injectivity for closed subgroups of a semisimple Lie group, as this property passes to closed subgroups. Since then, an abstraction of the Dirac–dual-Dirac method, explained in Section 4.4, has been used by Kasparov and Skandalis [KS03], to prove the injectivity of the assembly map for a large class of groups denoted by $\mathcal{C}$ in [Laf02b]. This class contains, for example, all locally compact groups acting continuously, properly and isometrically on a complete and simply connected Riemannian manifold of non-positive scalar curvature (see [Kas88]), or on a Bruhat–Tits affine building (for example, all p-adic groups, see [KS91]), all hyperbolic groups (see [KS03]). So the injectivity of the Baum–Connes assembly map has been proven for a huge class of groups.

The conjecture with coefficients has been proven for a large class of groups that includes all groups with the Haagerup property (e.g., $SL_2(\mathbf{R})$, $SO_0(n, 1)$, $SU(n, 1)$, and all free groups). For those groups the proof is due to Higson and Kasparov (see [HK01]) and it is also based on the "Dirac–dual-Dirac" method. This method cannot, however, be applied to non-compact groups having property (T), not even for the conjecture without coefficients: see Section 5.1 for more on the tension between the Haagerup and Kazhdan properties.

Nevertheless, as will be explained in Section 6.1, Lafforgue managed to prove the conjecture without coefficients for all semisimple Lie groups and for some of their discrete subgroups, precisely those having property (RD) (as defined in Section 6.1.4). For example, the conjecture without coefficients is true for all cocompact lattices in $SL_3(\mathbf{R})$ but it is still open for $SL_3(\mathbf{Z})$.[2]

On the other hand, the conjecture with coefficients has been proven for all hyperbolic groups (see [Laf12]), but it still open for higher rank semisimple Lie groups and their closed subgroups: see Sections 5.2 and 6.2.3 for more on that.

An example of a group for which, at the time of writing, μ_r is not known to be either injective or surjective is the free Burnside group $B(d, n)$, as soon as it is infinite.[3]

1.6 A great conjecture?

What makes a conjecture great? Here we should of course avoid the chicken-and-egg answer "*It's a great conjecture because it is due to great mathematicians.*" We should also be suspicious of the pure maths self-referential answer: "*It's a great conjecture because it implies several previous conjectures*": that an abstruse conjecture implies even more abstruse ones,[4] does not necessarily make it great.

We believe that the interest of a conjecture lies in the feeling of unity of mathematics that it entails. We hope that the reader, in particular the young expert, after glancing at the table of contents and the various subjects listed in Section 1.7 below, will not let her/himself be discouraged. Rather (s)he should take this as an incentive to learn new mathematics, and most importantly connections between them.

Judging by the amount of fields that it helps bridging (representation theory, geometric group theory, metric geometry, dynamics,...), we are convinced that yes, the Baum–Connes conjecture is indeed a great conjecture.

---

[2] In the case of $SL_3(\mathbf{Z})$, surjectivity of μ_r is the open problem; the LHS of the Baum–Connes conjecture was computed in [SG08].

[3] Recall that $B(d, n)$ is defined as the quotient of the non-abelian free group $\mathbf{F}_d$ by the normal subgroup generated by all n's powers in $\mathbf{F}_d$.

[4] Compare with Sections 2.5 and 4.5.

1.7 Which mathematics are needed?

We use freely the following concepts; for each we indicate one standard reference:

- locally compact groups (Haar measure, unitary representations): see [Dix96];
- semisimple Lie groups and symmetric spaces: see [Hel62];
- operator algebras (full and reduced group C^*-algebras, full and reduced crossed products): see [Ped79];
- K-theory for C^*-algebras (Bott periodicity, 6-term exact sequences, Morita equivalence): see [WO93];
- index theory: see [BBB13].

2 Birth of a conjecture

2.1 Elliptic (pseudo-) differential operators

Let M be a closed manifold, and let D be a (pseudo-) differential operator acting on smooth sections of some vector bundles E, F over M, so D maps $C^\infty(E)$ to $C^\infty(F)$. Let T^*M denote the cotangent bundle of M. The (principal) symbol is a bundle map $\sigma(D)$ from the pullback of E to the pullback of F on T^*M. Recall that D is said to be *elliptic* if $\sigma(D)$ is invertible outside of the zero section of T^*M. In this case standard elliptic theory guarantees that $\ker(D)$ and $\mathrm{coker}(D)$ are finite-dimensional, so that the (Fredholm) index of D is defined as

$$Ind(D) = \dim_{\mathbf{C}} \ker(D) - \dim_{\mathbf{C}} \mathrm{coker}(D) \in \mathbf{Z}.$$

The celebrated Atiyah–Singer theorem [AS68] then provides a topological formula for $Ind(D)$ in terms of topological invariants associated with M and $\sigma(D)$.

Now let $\tilde{M} \to M$ be a Galois covering of M, with group Γ, so that $M = \Gamma\backslash\tilde{M}$. Assume that D lifts to a Γ-invariant operator $\tilde{D}$ on $\tilde{M}$, between smooth sections of $\tilde{E}, \tilde{F}$, the vector bundles pulled back from E, F via the covering map.

- Assume first that Γ is finite, i.e., our covering has $n = |\Gamma|$ sheets. Then $\tilde{M}$ is a closed manifold, and the index of $\tilde{D}$ satisfies $Ind(\tilde{D}) = n \cdot Ind(D)$. Now we may observe that, in this case, there is a more refined analytical index, obtained by observing that $\ker(\tilde{D})$ and $\mathrm{coker}(\tilde{D})$ are finite-dimensional representation spaces of Γ, hence their formal difference makes sense in the additive group of the representation ring $R(\Gamma)$: we get an element $\Gamma - Ind(\tilde{D}) \in R(\Gamma)$; the character of this virtual representation of Γ, evaluated at $1 \in \Gamma$, gives precisely $Ind(\tilde{D})$.
- Assume now that Γ is infinite. Then the L^2-kernel and L^2-cokernel of $\tilde{D}$ are closed subspaces of the suitable space of L^2-sections, namely $L^2(\tilde{M}, \tilde{E})$ and $L^2(\tilde{M}, \tilde{F})$, and by Γ-invariance those spaces are representation spaces of Γ. The problem with these representations is that their classical dimension is infinite.

Atiyah's idea in [Ati76] is to measure the size of these spaces via the dimension theory of von Neumann algebras.

More precisely, the L^2-kernel of $\tilde{D}$ is Γ-invariant, so the orthogonal projection onto that kernel belongs to the algebra A of operators commuting with the natural Γ-representation on $L^2(\tilde{M}, \tilde{E})$. Choosing a fundamental domain for the Γ-action on $\tilde{M}$ allows to identify Γ-equivariantly $L^2(\tilde{M}, \tilde{E})$ with $\ell^2(\Gamma) \otimes L^2(M, E)$. So A becomes the von Neumann algebra $L(\Gamma) \otimes \mathcal{B}(L^2(M, E))$, where $L(\Gamma)$, the *group von Neumann algebra* of Γ, is generated by the right regular representation of Γ on $\ell^2(\Gamma)$. The canonical trace on $L(\Gamma)$ (defined by $\tau(a) = \langle a(\delta_e), \delta_e \rangle$ for $a \in L(\Gamma)$) provides a dimension function $\dim_\Gamma$ on the projections in A. Atiyah's *L^2-index theorem* [Ati76] states that

Theorem 2.1 *In the situation above:*

$$Ind(D) = Ind_\Gamma(\tilde{D}),$$

where the right-hand side is defined as

$$Ind_\Gamma(\tilde{D}) := \dim_\Gamma(\ker \tilde{D})) - \dim_\Gamma(\mathrm{coker}(\tilde{D})).$$

2.2 Square-integrable representations

Recall that, for G a locally compact unimodular group, a unitary irreducible representation π of G is said to be *square-integrable* if, for every two vectors ξ, η in the Hilbert space of the representation π, the coefficient function

$$g \mapsto \langle \pi(g)\xi, \eta \rangle$$

is square-integrable on G. Equivalently, π is a sub-representation of the left regular representation λ_G of G on $L^2(G)$ (see [Dix96], section 14.1, for the equivalence). The set of square-integrable representations of G is called the *discrete series* of G.

When G is a semisimple Lie group with finite center, we denote by $\hat{G}_r$ the *reduced dual*, or *tempered dual* of G: this is the set of (equivalence classes of) unitary irreducible representations of G weakly contained in λ_G; it may also be defined as the support of the Plancherel measure on the full dual $\hat{G}$ of G. A cornerstone of twentieth century mathematics is Harish-Chandra's explicit description of the Plancherel measure on semisimple Lie groups, and it turns out that the discrete series of G is exactly the set of atoms of the Plancherel measure.

Let us be more specific. Let K be a maximal compact subgroup of G, a connected semisimple Lie group with finite center. The first result of Harish-Chandra states that the discrete series of G is non-empty if and only if G and K have equal rank. This exactly means that a maximal torus of K is also a maximal torus of G. Let us assume that this holds, and let us fix a maximal torus T in K. Let $\mathfrak{g}_\mathbb{C}, \mathfrak{k}_\mathbb{C}, \mathfrak{t}_\mathbb{C}$

be the complexified Lie algebras of G, K, T respectively. Decomposing the adjoint representations of T on $\mathfrak{k}_{\mathbb{C}}$ and $\mathfrak{g}_{\mathbb{C}}$ respectively, we get two root systems Φ_c and Φ, with $\Phi_c \subset \Phi$: we say that Φ is the set of roots, while Φ_c is the set of *compact* roots. Correspondingly there are two Weyl groups $W(K) \subset W$. We denote by Λ the lattice of weights of T. An element of $\mathfrak{t}_{\mathbb{C}}$ is *regular* if its stabilizer in W is trivial. We denote by ρ half the sum of positive roots in Φ (with respect to a fixed set Ψ of positive roots), and by ρ_c half the sum of the positive compact roots. We have then Harish-Chandra's main result on existence and exhaustion of discrete series (see [Lip74], section I.B.2 for a nice summary of Harish-Chandra's theory):

Theorem 2.2 *To each regular element $\lambda \in \Lambda + \rho$ is naturally associated a square-integrable irreducible representation π_λ of G such that $\pi_\lambda|_K$ contains with multiplicity 1 the K-type with highest weight $\lambda + \rho - 2\rho_c$. Every discrete series representation of G appears in this way. If $\lambda, \mu \in \Lambda + \rho$, the representations π_λ, π_μ are unitarily equivalent if and only if λ and μ are in the same $W(K)$-orbit.*

Impressive as it is, Theorem 2.2 left open the question of constructing geometrically the discrete series representations π_λ. That question was solved by Atiyah and Schmid [AS77]. Assume that G has discrete series representations, which forces the symmetric space G/K to be even-dimensional. Assume moreover that G/K carries a G-invariant spin structure, meaning that the isotropy representation of K on $V := \mathfrak{g}/\mathfrak{k}$ lifts to the spin group of V; this can be ensured by replacing G by a suitable double cover. Then we have the two irreducible spinor representations S^+, S^- of $Spin(V)$, that we view as K-representations.[5] Fix a regular element λ in $\Lambda + \rho$; conjugating Ψ by some element of W, we may assume that λ is dominating for Ψ. Then $\mu := \lambda - \rho_c \in \Lambda$ is a weight dominating for $\Phi_c \cap \Psi$, and we denote by E_μ the irreducible representation of K with highest weight μ. Form the G-equivariant induced vector bundles $G \times_K (E_\mu \otimes S^\pm)$ over G/K, and let

$$D_\mu : C^\infty(G \times_K (E_\mu \otimes S^+)) \to C^\infty(G \times_K (E_\mu \otimes S^-))$$

be the corresponding Dirac operator with coefficients in μ. The main result of Atiyah and Schmid (see [AS77, 9.3]) is then:

Theorem 2.3 *Let $\lambda \in \Lambda + \rho$ be regular, with $\lambda = \mu + \rho_c$ as above. Then* $\mathrm{coker}(D_\mu^+) = 0$ *and the G-representation on* $\ker(D_\mu^+)$ *is the discrete series representation π_λ. If λ is not regular, then* $\ker(D_\mu^+) = \mathrm{coker}(D_\mu^+) = 0$.

It is interesting to observe that Atiyah's L^2-index theorem plays a role in the proof, as the authors need a torsion-free cocompact lattice Γ in G and apply the L^2-index theorem to the covering of the compact manifold $\Gamma\backslash G/K$ by G/K.

[5] G/K carries a G-invariant spin structure if and only if $\rho - \rho_c \in \Lambda$, see [AS77, 4.34]; the distinction between S^+ and S^- is made by requiring that $\rho - \rho_c$ is the highest weight for S^+, see [AS77, 3.13].

To summarize, *Dirac induction* (i.e., realizing G-representations by means of Dirac operators with coefficients in K-representations) sets up a bijection between a generic set of irreducible representations of K and all square-integrable representations of G. Suitably interpreted using K-theory of C^*-algebras, this principle paved the way towards the Connes–Kasparov conjecture, which was the first form of the Baum–Connes conjecture.

2.3 Enters K-theory for group C^*-algebras

The Atiyah–Schmid construction of the discrete series, served as a crucial motivation for Connes and Moscovici [CM82] in their study of the G-index for G-equivariant elliptic differential operators D on homogeneous spaces of the form G/K, where G is a unimodular Lie group with countably many connected components, and K is a compact subgroup. Their aim is to define the G-index of D intrinsically, i.e., without appealing to Atiyah's L^2-index theory (so, not needing an auxiliary cocompact lattice in G): D will not be Fredholm in the usual sense (unless G is compact), but $\ker(D)$ and $\mathrm{coker}(D)$ will have finite G-dimension in the sense of the Plancherel measure on $\hat{G}_r$. The formal difference of these, the G-index of D, is a real number shown to depend only on the class $[\sigma(D)]$ of the symbol of D in $\mathrm{K}_K(V^*)$, where K_K denotes equivariant K-theory with compact supports and V^* is the cotangent space to G/K at the origin. This G-index is computed in terms of the symbol of D, and this index formula is used to prove that $\ker(D)$ is a finite direct sum of square-integrable representations of G.

Crucial for our story is the final section of [CM82]. Indeed, there Connes and Moscovici sketch the construction of an index taking values in $\mathrm{K}_*(C_r^*(G))$, the topological K-theory of the reduced C^*-algebra of G. It goes as follows: let ρ be a finite-dimensional unitary representation of K on H_ρ, form the induced vector bundle $E_\rho := G \times_K H_\rho$ over G/K. Denote by $\Psi_G^*(G/K, E_\rho)$ be the norm closure of the space of 0-th order G-invariant pseudo-differential operators on G/K acting on sections of E_ρ: since such an operator acts by bounded operators on $L^2(G/K, E_\rho)$, we see that $\Psi_G^*(G/K, E_\rho)$ is a C^*-algebra on $L^2(G/K, E_\rho)$. The symbol map induces a $*$-homomorphism $\Psi_G^*(G/K, E_\rho) \to C_K(S(V^*), \mathcal{B}(H_\rho))$, where the latter is the algebra of K-invariant, $\mathcal{B}(H_\rho)$-valued continuous functions on $S(V^*)$, the unit sphere in V^*. It fits into a short exact sequence

$$0 \to C_G^*(G/K, E_\rho) \to \Psi_G^*(G/K, E_\rho) \to C_K(S(V^*), \mathcal{B}(H_\rho)) \to 0, \tag{2.1}$$

where the kernel $C_G^*(G/K, E_\rho)$ is the norm closure of G-invariant regularizing operators on G/K. When ρ is the left regular representation of K, Connes and Moscovici observe that $C_G^*(G/K, E_\rho)$ is canonically isomorphic to the reduced C^*-algebra $C_r^*(G)$ of G. If $D \in \Psi_G^*(G/K, E_\rho)$ is elliptic, then its symbol is invertible in $C_K(S(V^*), \mathcal{B}(H_\rho))$, so defines an element $[\sigma(D)] \in K_1(C_K(S(V^*)))$. The short exact sequence (2.1) defines a 6-term exact sequence in K-theory, and the connecting map $K_1(C_K(S(V^*))) \to K_0(C_r^*(G))$ allows to define $ind_G(D) \in K_0(C_r^*(G))$. So the K-theory $K_*(C_r^*(G))$ appears as a receptacle for indices of G-invariant elliptic pseudo-differential operators on manifolds of the form G/K, with K compact.

We quote the final lines of [CM82]: "*Of course, to obtain a valuable formula for the index map* ind_G, *one first has to compute* $K_0(C_r^*(G))$. *When G is simply connected and solvable, it follows from the Thom isomorphism in [Con81] that* $K_i(C^*(G)) \simeq K^{i+j}(point), i, j \in \mathbf{Z}_2$, *where j is the dimension* mod 2 *of G. The computation of the K-theory of* $C^*(G)$ *for an arbitrary Lie group G and the search for an "intrinsic" index formula certainly deserve further study.*" This served as a research program for the following years![6]

Let us end this section by mentioning that, since the framework in [CM82] is unimodular Lie groups with countably many connected components, it applies in particular to countable discrete groups Γ. In this case the canonical trace $\tau : C_r^*(\Gamma) \to \mathbf{C}$ defines a homomorphism $\tau_* : K_0(C_r^*(\Gamma)) \to \mathbf{R}$, and $\tau_*(ind_\Gamma(D)) = Ind_\Gamma(D)$, the Γ-index of D as in (2.1).

2.4 The Connes–Kasparov conjecture

Disclaimer: the Connes–Kasparov conjecture is not a conjecture anymore since 2003! After proofs of several particular cases, starting with the case of simply connected solvable groups established by Connes [Con81], and the cornerstone of semisimple groups being established first by Wassermann [Was87] by representation-theoretic methods then by Lafforgue [Laf02b] by geometric/analytical techniques, the general case was handled by Chabert–Echterhoff–Nest [CEN03] building on Lafforgue's method. Nevertheless the Connes–Kasparov conjecture was fundamental for the later formulation of the more general Baum–Connes conjecture.

Let G be a connected Lie group, and let K be a maximal compact subgroup (it follows from structure theory that K is unique up to conjugation). Set $V = \mathfrak{g}/\mathfrak{k}$; assume that G/K carries a G-invariant spin structure, i.e., that the adjoint representation of K on V lifts to $Spin(V)$. Let S^+, S^- be the spinor representations of $Spin(V)$ (with the convention $S^+ = S^-$ if $j = \dim G/K$ is odd), that we view as K-representations. Let ρ be a finite-dimensional representation of K, form the induced G-vector bundles $E_\rho^\pm = G \times_K (\rho \otimes S^\pm)$. Let $D_\rho : C^\infty(E_\rho^+) \to C^\infty(E_\rho^-)$ be the corresponding Dirac operator. Let $R(K)$ be the representation ring of K. Thanks to the previous section, we may define the *Dirac induction*

$$\mu_G : R(K) \to K_j(C_r^*(G)) : \rho \mapsto ind_G(D_\rho^+),$$

[6] We believe that Connes and Moscovici actually had $C_r^*(G)$, not $C^*(G)$, in mind when writing this.

a homomorphism of abelian groups. The *Connes–Kasparov conjecture* (see [BC00], section 5; [Kas87]; [Kas95], Conjecture 1) is the following statement:

Conjecture 1 (1st version) Let G be a connected Lie group, K a maximal compact subgroup, $j = \dim(G/K)$. Assume that G/K carries a G-invariant spin structure.

(1) The Dirac induction $\mu_G : R(K) \to K_j(C_r^*(G))$ is an isomorphism;
(2) $K_{j+1}(C_r^*(G)) = 0$

Remark 2.4 If G is semisimple with finite center, and π is a square-integrable representation of G, then π defines an isolated point of $\hat{G}_r$, so there is a splitting $C_r^*(G) = J_\pi \oplus \mathcal{K}$, where J_π is the C^*-kernel of π and $\mathcal{K}$ is the standard algebra of compact operators. Hence $K_0(C_r^*(G)) = K_0(J_\pi) \oplus \mathbf{Z}$, i.e., π defines a free generator $[\pi]$ of $K_0(C_r^*(G))$. In terms of the Connes–Kasparov conjecture, Theorem 2.3 expresses the fact that the Dirac induction μ_G induces an isomorphism between an explicit free abelian subgroup of $R(K)$ and the free abelian part of $K_0(C_r^*(G))$ associated with the discrete series.

Example 2.5 Take $G = SL_2(\mathbf{R})$, so that $K = T = SO(2)$. Then the set Λ of weights of T identifies with $\mathbf{Z}$, the set Φ of roots is $\{-2, 0, 2\}$ (so that $\rho = 1$ if $\Psi = \{2\}$), the set Φ_c of compact roots is $\{0\}$, and the Dirac induction consists in associating to $n > 0$ the holomorphic discrete series representation π_{n+1} (with minimal K-type $n + 1$), and to $n < 0$ the anti-holomorphic discrete series representation π_{n-1} (with minimal K-type $n - 1$). For the singular weight $n = 0$ (i.e., the trivial character of K), it follows from Theorem 2.3 that the corresponding Dirac operator D_0 has no kernel or cokernel. However, as prescribed by Conjecture 1, its image by μ_G provides the "missing" generator of $K_0(C_r^*(G))$. To understand this, let us dig further into the structure of $C_r^*(G)$: apart from discrete series representations, $\hat{G}_r$ comprises two continuous series of representations. To describe those, consider the subgroup B of upper triangular matrices and define two families of unitary characters (where $t \geq 0$):

$$\chi_{0,t} : B \to \mathbf{T} : \begin{pmatrix} a & b \\ 0 & a^{-1} \end{pmatrix} \mapsto |a|^{it}$$

$$\chi_{1,t} : B \to \mathbf{T} : \begin{pmatrix} a & b \\ 0 & a^{-1} \end{pmatrix} \mapsto sign(a) \cdot |a|^{it}$$

For $\epsilon = 0, 1$ and $t \geq 0$, denote by $\sigma_{\epsilon,t}$ the unitarily induced representation:

$$\sigma_{\epsilon,t} = Ind_B^G \chi_{\epsilon,t}.$$

The family $\{\sigma_{0,t} : t \geq 0\}$ (resp. $\{\sigma_{1,t} : t \geq 0\}$) is the *even principal series* (resp. *odd principal series*). For $t > 0$ or for $\epsilon = 0$, the representation $\sigma_{\epsilon,t}$ is irreducible. But $\sigma_{1,0}$ splits into two irreducible components σ_1^+, σ_1^- (sometimes called *mock discrete* representations), and $\hat{G}_r$ is the union of the discrete series, the even and the odd principal series of representations. The topology on the even principal series is the topology of $[0, +\infty[$, while the topology on the odd principal series is mildly non-Hausdorff: for $t \to 0$, the representation $\sigma_{1,t}$ converges simultaneously to σ_1^+ and σ_1^-. As a consequence, the direct summand of $C_r^*(G)$ corresponding to the even principal series is Morita equivalent to $C_0([0, +\infty[)$, and hence is trivial in K-theory, while the direct summand corresponding to the odd principal series is Morita equivalent to

$$\{f \in C_0([0, +\infty[, M_2(\mathbf{C})) : f(0) \text{ is diagonal}\},$$

that contributes a copy of $\mathbf{Z}$ to $K_0(C_r^*(G))$, generated by the image of the trivial character of K under Dirac induction. This description of $C_r^*(G)$ also gives $K_1(C_r^*(G)) = 0$ by direct computation.

Coming back to the general framework (G connected Lie group, K maximal compact subgroup), let us indicate how to modify the conjecture when G/K does *not* have a G-invariant spin structure. Then we may construct a double cover $\tilde{G}$ of G, with maximal compact subgroup $\tilde{K}$, such that $\tilde{G}/\tilde{K} = G/K$ carries a $\tilde{G}$-invariant spin structure. Let $\varepsilon \in Z(\tilde{G})$ be the non-trivial element of the covering map $\tilde{G} \to G$. Then $R(\tilde{K})$ splits into a direct sum

$$R(\tilde{K}) = R(\tilde{K})^0 \oplus R(\tilde{K})^1,$$

where $R(\tilde{K})^0$ (resp. $R(\tilde{K})^1$) is generated by those irreducible representations $\rho \in \hat{\tilde{K}}$ such that $\rho(\varepsilon) = 1$ (resp. $\rho(\varepsilon) = -1$). So $R(\tilde{K})^0$ identifies canonically with $R(K)$. Similarly $C_r^*(\tilde{G})$ splits into the direct sum of two ideals $C_r^*(\tilde{G}) = J^0 \oplus J^1$, where J^0 (resp. J^1) corresponds to those representations $\pi \in \widehat{(\tilde{G})}_r$ such that $\pi(\varepsilon) = 1$ (resp. $\pi(\varepsilon) = -1$); so J^0 identifies canonically with $C_r^*(G)$. Now we observe that the Dirac induction for $\tilde{G}$:

$$\mu_{\tilde{G}} : R(\tilde{K}) = R(\tilde{K})^0 \oplus R(\tilde{K})^1 \to K_j\left(C_r^*(\tilde{G})\right) = K_j(J^0) \oplus K_j(J^1)$$

interchanges the $\mathbf{Z}/2$-gradings: indeed the spin representations $S^\pm$ do not factor through K by assumption, but if ρ is in $R(\tilde{K})^1$, then $S^\pm \otimes \rho$ factors through K (as ε acts by the identity). Hence the second case of the Connes–Kasparov conjecture:

Conjecture 2 (2nd version) Let G be a connected Lie group, K a maximal compact subgroup, $j = \dim(G/K)$. Assume that G/K does not carry a G-invariant spin structure.

(1) The Dirac induction $\mu_{\tilde{G}} : R(\tilde{K})^1 \to K_j(C_r^*(G))$ is an isomorphism;
(2) $K_{j+1}(C_r^*(G)) = 0$

As we said before, the Connes–Kasparov conjecture was eventually proved for arbitrary connected Lie groups by Chabert et al. [CEN03], whose result is even more general as it encompasses almost connected groups, i.e., locally compact groups whose group of connected components is compact.

Theorem 2.6 *The Connes–Kasparov conjecture holds for almost connected groups.*

In the same paper [CEN03], Chabert–Echterhoff–Nest obtain a purely representation-theoretic consequence of Theorem 2.6:

Corollary 2.7 *Let G be a connected unimodular Lie group. Then all square-integrable factor representations of G are type I. Moreover, G has no square-integrable factor representations if* $\dim(G/K)$ *is odd.*

2.5 The Novikov conjecture

For discrete groups, an important motivation for the Baum–Connes conjecture was provided by the work of Mishchenko (see, e.g., [Mis74]) and Kasparov (see, e.g., [Kas95]) on the *Novikov conjecture*, whose statement we now recall.

For a discrete group Γ, denote by $B\Gamma$ "the" *classifying space* of Γ, a CW-complex characterized, up to homotopy, by the properties that its fundamental group is Γ and its universal cover $E\Gamma$ is contractible.[7] Alternatively, $B\Gamma$ is a $K(\Gamma, 1)$-space. As a consequence, group cohomology of Γ, defined algebraically, is canonically isomorphic to cellular cohomology of $B\Gamma$.

Let M be a smooth, closed, oriented manifold of dimension n, equipped with a map $f : M \to B\Gamma$. For $x \in H^*(B\Gamma, \mathbf{Q})$ (cohomology with rational coefficients), consider the *higher signature*

$$\sigma_x(M, f) = \langle f^*(x) \cup L(M), [M] \rangle \in \mathbf{Q},$$

where $L(M)$ is the L-class (a polynomial in the Pontryagin classes, depending on the smooth structure of M), and $[M]$ is the fundamental class of M. The Novikov conjecture states that these numbers are homotopy invariant (and so do not depend on the smooth structure of M):

Conjecture 3 (The Novikov conjecture on homotopy invariance of higher signatures) Let $h : N \to M$ be a homotopy equivalence; then for any $x \in H^*(B\Gamma, \mathbf{Q})$:

$$\sigma_x(M, f) = \sigma_x(N, f \circ h).$$

We say that *the Novikov conjecture holds for* Γ if Conjecture 3 holds for every $x \in H^*(B\Gamma, \mathbf{Q})$. We refer to the detailed survey paper [FRR95] for the history of this conjecture, and an explanation why it is important.

[7] As Connes once pointed out: "*$E\Gamma$ is a point on which Γ acts freely!*."

We summarize now Kasparov's approach from section 9 in [Kas95].[8] Keeping notations as in Conjecture 3, Kasparov considers the homology class $\mathcal{D}(M) = L(M) \cap [M] \in H_*(M, \mathbf{Q})$ which is Poincaré-dual to $L(M)$, and Conjecture 3 is equivalent to the homotopy invariance of the class $f_*(\mathcal{D}(M)) \in H_*(B\Gamma, \mathbf{Q})$.

Let $d : \Omega^p(M) \to \Omega^{p+1}(M)$ be the exterior derivative on differential forms. Up to crossing M with the circle S^1, we may assume that $n = \dim M$ is even. Fix an auxiliary Riemannian metric on M. This allows to define the adjoint $d^* : \Omega^p(M) \to \Omega^{p-1}(M)$: it satisfies $d^* = - \star d \star$, where $\star$ is the Hodge operator associated with the Riemannian structure.

Now consider $d + d^*$ acting on the space of all forms $\Omega(M) = \bigoplus_{p=0}^{n} \Omega^p(M)$. One way to consider this as a graded operator is the following: let τ be an involution on the space of all forms defined by:

$$\tau(\omega) = i^{p(p-1)+\frac{n}{2}} \star \omega \quad , \quad \omega \in \Omega^p(M).$$

It is verified that $d + d^*$ anti-commutes with τ: with this grading on forms, $d + d^*$ is the *signature operator* on M. As it is an elliptic operator, it defines an element $[d + d^*]$ in the group $K_0(M)$ of K-homology[9] of M. Note that, by connectedness of the space of Riemannian metrics on M, the element $[d + d^*] \in K_0(M)$ does not depend on the choice of a Riemannian metric. Using Hodge theory, it is classical to check that the index of $d + d^*$ is exactly the *topological signature* of M, i.e., the signature of the quadratic form given by cup product on the middle-dimensional cohomology $H^{\frac{n}{2}}(M, \mathbf{C})$. Now consider the *index pairing* between K-theory and K-homology of M:

$$K^0(M) \times K_0(M) \to \mathbf{Z} : (\xi, D) \mapsto Ind(D_\xi),$$

the index of the differential operator D_ξ, which is D with coefficients in the vector bundle ξ on M. In particular $Ind((d + d^*)_\xi)$ is the index of the signature operator with coefficients in ξ, i.e., acting on sections of $\Lambda^*(M) \otimes \xi$. It is given by the cohomological version of the Atiyah–Singer index theorem:

$$Ind((d + d^*)_\xi) = \langle Ch^*(\xi) \cup L(M), [M] \rangle, \tag{2.2}$$

where Ch^* denotes the Chern character in cohomology. Recall that, for every finite CW-complex X, we have Chern characters in cohomology and homology:

$$Ch^* : K^0(X) \to \bigoplus_{k=0}^{\infty} H^{2k}(X, \mathbf{Q});$$

[8] Although published only in 1995, the celebrated "*Conspectus*" was first circulated in 1981.

[9] K-homology is the homology theory dual to topological K-theory. It was shown by Atiyah [Ati70] that an elliptic (pseudo-)differential operator on M defines an element in $K_0(M)$.

$$Ch_* : K_0(X) \to \bigoplus_{k=0}^{\infty} H_{2k}(X, \mathbf{Q}),$$

which are rational isomorphisms, compatible with the index pairing and with the pairing between cohomology and homology. Equation 2.2 then implies that

$$Ch_*([d+d^*]) = L(M) \cap [M] = \mathcal{D}(M). \tag{2.3}$$

Assume for simplicity that $B\Gamma$ is a closed manifold,[10] which implies that Γ is torsion-free. Recall that Conjecture 3 is equivalent to homotopy invariance of $f_*(\mathcal{D}(M))$. By Equation 2.3 and functoriality of Ch_*, we have:

$$f_*(\mathcal{D}(M)) = f_*(Ch_*([d+d^*])) = Ch_*(f_*[d+d^*]).$$

By rational injectivity of Ch_*, we see that Conjecture 3 is equivalent to the homotopy invariance of $f_*[d+d^*]$ in $K_0(B\Gamma) \otimes_{\mathbf{Z}} \mathbf{Q}$.

In the final section of [Kas95], Kasparov defines a homomorphism $\beta : K_i(B\Gamma) \to K_i(C_r^*(\Gamma))$ that later was identified with the assembly map $\mu_r : K_i^\Gamma(\underline{E\Gamma}) \to K_i(C_r^*(\Gamma))$. Kasparov's β is defined as follows: keep the assumption that $B\Gamma$ is a finite complex. Form the induced vector bundle $\mathcal{L}_\Gamma = E\Gamma \times_\Gamma C_r^*(\Gamma)$ (where Γ acts on $C_r^*(\Gamma)$ by left translations). This is a vector bundle with fiber $C_r^*(\Gamma)$ over $B\Gamma$, sometimes called the *Mishchenko line bundle*. Its space $C(E\Gamma, C_r^*(\Gamma))^\Gamma$ of continuous sections, is a projective finite type module over $C(B\Gamma) \otimes C_r^*(\Gamma)$ (and as such it defines a K-theory element $[\mathcal{L}_\Gamma] \in K_0(C(B\Gamma) \otimes C_r^*(\Gamma))$). For a K-homology element $[D] \in K_0(B\Gamma)$ given by an elliptic (pseudo-)differential operator D over $B\Gamma$ we may form the operator $D_{\mathcal{L}_\Gamma}$ with coefficients in $\mathcal{L}_\Gamma$: its kernel and cokernel are projective finite type modules over $C_r^*(\Gamma)$, so their formal difference defines an element $\beta[D] \in K_0(C_r^*(\Gamma))$: this defines the desired homomorphism[11]

$$\beta : K_0(B\Gamma) \to K_0(C_r^*(\Gamma)).$$

Coming back to the Novikov conjecture, recall that it is equivalent to the homotopy invariance of $f_*[d+d^*]$ in $K_0(B\Gamma) \otimes_{\mathbf{Z}} \mathbf{Q}$. Now one of Kasparov's result in [Kas95] (Theorem 2 in the final section) is:

Theorem 2.8 *If M is an even-dimensional smooth, closed, oriented manifold and $f : M \to B\Gamma$ is a continuous map, then $\beta(f_*[d+d^*]) \in K_0(C_r^*(\Gamma))$ is a homotopy invariant of M.* □

---

[10] When $B\Gamma$ is a general CW-complex we must replace $K_0(B\Gamma)$ by $RK_0(B\Gamma) = \varinjlim_X K_0(X)$, where X runs along compact subsets of $B\Gamma$.

[11] In terms of Kasparov theory, to be defined in Chapter 3 below, this can be expressed using Kasparov product: $\beta[D] = [\mathcal{L}_\Gamma] \otimes_{C(B\Gamma)} [D]$.

As an immediate consequence of Theorem 2.8, we get the following result:

Corollary 2.9 *If the map β is rationally injective, then the Novikov conjecture (Conjecture 3) holds for Γ.*

The main result of Kasparov's Conspectus [Kas95] is the following:

Theorem 2.10 *If Γ is a discrete subgroup of a connected Lie group, then the map β is injective.*

Corollary 2.11 *The Novikov conjecture holds for any discrete subgroup of a connected Lie group.*

3 Index maps in K-theory: the contribution of Kasparov

3.1 Kasparov bifunctor

The powerful tool developed by Kasparov in his proof of the Novikov conjecture is the equivariant KK-theory. We refer to [Kas95] and [Kas88].

For any locally compact group G and A, B two $G - C^*$-algebras (i.e., C^*-algebras equipped with a strongly continuous action by automorphisms of the group G), Kasparov defines an abelian group $KK_G(A, B)$. The main tool in the theory is the cup product

$$KK_G(A, B) \times KK_G(B, C) \to KK_G(A, C) : (x, y) \mapsto x \otimes_B y.$$

In particular, if $\mathbf{C}$ is the field of complex numbers equipped with the trivial G-action, $KK_G(\mathbf{C}, \mathbf{C})$ is a ring, which turns out to be commutative. Moreover the homomorphisms

$$\tau_D : KK_G(A, B) \to KK_G(A \otimes D, B \otimes D)$$

defined by tensoring by a C^*-algebra D equip all the $KK_G(A, B)$'s with a structure of $KK_G(\mathbf{C}, \mathbf{C})$-modules.

One of the most important ingredients in G-equivariant KK-theory is the existence of descent maps: for all $G - C^*$algebras A and B there are group homomorphisms

$$j_{G,r} : KK_G(A, B) \to KK\left(C_r^*(G, A), C_r^*(G, B)\right)$$

$$j_{G,\max} : KK_G(A, B) \to KK\left(C_{\max}^*(G, A), C_{\max}^*(G, A)\right),$$

where $C_r^*(G, A)$ and $C_{\max}^*(G, A)$ denote respectively the reduced and the full crossed product.

The abelian group $KK_G(A, B)$ is defined as follows:

Definition 3.1 An (A, B)-Fredholm bimodule is given by:

(i) a B-Hilbert module E;
(ii) a covariant representation $(\pi, \rho(g))$ of (G, A) on the Hilbert module E;
(iii) an operator T on E, B-bounded and self-adjoint (i.e., $T = T^*$) and such that: for any a in A and g in G, the operators $(1 - T^2)\rho(a)$, $T\rho(a) - \rho(a)T$, and $T\pi(g) - \pi(g)T$ are B-compact operators; moreover the map $g \mapsto T\pi(g) - \pi(g)T$ is norm continuous.

Such a (A, B)-Fredholm module is also called odd (A, B)-Fredholm module. An even (A, B)-Fredholm module is given by a (A, B)-Fredholm module together with a $\mathbf{Z}/2$-grading on the module E, such that the covariant representation preserves the grading, and the operator T is odd with respect to the grading.

One defines a homotopy of (A, B)-Fredholm modules to be a $(A, B \otimes C[(0, 1])$-Fredholm module. An element of $KK_G(A, B)$ is defined as a homotopy class of even (A, B)-Fredholm modules. Addition is given by direct sum. The zero element is given by the class of degenerate modules, i.e., those where "compact" is replaced by "zero" in Definition 3.1. When necessary we use the notation $KK_G^j(A, B)$ with $j = 0$ (resp. 1) for the even (resp. odd case).

When there is no group acting, we simply write $KK(A, B)$. Ordinary K-theory for C^*-algebras is recovered by $K_*(B) = KK^*(\mathbf{C}, B)$, while K-homology corresponds to $K^*(A) = KK^*(A, \mathbf{C})$.

3.2 *Dirac induction in KK-theory*

In [Kas95], Kasparov gives an interpretation of the Dirac induction map from $K_*(C^*(K, A))$ to $K_*(C_r^*(G, A))$ in the framework of KK-theory. Here G is a semisimple Lie group with finite center and K a maximal compact subgroup. We assume that the adjoint representation of K on $V = \mathfrak{g}/\mathfrak{k}$ lifts to $Spin(V)$. The symmetric space $X = G/K$ then carries a G-invariant spin structure. Let D be the corresponding Dirac operator, a G-invariant elliptic operator defined on the sections of the spinor bundle S of X.

We define an element α of the group $KK_G^j(C_0(X), \mathbf{C})$ as the homotopy class of the $(C_0(X), \mathbf{C})$-Fredholm bimodule defined by:

(1) The Hilbert space $L^2(X, S)$ of L^2-sections of the spinor bundle S.
(2) The covariant action on $L^2(X, S)$ of the $G - C^*$-algebra $C_0(X)$ of continuous functions on X vanishing at infinity.
(3) The operator $F = D(1 + D^2)^{-1/2}$ obtained by functional calculus from the Dirac operator D.

Note that the bundle S is graded for j even, and trivially graded if j is odd. The above Fredholm module therefore defines an element $\alpha \in KK_G^j(C_0(G/K), \mathbf{C})$, where $j = \dim G/K \pmod 2$.

Now consider the following composition

$$KK_G(C_0(G/K), \mathbf{C}) \to KK_G(C_0(G/K) \otimes A, A) \to KK\left(C^*(K, A), C_r^*(G, A)\right),$$

where the first map is τ_A and the second is $j_{G,r}$, taking into account the Morita equivalence of $C_r^*(G, C_0(G/K) \otimes A)$ with $C^*(K, A)$. The image of α by the above composed map is an element of $KK^j(C^*(K, A), C_r^*(G, A))$ which defines a map

$$\tilde{\alpha}_A : K_{*+j}(C^*(K, A)) \to K_*\left(C_r^*(G, A)\right).$$

Note the two special cases:

(1) When $A = \mathbf{C}$, this is nothing but the Connes–Kasparov map $K_{*+j}(C^*(K)) \to K_*(C_r^*(G))$, see Conjecture 1.
(2) When Γ is a torsion-free discrete cocompact subgroup of G, and $M = \Gamma\backslash G/K$, this gives the map $\beta : K_{*+j}(C(M)) \to K_*(C_r^*(\Gamma))$, see section 2.5 where $M = B\Gamma$, the classifying space of Γ.

3.3 The dual-Dirac method and the γ-element

In order to construct the inverse map, Kasparov defines in [Kas95] the element

$$\beta \in KK_G^j(\mathbf{C}, C_0(X))$$

as the homotopy class of the following $(\mathbf{C}, C_0(X))$-Fredholm bimodule:

(1) The $C_0(X)$-Hilbert module $C_0(X, S)$ of sections of the spinor bundle S;
(2) the natural action of G on $C_0(X)$;
(3) the operator on $C_0(X)$ which is the Clifford multiplication by the vector field b on X defined as follows: let x_0 be the origin in X (i.e., the class of the identity in G/K), then the value of b at a point $x \in X$ is the vector tangent to the geodesic from x to x_0, and of length $\rho(1+\rho^2)^{-1/2}$ if ρ is the distance between x and x_0.

Similarly to what was done for α, the element $\beta \in KK_G^j(\mathbf{C}, C_0(G/K))$ gives rise to an element $\tilde{\beta}$ of $KK^j(C^*_{\mathrm{red}}(G, A), C^*(K, A))$ by applying to β the following maps:

$$KK_G((\mathbf{C}, C_0(G/K)) \to KK_G(A, C_0(G/K) \otimes A) \to KK\left(C_r^*(G, A), C^*(K, A)\right),$$

hence a map $K_*(C_r^*(G, A)) \to K_{*+j}(C^*(K, A))$ which is a candidate to be the inverse of the index map.

In other words, one would hope that the following equalities hold in KK-theory: $\alpha \otimes_{\mathbf{C}} \beta = 1$ in $KK_G(C_0(X), C_0(X))$ and $\beta \otimes_{C_0(X)} \alpha = 1$ in $KK_G(\mathbf{C}, \mathbf{C})$. However, such a dream is not fulfilled. Only the first statement is true in general.

Theorem 3.2 *One has* $\alpha \otimes_{\mathbf{C}} \beta = 1$ *in* $KK_G(C_0(X), C_0(X))$. *As a consequence,* $\gamma := \beta \otimes_{C_0(X)} \alpha$ *is an idempotent of the ring* $KK_G(\mathbf{C}, \mathbf{C})$, *i.e.,* $\gamma \otimes_{\mathbf{C}} \gamma = \gamma$.

This element γ plays a key role in the Baum–Connes conjecture. The main step in the proof of Theorem 3.2 is the following *rotation lemma*:

Lemma 3.3 $\alpha \otimes_{\mathbf{C}} \beta = \tau_{C_0(X)}(\beta \otimes_{C_0(X)} \alpha)$.

On the other hand, Kasparov shows

Lemma 3.4 $\mathrm{Rest}_K^G(\gamma) = 1$ *in* $R(K)$,

where

$$\mathrm{Rest}_K^G : KK_G(\mathbf{C}, \mathbf{C}) \to R(K)$$

is the natural restriction map. This is a K-equivariant version of the Bott periodicity. Namely, from the K-equivariant point of view, the space G/K can be replaced by its tangent space V at x_0. Then the Euclidean space V is equipped with a representation of K which factors though $Spin(V)$ and the Bott periodicity has an equivariant version, an isomorphism between $K_*(C^*(K, C_0(V)))$ and $R(K)$.

Corollary 3.5 $\tau_{C_0(G/K)}(\gamma) = 1$ *in* $KK_G(C_0(G/K), C_0(G/K))$.

This follows from the fact that $\tau_{C_0(G/K)} = \mathrm{Ind}_K^G \circ \mathrm{Rest}_K^G$, where the induction $\mathrm{Ind}_K^G : R(K) \to KK_G(C_0(G/K), C_0(G/K))$ is defined in [Kas88]. Theorem 3.2 follows by combining Lemma 3.3 with Corollary 3.5.

Since γ is an idempotent, the ring $KK_G(\mathbf{C}, \mathbf{C})$ is a direct sum of two subrings

$$KK_G(\mathbf{C}, \mathbf{C}) = \gamma KK_G(\mathbf{C}, \mathbf{C}) \oplus (1 - \gamma) KK_G(\mathbf{C}, \mathbf{C}).$$

Moreover, by Lemma 3.4 the restriction map $KK_G(\mathbf{C}, \mathbf{C}) \to KK_K(\mathbf{C}, \mathbf{C}) = R(K)$ is an isomorphism from $\gamma KK_G(\mathbf{C}, \mathbf{C})$ to $R(K)$, and vanishes on the complement $(1 - \gamma) KK_G(\mathbf{C}, \mathbf{C})$. More generally for any A, B as above,

$$KK_G(A, B) = \gamma KK_G(A, B) \oplus (1 - \gamma) KK_G(A, B),$$

the restriction map is an isomorphism from $\gamma KK_G(A, B)$ to $KK_K(A, B)$ and vanishes on $(1 - \gamma) KK_G(A, B)$.

The element γ acts on the K-theory of $C_r^*(G, A)$ by an idempotent map which can be described as follows: consider the composition of ring homomorphisms

$$\begin{aligned} KK_G(\mathbf{C}, \mathbf{C}) \to KK_G(A, A) \to KK\left(C_r^*(G, A), C_r^*(G, A)\right) \\ \to \mathrm{End}\left(K_*\left(C_r^*(G, A)\right)\right) \end{aligned}$$

and take the image of the idempotent γ by the above map:

$$\tilde{\gamma}_A \in \mathrm{End}\left(K_*\left(C_r^*(G,A)\right)\right).$$

The results of Kasparov [Kas95] [Kas88] can then be summarized as follows:

Theorem 3.6 *The map $\tilde{\alpha}_A$ is injective.*[12] *Its image in $K_*(C_r^*(G,A))$ is equal to the image of the idempotent map $\tilde{\gamma}_A$.*

Corollary 3.7 *The Connes–Kasparov conjecture with coefficients in A (i.e., the statement that $\tilde{\alpha}_A$ is an isomorphism) is equivalent to the equality $\tilde{\gamma}_A = \mathrm{Id}$.*

Corollary 3.8 *If $\gamma = 1$ in $KK_G(\mathbf{C},\mathbf{C})$, then the Connes–Kasparov conjecture with coefficients is true.*

3.4 From K-theory to K-homology

All the constructions above rest upon the assumption that the space $X = G/K$ carries a G-equivariant structure of a *spin manifold*, or equivalently that the representation of K on $V^* = T^*_{x_0}X$ is spinorial.

In the case of a general connected Lie group, this is not necessarily the case, and Kasparov's constructions have to be modified as follows: consider the cotangent bundle T^*X which has an almost-complex structure. There is therefore a Dirac operator on T^*X, which defines an element $\alpha \in KK_G(C_0(T^*X),\mathbf{C})$. Applying the same procedure as above yields an element of $KK(C^*(K, A\otimes C_0(V^*)), C_r^*(G,A))$ since $C^*(G, A\otimes C_0(T^*X))$ is Morita equivalent to $C^*(K, A\otimes C_0(V^*))$.

Therefore the element α defines a map

$$K_*\left(C^*\left(K, A\otimes C_0(V^*)\right)\right) \to K_*\left(C_r^*(G,A)\right).$$

Note that there is no dimension shift but that A is replaced by $A\otimes C_0(V^*)$. As usual, note the special cases $A = \mathbf{C}$ and $A = C(G/\Gamma)$

(1) $K_*(C^*(K, C_0(V^*))) \to K_*(C_r^*(G))$;
(2) $K^*(T^*M) \to K_*(C_r^*(\Gamma))$, where $M = \Gamma\backslash G/K$.

In the same way one can define a dual-Dirac element $\beta \in KK_G(\mathbf{C}, C_0(T^*X))$ and an element $\gamma \in KK_G(\mathbf{C},\mathbf{C})$. The same results as above do hold.

The role of the cotangent bundle T^*X or equivalently the representation of K on $V^* = T^*_{x_0}X$ is closely related to Poincaré duality in K-theory. The latter is conveniently formulated in Kasparov theory as follows: as we shall see, the left-hand side of the conjecture should in fact be interpreted, rather than a K-theory group, as

[12] The injectivity of $\tilde{\alpha}_A$ is responsible for the Novikov conjecture, Conjecture 3: see Section 4.5.1.

a K-homology group. The Dirac induction map appears rather as the composition of the assembly map with the Poincaré duality map.

Let us explain that point. In Kasparov theory, the K-homology $K^*(A)$ of a C^*-algebra is defined as the group $KK(A, \mathbf{C})$. There is a duality pairing

$$K_*(A) \otimes K^*(A) \to \mathbf{Z}$$

with the K-theory $K_*(A) = KK(\mathbf{C}, A)$, defined by the cup product

$$KK(\mathbf{C}, A) \otimes KK(A, \mathbf{C}) \to KK(\mathbf{C}, \mathbf{C}) = \mathbf{Z}.$$

For example, if M is a compact manifold, the K-homology group $K_*(M) = K^*(C(M))$ can be described, according to Atiyah [Ati70], as the group $\mathrm{Ell}(M)$ of classes of elliptic operators on the manifold M. The pairing $K^*(M) \otimes K_*(M) \to \mathbf{Z}$ associates to a vector bundle E and an elliptic operator D the index of the operator D_E with coefficients in E. Poincaré duality in K-theory is a canonical isomorphism

$$K^*(T^*M) \to K_*(M)$$

between the K-homology of M and the K-theory of the total space T^*M of its cotangent bundle. Such a map can be interpreted as follows: an element of $K^*(T^*M)$ is the homotopy class of an elliptic symbol on M. Its image in $K_*(M)$ is the class of an elliptic pseudo-differential operator associated to that symbol. In Kasparov theory, one can interpret Poincaré duality as the existence of two elements, respectively of $KK(C(M) \otimes C_0(T^*M), \mathbf{C})$ and of $KK(\mathbf{C}, C(M) \otimes C_0(T^*M))$, inverse to each other for the cup product. See the details in [Kas88].

This allows to reformulate the conjecture as follows: for the case of a torsion-free discrete cocompact subgroup Γ as above, the map $K^*(T^*M) \to K_*(C_r^*(\Gamma))$ becomes[13]

$$K_*(M) \to K_*\left(C_r^*(\Gamma)\right).$$

In general, one needs the G-equivariant version of Poincaré duality for the space $X = G/K$. There are two elements one of $KK_G(C_0(X) \otimes C_0(T^*X), \mathbf{C})$ and the other of $KK_G(\mathbf{C}, C_0(X) \otimes C_0(T^*X))$ that are inverse to each other.

Then for any $G - C^*$-algebra A, one has an isomorphism

$$KK^G\left(\mathbf{C}, C_0(T^*X) \otimes A\right) \to KK^G\left(C_0(X), A\right).$$

One can show that the first group is isomorphic to

$$KK\left(\mathbf{C}, C^*\left(G, C_0(T^*X) \otimes A\right)\right) = KK\left(\mathbf{C}, C^*\left(K, C_0(V^*) \otimes A\right)\right).$$

[13] This is actually the same map as the map β from Section 2.5.

The Dirac induction with coefficients in A can therefore be defined as a map

$$KK^G(C_0(X), A) \to K_*\left(C^*_{\mathrm{red}}(G, A)\right)$$

which in the case without coefficients can be written as $K^G_*(X) \to K_*(C^*_{\mathrm{red}}(G))$.

3.5 Generalization to the p-adic case

Shortly after the work of Kasparov, it became natural to investigate the analogue of the Kasparov Dirac–dual-Dirac method when real Lie groups are replaced by p-adic groups. According to the philosophy of Bruhat and Tits the p-adic analogue of the symmetric space is a building of affine type (see [BT72, Tit75]). It shares with symmetric spaces the property of unique geodesics between two points, and the fact that the stabilizers of vertices are maximal compact subgroups (note that there may be several conjugacy classes of such subgroups). In the rank one case, e.g., $SL(2, \mathbf{Q}_p)$, the Bruhat–Tits building is the Bass–Serre tree. Julg and Valette [JV88] have constructed an element γ for buildings using an operator on the Hilbert space $\ell^2(X)$ (the set X is seen as the set of objects of all dimensions) which may be seen as the "vector pointing to the origin," generalizing the Julg-Valette element for trees [JV84].

The question of an analogue of the Connes–Kasparov conjecture for p-adic groups has been considered by Kasparov and Skandalis in [KS91]. They met the following difficulty: the building is not a manifold, and it does not satisfy the Poincaré duality in the usual sense. However, if X is a simplicial complex, there is an algebra $\mathcal{A}_X$ which plays the role played by the algebra $C^*(TM) = C_0(T^*M)$ in the case of a manifold M. The algebra $\mathcal{A}_X$ is not commutative, it is in fact the algebra of a groupoid associated to the simplicial complex X. Moreover, $\mathcal{A}_X$ is Poincaré dual in K-theory to the commutative algebra $C_0(|X|)$ of continuous functions on the geometric realization of X: there is a canonical isomorphism

$$K_*(\mathcal{A}_X) \to K^*(C_0(|X|))$$

from the K-theory of the algebra A_X to the K-homology of the space $|X|$.

Let us now assume that X is the Bruhat–Tits building of a reductive linear group over a non-Archimedean local field (e.g., $\mathbf{Q}_p$). Then the above form of the Poincaré duality, in a G-equivariant way, shows the isomorphism

$$KK_G(C_0(|X|), A) = K_*\left(C^*(G, \mathcal{A}_X \otimes A)\right)$$

for any $G - C^*$-algebra A.

By analogy with the Lie group case, it was natural to construct a map from the group above to the K-theory group $K_*(C^*_r(G, A))$. Kasparov and Skandalis [KS91] construct a Dirac element $\alpha \in KK_G(\mathcal{A}_X, \mathbf{C})$ which defines as above maps in K-theory:

$$K_*(C^*(G, \mathcal{A}_X \otimes A)) \to K_*(C^*_r(G, A)).$$

The left-hand side can be computed by Morita equivalence from the K-theory of crossed products of A by the compact subgroups of G stabilizing the vertices of a simplex viewed as a fundamental domain. A special case is the Pimsner exact sequence for trees [Pim86].

Kasparov and Skandalis have shown the injectivity of the above map (which implies the Novikov conjecture for discrete subgroups of p-adic groups) by constructing a dual-Dirac element $\beta \in KK_G(\mathbf{C}, \mathcal{A}_X)$. They show that

$$\beta \otimes_{A_X} \alpha = \gamma \in KK_G(\mathbf{C}, \mathbf{C}),$$

the Julg-Valette element of [JV88]. A rotation trick shows that $\alpha \otimes_{\mathbf{C}} \beta = 1$.

At this point we note that the Lie group case and the p-adic group case can be unified by the K-homology formulation of the conjecture. If Z denotes the locally compact G-space which is the symmetric space G/K in the Lie case, the geometric realization $|X|$ of the Bruhat–Tits building in the p-adic case, the conjecture is that a certain map

$$KK_G(C_0(Z), A) \to K_*(C^*(G, A))$$

is an isomorphism. This will become more precise with the Baum–Connes–Higson formulation of the conjecture for general locally compact groups: the role of the symmetric spaces or Bruhat–Tits buildings will be clarified as classifying spaces for proper actions, see Sections 4.2 and 4.3. In both cases injectivity can be proved by a Dirac–dual-Dirac method, which hints to a general notion of γ-element, as explained in Section 4.4.

4 Towards the official version of the conjecture

4.1 *Time-dependent left-hand side*

There is a certain time-dependency in the left-hand side of the Baum–Connes conjecture, hence also in the assembly map. Let us first recall the fundamental concept of proper actions.

Definition 4.1

1. Let G be a locally compact group. A G-action on a locally compact space X is said to be *proper* if the action map

$$G \times X \to X : (g, x) \mapsto gx$$

is proper, i.e., the inverse image of a compact subset of X, is compact.
2. If X is a locally compact, proper G-space, then the quotient space $G\backslash X$ is locally compact, and X is said to be *G-compact* if $G\backslash X$ is compact.

In the original paper of Baum–Connes [BC00], the conjecture is formulated only for Lie groups—possibly with infinitely many connected components, so as to include discrete groups. However, the authors take great care in allowing coefficients, in the form of group actions on smooth manifolds. So if G is a Lie group (not necessarily connected) and M is a manifold, the goal is to identify the analytical object $K_*(C_r^*(G, C_0(M)))$ (the K-theory of the reduced crossed product C^*-algebra), with something of geometrical nature.

This is done in two steps. First, let Z be a proper G-manifold. Denote by $V_G^0(Z)$ the collection of all G-elliptic complexes of vector bundles (E_+, E_-, σ), where E_+, E_- are G-vector bundles over Z, and $\sigma : E_+ \to E_-$ is a G-equivariant vector bundle map, which is invertible outside of a G-compact set. One also defines $V_G^1(Z) = V_G^0(Z \times \mathbf{R})$, where G acts trivially on $\mathbf{R}$.

The second—and main—step is to consider an *arbitrary* G-manifold M and to "approximate" it by proper G-manifolds; here one can identify, in germ, the presence of the classifying space for G-proper actions that will come to the forefront in the "official" version of the conjecture in [BCH94]; see Section 4.3 below. In [BC00], a *K-cocycle* for M will be a triple (Z, f, ξ), where:

- Z is a proper, G-compact, G-manifold;
- $f : Z \to M$ is a G-map;
- $\xi \in V_G^*(T^*Z \oplus f^*T^*M)$.

We denote by $\Gamma(G, M)$ the set of K-cocycles for M. If (Z, f, ξ) and (Z', f', ξ') are two equivariant K-cycles for X, then their disjoint union is the equivariant K-cycle $(Z \coprod Z', f \coprod f', \xi \coprod \xi')$. It is assumed that manifolds are not necessarily connected, and their connected components do not always have the same dimension. The operation of disjoint union will give addition.

Suppose that the manifolds Z_1, Z_2, M and the G-maps f_1, f_2, g fit into a commutative diagram

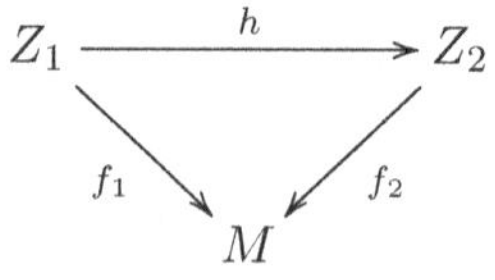

Then, using the Thom isomorphism, it is possible to construct a "*wrong way functoriality*" Gysin map

$$h_! : K_G^*\left(T^*Z_1 \oplus f_1^*T^*M\right) \to K_G^*\left(T^*Z_2 \oplus f_2^*T^*M\right).$$

Two K-cocycles (Z_1, f_1, ξ_1), (Z_2, f_2, ξ_2) are said to be equivalent[14] if there exists a K-cocycle $(\tilde{Z}, \tilde{f}, \tilde{\xi})$ and G-maps $h_1 : Z_1 \to \tilde{Z}$, $h_2 : Z_2 \to \tilde{Z}$ making the following diagram commutative:

$$\begin{array}{ccccc} Z_1 & \xrightarrow{h_1} & \tilde{Z} & \xleftarrow{h_2} & Z_2\,, \\ & {\scriptstyle f_1}\searrow & \downarrow{\scriptstyle \tilde{f}} & \swarrow{\scriptstyle f_2} & \\ & & M & & \end{array}$$

and such that $h_{1,!}(\xi_1) = \tilde{\xi} = h_{2,!}(\xi_2)$. Then we define $K^{top}(G, M)$ as the quotient of $\Gamma(G, M)$ by this equivalence relation.

To construct the assembly map $\mu_{r,M} : K^{top}(G, M) \to K_*(C_r^*(G, C_0(M)))$, the construction is roughly as follows: start from a K-cocycle $(Z, f, \xi) \in \Gamma(G, M)$. Observe that $f = p \circ i$, where $i : Z \to Z \times M : z \mapsto (z, f(z))$ and $p : Z \times M \to M$ is the projection onto the second factor. Replacing Z by $Z \times M$ and f by p, we may assume that f is a submersion. Let then τ be the cotangent bundle along the fibers of f. By the Thom isomorphism, the class $\xi \in V_G^*(T^*Z \oplus f^*T^*M)$ determines a unique class $\eta \in V_G^*(\eta)$. For $x \in M$, set $Z_x = f^{-1}(x)$. Then, restricting η to Z_x we get $\eta_x \in V^*(Z_x)$, which can be viewed as the symbol of some elliptic differential operator D_x on Z_x. Then the family $(D_x)_{x \in M}$ is a G-equivariant family of elliptic differential operators on M, so its G-index belongs to $K_*(C_r^*(G, C_0(M)))$ and we set:

$$\tilde{\mu}_{r,M}(Z, f, \xi) = Ind_G(D_x)_{x \in M}.$$

It is stated in Theorem 5 of [BC00] that this map $\tilde{\mu}_{r,M}$ is compatible with wrong way Gysin maps, so it descends to a homomorphism of abelian groups:

$$\mu_{r,M} : K^{top}(G, M) \to K_* \left(C_r^*(G, C_0(M))\right),$$

and the main conjecture in [BC00] is that $\mu_{r,M}$ is an isomorphism for every Lie group G and every G-manifold M.

4.2 *The classifying space for proper actions, and its K-homology*

In the paper [BCH94], Baum, Connes, and Higson consider the class of all 2nd countable, locally compact groups G. They make a systematic use of the classifying space for proper actions $\underline{EG}$, first introduced in this context in [10]. The G/K space

[14]The fact that it is indeed an equivalence relation does not appear in [BC00].

associated to a connected Lie group and the Bruhat–Tits building of a p-adic group are special cases of classifying space of proper actions as we mentioned already in Section 3.5.

Definition 4.2 Let G be a 2nd countable locally compact group. A classifying space for proper actions for G, is a proper G-space $\underline{EG}$ with the properties that, if X is any proper G-space, then there exists a G-map $X \to \underline{EG}$, and any two G-maps from X to $\underline{EG}$ are G-homotopic.

When Γ is a countable discrete group, we could also define $\underline{E\Gamma}$ as a Γ-CW-complex such that the fixed point set $\underline{E\Gamma}^H$ is empty whenever H is an infinite subgroup of Γ, and is contractible whenever H is a finite subgroup (in particular $\underline{E\Gamma}$ is itself contractible).

Back to the general case: even if we refer to $\underline{EG}$ as "the" universal space for proper actions of G, it is important to keep in mind that $\underline{EG}$ is only unique up to G-equivariant homotopy, and the definition of the left-hand side $K_*^{top}(G, A)$ will have to account for this ambiguity. So we define

$$K_*^{top}(G, A) = \lim_X KK_*^G(C_0(X), A),$$

where X runs in the directed set of closed, G-compact subsets of $\underline{EG}$. This is the left-hand side of the assembly map for $G \curvearrowright A$.

4.3 *The Baum–Connes–Higson formulation of the conjecture*

For any proper, G-compact G-space X, the space $C_0(X)$ is a module of finite type over the algebra $C^*(G, C_0(X))$ (which is both the full and the reduced one) whose class in $K_0(C^*(G, C_0(X))) = KK(\mathbf{C}, C^*(G, C_0(X)))$ will be denoted by e_X. Then for any $G - C^*$-algebra A, Kasparov's descent map

$$j_{G,r} : KK_G(C_0(X), A) \to KK\left(C^*(G, C_0(X)), C_r^*(G, A)\right)$$

can be composed with the left multiplication by e_X :

$$KK\left(C^*(G, C_0(X)), C_r^*(G, A)\right) \to KK\left(\mathbf{C}, C_r^*(G, A)\right)$$

to define a map $KK_G(C_0(X), A) \to K_*(C_r^*(G, A))$.

When X runs in the directed set of closed, G-compact subsets of $\underline{EG}$, those maps are compatible with the direct limit, hence define the *assembly map* or *index map*:

$$\mu_{A,r} : K_*^{top}(G, A) \to K_*\left(C_r^*(G, A)\right).$$

For $A = \mathbf{C}$, the map $\mu_{A,r}$ is simply denoted by μ_r. The Baum–Connes conjecture is then stated as follows, in its two classical versions:

Conjecture 4 (The Baum–Connes conjecture) For all locally compact, 2nd countable groups G the assembly map μ_r is an isomorphism.

Conjecture 5 (The Baum–Connes conjecture with coefficients) For all locally compact 2nd countable groups G and for all $G - C^*$-algebras A, the assembly map $\mu_{A,r}$ is an isomorphism.

Conjecture 5 has the advantage of being stable under passing to closed subgroups (see [BCH94]), and the disadvantage of being false in general: see Sections 7.2 and 9.3.3. If G is discrete, the classifying space BG classifies actions of G which are free and proper. By forgetting about freeness of the action we get a canonical map

$$\iota_G : K_*(BG) \to K_*^{top}(G)$$

which is rationally injective. The *Strong Novikov conjecture* for G is the rational injectivity of $\mu_r \circ \iota_G$.

Remark 4.3 If $p \in K_*(C^*(G, C_0(X))) = KK_*(\mathbf{C}, C^*(G; C_0(X)))$ is a fixed element, the Kasparov product $p \otimes_{C^*(G,C_0(X))} : x \mapsto p \otimes_{C^*(G,C_0(X))x}$ provides a map $KK_*(C^*(G, C_0(X)), C_r^*(G, A)) \to KK_*(\mathbf{C}, C_r^*(G, A))$. Observe that if p is given by an idempotent of $C^*(G, C_0(X))$, and $x = (E_+, E_-, F)$, with E_+, E_- Hilbert C^*-modules over $C_r^*(G, A)$ and $F \in \mathcal{B}_{C_r^*(G,A)}(E_+, E_-)$, then $p \otimes_{C^*(G,C_0(X))} x$ is described simply as (pE_+, pE_-, pFp). It turns out that e_X can be described by such an idempotent. Indeed, by properness and G-compactness, there exists a Bruhat function on X, i.e., a non-negative function $f \in C_c(X)$ such that $\int_G f(g^{-1}x)\, dg = 1$ for every $x \in X$. Set then $e(x, g) = \sqrt{f(x)f(g^{-1}x)}$. Recalling that the product in $C_c(X \times G)$ is given by $(a \star b)(x, g) = \int_G a(x, h)b(h^{-1}x, h^{-1}g)\, dh$, one sees immediately that $e^2 = e$. Since the set of Bruhat functions is clearly convex, we have a canonical K-theory class $[e_X] \in K_0(C^*(G, C_0(X)))$.

Remark 4.4 Assume that $A = \mathbf{C}$. Let $x = (E_+, E_-, F)$ be an element of $KK_0^G(C_0(X), \mathbf{C})$. Denote by $\pi_\pm$ the representation of $C_0(X)$ on $E_\pm$. Say that F is *properly supported* if for every $\phi \in C_c(X)$ there exists $\psi \in C_c(X)$ such that $\pi_-(\psi)F\pi_+(\phi) = F\pi_+(\phi)$. Replacing F by some homotopical operator (so not changing the K-homology class of (E_+, E_-, F), we may assume that F is properly supported. Consider then the linear subspaces $\pi_\pm(C_c(X))E_\pm$ of $E_\pm$: those are not Hilbert spaces in general, but these are $C_c(G)$-modules and F induces a G-intertwiner between them. These spaces carry the $C_c(G)$-valued scalar product:

$$\langle \xi, \eta \rangle (g) =: \langle \xi, \rho_\pm(g)\eta \rangle \ (\xi, \eta \in E_\pm),$$

where $\rho_\pm$ denotes the unitary representation of G on $E_\pm$. Completing those spaces into C^*-modules over $C_r^*(G)$, and extending F to the completion, we get a triple $\mu_r(x) = (\mathcal{E}_+, \mathcal{E}_-, \mathcal{F}) \in KK_*(\mathbf{C}, C_r^*(G)) = K_*(C_r^*(G))$, also called the G-index of F.

The two above approaches, for $A = \mathbf{C}$, were shown to be equivalent in Corollary 2.16 of Part 2 of [MV03].[15]

Remark 4.5 It was only in 2009 that Baum et al. [BHS10] reconciled the original approach of [BC00] with the Kasparov-based approach of [BCH94], in the case of discrete groups.

For general Lie groups (with arbitrarily many connected components), the equivalence between the approaches in [BC00] and [BCH94] has not been proved in print so far. However, for connected Lie groups both approaches reduce to the Connes–Kasparov conjecture so there is no problem.

Remark 4.6 There is also a homotopical approach to the Baum–Connes conjecture, developed by Davis and Lück [DL98]; it is valid for discrete groups only. It uses homotopy spectra over the orbit category. More precisely, let G be a group, and denote by $\mathbb{O}_{\mathcal{F}}(G)$ the category whose objects are homogeneous spaces G/H, with H a finite subgroup, and morphisms are G-equivariant maps. Equivariant K-homology is obtained by defining some functor from $\mathbb{O}_{\mathcal{F}}(G)$ to the category of Ω-spectra, extending it to a functor from G-spaces to Ω-spectra, and then applying the i-th homotopy group to get K_i^G (with $i \geq 0$). It turns out that the value of their functor on G/H, for every subgroup H on G, is $K_*(C_r^*(G))$. Hence the assembly map, in that framework, is the map functorially associated to the projection $\underline{EG} \to G/G = \{*\}$. The equivalence with the approach in [BCH94] was worked out by Hambleton and Pedersen [Hp04].

For the operator algebra inclined reader, we emphasize that the Davis-Lück approach, abstract as it may seem, allows for explicit computations of the left-hand side $K_*^{top}(G)$, for G discrete: this is due to the existence of an Atiyah–Hirzebruch spectral sequence relating Bredon homology $H_*^{\mathcal{F}}(\underline{EG}, R_{\mathbf{C}})$ to equivariant K-homology. In favorable circumstances (e.g., dim $\underline{EG} \leq 3$), there are exact sequences allowing one to compute exactly (i.e., integrally, not just rationally) $K_*^{top}(G)$ from Bredon homology (see [MV03], Theorem I.5.27). For specific classes of groups, the Baum–Connes conjecture can be checked by hand in this way (see, e.g., [FPV17] for the case of lamplighter groups $F \wr \mathbf{Z}$, with F a finite group).

4.4 Generalizing the γ-element method

4.4.1 The case of groups acting on bolic spaces

The general formulation of the Baum–Connes conjecture suggests the problem of generalizing the γ-element method, which was first elaborated in the realm of Riemannian symmetric spaces and of their p-adic analogues, Bruhat–Tits buildings.

[15] Note that the proof is given there only for discrete groups, but the proof goes over to locally compact group.

Kasparov and Skandalis [KS03] have explored the case of a combinatorial analogue of simply connected Riemannian manifold with non-positive curvature. The good framework is that of weakly bolic, weakly geodesic metric spaces of bounded coarse geometry (see the definition in their paper). They prove the following:

Theorem 4.7 *Let G be a group acting properly by isometries on a weakly bolic, weakly geodesic metric space of bounded coarse geometry. Then the Baum–Connes assembly map is injective.*

The proof involves analogues of the Dirac, dual-Dirac, and γ-elements. However, α and β should no more be thought as defining the Baum–Connes assembly map and the candidate for its inverse. They rather give maps imbedding the K-theory of arbitrary crossed products into the K-theory of crossed products by some proper G-algebras, for which the conjecture is known to be true:

Definition 4.8 Let X be a G-space. A $G - X - C^*$-algebra is a $G - C^*$-algebra B equipped with a G-equivariant homomorphism $C_0(X) \to Z(M(B))$, the center of the multiplier algebra of B. A $G - C^*$-algebra B is *proper* if there exists a proper G-space X such that B is a $G - X - C^*$ algebra.

The following was proved by Chabert et al. [CEM01][16]:

Theorem 4.9 *The Baum–Connes morphism with coefficients in a proper G-algebra is an isomorphism.*

In the case of a discrete group G acting properly by isometries on a weakly bolic, weakly geodesic metric space of bounded coarse geometry, Kasparov and Skandalis define a proper algebra B, Dirac and dual-Dirac elements $\alpha \in KK_G(B, \mathbf{C})$, $\beta \in KK_G(\mathbf{C}, B)$ and consider the product $\gamma = \beta \otimes_B \alpha \in KK_G(\mathbf{C}, \mathbf{C})$. In that case, it is no more the case that $\alpha \otimes_{\mathbf{C}} \beta$ is equal to 1 in $KK_G(B, B)$, and this is in fact not needed. However, one still has the fact that γ becomes 1 when restricted to finite subgroups. This is enough to prove injectivity of the assembly map for such a group G.

4.4.2 Tu's abstract gamma element

The Kasparov–Skandalis method has been formalized by Tu who defined a general notion of γ element for a locally compact group, such that the mere existence of $\gamma \in KK_G(\mathbf{C}, \mathbf{C})$ implies the injectivity of the Baum–Connes map, and that the surjectivity is equivalent to the fact that $\tilde{\gamma}_A = \mathrm{Id}$ with notations as in Theorem 3.6. The techniques use the representable KK-theory of Kasparov and can also be beautifully interpreted in the framework of equivariant KK-theory for groupoids

[16] See also Higson and Guentner [HG04, Theorem 2.19] and Kasparov and Skandalis [KS03]. The case where G is a connected Lie group and $B = C_0(X)$, where X is a proper G-space, was previously treated by Valette [Val88].

as introduced by Le Gall [LG99]. See Chapter 7 below for details on the groupoid framework.

Definition 4.10 A *γ-element* for G is an element γ of the ring $KK_G(\mathbf{C}, \mathbf{C})$ satisfying the following two conditions:

(1) there exists a proper $G - C^*-$algebra B and two elements $\alpha \in KK_G(B, \mathbf{C})$ and $\beta \in KK_G(\mathbf{C}, B)$ such that $\gamma = \beta \otimes_B \alpha \in KK_G(\mathbf{C}, \mathbf{C})$;
(2) for any compact subgroup K of G, the image of γ by the restriction map $KK_G(\mathbf{C}, \mathbf{C}) \to R(K)$ is the trivial representation 1_K.

Remark 4.11 The second condition is technically formulated as follows: for any proper G-space X, we have $p_*(\gamma) = 1$ in $RKK_G(X; \mathbf{C}, \mathbf{C})$ (where p_* denotes the induction homomorphism $KK_G(\mathbf{C}, \mathbf{C}) \to RKK_G(X; \mathbf{C}, \mathbf{C})$). The notations are as follows: for X a G-space, A and B two $G - X - C^*$-algebras, Kasparov defines $\mathcal{R}KK_G(X; A, B)$ as the set of homotopy classes of (A, B)-Fredhom bimodules equipped with a covariant action of the C^*-algebra $C_0(X)$, with the usual assumption of compactness of commutators. The beautiful language of groupoids allows to think of A and B as $\mathcal{G} - C^*$-algebras with $\mathcal{G} = X \rtimes G$ the groupoid given by the action of G on X. Then

$$\mathcal{R}KK_G(X; A, B) = KK_{\mathcal{G}}(A, B).$$

Now for two $G - C^*$-algebras A and B (no action of $C_0(X)$ is needed), Kasparov defines

$$\begin{aligned} RKK_G(A, B) &= \mathcal{R}KK_G(X; A \otimes C_0(X), B \otimes C_0(X)) \\ &= KK_{\mathcal{G}}(A \otimes C_0(X), B \otimes C_0(X)). \end{aligned}$$

In the definition of a γ-element, the map

$$p_* : KK_G(\mathbf{C}, \mathbf{C}) \to RKK_G(X; \mathbf{C}, \mathbf{C})$$

is the pullback by the groupoid homomorphism $p : \mathcal{G} = X \rtimes G \to G$. Note that if $X = G/K$ with K a compact subgroup, then $RKK_G(X; \mathbf{C}, \mathbf{C}) = R(K)$.

Tu has proved the following [Tu00]:

Proposition 4.12 *If an element γ exists, then it is unique. Moreover, it is an idempotent of the ring $KK_G(\mathbf{C}, \mathbf{C})$, namely $\gamma \otimes_{\mathbf{C}} \gamma = \gamma$.*

Observe that, if a γ-element does exist, then it acts as the identity on any group $K_*^{top}(G, A)$, for every $G - C^*$-algebra A. The relation with the Baum–Connes conjecture can be stated as follows

Theorem 4.13 (Theorems 4.2 and 4.4 [Tu99c]) *Let G be a locally compact group admitting a γ-element.*

(1) The map $\mu_{A,r}$ is injective for every $G - C^*$-algebra A.

(2) The map $\mu_{A,r}$ is surjective if and only if the map $\tilde{\gamma}_A$ (i.e., Kasparov product by $j_{G,r}(\tau_A(\gamma))$) is the identity on $K_*(C_r^*(G, A))$. This is in particular true if $\gamma = 1$.

Proof Let $\gamma = \beta \otimes_B \alpha$ be a γ-element, with B a proper $G - C^*$-algebra. Let A be any $G - C^*$-algebra. Then we have a commutative diagram:

$$\begin{array}{ccccc}
K^{top}(G, A) & \xrightarrow{\otimes_A \tau_A(\beta)} & K_*^{top}(G, A \otimes B) & \xrightarrow{\otimes_{A\otimes B}(\tau_A(\alpha))} & K^{top}(G, A) \\
\downarrow{\scriptstyle \mu_{A,r}} & & \simeq \downarrow{\scriptstyle \mu_{A\otimes B,r}} & & \downarrow{\scriptstyle \mu_{A,r}} \\
K_*(C_r^*(G, A)) & \xrightarrow{\otimes_{C_r^*(G,A)} j_G(\tau_A(\beta))} & K_*(C^*(G; A \otimes B)) & \xrightarrow{\otimes_{C^*(G,A\otimes B)} j_G(\tau(\alpha))} & K_*(C_r^*(G, A)),
\end{array}$$

with j_G the descent map as in Section 3.1. Since $A \otimes B$ is a proper $G - C^*$-algebra, the map $\mu_{A\otimes B,r}$ is an isomorphism, by Theorem 4.9. The assumption in (1) is that the composition of the two maps on the top row is the identity: this implies that $\mu_{A,r}$ is injective. The assumption in (2) is that moreover the composition of the two maps on the bottom row is the identity: this implies that $\mu_{A,r}$ is also surjective. □

Remark 4.14 The element γ initially defined by Kasparov in [Kas95] is of course a special case of γ-element in the sense of Tu. Note that if K is a maximal compact subgroup of a connected Lie group G, the element γ is simply characterized by the conditions (cf. Proposition 4.1 in [Tu00]) that it factorizes through a proper $G - C^*$-algebra and that the image of γ by the restriction map $KK_G(\mathbf{C}, \mathbf{C}) \to R(K)$ is the trivial representation 1_K.

4.4.3 Nishikawa's new approach

Very recently (March 2019), Nishikawa [Nis19] introduced a new idea in the subject, that amounts to constructing the γ element *without* having to construct the Dirac and dual-Dirac elements. We briefly explain his approach. The standing assumption is that the group G admits a cocompact model for $\underline{EG}$ (in particular $\underline{EG}$ is locally compact).

Definition 4.15 Let x be an element of $KK_G(\mathbf{C}, \mathbf{C})$. Say that x has property (γ) if it can be represented by a Fredhom module $KK_G(\mathbf{C}, \mathbf{C})$ such that:

1. For every compact subgroup K of G, x restricts to 1_K in $R(K)$.
2. The Hilbert space $\mathcal{H}$ carries a G-equivariant non-degenerate representation of $C_0(\underline{EG})$ such that, for every $f \in C_0(\underline{EG})$, the map $g \mapsto [g(f), T]$ is a norm continuous map vanishing at infinity on G, with values in the ideal of compact operators.

3. Moreover, the integral

$$\int_G g(c)Tg(c)dg - T = -\int_G g(c)[g(c), T]dg$$

is compact, where c is a compactly supported function on $\mathbf{E}G$ such that $\int_G g(c)^2 dg = 1$.

It is not known whether the technical condition 3 follows from condition 2 or is really needed. Nishikawa shows that such a Fredholm module allows to define, for every $G - C^*$-algebra A, a map $\nu_A^x : K_*(C_r^*(G, A)) \to K_*^{top}(G, A) = KK_G(C_0(\underline{EG}), A)$, which is a left inverse for the assembly map $\mu_{A,r}$. One has the following theorem:

Theorem 4.16 *Assume that there exists a Fredholm module $x = (\mathcal{H}, F)$ with property (Γ). Then:*

1. *For every $G - C^*$-algebra A, the map $\mu_{A,r}$ is injective.*
2. *For every $G - C^*$-algebra A, the map $\mu_{A,r}$ is surjective if and only if the element x defines the identity on $K_*(C_r^*(G, A))$. In particular, if $x = 1$ in $KK_G(\mathbf{C}, \mathbf{C})$, Conjecture 5 holds for G.*

Nishikawa also proves the following result:

Theorem 4.17

1. *If there exists an element x of $KK_G(\mathbf{C}, \mathbf{C})$ with property (γ), then it is unique and is an idempotent in $KK_G(\mathbf{C}, \mathbf{C})$.*
2. *If G admits a γ element in the sense of Tu, then $x = \gamma$ has the (γ) property.*

In particular, in the case of groups admitting an abstract γ element, any element with the (γ) property is in fact equal to γ.

Using this new approach, Nishikawa can reprove Conjecture 5 for Euclidean motion groups, as well as the injectivity of the Baum–Connes map with coefficients $\mu_{A,r}$ for G a semisimple Lie group. He also reproves the conjecture for groups acting properly on locally finite trees and announces a generalization (with Brodzki, Guentner, and Higson) to groups acting properly on $CAT(0)$ cubic complexes.

4.5 Consequences of the Baum–Connes conjecture

4.5.1 Injectivity: the Novikov conjecture

In Section 2.5, we already emphasized that the Novikov conjecture (Conjecture 3) on homotopy invariance of higher signature followed from the (rational) injectivity of Kasparov's map

$$\beta : K_0(B\Gamma) \to K_0\left(C_r^*(\Gamma)\right).$$

In the case of a cocompact, torsion-free lattice of a connected Lie group G, the map β coincides with the Dirac induction map

$$K_0(M) \to K_0\left(C_r^*(\Gamma)\right)$$

of Section 3.1. In general there is a natural injection group $\iota_\Gamma : K_0(B\Gamma) \to K_0^\Gamma(\underline{E\Gamma})$ and its composition with the assembly map μ_r gives β. That fact, taken for granted for a long time, was proved only fairly recently by Land [Lan15].

Therefore, the Novikov conjecture follows from the Strong Novikov conjecture, i.e., from the rational injectivity of the map $\mu_r \circ \iota_\Gamma$. In particular, the Novikov conjecture follows from the injectivity of the assembly map μ_r.

We must here mention the beautiful recent approach of Antonini et al. [AAS18] on K-theory with coefficients in the real numbers. They make use of von Neumann theory of II_1-factors. For such a factor N, the trace defines naturally an isomorphism from $K_0(N)$ to $\mathbf{R}$ whereas $K_1(N) = 0$. The KK-theory with real coefficients $KK^G_{\mathbf{R}}(A, B)$ is defined as the inductive limit: of the groups $KK_G(A, B \otimes N)$ for all N a II_1-factors N. Note that there is a map $KK^G(A, B) \otimes \mathbf{R} \to KK^G_{\mathbf{R}}(A, B)$ but it is in general not an isomorphism. Any trace on A defines an element of $KK_{\mathbf{R}}(A, \mathbf{C})$. In particular for Γ a discrete group, the canonical trace τ defines an element $[\tau]$ of $KK_{\mathbf{R}}(C_r^*(\Gamma), \mathbf{C}) = KK^\Gamma_{\mathbf{R}}(\mathbf{C}, \mathbf{C})$. The crucial remark of [AAS18] is the following:

Proposition 4.18 *The element $[\tau]$ is an idempotent of the ring $KK^\Gamma_{\mathbf{R}}(\mathbf{C}, \mathbf{C})$. Moreover for any proper and free space X, the identity $1_{C_0(X)}$ of the ring $KK^\Gamma_{\mathbf{R}}(\mathbf{C}, \mathbf{C})$ satisfies $1_{C_0(X)} \otimes [\tau] = 1_{C_0(X)}$.*

The authors define the $KK_{\mathbf{R}}$-groups localized at the identity as the products by the idempotent $[\tau]$, i.e., $KK^\Gamma_{\mathbf{R}}(A, B)_\tau = KK^\Gamma_{\mathbf{R}}(A, B) \otimes_{\mathbf{C}} [\tau]$. In particular the Baum–Connes map can be localized as

$$\mu_\tau : K^{top}_{*,\mathbf{R}}(\Gamma)_\tau \to K_{*,\mathbf{R}}\left(C_r^*(\Gamma)\right)_\tau,$$

where the right-hand side is nothing but $KK^\Gamma_{\mathbf{R}}(\mathbf{C}, C_r^*(\Gamma))_\tau$ and the left-hand side is $KK^\Gamma_{\mathbf{R}}(C_0(X), \mathbf{C})_\tau$ (assume for simplicity that $\underline{E}\Gamma$ is cocompact).

The results of [AAS18] can be summarized as follows

Theorem 4.19 *Let Γ be a discrete group.*

1. *If the Baum–Connes conjecture (with coefficients) holds for Γ, then μ_τ is an isomorphism.*
2. *If the map μ_τ is injective, then the Strong Novikov conjecture holds for Γ.*

The first point uses the Baum–Connes map with coefficients in any II_1-factor. The second point rests upon the observation that the map from $E\Gamma$ to $\underline{E}\Gamma$ induces an isomorphism from

$$K_*(B\Gamma) \otimes \mathbf{R} = KK^\Gamma_{\mathbf{R}}(C_0(E\Gamma), \mathbf{C}) \to K^{top}_{*,\mathbf{R}}(\Gamma)_\tau = KK^\Gamma_{\mathbf{R}}(C_0(\underline{E}\Gamma), \mathbf{C})_\tau.$$

In other words, the conjecture that μ_τ is an isomorphism is intermediate between the Baum–Connes conjecture (without coefficents) and the Strong Novikov conjecture.

4.5.2 Injectivity: the Gromov–Lawson–Rosenberg conjecture

Let M be a Riemannian manifold of dimension n. The *scalar curvature* is a smooth function $\kappa : M \to \mathbf{R}$ that, at a point $p \in M$, measures how fast the volume of small balls centered at p grows when compared to the volume of small balls of the same radius in Euclidean space $\mathbf{E}^n$. More precisely we expand the ratio $\frac{Vol\ B_M(p,r)}{Vol\ B_{\mathbf{E}^n}(0,r)}$ as a power series in r:

$$\frac{Vol\ B_M(p,r)}{Vol\ B_{\mathbf{E}^n}(0,r)} = 1 - \frac{\kappa(p)}{6(n+2)}r^2 + o(r^2);$$

so positive scalar curvature means that small balls in M grow more slowly than corresponding Euclidean balls.

Let M be now a closed spin manifold, and D the Dirac operator of M, the Atiyah–Singer index formula for D is

$$Ind(D) = \langle \hat{\mathbf{A}}(\mathbf{M}), [M] \rangle,$$

where $\hat{\mathbf{A}}(\mathbf{M})$ is a polynomial in the Pontryagin classes, and $[M]$ is the fundamental class of M; see [BBB13]. Let $\Gamma = \pi_1(M)$ be the fundamental group of M, and let $f : M \to B\Gamma$ be the classifying map. Fix $x \in H^*(B\Gamma, \mathbf{Q})$. The number $\langle \hat{\mathbf{A}}(\mathbf{M})[M] \rangle$ being called the $\hat{A}$-*genus*, it is natural to call the numbers

$$\hat{A}_x(M) =: \langle f^*(x) \cup \hat{\mathbf{A}}(\mathbf{M}), [M] \rangle$$

higher $\hat{A}$-*genera*, by analogy with higher signatures. The *Gromov–Lawson–Rosenberg conjecture* (GLRC) states:

Conjecture 6 (GLRC) Let M be a closed spin manifold M with fundamental group Γ. If M admits a Riemannian metric with positive scalar curvature, then all higher $\hat{A}$-genera do vanish: $\hat{A}_x(M) = 0$ for all $x \in H^*(B\Gamma, \mathbf{Q})$.

GLRC for manifolds with given fundamental group Γ, follows from injectivity of the assembly map for Γ, see Theorem 7.11 in [BCH94]. The fact that the usual $\hat{A}$-genus vanishes for a closed spin manifold with positive scalar curvature, is a famous result by Lichnerowicz.

See [RS95] for a lucid discussion of GLRC, together with speculations about a suitable converse: does the vanishing of a certain K-theory class in the real K-theory of $C_r^*(\Gamma)$ implies the existence of a metric with positive scalar curvature on M?

4.5.3 Surjectivity: the Kadison–Kaplansky conjecture

Let Γ be a discrete group. If $g \in \Gamma$ is a group element of finite order $n > 1$, then $e = \frac{1}{n}\sum_{k=0}^{n-1} g^k$ defines a non-trivial element in the complex group ring $\mathbf{C}\Gamma$ ("non-trivial" meaning: distinct from 0 and 1). When Γ is torsion-free, it is not clear that $\mathbf{C}\Gamma$ admits non-trivial idempotents, and around 1950, Kaplansky turned this into a conjecture:

Conjecture 7 If Γ is a torsion-free group, then $\mathbf{C}\Gamma$ has no non-trivial idempotent.

Around 1954, Kadison and Kaplansky conjectured that this should be even true by replacing $\mathbf{C}\Gamma$ by the larger reduced group C^*-algebra:

Conjecture 8 If Γ is a torsion-free group, then $C_r^*(\Gamma)$ has no non-trivial idempotent.

In contrast with the Novikov conjecture (Conjecture 3), Conjecture 8 is easy to state. It is interesting that it follows too from the Baum–Connes conjecture (Conjecture 4), actually from the surjectivity part.

Proposition 4.20 *Let Γ be a torsion-free group. If the assembly map μ_r is onto, then Conjecture 8 holds for Γ.*

The proof of Proposition 4.20 goes through an intermediate conjecture. To motivate this one, recall that any trace σ on a complex algebra A defines a homomorphism

$$\sigma_* : K_0(A) \to \mathbf{C} : [e] \mapsto (Tr_n \otimes \sigma)(e),$$

where $e = e^2 \in M_n(A)$ and $Tr_n : M_n(A) \to A$ is the usual trace. If A is a C^*-algebra and σ is a positive trace, then the image of σ_* is contained in $\mathbf{R}$. Consider now the canonical trace τ on $C_r^*(\Gamma)$. The following conjecture is known as *conjecture of integrality of the trace*.

Conjecture 9 If Γ is a torsion-free group, then the canonical trace τ_* maps $K_0(C_r^*(\Gamma))$ to $\mathbf{Z}$.

It is then easy to see that Conjecture 9 implies the Kadison–Kaplansky conjecture (Conjecture 8). Indeed, take $e = e^2 \in C_r^*(\Gamma)$. Since an idempotent in a unital C^*-algebra is similar to a projection, we may assume that $e = e^* = e^2$. As $0 \leq e \leq 1$ and τ is a positive trace, we have $0 \leq \tau(e) \leq 1$. But $\tau(e) \in \mathbf{Z}$ by Conjecture 9, so $\tau(e)$ is either 0 or 1. If $0 = \tau(e) = \tau(e^*e)$, then $e = 0$ by faithfulness of τ. Replacing e by $1 - e$, we see that if $\tau(e) = 1$, then $e = 1$.

Proof of Proposition 4.20 By the previous remarks, it is enough to see that, for a torsion-free group Γ such that μ_r is onto, Conjecture 9 holds. Actually we prove that, assuming Γ to be torsion-free, τ_* is always integer-valued on the image of μ_r in $K_0(C_r^*(\Gamma))$.

Thanks to Remark 4.5, the domain of μ_r, i.e., the left-hand side of the Baum–Connes conjecture, is the group $K_0(\Gamma, pt)$, whose cycles are of the form (Z, ξ) with

Z a proper Γ-compact manifold and $\xi \in V_\Gamma(T^*Z)$, and by Section 4.1 we have $\mu_r(Z, \xi) = Ind_\Gamma(\tilde{D})$, where $\tilde{D}$ is some Γ-invariant elliptic differential operator on Z. As Γ is torsion-free, any proper Γ-action is free and proper, so that the map $Z \to \Gamma\backslash Z$ is a Γ-covering and we may appeal to Atiyah's L^2-index theorem (Theorem 2.1): the operator $\tilde{D}$ descends to an elliptic operator on the compact manifold $\Gamma\backslash Z$ and

$$\tau_*(\mu_r(Z, \xi)) = Ind_\Gamma(\tilde{D}) = Ind(D).$$

Since $Ind(D) \in \mathbf{Z}$, this concludes the proof.[17] □

4.5.4 Surjectivity: vanishing of a topological Whitehead group

For a group Γ, denote by $\mathbf{Z}\Gamma$ its integral group ring, and let

$$K_1^{alg}(\mathbf{Z}\Gamma) =: \varinjlim GL_n(\mathbf{Z}\Gamma)/E_n(\mathbf{Z}\Gamma)$$

be the first algebraic K-theory group of $\mathbf{Z}\Gamma$, where $E_n(\mathbf{Z}\Gamma)$ is the subgroup of elementary matrices. We denote by $[\pm\Gamma]$ the subgroup of $K_1^{alg}(\mathbf{Z}\Gamma)$ generated by the image of the inclusion of $\Gamma \times \{\pm 1\}$ into $GL_1(\mathbf{Z}\Gamma)$. The *Whitehead group* $Wh(\Gamma)$ is then

$$Wh(\Gamma) = K_1^{alg}(\mathbf{Z}\Gamma)/[\pm\Gamma].$$

By analogy, using the inclusion of Γ in the unitary group of $C_r^*(\Gamma)$, we may define the *topological Whitehead group* as $Wh^{top}(\Gamma) =: K_1(C_r^*(\Gamma))/[\Gamma]$. So the vanishing of $Wh^{top}(\Gamma)$ is equivalent to the fact that every unitary matrix in $M_\infty(C_r^*(\Gamma))$ is in the same connected component as a diagonal matrix $diag(\gamma, 1, 1, 1, \ldots)$ for some $\gamma \in \Gamma$.

Conjecture 10 Assume that there is a 2-dimensional model for $B\Gamma$. Then $Wh^{top}(\Gamma) = 0$.

The following result appears in [BMV05]:

Proposition 4.21 *When Γ has a 2-dimensional model for $B\Gamma$, Conjecture 10 follows from the surjectivity of the assembly map μ_r.*

Proof Let Γ^{ab} denote the abelianization of Γ. The inclusion of Γ in the unitary group of $C_r^*(\Gamma)$ induces a map $\beta : \Gamma^{ab} \to K_1(C_r^*(\Gamma))$, as K_1 is an abelian group.

[17] For a nice proof of that result NOT appealing to Atiyah's L^2-index theorem, see lemma 7.1 in [MV03].

By lemma 7.5 in [BMV05], as $B\Gamma$ is 2-dimensional, the Chern character $Ch : K_1(B\Gamma) \to H_1(B\Gamma, \mathbf{Z})$ is an isomorphism. Of course we have $H_1(B\Gamma, \mathbf{Z}) = H_1(\Gamma, \mathbf{Z}) = \Gamma^{ab}$. Moreover we have a commutative diagram

$$\begin{array}{ccc} K_1(B\Gamma) & \xrightarrow{\mu_r} & K_1(C_r^*(\Gamma)) \, . \\ & {\scriptstyle Ch}\searrow \quad \nearrow{\scriptstyle \beta} & \\ & \Gamma^{ab} & \end{array}$$

So β is onto as soon as μ_r is, and this implies $Wh^{top}(\Gamma) = 0$ □

4.5.5 Surjectivity: discrete series of semisimple Lie groups

Let G be a semisimple connected Lie group with finite center and maximal compact subgroup K. As we shall see in Theorem 6.10 below, Lafforgue has given a proof of the Baum–Connes conjecture without coefficients for G which is independent of Harish-Chandra theory. On the other hand, let us present here a beautiful argument, also due to Lafforgue [Laf02a], showing that the surjectivity of the assembly map does say something on the representation theory: namely, surjectivity implies that the Dirac induction μ_G maps bijectively a subset of the dual $\hat{K}$ to the discrete series of G; compare with Remark 2.4.

Recall that semisimple groups are CCR, i.e., any unitary irreducible representation σ of G maps $C^*(G)$ onto the compact operators on $\mathcal{H}_\sigma$; so in K-theory σ induces a homomorphism $\sigma_* : K_0(C^*(G)) \to \mathbf{Z}$.

As the main ingredient for Lafforgue's observation, we just need to recall from Remark 2.4 that any discrete series π of G defines a K-theory class $[\pi] \in K_0(C_r^*(G))$ such that $\pi_*([\pi]) = 1$. In particular $[\pi] \neq 0$. Note that if G/K is odd dimensional, then the surjectivity part of the conjecture implies that $K_0(C_r^*(G)) = 0$ so that G has no discrete series, reproving a well-known fact in Harish-Chandra theory. We therefore now assume that G/K has even dimension.

Assume for simplicity that G/K has a G-invariant spin structure, i.e., the adjoint representation of K in $V = \mathfrak{g}/\mathfrak{k}$ lifts to $Spin(V)$. The Connes–Kasparov map μ_G then coincides with Kasparov's Dirac map $\tilde{\alpha} : R(K) = K_0(C^*(K)) \to K_0(C_r^*(G))$. The inverse of the map is Kasparov's dual-Dirac map $\tilde{\beta} : K_0(C_r^*(G)) \to R(K)$. Lafforgue's observation is the following duality:

Lemma 4.22 *For any discrete series π of G and any irreducible representation ρ of K, the following integers are equal:*

$$\pi_*(\tilde{\alpha}([\rho])) = \rho_*(\tilde{\beta}([\pi])).$$

Indeed, one can show that both are equal to the dimension of the intertwining space $\mathrm{Hom}_K(S \otimes V_\rho, H_\pi)$, where S is the spinor representation of K.

Fix π a discrete series of G. Viewing $R(K)$ as the free abelian group on $\hat{K}$, we may write

$$\tilde{\beta}([\pi]) = \sum_{\rho \in \hat{K}} n_{\pi,\rho}[\rho],$$

where $n_{\pi,\rho}$ is the integer defined in two different ways in Lemma 4.22. Now the assumed surjectivity of μ_G translates into $\tilde{\alpha} \circ \tilde{\beta} = \mathrm{Id}$, which implies the following decomposition in $K_0(C_r^*(G))$:

$$[\pi] = \sum_{\rho} n_{\pi,\rho} \tilde{\alpha}([\rho]).$$

Now the equality $\pi_*([\pi]) = 1$ and Lemma 4.22 yield:

$$1 = \sum_{\rho} n_{\pi,\rho} \pi_*(\tilde{\alpha}([\rho])) = \sum_{\rho} n_{\pi,\rho}^2.$$

So the integers $n_{\pi,\rho}$ satisfy $\sum_\rho n_{\pi,\rho}^2 = 1$, hence there is a unique ρ such that $n_{\pi,\rho} = \pm 1$, the others being zero. Then $\tilde{\alpha}([\rho]) = \pm[\pi]$, and the Dirac induction maps bijectively a subset of the dual $\hat{K}$ to the discrete series of G; in other words, we have recovered Theorem 2.3 in a qualitative way.

5 Full and reduced C^*-algebras

5.1 Kazhdan vs. Haagerup: property (T) as an obstruction

The assembly map could as well be constructed using maximal C^*-algebras instead of reduced. There is indeed a map

$$\mu_{A,\max} : K_*^G(\underline{EG}, A) \to K_*\left(C^*_{\max}(G, A)\right)$$

so that $\mu_{A,r}$ is the composition of $\mu_{A,\max}$ with the map λ_A^* obtained by functoriality in K-theory from the map

$$\lambda_A : C^*_{\max}(G, A) \to C_r^*(G, A).$$

In other words we have a commutative diagram

$$\begin{array}{ccc} sK_*^G(\underline{EG}, A) & \xrightarrow{\mu_{A,\max}} & K_*(C^*_{\max}(G, A)) \\ & \searrow^{\mu_{A,r}} & \downarrow \lambda_A^* \\ & & K_*(C_r^*(G, A))), \end{array}$$

The main difficulty in that the Baum–Connes conjecture is a conjecture about $\mu_{A,r}$, not $\mu_{A,\max}$. In order to understand that crucial point, it will be enlightening to consider two classes of groups: one for which both $\mu_{A,r}$ and $\mu_{A,\max}$ are isomorphisms, hence also λ_A^*.; another for which λ_A^* is not injective, $\mu_{A,\max}$ not surjective, and for which the conjectural bijectivity of $\mu_{A,r}$ is difficult and proved only in very few cases. We refer to [Jul98] for more details.

Definition 5.1 A locally compact second countable group G has the Haagerup property[18] if the following equivalent conditions are satisfied:

(i) There exists an action of G by affine isometries on a Hilbert space which is metrically proper.
(ii) There exists a unitary representation π of G on a Hilbert space $\mathcal{H}$, and a 1-cocycle (i.e., a map $b : G \to \mathcal{H}$ such that $b(gg') = b(g) + \pi(g)b(g')$) which is proper.
(iii) There exists a function of conditional negative type on G which is proper.

Definition 5.2 A locally compact second countable group G has Kazhdan's property (T) if the following equivalent conditions are satisfied:

(i) Any action of G by affine isometries on a Hilbert space admits a fixed point.
(ii) For any unitary representation π of G on a Hilbert space $\mathcal{H}$, any 1-cocycle is bounded.
(iii) Any function of conditional negative type on G is bounded.

Note that only compact groups are both Haagerup and Kazhdan. The above definitions can also be expressed in terms of the *almost invariant vectors* property: a unitary representation π of G on $\mathcal{H}$ almost admits invariant vectors if for any $\varepsilon > 0$ and any compact subset C of G, there is a unit vector $x \in \mathcal{H}$ such that $\|\pi(g)x - x\| \leq \varepsilon$ for any $g \in C$.

Proposition 5.3 *A locally compact group G has property (T) if and only if any unitary representation almost admitting invariant vectors has a non-zero invariant vector. It has the Haagerup property if and only if there exists a unitary representation with coefficients vanishing at infinity and almost admitting invariant vectors.*

The above characterization of property (T) is the original definition of Kazhdan. As to the characterization of the Haagerup property, it is due to Jolissaint and implies that all amenable groups have the Haagerup property. For examples of groups having Haagerup or Kazhdan property, we refer to [BdlHV08] and to [CCJ$^+$01]. Typical examples of non-amenable discrete groups with Haagerup property are the free groups $F_n (n \geq 2)$ or $SL_2(\mathbf{Z})$, whereas typical discrete groups having Kazhdan property are $SL_n(\mathbf{Z})$, $n \geq 3$.

[18]Or is a-(T)-menable, according to Gromov.

Let us now explain the link with the Baum–Connes conjecture. We begin with a C^*-algebraic characterization of property (T) (see [AW81]), in terms of the existence of a *Kazhdan projection.*

Proposition 5.4 *The locally compact group G has property (T) if and only if there exists an idempotent $e_G \in C^*_{\max}(G)$ such that, for every unitary representation π of G, the idempotent $\pi(e_G)$ is the orthogonal projector on the space of $\pi(G)$-fixed vectors in $\mathcal{H}_\pi$.*

From this we deduce a key observation made by Connes in the early 1980s: let us consider, for a locally compact group, the map $\lambda : C^*_{\max}(G) \to C^*_r(G)$ associated with the left regular representation of G.

Lemma 5.5 *If G is non-compact with property (T), the map induced in K-theory*

$$\lambda_* : K_*\left(C^*_{\max}(G)\right) \to K_*\left(C^*_r(G)\right)$$

is not injective: its kernel contains a copy of $\mathbf{Z}$ which is a direct summand in $K_0(C^*_{\max}(G))$.

Proof Because of property (T), we have a direct sum decomposition

$$C^*_{\max}(G) = \ker(\epsilon_G) \oplus \mathbf{C}e_G,$$

where ϵ_G is the trivial one-dimensional representation of G. So $K_0(C^*_{\max}(G)) = K_0(\ker(\epsilon_G)) \oplus \mathbf{Z}$. On the other hand, as G is not compact: $\lambda_*(e_G) = 0$, which ends the proof. □

Corollary 5.6 *Assume that G is non-compact with property (T), and admits a γ-element. Then $\mu_{\max}$ is not surjective. In particular, $\gamma \neq 1$ in $KK_G(\mathbf{C}, \mathbf{C})$.*

Proof We have $\mu_r = \lambda_* \circ \mu_{\max}$, and the injectivity of μ_r (see Theorem 4.13) trivially implies that a non-zero element of the kernel of λ_* cannot be in the image of $\mu_{\max}$. Moreover, if $\gamma = 1$, the Kasparov machine, which works also for full crossed products, shows that $\mu_{A,\max}$ is an isomorphism, a contradiction. □

On the other hand, Higson and Kasparov have proved in the 1990s the following beautiful result:

Theorem 5.7 *Let G be a locally compact group having the Haagerup property. Then G has a γ-element equal to 1. As a consequence, the three maps $\mu_{A,r}$, $\mu_{A,\max}$, and $(\lambda_A)_*$ are isomorphisms. In particular Conjecture 5 holds for G.*

For a proof (using E-theory instead of KK-theory) we refer to [HK01] and [Jul98]. We shall only explain how a locally compact proper G-space can be constructed from an affine action on a Hilbert space. Consider the space $Z = H \times [0, +\infty[$ equipped by the topology pulled back by the map $(x, t) \mapsto (x, \sqrt{\|x\|^2 + t^2})$ of the topology of the space $H_w \times [0, +\infty[$, where H_w is the space H with weak topology. The space Z is a locally compact space and carries a proper action defined by $g.(x, t) = (g.x, t)$ for $g \in G$. It is a representative of the classifying space of proper actions $\underline{EG}$. The space Z can also be defined as a projective limit of spaces $[0, +\infty[\times V$ over all affine subspaces V of H, with the maps $[0, +\infty[\times V' \to [0, +\infty[\times V$ (for all $V \subset V'$) combining the projection to V with the map $x \mapsto \sqrt{\|x\|^2 + t^2}$ on the vector subspace orthogonal to V in V'.

A locally compact group G is *K-amenable* (see, e.g., [JV84]) if, for any $G - C^*$-algebra A, the full crossed product $C^*_{\text{max}}(G, A)$ and the reduced crossed product $C^*_r(G, A)$ do have the same K-theory via the map $(\lambda_A)_*$. So Theorem 5.7 says that groups with the Haagerup property are K-amenable, while Corollary 5.6 says that non-compact groups with property (T), are not.

Remark 5.8 In a recent preprint, Gong et al. [GWY19] prove the Strong Novikov conjecture for discrete groups acting isometrically and metrically properly on a Hilbert–Hadamard manifold (i.e., an infinite-dimensional analogue of simply connected and non-positively curved manifold). This contains of course the case of groups with the Haagerup property, but also the case of geometrically discrete subgroups of the group of volume preserving diffeomorphisms of a compact smooth manifold. Their proof uses a generalization of the Higson–Kasparov construction, but also the techniques of Antonini et al. [AAS18].

5.2 A trichotomy for semisimple Lie groups

Let us now assume that G is a semisimple Lie group, connected with finite center. The conjecture without coefficients (Conjecture 4) for G is known to be true. There are now three completely distinct proofs of that fact. In 1984, Wassermann [Was87] (following the work of Penington and Plymen [PP83] and Valette [Val85, Val84]) proved the conjecture using the whole machinery of Harish-Chandra theory together with the work of Knapp-Stein and Arthur, allowing for a precise description of the reduced dual of such groups. The second proof, due to Lafforgue, only uses Harish-Chandra's Schwartz space, but appeals to the whole of his Banach KK-theory, sketched in Chapter 6 below. Another idea of proof had been suggested by Baum et al. [BCH94] following the idea of Mackey correspondence, i.e., of a very subtle correspondence between the reduced dual of a semisimple Lie group G and the dual of its Cartan motion group, i.e., the semidirect product $G_0 = \mathfrak{g}/\mathfrak{k} \rtimes K$, where K is a maximal compact subgroup of G. Very recently Afgoustidis [Afg16] has given such a proof using the notion of minimal K-types introduced by Vogan [Vog81].

But the most difficult problem arises when one is interested in the conjecture for a discrete subgroup Γ of G. Such groups inherit the geometry from G, but there is of course no hope to describe their reduced dual. However, the conjecture (with or without coefficients) for Γ follows from the conjecture *with coefficients* (Conjecture 5) for the Lie group G, a fact stated without proof in [BCH94] and first proved by Oyono-Oyono [OO01].

As a result the question of Baum–Connes for Γ can be summarized as follows, resulting from Kasparov's work:

(1) injectivity of the Baum–Connes assembly map for G holds with coefficients in any $G - C^*$-algebra, hence it also holds for the discrete group Γ.
(2) the question of surjectivity of the Baum–Connes assembly map for the discrete group Γ, or more generally the surjectivity of the Baum–Connes assembly map with coefficients in any A for the Lie group G, are difficult questions and can be considered as a crucial test for Conjecture 4.

We shall have to distinguish, among simple Lie groups, the real rank 1 and the higher rank cases. We need to recall the classification of real rank 1 simple Lie groups. Up to local isomorphism, the list is: $SO_0(n, 1)$, $SU(n, 1)$, $Sp(n, 1)$, $F_{4(-20)}$, i.e., the isometry group of the n-dimensional hyperbolic space over the division algebras **R**, **C**, **H** (the Hamilton quaternions), and **O** (the Cayley octonions); over **R**, we restrict to orientation-preserving isometries; over **O**, there is only $n = 2$.

Assume that G is locally isomorphic to a simple Lie group. There is the following trichotomy:

(1) *If G is (locally isomorphic to) one of the real rank one groups $SO_0(n, 1)$ or $SU(n, 1)$ ($n \geq 2$):* then G is known to have the Haagerup property. Therefore, by Theorem 5.7, G satisfies the Baum–Connes conjecture with coefficients (Conjecture 5), and so do all its discrete subgroups.

   However, it is worth noting that the $SO_0(n, 1)$ and $SU(n, 1)$ cases had been solved *before* the Higson–Kasparov theorem by more geometric methods in the works of Kasparov [Kas84], Chen [Che96] and Julg-Kasparov [JK95]. Indeed, the above authors have produced a construction of a representative of γ combining a complex on the flag manifold (which is the boundary of the symmetric space) and a Poisson transform, as explained in Section 5.3 below. Then a homotopy using the so-called complementary series yields the required equality $\gamma = 1$ in $KK_G(\mathbf{C}, \mathbf{C})$.
(2) *If G is (locally isomorphic to) one of the real rank one groups $Sp(n, 1)$ ($n \geq 2$) or $F_{4(-20)}$*: then by a result of Kostant, G has Kazhdan's property (T). This fact makes the Baum–Connes conjecture more difficult since the full and reduced crossed product do not have in general the same K-theory. The first deep result in that direction was obtained by Lafforgue in 1998 [Laf00] by combining the Banach analogue of the conjecture, explained in Chapter 6, with Jolissaint's rapid decay property (see Section 6.1.4 below): if Γ is a cocompact discrete subgroup of such a group G, then Γ satisfies Conjecture 4 (i.e., without coefficients).

   Moreover, Julg has been able to extend to those cases the method of flag manifolds and Poisson transforms, which gives again the construction of a Fredholm module representing γ. However, it is no longer possible to use the theory of unitary representations since the complementary series stays away from the trivial representation, in accordance with property (T). An idea proposed by Julg in 1994 is to use a family of uniformly bounded representations approaching the trivial representation. Such a family of uniformly bounded representations has been constructed by Cowling [Cow82]: see Section 6.2.2 for more details.

It should be possible to show that the element γ, though not equal to 1 in $KK_G(\mathbf{C}, \mathbf{C})$, still gives the identity map in $K_*(C_r^*(G, A))$ (but of course not in $K_*(C_{\max}^*(G, A))$. Technically the notion of uniformly bounded representations has to be extended to representations with an arbitrary slow exponential growth, following an idea of Higson explained in Section 6.2.1 below. The details of the proof announced by Julg [Jul02] have not yet been fully written, we refer to [Jul19].

On the other hand, there is a detailed proof of a similar result by Lafforgue [Laf12]: any Gromov hyperbolic group Γ satisfies Conjecture 5 (with coefficients). His proof uses the same idea of arbitrary slow exponential growth representations, see Section 6.2.3 below.

The result of Lafforgue and the announced result of Julg have in common the following important case, namely the case of a cocompact lattice Γ of $Sp(n, 1)$ $(n \geq 2)$ or $F_{4(-20)}$. Note, however, that Lafforgue's result applies to general Gromov hyperbolic groups (many do have property (T)), whereas Julg's claim would apply to all discrete subgroups of $Sp(n, 1)$ $(n \geq 2)$ or $F_{4(-20)}$, including non-cocompact lattices,[19] which also have property (T).

(3) *If G is a simple group of real rank greater or equal to 2*: this is the very difficult case. Actually Lafforgue found that for higher rank Lie groups an obstruction persists: they satisfy a stronger version of property (T), explained in Section 6.3, that prevents the use of representations of arbitrary small exponential growth to succeed (see [Laf08] and [Laf10]). In this case very few are known. The only results are for the cocompact discrete subgroups Γ of a simple Lie group G of rank 2 locally isomorphic to $SL_3(\mathbf{R})$, $SL_3(\mathbf{C})$, $SL_3(\mathbf{H})$, or $E_{6(-26)}$. The proof combines again Lafforgue's result on the Banach analogue of the Baum–Connes conjecture (see Chapter 6), and Jolissaint's (RD) property that we recall in 6.1.4.

5.3 Flag manifolds and KK-theory

Let G be a semisimple Lie group, connected with finite center. Kasparov [Kas84] has made the following remark: Let $P = MAN$ be the minimal parabolic (or Borel) subgroup. The flag manifold G/P is a compact G-space satisfying the following proposition:

Proposition 5.9 *An element of $KK_G(\mathbf{C}, \mathbf{C})$ which is in the image of the map $KK_G(C(G/P), \mathbf{C}) \to KK_G(\mathbf{C}, \mathbf{C})$ and restricts to 1 in $R(K)$ is equal to γ.*

---

[19] A concrete example of a non-cocompact lattice in $Sp(n, 1)$, is $Sp(n, 1)(\mathbf{H}(\mathbf{Z}))$, the group of points of the real algebraic group $Sp(n, 1)$ over the ring $\mathbf{H}(\mathbf{Z})$ of integral quaternions. For such a group Conjecture 5 is still open.

This result follows from the fact that the restriction of γ to the amenable connected Lie group P is equal to 1. Hence $(1-\gamma)KK_G(C(G/P), \mathbf{C}) = 0$.

A stronger statement is used by Julg-Kasparov[JK95] and Julg [Jul19]. Let us compactify the symmetric space $X = G/K$ by adding at infinity the flag manifold G/P. Consider $\bar{X} = G/K \cup G/P$. They prove the following:

Proposition 5.10 *An element of $KK_G(\mathbf{C}, \mathbf{C})$ which is in the image of the map $KK_G(C(\bar{X}), \mathbf{C}) \to KK_G(\mathbf{C}, \mathbf{C})$ and restricts to 1 in $R(K)$ is equal to γ.*

5.3.1 The BGG complex

An important object associated to flag manifolds is the so-called Bernstein–Gelfand–Gelfand (BGG) complex on G/P. The following construction is due to Čap et al. [cSS01].

Lemma 5.11 *The cotangent bundle T^*G/P carries a G-equivariant structure of Lie algebra bundle.*

Proof The group G acts transitively on the flag manifold G/P. Let us consider a point $x \in G/P$. Its stabilizer in G is a parabolic subgroup P_x, a conjugate of P. The tangent space at x is the quotient of Lie algebras $\mathfrak{g}/\mathfrak{p}_x$. The Killing form on G identifies the cotangent space $T_x^*G/P = (\mathfrak{g}/\mathfrak{p}_x)^*$ with the Lie algebra $\mathfrak{n}_x$ of the maximal nilpotent normal subgroup N_x of P_x. The Lie algebras $\mathfrak{n}_x$ form a Lie algebra bundle on G/P, which is, as a vector bundle, G-equivariantly isomorphic to T^*G/P. □

Let $\delta_x : \bigwedge^k \mathfrak{n}_x \to \bigwedge^{k-1} \mathfrak{n}_x$ be the boundary operator defining the *homology* of the Lie algebra $\mathfrak{n}_x$ for each $x \in G/P$. Recall the formula for δ_x:

$$\delta_x(X_1 \wedge \ldots \wedge X_k) = \sum_{i<j} (-1)^{i+j}[X_i, Xi_j] \wedge X_1 \wedge \ldots \wedge \hat{X}_i \wedge \ldots \wedge \hat{X}_j \wedge \ldots \wedge X_k.$$

Transporting δ_x from $\mathfrak{n}_x$ to T_x^*G/P defines a bundle map

$$\delta : \bigwedge^k T^*G/P \to \bigwedge^{k-1} T^*G/P.$$

Let $\Omega = \Omega(G/P)$ be the graded algebra of differential forms on the flag manifold G/P. We consider on Ω the two operators d and δ, respectively of degree 1 and -1. Since d^2 and δ^2 are both zero, the degree zero map $d\delta + \delta d$ commutes both with d and δ. In fact, as proved by Čap and Souček [cS07]:

Proposition 5.12 *The Casimir operator of G acting on Ω is equal to $-2(d\delta + \delta d)$.*

Let Ω_0 be kernel of the Casimir operator in Ω, which is a subcomplex of the de Rham complex.

Theorem 5.13

(1) $\Omega_0 = \ker(d\delta + \delta d) = \ker\delta \cap \ker\delta d$.
(2) The canonical injection $\Omega_0 \to \Omega$ *induces an isomorphism in cohomology.*
(3) The canonical map $\ker\delta \cap \ker\delta d \to \ker\delta/\mathrm{im}\delta$ *is a* G*-equivariant isomorphism from* Ω_0 *to the space of sections of the bundle* $\ker\delta/\mathrm{im}\delta$, *whose fibers are the homology groups* $H_k(\mathfrak{n}_x)$ *of the Lie algebra* $T_x^*G/P = \mathfrak{n}_x$.
(4) The complex D *transported from the complex* d *on* Ω_0 *is a complex of differential operators on the space of sections of the above bundle.*

Remark 5.14 The adjoint action of N_x on $\mathfrak{n}_x$ induces the identity on $H_k(\mathfrak{n}_x)$, a classical fact about Lie algebra homology. Therefore the BGG complex is defined on a space of sections of a bundle on G/P obtained from a representation of P *which is trivial on the nilpotent normal subgroup* N, i.e., factors through $P/N = MA$. In the language of representation theory, it means that the representation involved in the BGG complex are finite sums of (non-unitary) principal series of G.

5.3.2 The model: $SO_0(n, 1)$

Let us now explain Kasparov's proof [Kas88] of the Connes–Kasparov conjecture with coefficients for the Lorentz groups $G = SO_0(2n+1, 1)$. The flag manifold G/P is the sphere S^{2n}, which is the boundary of the hyperbolic space of dimension $2n + 1$. Because the nilpotent group N is abelian, the operator δ is zero and the BGG complex is nothing but the de Rham complex. Kasparov constructs a Fredholm module representing the element γ using the crucial fact that $G/P = S^{2n}$ carries a G-invariant *conformal structure*. Indeed, let us equip the sphere with its K-invariant metric. The action of $g \in G$ transforms the metric into its multiple by some scalar function λ_g^2.

(1) We make the action of G unitary by twisting the representation by a cocycle thanks to the conformal structure. More precisely, let

$$\pi(g)\alpha = \lambda_g^{n-k} g^{-1*}\alpha.$$

The representation π is unitary on the Hilbert space of L^2 forms of degree k.
(2) We make the operator d bounded by considering $F = d(1 + \Delta)^{-1/2}$, where $\Delta = dd^* + d^*d$ is the Laplace–Beltrami operator. The bounded complex F is no more G-invariant, but the natural action of $g \in G$ takes the zero order pseudo-differential operator F to $\lambda_g F$ plus a negative order pseudo-differential operator, as easily seen at the principal symbol level.
(3) Combining the two preceding items (and the fact that F maps k-forms to $(k+1)$-forms) we easily see that the conjugate $\pi(g)F\pi(g)^{-1}$ equals F plus a negative order pseudo-differential operator, hence the compactness of the commutator $[F, \pi(g)]$.

The Fredholm module thus obtained is not quite the good one, since its index is 2 (the Euler characteristic of S^{2n}). To solve the problem, Kasparov cuts the complex in the middle: the group acts on the sphere S^{2n} by conformal transformations and the Hodge $*$ operation on forms of degree n is therefore G-invariant. The half complex consists in taking the forms of degree 0 to $n-1$, and in degree n only half of them, those for which $* = i^n$. Then the index is 1. In the smallest dimension case $n = 1$ ($G = PSL(2, \mathbf{C})$) it amounts to take the $\bar{\partial}$ operator instead of the d operator on $S^2 = P^1(\mathbf{C})$. The G-Fredholm module thus obtained represents the element γ by Proposition 5.10.

In [Kas84], the case of $SO_0(2n, 1)$ was settled as a mere corollary of the case of $SO_0(2n+1, 1)$. Indeed $SO_0(2n, 1)$ is a subgroup of $SO_0(2n+1, 1)$ and the element γ restricts to closed subgroups. However, it was most interesting to treat the case of $SO_0(2n, 1)$ in itself before passing to the other rank one groups. The direct proof for $SO_0(2n, 1)$ has been treated by Chen in his thesis [Che96]. The G-equivariant de Rham complex on S^{2n-1} is again turned, thanks to the conformal structure, into a G-Fredholm module, but this time the index is 0 (the Euler characteristic of S^{2n-1}). To get a Fredholm module of index 1, something new is needed, which has no analogue in the $SO_0(2n+1, 1)$ case. One must use the L^2-cohomology of the hyperbolic space of dimension $2n$, i.e., the Hilbert space $\mathcal{H}^n$ of square-integrable harmonic forms (which are of degree n), which is a sum of two discrete series of G. The truncated module (with index 1) is obtained by considering forms of degree $\leq n-1$, and completing by a map from $\Omega^{n-1}(S^{2n-1})$ to $\mathcal{H}^n$. For $n = 1$, the map $\Omega^0(S^1) \to \mathcal{H}^1$ is just the composition of the classical Poisson transform with the de Rham differential. In general one must use Gaillard's Poisson transform for forms [Gai86]. One thus obtains an element of $KK_G(\mathbf{C}, \mathbf{C})$ which is equal to γ by Proposition 5.10.

5.3.3 Generalization to other rank one groups

The de Rham complex is replaced by the BGG complex on the flag manifold. This is done by Julg and Kasparov in [JK95] for $G = SU(n, 1)$ where the BGG complex is the so-called Rumin complex associated to the G-invariant contact structure on $G/P = S^{2n-1}$, and for $Sp(n, 1)$ or $F_{4(-20)}$ by Julg [Jul19]. In order to turn the BGG complex into a G-Fredholm module, one has to replace, in the above $SO_0(n, 1)$-picture, conformal structure by quasi-conformal structure: the tangent bundle has a G-equivariant subbundle E of codimension 1, 3, or 7 respectively for $G = SU(n, 1)$, $Sp(n, 1)$ or $F_{4(-20)}$, whose fiber E_x at any point $x \in G/P$ is defined as the subspace of $T_x G/P = \mathfrak{n}_x^*$ orthogonal to the subalgebra $[\mathfrak{n}_x, \mathfrak{n}_x]$ of the Lie algebra $\mathfrak{n}_x$. The K-invariant metric is no more conformal, but quasi-conformal in the following sense: consider the action of G on the subbundle E and on the quotient TM/E (note that there is no invariant supplementary subbundle), then under $g \in G$ the metric on E is multiplied by some scalar function λ_g^2, and on the quotient TM/E by λ_g^4. The action of G on forms is not conformal, but after passing to the δ-homology $H_*(\mathfrak{n}_x)$ splits into conformal components. Such a splitting is defined by the weight, i.e., the action of the abelian group $\mathbf{R}_+^*$ seen as a subgroup of the quotient P_x/N_x. This is closely related to the splitting of the representation in the BGG complex into (non-unitary) principal series of G mentioned in Remark 5.14. It follows that one can modify the action of G into a unitary representation $\pi(g)$ on the space of L^2 sections of the BGG complex.

Exploring the analytical properties of the complex D requires to replace the ordinary K-invariant Laplacian by the K-equivariant *sub-Laplacian* on G/P. Namely, $\Delta = -\sum X_i^2$ where the vector fields X_j form at each point $x \in G/P$ an orthonormal basis (for a K-equivariant metric) of the subspace of $T_x G/P = \mathfrak{n}_x^*$ orthogonal to the subalgebra $[\mathfrak{n}_x, \mathfrak{n}_x]$ of $\mathfrak{n}_x$. The operator Δ is not elliptic but hypoelliptic. It has a parametrix which is not a classical pseudo-differential operator, but belongs to a new pseudo-differential calculus in which Fourier analysis is replaced by representation theory of nilpotent Lie groups. Such calculi have been constructed in special cases by Beals and Greiner [BG88] or by Christ et al. [CGGP92]. However, what is needed here is the general construction, which seems to appear only in Melin's 1982 preprint [Mel82], unfortunately very difficult to find. It is worth to mention that noncommutative geometry has motivated a revival of work on the subject, in particular the groupoid approach. The groupoid adapted to the situation has been constructed by various authors: Ponge [Pon06], van Erp and Yuncken [vEY17b], see also [JvE18]. The most beautiful construction of the groupoid using the functoriality of the deformation to the normal cone can be found in the recent thesis of Mohsen [Moh18]. The link between the groupoid and the pseudo-differential calculus is discussed in [DS14] and [vEY17a].

The following theorem explains how to combine the sublaplacian and the weight grading to produce an element of $KK_G(C(G/P), \mathbf{C})$ out of the BGG complex. See [Rum99, Jul19] and [DH17].

Theorem 5.15 *Let $\Delta^{W/2}$ be the pseudo-differential operator equal to the power $\Delta^{w/2}$ on the w weight component of the BGG complex. Then the conjugate $F = \Delta^{W/2} D \Delta^{-W/2}$ is a bounded operator satisfying $F^2 = 0$ on the Hilbert space of L^2 sections of the BGG complex. The commutators $[F, f]$ and $[F, \pi(g)]$ are compact operators for any $f \in C(G/P)$ and $g \in G$. Moreover F admits a parametrix, i.e., a bounded operator Q such that $FQ + QF - 1$, Q^2, as well as $[Q, f]$ and $[Q, \pi(g)]$ for $f \in C(G/P)$ and $g \in G$ are compact operators.*

As above in the $SO_0(2n, 1)$ case, one has to modify the complex in order to get a truncated complex of index 1 in $R(K)$. Then Proposition 5.10 will ensure that its class in $KK_G(\mathbf{C}, \mathbf{C})$ is the element γ. Here again discrete series must be involved, namely those entering the L^2-cohomology of the symmetric space G/K, i.e., the Hilbert space $\mathcal{H}^m$ of harmonic L^2 forms of degree $m = \frac{\dim G/K}{2}$, namely $m = n, n, 2n$, and 8 respectively in the cases of $SO_0(2n, 1)$, $SU(n, 1)$, $Sp(n, 1)$, and $F_{4(-20)}$. Note that the Casimir operator vanishes on $\mathcal{H}^m$ (since the Casimir operator is equal to $-\Delta$ (the Laplace–Beltrami operator) on $L^2(\Omega(G/K))$, allowing to build an adequate Poisson transform [cHJ19] sending the BGG component in degree m to harmonic forms of degree m in G/K. The half complex is then obtained by taking the BGG complex up to degree $m-1$ and to complete by the composition of D with such a Poisson transform [Jul19].

5.3.4 Generalization to higher rank groups

More difficult is the case where G is a simple Lie group of real rank ≥ 2. So far only the case of $SL(3, \mathbf{C})$ has been treated, by Yuncken [Yun11] who has been able to produce a G-Fredholm module representing γ out of the BGG complex. Here the flag manifold G/P, where P is the minimal parabolic of G comes with two G-equivariant fibrations $G/P \rightarrow G/P_i$ ($i = 1, 2$) onto smaller flag manifolds coresponding to P_1 and P_2 the two other parabolics containing P. The operators in the BGG complex turn out to be longitudinally elliptic differential operators along the fibers. Considering a class of pseudo-differential operators on multifiltered manifolds, and making an unexpected use of Kasparov's technical lemma yields a Fredholm module representing an element of $KK_G(C(G/P), \mathbf{C})$. Its index can be taken as 1 in $R(K)$ if one considers the holomorphic BGG complex (as in the $SL(2, \mathbf{C})$ case of [Kas84], where d is replaced by $\bar{\partial}$). Its class in $KK_G(\mathbf{C}, \mathbf{C})$ is therefore γ by Proposition 5.10.

6 Banach algebraic methods

As Julg pointed out in [Jul97], once non-unitary representations appear, one can no longer work with C^*-algebras but with more general topological algebras, for instance, Banach algebras. Unfortunately, Kasparov's KK-theory is a purely C^*-algebraic tool. However, K-theory can be defined for all kind of topological algebras (see the appendix of [Bos90] for the notion of good topological algebras for which the K-theory can be defined); consequently, one has to be able to work in a more flexible framework whose foundations were laid by Lafforgue.

6.1 Lafforgue's approach

6.1.1 Banach KK-theory

In [Laf02b], Lafforgue defined a bi-equivariant KK-theory, denoted by KK^{ban}, for general Banach algebras. The basic idea to start with, was to define a group $KK_G^{\mathrm{ban}}(\mathbf{C}, \mathbf{C})$, in the same way as Kasparov defined $KK_G(\mathbf{C}, \mathbf{C})$, but where unitary representations on Hilbert spaces are replaced by isometric representations on Banach spaces and, therefore, replacing C^*-algebras by Banach algebras. More generally, Lafforgue defined a group $KK_{G,\ell}^{\mathrm{ban}}(\mathbf{C}, \mathbf{C})$ using representations on Banach spaces that are not necessarily isometric but for which the growth is exponentially controlled by a length function on the group G.

Knowing that the trivial representation is not isolated among representations on Banach spaces,[20] Lafforgue was able to prove for a class of groups called $\mathcal{C}'$ in [Laf02b], which is contained in the class $\mathcal{C}$ and hence for which a γ-element has been constructed, that such a γ is equal to 1 in $KK^{\mathrm{ban}}_{G,\ell}(\mathbf{C}, \mathbf{C})$. The class $\mathcal{C}'$ contains:

- semisimple real Lie groups and their closed subgroups;
- simple algebraic groups over non-Archimedean local fields, and their closed subgroups;
- hyperbolic groups.

The equality $\gamma = 1$ in $KK^{\mathrm{ban}}_{G,\ell}(\mathbf{C}, \mathbf{C})$ allowed Lafforgue to prove, for all groups in $\mathcal{C}'$, an analogue of the Baum–Connes conjecture where $C^*(G)$ is replaced by $L^1(G)$, which for general G is a conjecture of Bost. More precisely, Lafforgue used his equivariant KK^{ban} to define a morphism

$$\mu^A_{L^1} : K^{top}_*(G, A) \to K_*(L^1(G, A)),$$

for all locally compact groups G and all $G - C^*$-algebra A.

More precisely, Lafforgue used his equivariant KK^{ban} to define a morphism

$$\mu^A_{L^1} : K^{top}_*(G, A) \to K_*(L^1(G, A)),$$

for all locally compact groups G and all $G - C^*$-algebras A.

Let us recall the important features of Lafforgue's Banach KK-theory that allow one to define the morphism $\mu^A_{L^1}$. If A and B are two Banach algebras endowed with an action of a locally compact group G then there exists a descent map

$$j^{L^1}_G : KK^{\mathrm{ban}}_G(A, B) \to KK^{\mathrm{ban}}(L^1(G, A), L^1(G, B)).$$

Unlike Kasparov's bivariant theory, Banach KK-theory does not have a product but nevertheless, it still acts on K-theory, i.e., there is a morphism

$$K_0(A) \times KK^{\mathrm{ban}}(A, B) \to KK^{\mathrm{ban}}(\mathbf{C}, B)$$

and for every Banach algebra B, the group $KK^{\mathrm{ban}}(\mathbf{C}, B)$ is isomorphic to $K_0(B)$. Consequently, following the Baum–Connes–Higson formulation of the conjecture and hence the construction of the assembly map (see 4.3), one gets, without too much effort, the morphism $\mu^A_{L^1} : K^{top}_*(G, A) \to K_*(L^1(G, A))$.

Let us stress in addition that, unlike Hilbert spaces, Banach spaces are in general not self-dual; so to define the groups $KK^{\mathrm{ban}}(A, B)$ Lafforgue has to replace Hilbert

[20] See the discussion of strong property (T) in 6.3.

modules by pairs of Banach modules together with a duality condition. For details, see Chapter 10 in Valette's book [Val02].

6.1.2 Bost conjecture and unconditional completions

Bost's conjecture (with coefficients) is stated as follows

Conjecture 11 (Bost) For all locally compact groups G and for all $G - C^*$-algebras A the map $\mu_{L^1}^A$ is an isomorphism.

The moral is that when using representations with controlled growth on Banach spaces to construct a homotopy between a γ-element and 1, as one gets out of the C^*-algebraic context, the K-theory that we are able to compute is the K-theory of a Banach algebra. In the case of the Bost conjecture, the Banach algebra that we get is $L^1(G)$.

There are two good things about the Bost conjecture, the first one is that it is easier to prove than the Baum–Connes conjecture (meaning that it has been proven by Lafforgue for a wide class of groups containing all semisimple Lie groups as well as their lattices) and no counter-example to the Bost conjecture is known, to the best of our knowledge. Secondly, the original map $\mu_{A,r}$ of Baum–Connes–Higson factors through the map $\mu_{L^1}^B$ so that the following diagram is commutative:

$$\begin{array}{ccc} K_*^{\mathrm{top}}(G,B) & \xrightarrow{\mu_r^B} & K_*(C_r^*(G,B)) \\ & \searrow^{\mu_{L^1}^B} & \uparrow \iota_* \\ & & K_*(L^1(G,B)), \end{array}$$

where ι denotes inclusion $L^1(G, B) \to C_r^*(G, B)$. Therefore, if we take G to be a group for which the Bost conjecture has been proven, for example, a semisimple Lie group or a lattice in such a group, trying to prove the Baum–Connes conjecture for G amounts to prove that ι_* is an isomorphism, in other words that ι induces an isomorphism in K-theory.

Unfortunately, the usual criteria to prove that the continuous inclusion of $L^1(G)$ in $C_r^*(G)$ induces an isomorphism in K-theory, is not true for most of the locally compact groups. For example, the algebra $L^1(G)$ is not stable under holomorphic calculus if G is a non-compact semisimple Lie group [LP79]. To illustrate this, let us recall the usual criterion to determine whether an injective morphism of Banach algebras induces an isomorphism at the level of their K-theory groups.

Definition 6.1 Let A and B be two *unital* Banach algebras and $\phi : A \to B$ is a morphism of Banach algebras between them. Then ϕ is called spectral, if for every $x \in A$ the spectrum of x in A equals the spectrum of $\phi(x)$ in B. It is called dense if $\phi(A)$ is dense in B.

This terminology is taken from [Nic08]. When ϕ is injective, A can be considered as a subalgebra of B. In this case, A is said to be *"stable under holomorphic calculus in B,"* because, for every $x \in A$ and every holomorphic function f on a neighborhood of the spectrum of x in B, the element $f(x)$ constructed using holomorphic functional calculus in B belongs to A (see [Bos90]).

The theorem below is a classical result known as the *Density Theorem*; it is due to Swan and Karoubi (see [Swa77, Section 2.2 and 3.1], [Kar08, p. 109], [Con81, Appendix 3] and [Bos90, Théorème A.2.1]).

Theorem 6.2 *If A and B are two unital Banach algebras and $\phi : A \to B$ is dense and spectral morphism of Banach algebras then ϕ induces an isomorphism $\phi_* : K_*(A) \to K_*(B)$.*

What Bost noticed is that the condition of been spectral is, somehow, too strong: if ϕ is spectral it induces *strong isomorphisms* in K-theory:

Definition 6.3 ([Bos90]) An injective morphism between two unital Banach algebras $\phi : A \to B$, induces a strong isomorphism in K-theory if for every $n \geq 1$ the maps

$$M_n(\phi) : P_n(A) \to P_n(B) \quad \text{and} \quad GL_n(\phi) : GL_n(A) \to GL_n(B),$$

induced by ϕ are homotopy equivalences.

Here for an algebra A and for an integer n, we denote by $M_n(A)$ the set of $n \times n$ matrices with coefficients in A and $P_n(A) = \{p \in M_n(A) \mid p^2 = p\}$ is the set of idempotent matrices.

If the maps $M_n(\phi)$ and $GL_n(\phi)$ above are homotopy equivalences then the morphism induced by ϕ, say $\phi_* : \mathcal{P}(A) \to \mathcal{P}(B)$ is an isomorphism (where $\mathcal{P}(A)$ denotes the semi-group of isomorphism classes of projective A-modules of finite type). This is stronger than inducing an isomorphism in K-theory.

The next example shows that, because of this strength, it is not easy to pass from Bost conjecture to the Baum–Connes conjecture.

Example 6.4 Set $G = SL_2(\mathbf{R})$. Then G has two representations in the discrete series (i.e., square-integrable representations) that are not integrable (i.e., the matrix coefficients do not belong to $L^1(G)$) and are therefore not isolated in the dual space of $L^1(SL_2(\mathbf{R}))$. This implies that the idempotent of $C_r^*(SL_2(\mathbf{R}))$ associated to one of these discrete series (which is equal to a matrix coefficient) does not belong to $L^1(SL_2(\mathbf{R}))$; hence, if π_0 denotes the set of connected components, the map from $\pi_0\Big(P_1(L^1(SL_2(\mathbf{R}))\Big) \to \pi_0\Big(P_1(C_r^*(SL_2(\mathbf{R}))\Big)$, which is induced by ι, is not surjective. Therefore applying what is known for the Bost conjecture to the Baum–Connes conjecture is not in any case automatic.

Fortunately, Lafforgue's proof of the Bost conjecture can actually be used to compute the K-theory of a class of Banach algebras more general than $L^1(G)$ called *unconditional completions* of $C_c(G)$.

Definition 6.5 Let G be a locally compact group. A Banach algebra completion $\mathcal{B}(G)$ of $C_c(G)$ is called *unconditional* if the norm $\|f\|_{\mathcal{B}(G)}$ only depends on the map $g \mapsto |f(g)|$, i.e., for $f_1, f_2 \in C_c(G)$, $\|f_1\|_{\mathcal{B}(G)} \leq \|f_2\|_{\mathcal{B}(G)}$ if $|f_1(g)| \leq |f_2(g)|$ for all $g \in G$.

Example 6.6 For a locally compact group G, the algebra $L^1(G)$ is an unconditional completion of $C_c(G)$.

Example 6.7 If G is a connected semisimple Lie group and K is a maximal compact subgroup, let $t \in \mathbf{R}^+$ and let $\mathcal{S}_t(G)$ be the completion of $C_c(G)$ for the norm given by:

$$\|f\|_{\mathcal{S}_t(G)} = \sup_{g \in G} |f(g)|\phi(g)^{-1}(1 + d(g))^t,$$

where ϕ is the Harish-Chandra function on G (see Chapter 4 in [Kna01]) and for $g \in G$, $d(g)$ is the distance of gK to the origin in G/K. Then, for t large enough, $\mathcal{S}_t(G)$ is an unconditional completion (see Section 4 in [Laf02b]).

Another important example of unconditional completions appears in connexion with the rapid decay property, to be discussed in Section 6.1.4 below.

Inspired by the definitions of the algebras $L^1(G, A)$, one can define analogues of crossed products in the context of Banach algebras using unconditional completions as follows: if A is a $G - C^*$-algebra and $\mathcal{B}(G)$ is an unconditional completion of $C_c(G)$, we define the algebra $\mathcal{B}(G, A)$ as the completion of $C_c(G, A)$ for the norm

$$\|f\|_{\mathcal{B}(G,A)} = \left\| g \mapsto \|f(g)\|_A \right\|_{\mathcal{A}(G)}.$$

For all locally compact group G, all $G - C^*$-algebra A and all unconditional completions $\mathcal{B}(G)$ Lafforgue used his Banach KK-theory to construct a morphism

$$\mu^A_{\mathcal{B}(G)} : K^{top}_*(G, A) \to K_*(\mathcal{B}(G, A)).$$

He then obtained an analogue of the "Dirac–dual-Dirac method" in this context:

Theorem 6.8 (Lafforgue) *If the group G has a γ-element in $KK_G(\mathbf{C}, \mathbf{C})$ and if there exists a length function ℓ on G, such that, for all $s > 0$, $\gamma = 1$ in $KK^{\mathrm{ban}}_{G,s\ell}(\mathbf{C}, \mathbf{C})$, then $\mu^A_{\mathcal{B}(G)}$ is an isomorphism for all unconditional completions $\mathcal{B}(G)$ and for all G-algebras A.*

Lafforgue proved the equality $\gamma = 1$ in $KK^{\mathrm{ban}}_{G,s\ell}(\mathbf{C}, \mathbf{C})$ for all groups in the class $\mathcal{C}'$ (see [Laf02b, Introduction]). All real semisimple Lie groups and all p-adic reductive Lie groups as well as their closed subgroups, all discrete groups acting properly, cocompactly, continuously and by isometries on a $CAT(0)$ space and all hyperbolic groups belong to this class. For all these groups G and all G-algebras A the map $\mu^A_{\mathcal{B}(G)}$ is an isomorphism and hence the Bost conjecture holds (see [Laf02b]). For a nice expository explanation on how the homotopy between γ and

1 is constructed using Banach representations, see [Ska02] where the combinatorial case is explained in details, i.e., the case containing p-adic groups.

6.1.3 Application to the Baum–Connes conjecture

Let $\mathcal{B}(G)$ be an unconditional completion of $C_c(G)$ that embeds in $C_r^*(G)$. In that case, the Baum–Connes map μ_r factors through $\mu_{\mathcal{B}(G)}$ so that the following diagram is commutative:

$$\begin{array}{ccc} K_*^{\text{top}}(G) & \xrightarrow{\mu_{\mathcal{B}(G)}} & K_*(\mathcal{B}(G)) \\ & \searrow^{\mu_r} & \downarrow i_* \\ & & K_*(C_r^*(G)) \end{array},$$

where i_* is the inclusion map induced by the map $i : \mathcal{B}(G) \to C_r^*(G)$.

Proposition 6.9 *Let G be a group in Lafforgue's class $\mathcal{C}'$. Suppose there exists an unconditional completion $\mathcal{B}(G)$ which is a dense subalgebra stable under holomorphic calculus in $C_r^*(G)$. Then the Baum–Connes assembly map μ_r is an isomorphism.*

Using Example 6.7, we can state the first result of Lafforgue concerning connected Lie groups (see the discussion in Section 5.2 regarding those groups)

Theorem 6.10 (Lafforgue) *Let G be a connected semisimple Lie group. Then Conjecture 4 (without coefficients) is true for G.*

Proof For $t \in \mathbf{R}^+$ large enough, the algebra $\mathcal{S}^t(G)$ from Example 6.7 is an unconditional completion which is dense and stable under holomorphic calculus in $C_r^*(G)$ (cf. Section 4 in [Laf02b]).

□

As a matter of fact, Lafforgue's theorem is much more general. Let G be a locally compact group. A quadruplet (G, K, d, ϕ) is a *Harish-Chandra quadruplet* if G is unimodular with Haar measure denoted by dg, K is a compact subgroup endowed with his unique Haar measure of mass equal to 1, d is a length function on G such that $d(k) = 0$ for all $k \in K$ and $d(g^{-1}) = d(g)$ for all $g \in G$ and $\phi : G \to]0, 1]$ is a continuous function satisfying the following 5 properties:

1. $\phi(1) = 1$,
2. $\forall g \in G, \phi(g^{-1}) = \phi(g)$,
3. $\forall g \in G, \forall k, k' \in K, \phi(kgk') = \phi(g)$,
4. $\forall g, g' \in G, \int_K \phi(gkg')dk = \phi(g)\phi(g')$,
5. for all $t \in \mathbf{R}_+$ large enough, $\int_G \phi^2(g)(1 + d(g))^{-t} dg$ converges.

When one has a Harish-Chandra quadruplet, then one can define a Schwartz space on G following Example 6.7 : $\mathcal{S}_t(G)$ is the Banach space completion of $C_c(G)$ with respect to the norm given by

$$\|f\|_{\mathcal{S}_t(G)} = \sup_{g\in G} |f(g)|\phi(g)^{-1}(1+d(g))^t.$$

Lafforgue's result is then stated a follows:

Proposition 6.11 *Let (G, K, d, ϕ) be a Harish-Chandra quadruplet. Then, for $t \in \mathbf{R}_+$ large enough, $\mathcal{S}_t(G)$ is an unconditional completion of $C_c(G)$ which is a subalgebra of $C_r^*(G)$ dense and stable under holomorphic calculus.*

In Section 4 of [Laf02b], Lafforgue constructed a Harish-Chandra quadruplet for all linear reductive Lie groups on local fields.

Remark 6.12 The method of finding a Schwartz type unconditional completion dense and stable under holomorphic calculus in $C_r^*(G)$ like the algebra $\mathcal{S}_t(G)$ for semisimple Lie groups, does not work with coefficients (see the remark after the Proposition 4.8.2 of [Laf02b]). If Γ is a lattice in a semisimple Lie group G, we can define an algebra $\mathcal{S}_t(\Gamma)$ in the same manner as for G: it is the completion of $C_c(\Gamma)$ for the norm $\|f\|_{\mathcal{S}^t(\Gamma)} = \sup_{\gamma\in\Gamma} |f(\gamma)|(1+d(\gamma))^t\phi(\gamma)^{-1}$, where ϕ is the Harish-Chandra function of G and the d is the appropriate distance in G (see [Boy17] where this algebras are studied). Suppose now that Γ is cocompact. Then $\mathcal{S}_t(G, C(G/\Gamma))$ is not stable under holomorphic calculus in $C_r^*(G, C(G/\Gamma))$ as these algebras are Morita equivalent to $\mathcal{S}_t(\Gamma)$ and $C_r^*(\Gamma)$, respectively, and $\mathcal{S}_t(\Gamma)$ is not stable under holomorphic calculus in $C_r^*(\Gamma)$. Indeed, if $\gamma \in \Gamma$ is a hyperbolic element, since $d(\gamma^n)$ grows linearly in n if we denote by e_γ the corresponding Dirac function in $\mathbf{C}\Gamma$, its spectral radius as an element of $C_r^*(\Gamma)$ is 1 whereas its spectral radius in $\mathcal{S}_t(\Gamma)$ is > 1. To see this we use the following classical estimate on the Harish-Chandra ϕ-function (see Proposition 7.15 in [Vog81]): there are positive constants $C, \ell > 0$ such that for every $g \in G$:

$$\phi(g) \leq Ce^{-d(g)}(1+d(g))^\ell.$$

Hence

$$\|e_\gamma^n\|_{\mathcal{S}_t(\Gamma)} = \frac{(1+d(\gamma^n))^t}{\phi(\gamma^n)} \geq C^{-1}(1+d(\gamma^n))^{t-\ell}e^{d(\gamma^n)}.$$

Since $d(\gamma^n)$ grows linearly in n, we have for the spectral radius of e_γ in $\mathcal{S}_t(\Gamma)$:

$$\lim_{n\to\infty} \|e_\gamma^n\|_{\mathcal{S}_t(\Gamma)}^{1/n} > 1.$$

6.1.4 The rapid decay property

To state Lafforgue's results concerning lattices in connected Lie groups, and hence examples of discrete groups having property (T) and verifying the Baum–Connes conjecture (without coefficients), we need to introduce the property of rapid decay, denoted by (RD).

Recall that, for G a locally compact group, a continuous function $\ell : G \to \mathbf{R}^+$ is a *length function* if $\ell(1) = 0$ and $\ell(gh) \leq \ell(g) + \ell(h)$ for every $g, h \in G$.

Example 6.13 If Γ is a finitely generated group and S is a finite generating subset, then $\ell(g) = |g|_S$ (word length with respect to S) defines a length function on Γ.

The following definition is due to Jolissaint [Jol90].

Definition 6.14 Il ℓ is a length function on the locally compact group G, we say that G has the property of rapid decay with respect to ℓ (abridged property (RD)) if there exist positive constants C, k such that, for every $f \in C_c(G)$:

$$\|\lambda(f)\| \leq C \cdot \|f(1+\ell)^k\|_2.$$

In other words the norm of f in $C_r^*(G)$, i.e., the operator norm of f as a convolutor on $L^2(G)$, is bounded above by a weighted L^2-norm given by a polynomial in the length function.

The relevance of property (RD) regarding Baum–Connes comes from the following fact: If Γ is a discrete group with property (RD) with respect to a length function ℓ, then, for a real number s which is large enough, the space

$$H_\ell^s(\Gamma) = \left\{ f : \Gamma \to \mathbf{C} \mid \|f\|_{\ell,s} = \left(\sum_{\gamma \in \Gamma} |f(\gamma)|^2 (1+\ell(\gamma))^{2s} \right)^{\frac{1}{2}} < \infty \right\}$$

is a convolution algebra and an unconditional completion of $C_c(\Gamma)$ that is stable under holomorphic calculus in $C_r^*(\Gamma)$ (see, for example, [Val02], 8.15, Example 10.5). Note that functions in $H_\ell^s(\Gamma)$, with $s \gg 0$, are decaying fast at infinity on Γ, hence the name *rapid decay*.

We can now state Lafforgue's result concerning discrete groups (Corollaire 0.0.4 in [Laf02b]):

Theorem 6.15 *Let Γ be a group with property (RD) in Lafforgue's class $\mathcal{C}'$ (see Section 6.1.1). Then Conjecture 4 (without coefficients) for Γ is true.*

Jolissaint [Jol90] has shown that property (RD) holds for cocompact lattices in real rank 1 groups, a fact generalized in two directions:

- by de la Harpe [dlH88] to all Gromov hyperbolic groups;
- by Chatterji and Ruane [CR05] to *all* lattices in real rank 1 groups.

By Theorem 6.15, those groups do satisfy the Baum–Connes conjecture (without coefficients).

Remark 6.16 The first spectacular application of property (RD) was the proof of the Novikov conjecture for Gromov hyperbolic groups by Connes and Moscovici [CM90]. Their result is the following:

Theorem 6.17 *Assume that the group Γ satisfies both Jolissaint's (RD) property and the bounded cohomology property (i.e., that any group n-cocycle is cohomologous to a bounded one, for $n \geq 2$). Then Γ satisfies the Novikov conjecture.*

Sketch of proof Let $x \in H^n(\Gamma, \mathbf{Q})$ be a cohomology class. Let M be a closed, Spin manifold and $f : M \to B\Gamma$ a map; let $\tilde{M}$ be the pullback of $\widetilde{B\Gamma}$ via f. Let D be a Γ-invariant Dirac operator on $\tilde{M}$. Connes and Moscovici show that the index of D in $K_0(C_r^*(\Gamma))$ has a more refined version living in $K_0(\mathbf{C}(\Gamma) \otimes \mathcal{R})$, where $\mathcal{R}$ is the algebra of smoothing operators. They deduce a cohomological formula for the higher signature $\sigma_x(M, f)$ (defined in Section 2.5) by evaluating a cyclic cocycle τ_x associated with x on the index in $K_0(\mathbf{C}(\Gamma)\otimes\mathcal{R})$. The two assumptions of Theorem 6.17 ensure that the cocycle τ_x extends from $\mathbf{C}(\Gamma) \otimes \mathcal{R}$ to a subalgebra of the C^*-algebra $C_r^*(\Gamma) \otimes \mathcal{K}$ which is stable under holomorphic functional calculus. Therefore $\sigma_x(M, f)$ only depends on the index $\mu_r(f_*[D]) \in K_0(C_r^*(\Gamma))$, which is a homotopy invariant by Theorem 2.8. The hypothesis in the theorem holds in particular for Gromov's hyperbolic groups: the fact that they do satisfy the bounded cohomology property is a result stated by Gromov and proved by Mineyev [Min01]. □

In higher rank it can be proved that non-cocompact lattices do not satisfy property (RD). However, we have a conjecture by Valette (see p.74 in [FRR95]):

Conjecture 12 Let Γ be a group acting properly, isometrically, with compact quotient, either on a Riemannian symmetric space or on a Bruhat–Tits building. Then Γ has the (RD) property.

Valette's conjecture holds in higher rank for the following special cases, all in rank 2: assume G is locally isomorphic to $SL_3(\mathbf{R})$ or $SL_3(\mathbf{C})$: Lafforgue has shown that any cocompact lattice Γ of G satisfies property (RD). Chatterji has generalized this proof to $SL_3(\mathbf{H})$ and $E_{6(-26)}$, see [Cha03]. Their proofs are based on ideas of Ramagge, Robertson, and Steger for $SL_3(\mathbf{Q}_p)$ ([RRS98]). Conjecture 4 therefore follows for such lattices. As mentioned in Section 5.2 above, this gave the first examples of infinite discrete groups having property (T) and satisfying the Baum–Connes conjecture.

It is not known whether such a group Γ (or the Lie group G itself) satisfies the conjecture with coefficients. Moreover, nothing is known about the Baum–Connes conjecture for general discrete subgroups of G. In particular it is not known whether $SL_3(\mathbf{Z})$ satisfies Conjecture 4, or similarly whether $SL_3(\mathbf{R})$ satisfies Conjecture 5.

On the other hand, regarding lattices in another real rank 2 simple Lie group (e.g., the symplectic group $Sp_4(\mathbf{R})$), or in a simple group with real rank at least 3, absolutely nothing is known, in particular for lattices in $SL_n(\mathbf{R})$ or $SL_n(\mathbf{C})$ when $n \geq 4$.

Remark 6.18 The group $\Gamma = SL_3(\mathbf{Z})$ does not have property (RD) (see [Jol90]). Moreover, there is no unconditional completion $\mathcal{B}(\Gamma)$ that is a dense subalgebra of $C_r^*(\Gamma)$ stable under holomorphic calculus. The following argument is due to Lafforgue (see [Laf10]). Let us consider the action of $\mathbf{Z}$ on $\mathbf{Z}^2$ induced by the map from $\mathbf{Z}$ to $\mathbf{Z}^2$ that sends $n \in \mathbf{Z}$ to $\begin{pmatrix} 3 & 1 \\ 2 & 1 \end{pmatrix}^n$ and the semi-direct product $H := \mathbf{Z} \rtimes \mathbf{Z}^2$ constructed using this action. The group H is solvable, hence amenable, and can be embedded as a subgroup of $SL_3(\mathbf{Z})$ using the map: $\left(n, \begin{pmatrix} a \\ b \end{pmatrix}\right) \mapsto \begin{pmatrix} \begin{pmatrix} 3 & 1 \\ 2 & 1 \end{pmatrix}^n & \begin{pmatrix} a \\ b \end{pmatrix} \\ 0 & 1 \end{pmatrix}$.

Suppose by contradiction that there is an unconditional completion $\mathcal{B}(G)$ that is a subalgebra of $C_r^*(G)$. Then the algebra $\mathcal{B}(H) = \mathcal{B}(G) \cap C_r^*(H)$ is contained in $\ell^1(H)$ because as H is amenable, for every non-negative function f on H, one has $\|f\|_{C_r^*(H)} = \|f\|_{L^1(H)}$. However, $\ell^1(H)$ is not spectral in $C_r^*(H)$ (see [Jen69]).

6.2 *Back to Hilbert spaces*

The motto of this section is the following: in the case where property (T) imposes that $\gamma \neq 1$ in $KK_G(\mathbf{C}, \mathbf{C})$, the idea for showing that γ nevertheless acts by the identity in the K-theory groups $K_*(C_r^*(G, A))$ is to make the γ-element homotopic to the trivial representation in a weaker sense, getting out of the class of unitary representations, but staying within the framework of Hilbert spaces.

6.2.1 Uniformly bounded and slow growth representations

The idea of using uniformly bounded representations is a remark that Julg made in 1994. A uniformly bounded representation of a locally group G is a strongly continuous representation by bounded operators on a Hilbert space H, such that there is a constant C with $\|\pi(g)\| \leq C$ for any $g \in G$. Equivalently, it is a representation by isometries for a Banach norm equivalent to a Hilbert norm.

Following Kasparov [Kas95], let us denote $R(G) = KK_G(\mathbf{C}, \mathbf{C})$. Let $R_{\mathrm{ub}}(G)$ be the group of homotopy classes of G-Fredholm modules, with uniformly bounded representations replacing unitary representations, as in [Jul97].

Proposition 6.19 *For any $G - C^*$-algebra A, the Kasparov map*

$$R(G) \to \mathrm{End} K_* \left(C_r^*(G, A)\right)$$

factors through the map $R(G) \to R_{\mathrm{ub}}(G)$.

This follows from an easy generalization of the classical Fell lemma: indeed, if π is a uniformly bounded representation of a group G in a Hilbert space H, and λ is the left regular representation of G on $L^2(G)$, there exists a bounded invertible operator U on $H \otimes L^2(G)$, such that

$$\pi(g) \otimes \lambda(g) = U(1 \otimes \lambda(g))U^{-1},$$

when π is a unitary representation, U is of course a unitary operator.

To any Hilbert space H equipped with a uniformly bounded representation π, let us associate as in the construction of the map $j_{G,r}$ from [Kas95, Kas88], the Hilbert module $E = H \otimes C_r^*(G, A)$ and the covariant representation of (G, A) with values in $\mathcal{L}_{C_r^*(G,A)}(E)$ defined by:

$$a \mapsto 1 \otimes a, \;\; g \mapsto \pi(g) \otimes \lambda(g).$$

Then the representation $\pi_A : C_c(G, A) \to \mathcal{L}_{C_r^*(G,A)}(E)$ extending the above covariant representation factors through the reduced crossed product $C_r^*(G, A)$.

To a G-Fredholm module (H, π, T) we can therefore associate the triple $(H \otimes C_r^*(G, A), \pi_A, T_A)$, where $\pi_A : C_r^*(G, A) \to \mathcal{L}_{C^*_{\mathrm{red}}(G,A)}(E)$ is the Banach algebra homomorphism defined above, and $T_A = T \otimes 1 \in \mathcal{L}_{C^*_{\mathrm{red}}(G,A)}(E)$. The Banach G-Fredholm module thus obtained defines a map from the group $K_*(C^*_{\mathrm{red}}(G, A))$ to itself. Note that such a construction has no analogue for $C^*_{\max}(G, A)$ since it relies upon a specific feature of the regular representation.

As in the case of Lafforgue's Banach representation, it often happens that a family of representations can be deformed to a representation containing the trivial representation, but with a uniform boundedness constant tending to infinity. One must therefore use a more general class, as we now explain. Fix $\varepsilon > 0$. Let l be a length function on G.

Definition 6.20 We say that a representation π of G is *of ε-exponential type* if there is a constant C such that for any $g \in G$,

$$\|\pi(g)\| \le Ce^{\varepsilon l(g)}.$$

The following ideas come from a discussion between Higson, Julg, and Lafforgue in 1999. We define as above a G-Fredholm module of ε-exponential type, and similarly a homotopy of such modules. Let $R_\varepsilon(G)$ be the abelian group of homotopy classes. The obvious maps $R_\varepsilon(G) \to R_{\varepsilon'}(G)$ for $\varepsilon < \varepsilon'$ form a projective system and we consider the projective limit $\varprojlim R_\varepsilon(G)$ when $\varepsilon \to 0$.

We would like to have an analogue of the above proposition with the group $\varprojlim R_\varepsilon(G)$ instead of $R_{\mathrm{ub}}(G)$. In fact there is a slightly weaker result, due to Higson and Lafforgue (cf. [Laf12] Théorème 2.3) which is enough for our purpose. We assume now that G is a connected Lie group.

Theorem 6.21 *The kernel of the map*

$$R(G) \to \varprojlim R_\varepsilon(G)$$

is included in the kernel of the map

$$R(G) \to \mathrm{End} K_* \left(C_r^*(G, A)\right).$$

Let us sketch the proof following [Laf12]. As above, to any representation π of G is associated an algebra homomorphism

$$\pi_A : C_c(G, A) \to \mathcal{L}_{C_r^*(G,A)}(E)$$

where $E = H \otimes C^*_{\mathrm{red}}(G, A)$.

For all $\varepsilon > 0$ there is a Banach algebra C_ε which is a completion of $C_c(G, A)$ such that for any representation π of ε-exponential type, the above map π_A extends to a bounded homomorphism $C_\varepsilon \to \mathcal{L}_{C_r^*(G,A)}(E)$. The Banach Fredholm module thus obtained defines a map

$$R_\varepsilon(G) \to \mathrm{Hom}\left(K_*(C_\varepsilon), K_* \left(C_r^*(G, A)\right)\right).$$

This being done for each ε, we have a system of maps compatible with the maps $C_\varepsilon \to C_{\varepsilon'}$ for $\varepsilon' < \varepsilon$, so that there is a commutative diagram (cf. [Laf12] prop 2.5)

$$\begin{array}{ccc} R(G) & \longrightarrow & \mathrm{End} K_*(C_r^*(G, A)) \\ \downarrow & & \downarrow \\ \varprojlim R_\varepsilon(G) & \longrightarrow & \varprojlim \mathrm{Hom}(K_*(C_\varepsilon), K_*(C_r^*(G, A))). \end{array}$$

The theorem of Higson–Lafforgue then follows immediately, thanks to the following lemma:

Lemma 6.22 *The group $K_*(C_r^*(G, A))$ is the union of the images of the maps $K_*(C_\varepsilon) \to K_*(C_r^*(G, A))$.*

To prove the lemma, Higson and Lafforgue use the fact that the symmetric space $Z = G/K$ has finite asymptotic dimension. They give an estimate of the form (prop 2.6 in [Laf12])

$$\|f\|_{C_\varepsilon} \leq k_\varepsilon e^{\varepsilon(ar+b)} \|f\|_{C_r^*(G,A)}$$

for $f \in C_c(G, A)$ with support in a ball of radius r (for the length l).

The spectral radius formula in Banach algebras then implies for such an f,

$$\rho_{C_\varepsilon}(f) \le e^{\varepsilon a r} \rho_{C_r^*(G,A)}(f),$$

so that $\rho_{C_r^*(G,A)}(f) = \inf \rho_{C_\varepsilon}(f)$. This fact, by standard holomorphic calculus techniques, implies the lemma.

6.2.2 Cowling representations and γ

The beautiful work of Cowling and Haagerup on completely bounded multipliers of the Fourier algebras for rank one simple Lie groups [CH89] inspired Julg to use Cowling's strip of uniformly bounded representations to prove the Baum–Connes conjecture for such groups. Consider the Hilbert space $L^2(G/P)$ associated to a K-invariant measure on the flag manifold G/P. Let π_1 be the natural action of G, i.e., $\pi_1(g)f = f \circ g^{-1}$, and let π_0 be the unitary representation obtained by twisting π_1 by a suitable cocycle: $\pi_0(g) = \lambda_g^{N/2}\pi_1(g)$. One can interpolate between π_0 and π_1 by taking

$$\pi_s(g) = \lambda_g^{\frac{(1-s)N}{2}} \pi_1(g),$$

with s being a complex number. The result of Cowling [Cow82, ACDB04] is the following:

Theorem 6.23 *The representation* $(1+\Delta_E)^{(1-s)N/4}\pi_s(g)(1+\Delta_E)^{-(1-s)N/4}$ *is uniformly bounded for any* s *in the strip* $-1 < \Re s < 1$.

In particular this holds for $-1 < s < 1$. The important point is to compare with Kostant's result on the unitarizability of π_s. The representations π_s are by construction unitary if $\Re s = 0$. Otherwise they are unitarizable (i.e., admit an intertwining operator T_s such that $T_s^{-1}\pi_s(g)T_s$ is unitary) if and only if $-c < s < c$ for a certain $c \le 1$. This is the so-called complementary series. The critical value s is as follows:

(1) If $G = SO_0(n, 1)$ or $SU(n, 1)$, $c = 1$.
(2) If $G = Sp(n, 1)$, $c = \frac{2n-1}{2n+1}$.
(3) If $G = F_{4(-20)}$, $c = \frac{5}{11}$.

In case 1, G has the Haagerup property, and the complementary series approaches the trivial representation. In cases 2 and 3 one has $c < 1$ so that there is a gap between the complementary series and the trivial representation, as expected from property (T).

The above family π_s $(0 \le s < 1)$ and its generalizations to the other principal series are the tool for constructing a homotopy between γ and 1. Indeed the proofs of $\gamma = 1$ by Kasparov [Kas84], Chen [Che96] and Julg-Kasparov [JK95] rest upon the complementary series. In the general case, Julg [Jul19] constructs a similar homotopy involving Cowling uniformly bounded representations. Modulo some (not yet fully clarified) estimates, that would prove that γ is 1 in $R_\varepsilon(G)$ for all $\varepsilon > 0$ (with the above notations).

6.2.3 Lafforgue's result for hyperbolic groups

In 2012, in a very long and deep paper, Vincent Lafforgue has proved the following result.

Theorem 6.24 *Let G be a word hyperbolic group. Then G satisfies the Baum–Connes conjecture with coefficients (Conjecture 5).*

Remark 6.25 Lafforgue proves more generally the same result for G a locally compact group acting continuously, isometrically and properly on a metric space X which is hyperbolic, weakly geodesic and uniformly locally finite.

Let us sketch the main steps of Lafforgue's proof. The basic geometric object is the Rips complex $\Delta = P_R(G)$ of the group G seen as a metric space with respect to the word metric d_S associated with a set of generators S.

Definition 6.26 Let Y be a locally finite metric space (i.e., every ball in Y is finite). Fix $R \geq 0$. The *Rips complex* $P_R(Y)$ is the simplicial complex with vertex set Y, such that a subset F with $(n+1)$-elements spans a n-simplex if and only if $diam(F) \leq R$.

Because G is hyperbolic, one can choose the radius R big enough so that Δ is contractible. Let ∂ be the coboundary

$$\mathbf{C}[\Delta^0] \leftarrow \mathbf{C}[\Delta^1] \leftarrow \mathbf{C}[\Delta^2] \leftarrow \dots$$

of the Rips complex. Let us recall the formula for ∂:

$$\partial\delta_{g_0,g_1,\dots,g_k} = \sum_{i=0}^{k}(-1)^i\delta_{g_0,\dots,\hat{g}_i,\dots,g_k}$$

Contractibility of the Rips complex implies that the homology of the complex ∂ is zero in all degrees, except in degree 0 where it is one-dimensional. But a concrete contraction onto the origin x_0 of the graph gives rise to a parametrix, i.e., maps $h : \mathbf{C}[\Delta^k] \to \mathbf{C}[\Delta^{k+1}]$ such that $\partial h + h\partial = 1$ (except in degree zero where it is $1 - p_{x_0}$ where p_0 has image in $\mathbf{C}\delta_{x_0}$) and $h^2 = 0$. The prototype is the case of a tree, where $h\delta_x = \sum \delta_e$, the sum being extended to the edges on the geodesic from x_0 to x. The case of a hyperbolic group is more subtle, and the construction of h has to involve some averaging over geodesics. Suitable parametrices have been considered by Lafforgue in the Banach framework.

Kasparov and Skandalis in [KS91] have shown that hyperbolic groups admit a γ-element which can be represented by an operator on the space $\ell^2(\Delta)$. Lafforgue considers the following variant of the Kasparov–Skandalis construction. Let us conjugate the operator $\partial + h$ by a suitable function of the form $e^{t\rho}$, where ρ is the (suitably averaged) distance function to the point x_0. Then for t big enough, the operator $e^{t\rho}(\partial + h)e^{-t\rho}$, on the Hilbert space $\ell^2(\Delta)$ equipped with the even/odd grading and the natural representation π of G, represents the γ-element.

Lafforgue's *tour de force* is to modify the construction of the operator h and to construct Hilbert norms $\|.\|_\varepsilon$ on $\mathbf{C}[\Delta]$ such that the operators $e^{t\rho}(\partial+h)e^{-t\rho}$ become a homotopy between γ (for t big) and 1 (for $t = 0$), this homotopy being through ε-exponential representations. Let us give the precise statement:

Theorem 6.27 *Let G be a word hyperbolic group; let Δ and ∂ be as above. Fix $\varepsilon > 0$. There exists a suitable parametrix h satisfying the conditions above, a Hilbert completion H_ε of the space $\mathbf{C}[\Delta]$, and a distance function d on G differing from d_S by a bounded function such that:*

1. *the operator $F_t = e^{t\rho}(\partial + h)e^{-t\rho}$ (where ρ is the distance to the origin x_0) extends to a bounded operator on H_ε for any t,*
2. *the representation π of G extends to a representation on H_ε with estimates $\|\pi(g)\|_\varepsilon \leq Ce^{\varepsilon d(gx_0,x_0)}$,*
3. *the operators $[F_t, \pi(g)]$ are compact on H_ε.*

Let us give an idea of how the Hilbert norms $\|.\|_\varepsilon$ on $\mathbf{C}[\Delta]$ are constructed. It is most enlightening to consider the prototype case of trees. Let S^n denote the sphere of radius n, i.e., the set of vertices at distance n from the origin x_0 and B^n the ball of radius n, i.e., the set of vertices at distance $\leq n$ of x_0. Suppose that $f \in \mathbf{C}[\Delta^0]$ has support in S^n. Then

$$\|f\|_\varepsilon^2 = e^{2\varepsilon n} \sum_{z \in B^n} \left| \sum_{x \to z} f(x) \right|^2$$

where the last sum is over all $x \in S^n$ such that z lies on the path from x_0 to x. For general $f \in \mathbf{C}[\Delta^0]$, one defines $\|f\|_\varepsilon^2 = \sum_{n=0}^{\infty} \|f_n\|_\varepsilon^2$, where f is the restriction of f to S^n. A similar formula defines the norm $\|.\|_\varepsilon$ on $\mathbf{C}[\Delta^1]$. The way the norm $\|.\|_\varepsilon$ is constructed makes relatively easy to prove the continuity of the operator $e^{t\rho}(\partial+h)e^{-t\rho}$ for any t (and uniformly with respect to t). More subtle is the estimate for the action $\pi(g)$ of a group element g. Equivalently, it amounts to compare the norms $\|.\|_\varepsilon$ for two choices of x_0. Lafforgue establishes an inequality of the form

$$\|\pi(g)\|_\varepsilon \leq P(l(g))e^{\varepsilon l(g)}$$

with a certain polynomial P. In particular $\|\pi(g)\|_\varepsilon \leq Ce^{\varepsilon' l(g)}$ for any $\varepsilon' > \varepsilon$.

According to the philosophy of Gromov, the geometry of trees is a model for the geometry of general hyperbolic spaces. The implementation of that principle can, however, be technically hard. In our case, Lafforgue needs almost 200 pages of difficult calculations to construct the analogue of the norms $\|.\|_\varepsilon$ above and for all the required estimates. We refer to [Laf12] and [Pus14] for the details.

6.3 Strong property (T)

Theorem 6.24 yields examples of discrete groups with property (T) satisfying Conjecture 5. Indeed, many hyperbolic groups have property (T). On the other hand, as a by-product of his proof, Lafforgue shows that hyperbolic groups do not satisfy a certain strengthening of property (T), in which unitary representations are replaced by ε-exponential representations. To that effect, let us consider the representation π of G on the completion of $\mathbf{C}[\Delta^0]$ for the norm $\|.\|_\varepsilon$.

Lemma 6.28 *The representation π on H_ε has no non-zero invariant vector, whereas its contragredient $\check{\pi}$ does have non-zero invariant vectors.*

Proof The first fact is obvious since a constant function is not in H_ε. On the other hand, the G-invariant form $f \mapsto \sum_{g\in G} f(g)$ extends to a continuous form on H_ε. Let us explain that point in the case of a tree: it follows immediately from the definition of the norm $\|.\|_\varepsilon$ that any $f \in \mathbf{C}[\Delta^0]$ satisfies the inequality

$$\sum_{n=0}^{\infty} e^{2\varepsilon n} \left| \sum_{x\in S^n} f(x) \right|^2 \leq \|f\|_\varepsilon^2$$

hence by Cauchy–Schwarz inequality,

$$\left|\sum f(x)\right|^2 \leq \left(\sum_{n=0}^{\infty} e^{-2\varepsilon n}\right)\left(\sum_{n=0}^{\infty} e^{2\varepsilon n} \left|\sum_{x\in S^n} f(x)\right|^2\right) \leq \left(1-e^{-2\varepsilon}\right)^{-1} \|f\|_\varepsilon^2.$$

The identification of H_ε with its dual therefore gives a non-zero invariant vector for the contragredient representation $\check{\pi}$. □

Let G be a locally compact group, l a length function on G, and real numbers $\varepsilon > 0$, $K > 0$. Let $\mathcal{F}_{\varepsilon,K}$ the family of representations π of G on a Hilbert space satisfying $\|\pi(g)\| \leq K e^{\varepsilon l(g)}$, and let $\mathcal{C}_{\varepsilon,K}(G)$ be the Banach algebra defined as the completion of $C_c(G)$ for the norm $\sup \|\pi(f)\|$, where the supremum is taken over representations π in $\mathcal{F}_{\varepsilon,K}$.

Definition 6.29 A Kazhdan projection in the Banach algebra $\mathcal{C}_{\varepsilon,K}(G)$ is an idempotent element p satisfying the following condition: for any representation π belonging to $\mathcal{F}_{\varepsilon,K}$, on a Hilbert space H, the range of the idempotent $\pi(p)$ is the space H^π of G-invariant vectors.

Remark 6.30 The above definition is given in a more general setting by de la Salle [dlS16], whose Proposition 3.4 and Corollary 3.5 also show that, since the family $\mathcal{F}_{\varepsilon,K}$ is stable under contragredient, a Kazhdan projection is necessarily central, hence unique and self-adjoint.

The above lemma has the following consequence:

Corollary 6.31 *Let G be a hyperbolic group. Then for any $\varepsilon > 0$ there exists $K > 0$ such that the Banach algebra $\mathcal{C}_{\varepsilon,K}(G)$ has no Kazhdan projection.*

Indeed, assume there is such a projection p. By the above remark p is self-adjoint, so that $\pi(p)^* = \check{\pi}(p)$, where π is the representation of G in H_ε. But by the lemma, $\pi(p) = 0$ and $\check{\pi}(p) \neq 0$, a contradiction.

The following definition should be thought as a strengthening of the characterization of Kazhdan's property (T) by a Kazhdan projection in $C^*_{\max}(G)$, cf. Proposition 5.4.

Definition 6.32 The group G has strong property (T) for Hilbert spaces if for any length function l, there exists an $\varepsilon > 0$ such that for every K there is a Kazhdan projection in $\mathcal{C}_{\varepsilon,K}(G)$.

We thus conclude:

Theorem 6.33 *Gromov hyperbolic groups do not satisfy the strong property (T) for Hilbert spaces.*

On the other hand, it follows from the works of Lafforgue, Liao, de Laat and de la Salle (see [Laf08, Lia14, dlS18, dLdlS15]) that in higher rank the situation is completely different.

Theorem 6.34 *Let G be a simple connected Lie group of real rank ≥ 2 or a simple algebraic group of split rank ≥ 2 over a non archimedian local field. Then G has strong property (T) in Hilbert spaces. The same holds for any lattice in such a G.*

Lafforgue more generally defines strong property (T) for a given class $\mathcal{E}$ of Banach spaces. The theorem above also holds provided the class of Banach spaces $\mathcal{E}$ has a non-trivial type, i.e., if the Banach space ℓ^1 is not finitely representable in $\mathcal{E}$.

Strong property (T) had been introduced by Lafforgue [Laf10] to understand the obstruction, if not to the Baum–Connes conjecture, at least to the proofs considered so far. But in fact, he has been led to introduce the following variant of strong property (T). We consider a locally compact group G and a compact subgroup K. Let l be a K-biinvariant length function on G and $\varepsilon > 0$.

Definition 6.35 An ε-exponential K-biinvariant Schur multiplier is a K-biinvariant function c on G such that for any K-biinvariant function f on G with values in $C_c(G)$ and support in the ball of radius R for the length l,

$$\|cf\| \leq e^{\varepsilon R}\|f\|$$

where cf is the pointwise product on G and $\|.\|$ is the norm in the crossed product $C^*(G, C_0(G)) = \mathcal{K}(L^2(G))$.

Definition 6.36 The group G has Schur property (T) relative to the compact subgroup K if for any K-biinvariant length function l, there exists $\varepsilon > 0$ and a K-biinvariant function φ on G with non-negative values and vanishing at infinity satisfying the following property: any ε-exponential K-biinvariant Schur multiplier c has a limit c_∞ at infinity and satisfies $|c(g) - c_\infty| \leq \varphi(g)$ for any $g \in G$.

Lafforgue explains in [Laf10] that Schur property (T) for a group G relative to a compact subgroup K is an obstacle to the above attempts to prove the Baum–Connes conjecture. It contradicts the existence, for any $G - C^*$ algebra A and any $\varepsilon > 0$, of a Banach subalgebra $\mathcal{B}$ of the reduced crossed product $C_r^*(G, A)$ satisfying the inequality $\|f\|_{\mathcal{B}} \leq e^{\varepsilon R}\|f\|_{C_r^*(G,A)}$ for any $f \in C_c(G, A)$ supported in the ball of radius R. In particular, supposing that G admits a γ element, it is hopeless to try to prove the Baum–Connes conjecture with coefficients using a homotopy of γ to 1 through ε-exponential representations as suggested above. It is also shown in [Laf10] that $SL_3(\mathbf{R})$ and $SL_3(\mathbf{Q}_p)$ do satisfy Schur property (T) with respect to their maximal compact subgroups. Liao [Lia16] has a similar result for the group Sp_4 over a nonarchimedian local field of finite characteristic. It is very likely, but as far as we know not yet proved, that it is also the case for simple groups of higher rank and with finite center.

Remark 6.37 The logical link between strong property (T) and Schur property (T) is not completely clear. One would expect that Schur property (T) for G relative to some compact subgroup K implies strong property (T) for G. But as noted by Lafforgue, this is not quite the case. As suggested to us by M. de la Salle, there should be a natural strengthening of Schur property (T) implying strong property (T).

6.4 Oka principle in Noncommutative Geometry

As explained in the previous section, Lafforgue observed that the "Dirac–dual-Dirac"-like methods used so far, would probably not work to prove the Baum–Connes conjecture with arbitrary coefficients for simple Lie groups of higher rank, mainly because of the presence of a variant of strong property (T) (see Section 6.3). In [Laf10], he even gave a necessary condition for this kind of approach to work and proved that these methods would certainly not succeed, leaving very few hope in proving further cases of the conjecture using the classical techniques. Nonetheless, he indicates that Bost's ideas on Oka principle are still open and he leaves them as a path for investigating the problem of surjectivity.

6.4.1 Isomorphisms in K-theory

In analytic geometry, the reduction of holomorphic problems to topological problems is known as Oka principle, whose classical version is the so-called *Oka–Grauert principle*. In its simplest form, it states that the holomorphic classification of complex vector bundles over an analytic Stein space agrees with their topological classification. The case of line bundles was proven by Oka in 1939 and it was then generalized by Grauert in 1958 ([Gra94]; see also [Gro89] for a seminal paper on the theory and [FL11] for a survey). Let us state Grauert's Theorem regarding complex vector bundles.

Theorem 6.38 (Grauert) *Let X be an analytic Stein space. Then,*

1. *if E and F are two complex holomorphic vector bundles over X which are continuously isomorphic, then E and F are holomorphically isomorphic.*
2. *every continuous vector bundle over X carries a holomorphic vector bundle structure that is uniquely determined.*
3. *the inclusion* $\iota : \mathcal{O}(X, GL_n(\mathbf{C})) \hookrightarrow C(X, GL_n(\mathbf{C}))$ *of the space of all holomorphic maps* $X \to GL_n(\mathbf{C})$ *into the space of all continuous maps is a* weak homotopy equivalence *with respect to the compact-open topology, i.e.,* ι *induces isomorphisms of all homotopy groups:*

$$\pi_k(\iota) : \pi_k\big(\mathcal{O}(X, GL_n(\mathbf{C}))\big) \xrightarrow{\simeq} \pi_k\big(C(X, GL_n(\mathbf{C}))\big), \quad k = 0, 1, 2, \dots$$

Let us assume X is compact. Let $\mathcal{O}(X)$ be the set of all continuous functions on X which are holomorphic on the interior of X, as a Banach subalgebra of $C(X)$. Then the injection $\iota : \mathcal{O}(X) \to C(X)$ is a strong isomorphism in K-theory.

In [Bos90], Bost asks the following question: *Let A and B be two Banach algebras and* $\iota : A \to B$ *a continuous injective morphism with dense image. What can be said about the map* $\iota_* : K(A) \to K(B)$*? More precisely, under which conditions on* ι *is the map* ι_* *an isomorphism?* As we have already mentioned in Section 6.1.2, the most classical criteria for the map ι_* to be an isomorphism is the fact that A is a dense subalgebra stable under holomorphic calculus in B (see [Kar08, p. 209], [Swa77, 2.2 and 3.1]). The discussion from Section 6.1 makes it clear why having a good criteria to ensure that ι_* is an isomorphism, can be very helpful when trying to prove the Baum–Connes conjecture. We will see that a closer relation can be stated.

The link between Bost's question and Grauert's Theorem 6.38 can be philosophically thought as follows: we start from a Banach algebra B, e.g., a C^*-algebra that we may think as the algebra of continuous functions on some noncommutative space T. Assume that T can be imbedded in some neighborhood X which is homotopic to T and carries a (non-commutative analogue of) complex structure. The dense subalgebra A is the set of functions on T which extend to functions on X which are holomorphic. Then the injection $\iota : A \to B$ can be seen as the composition of the Banach space injection $A = \mathcal{O}(X) \subset C(X)$ and a restriction map $C(X) \to C(T) = B$. The first should be an isomorphism in K-theory by a noncommutative analogue of Oka–Grauert's principle, and the second by the homotopy invariance of K-theory.

More precisely, Bost considers the following situation: Let B be a Banach algebra endowed with a continuous action of $\mathbf{R}^n$ denoted by α. Let F be a compact and convex subset of $\mathbf{R}^n$ containing 0 and with non-empty interior. Then one defines $A = \mathcal{O}(B, \alpha, F)$ as the set of elements a in B such that the continuous map $t \mapsto \alpha_t(a)$ from $\mathbf{R}^n$ to B has a continuous extension on $\mathbf{R}^n + iF \subset \mathbf{C}^n$ which is holomorphic on $\mathbf{R}^n + i\mathring{F}$, where $\mathring{F}$ is the interior of F.

For $z \in \mathbf{R}^n + iF$, denote by $\alpha_z(a) \in B$ the value of the map that extends α at z. Then $A = \mathcal{O}(B, \alpha, F)$ is a Banach algebra endowed with the norm

$$\|a\|_F = \sup_{z \in \mathbf{R}^n + iF} \|\alpha_z(a)\|,$$

and the inclusion map $\iota : A = \mathcal{O}(B, \alpha, F) \to B$ is dense (see [Bos90, 3.1 and Corollaire 3.2.4]). As mentioned by Bost, the algebra $\mathcal{O}(B, \alpha, F)$ is not in general stable under holomorphic calculus in B (see [Bos90, 1.3.1]), but the map ι still induces a strong isomorphism in K-theory (see Definition 6.3, see also [Nic08] for other criteria on ι so that ι_* is an isomorphism):

Theorem 6.39 ([Bos90, Théorème 2.2.1]) *Let B be a complex Banach algebra endowed with an action of $\mathbf{R}^n$ denoted by α. For all compact and convex subset F of $\mathbf{R}^n$, containing 0 and with non-zero interior, the inclusion map $\iota : A = \mathcal{O}(B, \alpha, F) \to B$ induces a strong isomorphism in K-theory.*

The idea of the proof is the following. The map which to $a \in A$ associates the function $\tau \mapsto \alpha_{i\tau}(a)$ provides an isometric embedding of the Banach algebra $A = \mathcal{O}(B, \alpha, F)$ into $C(F, B)$. Bost's proof then imitates the proof of Theorem 6.38 to show that the canonical injection $A \to C(F, B)$ is a strong isomorphism in K-theory. Composing with the evaluation at 0 from $C(F, B)$ to B (which is also a strong isomorphism theorem by the usual homotopy argument) yields the result.

The following examples are the basic examples of [Bos90]. Example 6.40 is equivalent to Grauert's theorem for a corona $U = \{z \in \mathbf{C} \mid \rho_1 \leq |z| \leq \rho_2\}$:

Example 6.40 Let $\mathbf{S}^1 = \{z \in \mathbf{C} \mid |z| = 1\}$ denote the unit circle, and let B be the algebra $C(\mathbf{S}^1)$ of continuous functions on $\mathbf{S}^1$ with complex values. Let ρ_1 and ρ_2 be two real numbers such that $0 < \rho_1 < 1 < \rho_2$, consider the closed corona $U = \{z \in \mathbf{C} \mid \rho_1 \leq |z| \leq \rho_2\}$ and let A be the subalgebra of $C(U)$ of continuous functions $\phi : U \to \mathbf{C}$ which are holomorphic in $\overset{\circ}{U}$. The algebra A, endowed with the norm of uniform convergence, is closed in $C(U)$ and hence it is a Banach algebra. Then, Theorem 6.39 says that the inclusion map $\iota : A \to B$ induces an isomorphism in K-theory. Indeed, let $(\alpha_t f)(z) = f(e^{-it}z)$, then $(\alpha_t)_{t \in \mathbf{R}}$ defines a one parameter group of isometric algebra automorphisms of B and $\mathcal{O}(B, \alpha, I) = A$ for $I = [\log \rho_1, \log \rho_2] \subset \mathbf{R}$.

Example 6.41 Let B be the convolution algebra $l^1(\mathbf{Z})$. Let $R > 0$ be a real number and let $A = \left\{(a_n) \in \mathbf{C}^{\mathbf{Z}} \mid \sum_{n=-\infty}^{+\infty} e^{R|n|}|a_n| < +\infty\right\}$. Hence A endowed with the norm $\|(a_n)\|_R = \sum_{n=-\infty}^{+\infty} e^{R|n|}|a_n|$ is a Banach algebra which is densely embedded in B. Theorem 6.39 says that the inclusion map $\iota : A \hookrightarrow B$ induces an isomorphism in K-theory. In this case, the one parameter group of isometric automorphisms of B is defined by $(\alpha_t(a_n) = (e^{int}a_n)$, and if $I = [-R, R] \subset \mathbf{R}$, then $\mathcal{O}(B, \alpha, I) = A$.

Example 6.42 The previous example can be also considered with coefficients so that things can be formulated in a noncommutative way: if A is a Banach algebra and α is an action of $\mathbf{Z}$ by isometric automorphism of A, let $B := \ell^1(\mathbf{Z}, A)$ be the completion of the convolution algebra $C_c(\mathbf{Z}, A)$ given by $\|(b_n)_n\|_1 = \sum_{n\in\mathbf{Z}} \|b_n\|_A$, for $(b_n)_n \in C_c(\mathbf{Z}, A)$. The product in B is given by twisted convolution, i.e., $(bb')_n = \sum_{k\in\mathbf{Z}} b_k\alpha(k)(b_{n-k})$, for $b, b' \in C_c(\mathbf{Z}, A)$. For all $t \in \mathbf{R}$, set $\beta_t((b_n)_n) = (e^{-int}b_n)_n$ and

$$\mathcal{O}(B, \beta, I) = \left\{ (b_n)_n \in \ell^1(\mathbf{Z}, A) \,\middle|\, \sum_{n=-\infty}^{+\infty} e^{R|n|}\|a_n\|_A < +\infty \right\},$$

where $I = [-R, R]$. Then Theorem 6.39 applies and $\mathcal{O}(B, \beta, I) \hookrightarrow B = \ell^1(\mathbf{Z}, A)$ induces an isomorphism in K-theory.

Theorem 6.39 can be applied to more general crossed products algebras for which it states that a certain subalgebra defined using an exponential decay condition on $L^1(G)$ has the same K-theory as Ł$^1(G)$. For a general locally compact group G, a Banach G-algebra B and a continuous function $a : G \to \mathbf{R}^+$ such that $a(g_1g_2) \leq a(g_1) + a(g_2)$, for $g_1, g_2 \in G$, define a subspace $\mathrm{Exp}_a(G, B)$ of $L^1(G, B)$ by the following decay condition:

$$\phi \in \mathrm{Exp}_a(G, B) \quad \text{if and only if} \quad e^a\phi \in L^1(G, B).$$

Then, endowed with the norm given by $\|\phi\|_a = \|e^a\phi\|_1$, $\mathrm{Exp}_a(G, B)$ is a Banach dense subalgebra of $L^1(G, B)$. Bost proved that if G is an elementary abelian group, then $K_*(\mathrm{Exp}_a(G, B))$ is isomorphic to $K_*(L^1(G, B))$. Let us state his result more precisely,

Theorem 6.43 ([Bos90, Théorème 2.3.2]) *Let G be a locally compact group and B a Banach algebra endowed with an action of G. If G is an extension by a compact group of a group of the form $\mathbf{Z}^p \times \mathbf{R}^q$ (i.e., there is a compact group K and a short exact sequence $1 \to K \to G \to \mathbf{Z}^p \times \mathbf{R}^q \to 1$), then, for every subadditive function $a : G \to \mathbf{R}_+$, the inclusion morphism*

$$\mathrm{Exp}_a(G, B) \hookrightarrow L^1(G, B)$$

induces an isomorphism in K-theory.

6.4.2 Relation with the Baum–Connes conjecture

Since we are dealing with K-theoretic issues, we focus on the right-hand side of the assembly map and therefore we are interested in surjectivity: let G be a group for which injectivity of the Baum–Connes assembly map is known (take, for example, any group in Lafforgue's class $\mathcal{C}$), and let A be a $G - C^*$-algebra. Let $\rho : G \to \mathrm{End}(V)$ be a representation of G on a complex hermitian vector space V of finite dimension. Then the norm of $\rho(g)$ can be used as a weight to define exponential decay subalgebras of crossed product algebras. In the case of $L^1(G, A)$, these are easy to define: using the notation of the previous paragraph and taking $a(g) = \log \|\rho(g)\|$, denote by $\mathrm{Exp}_\rho(G, B) := \mathrm{Exp}_a(G, B)$ which is the completion of $C_c(G, A)$ for the norm

$$\|f\|_{1,\rho} = \int_G \|f(g)\|_A \left(1 + \|\rho(g)\|_{\mathrm{End}(V)}\right) dg.$$

Hence $\mathrm{Exp}_\rho(G, A)$ is a dense subalgebra of $L^1(G, A)$ and the representation ρ is used as a weight to define exponential decay subalgebras of L^1. An Oka principle applied to this case, would state that these two algebras have the same K-theory.

Notice that for all groups belonging to the class $\mathcal{C}'$, as the algebra $\mathrm{Exp}_\rho(G, \mathbf{C})$ is an unconditional completion, by Theorem 6.8 we know that

$$K_*\left(\mathrm{Exp}_\rho(G, A)\right) \simeq K_*\left(L^1(G, A)\right).$$

Furthermore, we can use ρ to define exponential decay subalgebras of any unconditional completion,

Definition 6.44 Let $\mathcal{B}(G)$ be an unconditional completion of $C_c(G)$ and A a $G - C^*$-algebra. Let $\mathcal{B}^\rho(G, A)$ be the completion of $C_c(G, A)$ for the norm

$$\|f\|_{\mathcal{B}^\rho} = \left\| g \mapsto \|f(g)\|_A \, \|\rho(g)\|_{\mathrm{End}(V)} \right\|_{\mathcal{B}(G)}.$$

When $\mathcal{B}(G) = L^1(G)$, if ρ satisfies the following growth condition:

$$\int_G \frac{1}{\|\rho(g)\|} dg < +\infty,$$

then $L^\rho(G, A)$ is embedded in $L^1(G, A)$.

In the case of the reduced (resp. maximal) C^*-crossed products, an algebra that we call *weighted crossed product* and denoted by $\mathcal{A}_r^\rho(G, A)$ (resp. $\mathcal{A}^\rho(G, A)$) was defined in [GA10] (for more details see 6.4.3 below). Taking ρ to be *very large* (meaning that $\int_G \frac{1}{\|\rho(g)\|} dg < +\infty$) this algebra plays the same role in $C_r^*(G, A)$ as $\mathrm{Exp}_\rho(G, A)$ in $L^1(G, A)$; they are constructed to be some kind of "exponential decay subalgebras" of $C_r^*(G, A)$. Suppose now that G is a group for which the Bost conjecture is known to be true, in other words, the map $\mu_{L^1}^A : K_*^{top}(G, A) \to K_*(L^1(G, A))$ is an isomorphism. We will see that taking ρ very large allows us to have a morphism $\iota : K_*(\mathcal{A}_r^\rho(G, A)) \to K_*(L^1(G, A))$ and hence a morphism $\varphi : K_*(\mathcal{A}_r^\rho(G, A)) \to K_*(C_r^*(G, A))$ (see Proposition 6.51 below), so that the following diagram is commutative:

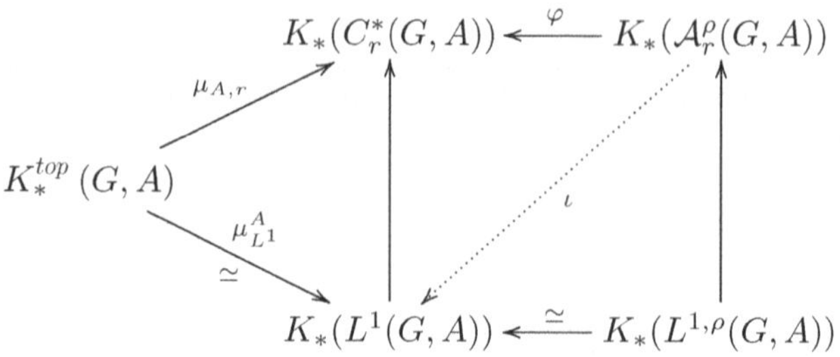

A suitable Oka principle applied to these crossed products states that the weighted group algebras $\mathcal{A}_r^\rho(G, A)$, have the same K-theory as $C_r^*(G, A)$, i.e., φ is an isomorphism. This would then imply the surjectivity of $\mu_{A,r}$ and hence the Baum–Connes conjecture with coefficients for G.

6.4.3 Weighted group algebras

In this section, we will recall the construction of weighted group algebras constructed in [GA10]. Let us first recall some definition and establish some notation.

Let G be a locally compact group and let dg a left Haar measure on G. Let Δ be the modular function on G (i.e., $dg^{-1} = \Delta(g)^{-1}dg$ for all $g \in G$).

Let A be a $G - C^*$-algebra. For all $g \in G$ and for all $a \in A$, let $g.a$, or $g(a)$, be the action of g on a. The space of continuous functions with compact support on G with values in A, denoted by $C_c(G, A)$, is endowed with a structure of involutive algebra where the multiplication and the involution are given, respectively, by the formulas:

$$(f_1 * f_2)(g) = \int_G f_1(g_1)g_1\left(f_2\left(g_1^{-1}g\right)\right)dg_1,$$

for $f_1, f_2 \in C_c(G, A)$ and

$$f^*(g) = g\left(f\left(g^{-1}\right)\right)^* \Delta\left(g^{-1}\right),$$

for $f \in C_c(G, A)$ and $g \in G$. In a general, we write every element f in $C_c(G, A)$ as the formal integral $\int_G f(g)e_g dg$, where e_g is a formal letter satisfying the following conditions:

$$e_g e_{g'} = e_{gg'}, \quad e_g^* = (e_g)^{-1} = e_{g^{-1}} \quad \text{and} \quad e_g a e_g^* = g.a,$$

for all $g, g' \in G$ and for all $a \in A$.

We denote by $C_{\max}^*(G, A)$ and $C_r^*(G, A)$ the maximal and the reduced crossed product of G and A, respectively. Moreover, we denote by

$$L^2(G, A) = \left\{ f \in C_c(G, A) \,\middle|\, \int_G f(g)^* f(g) dg \text{ converges in } A \right\},$$

and $\lambda_{G,A}$ the left regular representation of $C_c(G, A)$ on $L^2(G, A)$ which is given by the formula:

$$\lambda_{G,A}(f)(h)(t) = \int_G t^{-1}(f(s))h(s^{-1}t)ds,$$

for $f \in C_c(G, A)$, $h \in L^2(G, A)$ and $t \in G$. Recall that $\lambda_{G,A}$ induces a unique morphism of C^*-algebras from $C^*_{\max}(G, A)$ to $C^*_r(G, A)$; we also denote that morphism by $\lambda_{G,A}$, by abuse of notation.

Let (ρ, V) be a finite-dimensional representation of G. We then consider the map

$$C_c(G, A) \to C_c(G, A) \otimes \mathrm{End}(V)$$

$$\int_G f(g)e_g dg \mapsto \int_G f(g)e_g \otimes \rho(g)dg.$$

Definition 6.45 The reduced crossed product weighted by ρ of G and A, denoted by $\mathcal{A}^\rho_r(G, A)$, is the completion of $C_c(G, A)$ for the norm:

$$\| \int_G f(g)e_g dg \|_{A \rtimes^\rho G} = \| \int_G f(g)e_g \otimes \rho(g)dg \|_{C^*_r(G,A)\otimes \mathrm{End}(V)},$$

for $f \in C_c(G, A)$. If $A = \mathbf{C}$, we denote it by $\mathcal{A}^\rho_r(G) := \mathcal{A}^\rho_r(G, \mathbf{C})$.

It is then easy to prove that the reduced weighted crossed product $\mathcal{A}^\rho_r(G, A)$ is a Banach algebra. When ρ is an unitary representation of G then $\mathcal{A}^\rho_r(G, A) = C^*_r(G, A)$, up to norm equivalence.

Remark 6.46 In the same manner, we can define weighted maximal crossed products, however, we don't treat them here because of the discussion held in 5.1.

Example 6.47 Let $G = \mathbf{Z}$ and let $\rho : \mathbf{Z} \to \mathbf{C}^*$ be a character of $\mathbf{Z}$. Let $S^\rho := \{z \in \mathbf{C} \mid |z| = |\rho(1)|\}$ the circle of radius $|\rho(1)|$. Hence, $\mathcal{A}^\rho_r(G)$ is the algebra of continuous functions on S^ρ.

Example 6.48 Let $G = \mathbf{Z}$ and let $\rho_1 : \mathbf{Z} \to \mathbf{C}^*$ and $\rho_2 : \mathbf{Z} \to \mathbf{C}$ be two characters of $\mathbf{Z}$ such that $R_1 < R_2$, where $R_1 = |\rho_1(1)|$ and $R_2 = |\rho_2(1)|$. Then, $\mathcal{A}^{\rho_1 \oplus \rho_2}(G)$ is the algebra of continuous functions on the closed corona $U := \{z \in \mathbf{C} \mid |\rho_1(1)| \leq |z| \leq |\rho_2(1)|\}$ holomorphic on $\overset{\circ}{U}$. Indeed, we have the following diagram:

$$\begin{array}{ccc} \mathcal{A}^{\rho_1 \oplus \rho_2}(G) & \longrightarrow & C(\mathbf{S}^1, \mathrm{End}(\mathbf{C}^2)) \\ \uparrow & & \uparrow \\ \ell^{1,\rho_1 \oplus \rho_2}(G) & \longrightarrow & \ell^1(\mathbf{Z}, \mathrm{End}(\mathbf{C}^2)) \end{array}$$

where the vertical arrows are given by Fourier series and the norm in $\ell^{1,\rho_1\oplus\rho_2}(G)$ is given by $\|(a_n)_n\| = \sum_{n\in\mathbf{Z}} |a_n|\|(\rho_1 \oplus \rho_2)(n)\|$. It is then clear that the algebra $\ell^{1,\rho_1\oplus\rho_2}(G)$ can be identified with the algebra

$$A = \left\{ (a_n) \in \mathbf{C}^{\mathbf{Z}} \;\middle|\; \sum_{n=-\infty}^{+\infty} e^{|n|\log r}|a_n| < +\infty,\ \text{for all } r \in]R_1, R_2[\right\},$$

which is identified by Fourier series with the algebra of continuous functions on U holomorphic on $\overset{\circ}{U}$. Applying Theorem 6.39, taking $R_1 < 1 < R_2$, we get that the algebras $\mathcal{A}^{\rho_1\oplus\rho_2}(\mathbf{Z})$ and $C_r^*(\mathbf{Z})$ have the same K-theory.

In [GA10], a weighted version of the Baum–Connes morphism was constructed using Lafforgue's Banach KK-theory:

$$\mu^{\rho}_{r,A} : K^{top}(G, A) \to K(\mathcal{A}^{\rho}_r(G, A));$$

it computes the K-theory of this weighted algebras. Analogues of Kasparov's and Lafforgue's Dirac–dual-Dirac methods were proven in this context. We state them as the following two theorems.

Theorem 6.49 ([GA10]) *Let G be a locally compact group with a γ-element. Then, for every $G - C^*$-algebra A and every finite-dimensional representation ρ of G, the weighted morphism $\mu^{\rho}_{r,A}$ is injective. If moreover, $\gamma = 1$ in $KK_G(\mathbf{C}, \mathbf{C})$, then $\mu^{\rho}_{r,A}$ is surjective.*

Theorem 6.50 ([GA09]) *Let G be a locally compact group with a γ-element. If $\gamma = 1$ in $KK^{\mathrm{ban}}_{G,\ell}(\mathbf{C}, \mathbf{C})$ and there is an unconditional completion stable under holomorphic calculus in $C_r^*(G)$, then μ^{ρ}_r is an isomorphism for every finite-dimensional representation ρ of G.*

Hence the morphism $\mu^{\rho}_{r,A}$ is an isomorphism, for example, for all groups with the Haagerup property and more general, for all K-amenable groups, and when $A = \mathbf{C}$, the morphism μ^{ρ}_r is an isomorphism for all semisimple Lie groups and all cocompact lattices in a semisimple Lie group.

It is worth nothing to mention that, proving that the weighted map is an isomorphism is not easier than proving the Baum–Connes conjecture; one of the reasons is that, even though the algebras $\mathcal{A}^{\rho}_r(G, A)$ are in general not C^*-algebras, there are constructed in a very C^*-algebraic way. However, the following proposition shows that the weighted crossed products can be very small when the representation ρ is very large.

Proposition 6.51 ([GA10, Proposition 1.5]) *Let Γ be a discrete group and A a $\Gamma - C^*$-algebra. Let $\rho : \Gamma \to \mathrm{End}(V)$ a finite-dimensional representation of Γ such that $\sum_{\gamma\in\Gamma} \frac{1}{\|\rho(\gamma)\|}$ converges. Then $\mathcal{A}^{\rho}_r(\Gamma, A)$ embeds into $\ell^1(\Gamma, A)$.*

We then have the inclusions $\mathcal{A}_r^\rho(\Gamma, A) \hookrightarrow \ell^1(\Gamma, A) \hookrightarrow C_r^*(\Gamma, A)$ and hence, if we take a group Γ for which we know that $K_*^{top}(\Gamma) \simeq K_*(\ell^1(\Gamma, A))$, proving that $\mathcal{A}_r^\rho(\Gamma, A)$ and $C_r^*(\Gamma, A)$ have the same K-theory would prove the surjectivity of the Baum–Connes map with coefficients for Γ. These ideas also work for more general locally compact groups, but we don't always have a continuous map from $\mathcal{A}_r^\rho(G, A)$ to $C_r^*(G, A)$ (this map exists if and only if the regular representation λ_G is weakly contained in $\lambda_G \otimes \rho$). Nevertheless, thanks to the following proposition, we have a map at the level of K-theory:

Proposition 6.52 *Let G be a locally compact group and let $\rho : G \to \mathrm{End}(V)$ a finite-dimensional representation of G such that $\int_G \frac{1}{\|\rho(g)\|} dg$ converges. Then, if A is a $G - C^*$-algebra, $\mathcal{A}_r^\rho(G, A) \cap L^1(G, A)$ is relatively spectral in $\mathcal{A}_r^\rho(G, A)$.*

Definition 6.53 A morphism $\phi : A \to B$ between two algebras is *relatively spectral* if $\mathrm{sp}_B(\phi(x)) = \mathrm{sp}_A(x)$ for all x in some dense subalgebra X of A. It is a weaker condition than being stable under holomorphic calculus and it induces an isomorphism in K-theory (see [Nic08]).

As a result, we have a map from $K_*(\mathcal{A}_r^\rho(G, A))$ to $K_*(L^1(G, A))$ defined through $K(\mathcal{A}_r^\rho(G, A) \cap L^1(G, A))$ and we can prove that the following diagram is commutative:

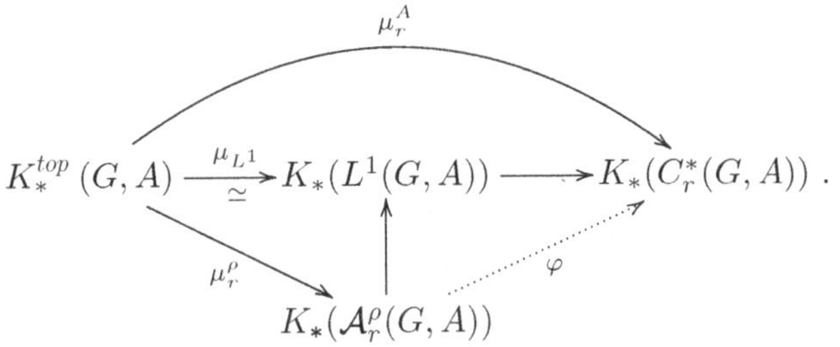

Hence, we get a morphism $\varphi : K_*(\mathcal{A}_r^\rho(G, A)) \to K_*(C_r^*(G, A))$. The following result is then straightforward:

Theorem 6.54 *Let G be a locally compact group with a γ-element and let (ρ, V) be a finite-dimensional representation of G such that $\int_G \frac{1}{\|\rho(g)\|} dg$ converges. If φ is an isomorphism then the Baum–Connes conjecture with coefficients in A is true for G.*

Let us give two examples of groups having a "very large" finite-dimensional representation.

Example 6.55

1. Let $G = \mathbf{R}$ and let $\rho : \mathbf{R} \to \mathrm{GL}_3(\mathbf{C})$, be the representation of G defined by $t \mapsto \mathrm{Exp}(tX)$, where $X = \begin{pmatrix} 0 & 1 & 0 \\ 0 & 0 & 1 \\ 0 & 0 & 0 \end{pmatrix}$. Then,

$$\mathrm{Exp}(tX) = 1 + tX + \frac{t^2}{2}X^2 = \begin{pmatrix} 1 & t & \frac{t^2}{2} \\ 0 & 1 & t \\ 0 & 0 & 1 \end{pmatrix}$$

and hence, $\|\mathrm{Exp}(tX)\| \geq \left(\frac{t^4}{4} + t^2 + 1\right)^{\frac{1}{2}} = 1 + \frac{t^2}{2}$. It follows that,

$$\int_{-\infty}^{+\infty} \frac{dt}{\|\mathrm{Exp}(tX)\|} \leq \int_{-\infty}^{+\infty} \frac{dt}{1 + \frac{t^2}{2}} < +\infty.$$

2. Take $G = SL_2(\mathbf{R})$. Set $K = SO(2)$, and let

$$A = \left\{ a_t = \begin{pmatrix} e^t & 0 \\ 0 & e^{-t} \end{pmatrix} : t \in \mathbf{R} \right\}$$

be the diagonal subgroup. Recall that the Haar measure in the Cartan decomposition $G = KA^+K$ is expressed as

$$\int_G f(g)\, dg = \int_K \int_0^\infty \int_K \sinh(2t) f(k_1 a_t k_2)\, dk_1\, dt\, dk_2$$

for $f \in C_c(G)$. Let ρ_n be the $(n+1)$-dimensional representation of G on homogeneous polynomials of degree n on $\mathbf{C}^2$. Then $\|\rho_n(a_t)\| = e^{nt}$ for $t \geq 0$, so that ρ_n is very large exactly when $n \geq 3$.

Accordingly, proving the Baum–Connes conjecture with coefficients for a group for which injectivity is known (for example, a semisimple Lie group or one of its closed subgroups) amounts to prove that the map φ is surjective. To illustrate the fact that proving the surjectivity of φ fits in the framework of Oka's principle as introduced by Bost in [Bos90], let us state the following proposition. The first point is a generalization of Theorem 6.43 concerning L^1 algebras; even though this result does not appear in [Bos90], the proof is due to Bost.

Proposition 6.56 *Let G be a locally compact group and let $\rho : G \to \mathrm{GL}_n(\mathbf{R})$ a representation of G.*

1. *If $\overline{\rho(G)}$ is amenable and $a(g) = \log(\|\rho(g)\|)$, then the map $K_*(\mathrm{Exp}_a(G, B)) \to K_*(L^1(G, B))$ is an isomorphism.*

2. *If $\overline{\rho(G)}$ is amenable and $\int_G \frac{1}{\|\rho(g)\|} dg$ converges then the map*

$$K_*\left(\mathcal{A}_r^\rho(G,B)\right) \to K_*\left(C_r^*(G,B)\right)$$

defined using Proposition 6.52 is an isomorphism.

The conditions that $\int_G \frac{1}{\|\rho(g)\|} dg$ converges and that $\overline{\rho(G)}$ is amenable imply that G is amenable. This is because the condition that $\int_G \frac{1}{\|\rho(g)\|} dg$ converges implies that ρ is proper. Hence, Theorem 6.56 does not give anything new apart from proving that the Baum–Connes conjecture is true for an amenable group. Yet, it seemed to us that this result gives a good idea of how Bost's version of Oka principle works, and therefore we give the main ideas of the proof below.

We will use the following properties of weighted algebras. Analogous properties are satisfied by $\mathrm{Exp}_\rho(G,B)$.

Lemma 6.57 *Let ρ, ρ', π, σ finite-dimensional representations of a locally compact group G.*

1. *If ρ' is either a sub-representation or a quotient of ρ, then $\mathcal{A}_r^\rho(G,B) \subset \mathcal{A}_r^{\rho'}(G,B)$.*
2. *If $\rho = \pi \otimes \sigma$ and σ is unitary, then $\mathcal{A}_r^\rho(G,B) = \mathcal{A}_r^\pi(G,B)$.*
3. *If $\rho = \bigoplus_k \rho_k$, then $\mathcal{A}_r^\rho(G,B) \subset \bigcap_k \mathcal{A}_r^{\rho_k}(G,B)$.*

Lemma 6.58 *Let $\rho : G \to \mathrm{GL}_n(\mathbf{R})$ be a representation of a locally compact group. If $\mathbf{R}^n$ has a G-invariant filtration of the form $0 = V_0 \subset V_1 \subset \cdots \subset V_r = \mathbf{R}^n$ and $\sigma_k : G \to \mathrm{End}(V_k/V_{k-1})$ is the corresponding representation on V_k/V_{k-1} and $\sigma = \bigoplus_k \sigma_k$ is its semi-simplification, then $\mathcal{A}_r^\rho(G,B) \subset \mathcal{A}_r^\sigma(G,B)$ and, moreover, $\mathcal{A}_r^\rho(G,B)$ is stable under holomorphic calculus in $\mathcal{A}_r^\sigma(G,B)$.*

If $\overline{\rho(G)}$ is amenable then the Zariski closure of $\rho(G)$ is also amenable by a result of Moore (see, for example, [Zim84, page 64]). Using Furstenberg's Lemma we may suppose that $\rho(G)$ is contain in the a subgroup of $GL_n(\mathbf{R})$ of the form

$$\begin{pmatrix} \mathbf{R}_+^* \times SO(n_1) & * & \cdots & * \\ 0 & \mathbf{R}_+^* \times SO(n_2) & * & \vdots \\ \vdots & & \ddots & * \\ 0 & \cdots & 0 & \mathbf{R}_+^* \times SO(n_k) \end{pmatrix}.$$

Hence, we may apply Lemma 6.57 with $\sigma_i = \chi_i \otimes u_i$, where χ_i is a character of $\mathbf{R}_+^*$ and u_i is an unitary representation of $SO(n_i)$. Using the fact that $\mathcal{A}_r^{\sigma_i}(G,B) = \mathcal{A}_r^{\chi_i}(G,B)$, we get a injective morphism

$$\mathcal{A}_r^\rho(G,B) \to \mathcal{A}_r^\pi(G,B)$$

where $\pi = \bigoplus_{k=1}^{m} \chi_k$ and this morphism is dense and stable under holomorphic calculus. Therefore,

$$K_*\big(\mathcal{A}_r^{\rho}(G,B)\big) \simeq K_*\big(\mathcal{A}_r^{\pi}(G,B)\big).$$

It remains to prove that the inclusion

$$\mathcal{A}_r^{\pi}(G,B) \to C_r^*(G,B)$$

induces an isomorphism in K-theory.

Let W be the space of real-valued functions on G defined as $W = \sum_k \mathbf{R} \log(\chi_k)$. We define an action of W on $C_r^*(G,B)$ by the formula $\alpha_\xi(f)(g) = f(g)e^{-i\xi(g)}$, for $f \in C_c(G,B)$ and $\xi \in W$. Then, we need to check that

$$\mathcal{A}_r^{\pi}(G,B) = \mathcal{O}(K, C_r^*(G,B), \alpha)$$

where K is the convex hull of $\{0, \log \chi_k\}$. We conclude by applying Theorem 6.39.

7 The Baum–Connes conjecture for groupoids

Let $\mathcal{G}$ be a locally compact, σ-compact, Hausdorff groupoid with Haar system and let $C_r^*(\mathcal{G})$ be its reduced C^*-algebra (see the definition below). The Baum–Connes conjecture for $\mathcal{G}$ states that a certain map

$$\mu_r : K_*^{top}(\mathcal{G}) \to K_*\left(C_r^*(\mathcal{G})\right)$$

is an isomorphism. Many important examples of operator algebras may be realized as the C^*-algebra associated to a groupoid. This is the case, for example, for C^*-algebras associated to a foliation, to an action of a group on a space as well as the C^*-algebra associated to a group. Therefore, a version of the Baum–Connes conjecture for groupoids allows to study the K-theory of all of these algebras in a very general framework; we will see that it is also the case for the coarse Baum–Connes conjecture developed in Chapter 8.

The Baum–Connes map μ_r for groupoid C^*-algebras appeared in the work of Baum and Connes on the Novikov conjecture for foliations (see [Con82] for a very nice survey on the subject). In [BC85], Baum and Connes gave a proof of the injectivity of μ_r in the case of groupoids coming from foliations that have negatively curved leaves which is based on the construction of a Dual-Dirac element following ideas of both Kasparov and Mishchenko. Using a construction of a Kasparov bivariant theory which is equivariant with respect to the action of a groupoid defined by Le Gall in [LG99], Tu stated in [Tu99c] the Dirac–dual-Dirac method in a very general context. He then proved injectivity of μ_r for a class of groupoids called bolic, generalizing Kasparov and Skandalis's work for groups, and that μ_r is an isomorphism for amenable groupoids generalizing the results of Higson and Kasparov (see [Tu99c, Tu99b]).

7.1 Groupoids and their C^*-algebras

In this section, we recall the definition of the C^*-algebras associated to groupoids and the Baum–Connes conjecture for those. It is mostly taken from the survey written by Tu on the subject [Tu00].

A groupoid is a small category in which all morphisms are invertible. More concretely, it is given by the following data:

1. the set of objects $\mathcal{G}^0$, also called the unit space,
2. the set of morphisms $\mathcal{G}$,
3. an inclusion $i : \mathcal{G}^0 \hookrightarrow \mathcal{G}$,
4. two maps "range" and "source" $r, s : \mathcal{G} \to \mathcal{G}^0$ such that $r \circ i = s \circ i = \mathrm{Id}$,
5. an involution $\mathcal{G} \to \mathcal{G}$, denoted by $g \mapsto g^{-1}$ such that $r(g) = s(g^{-1})$ for every $g \in \mathcal{G}$,
6. a partially defined product $\mathcal{G}^2 \to \mathcal{G}$, denoted by $(g, h) \mapsto gh$, where $\mathcal{G}^2 := \{(g, h) \in \mathcal{G} \times \mathcal{G} \,|\, s(g) = r(h)\}$ is the set of composable pairs.

It is assumed moreover that the product is associative (i.e., if $(g, h), (h, k) \in \mathcal{G}^2$ then the products $(gh)k$ and $g(hk)$ are defined and are equal), that for all $g \in \mathcal{G}$, $i(r(g))g = gi(s(g)) = g$ and for all $g \in \mathcal{G}$, $gg^{-1} = i(r(g))$.

A topological groupoid is a groupoid such that $\mathcal{G}$ and $\mathcal{G}^0$ are topological spaces and all maps appearing in the definitions are continuous. When a topological groupoid $\mathcal{G}$ is locally compact and Hausdorff, it is said to be

(a) *principal* if $(r, s) : \mathcal{G} \to \mathcal{G}^0 \times \mathcal{G}^0$ is injective,
(b) *proper* if $(r, s) : \mathcal{G} \mapsto \mathcal{G}^0 \times \mathcal{G}^0$ is proper,
(c) *étale*, or *r-discrete*, if the range map $r : \mathcal{G} \mapsto \mathcal{G}^0$ is local homeomorphism, i.e., if every $x \in \mathcal{G}$ admits an open neighborhood U such that $r(U)$ is an open subset of $\mathcal{G}^0$ and $r : U \mapsto r(U)$ is a homeomorphism

Before giving some examples of groupoids, let us introduce some notations: for all $x, y \in \mathcal{G}^0$, $\mathcal{G}_x := s^{-1}(x)$, $\mathcal{G}^x = r^{-1}(x)$, $\mathcal{G}^x_y = \mathcal{G}_x \cap \mathcal{G}^y$.

Example 7.1

1. *Groups and Spaces*. A group G is a groupoid with $G^0 = \{1\}$, the unit element. A space X is a groupoid where $\mathcal{G} = \mathcal{G}^0 = X$ and $r = s = \mathrm{Id}_X$.
2. An equivalence relation $R \subset X \times X$ on a set X can be endowed with a groupoid structure; the unit space is X, the range and source maps are $r(x, y) = x$, $s(x, y) = y$, respectively, composition is defined by $(x, y)(z, t) = (x, t)$ if $y = z$ and inverses by $(x, y)^{-1} = (y, x)$. In particular, the space $X \times X$ is a groupoid.

3. If a group Γ acts on the right on a space X, then one obtains a groupoid $\mathcal{G} = X \rtimes \Gamma$ by taking as a set $\mathcal{G} = X \times \Gamma$ as unit space $\mathcal{G}^0 = X \times \{1\} \simeq X$, $r(x, \gamma) = x$, $s(x, \gamma) = x\gamma, x, \gamma)^{-1} = (x\gamma, \gamma^{-1})$, $(x, \gamma)(x\gamma, \gamma') = (x, \gamma\gamma')$. If X is a topological space, Γ a topological group and the cation is continuous then $X \rtimes$ is a topological groupoid, which is Hausdorff if X and Γ are. In that case, if Γ is discrete, $X \rtimes \Gamma$ is étale and it is principal if the action is free.
4. Let X be topological space and take $\mathcal{G}$ to be the set of equivalence classes of paths $\varphi : [0, 1] \to X$ where two paths are equivalent if and only if they are homotopic with fixed endpoints. Then $\mathcal{G}^0 \simeq X$ is the set of equivalence classes of constant paths on X. If φ is a path on X and $g = [\varphi]$ is its class in $\mathcal{G}$, then $r(g) = \varphi(1)$, $s(g) = \varphi(0)$, $g^{-1} = [\varphi^{-1}]$, where $\varphi^{-1}(t) = \varphi(1 - t)$ and $[\varphi][\psi] = [\varphi * \psi]$, where $\varphi * \psi(t) = \varphi(2t)$ for $t \in \left[0, \frac{1}{2}\right]$ and $\varphi * \psi(t) = \psi(2t - 1)$ for $t \in \left[\frac{1}{2}, 1\right]$. $\mathcal{G}$ is called *the fundamental groupoid of* X.
5. Let (V, F) be a foliation. *The holonomy groupoid* $\mathcal{G}$ is the set of equivalence classes of paths whose support is contained in one leaf, where two paths are identified if they have the same end points and they define the same holonomy element. Composition and inverse are defined in the same way as for the fundamental groupoid. The space of units is V; if V is of dimension n and the foliation of codimension q then $\mathcal{G}$ is a differentiable groupoid of dimension $2n - q$. It is not Hausdorff in general. If T is a transversal that meets all leaves of the foliation, then the restriction of the holonomy groupoid to T is an étale groupoid equivalent to $\mathcal{G}$.

From now on let $\mathcal{G}$ be a locally compact Hausdorff groupoid. An action (on the right) of $\mathcal{G}$ on a space Z is given by a map $p : Z \to \mathcal{G}^0$, called the source map, and a continuous map from $Z \times_{\mathcal{G}^0} \mathcal{G} = \{(z, g) \mid p(z) = r(g)\}$ to Z, denoted by $(z, g) \mapsto zg$, such that $(zg)h = z(gh)$ whenever $p(z) = r(g)$ and $s(g) = r(h)$ and $zp(z) = z$. A space endowed with an action of $\mathcal{G}$ is called a $\mathcal{G}$-space.

We can then define a groupoid denoted by $Z \rtimes \mathcal{G}$ with underlying set $Z \times \mathcal{G}$, unit space $Z \simeq \{(z, p(z)) \mid z \in Z\}$, source and range maps $s(z, g) = zg$, $r(z, g) = z$, inverse $(z, g)^{-1} = (zg, g^{-1})$ and products $(z, g)(zg, h) = (z, gh)$. Note that $Z \rtimes \mathcal{G}$ is étale if $\mathcal{G}$ is. If Z and $\mathcal{G}$ are locally compact Hausdorff, the action of $\mathcal{G}$ on Z is free (resp. proper) if and only if the groupoid $Z \rtimes \mathcal{G}$ is principal (resp. proper). A $\mathcal{G}$-space Z is said to be $\mathcal{G}$-compact if the action is proper and the quotient $Z/\mathcal{G}$ is compact.

A $\mathcal{G}$-algebra is an algebra A endowed with an action of $\mathcal{G}$ i.e., A is a $C(\mathcal{G}^0)$-algebra and the action of $\mathcal{G}$ on A is given by an isomorphism of $C(\mathcal{G})$-algebras $\alpha : s^*A \to r^*A$ such that the morphisms $\alpha_g : A_{s(g)} \to A_{r(g)}$ satisfy the relation $\alpha_g \circ \alpha_h = \alpha_{gh}$. Recall that if X is a locally compact Hausdorff space, a $C(X)$-algebra is a C^*-algebra endowed with a $*$-homomorphism θ from $C_0(X)$ to the center $Z(M(A))$ of the multiplier algebra of A, such that $\theta(C_0(X))A = A$. If $p : X \to Y$ is a map between two locally compact Hausdorff spaces and A is a $C(X)$-algebra, then $p^*A = A \otimes_{C_0(X)} C_0(Y)$ is a $C(Y)$-algebra. If $x \in X$, the fiber A_x of A over x is defined by i^*A where $i_x : \{x\} \to X$ is the inclusion map.

Suppose $\mathcal{G}$ is σ-compact and has a Haar system $\lambda = \{\lambda_x \mid x \in \mathcal{G}^0\}$ (we can take, for example, $\mathcal{G}$ to be étale and then λ_x is the counting measure on $\mathcal{G}^x$). A *cutoff function* on $\mathcal{G}$ is a continuous function $c : \mathcal{G}^0 \to \mathbb{R}^+$ such that for every $x \in \mathcal{G}^0$, $\int_{g \in \mathcal{G}^x} c(s(g))d\lambda^x(g) = 1$, and for every compact $K \subset \mathcal{G}^0$, $\text{supp}(c) \cap s(G^K)$ is compact. Such a function exists if and only if $\mathcal{G}$ is proper [Tu99c, Propositions 6.10, 6.11].

Let A be a $\mathcal{G}$-algebra. The full and reduced crossed products of A by $\mathcal{G}$, denoted $C^*(\mathcal{G}, A)$ and $C^*_r(\mathcal{G}, A)$ respectively are defined in the following way: let $C_c(\mathcal{G}, r^*A)$ be the space of functions with compact support $g \mapsto \varphi(g) \in A_{r(g)}$ continuous in the sense of [LG99]. The product and adjoint are defined respectively by

$$\varphi * \psi(g) = \int_{h \in \mathcal{G}^{r(g)}} \varphi(h)\alpha_h(\psi(h^{-1}g))d\lambda^{r(g)}(h),$$

$$\varphi^*(g) = \alpha_g(\varphi(g^{-1}))^*.$$

Then, $L^1(\mathcal{G}, r^*A)$ denotes the completion of $C_c(\mathcal{G}, r^*A)$ for the norm

$$\|\varphi\| = \max(|\varphi|_1, |\varphi^*|_1),$$

where $|\varphi|_1 = \sup_{x \in \mathcal{G}^0} \int_{g \in \mathcal{G}^x} \|\varphi(g)\| d\lambda^x(g)$ and $C^*(\mathcal{G}, A)$ is the enveloping C^*-algebra of $L^1(\mathcal{G}, r^*A)$ and $C^*_r(\mathcal{G}, A)$ is the closure of $L^1(\mathcal{G}, r^*A)$ in $\mathcal{L}(L^2(\mathcal{G}, r^*A))$.

When the $\mathcal{G}$-algebra A is the algebra $C_0(\mathcal{G}^0)$ of continuous functions vanishing at infinity on the space of objects $\mathcal{G}^0$, the crossed products $C^*(\mathcal{G}, A)$ and $C^*_r(\mathcal{G}, A)$ will simply be denoted $C^*(\mathcal{G})$ and $C^*_r(\mathcal{G})$, and called groupoid full and reduced C^*-algebras.

In [LG97, LG99], for every pair (A, B) of graded $\mathcal{G}$-algebras, Le Gall defined a bifunctor $KK_{\mathcal{G}}(A, B)$ generalizing Kasparov's KK-bifuntor for groups (see Section 3.1) that has mostly the same features, in particular, there is an associative product $KK_{\mathcal{G}}(A, D) \times KK_{\mathcal{G}}(D, B) \to KK_{\mathcal{G}}(A, B)$ that satisfies the same naturality properties as in case of the non-equivariant KK-functor. The product of two elements $\alpha \in KK_{\mathcal{G}}(A, D)$, $\beta \in KK_{\mathcal{G}}(D, B)$ is denoted by $\alpha \otimes_D \beta$. And there are descent morphisms

$$j_{\mathcal{G}} : KK_{\mathcal{G}}(A, B) \to KK\left(C^*(\mathcal{G}, A), C^*(\mathcal{G}, B)\right),$$

$$j_{\mathcal{G},r} : KK_{\mathcal{G}}(A, B) \to KK\left(C^*_r(\mathcal{G}, A), C^*_r(\mathcal{G}, B)\right),$$

compatible with the product.

Suppose that $\mathcal{G}$ is proper and that $\mathcal{G}^0/\mathcal{G}$ is compact and let c be a cutoff function for $\mathcal{G}$. The function $g \mapsto \sqrt{c(r(g))c(s(g))}$, which is continuous with compact support, defines a projection in $C^*(\mathcal{G}) = C^*_r(\mathcal{G})$ whose homotopy class is independent of the choice of the cutoff function and hence defines a canonical element $\lambda_{\mathcal{G}} \in K_0(C^*(\mathcal{G}))$. If Z is a $\mathcal{G}$-compact proper space and B is a $\mathcal{G}$-algebra, the map

$$KK^*(C_0(Z), B) \xrightarrow{j_{\mathcal{G},r}} KK^*(C^*(Z \rtimes \mathcal{G}), C_r^*(\mathcal{G}, B)) \xrightarrow{\lambda_{Z \rtimes G} \otimes \cdot} K_*(C_r^*(\mathcal{G}, B))$$

induces the Baum–Connes map with coefficients

$$\mu_r^B : K_*^{top}(\mathcal{G}; B) = \lim_{\rightarrow} KK_{\mathcal{G}}^*(C_0(Z), B) \rightarrow K_*\left(C_r^*(\mathcal{G}, B)\right),$$

where the inductive limit is taken among all the Z subspace of $\underline{E}\mathcal{G}$ that are $\mathcal{G}$-compact and $\underline{E}\mathcal{G}$ is the classifying space for proper actions of $\mathcal{G}$. As shown in [Tu99c], one can take $\underline{E}\mathcal{G}$ to be the se of positive measures μ on $\mathcal{G}$ such that $s_*\mu$ is a Dirac measure on $\mathcal{G}^0$ and $|\mu| \in (\frac{1}{2}, 1]$.

The Baum–Connes conjecture with coefficients for groupoids can be stated as follows

Conjecture 13 For every locally compact Haussdorf groupoid with Haar system $\mathcal{G}$ and every $\mathcal{G}$-algebra, $\mu_r^B(\mathcal{G})$ is an isomorphism.

When $B = C_0(\mathcal{G}^0)$, we get the Baum–Connes map without coefficients:

$$\mu_r : K_*^{top}(\mathcal{G}) = K_*^{top}(\mathcal{G}; C_0(\mathcal{G})) = \lim KK_{\mathcal{G}}^*(C_0(Z), B) \rightarrow K_*\left(C_r^*(\mathcal{G})\right),$$

And the conjecture without coefficients states that $\mu_r(\mathcal{G})$ is an isomorphism.

Tu's general definition of the dual-Dirac method as discussed in Section 4.4 is stated in terms of groupoids as follows: let $\mathcal{G}$ be a locally compact, σ-compact groupoid with Haar system. Suppose there exists a proper $\mathcal{G}$-algebra A and elements

$$\eta \in KK_{\mathcal{G}}(C_0(\mathcal{G}^0, A), \quad D \in KK_{\mathcal{G}}(A, C_0(\mathcal{G}^0)),$$

$$\gamma \in KK_{\mathcal{G}}(C_0(\mathcal{G}^0), \mathcal{G}^0))$$

such that $\eta \otimes_A D = \gamma$ and $p^*\gamma = 1 \in KK_{\underline{E}\mathcal{G} \rtimes \mathcal{G}}(C_0(\underline{E}\mathcal{G}), C_0(\underline{E}\mathcal{G}))$, where $p : \underline{E}\mathcal{G} \rightarrow \mathcal{G}^0$ is the source map for the action of $\mathcal{G}$ on $\underline{E}\mathcal{G}$. Then this element is unique and $\mathcal{G}$ is said to have a γ-element. It is the same element as the one constructed by Kasparov for every connected locally compact group [Kas95] (see Section 3.3). Tu's result is stated as follows

Theorem 7.2 ([Tu99c, Proposition 5.23], [Tu99a, Theorem 2.2]) *If the groupoid $\mathcal{G}$ has a γ-element, then the Baum–Connes maps with coefficients μ and μ_r are split injective. Moreover, if $\gamma = 1$ in $KK_{\mathcal{G}}(C_0(\mathcal{G}^0), C_0(\mathcal{G}^0))$, then μ and μ_r with coefficients are isomorphisms and $\mathcal{G}$ is K-amenable.*

As explained by Tu in [Tu00], proofs of injectivity of μ_r based in Theorem 7.2 are constructive: they require explicit constructions of a proper C^*-algebra and the elements in $KK_{\mathcal{G}}$ appearing in the definition of a γ-element, and to do so one uses the existence of an action of the corresponding groupoid on some space with particular geometric properties.

Using Theorem 7.2 Tu proved that the assembly map μ_r is injective for bolic foliations (cf. [Tu99c], Définition 1.15) and that it is a isomorphism for groupoids satisfying the Haagerup property, for example, amenable groupoids (cf. [Tu99b]).

As an example, let us mention that Higson and Roe proved that a discrete group Γ has property A if and only if the groupoid $\beta\Gamma \rtimes \Gamma$ is amenable, where $\beta\Gamma$ is the Stone-Čech compactification of Γ (see Section 9.3.1 for a discussion on property A and [HR00]).

Higson also proved that if Γ has property A, then the Baum–Connes map with coefficients μ_r for Γ is injective and $C_r^*(\Gamma)$ is an exact C^*-algebra [Hig00].

On the other hand, Skandalis, Tu, and Yu proved in [STY02] that Γ can be coarsely embedded into a Hilbert space if and only if $\beta\Gamma \rtimes \Gamma$ has Haagerup property. If this is the case, then the Baum–Connes map with coefficients for Γ is injective.

We mention here that there is also a Banach version of the dual-Dirac technique for groupoids developed by Lafforgue in [Laf07]. He defined a KK-theory for Banach algebras that is equivariant with respect to the action of a groupoid and he used a notion of unconditional completion that he established in this context to prove the Baum–Connes conjecture with commutative coefficients for hyperbolic groups.

7.2 Counter-examples for groupoids

This section is based on sections 1 and 2 of [HLS02]. Let $\mathcal{G}$ be a locally compact, Hausdorff groupoid. Say that a closed subset F of the unit space $\mathcal{G}^0$ is *saturated* if every morphism with source in F has also range in F. Set $U = \mathcal{G} \backslash F$. Let $\mathcal{G}_F$ be the groupoid obtained by restricting $\mathcal{G}$ to F, and let $\mathcal{G}_U$ be the open subgroupoid of $\mathcal{G}$ comprising those morphisms with source and range in U. Then there is a short exact sequence at the level of maximal C^*-algebras:

$$0 \to C^*_{\max}(\mathcal{G}_U) \to C^*_{\max}(\mathcal{G}) \to C^*_{\max}(\mathcal{G}_F) \to 0,$$

but the corresponding sequence at the level of *reduced* C^*-algebras

$$0 \to C_r^*(\mathcal{G}_U) \to C_r^*(\mathcal{G}) \to C_r^*(\mathcal{G}_F) \to 0$$

may fail to be exact; in favorable circumstances this lack of exactness can even be detected at the level of K-theory. This can be exploited to produce counter-examples to the Baum–Connes conjecture.

Lemma 7.3 *Assume that the sequence*

$$K_0\left(C_r^*(\mathcal{G}_U)\right) \to K_0\left(C_r^*(\mathcal{G})\right) \to K_0\left(C_r^*(\mathcal{G}_F)\right) \tag{7.1}$$

is NOT exact in the middle term. If the assembly map $K_0^{top}(\mathcal{G}_F) \to K_0(C_r^*(\mathcal{G}_F))$ *is injective, then the assembly map* $K_0^{top}(\mathcal{G}) \to K_0(C_r^*(\mathcal{G}))$ *is NOT surjective.*

Proof By contrapositive, we assume that $K_0^{top}(\mathcal{G}) \to K_0(C_r^*(\mathcal{G}))$ is surjective, and prove that the sequence 7.1 is exact. For this we chase around the commutative diagram:

$$\begin{array}{ccccccccc}
 & & & & K_0^{top}(\mathcal{G}) & \longrightarrow & K_0^{top}(\mathcal{G}_F) & & \\
 & & & & \downarrow & & \downarrow & & \\
0 & \longrightarrow & K_0(C^*_{\max}(\mathcal{G}_U)) & \longrightarrow & K_0(C^*_{\max}(\mathcal{G})) & \longrightarrow & K_0(C^*_{\max}(\mathcal{G}_F)) & \longrightarrow & 0 \\
 & & \downarrow & & \downarrow & & \downarrow & & \\
0 & \longrightarrow & K_0(C^*_r(\mathcal{G}_U)) & \longrightarrow & K_0(C^*_r(\mathcal{G})) & \longrightarrow & K_0(C^*_r(\mathcal{G}_F) & \longrightarrow & 0
\end{array}$$

Let y be in the kernel of $K_0(C_r^*(\mathcal{G})) \to K_0(C_r^*(\mathcal{G}_F))$. By the assumed surjectivity of the assembly map for $\mathcal{G}$, we write y as the image of $x \in K_0^{top}(\mathcal{G})$. Then the image of x in $K_0^{top}(\mathcal{G}_F)$ is zero, by the assumed injectivity of the assembly map for $\mathcal{G}_F$. So $\mu_{\max}(x)$ is in the kernel of $K_0(C^*_{\max}(\mathcal{G})) \to K_0(C^*_{\max}(\mathcal{G}_F))$ and therefore in the image of $K_0(C^*_{\max}(\mathcal{G}_U))$, by exactness of the middle row. So $y = \mu_r(x)$ is in the image of $K_0(C_r^*(\mathcal{G}_U))$. □

Let us give a simple example where this happens.

Definition 7.4 A group Γ is *residually finite* if Γ admits a *filtration*, i.e., a decreasing sequence $(N_k)_{k>0}$ of finite index normal subgroups with trivial intersection.

We recall that finitely generated linear groups are residually finite, which provides a wealth of examples. If $(N_k)_{k>0}$ is a filtration of Γ, we denote by λ_{Γ/N_k} the representation of Γ obtained by composing the regular representation of Γ/N_k with the quotient map $\Gamma \to \Gamma/N_k$, and by λ^0_{Γ/N_k} the restriction of λ_{Γ/N_k} to the orthogonal of constants.

Definition 7.5 If $(N_k)_{k>0}$ is a filtration of Γ, the group Γ has *property* (τ) *with respect to the filtration* $(N_k)_{k>0}$ if the representation $\oplus_{k>0}\lambda^0_{\Gamma/N_k}$ does not almost admit invariant vectors.

It follows from Proposition 5.3 that a residually finite group with property (T) has property (τ) with respect to every filtration. For a group like the free group, this property depends crucially on the choice of a filtration.

Fix now a filtration $(N_k)_{k>0}$ in the residually finite[21] group Γ_∞, let $q_k : \Gamma_\infty \to \Gamma_k = \Gamma_\infty/N_k$ be the quotient homomorphism. Let $\overline{\mathbf{N}} = \mathbf{N} \cup \{\infty\}$ be the one-point compactification of $\mathbf{N}$, endow $\overline{\mathbf{N}} \times \Gamma_\infty$ with the following equivalence relation:

$$(m,g) \sim (n,h) \Leftrightarrow \begin{cases} \text{either } m = n = \infty \text{ and } & g = h \\ \text{or } \quad m = n \in \mathbf{N} \text{ and } & q_m(g) = q_m(h). \end{cases}$$

Let $\mathcal{G}$ be the groupoid with set of objects $\mathcal{G}^0 = \overline{\mathbf{N}}$, and with set of morphisms $\mathcal{G}^1 = (\overline{\mathbf{N}} \times \Gamma_\infty)/\sim$, with the quotient topology; observe that $\mathcal{G}$ is a Hausdorff groupoid, as $(N_k)_{k>0}$ is a filtration. We may view $\mathcal{G}$ as a continuous field of groups over $\overline{\mathbf{N}}$, with Γ_k sitting over $k \in \overline{\mathbf{N}}$. Set $F = \{\infty\}$ and $U = \mathbf{N}$.

Proposition 7.6 *Let Γ_∞ be an infinite, discrete subgroup of $SL_n(\mathbf{R})$. Assume that there exists a filtration $(N_k)_{k>0}$ such that Γ_∞ has property (τ) with respect to it. Let $\mathcal{G}$ be the groupoid construct above, associated with this filtration. The assembly map for $\mathcal{G}$ is not surjective.*

Proof We check the two assumptions of Lemma 7.3. First, $\mathcal{G}_F = \Gamma_\infty$. As the assembly map μ_r is injective for every closed subgroup of any connected Lie group (e.g., $SL_n(\mathbf{R})$), it is injective for $\mathcal{G}_F$. It remains to see that the sequence (7.1) is not exact in our case. For the representation $\pi = \oplus_{k>0}\lambda_{\Gamma_\infty/N_k}$ of Γ_∞, denote by $C^*_\pi(\Gamma_\infty)$ the completion of $\mathbf{C}\Gamma_\infty$ defined by π. Because of property (τ) there exists a Kazhdan projection $e_\pi \in C^*_\pi(\Gamma_\infty)$ that projects on the Γ_∞-invariant vectors[22] in every representation of $C^*_\pi(\Gamma_\infty)$.

Now $C^*_r(\mathcal{G})$ is the completion of $C_c(\mathcal{G}^1)$ for the norm

$$\|f\| = \sup_{k\in\overline{\mathbf{N}}} \|\lambda_{\Gamma_\infty/N_k}(f_k)\|,$$

where $f \in C_c(\mathcal{G}^1)$ and $f_k = f|_{\{k\}\times\Gamma_k}$.

Consider the homomorphism $\mathbf{C}\Gamma_\infty \to C_c(\mathcal{G}^1)$ which to $g \in \Gamma_\infty$ associates the characteristic function of the set of $(k,h) \in \overline{\mathbf{N}} \times \Gamma_\infty$ such that $h = q_k(g)$. It extends to a homomorphism $\alpha : C^*_\pi(\Gamma_\infty) \to C^*_r(\mathcal{G})$, as is easily checked. The projection $\alpha(e_\pi)$ is in the kernel of the map $C^*_r(\mathcal{G}) \to C^*_r(\mathcal{G}_F)$: as Γ_∞ is infinite, its regular representation has no non-zero invariant vector. Therefore the class $[\alpha(e_\pi)] \in K_0(C^*_r(\mathcal{G}))$ is in the kernel of the map $K_0(C^*_r(\mathcal{G})) \to K_0(C^*_r(\mathcal{G}_F))$.

On the other hand $\mathcal{G}_U = \coprod_{k>0}(\Gamma_\infty/N_k)$, so $C^*_r(\mathcal{G}_U) = \oplus_{k>0}C^*(\Gamma_\infty/N_k)$ (a C^*-direct sum) and $K_0(C^*_r(\mathcal{G}_U)) = \oplus_{k>0}K_0(C^*(\Gamma_\infty/N_k))$ (an algebraic direct sum). Considering now the natural homomorphism $\lambda_{\Gamma_\infty/N_k} : C^*_r(\mathcal{G}_U) \to C^*(\Gamma_\infty/N_k)$, we see in this way that $(\lambda_{\Gamma_\infty/N_k})_*(x) \neq 0$ for only finitely many k's if x lies in the

[21] Until the end of Proposition 7.6, we denote a countable group by Γ_∞ rather than Γ, as we view Γ_∞ as the limit of its finite quotients Γ_k.

[22] If Γ_∞ has property (T), e_π is the image in $C^*_\pi(\Gamma_\infty)$ of the Kazhdan projection $e_G \in C^*_{\max}(\Gamma_\infty)$ from Proposition 5.4.

image of $K_0(C_r^*(\mathcal{G}_U))$ in $K_0(C_r^*(\mathcal{G}))$, while $(\lambda_{\Gamma_\infty/N_k})_*[\alpha(e_\pi)] \neq 0$ for every $k \in \mathbf{N}$. This shows that $[\alpha(e_\pi)]$ is not in the image of $K_0(C_r^*(\mathcal{G}_U))$. □

Example 7.7 Explicit examples where Proposition 7.6 applies, are $SL_n(\mathbf{Z})$ with $n \geq 3$ and any filtration (because of property (T)), and $SL_2(\mathbf{Z})$ with a filtration by congruence subgroups (property (τ) is established in [Lub10]).

The paper [HLS02] by Higson–Lafforgue–Skandalis contains several other counter-examples to the Baum–Connes conjecture for groupoids:

- injectivity counter-examples for Hausdorff groupoids;
- injectivity counter-examples for (non-Hausdorff) holonomy groupoids of foliations;
- surjectivity counter-examples for semi-direct product groupoids $Z \rtimes \Gamma$, where Z is a suitable locally compact space carrying an action of a Gromov monster Γ (see Section 9.2 below for more on Gromov monsters). In terms of C^*-algebras, since $C_r^*(Z \rtimes \Gamma) = C_r^*(\Gamma, C_0(Z))$, this is a counter-example for the Baum–Connes conjecture with coefficients (Conjecture 5).

8 The coarse Baum–Connes conjecture (CBC)

We dedicate this section to the memory of John Roe (1959–2018)

The idea behind coarse, or large scale-geometry is very simple: ignore the local, small-scale features of a geometric space and concentrate on its large-scale, or long-term, structure. By doing so, trends or qualities may become apparent which are obscured by small-scale irregularities. For a metric space X, the *coarse Baum–Connes conjecture* postulates an isomorphism

$$\mu_X : KX_*(X) = \lim_{d\to\infty} K_*(P_d(X)) \xrightarrow{\simeq} K_*(C^*(X)),$$

where the actors only depend on large scale, or coarse structure of X. The right-hand side is the K-theory of a certain C^*-algebra, the *Roe algebra of* X—a noncommutative object; while the left-hand side is the limit of the K-homology groups of certain metric spaces (i.e., commutative objects), namely Rips complexes of X, see Definition 6.26, and the isomorphism should be given by a concrete map, the *coarse assembly map* μ_X. This way the analogy with the classical Baum–Connes conjecture (Conjecture 4) becomes apparent: both are in the spirit of bridging noncommutative geometry with classical topology and geometry. CBC has several applications, e.g., the Novikov conjecture (Conjecture 2.5) when $X = \Gamma$, a finitely generated group equipped with a word metric.

Let (X, d_X), (Y, d_Y) be metric spaces, and $f : X \to Y$ a map (not necessarily continuous). We say that f is *almost surjective* if there exists $C > 0$ such that Y is the C-neighborhood of $f(X)$. Recall that f is a *quasi-isometric embedding* if there exists $A > 0$ such that

$$\frac{1}{A}d_X(x,x') - A \le d_Y(f(x), f(x')) \le Ad_X(x,x') + A,$$

for every $x, x' \in X$, and that f is a *quasi-isometry* if f is a quasi-isometric embedding which is almost surjective. A weaker condition is provided by coarse embeddings, relevant for large-scale structure and corresponding to injections in the coarse category: f is a *coarse embedding* if there exist functions $\rho_+, \rho_- : \mathbf{R}^+ \to \mathbf{R}^+$ (called control functions) such that $\lim_{t\to\infty} \rho_\pm(t) = \infty$ and

$$\rho_-(d_X(x,x')) \le d_Y(f(x), f(x')) \le \rho_+(d_X(x,x'))$$

for every $x, x' \in X$. Finally, f is a *coarse equivalence* if f is a coarse embedding which is almost surjective; coarse equivalences are isomorphisms in the coarse category.

8.1 Roe algebras

8.1.1 Locality conditions on operators

Let (X, d_X) be a proper metric space. A *standard module over* $C_0(X)$ is a Hilbert space $\mathcal{H}_X$ carrying a faithful representation of $C_0(X)$, whose image meets the compact operators only in 0. Fix a bounded operator T on $\mathcal{H}_X$. A point $(x, x') \in X \times X$ is in *the complement of the support of* T if there exists $f, f' \in C_c(X)$, with $f(x) \neq 0 \neq f'(x')$ and $f'Tf = 0$.

Say that T is *pseudo-local* if the commutator $[T, f]$ is compact for every $f \in C_0(X)$, that T is *locally compact* if Tf and fT are compact operators for every $f, f' \in C_0(X)$. Say that T has *finite propagation* if the support of T is contained in a neighborhood of the diagonal in X of the form $\{(x,x') \in X \times X : d_X(x,x') \le R\}$.

Definition 8.1 The Roe algebra $C^*(X)$ is the norm closure of the set of locally compact operators with finite propagation on $\mathcal{H}_X$.

It can be shown that $C^*(X)$ does not depend on the choice of the standard module $\mathcal{H}_X$ over $C_0(X)$. The K-theory $K_*(C^*(X))$ will be the right-hand side of the CBC.

Example 8.2 If X is a uniformly discrete metric space (i.e., the distance between two distinct points is bounded below by some positive number), then we may take $\mathcal{H}_X = \ell^2(X) \otimes \ell^2(\mathbf{N})$, any operator $T \in \mathcal{B}(\mathcal{H}_X)$ can be viewed as a matrix $T = (T_{xy})_{x,y\in X}$. Then T is locally compact if and only if T_{xy} is compact for every $x, y \in X$, and T has finite propagation if and only if there is $R > 0$ such that $T_{xy} = 0$ for $d(x, y) > R$. In particular $\ell^\infty(X, \mathcal{K})$, acting diagonally on $\mathcal{H}_X$, is contained in $C^*(X)$.

Example 8.3 Let Γ be a finitely generated group, endowed with the word metric $d(x, y) = |x^{-1}y|_S$ associated with some finite generating set S of G. Let $|\Gamma|$ denote the underlying metric space, which is clearly uniformly discrete. Let ρ be the right regular representation of G on $\ell^2(G)$; observe that, because $d(xg, x) = |g|_S$, the operator $\rho(g) \otimes 1$ has finite propagation. Actually the Roe algebra in this case is $C^*(|\Gamma|) = \ell^\infty(\Gamma, \mathcal{K}) \rtimes_r \Gamma$, where Γ acts via ρ.

8.1.2 Paschke duality and the index map

Let X be a proper metric space and $\mathcal{H}_X$ a standard module over $C_0(X)$, as in the previous paragraph. Denote by $\Psi_0(X, \mathcal{H}_X)$ the set of pseudo-local operators, and by $\Psi_{-1}(X, \mathcal{H}_X)$ the set of locally compact operators. It follows from the definitions that $\Psi_0(X, \mathcal{H}_X)$ is a C^*-algebra containing $\Psi_{-1}(X, \mathcal{H}_X)$ as a closed 2-sided ideal.

The K-homology of X may be related to the K-theory of the quotient

$$\Psi_0(X, \mathcal{H}_X)/\Psi_{-1}(X, \mathcal{H}_X).$$

For $i = 0, 1$ there are maps

$$K_i(\Psi_0(X, \mathcal{H}_X)/\Psi_{-1}(X, \mathcal{H}_X)) \to K_{1-i}(X) \tag{8.1}$$

defined as follows: for $i = 0$, let p be a projection in $\Psi_0(X, \mathcal{H}_X)/\Psi_{-1}(X, \mathcal{H}_X)$ (or in a matrix algebra over $\Psi_0(X, \mathcal{H}_X)/\Psi_{-1}(X, \mathcal{H}_X)$), form the self-adjoint involution $f = 2p - 1$, let F be a self-adjoint lift of f in $\Psi_0(X, \mathcal{H}_X)$. Then the pair $(\mathcal{H}_X, F)$ is an odd Fredholm module over $C_0(X)$, in the sense of Definition 3.1, so it defines an element of the K-homology $K_1(X)$. For $i = 1$, let u be a unitary in $\Psi_0(X, \mathcal{H}_X)/\Psi_{-1}(X, \mathcal{H}_X)$ (or in a matrix algebra over it), let U be a lift of u in $\Psi_0(X, \mathcal{H}_X)$, form the self-adjoint operator

$$F = \begin{pmatrix} 0 & U \\ U & 0 \end{pmatrix}$$

on $\mathcal{H}_X \oplus \mathcal{H}_X$: then $(\mathcal{H}_X \oplus \mathcal{H}_X, F)$ is an even Fredholm module over $C_0(X)$, defining an element of the K-homology $K_0(X)$. Paschke [Pas81] proved that, when $\mathcal{H}_X$ is a standard module, the homomorphisms in 8.1 are isomorphisms: this is Paschke duality.

Now define $D^*(X, \mathcal{H}_X)$ as the norm closure of the pseudo-local, finite propagation operators. It is clear that $C^*(X)$ is a closed 2-sided ideal in $D^*(X, \mathcal{H}_X)$. It was proved by Higson and Roe (see [HR95], lemma 6.2), that the inclusion $D^*(X, \mathcal{H}_X) \subset \Psi_0(X, \mathcal{H}_X)$ induces an isomorphism $D^*(X, \mathcal{H}_X)/C^*(X) \simeq \Psi_0(X, \mathcal{H}_X)/\Psi_{-1}(X, \mathcal{H}_X)$ of quotient C^*-algebras. Now consider the 6-term exact sequence in K-theory associated with the short exact sequence

$$0 \to C^*(X) \to D^*(X, \mathcal{H}_X) \to D^*(X, \mathcal{H}_X)/C^*(X) \to 0;$$

the connecting maps $K_{1-i}(D^*(X, \mathcal{H}_X)/C^*(X)) \rightarrow K_i(C^*(X)$ $(i = 0, 1)$ can be seen as maps $K_{1-i}(\Psi_0(X, \mathcal{H}_X)/\Psi_{-1}(X, \mathcal{H}_X)) \rightarrow K_i(C^*(X))$. Applying Paschke duality, we get an *index map*

$$Ind_X : K_*(X) \rightarrow K_*(C^*(X)),$$

for every proper metric space X.

Example 8.4 If X is compact, then $C^*(X)$ is the C^*-algebra of compact operators, so $K_0(C^*(X)) = \mathbf{Z}$ and the map $Ind_X : K_0(X) \rightarrow \mathbf{Z}$ is the usual index map that associates its Fredholm index to an even Fredholm module over $C(X)$.

8.2 Coarse assembly map and Rips complex

8.2.1 The Rips complex and its K-homology

We now define the left-hand side of the assembly map, in terms of Rips complexes. Recall from Definition 6.26 that, for X a locally finite metric space (i.e., every ball in X is finite) and $d \geq 0$, the *Rips complex* $P_d(X)$ is the simplicial complex with vertex set X, such that a subset F with $(n+1)$-elements spans a n-simplex if and only if $diam(F) \leq d$. We define a metric on $P_d(X)$ by taking the maximal metric that restricts to the spherical metric on every n-simplex—the latter being obtained by viewing the n-simplex as the intersection of the unit sphere S^n with the positive octant in $\mathbf{R}^{n+1}$.

The *coarse K-homology* of X is then defined as:

$$KX_*(X) := \lim_{d\rightarrow\infty} K_*(P_d(X));$$

this will be the left-hand side of the CBC. Observe that, for every $d \geq 0$, the spaces X and $P_d(X)$ are coarsely equivalent. Then, taking K-theory, we see that $\lim_{d\rightarrow\infty} K_*(C^*(P_d(X)))$ is isomorphic to $K_*(C^*(X))$.

Example 8.5 If Γ is a finitely generated group and $X = |\Gamma|$, then $KX_*(X) = \lim_Y K_*(Y)$, where Y runs in the directed set of closed, Γ-compact subsets of the classifying space for proper actions $\underline{E}\Gamma$. This is to say that CBC can really be seen as a non-equivariant version of the Baum–Connes Conjecture 4.

8.2.2 Statement of the CBC

The index map $Ind_{P_d(X)}$ is compatible with the maps $K_*(P_d(X)) \rightarrow K_*(P_{d'}(X))$ and $K_*(C^*(P_d(X))) \rightarrow K_*(C^*(P_{d'}(X)))$ induced by the inclusion $P_d(X) \rightarrow P_{d'}(X)$ for $d < d'$. Passing to the limit for $d \rightarrow \infty$, we get the *coarse assembly map*

$$\mu_X : KX_*(X) \to K_*(C^*(X)).$$

Say that X has *bounded geometry* if, for every $R > 0$, the cardinality of balls of radius R is uniformly bounded over X. Here is now the statement of the *coarse Baum–Connes conjecture*.

Conjecture 14 (CBC) For every space X with bounded geometry, the coarse assembly map μ_X is an isomorphism.

8.2.3 Relation to the Baum–Connes conjecture for groupoids

It is a result of Yu [Yu95] that, if Γ is a finitely generated group, the CBC for the metric space $|\Gamma|$ is the usual Baum–Connes conjecture for Γ with coefficients in the C^*-algebra $\ell^\infty(\Gamma, \mathcal{K})$ (compare with Example 8.3). Skandalis et al. [STY02] generalize this by associating to every discrete metric space X with bounded geometry, a groupoid $\mathcal{G}(X)$ such that the coarse assembly map for X is equivalent to the Baum–Connes assembly map for $\mathcal{G}(X)$ with coefficients in the C^*-algebra $\ell^\infty(X, \mathcal{K})$.

Let us explain briefly the groupoid $\mathcal{G}(X)$. So let X be a countable metric space with bounded geometry. A subset E of $X \times X$ is called an *entourage* if d is bounded on E, i.e., if there exists $R > 0$ such that $\forall (x, y) \in E, \quad d(x, y) \leq R$.

Let

$$\mathcal{G}(X) = \bigcup_{E \text{ entourage}} \overline{E} \subset \beta(X \times X),$$

where $\beta(X \times X)$ is the Stone-Čech compactification of $X \times X$ and $\overline{E}$ is the closure of E in $\beta(X \times X)$. $\mathcal{G}(X)$ is the spectrum of the abelian C^*-subalgebra of $\ell^\infty(X \times X)$ generated by the characteristic functions χ_E of entourages E. Skandalis, Tu, and Yu proved that it can be endowed with a structure of groupoid extending the one on $X \times X$. Recall that $X \times X$ is endowed with a structure of groupoid where the source and range are defined by $s(x, y) = y$ and $r(x, y) = x$. These maps extend to maps from $\beta(X \times X)$ to βX, hence to maps from $\mathcal{G}(X)$ to βX so that $\mathcal{G}(X)$ is a groupoid whose unit space is βX and which is étale, locally compact, Hausdorff and principal (cf. [STY02], Proposition 3.2).

In the case where X is a finitely generated discrete group Γ with a word metric, the groupoid $\mathcal{G}(X)$ is $\beta\Gamma \rtimes \Gamma$. Skandalis, Tu, and Yu proved the following result.

Theorem 8.6 ([STY02]) *Let X be a discrete metric space with bounded geometry. Then X has property A(in the sense of Definition 9.4 below) if and only if $\mathcal{G}(X)$ is amenable. Moreover, X is coarsely embedded into a Hilbert space if and only if $\mathcal{G}(X)$ has Haagerup property.*

The coarse Baum–Connes conjecture can be put inside the framework of the conjecture for groupoids: let $C^*(X)$ be the Roe algebra associated to (X, d), see Definition 8.1. Then $C^*(X)$ is isomorphic to the reduced crossed product $C_r^*(\mathcal{G}(X), \ell^\infty(X, \mathcal{K}))$ and the coarse assembly map identifies with the Baum–Connes assembly map for the groupoid $\mathcal{G}(X)$ with coefficients in $\ell^\infty(X, \mathcal{K})$.

8.2.4 The descent principle

For a finitely generated group Γ, there is a "descent principle" saying that the CBC for $|\Gamma|$ implies the Novikov conjecture for Γ (see Theorem 8.4 in [Roe96])

Theorem 8.7 *Let Γ be a finitely generated group. Assume that Γ admits a finite complex as a model for its classifying space $B\Gamma$. If CBC holds for the underlying metric space $|\Gamma|$, then the assembly map μ_Γ is injective; in particular the Novikov conjecture (Conjecture 3) holds for Γ.*

8.3 Expanders

Expanders are families of sparse graphs which are ubiquitous in mathematics, from theoretical computer science to dynamical systems, to coarse geometry.

Let $X = (V, E)$ be a finite, connected, d-regular graph. The *combinatorial Laplace operator* on X is the operator Δ on $\ell^2(V)$ defined by

$$(\Delta f)(x) = d \cdot f(x) - \sum_{y \in V : y \sim x} f(y),$$

where $f \in \ell^2(V)$ and $\sim$ denotes the adjacency relation on X.

It is well known from algebraic graph theory (see, e.g., [Lub10, DSV03]) that, if X has n vertices, the spectrum of Δ consists of n eigenvalues (repeated according to multiplicity):

$$0 = \lambda_0 < \lambda_1 \leq \lambda_2 \leq \ldots \leq \lambda_{n-1} \in [0, 2d].$$

On the other hand, the *Cheeger constant*, or *isoperimetric constant* of X, is defined as

$$h(X) = \inf_{A \subset V} \frac{|\partial A|}{\min\{|A|, |V \backslash A|\}},$$

where ∂A is the *boundary* of A, i.e., the set of edges connecting A with $V \backslash A$. The Cheeger constant measures the difficulty of disconnecting X.

The *Cheeger–Buser inequality* says that $h(X)$ and $\lambda_1(X)$ essentially measure the same thing:

$$\frac{\lambda_1(X)}{2} \leq h(X) \leq \sqrt{2k\lambda_1(X)}.$$

Expanders are families of large graphs which are simultaneously sparse (i.e., they have few edges, a condition ensured by d-regularity, with d fixed) and hard to disconnect (a condition ensured by $h(X)$ being bounded away form 0).

Definition 8.8 A family $(X_k)_{k>0}$ of finite, connected, d-regular graphs is a family of expanders if $\lim_{k\to\infty} |V_k| = +\infty$ and there exists $\varepsilon > 0$ such that $\lambda_1(X_k) \geq \varepsilon$ for all k (equivalently: there exists $\varepsilon' > 0$ such that $h(X_k) \geq \varepsilon'$ for every k).

The tension between sparsity of X and $h(X)$ being bounded away from 0, makes the mere existence of expanders non-trivial. The first explicit construction, using property (T), is due to Margulis:

Theorem 8.9 *Let Γ be a discrete group with property (T), let $S = S^{-1}$ be a finite, symmetric, generating set of Γ. Assume that Γ admits a sequence of finite index normal subgroups $N_k \lhd \Gamma$ with $\lim_{k\to\infty}[\Gamma : N_k] = +\infty$. Then the sequence of Cayley graphs $(Cay(\Gamma/N_k, S))_{k>0}$ is a family of expanders.*

Example 8.10 Take $\Gamma = SL_d(\mathbf{Z})$, with $d \geq 3$, and $N_k = \Gamma(k)$ the congruence subgroup of level k, i.e., the kernel of the map of reduction modulo k:

$$\Gamma(k) = \ker(SL_d(\mathbf{Z}) \to SL_d(\mathbf{Z}/k\mathbf{Z})).$$

Coarse geometry prompts us to view a family $(X_k)_{k>0}$ of finite connected graphs as a single metric space. This is achieved by the *coarse disjoint union*: on the disjoint union $\coprod_{k>0} X_k$, consider a metric d such that the restriction of d to each component X_k is the graph metric, and $d(X_k, X_\ell) \geq diam(X_k) + diam(X_\ell)$ for $k \neq \ell$. Such a metric is unique up to coarse equivalence.

A favorite source of examples comes from *box spaces*, that we now define. Let Γ be a finitely generated, residually finite group, and let $(N_k)_{k>0}$ be a filtration in the sense of Definition 7.4. If S is a finite, symmetric, generating set of Γ, we may form the Cayley graph $Cay(\Gamma/N_k, S)$, as in Theorem 8.9.

Definition 8.11 The coarse disjoint union $\coprod_{k>0} Cay(\Gamma/N_k, S)$ is the box space of Γ associated with the filtration $(N_k)_{k>0}$.

It is clear that, up to coarse equivalence, it does not depend on the finite generating set S, so we simple write $\coprod_{k>0} \Gamma/N_k$. By Theorem 8.9, any box space of a residually finite group with property (T) is an expander. More generally, it is a result by Lubotzky and Zimmer [LZ89] that $\coprod_{k>0} Cay(\Gamma/N_k, S)$ is a family of expanders if and only if Γ has property (τ) with respect to the filtration $(N_k)_{k>0}$, in the sense of Definition 7.5.

For future reference (see Section 9.4.1), we give one more characterization of expanders:

Proposition 8.12 *Let $(X_k)_{k>0}$ be a sequence of finite, connected, d-regular graphs with $\lim_{k\to\infty} |V_k| = +\infty$. The family $(X_k)_{k>0}$ is a family of expanders if and only if there exists $C > 0$ such that, for every map f from $\coprod_{k>0} X_k$ to a Hilbert space $\mathcal{H}$, the following Poincaré inequality holds for every $k > 0$:*

$$\frac{1}{|V_k|^2} \sum_{x,y\in V_k} \|f(x) - f(y)\|^2 \leq \frac{C}{|V_k|} \sum_{x\sim y} \|f(x) - f(y)\|^2. \tag{8.2}$$

Proof

(1) Let $X = (V, E)$ be a finite connected graph. We first re-interpret the first non-zero eigenvalue λ_1 of Δ. Consider two quadratic forms on $\ell^2(V)$, both with kernel the constant functions: $\phi \mapsto \frac{1}{|V|^2} \sum_{x,y\in V} |\phi(x) - \phi(y)|^2$ and $\phi \mapsto \frac{1}{|V|} \sum_{x\sim y} |\phi(x) - \phi(y)|^2$. Then $\frac{1}{\lambda_1}$ is the smallest constant $K > 0$ such that[23]

$$\frac{1}{|V|^2} \sum_{x,y\in V} |\phi(x) - \phi(y)|^2 \leq \frac{K}{|V|} \sum_{x\sim y} |\phi(x) - \phi(y)|^2$$

for all $\phi \in \ell^2(V)$.

(2) By the first step, the sequence $(X_k)_{k>0}$ is an expander if and only if there exists a constant C such that, for every function ϕ on $\coprod_{k>0} X_k$, we have:

$$\frac{1}{|V_k|^2} \sum_{x,y\in V_k} |\phi(x) - \phi(y)|^2 \leq \frac{C}{|V_k|} \sum_{x\sim y} |\phi(x) - \phi(y)|^2.$$

(3) Taking a map $f : \coprod_{k>0} X_k \to \mathcal{H}$ and expanding in some orthonormal basis of $\mathcal{H}$, we immediately deduce inequality (8.2) from the 2nd step.

□

8.4 Overview of CBC

8.4.1 Positive results

The CBC was formulated by Roe in 1993, see [Roe93].

- Yu 2000: if a discrete metric space with bounded geometry that admits a coarse embedding into Hilbert space, then CBC holds for X, see [Yu00];

[23] The re-interpretation goes as follows: fix an auxiliary orientation on the edges of E, allowing one to define the *coboundary operator* $d : \ell^2(V) \to \ell^2(E) : \phi \mapsto d\phi$, where $d\phi(e) = \phi(e^+) - \phi(e^-)$. Observe that $\Delta = d^*d$, so that $\langle \Delta\phi, \phi\rangle = \|d\phi\|^2 = \frac{1}{2} \sum_{x\sim y} |\phi(x) - \phi(y)|^2$. By the Rayleigh quotient, $\frac{1}{\lambda_1}$ is the smallest constant $K > 0$ such that $\|\phi\|^2 \leq K\|d\phi\|^2$ for every $\phi \perp 1$. We leave the rest as an exercise.

- Kasparov and Yu 2006: if X is a discrete metric space with bounded geometry that coarsely embeds into a super-reflexive Banach space, then the coarse Novikov conjecture (i.e., the injectivity of μ_X) holds for X, see [KY06].

8.4.2 Negative results

- Yu 1998: the coarse assembly map is not injective for the coarse disjoint union $\coprod_{n>0} n \cdot S^{2n}$, where $n \cdot S^{2n}$ denotes the sphere of radius n in $(2n+1)$-Euclidean space, with induced metric, see [Yu98].
- Willett and Yu 2012: the coarse assembly map is not surjective for expanders with large girth, see [WY12].
- Higson, Lafforgue, and Skandalis 2001: the coarse assembly map is not surjective for box spaces of residually finite groups Γ which happen to be expanders, when Γ moreover satisfies injectivity of the assembly map with coefficients, see [HLS02].

Let us describe those counter-examples of Higson et al. [HLS02] more precisely. We first observe (building on Lemma 7.3) that any family of expanders provides a counter-example either to injectivity or to surjectivity of the Baum–Connes assembly map for suitable associated groupoids. To see this, let $(X_k)_{k>0}$ be a family of d-regular expanders, and let $X = \coprod_{k>0} X_k$ be their coarse disjoint union. Let $\mathcal{G}(X)$ be the groupoid associated to X, as in Section 8.2.2. Let $F = \beta(X)\backslash X$ be a saturated closed subset in the space of objects, and $U = X$ its complement.

Proposition 8.13 *Let X be the coarse disjoint union of a family of d-regular expanders. Let $\mathcal{G}(X)$ be the associated groupoid, set $F = \beta(X)\backslash X$. Either the assembly map is not injective for the groupoid $\mathcal{G}(X)_F$ or the coarse assembly map is not surjective for the space X. The same holds true for the assembly map with coefficients in $\ell^\infty(X, \mathcal{K})$.*

Sketch of proof In view of Lemma 7.3, we must check that

$$K_0\left(C_r^*(\mathcal{G}(X)_U)\right) \to K_0\left(C_r^*(\mathcal{G}(X))\right) \to K_0\left(C_r^*(\mathcal{G}(X)_F)\right)$$

is NOT exact in the middle term. Set $\mathcal{H}_X = \ell^2(X) \otimes \ell^2(\mathbf{N})$, fix some rank 1 projection $e \in \mathcal{K}(\ell^2(\mathbf{N}))$ on some unit vector ξ, let Δ_k denote the combinatorial Laplacian on X_k, and set $\Delta_X = \oplus_{k>0}(\Delta_k \otimes e)$. Then Δ_X a locally compact operator with finite propagation on $\mathcal{H}_X$, as such it defines an element of the Roe algebra $C^*(X)$. The fact that $(X_k)_{k>0}$ is a family of expanders exactly means that 0 is isolated in the spectrum of Δ_X. By functional calculus, the spectral projector p_X associated with $\{0\}$ is also in $C^*(X)$. Now the kernel of Δ_k on $\ell^2(X_k)$ is spanned by u_k, with $u_k = (1, 1, \dots, 1)$, so the restriction of p_X to $\ell^2(X_k)\otimes\ell^2(\mathbf{N})$ is $p_k\otimes(1-e)$, where p_k is the $|V_k| \times |V_k|$-matrix with all entries equal to $\frac{1}{|V_k|}$. In particular entries $(p_X)_{x,y}$ of p_X, go to 0 when $d(x, y) \to \infty$, so p_X is in the kernel of the map $C_r^*(\mathcal{G}(X))) \to C_r^*(\mathcal{G}(X)_F)$.

It remains to show that the class $[p_X]$ in $K_0(C_r^*(\mathcal{G}(X)))$ does not lie in the image of $K_0(C_r^*(\mathcal{G}(X)_U))$. To see this, first observe that $\mathcal{G}(X)_U$ is the groupoid with space of objects X and exactly one morphism between every two objects. So $C_r^*(\mathcal{G}(X)_U)$ is nothing but $\mathcal{K}(\ell^2(X))$. To proceed, for an operator T with finite propagation on X, denote by T_k the restriction of T to $X_k \times X_k$. If S, T are operators with finite propagation then, for k large enough, we have $(ST)_k = S_k T_k$: the reason is that, given $R > 0$, for $k \gg 0$ an R-neighborhood in X coincides with an R-neighborhood in X_k, as the X_k's are further and further apart. As a consequence, there exists a homomorphism

$$C^*(X) \to \left(\prod_{k>0} \mathcal{K}(\ell^2(X_k)) \otimes \mathcal{K})/(\oplus_{k>0}\mathcal{K}(\ell^2(X_k)) \otimes \mathcal{K}\right),$$

that factors through $C^*(X)/\mathcal{K}(\ell^2(X))$. To conclude, it is enough to show that the image of $[p_X]$ is non-zero in $K_0\big((\prod_{k>0} \mathcal{K}(\ell^2(X_k)) \otimes \mathcal{K})/(\oplus_{k>0}\mathcal{K}(\ell^2(X_k)) \otimes \mathcal{K})\big)$. For this observe that p_X lifts to a projector $\tilde{p}_X \in \prod_{k>0} \mathcal{K}(\ell^2(X_k)) \otimes \mathcal{K}$, and that projections on all factors define a homomorphism

$$K_0\left(\prod_{k>0} \mathcal{K}(\ell^2(X_k)) \otimes \mathcal{K}\right) \to \mathbf{Z}^{\mathbf{N}}$$

that maps $[\tilde{p}_X]$ to $(1, 1, 1, \dots) \in \mathbf{Z}^{\mathbf{N}}$. Since that homomorphism also maps the group $K_0(\oplus_{k>0}\mathcal{K}(\ell^2(X_k)) \otimes \mathcal{K})$ to $\mathbf{Z}^{(\mathbf{N})}$, we have shown that $[\tilde{p}_X]$ is not in the image of

$$K_0(\oplus_{k>0}\mathcal{K}(\ell^2(X_k)) \otimes \mathcal{K}) \to K_0\left(\prod_{k>0} \mathcal{K}(\ell^2(X_k)) \otimes \mathcal{K}\right),$$

so $[p_X] \neq 0$ in $K_0((\prod_{k>0} \mathcal{K}(\ell^2(X_k)) \otimes \mathcal{K})/(\oplus_{k>0}\mathcal{K}(\ell^2(X_k)) \otimes \mathcal{K}))$. □

By carefully choosing the family of expanders, we get actual counter-examples to surjectivity in the CBC. For this we need a group Γ exactly as in Proposition 7.6 (with explicit examples provided by Example 7.7), and a box space in the sense of Definition 8.11.

Theorem 8.14 *Let Γ be an infinite, discrete subgroup of $SL_n(\mathbf{R})$, endowed with a filtration $(N_k)_{k>0}$ such that Γ has property (τ) with respect to it. Then the coarse assembly map for the box space X associated with this filtration, is not surjective.*

Proof Because of property (τ), the space X is the coarse disjoint union of a family of expanders, and Proposition 8.13 will apply. Since by [STY02] the coarse assembly map for X is the Baum–Connes assembly map for the groupoid $\mathcal{G}(X)$ with coefficients in $\ell^\infty(X, \mathcal{K})$, by Lemma 7.3 it is enough to check that the assembly map for the groupoid $\mathcal{G}(X)_F$ is injective with coefficients in $\ell^\infty(X, \mathcal{K})$. Now, because X is a box space, $\mathcal{G}(X)_F$ identifies with the semi-direct product groupoid $(\beta(X)\backslash X) \rtimes \Gamma$. Since Γ is a discrete subgroup of $SL_n(\mathbf{R})$, the assembly map $\mu_{A,r}$ is injective for any coefficient C^*-algebra A: this proves the desired injectivity, so the coarse assembly map for X is not surjective by Proposition 8.13. □

8.5 Warped cones

Warped cones were introduced by Roe in 2005, see [Roe05]; he had the intuition that they might lead to counter-examples to CBC. Let (Y, d_Y) be a compact metric space. Let Γ be a finitely generated group, with a fixed finite generating set S. Assume that Γ acts on Y by Lipschitz homeomorphisms, not necessarily preserving d_Y. The *warped metric* d_S on Y is the largest metric $d_S \leq d_Y$ such that, for every $x \in Y, s \in S, d_S(sx, x) \leq 1$. It is given by

$$d_S(x, y)$$
$$= \inf\{n + \sum_{i=0}^{n} d_Y(x_i, y_i) : x_0 = x, y_n = y, x_i = s_i(y_{i-1}), s_i \in S \cup S^{-1}, n \in \mathbf{N}\}.$$

Intuitively, we modify the metric d_Y by introducing "group shortcuts," as two points $x, \gamma x$ will end at distance $d_S(x, \gamma x) \leq |\gamma|_S$, where $|.|_S$ denotes word length on Γ.

Form the "cone" $Y \times]1, +\infty[$, with the distance d given by:

$$d_{Cone}((y_1, t_1), (y_2, t_2)) =: |t_1 - t_2| + \min\{t_1, t_2\} \cdot d_Y(y_1, y_2).$$

Let Γ act trivially on the second factor. The *warped cone* $\mathcal{O}_\Gamma Y$ is the cone $Y \times]1, +\infty[$, with the warped metric obtained from d_{Cone}. To get an intuition of what the warped metric does on the level sets $Y \times \{t\}$: assume for a while that Y is a closed Riemannian manifold, fix a $\frac{1}{t}$-net on Y, and consider the Voronoi tiling of Y associated to this net (if y is a point in the net, the tile around y is the set of points of Y closer to y that to any other point in the net). Define a graph X_t whose vertices are closed Voronoi tiles, and two tiles T_1, T_2 are adjacent if there exists $s \in S \cup S^{-1} \cup \{1\}$ such that $s(\overline{T_1}) \cap \overline{T_2} \neq \emptyset$. Then the family of level sets $(Y \times \{t\})_{t>1}$ is uniformly quasi-isometric to the family of graphs $(X_t)_{t>1}$ (i.e., the quasi-isometry constants do not depend on t).

In 2015, Druţu and Nowak [DN17] made Roe's intuition more precise with the following conjecture. Assume that, on top of the above assumptions, Y carries a Γ-invariant probability measure ν such that the action $\Gamma \curvearrowright (Y, \nu)$ is ergodic. Assume that the measure ν is adapted to the metric d_Y in the sense that $\lim_{r \to 0} \sup_{y \in Y} \nu(B(y, r)) = 0$.

Conjecture 15 If the action of Γ on Y has a spectral gap (i.e., the Γ-representation on $L^2_0(Y, nu)$ does not have almost invariant vectors), then $\mathcal{O}_\Gamma Y$ violates CBC.

At the time of writing, warped cones are a hot topic:

- Nowak and Sawicki 2015: warped cones do not embed coarsely into a large class of Banach spaces (those with non-trivial type), containing in particular all L^p-spaces ($1 \leq p < +\infty$), see [NS17].
- Vigolo 2016: relates warped cones and expanders, therefore getting new families of expanders [Vig19].
- Sawicki 2017: the level sets $Y \times \{t\}$ of warped cones provide new examples of super-expanders, i.e., expanders not embedding coarsely into any Banach space with non-trivial type, see [Saw17b].
- de Laat and Vigolo 2017: those examples of super-expanders are different (i.e., not coarsely equivalent) to Lafforgue's super-expanders, see [dLV17].
- Fisher, Nguyen, and Van Limbeek 2017: there is a continuum of coarsely pairwise inequivalent super-expanders obtained from warped cones, see [FNVL17]. See Section 9.4.1 for super-expanders.

In 2017, Sawicki [Saw17c] confirmed Roe's intuition by proving the following form of Conjecture 15.

Theorem 8.15 *Let Γ having Yu's property A. Assume that Γ acts on Y by Lipschitz homeomorphisms, freely, and with a spectral gap. Set $A = \{2^n : n \in \mathbf{N}\} \subset]1, +\infty[$, let $\mathcal{O}'_\Gamma Y$ be the subspace $Y \times A \subset \mathcal{O}_\Gamma Y$, equipped with the warped cone metric. Then μ_{CBC} is not surjective for $\mathcal{O}'_\Gamma Y$.* □

By looking at actions on Cantor sets, Sawicki is even able to produce counter-examples to CBC which are NOT coarsely equivalent to any family of graphs.

9 Outreach of the Baum–Connes conjecture

The Baum–Connes conjecture and the coarse Baum–Connes conjecture prompted a surge of activity at the interface between operator algebras and other fields of mathematics, e.g., geometric group theory and metric geometry. Indeed results like the Higson–Kasparov theorem (see Theorem 5.7 above) are of the form "*groups (resp. spaces) in a given class satisfy the Baum–Connes (resp. coarse Baum–Connes) conjecture.*" This leads naturally to trying to extend the class of groups (resp. spaces) in question, as a way of enlarging the domain of validity of either conjecture. The study of a class of groups (resp. spaces) has two obvious counterparts: providing new examples, and studying permanence properties of the class. We sketch some of those developments below.

9.1 The Haagerup property

The 5-authors book [CCJ+01] was the first survey on the subject. Although motivated by Theorem 5.7, it barely mentions the Baum–Connes conjecture and focuses on new examples and stability properties. It was updated in the paper [Val18], which can serve as a guide to more recent literature. Here we mention some long-standing open questions on the Haagerup property, and partial results.

- Let B_n denote the braid group on n strands. Does B_n have the Haagerup property? Yes trivially for $B_2 \simeq \mathbf{Z}$, and yes easily for $B_3 \simeq \mathbf{F}_2 \rtimes \mathbf{Z}$. A recent result by Haettel [Hae19] shows that, if the general answer is affirmative, it will not be for a very geometric reason: for $n \geq 4$, the group B_n has no proper, cocompact isometric action on a CAT(0) cube complex.[24] Note that a fairly subtle proof of the Baum–Connes conjecture with coefficients for B_n, has been given by Schick [Sch07].
- Unlike amenability or property (T), the Haagerup property is *not* stable under extensions.[25] The standard examples to see this are $\mathbf{Z}^2 \rtimes SL_2(\mathbf{Z})$ and $\mathbf{R}^2 \rtimes SL_2(\mathbf{R})$, where the relative property (T) with respect to the non-compact normal subgroup, is an obstruction to the Haagerup property. However, the Haagerup property is preserved by some types of semi-direct products: e.g., Cornulier–Stalder–Valette [CSV12] proved that, if Γ, Λ are countable groups with the Haagerup property, then the wreath product $\Gamma \wr \Lambda = (\oplus_\Lambda \Gamma) \rtimes \Lambda$ has the Haagerup property. A probably difficult question: is G, N are locally compact groups with the Haagerup property and G acts continuously on N by automorphisms, under which conditions on the action $G \curvearrowright N$ does the semi-direct product $N \rtimes G$ have the Haagerup property? When G, N are σ-compact and N is *abelian*, the answer was provided by Cornulier–Tessera (Theorem 4 in [CT11]): $N \rtimes G$ has the Haagerup property if and only if there exists a net $(\mu_i)_{i \in I}$ of Borel probability measures on the Pontryagin dual $\hat{N}$, such that there is a weak-* convergence $\mu_i \to \delta_1$, and $\mu_i\{1\} = 0$ for every $i \in I$, and $\|g \cdot \mu_i - \mu_i\| \to 0$ uniformly on compact subsets of G, and finally the Fourier transform $\widehat{\mu_i}$ is a C_0 function on N for every $i \in I$.
- The behavior of the Haagerup property under central extensions is a widely open question. More precisely: if Z is a closed central subgroup in the locally compact group G, is it true that G has the Haagerup property if and only if G/Z has it? Both implications are open. See Proposition 4.2.14 and Section 7.3.3 in [CCJ+01] for partial results on lifting the Haagerup property from G/Z to G, in particular from $SU(n, 1)$ to $\widetilde{SU(n, 1)}$.

[24] Recall that a group acting properly isometrically on a CAT(0) cube complex, has the Haagerup property, see, e.g., Corollary 1 in [Val18].

[25] Amenability (resp. property (T)) can be defined by a fixed point property: existence of a fixed point for affine actions on compact convex sets (resp. affine isometric actions on Hilbert spaces). This makes clear that it is preserved under extensions.

- The Haagerup property for discrete groups is stable under free products or more generally amalgamated products over finite groups, by Proposition 6.2.3(1) of [CCJ$^+$01]. In general, it is *not* true that, if A, B have the Haagerup property and C is a common subgroup, then $A *_C B$ has the Haagerup property: see section 4.3.3 in [Val18] for an example with $C = \mathbf{Z}^2$. An open question concerns the permanence of the Haagerup property for amalgamated products $A *_C B$ with C virtually cyclic; The first positive result was obtained recently by Carette et al. [CWW17]: recall that if a group G acts by isometries on a metric space (X, d), the action of G on X is said to be semisimple if, for every $g \in G$, the infimum $\inf_{x\in X} d(gx, x)$ is actually a minimum. They proved that, if A, B are groups acting properly and semisimply on some real hyperbolic space $\mathbb{H}^n(\mathbf{R})$, and C is a cyclic subgroup common to A and B, then the amalgamated product $A *_C B$ has the Haagerup property.

9.2 *Coarse embeddings into Hilbert spaces*

In 2000, Yu [Yu00] opened a new direction in mathematics by uniting the fields of K-theory for C^*-algebras and of metric embeddings into Hilbert space. Indeed he proved that if a metric space X with bounded geometry coarsely embeds into Hilbert space, then X satisfies the CBC. Using the descent principle (Theorem 8.7), this implies that if some Cayley graph $|\Gamma|$ of a finitely generated group Γ coarsely embeds into Hilbert space, then the Baum–Connes assembly map for Γ is injective,[26] i.e., the assembly map μ embeds the K-homology of the classifying space $B\Gamma$ into the K-theory of the reduced C^*-algebra of Γ. This implies the Novikov conjecture on the homotopy invariance of the higher signatures for Γ. This was a stunning result, as a strong topological conclusion resulted from a weak metric assumption.

Finitely generated groups with the Haagerup property coarsely embed into Hilbert space. Indeed if α is a proper isometric action of Γ on $\mathcal{H}$, then for every $x \in \mathcal{H}$ the orbit map $g \mapsto \alpha(g)x$ is a coarse embedding.

Using their groupoid approach, Skandalis, Tu, and Yu (Theorem 6.1 in [STY02]) proved the following:

Theorem 9.1 *Let Γ be a finitely generated group that admits a coarse embedding into Hilbert space. Then the assembly map $\mu_{A,r}$ is injective for every $\Gamma - C^*$-algebra A.*

Lots of finitely generated groups embed coarsely into Hilbert space, as they satisfy the stronger property A (see Section 9.3.1 below). Actually it is not even easy to find a *bounded geometry space* not embedding coarsely. The most famous

[26]Under the assumption that $|\Gamma|$ coarsely embeds into Hilbert space, the assumption that $B\Gamma$ is a finite complex was removed by Skandalis et al. [STY02], using their groupoid approach to CBC.

example is due to Matousek [Mat97], and was popularized by Gromov [Gro03]; we will give a proof of a stronger result in Proposition 9.17:

Proposition 9.2 *Let X be the coarse disjoint union of a family of expanders. Then X does not coarsely embed into Hilbert space.*

In [Gro03], Gromov sketched the construction of families of groups containing families of expanders coarsely embedded in their Cayley graphs, which therefore do not embed coarsely into Hilbert space. These are called Gromov's *random groups*, or *Gromov monsters*. Details of their construction were supplied by Arzhantseva and Delzant [AD08]. It was shown by Higson et al. [HLS02] that those groups provide counter-examples to the Baum–Connes conjecture with coefficients (Conjecture 5).

Theorem 9.3 *Let Γ be a Gromov monster. Consider the commutative C^*-algebra $A = \ell^\infty(\mathbf{N}, c_0(\Gamma))$, with the natural Γ-action. Then the Baum–Connes conjecture with coefficients fails for Γ and A, in the sense that $\mu_{A,r}$ is not onto.*

We will come back on those groups in Section 9.3.1, and explain what exactly is needed to get counter-examples to Conjecture 5.

9.3 Yu's property A: a polymorphous property

One of the crucial new invariants of metric spaces introduced by Yu [Yu00] is property A, a non-equivariant form of amenability. Like standard amenability, it has several equivalent definitions. In particular we will see that three concepts from different areas (property A for discrete spaces, boundary amenability from topological dynamics, and exactness from C^*-algebra theory) provide one and the same concept when applied to finitely generated groups.

9.3.1 Property A

Definition 9.4 Let (X, d) be a discrete metric space. The space X has property A if there exists a sequence $\Phi_n : X \times X \to \mathbf{C}$ of normalized, positive-definite kernels on X such that Φ_n is supported in some entourage,[27] and $(\Phi_n)_{n>0}$ converges to 1 uniformly on entourages for $n \to \infty$.

This is inspired by the following characterization of amenability for a countable group Γ: the group Γ is amenable if and only if there exists a sequence $\varphi_n : \Gamma \to \mathbf{C}$ of normalized, finitely supported, positive-definite functions on Γ such that $(\varphi_n)_{n>0}$ converges to 1 for $n \to \infty$. If this happens and if Γ is finitely generated, then $\Phi_n(s,t) = \varphi_n(s^{-1}t)$ witnesses that $|\Gamma|$ has property A. However, there are many

[27] Recall from Section 8.2.3 that an entourage is a subset of $X \times X$ on which $d(.,.)$ is bounded.

more examples of finitely generated groups with property A. Other natural examples are provided by linear groups, i.e., subgroups of the group $GL_n(F)$ for some field F, this is a result by Guentner et al. [GHW05]; this class includes many groups with property (T). The list of classes of groups that satisfy property A also includes one-relator groups, Coxeter groups, groups acting on finite-dimensional CAT(0) cube complexes, and many more.

Theorem 9.5 (see Theorem 2.2 in [Yu00]) *A discrete metric space with property A admits a coarse embedding into Hilbert space.*

The converse is false: endow $\{0,1\}^n$ with the Hamming distance; then the coarse disjoint union $\coprod_n \{0,1\}^n$ coarsely embeds into Hilbert space but does not have property A, as proved by Nowak [Now07]; however, this space does not have bounded geometry. For a while, an unfortunate situation was that the only way of disproving property A for a space X, was to prove that X has no coarse embedding into Hilbert space (see Section 9.2). The situation began to evolve with a paper of Willett [Wil11] containing a nice result addressing property A directly: the coarse disjoint union of a sequence of finite regular graphs with girth tending to infinity (i.e., graphs looking more and more like trees), does not have property A. On the other hand some of them can be coarsely embedded into Hilbert space, as was shown by Arzhantseva et al. [AGv12] using box spaces of the free group. For every group G, denote by $G^{(2)}$ the normal subgroup generated by squares in G, and define a decreasing sequence of subgroups in G by $G_0 = G$ and $G_n = G_{n-1}^{(2)}$. The main result of [AGv12] is:

Theorem 9.6 *For the free group $\mathbf{F}_k$ of rank $k \geq 2$, with $(\mathbf{F}_k)_n$ defined as above, the box space $\coprod_{n>0} \mathbf{F}_k/(\mathbf{F}_k)_n$ does not have property A but is coarsely embeddable into Hilbert space.*

To summarize the above discussion, we have a square of implications, for finitely generated groups (where CEH stands for *coarse embeddability into Hilbert space*):

$$\begin{array}{ccc} \text{amenable} & \Longrightarrow & \text{property A} \\ \Downarrow & & \Downarrow \\ \text{Haagerup property} & \Longrightarrow & \text{CEH} \end{array}$$

Let us observe:

- The top horizontal and the left vertical implications cannot be reversed: indeed a non-abelian free group enjoys both property A and the Haagerup property, but is not amenable.
- The bottom horizontal implication cannot be reversed: $SL_3(\mathbf{Z})$ has CEH but, because of property (T), it does not have the Haagerup property. The same example shows that property A does *not* imply the Haagerup property.

This leaves possibly open the implications "*CEH $\Rightarrow$ property A*" (which was known to be false for spaces, by Theorem 9.6), and the weaker implication

"*Haagerup property ⇒ property A*." The latter was disproved by Osajda [Osa14]: he managed, using techniques of graphical small cancellation, to embed sequences of graphs isometrically into Cayley graphs of suitably constructed groups. This way he could prove:

Theorem 9.7 *There exists a finitely generated group not having property A, but admitting a proper isometric action on a CAT(0) cube complex (and therefore having the Haagerup property).*

We refer to [Khu18] for a nice survey of that work.

9.3.2 Boundary amenability

Let Γ be a countable group; we denote by $Prob(\Gamma)$ the set of probability measures on Γ, endowed with the topology of pointwise convergence.

Definition 9.8

1. Let X be a compact space on which Γ acts by homeomorphisms. We say that the action $\Gamma \curvearrowright X$ is topologically amenable if there exists a sequence of continuous maps $\mu_n : X \to Prob(\Gamma)$ which are almost Γ-equivariant, i.e.,

$$\lim_{n\to\infty} \sup_{x\in X} \|\mu_n(gx) - g\mu_n(x)\|_1 = 0.$$

2. The group Γ is boundary amenable if Γ admits a topologically amenable on some compact space.

For example, the action of Γ on a point is topologically amenable if and only if Γ is amenable, so boundary amenability is indeed a generalization of amenability. We will see in Theorem 9.9 below that, for finitely generated group, boundary amenability is equivalent to property A. Boundary amenability attracted the attention of low-dimensional topologists, so that the following groups were shown to verify it:

- Mapping class groups, see [Ham09, Kid08];
- $Out(\mathbf{F}_n)$, the outer automorphism group of the free group, see [BHG17].

9.3.3 Exactness

For C^*-algebras A, B, denote by $A \otimes_{\min} B$ (resp. $A \otimes_{\max} B$) the minimal (resp. maximal) tensor product. Recall that A is *nuclear* if the canonical map $A \otimes_{\max} B \to A \otimes_{\min} B$ is an isomorphism for every C^*-algebra B, and that A is *exact* if the minimal tensor product with A preserves short exact sequences of C^*-algebras. As the maximal tensor product preserves short exact sequences, nuclear implies exact.

A classical result of Lance says that, for discrete groups, a group Γ is amenable if and only if $C_r^*(\Gamma)$ is nuclear. It turns out that, for exactness we have an analogous result merging this section with Sections 9.3.1 and 9.3.2; it is a combination of results by Anantharaman-Delaroche and Renault [AR01], Guentner and Kaminker [GK02], Higson and Roe [HR00], and Ozawa [Oza00].

Theorem 9.9 *For a finitely generated group Γ, the following are equivalent:*

1. *Γ has property A;*
2. *Γ is boundary amenable;*
3. *$C_r^*(\Gamma)$ is exact.*

Combining with Theorems 9.5 and 9.1, we get immediately:

Corollary 9.10 *If Γ is a finitely generated group with property A, then for every $\Gamma - C^*$-algebra A the assembly map $\mu_{A,r}$ is injective.*

As a consequence of Theorem 9.9, for a finitely generated group Γ, nuclearity and exactness of $C_r^*(\Gamma)$ are quasi-isometry invariants (which is by no means obvious on the analytical definitions). An interesting research question is: which other properties of $C_r^*(\Gamma)$ are quasi-isometry invariants of Γ?

We now explain how the lack of exactness of $C_r^*(\Gamma)$, when detected at the level of K-theory, leads to counter-examples to Conjecture 5.

Definition 9.11 A C^*-algebra C is *half-K-exact* if for any short exact sequence $0 \to J \to A \to B \to 0$ of C^*-algebras, the sequence

$$K_*(J \otimes_{\min} C) \to K_*(A \otimes_{\min} C) \to K_*(B \otimes_{\min} C)$$

is exact in the middle term.

The following statement is an unpublished result by Ozawa (see, however, Theorem 5.2 in [Oza01]).

Theorem 9.12 *Gromov monsters are not half-K-exact.*

Proof Let Γ be a Gromov monster. So there is a family $(X_k)_{k>0}$ of d-regular expanders which coarsely embeds in Γ, i.e., there exists a family of maps $f_k : X_k \to \Gamma$ such that, for $x_k, y_k \in X_k$, we have $d_{X_k}(x_k, y_k) \to +\infty \iff d_\Gamma(f_k(x_k), f_k(y_k)) \to +\infty$. We will need below a consequence of this fact: there exists a constant $K > 0$ such that the fiber $f_k^{-1}(g)$ has cardinality at most K, for every $k > 0$ and every $g \in \Gamma$. (Indeed, first observe that, as a consequence of the coarse embedding, there exists $R > 0$ such that, for every k and g, we have $d_{X_k}(x, y) \leq R$ for every $x, y \in f_k^{-1}(g)$; then use the bounded geometry of the family $(X_k)_{k>0}$: we may, for example, take for K the cardinality of a ball of radius R in the d-regular tree.)

We now start the proof really. Denote by n_k the number of vertices of X_k, and form the product of matrix algebras $M = \prod_{k>0} M_{n_k}(\mathbf{C})$ together with its ideal $J = \oplus_{k>0} M_{n_k}(\mathbf{C})$. We are going to show that the sequence

$$K_0(J \otimes_{\min} C_r^*(\Gamma)) \to K_0(M \otimes_{\min} C_r^*(\Gamma)) \to K_0((M/J) \otimes_{\min} C_r^*(\Gamma))$$

is not exact at its middle term. Let us identify $M_{n_k}(\mathbf{C})$ with $End(\ell^2(X_k)$ via the canonical basis.

We first define an injective homomorphism $\iota_k : M_{n_k}(\mathbf{C}) \to M_{n_k}(\mathbf{C}) \otimes C_r^*(\Gamma)$ by $\iota_k(E_{xy}) = E_{xy} \otimes f_k(x)^{-1} f_k(y)$, where E_{xy} is the standard set of matrix units in $End(\ell^2(X_k))$. We then use an idea similar to the one in the proof of Proposition 8.13. Let Δ_k be the combinatorial Laplace operator on X_k, let p_k be the projection on its 1-dimensional kernel: recall that $(p_k)_{xy} = \frac{1}{n_k}$ for every $x, y \in X_k$. Then $\Delta := (\iota_k(\Delta_k))_{k>0} \in M \otimes_{\min} C_r^*(\Gamma)$ has 0 as an isolated point in its spectrum, as the X_k's are a family of expanders. The spectral projection associated with 0 is $q = (\iota_k(p_k))_{k>0}$. The class $[q] \in K_0(M \otimes_{\min} C_r^*(\Gamma))$ will witness the desired non-exactness.

Let $\pi : M \to M/J$ denote the quotient map. To show that q is in the kernel of $\pi \otimes_{\min} Id$, consider the conditional expectation $\mathbb{E}_M = Id_M \otimes \tau : M \otimes_{\min} C_r^*(\Gamma) \to M$, where τ denotes the canonical trace on $C_r^*(\Gamma)$. We have

$$(Id_{n_k} \otimes \tau)(\iota_k(p_k))_{xy} = \begin{cases} \frac{1}{n_k} \; if \; f_k(x) = f_k(y) \\ 0 \; if \; f_k(x) \neq f_k(y) \end{cases}$$

So the operator norm of $(Id_{n_k} \otimes \tau)(\iota_k(p_k))$ satisfies:

$$\|(Id_{n_k} \otimes \tau)(\iota_k(p_k))\| \leq \frac{1}{n_k} \cdot \max_{x \in X_k} \left| f_k^{-1}(f_k(x)) \right| \leq \frac{K}{n_k},$$

where K is the constant introduced at the beginning of the proof. As a consequence $\mathbb{E}_M(q)$ belongs to J and

$$0=\pi(\mathbb{E}_M(q))=\pi((Id_M \otimes \tau)(q))=(Id_{M/J} \otimes \tau)((\pi \otimes_{\min} Id)(q))=\mathbb{E}_{M/J}((\pi \otimes_{\min} Id)(q));$$

by faithfulness of $\mathbb{E}_{M/J}$ we get $(\pi \otimes_{\min} Id)(q) = 0$.

It remains to show that $[q]$ is not in the image of $K_0(J \otimes_{\min} C_r^*(\Gamma))$ in $K_0(M \otimes_{\min} C_r^*(\Gamma)$. For this, denote by $\sigma_k : M \otimes_{\min} C_r^*(\Gamma) \to M_{n_k}(\mathbf{C}) \otimes C_r^*(\Gamma)$ the projection on the k-th factor. Because $K_0(J \otimes_{\min} C_r^*(\Gamma)) = \oplus_{k>0} K_0(M_{n_k}(\mathbf{C}) \otimes C_r^*(\Gamma))$, for every $x \in K_0(J \otimes_{\min} C_r^*(\Gamma))$ we have $(\sigma_k \otimes \tau)(x) = 0$ for k large enough. On the other hand $(\sigma_k \otimes \tau)(q) > 0$ for every $k > 0$. □

The following result may be extracted from [HLS02], where it is not stated explicitly.

Theorem 9.13 *Let Γ be a countable group. If Γ is not half-K-exact, then there is a C^*-algebra C with trivial Γ-action such that the assembly map*

$$\mu_{C,r} : K_*^{top}(\Gamma, C) \to K_* \left(C_r^*(\Gamma, C)\right)$$

is NOT onto.

The proof will be given below. Combining with Theorem 9.12 and its proof, we immediately get the following:

Corollary 9.14 *If Γ is a Gromov monster, there exists a noncommutative C^*-algebra C with trivial Γ-action such that the assembly map*

$$\mu_{C,r} : K_*^{top}(\Gamma, C) \to K_*\left(C_r^*(\Gamma, C)\right)$$

is NOT onto. □

It seems this is as close as one can get to a counter-example to the Baum–Connes conjecture without coefficients (Conjecture 4).

To prove Theorem 9.13, we start by some recalls about mapping cones.

Definition 9.15 Let $\beta : A \to B$ be a homomorphism of $C*$-algebras. The *mapping cone* of β is the C^*-algebra $C(\beta) = \{(a, f) \in A \oplus C([0,1], B) : f(0) = \beta(a), f(1) = 0\}$.

Consider now the following situation, with three C^*-algebras J, A, B and homomorphisms:

- $\alpha : J \to A$, injective;
- $\beta : A \to B$, surjective, such that $\beta \circ \alpha = 0$.

We then have an inclusion $\gamma : J \to C(\beta) : j \mapsto (\alpha(j), 0)$.

Lemma 9.16

1. *If $Im(\alpha) = \ker(\beta)$, i.e., the sequence $0 \to J \to A \to B \to 0$ is exact, then $\gamma_* : K_*(J) \to K_*(C(\beta))$ is an isomorphism.*
2. *If γ_* is an isomorphism, then the sequence $K_*(J) \stackrel{\alpha_*}{\to} K_*(A) \stackrel{\beta_*}{\to} K_*(B)$ is exact.*
3. *γ_* is an isomorphism if and only if $K_*(C(\gamma)) = 0$.*

Proof of lemma 9.16 1. See Exercise 6.N in [WO93].

2. Set $I = \ker(\beta)$ and $\tilde{\gamma} : I \to C(\beta) : x \mapsto (x, 0)$, so that $\gamma = \tilde{\gamma} \circ \alpha$. Since $\tilde{\gamma}_*$ is an isomorphism by the previous point, and γ_* is an isomorphism by assumption, we get that $\alpha_* : K_*(J) \to K_*(I)$ is an isomorphism. Since the sequence $K_*(I) \longrightarrow K_*(A) \stackrel{\beta_*}{\longrightarrow} K_*(B)$ is exact, so is the sequence

$$K_*(J) \xrightarrow{\alpha_*} K_*(A) \xrightarrow{\beta_*} K_*(B).$$

3. Since γ is injective, we may identify the mapping cone $C(\gamma)$ with $\{f \in C([0,1], C(\beta)) : f(0) \in \gamma(J), f(1) = 0\}$. By evaluation at 0, we get a short exact sequence

$$0 \longrightarrow C_0]0,1[\otimes C(\beta) \longrightarrow C(\gamma) \longrightarrow J \longrightarrow 0.$$

In the associated 6-term exact sequence in K-theory, the use of Bott periodicity to identify $K_*(C_0]0,1[\otimes C(\beta))$ with $K_*(C(\beta))$ allows to identify the connecting maps with γ_*, so the result follows.

□

Proof of Theorem 9.13 Since Γ is not half-K-exact, we find a short exact sequence

$$0 \longrightarrow J \xrightarrow{\alpha} A \xrightarrow{\beta} B \longrightarrow 0 \quad \text{such that}$$

$$K_*\left(J \otimes_{\min} C_r^*(\Gamma)\right) \xrightarrow{(\alpha\otimes_{\min} Id)_*} K_*\left(A \otimes_{\min} C_r^*(\Gamma)\right) \xrightarrow{(\beta\otimes_{\min} Id)_*} K_*\left(B \otimes_{\min} C_r^*(\Gamma)\right) \tag{9.1}$$

is not exact in the middle term. As above, define the mapping cone $C(\beta)$ and the inclusion $\gamma : J \to C(\beta)$. Set $C = C(\gamma)$, with trivial Γ-action. We prove in three steps that the assembly map $\mu_{C,r}$ with coefficients in C, is not onto.

- $K_*(C \otimes_{\min} C_r^*(\Gamma)) = K_*(C(\gamma \otimes_{\min} Id))$ is non-zero: this follows from non-exactness of the sequence (9.1) together with the two last statements of Lemma 9.16.
- $K_*(C \otimes_{\max} C_{\max}^*(\Gamma)) = K_*(C(\gamma \otimes_{\max} Id))$ is zero: this follows from exactness of

$$0 \to J \otimes_{\max} C_{\max}^*(\Gamma) \to A \otimes_{\max} C_{\max}^*(\Gamma) \to B \otimes_{\max} C_{\max}^*(\Gamma) \to 0$$

  together with the first and last statements of Lemma 9.16.
- The assembly map $\mu_{C,r} : K_*^{top}(\Gamma, C) \to K_*(C_r^*(\Gamma, C)) = K_*(C \otimes_{\min} C_r^*(\Gamma))$ is zero, and therefore is not onto: this is because, as explained in the beginning of Section 5.1, $\mu_{C,r}$ factors through

$$\mu_{C,\max} : K_*^{top}(\Gamma, C) \to K_*\left(C_{\max}^*(\Gamma, C)\right) = K_*\left(C \otimes_{\max} C_{\max}^*(\Gamma)\right),$$

  and this is the zero map.

□

9.4 *Applications of strong property (T)*

9.4.1 Super-expanders

A Banach space is *super-reflexive* if it admits an equivalent norm making it uniformly convex. As mentioned in Section 8.4.1 Kasparov and Yu proved in

[KY06] that if a discrete metric space with bounded geometry coarsely embeds into some super-reflexive space, then the coarse assembly map μ_X is injective. Since families of expanders do not embed coarsely into Hilbert space, by Proposition 9.2, it is natural to ask: *is there a family of expanders that admits a coarse embedding into some super-reflexive Banach space?* This is a very interesting open question. However, certain families of expanders are known *not* to embed coarsely into any super-reflexive Banach space, and we wish to explain the link with strong property (T) from Section 6.3.

Let $(X_k = (V_k, E_k))_{k>0}$ be a family of finite, connected, d-regular graphs with $\lim_{k\to\infty} |V_k| = +\infty$, and let B be a Banach space. We say that $(X_k)_{k>0}$ *satisfies a Poincaré inequality with respect to* B if there exists $C = C(B) > 0$ such that for every map $f : \coprod_{k>0} X_k \to B$ we have:

$$\frac{1}{|V_k|^2}\sum_{x,y\in V_k}\|f(x)-f(y)\|_B^2 \le \frac{C}{|V_k|}\sum_{x\sim y}\|f(x)-f(y)\|_B^2. \tag{9.2}$$

Compare with inequality (8.2), which is the Poincaré inequality with respect to Hilbert spaces. In view of Proposition 8.12, the following result implies Proposition 9.2.

Proposition 9.17 *Assume that the family $(X_k)_{k>0}$ satisfies a Poincaré inequality with respect to the Banach space B. Then the coarse disjoint union X of the X_k's, admits no coarse embedding into B.*

Proof Suppose by contradiction that there exists a coarse embedding $f : X \to B$, with control functions $\rho_\pm$. Then, using $\|f(x) - f(y)\|_B \le \rho_+(1)$ for $x \sim y$ in any X_k, we get for every $k > 0$:

$$\begin{aligned}\frac{1}{|V_k|^2}\sum_{x,y\in V_k}\rho_-(d(x,y))^2 &\le \frac{1}{|V_k|^2}\sum_{x,y\in V_k}\|f(x)-f(y)\|_B^2\\ &\le \frac{C}{|V_k|}\sum_{x\sim y}\|f(x)-f(y)\|_B^2\\ &\le \frac{2C|E_k|\rho_+(1)^2}{|V_k|} = dC\rho_+(1)^2,\end{aligned}$$

where the second inequality is the Poincaré inequality and the final equality is $|E_k| = \frac{d|V_k|}{2}$. Set $M = dC\rho_+(1)^2$; since the mean of the quantities $\rho_-(d(x,y))^2$ is at most M, this means that for at least half of the pairs $(x, y) \in V_k \times V_k$, we have $\rho_-(d(x,y))^2 \le 2M$, for every $k > 0$. Since $\lim_{t\to\infty}\rho_-(t) = +\infty$, we find a constant $N > 0$ such that, for every $k > 0$ and at least half of the pairs $(x, y) \in V_k \times V_k$, we have $d(x, y) \le N$. But as X_k is d-regular, the cardinality of a ball of radius N is at most $(d + 1)^N$, so the cardinality of the set of pairs

$(x, y) \in V_k \times V_k$ with $d(x, y) \leq N$, is at most $|V_k|(d+1)^N$. For $k \gg 0$, this is smaller than $\frac{|V_k|^2}{2}$, and we have reached a contradiction. □

Definition 9.18 A sequence $(X_k)_{k>0}$ of finite, connected, d-regular graphs with $\lim_{k\to\infty} |X_k| = +\infty$, is a family of super-expanders if, for any super-reflexive Banach space B, the sequence $(X_k)_{k>0}$ satisfies the Poincaré inequality (9.2) with respect to B.

It follows from Proposition 8.12 that, assuming they do exist, super-expanders are expanders, and from Proposition 9.17 that super-expanders do not admit a coarse embedding into *any* super-reflexive Banach space. Lafforgue's construction of super-expanders in [Laf08, Laf09], following a suggestion by Naor, answered a question from [KY06]:

Theorem 9.19 *Let F be a non-Archimedean local field, let G be a simple algebraic group of higher rank defined over F, and let $G(F)$ be the group of F-rational points of G. Let Γ be a lattice in $G(F)$, fix any filtration $(N_k)_{k>0}$ of Γ. Then the box space $\coprod_{k>0} Cay(\Gamma/N_k, S)$ (see Definition 8.11) is a family of super-expanders.*

Proof Write $X_k =: Cay(\Gamma/N_k, S)$. Let B be a super-reflexive Banach space. The goal is to show that the Poincaré inequality (9.2) is satisfied.

1. Let B_k be the space of functions $X_k \to B$, with norm $\|f\|^2_{B_k} = \frac{1}{|X_k|}\sum_{x\in X_k} \|f(x)\|^2_B$. For $f \in B_k$, set $m_f = \frac{1}{|X_k|}\sum_{x\in X_k} f(x) \in B$. Then[28]
$$\frac{1}{|X_k|^2}\sum_{x,y\in X_k} \|f(x) - f(y)\|^2_B \leq \frac{4}{|X_k|}\sum_{x\in X_k} \|f(x) - m_f\|^2_B. \tag{9.3}$$
To see this: by translation we may assume $m_f = 0$. Then by the triangle inequality:
$$\|f(x) - f(y)\|^2_B \leq (\|f(x)\|_B + \|f(y)\|_B)^2 \leq 2\left(\|f(x)\|^2_B + \|f(y)\|^2_B\right),$$
and inequality 9.3 follows by averaging over $X_k \times X_k$.
2. Let π_k be the natural isometric representation of Γ on B_k. As Γ acts transitively on $X - K$, the fixed point space of Γ in B_k is the space of constant functions. Now strong property (T) for representations in a Banach space is defined by analogy with Definition 6.32, by replacing Hilbert space by a suitable class of Banach spaces: it posits the existence of a Kazhdan projection projecting onto the fixed point space, for any representation in a suitable class. It turns out that the lattice Γ has strong property (T) for isometric representations in super-reflexive Banach spaces: this is due to Lafforgue [Laf08, Laf09] when $G(F)$

[28] Note typos regarding inequality 9.3 in Proposition 5.2 of [Laf08] and in Proposition 5.5 of [Laf09]: $\leq \frac{4}{|X_k|}$ is erroneously written as $= \frac{2}{|X_k|}$.

contains $SL_3(F)$, and to Liao [Lia14] in general. So, denoting by $\mathcal{C}_{0,1}(\Gamma)$ the Banach algebra completion of $\mathbf{C}\Gamma$ with respect to isometric Γ-representations in the spaces $(B_k)_{k>0}$, there exists an idempotent $p \in \mathcal{C}_{0,1}(\Gamma)$ such that in particular $\pi_k(p)f = m_f$ for every $f \in B_k$. Inequality 9.3 is then reformulated

$$\frac{1}{|X_k|^2} \sum_{x,y \in X_k} \|f(x) - f(y)\|_B^2 \leq 4\|f - \pi_k(p)f\|_{B_k}^2. \tag{9.4}$$

3. Let $q \in \mathbf{C}\Gamma$ be an element such that $\|p - q\|_{\mathcal{C}_{0,1}(\Gamma)} < \frac{1}{2}$ and $\sum_\gamma q(\gamma) = 1$. Then

$$\begin{aligned}\|\pi_k(p)f - \pi_k(q)f\|_{B_k} &= \|(\pi_k(p) - \pi_k(q))(f - m_f)\|_{B_k} \\ &\leq \frac{1}{2}\|f - m_f\|_{B_k} = \frac{1}{2}\|f - \pi_k(p)f\|_{B_k};\end{aligned}$$

   but $\|f - \pi_k(p)f\|_{B_k} \leq \|f - \pi_k(q)f\|_{B_k} + \|\pi_k(q)f - \pi_k(p)f\|_{B_k}$ by the triangle inequality, so

$$\|f - \pi_k(p)f\|_{B_k} \leq 2\|f - \pi_k(q)f\|_{B_k},$$

   that we plug in (9.4).
4. Finally it is easy to see that there exists a constant $C_1 > 0$, only depending on q, such that for every $k > 0$:

$$\|f - \pi_k(q)f\|_{B_k}^2 \leq \frac{C_1}{|X_k|} \sum_{x \sim y} \|f(x) - f(y)\|_B^2.$$

□

Later on, other constructions of super-expanders were provided:

- by Mendel and Naor [MN14], using zig-zag products;
- independently by Sawicki [Saw17a] and by de Laat and Vigolo [dLV18], using warped cones, as defined in Section 8.5: the constructions appeal to actions on manifolds of groups with strong property (T).

9.4.2 Zimmer's conjecture

A striking, unexpected application of Lafforgue's strong property (T) from Section 6.3 is the recent solution of Zimmer's conjecture on actions of higher rank lattices on manifolds. Roughly speaking, Zimmer's conjecture claims that a lattice Γ in a higher rank simple Lie group G, has only finite actions on manifolds of dimension small enough (relative to data only associated with G). Somewhat more precisely, in this section:

- *higher rank* means that the real rank of G is at least 2 (think of $G = SL_n(\mathbf{R})$, for $n \geq 3$; or $G = Sp_{2n}(\mathbf{R})$, for $n \geq 2$);
- *manifold* means a smooth closed manifold M;
- *action of* Γ *on* M means an action by diffeomorphisms of class at least C^2;
- *a finite action of* Γ is one that factors through a finite quotient of Γ.

It remains to explain "*dimension small enough*" and for this we will restrict to $G = SL_n(\mathbf{R}), n \geq 3$. For the general case, we refer to Conjecture 1.2 in [BFHS16]. For the original statement by Zimmer, see [Zim87].

If Γ is a lattice in $SL_n(\mathbf{R})$, we may let it act linearly on $\mathbf{R}^n$. So we get an infinite action of Γ on the $(n-1)$-dimensional projective space $P^{n-1}(\mathbf{R})$; we observe that this action has no invariant volume form. On the other hand, $\Gamma = SL_n(\mathbf{Z})$ has an infinite action on the n-dimensional torus $\mathbf{T}^n = \mathbf{R}^n/\mathbf{Z}^n$, this one clearly preserving a volume form. Zimmer's conjecture basically claims that those examples are of minimal dimension among non-finite actions. Precisely, Zimmer's conjecture for cocompact lattices in $SL_n(\mathbf{R})$, is now the following result by Brown et al. (Theorem 1.1 in [BFHS16]):

Theorem 9.20 *Let* Γ *be a cocompact lattice in* $SL_n(\mathbf{R}), n \geq 3$.

1. *If* $\dim M < n-1$, *any action of* Γ *on* M *is finite.*
2. *If* $\dim M < n$, *any volume-preserving action of* Γ *on* M *is finite.*

Let us give a rough sketch, in 3 steps, of the proof of the first statement in Theorem 9.20. So we consider $\alpha : \Gamma \to Diff(M)$, with $\dim M < n-1$, we must show that α is finite.

- Let $\alpha : \Gamma \to Diff^{\infty}(M)$ be a homomorphism (for simplicity we assume that Γ acts by C^{∞} diffeomorphisms). Fix any Riemannian structure on M. For $x \in M, \gamma \in \Gamma$, denote by $D_x\alpha(\gamma)$ the differential of $\alpha(\gamma)$ at x. Then α has *uniform subexponential growth of derivatives*, i.e., for every $\varepsilon > 0$, there exists $C \geq 1$ such that for every $\gamma \in \Gamma$:

$$\sup_{x \in M} \|D_x\alpha(\gamma)\| \leq Ce^{\varepsilon \ell(\gamma)}, \tag{9.5}$$

  where ℓ denotes the word length with respect to a fixed finite generating set of Γ. Morally, this means that generators of Γ are close to being isometries of M.
- A Riemannian structure of class C^k on M is a C^k section of the symmetric square $S^2(TM)$ of the tangent bundle TM of M. Via α, the group Γ acts on C^k Riemannian structures on M and this defines a homomorphism $\alpha_\sharp$ from Γ to the group of invertibles in the algebra $\mathcal{B}(C^k(S^2(TM)))$ of bounded operators on $C^k(S^2(TM))$. At this point we introduce the Hilbert space $\mathcal{H}^k$ which is the Sobolev space of sections of $S^2(TM)$ with weak k-th derivative being L^2. By the Sobolev embedding theorem, we have $\mathcal{H}^k \subset C^\ell(S^2(TM))$ for $k \gg \ell$. If α satisfies (9.5), then $\alpha_\sharp$ has slow exponential growth: for all $\varepsilon > 0$, there exists $C \geq 1$ such that for all $g \in G$:

$$\|\alpha_\sharp(g)\|_{\mathcal{H}^k\to\mathcal{H}^k} \leq Ce^{\varepsilon\ell(g)}.$$

It is here that strong property (T) enters the game; it is, however, needed in a form both stronger and more precise than in Definition 6.32, namely there exists a constant $\delta > 0$ and a sequence μ_n of probability measures supported in the balls $B_\ell(n)$ of radius n in Γ, such that for all $C > 0$ and any representation π on a Hilbert space with $\|\pi(g)\| \leq Ce^{\delta\ell(g)}$, the operators $(\pi(\mu_n))_{n>0}$ converge exponentially quickly to a projection P_∞ onto the space of invariant vectors. That is, there exists $K > 0$ and $0 < \lambda < 1$, independent of π, such that $\|\pi(\mu_i) - P_\infty\| < K\cdot\lambda^i$. Theorem 6.3 in [BFHS16] explains how to deduce the extra desired features (exponentially fast convergence and approximation by positive measures rather than signed measures) from the *proofs* of Theorem 6.34 by Lafforgue, de Laat and de la Salle [Laf08, dlS18, dLdlS15].[29]

Coming back to our sketch of proof of Theorem 9.20:

Proposition 9.21 *$\alpha(\Gamma)$ preserves some C^ℓ Riemannian structure on M.*

Proof We will apply the above form of strong property (T) to the representation $\alpha_\sharp$. Let $(\mu_n)_{n>0}$ be the sequence of probability measures as above, set $P_n = \alpha_\sharp(\mu_n)$, so that $\|P_i - P_\infty\|_{\mathcal{H}^k\to\mathcal{H}^k} < K\cdot\lambda^i$.

We start with any smooth Riemannian metric g on M, view it as an element in $\mathcal{H}^k$, and apply the averaging operators P_i: then $g_i =: P_i(g)$. We set $g_\infty = \lim_{i\to\infty} g_i$, so that g_∞ is $\alpha_\sharp(\Gamma)$-invariant in $\mathcal{H}^k$, hence also in $C^\ell(S^2(TM))$. We have $g_\infty(v,v) \geq 0$ for every $v \in TM$, as g_∞ is a limit of positive-definite forms, but we must show that g_∞ is positive-definite, i.e., $g_\infty(v,v) > 0$ for every unit vector $v \in TM$. By the previous point (subexponential growth of derivatives), taking $e^\varepsilon = \lambda^{-\frac{1}{3}}$, we have for every $\gamma \in \Gamma$:

$$C^2\lambda^{-\frac{2\ell(\gamma)}{3}} \geq \|D\alpha(\gamma^{-1})\|^2 = \sup_{u\in TM} \frac{g(u,u)}{g(D\alpha(\gamma)(u), D\alpha(\gamma)(u))} \geq \frac{1}{g(D\alpha(\gamma)(v), D\alpha(\gamma)(v))}$$

hence, if $\ell(\gamma) \leq i$:

$$g(D\alpha(\gamma)(v), D\alpha(\gamma)(v)) \geq \frac{1}{C^2}\cdot\lambda^{\frac{2\ell(\gamma)}{3}} \geq \frac{1}{C^2}\cdot\lambda^{\frac{2i}{3}}$$

Since μ_i is supported in the ball of radius i of Γ, we have

[29] The subtlety here is that, as lucidly explained in [dlS16], Definition 6.32 for an arbitrary finitely generated group is equivalent to the existence of a sequence of *signed* probability measures as above.

$$g_i(v,v) \geq \frac{1}{C^2} \cdot \lambda^{\frac{2i}{3}}.$$

On the other hand $|g_\infty(v,v) - g_i(v,v)| \leq K \cdot \lambda^i$, hence

$$g_\infty(v,v) \geq g_i(v,v) - K \cdot \lambda^i \geq \frac{1}{C^2} \cdot \lambda^{\frac{2i}{3}} - K \cdot \lambda^i,$$

which is positive for $i \gg 0$. □

- Set $m = \dim M$. Let g be an $\alpha(\Gamma)$-invariant C^ℓ metric on M, so that $\alpha(\Gamma)$ is a subgroup of the isometry group $K =: Isom(M, g)$. Now K is a compact Lie group, of dimension at most $\frac{m(m+1)}{2}$. Assuming by contradiction that $\alpha(\Gamma)$ is infinite, a suitable version of Margulis' super-rigidity says that the Lie algebra $\mathfrak{su}_n$, which is the compact real form of $\mathfrak{sl}_n(\mathbf{R})$, must embed into the Lie algebra of K. Counting dimensions we get

$$n^2 - 1 = \dim \mathfrak{su}_n \leq \dim K \leq \frac{m(m+1)}{2},$$

contradicting the assumption $m < n - 1$. So α is finite.

More recently in [BFHS17], Brown, Fisher, and Hurtado verified Zimmer's conjecture for $SL_3(\mathbf{Z})$. For this they had to appeal to de la Salle's result [dlS18] that strong property (T) holds for arbitrary lattices in higher rank simple Lie groups.

It is expected that in 2019, Brown, Fisher, and Hurtado, with the help of D. Witte-Morris, will complete a proof of Zimmer's conjecture for any lattice in any higher rank simple Lie group.

Acknowledgement Thanks are due to J.-B. Bost, R. Coulon, N. Higson, V. Lafforgue, P.-Y. Le Gall, H. Oyono-Oyono, N. Ozawa, and M. de la Salle for useful conversations and exchanges.

References

[AAS18] P. Antonini, S. Azzali, and G. Skandalis. The Baum-Connes conjecture localized at the unit element of a discrete group. Preprint 2018, `arXiv:1807.05892v1`, 2018.

[ACDB04] F. Astengo, M. Cowling, and B. Di Blasio. The Cayley transform and uniformly bounded representations. *J. Funct. Anal.*, 213(2):241–269, 2004.

[AD08] G. Arzhantseva and T. Delzant. Examples of random groups. Preprint available on http://irma.math.unistra.fr/~delzant/random.pdf, 2008.

[Afg16] A. Afgoustidis. On the analogy between real reductive groups and Cartan motion groups. iii : A proof of the Connes- Kasparov isomorphism. arXiv :1602.08891, 2016.

[AGv12] G. Arzhantseva, E. Guentner, and J. Špakula. Coarse non-amenability and coarse embeddings. *Geom. Funct. Anal.*, 22(1):22–36, 2012.

[AR01] C. Anantharaman and J. Renault. Amenable groupoids. In *Groupoids in analysis, geometry, and physics (Boulder, CO, 1999)*, volume 282 of *Contemp. Math.*, pages 35–46. Amer. Math. Soc., Providence, RI, 2001.

[AS68] M. F. Atiyah and I. M. Singer. The index of elliptic operators. I. *Ann. of Math. (2)*, 87:484–530, 1968.

[AS77] M. F. Atiyah and W. Schmid. A geometric construction of the discrete series for semisimple Lie groups. *Invent. Math.*, 42:1–62, 1977.

[Ati70] M. F. Atiyah. Global theory of elliptic operators. In *Proc. Internat. Conf. on Functional Analysis and Related Topics (Tokyo, 1969)*, pages 21–30. Univ. of Tokyo Press, Tokyo, 1970.

[Ati76] M. F. Atiyah. Elliptic operators, discrete groups and von Neumann algebras. pages 43–72. Astérisque, No. 32–33, 1976.

[AW81] C. Akemann and M. Walter. Unbounded negative definite functions. *Canad. J. Math.*, 33(4):862–871, 1981.

[BBB13] D. Bleecker and B. Booß-Bavnbek. *Index theory—with applications to mathematics and physics*. International Press, Somerville, MA, 2013.

[BC85] P. Baum and A. Connes. Leafwise homotopy equivalence and rational Pontrjagin classes. In *Foliations (Tokyo, 1983)*, volume 5 of *Adv. Stud. Pure Math.*, pages 1–14. North-Holland, Amsterdam, 1985.

[BC00] P. Baum and A. Connes. Geometric K-theory for Lie groups and foliations. *Enseign. Math. (2)*, 46(1-2):3–42, 2000.

[BCH94] P. Baum, A. Connes, and N. Higson. Classifying space for proper actions and K-theory of group C^*-algebras. In *C^*-algebras: 1943–1993 (San Antonio, TX, 1993)*, volume 167 of *Contemp. Math.*, pages 240–291. Amer. Math. Soc., Providence, RI, 1994.

[BdlHV08] B. Bekka, P. de la Harpe, and A. Valette. *Kazhdan's property (T)*, volume 11 of *New Mathematical Monographs*. Cambridge University Press, Cambridge, 2008.

[BFHS16] A. Brown, D. Fisher, and S. Hurtado-Salazar. Zimmer's conjecture: Subexponential growth, measure rigidity, and strong property (T). Preprint `arXiv:1608.04995`, 2016.

[BFHS17] A. Brown, D. Fisher, and S. Hurtado-Salazar. Zimmer's conjecture for actions of $\mathrm{SL}(m, \mathbb{Z})$. Preprint `arXiv:1710.02735` , 2017.

[BG88] R. Beals and P. Greiner. *Calculus on Heisenberg manifolds*, volume 119 of *Annals of Mathematics Studies*. Princeton University Press, Princeton, NJ, 1988.

[BHG17] M. Bestvina, C. Horbez, and V. Guirardel. Boundary amenability of $Out(F_n)$. Preprint 2017, `arXiv:1705.07017`, 2017.

[BHS10] P. Baum, N. Higson, and T. Schick. A geometric description of equivariant K-homology for proper actions. In *Quanta of maths*, volume 11 of *Clay Math. Proc.*, pages 1–22. Amer. Math. Soc., Providence, RI, 2010.

[BKKO17] E. Breuillard, M. Kalantar, M. Kennedy, and N. Ozawa. C^*-simplicity and the unique trace property for discrete groups. *Publ. Math. Inst. Hautes Études Sci.*, 126:35–71, 2017.

[BMV05] H. Bettaieb, M. Matthey, and A. Valette. Unbounded symmetric operators in K-homology and the Baum-Connes conjecture. *J. Funct. Anal.*, 229(1):184–237, 2005.

[Bos90] J.-B. Bost. Principe d'Oka, K-théorie et systèmes dynamiques non commutatifs. *Invent. Math.*, 101:261–333, 1990.

[Boy17] A. Boyer. Harish-Chandra's Schwartz algebras associated with discrete subgroups of semisimple Lie groups. *J. Lie Theory*, 27(3):831–844, 2017.

[BT72] F. Bruhat and J. Tits. Groupes réductifs sur un corps local. *Inst. Hautes Études Sci. Publ. Math.*, (41):5–251, 1972.

[CCJ+01] P.-A. Cherix, M. Cowling, P. Jolissaint, P. Julg, and A. Valette. *Groups with the Haagerup property*, volume 197 of *Progress in Mathematics*. Birkhäuser Verlag, Basel, 2001.

[CEM01] J. Chabert, S. Echterhoff, and R. Meyer. Deux remarques sur l'application de Baum-Connes. *C. R. Acad. Sci. Paris Sér. I Math.*, 332(7):607–610, 2001.

[CEN03] J. Chabert, S. Echterhoff, and R. Nest. The Connes-Kasparov conjecture for almost connected groups and for linear p-adic groups. *Publ. Math. Inst. Hautes Études Sci.*, (97):239–278, 2003.

[CGGP92] M. Christ, D. Geller, P. Głowacki, and L. Polin. Pseudodifferential operators on groups with dilations. *Duke Math. J.*, 68(1):31–65, 1992.

[CH89] M. Cowling and U. Haagerup. Completely bounded multipliers of the Fourier algebra of a simple Lie group of real rank one. *Invent. Math.*, 96(3):507–549, 1989.

[Cha03] I. Chatterji. Property (RD) for cocompact lattices in a finite product of rank one Lie groups with some rank two Lie groups. *Geom. Dedicata*, 96:161–177, 2003.

[Che96] Z. Chen. Séries complémentaires des groupes de Lorentz et KK-théorie. *J. Funct. Anal.*, 137(1):76–96, 1996.

[cHJ19] A. Čap, C. Harrach, and P. Julg. A Poisson transform adapted to the Rumin complex. Preprint 2019, `arXiv:1904.00635`, 2019.

[CM82] A. Connes and H. Moscovici. The L^2-index theorem for homogeneous spaces of Lie groups. *Ann. of Math. (2)*, 115(2):291–330, 1982.

[CM90] A. Connes and H. Moscovici. Cyclic cohomology, the Novikov conjecture and hyperbolic groups. *Topology*, 29(3):345–388, 1990.

[Con81] A. Connes. An analogue of the Thom isomorphism for crossed products of a C^*-algebra by an action of **R**. *Adv. in Math.*, 39(1):31–55, 1981.

[Con82] A. Connes. A survey of foliations and operator algebras. In *Operator algebras and applications, Part I (Kingston, Ont., 1980)*, volume 38 of *Proc. Sympos. Pure Math.*, pages 521–628. Amer. Math. Soc., Providence, R.I., 1982.

[Cow82] M. Cowling. Unitary and uniformly bounded representations of some simple Lie groups. In *Harmonic analysis and group representations*, pages 49–128. Liguori, Naples, 1982.

[CR05] I. Chatterji and K. Ruane. Some geometric groups with rapid decay. *Geom. Funct. Anal.*, 15(2):311–339, 2005.

[cS07] A. Čap and V. Souček. Curved Casimir operators and the BGG machinery. *SIGMA Symmetry Integrability Geom. Methods Appl.*, 3:Paper 111, 17, 2007.

[cSS01] A. Čap, J. Slovák, and V. Souček. Bernstein-Gelfand-Gelfand sequences. *Ann. of Math. (2)*, 154(1):97–113, 2001.

[CSV12] Y. Cornulier, Y. Stalder, and A. Valette. Proper actions of wreath products and generalizations. *Trans. Amer. Math. Soc.*, 364(6):3159–3184, 2012.

[CT11] Yves Cornulier and Romain Tessera. A characterization of relative Kazhdan property T for semidirect products with abelian groups. *Ergodic Theory Dynam. Systems*, 31(3):793–805, 2011.

[CWW17] Mathieu Carette, Daniel T. Wise, and Daniel J. Woodhouse. The A-T-menability of some graphs of groups with cyclic edge groups. *Math. Proc. Cambridge Philos. Soc.*, 163(1):145–159, 2017.

[DH17] S. Dave and S. Haller. Graded hypoellipticity of BGG sequences. `arXiv:1705.01659`, 2017.

[Dix96] J. Dixmier. *Les C^*-algèbres et leurs représentations*. Les Grands Classiques Gauthier-Villars. [Gauthier-Villars Great Classics]. Éditions Jacques Gabay, Paris, 1996. Reprint of the second (1969) edition.

[DL98] J. Davis and W. Lück. Spaces over a category and assembly maps in isomorphism conjectures in K- and L-theory. *K-Theory*, 15(3):201–252, 1998.

[dLdlS15] T. de Laat and M. de la Salle. Strong property (T) for higher-rank simple Lie groups. *Proc. Lond. Math. Soc. (3)*, 111(4):936–966, 2015.

[dlH88] P. de la Harpe. Groupes hyperboliques, algèbres d'opérateurs et un théorème de Jolissaint. *C. R. Acad. Sci. Paris Sér. I Math.*, 307(14):771–774, 1988.

[dlS16] M. de la Salle. A local characterization of Kazhdan projections and applications. to appear in *Comment. Math. Helvet.* `arXiv:1604.01616`, 2016.

[dlS18] M. de la Salle. Strong property (T) for higher rank lattices. Preprint `arXiv:1711.01900`, 2018.

[dLV17] T. de Laat and F. Vigolo. Superexpanders from group actions on compact manifolds. `arXiv:1707.01399`, 2017.

[dLV18] T. de Laat and F. Vigolo. Superexpanders from group actions on compact manifolds. Preprint 2018, to appear in Geometriae Dedicata `arXiv:1707.01399v2`, 2018.

[DN17] C. Druţu and P. Novak. Kazhdan projections, random walks and ergodic theorems. *Journal für die reine und angewandte Mathematik*, 2017. https://doi.org/10.1515/crelle-2017-0002.

[DS14] C. Debord and G. Skandalis. Adiabatic groupoid, crossed product by $\mathbb{R}_+^*$ and pseudodifferential calculus. *Adv. Math.*, 257:66–91, 2014.

[DSV03] G. Davidoff, P. Sarnak, and A. Valette. *Elementary number theory, group theory, and Ramanujan graphs*, volume 55 of *London Mathematical Society Student Texts*. Cambridge University Press, Cambridge, 2003.

[FL11] F. Forstnerič and F. Lárusson. Survey of Oka theory. *New York J. Math.*, 17A:11–38, 2011.

[FNVL17] D. Fisher, T Nguyen, and W. Van Limbeek. Rigidity of warped cones and coarse geometry of expanders. Preprint, `arXiv:1710.03085`, 2017.

[FPV17] R. Flores, S. Pooya, and A. Valette. K-homology and K-theory for the lamplighter groups of finite groups. *Proc. Lond. Math. Soc. (3)*, 115(6):1207–1226, 2017.

[FRR95] S. Ferry, A. Ranicki, and J. Rosenberg. A history and survey of the Novikov conjecture. In *Novikov conjectures, index theorems and rigidity, Vol. 1 (Oberwolfach, 1993)*, volume 226 of *London Math. Soc. Lecture Note Ser.*, pages 7–66. Cambridge Univ. Press, Cambridge, 1995.

[GA09] M. P. Gomez Aparicio. Représentations non unitaires, morphisme de Baum-Connes et complétions inconditionnelles. *J. Noncommut. Geom.*, 3(3):419–446, 2009.

[GA10] M. P. Gomez-Aparicio. Morphisme de Baum-Connes tordu par une représentation non unitaire. *J. K-Theory*, 6(1):23–68, 2010.

[Gai86] P.-Y. Gaillard. Transformation de Poisson de formes différentielles. Le cas de l'espace hyperbolique. *Comment. Math. Helv.*, 61(4):581–616, 1986.

[GHW05] E. Guentner, N. Higson, and S. Weinberger. The Novikov conjecture for linear groups. *Publ. Math. Inst. Hautes Études Sci.*, (101):243–268, 2005.

[GK02] E. Guentner and J. Kaminker. Exactness and the Novikov conjecture. *Topology*, 41(2):411–418, 2002.

[Gra94] H. Grauert. *Selected papers. Vol. I, II*. Springer-Verlag, Berlin, 1994. With commentary by Y. T. Siu et al.

[Gro89] M. Gromov. Oka's principle for holomorphic sections of elliptic bundles. *J. Amer. Math. Soc.*, 2(4):851–897, 1989.

[Gro03] M. Gromov. Random walk in random groups. *Geom. Funct. Anal.*, 13(1):73–146, 2003.

[GWY19] S. Gong, J. Wu, and G. Yu. The Novikov conjecture, the group of volume preserving diffeomorphisms and Hilbert-Hadamard spaces. Preprint 2019, `arXiv:1811.02086v2`, 2019.

[Hae19] T. Haettel. Virtually cocompactly cubulated artin-tits groups. Preprint 2019, `arXiv:1509.08711v4`, march 2019.

[Hp04] I. Hambleton and E. K. Pedersen. Identifying assembly maps in K- and L-theory. *Math. Ann.*, 328(1–2):27–57, 2004.

[Ham09] U. Hamenstädt. Geometry of the mapping class groups. I. Boundary amenability. *Invent. Math.*, 175(3):545–609, 2009.

[Hel62] S. Helgason. *Differential geometry and symmetric spaces*. Pure and Applied Mathematics, Vol. XII. Academic Press, New York-London, 1962.

[HG04] N. Higson and E. Guentner. Group C^*-algebras and K-theory. In *Noncommutative geometry*, volume 1831 of *Lecture Notes in Math.*, pages 137–251. Springer, Berlin, 2004.

[Hig00] N. Higson. Bivariant K-theory and the Novikov conjecture. *Geom. Funct. Anal.*, 10(3):563–581, 2000.

[HK01] N. Higson and G. G. Kasparov. E-theory and KK-theory for groups which act properly and isometrically on Hilbert space. *Invent. Math.*, 144(1):23–74, 2001.

[HLS02] N. Higson, V. Lafforgue, and G. Skandalis. Counterexamples to the Baum-Connes conjecture. *Geom. Funct. Anal.*, 12(2):330–354, 2002.

[HR95] N. Higson and J. Roe. On the coarse Baum-Connes conjecture. In *Novikov conjectures, index theorems and rigidity, Vol. 2 (Oberwolfach, 1993)*, volume 227 of *London Math. Soc. Lecture Note Ser.*, pages 227–254. Cambridge Univ. Press, Cambridge, 1995.

[HR00] N. Higson and J. Roe. Amenable group actions and the Novikov conjecture. *J. Reine Angew. Math.*, 519:143–153, 2000.

[Jen69] J. W. Jenkins. An amenable group with a non-symmetric group algebra. *Bull. Amer. Math. Soc.*, 75:357–360, 1969.

[JK95] P. Julg and G. G. Kasparov. Operator K-theory for the group $\mathrm{SU}(n, 1)$. *J. Reine Angew. Math.*, 463:99–152, 1995.

[Jol90] P. Jolissaint. Rapidly decreasing functions in reduced C^*-algebras of groups. *Trans. Amer. Math. Soc.*, 317(1):167–196, 1990.

[Jul97] P. Julg. Remarks on the Baum-Connes conjecture and Kazhdan's property T. In *Operator algebras and their applications (Waterloo, ON, 1994/1995)*, volume 13 of *Fields Inst. Commun.*, pages 145–153. Amer. Math. Soc., Providence, RI, 1997.

[Jul98] P. Julg. Travaux de N. Higson et G. Kasparov sur la conjecture de Baum-Connes. *Astérisque*, (252):Exp. No. 841, 4, 151–183, 1998. Séminaire Bourbaki. Vol. 1997/98.

[Jul02] P. Julg. La conjecture de Baum-Connes à coefficients pour le groupe $\mathrm{Sp}(n, 1)$. *C. R. Math. Acad. Sci. Paris*, 334(7):533–538, 2002.

[Jul19] P. Julg. How to prove the Baum-Connes conjecture for $Sp(n, 1)$? *Journal of Geometry and Physics*, https://doi.org/10.1016/j.geomphys.2019.02.009, 2019.

[JV84] P. Julg and A. Valette. K-theoretic amenability for $\mathrm{SL}_2(\mathbf{Q}_p)$, and the action on the associated tree. *J. Funct. Anal.*, 58(2):194–215, 1984.

[JV88] P. Julg and A. Valette. Fredholm modules associated to Bruhat-Tits buildings. In *Miniconferences on harmonic analysis and operator algebras (Canberra, 1987)*, volume 16 of *Proc. Centre Math. Anal. Austral. Nat. Univ.*, pages 143–155. Austral. Nat. Univ., Canberra, 1988.

[JvE18] P. Julg and E. van Erp. The geometry of the osculating nilpotent group structures of the Heisenberg calculus. *J. Lie Theory*, 28(1):107–138, 2018.

[Kar08] M. Karoubi. *K-theory*. Classics in Mathematics. Springer-Verlag, Berlin, 2008. An introduction, Reprint of the 1978 edition, With a new postface by the author and a list of errata.

[Kas84] G. G. Kasparov. Lorentz groups: K-theory of unitary representations and crossed products. *Dokl. Akad. Nauk SSSR*, 275(3):541–545, 1984.

[Kas87] G. G. Kasparov. Operator K-theory and its applications. *Journal of Soviet Mathematics*, 37(6):1373–1396, Jun 1987.

[Kas88] G. G. Kasparov. Equivariant KK-theory and the Novikov conjecture. *Invent. Math.*, 91(1):147–201, 1988.

[Kas95] G. G. Kasparov. K-theory, group C^*-algebras, and higher signatures (conspectus). In *Novikov conjectures, index theorems and rigidity, Vol. 1 (Oberwolfach, 1993)*, volume 226 of *London Math. Soc. Lecture Note Ser.*, pages 101–146. Cambridge Univ. Press, Cambridge, 1995.

[Khu18] A. Khukhro. Espaces et groupes non exacts admettant un plongement grossier dans un espace de Hilbert, d'après Arzhantseva-Guentner-Špakula, Arzhantseva-Osajda, Osajda, et al. In *Séminaire Bourbaki, Vol. 1154, 2018*. 2018.

[Kid08] Y. Kida. The mapping class group from the viewpoint of measure equivalence theory. *Mem. Amer. Math. Soc.*, 196(916):viii+190, 2008.

[Kna01] A. W. Knapp. *Representation theory of semisimple groups*. Princeton Landmarks in Mathematics. Princeton University Press, Princeton, NJ, 2001. An overview based on examples, Reprint of the 1986 original.

[KS91] G. G. Kasparov and G. Skandalis. Groups acting on buildings, operator K-theory, and Novikov's conjecture. *K-Theory*, 4(4):303–337, 1991.

[KS03] G. G. Kasparov and G. Skandalis. Groups acting properly on "bolic" spaces and the Novikov conjecture. *Ann. of Math. (2)*, 158(1):165–206, 2003.

[KY06] G. G. Kasparov and G. Yu. The coarse geometric Novikov conjecture and uniform convexity. *Adv. Math.*, 206(1):1–56, 2006.

[Laf00] V. Lafforgue. A proof of property (RD) for cocompact lattices of $SL(3, \mathbf{R})$ and $SL(3, \mathbf{C})$. *J. Lie Theory*, 10(2):255–267, 2000.

[Laf02a] V. Lafforgue. Banach KK-theory and the Baum-Connes conjecture. In *Proceedings of the International Congress of Mathematicians, Vol. II (Beijing, 2002)*, pages 795–812, Beijing, 2002. Higher Ed. Press.

[Laf02b] V. Lafforgue. K-théorie bivariante pour les algèbres de Banach et conjecture de Baum-Connes. *Invent. Math.*, 149(1):1–95, 2002.

[Laf07] V. Lafforgue. K-théorie bivariante pour les algèbres de Banach, groupoïdes et conjecture de Baum-Connes. Avec un appendice d'Hervé Oyono-Oyono. *J. Inst. Math. Jussieu*, 6(3), 2007.

[Laf08] V. Lafforgue. Un renforcement de la propriété (T). *Duke Math. J.*, 143(3):559–602, 2008.

[Laf09] V. Lafforgue. Propriété (T) renforcée banachique et transformation de Fourier rapide. *J. Topol. Anal.*, 1(3):191–206, 2009.

[Laf10] V. Lafforgue. Propriété (T) renforcée et conjecture de Baum-Connes. In *Quanta of maths*, volume 11 of *Clay Math. Proc.*, pages 323–345. Amer. Math. Soc., Providence, RI, 2010.

[Laf12] V. Lafforgue. La conjecture de Baum-Connes à coefficients pour les groupes hyperboliques. *J. Noncommut. Geom.*, 6(1):1–197, 2012.

[Lan15] M. Land. The analytical assembly map and index theory. *J. Noncommut. Geom.*, 9(2):603–619, 2015.

[LG97] P.-Y. Le Gall. Théorie de Kasparov équivariante et groupoïdes. *C. R. Acad. Sci. Paris Sér. I Math.*, 324(6):695–698, 1997.

[LG99] P.-Y. Le Gall. Théorie de Kasparov équivariante et groupoïdes. I. *K-Theory*, 16(4):361–390, 1999.

[Lia14] B. Liao. Strong Banach property (T) for simple algebraic groups of higher rank. *J. Topol. Anal.*, 6(1):75–105, 2014.

[Lia16] B. Liao. About the obstacle to proving the Baum-Connes conjecture without coefficient for a non-cocompact lattice in Sp_4 in a local field. *J. Noncommut. Geom.*, 10(4):1243–1268, 2016.

[Lip74] R. Lipsman. *Group representations*. Lecture Notes in Mathematics, Vol. 388. Springer-Verlag, Berlin-New York, 1974. A survey of some current topics.

[LP79] H. Leptin and D. Poguntke. Symmetry and nonsymmetry for locally compact groups. *J. Funct. Anal.*, 33(2):119–134, 1979.

[Lub10] A. Lubotzky. *Discrete groups, expanding graphs and invariant measures*. Modern Birkhäuser Classics. Birkhäuser Verlag, Basel, 2010. With an appendix by Jonathan D. Rogawski, Reprint of the 1994 edition.

[LZ89] A. Lubotzky and R.J. Zimmer. Variants of Kazhdan's property for subgroups of semisimple groups. *Israel J. Math.*, 66(1-3):289–299, 1989.

[Mat97] J. Matoušek. On embedding expanders into l_p spaces. *Israel J. Math.*, 102:189–197, 1997.

[Mel82] A. Melin. Lie filtrations and pseudo-differential operators. Preprint, scan available on request from the 2nd author, 1982.

[Min01] I. Mineyev. Straightening and bounded cohomology of hyperbolic groups. *Geom. Funct. Anal.*, 11(4):807–839, 2001.

[Mis74] A. S. Miscenko. Infinite-dimensional representations of discrete groups, and higher signatures. *Izv. Akad. Nauk SSSR Ser. Mat.*, 38:81–106, 1974.

[MN14] M. Mendel and A. Naor. Nonlinear spectral calculus and super-expanders. *Publ. Math. Inst. Hautes Études Sci.*, 119:1–95, 2014.

[Moh18] O. Mohsen. On the deformation groupoid of the inhomogeneous pseudo-differential calculus. arXiv:1806.08585, 2018.

[MV03] G. Mislin and A. Valette. *Proper group actions and the Baum-Connes conjecture*. Advanced Courses in Mathematics. CRM Barcelona. Birkhäuser Verlag, Basel, 2003.

[Nic08] B. Nica. Relatively spectral morphisms and applications to K-theory. *J. Funct. Anal.*, 255(12):3303–3328, 2008.

[Nis19] S. Nishikawa. Direct splitting method for the Baum-Connes conjecture. Preprint 2019, arXiv:1808.08298v2, 2019.

[Now07] P. Nowak. Coarsely embeddable metric spaces without Property A. *J. Funct. Anal.*, 252(1):126–136, 2007.

[NS17] P. W. Nowak and D. Sawicki. Warped cones and spectral gaps. *Proc. Amer. Math. Soc.*, 145(2):817–823, 2017.

[OO01] H. Oyono-Oyono. Baum-Connes conjecture and group actions on trees. *K-Theory*, 24(2):115–134, 2001.

[Osa14] D. Osajda. Small cancellation labellings of some infinite graphs and applications. Preprint 2014, arXiv:1406.5015, 2014.

[Oza00] N. Ozawa. Amenable actions and exactness for discrete groups. *C. R. Acad. Sci. Paris Sér. I Math.*, 330(8):691–695, 2000.

[Oza01] N. Ozawa. *Local theory and local reflexivity for operator spaces*. PhD thesis, Texas A&M University, 2001.

[Pas81] William L. Paschke. K-theory for commutants in the Calkin algebra. *Pacific J. Math.*, 95(2):427–434, 1981.

[Ped79] G. K. Pedersen. *C^*-algebras and their automorphism groups*, volume 14 of *London Mathematical Society Monographs*. Academic Press Inc. [Harcourt Brace Jovanovich Publishers], London, 1979.

[Pim86] M. Pimsner. KK-groups of crossed products by groups acting on trees. *Invent. Math.*, 86(3):603–634, 1986.

[Pon06] R. Ponge. The tangent groupoid of a Heisenberg manifold. *Pacific J. Math.*, 227(1):151–175, 2006.

[PP83] M. G. Penington and R. J. Plymen. The Dirac operator and the principal series for complex semisimple Lie groups. *J. Funct. Anal.*, 53(3):269–286, 1983.

[Pus14] M. Puschnigg. The Baum-Connes conjecture with coefficients for word-hyperbolic groups (after Vincent Lafforgue). *Astérisque*, (361):Exp. No. 1062, vii, 115–148, 2014.

[PV82] M. Pimsner and D. Voiculescu. K-groups of reduced crossed products by free groups. *J. Operator Theory*, 8(1):131–156, 1982.

[Roe93] J. Roe. Coarse cohomology and index theory on complete Riemannian manifolds. *Mem. Amer. Math. Soc.*, 104(497):x+90, 1993.

[Roe96] J. Roe. *Index theory, coarse geometry, and topology of manifolds*, volume 90 of *CBMS Regional Conference Series in Mathematics*. Published for the Conference Board of the Mathematical Sciences, Washington, DC; by the American Mathematical Society, Providence, RI, 1996.

[Roe05] J. Roe. Warped cones and property A. *Geom. Topol.*, 9:163–178, 2005.

[RRS98] J. Ramagge, G. Robertson, and T. Steger. A Haagerup inequality for $\widetilde{A}_1 \times \widetilde{A}_1$ and $\widetilde{A}_2$ buildings. *Geom. Funct. Anal.*, 8(4):702–731, 1998.

[RS95] J. Rosenberg and S. Stolz. A "stable" version of the Gromov-Lawson conjecture. In *The Čech centennial (Boston, MA, 1993)*, volume 181 of *Contemp. Math.*, pages 405–418. Amer. Math. Soc., Providence, RI, 1995.

[Rum99] M. Rumin. Differential geometry on C-C spaces and application to the Novikov-Shubin numbers of nilpotent Lie groups. *C. R. Acad. Sci. Paris Sér. I Math.*, 329(11):985–990, 1999.

[Saw17a] D. Sawicki. Super-expanders and warped cones. Preprint 2017, `arXiv:1704.03865`, 2017.

[Saw17b] D. Sawicki. Warped cones, (non-)rigidity, and piecewise properties (with a joint appendix with Dawid Kielak). Preprint, `arXiv:1707.02960`, 2017.

[Saw17c] D. Sawicki. Warped cones violate the Baum-Connes conjecture. Preprint available on https://www.impan.pl/ `~dsawicki/files/counterexamples.pdf`, 2017.

[Sch07] T. Schick. Finite group extensions and the Baum-Connes conjecture. *Geom. Topol.*, 11:1767–1775, 2007.

[SG08] R. Sánchez-García. Bredon homology and equivariant K-homology of $SL(3, \mathbb{Z})$. *J. Pure Appl. Algebra*, 212(5):1046–1059, 2008.

[Ska02] G. Skandalis. Progrès récents sur la conjecture de Baum-Connes. Contribution de Vincent Lafforgue. *Astérisque*, (276):105–135, 2002. Séminaire Bourbaki, Vol. 1999/2000.

[STY02] G. Skandalis, J. L. Tu, and G. Yu. The coarse Baum-Connes conjecture and groupoids. *Topology*, 41(4):807–834, 2002.

[Swa77] R. Swan. Topological examples of projective modules. *Trans. Amer. Math. Soc.*, 230:201–234, 1977.

[Tit75] J. Tits. On buildings and their applications. In *Proceedings of the International Congress of Mathematicians (Vancouver, B. C., 1974), Vol. 1*, pages 209–220. Canad. Math. Congress, Montreal, Que., 1975.

[Tu99a] J.-L. Tu. The Baum-Connes conjecture and discrete group actions on trees. *K-Theory*, 17(4):303–318, 1999.

[Tu99b] J.-L. Tu. La conjecture de Baum-Connes pour les feuilletages moyennables. *K-Theory*, 17(3):215–264, 1999.

[Tu99c] J.-L. Tu. La conjecture de Novikov pour les feuilletages hyperboliques. *K-Theory*, 16(2):129–184, 1999.

[Tu00] J.-L. Tu. The Baum-Connes conjecture for groupoids. In *C^*-algebras (Münster, 1999)*, pages 227–242. Springer, Berlin, 2000.

[Val84] A. Valette. K-theory for the reduced C^*-algebra of a semisimple Lie group with real rank 1 and finite centre. *Quart. J. Math. Oxford Ser. (2)*, 35(139):341–359, 1984.

[Val85] A. Valette. Dirac induction for semisimple Lie groups having one conjugacy class of Cartan subgroups. In *Operator algebras and their connections with topology and ergodic theory (Buşteni, 1983)*, volume 1132 of *Lecture Notes in Math.*, pages 526–555. Springer, Berlin, 1985.

[Val88] A. Valette. Le rôle des fibrés de rang fini en théorie de Kasparov équivariante. *Acad. Roy. Belg. Cl. Sci. Mém. Collect. 8° (2)*, 45(6), 1988.

[Val02] A. Valette. *Introduction to the Baum-Connes conjecture*. Lectures in Mathematics ETH Zürich. Birkhäuser Verlag, Basel, 2002. From notes taken by Indira Chatterji, With an appendix by Guido Mislin.

[Val18] A. Valette. Proper isometric actions on hilbert spaces: a-(T)-menability and Haagerup property. handbook of group actions. *Adv. Lect. Math. (ALM), Int. Press, Somerville, MA*, IV(41), 2018.

[vEY17a] E. van Erp and R. Yuncken. A groupoid approach to pseudodifferential operators. *J. Reine Agnew. Math.* https://doi.org/10.1515/crelle-2017-0035, 2017.

[vEY17b] E. van Erp and R. Yuncken. On the tangent groupoid of a filtered manifold. *Bull. Lond. Math. Soc.*, 49(6):1000–1012, 2017.

[Vig19] F. Vigolo. Measure expanding actions, expanders and warped cones. *Trans. Amer. Math. Soc.*, 371(3):1951–1979, 2019.

[Vog81] D. Vogan, Jr. *Representations of real reductive Lie groups*, volume 15 of *Progress in Mathematics*. Birkhäuser, Boston, Mass., 1981.

[Was87] A. Wassermann. Une démonstration de la conjecture de Connes-Kasparov pour les groupes de Lie linéaires connexes réductifs. *C. R. Acad. Sci. Paris Sér. I Math.*, 304(18):559–562, 1987.

[Wil11] R. Willett. Property A and graphs with large girth. *J. Topol. Anal.*, 3(3):377–384, 2011.

[WO93] N. E. Wegge-Olsen. *K-theory and C^*-algebras*. Oxford Science Publications. The Clarendon Press, Oxford University Press, New York, 1993. A friendly approach.

[WY12] R. Willett and G. Yu. Higher index theory for certain expanders and Gromov monster groups, I. *Adv. Math.*, 229(3):1380–1416, 2012.

[Yu95] G. Yu. Coarse Baum-Connes conjecture. *K-Theory*, 9(3):199–221, 1995.

[Yu98] G. Yu. The Novikov conjecture for groups with finite asymptotic dimension. *Ann. of Math. (2)*, 147(2):325–355, 1998.

[Yu00] G. Yu. The coarse Baum-Connes conjecture for spaces which admit a uniform embedding into Hilbert space. *Invent. Math.*, 139(1):201–240, 2000.

[Yun11] R. Yuncken. The Bernstein-Gelfand-Gelfand complex and Kasparov theory for $SL(3, \mathbb{C})$. *Adv. Math.*, 226(2):1474–1512, 2011.

[Zim84] R. Zimmer. *Ergodic theory and semisimple groups*, volume 81 of *Monographs in Mathematics*. Birkhäuser Verlag, Basel, 1984.

[Zim87] R. Zimmer. Actions of semisimple groups and discrete subgroups. In *Proceedings of the International Congress of Mathematicians, Vol. 1, 2 (Berkeley, Calif., 1986)*, pages 1247–1258. Amer. Math. Soc., Providence, RI, 1987.

Lie groupoids, pseudodifferential calculus, and index theory

Claire Debord and Georges Skandalis

Abstract Alain Connes introduced the use of Lie groupoids in noncommutative geometry in his pioneering work on the index theory of foliations. In the present paper, we recall the basic notion involved: groupoids, their C^*-algebras, their pseudodifferential calculus, etc. We review several recent and older advances on the involvement of Lie groupoids in noncommutative geometry. We then propose some open questions and possible developments of the subject.

1 Introduction

Groupoids, and especially smooth ones, appear naturally in various areas of modern mathematics. One can find a recent overview on Lie groupoids, with a historical introduction in [23].

Our aim in this paper is to review some advances in the study of Lie groupoids as objects of noncommutative geometry. This theory of Lie groupoids is very much linked with various index problems. A main tool for this index theory is the corresponding pseudodifferential calculus. On the other hand, index theory and pseudodifferential calculus are strongly linked with deformation groupoids.

Groupoids first appeared in the theory of operator algebras in the measurable—von Neumann algebra—setting. They were natural generalizations of actions of groups on spaces. These crossed product operator algebras go back to the "group measure space construction" of Murray von Neumann [94], who used it in order to construct factors of all different types. It often happens that two group actions give rise to isomorphic groupoids (especially in the world of measurable actions—i.e., the von Neumann algebra case). It was noticed in [52] that the corresponding operator algebras only depend on the corresponding groupoid. A recent survey of this measurable point of view can be found in [53].

C. Debord · G. Skandalis (✉)
UPMC Paris 06 CNRS, IMJ-PRG UFR de Mathématiques, Université Paris Diderot, Sorbonne Paris Cité Sorbonne Universités, Paris, France
e-mail: cdebord@math.univ-paris-diderot.fr; skandalis@math.univ-paris-diderot.fr

A. Chamseddine et al. (eds.), *Advances in Noncommutative Geometry*,
https://doi.org/10.1007/978-3-030-29597-4_4

Two almost simultaneous major contributions forced groupoids into the topological noncommutative world, i.e., C^*-algebras:

- The construction by Jean Renault of the C^*-algebra of a locally compact groupoid, its representation theory [102],
- The construction by Alain Connes of the von Neumann algebra and the C^*-algebra of a foliation based on its holonomy groupoid [28, 30, 32].

Moreover, as a motivation for Connes was the generalization of the Atiyah–Singer index theorem [8, 10], he used the smooth longitudinal structure in order to construct the associated pseudodifferential calculus and the C^*-algebraic exact sequence of pseudodifferential operators. This allowed him to construct the analytic index, and to prove a measured index theorem [28]. Very soon after, he constructed a topological index with values in the C^*-algebra of the foliation [30]. The corresponding index theorem was proved in [36].

This pseudodifferential calculus on groupoids and the construction of the analytic index in the groupoid C^*-algebra was then easily generalized to all Lie groupoids [91, 95]. They gave rise to several index theorems—cf. [43, 22, 42].

Another very nice construction of Connes (see [32]) gave a geometric insight on this generalized index theory: the construction of the *tangent groupoid*. This tangent groupoid allowed to construct the analytic index of (pseudo)differential operators without pseudodifferential calculus. It was also used to give beautiful alternate proofs of the Atiyah–Singer index theorem [32, 43].

Connes' tangent groupoid was an inspiration for many papers (cf. [61, 91, 95]...) where this idea was generalized to various geometric contexts. Its natural setting is the *deformation to the normal cone* (DNC) construction. Since DNC is functorial, Connes' construction can be extended to any case of a sub-Lie groupoid of a Lie groupoid (see [46]). Moreover, this construction is immediately related to the Connes–Higson E-theory (cf. [33]). It therefore opened a whole world of deformation groupoids that are useful in many situations and gave rise to many interesting K-theoretic constructions and computations.

One also sees that the C^*-algebra extension of the pseudodifferential operators on a groupoid is directly related to the one naturally associated with the DNC construction and the canonical action of $\mathbb{R}_+^*$ on it [1, 45]. There is a well-defined Morita equivalence between these exact sequences, and the corresponding bimodule gives an alternative definition of the pseudodifferential calculus on a groupoid (cf. [45])—which in turn should be used to various contexts.

In the present survey, we recall definitions and several examples of Lie groupoids and describe their C^*-algebras. Next, we study the pseudodifferential operators associated with Lie groupoids from various view points. Examples of various constructions of groupoids giving rise to interesting K-theoretic computations are then outlined. We end with a few remarks and several natural questions concerning groupoids, deformations, and applications.

2 Lie groupoids and their operator algebras

We refer to [78, 80] for the classical definitions and constructions related to groupoids and their Lie algebroids. The construction of the C^*-algebra of a groupoid is due to Jean Renault [102], one can look at his course [103] which is mainly devoted to locally compact groupoids.

2.1 Generalities on Lie groupoids

A *groupoid* is a small category in which every morphism is an isomorphism. Thus a groupoid G is a pair $(G^{(0)}, G^{(1)})$ of sets together with structural morphisms:

Units and arrows The set $G^{(0)}$ denotes the set of *objects* (or *units*) of the groupoid, whereas the set $G^{(1)}$ is the set of *morphisms* (or *arrows*). The *unit map* $u : G^{(0)} \to G^{(1)}$ is the injective map which assigns to any object of G its identity morphism.

The source and range maps $s, r : G^{(1)} \to G^{(0)}$ are (surjective) maps equal to identity in restriction to $G^{(0)}$: $s \circ u = r \circ u = Id$.

The inverse $\iota : G^{(1)} \to G^{(1)}$ is an involutive map which exchanges the source and range:

$$\text{for } \alpha \in G,\ (\alpha^{-1})^{-1} = \alpha \text{ and } s(\alpha^{-1}) = r(\alpha),\ \text{ where } \alpha^{-1} \text{ denotes } \iota(\alpha)$$

The partial multiplication $m : G^{(2)} \to G^{(1)}$ is defined on the set of *composable pairs* $G^{(2)} = \{(\alpha, \beta) \in G^{(1)} \times G^{(1)} \mid s(\alpha) = r(\beta)\}$. It satisfies for any $(\alpha, \beta) \in G^{(2)}$:

$$r(\alpha\beta) = r(\alpha),\ s(\alpha\beta) = s(\beta),\ \alpha u(s(\alpha)) = u(r(\alpha))\alpha = \alpha,\ \alpha^{-1}\alpha = u(s(\alpha))$$

where $\alpha\beta$ stands for $m(\alpha, \beta)$. Moreover the product is associative, if $\alpha, \beta, \gamma \in G$:

$$(\alpha\beta)\gamma = \alpha(\beta\gamma) \text{ when } s(\alpha) = r(\beta) \text{ and } s(\beta) = r(\gamma)$$

We often identify $G^{(0)}$ with its image in $G^{(1)}$ and make the confusion between G and $G^{(1)}$. A groupoid $G = (G^{(0)}, G^{(1)}, s, r, u, \iota, m)$ will be simply denoted $G \overset{r,s}{\rightrightarrows} G^{(0)}$ or just $G \rightrightarrows G^{(0)}$.

Notation For any maps $f : A \to G^{(0)}$ and $g : B \to G^{(0)}$, define

$$G^f = \{(x, \gamma) \in A \times G;\ r(\gamma) = f(x)\},\ G_g = \{(\gamma, x) \in G \times B;\ s(\gamma) = g(x)\}$$

and

$$G_g^f = \{(x, \gamma, y) \in A \times G \times B;\ r(\gamma) = f(x),\ s(\gamma) = g(y)\}\,.$$

In particular for $A, B \subset G^{(0)}$, we put $G^A = \{\gamma \in G;\ r(\gamma) \in A\}$ and $G_A = \{\gamma \in G;\ s(\gamma) \in A\}$; we also put $G_A^B = G_A \cap G^B$.

Remark 2.1 For any $x \in G^{(0)}$, G_x^x is a group with unit x, called the *isotropy group* at x. It acts by left (resp. right) multiplication on G^x (resp. G_x) and the quotient identifies with $s(G^x) = r(G_x) \subset G^{(0)}$ which is called the *orbit* of G passing through x. Thus a groupoid acts on its set of units.

Note that A is a *saturated* subset of $G^{(0)}$ (for the action of G) if and only if $G_A = G^A = G_A^A$.

In order to construct the C^*-algebra of a groupoid, we will assume that it is *locally compact*. This means that $G^{(0)}$ and G are endowed with topologies for which

- $G^{(0)}$ is a locally compact Hausdorff space,
- G is second countable, and locally compact locally Hausdorff, i.e., each point γ in G has a compact (Hausdorff) neighborhood;
- all structural maps (s, r, u, ι, m) are continuous and s is open.

In this situation the map r is open and the s-fibers of G are Hausdorff.
In order to study differential operators and index theories, we will assume our groupoid to be smooth. The groupoid $G \rightrightarrows G^{(0)}$ is *Lie* or *smooth* when G and $G^{(0)}$ are second countable smooth manifolds with $G^{(0)}$ Hausdorff, s is a smooth submersion (hence $G^{(2)}$ is a manifold), and the structural morphisms are smooth.

The Lie groupoid G is said to be *s-connected* when for any $x \in G^{(0)}$, the s-fiber of G over x is connected. The s-connected component of a groupoid G is $\cup_{x \in G^{(0)}} CG_x$ where CG_x is the connected component of the s-fiber G_x which contains the unit $u(x)$. The groupoid CG is the smallest open subgroupoid of G containing its units.

Example 2.2

(a) A space M is a groupoid over itself with $s = r = u = \mathrm{Id}$. Thus, a manifold is a Lie groupoid.
(b) A group $H \rightrightarrows \{e\}$ is a groupoid over its unit e, with the usual product and inverse map. A Lie group is a Lie groupoid!
(c) A group bundle: $\pi : E \to M$ is a groupoid $E \rightrightarrows M$ with $r = s = \pi$ and algebraic operations given by the group structure of each fiber E_x, $x \in M$. In particular, a smooth vector bundle over a manifold gives thus rise to a Lie groupoid.
(d) If $\mathcal{R}$ is an equivalence relation on a space M, then the graph of $\mathcal{R}$, $G_{\mathcal{R}} := \{(x, y) \in M \times M \mid x\mathcal{R}y\}$, admits a structure of groupoid over M, which is given by:

$$u(x) = (x, x) \, , \; s(x, y) = y \, , \; r(x, y) = x \, ,$$

$$(x, y)^{-1} = (y, x) \, , \; (x, y) \cdot (y, z) = (x, z)$$

for $x, \; y, \; z$ in M. Notice that the orbits of the groupoid $G_{\mathcal{R}}$ are precisely the orbits of the equivalence relation $\mathcal{R}$. If M is a manifold, $G_{\mathcal{R}}$ is a smooth submanifold of $M \times M$ and s restricts to a submersion, it is a Lie groupoid.

When $x\mathcal{R}y$ for any $x, \; y$ in M, $G_{\mathcal{R}} = M \times M \rightrightarrows M$ is called the *pair groupoid*. If M is a manifold, the pair groupoid $M \times M$ is a Lie groupoid.

Without entering too much in the details, let us say that a smooth regular foliation on a manifold M(of dimension n) is an equivalence relation on M whose orbits, called the leaves, are immersed connected submanifolds of M (of dimension p). The corresponding groupoid $G_{\mathcal{R}}$ does not have a smooth structure, but there is a "smallest" Lie groupoid of dimension $n + p$, called the *holonomy groupoid of the foliation*, whose orbits are the leaves [118, 101, 55, 80]. The holonomy appears to be exactly the obstruction for $G_{\mathcal{R}}$ to be smooth!

(e) If H is a group acting on a space M, the *groupoid of the action* is $H \ltimes M \rightrightarrows M$ with the following structural morphisms:

$$u(x) = (e, x) \, , \; s(g, x) = x \, , \; r(g, x) = g \cdot x \, ,$$
$$(g, x)^{-1} = (g^{-1}, g \cdot x) \, , \; (h, g \cdot x) \cdot (g, x) = (hg, x) \, ,$$

for x in M and $g, \; h$ in H. Once again, notions of isotropy groups, and orbits of the groupoid coincide with the corresponding notions for the group action.

If H is a Lie group, M is a smooth manifold and the action is smooth, then $H \ltimes M$ is a Lie groupoid.

(f) Let M be a smooth manifold of dimension n. The *Poincaré groupoid* of M is

$$\Pi(M) := \{\bar{\gamma} \mid \gamma : [0, 1] \to M \text{ a continuous path}\} \rightrightarrows M$$

where $\bar{\gamma}$ denotes the homotopy class (with fixed endpoints) of γ. For $x \in M$, $u(x)$ will be the (class of the) constant path at x, $s(\bar{\gamma}) = \gamma(0)$, $r(\bar{\gamma}) = \gamma(1)$, the product comes from the concatenation product of paths.

The groupoid $\Pi(M)$ is naturally endowed with a smooth structure (of dimension $2n$). For any $x \in M$, the isotropy group $\Pi(M)_x^x$ is the fundamental group of M with base point x and $\Pi(M)_x$ the corresponding universal covering.

(g) If $G \rightrightarrows M$ is a groupoid and $f : N \to M$ a map, $G_f^f \rightrightarrows N$ is again a groupoid:

$$u(x) = (x, f(x), x), \; s(x, \alpha, y) = y,$$

$$(x, \alpha, y)^{-1} = (y, \alpha^{-1}, x), \; (x, \alpha, y)(y, \beta, z) = (x, \alpha\beta, z)$$

where α, β are in G, $x, \; y, \; z$ in N and $f(x) = r(\alpha)$, $f(y) = s(\alpha) = r(\beta)$ and $f(z) = s(\beta)$.

When G is a Lie groupoid, and f a smooth map transverse to G (see Definition 2.4), $G_f^f \rightrightarrows N$ is a Lie groupoid.

The infinitesimal object associated with a Lie groupoid is its *Lie algebroid*:

Definition 2.3 A *Lie algebroid* $\mathcal{A}$ over a manifold M is a vector bundle $\mathcal{A} \to M$, together with a Lie algebra structure on the space $\Gamma(\mathcal{A})$ of smooth sections of $\mathcal{A}$ and a bundle map $\varrho : \mathcal{A} \to TM$, called the *anchor*, whose extension to sections of these bundles satisfies

(i) $\varrho([X, Y]) = [\varrho(X), \varrho(Y)]$, and
(ii) $[X, fY] = f[X, Y] + (\varrho(X)f)Y$,

for any smooth sections X and Y of $\mathcal{A}$ and any smooth function f on M.

Now, let $G \overset{s,r}{\rightrightarrows} G^{(0)}$ be a Lie groupoid.

For any α in G, let $R_\alpha : G_{r(\alpha)} \to G_{s(\alpha)}$ be the right multiplication by α. A tangent vector field Z on G is *right invariant* if it satisfies,

- Z is s-vertical, namely $Ts(Z) = 0$, i.e., for every $\alpha \in G$, $Z(\alpha)$ is tangent to the fiber $G_{s(\alpha)}$.
- For all (α, β) in $G^{(2)}$, $Z(\alpha\beta) = TR_\beta(Z(\alpha))$.

The Lie algebroid $\mathfrak{A}G$ of the Lie groupoid G is defined as follows [78],

- The fiber bundle $\mathfrak{A}G \to G^{(0)}$ is the restriction of the kernel of the differential Ts of s to $\mathcal{G}^{(0)}$. In other words, $\mathfrak{A}G = \cup_{x \in G^{(0)}} T_x G_x$ is the union of the tangent spaces to the s-fibers at the corresponding unit.
- The anchor $\varrho : \mathfrak{A}G \to T\mathcal{G}^{(0)}$ is the restriction of the differential Tr of r to $\mathfrak{A}G$.
- If $Y : U \to \mathfrak{A}G$ is a local section of $\mathfrak{A}G$, where U is an open subset of $G^{(0)}$, we define the local *right invariant vector field* Z_Y *associated* with Y by

$$Z_Y(\alpha) = TR_\alpha(Y(r(\alpha))) \text{ for all } \alpha \in G^U .$$

The Lie bracket is then defined by

$$\begin{array}{rcl} [\ ,\ ] : \Gamma(\mathfrak{A}G) \times \Gamma(\mathfrak{A}G) & \longrightarrow & \Gamma(\mathfrak{A}G) \\ (Y_1, Y_2) & \mapsto & [Z_{Y_1}, Z_{Y_2}]_{G^{(0)}} \end{array}$$

where $[Z_{Y_1}, Z_{Y_2}]_{G^{(0)}}$ is the restriction of the usual bracket $[Z_{Y_1}, Z_{Y_2}]$ to $G^{(0)}$.

Notice that $\mathfrak{A}G$ identifies with the normal bundle $\mathcal{N}^G_{G^{(0)}}$ of the inclusion $u : G^{(0)} \hookrightarrow G$.

Definition 2.4 A smooth map $f : N \to M$ is *transverse* to a Lie groupoid $G \rightrightarrows M$ when for all $x \in N$: $Tf(T_x N) + \varrho(\mathfrak{A}G)_{f(x)} = T_{f(x)}M$.

Lie theory for groupoids is much trickier than for groups: a Lie algebroid does not always integrate into a Lie groupoid [2].

Nevertheless, when the anchor of a Lie algebroid $\mathcal{A}$ is injective in restriction to a dense open subset it is integrable and there is a "smallest" s-connected Lie groupoid integrating it [38]. This situation is often encountered in index theory where such a Lie algebroid of vector fields is naturally associated with the studied geometrical object (e.g., manifolds with corners, conical singularities, etc.). See [37] for a complete answer to this integrability problem.

Example 2.5

(a) The Lie algebroid of a Lie group is the Lie algebra of the group.
(b) The Lie algebroid of the pair groupoid $M \times M \rightrightarrows M$ on a smooth manifold M is TM with identity as anchor.
(c) If $f : M \to B$ is a smooth submersion, the Lie groupoid of the equivalence relation on M "being on the same fiber of f" is $B_f^f = M \times_f M \rightrightarrows M$ and its Lie algebroid is the kernel of Tf with anchor the inclusion.
(d) More generally, if $\mathcal{F}$ is a regular foliation on a manifold M, $T\mathcal{F}$ with inclusion as anchor, defines a Lie algebroid over M and the *holonomy groupoid* is the smallest Lie groupoid which integrates $T\mathcal{F}$ [118, 101].
(e) Let M be a manifold and V an hypersurface cutting M into two pieces. The module of smooth vector fields on M that are tangent to V was considered by Melrose for the study of *b-operators* for manifold with boundary [83]. This module is the module of sections of a Lie algebroid $\mathfrak{A}_b$ over M which integrates into the b-groupoid G_b [90].

    If $M = V \times \mathbb{R}$ and $V = V \times \{0\}$ (which is always locally the case around V up to a diffeomorphism), then $\mathfrak{A}_b = TV \times T\mathbb{R}$ with anchor $\varrho(x, U, t, \xi) \mapsto (x, U, t, t\xi)$ and $G_b = (V \times V) \times (\mathbb{R} \rtimes \mathbb{R}_+^*)$ is the product of the pair groupoid on V with the groupoid of the multiplicative action of $\mathbb{R}_+^*$ on $\mathbb{R}$.

Remark 2.6 The unit spaces of many interesting groupoids have boundaries or corners. In (almost) all the situations, these groupoids sit naturally inside Lie groupoids *without boundaries* as restrictions to closed *saturated* subsets. This means that the object under study is a subgroupoid $G_V^V = G_V$ of a Lie groupoid $G \rightrightarrows G^{(0)}$ where V is a closed saturated subset of $G^{(0)}$. Such groupoids have a natural algebroid, adiabatic deformation, pseudodifferential calculus, etc. that are restrictions to V and G_V of the corresponding objects on $G^{(0)}$ and G. We chose to give definitions and constructions for Lie groupoids; the case of a longitudinally smooth groupoid over a manifold with corners is a straightforward generalization using a convenient restriction.

In [92, 97] are treated slightly more singular cases. There, the authors deal with foliations and groupoids that are only smooth in the direction of the orbits. One can perform with some effort the above constructions in these generalized cases too.

2.1.1 Morita equivalence of Lie groupoids

An important feature of noncommutativity is a nontrivial notion of Morita equivalence: indeed a Morita equivalence of commutative algebras is just an isomorphism.

In the same way in the world of groupoids there is an interesting notion of Morita equivalence, although this notion reduces to isomorphism both for spaces and for groups.

Definition 2.7 Two Lie groupoids $G_1 \overset{r,s}{\rightrightarrows} M_1$ and $G_2 \overset{r,s}{\rightrightarrows} M_2$ are *Morita equivalent* if there exists a groupoid $G \overset{r,s}{\rightrightarrows} M$ and smooth maps $f_i : M_i \to M$ transverse to G such that the pull-back groupoids $G_{f_i}^{f_i}$ identify to G_i and $f_i(M_i)$ meets all the orbits of G.

More precisely, a Morita equivalence is given by a linking manifold X with extra data: surjective smooth submersions $r : X \to M_1$ and $s : X \to M_2$ and compositions $G_1 \times_{s,r} X \to X$, $X \times_{s,r} G_2 \to X$, $X \times_{r,r} X \to G_2$ and $X \times_{s,s} X \to G_1$ with natural associativity conditions (see [93] for details). In the above situation, X is the manifold $G_{f_2}^{f_1}$ and the extra data are the range and source maps and the composition rules of the groupoid $G_{f_1 \sqcup f_2}^{f_1 \sqcup f_2} \rightrightarrows M_1 \sqcup M_2$ (see [93]).

Example 2.8 There are many interesting Morita equivalences of Lie groupoids.

(a) Given a surjective submersion $f : M \to B$, the subgroupoid $G = \{(x, y) \in M \times M;\ f(x) = f(y)\}$ of the pair groupoid $M \times M$ is Morita equivalent to the space B, i.e., the groupoid $B \rightrightarrows B$ (cf. Example 2.5c).
(b) More generally, if $G \rightrightarrows B$ is a Lie groupoid and $f : M \to B$ is a submersion[1] whose image meets all G orbits, then the groupoid G_f^f is a Lie groupoid Morita equivalent to G (cf. Example 2.5d).
(c) If M is a connected manifold, its Poincaré groupoid $\Pi(M)$ is Morita equivalent to the fundamental group $\pi_1(M)$ (cf. Example 2.2f).

2.2 *C*-algebra of a Lie groupoid*

2.2.1 Convolution*-algebra of smooth functions with compact support

Recall that on a Lie group H, the *convolution product formula* is given for f and g in $C_c^\infty(H)$ (i.e., smooth functions with compact supports on H) by

$$f * g(x) = \int_H f(y)g(y^{-1}x)dy$$

[1] We may just assume that f satisfies the transversality condition.

For M a manifold, the *convolution product formula of kernels* is given for f and g in $C_c^\infty(M \times M)$ by

$$f * g(x, y) = \int_M f(x, z)g(z, y)dz$$

Both these convolution products are special cases of convolution product on a Lie groupoid, the first one is for the Lie group viewed as a Lie groupoid $H \rightrightarrows \{e\}$ and the second one corresponds to the pair groupoid $M \times M \rightrightarrows M$.

Let us assume now that $G \rightrightarrows G^{(0)}$ is a Lie groupoid with source s and range r. We define a *convolution algebra* structure on $C_c^\infty(G)$ in the following way:

Convolution product. For $f,\ g \in C_c^\infty(G)$ and $\gamma \in G$ one wants to define a convolution formula of the following form:

$$f * g(\gamma) = \int_{(\alpha,\beta)\in G^{(2)};\ \alpha\beta=\gamma} f(\alpha)g(\beta)$$

In order to define the previous integral, one can choose a smooth *Haar system* on G that is

- a smooth family of Lebesgue measure ν^x on every G^x, $x \in G^{(0)}$,
- with left invariance property: for every $\alpha \in G$, the diffeomorphism $\beta \mapsto \alpha \cdot \beta$ from $G^{s(\alpha)}$ with $G^{r(\alpha)}$ sends the measure $\nu^{s(\alpha)}$ to the measure $\nu^{r(\alpha)}$.

Now the convolution formula becomes:

$$f * g(\gamma) = \int_{G^{r(\gamma)}} f(\alpha)g(\alpha^{-1}\gamma)\, d\nu^{r(\gamma)}(\alpha).$$

The convolution product is associative by invariance of the Haar system (and Fubini).

Adjoint. For $f \in C_c^\infty(G)$, its adjoint is the function $f^* : \alpha \mapsto \overline{f(\alpha^{-1})}$.

Remarks 2.9 Let us mention two very useful constructions that appeared in [30].

(a) In order to have intrinsic formulas for the convolution and adjoint, it is suitable to replace the space $C_c^\infty(G)$ by the space of sections of a bundle of (half) densities on the groupoid, more precisely sections of the vector bundle $\Omega^{1/2} = |\Lambda|^{1/2}(\ker Ts \times \ker Tr)$ of half densities of the bundle $\ker Ts \times \ker Tr$. Note that the Haar system is just an invariant section of the bundle $|\Lambda|^1(\ker Tr)$ of 1-densities of the bundle $\ker(\ker Tr)$ and can be used to trivialize the bundle $|\Lambda|^{1/2}(\ker Ts \times \ker Tr)$.

(b) Also in [30] Connes explains how to naturally define the convolution algebra "$C_c^\infty(G)$" when the groupoid G is not assumed to be Hausdorff (but is still a *locally* Hausdorff manifold).

2.3 Norm and C*-algebra

For $f \in C_c^\infty(G)$, define

$$\|f\|_I = \max\left(\sup_{x\in G^{(0)}} \int_{G^x} |f(\gamma)| d\nu^{r(\gamma)}(\alpha) \,;\, \sup_{x\in G^{(0)}} \int_{G^x} |f(\gamma^{-1})| d\nu^{r(\gamma)}(\alpha) \right)$$

A $*$-representation $\pi : C_c^\infty(G) \longrightarrow \mathcal{B}(\mathcal{H})$, where $\mathcal{B}(\mathcal{H})$ is the algebra of bounded operator on the separable Hilbert space $\mathcal{H}$, is *bounded* when it satisfies: $\|\pi(f)\| \leq \|f\|_I$ for any $f \in C_c^\infty(G)$.

We define the *maximal norm* of $f \in C_c^\infty(G)$ by:

$$\|f\|_{max} = \sup_{\pi \text{ bounded}} \|\pi(f)\|_{\mathcal{B}(\mathcal{H})}$$

For any $x \in G^{(0)}$ the map $\pi^x : C_c^\infty(G) \to \mathcal{B}(L^2(G_x))$ defined by the formula:

$$\pi^x(f)(\tau)(\gamma) = \int_{G^{r(\gamma)}} f(\alpha)\tau(\alpha^{-1}\gamma)\, d\nu^{r(\gamma)}(\alpha).$$

where $f \in C_c^\infty(G)$, $\tau \in L^2(G_x)$ and $\gamma \in G$, is a bounded representation.

We define the *minimal norm* of $f \in C_c^\infty(G)$ by:

$$\|f\|_{min} = \sup_{x\in G^{(0)}} \|\pi^x(f)\|_{\mathcal{B}(L^2(G_x))}$$

The *reduced C^*-algebra* of G is the completion of $C_c^\infty(G)$ with respect to the minimal norm: $C_r^*(G) = \overline{C_c^\infty(G)}^{min}$. The *maximal C^*-algebra* of G is the completion of $C_c^\infty(G)$ with respect to the maximal norm: $C^*(G) = \overline{C_c^\infty(G)}^{max}$.

The identity induces a surjective morphism from $C^*(G)$ to $C_r^*(G)$. This morphism is an isomorphism when the groupoid G is *amenable* (see [4] for a discussion of the amenability of locally compact groupoids).

The C^*-completions $C^*(G)$ and $C_r^*(G)$ have both advantages and disadvantages. Some properties hold for one of them and not necessarily for the other one. The celebrated Baum–Connes conjecture [11, 12] is a statement for the K-theory of the reduced one—and Kaszdan's property T shows easily that it cannot hold for the maximal one (see [112] for a very nice discussion on the Baum–Connes conjecture).

On the other hand, let $G \rightrightarrows M$ be a Lie groupoid and $X \subset M$ a closed subset of M *saturated* for G (i.e., if for $\alpha \in G$, $s(\alpha) \in X$ if and only if $r(\alpha) \in X$). Then we have an exact sequence:

$$0 \to C^*(G_{M\setminus X}) \to C^*(G) \to C^*(G_X) \to 0$$

of maximal C^*-algebras. The corresponding sequence at the level of reduced C^*-algebras is *not* always exact in its middle term. This nonexactness is responsible for counterexamples to the Baum–Connes conjecture in [56]. See [13, 20, 21] for a possible solution to this lack of exactness.

2.4 Deformation to the normal cone and blowup groupoids

2.4.1 Deformation to the normal cone groupoid

Let us first recall the standard *deformation to the normal cone* construction. Let $V \subset M$ be a submanifold of a manifold M and denote $\mathcal{N}_V^M$ the normal bundle. The *deformation to the normal cone* of V in M is:

$$DNC(M, V) = (M \times \mathbb{R}^*) \sqcup (\mathcal{N}_V^M \times \{0\})$$

It is equipped with the natural smooth structure generated by the following constraints:

- the map $\varphi : DNC(M, V) \to M \times \mathbb{R}$ given by $(x, t) \in M \times \mathbb{R}^* \mapsto (x, t)$ and $(x, \xi, 0) \in \mathcal{N}_V^M \times \{0\} \mapsto (x, 0)$ is smooth.
- if $f : M \to \mathbb{R}$ is any smooth function that vanishes on V, the function f^{dnc} : $DNC(M, V) \to \mathbb{R}$ given by $(x, t) \in M \times \mathbb{R}^* \mapsto \frac{f(x)}{t}$ and $(x, \xi, 0) \in \mathcal{N}_V^M \times \{0\} \mapsto df(\xi)$ is smooth.

One can also define the smooth structure with the choice of an exponential map $\theta : U' \subset \mathcal{N}_V^M \to U \subset M$, by requiring the map

$$\Theta : (x, \xi, t) \mapsto \begin{cases} (\theta(x, t\xi), t) \; for \; t \neq 0 \\ (x, \xi, 0) \; for \; t = 0 \end{cases}$$

to be a diffeomorphism from the open neighborhood $W' = \{(x, \xi, t) \in \mathcal{N}_V^M \times \mathbb{R} \mid (x, t\xi) \in U'\}$ of $\mathcal{N}_V^M \times \{0\}$ in $\mathcal{N}_V^M \times \mathbb{R}$ on its image.

Note that describing the smooth structure thanks to the choice of an exponential map ensures that such a structure exists, while the description thanks to the characterization of the smooth functions ensures that this structure is independent of choices.

Remark 2.10 By the characterization of the smooth functions, it follows that the deformation to the normal cone construction is functorial.

A consequence of this remak is

Corollary 2.11 *If G is a Lie groupoid and H is a subgroupoid, $DNC(G, H) \rightrightarrows DNC(G^{(0)}, H^{(0)})$ is a Lie groupoid. Moreover the Lie algebroid of $DNC(G, H)$ is $DNC(\mathfrak{A}G, \mathfrak{A}H)$.*

At the set level $DNC(G, H) = G \times \mathbb{R}^* \sqcup \mathcal{N}_H^G \times \{0\}$. The normal space $\mathcal{N}_H^G$ is equipped with a Lie groupoid structure with units $\mathcal{N}_{H^{(0)}}^{G^{(0)}}$ called the *normal groupoid* of the inclusion of H in G. Its Lie algebroid is $\mathcal{N}_{\mathfrak{A}H}^{\mathfrak{A}G}$.

Example 2.12 The first and most famous example of groupoid resulting from the DNC construction is the *tangent groupoid* of Connes [32]: let M be a smooth manifold diagonally embedded in the pair groupoid $M \times M$, perform $DNC(M \times M, M)$, and restrict it to $M \times [0, 1]$:

$$G_T(M) = DNC(M \times M, M)|_{M\times[0,1]} = M \times M \times (0, 1] \sqcup TM \times \{0\} \rightrightarrows M \times [0, 1]$$

Similarly, the *adiabatic groupoid* of a Lie groupoid $G \rightrightarrows G^{(0)}$ is the restriction over $G^{(0)} \times [0, 1]$ of $DNC(G, G^{(0)})$ [91, 95]:

$$G_{ad} = G \times (0, 1] \sqcup \mathfrak{A}G \times \{0\} \rightrightarrows G^{(0)} \times [0, 1]$$

2.4.2 Blowup groupoid

We keep the notation of the previous section: $V \subset M$ is a submanifold of a manifold M and $DNC(M, V)$ is the deformation to the normal cone of V in M.

Recall that $\varphi : DNC(M, V) = M \times \mathbb{R}^* \sqcup \mathcal{N}_V^M \times \{0\} \to M \times \mathbb{R}$ is the natural (smooth) map. We will consider the manifold with boundary $DNC_+(M, V) = \varphi^{-1}(M \times \mathbb{R}_+) = M \times \mathbb{R}_+^* \sqcup \mathcal{N}_V^M$.

The *scaling action* of $\mathbb{R}_+^*$ on $M \times \mathbb{R}^*$ extends to the *zooming action* of $\mathbb{R}_+^*$ on $DNC_+(M, V)$:

$$\begin{array}{ccc} DNC_+(M, V) \times \mathbb{R}^* & \longrightarrow & DNC_+(M, V) \\ (z, t, \lambda) & \mapsto & (z, \lambda t) \ for\ t \neq 0 \\ (x, X, 0, \lambda) & \mapsto & \left(x, \frac{1}{\lambda}X, 0\right) \ for\ t = 0 \end{array}$$

By functoriality, the manifold $V \times \mathbb{R}_+ = DNC_+(V, V)$ embeds in $DNC_+(M, V)$. The zooming action is free and proper on the open subset $DNC_+(M, V) \setminus V \times \mathbb{R}_+$ of $DNC_+(M, V)$. We let the *spherical blowup* of V in M be:

$$SBlup(M, V) = \big(DNC_+(M, V) \setminus V \times \mathbb{R}_+\big)/\mathbb{R}_+^* = M \setminus V \cup \mathbb{S}(N_V^M)\,.$$

Remark 2.13 The spherical blowup construction is functorial "wherever it is defined". Precisely, suppose that we have a commutative diagram $\begin{array}{ccc} V & \hookrightarrow & M \\ {\scriptstyle f|_V}\downarrow & & \downarrow{\scriptstyle f} \\ V' & \hookrightarrow & M' \end{array}$ where the horizontal arrows are embeddings of submanifolds. Functoriality of the deformation to the normal cones constructions yields a smooth map $DNC(f)$: $DNC(M, V) \to DNC(M', V')$. This smooth map restricts to the map $DNC(f)_+$: $DNC_+(M, V) \to DNC_+(M', V')$ which is equivariant under the zooming action.

Let $SU_f(M, V) = DNC_+(M, V) \setminus DNC(f)^{-1}(V' \times \mathbb{R}_+)$ and define

$$SBlup_f(M, V) = SU_f/\mathbb{R}_+^* \subset SBlup(M, V)$$

Then $DNC(f)_+$ passes to the quotient:

$$SBlup(f) : SBlup_f(M, V) \to SBlup(M', V')$$

If $\begin{array}{ccc} V & \hookrightarrow & M \\ {\scriptstyle g|_V}\downarrow & & \downarrow{\scriptstyle g} \\ V' & \hookrightarrow & M' \end{array}$ is another smooth *map of embeddings*, we will denote $SU_{f,g}(M, V) = DNC_+(M, V) \setminus (DNC(f)^{-1}(V' \times \mathbb{R}_+) \cup DNC(g)^{-1}(V' \times \mathbb{R}_+))$ and $SBlup_{f,g}(M, V)$ its quotient under the zooming action.

A consequence of the preceding remak is

Corollary 2.14 *If* $G \overset{r,s}{\rightrightarrows} G^{(0)}$ *is a Lie groupoid and* H *is a subgroupoid,* $SBlup_{r,s}(G, H) \rightrightarrows SBlup(G^{(0)}, H^{(0)})$ *is a Lie groupoid. Moreover the Lie algebroid of* $SBlup_{r,s}(G, \Gamma)$ *is* $SBlup_{r,s}(\mathfrak{A}G, \mathfrak{A}\Gamma)$.

Example 2.15

(a) If $G \rightrightarrows G^{(0)}$ is a Lie groupoid, define $\mathbb{G} = G \times \mathbb{R} \times \mathbb{R} \rightrightarrows G^{(0)} \times \mathbb{R}$, the product of G with the pair groupoid on $\mathbb{R}$. One can check that

$$SBlup_{r,s}(\mathbb{G}, \mathbb{G}^{(0)} \times \{(0, 0)\}) = DNC_+(G, G^{(0)}) \rtimes \mathbb{R}_+^* \rightrightarrows G^{(0)} \times \mathbb{R}$$

and recover the *Gauge adiabatic groupoid* of [45].

(b) Let $V \subset M$ be a hypersurface. The blowup procedure enables to recover groupoids and spaces involved in the pseudodifferential calculus on manifold with boundary. In particular,

$$\underbrace{G_b = SBlup_{r,s}(M \times M, V \times V)}_{\text{The b-calculus groupoid}} \subset \underbrace{SBlup(M \times M, V \times V)}_{\text{Melrose's b-space}}$$

$$\underbrace{G_0 = SBlup_{r,s}(M \times M, \Delta(V))}_{\text{The 0-calculus groupoid}} \subset \underbrace{SBlup(M \times M, \Delta(V))}_{\text{Mazzeo-Melrose's 0-space}}$$

(c) One can iterate these constructions to go to the study of manifolds with corners, or consider a foliation with no holonomy on V, or define the holonomy groupoid of a manifold with iterated fibered corners, etc.

3 Pseudodifferential calculus on Lie groupoids

The "classical" pseudodifferential calculus was developed in the 1960s and was crucial in the Atiyah–Singer index theorem [8].

Pseudodifferential operators appear naturally when trying to solve (elliptic) differential equations. Using Fourier transform, one associates canonically to a differential operator a polynomial function—its symbol. The composition of differential operators is not commutative and therefore it does not induce just the product of these polynomials, but at least the leading term of the symbol of the product is the product of the leading terms of the symbols. When trying to solve such an equation, one then naturally tries to invert this symbol. This inverse is no longer a polynomial of course, but one can still associate to it an operator—a pseudodifferential operator.

The pseudodifferential operators are used in many different parts of mathematics. Information on pseudodifferential operators and much more can be found in the classical books [66, 67, 68, 69, 107, 109, 110, 111].

Here we will concentrate to the use of pseudodifferential operators in connection with Lie groupoids and noncommutative geometry.

In [28], Alain Connes, in order to generalize the Atiyah–Singer index theorem for families [10] to the case of general foliations, considered the C^*-algebra of the holonomy groupoid as a noncommutative generalization of the space of parameters and studied index problems with values in this algebra. He therefore introduced the pseudodifferential calculus on the holonomy groupoid of a foliation.

This pseudodifferential calculus was easily extended to general Lie groupoids (see [91, 95]). In this way one constructs an analytic index map $K_*(C_0(\mathfrak{A}^*G)) \to K_*(C^*(G))$ for every Lie groupoid G.

Alain Connes made another beautiful observation. His tangent groupoid that we described in Section 2.4.1 can be used in order to construct the analytic index of elliptic operators in a differential and pseudodifferential free way. The fact that this indeed coincides with the analytic index of elliptic operators is just a consequence of the existence of a pseudodifferential calculus on every Lie groupoid.

In this section, we will discuss various constructions of this pseudodifferential calculus on groupoids and the construction of the index.

3.1 Distributions on G conormal to $G^{(0)}$

The point of view developed by Connes is the following: locally the foliation looks like a fibration. On a *foliation chart* $\Omega_i \simeq U_i \times T_i$, where T_i is the local transversal and U_i represents the leaf direction, a pseudodifferential operator P_i is a family indexed by T_i of operators on U_i (in the sense of [10]). Connes then defines a pseudodifferential operator on the foliation as a finite sum $f + \sum_i P_i$ of such local pseudodifferential families P_i (with compact support) and $f \in C_c^\infty(G)$.

It was then quite easy to extend this calculus to a general Lie groupoid and this was done independently in [91] and [95]. There, pseudodifferential operators appear as G-invariant pseudodifferential families acting on the source fibers of G.

In fact, it is probably easier and more natural to consider pseudodifferential operators on Lie groupoids as distributions on G which are conormal to $G^{(0)}$. This point of view appears in various calculi of Melrose (on some spaces that contain Lie groupoids as dense open subsets—see, e.g., [85]) and in [6]. It is explained and developed in [75, 76] where the interested reader will have all details and complete description of the pseudodifferential calculus on a Lie groupoid.

Given a manifold M and a (locally) closed submanifold V, conormal distributions are some particular distributions on M, with singular support in V.

A remak on densities As the elements of the convolution algebra of a groupoid are sections of a density bundle $\Omega^{1/2}$ rather than functions, the distributions that we consider are generalized sections of the same density bundle. The distribution associated with a smooth section of a bundle E over a manifold M is a continuous linear mapping on the topological vector space $C_c^\infty(M; \Omega^1(M) \otimes E^*)$ of smooth sections with compact support of the tensor product of the bundle of one densities on M with the dual bundle of E. In order to simplify our exposition we will drop all these trivial bundles and just consider functions—although this issue is not completely trivial. We will in fact assume that coherent choices of sections of these bundles are made.

A guiding principle is that symbols (of scalar operators) are indeed functions.

3.1.1 Symbols and conormal distributions

Symbols and conormal distributions on $\mathbb{R}^n$ Let $\alpha = (\alpha_1, \dots, \alpha_n) \in \mathbb{N}^n$. Put $|\alpha| = \sum \alpha_i$. The map $D_\alpha : f \mapsto \frac{\partial^{|\alpha|} f}{(\partial x_1)^{\alpha_1} \dots (\partial x_n)^{\alpha_n}}(0)$ is a distribution on $\mathbb{R}^n$ with (singular) support 0. In Fourier terms it can be written as $D_\alpha(f) = \frac{1}{(2\pi)^n} \int_{\mathbb{R}^n} (i\xi)^\alpha \hat{f}(\xi)\, d\xi$.

A *classical symbol* on $\mathbb{R}^n$ of order $m \in \mathbb{Z}$ (or in $\mathbb{C}$) is a function a on $\mathbb{R}^n$ that can be written as

$$a(\xi) \sim \sum_{k=0}^{+\infty} a_{m-k}(\xi),$$

where a_j is a smooth function on $\mathbb{R}^n \setminus \{0\}$ homogeneous of degree j, i.e., such that, for $t \in \mathbb{R}_+^*$ and $\xi \in \mathbb{R}^n \setminus \{0\}$ we have $a_j(t\xi) = t^j a_j(\xi)$.

The notation $\sim$ means that for every $k \in \mathbb{N}$, and every $\alpha \in \mathbb{N}^n$, there is a constant $M_{k,\alpha}$ such that, for $\|\xi\| \geq 1$, we have

$$\frac{\partial^{|\alpha|}(a - \sum_{j=0}^{k-1} a_{m-j})}{(\partial \xi_1)^{\alpha_1} \dots (\partial \xi_n)^{\alpha_n}}(\xi) \leq M_{k,\alpha} \|\xi\|^{m-k-|\alpha|}.$$

Such a symbol a gives rise to a distribution using the formula

$$P_a(f) = \frac{1}{(2\pi)^n} \int_{\mathbb{R}^n} a(\xi) \hat{f}(\xi)\, d\xi.$$

- The support of this distribution is now $\mathbb{R}^n$, but its *singular support* is 0: For every neighborhood V of 0, one can write $P_a = Q_a + \kappa$ where Q_a has support in V and κ is a smooth (Schwartz) function.
- The function a is called the *total symbol* of P_a. It is the Fourier transform of a since $a(\xi) = P_a(h_\xi)$ where $h_\xi(x) = e^{i\langle x|\xi\rangle}$.
- The homogeneous function a_m is called the *principal symbol* of P_a. Note that $a_m(\xi) = \lim_{t\to+\infty} t^{-m} a(t\xi)$—and therefore a_m only depends on P_a.

Symbols on a vector bundle Let now $p : E \to B$ be a real vector bundle over a manifold B. We consider symbols on E as being families—indexed by B of symbols on the fibers $(E_x)_{x\in B}$. Such a symbol is then a function $a : E^* \to \mathbb{C}$ where E^* is the dual vector bundle such that

$$a(x,\xi) \sim \sum_{k=0}^{+\infty} a_{m-k}(x,\xi),$$

where a_j is a smooth function on $E^* \setminus B$ (where $B \subset E^*$ is the 0 section of the bundle E^*) homogeneous of degree j in ξ, i.e., such that, for $t \in \mathbb{R}_+^*$, $x \in B$ and $\xi \in E_x^* \setminus \{0\}$ we have $a_j(x, t\xi) = t^j a_j(x,\xi)$.

The writing $\sim$ means here that, in local coordinates, putting $B = \mathbb{R}^p$ and $E = \mathbb{R}^p \times \mathbb{R}^n$, for every (k, α, β, K) where $k \in \mathbb{N}$, $\alpha \in \mathbb{N}^n$, $\beta \in \mathbb{N}^p$ and $K \subset B$ is a compact subset, there is a constant $M_{k,\alpha,\beta,K}$ such that, for $x \in K$ and $\|\xi\| \geq 1$, we have

$$\frac{\partial^{|\alpha|+|\beta|}\left(a - \sum_{j=0}^{k-1} a_{m-j}\right)}{(\partial x_1)^{\beta_1}\dots(\partial x_p)^{\beta_p}\,\partial\xi_1)^{\alpha_1}\dots(\partial\xi_n)^{\alpha_n}}(x,\xi) \leq M_{k,\alpha,\beta,K}\|\xi\|^{m-k-|\alpha|}.$$

Remark 3.1 Note that giving a symbol a of order m on E^* is equivalent to a homogeneous smooth function b of order m on $E^* \times \mathbb{R}_+ \setminus B \times \{0\}$.

- given a homogeneous smooth function $b : E^* \times \mathbb{R}_+ \setminus B \times \{0\} \to \mathbb{R}$, put $a(x,\xi) = b(x,\xi,1)$;
- given a symbol a, put $b(x,\xi,t) = t^{-m} a\left(x, \frac{\xi}{t}\right)$ for $t \neq 0$ and $b(x,\xi,0) = a_m(u,\xi)$ (where a_m is the principal symbol of a).

The expansion $a(x, \xi) \sim \sum_j a_{m-j}(x, \xi)$ corresponds to the Taylor expansion of b at $t = 0$: we have $b(x, \xi, t) \sim \sum_j t^j a_{m-j}(x, \xi)$.

Associated conormal distributions To such a symbol, we may still associate a distribution P_a given by the formula

$$P_a(f) = \frac{1}{(2\pi)^n} \int_{E^*} a(x, \xi) \hat{f}(x, \xi)\, dx\, d\xi$$

(where $\hat{f}(x, \xi) = \int_{E_x} e^{-i\langle u|\xi\rangle} f(x, u)\, du$).

- The *singular support* of this distribution is contained in $B \subset E$.
- The function a is called the *total symbol* of P_a. The homogeneous function a_m is called the *principal symbol* of P_a.

Symbols and conormal distributions on a manifold Let now M be a manifold and $B \subset M$ a closed submanifold of M. The tubular neighborhood construction provides us with a neighborhood U of B in M and a diffeomorphism $\varphi : N \to U$ where $N = \mathcal{N}_B^M$ is the normal bundle: for $x \in B$, we have $N_x = T_x M / T_x B$. The requirement for such a diffeomorphism is $\varphi(b) = b$ for every $b \in B$ and, for every x in B, the differential $Tf : T_x N \to T_x M$ satisfies $p_x \circ Tf_x(\xi) = \xi$ for $\xi \in N_x \subset T_x(N)$ where $p_x : T_x M \to N_x = T_x M / T_x B$ is the projection.

Using φ, we obtain a family of distributions on M: those that are a sum $Q = \varphi_*(P_a) + \kappa$ of a smooth function κ on M (with compact support) and a conormal distribution P_a on N where a is a symbol on N^*: we write

$$Q(f) = \int_M \kappa(y) f(y)\, dy + \int_{N^*} a(x, \xi) \Big(\int_{u \in N_x} e^{-i\langle u|\xi\rangle} f \circ \varphi(x, u)\, du \Big)\, dx\, d\xi.$$

Diffeomorphism invariance of conormal distributions and principal symbol It turns out that

- the space of distributions on M of the form $\varphi_*(P_a) + \kappa$ does not depend on the partial diffeomorphism $\varphi : N_B^M \to M$;
- the principal symbol a_m of $\varphi_*(P_a) + \kappa$ does not depend on φ either.

Remark We can also write

$$Q(f) = \lim_{R \to \infty} \int_M \kappa_R(y) f(y)\, dy \quad \text{with}$$

$$\kappa_R(y) = \kappa(y) + \chi(y) \int_{\xi \in N^*_{p(y)},\ \|\xi\| \le R} a(p(y), \xi) e^{-i\langle \theta(y)|\xi\rangle}\, d\xi,$$

where $\chi \in C^\infty(M)$ is a smooth function which is equal to 0 outside U and to 1 for $y \in B$ (taking into account the Jacobian of φ), and where we have written $\varphi^{-1}(y) = (p(y), \theta(y))$ with $p(y) \in B$ and $\theta(y) \in (\mathcal{N}_B^M)_{p(y)}$.

Conormal distributions We denote by $\mathcal{I}_c^m(M, B)$ the space of such conormal distributions with compact support. The map $\sigma_m : \varphi_*(P_a)+\kappa \mapsto a_m$ is the principal symbol map.

Note that we have a natural exact sequence $0 \to \mathcal{I}_c^{m-1}(M, B) \to \mathcal{I}_c^m(M, B) \xrightarrow{\sigma_m} C_m^\infty((\mathcal{N}_B^M)^* \setminus B) \to 0$ where we denoted by $C_m^\infty((\mathcal{N}_B^M)^* \setminus B)$ the space of smooth functions on $(\mathcal{N}_B^M)^* \setminus B$ which are homogeneous of degree m.

Definition 3.2 A (classical) pseudodifferential operator of order m on a Lie groupoid $G \rightrightarrows M$ is a conormal distribution $P \in \mathcal{I}_c^m(G, M)$.

3.1.2 Convolution

Let $G \rightrightarrows M$ be a Lie groupoid.

The convolution for $C_c^\infty(G)$ can be understood in the following way. Let $G^{(2)} = \{(\alpha, \beta) \in G \times G;\ s(\alpha) = r(\beta)\}$ be the set of composable elements, and let $p_1, p_2, m : G^{(2)} \to G$ be the maps defined by $p_1(\alpha, \beta) = \alpha$, $p_2(\alpha, \beta) = \beta$ and $m(\alpha, \beta) = \alpha\beta$.

Take $f_1, f_2 \in C_c^\infty(G)$. We then may write $f_1 * f_2 = m_!(p_1^*(f_1).p_2^*(f_2))$, where:

- $p_i^* : C^\infty(G) \to C^\infty(G^{(2)})$ is given by $p_i^*(f_i) = f_i \circ p_i$;
- $p_1^*(f_1).p_2^*(f_2)$ is just the pointwise product of the functions $p_i^*(f_i)$;
- $m_! : C_c^\infty(G^{(2)}) \to C_c^\infty(G)$ is the integration along the fibers of the submersion m:

$$m_!(f)(\gamma) = \int_{\alpha\beta=\gamma} f(\alpha, \beta)\, d\nu = m_!(f)(\gamma) = \int_{G^{r(\gamma)}} f(\alpha, \alpha^{-1}\gamma) d\nu^{r(\gamma)}(\alpha).$$

We wish to extend these three operations to the case when f_1 and f_2 are conormal distributions.

Push-forward, pull-back, product of distributions

Push-forward Let $\varphi : M \to M'$ be a smooth map. Dual to $\varphi^* : C^\infty(M') \to C^\infty(M)$ is a map $\varphi_* : C_c^{-\infty}(M) \to C_c^{-\infty}(M')$ given by $(\varphi_*(P))(f) = P(\varphi^*(f))$ for $P \in C_c^{-\infty}(M)$ and $f \in C^\infty(M')$.

Let $V \subset M$ be a submanifold. Assume that φ is a submersion and that the restriction of φ to V is a diffeomorphism. Then, the image φ_* of $P \in \mathcal{I}_c^m(M, V)$ is a smooth distribution: $\varphi_*(P) \in C_c^\infty(M') \subset C_c^{-\infty}(M')$.

We may indeed, by restricting to a neighborhood of V in M, assume that $\varphi : M \to V \simeq M'$ is a vector bundle projection and that $P = P_a$ where a is a symbol on the dual bundle M^*. Then P is in fact a family of pseudodifferential operators $(P_x)_{x \in V}$ and

$$\varphi_*(P)(f) = \int_V f(x)P_x(1)\,dx = \int_V f(x)a(x,0)\,dx.$$

In other words, $\varphi_*(P)$ is the distribution associated with the function $x \mapsto a(x, 0)$.

Pull-back by a submersion Let $p : M' \to M$ be a submersion. Dual to the integration along the fibers $p_! : C_c^\infty(M') \to C_c^\infty(M)$ is a map $p^! : C^{-\infty}(M) \to C^{-\infty}(M')$ given by $(p^!(P))(f) = P(p_!(f))$ for $P \in C^{-\infty}(M)$ and $f \in C_c^\infty(M')$. The map $p^!$ extends to distributions the map $p^* : C^\infty(M) \to C^\infty(M')$.

Proposition 3.3 *Let $V \subset M$ be a closed submanifold. Put $V' = p^{-1}(V)$.*

(a) *The normal bundle of V' in M' identifies with the pull back of the normal bundle of V in M.*
(b) *Assume that $P \in \mathcal{I}_c^m(M, V)$. Then $p^!(P) \in \mathcal{I}_c^m(M', V')$. Under the identification of the normal bundle N' of V' in M' with $p^*\mathcal{N}_V^M$, the principal symbol of p^*P is given by $\sigma_m(p^*P) = \sigma_m(P) \circ p$.*

Proof The first statement is obvious. Thanks to it, we may assume that M is a vector bundle $p : E \to V$ and M' is the pull-back vector bundle $E' = E \times_V V' \to V'$. The second statement is then immediate too. □

Products of conormal distributions

(a) One can extend the (pointwise) product of functions to the case where one of them is a distribution. The pointwise product of two distributions is not always well defined.

    Already in this way, we may define the convolution of a classical pseudodifferential operator P with an element $f \in C_c^\infty(G)$: we have $P * f = m_*(p_1^!(P).p_2^*(f_2))$. Then $p_1^!(P).p_2^*(f_2) \in \mathcal{I}_c(G^{(2)}, p_1^{-1}G^{(0)})$ and the submersion $m : G^{(2)} \to G$ induces a diffeomorphism $p_1^{-1}(G^{(0)}) \to G$, whence $P * f \in C_c^\infty(G)$ and in the same way, $f * P \in C_c^\infty(G)$.

    From this it follows that pseudodifferential operators define *multipliers* of $C_c^\infty(G)$.
(b) To explain the convolution of two pseudodifferential operators we use the two following facts which reduce to linear algebra.

    (i) Let M be a manifold, V_1, V_2 two closed submanifolds of M that are transverse to each other. This means that, for every $x \in V_1 \cap V_2$ we have $T_xM = T_xV_1 + T_xV_2$ (we do not assume that this sum is a direct sum). Then if $Q_1 \in \mathcal{I}_c^{\ell_1}(M, V_1)$ and $Q_2 \in \mathcal{I}_c^{\ell_2}(M, V_2)$, then the distribution $Q_1.Q_2$ makes sense.
    (ii) If moreover $Q_1.Q_2$ has compact support and $m : M \to M'$ is a submersion whose restriction to both V_1 and V_2 is a diffeomorphism $V_i \to M'$, then $m_*(Q_1.Q_2) \in \mathcal{I}_c^{\ell_1+\ell_2}(M', m(V_1 \cap V_2))$ and its principal symbol is the product of the symbols of Q_1 and Q_2 under the natural identification of the normal bundles:

- the restriction $(\mathcal{N}_{V_1}^{M})_{|V_1\cap V_2}$ of $\mathcal{N}_{V_1}^{M}$ to $V_1 \cap V_2$ identifies with $\mathcal{N}_{V_1\cap V_2}^{V_2}$;
- the restriction $(\mathcal{N}_{V_2}^{M})_{|V_1\cap V_2}$ of $\mathcal{N}_{V_1}^{M}$ to $V_1 \cap V_2$ identifies with $\mathcal{N}_{V_1\cap V_2}^{V_1}$;
- finally, using the map m, we identify $\mathcal{N}_{V_1\cap V_2}^{V_1}$ and $\mathcal{N}_{V_1\cap V_2}^{V_2}$ identify with $\mathcal{N}_{m(V_1\cap V_2)}^{M'}$.

Given $P_1 \in \mathcal{I}_c^{\ell_1}(G, G^{(0)})$ and $P_2 \in \mathcal{I}_c^{\ell_2}(G, G^{(0)})$ we then put

- $M = G^{(2)} = \{(\alpha, \beta) \in G \times G;\ s(\alpha) = r(\beta)\}$;
- $Q_i = p_i^!(P_i)$ where $p_i : G^{(2)} \to G$ are the submersions $(\gamma_1, \gamma_2) \mapsto \gamma_i$;
- $M' = G$ and $m : (\alpha, \beta) \mapsto \alpha\beta$ is the composition.

It follows that $P_1 P_2 \in \mathcal{I}_c^{\ell_1+\ell_2}(G, G^{(0)})$ with principal symbol $\sigma_{\ell_1}(P_1)\sigma_{\ell_2}(P_2)$.

Formal adjoint One can also define the adjoint of a pseudodifferential operator P by setting $P^* = j_*(\overline{P})$, where $j : G \to G$ is the diffeomorphism $\gamma \mapsto \gamma^{-1}$.

3.1.3 Pseudodifferential operators of order ≤ 0

Pseudodifferential operators with compact supports on a Lie groupoid $G \rightrightarrows M$ appear as multipliers of $C_c^\infty(G)$.

Proposition 3.4 *Pseudodifferential operators with compact support of order ≤ 0 extend to multipliers of $C^*(G)$; pseudodifferential operators of order < 0 are in fact elements of $C^*(G)$.*

The first statement means that if P is a pseudodifferential operator with compact support in G and of order ≤ 0, then there exists a constant c such that, for all $f \in C_c^\infty(G)$, we have $\|P * f\| \leq c\|f\|$ and $\|f * P\| \leq c\|f\|$ (this is true for both the maximal and the reduced C^*-norm of G).

Proof To establish this statement, first assume that P is of order $< -p$ where $p = \dim G - \dim M$ is the dimension of the algebroid. Note that if a is a symbol of order $\leq -p$, then P_a is a continuous function. Therefore, P is a continuous function with compact support on G, and thus an element of $C^*(G)$.

If P is of order $< -p/2$, then $\|P * f\|^2 = \|f^* * P^* * P * f\|$ (and $\|f * P\|^2 = \|f * P * P^* * f^*\|$) and as $P^* * P$ is of order $< -p$, it is in $C^*(G)$ and thus $\|P * f\|^2 \leq \|P^* * P\|\|f\|^2$. It follows that P is a multiplier, and as $P^* * P \in C^*(G)$ we find $P \in C^*(G)$.

If P is of negative order, $(P^*P)^{2^k} \in C^*(G)$ for some $k \in \mathbb{N}$, and by induction in k, $P \in C^*(G)$.

Let P be a pseudodifferential operator of order 0.

Note first that every smooth function $q \in C_c^\infty(M)$ is a pseudodifferential operator of order 0 with principal symbol $\sigma_q : (x, \xi) \mapsto q(x)$—and of course a bounded multiplier: we have $(q * f)(\gamma) = q(r(\gamma))f(\gamma)$ and $(f * q)(\gamma) = f(\gamma)q(s(\gamma))$.

Let $q \in C_c(M)$ which is equal to 1 on the support of σ_P—i.e., the projection on M of the closure of $\{(x,\xi);\ \sigma_q(x,\xi) \neq 0\}$ (which is assumed to be compact in the space of half lines of the bundle $\mathfrak{A}^*$). Let $c \in \mathbb{R}_+$ with $c > \sigma_P(x,\xi)$ for all (x,ξ). Put $b(x,\xi) = q(x)\sqrt{c^2+1-|\sigma_q(x,\xi)|^2}$, and let Q be a pseudodifferential operator with principal symbol b. Then P^*P+Q^*Q which has symbol $(1+c^2)|q|^2$ is of the form $(1+c^2)|q|^2+R$ where R is of negative order and therefore P^*P+Q^*Q is bounded.

For all $f \in C_c(G)$, $\|Pf\|^2 = \|f^*P^*Pf\| \leq \|f^*P^*Pf+f^*Q^*Qf\| \leq \|P^*P+Q^*Q\|\|f\|^2$, and thus $f \mapsto Pf$ is bounded.

In the same way $f \mapsto fP$ is bounded. □

As a consequence, one gets:

Theorem 3.5 *We have a short exact sequence of C^*-algebras*

$$0 \to C^*(G) \to \Psi^*(G) \overset{\sigma}{\to} C_0(S\mathfrak{A}^*) \to 0$$

where $\Psi^*(G)$ is the closure of the algebra of order 0 *pseudodifferential operators in the multiplier algebra of $C^*(G)$ and $S\mathfrak{A}^*$ is the sphere bundle (the set of half lines) of the dual $\mathfrak{A}^*$ of the algebroid $\mathfrak{A}$ of G.*

Note that this statement is true for the full groupoid C^*-algebra as well as for the reduced one.

Proof First note that as $\Psi^*(G)$ contains $C_c^\infty(G)$, it contains its closure $C^*(G)$.

The only statement which does not follow from Proposition 3.4 is that the principal symbol map is well defined, i.e., if $\sigma(P) \neq 0$, then $P \notin C_r^*(G)$ (it is enough to check this for the reduced algebra).

In fact, one shows that for every pseudodifferential operator P of order 0 every $x \in M$ and a nonzero $\xi \in \mathfrak{A}_x^*$, we have $\sigma_P(x,\xi) = \lim_{n\to\infty}\langle \varphi_n, \lambda_x(P)\varphi_n\rangle$ where φ_n is a function on G_x of L^2-norm 1 whose support is concentrated around x and whose Fourier transform is concentrated in $\mathbb{R}_+^*\xi$: we may take, in local coordinates, $\varphi_n(y) = (2n)^{p/4}e^{-n\pi\|x-y\|^2-in\langle(y-x)|\xi\rangle}$. Here, λ_x is the representation of $C^*(G)$ on $L^2(G_x)$ by left convolution—extended to the multipliers. On the other hand, for $f \in C_c^\infty(G)$, we have $\lim_{n\to\infty}\langle \varphi_n, \lambda_x(f)\varphi_n\rangle = 0$ and by continuity the same is true for $f \in C_r^*(G)$. □

3.1.4 Analytic index

The connecting map of the exact sequence of Theorem 3.5 is the *analytic index* of the Lie groupoid

$$\partial_G : K_{i+1}(C_0(S\mathfrak{A}^*)) \to K_i(C^*(G)).$$

This analytic index can be improved a little by taking vector bundles into account. Indeed, the starting point of an index problem is often a pair of bundles $E^{\pm}$ over M together with a symbol of order 0 which gives a smooth family of isomorphisms $a(x,\xi) : E_x^+ \to E_x^-$.

Such a symbol defines an element in the *relative K-theory* of the morphism $\mu : C_0(M) \to C_0(S\mathfrak{A}^*)$, in other words an element of the K-theory of the *mapping cone*

$$C_\mu = \{(f,g) \in C_0(M) \times C_0(S\mathfrak{A}^* \times \mathbb{R}_+);\ \forall (x,\xi) \in S\mathfrak{A}^*,\ g(x,\xi,0) = f(x)\}$$

of μ. This mapping cone is naturally isomorphic to $C_0(\mathfrak{A}^*)$ using the map $(x,\xi,t) \mapsto (x,t\xi)$.

Consider the morphism $\psi : C_0(M) \to \Psi^*(G)$ which associates to a (smooth) function f the order 0 (pseudo)differential operator multiplication by f. Note that we have $\mu = \sigma \circ \psi$. Using the commutative diagram

$$\begin{array}{ccc} C_0(M) & \xrightarrow{\mu} & C_0(S\mathfrak{A}^*) \\ \downarrow{\scriptstyle\psi} & & \| \\ \Psi^*(G) & \xrightarrow{\sigma} & C_0(S\mathfrak{A}^*) \end{array}$$

we obtain a morphism $\tilde{\psi} : C_0(\mathfrak{A}^*) = C_\mu \to C_\psi$. Now, we also have an exact sequence

$$0 \to C^*(G) \xrightarrow{e_G} C_\psi \longrightarrow C_0(S\mathfrak{A}^* \times \mathbb{R}_+) \to 0.$$

As the algebra $C_0(S\mathfrak{A}^* \times \mathbb{R}_+)$ is contractible (and nuclear), the *excision map e_G* is a KK-equivalence.

Definition 3.6 The *analytic index map* of the Lie groupoid G is the composition

$$\operatorname{ind}_G = [e_G]^{-1} \circ [\tilde{\psi}] : K^j(\mathfrak{A}^*) = K_j(C_0(\mathfrak{A}^*)) \to K_j(C^*(G)).$$

The index ∂_G is the composition of the morphism $K_{j+1}(C_0(S\mathfrak{A}^*)) \to K_j(C_0(\mathfrak{A}^*))$ induced by the inclusion $S\mathfrak{A}^* \times \mathbb{R}_+^* \to \mathfrak{A}^*$ with the index map ind_G).

3.2 Classical examples

The analytic index for groupoids recovers many classical situations.

(a) Assume that $G = M \times M$ is just the pair groupoid. Then the corresponding index is the classical Atiyah–Singer index $K^0(T^*M) \to K_0(\mathcal{K}) = \mathbb{Z}$ of (pseudo)differential operators on M, i.e., the one constructed and computed in [8].

(b) Assume that G is the groupoid $M \times_Y M$ associated with a smooth fibration $\pi : M \to Y$. The corresponding index is the Atiyah–Singer index $K^j(T_F^*M) \to K_j(C^*(G)) = K^j(Y)$ of families of (pseudo)differential operators on the fibers of π, i.e., the one constructed and computed in [10].
(c) Assume that G is the groupoid $G = (\widetilde{M} \times \widetilde{M})/\Gamma$ where Γ is a countable group acting freely and properly on a manifold $\widetilde{M}$. The corresponding index $K^j(T^*M) \to K_0(C^*(G)) = K_j(C^*(\Gamma))$. This situation was introduced by Atiyah in [7] where it is shown that the von Neumann dimension of the index (for $j = 0$) is in fact the index $K^0(T^*M) \to \mathbb{Z}$. In the C^*-context, it was studied in [87] and has been since then studied in many, many papers.

3.3 Analytic index via deformation groupoids

The deformation groupoids allow to construct the analytic index without use of pseudodifferential operators. This was Connes' main motivation for introducing them.

Let $G \rightrightarrows M$ be a Lie groupoid and let $G_{ad} = \mathfrak{A} \times \{0\} \cup G \times (0,1]$ be its adiabatic groupoid. Consider the evaluation maps $ev_0 : C^*(G_{ad}) \to C^*(\mathfrak{A}^*)$ and $ev_t : C^*(G_{ad}) \to C^*(G)$ for $t \neq 0$.

As the sequence

$$0 \to C^*(G \times (0,1]) \longrightarrow C^*(G_{ad}) \xrightarrow{ev_0} C^*(\mathfrak{A}^*) \to 0$$

is exact and $C^*(G \times (0,1])$ is contractible, the evaluation ev_0 is K-invertible. We then have the following important theorem which is in a sense just an observation.

Theorem 3.7 (Connes. cf. [32, 91, 95, 42]) *The analytic index is the composition* $\mathrm{ind}_G = [ev_1] \circ [ev_0]^{-1}$.

The proof of this theorem reduces to the following two observations:

(a) (*Naturality of the analytic index.*) Let $G_1 \rightrightarrows M_1$ be a Lie groupoid and let $M_2 \subset M_1$ be a closed submanifold which is *saturated* for G_1 (i.e., for $\gamma \in G_1$ we have $r(\gamma) \in M_2$ if and only if $s(\gamma) \in M_2$). Denote by $G_2 \rightrightarrows M_2$ the Lie groupoid $\{\gamma \in G_1;\ s(\gamma) \in M_2\}$. The algebroid $\mathfrak{A}_2$ of G_2 is the restriction to M_2 of the algebroid $\mathfrak{A}_1$ of G_1. We have restriction maps $r_G : C^*(G_1) \to C^*(G_2)$ and $r_{\mathfrak{A}^*} : C_0(\mathfrak{A}_1^*) \to C_0(\mathfrak{A}_2^*)$.

Then the diagram $$\begin{array}{ccc} K_j(C_0(\mathfrak{A}_1^*)) & \xrightarrow{\mathrm{ind}_{G_1}} & K_j(C^*(G_1)) \\ \downarrow r_{\mathfrak{A}^*} & & \downarrow r_G \\ K_j(C_0(\mathfrak{A}_2^*)) & \xrightarrow{\mathrm{ind}_{G_2}} & K_j(C^*(G_2)) \end{array}$$ is commutative.

(b) If $E \to B$ is a vector bundle considered as a Lie groupoid, then

$$\mathrm{ind}_E : K_j(C_0(E^*)) \to K_j(C^*(E)) = K_j(C_0(E^*))$$

is the identity.

Remarks 3.8

(a) One can also use the deformation groupoids in order to prove index theorems. In [32], Alain Connes gives a beautiful proof of the Atiyah–Singer index theorem in K-theory—based on the analogue of the Thom isomorphism for crossed products by $\mathbb{R}^n$ of [29]. A different proof was given in [43]. These proofs naturally generalize to more general index theorems.
(b) An extra step is also taken in his lectures at the Collège de France (see also, e.g., [50]) where the asymptotics of the deformation groupoid are used in order to derive the cohomological formula of the Atiyah–Singer index theorem [9]. This is a very important issue for index theory, but we will not discuss it any further here.

3.4 *Deformation to the normal cone, zooming action and PDO*

The deformation groupoids as we saw give some insight to the pseudodifferential calculus on a groupoid. We will see in fact that the deformation groupoids allow to recover the pseudodifferential calculus itself, using the natural "zooming" action (called *gauge action* in [45]).

3.4.1 The zooming action of $\mathbb{R}_+^*$ on a deformation to the normal cone

Let M be a smooth manifold and V a closed submanifold. Denote by $N = \mathcal{N}_V^M$ the normal bundle of V in M. The group $\mathbb{R}_+^*$ acts smoothly on $DNC(M, V)$: for $t \in \mathbb{R}_+^*$ put $\alpha_t(z, \lambda) = (z, t\lambda)$ for $z \in M$ and $\lambda \in \mathbb{R}^*$ and $\alpha_t(x, U, 0) = (x, \frac{U}{t}, 0)$ for $x \in V$ and $U \in N_x$.

This zooming action gives two interpretations of conormal distributions.

3.4.2 Integrals of smooth functions

Define first the space $\mathcal{S}(M, V)$ of Schwartz functions on $DNC_+(M, V)$: notice that $DNC_+(M, V)$ is an open dense subset of the blowup $SBlup(M \times \mathbb{R}, V \times \{0\})$. The space $\mathcal{S}(M, V)$ is the set of smooth functions with compact support on $SBlup(M \times \mathbb{R}, V \times \{0\})$ which vanish at infinite order on the complement of $DNC_+(M, V)$.

The subspace $\mathcal{J}(M, V)$ An element $k \in \mathcal{S}(M, V)$ defines a family $(k_t)_{t \neq 0}$ of smooth function on M. We define then the subspace $\mathcal{J}(M, V) \subset \mathcal{S}(M, V)$ to be the set of elements $k \in \mathcal{S}(M, V)$ such that $(k_t)_{t \neq 0}$ vanishes at infinite order at 0 as a distribution on M, i.e., such that for every $f \in C^\infty(M)$, the function $t \mapsto \langle k_t | f \rangle = \int_M k_t(x) f(x)\, dx$ extends to a smooth function on $\mathbb{R}$ which vanishes at infinite order when $t \to 0$.

In local coordinates, i.e., if M is a vector bundle over V, identifying $DNC(M, V)$ with $M \times \mathbb{R}$, an element $k \in \mathcal{S}(M, V)$ is in $\mathcal{J}(M, V)$ if and only if $\hat{k}$ vanishes at infinite order on $V \times \{0\} \subset M^* \times \mathbb{R}$. Indeed, in that case, $\langle k_t | f \rangle = \int_V du \Big(\int_{M_u^*} \hat{k}_t(u, t\xi) \hat{f}(u, \xi)\, d\xi \Big)$.

It is then easy to see—using local coordinates:

Theorem 3.9 ([45]) *For $m \in \mathbb{C}$, $\mathcal{I}_c^m(M, V)$ is the set of distributions of the form $\int_{\mathbb{R}+} k_t\, t^{-1-m}\, dt$ where k runs over $\mathcal{J}(M, V)$.*

Proof We may assume that M is a vector bundle over V. We need then to write a symbol $a(u, \xi) \sim \sum a_{m-j}(u, \xi)$ on the dual bundle M^* as an integral $a(u, \xi) = \int_0^{+\infty} g(u, t\xi, t)\, t^{-1-m}\, dt$. Such a g will have a Taylor expansion at $t = 0$ of the form $g(u, \xi, t) \sim \sum\limits_{j=0}^{+\infty} t^n g_j(x, \xi)$. Then

$$\int_0^{+\infty} g(u, t\xi, t)\, t^{-1-m}\, dt \sim \sum_{j=0}^{+\infty} b_{m-j}(u, \xi)$$

where $b_{m-j}(u, \xi) = \int_0^{+\infty} g_j(u, t\xi)\, t^{n-1-m}\, dt$ is homogeneous in ξ of order $m - j$. Using Borel's theorem, one easily finds a smooth function g whose Taylor expansion g_j yields $b_{m-j} = a_{m-j}$. The theorem follows. □

3.4.3 Almost equivariant distributions

In [114], Erik van Erp and Robert Yuncken presented another point of view on pseudodifferential calculus on groupoids. Their construction can be carried to conormal distribution in the following way:

Theorem 3.10 *A conormal distribution $P \in \mathcal{I}_c^m(M, V)$ is a distribution on M with compact support such that there exists a distribution Q on $DNC(M, V)$ given by a smooth family $(P_t)_{t \in \mathbb{R}^*}$, (i.e., such that $\langle Q | f \rangle = \int_{\mathbb{R}} \langle P_t | f_t \rangle\, dt$) which satisfies $P_1 = P$ and $\alpha_\lambda Q - \lambda^m Q \in C^\infty(DNC(M, V))$ for every $\lambda \in \mathbb{R}_+^*$.*

Proof We may of course assume that M is the total space of a vector bundle $E \to V$ (and V is the 0-section). As Q has compact support, we may write $\langle Q | f \rangle = \int_{\mathbb{R}} \langle \widehat{P_t} | \widehat{f_t} \rangle\, dt$ where $\widehat{P_t}$ is a smooth function on E^*. The family $(\widehat{P_t})_{t \in \mathbb{R}_+}$ is then a smooth function F on $E^* \times \mathbb{R}_+$ such that, for every $\lambda \in \mathbb{R}_+^*$, the function $(x, \xi, t) \mapsto \lambda^m F(x, \xi, t) - F(x, \lambda x, \lambda t)$ has compact support.

If F is of that form, then $(x, \xi) \mapsto F(x, \xi, 1)$ is a symbol; if a is a symbol, just put $F(x, \xi, t) = \chi(\|\xi\|^2 + t^2) t^{-m} a\left(x, \frac{\xi}{t}\right)$ for $t \neq 0$ and $F(x, \xi, 0) = \chi(\|\xi\|^2) a_m(x, \xi)$. □

Remark: Equivariant linear forms on $\mathcal{J}(M, V)$ One can modify a little bit this construction. The distributions used here are linear forms on $\mathcal{S}(M, V)$. Such linear forms can be exactly equivariant for $m \in \mathbb{N}$—and in this case we obtain only differential operators. If instead we take linear forms only defined on the subspace $\mathcal{J}(M, V) \subset \mathcal{S}(M, V)$ of [45], we probably construct an exactly equivariant family of linear forms on $\mathcal{J}(M, V)$ extending any element of $\mathcal{I}_c^m(M, V)$.

3.5 *Some generalizations*

We will just cite here, without going too much into the details, some more general distributions which were constructed and used in operator algebras.

3.5.1 More general families of pseudodifferential operators

One can associate useful distributions to much more general symbols. In [61] appear symbols of type (ρ, δ) and the associated pseudodifferential operators.

Let $\rho, \delta \in [0, 1]$. Let $E \to M$ be a Euclidean vector bundle. A symbol of order m and type (ρ, δ) is a function $a : E \to \mathbb{C}$ such that, in local coordinates, for every multiindices α, β, and every compact subset K in M, there exists $C \in \mathbb{R}_+$ such that

$$\left| \frac{\partial^{|\alpha|+|\beta|} a}{\partial^\alpha x \, \partial^\beta \xi}(x, \xi) \right| \leq C(\|\xi\| + 1)^{m-\rho|\beta|+\delta|\alpha|}$$

for every $x \in K$ and $\xi \in E_x$ [62].

Polyhomogeneous symbols, i.e., the ones considered above, are particular cases of symbols of type $(1, 0)$.

These symbols of type (ρ, δ) were used in [61] in order to construct *holonomy almost invariant* transversally elliptic operators on any foliation, i.e., holonomy invariant up to lower order. Restricting to a transversal this amounts to finding operators on a manifold almost invariant under the action of a (pseudo)group Γ. Thanks to the work of Connes [31], one may assume that the (tangent) bundle E has an invariant subbundle F which has an invariant Euclidean metric as well as the quotient E/F. In [61] were constructed pseudodifferential operators of order 0 that "differentiate more" along the direction F and were used to construct almost invariant Dirac type operators.

3.5.2 Inhomogeneous calculus

Connes–Moscovici [34, 35], in order to write formulae in cyclic cohomology, used a more specific and unbounded analogue of [61]. This was a differential operator which is of second order in the direction tangent to F and of order 1 in the complementary direction. This falls into a construction of an inhomogeneous pseudodifferential calculus modelled on nilpotent groups studied by many authors—see the book [14].

In order to understand better this calculus and the corresponding index map, Choi-Ponge [25, 27, 26] and van Erp-Yuncken [115] constructed independently a deformation Lie groupoid of the form $(M \times M \times \mathbb{R}^*) \sqcup \mathcal{N} \times \{0\}$.

Let us briefly describe the very general and nice setting for this inhomogeneous calculus.

Let M be a smooth manifold and let $\{0\} \subset H^1 \subset H^2 \subset \ldots \subset H^k = TM$ be a filtration of TM by subbundles. We assume that if X is a smooth section of H^i and Y is a smooth section of H^j then $[X, Y]$ is a section of H^{i+j} (we put $H^\ell = TM$ for $\ell \geq k$). We declare then that a vector field which is a section of H^i is a differential operator of degree i.

Note that the bundle $\mathfrak{N} = \bigoplus_{i=1}^{n} H^i$ is naturally equipped with a nilpotent Lie algebra structure, thanks to the Lie brackets of sections: Let $x \in M$. If X is a smooth section of H^i and Y is a smooth section of H^j, then the class of $[X, Y]_x$ is H_x^{i+j}/H_x^{i+j-1} only depends on the class of X_x in H_x^i/H_x^{i-1} and of Y_x in H_x^j/H_x^{j-1}. One then has an associated nilpotent Lie group bundle $\mathcal{N}$ over M, which as a set in $\mathfrak{N}$ and the product is constructed, thanks to the Baker–Campbell–Hausdorff formula. An important feature of this is that there is a natural action α of the group $\mathbb{R}_+^*$ on $\mathfrak{N}$ and $\mathcal{N}$ given by $\alpha_\lambda(X) = \lambda^i X$ if $X \in H^i$.

Constructions as those explained in Section 3.4 (should) naturally allow to recover the associated inhomogeneous pseudodifferential calculus out of this deformation groupoid [115].

Mohsen [88] gave a very nice construction of this groupoid based on deformations to the normal cone. We will come back to Mohsen's construction of this deformation groupoid in Section 4.2.2.

3.5.3 Fourier integral operators

A "classical" family of operators generalizing pseudodifferential calculus is that of Fourier integral operators (cf. [69]). These were constructed by Hörmander [63] in order to better understand the propagation of singularities for some strictly hyperbolic operators as the wave equation. These operators were studied by several authors and were very useful in local analysis [49, 64, 65, 69, 48, 107, 111]. Recently, an index theory based on Fourier integral operators was developed (cf. [105]).

In [76], Lescure and Vassout show how to define Fourier integral operators on Lie groupoids. Fourier integral operators with proper support define multipliers of the convolution algebra $C_c^\infty(G)$, those of order 0 define multipliers of the C^*-algebra of the groupoid, and negative order ones define elements of the C^*-algebra.

4 Constructions based on Lie groupoids and their deformations

We briefly discuss here some constructions where deformation groupoids are naturally obtained and used. Such groupoids give a geometric description of important pseudodifferential calculi. Some others are used to construct elements of Kasparov's KK-theory [72] such as index maps, or Poincaré duality.

4.1 The associated index map

Let $G \rightrightarrows M$ be a Lie groupoid and $\Gamma \rightrightarrows V$ a sub-Lie groupoid, i.e., a submanifold and a subgroupoid of G. The groupoid $DNC(G,\Gamma)$ when restricted to the interval $[0,1]$ gives rise to a diagram:

$$\begin{array}{ccccccccc} 0 & \longrightarrow & C^*(G\times(0,1]) & \longrightarrow & C^*(DNC(G,\Gamma)_{[0,1]}) & \xrightarrow{ev_0} & C^*(\mathcal{N}_\Gamma^G) & \longrightarrow & 0 \\ & & & & \downarrow{\scriptstyle ev_1} & \swarrow{\scriptstyle \partial_\Gamma^G} & & & \\ & & & & C^*(G) & & & & \end{array}$$

where the top line is exact—at least in the full C^*-algebra level.

As $C^*(G\times(0,1])$ is contractible, the map ev_0 is invertible in E-theory of Connes–Higson [33]—and in KK-theory if this exact sequence admits a completely positive splitting—which is the case if the groupoid $\mathcal{N}_\Gamma^G$ is amenable.

We thus obtain an index element $\partial_\Gamma^G = [ev_1] \otimes [ev_0]^{-1} \in E(C^*(\mathcal{N}_\Gamma^G), C^*(G))$.

Let us see some examples:

4.1.1 The "Dirac element" of a Lie group

Consider an inclusion $H \subset G$ of Lie groups. Note that the groupoid $\mathcal{N}_H^G$ is actually a group. This group is immediately seen to be the semidirect product $H \ltimes (\mathfrak{G}/\mathfrak{H})$—where H acts on the lie algebra $\mathfrak{G}$ of G via the adjoint representation of G and fixes the Lie algebra $\mathfrak{H}$ of H.

Assume G is a (almost) connected Lie group and K is its maximal compact subgroup. The K theory of the group $C^*(\mathcal{N}_K^G)$ is a twisted K theory of $C^*(G)$ and the map $\partial_K^G : K_0(C^*(\mathcal{N}_K^G)) \to C_r^*(G)$ identifies with the "Dirac element"—i.e., the Connes–Kasparov map.

4.1.2 Foliation and shriek map for immersions

Let (M_1, F_1) and (M_2, F_2) be smooth (regular) foliations. In [61] is considered a notion of maps between leaf spaces $f : M_1/F_1 \to M_2/F_2$. The goal of that paper is to construct wrong way functoriality maps $f! : K(C^*(M_1, F_1)) \to K(C^*(M_2, F_2))$ generalizing constructions of [36].

As in [36], writing f as a composition $p \circ i$ where $i : M_1/F_1 \to M_1/F_1 \times M_2/F_2$ is (somewhat loosely speaking) $\ell \mapsto (\ell, f(\ell))$—where ℓ is a leaf of (M_1, F_1) and p is the projection $M_1/F_1 \times M_2/F_2 \to M_2/F_2$ the problem is reduced to the case of immersions and submersions.

Following Connes' construction of the tangent groupoid, a deformation groupoid was used in [61] in order to construct the wrong way functoriality map for immersions between spaces of leaves, in the following way:

- Using a Morita equivalence, one may reduce to transversals in order to understand an immersion $M_1/F_1 \to M_2/F_2$ to be an inclusion $f : G_1 \hookrightarrow G_2$ where M_1 is a submanifold of M_2 which is saturated and G_1 is the restriction of G_2 to M_1—i.e., for $\gamma \in G_2$ we have the equivalences: $s(\gamma) \in M_1 \iff r(\gamma) \in M_1 \iff \gamma \in G_1$.
- Then, the DNC construction is used in order to obtain a wrong way functoriality element $f! \in E(C^*(G_1), C^*(G_2))$.[2] This element is the Kasparov product $[th] \otimes \partial_{G_1}^{G_2}$ of a Thom isomorphism element $[th] \in KK(C^*(G_1), C^*(\mathcal{N}_{G_1}^{G_2}))$ with the index element $\partial_{G_1}^{G_2} \in E(C^*(\mathcal{N}_{G_1}^{G_2}), C^*(G_2))$.

Remark 4.1 In order to construct the Thom element $[th]$ in [61] one of course has to assume a K-orientation. Moreover, is used the fact that the groupoid G_1 acts naturally on the normal bundle $\mathcal{N}_{M_1}^{M_2}$.

Question 1 Can one construct the Thom element when the groupoid G_1 does not act on the bundle $\mathcal{N}_{M_1}^{M_2}$? What is the right condition of K-orientation for $\mathcal{N}_{G_1}^{G_2}$?

Remark 4.2 It is mentioned also in [61] that one could use a deformation groupoid to construct $f!$ for submersions.

4.1.3 On the computation of the index map in some cases

In [47], the index map index element $\partial_\Gamma^G = [ev_1] \otimes [ev_0]^{-1} \in E(C^*(\mathcal{N}_\Gamma^G), C^*(G))$ associated with the inclusion of groupoids is computed in some situations. In particular, when Γ is just a space $V \subset M$, the C^*-algebra of the groupoid $\mathcal{N}_V^G$ has the same K theory as the space $\mathcal{N}_V^G$. We have an embedding $j : \mathcal{N}_V^G \to \mathfrak{A}G$ of this space, via a tubular neighborhood construction, as an open subset of the Lie algebroid of G. The index ∂_Γ^G is the composition:

$$K_*\left(C^*\left(\mathcal{N}_V^G\right)\right) \simeq K_*\left(C_0\left(\mathcal{N}_V^G\right)\right) \xrightarrow{j} K_*(C_0(\mathfrak{A}G)) \xrightarrow{\mathrm{ind}_G} K_0(C^*(G))$$

(and this is actually true for KK-elements instead of morphisms of K-theory).

Also, one can compare it to the connecting map $\tilde{\partial}_\Gamma^G$ of the exact sequence

$$0 \to C^*(\mathring{G}) \to C^*(SBlup(G, \Gamma)) \to C^*\left(S\mathcal{N}_\Gamma^G\right) \to 0,$$

[2] In [61] this element is just a morphism of K-groups since E-theory of Connes–Higson was defined later.

associated with the open saturated subset $\mathring{M} = M \setminus V \subset SBlup(M, V)$. Here, $\mathring{G}$ is the groupoid $G_{\mathring{M}}^{\mathring{M}}$.

We then have a commutative diagram[3]

$$\begin{array}{ccc}
K_j(C^*(\mathring{G})) & \xrightarrow{\tilde{\partial}_\Gamma^G} & K_{j+1}(C^*(S\mathcal{N}_\Gamma^G)) \\
\downarrow{\scriptstyle j} & & \downarrow{\scriptstyle th} \\
K_j(C^*(G)) & \xrightarrow{\partial_\Gamma^G} & K_j(C^*(\mathcal{N}_\Gamma^G))
\end{array}$$

where th is a map based on the Connes analogue of the Thom isomorphism (cf. [29], see also [51] for its construction in Kasparov's bivariant groups).

It is then easily seen that:

- if no G orbit is contained in V, then the inclusion $C^*(\mathring{G}) \to C^*(G)$ is a Morita equivalence and therefore $j : K_*(C^*(\mathring{G})) \to K_*(C^*(G))$ is an isomorphism;
- if for every $x \in V$, the tangent to the G orbit through x, i.e., the image by the anchor map $\varrho_x : \mathfrak{A}G_x \to T_xM$ is not contained in T_xV, then the Thom element th is also an isomorphism.

4.1.4 Full index

Another natural question, involving blowup groupoids, appears: Let P be an elliptic operator on $SBlup_{r,s}(G, \Gamma)$. When is it invertible modulo $C^*(\mathring{G})$? If this is the case, can one compute its index as an element in $K_*(C^*(\mathring{G}))$? There is a particular interest when $\mathring{G} = \mathring{M} \times \mathring{M}$, in which case the index is in $\mathbb{Z}$.

We have a commutative diagram

$$\begin{array}{ccccccccc}
 & & 0 & & 0 & & 0 & & \\
 & & \downarrow & & \downarrow & & \downarrow & & \\
0 & \to & C^*(\mathring{G}) & \to & \Psi^*(\mathring{G}) & \to & C_0(\mathbb{S}^*\mathfrak{A}\mathring{G}) & \to & 0 \\
 & & \downarrow & & \downarrow & & \downarrow & & \\
0 & \to & C^*(SBlup_{r,s}(G,\Gamma)) & \to & \Psi^*(SBlup_{r,s}(G,\Gamma)) & \xrightarrow{\sigma} & C_0(\mathbb{S}^*\mathfrak{A}SBlup_{r,s}(G,\Gamma)) & \to & 0 \\
 & & \downarrow & & \downarrow{\scriptstyle q} & & \downarrow & & \\
0 & \to & C^*(S\mathcal{N}_\Gamma^G) & \to & \Psi^*(S\mathcal{N}_\Gamma^G) & \to & C_0(\mathbb{S}^*\mathfrak{A}S\mathcal{N}_\Gamma^G) & \to & 0 \\
 & & \downarrow & & \downarrow & & \downarrow & & \\
 & & 0 & & 0 & & 0 & &
\end{array}$$

where lines and columns are exact.

[3]The arrows are in fact E-theory elements or even KK-theory elements.

- The columns represent the exact sequence corresponding to the partition of the groupoid $SBlup_{r,s}(G, \Gamma)$ into the open subgroupoid $\overset{\circ}{G}$ and the closed subgroupoid $S\mathcal{N}_\Gamma^G$ at the level of C^*-algebras, order 0 pseudodifferential operators and principal—0-homogeneous symbol.
- The lines are the exact sequences of zero order pseudodifferential operators of the groupoids $\overset{\circ}{G}$, $SBlup_{r,s}(G, \Gamma)$, and $S\mathcal{N}_\Gamma^G$ (see Theorem 3.5).

It follows that a pseudodifferential operator, i.e., an element $P \in \Psi^*(SBlup_{r,s}(G, \Gamma))$ is invertible modulo $C^*(\overset{\circ}{G})$ if and only if its *classical symbol* $\sigma(P)$ and its *noncommutative symbol*, i.e., its restriction $q(P)$ to the "singular part" $S\mathcal{N}_\Gamma^G$ are both invertible.

In other words, we have a *full symbol algebra* $\Sigma_\Gamma^G = C_0(\mathbb{S}^*\mathfrak{A}SBlup_{r,s}(G, \Gamma)) \times_{C_0(\mathbb{S}^*\mathfrak{A}S\mathcal{N}_\Gamma^G)} \Psi^*(S\mathcal{N}_\Gamma^G)$ and an exact sequence

$$0 \to C^*(\overset{\circ}{G}) \to \Psi^*(SBlup_{r,s}(G, \Gamma)) \to \Sigma_\Gamma^G \to 0.$$

One can compute in some cases the K-theory of Σ_Γ^G and the connecting map. For instance, if—as above Γ is just a submanifold $V \subset M$ and if we assume that the anchor map $\varrho_x : \mathfrak{A}G_x \to T_xM$ is not contained in T_xV, then Σ_V^G is K-equivalent to the algebra of pseudodifferential operators on G whose symbol is "trivial" when restricted to V—i.e., a function on V, and the index map is the restriction to this subalgebra of $\Psi^*(G)$ to the index map of G.

4.2 *Groupoids using deformation constructions*

As Lie groupoids are useful in defining natural pseudodifferential calculi and in several K-theoretic constructions, many authors have introduced interesting groupoids using various techniques: gluing, integration of algebroids, etc. Recognizing some of these groupoids as deformations or blowups often simplifies their construction, may help their understanding and give a geometric insight on their properties.

In this section we outline some of these natural and useful constructions of deformation or blowup Lie groupoids.

4.2.1 Pseudodifferential calculi on singular manifolds

Let M be a singular manifold. This can be a manifold with boundary, or with corners, or even a stratified manifold. Many natural pseudodifferential calculi were constructed by analysts—especially in the school of Richard Melrose—in order to take into account, sometimes in a very fine way, the behavior of the operators near the boundary (cf. [85, 80, 79, 81, 86]).

Some of these calculi were already constructed using blowup constructions. Some others just putting some conditions on the Riemannian metric in the regular part $\overset{\circ}{M} \subset M$ degenerating near the boundary. In many cases, when this metric is complete, this metric actually corresponds to an algebroid. More precisely, the space of bounded vector fields with respect to this metric is the module of sections of a Lie algebroid on M. It follows from [38] that this algebroid integrates to a Lie groupoid $G \rightrightarrows M$.

Of course knowing this groupoid will certainly not solve at once all the questions for which the corresponding calculus was constructed! It may however help understanding some of its properties: the decomposition of $C^*(G)$ into ideals that can often be seen geometrically can simplify the study of conditions for Fredholmness of an associated (pseudo)differential operator; it is also relevant for various index computations.

Let us outline specific examples.

The groupoid of the b-calculus Let M be manifold with boundary. Melrose constructs the b space which is the blowup of $M \times M$ along its corner $\partial M \times \partial M$ (cf. [83, 84, 86, 85]). The pseudodifferential operators of the corresponding b-calculus consist of operators which are distributions on the b-space that are conormal along the diagonal M and have a specific decay near the boundary components $M \times \partial M$ and $\partial M \times M$.

Monthubert [89, 90] constructed the associated b-groupoid which is nothing else than the dense open subspace $G_b = SBlup_{r,s}(M \times M, \partial M \times \partial M)$ of the b-space $SBlup(M \times M, \partial M \times \partial M)$.

Let us note that in fact all these constructions were also performed in the more general case of manifolds with corners.

Fibered corners We restrict again to the case of a manifold with boundary, although the constructions below extend to more general settings of manifolds with corners.

Let M be manifold with boundary ∂M and let $p : \partial M \to B$ be a fibration. Mazzeo [79] studied the *edge calculus* in this situation. This corresponds to the blowup $SBlup(M \times M, \partial M \times_B \partial M)$ and of course to the corresponding groupoid $G_e = SBlup_{r,s}(M \times M, \partial M \times_B \partial M)$.

Later, [81] Mazzeo and Melrose introduced and studied in the same situation the Φ calculus which correspond to the algebroid of vector fields that are tangent to the fibers at the boundary but also whose derivative is tangent to ∂M: these are vector fields of the form $X + tY + t^2N$—where t is a defining function of the boundary, X is a vector field along the fibration (extended near the boundary), Y is tangent to the boundary, and N is normal to the boundary. Piazza and Zenobi [99] actually realized that the groupoid constructed in [44] integrating this algebroid can be obtained via a double blowup construction. See also [119] for topological aspects of indices in this context.

Question 2 A natural question is also to try to understand the noncomplete case too in terms of deformation groupoids.

4.2.2 Inhomogeneous pseudodifferential calculus

The "inhomogeneous" pseudodifferential calculus (also called "filtered" or "Carnot") deals with manifolds M whose tangent bundle is endowed with a filtration $(H_i)_{0\leqslant 1\leqslant k}$ (with $H_0 = 0$ and $H_k = TM$) satisfying $[\Gamma(H_i), \Gamma(H_j)] \subset \Gamma(H_{i+j})]$. A natural pseudodifferential calculus has been constructed in this framework, generalizing the case of contact manifolds (cf. [14]). In this sub-elliptic or Carnot calculus, the vector fields that are sections of the bundle H_i are considered as differential operators of order i.

A deformation groupoid taking into account this inhomogeneous calculus was constructed independently in [100, 25, 27, 26] and [113, 70, 115]. Both these constructions are rather technical and based on higher jets. This groupoid plays the role of Connes' tangent groupoid in this setting. It allows to recover the inhomogeneous pseudodifferential calculus (cf. [115]).

Omar Mohsen presents in [88] a very elegant construction of this deformation groupoid in terms of successive deformations to the normal cone. In the case where there is only one subbundle $H \subset TM$, one just considers the inclusion of $H \times \{0\}$ into Connes' tangent groupoid $(M \times M) \times \mathbb{R}^* \sqcup TM \times \{0\}$. The crucial fact that the object built is canonically a groupoid is clear in this construction—while it leads to relatively sophisticated computations in the works cited above. The general case is treated by induction.

In addition, this construction has the advantage of being very flexible and generalizing immediately, for example, in the context of an inhomogeneous a pseudodifferential calculus transverse to a foliation—as the one appearing in the work of Connes–Moscovici [34, 35].

4.2.3 Poincaré dual of a stratified manifold

K-duality Kasparov in [72] defines a formal KK-duality of C^*-algebras. If A and B are K-dual C^*-algebras, the K-homology of A is isomorphic to the K-theory of B.

Let M be a smooth compact manifold. The algebra $C(M)$ has naturally a K-dual which is $C_0(T^*M)$ (cf. [71, 72, 36]). The corresponding duality map associates to the K-homology class of an elliptic (pseudo)differential the K-theory class of its symbol.

Several generalizations to manifolds with singularities have been given by various authors: manifolds with boundary [36], non Hausdorff manifolds [73], manifolds with conic singularity [40, 22], stratified manifolds [41]. Many of them use naturally Lie groupoids.

Manifolds with a conic singularity Let us outline the construction of [40].

Let M be a compact manifold with boundary ∂M. Denote by $\mathring{M}$ the open subset $M \setminus \partial M$. The one point compactification M_+ of $\mathring{M}$ is the quotient of M by the equivalence relation which identifies all the points of the boundary ∂M. It is a *manifold with conic singularity*.

Let $G_b = Blup_{r,s}(M \times M, \partial M \times \partial M) = \overset{\circ}{M} \times \overset{\circ}{M} \sqcup \partial M \times \partial M \times \mathbb{R}_+^*$ be the groupoid of the b-calculus of the manifold with boundary M. The Poincaré dual of M_+ constructed in [40] is a closed subgroupoid $\mathcal{G}$ of the "adiabatic" groupoid $DNC(G_b, G_b^{(0)})$ of G_b: it is $\mathcal{G} = DNC(G_b, G_b^{(0)}) \setminus \overset{\circ}{M} \times \overset{\circ}{M} \times \mathbb{R}_+^*$ [40]. Note that $\mathcal{G}$ is the union of the algebroid $\mathfrak{A}G_b = TM = T\overset{\circ}{M} \sqcup T\partial M \times \mathbb{R}$ of G_b with $\partial M \times \partial M \times \mathbb{R}^2$.

The *Poincaré duality* element is an element $\psi \in KK(C(M_+) \otimes C^*(\mathcal{G}), \mathbb{C})$. It is obtained as follows:

- first $C(M_+)$ sits naturally in the center of the multiplier algebra of $C^*(\mathcal{G})$, by extending in a unital way the map $C_0(\overset{\circ}{M}) \to \mathcal{M}(C_0(T^*\overset{\circ}{M})) = \mathcal{M}(C^*(TM))$. We thus have a morphism of C^*-algebras $m : C(M_+) \otimes C^*(\mathcal{G}) \to C^*(\mathcal{G})$.
- As $C^*(\overset{\circ}{M} \times \overset{\circ}{M} \times \mathbb{R}_+^*) \simeq \mathcal{K} \otimes \mathbb{C}_0(\mathbb{R}_+^*)$, the extension

$$0 \to C^*\left(\overset{\circ}{M} \times \overset{\circ}{M} \times \mathbb{R}_+^*\right) \to C^*\left(DNC\left(G_b, G_b^{(0)}\right)\right) \to C^*(\mathcal{G}) \to 0$$

  gives rise to an element $d \in KK^1(C^*(\mathcal{G}), C_0(\mathbb{R}_+^*)) = KK(C^*(\mathcal{G}), \mathbb{C})$.

Put then $\psi = m^*(d)$.

Question 3 It would be nice to understand this Poincaré duality by describing as precisely as possible which operators on $\overset{\circ}{M}_+$ correspond to symbols on $C^*(\mathcal{G})$. In particular, what is the symbol class of a Fuchs type operator studied in [74]? This question is certainly linked with Question 2.

Stratified manifolds One can generalize immediately this construction to manifolds with a fibered boundary: given a fibration $p : \partial M \to B$ one can form the space $M/\sim$, where $\sim$ is the equivalence relation on M given by

$$x \sim y \iff \begin{cases} x = y \text{ or} \\ x, y \in \partial M \text{ and } p(x) = p(y). \end{cases}$$

One just replaces the groupoid G_b by the groupoid G_e of the edge calculus or by the groupoid G_Φ of the Φ calculus.

This construction was extended in [41], using an induction process, to describe in a similar way the Poincaré dual of general stratified manifolds X. The construction can in fact be obtained by use of several blowups: one blows up inductively all the strata to obtain a groupoid G_X generalizing G_b; then, one uses as above the adiabatic deformation of G_X in order to construct the dual groupoid $\mathcal{G}_X = DNC(G_X, G_X^{(0)}) \setminus \overset{\circ}{X} \times \overset{\circ}{X} \times \mathbb{R}_+^*$ and the Poincaré duality element $\psi_X \in KK(C(X) \otimes C^*(\mathcal{G}_X), \mathbb{C})$. Details will appear in [39].

5 Related topics and further questions

In this section, we examine some topics that are not a priori based on Lie groupoids, but are actually linked to our discussion.

5.1 Relation to Roe algebras

Starting from a finite propagation speed principle for differential operators of order one [24], John Roe developed a very beautiful theory of *coarse* spaces and algebras [104]. The K-theory of Roe algebras has been used as a receptacle for various index problems [60].

We will not develop here this theory. Let us just outline some links with groupoids that have been made in the literature.

- In [108], a groupoid is constructed out of a coarse space and it is shown that the Roe C^*-algebra of locally compact functions with finite propagation on this coarse space is the C^*-algebra of this groupoid
- Out of a Lie groupoid one constructs Roe type algebras (see, e.g., [15]). A natural example is the case of a covering space $\widetilde{M} \to M$ with group Γ. In that case, one may also define naturally the index with values in the K-theory $C^*(\Gamma)$ for Γ-invariant elliptic operators on $\widetilde{M}$ using coarse techniques. In particular, there is an exact sequence of Roe algebras [57, 58, 59, 98]

$$0 \to \mathcal{C}^*\left(\widetilde{M}\right)^{\Gamma} \longrightarrow \mathcal{D}^*\left(\widetilde{M}\right)^{\Gamma} \longrightarrow \mathcal{D}^*\left(\widetilde{M}\right)^{\Gamma} / \mathcal{C}^*\left(\widetilde{M}\right)^{\Gamma} \to 0.$$

  The algebra $\mathcal{C}^*(\widetilde{M})^{\Gamma}$ is Morita equivalent to $C^*(\Gamma)$ and that of $\mathcal{D}^*(\widetilde{M})^{\Gamma}/\mathcal{C}^*(\widetilde{M})^{\Gamma}$ is the K-homology of M (Paschke duality—[96]).
- In this precise case, Zenobi [120] identified the K-theory exact sequence of the Higson–Roe sequence with the adiabatic groupoid exact sequence of the groupoid $(\widetilde{M} \times \widetilde{M})/\Gamma \rightrightarrows M$.

Question 4 How far can one push this parallel between groupoid C^*-algebras and Roe algebras?

5.2 *Singular foliations and "singular Lie groupoids"*

Let us just say a few words on very singular Lie groupoids associated with singular foliations in [5].

A singular foliation on a compact manifold M is a submodule $\mathcal{F}$ of the $C^\infty(M)$ module of vector fields $\Gamma(TM)$ which is finitely generated and *involutive*—i.e., closed under Lie brackets: $[\mathcal{F}, \mathcal{F}] \subset \mathcal{F}$.

The module $\mathcal{F}$ tells us which differential operators are "longitudinal." In [5] is constructed the holonomy groupoid and the C^*-algebra of such a singular foliation. In [6] is also constructed a pseudodifferential calculus and an analytic index for a singular foliations as well as a deformation holonomy groupoid, which gives rise to an alternate way of defining the analytic index.

Question 5 Let $(M, \mathcal{F})$ be a singular foliation in the sense of [5]. It is sometimes possible by blowing up singular leaves to obtain a less singular foliation: one for which the holonomy groupoid is a Lie groupoid.

For instance, take the singular foliation of $\mathbb{R}^3$ whose leaves are the spheres of center 0—given by the action of $SO(3)$—has a singular holonomy groupoid $\{(x, y) \in \mathbb{R}^3 \setminus \{0\};\ \|x\| = \|y\|\} \sqcup \{0\} \times SO(3)$. Blowing up the singular leaf $\{0\}$, we obtain the regular foliation on $\mathbb{R}_+ \times \mathbb{S}^2$ with leaves $\{r\} \times \mathbb{S}^2$.

This poses several questions

(a) How general/canonical can this procedure be?
(b) When such a blowup is possible, what is the precise relation between the C^*-algebra of the foliation in the sense of [5] and that of the Lie groupoid obtained by blowing up?

Question 6 The holonomy groupoid $Hol(M, \mathcal{F})$ of a singular foliation is a "bad" topological space. On the other hand, there is a natural notion of a smooth map $V \to Hol(M, \mathcal{F})$—and also of a smooth submersion—for a manifold V (this notion is called a *bisubmersion* in [5]). Is there a natural structure to express this in a functorial way?

5.3 Computations using cyclic cohomology

A classical way to make K-theoretic computations for spaces is to use the Chern isomorphism with (co)homology. In the case of manifolds, one naturally uses the de Rham cohomology. The de Rham cohomology extends to the noncommutative setting, thanks to Connes' cyclic cohomology. On the other hand, de Rham cohomology uses differentiation and is not well suited for continuous functions, but rather smooth functions. So, starting with Connes (cf. [31, 32]) and then many others, one constructs natural cyclic cocycles defined on the algebra $C_c^\infty(G)$ of smooth functions with compact support on a Lie groupoid G—and then we need to extend them to a larger algebra that has the same K-theory as $C^*(G)$. Such an extension was performed in [31] using a notion of n-traces—which are "well behaved cyclic cocycles".

Question 7 Extend natural cocycles so that they pair with the K-theory of the C^*-algebra of the groupoid and compute this pairing.

5.4 *Behavior of the resolvent of an elliptic operator*

Let us start with say a positive Laplacian Δ on a groupoid G with compact $G^{(0)}$. As Δ is elliptic, self-adjoint, and positive, the operator $1 + \Delta$ is invertible and its inverse is in $C^*(G)$ (cf. [116]).

Question 8 What is the behavior of $(1 + \Delta)^{-1}$? Is it a smooth function outside $G^{(0)}$? What kind of decay at infinity does it have?

Just a few cases have been worked out (cf. [85, 80]).

Let us make a comment about these problems. The elements of $C_r^*(G)$ define distributions on G. In other words there is an injective map $C_r^*(G) \to C^{-\infty}(G)$. So it is a natural question to ask what kind of distributions are elements like $(1+\Delta)^{-1}$. The distribution associated with an element of $C^*(G)$ only depends on its image in $C_r^*(G)$. So that these problems a priori concern the reduced C^*-algebra of G.

5.5 *Relations with the Boutet de Monvel calculus*

This was a motivation for us from the beginning.

Let us very briefly say a few words on the Boutet de Monvel calculus. Details can be found in [18, 19, 54, 106].

Let M be a manifold with boundary. We consider M as included in a manifold $\widetilde{M}$ without boundary in which a smooth hypersurface ∂M of $\widetilde{M}$ separates $\widetilde{M}$ into two open subsets $\overset{\circ}{M}$ and M_-.

Denote by χ_M the characteristic function of M. Boutet de Monvel defines:

Definition 5.1 A pseudodifferential operator Φ (with compact support) on $\widetilde{M}$ is said to have the *transmission property* if for every smooth function $\tilde{f}$ on $\widetilde{M}$, the function $\Phi(\chi_M \tilde{f})$ coincides on M with a smooth function on $\widetilde{M}$.[4]

Of course, a smoothing operator satisfies the transmission property, and thus this property can be described in terms of the (restriction to ∂M of the) total symbol of Φ. This condition is explicitly given in local coordinates.

Assume that Φ satisfies the transmission property, and let $\Phi_+(f)$ be the restriction to M of the smooth function which coincides with $\Phi(\chi_M \tilde{f})$ on M.

The operators Φ_+, *do not form an algebra* since $\Phi_+\Psi_+ \neq (\Phi\Psi)_+$, and the difference $\Phi_+\Psi_+ - (\Phi\Psi)_+$ is not a pseudodifferential operator. On the other hand, this difference belongs to a new class of operators—described again precisely in local coordinates—called *singular Green* operators.

[4]We actually have to assume that the same holds also for the adjoint of Φ.

The set of operators of the form $\Phi_+ + S$ where P is a pseudodifferential operator with the transmission property and S a singular Green operator is an algebra. Call it $\mathcal{P}_{BM}(M)$.

Boutet de Monvel moreover defines *singular Poisson* (or *Potential*) operators mapping functions on ∂M to functions on M and *singular Trace* operators which map functions on M to functions on ∂M. Singular Poisson operators and singular Trace operators are adjoint of each other.

They form bimodules yielding a Morita equivalence between singular Green operators on M and ordinary pseudodifferential operators on its boundary.

We thus obtain the *Boutet de Monvel algebra* which consists of matrices of the form $\begin{pmatrix} \Phi_+ + S & P \\ T & Q \end{pmatrix}$ where

- Φ is a pseudodifferential operator on $\widetilde{M}$ with the transmission property, and Φ_+ the corresponding operator on smooth functions on M;
- S is a singular Green operator acting on M;
- P is a singular Poisson operator mapping functions on ∂M to functions on M;
- T is a singular trace operator mapping functions on M to functions on ∂M;
- Q is a pseudodifferential operator on ∂M.

All these constructions are generalized to the case of families of manifolds and more generally to a Lie groupoid $G \rightrightarrows M$ on a manifold with boundary M assuming that G is *transverse* to the boundary M (cf. [82, 77, 16, 17]).

When taking the closure of the *bounded* singular Green operators, we find an exact sequence

$$0 \to \mathcal{K}(L^2(M)) \longrightarrow \text{Green} \xrightarrow{\sigma_g} \Sigma_M^G \to 0, \qquad \text{(Green)}$$

where σ_M^G is the *noncommutative symbol* of singular Green operators with values in the algebra $\Sigma_M = C_0(S^*\partial M) \otimes \mathcal{K}$. This exact sequence can be compared with the exact sequence

$$0 \to \mathcal{K}(L^2(\partial M)) \longrightarrow \Psi^*(\partial M) \xrightarrow{\sigma_{\partial M}} C_0(S^*\partial M) \to 0, \qquad (\Psi_{\partial M})$$

of pseudodifferential operators on ∂M. Bounded singular Poisson operators and singular trace operators form bimodules yielding a Morita equivalence of these sequences.

Now, the group $\mathbb{R}_+^*$ naturally acts on Connes tangent groupoid $T\partial M \times \{0\} \sqcup (\partial M \times \partial M) \times \mathbb{R}_+^*$ of the manifold ∂M via the zooming action. The corresponding crossed product groupoid $\mathcal{G}$ gives naturally rise to an exact sequence

$$0 \to \mathcal{K}\left(L^2(\partial M) \times \mathbb{R}_+^*\right) \longrightarrow C^*(\mathcal{G}) \xrightarrow{\text{ev}_0} C_0(T^*\partial M) \rtimes \mathbb{R}_+^* \to 0. \qquad (\mathcal{G})$$

Note that $C_0(S^*\partial M) \otimes \mathcal{K}$ naturally sits in $C_0(T^*\partial M) \rtimes \mathbb{R}_+^*$ as an ideal. In [1], the exact sequence $(\mathcal{G})$ restricted to this ideal was shown to coincide with (Green)—by showing that they define the same KK-element and then using Voiculescu's theorem [117].

This construction was generalized in [45], where for any Lie groupoid $G \rightrightarrows V$ is directly constructed a (sub)-Morita equivalence relating the pseudodifferential exact sequence

$$0 \to C^*(G) \longrightarrow \Psi^*(G) \xrightarrow{\sigma_G} C_0(S^*\mathfrak{A}G) \to 0, \qquad (\Psi_G)$$

of the groupoid G and the exact sequence

$$0 \to C^*(G) \otimes \mathcal{K}\left(L^2\left(\mathbb{R}_+^*\right)\right) \longrightarrow C^*(\mathcal{G}_G) \xrightarrow{\mathrm{ev}_0} C_0(\mathfrak{A}^*G) \rtimes \mathbb{R}_+^* \to 0, \qquad (\mathcal{G}_G)$$

where $\mathcal{G}_G$ is the "gauge adiabatic groupoid" obtained as the crossed product by the natural scaling action of $\mathbb{R}_+^*$ on the adiabatic groupoid $G_{ad} = \mathfrak{A}G \times \{0\} \sqcup G \times \mathbb{R}_+^*$.

In [46], the construction of the "gauge adiabatic groupoid" is generalized using the blowup construction for groupoids discussed above (cf. Section 2.4.2). If $V \subset M$ is a hypersurface which is transverse to the groupoid G, the groupoid $Blup_{r,s}(G, V) \to Blup(M, V) \simeq M$ is the gauge adiabatic groupoid of G_V^V. More generally, if V is *any* submanifold of M which is transverse to G, one still constructs a Boutet de Monvel type calculus.

Question 9 It is natural to try to show that the closure of the algebra of bounded elements of the Boutet de Monvel algebra coincides with the one obtained in this way. We have quite well understood this and should write it precisely. Actually, the transmission property gives rise to a small difference between them.

5.6 *Algebroids and integrability*

Recall from Definition 2.3 that an algebroid over a manifold M is a smooth bundle A over M endowed with the following structure:

- a Lie algebra bracket on the space of smooth sections of A;
- a bundle morphism $\varrho : A \to TM$.

These are assumed to satisfy: $[X, fY] = \varrho(X)(f)Y + f[X, Y]$.

The integrability problem for algebroids is not "trivial". There are indeed algebroids that are not associated with groupoids (cf. [3]—se also [37] where necessary and sufficient conditions for integrability are given).

On the other hand, one can define differential operators on an algebroid and even pseudodifferential operators locally using local integration, which is also possible. These operators act naturally on $C^\infty(M)$ (or $L^2(M)$ when they are bounded). When the algebroid is integrated to a Lie groupoid G, the C^*-algebra $C^*(G)$ is generated by resolvents of elliptic operators. Also, in that case, we find more representations where these operators act.

Question 10 How much of a groupoid C^*-algebra can one construct out of just its algebroid? Is there a construction of an (s-simply connected) Lie groupoid C^*-algebra which makes sense even for nonintegrable algebroids?

Remark 5.2 Let us remak that we can also restrict this question to the case of Poisson manifolds which are particular cases of Lie algebroids.

5.7 Fourier integral operators

Fourier integral operators [64, 65, 49] form a class of operators, larger than pseudodifferential operators. They are very useful in order to understand hyperbolic differential operators.

In [76], Fourier integral operators on a Lie groupoid are defined and studied. We do not wish to say much on these. Let us just ask a question.

Question 11 Is there a construction of Fourier integral operators in the spirit of [45] or [114] (See Section 3.4 above)?

Acknowledgements The authors were partially supported by ANR-14-CE25-0012-01 (SINGSTAR).

References

1. Johannes Aastrup, Severino T. Melo, Bertrand Monthubert, and Elmar Schrohe, *Boutet de Monvel's calculus and groupoids I.*, J. Noncommut. Geom. **4 no. 3** (2010), 313–329.
2. Rui Almeida and Pierre Molino, *Suites d'Atiyah et feuilletages transversalement complets*, C. R. Acad. Sci. Paris Sér. I Math. **300** (1985), no. 1, 13–15. MR 778785
3. ———, *Flots riemanniens sur les* 4*-variétés compactes*, Tohoku Math. J. (2) **38** (1986), no. 2, 313–326. MR 843815
4. C. Anantharaman-Delaroche and J. Renault, *Amenable groupoids*, Monographies de L'Enseignement Mathématique [Monographs of L'Enseignement Mathématique], vol. 36, L'Enseignement Mathématique, Geneva, 2000, With a foreword by Georges Skandalis and Appendix B by E. Germain. MR 1799683
5. Iakovos Androulidakis and Georges Skandalis, *The holonomy groupoid of a singular foliation*, J. Reine Angew. Math. **626** (2009), 1–37.
6. ———, *Pseudodifferential calculus on a singular foliation*, J. Noncommut. Geom. **5** (2011), no. 1, 125–152.
7. Michael F. Atiyah, *Elliptic operators discrete groups and von Neumann algebras*, Astérisque **32–33** (1976), 43–72.
8. Michael F. Atiyah and Singer Isadore M., *The index of elliptic operators. I*, Ann. of Math. (2) **87** (1968), 484–530.
9. ———, *The index of elliptic operators. III*, Ann. of Math. (2) **87** (1968), 546–604. MR 0236952

10. ______, *The index of elliptic operators. IV*, Ann. of Math. (2) **93** (1971), 119–138.
11. Paul Baum and Alain Connes, *Geometric K-theory for Lie groups and foliations*, Enseign. Math. (2) **46** (2000), no. 1-2, 3–42. MR 1769535
12. Paul Baum, Alain Connes, and Nigel Higson, *Classifying space for proper actions and K-theory of group C^*-algebras*, C^*-algebras: 1943–1993 (San Antonio, TX, 1993), Contemp. Math., vol. 167, Amer. Math. Soc., Providence, RI, 1994, pp. 240–291. MR 1292018
13. Paul Baum, Erik Guentner, and Rufus Willett, *Expanders, exact crossed products, and the Baum-Connes conjecture*, Ann. K-Theory **1** (2016), no. 2, 155–208. MR 3514939
14. Richard Beals and Peter Greiner, *Calculus on Heisenberg manifolds*, Annals of Mathematics Studies, vol. 119, Princeton University Press, Princeton, NJ, 1988. MR 953082
15. Moulay Tahar Benameur and Indrava Roy, *The Higson-Roe sequence for étale groupoids. I. dual algebras and compatibility with the BC map*, arXiv:1801.06040.
16. Karsten Bohlen, *Boutet de Monvel operators on singular manifolds*, C. R. Math. Acad. Sci. Paris **354** (2016), no. 3, 239–243. MR 3463018
17. ______, *Boutet de Monvel operators on Lie manifolds with boundary*, Adv. Math. **312** (2017), 234–285. MR 3635812
18. Louis Boutet de Monvel, *Boundary problems for pseudo-differential operators*, Acta Math. **126** (1971), no. 1-2, 11–51.
19. ______, *A course on pseudo differential operators and their applications*, Mathematics Department, Duke University, Durham, N.C., 1976, Duke University Mathematics Series, No. II.
20. Alcides Buss, Siegfried Echterhoff, and Rufus Willett, *Exotic crossed products*, Operator algebras and applications—the Abel Symposium 2015, Abel Symp., vol. 12, Springer, [Cham], 2017, pp. 67–114. MR 3837592
21. ______, *Exotic crossed products and the Baum-Connes conjecture*, J. Reine Angew. Math. **740** (2018), 111–159. MR 3824785
22. P. Carrillo Rouse, J. M. Lescure, and B. Monthubert, *A cohomological formula for the Atiyah-Patodi-Singer index on manifolds with boundary*, J. Topol. Anal. **6** (2014), no. 1, 27–74. MR 3190137
23. Pierre Cartier, *Groupoïdes de Lie et leurs algébroïdes*, Astérisque (2009), no. 326, Exp. No. 987, viii, 165–196 (2010), Séminaire Bourbaki. Vol. 2007/2008. MR 2605322
24. Paul R. Chernoff, *Essential self-adjointness of powers of generators of hyperbolic equations*, J. Functional Analysis **12** (1973), 401–414. MR 0369890
25. Woocheol Choi and Raphael Ponge, *Tangent maps and tangent groupoid for Carnot manifolds*, arXiv:1510.05851, 2015.
26. ______, *Privileged coordinates and nilpotent approximation for Carnot manifolds, II. Carnot coordinates*, arXiv:1703.05494, 2017.
27. ______, *Privileged coordinates and nilpotent approximation of Carnot manifolds, I. General results*, arXiv:1709.09045, 2017.
28. Alain Connes, *Sur la théorie non commutative de l'intégration*, Algèbres d'opérateurs (Sém., Les Plans-sur-Bex, 1978), Lecture Notes in Math., vol. 725, Springer, Berlin, 1979, pp. 19–143.
29. ______, *An analogue of the Thom isomorphism for crossed products of a C^*-algebra by an action of* **R**, Adv. in Math. **39** (1981), no. 1, 31–55.
30. ______, *A survey of foliations and operator algebras*, Operator algebras and applications, Part I (Kingston, Ont., 1980), Proc. Sympos. Pure Math., vol. 38, Amer. Math. Soc., Providence, R.I., 1982, pp. 521–628.
31. ______, *Cyclic cohomology and the transverse fundamental class of a foliation*, Geometric methods in operator algebras (Kyoto, 1983), Pitman Res. Notes Math. Ser., vol. 123, Longman Sci. Tech., Harlow, 1986, pp. 52–144. MR 866491 (88k:58149)
32. ______, *Noncommutative geometry*, Academic Press Inc., San Diego, CA, 1994.
33. Alain Connes and Nigel Higson, *Déformations, morphismes asymptotiques et K-théorie bivariante*, C. R. Acad. Sci. Paris Sér. I Math. **311** (1990), no. 2, 101–106. MR 1065438

34. Alain Connes and Henri Moscovici, *The local index formula in noncommutative geometry*, Geom. Funct. Anal. **5** (1995), no. 2, 174–243. MR 1334867
35. ———, *Hopf algebras, cyclic cohomology and the transverse index theorem*, Comm. Math. Phys. **198** (1998), no. 1, 199–246. MR 1657389
36. Alain Connes and Georges Skandalis, *The longitudinal index theorem for foliations*, Publ. Res. Inst. Math. Sci. **20** (1984), no. 6, 1139–1183. MR 775126
37. Marius Crainic and Rui Loja Fernandes, *Integrability of Lie brackets*, Ann. of Math. (2) **157** (2003), no. 2, 575–620. MR 1973056
38. Claire Debord, *Holonomy groupoids of singular foliations*, J. Differential Geom. **58** (2001), no. 3, 467–500. MR 1906783
39. Claire Debord and Jean-Marie Lescure, In preparation.
40. ———, *K-duality for pseudomanifolds with isolated singularities*, J. Funct. Anal. **219** (2005), no. 1, 109–133.
41. ———, *K-duality for stratified pseudomanifolds*, Geom. Topol. **13** (2009), no. 1, 49–86. MR 2469513
42. ———, *Index theory and groupoids*, Geometric and topological methods for quantum field theory, Cambridge Univ. Press, Cambridge, 2010, pp. 86–158.
43. Claire Debord, Jean-Marie Lescure, and Victor Nistor, *Groupoids and an index theorem for conical pseudo-manifolds*, J. Reine Angew. Math. **628** (2009), 1–35. MR 2503234
44. Claire Debord, Jean-Marie Lescure, and Frédéric Rochon, *Pseudodifferential operators on manifolds with fibred corners*, Ann. Inst. Fourier (Grenoble) **65** (2015), no. 4, 1799–1880.
45. Claire Debord and Georges Skandalis, *Adiabatic groupoid, crossed product by $\mathbb{R}_+^*$ and pseudodifferential calculus*, Adv. Math. **257** (2014), 66–91.
46. ———, *Blowup constructions for Lie groupoids and a Boutet de Monvel type calculus*, arXiv:1705.09588, 2017.
47. ———, *Lie groupoids, exact sequences, Connes-Thom elements, connecting maps and index maps*, Preprint (part of arXiv:1705.09588), 2017.
48. J. J. Duistermaat, *Fourier integral operators*, Modern Birkhäuser Classics, Birkhäuser/Springer, New York, 2011, Reprint of the 1996 edition [MR1362544], based on the original lecture notes published in 1973 [MR0451313]. MR 2741911
49. J. J. Duistermaat and L. Hörmander, *Fourier integral operators. II*, Acta Math. **128** (1972), no. 3-4, 183–269. MR 0388464
50. George A. Elliott, Toshikazu Natsume, and Ryszard Nest, *The Atiyah-Singer index theorem as passage to the classical limit in quantum mechanics*, Comm. Math. Phys. **182** (1996), no. 3, 505–533. MR 1461941
51. Thierry Fack and Georges Skandalis, *Connes' analogue of the Thom isomorphism for the Kasparov groups*, Invent. Math. **64** (1981), no. 1, 7–14.
52. Jacob Feldman and Calvin C. Moore, *Ergodic equivalence relations, cohomology, and von Neumann algebras. I*, Trans. Amer. Math. Soc. **234** (1977), no. 2, 289–324. MR 0578656
53. Damien Gaboriau, *Orbit equivalence and measured group theory*, Proceedings of the International Congress of Mathematicians. Volume III, Hindustan Book Agency, New Delhi, 2010, pp. 1501–1527. MR 2827853
54. Gerd Grubb, *Functional calculus of pseudodifferential boundary problems*, second ed., Progress in Mathematics, vol. 65, Birkhäuser Boston Inc., 1996.
55. André Haefliger, *Groupoïdes d'holonomie et classifiants*, Astérisque (1984), no. 116, 70–97, Transversal structure of foliations (Toulouse, 1982). MR 755163
56. Nigel Higson, Vincent Lafforgue, and Georges Skandalis, *Counterexamples to the Baum-Connes conjecture*, Geom. Funct. Anal. **12** (2002), no. 2, 330–354. MR 1911663 (2003g:19007)
57. Nigel Higson and John Roe, *Mapping surgery to analysis. I. Analytic signatures*, *K*-Theory **33** (2005), no. 4, 277–299. MR 2220522
58. ———, *Mapping surgery to analysis. II. Geometric signatures*, *K*-Theory **33** (2005), no. 4, 301–324. MR 2220523

59. ———, *Mapping surgery to analysis. III. Exact sequences*, K-Theory **33** (2005), no. 4, 325–346. MR 2220524
60. ———, *K-homology, assembly and rigidity theorems for relative eta invariants*, Pure and Applied Mathematics Quarterly **6** (2010), no. 2, 555–601.
61. Michel Hilsum and Georges Skandalis, *Morphismes K-orientés d'espaces de feuilles et fonctorialité en théorie de Kasparov (d'après une conjecture d'A. Connes)*, Ann. Sci. École Norm. Sup. (4) **20** (1987), no. 3, 325–390.
62. Lars Hörmander, *Pseudo-differential operators and hypoelliptic equations*, Singular integrals (Proc. Sympos. Pure Math., Vol. X, Chicago, Ill., 1966), Amer. Math. Soc., Providence, R.I., 1967, pp. 138–183. MR 0383152
63. ———, *The spectral function of an elliptic operator*, Acta Math. **121** (1968), 193–218. MR 0609014
64. ———, *The calculus of Fourier integral operators*, Prospects in mathematics (Proc. Sympos., Princeton Univ., Princeton, N.J., 1970), Princeton Univ. Press, Princeton, N.J., 1971, pp. 33–57. Ann. of Math. Studies, No. 70. MR 0341193
65. ———, *Fourier integral operators. I*, Acta Math. **127** (1971), no. 1-2, 79–183. MR 0388463
66. ———, *The analysis of linear partial differential operators. I*, Classics in Mathematics, Springer-Verlag, Berlin, 2003, Distribution theory and Fourier analysis, Reprint of the second (1990) edition [Springer, Berlin; MR1065993 (91m:35001a)]. MR 1996773
67. ———, *The analysis of linear partial differential operators. II*, Classics in Mathematics, Springer-Verlag, Berlin, 2005, Differential operators with constant coefficients, Reprint of the 1983 original. MR 2108588
68. ———, *The analysis of linear partial differential operators. III*, Classics in Mathematics, Springer, Berlin, 2007, Pseudo-differential operators, Reprint of the 1994 edition. MR 2304165
69. ———, *The analysis of linear partial differential operators. IV*, Classics in Mathematics, Springer-Verlag, Berlin, 2009, Fourier integral operators, Reprint of the 1994 edition. MR 2512677
70. Pierre Julg and Erik van Erp, *The geometry of the osculating nilpotent group structures of the Heisenberg calculus*, J. Lie Theory **28** (2018), no. 1, 107–138.
71. Gennadi G. Kasparov, *Topological invariants of elliptic operators I: K-homology*, Izv. Akad. Nauk. S.S.S.R. Ser. Mat. **39** (1975), 796–838.
72. ———, *The operator K-functor and extensions of C^*-algebras*, Izv. Akad. Nauk SSSR Ser. Mat. **44** (1980), no. 3, 571–636.
73. Gennadi G. Kasparov and Georges Skandalis, *Groups acting on buildings, operator K-theory and Novikov's conjecture*, K-theory **4** (1991), 303–337.
74. Jean-Marie Lescure, *Triplets spectraux pour les variétés à singularité conique isolé e*, Bull. Soc. math. France **129** (2001), no. 4, 593–623.
75. Jean-Marie Lescure, Dominique Manchon, and Stéphane Vassout, *About the convolution of distributions on groupoids*, J. Noncommut. Geom. **11** (2017), no. 2, 757–789. MR 3669118
76. Jean-Marie Lescure and Stéphane Vassout, *Fourier integral operators on Lie groupoids*, Adv. Math. **320** (2017), 391–450. MR 3709110
77. Pedro T. P. Lopes and Severino T. Melo, *K-theory of the Boutet de Monvel algebra with classical SG-symbols on the half space*, Math. Nachr. **287** (2014), no. 16, 1804–1827. MR 3274491
78. Kirill C. H. Mackenzie, *General theory of Lie groupoids and Lie algebroids*, London Mathematical Society Lecture Note Series, vol. 213, Cambridge University Press, Cambridge, 2005.
79. Rafe Mazzeo, *Elliptic theory of differential edge operators. I*, Comm. Partial Differential Equations **16** (1991), no. 10, 1615–1664. MR 1133743
80. Rafe R. Mazzeo and Richard B. Melrose, *Meromorphic extension of the resolvent on complete spaces with asymptotically constant negative curvature*, J. Funct. Anal. **75** (1987), no. 2, 260–310.

81. ______, *Pseudodifferential operators on manifolds with fibred boundaries*, Asian J. Math. **2** (1998), no. 4, 833–866, Mikio Sato: a great Japanese mathematician of the twentieth century.
82. Severino T. Melo, Thomas Schick, and Elmar Schrohe, *Families index for Boutet de Monvel operators*, Münster J. Math. **6** (2013), no. 2, 343–364. MR 3148215
83. Richard B. Melrose, *Pseudodifferential operators, corners and singular limits*, Proceedings of the International Congress of Mathematicians, Vol. I, II (Kyoto, 1990), Math. Soc. Japan, Tokyo, 1991, pp. 217–234. MR 1159214
84. ______, *Calculus of conormal distributions on manifolds with corners*, Internat. Math. Res. Notices (1992), no. 3, 51–61. MR 1154213
85. ______, *The Atiyah-Patodi-Singer index theorem*, Research Notes in Mathematics, vol. 4, A K Peters Ltd., Wellesley, MA, 1993.
86. Richard B. Melrose and Paolo Piazza, *Analytic K-theory on manifolds with corners*, Adv. Math. **92** (1992), no. 1, 1–26. MR 1153932
87. A. S. Mishchenko and A. T. Fomenko, *The index of elliptic operators over C^*-algebras*, Izv. Akad. Nauk. S.S.S.R. Ser. Mat. **43** (1979), no. 4, 831–859.
88. Omar Mohsen, *On the deformation groupoid of the inhomogeneous pseudo-differential calculus*, arXiv:1806.08585, 2018.
89. Bertrand Monthubert, *Pseudodifferential calculus on manifolds with corners and groupoids*, Proc. Amer. Math. Soc. **127** (1999), no. 10, 2871–2881.
90. ______, *Groupoids and pseudodifferential calculus on manifolds with corners*, J. Funct. Anal. **199** (2003), no. 1, 243–286.
91. Bertrand Monthubert and François Pierrot, *Indice analytique et groupoïdes de Lie*, C. R. Acad. Sci. Paris Sér. I Math. **325** (1997), no. 2, 193–198.
92. Calvin C. Moore and Claude L. Schochet, *Global analysis on foliated spaces*, second ed., Mathematical Sciences Research Institute Publications, vol. 9, Cambridge University Press, New York, 2006. MR 2202625
93. Paul S. Muhly, Jean N. Renault, and Dana P. Williams, *Equivalence and isomorphism for groupoid C^*-algebras*, J. Operator Theory **17** (1987), no. 1, 3–22.
94. F. J. Murray and J. Von Neumann, *On rings of operators*, Ann. of Math. (2) **37** (1936), no. 1, 116–229. MR 1503275
95. Victor Nistor, Alan Weinstein, and Ping Xu, *Pseudodifferential operators on differential groupoids*, Pacific J. Math. **189** (1999), no. 1, 117–152.
96. William L. Paschke, *K-theory for commutants in the Calkin algebra*, Pacific J. Math. **95** (1981), no. 2, 427–434. MR 632196
97. Alan L. T. Paterson, *Continuous family groupoids*, Homology Homotopy Appl. **2** (2000), 89–104. MR 1782594
98. Paolo Piazza and Thomas Schick, *The surgery exact sequence, K-theory and the signature operator*, Ann. K-Theory **1** (2016), no. 2, 109–154. MR 3514938
99. Paolo Piazza and Vito Felice Zenobi, *Singular spaces, groupoids and metrics of positive scalar curvature*, arXiv:1803.02697, 2018.
100. Raphaël Ponge, *The tangent groupoid of a Heisenberg manifold*, Pacific J. Math. **227** (2006), no. 1, 151–175.
101. Jean Pradines, *Feuilletages: holonomie et graphes locaux*, C. R. Acad. Sci. Paris Sér. I Math. **298** (1984), no. 13, 297–300. MR 765427
102. Jean Renault, *A groupoid approach to C^*-algebras*, Lecture Notes in Mathematics, vol. 793, Springer, Berlin, 1980.
103. ______, *C^*-algebras and dynamical systems*, Publicações Matemáticas do IMPA. [IMPA Mathematical Publications], Instituto Nacional de Matemática Pura e Aplicada (IMPA), Rio de Janeiro, 2009, 27o Colóquio Brasileiro de Matemática. [27th Brazilian Mathematics Colloquium]. MR 2536186
104. John Roe, *Lectures on coarse geometry*, University Lecture Series, vol. 31, American Mathematical Society, 2003.
105. Anton Savin, Elmar Schrohe, and Sternin Boris, *Elliptic operators associated with groups of quantized canonical transformations*, arXiv:1612.02981.

106. Elmar Schrohe, *A short introduction to Boutet de Monvel's calculus*, Approaches to singular analysis (Berlin, 1999), Oper. Theory Adv. Appl., vol. 125, Birkhäuser, Basel, 2001, pp. 85–116.
107. M. A. Shubin, *Pseudodifferential operators and spectral theory*, second ed., Springer-Verlag, Berlin, 2001, Translated from the 1978 Russian original by Stig I. Andersson. MR 1852334
108. Georges Skandalis, Jean-Louis Tu, and Guoliang Yu, *The coarse Baum-Connes conjecture and groupoids*, Topology **41** (2002), 807–834.
109. Michael E. Taylor, *Pseudodifferential operators*, Princeton Mathematical Series, vol. 34, Princeton University Press, Princeton, N.J., 1981. MR 618463
110. François Trèves, *Introduction to pseudodifferential and Fourier integral operators. Vol. 1*, Plenum Press, New York-London, 1980, Pseudodifferential operators, The University Series in Mathematics. MR 597144
111. ———, *Introduction to pseudodifferential and Fourier integral operators. Vol. 2*, Plenum Press, New York-London, 1980, Fourier integral operators, The University Series in Mathematics. MR 597145
112. Alain Valette, *Introduction to the Baum-Connes conjecture*, Lectures in Mathematics ETH Zürich, Birkhäuser Verlag, Basel, 2002, From notes taken by Indira Chatterji, With an appendix by Guido Mislin. MR 1907596
113. Erik van Erp, *The Atiyah-Singer index formula for subelliptic operators on contact manifolds. Part I*, Ann. of Math. (2) **171** (2010), no. 3, 1647–1681. MR 2680395
114. Erik van Erp and Robert Yuncken, *A groupoid approach to pseudodifferential operators*, J. Reine Agnew. Math. **2019** (2019), no. 756, 151–182.
115. ———, *On the tangent groupoid of a filtered manifold*, Bull. Lond. Math. Soc. **49** (2017), no. 6, 1000–1012.
116. Stéphane Vassout, *Unbounded pseudodifferential calculus on Lie groupoids*, J. Funct. Anal. **236** (2006), no. 1, 161–200. MR 2227132
117. Dan Voiculescu, *A non-commutative Weyl-von Neumann theorem*, Rev. Roumaine Math. Pures Appl. **21** (1976), no. 1, 97–113. MR 0415338
118. H. E. Winkelnkemper, *The graph of a foliation*, Ann. Global Anal. Geom. **1** (1983), no. 3, 51–75. MR 739904
119. Mayuko Yamashita, *A topological approach to indices of geometric operators on manifolds with fibered boundaries*, arXiv:1902.03767.
120. Vito Felice Zenobi, *The adiabatic groupoid and the Higson-Roe exact sequence*, arXiv:1901.05081.

Cyclic homology in a special world

Bjørn Ian Dundas

On the occasion of Alain Connes' 70 year celebration.

Abstract In work of Connes and Consani, Γ-spaces have taken a new importance. Segal introduced Γ-spaces in order to study stable homotopy theory, but the new perspective makes it apparent that also information about the *unstable* structure should be retained. Hence, the question naturally presents itself: to what extent are the commonly used invariants available in this context? We offer a quick survey of (topological) cyclic homology and point out that the categorical construction is applicable also in an $\mathbb{N}$-algebra (aka. semi-ring or rig) setup.

Keywords Cyclic homology · Ring spectra · Topological cyclic homology · Special gamma spaces · Unstable homology · Group completion

Mathematical Subject Classification (2010) Primary: 13D03; Secondary: 18G60, 19D55, 55P92

Alain Connes introduced cyclic homology in 1981 as a generalization of de Rham homology suitable for non-commutative geometry. Boris Tsygan reintroduced it in 1983 as an "additive" version of algebraic K-theory (see Section 1 for a brief overview with citations of the part of the theory relevant for our considerations). Almost immediately it became apparent that cyclic homology was a very good invariant for studying K-theory, at least rationally. However, for torsion information one needed to extend the construction from rings to so-called $\mathbb{S}$-algebras (i.e., replacing the ring $\mathbb{Z}$ of integers with the sphere spectrum $\mathbb{S}$), resulting in Bökstedt, Hsiang, and Madsen's topological cyclic homology TC. A possible framework for extending cyclic homology in this direction is Segal's category of Γ-spaces,

B. I. Dundas (✉)
University of Bergen, Bergen, Norway
e-mail: dundas@math.uib.no

A. Chamseddine et al. (eds.), *Advances in Noncommutative Geometry*,
https://doi.org/10.1007/978-3-030-29597-4_5

generalizing the concept of abelian groups in a way that also allows objects where the axioms of an abelian group are perhaps only true up to some notion of equivalence—see Section 2.1 for an elementary introduction to Γ-spaces where we try to explain why the structure is virtually forced upon us from the algebraic origins.

Actually, in many of these examples there is one axiom that does not hold at all—the existence of negatives. For instance, at the outset there are no sets with a negative number of elements or vector spaces of negative dimension. However, experience—starting in elementary school—has taught us that we get a much more effective theory if we adjoin negatives.

Stable homotopy theory is the study of such examples after adjoining negatives by a process dubbed stabilization. This is an enormously successful theory: the sphere spectrum sees much more combinatorial data than the ring of integers does, and some of this combinatorial data is reflected in the number theory revealed by algebraic K-theory. In fact, Thomason [60] showed that algebraic K-theory can be viewed as a localization from the category $\mathcal{SMC}$ of small symmetric monoidal categories to the homotopy category (with respect to stable equivalences) of Γ-spaces.

Example 0.0.1 One crucial difference between abelian groups and stable homotopy theory is how symmetries behave. For instance, if A is an $\mathbb{S}$-algebra we can consider the smash $A \wedge A$—the analog of the tensor product or, in algebraic geometry, a product $X \times X$. The cyclic group of order two acts and we can consider the fixed points $(A \wedge A)^{C_2}$—analog to the symmetric product $(X \times X)/C_2$.

What does *not* have an analog in the algebraic or geometric situation is that in stable homotopy theory $(A \wedge A)^{C_2}$ is much like a form of Witt vectors: there is a "restriction" map $(A \wedge A)^{C_2} \to A$ (it is not any sort of multiplication! At the level of path components it is the extension of the group $\pi_0 A$ by $(\pi_0 A \otimes \pi_0 A)/C_2$ by the cocycle $(x, y) \mapsto -x \otimes y$) which often is the first step of a lift, either from finite characteristic to infinite characteristic or higher up in the so-called chromatic tower of stable homotopy theory. The restriction map is essential for the construction of TC in Section 1.5 and can be viewed as the source of most of the verifications we know of the so-called "red-shift conjecture" in algebraic K-theory.

The restriction map reappears below in a special context as a composite of the geometric diagonal of Section 2.3.4 and isotropy separation in Section 2.2.1.

However, algebraic K-theory kills much information one might be interested in. For a ring A, the Grothendieck group $K_0(A)$ is obtained from the isomorphism classes of finitely generated projective A-modules by introducing objects of "negative rank." For many situations this is a rather innocent operation (from the natural numbers one obtains the integers), but in other situations group completion can drastically alter the object at hand.

Example 0.0.2 If k is a field, consider the category $\mathrm{Vect}_k^{\mathrm{count}}$ of all k-vector spaces of *countable* dimension. Then $k^n \oplus k^\infty \cong k^\infty$ for all $n \leq \infty$, and if you group complete with respect to sum—essentially introducing negative dimensions—you have cancellation, leaving you with the rather uninformative trivial group. This sorry state of affairs is sometimes referred to as the *Eilenberg swindle*. This is in stark contrast with the situation where you only consider *finite* dimensional vector spaces which leads to the usual algebraic K-theory, which is far from trivial and (the higher homotopy) contains much information about the field.

For all its categorical defects, the category $\mathcal{SMC}$ of small symmetric monoidal categories is in many ways the natural philosophical relaxation of the category of abelian groups. We must perhaps live without negatives and that laws like commutativity only hold up to isomorphism. While we want to retain as much information about $\mathcal{SMC}$ as possible, in order to obtain a situation we can calculate with, some localization seems necessary.

One choice is to study the localization of $\mathcal{SMC}$ with respect to the *unstable equivalences*: a symmetric monoidal functor $f: c \to d$ in $\mathcal{SMC}$ is a weak equivalence if the map of nerves $Nc \to Nd$ is a weak equivalence in spaces. Mandell [46] improves on Thomason's result by showing that the localization of $\mathcal{SMC}$ with respect to unstable equivalences is equivalent to the localization of the category of Γ-spaces with respect to the *special* equivalences (we will discuss these in Section 2.2.3).

Forty years after Segal's discovery, Γ-spaces reappear in work of Connes and Consani [18] where it becomes clear that this generalization of abelian groups fits as a common framework for many of the current efforts of understanding the "field with one element." The rôle of the field with one element is taken by the sphere spectrum $\mathbb{S}$ and the rôle of the tensor product is taken by the smash product $\wedge$, see Section 2 for further details. However, for Connes and Consani it is vital that we do *not* adjoin negative elements; we are no longer in the realm of stable homotopy theory and many of the identifications we are used to no longer hold. A priori this has serious consequences for invariants—we may have used identifications that only hold after having adjoined negatives.

What follows is a tentative study of to what extent we can hope to extend invariants to a context that handles symmetric monoidal categories well without adjoining negatives by treating (successfully) the case of topological cyclic homology, see Section 3. Central to this is that the underlying machinations of the restriction map of Example 0.0.1 carry through.

The reader should be aware of the fact that this is only a tentative study: ultimately we are after a theory that better reflects the intuition of how modules over the field with one element should behave. In order to achieve this, there are reasons to not take spaces as our primitive notion, but rather quasi-categories (i.e., the Joyal model structure on the category of spaces). However, this theory is more technical and not at all suited for a survey-type paper of this sort. While most things follow a path very similar to the one sketched below, we have not yet written down all the details and hope to return to it in a future paper.

Overview

In Section 1 we give a quick overview of the history and some results pertaining to cyclic homology that are relevant to our discussion.

In Section 2 we study the equivariant theory you get from Γ-spaces when you stop short of group completing. This is the so-called *special* model and the most important output is that the categorical model adapts to the current situation. The category of Γ-spaces is in many ways a much less friendly world than most of its competitors modeling stable homotopy theory, but it is almost finitely generated (a technical term of Voevodsky's) which means that we retain just enough control also over the special situation.

In Section 3 we see that the equivariant control we obtained in Section 2 is exactly what is needed in order to set up TC in the special situation.

Lastly we collect some results on modules and monoids that are of interest, but require input that did not fit with the equivariant focus of the rest of the paper. In particular, one of the examples Connes and Consani pay special attention to is the so-called Boolean algebra $\mathbf{B} = \{0, 1\}$ with $1+1 = 1$. In Section 4 we show that $\mathbf{B}$ is "specially solid": the multiplication map $\mathbf{B}\wedge^L\mathbf{B} \to \mathbf{B}$ is a special equivalence. This can be seen as a disappointment: although we have devised a theory that avoids the Scylla of group completing the *monoidal structure* of a symmetric monoidal category, we still must deal with the Charybdis of weak equivalences which is akin to inverting the *morphisms*. In this example these processes are much the same (not quite: otherwise everything would be zero). One fix is to consider the Joyal structure referred to above, but this is as mentioned postponed to another day.

Notational conventions

1. The category of symmetric monoids is symmetric monoidal with unit $\mathbb{N} = \{0, 1, 2, \dots\}$ and tensor $\otimes_{\mathbb{N}}$ (defined exactly as the usual tensor product). To avoid the rig/semi-ring controversy (we find neither alternative particularly attractive, but we *really* dislike "semi-rings") we call the monoids with respect to the tensor in symmetric monoids $\mathbb{N}$-*algebras*.
2. If C is a category and c, c' are objects, then $C(c, c')$ is the set (or space according to flavor) of morphisms $c \to c'$ in C. The functor $c' \mapsto C(c, c')$ is denoted $C(c, -)$.
3. If X and Y are pointed sets, then the *wedge* $X \vee Y \subseteq X \times Y$ is the subset where one of the coordinates is the base point and the *smash product* is the quotient $X\wedge Y = X \times Y/X \vee Y$.
4. We use "k_+" as shorthand for the set $\{0, 1, \dots, k\}$ pointed at 0 and $[k]$ for the ordered set $\{0 < 1 < \cdots < k\}$. The category of pointed finite sets is called Γ^o and the category of nonempty totally ordered finite sets is called Δ.
5. Objects in the category $\mathcal{S}_*$ of pointed simplicial sets will be referred to as *spaces*. A Γ-*space* is a pointed functor from Γ^o to $\mathcal{S}_*$.
6. If X is an object on which a group G acts, then $X^G \to X$ is the "inclusion of fixed points" (as defined e.g., by a categorical limit over G).

1 Cyclic homology

1.1 Prehistory

The connection between algebraic K-theory and de Rham cohomology was pointed out already in the early days of higher algebraic K-theory (ca. 1972). For a commutative ring A, Gersten provided a map to the Kähler differentials

$$\mathrm{d}\log\colon K_*A \to \Omega^*_A$$

and Bloch [5] proved that the "tangent space" $TK_n(A) = \ker\{K_n(A[\epsilon]/\epsilon^2) \to K_n(A)\}$ contains Ω^{n-1}_A as a split summand when A is local, $1/2 \in A$ and $n > 0$. Also other connections between algebraic K-theory and homological theories were investigated, for instance the Dennis trace map to Hochschild homology $\mathrm{HH}(A)$.

1.2 Cyclic homology

In 1980 Alain Connes was searching for a cohomology theory of de Rham type [14] suitable for non-commutative algebras and introduced cyclic homology the year after [15, 17]. A few years later, Tsygan [61] rediscovered cyclic homology and in parallel with Loday and Quillen [42] proved that in characteristic zero, cyclic homology $\mathrm{HC}_{*-1}(A)$ is isomorphic to the primitive part of the homology of the Lie algebra $\mathfrak{gl}(A)$. Goodwillie completed the picture by showing [31] that in the context of a nilpotent extension of associative rings, the relative algebraic K-theory agrees, rationally and up to a shift in grading, with relative cyclic homology.

It is tempting to think of the map used by Goodwillie between relative K-theory and cyclic homology as a "logarithm" from the general linear group $\mathrm{GL}(A)$ to its "tangent space," the Lie algebra $\mathfrak{gl}(A)$. In this interpretation the rationality assumption is necessary for the coefficients in the Taylor expansion of the logarithm to be defined, the nilpotence assures convergence and finally the need for taking primitives stems from the correspondence in rational stable homotopy between homology and homotopy. See [19] and [63] for ideas along this line.

Connes demonstrated [16] that the Hochschild homology is a *cyclic* object and its associated spectrum $\mathrm{HH}(A)$ comes with an action by the circle $\mathbb{T}$ (see Section 3.1 for more on cyclic objects). In this interpretation, cyclic homology corresponds to the homotopy *orbits* $\mathrm{HC}(A) = \mathrm{HH}(A)_{h\mathbb{T}}$ (the double complex is a concrete algebraic representation of the fact that the classifying space $B\mathbb{T} \cong \mathbb{C}\mathrm{P}^\infty$ of the circle $\mathbb{T}$ has a single cell in each even dimension). Goodwillie (and Jones [38]) showed that the Dennis trace factors through the homotopy *fixed points* $\mathrm{HH}(A)^{h\mathbb{T}}$ (which was dubbed "negative cyclic homology"). The difference between the homotopy orbits and fixed points is measured by the "norm map" $N\colon \Sigma\mathrm{HH}(A)_{h\mathbb{T}} \to \mathrm{HH}(A)^{h\mathbb{T}}$ (note the suspension which is responsible for the above observed shift in grading) which is part of a fiber sequence

$$\Sigma \mathbb{HH}(A)_{h\mathbb{T}} \to \mathbb{HH}(A)^{h\mathbb{T}} \to \mathbb{HH}(A)^{t\mathbb{T}}.$$

The last term—the "Tate-construction" on $\mathbb{HH}(A)$ and whose homotopy groups are referred to as periodic homology—is defined by this sequence and vanishes in certain key situations. Most notably, in the rational nilpotent situation the relative periodic homology vanishes.

1.3 The "topological" version

However, as Goodwillie and Waldhausen pointed out, Hochschild homology in itself contains much too little information to be a useful starting point for measuring algebraic K-theory and they conjectured the existence of a version built on the sphere spectrum $\mathbb{S}$ instead of the integers $\mathbb{Z}$ and the smash product $\wedge$ instead of the tensor product $\otimes$. This idea was realized by Bökstedt and dubbed "topological Hochschild homology," $\mathbb{THH}(A)$ or—emphasizing that this is nothing but Hochschild homology over $\mathbb{S}$—$\mathbb{HH}^{\mathbb{S}}(A)$.

Topological Hochschild homology has a richer inner life than Hochschild homology over the integers, and Bökstedt et al. [10] used this to define topological cyclic homology and prove an algebraic K-theory version of the Novikov conjecture. Although predating the first fully adequate setups for $\mathbb{S}$-algebras, their approach critically used the ability to move freely between $\mathbb{S}$ and $\mathbb{Z}$ as ground rings and that rationally the difference is very small.

Topological Hochschild homology gained further credibility from the fact [62, 56, 27, 26] that it agreed with stable K-theory as predicted by Goodwillie and Waldhausen. Stable K-theory is a version of Bloch's tangent space, where the dual numbers $A \ltimes \epsilon A = A[\epsilon]/\epsilon^2$ is replaced by a square zero extension $A \ltimes M$ where the connectivity of M is allowed to tend to infinity—it is the differential of algebraic K-theory in a way made precise by Goodwillie's calculus of functors. Related to early ideas of Goodwillie, Lindenstrauss, and McCarthy [41] show that it actually is (relatively) fair to think of $\mathbb{TC}$ as the Taylor tower of K-theory. This also sheds light on the nature of the action by cyclic group $C_n \subseteq \mathbb{T}$: it is a remnant of the action by the symmetric group hiding behind the denominator $n!$ in the usual Taylor series. Much of this insight was clear already at the time of [47].

It is not only the connection to algebraic K-theory that makes topological cyclic homology and its relatives interesting. Topological cyclic homology carries interesting information from an algebro-geometric and number theoretic point of view, as a theory with close connections to motivic, étale, crystalline and de Rham cohomology. Some of this was clear from the very start, but some aspects have become apparent more recently, see e.g., [33] and [34].

1.4 The cyclotomic trace

The cyclotomic trace $K \to \mathrm{TC}$, first defined by Bökstedt et al. [10], refined in [22] and beautifully pinned down in [6] and [7], is of crucial importance for two reasons:

1. TC has surprisingly often been possible to calculate
2. the homotopy fiber K^{inv} of the cyclotomic trace $K \to \mathrm{TC}$ is very well behaved.

The starting point for many applications is that the K-theory of finite fields is known by Quillen [50] and that topological cyclic homology is possible to calculate in a number of difficult situations. From there the following omnibus theorem will take you a long way.

Theorem 1.4.1 *Let $A \to B$ be a map of connective $\mathbb{S}$-algebras such that the map $\pi_0 A \to \pi_0 B$ is a surjection with kernel I.*

Locally constant *If I is nilpotent, then the map $K^{\mathrm{inv}} A \to K^{\mathrm{inv}}(B)$ is a stable equivalence [22]*

Rigidity *If $\pi_0 A$ and $\pi_0 B$ are commutative and $(\pi_0 A, I)$ is a Henselian pair, then the map $K^{\mathrm{inv}}(A) \to K^{\mathrm{inv}}(B)$ is a stable equivalence with mod-n coefficients for $n \in \mathbb{N}$ [13]*

Closed excision *If $C \to B$ is a map of connective $\mathbb{S}$-algebras and $D = A \times^h_B C$ the homotopy pullback, then*

$$\begin{array}{ccc} K^{\mathrm{inv}} D & \longrightarrow & K^{\mathrm{inv}} C \\ \downarrow & & \downarrow \\ K^{\mathrm{inv}} A & \longrightarrow & K^{\mathrm{inv}} B \end{array}$$

is a homotopy pullback square [40].

□

Closed excision was proved rationally by Cortiñas [20], after completion by Geisser and Hesselholt [28] for rings and in general by Dundas–Kittang [23]. Land and Tamme's preprint [40] removes an unnecessary surjectivity condition from the integral result of [24].

The combined outcome of the calculations of TC and Theorem 1.4.1 has been that a vast range of calculations in algebraic K-theory has become available, at least after profinite completion, but also integrally when coupled with motivic calculations. Even a somewhat random and very inadequate listing of results would include [10, 9, 35, 52, 4, 53, 54, 51, 36, 29, 30, 3, 49, 32, 2, 1, 43]. See [45] and [22] for more background on these methods.

1.5 The original construction of TC

Bökstedt et al. [10] relied on equivariant stable homotopy theory to produce a "naively invariant" theory out of categorical fixed points with respect to the finite cyclic subgroups C_m of the circle. In particular, if $m|n$ the model for THH provided by Bökstedt comes with a hands-on *restriction map*

$$R_n^m : \mathrm{THH}(A)^{C_n} \to \mathrm{THH}(A)^{C_m}$$

related to the restriction map in Example 0.0.1. The restriction map has very good homotopical properties; for instance, if p is a prime, it fits into the "fundamental cofibration sequence"

$$\mathrm{THH}(A)_{hC_{p^n}} \longrightarrow \mathrm{THH}(A)^{C_{p^n}} \xrightarrow{R_{p^n}^{p^{n-1}}} \mathrm{THH}(A)^{C_{p^{n-1}}},$$

i.e., the homotopy fiber of the restriction map $R_{p^n}^{p^{n-1}}$ is naturally equivalent to the homotopy orbits $\mathrm{THH}(A)_{hC_{p^n}}$. The algebraic analog of the restriction map is the restriction map of truncated Witt vectors, and the inclusion of fixed points

$$F_n^m : \mathrm{THH}(A)^{C_n} \to \mathrm{THH}(A)^{C_m}$$

turns out to mirror the Frobenius. Focusing on one prime p, one defines

$$\mathrm{TC}(A, p) = \operatorname*{holim}_{\overleftarrow{F,R}} \mathrm{THH}(A)^{C_{p^n}}.$$

Note that one has full homotopic control of this construction. For instance, if a map $A \to B$ induces an equivalence $\mathrm{THH}(A) \to \mathrm{THH}(B)$, the fundamental cofibration sequences guarantee that the same is true for all C_{p^n}-fixed points and ultimately $\mathrm{TC}(A, p) \to \mathrm{TC}(B, p)$ is an equivalence too.

After p-completion, the inclusion of the p-power roots of unity induces an equivalence of classifying spaces $\lim_{\overrightarrow{n}} BC_{p^n} = BC_{p^\infty} \to B\mathbb{T}$, and so the target of the natural map

$$\mathrm{TC}(A, p) \to \operatorname*{holim}_{\overleftarrow{F}} \mathrm{THH}(A)^{C_{p^n}} \to \operatorname*{holim}_{\overleftarrow{F}} \mathrm{THH}(A)^{hC_{p^n}}$$

(given by restricting to the Frobenius maps and mapping the fixed points to the homotopy fixed points) is equivalent after p-completion to $\mathrm{THH}(A)^{h\mathbb{T}}$, and one defines integral topological cyclic homology by the pullback

$$\begin{array}{ccc} \mathrm{TC}(A) & \longrightarrow & \mathrm{THH}(A)^{h\mathbb{T}} \\ \downarrow & & \downarrow \\ \prod_{p \text{ prime}} \mathrm{TC}(A,p)_p & \longrightarrow & \prod_{p \text{ prime}} \mathrm{THH}(A)_p^{h\mathbb{T}}. \end{array}$$

Hesselholt and Madsen packaged in [35] the information about the nature of the restriction map in the language of stable equivariant homotopy theory via their notion of *cyclotomic spectra* by focusing on the so-called *geometric fixed points* Φ^N (which we will discuss more extensively in Sections 2.2 and 2.3) and one way of stating this is that there is an equivalence between $\mathbb{THH}(A)$ and its C_p-geometric fixed points $\Phi^{C_p}\mathbb{THH}(A)$, see Lemma 3.3.1.

1.6 The Nikolaus–Scholze approach

The fundamental cofibration sequence/cyclotomic structure implies that the categorical fixed points of topological Hochschild homology is a homotopy invariant (for instance, $\mathbb{THH}(A)^{C_p}$ is the homotopy pullback of a diagram of the form $\mathbb{THH}(A)^{hC_p} \to \mathbb{THH}(A)^{tC_p} \leftarrow \mathbb{THH}(A)$). Nikolaus and Scholze [48] showed that this gives rise to an extremely elegant formula expressing topological cyclic homology in terms of functors that are manifestly homotopy invariant; namely as the homotopy fiber of a certain map

$$\mathbb{THH}(A)^{h\mathbb{T}} \to \mathbb{THH}(A)^{t\mathbb{T}^\wedge}$$

from the homotopy fixed points to the profinite completion of the Tate-construction of topological Hochschild homology.

2 The special version

We have seen that there are many reasons to consider Γ-spaces. If we are especially careful (as we will be) it models symmetric monoidal categories very faithfully but still has very good algebraic properties and is a common framework for various points of view of the "vector spaces over the field with one element." In what follows, we explore how we can formulate some important invariants in this special context.

2.1 Γ-spaces as a generalization of symmetric monoids

Graeme Segal introduced Γ-spaces as an infinite delooping machine in [58], and Manos Lydakis [44] realized that this very down-to-earth approach actually possessed very good properties. Other useful sources for the properties of Γ-spaces are Bousfield and Friedlander [11] and Schwede [57]. As we try to elucidate below, apart from being very concrete, one of the benefits of Γ-spaces is that their algebraic origin is very clear.

A *symmetric monoid* is a set M together with a multiplication and a unit element so that any two maps $M^{\times j} \to M$ obtained by composing maps in the diagram

$$* \xrightarrow{\text{unit}} M \begin{array}{c} \xrightarrow{m\mapsto(1,m)} \\ \xleftarrow{\text{multiplication}} \\ \xrightarrow{m\mapsto(m,1)} \end{array} \overset{\overset{\text{twist}}{\circlearrowleft}}{M\times M} \begin{array}{c} \xleftarrow{(m_1m_2,m_3)\leftarrow\!\shortmid(m_1,m_2,m_3)} \\ \xleftarrow{(m_1,m_2m_3)\leftarrow\!\shortmid(m_1,m_2,m_3)} \end{array} M\times M\times M$$

are equal. The diagram is mirrored by the diagram of sets

$$\emptyset \longrightarrow \{1\} \begin{array}{c} \xrightarrow{1\mapsto 2} \\ \xleftarrow{1\leftarrow\!\shortmid 1,2} \\ \xrightarrow{1\mapsto 1} \end{array} \overset{\overset{1\leftrightarrow 2}{\circlearrowleft}}{\{1,2\}} \begin{array}{c} \xleftarrow{1\leftarrow\!\shortmid 1\ \ 2\leftarrow\!\shortmid 2,3} \\ \xleftarrow{1\leftarrow\!\shortmid 1,2\ \ 2\leftarrow\!\shortmid 3} \end{array} \{1,2,3\}.$$

We will need to encode the two projections $M \times M \to M$ as well, and for this purpose we add a basepoint and consider the category Γ^o of finite pointed sets (the functions must preserve the base point), so that the diagram governing the axioms of a monoid looks like

$$0_+ \longrightarrow 1_+ \begin{array}{c} \longrightarrow \\ \longleftarrow \\ \longrightarrow \end{array} \overset{\circlearrowleft}{2_+} \begin{array}{c} \longleftarrow \\ \longleftarrow \end{array} 3_+,$$

where $k_+ = \{0, 1, \dots, k\}$. Segal realized that if one wants to relax the axioms for symmetric monoids so that they only are true up to some sort of equivalence (as for instance is the case for symmetric monoidal categories) it is fruitful to extend this diagram to all finite pointed sets: send k_+ to $HM(k_+) = M^{\times k}$ and a pointed function $\phi\colon k_+ \to l_+$ to

$$\phi_*\colon M^{\times k} \to M^{\times l}, \qquad \phi_*(m_1, \dots, m_k) = \left(\prod_{\phi(j)=1} m_j, \dots, \prod_{\phi(j)=k} m_j \right).$$

This is the so-called *Eilenberg–Mac Lane* construction which identifies the category of symmetric monoids with a combinatorially easily recognizable subcategory of the category of Γ*-sets* (pointed functors from the category Γ^o of finite pointed sets to pointed sets): we get an isomorphism between the categories of symmetric monoids and of the full category of Γ-sets sending $\vee$ to $\times$ strictly (e.g., $3_+ = 1_+ \vee 1_+ \vee 1_+$ must be sent to the triple product of the values at 1_+). The projections $HM(k_+) = M^{\times k} \to M = HM(1_+)$ are given by the characteristic functions

$$\delta^i : k_+ \to 1_+, \qquad \delta^i(j) = \begin{cases} 1 & i = j \\ 0 & i \neq j \end{cases}$$

for $i = 1, \dots k$ and the multiplication $M \times M \to M$ is given by $\nabla\colon 2_+ \to 1_+$ with $\nabla(1) = \nabla(2) = 1$.

We want to be able to handle not only symmetric monoids but also symmetric monoidal *categories*, so we allow a simplicial direction to harbor morphisms: Let

- $\mathcal{S}_* =$ pointed simplicial sets ("spaces"),
- $\Gamma^o =$ pointed finite sets,
- $\Gamma\mathcal{S}_* =$ "Γ-spaces" = pointed functors $\Gamma^o \to \mathcal{S}_*$.

Note that (up to natural isomorphism) it is enough to specify a Γ-space on the skeletal subcategory containing the objects of the form k_+ only.

2.1.1 Smash as a generalization of tensor

The *smash product*—even more than its sibling the tensor product—is often shrouded in mystery, but I insists it is a natural object forced on us by bilinearity and can be motivated as follows: Fiddling with the functoriality of the Eilenberg–Mac Lane construction

$$H: s\mathcal{M}on \to \Gamma\mathcal{S}_*, \qquad M \mapsto HM = \{k_+ \mapsto M^{\times k}\}$$

defined above we see that a transformation $HM \to HN$ is uniquely given by its value on 1_+: it comes from a unique homomorphism $M \to N$, and the canonical map $s\mathcal{M}on(M, N) \to \Gamma\mathcal{S}_*(HM, HN)$ is an isomorphism (actually, of spaces, but you may ignore this enrichment if you just want to understand the smash of Γ-sets). However, $s\mathcal{M}on(M, N)$ is obviously itself a symmetric (simplicial) monoid and

$$\begin{aligned} Hs\mathcal{M}on(M, N)(k_+) &\cong s\mathcal{M}on\left(M, N^{\times k}\right) \cong \Gamma\mathcal{S}_*\left(HM, HN^{\times k}\right) \\ &\cong \Gamma\mathcal{S}_*(HM, HN(-\wedge k_+)) \end{aligned}$$

(where $k'_+ \wedge k_+ \cong k'k_+$ is the smash of finite pointed sets), so if we define the internal morphism object by $\underline{\Gamma\mathcal{S}_*}(X, Y) = \{k_+ \mapsto \Gamma\mathcal{S}_*(X, Y(-\wedge k_+))\} \in \Gamma\mathcal{S}_*$ for arbitrary $X, Y \in \Gamma\mathcal{S}_*$ we get a natural isomorphism of Γ-spaces

$$Hs\mathcal{M}on(M, N) \cong \underline{\Gamma\mathcal{S}_*}(HM, HN).$$

Now, we want the smash product to be the adjoint:

$$\underline{\Gamma\mathcal{S}_*}(X\wedge Y, Z) \cong \underline{\Gamma\mathcal{S}_*}(Y, \underline{\Gamma\mathcal{S}_*}(X, Z)),$$

and the usual Yoneda yoga "solving the equation with respect to $X\wedge Y$" gives us the smash product by means of a concrete coend formula

$$X\wedge Y = \int^{m_+, n_+ \in \Gamma^o} \Gamma^o\left(m_+ \wedge n_+, -\right) \wedge X\left(m_+\right) \wedge Y\left(n_+\right)$$

i.e., as the "weighted average of all pointwise smash products." Even more concretely, we have an identification between maps $X\wedge Y \to Z \in \Gamma\mathcal{S}_*$ and transformations $X(m_+) \wedge Y(n_+) \to Z(m_+ \wedge n_+)$ natural in $m_+, n_+ \in \Gamma^o$, specifying $X \wedge Y$ up to unique isomorphism.

This affords $\Gamma\mathcal{S}_*$ the structure of a closed symmetric monoidal category. This categorical construction is a special case of the Day construction known since the 1970s but it was Lydakis who realized that it actually was the relevant construction for stable homotopy theory [44].

The unit for the smash is the inclusion $\Gamma^o \subseteq \mathcal{S}_*$ denoted either by $\mathbb{S}$ or $\Gamma^o(1_+, -)$ and often referred to as the *sphere spectrum* (since under the equivalence between the stable homotopy categories of Γ-spaces and connective spectra $\mathbb{S}$ corresponds to the actual sphere spectrum).

Hence it makes sense to talk about monoids with respect to the smash products, and we refer to these as $\mathbb{S}$-algebras. By design, the Eilenberg–Mac Lane construction is lax symmetric monoidal from $(s\mathcal{M}on, \otimes, \mathbb{N})$ to $(\Gamma\mathcal{S}_*, \wedge, \mathbb{S})$ and so takes $\mathbb{N}$-algebras (aka. rigs or semi-rings—they do not necessarily have additive inverses but otherwise satisfy the axioms of rings) to $\mathbb{S}$-algebras.

2.1.2 Special Γ-spaces

Although simplicial monoids are too restrictive for our purposes, some Γ-spaces are more important than others (in particular those that arise from symmetric monoidal categories) and we consider Segal's "up to homotopy" notion.

A Γ-space $X \in \Gamma\mathcal{S}_*$ is isomorphic to the Eilenberg–Mac Lane construction of a symmetric monoid if and only if

$$\delta_k \colon X(k_+) \to X(1_+)^{\times k}, \qquad \delta_k(x) = \left(\delta_*^1 x, \ldots, \delta_*^k x\right)$$

is an isomorphism for all $k \geq 0$. The "up to homotopy" notion is the following

Definition 2.1.3 A Γ-space X is *special* if $\delta_k \colon X(k_+) \to X(1_+)^{\times k}$ is a weak equivalence for all k.

An equivalent, and for our purposes better, way of expressing this is as follows. For $k_+ \in \Gamma^o$, consider the inclusion

$$s_k \colon \Gamma^o(1_+, -) \wedge k_+ = \Gamma^o(1_+, -)^{\vee k} \subseteq \Gamma^o(1_+, -)^{\times k} = \Gamma^o(k_+, -).$$

Under the Yoneda isomorphisms $X(k_+) \cong \Gamma\mathcal{S}_*(\Gamma^o(k_+, -), X)$ and $X(1_+)^{\times k} \cong \Gamma\mathcal{S}_*(\Gamma^o(1_+, -) \wedge k_+, X)$ we see that δ_k corresponds to $s_k^* \colon \Gamma\mathcal{S}_*(\Gamma^o(k_+, -), X) \to \Gamma\mathcal{S}_*(\Gamma^o(1_+, -) \wedge k_+, X)$. Let

$$\mathcal{L} = \{s_k \mid k_+ \in \Gamma^o\}.$$

Example 2.1.4 Among examples of special Γ-spaces we have those that arise from symmetric monoidal categories: The Eilenberg–Mac Lane construction extends from symmetric monoids to symmetric monoidal categories—and in this guise it is often referred to as algebraic K-theory

$$H\colon \mathcal{SMC} \to \Gamma\mathcal{S}_*$$

by incorporating automorphisms into the construction (functorially rectifying the pseudo-functor you get by taking the formula for monoids either through one of the standard machines or by hand), and all special Γ-spaces are unstably equivalent to something in its image [46].

This is yet another manifestation of the idea that $\Gamma\mathcal{S}_*$ represents the categorification of the category of symmetric monoids. Driving home this message, de Brito and Moerdijk [8] prove a special refinement of the famous Barratt–Priddy–Quillen theorem: "the canonical map $\mathbb{S} \to H\Sigma$ is a fibrant replacement in the special structure," where Σ is the category of finite sets and isomorphisms.

The problem is that standard operations of special Γ-spaces give output that is not special (case in point: the smash product), and the standard remedy is to allow for *all* Γ-spaces, but localize with respect to $\mathcal{L}$. We will need to do this in the presence of extra symmeries, so we will bake this into the presentation from the start.

2.2 Symmetries on Γ-spaces

From now on, let G be a finite group. A pointed G-set is a pointed set together with an action of G preserving the base point. Let (deleting "the category of" for convenience)

- Γ^o_G: finite pointed G-sets and *all* (not necessarily equivariant) pointed maps
- $\mathcal{S}_G$: pointed simplicial G-sets and *all* (not necessarily equivariant) pointed maps;
  $G\mathcal{S}_*$: pointed simplicial G-sets and pointed G-equivariant maps
- $\Gamma_G\mathcal{S}_G$: pointed G-functors $\Gamma^o_G \to \mathcal{S}_G$ and G-natural transformations; $\Gamma G\mathcal{S}_*$: pointed functors $\Gamma^o \to G\mathcal{S}_*$ and natural transformations; in other words, G-objects in $\Gamma\mathcal{S}_*$.

To elucidate the distinctions, let us list some functors connecting these (see Shimakawa [59])

- the inclusion $\nu\colon \Gamma^o \subseteq \Gamma^o_G$ giving a finite pointed set the trivial G-action is an equivalence of categories with retraction $\Gamma^o_G \to \Gamma^o$ the forgetful functor.
- sending $X \in \Gamma_G\mathcal{S}_G$ to $\{k_+ \to X(\nu k_+)\} \in G\mathcal{S}_*$ induces an equivalence $\nu^*\colon \Gamma_G\mathcal{S}_G \to \Gamma G\mathcal{S}_*$ with inverse $\Gamma G\mathcal{S}_* \to \Gamma_G\mathcal{S}_G$ sending $Y \in \Gamma G\mathcal{S}_*$ to $\{A \mapsto Y(A) = \int^{k_+} \Gamma^o(k_+, A)\wedge Y(k_+)\} \in \Gamma_G\mathcal{S}_G$ (with G acting diagonally on $\Gamma^o(k_+, A)\wedge Y(k_+)$).

Analogous to the set of maps $\mathcal{L}$ determining the special Γ-spaces we have the set $\mathcal{L}_G$ of inclusions

$$s_A\colon \Gamma^o(1_+,-)\wedge A \subseteq \Gamma^o(A,-) \in \Gamma_G\mathcal{S}_G$$

(where A is a G-set which for the sake of keeping $\mathcal{L}_G$ a set is of the form k_+ for some $k \in \mathbb{N}$ and some homomorphism $G \to \Sigma_k$) and we say that $X \in \Gamma_G\mathcal{S}_G$ is *special* if the maps $\Gamma_G\mathcal{S}_G(s_A, X) \in G\mathcal{S}_*$ are G-equivalences (i.e., for every subgroup $H \subseteq G$, the map of H-fixed points $\Gamma_G\mathcal{S}_G(s_A, X)^H = \Gamma\mathcal{S}_*(s_A, X)^H$ is a weak equivalence

$$X(A)^H \xrightarrow{\sim} \mathcal{S}_*(A, X(1_+))^H$$

of simplicial sets).

2.2.1 Fixed points

If $f\colon G \twoheadrightarrow J$ is a surjective group homomorphism with kernel N we let

$$\lfloor - \rfloor^N, \Phi^N\colon \Gamma_G\mathcal{S}_G \to \Gamma_J\mathcal{S}_J$$

be the *categorical* and *geometric* fixed point functors sending $X \in \Gamma_G\mathcal{S}_G$ to the objects in $\Gamma_J\mathcal{S}_J$ sending $V \in \Gamma^o_J$ to

$$X^N(V) = [X(f^*V)]^N,$$

$$\Phi^N X(V) = \operatorname{coeq}\left\{ \bigvee_{W\in\Gamma^o_G} \Gamma^o\left(W^N, V\right)\wedge[X(W)]^N \right.$$

$$\left. \leftleftarrows \bigvee_{W,W'\in\Gamma^o_G} \Gamma^o\left(W^N, V\right)\wedge\left[\Gamma^o(W', W)\wedge X(W')\right]^N \right\}$$

(with the two maps in the coequalizer given by functoriality $\Gamma^o(W', W)\wedge X(W') \to X(W)$ and composition $\Gamma^o(W^N, V)\wedge\Gamma^o(W', W)^N \to \Gamma^o([W']^N, V)$). Although weird-looking when presented like this without any motivation, the geometric fixed points are in many ways more convenient. In particular, Φ^N preserves much structure, like colimits and smash; a fact that becomes particularly potent when coupled with the isomorphism

$$\Phi^N(\Gamma^o(A,-)\wedge K) \cong \Gamma^o\left(A^N,-\right)\wedge K^N$$

(for $A \in \Gamma^o_G$ and $K \in \mathcal{S}_G$) obtained from the dual Yoneda lemma plus the fact that for G-spaces fixed points commute with smash. This isomorphism is the Γ-space version of "geometric fixed points commute with forming the suspension spectrum."

By writing out the definitions we see that Φ^N has a very special effect on the elements of $\mathcal{L}_G$:

Lemma 2.2.2 *If $f\colon G \twoheadrightarrow J$ is a surjective group homomorphism with kernel N and $A \in \Gamma_G^o$, then*

$$\begin{array}{ccc}
\Phi^N\Gamma^o(A,-) & \xrightarrow[\cong]{\text{dual Yoneda}} & \Gamma^o(A^N,-) \\
\Big\downarrow{\scriptstyle \Phi^N s_A} & & \Big\downarrow{\scriptstyle s_{A^N}} \\
\Phi^N(\Gamma^o(1_+,-)\wedge A) & \xrightarrow[\cong]{\text{dual Yoneda}} & \Gamma^o(1_+,-)\wedge A^N
\end{array}$$

commutes.

Note that since $[f^*V]^N = V$, we have a canonical *isotropy separation* map

$$X^N \to \Phi^N X.$$

2.2.3 Model structures

We record a minimum of the model theoretic properties that we need. Readers unfamiliar with this technology can for a large part ignore this and the next section at the price of accepting as black boxes the special equivalences and the few references to (co)fibrant replacements occurring later (in particular to Lemma 2.2.9).

The *projective* model structure on $\Gamma_G\mathcal{S}_G$ is the one where a map $X \to Y$ is a fibration (resp. weak equivalence) if for every subgroup $H \subseteq G$ and $V \in \Gamma_G^o$, the induced map $X(V)^H \to Y(V)^H$ is a (Kan) fibration (resp. weak equivalence) in $\mathcal{S}_*$.

As sets of generating cofibrations and generating acyclic cofibrations for the projective structure on $\Gamma_G\mathcal{S}_G$ we may choose

$$I_G = \{\Gamma_G^o(A,-)\wedge(G/H \times \partial\Delta[n])_+ \to \Gamma_G^o(A,-)\wedge(G/H \times \Delta[n])_+\}_{A,H,0\leq n}$$

$$J_G = \{\Gamma_G^o(A,-)\wedge\left(G/H \times \Lambda_k^n\right)_+ \to \Gamma_G^o(A,-)\wedge(G/H \times \Delta[n])_+\}_{A,H,0<n,0\leq k\leq n},$$

where A varies over Γ_G^o and H over the subgroups of G and $\Lambda_k^n \subseteq \Delta[n]$ is the k-th horn in the n-simplex. The source and targets of the maps in I_G and J_G are finitely presented, and so the projective structure is finitely generated. The internal morphism object is

$$\underline{\Gamma_G\mathcal{S}_G}(C,Z) = \{V \mapsto \Gamma_G\mathcal{S}_G(C, Z(V\wedge -))\} \in \Gamma_G\mathcal{S}_G.$$

Cell induction implies that smashing with a cofibrant object preserves projective equivalences.

Definition 2.2.4 The *special* model structure on $\Gamma_G\mathcal{S}_G$ is the one obtained from the projective model structure by (left Bousfield) localizing with respect to $\mathcal{L}_G$. The weak equivalences and fibrations in the special structure are referred to as special equivalences and special fibrations, whereas—since the cofibrations are the same in the projective structure and its localizations—we refer to the cofibrations simply as *cofibrations* without any qualifications.

Note 2.2.5 Even if we started with the Joyal structure, the cofibrations would remain the same.

Explicitly, a map $A \to B \in \Gamma_G\mathcal{S}_G$ with A, B cofibrant is a special equivalence if and only if for all specially fibrant (special and projectively fibrant) $Z \in \Gamma_G\mathcal{S}_G$, the induced map

$$\Gamma_G\mathcal{S}_G(B, Z) \to \Gamma_G\mathcal{S}_G(A, Z)$$

is a weak equivalence on all fixed points. In general, a map is a special equivalence if its (projective) cofibrant replacement is.

Lemma 2.2.6 *Smashing with a cofibrant object preserves special equivalences.*

Proof Let $C \in \Gamma_G\mathcal{S}_G$ be cofibrant. Since smashing with cofibrant objects preserves *projective* equivalences, we may consider the case of a special equivalence $A \to B$ with A and B cofibrant. If Z is specially fibrant and $V \in \Gamma^o_G$, then the map $Z(V\wedge-) \to \mathcal{S}_*(V, Z) \in \Gamma_G\mathcal{S}_G$ is a projective equivalence and so the internal morphism object $\underline{\Gamma_G\mathcal{S}_G}(C, Z)$ is specially fibrant. By the adjointness of smash and internal morphism object, this implies that $A\wedge C \to B\wedge C$ is a special equivalence. □

2.2.7 Special fibrant replacements and geometric fixed points

We need some control over special fibrant replacements in $\Gamma_G\mathcal{S}_G$, so for the moment we allow ourselves to be a bit technical. For $s_A \colon \Gamma^o(1_+, -)\wedge A \to \Gamma^o(A, -) \in \mathcal{L}_G$, let

$$\tilde{s}_A \colon \Gamma^o(1_+, -)\wedge A \rightarrowtail M_A$$

be the result of applying the simplicial mapping cylinder construction to s_A, so that $\tilde{s}_A$ is a cofibration while localizing with respect to $\tilde{\mathcal{L}}_G = \{\tilde{s}_A\}$ still gives the special structure on $\Gamma_G\mathcal{S}_G$. Finally, we let $\Lambda(\mathcal{L}_G) = \{\tilde{s}_A\Box i \mid \tilde{s}_A \in \tilde{\mathcal{L}}_G, i \in I_G\}$. Here $\Box$ is the "pushout product": if $f : X \to X'$ and $g : Y \to Y'$, then $f\Box g$ is the universal map form the pushout to the final vertex $X' \wedge Y'$ in

$$\begin{array}{ccc} X\wedge Y & \xrightarrow{f\wedge \mathrm{id}} & X'\wedge Y \\ \downarrow{\scriptstyle \mathrm{id}\wedge g} & & \downarrow{\scriptstyle \mathrm{id}\wedge g} \\ X\wedge Y' & \xrightarrow{f\wedge \mathrm{id}} & X'\wedge Y'. \end{array}$$

The following results show that we have good control over the specially fibrant objects.

Lemma 2.2.8 *Consider a map $f\colon X \to Y \in \Gamma_G\mathcal{S}_G$ with Y specially fibrant. Then f is a special fibration if and only if f has the right lifting property with respect to $\Lambda(\mathcal{L}_G) \cup J_G$. In particular, X is specially fibrant if $X \to *$ has the right lifting property with respect to $\Lambda(\mathcal{L}_G) \cup J_G$.*

By the small object argument, we construct a specially fibrant replacement functor $X \to X^{fG}$ as a relative $(\Lambda(\mathcal{L}_G) \cup J_G)$-cell. A cell induction using Lemma 2.2.2 then gives that

Lemma 2.2.9 *If $f\colon G \twoheadrightarrow J$ is a surjection of groups with kernel N and $X \in \Gamma_G\mathcal{S}_G$ then the geometric N-fixed points applied to the specially fibrant replacement, $\Phi^N(X) \to \Phi^N(X^{fG})$, is a special equivalence in $\Gamma_J\mathcal{S}_J$.*

Note 2.2.10 There is a slight variant that is occasionally useful. Note that the source and targets in $\Lambda(\mathcal{L}_G) \cup J_G$ are cofibrant, so smashing one of these with a projective equivalence $X \xrightarrow{\sim} Y \in \Gamma_G\mathcal{S}_G$ gives a projective equivalence. Since cofibrant replacements are projective equivalences we get that all the maps in $(\Lambda(\mathcal{L}_G) \cup J_G) \wedge \Gamma_G\mathcal{S}_G$ are special equivalences. Applying this to the construction in [25, 3.3.2] we get a fibrant replacement $\Gamma_G\mathcal{S}_G$-functor

$$\mathrm{id} \to R_G.$$

In particular, we get an induced map of internal morphism objects

$$R_G\colon \underline{\Gamma_G\mathcal{S}_G}(X, Y) \to \underline{\Gamma_G\mathcal{S}_G}(R_G X, R_G Y) \in \Gamma_G\mathcal{S}_G,$$

$R_G X$ is specially fibrant and $X \to R_G X$ is a special equivalence (it may not be a cofibration).

2.3 Fixed points of smash powers

The (co)domains of the generating cofibrations behave nicely with respect to the smash product:

Lemma 2.3.1 *If $A, A' \in \Gamma^o_G$ and $K, K' \in \mathcal{S}_G$, then the smash*

$$\Gamma^o_G(A, B) \wedge \Gamma^o_G(A', B') \to \Gamma^o_G(A \wedge A', B \wedge B')$$

(for $B, B' \in \Gamma^o_G$) induces an isomorphism

$$\left(\Gamma^o_G(A, -) \wedge K\right) \wedge \left(\Gamma^o_G(A', -) \wedge K'\right) \cong \Gamma^o_G\left(A \wedge A', -\right) \wedge K \wedge K'.$$

2.3.2 Smash powers

If $X \in \Gamma_G \mathcal{S}_G$ we can form smash indexed over arbitrary finite sets S:

$$\bigwedge_S X = X \wedge \ldots \wedge X$$

(either by choosing orderings on every S and coherently sticking to these choices or defining the S-fold smash in a symmetric fashion from scratch as we did for $S = \{1, 2\}$). This will at the outset only be functorial with respect to bijections of sets, but if X has more structure (if for instance X is an $\mathbb{S}$-algebra) then we obtain more functoriality as in [12]. However, the functoriality in mere bijections means that all the symmetries of S are present in the smash: $\bigwedge_S X$ is an $\mathrm{Aut}(S)$-object in $\Gamma_G \mathcal{S}_G$, or equivalently, an element in $\Gamma^o(\mathrm{Aut}(S) \times G)\mathcal{S}_* \simeq \Gamma_{\mathrm{Aut}(S)\times G}\mathcal{S}_{\mathrm{Aut}(S)\times G}$.

For some applications one may want to consider cases where S varies over sets with some prescribed group interacting with G. However, for our current purposes, it even suffices to focus on the symmetries of S, not on the symmetries of the incoming Γ-spaces. So, for simplicity we will start with $X \in \Gamma\mathcal{S}_*$ and since then "G" is freed from its duties and is such a good letter for a group, we let G be a group acting on S and view the S-fold smash as a functor

$$\bigwedge_S : \Gamma\mathcal{S}_* \to \Gamma_G \mathcal{S}_G.$$

Note that

$$\bigwedge_S \left(\Gamma^o(A, -) \wedge K\right) \cong \Gamma^o\left(A^{\wedge S}, -\right) \wedge K^{\wedge S} = \Gamma^o_G\left(A^{\wedge S}, -\right) \wedge K^{\wedge S}$$

is cofibrant (where $A^{\wedge S}$ is considered as an object in Γ^o_G and $K^{\wedge S}$ an object in $\mathcal{S}_G$) and a cell induction yields

Lemma 2.3.3 *If S is a finite G-set, then the S-fold smash $\bigwedge_S : \Gamma\mathcal{S}_* \to \Gamma_G\mathcal{S}_G$ preserves cofibrations.*

2.3.4 The Geometric diagonal

The geometric fixed points treat smash powers of cofibrant objects like fixed points of sets treat Cartesian power. The beginning of the induction needed to show this is

Lemma 2.3.5 *If $f : G \twoheadrightarrow J$ is a surjection of groups with kernel N, then the dual Yoneda lemma gives isomorphisms*

$$\begin{aligned}\Phi^N\left(\bigwedge_S(\Gamma^o(A,-)\wedge K)\right) &\cong\Gamma^o\left([A^{\wedge S}]^N,-\right)\wedge[K^{\wedge S}]^N\\ &\cong\Gamma^o_J\left(A^{\wedge S/N},-\right)\wedge K^{\wedge S/N}\\ &\cong\bigwedge_{S/N}\left(\Gamma^o(A,-)\wedge K\right).\end{aligned}$$

Inspired by the observation 2.3.5 we define, following the pattern laid out in e.g., [12, 37, 39], a chain of natural (in $X\in\Gamma\mathcal{S}_*$) transformations connecting $\bigwedge_{S/N}X$ and $\Phi^N\bigwedge_S X$, which in the case when X is cofibrant(!) gives an *isomorphism*

$$\Phi^N\bigwedge_S X\cong\bigwedge_{S/N}X$$

called the *geometric diagonal*. The tricky part is the functoriality in S. For Example 0.0.1 there is no requirement, and for topological Hochschild homology, as discussed in Section 3.3, when X is an $\mathbb{S}$-algebra and G is a cyclic group we only need functoriality with respect to the structure maps in the (subdivisions of the simplicial) circle. For commutative $\mathbb{S}$-algebras this is much more demanding since we have to be more careful with our cofibration hypotheses and typically we want functoriality with respect to a wide range of functions of finite sets.

3 TC in a special world

It is relatively straightforward to express (topological) Hochschild homology in Γ-spaces: you simply do exactly as Goodwillie and Walhausens envisioned: in the standard complex replace the tensor with the smash (tensor over $\mathbb{S}$). Just as in the algebraic case there are flatness concerns, but that is all you need to worry about (and taken care of by the unproblematic demand that the input being cofibrant).

However, if you want to make further refinements like cyclic homology you need to take a right derived version (aka a fibrant replacement). Magically, Bökstedt's topological Hochschild homology is an explicit version of such a right derived version: *its very construction has built in deloopings with respect to all finite subgroups of the circle*. This extremely fortunate state of affairs is crucially used in [10] for the definition of topological cyclic homology; most importantly the restriction map is simply obtained by restricting an equivariant map to the fixed points.

Since we want to avoid group completion we do *not* want to deloop, but we *do* want to retain homotopical control. Luckily, the categorical approach works wonderfully, as we now will sketch.

3.1 Cyclic objects

Connes' cyclic category Λ and its variants Λ_a (for $a = 1, 2, \ldots, \infty$ with $\Lambda = \Lambda_1$) can be obtained as follows. Fixing a there is an object $[n]_a \in \Lambda_a$ for each $n = 0, 1, \ldots$. For fixed m and n the set of morphisms $\Lambda_\infty([m]_\infty, [n]_\infty)$ is the set of order preserving functions $f\colon \frac{1}{m+1}\mathbb{Z} \to \frac{1}{n+1}\mathbb{Z}$ with $f\left(\frac{i}{m+1} + 1\right) = f\left(\frac{i}{n+1}\right) + 1$ for all i. Fixing $a < \infty$, we let $\Lambda_a([m]_a, [n]_a)$ be the quotient of $\Lambda_\infty([m]_\infty, [n]_\infty)$ by the equivalence relation generated by $f \sim f + a$. Composition in Λ_a is composition of functions.

The cyclic group C_a of order a acts on Λ_a by the identity on objects and by $f \mapsto f - 1$ on $\Lambda_a([m]_a, [n]_a)$. This means that functors from Λ_a comes with a natural C_a-action. The group of automorphisms $\mathrm{Aut}_{\Lambda_a}([n]_a)$ is cyclic of order $a(n+1)$ generated by the class $t_{a,n}$ of the function $\frac{1}{n+1}\mathbb{Z} \to \frac{1}{n+1}\mathbb{Z}$ given by $i \mapsto i + \frac{1}{n+1}$. The faithful inclusion $j_a\colon \Delta \to \Lambda_a$ is given by $j[n] = [n]_a$ and by sending $\phi \in \Delta([m], [n])$ to the class of the function $\frac{1}{m+1}\mathbb{Z} \to \frac{1}{n+1}\mathbb{Z}$ with $\frac{i}{m+1} \mapsto \frac{\phi(i)}{n+1}$ for $0 \le i \le m$.

A functor X from $\Lambda^o = \Lambda_1^o$ to some category is called a *cyclic object* in that category and the composite $j^*X = Xj = Xj_1$ is referred to as the underlying simplicial object.

A particularly important example is the cyclic set $S^1 = \Lambda[0]$ modeling the circle. An element in $S^1_n = \Lambda([n], [0])$ can be composed uniquely into an automorphism of $[n]$ followed by the unique map $[n] \to [0]$ coming from Δ. Hence, S^1_n is identified with the cyclic group $\mathrm{Aut}_\Lambda([n])$ of order $n+1$. Restricting to Δ^{op} we have the usual simplicial circle: $j^*S^1 = \Delta[1]/\partial\Delta[1]$.

3.2 Edgewise subdivision

Essentially because $|S^1|$ is homeomorphic to the circle, the geometric realization of cyclic object comes equipped with an action by the circle group $\mathbb{T} = |S^1|$. Bökstedt, Hsiang, and Madsen [10] introduced the edgewise subdivision as a way of making the action of the finite cyclic subgroups of $\mathbb{T}$ combinatorial. Let $sd^r\colon \Delta \to \Delta$ be the r-fold concatenation $S \mapsto S \sqcup \cdots \sqcup S$. Note that $sd^r[k-1] = [kr-1]$ and that $sd^r sd^s = sd^{rs}$. This extends to the cyclic situation

$$\begin{array}{ccc} \Delta & \xrightarrow{sd^r} & \Delta \\ \Big\downarrow{\scriptstyle j_{ar}} & & \Big\downarrow{\scriptstyle j_a} \\ \Lambda_{ar} & \xrightarrow{sd^r} & \Lambda_a \end{array}$$

by setting $sd^r(t_{ar}) = t_a$. Precomposing any cyclic object X with sd^r gives $sd_r X = X \circ sd^r$, the *r-fold edgewise subdivision of* X, giving us a functor from cyclic objects to Λ_r-objects. We note that $(sd_r X)_{k-1} = X_{kr-1}$ and that

$$(sd_{mn}S^1)/C_n \cong sd_m S^1.$$

From [10] we know that there is a natural C_r-equivariant homeomorphism $D\colon |sd_{C_r}X| \cong |X|$, where the C_r-action on $|sd_{C_r}X|$ comes from the C_r-action on $sd_{C_r}X$, and the action on $|X|$ comes from the cyclic structure on X. The resulting homeomorphism $|sd_{C_r}X^{C_r}| \cong |X|^{C_r}$ is $\mathbb{T}$-equivariant if we let $\mathbb{T}$ act on $|sd_{C_r}X^{C_r}|$ via the cyclic structure, and on $|X|^{C_r}$ through the isomorphism $\mathbb{T} \cong \mathbb{T}/C_r$.

3.3 (Topological) Hochschild homology

Topological cyclic homology makes sense in the special world. This is not obvious since the classical construction relies on various objects being equivalent, and when the meaning of "equivalent" is changed not all constructions translate. There is much to be said, for instance in regard to compatibility, but we present only what is needed for setting up the framework.

(Topological) Hochschild homology for $\mathbb{S}$-algebras is defined exactly as ordinary Hochschild homology, with $(\mathcal{A}b, \otimes, \mathbb{Z})$ replaced by $(\Gamma\mathcal{S}_*, \wedge, \mathbb{S})$: if A is an $\mathbb{S}$-algebra then $\mathrm{HH}^{\mathbb{S}}(A)$ is the cyclic Γ-space

$$[q] \mapsto A^{\wedge(q+1)} = A \wedge \ldots \wedge A,$$

with face maps induced by multiplication, degeneracy maps by insertion of identities, and the cyclic operator acting by cyclic permutation. As in the algebraic case where the analogous definition is problematic unless the ring is flat, we really only ever use this definition for sufficiently flat A—being cofibrant is more than enough. We have chosen to use the notation $\mathrm{HH}^{\mathbb{S}}$ rather than THH to emphasize that we are using the categorical smash powers.

Recall the discussion of the smash powers and geometric fixed points from Section 2.3. When A is a cofibrant, then the geometric diagonal $\Phi^{C_p} \bigwedge_{S \sqcup p} A \cong \bigwedge_S A$ is an isomorphism 2.3.4 which is natural in A and natural enough in S to give an isomorphism on the level of Hochschild homology:

Lemma 3.3.1 *If A is a cofibrant $\mathbb{S}$-algebra, then the geometric diagonal yields an isomorphism*

$$\Delta\colon \Phi^{C_p} sd_{p^{n+1}} \mathrm{HH}^{\mathbb{S}}(A) \cong sd_{p^n} \mathrm{HH}^{\mathbb{S}}(A).$$

Hence the considerations of Section 2.2.1 give an isotropy separation or "restriction" map

$$|\mathrm{HH}^{\mathbb{S}}(A)|^{C_p} \cong |sd_p \mathrm{HH}^{\mathbb{S}}(A)^{C_p}| \to |\Phi^{C_p} sd_p \mathrm{HH}^{\mathbb{S}}(A)| \cong |\mathrm{HH}^{\mathbb{S}}(A)|$$

and an inclusion of fixed points "Frobenius" $|\mathrm{HH}^{\mathbb{S}}(A)|^{C_p} \subseteq |\mathrm{HH}^{\mathbb{S}}(A)|$ and we want to build the theory from here.

Note 3.3.2 There are technicalities regarding fibrant replacements that we for the sake of exposition have glossed over, but which can be handled as follows. If X is a C_{p^n}-Γ-space (simplicial or topological) we let $X \to X^{f_n}$ be the specially C_{p^n}-fibrant replacement. Note that if $i\colon C_{p^k} \subseteq C_{p^n}$, then $i^*X \to i^*(X^{f_n})$ is a special C_{p^k}-fibrant replacement, and so naturally equivalent (but not equal) to $i^*X \to i^*(X)^{f_k}$. In all honesty, the "restriction map" is the chain

$$|\mathrm{HH}^{\mathbb{S}}A|^{f_n C_{p^n}} \dashrightarrow |\mathrm{HH}^{\mathbb{S}}A|^{f_{n-1} C_{p^{n-1}}}$$

given by composing the map

$$|\mathrm{HH}^{\mathbb{S}}A|^{f_n C_{p^n}} = [|\mathrm{HH}^{\mathbb{S}}A|^{f_n C_p}]^{C_{p^n}/C_p} \xrightarrow{\sim} |\mathrm{HH}^{\mathbb{S}}A|^{f_n C_p\, f_{n-1} C_{p^{n-1}}}$$

induced by fibrant replacement with $(-)^{f_{n-1} C_{p^{n-1}}}$ of

$$\begin{array}{ccccc}
|\mathrm{HH}^{\mathbb{S}}A|^{f_n C_p} & \xrightarrow[\cong]{D_n} & |sd_{p^n}\mathrm{HH}^{\mathbb{S}}(A)|^{f_n C_p} & \xrightarrow[\sim]{} & |sd_{p^n}\mathrm{HH}^{\mathbb{S}}(A)^{f_n}|^{f_n C_p} \\
 & & & & \uparrow{\scriptstyle \sim} \\
|\Phi^{C_p}(sd_{p^n}\mathrm{HH}^{\mathbb{S}}(A)^{f_n})| & \xleftarrow[\text{separation}]{\text{isotropy}} & |sd_{p^n}\mathrm{HH}^{\mathbb{S}}(A)^{f_n C_p}| & \xrightarrow{\cong} & |sd_{p^n}\mathrm{HH}^{\mathbb{S}}(A)^{f_n}|^{C_p} \\
{\scriptstyle \text{Lemma } 2.2.9}\uparrow{\scriptstyle \sim} & & & & \\
|\Phi^{C_p}(sd_{p^n}\mathrm{HH}^{\mathbb{S}}(A))| & \xrightarrow[\cong]{\Delta} & |sd_{p^{n-1}}\mathrm{HH}^{\mathbb{S}}A| & \xleftarrow[\cong]{D_{n-1}} & |\mathrm{HH}^{\mathbb{S}}A|
\end{array}$$

(the unmarked equivalences are annoying but innocent jugglings with fibrant replacements written out in its most primitive form), whereas the Frobenius is the inclusion of fixed points (and change of fibrant replacement)

$$|\mathrm{HH}^{\mathbb{S}}A|^{f_n C_{p^n}} \subseteq |\mathrm{HH}^{\mathbb{S}}A|^{f_n C_{p^{n-1}}} \xrightarrow{\sim} |\mathrm{HH}^{\mathbb{S}}A|^{f_n f_{n-1} C_{p^{n-1}}} \xleftarrow{\sim} |\mathrm{HH}^{\mathbb{S}}A|^{f_{n-1} C_{p^{n-1}}}.$$

3.4 Topological cyclic homology

We define $\mathrm{TC}(A;p)$ as the homotopy limit over the R and F-maps (which makes sense since the arrows pointing in the "wrong" directions are equivalences and we can choose an explicit model taking this into account).

Note that for any $\mathbb{T}$-space X, there is a chain $X^{f_n C_{p^n}} \to X^{f_n\, hC_{p^n}} \xleftarrow{\sim} X^{hC_{p^n}}$ compatible with the inclusion of fixed points (we use $E\mathbb{T}$ for all the EC_{p^n}s occurring in the homotopy fixed points), so that we get a map

$$\operatorname{holim}_{\overleftarrow{F}} |\mathrm{HH}^{\mathbb{S}} A|^{f_n C_{p^n}} \to \operatorname{holim}_{\leftarrow} |\mathrm{HH}^{\mathbb{S}} A|^{hC_{p^n}},$$

where after p-completion the latter object is naturally equivalent to the homotopy $\mathbb{T}$-fixed points $|\mathrm{HH}^{\mathbb{S}} A|^{h\mathbb{T}}$. Ultimately, this leads us to the same definition for $\mathrm{TC}(A)$ as in the stable case.

Definition 3.4.1 Let A be a cofibrant $\mathbb{S}$-algebra. Then the topological cyclic homology $\mathrm{TC}(A)$ is the homotopy pullback of

$$\prod_p \mathrm{TC}(A; p)_p \to \prod_p |\mathrm{HH}^{\mathbb{S}}(A)|_p^{h\mathbb{T}} \leftarrow |\mathrm{HH}^{\mathbb{S}}(A)|^{h\mathbb{T}}.$$

Note 3.4.2 I do not know whether the setup of Nikolaus and Scholze of Section 1.6 translates well to the special situation since the nature of the Tate-construction is somewhat mysterious in this case.

Note 3.4.3 The extension from $\mathbb{S}$-algebras to categories enriched in Γ-spaces is straightforward and left to the reader.

4 On modules and monoids

We end by discussing some algebraic properties. In particular we show that if **B** is the Boolean $\mathbb{N}$-algebra, then $H\mathbf{B}$-modules are specially homotopy discrete. For this purpose we first give a more concrete characterization of special equivalences. Recall that a map $A \to B$ of cofibrant Γ-spaces is a special equivalence if for all special and projectively fibrant $Z \in \Gamma\mathcal{S}_*$ the induced map $\Gamma\mathcal{S}_*(B, Z) \to \Gamma\mathcal{S}_*(A, Z) \in \mathcal{S}_*$ is a weak equivalence. In general, a map $A \to B$ is a special equivalence if its cofibrant replacement is a special equivalence.

Using the fibrant replacement $\Gamma\mathcal{S}_*$-functor of 2.2.10 (we use the enriched fibrant replacement in order to apply it to modules) we can simplify this to the statement that $A \to B$ is a special equivalence if and only if $RA \to RB$ is a projective equivalence, which in view of the fact that RA and RB are special is the same as saying that $RA(1_+) \to RB(1_+)$ is a weak equivalence of simplicial sets.

4.1 Linearization

The Eilenberg–Mac Lane construction has a left adjoint $L\colon \Gamma\mathcal{S}_* \to s\mathcal{M}on$ with LX given as the coequalizer in $s\mathcal{M}on$ of the two maps

$$\delta_*^1 + \delta_*^2, \nabla_*\colon \mathbb{N}[X(2_+)] \rightrightarrows \mathbb{N}[X(1_+)],$$

where $\mathbb{N}[-]$ is free functor adjoint to the forgetful functor from symmetric monoids to pointed sets. Adapting the argument in [57, Lemma 1.2] we get

Lemma 4.1.1 *The unit of adjunction $A \to LHA$ is an isomorphism. The adjunction is enriched in the sense that it extends to a natural isomorphism*

$$\underline{\Gamma\mathcal{S}_*}(X, HA) \cong Hs\mathcal{M}on(LX, A) \in \Gamma\mathcal{S}_*$$

and L is strong symmetric monoidal: the maps induced by the enriched adjunction are isomorphisms $\mathbb{N} \cong L(\mathbb{S})$, $LX \otimes_{\mathbb{N}} LY \cong L(X \wedge Y)$. Furthermore, L preserves finite products.

Lemma 4.1.2 *The Eilenberg–Mac Lane functor $H\colon s\mathcal{M}on \to \Gamma\mathcal{S}_*$ is a right Quillen map, both with respect to the projective and the special structures on $\Gamma\mathcal{S}_*$.*

Proof It is enough to show that H preserves acyclic fibrations and fibrations between fibrant objects (see [21]). Since finite products preserve fibrations and equivalences, H sends fibrations/weak equivalences to projective fibrations/equivalences. Since acyclic fibrations in the projective and special structures coincide, it is enough to show that if $M \twoheadrightarrow N \in s\mathcal{M}on$ is a fibration between fibrant objects then $HM \to HN$ is a special fibration. This follows since HM, HN are specially fibrant and $HM \to HN$ projectively fibrant. □

Note that, contrary to what is the case in other formalisms the Eilenberg–Mac Lane functor very rarely takes cofibrant values.

Lemma 4.1.3 *If M is a nontrivial simplicial symmetric monoid, then $HM \in \Gamma\mathcal{S}_*$ is not cofibrant.*

Proof If $HM \in \Gamma\mathcal{S}_*$ is cofibrant, then $LHM \cong M$ is cofibrant in $s\mathcal{M}on$ with the projective structure ((L, H) is a Quillen pair), which is equivalent to M being a retraction of a free (in the sense of Quillen) simplicial symmetric monoid. Hence, it is enough to consider the case where M is free simplicial symmetric monoid. In that case, if n is the smallest dimension in which M_n is nontrivial (here we use that $M \neq 0$), then M_n is actually a nontrivial free symmetric monoid, and so contains $\mathbb{N}$ as a retract. By Schwede [57, A3], if HM were cofibrant then HM_n—and hence $H\mathbb{N}$—would be a wedge of representables.

However, $H\mathbb{N}$ has no proper retracts: if $X \subseteq H\mathbb{N} \to X$ is a retract, then $LX \subseteq \mathbb{N} \to LX$ is a retract (of symmetric monoids), implying that either $LX = 0$ or $LX = \mathbb{N}$. In the first case, the inclusion $X \subseteq H\mathbb{N}$ factors over $HLX = 0$, so that $X = 0$, and in the second case the composite $H\mathbb{N} \to X \to HLX$ is an isomorphism implying that the surjection $H\mathbb{N} \to X$ is an injection too.

Combining this information, we get that if $H\mathbb{N}$ were cofibrant, $H\mathbb{N}$ would be representable, which is nonsense given that representables are finite. □

4.1.4 The special path monoid

The special analog of the set of path components is the following.

Definition 4.1.5 If X is a Γ-space, then the *special path monoid* of X is the symmetric monoid $\pi_0^{\text{special}} X = \pi_0 RX(1_+)$.

It can alternatively be seen as the monoid of all maps $\mathbb{S} \to X$ in the special homotopy category, but for our purposes the characterization in terms of linearization is more useful.

Lemma 4.1.6 *A special equivalence $X \to Y$ (of not necessarily cofibrant Γ-spaces) induces an isomorphism $\pi_0 LX \cong \pi_0 LY$. If Z is special, then the map $\pi_0 Z(1_+) \to \pi_0 LZ$ induced by the unit $Z \to HLZ$ is an isomorphism. Hence, for any $X \in \Gamma\mathcal{S}_*$ we have a chain of natural isomorphisms*

$$\begin{aligned}\pi_0^{\text{special}} X = \pi_0 RX(1_+) &\xrightarrow{\cong} \pi_0 LRX \xleftarrow{\cong} \pi_0 LX \\ &\cong \operatorname{coeq}\{\mathbb{N}\pi_0 X(2_+) \rightrightarrows \mathbb{N}\pi_0 X(1_+)\}.\end{aligned}$$

Proof Since L is a left Quillen functor, it sends special equivalences between cofibrant objects to weak equivalences, so we need to show that $\pi_0 L$ sends projective equivalences to isomorphisms. This is true since π_0 commutes with colimits.

Likewise, if Z is special, then $\pi_0 Z(1_+)$ inherits a monoid structure and $\pi_0 LZ$ is a coequalizer of a diagram

$$\mathbb{N}\pi_0 Z(1_+) \otimes \mathbb{N}\pi_0 Z(1_+) \cong \mathbb{N}\pi_0 Z(2_+) \rightrightarrows \mathbb{N}\pi_0 Z(1_+)$$

exactly recovering the generators $\pi_0 Z(1_+) \subseteq \mathbb{N}\pi_0 Z(1_+)$ □

Corollary 4.1.7 *The isomorphism of Lemma 4.1.6 and the monoidality of L of Lemma 4.1.1 induce isomorphisms*

$$\begin{aligned}\pi_0^{\text{special}}(X \wedge Y) &\cong \pi_0 L(X \wedge Y) \cong \pi_0(LX \otimes_{\mathbb{N}} LY) \cong \pi_0 LX \otimes_{\mathbb{N}} \pi_0 LY \\ &\cong \pi_0^{\text{special}} X \otimes_{\mathbb{N}} \pi_0^{\text{special}} Y.\end{aligned}$$

In particular, if $A, B \in s\mathcal{M}on$, then $\pi_0^{\text{special}}(HA \wedge^L HB) \cong \pi_0^{\text{special}}(HA \wedge HB) \cong \pi_0 A \otimes_{\mathbb{N}} \pi_0 B$.

The superscript L signifies the derived smash, where the factors are functorially replaced by cofibrant objects. For instance we could use the standard simplicial replacement (symmetric version, Hochschild-style structure maps) for $X \wedge^L Y$ with Γ-space of q-simplices

$$\bigvee_{a_0,\dots,a_q,b_0,\dots,b_q\in\Gamma^o} X(a_q)\wedge Y(b_q)\wedge\Gamma^o(a_0\wedge b_0,-)\wedge \bigwedge_{j=1^q}\Gamma^o(a_i,a_{i-1})\wedge\Gamma^o(b_i,b_{i-1}).$$

Generally, $X\wedge Y$ and $X\wedge^L Y$ are not specially equivalent but, as we see, their path monoids coincide. On the other hand, allowing a simplicial direction offers no real advantage (or disadvantage) when considering $\mathbf{B}$-modules (or for that matter modules over other rigs where $1+\cdots+1=1$).

Example 4.1.8 Consider the Boolean algebra $\mathbf{B}=\{0,1\}$ (0 is "false" and 1 is "true") with the operation $+$ being "or" (($\mathbf{B}$, +) is isomorphic to $\mathbb{Z}/2^\times$, the integers mod 2 under multiplication) and $\cdot$ being "and". This is the $\mathbb{N}$-algebra on $\{0,1\}$ defined by $1+1=1$. If we force all elements to have negatives, then we can cancel 1 on each side of the expression $1+1=1$ resulting in $1=0$: the group completion is the trivial group.

The advantage of the special world is that we do not group complete and so the theory is not trivialized, but the special theory of $H\mathbf{B}$-modules is of an essentially discrete nature unless we change the underlying weak equivalences on simplicial sets (e.g., by using quasi-categories instead of Kan complexes as our fibrant objects).

The fact that $H\mathbf{B}$-modules are specially homotopy discrete can be seen as follows. It is enough to consider specially fibrant $H\mathbf{B}$-modules M and show that the map

$$M\to HLM\to H\pi_0 LM\cong H\pi_0^{\text{special}}M$$

is a projective equivalence, or equivalently, that $M(1_+)\to\pi_0 M(1_+)$ is a weak equivalence. Consider the part of the functoriality of the multiplication map $H\mathbf{B}\wedge M\to M$ expressed in the diagram

$$\begin{array}{ccccc}
(\mathbf{B}\wedge M(1_+))\times(\mathbf{B}\wedge M(1_+)) & \xleftarrow[(\mathrm{pr}_1\wedge\mathrm{id},\mathrm{pr}_2\wedge\mathrm{id})]{} & (\mathbf{B}\times\mathbf{B})\wedge M(1_+) & \xrightarrow[+\wedge\mathrm{id}]{} & \mathbf{B}\wedge M(1_+)\\
\downarrow\cong & & \downarrow & & \downarrow\cong\\
M(1_+)\times M(1_+) & \xleftarrow[(\delta^1_*,\delta^2_*)]{} & M(2_+) & \xrightarrow[\nabla_*]{} & M(1_+).
\end{array}$$

Choosing a basepoint $x\in M(1_+)$ we see that for $i>0$ the lower row defines a group homomorphism $+\colon\pi_i(M(1_+),x)\times\pi_i(M(1_+),x)\to\pi_i(M(1_+),x)$. If $\alpha\in\pi_i(M(1_+),x)$ and $0\in\pi_i(M(1_+),x)$ is the constant loop, then tracing through the diagram we see that $\alpha=0+\alpha=\alpha+0$, so that Eckmann–Hilton forces $+$ to be the usual group operation in $\pi_i(M(1_+),x)$, but also $\alpha+\alpha=\alpha$, which means that $\pi_i(M(1_+),x)=0$.

This is somewhat disappointing. For instance, it means that multiplication $\mathbf{B}\wedge^L\mathbf{B}\to\mathbf{B}$ is a special equivalence. Since topological Hochschild homology $\mathrm{HH}^{\mathbb{S}}(\mathbf{B})$ is a $\mathbf{B}$-module it also means that $\mathrm{HH}^{\mathbb{S}}(\mathbf{B})\to\mathbf{B}$ is a special equivalence. This is a good motivation for not only to moving from the stable to the special world, but also from Kan simplicial sets to quasi-categories (where not all paths have homotopy inverses, and so the argument about homotopy discreteness fails).

If A is an $\mathbb{S}$-algebra, then the fixed points of the smash powers and $\mathrm{TC}(A)$ are generally not an A-module since it is built out of non-split extensions (for instance, $\pi_0^{\text{special}}(H\mathbb{F}_2 \wedge^L H\mathbb{F}_2)^{C_2} \cong \mathbb{Z}/4\mathbb{Z}$ is not an $\mathbb{F}_2$-algebra), but Example 4.1.8 still makes it clear that the presence of elements killed by group completion puts severe restrictions on the theory. In addition, even when such a theory is set up, it is by no means clear that is has any of the calculational power that the original setup of Bökstedt, Hsiang, and Madsen had.

Note 4.1.9 Example 4.1.8 showed that if $1+1 = 1$ in an $\mathbb{N}$-algebra, then its modules are specially discrete. It should however be noted that the $\infty+1 = \infty$ encountered in Example 0.0.2 for the category $\mathrm{Vect}_k^{\text{count}}$ of countable vector spaces is less dramatic. The associated Γ-space (wrt. $\oplus$) is specially fibrant with value at 1_+ (the nerve of) the groupoid $\mathrm{Vect}_k^{\text{count}}$. This groupoid has a lot of automorphisms and so is not homotopy discrete.

Acknowledgements Apart from obvious input from Alain Connes this tribute has benefitted from enlightening conversations with M. Brun, K. Consani, L. Hesselholt, M. Hill, and C. Schlichtkrull. Also, the preprint [8] by de Brito and Moerdijk and the papers of Santhanam [55] and Mandell [46] were inspirational. The author also wants to thank an anonymous referee for correcting an unfortunate misconception about events in the early 1980s.

References

1. Vigleik Angeltveit. On the K-theory of truncated polynomial rings in non-commuting variables. *Bull. Lond. Math. Soc.*, 47(5):731–742, 2015.
2. Vigleik Angeltveit, Teena Gerhardt, Michael A. Hill, and Ayelet Lindenstrauss. On the algebraic K-theory of truncated polynomial algebras in several variables. *J. K-Theory*, 13(1):57–81, 2014.
3. Christian Ausoni. On the algebraic K-theory of the complex K-theory spectrum. *Invent. Math.*, 180(3):611–668, 2010.
4. Christian Ausoni and John Rognes. Algebraic K-theory of topological K-theory. *Acta Math.*, 188(1):1–39, 2002.
5. S. Bloch. On the tangent space to Quillen K-theory. In *Algebraic K-theory, I: Higher K-theories (Proc. Conf., Battelle Memorial Inst., Seattle, Wash., 1972)*, pages 205–210. Lecture Notes in Math., Vol. 341. Springer, Berlin, 1973.
6. Andrew J. Blumberg, David Gepner, and Gonçalo Tabuada. A universal characterization of higher algebraic K-theory. *Geom. Topol.*, 17(2):733–838, 2013.
7. Andrew J. Blumberg, David Gepner, and Gonçalo Tabuada. Uniqueness of the multiplicative cyclotomic trace. *Adv. Math.*, 260:191–232, 2014.
8. P. Boavida de Brito and I. Moerdijk. Dendroidal spaces, Γ-spaces and the special Barratt-Priddy-Quillen theorem. *ArXiv e-prints*, January 2017.
9. M. Bökstedt, G. Carlsson, R. Cohen, T. Goodwillie, W. C. Hsiang, and I. Madsen. On the algebraic K-theory of simply connected spaces. *Duke Math. J.*, 84(3):541–563, 1996.
10. M. Bökstedt, W. C. Hsiang, and I. Madsen. The cyclotomic trace and algebraic K-theory of spaces. *Invent. Math.*, 111(3):465–539, 1993.

11. A. K. Bousfield and E. M. Friedlander. Homotopy theory of Γ-spaces, spectra, and bisimplicial sets. In *Geometric applications of homotopy theory (Proc. Conf., Evanston, Ill., 1977), II*, volume 658 of *Lecture Notes in Math.*, pages 80–130. Springer, Berlin, 1978.
12. M. Brun, B. I. Dundas, and M. Stolz. Equivariant Structure on Smash Powers. *ArXiv e-prints*, April 2016.
13. D. Clausen, A. Mathew, and M. Morrow. K-theory and topological cyclic homology of henselian pairs. *ArXiv e-prints*, March 2018.
14. Alain Connes. C^* algèbres et géométrie différentielle. *C. R. Acad. Sci. Paris Sér. A-B*, 290(13):A599–A604, 1980.
15. Alain Connes. Spectral sequence and homology of currents for operator algebras. *Math. Forschungsinstitut Oberwolfach Tagungsbericht*, 42/81, 1981.
16. Alain Connes. Cohomologie cyclique et foncteurs Ext^n. *C. R. Acad. Sci. Paris Sér. I Math.*, 296(23):953–958, 1983.
17. Alain Connes. Noncommutative differential geometry. *Inst. Hautes Études Sci. Publ. Math.*, (62):257–360, 1985.
18. Alain Connes and Caterina Consani. Absolute algebra and Segal's Γ-rings: au dessous de $\overline{\mathrm{Spec}(\mathbb{Z})}$. *J. Number Theory*, 162:518–551, 2016.
19. Alain Connes and Max Karoubi. Caractère multiplicatif d'un module de Fredholm. *C. R. Acad. Sci. Paris Sér. I Math.*, 299(19):963–968, 1984.
20. Guillermo Cortiñas. The obstruction to excision in K-theory and in cyclic homology. *Invent. Math.*, 164(1):143–173, 2006.
21. Daniel Dugger. Replacing model categories with simplicial ones. *Trans. Amer. Math. Soc.*, 353(12):5003–5027, 2001.
22. Bjørn Ian Dundas, Thomas G. Goodwillie, and Randy McCarthy. *The local structure of algebraic K-theory*, volume 18 of *Algebra and Applications*. Springer-Verlag London, Ltd., London, 2013.
23. Bjørn Ian Dundas and Harald Øyen Kittang. Excision for K-theory of connective ring spectra. *Homology Homotopy Appl.*, 10(1):29–39, 2008.
24. Bjørn Ian Dundas and Harald Øyen Kittang. Integral excision for K-theory. *Homology Homotopy Appl.*, 15(1):1–25, 2013.
25. Bjørn Ian Dundas, Oliver Röndigs, and Paul Arne Østvær. Enriched functors and stable homotopy theory. *Doc. Math.*, 8:409–488, 2003.
26. Bjørn Ian Dundas. Relative K-theory and topological cyclic homology. *Acta Math.*, 179(2):223–242, 1997.
27. Bjørn Ian Dundas and Randy McCarthy. Stable K-theory and topological Hochschild homology. *Ann. of Math. (2)*, 140(3):685–701, 1994.
28. Thomas Geisser and Lars Hesselholt. Bi-relative algebraic K-theory and topological cyclic homology. *Invent. Math.*, 166(2):359–395, 2006.
29. Thomas Geisser and Lars Hesselholt. On the K-theory and topological cyclic homology of smooth schemes over a discrete valuation ring. *Trans. Amer. Math. Soc.*, 358(1):131–145, 2006.
30. Thomas Geisser and Lars Hesselholt. On the vanishing of negative K-groups. *Math. Ann.*, 348(3):707–736, 2010.
31. Thomas G. Goodwillie. Relative algebraic K-theory and cyclic homology. *Ann. of Math. (2)*, 124(2):347–402, 1986.
32. Lars Hesselholt. On the K-theory of planar cuspical curves and a new family of polytopes. In *Algebraic topology: applications and new directions*, volume 620 of *Contemp. Math.*, pages 145–182. Amer. Math. Soc., Providence, RI, 2014.
33. Lars Hesselholt. The big de Rham-Witt complex. *Acta Math.*, 214(1):135–207, 2015.
34. Lars Hesselholt. Topological Hochschild homology and the Hasse-Weil zeta function. In *An alpine bouquet of algebraic topology*, volume 708 of *Contemp. Math.*, pages 157–180. Amer. Math. Soc., Providence, RI, 2018.
35. Lars Hesselholt and Ib Madsen. On the K-theory of finite algebras over Witt vectors of perfect fields. *Topology*, 36(1):29–101, 1997.

36. Lars Hesselholt and Ib Madsen. On the K-theory of local fields. *Ann. of Math. (2)*, 158(1):1–113, 2003.
37. Michael A. Hill, Michael J. Hopkins, and Douglas C. Ravenel. On the non-existence of elements of Kervaire invariant one. In *Proceedings of the International Congress of Mathematicians—Seoul 2014. Vol. II*, pages 1219–1243. Kyung Moon Sa, Seoul, 2014.
38. John D. S. Jones. Cyclic homology and equivariant homology. *Invent. Math.*, 87(2):403–423, 1987.
39. T. A. Kro. Involutions on $S[\Omega M]$. *ArXiv Mathematics e-prints*, October 2005.
40. M. Land and G. Tamme. On the K-theory of pullbacks. *ArXiv e-prints*, 2018.
41. Ayelet Lindenstrauss and Randy McCarthy. On the Taylor tower of relative K-theory. *Geom. Topol.*, 16(2):685–750, 2012.
42. Jean-Louis Loday and Daniel Quillen. Cyclic homology and the Lie algebra homology of matrices. *Comment. Math. Helv.*, 59(4):569–591, 1984.
43. Wolfgang Lück, Holger Reich, John Rognes, and Marco Varisco. Algebraic K-theory of group rings and the cyclotomic trace map. *Adv. Math.*, 304:930–1020, 2017.
44. Manos Lydakis. Smash products and Γ-spaces. *Math. Proc. Cambridge Philos. Soc.*, 126(2):311–328, 1999.
45. Ib Madsen. Algebraic K-theory and traces. In *Current developments in mathematics, 1995 (Cambridge, MA)*, pages 191–321. Internat. Press, Cambridge, MA, 1994.
46. Michael A. Mandell. An inverse K-theory functor. *Doc. Math.*, 15:765–791, 2010.
47. Randy McCarthy. Relative algebraic K-theory and topological cyclic homology. *Acta Math.*, 179(2):197–222, 1997.
48. T. Nikolaus and P. Scholze. On topological cyclic homology. *ArXiv e-prints*, July 2017.
49. Crichton Ogle. On the homotopy type of $A(\Sigma X)$. *J. Pure Appl. Algebra*, 217(11):2088–2107, 2013.
50. Daniel Quillen. Cohomology of groups. In *Actes du Congrès International des Mathématiciens (Nice, 1970), Tome 2*, pages 47–51. Gauthier-Villars, Paris, 1971.
51. J. Rognes and C. Weibel. Two-primary algebraic K-theory of rings of integers in number fields. *J. Amer. Math. Soc.*, 13(1):1–54, 2000. Appendix A by Manfred Kolster.
52. John Rognes. Algebraic K-theory of the two-adic integers. *J. Pure Appl. Algebra*, 134(3): 287–326, 1999.
53. John Rognes. Two-primary algebraic K-theory of pointed spaces. *Topology*, 41(5):873–926, 2002.
54. John Rognes. The smooth Whitehead spectrum of a point at odd regular primes. *Geom. Topol.*, 7:155–184 (electronic), 2003.
55. Rekha Santhanam. Units of equivariant ring spectra. *Algebr. Geom. Topol.*, 11(3):1361–1403, 2011.
56. Roland Schwänzl, Ross Staffeldt, and Friedhelm Waldhausen. Stable K-theory and topological Hochschild homology of A_∞ rings. In *Algebraic K-theory (Poznań, 1995)*, volume 199 of *Contemp. Math.*, pages 161–173. Amer. Math. Soc., Providence, RI, 1996.
57. Stefan Schwede. Stable homotopical algebra and Γ-spaces. *Math. Proc. Cambridge Philos. Soc.*, 126(2):329–356, 1999.
58. Graeme Segal. Categories and cohomology theories. *Topology*, 13:293–312, 1974.
59. Kazuhisa Shimakawa. A note on Γ_G-spaces. *Osaka J. Math.*, 28(2):223–228, 1991.
60. R. W. Thomason. Symmetric monoidal categories model all connective spectra. *Theory Appl. Categ.*, 1:No. 5, 78–118 (electronic), 1995.
61. B. L. Tsygan. Homology of matrix Lie algebras over rings and the Hochschild homology. *Uspekhi Mat. Nauk*, 38(2(230)):217–218, 1983.
62. Friedhelm Waldhausen. Algebraic K-theory of spaces, concordance, and stable homotopy theory. In *Algebraic topology and algebraic K-theory (Princeton, N.J., 1983)*, volume 113 of *Ann. of Math. Stud.*, pages 392–417. Princeton Univ. Press, Princeton, NJ, 1987.
63. Charles A. Weibel. Nil K-theory maps to cyclic homology. *Trans. Amer. Math. Soc.*, 303(2):541–558, 1987.

Curvature in noncommutative geometry

Farzad Fathizadeh and Masoud Khalkhali

Dedicated to Alain Connes with admiration, affection, and much appreciation

Abstract Our understanding of the notion of curvature in a noncommutative setting has progressed substantially in the past 10 years. This new episode in noncommutative geometry started when a Gauss-Bonnet theorem was proved by Connes and Tretkoff for a curved noncommutative two torus. Ideas from spectral geometry and heat kernel asymptotic expansions suggest a general way of defining local curvature invariants for noncommutative Riemannian type spaces where the metric structure is encoded by a Dirac type operator. To carry explicit computations however one needs quite intriguing new ideas. We give an account of the most recent developments on the notion of curvature in noncommutative geometry in this paper.

1 Introduction

Broadly speaking, the progress of *noncommutative geometry* in the last four decades can be divided into three phases: *topological, spectral,* and *arithmetical.* One can also notice the pervasive influence of quantum physics in all aspects of the subject. Needless to say, each of these facets of the subject is still evolving, and there are many deep connections among them.

In its topological phase, noncommutative geometry was largely informed by index theory and a real need to extend index theorems beyond their classical realm of smooth manifolds, to what we collectively call *noncommutative spaces*. Thus

F. Fathizadeh
Department of Mathematics, Computational Foundry, Swansea University, Swansea, UK
e-mail: farzad.fathizadeh@swansea.ac.uk

M. Khalkhali (✉)
Department of Mathematics, University of Western Ontario, London, ON, Canada
e-mail: masoud@uwo.ca

A. Chamseddine et al. (eds.), *Advances in Noncommutative Geometry*,
https://doi.org/10.1007/978-3-030-29597-4_6

K-theory, K-homology, and KK-theory in general were brought in and with the discovery of cyclic cohomology by Connes [10, 11], a suitable framework was created by him to formulate noncommutative index theorems. With the appearance of the groundbreaking and now classical paper of Connes [12], results of which were already announced in Oberwolfach in 1981 [10], this phase of the theory was essentially completed. In particular a noncommutative Chern-Weil theory of characteristic classes was created with Chern character maps for both K-theory and K-homology with values in cyclic (co)homology. To define all these a notion of Fredholm module (bounded or unbounded, finitely summable or theta summable) was introduced which essentially captures and brings in many aspects of smooth manifolds into the noncommutative world. These results were applied to noncommutative quotient spaces such as the space of leaves of a foliation, or the unitary dual of noncompact and nonabelian Lie groups. Ideas and tools from global analysis, differential topology, operator algebras, representation theory, and quantum statistical mechanics were crucial. One of the main applications of this resulting noncommutative index theory was to settle some long-standing conjectures such as the Novikov conjecture and the Baum-Connes conjecture for specific and large classes of groups.

Next came the study of the geometry of noncommutative spaces and the impact of *spectral geometry*. Geometry, as we understand it here, historically has dealt with the study of spaces of increasing complexity and metric measurements within such spaces. Thus in classical differential geometry one learns how to measure distances and volumes, as well as various types of curvature of Riemannian manifolds of arbitrary dimension. One can say the two notions of Riemannian metric and the Riemann curvature tensor are hallmarks of classical differential geometry in general. This should be contrasted with topology where one studies spaces only from a rather soft homotopy theoretic point of view. A similar division is at work in noncommutative geometry. Thus, as we mentioned briefly above, while in its earlier stage of development noncommutative geometry was mostly concerned with the development of topological invariants like cyclic cohomology, Connes-Chern character maps, and index theory, starting in about 10 years ago noncommutative geometry entered a new truly geometric phase where one tries to seriously understand what a *curved noncommutative space* is and how to define and compute curvature invariants for such a noncommutative space.

This episode in noncommutative geometry started when a Gauss-Bonnet theorem was proved by Connes and Cohen for a *curved noncommutative torus* in [22] (see also the MPI preprint [8] where many ideas are already laid out). This paper was immediately followed in [30] where the Gauss-Bonnet was proved for general conformal structures. The metric structure of a noncommutative space is encoded in a (twisted) spectral triple. Giving a state-of-the-art report on developments following these works, and on the notion of curvature in noncommutative geometry, is the purpose of our present review.

Classically, geometric invariants are usually defined explicitly and algebraically in a local coordinate system, in terms of a metric tensor or a connection on the given manifold. However, methods based on local coordinates, or algebraic methods

This is not a quantum curved torus

This is not a quantum curved torus

based on commutative algebra, have no chance of being useful in a noncommutative setting, in general. But other methods, more analytic and more subtle, based on ideas of spectral geometry are available. In fact, thanks to spectral geometry, we know that there are intricate relations between Riemannian invariants and spectra of naturally defined elliptic operators like Laplace or Dirac operators on the given manifold. A prototypical example is the celebrated *Weyl's law* on the asymptotic distribution of eigenvalues of the Laplacian of a closed Riemannian manifold M^n in terms of its volume:

$$N(\lambda) \sim \frac{\omega_n \mathrm{Vol}(M)}{(2\pi)^n} \lambda^{\frac{n}{2}} \qquad \lambda \to \infty. \tag{1}$$

Here $N(\lambda)$ is the number of eigenvalues of the Laplacian in the interval $[0, \lambda]$ and ω_n is the volume of the unit ball in $\mathbb{R}^n$. In the spirit of Marc Kac's article [39], one says one can hear the volume of a manifold. But one can ask what else about a Riemannian manifold can be heard? Or even we can ask: what can we learn by listening to a noncommutative manifold? Results so far indicate that one can effectively define and compute, not only the volume, but in fact the scalar and Ricci curvatures of noncommutative curved spaces, at least in many examples.

In his Gibbs lecture of 1948, *Ramifications, old and new, of the eigenvalue problem*, Hermann Weyl had this to say about possible extensions of his asymptotic law (1): *I feel that these informations about the proper oscillations of a membrane, valuable as they are, are still very incomplete. I have certain conjectures on what a complete analysis of their asymptotic behavior should aim at; but since for more than 35 years I have made no serious attempt to prove them, I think I had better keep them to myself.*

One of the most elaborate results in spectral geometry is Gilkey's theorem that gives the first four non-zero terms in the asymptotic expansion of the heat kernel of Laplace type operators in terms of covariant derivatives of the metric tensor and the Riemann curvature tensor [36]. More precisely, if P is a Laplace type operator, then the heat operator e^{-tP} is a smoothing operator with a smooth kernel $k(t, x, y)$, and there is an asymptotic expansion near $t = 0$ for the heat kernel restricted to the diagonal of $M \times M$:

$$k(t, x, x) \sim \frac{1}{(4\pi t)^{m/2}}(a_0(x, P) + a_2(x, P)t + a_4(x, P)t^2 + \cdots),$$

where $a_i(x, P)$ are known as the Gilkey-Seeley-DeWitt coefficients. The first term $a_0(x, P)$ is a constant. It was first calculated by Minakshisundaram and Pleijel [51] for $P = \Delta$ the Laplace operator. Using Karamata's Tauberian theorem, one immediately obtains Weyl's law for closed Riemannian manifolds. Note that Weyl's original proof was for bounded domains with a regular boundary in Euclidean space and does not extend to manifolds in general. The next term $a_2(x, P)$, for $P = \Delta$, was calculated by MacKean and Singer [50] and it was shown that it gives the scalar curvature:

$$a_2(x, \Delta) = \frac{1}{6}S(x).$$

This immediately shows that the scalar curvature has a spectral nature and in particular the total scalar curvature is a spectral invariant. This result, or rather its localized version to be recalled later, is at the heart of the noncommutative geometry approach to the definition of scalar curvature. The expressions for $a_{2k}(x, P)$ get rapidly complicated as k grows, although in principle they can be recursively computed in a normal coordinate chart. They are reproduced up to term a_6 in the next section.

It is this analytic point of view on geometric invariants that play an important role in understanding the geometry of curved noncommutative spaces. The algebraic approach almost completely breaks down in the noncommutative case. Our experience so far in the past few years has been that in the noncommutative case spectral and hard analytic methods based on pseudodifferential operators yield results that are in no way possible to guess or arrive at from their commutative counterparts by algebraic methods. One just needs to take a look at our formulas for scalar, and now Ricci curvature, in dimensions two, three, and four, in later sections to believe in this statement. The fact that in the first step we had to rely on heavy symbolic computer calculations to start the analysis shows the formidable nature of this material. Surely computations, both symbolic and analytic, are quite hard and are done on a case-by-case basis, but the surprising end results totally justify the effort.

The spectral geometry of a *curved noncommutative two torus* has been the subject of intensive studies in recent years. As we said earlier, this whole episode started when a Gauss-Bonnet theorem was proved by Connes and Tretkoff (formerly Cohen) in [22] (see also [8] for an earlier version), and for general conformal structures in [30]. A natural question then was to define and compute the scalar curvature of a curved noncommutative torus. This was done, independently, by Connes-Moscovici [21] and Fathizadeh-Khalkhali [31]. The next term in the expansion, namely the term a_4, which in the classical case contains explicit information about the analogue of the Riemann tensor, is calculated and studied in [17]. A version of the Riemann-Roch theorem is proven in [41] and the study of local spectral invariants is extended to all finite projective modules on noncommutative two tori in [47].

A key idea to define a curved noncommutative space in the above works is to conformally perturb a flat spectral triple by introducing a noncommutative Weyl factor. The complex geometry of the noncommutative two torus, on the other hand, provides a Dirac operator which, in analogy with the classical case, originates from the Dolbeault complex. By perturbing this spectral triple, one can construct a (twisted) spectral triple that can be used to study the geometry of the conformally perturbed flat metric on the noncommutative two torus. Then, using the pseudodifferrential operator theory for C^*-dynamical systems developed by Connes in [9], the computation is performed and explicit formulas are obtained. The spectral geometry and the study of scalar curvature of noncommutative tori have been pursued further in [23, 32, 28].

Finally, for the latest on interactions between noncommutative geometry, number theory, and arithmetic algebraic geometry, the reader can start with the article by Connes and Consani [16] in this volume and the references therein.

2 Curvature in noncommutative geometry

This section is of an introductory nature and is meant to set the stage for later sections and to motivate the evolution of the concept of curvature in noncommutative geometry from its beginnings to its present form. Clearly we have no intention of giving even a brief sketch of the history of the development of the curvature concept in differential geometry. That would require a separate long article, if not a book. We shall simply highlight some key concepts that have impacted the development of the idea of curvature in noncommutative geometry.

2.1 A brief history of curvature

Curvature, as understood in classical differential geometry, is one of the most important features of a geometric space. It is here that geometry and topology differ in the ways they probe a space. To talk about curvature we need more than just topology or smooth structure on a space. The extra piece of structure is usually encoded in a (pseudo-)Riemannian metric, or at least a connection on the tangent bundle, or on a principal G-bundle. It is remarkable that Greek geometers missed the curvature concept altogether, even for simple curves like a circle, which they studied so intensely. The earliest quantitative understanding of curvature, at least for circles, is due to Nicole Oresme in the fourteenth century. In his treatise, *De configurationibus*, he correctly gives the inverse of radius as the curvature of a circle. The concept had to wait for Descartes' analytic geometry and the Newton-Leibniz calculus before to be developed and fully understood. In fact the first definitions of the (signed) curvature κ of a plane curve $y = y(x)$ are due to Newton, Leibniz, and Huygens in the seventeenth century:

$$\kappa = \frac{y''}{(1 + y'^2)^{3/2}}.$$

It is important to note that this is not an intrinsic concept. Intrinsically any one-dimensional Riemannian manifold is locally isometric to $\mathbb{R}$ with its flat Euclidean metric and hence its intrinsic curvature is zero.

Thus the first major case to be understood was the curvature of a surface embedded in a three-dimensional Euclidean space with its induced metric. In his magnificent paper of 1828 entitled *disquisitiones generales circa superficies curvas,* Gauss first defines the curvature of a surface in an *extrinsic* way, using the Gauss map and then he proves his *theorema egregium*: the curvature so defined is in fact an *intrisic* concept and can solely be defined in terms of the first fundamental form. That is the Gaussian curvature is an isometry invariant, or in Gauss' own words:

> Thus the formula of the preceding article leads itself to the remarkable Theorem. If a curved surface is developed upon any other surface whatever, the measure of curvature in each point remains unchanged.

Now the first fundamental form is just the induced Riemannian metric in more modern language. As we shall see, in the hands of Riemann, Theorema Egregium opened the way for the idea of intrinsic geometry of spaces in general. Surfaces, and manifolds in general, have an intrinsic geometry defined solely by metric relations within the space itself, independent of any ambient space.

If $g = e^h(dx^2 + dy^2)$ is a locally conformally flat metric, then its Gaussian curvature is given by

$$K = -\frac{1}{2}e^{-h}\Delta h,$$

where Δ is the flat Laplacian. We shall see later in this paper that the analogous formula in the noncommutative case, first obtained in [21, 31], takes a much more complicated form, with remarkable similarities and differences.

Another major result of Gauss' surface theory was his *local uniformization theorem*, which amounts to existence of *isothermal coordinates*: any analytic

Riemannian metric in two dimensions is locally conformally flat. The result holds for all smooth metrics in two dimensions, but Gauss' proof only covers analytic metrics. Since conformal classes of metrics on a two torus are parametrized by the upper half plane modulo the action of the modular group, this justifies the initial choice of metrics for noncommutative tori by Connes and Cohen in their Gauss-Bonnet theorem in [22], and for general conformal structures in our paper [30]. By all chances, in the noncommutative case one needs to go beyond the class of locally conformally flat metrics. For recent results in this direction, see [35].

A third major achievement of Gauss in differential geometry is his local *Gauss-Bonnet theorem*: for any geodesic triangle drawn on a surface with interior angles α, β, γ, we have

$$\alpha + \beta + \gamma - \pi = \int K dA,$$

where K denotes the Gauss curvature and dA is the surface area element. By using a geodesic triangulation of the surface, one can then easily prove the global Gauss-Bonnet theorem for a closed Riemannian surface:

$$\frac{1}{2\pi} \int_M K dA = \chi(M),$$

where $\chi(M)$ is the Euler characteristic of the closed surface M. It is hard to overemphasize the importance of this result which connects geometry with topology. It is the first example of an index theorem and the theory of characteristic classes.

To find a true analogue of the Gauss-Bonnet theorem in a noncommutative setting was the motivation for Connes and Tretkoff in their groundbreaking work [22]. After conformally perturbing the flat metric of a noncommutative torus, they noticed that while the above classical formulation has no clear analogue in the noncommutative case, its spectral formulation

$$\zeta(0) + 1 = \frac{1}{12\pi} \int_M K dA = \frac{1}{6} \chi(M),$$

makes perfect sense. Here

$$\zeta(s) = \sum \lambda_j^{-s}, \qquad \mathrm{Re}(s) > 1, \tag{2}$$

is the *spectral zeta function* of the scalar Laplacian $\Delta_g = d^*d$ of (M, g). The spectral zeta function has a meromorphic continuation to $\mathbb{C}$ with a unique (simple) pole at $s = 1$. In particular $\zeta(0)$ is defined. Thus $\zeta(0)$ is a topological invariant, and, in particular, it remains invariant under the conformal perturbation $g \to e^h g$ of the metric. This result was then extended to all conformal classes in the upper half plane in our paper [30].

After the work of Gauss, a decisive giant step was taken by Riemann in his epoch-making paper *Ueber die Hypothesen, welche der Geometrie zu Grunde liegen*, which is a text of his *Habilitationsvortrag* of June 1854. The notion of *space*, as an entity that exists on its own, without any reference to an ambient space or external world, was first conceived by Riemann. Riemannian geometry is intrinsic from the beginning. In Riemann's conception, a space, which he called a *mannigfaltigkeit*, manifold in English, can be discrete or continuous, finite or infinite dimensional. The idea of a geometric space as an abstract *set* endowed with some extra structure was born in this paper of Riemann. Local coordinates are just labels without any intrinsic meaning, and thus one must always make sure that the definitions are independent of the choice of coordinates. This is the *general principle of relativity*, which later came to be regarded as a cornerstone of modern theories of spacetime and Einstein's theory of gravitation. This idea quickly led to the development of tensor calculus, also known as the *absolute differential calculus*, by the Italian school of Ricci and his brilliant student Levi-Civita.

Riemann also introduced the idea of a Riemannian metric and understood that to define the bending or curvature of a space one just needs a Riemannian metric. This was of course directly inspired by Gauss' theorema egregium. In fact he gave two definitions for curvature. His *sectional* curvature is defined as the Gaussian curvature of two-dimensional submanifolds defined via the geodesic flow for each two-dimensional subspace of the tangent space at each point. For his second definition he introduced the geodesic coordinate systems and considered the Taylor expansion of the metric components $g_{ij}(x)$ in a geodesic coordinate. Let

$$c_{ij,kl} = \frac{1}{2} \frac{\partial^2 g_{ij}}{\partial x^k \partial x^l}.$$

He shows that sectional curvature is determined by the components $c_{ij,kl}$, and vice versa. Also, one knows that the components $c_{ij,kl}$ are closely related to Riemann curvature tensor.

The Riemann curvature tensor, in modern notation, is defined as

$$R(X, Y) = \nabla_X \nabla_Y - \nabla_Y \nabla_X - \nabla_{[X,Y]},$$

where ∇ is the Levi-Civita connection of the metric, and X and Y are vector fields on the manifold. The analogue of this curvature tensor of rank four is still an illusive concept in the noncommutative case. However, the components of the Riemann tensor appear in the term a_4 in the small time heat kernel expansion of the Laplacian of the metric, the analogue of which was calculated and studied in [17] for noncommutative two tori and for noncommutative four tori with product geometries.

It is hard to exaggerate the importance of the *Ricci curvature* in geometry and physics. For example, it plays an indispensable role in Einstein's theory of gravity and Einstein field equations. In particular, it directly leads, thanks to Schwarzschild solution, to the prediction of black holes. It is also fundamental for the Ricci flow. Ricci curvature can be formulated in spectral terms and this opened up the possibility of defining it in noncommutative settings [34]. The reader should consult later sections in this survey for more on this.

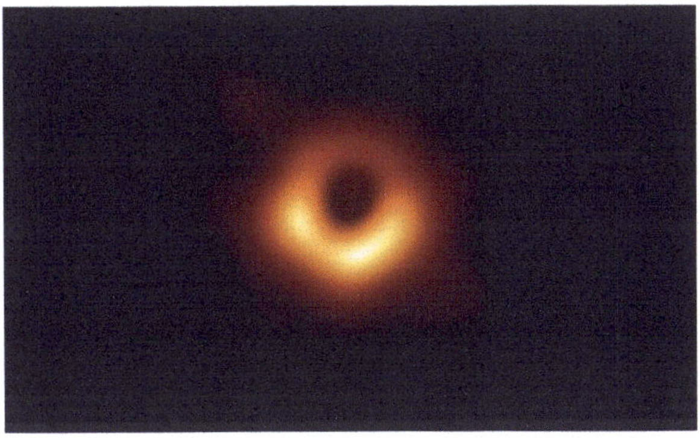

The first black hole image by Event Horizon Telescope, April 2019

Although they won't be a subject for the present exposition, let us briefly mention some other aspects of curvature that have found their analogues in noncommutative settings. These are mostly *linear* aspects of curvature, and have much to do with representation theory of groups. They include Chern-Weil theory of characteristic classes and specially the Chern-Connes character maps for both K-theory and K-homology, Chern-Simons theory, and Yang-Mills theory. Riemannian curvature, whose noncommutative analogue we are concerned with here, is a *nonlinear* theory and from our point of view that is why it took so long to find its proper formulation and first calculations in a noncommutative setting.

2.2 Laplace type operators and Gilkey's theorem

At the heart of spectral geometry, Gilkey's theorem [36] gives the most precise information on asymptotic expansion of heat kernels for a large class of elliptic PDEs. Since this result and its noncommutative analogue play such an important role in defining and computing curvature invariants in noncommutative geometry, we shall explain it briefly in this section. Let M be a smooth closed manifold with a Riemannian metric g and a vector bundle V on M. An operator $P : \Gamma(M, V) \to \Gamma(M, V)$ on smooth sections of V is called a Laplace type operator if in local coordinates it looks like

$$P = -g^{ij} \partial_i \partial_j + \text{lower orders.}$$

Examples of Laplace type operators include Laplacian on forms

$$\Delta = (d + d^*)^2 : \Omega^p(M) \to \Omega^p(M),$$

and the Dirac Laplacians $\Delta = D^*D$, where $D : \Gamma(S) \to \Gamma(S)$ is a generalized Dirac operator.

Now if P is a Laplace type operator, then there exist a unique connection ∇ on the vector bundle V and an endomorphism $E \in \mathrm{End}(V)$ such that

$$P = \nabla^*\nabla - E.$$

Here $\nabla^*\nabla$ is the connection Laplacian which is locally given by $-g^{ij}\nabla_i\nabla_j$. For example, the Lichnerowicz formula for the Dirac operator, $D^2 = \nabla^*\nabla - \frac{1}{4}R$, gives

$$E = \frac{1}{4}R,$$

where R is the scalar curvature. Now e^{-tP} is a smoothing operator with a smooth kernel $k(t, x, y)$. There is an asymptotic expansion near $t = 0$

$$k(t, x, x) \sim \frac{1}{(4\pi t)^{m/2}}(a_0(x, P) + a_2(x, P)t + a_4(x, P)t^2 + \cdots),$$

where $a_{2k}(x, P)$ are known as the Gilkey-Seeley-De Witt coefficients. Gilkey's theorem asserts that $a_{2k}(x, P)$ can be expressed in terms of universal polynomials in the metric g and its covariant derivatives. Gilkey has computed the first four non-zero terms and they are as follows:

$$a_0(x, P) = tr(1),$$

$$a_2(x, P) = tr\left(E - \frac{1}{6}R\right),$$

$$a_4(x, P) = \frac{1}{360}tr\Big(\big(-12R_{;kk} + 5R^2 - 2R_{jk}R_{jk} + 2R_{ijkl}R_{ijkl}\big) -60RE + 180E^2 + 60E_{;kk} + 30\Omega_{ij}\Omega_{ij}\Big).$$

$$\begin{aligned} a_6(x, P) = tr\Big\{\frac{1}{7!}\Big(&-18R_{;kkll} + 17R_{;k}R_{;k} - 2R_{jk;l}R_{jk;l} - 4R_{jk;l}R_{jl;k} \\ &+ 9R_{ijku;l}R_{ijku;l} + 28RR_{;ll} - 8R_{jk}R_{jk;ll} + 24R_{jk}R_{jl;kl} \\ &+ 12R_{ijkl}R_{ijkl;uu}\Big) \\ + \frac{1}{9\cdot 7!}\Big(&-35R^3 + 42RR_{lp}R_{lp} - 42RR_{klpq}R_{klpq} + 208R_{jk}R_{jl}R_{kl} \\ &- 192R_{jk}R_{ul}R_{jukl} + 48R_{jk}R_{julp}R_{kulp} - 44R_{ijku}R_{ijlp}R_{kulp} \\ &- 80R_{ijku}R_{ilkp}R_{jlup}\Big) \end{aligned}$$

$$+\frac{1}{360}\Big(8\Omega_{ij;k}\Omega_{ij;k}+2\Omega_{ij;j}\Omega_{ik;k}+12\Omega_{ij}\Omega_{ij;kk}-12\Omega_{ij}\Omega_{jk}\Omega_{ki}$$

$$-6R_{ijkl}\Omega_{ij}\Omega_{kl}+4R_{jk}\Omega_{jl}\Omega_{kl}-5R\Omega_{kl}\Omega_{kl}\Big)$$

$$+\frac{1}{360}\Big(6E_{;iijj}+60EE_{;ii}+30E_{;i}E_{;i}+60E^3+30E\Omega_{ij}\Omega_{ij}$$

$$-10RE_{;kk}-4R_{jk}E_{;jk}-12R_{;k}E_{;k}-30RE^2-12R_{;kk}E$$

$$+5R^2E-2R_{jk}R_{jk}E+2R_{ijkl}R_{ijkl}E\Big)\Big\}.$$

Here R_{ijkl} is the Riemann curvature tensor, R is the scalar curvature, Ω is the curvature matrix of two forms, and ; denotes the covariant derivative operator.

As we shall later see in this survey, the first two terms in the above list allow us to define the scalar and Ricci curvatures in terms of heat kernel coefficients and extend them to noncommutative settings.

Alternatively, one can use spectral zeta functions to extract information from the spectrum. Heat trace and spectral zeta functions are related via Mellin transform. For a concrete example, let $\triangle$ denote the Laplacian on functions on an m-dimensional closed Riemannian manifold. Define

$$\zeta_{\triangle}(s)=\sum\lambda_i^{-s}\qquad \mathrm{Re}(s)>\frac{m}{2}.$$

The spectral invariants a_i in the heat trace asymptotic expansion

$$\mathrm{Trace}(e^{-t\triangle})\sim(4\pi t)^{\frac{-m}{2}}\sum_{j=0}^{\infty}a_jt^j\qquad\qquad (t\to 0^+)$$

are related to residues of spectral zeta function by

$$\mathrm{Res}_{s=\alpha}\zeta_{\triangle}(s)=(4\pi)^{-\frac{m}{2}}\frac{a_{\frac{m}{2}-\alpha}}{\Gamma(\alpha)},\qquad \alpha=\frac{m}{2}-j>0.$$

To get to the local invariants like scalar curvature we can consider localized zeta functions. Let $\zeta_f(s):=\mathrm{Tr}\,(f\triangle^{-s}),\ f\in C^\infty(M)$. Then we have

$$\mathrm{Res}\,\zeta_f(s)|_{s=\frac{m}{2}-1}=\frac{(4\pi)^{-m/2}}{\Gamma(m/2-1)}\int_M f(x)R(x)dvol_x,\qquad m\geq 3,$$

$$\zeta_f(s)|_{s=0}=\frac{1}{4\pi}\int_M f(x)R(x)dvol_x-\mathrm{Tr}(fP),\qquad m=2,$$

where P is projection onto zero eigenmodes of $\triangle$. Thus the scalar curvature R appears as the density function for the localized spectral zeta function.

2.3 Noncommutative Chern-Weil theory

Although it is not our intention to review this subject in the present survey, we shall nevertheless explain some ideas of noncommutative Chern-Weil theory here. Many aspects of Chern-Weil theory of characteristic classes for vector bundles and principal bundles over smooth manifolds can be cast in an algebraic formalism and as such is even used in commutative algebra and algebraic geometry [5]. Thus one can formulate notions like de Rham cohomology, connection, curvature, Chern classes, and Chern character, over a commutative algebra and then for a scheme. This is a commutative theory which is more or less straightforward in the characteristic zero case. But there seemed to be no obvious extension of de Rham theory and the rest of Chern-Weil theory to the noncommutative case.

In [9] Connes realized that many aspects of Chern-Weil theory can be implemented in a noncommutative setting. The crucial ingredient was the discovery of cyclic cohomology that replaces de Rham homology of currents in a noncommutative setting [11, 12]. Let A be a not necessarily commutative algebra over the field of complex numbers. By a *noncommutative differential calculus* on A we mean a triple (Ω, d, ρ) such that (Ω, d) is a differential graded algebra and $\rho : A \to \Omega^0$ is an algebra homomorphism. Given a right A-module $\mathcal{E}$, a *connection* on $\mathcal{E}$ is a $\mathbb{C}$-linear map $\nabla : \mathcal{E} \longrightarrow \mathcal{E} \otimes_A \Omega^1$ satisfying the Leibniz rule $\nabla(\xi a) = \nabla(\xi)a + \xi \otimes da$, for all $\xi \in \mathcal{E}$ and $a \in A$. Let $\hat{\nabla} : \mathcal{E} \otimes_A \Omega^\bullet \to \mathcal{E} \otimes_A \Omega^{\bullet+1}$ be the (necessarily unique) extension of ∇ which satisfies the graded Leibniz rule $\hat{\nabla}(\xi\omega) = \hat{\nabla}(\xi)\omega + (-1)^{\deg \xi} \xi d\omega$ with respect to the right Ω-module structure on $\mathcal{E} \otimes_A \Omega$. The *curvature* of ∇ is the operator of degree 2, $\hat{\nabla}^2 : \mathcal{E} \otimes_A \Omega^\bullet \to \mathcal{E} \otimes_A \Omega^\bullet$, which can be easily checked to be Ω-linear.

Now to obtain Connes' Chern character pairing between K-theory and cyclic cohomology, $K_0(A) \otimes HC^{2n}(A) \to \mathbb{C}$, one can proceed as follows. Given a finite projective A-module $\mathcal{E}$, one can always equip $\mathcal{E}$ with a connection over the universal differential calculus ΩA. An element of $HC^{2n}(A)$ can be represented by a closed graded trace τ on $\Omega^{2n} A$. The value of the pairing is then simply $\tau(\hat{\nabla}^{2n})$. Here we used the same symbol τ to denote the extension of τ to the ring $End_{\Omega^\bullet}(\mathcal{E} \otimes_A \Omega^\bullet)$. One checks that this definition is independent of all choices that we made [12]. Connes in fact initially developed the more sophisticated Chern-Connes pairing in K-homology with explicit formulas that do not have a commutative counterpart. For all this and more, the reader should check Connes' book and his above cited article [12, 14] as well as the book [40].

2.4 From spectral geometry to spectral triples

The very notion of Riemannian manifold itself is now subsumed and vastly generalized through Connes' notion of *spectral triples*, which is a centerpiece of noncommutative geometry and applications of noncommutative geometry to particle physics.

Let us first motivate the definition of a spectral triple. During the course of their heat equation proof of the index theorem, it was discovered by Atiyah-Bott-Patodi [2] that it is enough to prove the theorem for Dirac operators twisted by vector bundles. The reason is that these twisted Dirac operators in fact generate the whole K-homology group of a spin manifold and thus it suffices to prove the theorem only for these first order elliptic operators. This indicates the preeminence of Dirac operators in topology. As we shall see below, Dirac operators also encode metric information of a Riemannian manifold in a succinct way. Broadly speaking, spectral triples, suitably enhanced, are noncommutative spin manifolds and form a backbone of noncommutative geometry, specially its metric aspects. One precise formulation of this idea is *Connes' reconstruction theorem* [15] which states that a commutative spectral triple satisfying some natural conditions is in fact the standard spectral triple of a spin^c manifold described below.

Recall that the Dirac operator D on a compact Riemannian spin^c manifold acts as an unbounded selfadjoint operator on the Hilbert space $L^2(M, S)$ of L^2-spinors on M. If we let $C^\infty(M)$ act on $L^2(M, S)$ by multiplication operators, then one can check that for any smooth function f, the commutator $[D, f] = Df - fD$ extends to a bounded operator on $L^2(M, S)$. The metric d on M, that is the geodesic distance of M, can be recovered, thanks to the *distance formula* of Connes [14]:

$$d(p, q) = \text{Sup}\{|f(p) - f(q)|;\ \| [D, f] \| \leq 1\}.$$

The triple $(C^\infty(M), L^2(M, S), \not{D})$ is a commutative example of a spectral triple.

The general definition of a spectral triple, in the odd case, is as follows.

Definition 2.1 *Let A be a unital algebra. An odd spectral triple on A is a triple $(A, \mathcal{H}, D)$ consisting of a Hilbert space $\mathcal{H}$, a selfadjoint unbounded operator $D: Dom(D) \subset \mathcal{H} \to \mathcal{H}$ with compact resolvent, i.e., $(D-\lambda)^{-1} \in \mathcal{K}(\mathcal{H})$, for all $\lambda \notin \mathbb{R}$, and a representation $\pi : A \to \mathcal{L}(\mathcal{H})$ of A such that for all $a \in A$, the commutator $[D, \pi(a)]$ is defined on $Dom(D)$ and extends to a bounded operator on $\mathcal{H}$.*

A spectral triple is called *finitely summable* if for some $n \geq 1$

$$|D|^{-n} \in \mathcal{L}^{1,\infty}(\mathcal{H}).$$

Here $\mathcal{L}^{1,\infty}(\mathcal{H})$ is the Dixmier ideal. It is an ideal of compact operators which is slightly bigger than the ideal of trace class operators and is the natural domain of the Dixmier trace. Spectral triples provide a refinement of Fredholm modules. Going from Fredholm modules to spectral triples is similar to going from the conformal class of a Riemannian metric to the metric itself. Spectral triples simultaneously provide a notion of *Dirac operator* in noncommutative geometry, as well as a Riemannian type *distance function* for noncommutative spaces. In later sections we shall define and work with concrete examples of spectral triples and their conformal perturbations.

3 Pseudodifferential calculus and heat expansion

In this section we discuss the classical pseudodifferential calculus on the Euclidean space and will then provide practical details of the pseudodifferential calculus of [9] that we use for heat kernel calculations on noncommutative tori.

3.1 *Classical pseudodifferential calculus*

In the Euclidean case we follow the notations and conventions of [36] as follows. For any multi-index $\alpha = (\alpha_1, \dots, \alpha_m)$ of non-negative integers and coordinates $x = (x_1, \dots, x_m) \in \mathbb{R}^m$ we set:

$$|\alpha| = \alpha_1 + \cdots + \alpha_m, \qquad \alpha! = \alpha_1! \cdots \alpha_m!, \qquad x^\alpha = x_1^{\alpha_1} \cdots x_m^{\alpha_m},$$

$$\partial_x^\alpha = \left(\frac{\partial}{\partial x_1}\right)^{\alpha_1} \cdots \left(\frac{\partial}{\partial x_m}\right)^{\alpha_m}, \qquad D_x^\alpha = (-i)^{|\alpha|} \partial_x^\alpha.$$

Also we normalize the Lebesgue measure on $\mathbb{R}^m$ by a multiplicative factor of $(2\pi)^{-m/2}$ and still denote it by dx. Therefore we have:

$$\int_{\mathbb{R}^m} \exp\left(-\frac{1}{2}|x|^2\right) dx = 1.$$

The main idea behind pseudodifferential calculus is that it uses the *Fourier transform* to turn a differential operator into multiplication by a function, namely the *symbol* of the differential operator. The Fourier transform $\hat{f}$ of a *Schwartz function* f on $\mathbb{R}^m$ is defined by the following integration:

$$\hat{f}(\xi) = \int_{\mathbb{R}^m} e^{-ix\cdot\xi} f(x)\, dx, \qquad \xi \in \mathbb{R}^m.$$

This integral is convergent because, by definition, the set of Schwartz functions $\mathcal{S}(\mathbb{R}^m)$ consists of all complex-valued smooth functions f on the Euclidean space such that for any multi-indices α and β of non-negative integers

$$\sup_{x\in\mathbb{R}^m} |x^\alpha D^\beta f(x)| < \infty.$$

It turns out that the Fourier transform preserves the L^2-norm, hence it extends to a unitary operator on $L^2(\mathbb{R}^m)$.

The differential operator D_x^α turns in the Fourier mode to multiplication by the monomial ξ^α, in the sense that:

$$\widehat{(D_x^\alpha f)}(\xi) = \xi^\alpha \hat{f}(\xi).$$

The monomial ξ^α is therefore called the symbol of the differential operator D_x^α. Then, the *Fourier inversion formula,*

$$f(x) = \int_{\mathbb{R}^m} e^{i\xi\cdot x} \hat{f}(\xi)\, d\xi, \qquad f \in \mathcal{S}(\mathbb{R}^m),$$

implies that

$$D_x^\alpha f(x) = \int_{\mathbb{R}^m} e^{ix\cdot\xi} \xi^\alpha \hat{f}(\xi)\, d\xi = \int_{\mathbb{R}^m}\int_{\mathbb{R}^m} e^{i(x-y)\cdot\xi} \xi^\alpha f(y)\, dy\, d\xi. \tag{3}$$

It is now clear from the above facts that the symbol of any differential operator, given by a finite sum of the form $\sum a_\alpha(x) D_x^\alpha$, is the polynomial in ξ of the form $\sum a_\alpha(x)\xi^\alpha$, whose coefficients are the functions $a_\alpha(x)$ (which we assume to be smooth). Using the notation $\sigma(\cdot)$ for the symbol it is an easy exercise to see that given two differential operators P_1 and P_2, the symbol of their composition $\sigma(P_1 \circ P_2)$ is given by the following expression:

$$\sum_{\alpha\in\mathbb{Z}^m_{\geq 0}} \frac{1}{\alpha!} \partial_\xi^\alpha \sigma(P_1)\, D_x^\alpha \sigma(P_2), \tag{4}$$

which is a finite sum because only finitely many of the summands are non-zero.

By considering a wider family of symbols, one obtains a larger family of operators which are called *pseudodifferential operators*. A smooth function $p : \mathbb{R}^m \times \mathbb{R}^m \to \mathbb{C}$ is *a pseudodifferential symbol of order* $d \in \mathbb{R}$ if it satisfies the following conditions:

- $p(x, \xi)$ has compact support in x,
- for any multi-indices $\alpha, \beta \in \mathbb{Z}^m_{\geq 0}$, there exists a constant $C_{\alpha,\beta}$ such that

$$|\partial_\xi^\beta \partial_x^\alpha p(x, \xi)| \leq C_{\alpha,\beta}(1 + |\xi|)^{d-|\beta|}. \tag{5}$$

Clearly the space of pseudodifferential symbols possesses a filtration because, denoting the space of symbols of order d by S^d, we have:

$$d_1 \leq d_2 \implies S^{d_1} \subset S^{d_2}.$$

Existence of symbols of arbitrary orders can be assured by observing that for any $d \in \mathbb{R}$ and any compactly supported function f_0, the function $p(x, \xi) = f_0(x)(1 + |\xi|^2)^{d/2}$ belongs to S^d.

Given a symbol $p \in S^d$, inspired by formula (3), the corresponding pseudodifferential operator P is defined by

$$Pf(x) = \int_{\mathbb{R}^m} e^{ix\cdot\xi} p(x, \xi) \hat{f}(\xi)\, d\xi, \qquad f \in \mathcal{S}(\mathbb{R}^m). \tag{6}$$

The space of pseudodifferential operators associated with symbols of order d is denoted by $\Psi^d(\mathbb{R}^m)$. Searching for an analog of formula (4) for general pseudodifferential operators leads to a complicated analysis which, at the end, gives *an asymptotic expansion* for the symbol of the composition of such operators. The formula is written as

$$\sigma\,(P_1 P_2) \sim \sum_{\alpha \in \mathbb{Z}^m_{\geq 0}} \frac{1}{\alpha!} \partial_\xi^\alpha \sigma(P_1)\, D_x^\alpha \sigma(P_2). \tag{7}$$

It is important to put in order some explanations about this formula. If $\sigma(P_1) \in S^{d_1}$ and $\sigma(P_2) \in S^{d_2}$, then there is a symbol in $S^{d_1+d_2}$ that gives $P_1 \circ P_2$ via formula (6). However $\sigma(P_1 \circ P_2)$ has a complicated formula which involves integrals, which can be seen by writing the formulas directly. The trick is then to use Taylor series and to perform analytic manipulations on the closed formula for $\sigma(P_1 \circ P_2)$ to derive the expansion (7). The error terms in the Taylor series that one uses in the manipulations are responsible for having an asymptotic expansion rather than a strict identity. The precise meaning of this expansion is that given any $d \in \mathbb{R}$, there exists a positive integer N such that

$$\sigma\,(P_1 P_2) - \sum_{|\alpha| \leq N} \frac{1}{\alpha!} \partial_\xi^\alpha \sigma(P_1)\, D_x^\alpha \sigma(P_2) \in S^d.$$

Therefore, as one subtracts the terms $\frac{1}{\alpha!} \partial_\xi^\alpha \sigma(P_1)\, D_x^\alpha \sigma(P_2)$ from $\sigma(P_1 \circ P_2)$, the orders of the resulting symbols tend to $-\infty$. Regarding this, it is convenient to introduce the space $S^{-\infty} = \bigcap_{d \in \mathbb{R}} S^d$ of the *infinitely smoothing pseudodifferential symbols*. For example, for any compactly supported function f_0, the symbol $p(x,\xi) = f_0(x) e^{-|\xi|^2}$ belongs to $S^{-\infty}$.

The composition rule (7) is a very useful tool. For instance, it can be used to find a *parametrix* for *elliptic* pseudodifferential operators. Important geometric operators such as Laplacians are elliptic, and by finding a parametrix, as we shall explain, one finds an approximation of the fundamental solution of the partial differential equation defined by such an important operator. Intuitively, a pseudodifferential symbol $p(x,\xi)$ of order $d \in \mathbb{R}$ is *elliptic* if it is non-zero when ξ is away from the origin (or invertible in the case of matrix-valued symbols), and $|p(x,\xi)^{-1}|$ is bounded by a constant times $(1 + |\xi|)^{-d}$ as $\xi \to \infty$. For our purposes, it suffices to know that a differential operator $D = \sum a_\alpha(x) D_x^\alpha$ of order $d = \max_\alpha |\alpha|$ is elliptic if its *leading symbol*,

$$\sigma_L(D) = \sum_{|\alpha|=d} a_\alpha(x) \xi^\alpha,$$

is non-zero (or invertible) for $\xi \neq 0$. Given such an elliptic differential operator one can use formula (7) to find an inverse for D, called a *parametrix*, in the quotient $\Psi/\Psi^{-\infty}$ of the algebra of pseudodifferential operators Ψ by infinitely smoothing operators $\Psi^{-\infty}$. This process can be described as follows. One makes the natural assumption that the symbol of the parametrix has an expansion starting with a leading term of order $-d$ and other terms whose orders descend to $-\infty$, namely terms of orders $-d-1, -d-2, \ldots$, and one continues as follows. The formula given by (7) can be used to find these terms recursively and thereby find a parametrix R such that

$$DR - I \sim RD - I \sim 0.$$

We will illustrate this carefully in Section 3.2 in a slightly more complicated situation, where a parameter λ and a parametric pseudodifferential calculus are involved in deriving heat kernel expansions. We just mention that invertibility of $\sigma_L(D)$ is the crucial point that allows one to start the recursive process, and to continue on to find the parametrix R.

3.2 Small-time heat kernel expansion

For simplicity and practical purposes we assume that P is a positive elliptic differential operator of order 2 with

$$\sigma(P) = p_2(x, \xi) + p_1(x, \xi) + p_0(x, \xi),$$

where each p_j is (homogeneous) of order j in ξ. We know that $p_2(x, \xi)$ is non-zero (or invertible) for non-zero ξ. The first step in deriving a small time asymptotic expansion for $\mathrm{Tr}(\exp(-tP))$ as $t \to 0^+$ is to use the Cauchy integral formula to write

$$e^{-tP} = \frac{1}{2\pi i} \int_\gamma e^{-t\lambda} (P - \lambda)^{-1} \, d\lambda, \tag{8}$$

where the contour γ goes clockwise around the non-negative real axis, where the eigenvalues of P are located. The term $(P - \lambda)^{-1}$ in the above integral can now be approximated by pseudodifferential operators as follows. We look for an approximation R_λ of $(P - \lambda)^{-1}$ such that

$$\sigma(R_\lambda) \sim r_0(x, \xi, \lambda) + r_1(x, \xi, \lambda) + r_2(x, \xi, \lambda) + \cdots,$$

where each r_j is a symbol of order $-2-j$ in the parametric sense which we will elaborate on later. For now one can use formula (7) to find the r_j recursively out of the equation

$$R_\lambda (P - \lambda) \sim I.$$

This means that the terms r_j in the expansion should satisfy

$$\sum_j r_j \circ ((p_2 - \lambda) + p_1 + p_0) \sim 1, \tag{9}$$

where the composition $\circ$ is given by (7). By writing the expansion one can see that there is only one leading term, which is of order 0, namely $r_0(p_2 - \lambda)$ and needs to be set equal to 1 so that it matches the corresponding (and the only term) on the right-hand side of the Equation (9). Therefore the leading term r_0 is found to be

$$r_0 = (p_2 - \lambda)^{-1}. \tag{10}$$

Here the ellipticity plays an important role, because we need to be ensured that the inverse of $p_2 - \lambda$ exists. Since, in our examples, P will be a Laplace type operator, the leading term p_2 is a positive number (or a positive invertible matrix in the vector bundle case) for any $\xi \neq 0$. Therefore for any λ on the contour γ, we know that $p_2 - \lambda$ is invertible. One can then proceed by considering the term that is homogeneous of order -1 in the expansion of the left-hand side of (9) and set it equal to 0 since there is no term of order -1 on the right-hand side. This will yield a formula for the next term r_1. By continuing this process one finds recursively that for $n = 1, 2, 3, \dots$, we have

$$r_n = -\left(\sum_{\substack{|\alpha|+j+2-k=n, \\ 0 \le j < n,\, 0 \le k \le 2}} \frac{1}{\alpha!} \partial_\xi^\alpha r_j \, D_x^\alpha p_k \right) r_0. \tag{11}$$

It turns out that the r_n calculated by this formula have the following homogeneity property:

$$r_n(x, t\xi, t^2\lambda) = t^{-2-n} r_n(t, \xi, \lambda).$$

Having an approximation of the resolvent $R_\lambda \sim (P - \lambda)^{-1}$ via the symbols r_n, one can use the formulas (8) and (6) to approximate the kernel K_t of the operator e^{-tP}, namely the unique smooth function such that

$$e^{-tP} f(x) = \int K_t(x, y) f(y) \, dy, \qquad f \in \mathcal{S}(\mathbb{R}^m).$$

Since $\mathrm{Tr}(e^{-tP})$ can be calculated by integrating the kernel on the diagonal,

$$\mathrm{Tr}\left(e^{-tP}\right) = \int K_t(x, x) \, dx,$$

the integration of the approximation of the kernel obtained by going through the procedure described above leads to an asymptotic expansion of the following form:

$$\operatorname{Tr}\left(e^{-tP}\right) \sim_{t\to 0^+} t^{-m/2} \sum_{n=0}^{\infty} a_{2n}(P)\, t^n, \tag{12}$$

where each coefficient a_{2n} is the integral of a density $a_{2n}(x, P)$ given by

$$a_{2n}(x, P) = \frac{1}{2\pi i} \int \int_{\gamma} e^{-\lambda} \operatorname{tr}(r_{2n}(x, \xi, \lambda))\, d\lambda\, d\xi.$$

In this integrand, the tr denotes the matrix trace which needs to be considered in the case of vector bundles.

It is a known fact that when P is a geometric operator such as the Laplacian of a metric, each $a_{2n}(x, P)$ can be written in terms of the Riemann curvature tensor, its contractions, and covariant derivatives, see, for example, [18]. However, in practice, as n grows, these terms become so complicated rapidly. One can refer to [36] for the formulas for the terms up a_6 derived using invariant theory.

3.3 *Pseudodifferential calculus and heat kernel expansion for noncommutative tori*

Now that we have illustrated the derivation of the heat kernel expansion (12), we explain briefly in this subsection that using the pseudodifferential calculus developed in [9] for C^*-dynamical systems, heat kernel expansions of Laplacians on noncommutative tori can be derived by taking a parallel approach. We note that, in [48], for *toric manifolds,* the Widom pseudodifferential calculus is adapted to their noncommutative deformations and it is used for the derivation of heat kernel expansions.

We first recall the pseudodifferential calculus on the algebra of noncommutative m-torus. A *pseudodifferential symbol* of order $d \in \mathbb{Z}$ on $\mathbb{T}^m_\Theta$ is a smooth mapping $\rho : \mathbb{R}^m \to C^\infty(\mathbb{T}^m_\Theta)$ such that for any multi-indices α and β of non-negative integers, there exists a constant $C_{\alpha,\beta}$ such that

$$||\partial_\xi^\beta \delta^\alpha \rho(\xi)|| \le C_{\alpha,\beta}(1 + |\xi|)^{d-|\beta|}.$$

Here $|| \cdot ||$ denotes the C^*-algebra norm, which is the equivalent of the supremum norm in the commutative setting. Therefore this definition is the noncommutative analog of the definition given by (5) in the classical case. A symbol of order d is *elliptic* if $\rho(\xi)$ is invertible for large enough ξ and there exists a constant $C_\rho > 0$ such that

$$||\rho(\xi)^{-1}|| \le C_\rho(1 + |\xi|)^{-d}.$$

Given a pseudodifferential symbol on $\mathbb{T}^m_\Theta$ the corresponding pseudodifferential operator $P_\rho : C^\infty(\mathbb{T}^m_\Theta) \to C^\infty(\mathbb{T}^m_\Theta)$ is defined in [9] by the oscillatory integral

$$P_\rho(a) = \iint e^{-is\cdot\xi} \rho(\xi)\, \alpha_s(a)\, ds\, d\xi, \qquad a \in C^\infty(\mathbb{T}^m_\Theta), \tag{13}$$

where α_s is the dynamics given by

$$\alpha_s(U^\alpha) = e^{is\cdot\alpha} U^\alpha.$$

For example, the symbol of a differential operator of the form $\sum_{|\alpha|\le d} a_\alpha \delta^\alpha$, $a_\alpha \in C^\infty(\mathbb{T}^m_\Theta)$ is $\sum_{|\alpha|\le d} a_\alpha \xi^\alpha$.

Given a positive elliptic operator P of order 2 acting on $C^\infty(\mathbb{T}^m_\Theta)$, such as the Laplacian of a metric, in order to derive an asymptotic expansion for $\mathrm{Tr}(e^{-tP})$ one can start by writing the Cauchy integral formula as we did in formula (8). However now one has to use the pseudodifferential calculus given by (13) to write $P - \lambda$ in terms of its symbol and thereby approximate its inverse. In this calculus, if ρ_1 and ρ_2 are, respectively, symbols of orders d_1 and d_2, then the composition $P_{\rho_1} P_{\rho_2}$ has a symbol of order $d_1 + d_2$ with the following asymptotic expansion:

$$\sigma\left(P_{\rho_1} P_{\rho_2}\right) \sim \rho_1 \circ \rho_2 := \sum_{\alpha \in \mathbb{Z}^m_{\ge 0}} \frac{1}{\alpha!} \partial_\xi^\alpha \rho_1\, \delta^\alpha \rho_2. \tag{14}$$

Having these tools available, one can then perform calculations as in the process illustrated in Section 3.2 to derive an asymptotic expansion for $\mathrm{Tr}(e^{-tP})$. That is, one writes $\sigma(P) = p_2 + p_1 + p_0$, where each p_j is homogeneous of order j, and finds recursively the terms r_j, $j = 0, 1, 2, \ldots$, that are homogeneous of order $-2 - j$ and

$$\sum_j r_j \circ ((p_2 - \lambda) + p_1 + p_0) \sim 1.$$

This means that we are using the composition rule (14) to approximate the inverse of $P - \lambda$. The result of this process is a recursive formula similar to the one given by (10) and (11). That is, one finds that

$$r_0 = (p_2 - \lambda)^{-1}. \tag{15}$$

and for $n = 1, 2, 3, \ldots$,

$$r_n = -\left(\sum_{\substack{|\alpha|+j+2-k=n, \\ 0\le j<n,\, 0\le k\le 2}} \frac{1}{\alpha!} \partial_\xi^\alpha r_j\, \delta^\alpha p_k \right) r_0. \tag{16}$$

Then one finds the small asymptotic expansion

$$\mathrm{Tr}(e^{-tP}) \sim_{t\to 0^+} t^{-m/2} \sum_{n=0}^{\infty} \varphi_0(a_{2n}) t^n,$$

where φ_0 is the canonical trace

$$\varphi_0 \left(\sum_{\alpha \in \mathbb{Z}^m} a_\alpha U^\alpha \right) = a_0$$

providing us with integration on the noncommutative torus $\mathbb{T}^m_\Theta$. The terms $a_{2n} \in C^\infty(\mathbb{T}^m_\Theta)$ can be calculated using (15) and (16) as follows:

$$a_{2n} = \frac{1}{2\pi i} \int_{\mathbb{R}^m} \int_\gamma e^{-\lambda} r_{2n}(\xi, \lambda)\, d\lambda\, d\xi. \tag{17}$$

We shall see in Section 4 that in order to perform this type of integrals in the noncommutative setting one encounters noncommutative features which will lead to the appearance of a functional calculus with a modular automorphism in the outcome of the integrals.

4 Gauss-Bonnet theorem and curvature for noncommutative 2-tori

The Gauss-Bonnet theorem for smooth oriented surfaces is a fundamental result that establishes a bridge between topology and differential geometry of surfaces. Given a surface, its Euler characteristic is a topological invariant which can be calculated by choosing an arbitrary triangulation on the surface and forming an alternating summation on the number of its vertices, edges, and faces. It is quite remarkable that the Euler characteristic is independent of the choice of triangulation and depends only on the genus of the surface. Clearly, under a diffeomorphism, or roughly speaking under changes on the surface that do not change the genus, the Euler characteristic remains unchanged. However the scalar curvature of the surface changes under such changes by diffeomorphisms, say when the surface is embedded in the three-dimensional Euclidean space and has inherited the metric of the ambient space. However, the striking fact, namely the statement of the Gauss-Bonnet theorem, is that the change of curvature on the surface occurs in a way that the increase and decrease of curvature over the surface compensate for each other to the effect that the curvature integrates to the Euler characteristic, up to multiplication by a universal constant that is independent of the surface. Hence, the total curvature, namely the integral of the scalar curvature over the surface, is a topological invariant.

4.1 Scalar curvature and Gauss-Bonnet theorem for $\mathbb{T}_\theta^2$

In noncommutative geometry, the analog of the Gauss-Bonnet theorem has been investigated for the noncommutative two torus. In this setting, the flat geometry of $\mathbb{T}_\theta^2$ was conformally perturbed by means of a conformal factor e^{-h}, where h is a selfadjoint element in $C^\infty(\mathbb{T}_\theta^2)$. In the late 1980s, a heavy calculation was performed by P. Tretkoff and A. Connes to find an expression for the analog of the total curvature of the perturbed metric on $\mathbb{T}_\theta^2$. The expression had a heavy dependence on the element h used for changing the metric, therefore it was not clear whether the analog of the Gauss-Bonnet theorem holds for $\mathbb{T}_\theta^2$, and they just recorded the result of their calculations in an MPI preprint [8]. However, following calculations for the spectral action in the presence of a dilaton [7] and developments in the theory of twisted spectral triples [20], there were indications that the complicated expression for the total curvature has to be independent of the element h. By further calculations, simplifications and using symmetries in the result, it was shown in [22] that the terms in the complicated expression for the total curvature indeed cancel each other out to 0, hence the analog of the Gauss-Bonnet theorem for $\mathbb{T}_\theta^2$. The conformal class of metrics that was used in [22] is associated with the simplest translation-invariant complex structure on $\mathbb{T}_\theta^2$, namely the complex structure associated with $i = \sqrt{-1}$. The Gauss-Bonnet theorem for $\mathbb{T}_\theta^2$ for the complex structure associated with an arbitrary complex number τ in the upper-half plane was established in [30].

After considering a general complex number τ in the upper half-plane to induce a complex structure and thereby a conformal structure on $\mathbb{T}_\theta^2$, and by conformally perturbing the flat metric in this class by a fixed conformal factor e^{-h}, $h = h^* \in C^\infty(\mathbb{T}_\theta^2)$, the Laplacian of the curved metric is shown [22, 30] to be anti-unitarily equivalent to the operator

$$\triangle_{\tau,h} = e^{h/2} \triangle_{\tau,0} \, e^{h/2},$$

where

$$\triangle_{\tau,0} = \delta_1^2 + 2\tau_1 \delta_1 \delta_2 + |\tau|^2 \delta_2^2$$

is the Laplacian of the flat metric in the conformal class determined by $\tau = \tau_1 + i\tau_2$ in the upper half-plane. The pseudodifferential symbol of $\triangle_{\tau,h}$ is the sum of the following homogeneous components of order 2, 1, and 0, in which we use $k = h/2$ for simplicity:

$$p_2(\xi) = \xi_1^2 k^2 + |\tau|^2 \xi_2^2 k^2 + 2\tau_1 \xi_1 \xi_2 k^2,$$

$$p_1(\xi) = 2\xi_1 k \delta_1(k) + 2|\tau|^2 \xi_2 k \delta_2(k) + 2\tau_1 \xi_1 k \delta_2(k) + 2\tau_1 \xi_2 k \delta_1(k),$$

$$p_0(\xi) = k\delta_1^2(k) + |\tau|^2 k \delta_2^2(k) + 2\tau_1 k \delta_1 \delta_2(k).$$

The analog of the scalar curvature is then the term $a_2(\Delta_{\tau,h}) \in C^\infty(\mathbb{T}^2_\theta)$ appearing in the small time ($t \to 0^+$) asymptotic expansion

$$\mathrm{Tr}(ae^{-t\Delta_{\tau,h}}) \sim t^{-1} \sum_{n=0}^{\infty} \varphi_0\left(a\, a_{2n}(\Delta_{\tau,h})\right) t^n, \qquad a \in C^\infty(\mathbb{T}^2_\theta). \tag{18}$$

By going through the process illustrated in Section 3.3 one can calculate a_2. However, there is a purely noncommutative obstruction for the calculation of the involved integrals in formula (17), namely one encounters integration of C^*-algebra valued functions defined on the Euclidean space, $\mathbb{R}^2$ in this case. By passing to a suitable variation of the polar coordinates, the angular integration can be performed easily, and the main obstruction remains in the radial integration which can be overcome by the following rearrangement lemma [22, 3, 21, 46]:

Lemma 4.1 *For any tuple* $m = (m_0, m_1, \ldots, m_\ell) \in \mathbb{Z}^{\ell+1}_{>0}$ *and elements* $\rho_1, \ldots, \rho_\ell \in C^\infty(\mathbb{T}^2_\theta)$, *one has*

$$\int_0^\infty \frac{u^{|m|-2}}{(e^h u+1)^{m_0}} \prod_1^\ell \rho_j (e^h u+1)^{-m_j}\, du = e^{-(|m|-1)h} F_m(\Delta_{(1)}, \ldots, \Delta_{(\ell)})\Big(\prod_1^\ell \rho_j\Big),$$

where

$$F_m(u_1, \ldots, u_\ell) = \int_0^\infty \frac{x^{|m|-2}}{(x+1)^{m_0}} \prod_1^\ell \Big(x \prod_1^j u_h + 1\Big)^{-m_j} dx,$$

and Δ *is the modular automorphism*

$$\Delta(a) = e^{-h} a e^h, \qquad a \in C(\mathbb{T}^2_\theta).$$

After applying this lemma to the numerous integrands with the help of computer programming, the result for the scalar curvature $a_2(\Delta_{\tau,h})$ was calculated in [21, 31] (Figure 1):

Theorem 4.1 *The scalar curvature* $a_2(\Delta_{\tau,h}) \in C^\infty(\mathbb{T}^2_\theta)$ *of a general metric in the conformal class associated with a complex number* $\tau = \tau_1 + i\tau_2$ *in the upper half-plane is given by*

$$\begin{aligned} a_2(\Delta_{\tau,h}) = K(\nabla)&\left(\delta_1^2\left(\frac{h}{2}\right) + |\tau|^2\delta_2^2\left(\frac{h}{2}\right) + 2\tau_1\delta_1\delta_2\left(\frac{h}{2}\right)\right) \\ &+H(\nabla,\nabla)\left(\delta_1\left(\frac{h}{2}\right)\delta_1\left(\frac{h}{2}\right) + |\tau|^2\delta_2\left(\frac{h}{2}\right)\delta_2\left(\frac{h}{2}\right)\right. \\ &\left. + \tau_1\delta_1\left(\frac{h}{2}\right)\delta_2\left(\frac{h}{2}\right) + \tau_1\delta_2\left(\frac{h}{2}\right)\delta_1\left(\frac{h}{2}\right)\right), \end{aligned}$$

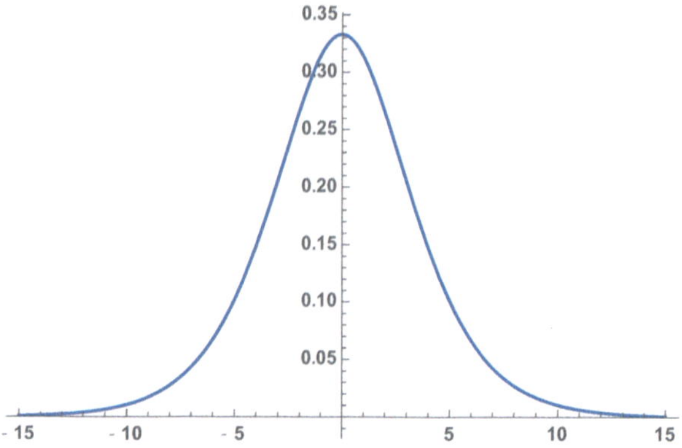

Fig. 1 Graph of K given by (19)

where

$$K(x) = \frac{2e^{x/2}(2 + e^x(-2 + x) + x)}{(-1 + e^x)^2 x}, \tag{19}$$

and

$$H(s, t) = -\frac{\begin{array}{c}-t(s+t)\cosh s + s(s+t)\cosh t - (s-t) \\ (s+t+\sinh s + \sinh t - \sinh(s+t))\end{array}}{\begin{array}{c}st(s+t)\sinh(s/2) \\ \sinh(t/2)\sinh^2((s+t)/2)\end{array}}. \tag{20}$$

Here the flat metric is conformally perturbed by e^{-h}, where $h = h^* \in C^\infty(\mathbb{T}_\theta^2)$, and ∇ is the logarithm of the modular automorphism $\Delta(a) = e^{-h}ae^h$, hence the derivation given by taking commutator with $-h$.

Using the symmetries of these functions describing the term $a_2(\Delta_{\tau,h})$ integrates to 0, hence the analog of the Gauss-Bonnet theorem. This result was proved in [22, 30] in a kind of simpler manner as by exploiting the trace property of φ_0 from the beginning of the symbolic calculations, only a one variable function was necessary to describe $\varphi_0(a_2(\Delta_{\tau,h}))$. However, for the description of a_2 one needs both one and two variable functions, which are given by (19) and (20). So we can state the Gauss-Bonnet theorem for $\mathbb{T}_\theta^2$ from [22, 30] as follows (Figure 2).

Theorem 4.2 *For any choice of the complex number τ in the upper half-plane and any conformal factor e^{-h}, where $h = h^* \in C^\infty(\mathbb{T}_\theta^2)$, one has*

$$\varphi_0(a_2(\Delta_{\tau,h})) = 0.$$

Hence the total curvature of $\mathbb{T}_\theta^2$ is independent of τ and h defining the metric.

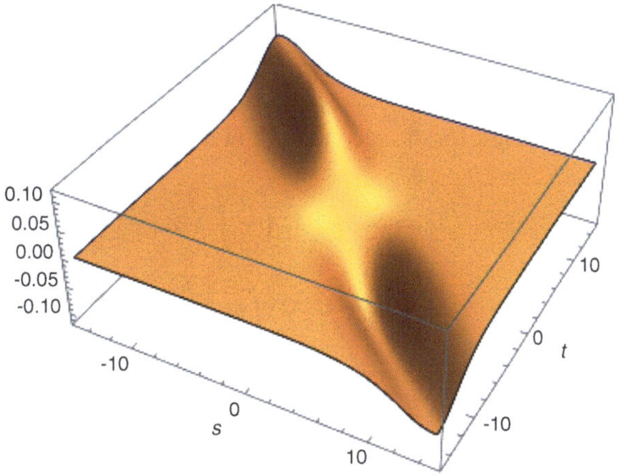

Fig. 2 Graph of H given by (20)

As we mentioned earlier, the validity of the Gauss-Bonnet theorem for $\mathbb{T}_\theta^2$ was suggested by developments on the spectral action in the presence of a dilaton [6] and also studies on twisted spectral triples [20]. In harmony with these developments, in fact a non-computational proof of the Gauss-Bonnet theorem can be given, as written in [21], in the spirit of conformal invariance of the value at the origin of the spectral zeta function of conformally covariant operators [4]. The argument is based on a variational technique: one can write a formula for the variation of the heat coefficients as one varies the metric conformally with e^{-sh}, where h is a dilaton, and the real parameter s goes from 0 to 1. However, the non-computational proof does not lead to an explicit formula for the curvature term $a_2(\Delta_{\tau,h})$. Hence the remarkable achievements in [22, 30, 21, 31] after heavy computer aided calculations include the explicit expression for the scalar curvature of $\mathbb{T}_\theta^2$ and the fact that the analog of the Gauss-Bonnet theorem holds for it.

4.2 *The Laplacian on* $(1,0)$*-forms on* $\mathbb{T}_\theta^2$ *with curved metric*

The analog of the Laplacian on $(1,0)$-forms is also considered in [21, 31] and the second term in its small time heat kernel expansion is calculated. The operator is anti-unitarily equivalent to the operator $\Delta_{\tau,h}^{(1,0)} = \bar{\partial} e^h \partial$, where $\partial = \delta_1 + \bar{\tau}\delta_2$ and $\bar{\partial} = \delta_1 + \tau\delta_2$. The symbol of this Laplacian is equal to $c_2(\xi) + c_1(\xi)$ where

$$c_2(\xi) = \xi_1^2 k^2 + 2\tau_1 \xi_1 \xi_2 k^2 + |\tau|^2 \xi_2^2 k^2,$$
$$c_1(\xi) = (\delta_1(k^2) + \tau\delta_2(k^2))\xi_1 + (\bar{\tau}\delta_1(k^2) + |\tau|^2 \delta_2(k^2))\xi_2.$$

Therefore by using the same strategy of using computer aided symbol calculations one can calculate the terms appearing in the following heat kernel expansion:

$$\mathrm{Tr}\left(ae^{-t\Delta_{\tau,h}^{(1,0)}}\right) \sim t^{-1}\sum_{n=0}^{\infty}\varphi_0\left(a\,a_{2n}(\Delta_{\tau,h}^{(1,0)})\right)t^n, \qquad a\in C^\infty(\mathbb{T}_\theta^2).$$

The result for the second term in this expansion is that [21, 31]

$$\begin{aligned}
a_2(\Delta_{\tau,h}^{(1,0)}) = S(\nabla)&\left(\delta_1^2\left(\frac{h}{2}\right)+|\tau|^2\delta_2^2\left(\frac{h}{2}\right)+2\tau_1\delta_1\delta_2\left(\frac{h}{2}\right)\right)\\
&+T(\nabla,\nabla)\left(\delta_1\left(\frac{h}{2}\right)\delta_1\left(\frac{h}{2}\right)+|\tau|^2\delta_2\left(\frac{h}{2}\right)\delta_2\left(\frac{h}{2}\right)\right.\\
&\left.+\,\tau_1\delta_1\left(\frac{h}{2}\right)\delta_2\left(\frac{h}{2}\right)+\tau_1\delta_2\left(\frac{h}{2}\right)\delta_1\left(\frac{h}{2}\right)\right)\\
&-i\tau_2 W(\nabla,\nabla)\left(\delta_1\left(\frac{h}{2}\right)\delta_2\left(\frac{h}{2}\right)-\delta_2\left(\frac{h}{2}\right)\delta_1\left(\frac{h}{2}\right)\right),
\end{aligned}$$

where

$$S(x) = -\frac{4e^x(-x+\sinh x)}{(-1+e^{x/2})^2(1+e^{x/2})^2x},$$

$$T(s,t) = -\cosh((s+t)/2)\times\frac{\frac{-t(s+t)\cosh s+s(s+t)\cosh t-(s-t)(s+t+\sinh s+\sinh t-\sinh(s+t))}{st(s+t)\sinh(s/2)\sinh(t/2)}}{\sinh^2((s+t)/2)},$$

and

$$W(s,t) = \frac{-s-t+t\cosh s+s\cosh t+\sinh s+\sinh t-\sinh(s+t)}{st\sinh(s/2)\sinh(t/2)\sinh((s+t)/2)}.$$

Using a simple iso-spectrality argument for the operators $\Delta_{\tau,h}$ and $\Delta_{\tau,h}^{(1,0)}$ one can argue that $\varphi_0\left(a_2(\Delta_{\tau,h}^{(1,0)})\right)=0$, based on the Gauss-Bonnet theorem proved in [22, 30]. However, one can also use properties of the functions S, T, W to prove this directly (Figure 3).

5 Noncommutative residues for noncommutative tori and curvature of noncommutative 4-tori

In this section we discuss noncommutative residues and illustrate an application of a noncommutative residue defined for noncommutative tori in calculating the

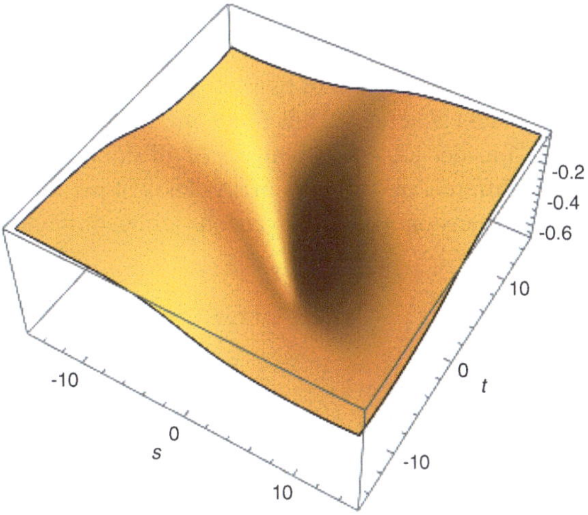

Fig. 3 Graph of W

scalar curvature of the noncommutative 4-torus in a convenient way with certain advantages.

5.1 *Noncommutative residues*

Noncommutative residues are trace functionals on algebras of pseudodifferential operators, which were first discovered by Adler and Manin in dimension 1 [1, 49]. In order to illustrate their construction in dimension 1 we consider the algebra $C^\infty(\mathbb{S}^1)$ of smooth functions on the circle $\mathbb{S}^1 = \mathbb{R}/\mathbb{Z}$, and the differentiation $(-i)d/dx$, whose pseudodifferential symbol is $\sigma(\xi) = \xi$. We then consider the algebra of pseudodifferential symbols of the form

$$\sum_{n=-\infty}^{N} a_n(x)\xi^n, \qquad a_n(x) \in C^\infty(\mathbb{S}^1), \qquad N \in \mathbb{Z}.$$

The product rule of this algebra can be deduced from the following relations:

$$\xi a(x) = a(x)\xi + a'(x), \qquad a_n(x) \in C^\infty(\mathbb{S}^1),$$

which are dictated by the Leibniz property of differentiation. The Adler-Manin trace is the linear functional defined by

$$\sum_{n=-\infty}^{N} a_n(x)\xi^n \mapsto \int_{\mathbb{S}^1} a_{-1}(x)\, dx,$$

which is shown to be a trace functional on the algebra of pseudodifferential symbols on the circle [1, 49]. A twisted version of this trace was worked out in [27], motivated by the notion of twisted spectral triples [20].

Wodzicki generalized this functional, in a remarkable work, to higher dimensions [55]. Consider a closed manifold M of dimension m and the algebra of classical pseudodifferential operators M. A classical pseudodifferential symbol σ of order d has an expansion with homogeneous terms, of the form

$$\sigma(x,\xi) \sim \sum_{j=0}^{\infty} \sigma_{d-j}(x,\xi),$$

where $\sigma_{d-j}(x, t\xi) = t^{d-j}\sigma_{d-j}(x,\xi)$ for any $t > 0$. The composition rule of this algebra is induced by the composition rule for the symbol of pseudodifferential operators:

$$\sigma_{P_1 P_2}(x,\xi) \sim \sum_{\alpha \in \mathbb{Z}^m_{\geq 0}} \frac{(-i)^{|\alpha|}}{\alpha!} \partial_\xi^\alpha \sigma_{P_1}(x,\xi)\, \partial_x^\alpha \sigma_{P_2}(x,\xi),$$

which we mentioned and used in Section 3 as well. Wodzicki's noncommutative residue WRes is the linear functional defined on the algebra of classical pseudodifferential symbols by

$$\mathrm{WRes}\left(\sum_{j=0}^{\infty} \sigma_{d-j}(x,\xi)\right) = \int_{S^*M} \mathrm{tr}(\sigma_{-m}(x,\xi))\, d^{m-1}\xi\, d^m x, \tag{21}$$

where S^*M is the cosphere bundle of the manifold with respect to a Riemannian metric. We stress that in this formula m is the dimension of the manifold M. It is proved that WRes is the unique trace functional on the algebra of classical pseudodifferential symbols on M [55].

The noncommutative residue has a spectral formulation as well. That is, one can fix a Laplacian $\triangle$ on M and define the noncommutative residue of a pseudodifferential operator P_σ to be the residue at $s = 0$ of the meromorphic extension of the zeta function defined, for complex numbers s with large enough real parts, by

$$s \mapsto \mathrm{Tr}(P_\sigma \triangle^{-s}).$$

This formulation is used in noncommutative geometry, when one works with the algebra of pseudodifferential operators associated with a spectral triple [19].

For noncommutative tori, the analog of formula (21) can be written and it was shown in [33] that it gives the unique *continuous* trace functional on the algebra of classical pseudodifferential operators on the noncommutative 2-torus. Although the argument written in [33] is for dimension 2, but it is general enough that works for any dimension, see, for example, [32] for the illustration in dimension 4. Given a classical pseudodifferential symbol $\rho : \mathbb{R}^m \to C^\infty(\mathbb{T}^m_\Theta)$ of order d on the noncommutative m-torus, by definition, there is an asymptotic expansion for $\xi \to \infty$ of the form

$$\rho(\xi) \sim \sum_{j=0}^{\infty} \rho_{d-j}(\xi),$$

where each ρ_{d-j} is positively homogeneous of order $d - j$. One can define the noncommutative residue Res of the corresponding pseudodifferential symbol as

$$\mathrm{Res}(P_\rho) = \int_{\mathbb{S}^{m-1}} \varphi_0 \left(\rho_{-m}\right) d\Omega, \tag{22}$$

where φ_0 is the canonical trace on $C(\mathbb{T}^m_\Theta)$ and $d\Omega$ is the volume form of the round metric on the $(m - 1)$-dimensional sphere in $\mathbb{R}^m$. The same argument as the one given in [33] shows that Res is the unique continuous trace on the algebra of classical pseudodifferential symbols on $\mathbb{T}^m_\Theta$.

5.2 *Scalar curvature of the noncommutative 4-torus*

The Laplacian associated with the flat geometry of the noncommutative four torus $\mathbb{T}^4_\Theta$ is simply given by the sum of the squares of the canonical derivatives, namely:

$$\triangle_0 = \delta_1^2 + \delta_2^2 + \delta_3^2 + \delta_4^2.$$

After conformally perturbing the flat metric on $\mathbb{T}^4_\Theta$ by means of a conformal factor e^{-h}, for a fixed $h = h^* \in C^\infty(\mathbb{T}^4_\Theta)$, the perturbed Laplacian is shown in [32] to be anti-unitarily equivalent to the operator

$$\triangle_h = e^h \bar{\partial}_1 e^{-h} \partial_1 e^h + e^h \partial_1 e^{-h} \bar{\partial}_1 e^h + e^h \bar{\partial}_2 e^{-h} \partial_2 e^h + e^h \partial_2 e^{-h} \bar{\partial}_2 e^h,$$

where

$$\partial_1 = \delta_1 - i\delta_3, \qquad \partial_2 = \delta_2 - i\delta_4,$$

$$\bar{\partial}_1 = \delta_1 + i\delta_3, \qquad \bar{\partial}_2 = \delta_2 + i\delta_4.$$

The latter are the analogues of the Dolbeault operators.

The scalar curvature of the metric on $\mathbb{T}^4_\Theta$ encoded in $\triangle_h$ is the term $a_2 \in C^\infty(\mathbb{T}^4_\Theta)$ appearing in the following small time asymptotic expansion:

$$\mathrm{Tr}(ae^{-t\Delta_h}) \sim t^{-2} \sum_{n=0}^{\infty} \varphi_0(a\, a_{2n}) t^n, \qquad a \in C^\infty(\mathbb{T}^4_\Theta).$$

The curvature term $a_2 \in C^\infty(\mathbb{T}^4_\Theta)$ was calculated in [32] by going through the procedure explained in Section 3.3. As we explained earlier, there is a purely noncommutative obstruction in this procedure that needs to be overcome by Lemma 4.1, the so-called rearrangement lemma. That is, one encounters integration over the Euclidean space of C^*-algebra valued functions. For this type of integrations, one can pass to polar coordinates and take care of the angular integrations with no problem. However, the redial integration brings forth the necessity of the rearrangement lemma.

Striking is the fact that after applying the rearrangement lemma to hundreds of terms, each of which involves a function from this lemma to appear in the calculations, the final formula for the curvature simplifies significantly with computer aid. In [28], by using properties of the noncommutative residue (22), it was shown that the curvature term $a_2 \in C^\infty(\mathbb{T}^4_\Theta)$ can be calculated as the integral over the 3-sphere of a homogeneous symbol. Therefore, with this method, the calculation of a_2 does not require radial integration, hence the calculation without using the rearrangement lemma and clarification of the reason for the significant simplifications. In fact, in [28], the term is shown to be a scalar multiple of $\int_{\mathbb{S}^3} b_2(\xi)\, d\Omega$, where b_2 is the homogeneous term of order -4 in the expansion of the symbol of the parametrix of Δ_h. The result, in agreement with the calculation of [32], is that

$$a_2 = e^{-h} K(\nabla)\Big(\sum_{i=1}^{4} \delta_i^2(h)\Big) + e^{-h} H(\nabla, \nabla)\Big(\sum_{i=1}^{4} \delta_i(h)^2\Big) \in C^\infty(\mathbb{T}^4_\Theta), \quad (23)$$

where $\nabla = [-h, \cdot]$, and

$$K(x) = \frac{1-e^{-x}}{2x},$$
$$H(s,t) = -\frac{e^{-s-t}\left((-e^s-3)\, s\left(e^t-1\right) + (e^s-1)\left(3e^t+1\right)t\right)}{4st(s+t)}. \quad (24)$$

The simplicity of this calculation also revealed in [28] the following functional relation between the functions K and H (Figure 4).

Theorem 5.1 *Let $\tilde{K}(s) = e^s K(s)$ and $\tilde{H}(s,t) = e^{s+t} H(s,t)$, where the function K and H are given by* (24). *Then*

$$\tilde{H}(s,t) = 2\frac{\tilde{K}(s+t) - \tilde{K}(s)}{t} + \frac{3}{2}\tilde{K}(s)\tilde{K}(t).$$

Another important result that we wish to recall from [32] is about the extrema of the analog of the Einstein-Hilbert action for $\mathbb{T}^4_\Theta$, namely $\varphi_0(a_2)$:

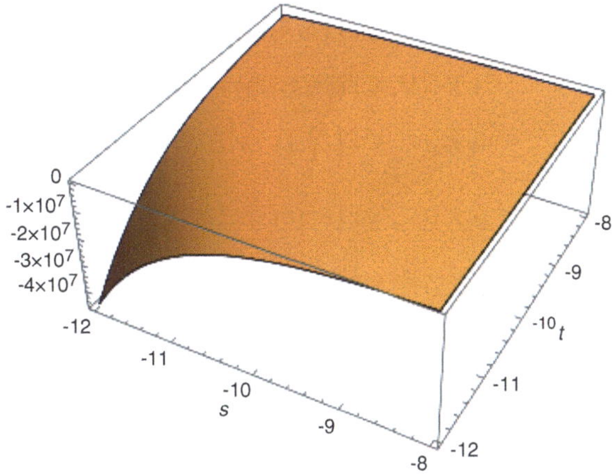

Fig. 4 Graph of H given by (24)

Theorem 5.2 *For any conformal factor e^{-h}, where $h = h^* \in C^\infty(\mathbb{T}^4_\Theta)$,*

$$\varphi_0(a_2) \leq 0,$$

where $a_2 \in C^\infty(\mathbb{T}^4_\Theta)$ is the scalar curvature given by (23). Moreover, we have $\varphi_0(a_2) = 0$ if and only if h is a scalar.

6 The Riemann curvature tensor and the term a_4 for noncommutative tori

The Riemann curvature tensor appears in the term a_4 in the heat kernel expansion for the Laplacian of any closed Riemannian manifold M. That is, if Δ_g is the Laplacian of a Riemannian metric g, which acts on $C^\infty(M)$, then

$$a_4(x, \Delta_g) = (4\pi)^{-1}(1/360)(-12\Delta_g R(x) + 5R(x)^2 - 2|Ric(x)|^2 + 2|Riem(x)|^2).$$

In this section we recall from [17] the formula obtained for the analog of the term a_4 in a noncommutative setting. Recall that in Section 4.1, we discussed the term a_2, namely the analog of the scalar curvature, for the noncommutative two torus when the flat metric is perturbed by a positive invertible element $e^{-h} \in C^\infty(\mathbb{T}^2_\theta)$, where $h = h^*$. These geometric terms appear in the expansion given by (18). Setting,

$$\ell = \frac{h}{2}$$

for the simplest conformal class (associated with $\tau = i$), the main calculation of [17] gives the term a_4 by a formula of the following form:

$$
\begin{aligned}
a_4(h) = -e^{2\ell}\Big(& K_1(\nabla)\left(\delta_1^2\delta_2^2(\ell)\right) + K_2(\nabla)\left(\delta_1^4(\ell)+\delta_2^4(\ell)\right) \\
&+K_3(\nabla,\nabla)\left((\delta_1\delta_2(\ell))\cdot(\delta_1\delta_2(\ell))\right) \\
&+K_4(\nabla,\nabla)\left(\delta_1^2(\ell)\cdot\delta_2^2(\ell)+\delta_2^2(\ell)\cdot\delta_1^2(\ell)\right) \\
&+K_5(\nabla,\nabla)\left(\delta_1^2(\ell)\cdot\delta_1^2(\ell)+\delta_2^2(\ell)\cdot\delta_2^2(\ell)\right) \\
&+K_6(\nabla,\nabla)\Big(\delta_1(\ell)\cdot\delta_1^3(\ell)+\delta_1(\ell)\cdot\left(\delta_1\delta_2^2(\ell)\right)+\delta_2(\ell)\cdot\delta_2^3(\ell) \\
&\qquad +\delta_2(\ell)\cdot\left(\delta_1^2\delta_2(\ell)\right)\Big) \\
&+K_7(\nabla,\nabla)\Big(\delta_1^3(\ell)\cdot\delta_1(\ell)+\left(\delta_1\delta_2^2(\ell)\right)\cdot\delta_1(\ell)+\delta_2^3(\ell)\cdot\delta_2(\ell) \\
&\qquad +\left(\delta_1^2\delta_2(\ell)\right)\cdot\delta_2(\ell)\Big) \\
&+K_8(\nabla,\nabla,\nabla)\left(\delta_1(\ell)\cdot\delta_1(\ell)\cdot\delta_2^2(\ell)+\delta_2(\ell)\cdot\delta_2(\ell)\cdot\delta_1^2(\ell)\right) \\
&+K_9(\nabla,\nabla,\nabla)\left(\delta_1(\ell)\cdot\delta_2(\ell)\cdot(\delta_1\delta_2(\ell))+\delta_2(\ell)\cdot\delta_1(\ell)\cdot(\delta_1\delta_2(\ell))\right) \\
&+K_{10}(\nabla,\nabla,\nabla)\left(\delta_1(\ell)\cdot(\delta_1\delta_2(\ell))\cdot\delta_2(\ell)+\delta_2(\ell)\cdot(\delta_1\delta_2(\ell))\cdot\delta_1(\ell)\right) \\
&+K_{11}(\nabla,\nabla,\nabla)\left(\delta_1(\ell)\cdot\delta_2^2(\ell)\cdot\delta_1(\ell)+\delta_2(\ell)\cdot\delta_1^2(\ell)\cdot\delta_2(\ell)\right) \\
&+K_{12}(\nabla,\nabla,\nabla)\left(\delta_1^2(\ell)\cdot\delta_2(\ell)\cdot\delta_2(\ell)+\delta_2^2(\ell)\cdot\delta_1(\ell)\cdot\delta_1(\ell)\right) \\
&+K_{13}(\nabla,\nabla,\nabla)\left((\delta_1\delta_2(\ell))\cdot\delta_1(\ell)\cdot\delta_2(\ell)+(\delta_1\delta_2(\ell))\cdot\delta_2(\ell)\cdot\delta_1(\ell)\right) \\
&+K_{14}(\nabla,\nabla,\nabla)\left(\delta_1^2(\ell)\cdot\delta_1(\ell)\cdot\delta_1(\ell)+\delta_2^2(\ell)\cdot\delta_2(\ell)\cdot\delta_2(\ell)\right) \\
&+K_{15}(\nabla,\nabla,\nabla)\left(\delta_1(\ell)\cdot\delta_1(\ell)\cdot\delta_1^2(\ell)+\delta_2(\ell)\cdot\delta_2(\ell)\cdot\delta_2^2(\ell)\right) \\
&+K_{16}(\nabla,\nabla,\nabla)\left(\delta_1(\ell)\cdot\delta_1^2(\ell)\cdot\delta_1(\ell)+\delta_2(\ell)\cdot\delta_2^2(\ell)\cdot\delta_2(\ell)\right) \\
&+K_{17}(\nabla,\nabla,\nabla,\nabla)\left(\delta_1(\ell)\cdot\delta_1(\ell)\cdot\delta_2(\ell)\cdot\delta_2(\ell)\right. \\
&\qquad\qquad \left.+\delta_2(\ell)\cdot\delta_2(\ell)\cdot\delta_1(\ell)\cdot\delta_1(\ell)\right) \\
&+K_{18}(\nabla,\nabla,\nabla,\nabla)\left(\delta_1(\ell)\cdot\delta_2(\ell)\cdot\delta_1(\ell)\cdot\delta_2(\ell)\right. \\
&\qquad\qquad \left.+\delta_2(\ell)\cdot\delta_1(\ell)\cdot\delta_2(\ell)\cdot\delta_1(\ell)\right) \\
&+K_{19}(\nabla,\nabla,\nabla,\nabla)\left(\delta_1(\ell)\cdot\delta_2(\ell)\cdot\delta_2(\ell)\cdot\delta_1(\ell)\right. \\
&\qquad\qquad \left.+\delta_2(\ell)\cdot\delta_1(\ell)\cdot\delta_1(\ell)\cdot\delta_2(\ell)\right) \\
&+K_{20}(\nabla,\nabla,\nabla,\nabla)\left(\delta_1(\ell)\cdot\delta_1(\ell)\cdot\delta_1(\ell)\cdot\delta_1(\ell)\right. \\
&\qquad\qquad \left.+\delta_2(\ell)\cdot\delta_2(\ell)\cdot\delta_2(\ell)\cdot\delta_2(\ell)\right)\Big).
\end{aligned} \tag{25}
$$

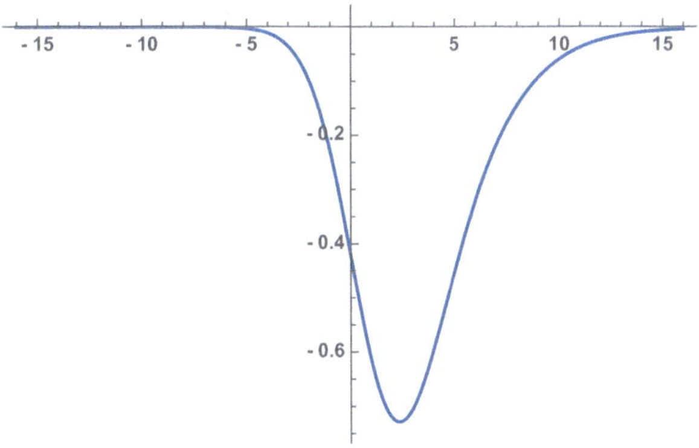

Fig. 5 Graph of K_1

We provide the explicit formulas for a few of the functions appearing in (25), and we refer the reader to [17] for the remaining functions, most of which have lengthy expressions. We have, for example, (Figure 5):

$$K_1(s_1) = -\frac{4\pi e^{\frac{3s_1}{2}} \left(\left(4e^{s_1} + e^{2s_1} + 1\right) s_1 - 3e^{2s_1} + 3\right)}{\left(e^{s_1} - 1\right)^4 s_1}, \tag{26}$$

and

$$K_3(s_1, s_2) = \frac{K_3^{\text{num}}(s_1, s_2)}{\left(e^{s_1} - 1\right)^2 \left(e^{s_2} - 1\right)^2 \left(e^{s_1+s_2} - 1\right)^4 s_1 s_2 \left(s_1 + s_2\right)}, \tag{27}$$

where the numerator is given by

$$\begin{aligned}
K_3^{\text{num}}(s_1, s_2) = 16\, e^{\frac{3s_1}{2} + \frac{3s_2}{2}} \pi \Big[& \left(e^{s_1} - 1\right) \left(e^{s_2} - 1\right) \left(e^{s_1+s_2} - 1\right) \\
& \times \Big\{ \Big(-5e^{s_1} - e^{s_2} + 6e^{s_1+s_2} - e^{2s_1+s_2} - 5e^{s_1+2s_2} \\
& \quad +3e^{2s_1+2s_2} + 3\Big) s_1 \\
& + \left(e^{s_1} + 5e^{s_2} - 6e^{s_1+s_2} + 5e^{2s_1+s_2} + e^{s_1+2s_2} - 3e^{2s_1+2s_2} - 3\right) s_2 \Big\} \\
& -2 \left(e^{s_1} - e^{s_2}\right) \left(e^{s_1+s_2} - 1\right) \\
& \times \left(-e^{s_1} - e^{s_2} - e^{2s_1+s_2} - e^{s_1+2s_2} + 2e^{2s_1+2s_2} + 2\right) s_1 s_2 \\
& +2e^{s_1} \left(e^{s_2} - 1\right)^3 \left(e^{s_1} - e^{s_1+s_2} + 2e^{2s_1+s_2} - 2\right) s_1^2 \\
& -2e^{s_2} \left(e^{s_1} - 1\right)^3 \left(e^{s_2} - e^{s_1+s_2} + 2e^{s_1+2s_2} - 2\right) s_2^2 \Big].
\end{aligned}$$

6.1 Functional relations

One of the main results of [17] is the derivation of a family of conceptually predicted functional relations among the functions $K_1, \ldots, K_{20}$ appearing in (25). As we shall see shortly the functional relations are highly nontrivial. There are two main reasons for the derivation of the relations, both of which are extremely important. First, the calculation of the term a_4 involves a really heavy computer aided calculation, hence, the need for a way of confirming the validity of the outcome by checking that the expected functional relations are satisfied. Second, patterns and structural properties of the relations give significant information that help one to obtain conceptual understandings about the structure of the complicated functions appearing in the formula for a_4. In order to present the relations, we need to consider the modification of each function K_j in (25) to a new function $\widetilde{K}_j$ by the formula

$$\widetilde{K}_j(s_1, \ldots, s_n) = \frac{1}{2^n} \frac{\sinh\left((s_1 + \cdots + s_n)/2\right)}{(s_1 + \cdots + s_n)/2} K_j(s_1, \ldots, s_n),$$

where $n \in \{1, 2, 3, 4\}$ is the number of variables, on which K_j depends. We also need to introduce the restriction of the functions K_j to certain hyperplanes by setting

$$k_j(s_1, \ldots, s_{n-1}) = K_j(s_1, \ldots, s_{n-1}, -s_1 - \cdots - s_{n-1}).$$

We shall explain shortly how these functional relations are predicted, using fundamental identities and lemmas [21, 17] (Figure 6).

Let us first list a few of the functional relations in which some auxiliary functions $G_n(s_1, \ldots, s_n)$ appear. These functions are mainly useful for relating the derivatives of e^h and those of h and we recall from [17] their recursive formula:

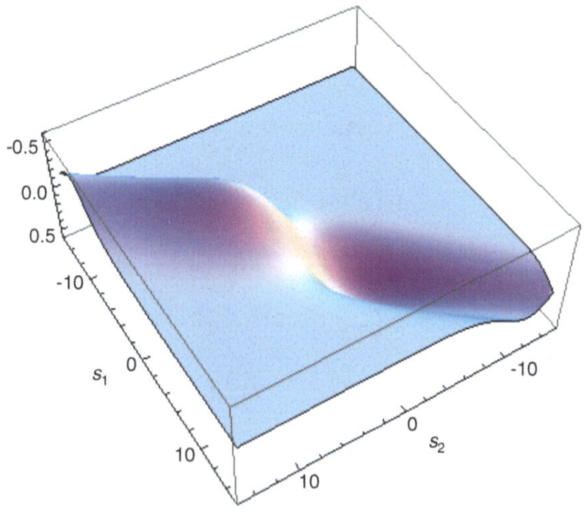

Fig. 6 Graph of K_3

Lemma 6.1 *The functions $G_n(s_1, \dots, s_n)$ are given recursively by*

$$G_0 = 1,$$

and

$$G_n(s_1, \dots, s_n) = \int_0^1 r^{n-1} e^{s_1 r}\, G_{n-1}(rs_2, rs_3, \dots, rs_n)\, dr.$$

Explicitly, for $n = 1, 2, 3$, one has:

$$G_1(s_1) = \frac{e^{s_1} - 1}{s_1}, \tag{28}$$

$$G_2(s_1, s_2) = \frac{e^{s_1}\left((e^{s_2} - 1)\, s_1 - s_2\right) + s_2}{s_1 s_2\, (s_1 + s_2)},$$

$$G_3(s_1, s_2, s_3) = \frac{\begin{array}{c} e^{s_1}\left(e^{s_2+s_3} s_1 s_2 (s_1+s_2) + (s_1+s_2+s_3)\right. \\ \left.((s_1+s_2)s_3 - e^{s_2} s_1 (s_2+s_3))\right) - s_2 s_3 (s_2+s_3) \end{array}}{\begin{array}{c} s_1 s_2 (s_1+s_2) s_3 \\ (s_2+s_3)(s_1+s_2+s_3) \end{array}}.$$

We can now write the relations. The functional relation associated with the function K_1 is given by

$$\begin{aligned} \widetilde{K}_1(s_1) = &-\frac{1}{15}\pi G_1(s_1) + \frac{1}{4} e^{s_1} k_3(-s_1) + \frac{1}{4} k_3(s_1) \\ &+ \frac{1}{2} e^{s_1} k_4(-s_1) + \frac{1}{2} k_4(s_1) - \frac{1}{2} e^{s_1} k_6(-s_1) \\ &- \frac{1}{2} k_6(s_1) - \frac{1}{2} e^{s_1} k_7(-s_1) - \frac{1}{2} k_7(s_1) - \frac{\pi\,(e^{s_1} - 1)}{15 s_1}. \end{aligned} \tag{29}$$

It is quite remarkable that such a nontrivial relation should exist among the functions, and it gets even more interesting when one looks at the case associated with a 2-variable function. For K_3 one finds the associated relation to be:

$$\begin{aligned} \widetilde{K}_3(s_1, s_2) = &\frac{1}{15}(-4)\pi G_2(s_1, s_2) + \frac{1}{2} k_8(s_1, s_2) + \frac{1}{4} k_9(s_1, s_2) \\ &- \frac{1}{4} e^{s_1+s_2} k_9(-s_1 - s_2, s_1) \\ &- \frac{1}{4} e^{s_1} k_9(s_2, -s_1 - s_2) - \frac{1}{4} k_{10}(s_1, s_2) - \frac{1}{4} e^{s_1+s_2} k_{10}(-s_1 - s_2, s_1) \\ &+ \frac{1}{4} e^{s_1} k_{10}(s_2, -s_1 - s_2) \end{aligned} \tag{30}$$

$$+\frac{1}{2}e^{s_1}k_{11}(s_2,-s_1-s_2)+\frac{1}{2}e^{s_1+s_2}k_{12}(-s_1-s_2,s_1)$$
$$-\frac{1}{4}k_{13}(s_1,s_2)+\frac{1}{4}e^{s_1+s_2}k_{13}(-s_1-s_2,s_1)-\frac{1}{4}e^{s_1}k_{13}(s_2,-s_1-s_2)$$
$$+\frac{1}{4}e^{s_2}G_1(s_1)k_3(-s_2)+\frac{1}{4}G_1(s_1)k_3(s_2)-G_1(s_1)k_6(s_2)$$
$$-e^{s_2}G_1(s_1)k_7(-s_2)$$
$$+\frac{\left(e^{s_1+s_2}-1\right)k_3(s_1)}{4(s_1+s_2)}+\frac{k_3(s_2)-k_3(s_1+s_2)}{4s_1}$$
$$+\frac{k_3(s_1+s_2)-k_3(s_1)}{4s_2}$$
$$+\frac{k_6(s_1)-k_6(s_1+s_2)}{s_2}+\frac{k_6(s_1+s_2)-k_6(s_2)}{s_1}$$
$$+\frac{e^{s_1}(k_7(-s_1)-e^{s_2}k_7(-s_1-s_2))}{s_2}$$
$$+\frac{e^{s_2}(e^{s_1}k_7(-s_1-s_2)-k_7(-s_2))}{s_1}$$
$$-\frac{e^{s_2}(e^{s_1}k_3(-s_1-s_2)-k_3(-s_2))}{4s_1}$$
$$-\frac{e^{s_1}(k_3(-s_1)-e^{s_2}k_3(-s_1-s_2))}{4s_2}$$
$$-\frac{e^{s_1}(k_3(-s_1)+e^{s_2}k_3(s_1)-e^{s_2}k_3(-s_2)-k_3(s_2))}{4(s_1+s_2)}.$$

The rapid pace in growing complexity of the functional relations can be seen in the higher variable cases as, for example, the functional relation corresponding to the 3-variable function K_8 is the following expression:

$$\widetilde{K}_8(s_1,s_2,s_3)=\frac{1}{15}(-2)\pi G_3(s_1,s_2,s_3)+\frac{1}{2}e^{s_3}G_2(s_1,s_2)k_4(-s_3) \tag{31}$$
$$-\frac{e^{s_3}\left(e^{s_2}s_1k_4(-s_2-s_3)+e^{s_2}s_2k_4(-s_2-s_3)-e^{s_1+s_2}s_2k_4(-s_1-s_2-s_3)-s_1k_4(-s_3)\right)}{2s_1s_2(s_1+s_2)}$$
$$+\frac{1}{2}G_2(s_1,s_2)k_4(s_3)+\frac{G_1(s_1)(k_4(s_3)-k_4(s_2+s_3))}{2s_2}$$
$$+\frac{s_1k_4(s_3)-s_1k_4(s_2+s_3)-s_2k_4(s_2+s_3)+s_2k_4(s_1+s_2+s_3)}{2s_1s_2(s_1+s_2)}$$
$$-\frac{1}{2}G_2(s_1,s_2)k_6(s_3)$$

$$+\frac{G_1(s_1)(k_6(s_2)-k_6(s_2+s_3))}{4s_3}+\frac{\begin{array}{c}k_6(s_2)-k_6(s_1+s_2)-k_6(s_2+s_3)\\+k_6(s_1+s_2+s_3)\end{array}}{4s_1s_3}$$

$$+\frac{\begin{array}{c}-s_3k_6(s_1)+s_2k_6(s_1+s_2)+s_3k_6(s_1+s_2)\\-s_2k_6(s_1+s_2+s_3)\end{array}}{4s_2s_3(s_2+s_3)}$$

$$+\frac{\begin{array}{c}-s_1k_6(s_3)+s_1k_6(s_2+s_3)+s_2k_6(s_2+s_3)\\-s_2k_6(s_1+s_2+s_3)\end{array}}{2s_1s_2(s_1+s_2)}$$

$$+\frac{e^{s_2}G_1(s_1)(k_7(-s_2)-e^{s_3}k_7(-s_2-s_3))}{4s_3}$$

$$-\frac{\begin{array}{c}e^{s_1}\Big(s_3k_7(-s_1)-e^{s_2}s_2k_7(-s_1-s_2)-e^{s_2}s_3k_7(-s_1-s_2)\\+e^{s_2+s_3}s_2k_7(-s_1-s_2-s_3)\Big)\end{array}}{4s_2s_3(s_2+s_3)}$$

$$-\frac{\begin{array}{c}e^{s_2}\Big(e^{s_1}k_7(-s_1-s_2)-k_7(-s_2)+e^{s_3}k_7(-s_2-s_3)\\-e^{s_1+s_3}k_7(-s_1-s_2-s_3)\Big)\end{array}}{4s_1s_3}$$

$$+\frac{e^{s_3}G_1(s_1)(e^{s_2}k_7(-s_2-s_3)-k_7(-s_3))}{2s_2}$$

$$-\frac{1}{2}e^{s_3}G_2(s_1,s_2)k_7(-s_3)$$

$$+\frac{\begin{array}{c}e^{s_3}\Big(e^{s_2}s_1k_7(-s_2-s_3)+e^{s_2}s_2k_7(-s_2-s_3)\\-e^{s_1+s_2}s_2k_7(-s_1-s_2-s_3)-s_1k_7(-s_3)\Big)\end{array}}{2s_1s_2(s_1+s_2)}$$

$$+\frac{\left(-1+e^{s_1+s_2+s_3}\right)k_8(s_1,s_2)}{8(s_1+s_2+s_3)}+\frac{k_8(s_1,s_2+s_3)-k_8(s_1,s_2)}{8s_3}$$

$$-\frac{1}{8}e^{s_2+s_3}G_1(s_1)k_8(-s_2-s_3,s_2)$$

$$+\frac{e^{s_1+s_2+s_3}(k_8(-s_1-s_2-s_3,s_1)-k_8(-s_1-s_2-s_3,s_1+s_2))}{8s_2}$$

$$+\frac{1}{8}G_1(s_1)k_9(s_2,s_3)$$

$$+\frac{k_9(s_2,s_3)-k_9(s_1+s_2,s_3)}{8s_1}$$

$$+\frac{k_9(s_1+s_2,s_3)-k_9(s_1,s_2+s_3)}{8s_2}$$

$$
\begin{aligned}
&+\frac{1}{8}e^{s_2}G_1(s_1)k_{10}(s_3,-s_2-s_3)\\
&+\frac{e^{s_2}\left(k_{10}(s_3,-s_2-s_3)-e^{s_1}k_{10}(s_3,-s_1-s_2-s_3)\right)}{8s_1}\\
&+\frac{e^{s_1}\left(e^{s_2}k_{10}(s_3,-s_1-s_2-s_3)-k_{10}(s_2+s_3,-s_1-s_2-s_3)\right)}{8s_2}\\
&-\frac{1}{8}G_1(s_1)k_{11}(s_2,s_3)\\
&+\frac{k_{11}(s_1,s_2+s_3)-k_{11}(s_1+s_2,s_3)}{8s_2}\\
&+\frac{k_{11}(s_1+s_2,s_3)-k_{11}(s_2,s_3)}{8s_1}\\
&-\frac{1}{8}e^{s_2}G_1(s_1)k_{12}(s_3,-s_2-s_3)\\
&+\frac{1}{8}e^{s_2+s_3}G_1(s_1)k_{13}(-s_2-s_3,s_2)\\
&+\frac{e^{s_2+s_3}\left(k_{13}(-s_2-s_3,s_2)-e^{s_1}k_{13}(-s_1-s_2-s_3,s_1+s_2)\right)}{8s_1}\\
&-\frac{1}{16}k_{17}(s_1,s_2,s_3)-\frac{1}{16}e^{s_1+s_2}k_{17}(s_3,-s_1-s_2-s_3,s_1)\\
&-\frac{1}{16}e^{s_1}k_{19}(s_2,s_3,-s_1-s_2-s_3)\\
&-\frac{1}{16}e^{s_1+s_2+s_3}k_{19}(-s_1-s_2-s_3,s_1,s_2)\\
&-\frac{e^{s_2+s_3}\left(k_8(-s_2-s_3,s_2)-e^{s_1}k_8(-s_1-s_2-s_3,s_1+s_2)\right)}{8s_1}\\
&-\frac{e^{s_2}\left(k_{12}(s_3,-s_2-s_3)-e^{s_1}k_{12}(s_3,-s_1-s_2-s_3)\right)}{8s_1}\\
&-\frac{e^{s_3}G_1(s_1)\left(e^{s_2}k_4(-s_2-s_3)-k_4(-s_3)\right)}{2s_2}\\
&-\frac{G_1(s_1)\left(k_6(s_3)-k_6(s_2+s_3)\right)}{2s_2}\\
&-\frac{e^{s_1}\left(e^{s_2}k_{12}(s_3,-s_1-s_2-s_3)-k_{12}(s_2+s_3,-s_1-s_2-s_3)\right)}{8s_2}\\
&-\frac{\substack{e^{s_1+s_2+s_3}(k_{13}(-s_1-s_2-s_3,s_1)\\ -k_{13}(-s_1-s_2-s_3,s_1+s_2))}}{8s_2}
\end{aligned}
$$

$$-\frac{e^{s_1}\left(k_{11}\left(s_2,-s_1-s_2\right)-k_{11}\left(s_2+s_3,-s_1-s_2-s_3\right)\right)}{8s_3}$$

$$-\frac{e^{s_1+s_2}\left(k_{12}\left(-s_1-s_2,s_1\right)-e^{s_3}k_{12}\left(-s_1-s_2-s_3,s_1\right)\right)}{8s_3}$$

$$-\frac{e^{s_1+s_2+s_3}\left(k_8\left(s_1,s_2\right)-k_8\left(-s_2-s_3,s_2\right)\right)}{8\left(s_1+s_2+s_3\right)}$$

$$-\frac{e^{s_1}\left(k_{11}\left(s_2,-s_1-s_2\right)-k_{11}\left(s_2,s_3\right)\right)}{8\left(s_1+s_2+s_3\right)}$$

$$-\frac{e^{s_1+s_2}\left(k_{12}\left(-s_1-s_2,s_1\right)-k_{12}\left(s_3,-s_2-s_3\right)\right)}{8\left(s_1+s_2+s_3\right)}.$$

The interested reader can refer to [17] to see that the functional relations of the 4-variable functions get even more complicated. The main point, which will be elaborated further, is that all these functional relations are derived conceptually, and by checking that our calculated functions $K_1, \ldots, K_{20}$ satisfy these relations, the validity of the calculations and their outcome, such as the explicit formulas (26), (27), is confirmed.

6.2 A partial differential system associated with the functional relations

When one takes a close look at the functional relations, one notices that there are terms in the right-hand sides (in the finite difference expressions) with $s_1+\cdots+s_n$ in their denominators. For example, in (30) one can see that there is a term with s_1+s_2 in the denominator. The question answered in [17], which leads to a differential system with interesting properties, is what happens when one restricts the functional relations to the hyperplanes $s_1+\cdots+s_n=0$ by setting $s_1+\cdots+s_n=\varepsilon$ and letting $\varepsilon\to 0$. For example, the restriction of the functional relation (30) to the hyperplane $s_1+s_2=0$ yields:

$$\begin{aligned}\frac{1}{4}e^{s_1}k_3'(-s_1)-\frac{1}{4}k_3'(s_1)=\frac{1}{60s_1}\Big(&16\pi s_1G_2(s_1,-s_1)-30s_1k_8(s_1,-s_1)\\&+15s_1k_9(0,s_1)+15e^{s_1}s_1k_9(-s_1,0)\\&-15s_1k_9(s_1,-s_1)+15s_1k_{10}(0,s_1)\\&-15e^{s_1}s_1k_{10}(-s_1,0)+15s_1k_{10}(s_1,-s_1)\\&-30e^{s_1}s_1k_{11}(-s_1,0)-30s_1k_{12}(0,s_1)\end{aligned} \tag{32}$$

$$
\begin{aligned}
&-15s_1k_{13}(0,s_1)+15e^{s_1}s_1k_{13}(-s_1,0)\\
&+15s_1k_{13}(s_1,-s_1)-15s_1G_1(s_1)k_3(-s_1)\\
&-15e^{-s_1}s_1G_1(s_1)k_3(s_1)+60s_1G_1(s_1)k_6(-s_1)\\
&+60e^{-s_1}s_1G_1(s_1)k_7(s_1)-15e^{s_1}k_3(-s_1)\\
&-15k_3(-s_1)-15e^{-s_1}k_3(s_1)-15k_3(s_1)\\
&+60k_6(-s_1)+60k_6(s_1)+60e^{s_1}k_7(-s_1)\\
&+60e^{-s_1}k_7(s_1)+60k_3(0)\\
&-120k_6(0)-120k_7(0)\Big).
\end{aligned}
$$

The restriction of the functional relation (31) to the hyperplane $s_1+s_2+s_3=0$ yields

$$
\begin{aligned}
&\frac{1}{8}e^{s_1}\partial_2k_{11}\left(s_2,-s_1-s_2\right)-\frac{1}{8}e^{s_1+s_2}\partial_2k_{12}\left(-s_1-s_2,s_1\right)\\
&\quad-\frac{1}{8}\partial_1k_8\left(s_1,s_2\right)+\frac{1}{8}e^{s_1+s_2}\partial_1k_{12}\left(-s_1-s_2,s_1\right)\\
&\quad=-\Big(\widetilde{K}_8(s_1,s_2,s_3)-\widetilde{K}_{8,\,\mathrm{s}}(s_1,s_2,s_3)\Big)\,\Big|_{s_3=-s_1-s_2},
\end{aligned}
\tag{33}
$$

where

$$
\begin{aligned}
\widetilde{K}_{8,\,\mathrm{s}}(s_1,s_2,s_3)=&\frac{1}{8\left(s_1+s_2+s_3\right)}\Big(-k_8\left(s_1,s_2\right)+e^{s_1+s_2+s_3}k_8\left(-s_2-s_3,s_2\right)\\
&-e^{s_1}k_{11}\left(s_2,-s_1-s_2\right)+e^{s_1}k_{11}\left(s_2,s_3\right)\\
&-e^{s_1+s_2}k_{12}\left(-s_1-s_2,s_1\right)+e^{s_1+s_2}k_{12}\left(s_3,-s_2-s_3\right)\Big).
\end{aligned}
$$

In order to see the general structure in a 4-variable case, we just mention that the restriction to the hyperplane $s_1+s_2+s_3+s_4=0$ of the functional relation corresponding to the function $\widetilde{K}_{17}$ gives

$$
\begin{aligned}
&-\frac{1}{16}e^{s_1+s_2}\partial_3k_{17}\left(s_3,-s_1-s_2-s_3,s_1\right)\\
&+\frac{1}{16}e^{s_1}\partial_3k_{19}\left(s_2,s_3,-s_1-s_2-s_3\right)\\
&+\frac{1}{16}e^{s_1+s_2}\partial_2k_{17}\left(s_3,-s_1-s_2-s_3,s_1\right)\\
&-\frac{1}{16}e^{s_1+s_2+s_3}\partial_2k_{19}\left(-s_1-s_2-s_3,s_1,s_2\right)
\end{aligned}
\tag{34}
$$

$$
\begin{aligned}
&-\frac{1}{16}\partial_1 k_{17}(s_1,s_2,s_3)+\frac{1}{16}e^{s_1+s_2+s_3}\partial_1 k_{19}(-s_1-s_2-s_3,s_1,s_2)\\
&=-\Big(\widetilde{K}_{17}(s_1,s_2,s_3,s_4)-\widetilde{K}_{17,\mathrm{s}}(s_1,s_2,s_3,s_4)\Big)\,\Big|_{s_4=-s_1-s_2-s_3},
\end{aligned}
$$

where

$$
\begin{aligned}
\widetilde{K}_{17,\mathrm{s}}(s_1,s_2,s_3,s_4)=\;&\frac{1}{16\,(s_1+s_2+s_3+s_4)}\\
&\Big(-k_{17}(s_1,s_2,s_3)\\
&-e^{s_1+s_2}k_{17}(s_3,-s_1-s_2-s_3,s_1)\\
&+e^{s_1+s_2}k_{17}(s_3,s_4,-s_2-s_3-s_4)\\
&+e^{s_1+s_2+s_3+s_4}k_{17}(-s_2-s_3-s_4,s_2,s_3)\\
&-e^{s_1}k_{19}(s_2,s_3,-s_1-s_2-s_3)+e^{s_1}k_{19}(s_2,s_3,s_4)\\
&-e^{s_1+s_2+s_3}k_{19}(-s_1-s_2-s_3,s_1,s_2)\\
&+e^{s_1+s_2+s_3}k_{19}(s_4,-s_2-s_3-s_4,s_2)\Big).
\end{aligned}
$$

6.3 Action of cyclic groups in the differential system, invariant expressions and simple flow of the system

In the partial differential system of the form given by (32), (33), (34) the action of the cyclic groups $\mathbb{Z}/2\mathbb{Z}$, $\mathbb{Z}/3\mathbb{Z}$, $\mathbb{Z}/4\mathbb{Z}$ is involved. For example, in (32) one can see very easily that $\mathbb{Z}/2\mathbb{Z}$ is acting by

$$T_2(s_1)=-s_1, \qquad s_1\in\mathbb{R}.$$

Using this fact, in [17], symmetries of some lengthy expressions are explored, which we recall in this subsection.

Theorem 6.1 *For any integers j_0, j_1 in $\{3,4,5,6,7\}$,*

$$e^{-\frac{s_1}{2}}\Big(-(k'_{j_0}(s_1)+k'_{j_1}(s_1))+e^{s_1}\Big(k'_{j_0}(-s_1)+k'_{j_1}(-s_1)\Big)\Big),$$

is in the kernel of $1+T_2$. Moreover, considering the finite difference expressions in the differential system corresponding to the following cases, one can find explicitly finite differences of the k_j that are in the kernel of $1+T_2$:

(1) *When $(j_0, j_1) = (3, 3)$.*
(2) *When $(j_0, j_1) = (4, 4)$.*

(3) *When* $(j_0, j_1) = (5, 5)$.
(4) *When* $(j_0, j_1) = (6, 7)$.

In (33), the action of $\mathbb{Z}/3\mathbb{Z}$ is involved as we have the following transformation acting on the variables:

$$T_3(s_1, s_2) = (-s_1 - s_2, s_1). \tag{35}$$

Using the latter, symmetries of more complicated expressions are discovered in [17]:

Theorem 6.2 *For any integers* j_0, j_1, j_2 *in* $\{8, 9, \ldots, 16\}$,

$$e^{-\frac{2s_1}{3}-\frac{s_2}{3}}\Big(- \partial_1(k_{j_0} + k_{j_1} + k_{j_2})(s_1, s_2) \\ -e^{s_1+s_2}(\partial_2 - \partial_1)(k_{j_0} + k_{j_1} + k_{j_2})(-s_1 - s_2, s_1) \\ +e^{s_1}\partial_2(k_{j_0} + k_{j_1} + k_{j_2})(s_2, -s_1 - s_2)\Big)$$

is in the kernel of $1 + T_3 + T_3^2$. *Also there are finite differences of the functions* k_j *associated with the following cases that are in the kernel of* $1 + T_3 + T_3^2$:

(1) *When* $(j_0, j_1, j_2) = (8, 12, 11)$.
(2) *When* $(j_0, j_1, j_2) = (9, 13, 10)$.
(3) *When* $(j_0, j_1, j_2) = (14, 16, 15)$.

The action of $\mathbb{Z}/4\mathbb{Z}$ in the partial differential system can be seen in (34) since the following transformation is involved:

$$T_4(s_1, s_2, s_3) = (-s_1 - s_2 - s_3, s_1, s_2).$$

The symmetries of the functions with respect to this action are also analyzed in [17]:

Theorem 6.3 *For any pair of integers* j_0, j_1 *in* $\{17, 18, 19, 20\}$,

$$e^{-\frac{3s_1}{4}-\frac{s_2}{2}-\frac{s_3}{4}}\Big(- \partial_1(k_{j_0} + k_{j_1})(s_1, s_2, s_3) \\ -e^{s_1+s_2+s_3}(\partial_2 - \partial_1)(k_{j_0} + k_{j_1})(-s_1 - s_2 - s_3, s_1, s_2) \\ -e^{s_1+s_2}(\partial_3 - \partial_2)(k_{j_0} + k_{j_1})(s_3, -s_1 - s_2 - s_3, s_1) \\ +e^{s_1}\partial_3(k_{j_0} + k_{j_1})(s_2, s_3, -s_1 - s_2 - s_3)\Big)$$

is in the kernel of $1 + T_4 + T_4^2 + T_4^3$. *Moreover, there are expressions given by finite differences of the* k_j *corresponding to the following cases that are in the kernel of* $1 + T_4 + T_4^2 + T_4^3$:

(1) *When* $(j_0, j_1) = (17, 19)$.
(2) *When* $(j_0, j_1) = (18, 18)$.
(3) *When* $(j_0, j_1) = (20, 20)$.

Moreover, in [17], it is shown that a very simple flow defined by

$$(s_1, s_2, \dots, s_n) \mapsto (s_1 + t, s_2, \dots, s_n),$$

combined with the action of the cyclic groups as described above, can be used to write the differential part of the partial differential system. In order to illustrate the idea, we just mention that, for example, in the case that the action of $\mathbb{Z}/3\mathbb{Z}$ is involved via the transformation (35), one defines the orbit $\mathcal{O}k$ of any 2-variable function k by

$$\mathcal{O}k(s_1, s_2) = (k(s_1, s_2), k(-s_1 - s_2, s_1), k(s_2, -s_1 - s_2)) .$$

Then one has to use the auxiliary function

$$\alpha_2(s_1, s_2) = e^{-\frac{2s_1}{3} - \frac{s_2}{3}},$$

to write

$$\left(\frac{d}{dt} \Big|_{t=0} \mathcal{O}k(s_1 + t, s_2) \right) \cdot (\mathcal{O}\alpha_2(s_1, s_2))$$

as a finite difference expression when $k = k_{j_0} + k_{j_1} + k_{j_2}$ and (j_0, j_1, j_2) is either $(8, 12, 11)$, $(9, 13, 10)$, or $(14, 16, 15)$. One can refer to §4.3 of [17] for more details and to see the treatment of all cases in detail.

6.4 Gradient calculations leading to functional relations

Here we explain how the functional relations written in Section 6.1 were derived in [17]. In fact the idea comes from [21], where a fundamental identity was proved and by means of a functional relation, the 2-variable function of the scalar curvature term a_2 was written in terms of its 1-variable function. The main identity to use from [21] is that, if we consider the conformally perturbed Laplacian,

$$\Delta_h = e^{h/2} \Delta e^{h/2}.$$

then for the spectral zeta function defined by

$$\zeta_h(a, s) = \mathrm{Tr}(a\, \Delta_h^{-s}), \qquad s \in \mathbb{C}, \ \Re(s) \gg 0,$$

one has

$$\frac{d}{d\varepsilon} \Big|_{\varepsilon=0} \zeta_{h+\varepsilon a}(1, s) = -\frac{s}{2} \zeta_h(\widetilde{a}, s), \tag{36}$$

where

$$\widetilde{a} = \int_{-1}^{1} e^{uh/2} a e^{-uh/2} \, du.$$

One can then see that

$$\zeta_h(a, -1) = -\varphi_0(a \, a_4(h)), \qquad a \in C^\infty(\mathbb{T}_\theta^2), \; h = h^* \in C^\infty(\mathbb{T}_\theta^2).$$

Therefore, it follows from (36) that

$$\frac{d}{d\varepsilon}\Big|_{\varepsilon=0} \varphi_0(a_4(h + \varepsilon a)) = -\frac{1}{2}\zeta_h(\widetilde{a}, -1) = \frac{1}{2}\varphi_0(\widetilde{a} \, a_4(h)) = -\varphi_0\left(a e^h \, \widetilde{a}_4(h)\right). \tag{37}$$

where $\widetilde{a}_4(h)$ is given by the same formula as (25) when the functions $K_j(s_1, \ldots, s_n)$ are replaced by

$$\widetilde{K}_j(s_1, \ldots, s_n) = \frac{1}{2^n} \frac{\sinh((s_1 + \cdots + s_n)/2)}{(s_1 + \cdots + s_n)/2} K_j(s_1, \ldots, s_n).$$

Hence, the gradient $\frac{d}{d\varepsilon}\big|_{\varepsilon=0} \varphi_0(a_4(h + \varepsilon a))$ can be calculated mainly by using the important identity (37).

There is a second way of calculating the gradient $\frac{d}{d\varepsilon}\big|_{\varepsilon=0} \varphi_0(a_4(h + \varepsilon a))$ which yields finite difference expressions. For this approach a series of lemmas were necessary as proved in [17], which are of the following type.

Lemma 6.2 *For any smooth function $L(s_1, s_2, s_3)$ and any elements x_1, x_2, x_3, x_4 in $C(\mathbb{T}_\theta^2)$, under the trace φ_0, one has:*

$$\begin{aligned}
e^h & \left(\frac{d}{d\varepsilon}\Big|_{\varepsilon=0} L(\nabla_\varepsilon, \nabla_\varepsilon, \nabla_\varepsilon)(x_1 \cdot x_2 \cdot x_3) \right) x_4 \\
&= a e^h L_{3,1}^\varepsilon(\nabla, \nabla, \nabla, \nabla)(x_1 \cdot x_2 \cdot x_3 \cdot x_4) \\
&\quad + a e^h L_{3,2}^\varepsilon(\nabla, \nabla, \nabla, \nabla)(x_2 \cdot x_3 \cdot x_4 \cdot x_1) \\
&\quad + a e^h L_{3,3}^\varepsilon(\nabla, \nabla, \nabla, \nabla)(x_3 \cdot x_4 \cdot x_1 \cdot x_2) \\
&\quad + a e^h L_{3,4}^\varepsilon(\nabla, \nabla, \nabla, \nabla)(x_4 \cdot x_1 \cdot x_2 \cdot x_3),
\end{aligned}$$

where

$$L_{3,1}^\varepsilon(s_1, s_2, s_3, s_4) := e^{s_1+s_2+s_3+s_4} \frac{L(-s_2 - s_3 - s_4, s_2, s_3) - L(s_1, s_2, s_3)}{s_1 + s_2 + s_3 + s_4},$$

$$L_{3,2}^\varepsilon(s_1, s_2, s_3, s_4) := e^{s_1+s_2+s_3} \frac{L(s_4, -s_2 - s_3 - s_4, s_2) - L(-s_1 - s_2 - s_3, s_1, s_2)}{s_1 + s_2 + s_3 + s_4},$$

$$L_{3,3}^{\varepsilon}(s_1,s_2,s_3,s_4) := e^{s_1+s_2}\frac{L(s_3,s_4,-s_2-s_3-s_4)-L(s_3,-s_1-s_2-s_3,s_1)}{s_1+s_2+s_3+s_4},$$

$$L_{3,4}^{\varepsilon}(s_1,s_2,s_3,s_4) := e^{s_1}\frac{L(s_2,s_3,s_4)-L(s_2,s_3,-s_1-s_2-s_3)}{s_1+s_2+s_3+s_4}.$$

Also, in order to perform necessary manipulations in the second calculation of the gradient $\frac{d}{d\varepsilon}\big|_{\varepsilon=0}\varphi_0(a_4(h+\varepsilon a))$, one needs lemmas of this type:

Lemma 6.3 *For any smooth function $L(s_1,s_2,s_3)$ and any elements x_1,x_2,x_3 in $C(\mathbb{T}_\theta^2)$, one has:*

$$\begin{aligned}
&\delta_j\left(L(\nabla,\nabla,\nabla)(x_1\cdot x_2\cdot x_3)\right)\\
&\quad= L(\nabla,\nabla,\nabla)(\delta_j(x_1)\cdot x_2\cdot x_3)+L(\nabla,\nabla,\nabla)(x_1\cdot\delta_j(x_2)\cdot x_3)\\
&\qquad+L(\nabla,\nabla,\nabla)(x_1\cdot x_2\cdot\delta_j(x_3))+L_{3,1}^{\delta}(\nabla,\nabla,\nabla,\nabla)(\delta_j(h)\cdot x_1\cdot x_2\cdot x_3)\\
&\qquad+L_{3,2}^{\delta}(\nabla,\nabla,\nabla,\nabla)(x_1\cdot\delta_j(h)\cdot x_2\cdot x_3)\\
&\qquad+L_{3,3}^{\delta}(\nabla,\nabla,\nabla,\nabla)(x_1\cdot x_2\cdot\delta_j(h)\cdot x_3)\\
&\qquad+L_{3,4}^{\delta}(\nabla,\nabla,\nabla,\nabla)(x_1\cdot x_2\cdot x_3\cdot\delta_j(h)),
\end{aligned}$$

where

$$L_{3,1}^{\delta}(s_1,s_2,s_3,s_4) := \frac{L(s_2,s_3,s_4)-L(s_1+s_2,s_3,s_4)}{s_1},$$

$$L_{3,2}^{\delta}(s_1,s_2,s_3,s_4) := \frac{L(s_1+s_2,s_3,s_4)-L(s_1,s_2+s_3,s_4)}{s_2},$$

$$L_{3,3}^{\delta}(s_1,s_2,s_3,s_4) := \frac{L(s_1,s_2+s_3,s_4)-L(s_1,s_2,s_3+s_4)}{s_3},$$

$$L_{3,4}^{\delta}(s_1,s_2,s_3,s_4) := \frac{L(s_1,s_2,s_3+s_4)-L(s_1,s_2,s_3)}{s_4}.$$

After performing the second gradient calculation in [17], and comparing it with the first calculation based on (37), the functional relations were derived conceptually.

6.5 *The term a_4 for non-conformally flat metrics on noncommutative four tori*

It was illustrated in [17] that, having the calculation of the term a_4 for the noncommutative two torus in place, one can conveniently write a formula for the term a_4 of a non-conformally flat metric on the noncommutative four torus that is the product of two noncommutative two tori. The metric is the noncommutative analog of the following metric. Let $(x_1, y_1, x_2, y_2) \in \mathbb{T}^4 = (\mathbb{R}/2\pi\mathbb{Z})^4$ be the coordinates of the ordinary four torus and consider the metric

$$g = e^{-h_1(x_1,y_1)}\left(dx_1^2 + dy_1^2\right) + e^{-h_2(x_2,y_2)}\left(dx_2^2 + dy_2^2\right),$$

where h_1 and h_2 are smooth real valued functions. The Weyl tensor is conformally invariant, and one can assure that the above metric is not conformally flat by calculating the components of its Weyl tensor and observing that they do not all vanish. The non-vanishing components are:

$$\begin{aligned} C_{1212} &= \frac{1}{6} e^{-h_1(x_1,y_1)} \partial_{y_1}^2 h_1(x_1, y_1) \\ &+ \frac{1}{6} e^{h_2(x_2,y_2)-2h_1(x_1,y_1)} \partial_{y_2}^2 h_2(x_2, y_2) \\ &+ \frac{1}{6} e^{-h_1(x_1,y_1)} \partial_{x_1}^2 h_1(x_1, y_1) \\ &+ \frac{1}{6} e^{h_2(x_2,y_2)-2h_1(x_1,y_1)} \partial_{x_2}^2 h_2(x_2, y_2), \end{aligned}$$

$$C_{1313} = -\frac{1}{2} e^{-h_2(x_2,y_2)+h_1(x_1,y_1)} C_{1212},$$

$$C_{2424} = C_{2323} = C_{1414} = C_{1313},$$

$$C_{3434} = e^{-2h_2(x_2,y_2)+2h_1(x_1,y_1)} C_{1212}.$$

Now, one can consider a noncommutative four torus of the form $\mathbb{T}^2_{\theta'} \times \mathbb{T}^2_{\theta''}$ that is the product of two noncommutative two tori. Its algebra has four unitary generators U_1, V_1, U_2, V_2 with the following relations: each element of the pair (U_1, V_1) commutes with each element of the pair (U_2, V_2), and there are fixed irrational real numbers θ' and θ'' such that:

$$V_1 U_1 = e^{2\pi i\theta'} U_1 V_1, \qquad V_2 U_2 = e^{2\pi i\theta''} U_2 V_2.$$

One can then choose conformal factors $e^{-h'}$ and $e^{-h''}$, where h' and h'' are selfadjoint elements in $C^\infty(\mathbb{T}^2_{\theta'})$ and $C^\infty(\mathbb{T}^2_{\theta''})$, respectively, and use them to perturb the flat metric of each component conformally. Then the Laplacian of the product geometry is given by

$$\Delta_{\varphi',\varphi''} = \Delta_{\varphi'} \otimes 1 + 1 \otimes \Delta_{\varphi''},$$

where $\triangle_{\varphi'}$ and $\triangle_{\varphi''}$ are, respectively, the Laplacians of the perturbed metrics on $\mathbb{T}^2_{\theta'}$ and $\mathbb{T}^2_{\theta''}$. Now one can use a simple Kunneth formula to find the term a_4 in the asymptotic expansion

$$\mathrm{Tr}(a \exp(-t\triangle_{\varphi',\varphi''})) \sim t^{-2}\Big((\varphi_0' \otimes \varphi_0'')(a\, a_0) + (\varphi_0' \otimes \varphi_0'')(a\, a_2)\, t \\ +(\varphi_0' \otimes \varphi_0'')(a\, a_4)\, t^2 + \cdots\Big) \tag{38}$$

in terms of the known terms appearing in the following expansions:

$$\mathrm{Tr}(a' \exp(-t\triangle_{\varphi'})) \sim t^{-1}\left(\varphi_0'(a'\, a_0') + \varphi_0'(a'\, a_2')\, t + \varphi_0'(a'\, a_4')\, t^2 + \cdots\right),$$

$$\mathrm{Tr}(a'' \exp(-t\triangle_{\varphi''})) \sim t^{-1}\left(\varphi_0''(a''\, a_0'') + \varphi_0''(a''\, a_2'')\, t + \varphi_0''(a''\, a_4'')\, t^2 + \cdots\right).$$

The general formula is

$$a_{2n} = \sum_{i=0}^{n} a_{2i}' \otimes a_{2(n-i)}'' \in C^\infty(\mathbb{T}^2_{\theta'} \times \mathbb{T}^2_{\theta''}),$$

hence an explicit formula for a_4 of the noncommutative four torus with the product geometry explained above since there are explicit formulas for its ingredients.

In this case of the non-conformally flat metric on the product geometry, two modular automorphisms are involved in the formulas for the geometric invariants and this motivates further systematic research on *twistings* that involve two-dimensional modular structures, cf. [13].

7 Twisted spectral triples and Chern-Gauss-Bonnet theorem for ergodic C^*-dynamical systems

This section is devoted to the notion of twisted spectral triples and some details of their appearance in the context of noncommutative conformal geometry. In particular we explain construction of twisted spectral triples for ergodic C^*-dynamical systems and the validity of the Chern-Gauss-Bonnet theorem in this vast setting.

7.1 *Twisted spectral triples*

The notion of twisted spectral triples was introduced in [20] to incorporate the study of type III algebras using noncommutative differential geometric techniques. In the definition of this notion, in addition to a triple (A, H, D) of a $*$-algebra A, a

Hilbert space H, and an unbounded operator D on H which plays the role of the Dirac operator, one has to bring into the picture an automorphism σ of A which interacts with the data as follows. Instead of the ordinary commutators $[D, a]$ as in the definition of an ordinary spectral triple, in the twisted case one asks for the boundedness of the twisted commutators $[D, a]_\sigma = Da - \sigma(a)D$. More precisely, here also one assumes a representation of A by bounded operators on H such that the operator $Da - \sigma(a)D$ is defined on the domain of D for any $a \in A$, and that it extends by continuity to a bounded operator on H.

This twisted notion of a spectral triple is essential for type III examples as this type of algebras do not possess non-zero trace functionals, and ordinary spectral triples with suitable properties cannot be constructed over them for the following reason [20]. If (A, H, D) is an m^+-summable ordinary spectral triple then the following linear functional on A defined by

$$a \mapsto \mathrm{Tr}_\omega(a|D|^{-m})$$

gives a trace. The main reason for this is that the kernel of the Dixmier trace Tr_ω is a large kernel that contains all operators of the form $|D|^{-m}a - a|D|^{-m}$, $a \in A$, if the ordinary commutators are bounded. In fact we are using the *regularity* assumption on the spectral triple, which in particular requires boundedness of the commutators of elements of A with $|D|$ as well as with D (indeed this is a natural condition and the main point is that one is using ordinary commutators). Hence, trace-less algebras cannot fit into the paradigm of ordinary spectral triples.

It is quite amazing that in [20], examples are provided where one can obtain boundedness of twisted commutators $Da - \sigma(a)D$ and $|D|a - \sigma(a)|D|$ for all elements a of the algebra by means of an algebra automorphism σ, where the Dirac operator D has the m^+-summability property. Then they use the boundedness of the twisted commutators to show that operators of the form $|D|^{-m}a - \sigma^{-m}(a)|D|^{-m}$ are in the kernel of the Dixmier trace and the linear functional $a \mapsto \mathrm{Tr}_\omega(a|D|^{-m})$ yields a twisted trace on A.

7.2 *Conformal perturbation of a spectral triple*

One of the main examples in [20] that demonstrates the need for the notion of twisted spectral triples in noncommutative geometry is related to conformal perturbation of Riemannian metrics. That is, if D is the Dirac operator of a spin manifold equipped with a Riemannian metric g, then, after a conformal perturbation of the metric to $\tilde{g} = e^{-4h}g$ by means of a smooth real valued function h on the manifold, the Dirac operator of the perturbed metric $\tilde{g}$ is unitarily equivalent to the operator

$$\tilde{D} = e^h D e^h.$$

So this suggests that given an ordinary spectral triple (A, H, D) with a noncommutative algebra A, since the metric is encoded in the analog D of the Dirac operator, one can implement conformal perturbation of the metric by fixing a self-adjoint element $h \in A$ and by then replacing D with $\tilde{D} = e^h D e^h$. However, it turns out that the triple $(A, H, \tilde{D})$ is not necessarily a spectral triple any more, since, because of noncommutativity of A, the commutators $[\tilde{D}, a]$, $a \in A$, are not necessarily bounded operators. Despite this, interestingly, the remedy brought forth in [20] is to introduce the automorphism

$$\sigma(a) = e^{2h} a e^{-2h}, \qquad a \in A,$$

which yields the bounded twisted commutators

$$[\tilde{D}, a]_\sigma = \tilde{D}a - \sigma(a)\tilde{D}, \qquad a \in A.$$

7.3 *Conformal perturbation of the flat metric on* $\mathbb{T}_\theta^2$

Another important example, which is given in [22], shows that twisted spectral triples can arise in a more intrinsic manner, compared to the example we just illustrated, when a conformal perturbation is implemented. In [22], the flat geometry of $\mathbb{T}_\theta^2$ is perturbed by a fixed conformal factor e^{-h}, where $h = h^* \in C^\infty(\mathbb{T}_\theta^2)$. This is done by replacing the canonical trace φ_0 on $C(\mathbb{T}_\theta^2)$ (playing the role of the volume form) by the tracial state $\varphi(a) = \varphi_0(ae^{-h})$, $a \in C(\mathbb{T}_\theta^2)$. In order to represent the opposite algebra of $C(\mathbb{T}_\theta^2)$ on the Hilbert space $\mathcal{H}_\varphi$, obtained from $C(\mathbb{T}_\theta^2)$ by the GNS construction, one has to modify the ordinary action induced by right multiplication. That is, one has to consider the action defined by

$$a^{op} \cdot \xi = \xi e^{-h/2} a e^{h/2}.$$

It then turns out that with the new action, the ordinary commutators $[D, a]$, $a \in C^\infty(\mathbb{T}_\theta^2)^{op}$, are not bounded any more, where D is the Dirac operator

$$D = \begin{pmatrix} 0 & \partial_\varphi^* \\ \partial_\varphi & 0 \end{pmatrix} : \mathcal{H} \to \mathcal{H}.$$

Here,

$$\partial_\varphi = \delta_1 + i\delta_2 : \mathcal{H}_\varphi \to \mathcal{H}^{(1,0)},$$

where $\mathcal{H}^{(1,0)}$, the analogue of $(1,0)$-forms, is the Hilbert space completion of finite sums $\sum a\partial(b)$, $a, b \in A_\theta^\infty$, with respect to the inner product

$$(a\partial b, c\partial d) = \varphi_0(c^* a(\partial b)(\partial d)^*),$$

and

$$\mathcal{H} = \mathcal{H}_\varphi \oplus \mathcal{H}^{(1,0)}.$$

The remedy for obtaining bounded commutators is to use a twist given by the automorphism

$$\sigma(a^{op}) = e^{-h/2} a e^{h/2},$$

which leads to bounded twisted commutators [22]

$$[D, a^{op}]_\sigma = Da^{op} - \sigma(a^{op})D, \qquad a \in C^\infty(\mathbb{T}_\theta^2)^{op}.$$

7.4 *Conformally twisted spectral triples for C^*-dynamical systems*

The example in Section 7.3 inspired the construction of twisted spectral triples for general ergodic C^*-dynamical systems in [29]. The Dirac operator used in this work, following more closely the geometric approach taken originally in [9], is the analog of the de Rham operator. An important reason for this choice is that an important goal in [29] was to confirm the validity of the analog of the Chern-Gauss-Bonnet theorem in the vast setting of ergodic C^*-dynamical systems.

In this subsection we consider a C^*-algebra $\mathcal{A}$ with a strongly continuous ergodic action α of a compact Lie group G of dimension n, and we let $\mathcal{A}^\infty$ denote the smooth subalgebra of $\mathcal{A}$, which is defined as:

$$\mathcal{A}^\infty = \{a \in \mathcal{A} : \text{the map } g \mapsto \alpha_g(a) \text{ is in } C^\infty(G, A)\}.$$

Following closely the construction in [9], we can define a space of *differential forms* on $\mathcal{A}$ by using the exterior powers of $\mathfrak{g}^*$, namely that for $k = 0, 1, 2, \ldots, n$, we set:

$$\Omega^k(\mathcal{A}, G) = \mathcal{A} \otimes \bigwedge^k \mathfrak{g}^*, \tag{39}$$

where $\mathfrak{g}^*$ is the linear dual of the Lie algebra $\mathfrak{g}$ of the Lie group G. We consider the inner product $\langle \cdot, \cdot \rangle$ on $\mathfrak{g}^*$ induced by the Killing form, and extend it to an inner product on $\bigwedge^k \mathfrak{g}^*$ by setting

$$\langle v_1 \wedge \cdots \wedge v_k, w_1 \wedge \cdots \wedge w_k \rangle = \det(\langle v_i, w_j \rangle).$$

After fixing an orthonormal basis $(\omega_j)_{j=1,\dots,n}$ for $\mathfrak{g}^*$, we equip the above differential forms with an exterior derivative $d : \Omega^k(\mathcal{A}, G) \to \Omega^{k+1}(\mathcal{A}, G)$ given by

$$d(a \otimes \omega_{i_1} \wedge \omega_{i_2} \wedge \cdots \wedge \omega_{i_k}) = \sum_{j=1}^{n} \partial_j(a) \otimes \omega_j \wedge \omega_{i_1} \wedge \omega_{i_2} \wedge \cdots \wedge \omega_{i_k} \tag{40}$$
$$- \frac{1}{2} \sum_{j=1}^{k} \sum_{\alpha,\beta} (-1)^{j+1} c^{i_j}_{\alpha\beta} a \otimes \omega_\alpha \wedge \omega_\beta$$
$$\wedge\, \omega_{i_1} \wedge \cdots \wedge \omega_{i_{j-1}} \wedge \omega_{i_{j+1}} \wedge \cdots \wedge \omega_{i_k},$$

where the coefficients $c^i_{\alpha\beta}$ are the *structure constants* of the Lie algebra $\mathfrak{g}$ determined by the relations

$$[\partial_\alpha, \partial_\beta] = \sum_{i=1}^{n} c^i_{\alpha\beta} \partial_i$$

for the predual $(\partial_j)_{j=1,\dots,n}$ of $(\omega_j)_{j=1,\dots,n}$. This exterior derivative satisfies $d \circ d = 0$ on $\Omega^\bullet(\mathcal{A}, G)$, therefore we have a complex $(\Omega^\bullet(\mathcal{A}, G), d)$. This complex is called the Chevalley-Eilenberg cochain complex with coefficients in $\mathcal{A}$, one can refer to [44] for more details.

We now define an inner product on $\Omega^k(\mathcal{A}, G)$, for which we make use of the unique G-invariant tracial state φ_0 on $\mathcal{A}$, see [37]. The inner product is defined by

$$(a \otimes v_1 \wedge \cdots \wedge v_k, a' \otimes w_1 \wedge \cdots \wedge w_k)_0 = \varphi_0(a^* a') \det(\langle v_i, w_j \rangle). \tag{41}$$

We denote the Hilbert space completion of $\Omega^k(\mathcal{A}, G)$ with respect to this inner product by $\mathcal{H}_{k,0}$.

In order to implement a conformal perturbation, we fix a selfadjoint element $h \in \mathcal{A}^\infty$, define the following new inner product on $\Omega^k(\mathcal{A}, G)$:

$$(a \otimes v_1 \wedge \cdots \wedge v_k, a' \otimes w_1 \wedge \cdots \wedge w_k)_h = \varphi_0(a^* a' e^{(n/2-k)h}) \det(\langle v_i, w_j \rangle), \tag{42}$$

and denote the associated Hilbert space by $\mathcal{H}_{k,h}$.

One of the goals is to construct ordinary and twisted spectral triples by using the unbounded operator $d + d^*$, the analog of the de Rham operator, acting on the direct sum of all $\mathcal{H}_{k,h}$. Here the adjoint d^* of d is of course taken with respect to the conformally perturbed inner product $(\cdot, \cdot)_h$. The Hilbert spaces are simply related by the unitary maps $U_k \colon \mathcal{H}_{k,0} \to \mathcal{H}_{k,h}$ given on degree k forms by:

$$U_k(a \otimes v_1 \wedge \cdots \wedge v_k) = a e^{-(n/2-k)h/2} \otimes v_1 \wedge \cdots \wedge v_k.$$

Therefore, for simplicity, we use these unitary maps to transfer the operator $d + d^*$ to an unbounded operator D acting on the Hilbert space $\mathcal{H}$ that is the direct sum of all $\mathcal{H}_{k,0}$. We can now state the following result from [29].

Theorem 7.1 *Consider the above constructions associated with a C^*-algebra $\mathcal{A}$ with an ergodic action of an n-dimensional Lie group G. The operator D has a selfadjoint extension which is n^+-summable. With the representation of $\mathcal{A}^\infty$ on $\mathcal{H} = \bigoplus_k \mathcal{H}_{k,0}$ induced by left multiplication, the triple $(\mathcal{A}^\infty, \mathcal{H}, D)$ is an ordinary spectral triple. However, when one represents the opposite algebra of $\mathcal{A}^\infty$ on $\mathcal{H}$ using multiplication from right, one obtains a twisted spectral triple with respect to the automorphism defined by $\beta(a^{op}) = e^h a e^{-h}$.*

The spectral triples described in the above theorem can in fact be equipped with the grading operator given by

$$\gamma(a \otimes v_1 \wedge \cdots \wedge v_k) = (-1)^k (a \otimes v_1 \wedge \cdots \wedge v_k).$$

Related to this grading, it is interesting to study the Fredholm index of the operator D, which is unitarily equivalent to $d + d^*$, when viewed as an operator from the direct sum of all even differential forms to the direct sum of all odd differential forms. We shall discuss this issue shortly.

7.5 The Chern-Gauss-Bonnet theorem for C*-dynamical systems

In Section 4 we briefly discussed the Gauss-Bonnet theorem for surfaces, which states that for any closed oriented two-dimensional Riemannian manifold Σ with scalar curvature R, one has

$$\frac{1}{4\pi} \int_\Sigma R = \chi(\Sigma),$$

where $\chi(\Sigma)$ is the Euler characteristic of Σ. The Chern-Gauss-Bonnet theorem generalizes this result to higher even dimensional manifolds. That is, in higher dimensions as well, the Euler characteristic, which is a topological invariant, coincides with the integral of a certain geometric invariant, namely the *Pfaffian* of the curvature form. Given a closed oriented Riemannian manifold M of even dimension n, consider the Levi-Civita connection, which is the unique torsion-free metric-compatible connection on the tangent bundle TM. Let us denote the matrix of local 2-forms representing the curvature of this connection by Ω. The Chern-Gauss-Bonnet theorem states that the Pfaffian of Ω (the square root of the determinant defined on the space of anti-symmetric matrices) integrates over the manifold to the Euler characteristic of the manifold, up to multiplication by a universal constant:

$$\frac{1}{(2\pi)^n}\int_M \mathrm{Pf}(\Omega) = \chi(M).$$

Interestingly, there is a spectral way of interpreting such relations between local geometry and topology of manifolds. Relevant to our discussion is indeed the Fredholm index of the de Rham operator $d + d^*$, where d is the de Rham exterior derivative and d^* is its adjoint with respect to the metric on the differential forms induced by the Riemannian metric. The Fredholm index of $d + d^*$ should be calculated when the operator is viewed as a map from the direct sum of all even differential forms to the direct sum of all odd differential forms on M:

$$d + d^* : \Omega^{even} M = \bigoplus \Omega^{2k} M \to \Omega^{odd} M = \bigoplus \Omega^{2k+1} M$$

The index of this operator is certainly an important geometric quantity since the adjoint d^* of d heavily depends on the choice of metric on the manifold. Amazingly, using the *Hodge decomposition theorem*, one can find a canonical identification of the de Rham cohomology group $H^k(M)$ with the kernel of the Laplacian $\Delta_k = d^*d + dd^* : \Omega^k M \to \Omega^k M$. This can then be used to show that the index of $d + d^*$ is equal to the Euler characteristic of M. Moreover, one can appeal to the *McKean-Singer index theorem* to find curvature related invariants appearing in small time heat kernel expansions associated with $d + d^*$ to see that the index is given by the integral of curvature related invariants.

In [29], this spectral approach is taken to show that the analog of the Chern-Gauss-Bonnet theorem can be established for ergodic C^*-dynamical systems. Let us consider the setup and the constructions presented in Section 7.4 for a C^*-algebra $\mathcal{A}$ with an ergodic action of a compact Lie group G of dimension n. Then, one of the main results proved in [29] is the following statement. Here, d is given by (40), $h = h^* \in \mathcal{A}^\infty$ is the element that was used to implement with e^h a conformal perturbation of the metric, and the Hilbert space $\mathcal{H}_{k,h}$ is the completion of the k-differential forms $\Omega^k(\mathcal{A}, G)$ with respect to the perturbed metric.

Theorem 7.2 *The Fredholm index of the operator*

$$d + d^* : \bigoplus_k \mathcal{H}_{2k,h} \to \bigoplus_k \mathcal{H}_{2k+1,h}$$

is equal to the Euler characteristic $\chi(A, G)$ of the complex $(\Omega^\bullet(A, G), d)$. Since $\chi(A, G) = \sum_k (-1)^k dim\left(H^k(A, G)\right)$ is the alternating sum of the dimensions of the cohomology groups, the index of $d + d^*$ is independent of the conformal factor e^h used for perturbing the metric.

8 The Ricci curvature

Classically, scalar curvature is only a deem shadow of the full Riemann curvature tensor. In fact there is no evidence that Riemann considered anything else but the full curvature tensor, and, equivalently, the sectional curvature. Both were defined by him for a Riemannian manifold. The Ricci and scalar curvatures were later defined by contracting the Riemann curvature tensor with the metric tensor. Once the metric is given in a local coordinate chart, all three curvature tensors can be computed explicitly via algebraic formulas involving only partial derivatives of the metric tensor. This is a purely algebraic process, with deep geometric and analytic implications. It is also a top-down process, going from the full Riemann curvature tensor, to Ricci curvature, and then to scalar curvature.

The situation in the noncommutative case is reversed and we have to move up the ladder, starting from the scalar curvature first, which is the easiest to define spectrally, being given by the second term of the heat expansion for the scalar Laplacian, the square of the Dirac operator in general. Thus after treating the scalar curvature, which we recalled in previous sections together with examples, one should next try to define and possibly compute, in some cases, a Ricci curvature tensor. But how? In [34] and motivated by the local formulas for the asymptotic expansion of heat kernels in spectral geometry, the authors propose a definition of Ricci curvature in a noncommutative setting. One necessarily has to use the asymptotic expansion of Laplacians on functions and 1-forms and a version of the Weitzenböck formula.

As we shall see in this section, the Ricci operator of an oriented closed Riemannian manifold can be realized as a spectral functional, namely the functional defined by the zeta function of the full Laplacian of the de Rham complex, localized by smooth endomorphisms of the cotangent bundle and their trace. In the noncommutative case, this Ricci functional uniquely determines a density element, called the Ricci density, which plays the role of the Ricci operator. The main result of [34] provides a general definition and an explicit computation of the Ricci density when the conformally flat geometry of the curved noncommutative two torus is encoded in the de Rham spectral triple. In a follow-up paper [24], the Ricci curvature of a noncommutative curved three torus is computed. In this section we explain these recent developments in more detail.

8.1 A Weitzenböck formula

The Weitzenböck formula

$$\textbf{Hodge} - \textbf{Bochner} = \textbf{Ricci}$$

in conjunction with Gilkey's asymptotic expansion gives an opening to define the Ricci curvature in spectral terms. Let M be a closed oriented Riemannian manifold. Consider the de Rham spectral triple

$$(C^\infty(M), L^2(\Omega^{ev}(M)) \oplus L^2(\Omega^{odd}(M)), d + \delta, \gamma),$$

which is the even spectral triple constructed from the de Rham complex. Here d is the exterior derivative, δ is its adjoint acting on the exterior algebra, and γ is the $\mathbb{Z}_2$-grading on forms. The eigenspaces for eigenvalues 1 and -1 of γ are even and odd forms, respectively. The full Laplacian on forms $\Delta = d\delta + \delta d$ is the Laplacian of the Dirac operator $d+\delta$, and is the direct sum of Laplacians on p-forms, $\Delta = \oplus \Delta_p$. As a Laplace type operator, Δ can be written as $\nabla^*\nabla - E$ by Weitzenböck formula, where ∇ is the Levi-Civita connection extended to all forms and

$$E = -\frac{1}{2} c(dx^\mu)c(dx^\nu)\Omega(\partial_\mu, \partial_\nu).$$

Here c denotes the Clifford multiplication and Ω is the curvature operator of the Levi-Civita connection acting on exterior algebra. The restriction of E to one forms gives the Ricci operator.

8.2 *Ricci curvature as a spectral functional*

The Ricci curvature of a Riemannian manifold (M^m, g) is originally defined as follows. Let ∇ be the Levi-Civita connection of the metric g. The Riemannian operator and the curvature tensor are defined for vector fields X, Y, Z, W by

$$Riem(X, Y) := \nabla_X \nabla_Y - \nabla_Y \nabla_X - \nabla_{[X,Y]},$$
$$\mathrm{Riem}(X, Y, Z, W) := g(Riem(X, Y)Z, W).$$

With respect to the coordinate frame $\partial_\mu = \frac{\partial}{\partial x^\mu}$, the components of the curvature tensor are denoted by

$$\mathrm{Riem}_{\mu\nu\rho\epsilon} := \mathrm{Riem}(\partial_\mu, \partial_\nu, \partial_\rho, \partial_\epsilon).$$

The components of the Ricci tensor Ric and scalar curvature R are given by

$$\mathrm{Ric}_{\mu\nu} := g^{\rho\epsilon}\mathrm{Riem}_{\mu\rho\epsilon\nu},$$
$$R := g^{\mu\nu}\mathrm{Ric}_{\mu\nu} = g^{\mu\nu}g^{\rho\epsilon}\mathrm{Riem}_{\mu\rho\epsilon\nu}.$$

Now these algebraic formulas have no chance to be extended to a noncommutative setting in general. One must thus look for a spectral alternative reformulation.

Let $P : C^\infty(V) \to C^\infty(V)$ be a positive elliptic differential operator of order two acting on the sections of a smooth Hermitian vector bundle V over M. The heat trace $\mathrm{Tr}(e^{-tP})$ has a short time asymptotic expansion of the form

$$\mathrm{Tr}(e^{-tP}) \sim \sum_{n=0}^{\infty} a_n(P) t^{\frac{n-m}{2}}, \qquad t \to 0^+,$$

where $a_n(P)$ are integrals of local densities

$$a_n(P) = \int \mathrm{tr}(a_n(x, P)) dx.$$

Here $dx = dvol_x$ is the Riemannian volume form and tr is the fiberwise matrix trace. The endomorphism $a_n(x, P)$ can be uniquely determined by localizing the heat trace by an smooth endomorphism F of V. It is easy to see that the asymptotic expansion of the localized heat trace $\mathrm{Tr}(Fe^{-tP})$ is of the form

$$\mathrm{Tr}(Fe^{-tP}) \sim \sum_{n=0}^{\infty} a_n(F, P) t^{\frac{n-m}{2}}, \tag{43}$$

with

$$a_n(F, P) = \int \mathrm{tr}\big(F(x) a_n(x, P)\big) dx. \tag{44}$$

If P is a Laplace type operator i.e., its leading symbol is given by the metric tensor, then the densities $a_n(x, P)$ can be expressed in terms of the Riemannian curvature, an endomorphism E, and their derivatives. The endomorphism E measures how far the operator P is from being the Laplacian $\nabla^*\nabla$ of a connection ∇ on V, that is

$$E = \nabla^*\nabla - P. \tag{45}$$

The first two densities of the heat equation for such P are given by [36, Theorem 3.3.1]

$$a_0(x, P) = (4\pi)^{-m/2} \mathrm{I}, \tag{46}$$

$$a_2(x, P) = (4\pi)^{-m/2} \left(\frac{1}{6} R(x) + E \right). \tag{47}$$

For the scalar Laplacian $\triangle_0$, the connection is the de Rham differential $d : C^\infty(M) \to \Omega^1(M)$, and $E = 0$. Hence the first two first terms of the heat kernel of $\triangle_0$ are given by

$$a_0(x, \Delta_0) = (4\pi)^{-m/2}, \tag{48}$$

$$a_2(x, \Delta_0) = (4\pi)^{-m/2}\frac{1}{6}R(x). \tag{49}$$

For Laplacian on one forms $\Delta_1 : \Omega^1(M) \to \Omega^1(M)$, the Hodge-de Rham Laplacian, the connection in (45) is the Levi-Civita connection on the cotangent bundle. The endomorphism E is the negative of the Ricci operator, $E = -\mathrm{Ric}$, on the cotangent bundle, which is defined by raising the first index of the Ricci tensor (denoted by Ric as well),

$$\mathrm{Ric}_x(\alpha^\sharp, X) = \mathrm{Ric}_x(\alpha)(X), \quad \alpha \in T_x^*M,\ X \in T_xM.$$

Therefore, one has

$$a_0(x, \Delta_1) = (4\pi)^{-m/2}\mathrm{I}, \tag{50}$$

$$a_2(x, \Delta_1) = (4\pi)^{-m/2}\left(\frac{1}{6}R(x) - \mathrm{Ric}_x\right). \tag{51}$$

We can use the function $\mathrm{tr}(F)$ to localize the heat trace of the scalar Laplacian Δ_0 and get the identity

$$a_2(\mathrm{tr}(F), \Delta_0) - a_2(F, \Delta_1) = (4\pi)^{-m/2}\int_M \mathrm{tr}\big(F(x)\mathrm{Ric}_x\big)dx. \tag{52}$$

This motivates the following definition.

Definition 8.1 ([34]) *The Ricci functional of the closed Riemannian manifold (M, g) is the functional on $C^\infty(\mathrm{End}(T^*M))$ defined as*

$$\mathcal{R}ic(F) = a_2(\mathrm{tr}(F), \Delta_0) - a_2(F, \Delta_1).$$

Proposition 8.1 *For a closed Riemannian manifold M of dimension m, we have the short time asymptotics*

$$Tr\left(\mathrm{tr}(F)e^{-t\Delta_0}\right) - Tr\left(Fe^{-t\Delta_1}\right) \sim \mathcal{R}ic(F)\,t^{1-\frac{m}{2}}.$$

Proof By (46) and (44), we have $\mathrm{tr}(F)a_0(x, \Delta_0) = \mathrm{tr}(F(x)a_0(x, \Delta_1))$. This implies that

$$a_0(\mathrm{tr}(F), \Delta_0) = a_0(F, \Delta_1), \qquad F \in C^\infty(\mathrm{End}(T^*M)). \tag{53}$$

The asymptotic expansion of the localized heat kernel (43) then shows that the first terms will cancel each other. The difference of the second terms, which are multiples of $t^{1-\frac{m}{2}}$, will become the first term in the asymptotic expansion of the differences of localized heat kernels. □

8.3 *Spectral zeta function and the Ricci functional*

The spectral zeta function of a positive elliptic operator P is defined as

$$\zeta(s, P) = \mathrm{Tr}(P^{-s}(\mathrm{I} - Q)), \quad \Re(s) \gg 0,$$

where Q is the projection on the kernel of P. Its localized version is $\zeta(s, F, P) = \mathrm{Tr}(FP^{-s}(\mathrm{I} - Q))$. These function have a meromorphic extension to the complex plane with isolated simple poles. Using the Mellin transform, one finds explicit relation between residue at the poles and coefficients of the heat kernel. This leads to the following expression for the Ricci functional in terms of zeta functions.

Proposition 8.2 *For a closed Riemannian manifold M of dimension $m > 2$, we have*

$$\mathcal{R}ic(F) = \Gamma\left(\frac{m}{2} - 1\right) \mathrm{res}_{\mathrm{s}=\frac{m}{2}-1}\Big(\zeta(\mathrm{s}, \mathrm{tr(F)}, \triangle_0) - \zeta(\mathrm{s}, \mathrm{F}, \triangle_1)\Big). \tag{54}$$

If M is two-dimensional, then

$$\mathcal{R}ic(F) = \zeta(0, \mathrm{tr}(F), \triangle_0) - \zeta(0, F, \triangle_1) + Tr(\mathrm{tr}(F)Q_0) - Tr(FQ_1), \tag{55}$$

where Q_j is the projection on the kernel of Laplacian $\triangle_j$, $j = 0, 1$.

It follows that the difference of zeta functions $\zeta(s, \mathrm{tr}(F), \triangle_0) - \zeta(s, F, \triangle_1)$ is regular at $m/2$, and its first pole is located at $s = m/2 - 1$.

To work with the Laplacian on one forms, we will use smooth endomorphisms F of the cotangent bundle. The smearing endomorphism $\tilde{F} = \mathrm{tr}(F)\mathrm{I}_0 \oplus F \in C^\infty(\mathrm{End}(\bigwedge^\bullet M))$, where I_0 denotes the identity map on functions, can be used to localize the heat kernel of the full Laplacian and

$$\mathcal{R}ic(F) = a_2(\gamma\tilde{F}, \triangle). \tag{56}$$

With the above notation, one can express the Ricci functional as special values of the (localized) spectral zeta functions

$$\mathcal{R}ic(F) = \begin{cases} \Gamma(\frac{m}{2} - 1)\mathrm{res}_{\mathrm{s}=\frac{m}{2}-1}\zeta(\mathrm{s}, \tilde{\mathrm{F}}\gamma, \triangle) & m > 2, \\ \zeta(0, \gamma\tilde{F}, \triangle) + \mathrm{Tr}(\mathrm{tr}(F)Q_0) - \mathrm{Tr}(FQ_1) & m = 2. \end{cases} \tag{57}$$

The flat de Rham spectral triple of the noncommutative two torus can be perturbed by a Weyl factor e^{-h} with $h \in A_\theta^\infty$ a self adjoint element. This procedure gives rise to the de Rham spectral triple of a curved noncommutative torus. The Ricci functional is defined in a similar fashion as above, and it can be shown that there exists an element $\mathbf{Ric} \in A_\theta^\infty \otimes M_2(\mathbb{C})$, called the Ricci density, such that

$$\mathcal{R}ic(F) = \frac{1}{\Im(\tau)}\varphi(\mathrm{tr}(F\mathbf{Ric})e^{-h}), \quad F \in A_\theta^\infty \otimes M_2(\mathbb{C}).$$

8.4 The de Rham spectral triple for the noncommutative two torus

In this section, we describe the de Rham spectral triple of a noncommutative two torus A_θ equipped with a complex structure. This is a deformation of the Dolbeault complex that we used in Section. Consider the vector space $W = \mathbb{R}^2$, and let τ be a complex number in the upper half plane. Let g_τ be the positive definite symmetric bilinear form on W given by

$$g_\tau = \frac{1}{\Im(\tau)^2}\begin{pmatrix} |\tau|^2 & -\Re(\tau) \\ -\Re(\tau) & 1 \end{pmatrix}. \tag{58}$$

The inverse $g_\tau^{-1} = \begin{pmatrix} 1 & \Re(\tau) \\ \Re(\tau) & |\tau|^2 \end{pmatrix}$ of g_τ is a metric on the dual of W. The entries of g_τ^{-1} will be denoted by g^{jk}.

Let $\bigwedge^\bullet W_\mathbb{C}^*$ be the exterior algebra of $W_\mathbb{C}^* = (W \otimes \mathbb{C})^*$. The algebra $A_\theta^\infty \otimes \bigwedge^\bullet W_\mathbb{C}^*$ is the algebra of differential forms on the noncommutative two torus A_θ. In this framework, the Hilbert space of functions, denoted $\mathcal{H}^{(0)}$, is simply the Hilbert space given by the GNS construction of A_θ^∞ with respect to $\frac{1}{\Im(\tau)}\varphi$. Additionally, the Hilbert space of one forms, denoted $\mathcal{H}^{(1)}$, is the space $\mathcal{H}_0 \otimes (\mathbb{C}^2, g_\tau^{-1})$ with inner product given by

$$\langle a_1 \oplus a_2, b_1 \oplus b_2\rangle = \frac{1}{\Im(\tau)}\sum_{j,k} g^{jk}\varphi(b_k^* a_j), \quad a_i,\ b_i \in A_\theta^\infty, \tag{59}$$

while the Hilbert space of two forms, denoted $\mathcal{H}^{(2)}$, is given by the GNS construction of A_θ^∞ with respect to $\Im(\tau)\,\varphi$.

The exterior derivative on elements of A_θ^∞ is given by

$$a \mapsto i\delta_1(a) \oplus i\delta_2(a), \quad a \in A_\theta^\infty. \tag{60}$$

It will be denoted by d_0, when considered as a densely defined operator from $\mathcal{H}^{(0)}$ to $\mathcal{H}^{(1)}$. The operator $d_1 : \mathcal{H}^{(1)} \to \mathcal{H}^{(2)}$ is defined on the elements of $A_\theta^\infty \oplus A_\theta^\infty$ as

$$a \oplus b \mapsto i\delta_1(b) - i\delta_2(a), \qquad a, b \in A_\theta^\infty. \tag{61}$$

The adjoints of the operators $d_0 : \mathcal{H}^{(0)} \to \mathcal{H}^{(1)}$ and $d_1 : \mathcal{H}^{(1)} \to \mathcal{H}^{(2)}$ are then given by

$$\begin{aligned} d_0^*(a \oplus b) &= -i\delta_1(a) - i\Re(\tau)\delta_2(a) - i\Re(\tau)\delta_1(b) - i|\tau|^2\delta_2(b), \\ d_1^*(a) &= (i|\tau|^2\delta_2(a) + i\Re(\tau)\delta_1(a)) \oplus (-i\Re(\tau)\delta_2(a) - i\delta_1(a)), \end{aligned}$$

for all $a, b \in A_\theta^\infty$.

Definition 8.2 *The (flat) de Rham spectral triple of A_θ is the even spectral triple $(A_\theta, \mathcal{H}, D)$, where $\mathcal{H} = \mathcal{H}^{(0)} \oplus \mathcal{H}^{(2)} \oplus \mathcal{H}^{(1)}$, $D = \begin{pmatrix} 0 & d^* \\ d & 0 \end{pmatrix}$, and $d = d_0 + d_1^*$.*

Note that the operator d and its adjoint $d^* = d_1 + d_0^*$ act on $A_\theta^\infty \oplus A_\theta^\infty$ as

$$d = \begin{pmatrix} i\delta_1 & i|\tau|^2\delta_2 + i\Re(\tau)\delta_1 \\ i\delta_2 & -i\Re(\tau)\delta_2 - i\delta_1 \end{pmatrix}, \quad d^* = \begin{pmatrix} -i\delta_1 - i\Re(\tau)\delta_2 & -i\Re(\tau)\delta_1 - i|\tau|^2\delta_2 \\ -i\delta_2 & i\delta_1 \end{pmatrix}. \tag{62}$$

Note also that the de Rham spectral triple introduced in Definition 8.2 is isospectral to the de Rham complex spectral triple of the flat torus $\mathbb{T}^2$ with the metric given by (58).

8.5 The twisted de Rham spectral triple

The conformal perturbation of the metric on the noncommutative two torus is implemented by changing the tracial state φ by a noncommutative Weyl factor e^{-h}, where the dilaton h is a selfadjoint smooth element of the noncommutative two torus, $h = h^* \in A_\theta^\infty$. The conformal change of the metric by the Weyl factor e^{-h} will change the inner product on functions and on two forms as follows. On functions, the Hilbert space given by GNS construction of A_θ with respect to the positive linear functional $\varphi_0(a) = \frac{1}{\Im(\tau)}\varphi(ae^{-h})$ will be denoted by $\mathcal{H}_h^{(0)}$. Therefore the inner product of $\mathcal{H}_h^{(0)}$ is given by

$$\langle a, b\rangle_0 = \frac{1}{\Im(\tau)}\varphi(b^*ae^{-h}), \quad a, b \in A_\theta.$$

On one forms, the Hilbert space will stay the same as in (59), and will be denoted by $\mathcal{H}_h^{(1)}$. On the other hand, the Hilbert space of two forms, denoted by $\mathcal{H}_h^{(2)}$, is the Hilbert space given by the GNS construction of A_θ with respect to $\varphi_2(a) = \Im(\tau)\varphi(ae^h)$. Hence its inner product is given by

$$\langle a, b\rangle_2 = \Im(\tau)\varphi(b^* a e^h), \quad a, b \in A_\theta.$$

The positive functional $a \mapsto \varphi(ae^{-h})$, called the conformal weight, is a twisted trace of which modular operator is given by

$$\Delta(a) = e^{-h} a e^h, \quad a \in A_\theta.$$

The logarithm $\log \Delta$ of the modular operator will be denoted by ∇, and is given by $\nabla(a) = -[h, a]$. For more details the reader can check the previous sections.

The exterior derivatives are defined in the same way they were defined in the flat case (60) and (61). However, to emphasize that they are acting on different Hilbert spaces, we will denote them by $d_{h,0} : \mathcal{H}_h^{(0)} \to \mathcal{H}_h^{(1)}$ and $d_{h,1} : \mathcal{H}_h^{(1)} \to \mathcal{H}_h^{(2)}$.

Next, we consider the Hilbert spaces $\mathcal{H}_h^+ = \mathcal{H}_h^{(0)} \oplus \mathcal{H}_h^{(2)}$ and $\mathcal{H}_h^- = \mathcal{H}_h^{(1)}$, and the operator $d_h : \mathcal{H}_h^+ \to \mathcal{H}_h^-$, $d_h = d_{h,0} + d_{h,1}^*$. Therefore

$$d_h = \begin{pmatrix} i\delta_1 & \big(i|\tau|^2\delta_2 + i\Re(\tau)\delta_1\big) \circ R_{k^2} \\ i\delta_2 & \big(-i\Re(\tau)\delta_2 - i\delta_1\big) \circ R_{k^2} \end{pmatrix},$$

and its adjoint is given by

$$d_h^* = \begin{pmatrix} R_{k^2} \circ \big(i\delta_1 - i\Re(\tau)\delta_2\big) & R_{k^2} \circ \big(-i\Re(\tau)\delta_1 - i|\tau|^2\delta_2\big) \\ -i\delta_2 & i\delta_1 \end{pmatrix}.$$

We also consider the operator

$$D_h = \begin{pmatrix} 0 & d_h^* \\ d_h & 0 \end{pmatrix},$$

which acts on $\mathcal{H}_h = \mathcal{H}_h^+ \oplus \mathcal{H}_h^-$. Define the Hilbert space $\mathcal{H} = \mathcal{H}_0 \oplus \mathcal{H}_0 \oplus \mathcal{H}_0 \oplus \mathcal{H}_0$ and the unitary operator $W : \mathcal{H} \to \mathcal{H}_h$,

$$W = R_k \oplus R_{k^{-1}} \oplus \mathrm{I}_{\mathcal{H}_0 \oplus \mathcal{H}_0}.$$

The operator D_h can be transferred to an operator $\tilde{D}_h$ on $\mathcal{H}$ by the inner perturbation

$$\tilde{D}_h := W^* D_h W = \begin{pmatrix} 0 & R_k \circ d^* \\ d \circ R_k & 0 \end{pmatrix}.$$

In order to define the twisted, or modular, de Rham spectral triple for the noncommutative two torus, we employ the following constructions from [21]. Let

$(\mathcal{A}, \mathcal{H}^+ \oplus \mathcal{H}^-, D)$ be an even spectral triple with grading operator γ, where $D = \begin{pmatrix} 0 & T^* \\ T & 0 \end{pmatrix}$ and $T : \mathcal{H}^+ \to \mathcal{H}^-$ is an unbounded operator with adjoint T^*. If $f \in \mathcal{A}$ is positive and invertible, then $(\mathcal{A}, \mathcal{H}, D_{(f,\gamma)})$ is a modular spectral triple with respect to the inner automorphism $\sigma(a) = faf^{-1}$, $a \in \mathcal{A}$ [21, Lemma 1.1], where the Dirac operator is given by

$$D_{(f,\gamma)} = \begin{pmatrix} 0 & fT^* \\ Tf & 0 \end{pmatrix}.$$

On the other hand, any modular spectral triple $(\mathcal{A}, \mathcal{H}, D)$ with an automorphism σ admits a transposed modular spectral triple $(\mathcal{A}^{\mathrm{op}}, \bar{\mathcal{H}}, D^t)$ [21, Proposition 1.3], where $\mathcal{A}^{\mathrm{op}}$ is the opposite algebra of $\mathcal{A}$, $\bar{\mathcal{H}}$ is the dual Hilbert space, the action of $\mathcal{A}^{\mathrm{op}}$ on $\bar{\mathcal{H}}$ is the transpose of the action of $\mathcal{A}$ on $\mathcal{H}$, D^t is the transpose of D, and σ' is the automorphism of $\mathcal{A}^{\mathrm{op}}$ given by $\sigma'(a^{\mathrm{op}}) = (\sigma^{-1}(a))^{\mathrm{op}}$.

Proposition 8.3 *Let $k = e^{h/2}$, where $h = h^* \in A_\theta^\infty$. The triple $(A_\theta^{\mathrm{op}}, \mathcal{H}, \tilde{D}_h)$ is a modular spectral triple, where the automorphism of A_θ^{op} is given by*

$$a^{\mathrm{op}} \mapsto (k^{-1}ak)^{\mathrm{op}}, \quad a \in A_\theta^\infty,$$

and the representation of A_θ^{op} on $\mathcal{H}$ is given by the right multiplication of A_θ on $\mathcal{H}$. Moreover, the transposed of the modular spectral triple $(A_\theta^{\mathrm{op}}, \mathcal{H}, \tilde{D}_h)$ is isomorphic to the perturbed spectral triple

$$(A_\theta, \mathcal{H}, \bar{D}_h), \quad \bar{D}_h = \begin{pmatrix} 0 & kd \\ d^*k & 0 \end{pmatrix}, \tag{63}$$

where the operators d and d^* are as in (62).

Definition 8.3 *The modular spectral triple $(A_\theta, \mathcal{H}, \bar{D}_h)$ in* (63) *will be called the modular de Rham spectral triple of the noncommutative two torus with dilaton h.*

8.6 *Ricci functional and Ricci curvature for the curved noncommutative torus*

Using the pseudodifferential calculus with symbols in $A_\theta^\infty \otimes M_4(\mathbb{C})$, one shows that the localized heat trace of $\bar{D}_h^2$ has an asymptotic expansion with coefficients of the form

$$a_n(E, \bar{D}_h^2) = \varphi \circ \mathrm{tr}\left(E\, c_n(\bar{D}_h^2)\right), \qquad E \in A_\theta^\infty \otimes M_4(\mathbb{C}),$$

where $c_n(\bar{D}_h^2) \in A_\theta^\infty \otimes \big(M_2(\mathbb{C}) \oplus M_2(\mathbb{C})\big)$ and tr is the matrix trace. The Ricci functional can now be defined:

Definition 8.4 ([34]) *The Ricci functional of the modular de Rham spectral triple* $(A_\theta, \mathcal{H}, \bar{D}_h)$ *is the functional on* $A_\theta \otimes M_2(\mathbb{C})$ *defined as*

$$\mathcal{R}ic(F) = a_2(\gamma \tilde{F}, \bar{D}^2) = \zeta(0, \gamma \tilde{F}, \bar{D}_h^2) + Tr(\mathrm{tr}(F) Q_0) - Tr(F Q_1),$$

where $\tilde{F} = \mathrm{tr}(F) \oplus 0 \oplus F$, *and* Q_j *is the orthogonal projection on the kernel of* $\triangle_{h,j}$, *for* $j = 0, 1$.

Lemma 8.1 *There exists an element* $\mathbf{Ric} \in A_\theta^\infty \otimes M_2(\mathbb{C})$ *such that for all* $F \in A_\theta^\infty \otimes M_2(\mathbb{C})$

$$\mathcal{R}ic(F) = \frac{1}{\Im(\tau)} \varphi(\mathrm{tr}(F \mathbf{Ric}) e^{-h}).$$

Proof For any such F we have

$$a_2(\gamma \tilde{F}, \bar{D}^2) = a_2(\mathrm{tr}(F), \triangle_{h,0}) - a_2(F, \triangle_{h,1}).$$

Now $\mathrm{tr}(F) e^{-t\triangle_{h,0}} = \mathrm{tr}(F e^{-t\triangle_{h,0} \otimes \mathrm{I}_2})$, and thus

$$a_2(\mathrm{tr}(F), \triangle_{h,0}) = a_2(F, \triangle_{h,0} \otimes I_2).$$

As a result, we have

$$\begin{aligned} \mathbf{Ric}(F) &= a_2(\mathrm{tr}(F), \triangle_{h,0}) - a_2(F, \triangle_{h,1}) \\ &= \varphi\left(\mathrm{tr}\Big(F\big(c_2(\triangle_{h,0}) \otimes \mathrm{I}_2 - c_2(\triangle_{h,1})\big)\Big)\right) \\ &= \frac{1}{\Im(\tau)} \varphi\left(\Im(\tau) \mathrm{tr}\Big(F\big(c_2(\triangle_{h,0}) \otimes \mathrm{I}_2 - c_2(\triangle_{h,1})\big)\Big) e^h e^{-h}\right). \end{aligned}$$

Hence,

$$\mathbf{Ric} = \Im(\tau)\Big(c_2(\triangle_{h,0}) \otimes \mathrm{I}_2 - c_2(\triangle_{h,1})\Big) e^h.$$

□

Definition 8.5 *The element* **Ric** *is called the Ricci density of the curved noncommutative torus with dilaton h.*

The terms $c_2(\triangle_{h,j})$ can be computed by integrating the symbol of the parametrix of $\triangle_{h,j}$. Since the operator $\triangle_{h,1}$ is a first order perturbation of $\triangle_\varphi^{(0,1)}$, we will only need to compute the difference $c_2(\triangle_{h,1}) - c_2(\triangle_\varphi^{(0,1)}) \otimes \mathrm{I}_2$. The terms

$c_2(\Delta_{h,0}) = c_2(k\Delta_0 k)$ and $c_2(\Delta_\varphi^{(0,1)})$ are computed previously in two places by Connes-Moscovici and Fathizadeh-Khalkhali, and their difference is given by

$$\begin{aligned} R^\gamma &= \big(c_2(k\Delta k)\otimes \mathrm{I}_2 - c_2(\Delta_\varphi^{(0,1)})\big)e^h \\ &= -\frac{\pi}{\Im(\tau)}\Big(K_\gamma(\nabla)(\Delta_0(\log k)) + H_\gamma\big(\nabla_1,\nabla_2\big)\left(\square_{\Re}(\log k)\right) \\ &\qquad + S(\nabla_1,\nabla_2)(\square_{\Im}(\log k))\Big)e^h. \end{aligned}$$

Here,

$$\square_{\Re}(\ell) = (\delta_1(\ell))^2 + \Re(\tau)\left(\delta_1(\ell)\delta_2(\ell)+\delta_2(\ell)\delta_1(\ell)\right) + |\tau|^2(\delta_2(\ell))^2,$$
$$\square_{\Im}(\ell) = i\Im(\tau)(\delta_1(\ell)\delta_2(\ell)-\delta_2(\ell)\delta_1(\ell))$$

with $\ell = \log k$. Moreover,

$$K_\gamma(u) = \frac{\frac{1}{2}+\frac{\sinh(u/2)}{u}}{\cosh^2(u/4)},$$

$$H_\gamma(s,t) = \big(1-\cosh((s+t)/2)\big) \times \frac{\frac{t(s+t)\cosh(s)-s(s+t)\cosh(t)+(s-t)}{(s+t+\sinh(s)+\sinh(t)-\sinh(s+t))}}{st(s+t)\sinh\left(\frac{s}{2}\right)\sinh\left(\frac{t}{2}\right)\sinh\left(\frac{s+t}{2}\right)^2},$$

$$S(s,t) = \frac{(s+t-t\,\cosh(s)-s\,\cosh(t)-\sinh(s)-\sinh(t)+\sinh(s+t))}{s\,t\left(\sinh\left(\frac{s}{2}\right)\sinh\left(\frac{t}{2}\right)\sinh\left(\frac{s+t}{2}\right)\right)}.$$

The term S coincides with the function S found in [21, 31] for scalar curvature.

Now the main result of [34] can be stated as follows. It computes the Ricci curvature density of a curved noncommutative two torus with a conformally flat metric. The proof of this theorem is quite long and complicated and will not be reproduced here.

Theorem 8.1 ([34]) *Let $k = e^{h/2}$ with $h \in A_\theta^\infty$ a selfadjoint element. Then the Ricci density of the modular de Rham spectral triple with dilaton h is given by*

$$\mathbf{Ric} = \frac{\Im(\tau)}{4\pi^2}R^\gamma \otimes \mathrm{I}_2 - \frac{1}{4\pi}S(\nabla_1,\nabla_2)\big([\delta_1(\log k),\delta_2(\log k)]\big)e^h \otimes \begin{pmatrix} i\Im(\tau) & \Im(\tau)^2 \\ -1 & i\Im(\tau)\end{pmatrix}.$$

It is important to check the classical limit for consistency. In the commutative limit the Ricci density **Ric** is retrieved as $\lim_{(s,t)\to(0,0)}$ **Ric**. Since (cf. [21] for a proof)

$$\lim_{(s,t)\to(0,0)} R^{\gamma} = -\frac{\pi}{\Im(\tau)}\Delta_0(\log k),$$

and $[\delta_1(\log k), \delta_2(\log k)] = 0$, we have

$$\mathbf{Ric}|_{\theta=0} = \frac{-1}{4\pi}\Delta_0(\log k)e^h \otimes \mathrm{I}_2.$$

If we take into account the normalization of the classical case that comes from the heat kernel coefficients, this gives the formula for the Ricci operator in the classical case.

Unlike the commutative case, the Ricci density **Ric** in the noncommutative case is not a symmetric matrix. Indeed, it has non-zero off diagonal terms, which are multiples of $S(\nabla_1, \nabla_2)([\delta_1(\log k), \delta_2(\log k)])$. This phenomenon, observed in [34] for the first time, is obviously a consequence of the noncommutative nature of the space. It is an interesting feature of noncommutative geometry that, contrary to the commutative case, the Ricci curvature is not a multiple of the scalar curvature even in dimension two. This manifests itself in the existence of off diagonal terms in the Ricci operator **Ric** above.

It is clear that one can define in a similar fashion a Ricci curvature operator for higher dimensional noncommutative tori, as well as for noncommutative toric manifolds. Its computation in these cases poses an interesting problem. This problem now is completely solved for noncommutative three tori in [24]. It would also be interesting to find the analogue of the Ricci flow based on our definition of Ricci curvature functional. It should be noted that for noncommutative two tori a definition of Ricci flow, without a notion of Ricci curvature, is proposed in [3].

9 Beyond conformally flat metrics and beyond dimension four

In the study of spectral geometry of noncommutative tori one is naturally interested in going beyond conformally flat metrics and beyond dimension four. Even in the case of noncommutative two torus it is important to consider metrics which are not conformally flat. In fact while by uniformization theorem we know that any metric on the two torus is conformally flat, there is strong evidence that this is not so in the noncommutative case. This is closely related to the problem of classification of complex structures on the noncommutative two torus via positive Hochschild cocycles, which is still unsolved.

As far as higher dimensions go, our original methods do not allow us to treat the dimension as a variable in the calculations and obtain explicit formulas in all dimensions in a uniform manner. This is in sharp contrast with the classical case where formulas work in a uniform manner in all dimensions. In this section we report on a very recent development [35] where progress has been made on both fronts.

In the recent paper [35], using a new strategy based on Newton divided differences, it is shown how to consider non-conformal metrics and how to treat all higher dimensional noncommutative tori in a uniform way. In fact based on older methods it was not clear how to extend the computation of the scalar curvature to a general higher dimensional case. The class of non-conformal metrics introduced in [35] is quite large and leads to beautiful combinatorial identities for the curvature via divided differences. In this section we shall briefly sketch the results obtained in [35], following closely its organization of material.

9.1 Rearrangement lemma revisited

To compute and effectively work with integrals of the form

$$\int_0^\infty (uk^2+1)^{-m} b (uk^2+1) u^m du,$$

the rearrangement lemma was proved by Connes and Tretkoff in [22]. Here $k = e^{h/2}$, $h, b \in C^\infty(\mathbb{T}^2_\theta)$ and h is selfadjoint. The problem stems from the fact that h and b need not commute. Later on this lemma was generalized, for the sake of curvature calculations, for more than one b in [21, 31]. A detailed study of this lemma for more general integrands of the form

$$\int_0^\infty f_0(u,k) b_1 f_1(u,k) \rho_2 \cdots b_n f_n(u,k) du,$$

was given by M. Lesch in [46], with a new proof and a new point of view. This approach uses the multiplication map

$$\mu : a_1 \otimes a_2 \otimes \cdots \otimes a_n \mapsto a_1 a_2 \cdots a_n$$

from the projective tensor product $A^{\otimes_\gamma n}$ to A. The above integral is expressed as the contraction of the product of an element $F(k_{(0)}, \cdots, k_{(n)})$ of $A^{\otimes_\gamma (n+1)}$, with the element $b_1 \otimes b_2 \otimes \cdots b_n \otimes 1$ which is

$$\mu\Big(F(k_{(0)}, \cdots, k_{(n)})(b_1 \otimes b_2 \otimes \cdots b_n \otimes 1)\Big).$$

The above element is usually written in the so-called *contraction form*

$$F(k_{(0)}, \cdots, k_{(n)})(b_1 \cdot b_2 \cdot \cdots \cdot b_n). \tag{64}$$

The following version of the rearrangement lemma is stated in [35] with the domain of integration changed from $[0, \infty)$ to any domain in $\mathbb{R}^N$.

Lemma 9.1 (Rearrangement Lemma [35]) *Let A be a unital C^*-algebra, $h \in A$ be a selfadjoint element, and Λ be an open neighborhood of the spectrum of h in $\mathbb{R}$. For a domain U in $\mathbb{R}^N$, let $f_j : U \times \Lambda \to \mathbb{C}$, $0 \le j \le n$, be smooth functions such that $f(u, \lambda) = \prod_{j=0}^{n} f_j(u, \lambda_j)$ satisfies the following integrability condition: for any compact subset $K \subset \Lambda^{n+1}$ and every given multi-index α we have*

$$\int_U \sup_{\lambda \in K} |\partial_\lambda^\alpha f(u, \lambda)| du < \infty.$$

Then,

$$\int_U f_0(u, h) b_1 f_1(u, h) \cdots b_n f_n(u, h) du = F(h_{(0)}, h_{(1)}, \cdots, h_{(n)})(b_1 \cdot b_2 \cdots b_n), \tag{65}$$

where $F(\lambda) = \int_U f(u, \lambda) du$. □

In particular it follows that every expression in the contraction form with a Schwartz function $F \in \mathcal{S}(\mathbb{R}^{n+1})$ used in the operator part can be written as an integral. In fact if we set

$$f_n(\xi, \lambda) = \hat{f}(\xi) e^{i\xi_n \lambda}, \quad f_j(\xi, \lambda) = e^{i\xi_j \lambda}, \quad 0 \le j \le n-1,$$

and $f(\xi, \lambda_0, \cdots, \lambda_n) = \prod_{j=0}^{n} f_j(\xi, \lambda_j)$, by the Fourier inversion formula, we have $F(\lambda) = \int f(\xi, \lambda) d\xi$. Then, Lemma 9.1 gives the equality

$$F(h_{(0)}, \cdots, h_{(n)})(b_1 \cdot b_2 \cdots b_n) = \int_{\mathbb{R}^n} e^{i\xi_0 h} b_1 e^{i\xi_1 h} b_2 \cdots b_n e^{i\xi_n h} \hat{f}(\xi) d\xi. \tag{66}$$

This is crucial for calculations in [35].

9.2 A new idea

As we saw in previous sections, to prove the Gauss-Bonnet theorem and to compute the scalar curvature of a curved noncommutative two torus in [22, 30] and [21, 31], the second density of the heat trace of the Laplacian D^2 of the Dirac operator had to be computed. First, the symbol of the parametrix of D^2 was computed, next a contour integral coming from Cauchy's formula for the heat operator had to be computed, and finally one had to integrate out the momentum variables. It was for this last step that the rearrangement lemma played an important role. Luckily, the contour integral could be avoided using a homogeneity argument.

A key observation in [35] is that one need not wait till the last step to have elements in the contraction form. It is just enough to start off with operators whose symbol is written in the contraction form

$$F(h_{(0)}, \cdots, h_{(n)})(b_1 \cdot b_2 \cdots b_n).$$

It is further noted that the symbol calculus can be effectively applied to differential operators whose symbols can be written in the contraction form. These operators are called *h-differential operators* in [35]. This is a new and larger class of differential operators that lends itself to precise spectral analysis. It is strictly larger than the class of Dirac Laplacians for conformally flat metrics on noncommutative tori which has been the subject of intensive studies lately.

Next, the Newton divided difference calculus was brought in to find the action of derivations on elements in contracted form (Theorem 9.1 below). For example, one has

$$\begin{aligned}\delta_j\big(f(h_{(0)}, h_{(1)})(b_1)\big) = f(h_{(0)}, h_{(1)})(\delta_j(b_1)) + \big[h_{(0)}, h_{(1)}; f(\cdot, h_{(2)})(\delta_j(h) \cdot b)\big] \\ +\big[h_{(1)}, h_{(2)}; f(h_{(0)}, \cdot)(b \cdot \delta_j(h))\big].\end{aligned}$$

Using this fact, and applying the pseudodifferential calculus, one can compute the spectral densities of positive h-differential operators whose principal symbol is given by a functional metric. These operators are called *Laplace type h-differential operator* in [35].

This change in order of the computations, i.e. writing symbols in the contraction form first, led to a smoother computation symbolically, and played a fundamental in computing with more general functional metrics. It also paved the way for calculating the curvature in all higher dimensions for conformally flat and twisted product of flat metrics.

9.3 Newton divided differences

A nice application of the rearrangement lemma is to find a formula for the differentials of a smooth element written in contraction form. To this end, Newton divided differences were used in [35].

Let $x_0, x_1, \cdots, x_n$ be distinct points in an interval $I \subset \mathbb{R}$ and let f be a function on I. The nth-*order Newton divided difference* of f, denoted by $[x_0, x_1, \cdots, x_n; f]$, is the coefficient of x^n in the interpolating polynomial of f at the given points. In other words, if the interpolating polynomial is $p(x)$, then

$$p(x) = p_{n-1}(x) + [x_0, x_1, \cdots, x_n; f](x - x_0) \cdots (x - x_{n-1}),$$

where $p_{n-1}(x)$ is a polynomial of degree at most $n - 1$. There is a recursive formula for the divided difference which is given by

$$[x_0; f] = f(x_0)$$
$$[x_0, x_1, \cdots, x_n; f] = \frac{[x_1, \cdots, x_n; f] - [x_0, x_1, \cdots, x_{n-1}; f]}{x_n - x_0}.$$

There is also an explicit formula for the divided difference:

$$[x_0, x_1, \cdots, x_n; f] = \sum_{j=0}^{n} \frac{f(x_j)}{\prod_{j \neq l}(x_j - x_l)}.$$

The *Hermite-Genocchi formula* gives an integral representation for the divided differences of an n times continuously differentiable function f as an integral over the standard simplex:

$$[x_0, \cdots, x_n; f] = \int_{\Sigma_n} f^{(n)}\Big(\sum_{j=0}^{n} s_j x_j\Big) ds. \tag{67}$$

Let δ be a densely defined, unbounded and closed derivation on a C^*-algebra A. If $a \in \mathrm{Dom}(\delta)$, then $e^{za} \in \mathrm{Dom}(\delta)$ for any $z \in \mathbb{C}$, and one has

$$\delta(e^{za}) = z \int_0^1 e^{zsa} \delta(a) e^{z(1-s)a} ds. \tag{68}$$

Using the rearrangement lemma, one can now express the differential of a smooth element given in contraction form. This result generalizes the expansional formula, also known as Feynman-Dyson formula, for e^{A+B}, and not only for elements of the form $f(h)$, but also for any element written in the contraction form.

Theorem 9.1 *Let δ be a closed derivation of a C^*-algebra A and $h \in \mathrm{Dom}(\delta)$ be a selfadjoint element. Let $b_j \in \mathrm{Dom}(\delta)$, $1 \leq j \leq n$, and let $f : \mathbb{R}^{n+1} \to \mathbb{C}$ be a smooth function. Then $f(h_{(0)}, \cdots, h_{(n)})(b_1 \cdot b_2 \cdot \cdots \cdot b_n)$ is in the domain of δ and*

$$\begin{aligned}
&\delta(f(h_{(0)}, \cdots, h_{(n)})(b_1 \cdot b_2 \cdots b_n)) \\
&= \sum_{j=1}^{n} f(h_{(0)}, \cdots, h_{(n)})\big(b_1 \cdots b_{j-1} \cdot \delta(b_j) \cdot b_{j+1} \cdots b_n\big) \\
&\quad + \sum_{j=0}^{n} f_j(h_{(0)}, \cdots, h_{(n+1)})\big(b_1 \cdots b_j \cdot \delta(h) \cdot b_{j+1} \cdots b_n\big),
\end{aligned}$$

where $f_j(t_0, \cdots, t_{n+1})$, which we call the partial divided difference, is defined as

$$f_j(t_0, \cdots, t_{n+1}) = \big[t_j, t_{j+1}; t \mapsto f(t_0, \cdots, t_{j-1}, t, t_{j+2}, \cdots, t_n)\big].$$

9.4 Laplace type h-differential operators and asymptotic expansions

Let us first recall this class of differential operators which is introduced in [35]. It extends the previous classes of differential operators on noncommutative tori, in particular Dirac Laplacians of conformally flat metrics.

Definition 9.1 ([35]) *Let $h \in C^\infty(\mathbb{T}_\theta^d)$ be a smooth selfadjoint element.*

(i) *An h-differential operator on $\mathbb{T}_\theta^d$ is a differential operator $P = \sum_\alpha p_\alpha \delta^\alpha$, with $C^\infty(\mathbb{T}_\theta^d)$-valued coefficients p_α which can be written in the contraction form*

$$p_\alpha = P_{\alpha,\alpha_1,\cdots,\alpha_k}(h_{(0)},\cdots,h_{(k)})(\delta^{\alpha_1}(h)\cdots\delta^{\alpha_k}(h)).$$

(ii) *A second order h-differential operator P is called a Laplace type h-differential operator if its symbol is a sum of homogeneous parts p_j of the form*

$$\begin{aligned}
p_2 &= P_2^{ij}(h)\xi_i\xi_j,\\
p_1 &= P_1^{ij}(h_{(0)},h_{(1)})(\delta_i(h))\xi_j,\\
p_0 &= P_{0,1}^{ij}(h_{(0)},h_{(1)})(\delta_i\delta_j(h)) + P_{0,2}^{ij}(h_{(0)},h_{(1)},h_{(2)})(\delta_i(h)\cdot\delta_j(h)),
\end{aligned}$$

where the principal symbol $p_2 : \mathbb{R}^n \to C^\infty(\mathbb{T}_\theta^d)$ is a $C^\infty(\mathbb{T}_\theta^d)$-valued quadratic form such that $p_2(\xi) > 0$ for all $\xi \in \mathbb{R}$.

We can allow the symbols to be matrix valued, that is $p_j : C^\infty(\mathbb{R}^d) \to C^\infty(\mathbb{T}_\theta^d) \otimes M_n(\mathbb{C})$, provided that all $p_2(\xi) \in C^\infty(\mathbb{T}_\theta^d) \otimes \mathrm{I}_n$ for all non-zero $\xi \in \mathbb{R}$.

Many of the elliptic second order differential operators on noncommutative tori which were studied in the literature are Laplace type h-differential operators. For instance, the two differential operators on $\mathbb{T}_\theta^2$ whose spectral invariants are studied in [21, 31] are indeed Laplace type h-differential operator. In fact with $k = e^{h/2}$, these operators are given by

$$k\triangle k = k\delta\delta^* k, \qquad \triangle_\varphi^{(0,1)} = \delta^* k^2 \delta,$$

where $\delta = \delta_1 + \bar{\tau}\delta_2$ and $\delta^* = \delta_1 + \tau\delta_2$ for some complex number τ in the upper half plane.

Let P be a positive Laplace type h-differential operator. Using the Cauchy integral formula, one has

$$e^{-tP} = \frac{-1}{2\pi i}\int_\gamma e^{-t\lambda}(P-\lambda)^{-1}d\lambda, \quad t > 0,$$

for a suitable contour γ. Expanding the symbol of the parametrix $\sigma((P-\lambda)^{-1})$, one obtains a short time asymptotic expansion for localized heat trace for any $a \in C^\infty(\mathbb{T}^d_\theta)$:

$$\mathrm{Tr}(ae^{-tP}) \sim \sum_{n=0}^{\infty} c_n(a) t^{(n-d)/2}.$$

Here, $c_n(a) = \varphi(ab_n)$ with

$$b_n = \frac{1}{(2\pi)^d} \int_{\mathbb{R}^d} \frac{-1}{2\pi i} \int_\gamma b_n(\xi, \lambda) d\lambda d\xi. \tag{69}$$

Using the rearrangement lemma (Lemma 9.1) and the fact that the contraction map and integration commute, one obtains

$$\begin{aligned} b_2 = &\left(\frac{1}{(2\pi)^d} \int_{\mathbb{R}^d} \frac{-1}{2\pi i} \int_\gamma B^{ij}_{2,1}(\xi, \lambda, h_{(0)}, h_{(1)}) e^{-\lambda} d\lambda d\xi \right) \big(\delta_i \delta_j(h)\big) \\ &+ \left(\frac{1}{(2\pi)^d} \int_{\mathbb{R}^d} \frac{-1}{2\pi i} \int_\gamma B^{ij}_{2,2}(\xi, \lambda, , h_{(0)}, h_{(1)}, , h_{(2)}) e^{-\lambda} d\lambda d\xi \right) \\ &\times \big(\delta_i(h) \cdot \delta_j(h)\big). \end{aligned}$$

The dependence of $B^{ij}_{2,k}$ on λ comes only from different powers of B_0 in its terms, while its dependence on ξ_j's is the result of appearance of ξ_j as well as of B_0 in the terms. Therefore, the contour integral will only contain $e^{-\lambda}$ and product of powers of $B_0(t_j)$. Hence, we need to deal with a certain kind of contour integral for which we shall use the following notation and will call them *T-functions*:

$$T_{\boldsymbol{n};\boldsymbol{\alpha}}(t_0, \cdots, t_n) := \frac{-1}{\pi^{d/2}} \int_{\mathbb{R}^d} \xi_{n_1} \cdots \xi_{n_{2|\alpha|-4}} \frac{1}{2\pi i} \int_\gamma e^{-\lambda} B_0^{\alpha_0}(t_0) \cdots B_0^{\alpha_n}(t_n) d\lambda d\xi, \tag{70}$$

where $\boldsymbol{n} = (n_1, \cdots, n_{2|\alpha|-4})$ and $\boldsymbol{\alpha} = (\alpha_0, \cdots, \alpha_n)$. We recall the T-functions and their properties a bit later. There is an explicit formula for $b_2(P)$ which we now recall from [35]:

Proposition 9.1 *For a positive Laplace type h-differential operator P with the symbol given by* (9.1)*, the term $b_2(P)$ in the contraction form is given by*

$$\begin{aligned} b_2(P) = (4\pi)^{-d/2} \Big(&B^{ij}_{2,1}(h_{(0)}, h_{(1)}) \big(\delta_i \delta_j(h)\big) \\ &+ B^{ij}_{2,2}(h_{(0)}, h_{(1)}, h_{(2)}) \big(\delta_i(h) \cdot \delta_j(h)\big) \Big), \end{aligned}$$

where the functions are defined by

$$\begin{aligned} B^{ij}_{2,1}(t_0,t_1) = &- T_{;1,1}(t_0,t_1) P^{ij}_{0,1}(t_0,t_1) + 2T_{k\ell;2,1}(t_0,t_1) P^{ik}_2(t_0) P^{j\ell}_1(t_0,t_1) \\ &+ T_{k\ell;2,1}(t_0,t_1)\, P^{ij}_2(t_0) \left[t_0,t_1;\, P^{k\ell}_2\right] \\ &- 4T_{k\ell mn;3,1}(t_0,t_1) P^{ik}_2(t_0) P^{j\ell}_2(t_0) \left[t_0,t_1;\, P^{mn}_2\right], \end{aligned}$$

and

$$\begin{aligned} B^{ij}_{2,2}(t_0,t_1,t_2) = &- T_{;1,1}(t_0,t_2) P^{ij}_{0,2}(t_0,t_1,t_2) \\ &+ T_{k\ell;1,1,1}(t_0,t_1,t_2)\, P^{ik}_1(t_0,t_1) P^{j\ell}_1(t_1,t_2) \\ &- 2T_{k\ell mn;2,1,1}(t_0,t_1,t_2)\, P^{im}_2(t_0) \left[t_0,t_1;\, P^{k\ell}_2\right] P^{jn}_1(t_1,t_2) \\ &+ 2T_{k\ell;2,1}(t_0,t_2)\, P^{ik}_2(t_0) \left[t_0,t_1;\, P^{j\ell}_1(\cdot,t_2)\right] \\ &+ 2T_{k\ell;2,1}(t_0,t_2)\, P^{jk}_2(t_0) \left[t_1,t_2;\, P^{i\ell}_1(t_0,\cdot)\right] \\ &+ T_{k\ell;1,1,1}(t_0,t_1,t_2)\, P^{ij}_1(t_0,t_1) \left[t_1,t_2;\, P^{k\ell}_2\right] \\ &- 2T_{k\ell mn;2,1,1}(t_0,t_1,t_2)\, P^{j\ell}_2(t_0) P^{ik}_1(t_0,t_1) \left[t_1,t_2;\, P^{mn}_2\right] \\ &- 2T_{k\ell mn;1,2,1}(t_0,t_1,t_2)\, P^{ik}_1(t_0,t_1) P^{j\ell}_2(t_1) \left[t_1,t_2;\, P^{mn}_2\right] \\ &- 2T_{k\ell mn;2,1,1}(t_0,t_1,t_2)\, P^{ij}_2(t_0) \left[t_0,t_1;\, P^{\ell m}_2\right] \left[t_1,t_2;\, P^{kn}_2\right] \\ &- 4T_{k\ell mn;2,1,1}(t_0,t_1,t_2)\, P^{ik}_2(t_0) \left[t_0,t_1;\, P^{j\ell}_2\right] \left[t_1,t_2;\, P^{mn}_2\right] \\ &+ 8T_{k\ell mnpq;3,1,1}(t_0,t_1,t_2)\, P^{ik}_2(t_0) P^{jn}_2(t_0) \left[t_0,t_1;\, P^{\ell m}_2\right] \\ &\quad \times \left[t_1,t_2;\, P^{pq}_2\right] \\ &+ 4T_{k\ell mnpq;2,2,1}(t_0,t_1,t_2) P^{ik}_2(t_0) \left[t_0,t_1;\, P^{\ell m}_2\right] P^{jn}_2(t_1) \\ &\quad \times \left[t_1,t_2;\, P^{pq}_2\right] \\ &+ 2T_{k\ell;2,1}(t_0,t_2)\, P^{ij}_2(t_0) \left[t_0,t_1,t_2;\, P^{k\ell}_2\right] \\ &- 8T_{k\ell mn;3,1}(t_0,t_2)\, P^{ik}_2(t_0) P^{j\ell}_2(t_0) [t_0,t_1,t_2;\, P^{mn}_2]. \end{aligned}$$

□

The computation of the higher heat trace densities for a Laplace type h-operator can be similarly carried out, expecting many more terms in the results. This would give a way to generalize results obtained for the conformally flat noncommutative two torus in [17] where b_4 of the Laplacian D^2 of the Dirac operator D is computed. This problem won't be discussed further in this paper, but is certainly an interesting problem.

Evaluating T-functions (70), the only parts of formulas for $B_{2,1}^{i,j}$ and $B_{2,2}^{i,j}$ that need to be evaluated, is not always an easy task. In [35] a concise integral formula for T-functions is given and their properties are studied. For the contour integral in (70). it is clear that there are functions $f_{\alpha_1,\cdots,\alpha_n}$ such that

$$\frac{-1}{2\pi i}\int_{\gamma} e^{-\lambda} B_0^{\alpha_0}(t_0)\cdots B_0^{\alpha_n}(t_n)d\lambda = f_{\alpha_0,\cdots,\alpha_n}\left(\|\xi\|_{t_0}^2,\cdots,\|\xi\|_{t_n}^2\right).$$

Here, we denoted $P_2^{ij}(t_k)\xi_i\xi_j$ by $\|\xi\|_{t_k}^2$. Examples of such functions are

$$f_{1,1}(x_0,x_1) = -\frac{e^{-x_0}}{x_0-x_1} - \frac{e^{-x_1}}{x_1-x_0},$$

$$f_{2,1}(x_0,x_1) = -\frac{e^{-x_0}}{x_0-x_1} - \frac{e^{-x_0}}{(x_0-x_1)^2} + \frac{e^{-x_1}}{(x_1-x_0)^2}.$$

Lemma 9.2 ([35]) *Let $P_2(t)$ be a positive definite $d \times d$ matrix of smooth real functions. Then*

$$T_{n;\boldsymbol{\alpha}}(t_0,\cdots,t_n) = \frac{1}{2^{|\alpha|-2}\boldsymbol{\beta}!}\int_{\Sigma_n}\prod_{j=0}^{n} s_j^{\alpha_j-1}\frac{\sum\prod_{\boldsymbol{n}} P^{-1}(s)_{n_i n_{\sigma(i)}}}{\sqrt{\det P(s)}}ds,$$

where $P(s)=\sum_{j=0}^{n} s_j P_2(t_j)$ and $\boldsymbol{\beta}=(\alpha_0-1,\cdots,\alpha_n-1)$. □

9.5 Functional metrics and scalar curvature

A natural question is if there exists a large class of noncommutative metrics whose Laplacians are h-differential operators and hence amenable to the spectral analysis developed in the last section. As we saw, conformally flat metrics on noncommutative tori is such a class. But there are more. One of the interesting concepts developed in [35] is the notion of a *functional metric* which is a much larger class than conformally flat metrics and whose Laplacian is still an h-differential operator. In this section we shall first recall this concept and reproduce the scalar curvature formula for these metrics developed in [35].

Definition 9.2 *Let h be a selfadjoint smooth element of a noncommutative d-torus and let $g_{ij}:\mathbb{R}\to\mathbb{R}$, $1\le i,j\le d$, be smooth functions such that the matrix $(g_{ij}(t))$ is a positive definite matrix for every t in a neighborhood of the spectrum of h. We shall refer to $g_{ij}(h)$ as a functional metric on A_θ^d.*

The construction of the Laplacian on functions on $\mathbb{T}_\theta^d$ equipped with a functional metric $g=g_{ij}(h)$ follows the same pattern as in previous sections. Details can be found in [35], where the following crucial result is also proved. The Laplacian $\delta^*\delta : \mathcal{H}_{0,g} \to \mathcal{H}_{0,g}$ on elements of $C^\infty(\mathbb{T}^d_\theta)$ is given by

$$\delta_j(a)g^{jk}(h)\delta_k(|g|^{\frac{1}{2}}(h))|g|^{-\frac{1}{2}}(h) + \delta_j(a)\delta_k(g^{jk}(h)) + i\delta_k(\delta_j(a))g^{jk}(h).$$

To carry the spectral analysis of the Laplacian $\delta^*\delta : \mathcal{H}_{0,g} \to \mathcal{H}_{0,g}$, we switch to the antiunitary equivalent setting as follows. Let $\mathcal{H}_0$ be the Hilbert space obtained by the GNS construction from A^d_θ using the nonperturbed tracial state φ.

Proposition 9.2 *The operator $\delta^*\delta : \mathcal{H}_{0,g} \to \mathcal{H}_{0,g}$ is antiunitary equivalent to a Laplace type h-differential operator $\Delta_{0,g} : \mathcal{H}_0 \to \mathcal{H}_0$ whose symbol, when expressed in the contraction form, has the functional parts given by*

$$\begin{aligned}
P_2^{jk}(t_0) &= g^{jk}(t_0),\\
P_1^{jk}(t_0,t_1) &= |g|^{-\frac{1}{4}}(t_0)\big[t_0,t_1; |g|^{\frac{1}{4}}\big]g^{jk}(t_1) + \big[t_0,t_1; g^{jk}\big]\\
&\quad + |g|^{\frac{1}{4}}(t_0)g^{jk}(t_0)\big[t_0,t_1; |g|^{-\frac{1}{4}}\big],\\
P_{0,1}^{jk}(t_0,t_1) &= |g|^{\frac{1}{4}}(t_0)g^{jk}(t_0)\big[t_0,t_1; |g|^{-\frac{1}{4}}\big],\\
P_{0,2}^{jk}(t_0,t_1,t_2) &= |g|^{-\frac{1}{4}}(t_0)\big[t_0,t_1; g^{jk}|g|^{\frac{1}{2}}\big]\big[t_1,t_2; |g|^{-\frac{1}{4}}\big]\\
&\quad + 2|g|^{\frac{1}{4}}(t_0)g^{jk}(t_0)\big[t_0,t_1,t_2; |g|^{-\frac{1}{4}}\big].
\end{aligned}$$

An important case of the functional metric is the conformally flat metric

$$g_{ij}(t) = f(t)^{-1}g_{ij}, \tag{71}$$

where f is a positive smooth function and g_{ij}'s are the entries of a constant metric on $\mathbb{R}^d$. The functions given by Proposition 9.2, for the conformally flat metrics, give us the following:

$$\begin{aligned}
P_2^{jk}(t_0) &= g^{jk}f(t_0), \\
P_1^{jk}(t_0,t_1) &= g^{jk}\Big(f(t_0)^{\frac{d}{4}}\big[t_0,t_1; f^{1-\frac{d}{4}}\big] + f(t_0)^{1-\frac{d}{4}}\big[t_0,t_1; f^{\frac{d}{4}}\big]\Big),\\
P_{0,1}^{jk}(t_0,t_1) &= g^{jk}f(t_0)^{1-\frac{d}{4}}\big[t_0,t_1; f^{\frac{d}{4}}\big],\\
P_{0,2}^{jk}(t_0,t_1,t_2) &= g^{jk}\Big(f(t_0)^{\frac{d}{4}}\big[t_0,t_1; f^{1-\frac{d}{2}}\big]\big[t_1,t_2; f^{\frac{d}{4}}\big]\\
&\quad + 2f(t_0)^{1-\frac{d}{4}}\big[t_0,t_1,t_2; f^{\frac{d}{4}}\big]\Big).
\end{aligned} \tag{72}$$

A careful examination of formula (72) shows that for any function $P_\bullet^{ij}$ there exists a function $P_\bullet$ such that $P_\bullet^{ij} = g^{ij}P_\bullet$. We have similar situation with the T-functions for conformally flat metrics.

Lemma 9.3 ([35]) *Let $\boldsymbol{\alpha}$ and $\boldsymbol{n} = (n_1, \cdots, n_{2|\boldsymbol{\alpha}|-4})$ be two multi-indices. Then the T-function $T_{\boldsymbol{n},\boldsymbol{\alpha}}$ for the conformally flat metric* (71) *is of the form*

$$T_{\boldsymbol{n},\boldsymbol{\alpha}}(t_0, \cdots, t_n) = \sqrt{|g|}\sum_{\boldsymbol{n}} \prod g_{n_i n_{\sigma(i)}} T_{\boldsymbol{\alpha}}(t_0, \cdots, t_n).$$

The function $T_{\boldsymbol{\alpha}}$ in dimension $d \neq 2$ is given by

$$T_{\boldsymbol{\alpha}}(t_0, \cdots, t_n) = \frac{(-1)^{|\boldsymbol{\alpha}|-1}\Gamma\left(\frac{d}{2}-1\right)}{\Gamma\left(\frac{d}{2}+|\boldsymbol{\alpha}|-2\right)} \partial_x^{\boldsymbol{\beta}}\left[x_0, \cdots, x_n; u^{1-\frac{d}{2}}\right]\Big|_{x_j=f(t_j)}, \tag{73}$$

where $\boldsymbol{\beta} = (\alpha_0 - 1, \cdots, \alpha_n - 1)$.

As an example, we have

$$\begin{aligned} T_{\alpha,1}(t_0, t_1) = &\frac{(-1)^{\alpha}\Gamma(\frac{d}{2}-1)}{2^{\alpha-1}\Gamma(\frac{d}{2}+\alpha-1)} \\ &\times \Big(\frac{f(t_1)^{1-\frac{d}{2}}}{(f(t_1)-f(t_0))^{\alpha}} \\ &\quad - \sum_{m=0}^{\alpha-1} \frac{(-1)^m\Gamma(\frac{d}{2}+m-1)}{\Gamma(\frac{d}{2}-1)m!} \frac{f(t_0)^{-\frac{d}{2}-m+1}}{(f(t_1)-f(t_0))^{\alpha-m}}\Big). \end{aligned}$$

Note that for dimension two, $T_{\alpha,1}(t_0, t_1)$ can be obtained by taking the limit of (73) as d approaches 2. When $f(t) = t$, we have

$$T_{\alpha,1}(t_0, t_1) = \frac{(-1)^{\alpha-1}}{2^{\alpha-1}\Gamma(\alpha)^2}\partial_{t_0}^{\alpha-1}\big[t_0, t_1; \log(u)\big].$$

Recall that the *scalar curvature density* of a given functional metric is defined by

$$R = (4\pi)^{\frac{d}{2}} b_2(\Delta_{0,g}).$$

This scalar curvature density is computed for two classes of examples in all dimensions: conformally flat metrics and twisted products of conformally flat metrics. Let us recall this result:

Theorem 9.2 ([35]) *The scalar curvature of the d-dimensional noncommutative tori $\mathbb{T}_\theta^d$ equipped with the metric $f(h)^{-1}g_{ij}$ is given by*

$$R = \sqrt{|g|}\Big(K_d(h_{(0)}, h_{(1)})(\Delta(h)) + H_d(h_{(0)}, h_{(1)}, h_{(2)})(\Box(h))\Big),$$

where $\triangle(h) = g^{ij}\delta_i\delta_j(h)$, $\square(h) = g^{ij}\delta_i(h)\cdot\delta_j(h)$. *The functions* K_d *and* H_d *are given by*

$$\begin{aligned} K_d(t_0,t_1) =&K_d^t(f(t_0),f(t_1))\big[t_0,t_1;f\big],\\ H_d(t_0,t_1,t_2) =&H_d^t(f(t_0),f(t_1),f(t_2))\big[t_0,t_1;f\big]\big[t_1,t_2;f\big]\\ &+2K_d^t(f(t_0),f(t_2))\big[t_0,t_1,t_2;f\big], \end{aligned} \tag{74}$$

where K_d^t *and* H_d^t *are the functions* K_d *and* H_d *when* $f(t) = t$. *For* $d \neq 2$, *they can be computed to be*

$$\begin{aligned} K_d^t(x,y) = &\frac{4x^{2-\frac{3d}{4}}y^{2-\frac{3d}{4}}}{d(d-2)(x-y)^3}\\ &\left((d-1)x^{\frac{d}{2}}y^{\frac{d}{2}-1}-(d-1)x^{\frac{d}{2}-1}y^{\frac{d}{2}}-x^{d-1}+y^{d-1}\right), \end{aligned}$$

and

$$\begin{aligned} H_d^t(x,y,z) = &\frac{2x^{-\frac{3d}{4}}y^{-d}z^{-\frac{3d}{4}}}{(d-2)d(x-y)^2(x-z)^3(y-z)^2}\\ &\times\left(x^dy^dz^2(x-y)\left(3x^2y-2x^2z-4xy^2+4xyz-2xz^2+yz^2\right)\right.\\ &+x^dy^{\frac{d}{2}+1}z^{\frac{d}{2}+1}(x-z)^2(z-y)(dx+(1-d)y)\\ &+x^dy^3z^d(z-x)^3+x^{\frac{d}{2}+1}y^{\frac{3d}{2}}z^2(x-y)(x-z)^2\\ &+2(d-1)x^{\frac{d}{2}+1}y^dz^{\frac{d}{2}+1}(x-y)(x-z)(z-y)(x-2y+z)\\ &-x^{\frac{d}{2}+1}y^{\frac{d}{2}+1}z^d(x-y)(x-z)^2((1-d)y+dz)\\ &-x^2y^{\frac{3d}{2}}z^{\frac{d}{2}+1}(x-z)^2(z-y)\\ &\left.+x^2y^dz^d(y-z)\left(x^2y-2x^2z+4xyz-2xz^2-4y^2z+3yz^2\right)\right). \end{aligned}$$

These functions for the dimension two are given by

$$\begin{aligned} K_2^t(x,y) = &-\frac{\sqrt{x}\sqrt{y}}{(x-y)^3}((x+y)\log(x/y)+2(y-x)),\\ H_2^t(x,y,z) = &\frac{2\sqrt{x}\sqrt{z}}{(x-y)^2(x-z)^3(y-z)^2}\\ &\times\Big(-(x-y)(x-z)(y-z)(x-2y+z)+y(x-z)^3\log(y) \end{aligned}$$

$$+ (y-z)^2(-2x^2+xy+yz)\log(x) \\ - (x-y)^2(xy+zy-2z^2)\log(z)\Big)\Big).$$

Note that the Function $K_d^t(x, y)$ is the symmetric part of the function

$$\frac{8x^{2-\frac{d}{4}}y^{2-\frac{3d}{4}}}{d(d-2)(x-y)^3}\left((d-1)y^{\frac{d}{2}-1}-x^{\frac{d}{2}-1}\right).$$

Similarly, $H_d^t(x, y, z)$ is equal to $(F_d(x, y, z) + F_d(z, y, x))/2$ where

$$\begin{aligned} F_d(x,y,z) =& \frac{4x^{-\frac{3d}{4}}y^{-d}z^{-\frac{3d}{4}}}{d(d-2)(x-y)^2(x-z)^3(y-z)^2} \\ &\times \Big(x^d y^d z^2(x-y)(3x^2y-2x^2z-4xy^2+4xyz-2xz^2+yz^2) \\ &+ x^d y^{\frac{d}{2}+1} z^{\frac{d}{2}+1}(x-z)^2(z-y)(dx+(1-d)y) \\ &+ \frac{1}{2}x^d y^3 z^d (z-x)^3 \\ &+ x^{\frac{d}{2}+1} y^{\frac{3d}{2}} z^2 (x-y)(x-z)^2 \\ &+ 2(d-1)x^{\frac{d}{2}+1} y^d z^{\frac{d}{2}+1}(x-y)^2(x-z)(z-y)\Big). \end{aligned}$$

As we recalled in earlier sections, in low dimensions two, three, and four, the curvature of the conformally flat metrics was computed in [21, 31, 28, 32, 24]. It is shown in [35] that the above general formula reproduces those results. We should first note that the functions found in all the aforementioned works are written in terms of the commutator $[h, \cdot]$, denoted by Δ. To produce those functions from our result, a linear substitution of the variables t_j in terms of new variables s_j is needed. On the other hand, it is important to note that the functions $K_d^t(x, y)$ and $H_d^t(x, y, z)$ are homogeneous rational functions of order $-\frac{d}{2}$ and $-\frac{d}{2} - 1$, respectively. Using formula (74), it is clear that the functions $K_d(t_0, t_1)$ and $H_d(t_0, t_1, t_2)$ are homogeneous of order $1 - \frac{d}{2}$ in $f(t_j)$'s. This is the reason that for function $f(t) = e^t$ and a linear substitution such as $t_j = \sum_{m=0}^{j} s_m$, a factor of some power of e^{s_0} comes out. This term can be replaced by a power of e^h multiplied from the left to the final outcome. This explains how the functions in the aforementioned papers have one less variable than our functions. In other words, we have

$$K_d(s_0, s_0+s_1) = e^{(1-\frac{d}{2})s_0}K_d(s_1), \quad H_d(s_0, s_0+s_1, s_0+s_1+s_2) = e^{(1-\frac{d}{2})s_0}H_d(s_1, s_2).$$

For instance, function $K_d(s)$ is given by

$$K_d(s_1) = \frac{8e^{\frac{d+2}{4}s_1}\left((d-1)\sinh\left(\frac{s_1}{2}\right) + \sinh\left(\frac{(1-d)s_1}{2}\right)\right)}{d(d-2)d\left(e^{s_1}-1\right)^2 s_1}.$$

Now, we can obtain functions in dimension two:

$$H_2(s_1) = -\frac{e^{\frac{s_1}{2}}\left(e^{s_1}(s_1-2)+s_1+2\right)}{\left(e^{s_1}-1\right)^2 s_1},$$

$$\begin{aligned} K_2(s_1,s_2) = \Big(&s_1(s_1+s_2)\cosh(s_2) - (s_1-s_2) \\ &\times (s_1+s_2+\sinh(s_1)+\sinh(s_2)-\sinh(s_1+s_2)) \\ &-s_2(s_1+s_2)\cosh(s_1)\Big) \\ &\times \operatorname{csch}\left(\frac{s_1}{2}\right)\operatorname{csch}\left(\frac{s_2}{2}\right)\operatorname{csch}^2\left(\frac{s_1+s_2}{2}\right)/(4s_1s_2(s_1+s_2)). \end{aligned}$$

we have $-4H_2 = H$ and $-2K_2 = K$ where K and H are the functions found in [21, 31]. The difference is coming from the fact that the noncommutative parts of the results in [28, Section 5.1] are $\Delta(\log(e^{h/2})) = \frac{1}{2}\Delta(h)$ and $\Box(\log(e^{h/2})) = \frac{1}{4}\Box(h)$.

The functions for dimension four, with the same conformal factor $f(t) = e^t$ and substitution $t_j = \sum_{m=0}^{j} s_m$, gives the following which up to a negative sign are in complete agreement with the results from our papers [32, 28]:

$$H_4(s_1) = \frac{1-e^{s_1}}{2e^{s_1}s_1}, \quad K_4(s_1,s_2) = \frac{\left(e^{s_1}-1\right)\left(3e^{s_2}+1\right)s_2 - \left(e^{s_1}+3\right)\left(e^{s_2}-1\right)s_1}{4e^{s_1+s_2}s_1s_2\left(s_1+s_2\right)}.$$

To recover the functions for curvature of a noncommutative three torus equipped with a conformally flat metric obtained in [42, 24], we need to set $f(t) = e^{2t}$ and $t_0 = s_0$, $t_1 = s_0 + s_1/3$ and $t_2 = s_0 + (s_1+s_2)/3$. Then up to a factor of e^{-s_0}, we have

$$H_3(s_1) = \frac{4-4e^{\frac{s_1}{3}}}{e^{\frac{s_1}{6}}\left(s_1e^{\frac{s_1}{3}}+1\right)}, \quad K_3(s_1,s_2) = \frac{\begin{array}{c}6(e^{\frac{s_1}{3}}-1)(3e^{\frac{s_2}{3}}+1)s_2 \\ -6(e^{\frac{s_1}{3}}+3)(e^{\frac{s_2}{3}}-1)s_1\end{array}}{e^{\frac{s_1+s_2}{6}}(e^{\frac{s_1+s_2}{3}}+1)s_1s_2(s_1+s_2)}.$$

Finally, one needs to check the classical limit of these formulas as $\theta \to 0$. In the commutative case, the scalar curvature of a conformally flat metric $\tilde{g} = e^{2h}g$ on a d-dimensional space reads

$$\tilde{R} = -2(d-1)e^{-2h}g^{jk}\partial_j\partial_k(h) - (d-2)(d-1)e^{-2h}g^{jk}\partial_j(h)\partial_k(h).$$

For $f(t) = e^{-2t}$, the limit is

$$\lim_{t_0,t_1\to t} K_d(t_0,t_1) = \frac{1}{3}(d-1)e^{(d-2)t}, \quad \lim_{t_0,t_1,t_2\to t} H_d(t_0,t_1,t_2) = \frac{1}{6}(d-2)(d-1)e^{(d-2)t}.$$

We should also add that since $\delta_j \to -i\partial_j$ as $\theta \to 0$, we have $\Delta(h) \to -g^{jk}\partial_j\partial_k(h)$ and $\Box(h) \to -g^{jk}\partial_j(h)\partial_k(h)$. Therefore, these results recover the classical result up to a factor of $\sqrt{|g|}e^{dh}/6$. The factor $\sqrt{|g|}e^{dh}$ represents the volume form in the scalar curvature density and the factor $1/6$ is due to the choice of normalization in (9.5).

9.6 Twisted product, warped product, and scalar curvature

In this section, following [35], we shall recall the computation of the scalar curvature density of a noncommutative d-torus equipped with a class of functional metrics, which is called a twisted product metric.

Definition 9.3 ([35]) *Let g be an $r \times r$ and $\tilde{g}$ be a $(d-r)\times(d-r)$ positive definite real symmetric matrices and assume f is a positive function on the real line. The functional metric*

$$G = f(t)^{-1}g \oplus \tilde{g}, \tag{75}$$

is called a twisted product functional metric with the twisting element $f(h)^{-1}$.

Some examples of the twisted product metrics on noncommutative tori were already studied. The asymmetric two torus whose Dirac operator and spectral invariants are studied in [23] is a twisted product metric for $r = 1$. The scalar and Ricci curvature of noncommutative three torus of twisted product metrics with $r = 2$ are studied in [24]. It is worth mentioning that conformally flat metrics as well as warped metrics are two special cases of twisted product functional metrics. The following theorem is proved in [35].

Theorem 9.3 *The scalar curvature density of the d-dimensional noncommutative tori $\mathbb{T}^d_\theta$ equipped with the twisted product functional metric* (75) *with the twisting element $f(h)^{-1}$ is given by*

$$R = \sqrt{|g||\tilde{g}|}\Big(K_r(h_{(0)},h_{(1)})(\Delta(h)) + H_r(h_{(0)},h_{(1)},h_{(2)})(\Box(h))$$
$$+ \tilde{K}_r(h_{(0)},h_{(1)})(\tilde{\Delta}(h)) + \tilde{H}_r(h_{(0)},h_{(1)},h_{(2)})(\tilde{\Box}(h))\Big),$$

where $\tilde{\Delta}(h) = \sum_{r<i,j}\tilde{g}^{ij}\delta_i\delta_j(h)$ and $\tilde{\Box}(h) = \sum_{r<i,j}\tilde{g}^{ij}\delta_i(h)\delta_j(h)$ and Δ, $\Box$, K_r and H_r are given by Theorem 9.2. The functions $\tilde{K}_r$ and $\tilde{H}_r$ for $r \neq 2, 4$ are given by

$$\tilde{K}_r(t_0,t_1) = \tilde{K}^t_r(f(t_0), f(t_1))\big[t_0,t_1; f\big],$$

$$\tilde{H}_r(t_0, t_1, t_2) = \tilde{H}_r^t(f(t_0), f(t_1), f(t_2))\big[t_0, t_1; f\big]\big[t_1, t_2; f\big]$$
$$+ 2\tilde{K}_r^t(f(t_0), f(t_2))\big[t_0, t_1, t_2; f\big].$$

The functions $\tilde{K}_r^t$ *and* $\tilde{H}_r^t$ *are*

$$\tilde{K}_r^t(x, y) = \frac{(2r-4)(x^2-y^2)x^{\frac{r}{2}}y^{\frac{r}{2}} + 4x^2y^r - 4x^ry^2}{(r-4)(r-2)x^{\frac{3}{4}r}y^{\frac{3}{4}r}(x-y)^3},$$

and

$$\begin{aligned}
\tilde{H}_r^t(x, y, z) = {} & \frac{2x^{-\frac{3r}{4}}y^{-r}z^{-\frac{3r}{4}}}{(r-4)(r-2)(x-y)^2(x-z)^3(y-z)^2} \\
& \times \Big(x^ry^rz^2(x-y)\left(x^2 + 2x(y-2z) - 4y^2 + 6yz - z^2\right) \\
& + x^ry^{\frac{r}{2}}z^{\frac{r}{2}}(x-z)^2(y-z)\big((r-3)yz - x((r-3)z + y)\big) \\
& - x^ry^2z^r(x-z)^3 + x^{\frac{r}{2}}y^{\frac{3r}{2}}z^2(x-y)(x-z)^2 \\
& - x^{\frac{r}{2}}y^rz^{\frac{r}{2}}(x-y)(x-z)(y-z)\big((r-3)x^2 - 2x((r-2)y + (1-r)z) \\
& \quad + z((r-3)z - 2(r-2)y)\big) \\
& + x^{\frac{r}{2}}y^{\frac{r}{2}}z^r(y-x)(x-z)^2(yz - (r-3)x(y-z)) \\
& + x^2y^{\frac{3r}{2}}z^{\frac{r}{2}}(x-z)^2(y-z) - x^2y^rz^r(y-z) \\
& \quad \times \left(x^2 - 6xy + 4xz + 4y^2 - 2yz - z^2\right)\Big).
\end{aligned}$$

When the selfadjoint element $h \in A_\theta^d$ has the property that $\delta_j(h) = 0$ for $1 \leq j \leq r$, we call the twisted product functional metric (75) a *warped functional metric* with the warping element $1/f(h)$.

Corollary 9.1 *The scalar curvature density of a warped product of* $\tilde{g}$ *and* g *with the warping element* $1/f(h)$ *is given by*

$$R = \sqrt{|g||\tilde{g}|}\Big(\tilde{K}_r(h_{(0)}, h_{(1)})(\tilde{\Delta}(h)) + \tilde{H}_r(h_{(0)}, h_{(1)}, h_{(2)})(\tilde{\Box}(h))\Big).$$

Proof It is enough to see that $\Delta(h)$ and $\Box(h)$ vanish for the warped metric. □

For $r = 2$ and $r = 4$, functions $\tilde{H}_r$ and $\tilde{K}_r$ are the limit of the functions given in Theorem 9.3 as r approaches 2 or 4. This is because of the fact that for these values of r, some of T_α^k functions are the limit case of formulas found earlier. For $r = 2$ we have

$$\tilde{K}_2(x,y) = \frac{-x^2+y^2+2xy\log(\frac{x}{y})}{\sqrt{xy}(x-y)^3},$$

and

$$\begin{aligned}\tilde{H}_2(x,y,z) = &\frac{1}{2y\sqrt{xz}(x-y)^2(x-z)^3(y-z)^2} \\ &\times\Big(-y(x+y)(x-z)^3(y+z)\log(y) \\ &-z(x-y)^2\left(-3x^2y+x^2z-8xy^2\right. \\ &\left.+10xyz-2xz^2+yz^2+z^3\right)\log(z) \\ &+x(y-z)^2\left(x^3+x^2y-2x^2z\right. \\ &\quad\left.+10xyz+xz^2-8y^2z-3yz^2\right)\log(x) \\ &+2y(y-x)(x-z)(x+z)(z-y)(x-2y+z)\Big).\end{aligned}$$

For $r=4$, we have

$$\tilde{K}_4(x,y) = \frac{x^2-y^2-(x^2+y^2)\log(\frac{x}{y})}{xy(x-y)^3},$$

and

$$\begin{aligned}\tilde{H}_4(x,y,z) = &\frac{1}{2x(x-y)^2y^2(x-z)^3(y-z)^2z} \\ &\times\Big(\left(x^2+y^2\right)(x-z)^3\left(y^2+z^2\right)\log(y) \\ &+\log(x)(y-z)^2\left(x^4y+x^4z-6x^3y^2-2x^3yz-2x^3z^2+3x^2y^3\right. \\ &\quad\left.+x^2y^2z+x^2yz^2+x^2z^3+2xy^3z-4xy^2z^2+3y^3z^2+y^2z^3\right) \\ &-\log(z)(x-y)^2\left(x^3y^2+x^3z^2+3x^2y^3-4x^2y^2z+x^2yz^2\right. \\ &\quad-2x^2z^3+2xy^3z+xy^2z^2-2xyz^3+xz^4+3y^3z^2 \\ &\quad\left.-6y^2z^3+yz^4\right) \\ &-2(x-y)(x-z)(y-z)\left(x^3z+x^2y^2-2x^2z^2-2xy^3+2xy^2z\right. \\ &\quad\left.+xz^3-2y^3z+y^2z^2\right)\Big).\end{aligned}$$

In [24, section 4.1], the scalar curvature density of twisted product functional metric on noncommutative three torus for $f(t) = e^{2t}$ and $r = 2$ is found. This result can be recovered from our formulas given in Theorem 9.3 by setting $t_0 = s_0$, $t_1 = s_0 + s_1/2$ and $t_2 = s_0 + s_1/2 + s_2/2$.

9.7 *Dimension two and Gauss-Bonnet theorem*

The following result which is proved in [35] shows that the total scalar curvature of a noncommutative two torus equipped with a functional metric g is independent of g. This result extends the Gauss-Bonnet theorem of [22, 30] earlier proved for conformally flat metrics. This is done by a careful study of the functions F_S^{ij} in dimension two, where it is shown that these functions vanish for the noncommutative two torus equipped with a functional metric g. This means that the total scalar curvature of $(\mathbb{T}_\theta^2, g)$ is independent of g. Similar to the case of conformally flat metrics, we call this result the Gauss-Bonnet theorem for functional metrics.

Theorem 9.4 (Gauss-Bonnet Theorem [35]) *The total scalar curvature $\varphi(R)$ of the noncommutative two tori equipped with a functional metric vanishes, hence it is independent of the metric.*

Let us summarize the results obtained in [35] where a new family of metrics, called functional metrics, on noncommutative tori is introduced and their spectral geometry is studied. A class of Laplace type operators for these metrics is introduced and their spectral invariants are obtained from the heat trace asymptotics. A formula for the second density of the heat trace is also obtained. In particular, the scalar curvature density and the total scalar curvature of functional metrics are explicitly computed in all dimensions for certain classes of metrics including conformally flat metrics and twisted product of flat metrics. Finally a Gauss-Bonnet type theorem for a noncommutative two torus equipped with a general functional metric is proved.

10 Matrix Gauss-Bonnet

As we emphasized in the previous section, it is quite important to go beyond conformally flat metrics, go beyond noncommutative tori, and beyond dimension four. For example, one naturally needs to consider noncommutative algebras that would represent higher genus noncommutative curves and other noncommutative manifolds. As far as noncommutative higher genus curves go, there is as yet no satisfactory theory, even at a topological level, and much less at a metric or spectral level. This is a largely uninvestigated area and we expect new methods and ideas will be needed to make further progress with these objects.

A reasonable class of noncommutative algebras are algebras of matrix valued functions on a smooth manifold. Now topologically they are Morita equivalent to commutative algebras and not so interesting, but their spectral geometry poses interesting questions. A first step was taken in [43] to address this question. In this paper a new class of noncommutative algebras that are amenable to spectral analysis, namely algebras of matrix valued functions on a Riemann surface of arbitrary genus, are studied. The Dirac operator is conformally rescaled by a diagonalizable matrix and a Gauss-Bonnet theorem is proved for them. This is the matrix Gauss-Bonnet in the title of this section. When the surface has genus one, scalar curvature is explicitly computed. We shall briefly sketch these results in this section.

Let M be a two-dimensional closed spin Riemannian manifold and consider the algebra of smooth matrix valued functions on M:

$$\mathcal{A} = C^{\infty}(M, M_n(\mathbb{C})).$$

The Dirac operator of M, $D : L^2(S) \to L^2(S)$ acts on the Hilbert space of spinors. The algebra $\mathcal{A}$ acts diagonally on the Hilbert space $\mathcal{H} = L^2(S) \otimes \mathbb{C}^n$ and we have a spectral triple.

Let $h \in \mathcal{A}$ be a positive element. We use h to perturb the spectral triple of $\mathcal{A}$ in the following way. Consider the operator $D_h = hDh$ as a conformally rescaled Dirac operator. Now D_h does not have bounded commutators with the elements of $\mathcal{A}$, but we still have a twisted spectral triple. This is similar to the situation with curved noncommutative tori. The question is if the Gauss-Bonnet theorem holds for D_h. One is also interested in knowing if the scalar curvature can be computed explicitly. The answer is positive as we sketch now.

To simplify the matters a bit, it is assumed that the Weyl conformal factor h is diagonalizable, that is $h = UHU^*$, where U is unitary and H is diagonal. Then we have

$$hDh = UHU^*DUHU^* = U\left(H\left(D + U^*[D, U]\right)H\right)U^*,$$

which shows that the spectrum of D_h and $D_{A,H} = H(D + A)H$ are equal. Here $A = U^*[D, U]$ is a matrix valued one-form on M and $D + A$ represents a fluctuation of the geometry represented by D. It is shown in [43] that the Gauss-Bonnet theorem holds for the family of conformally rescaled Dirac operators with possible fluctuations $D_{A,H} = H(D + A)H$ as above. Local expressions for the scalar curvature are computed as well. The results demonstrate that unlike the case of higher residues in [38], the expressions for the value of the ζ function at 0 are complicated also in the matrix case.

Let us consider first the canonical spectral triple for a flat torus $M = \mathbb{R}^2/\mathbb{Z}^2$. Its spin structure is defined by the Pauli spin matrices σ^1, σ^2 and its Dirac operator is

$$D = \sigma^1\delta_1 + \sigma^2\delta_2.$$

Here δ_1, δ_2 are the partial derivatives $\frac{1}{i}\frac{\partial}{\partial x}$ and $\frac{1}{i}\frac{\partial}{\partial y}$. To compute the resolvent kernel we work in the algebra of matrix valued pseudodifferential operators obtained by tensoring the algebra Ψ of pseudodifferential operators on a smooth manifold M by the algebra of n by n matrices.

The Resolvent The symbol of the Bochner Laplacian $D^2_{A,H} = H(D+A)H^2(D+A)H$ is given by $\sigma_{D^2_{A,H}} = a_2 + a_1 + a_0$, where

$$
\begin{aligned}
a_2 &= H^4\xi^2,\\
a_1 &= i\epsilon_{ij}\sigma^3 2H^3\delta_i(H)\xi^j + 4H^3\delta_i(H)\xi^i - i\epsilon_{ij}\sigma^3 H^3 A_i H\xi^j\\
&\quad + H^3 A_i H\xi^i + i\epsilon_{ij}\sigma^3 H A_i H^3\xi^j + H A_i H^3\xi^i,\\
a_0 &= H^4(\Delta H) + H^3 A_j\delta_i(H) - H^3 i\sigma^3\epsilon_{ij}A_i\delta_j(H) + H^3\delta_i(A_i)H\\
&\quad + i\sigma^3 H^3\epsilon_{ij}\delta_j(A_i)H + 2H^2\delta_i(H)\delta_i(H) + 2H^2\delta_i(H)A_i H\\
&\quad + 2HA_i H^2\delta_i(H) + 2i\sigma^3 H^2\epsilon_{ij}\delta_i(H)A_j H\\
&\quad + i\sigma^3\epsilon_{ij}HA_i H^2\delta_i(H) + i\sigma^3\epsilon_{ij}HA_i H^2 A_j + HA_i H^2 A_i H.
\end{aligned}
$$

The first three terms of the symbols of $(D_{H,a})^{-2} = b_0 + b_1 + b_2 + \cdots$ are:

$$
\begin{aligned}
b_0 &= (a_2+1)^{-1},\\
b_1 &= -\left(b_0 a_1 + \partial_k(b_0)\delta_k(a_2)\right) b_0,\\
b_2 &= -\left(b_1 a_1 + b_0 a_0 + \partial_k(b_0)\delta_k(a_1) + \partial_k(b_1)\delta_k(a_2) + \frac{1}{2}\partial_k\partial_j(b_0)\delta_k\delta_j(a_2)\right) b_0.
\end{aligned}
$$

10.1 Matrix curvature

Let us call a matrix-valued function $R : \mathbb{T}^2 \to M_n(\mathbb{C})$ the scalar curvature if for any matrix valued function $f \in \mathcal{A}$ we have:

$$\zeta_{f,D}(0) = \int_{\mathbb{T}^2} \operatorname{Tr} fR,$$

where the localized spectral zeta function is defined by

$$\zeta_{f,D}(s) = \operatorname{Tr} f|D|^{-s}.$$

It is found that four terms contribute to the scalar curvature R [43]:

Terms not Depending on A They depend only on H and derivatives of H, and since they commute with each other, they can be computed as in the classical case.

$$\begin{aligned}b_2(H,\xi) =&96\, b_0^5\delta_i(H)\delta_i(H)H^{14}(\xi^2)^3 - 136\, b_0^4\delta_i(H)\delta_i(H)H^{10}(\xi^2)^2\\&+ 46\, b_0^3\delta_i(H)\delta_i(H)H^6(\xi^2) - 2\, b_0^2\delta_i(H)\delta_i(H)H^2\\&- 8\, b_0^4\Delta(H)H^{11}(\xi^2)^2 + 8\, b_0^3\Delta(H)H^7(\xi^2) - b_0^2\Delta(H)H^3,\end{aligned} \tag{76}$$

To integrate over the ξ space, we can use the formula

$$\int_0^\infty \frac{r^{2k+1}dr}{(1+a^2r^2)^{2k+3}} = \frac{1}{2(k+1)a^{2(k+1)}},$$

and obtain

$$R(H) = -\pi\left(\frac{1}{3}H^{-2}\delta_i(H)\delta_i(H) + \frac{1}{3}H^{-1}\Delta(H)\right). \tag{77}$$

To continue, we use the rearrangement lemma of [46]. Let

$$\Delta(x) = H^{-4}xH^4.$$

Terms Linear in A We have:

$$\begin{aligned}b_2^{(1)}(H,A) = &-b_0hA_ib_0\delta_i(H)H^2 + 5b_0hA_ib_0^2\delta_i(H)H^6\xi^2\\&- 4b_0hA_ib_0^3\delta_i(H)H^{10}(\xi^2)^2\\&- b_0H^3A_ib_0\delta_i(H) + 7b_0H^3A_ib_0^2\delta_i(H)H^4\xi^2\\&- 4b_0H^3A_ib_0^3\delta_i(H)H^8(\xi^2)^2\\&+ 3b_0^2H^5A_ib_0\delta_i(H)H^2\xi^2 - 4b_0^2H^5A_ib_0^2\delta_i(H)H^6(\xi^2)^2\\&+ b_0^2H^7A_ib_0\delta_i(H)\xi^2 - 4b_0^2H^7A_ib_0^2\delta_i(H)H^4(\xi^2)^2\end{aligned}$$

and

$$\begin{aligned}b_2^{(1)}(H,A) = &-2b_0\delta_i(H)H^2A_ib_0H + 2b_0^2\delta_i(H)H^4A_ib_0H^3\xi^2\\&+ 6b_0^2\delta_i(H)H^6A_ib_0H\xi^2 - 4b_0^3\delta_i(H)H^8A_ib_0H^3(\xi^2)^2\\&- 4b_0^3\delta_i(H)H^{10}A_ib_0H(\xi^2)^2.\end{aligned}$$

Explicit computations give

$$R^{(1)}(H,A)=\sum_{i=1,2} 2\pi HG(\Delta)(A_i)\delta_i(H),$$

where G is the following function:

$$G(s)=\frac{(1+\sqrt{s})\sqrt{s}}{(s-1)^3}\big((s+1)\ln(s)-2(s-1)\big),$$

and a second term,

$$R^{(2)}(H,A)=\sum_{i=1,2} -2\pi H^{-2}\delta_i(H)G(\Delta)(A_i)H,$$

with the same function $G(s)$. After taking the trace these terms cancel each other, and we get

$$\mathrm{Tr}\left(R^{(1)}(H,A)+R^{(2)}(H,A)\right)=0.$$

Terms Linear in $\delta_i(A_i)$ In this case we have:

$$\begin{aligned}b_2(H,\delta_i(A_j)) &= -b_0H^3\delta_i(A_i)b_0H+b_0^2H^5\delta_i(A_i)b_0H^3\xi^2\\ &\quad+b_0^2H^7\delta_i(A_i)b_0H\xi^2,\end{aligned}$$

and integrating out the ξ variables, we get $\pi H^{-1}F(\Delta)(\delta_i(A_i))H$, where

$$F=-\frac{(1+\sqrt{s})\sqrt{s}}{(s-1)^2}\ln(s)+\frac{\sqrt{s}+1}{s-1}.$$

Again, it is not difficult to check that $F(1)=0$ and the expression vanishes after taking the trace:

$$\mathrm{Tr}\left(R(H,\delta_i(A_j))\right)=0.$$

Quadratic Terms in A_i We have:

$$\begin{aligned}b_2(H,A^2) &= -b_0HA_iH^2A_ib_0H+b_0HA_ib_0H^6A_ib_0H\xi^2\\ &\quad+b_0H^3A_ib_0H^2A_ib_0H^3\xi^2.\end{aligned}$$

Integrating over ξ we obtain:

$$R(H,A^2)=-\pi H^{-1}Q(\Delta^{(1)},\Delta^{(2)})(A_i\cdot A_i)H$$

where

$$Q(s,t) = \frac{\sqrt{s}(\sqrt{t}+s)}{(s-1))(s-t)} \ln s - \frac{\sqrt{s}\sqrt{s}}{(s-t)\sqrt{t}} \ln t.$$

To compute the trace, let

$$F(s) = Q(s,1) = \frac{(s+1)\ln s + 2(1-s)}{(s-1)^2},$$

and observe that due to the trace property:

$$\begin{aligned}\mathrm{Tr}\left(H^{-1}F(\Delta)(A_i)A_iH\right) &= \mathrm{Tr}\,(A_iF(\Delta)(A_i)) \\ &= \mathrm{Tr}\left(F(\Delta^{-1})(A_i)A_i\right).\end{aligned}$$

Now since

$$\begin{aligned}F(\frac{1}{s}) &= \frac{-(\frac{1}{s}+1)\ln s + 2(1-\frac{1}{s})}{(\frac{1}{s}-1)^2} \\ &= \frac{-(s+1)\ln s - 2(1-s)}{(s-1)^2} \\ &= -F(s),\end{aligned}$$

one gets

$$\mathrm{Tr}\, R(H, A^2) = 0,$$

and so the quadratic term vanishes as well.

10.2 The Gauss-Bonnet theorem

The term which does not depend on A is a total derivative term:

$$\frac{1}{3}\delta_i\left(H^{-1}\delta_i(H)\right).$$

Using Stokes theorem, this term is seen to vanish as well after integration. Putting it all together, one thus obtains:

Proposition 10.1 *For the matrix conformally rescaled Dirac operator on the two-dimensional torus, $D_h = hDh$, where h is a globally diagonalizable positive matrix, the Gauss-Bonnet theorem holds:*

$$\zeta_{D_h}(0) = \zeta_D(0).$$

10.3 Higher genus matrix Gauss-Bonnet

Now we look at the general case where M is a closed Riemann surface with a spin structure and a Dirac operator D. Consider the operator

$$D_{H,A} = H(D + A)H,$$

for H a diagonal matrix valued function on M and A a matrix-valued one-form, identified here with its Clifford image.

We must now compute the value of $\zeta_{D^2_{H,A}}(0)$ using methods of pseudodifferential calculus. Let us denote the symbols of D^2_H as:

$$D^2_H = (HDH)^2 = a^H_2 + a^H_1 + a^H_0,$$

and the symbol of D^2 as

$$D^2 = a^o_2 + a^o_1 + a^o_0.$$

As in the case of the torus, the computation is divided into the cases of terms not depending on A, linear in A and quadratic in A.

Terms Independent of A Since H is globally diagonalizable, we can assume it is scalar. Thus we are reduced to a conformal rescaling of the classical Dirac operator. Since in this case the Gauss-Bonnet theorem holds, it remains only to see that the contribution to the Gauss-Bonnet term from the linear and quadratic terms in A vanishes.

Terms Linear in A Linear terms do arise in b_2 from the following terms:

$$\begin{aligned} &b_0 a^H_1 b_0 a_1(A) b_0 + \partial^\xi_k(b_0)\partial^x_k(a^H_2) b_0 a_1(A) b_0 + b_0 a_1(A) b_0 a^H_1 b_0 \\ &- b_0 a_0(A) b_0 - \partial^\xi_k(b_0)\partial^x_k(a_1(A)) b_0 - \partial^\xi_k(b_0 a_1(A) b_0)\partial^x_k(a^H_2) b_0. \end{aligned}$$

where $a_1(A)$, $a_0(A)$ denote terms linear in A. Now, one can use normal coordinates at a given point x of M. The terms without derivatives reduce easily to the torus case. The only difficulty arises from terms with derivatives in x, that is, $\partial^x_k(a^H_2)$. and $\partial^x_k(a_1(A))$. Since $a^H_2 = H^4 g_{ij}\xi^i\xi^j$, and in normal coordinates the first derivatives of the metric vanish at the point x, we see that the only remaining term would be with the derivative of H^4, and again this term would be reduced to the term linear in A from the torus case.

Similar argument works also for the other term, $a_1(A)$, which is

$$\left(H^3 A_i H + H A_k H^3\right) \sigma^k \sigma^i \xi_i,$$

and since in $\partial_k^\xi(b_0)\partial_k^x(a_1(A))b_0$ there are no further σ matrices, one can compute first the trace over the Clifford algebra and write it as

$$\frac{1}{2}\left(H^3 A_i H + H A_k H^3\right) g^{ki} \xi_i.$$

Thus, in normal coordinates around x the expression is identical to the one for the flat torus. Therefore, the integration over ξ would yield the same result, and the density of the linear A contribution to the trace of b_2 vanishes at x. Consequently, the contribution to the Gauss-Bonnet term linear in A also vanishes.

Quadratic Terms The quadratic terms in A are

$$b_0 a_1(A) b_0 a_1(A) b_0 - b_0 a_0(A^2) b_0.$$

It is easy to see that in normal coordinates we have:

$$a_1(A) = \left(H^3 \sigma^j \xi_j (\sigma^i A_i) H + \sigma^i H A_i H^3 \sigma^j \xi_j\right),$$

$$a_0(A) = (\sigma^i H A_i H)(\sigma^k H A_k H).$$

Using normal coordinates, one reduces the ξ -integral to the situation already considered for the torus. Hence the density of Gauss-Bonnet term with quadratic contributions from A identically vanishes as well. This finishes the proof of Gauss-Bonnet for higher genus matrix valued functions with a general Dirac operator with fluctuation. This result was obtained in [43].

11 Curvature of the determinant line bundle

It would be interesting to know how far our hard analytic methods like pseudodifferential operators, spectral analysis and heat equation techniques, can be pushed in the noncommutative realm, at least for noncommutative tori and toric manifolds. So far we have seen that these analytic techniques, suitably modified and enhanced, has been quite successful in dealing with scalar and Ricci curvature. Along this idea, in [25] the curvature of the determinant line bundle on a family of Dirac operators for a noncommutative two torus is computed. Following Quillen's original construction for Riemann surfaces [53] and using zeta regularized determinant of Laplacians, the determinant line bundle is endowed with a natural Hermitian metric. By defining an analogue of Kontsevich-Vishik canonical trace, defined on Connes' algebra of classical pseudodifferential symbols for the noncommutative two torus, the curvature form of the determinant line bundle is computed through the second variation $\delta_w \delta_{\bar{w}} \log \det(\Delta)$. Calculus of symbols and the canonical trace were effectively used to bypass local calculations involving Green functions in [53] which is not applicable in the noncommutative case. In a sequel paper [26], the spectral eta function for certain families of Dirac operators on noncommutative 3-torus is studied and its regularity at zero is proved. By using variational techniques, it is shown that the eta function $\eta_D(0)$ is a conformal invariant. By studying the Laurent expansion at zero of $\mathrm{TR}(|D|^{-z})$, the conformal invariance of $\zeta'_{|D|}(0)$ for noncommutative 3-torus is proved. Finally, for the coupled Dirac operator, a local formula for the variation $\partial_A \eta_{D+A}(0)$ is derived which is the analogue of the so-called induced Chern-Simons term in quantum field theory literature.

In this section we shall recall and comment on results obtained in [25] on the curvature of the determinant line bundle on a noncommutative torus.

11.1 The determinant line bundle

Let $\mathcal{F} = \mathrm{Fred}(\mathcal{H}_0, \mathcal{H}_1)$ denote the set of Fredholm operators between Hilbert spaces $\mathcal{H}_0$ and $\mathcal{H}_1$. It is an open subset, in norm topology, in the complex Banach space of all bounded linear operators between $\mathcal{H}_0$ and $\mathcal{H}_1$. The index map $index : \mathcal{F} \to \mathbb{Z}$ is a homotopy invariant and in fact defines a bijection between connected components of $\mathcal{F}$ and the set of integers $\mathbb{Z}$. It is well known that $\mathcal{F}$ is a classifying space for K-theory (Atiyah-Janich): for any compact space X we have a natural ring isomorphism $K^0(X) = [X, \mathcal{F}]$ between the K-theory of X and the set of homotopy classes of continuous maps from X to $\mathcal{F}$.

In [53] Quillen defines a holomorphic line bundle $\mathrm{DET} \to \mathcal{F}$ over the space of Fredholm operators such that for any $T \in \mathcal{F}$

$$\mathrm{DET}_T = \Lambda^{max}(\ker(T))^* \otimes \Lambda^{max}(\mathrm{coker}(T)).$$

This is remarkable if we notice that $\ker(T)$ and $\mathrm{coker}(T)$ are not vector bundles due to discontinuities in their dimensions as T varies within $\mathcal{F}$.

It is tempting to think that since $c_1(\mathrm{DET})$ is the generator of $H^2(\mathcal{F}_0, \mathbb{Z}) \cong \mathbb{Z}$, $\mathcal{F}_0$ being the index zero operators, there might exist a natural Hermitian metric on DET whose curvature 2-form would be a representative of this generator. One problem is that the induced metric from $\ker(T)$ and $\ker(T^*)$ on DET is not even continuous. In [53] Quillen shows that for families of Cauchy-Riemann operators on a Riemann surface one can correct the Hilbert space metric by multiplying it by zeta regularized determinant and in this way one obtains a smooth Hermitian metric on the induced determinant line bundle. In [25] a similar construction for the noncommutative two torus is given as we explain later in this section.

11.2 The canonical trace and noncommutative residue

To carry the calculations, an analogue of the canonical trace of [45] for the noncommutative torus is constructed in [25]. First we need to extend our original algebra of pseudodifferential operators to *classical* pseudodifferential operators.

A smooth map $\sigma : \mathbb{R}^2 \to \mathcal{A}_\theta$ is called a classical symbol of order $\alpha \in \mathbb{C}$ if for any N and each $0 \leq j \leq N$ there exist $\sigma_{\alpha-j} : \mathbb{R}^2\backslash\{0\} \to \mathcal{A}_\theta$ positive homogeneous of degree $\alpha - j$ and a symbol $\sigma^N \in \mathcal{S}^{\Re(\alpha)-N-1}(\mathcal{A}_\theta)$, such that

$$\sigma(\xi) = \sum_{j=0}^{N} \chi(\xi)\sigma_{\alpha-j}(\xi) + \sigma^N(\xi) \quad \xi \in \mathbb{R}^2. \tag{78}$$

Here χ is a smooth cut-off function on $\mathbb{R}^2$ which is equal to zero on a small ball around the origin, and is equal to one outside the unit ball. It can be shown that the homogeneous terms in the expansion are uniquely determined by σ. We denote the set of classical symbols of order α by $\mathcal{S}^\alpha_{cl}(\mathcal{A}_\theta)$ and the associated classical pseudodifferential operators by $\Psi^\alpha_{cl}(\mathcal{A}_\theta)$.

The space of classical symbols $\mathcal{S}_{cl}(\mathcal{A}_\theta)$ is equipped with a Fréchet topology induced by the semi-norms

$$p_{\alpha,\beta}(\sigma) = \sup_{\xi\in\mathbb{R}^2} (1 + |\xi|)^{-m+|\beta|} ||\delta^\alpha \partial^\beta \sigma(\xi)||. \tag{79}$$

The analogue of the Wodzicki residue for classical pseudodifferential operators on the noncommutative torus is defined in [33].

Definition 11.1 *The Wodzicki residue of a classical pseudodifferential operator* P_σ *is defined as*

$$\mathrm{Res}(\mathrm{P}_\sigma) = \varphi_0\left(\mathrm{res}(\mathrm{P}_\sigma)\right),$$

where $\mathrm{res}(\mathrm{P}_\sigma) := \int_{|\xi|=1} \sigma_{-2}(\xi) d\xi$.

It is evident from its definition that Wodzicki residue vanishes on differential operators and on non-integer order classical pseudodifferential operators.

To define the analogue of the canonical trace on non-integer order pseudodifferential operators on the noncommutative torus, one needs the existence of the so-called cut-off integral for classical symbols.

Proposition 11.1 *Let $\sigma \in S^{\alpha}_{cl}(\mathcal{A}_\theta)$ and $B(R)$ be a disk of radius R around the origin. One has the following asymptotic expansion as $R \to \infty$*

$$\int_{B(R)} \sigma(\xi) d\xi \sim \sum_{j=0, \alpha-j+2\neq 0}^{\infty} \alpha_j(\sigma) R^{\alpha-j+2} + \beta(\sigma) \log R + c(\sigma),$$

where $\beta(\sigma) = \int_{|\xi|=1} \sigma_{-2}(\xi) d\xi$ and the constant term in the expansion, $c(\sigma)$, is given by

$$\int_{\mathbb{R}^n} \sigma^N + \sum_{j=0}^{N} \int_{B(1)} \chi(\xi) \sigma_{\alpha-j}(\xi) d\xi - \sum_{j=0, \alpha-j+2\neq 0}^{N} \frac{1}{\alpha-j+2} \int_{|\xi|=1} \sigma_{\alpha-j}(\omega) d\omega. \tag{80}$$

Definition 11.2 *The cut-off integral of a symbol $\sigma \in S^{\alpha}_{cl}(\mathcal{A}_\theta)$ is defined to be the constant term in the above asymptotic expansion, and we denote it by $⨍ \sigma(\xi) d\xi$.*

The cut-off integral of a symbol is independent of the choice of N. It is also independent of the choice of the cut-off function χ.

Definition 11.3 *The canonical trace of a classical pseudodifferential operator $P \in \Psi^{\alpha}_{cl}(\mathcal{A}_\theta)$ of non-integral order α is defined as*

$$\mathrm{TR}(P) := \varphi_0 \left(⨍ \sigma_P(\xi) d\xi \right).$$

Note that any pseudodifferential operator P of order less than -2 is a trace-class operator on $\mathcal{H}_0$ and its trace is given by

$$\mathrm{Tr}(P) = \varphi_0 \left(\int_{\mathbb{R}^2} \sigma_P(\xi) d\xi \right).$$

On the other hand, for such operators the symbol is integrable and we have

$$⨍ \sigma_P(\xi) d\xi = \int_{\mathbb{R}^2} \sigma_P(\xi) d\xi. \tag{81}$$

Therefore, the TR-functional and operator trace coincide on classical pseudodifferential operators of order less than -2.

The canonical trace TR is an analytic continuation of the operator trace and using this fact one can prove that it is actually a trace.

Proposition 11.2 *Given a holomorphic family $\sigma(z) \in S^{\alpha(z)}_{cl}(\mathcal{A}_\theta)$, $z \in W \subset \mathbb{C}$, the map*

$$z \mapsto ⨍ \sigma(z)(\xi) d\xi,$$

is meromorphic with at most simple poles. Its residues are given by

$$\mathrm{Res}_{z=z_0} ⨍ \sigma(z)(\xi)d\xi = -\frac{1}{\alpha'(z_0)} \int_{|\xi|=1} \sigma(z_0)_{-2} d\xi.$$

Using the above result one can show that if $A \in \Psi_{cl}^{\alpha}(\mathcal{A}_\theta)$ is of order $\alpha \in \mathbb{Z}$ and Q is a positive elliptic classical pseudodifferential operator of positive order q, then

$$\mathrm{Res}_{z=0}\mathrm{TR}(AQ^{-z}) = \frac{1}{q}\mathrm{Res}(A).$$

Using this and the uniqueness of analytic continuation one can prove the trace property of TR. That is, $\mathrm{TR}(AB) = \mathrm{TR}(BA)$ for any $A, B \in \Psi_{cl}(\mathcal{A}_\theta)$, provided that $\mathrm{ord}(A) + \mathrm{ord}(B) \notin \mathbb{Z}$.

11.3 Log-polyhomogeneous symbols

In general, z-derivatives of a classical holomorphic family of symbols is not classical anymore and therefore one needs to introduce log-polyhomogeneous symbols which include the z-derivatives of the symbols of the holomorphic family $\sigma(AQ^{-z})$.

Definition 11.4 *A symbol σ is called a log-polyhomogeneous symbol if it has the following form*

$$\sigma(\xi) \sim \sum_{j\geq 0} \sum_{l=0}^{\infty} \sigma_{\alpha-j,l}(\xi) \log^l |\xi| \quad |\xi| > 0, \tag{82}$$

with $\sigma_{\alpha-j,l}$ positively homogeneous in ξ of degree $\alpha - j$.

A prototypical example of an operator with such a symbol is $\log Q$ where $Q \in \Psi_{cl}^{q}(\mathcal{A}_\theta)$ is a positive elliptic pseudodifferential operator of order $q > 0$. The logarithm of Q can be defined by

$$\log Q = Q \left.\frac{d}{dz}\right|_{z=0} Q^{z-1} = Q \left.\frac{d}{dz}\right|_{z=0} \frac{i}{2\pi} \int_C \lambda^{z-1}(Q-\lambda)^{-1} d\lambda.$$

For an operator A with log-polyhomogeneous symbol as (82) we define

$$\mathrm{res}(A) = \int_{|\xi|=1} \sigma_{-2,0}(\xi) d\xi.$$

The following result can be proved along the lines of its classical counterpart in [52].

Proposition 11.3 *Let $A \in \Psi_{cl}^{\alpha}(\mathcal{A}_\theta)$ and Q be a positive, in general an admissible, elliptic pseudodifferential operator of positive order q. If $\alpha \in P$, then 0 is a possible simple pole for the function $z \mapsto \mathrm{TR}(AQ^{-z})$ with the following Laurent expansion around zero.*

$$
\begin{aligned}
\mathrm{TR}(AQ^{-z}) &= \frac{1}{q}\mathrm{Res(A)}\frac{1}{z} \\
&+ \varphi_0\left(⨍ \sigma(A)(\xi)\mathrm{d}\xi - \frac{1}{q}\mathrm{res}(A\log Q)\right) - Tr(A\Pi_Q) \\
&+ \sum_{k=1}^{K}(-1)^k\frac{(z)^k}{k!} \\
&\times \left(\varphi_0\left(⨍ \sigma(A(\log Q)^k)(\xi)d\xi - \frac{1}{q(k+1)}\mathrm{res}(A(\log Q)^{k+1})\right)\right. \\
&\left. - Tr(A(\log Q)^k\Pi_Q)\right) + o(z^K).
\end{aligned}
$$

where Π_Q is the projection on the kernel of Q.

For operators A and Q as in the previous Proposition, the *generalized zeta function* is defined by

$$
\zeta(A, Q, z) = \mathrm{TR}(AQ^{-z}). \tag{83}
$$

From Proposition 11.2, it follows that $\zeta(A, Q, z)$ is a meromorphic function with simple poles. Moreover, $\zeta(A, Q, z)$ is the analytic continuation of the spectral zeta function $\mathrm{Tr}(AQ^{-z})$. If A is a differential operator, the zeta function (83) is regular at $z = 0$ with a value

$$
\varphi_0\left(⨍ \sigma(A)(\xi)d\xi - \frac{1}{q}\mathrm{res}(A\log Q)\right) - \mathrm{Tr}(A\Pi_Q).
$$

11.4 Cauchy-Riemann operators on noncommutative tori

As we did before, we can fix a complex structure on A_θ by a complex number τ in the upper half plane. Consider the spectral triple

$$
\left(A_\theta, \mathcal{H}_0 \oplus \mathcal{H}^{0,1}, D_0 = \begin{pmatrix} 0 & \bar\partial^* \\ \bar\partial & 0 \end{pmatrix}\right), \tag{84}
$$

where $\bar{\partial} : A_\theta \to A_\theta$ is given by $\bar{\partial} = \delta_1 + \tau\delta_2$. The Hilbert space $\mathcal{H}_0$ is defined by the GNS construction from A_θ using the trace φ_0 and $\bar{\partial}^*$ is the adjoint of the operator $\bar{\partial}$.

As in the classical case, the Cauchy-Riemann operator on A_θ is the positive part of the twisted Dirac operator. All such operators define spectral triples of the form

$$\left(A_\theta, \mathcal{H}_0 \oplus \mathcal{H}^{0,1}, D_A = \begin{pmatrix} 0 & \bar{\partial}^* + \alpha^* \\ \bar{\partial} + \alpha & 0 \end{pmatrix}\right),$$

where $\alpha \in A_\theta$ is the positive part of a selfadjoint element

$$A = \begin{pmatrix} 0 & \alpha^* \\ \alpha & 0 \end{pmatrix} \in \Omega^1_{D_0}(A_\theta).$$

We recall that $\Omega^1_{D_0}(A_\theta)$ is the space of quantized one forms consisting of the elements $\sum a_i[D_0, b_i]$ where $a_i, b_i \in A_\theta$ [14]. Note that in this case the space $\mathcal{A}$ of Cauchy-Riemann operators is the space of (0, 1)-forms on A_θ.

11.5 The curvature of the determinant line bundle for A_θ

For any $\alpha \in \mathcal{A}$, the Cauchy-Riemann operator

$$\bar{\partial}_\alpha = \bar{\partial} + \alpha : \mathcal{H}_0 \to \mathcal{H}^{0,1}$$

is a Fredholm operator. We pull back the determinant line bundle DET on the space of Fredholm operators Fred$(\mathcal{H}_0, \mathcal{H}^{0,1})$, to get a line bundle $\mathcal{L}$ on $\mathcal{A}$. Following Quillen [53], one can define a Hermitian metric on $\mathcal{L}$ and the problem is to compute its curvature. Let us define a metric on the fiber

$$\mathcal{L}_\alpha = \Lambda^{max}(\ker \bar{\partial}_\alpha)^* \otimes \Lambda^{max}(\ker \bar{\partial}^*_\alpha)$$

as the product of the induced metrics on $\Lambda^{max}(\ker \bar{\partial}_\alpha)^*$ and $\Lambda^{max}(\ker \bar{\partial}^*_\alpha)$, with the zeta regularized determinant $e^{-\zeta'_{\Delta_\alpha}(0)}$. Here we define the Laplacian as $\Delta_\alpha = \bar{\partial}^*_\alpha \bar{\partial}_\alpha : \mathcal{H}_0 \to \mathcal{H}_0$, and its zeta function by

$$\zeta(z) = \mathrm{TR}(\Delta_\alpha^{-z}).$$

It is a meromorphic function and is regular at $z = 0$. Similar proof as in [53] shows that this defines a smooth Hermitian metric on the determinant line bundle $\mathcal{L}$.

On the open set of invertible operators each fiber of $\mathcal{L}$ is canonically isomorphic to $\mathbb{C}$ and the non-zero holomorphic section $\sigma = 1$ gives a trivialization. Also, according to the definition of the Hermitian metric, the norm of this section is given by

$$\|\sigma\|^2 = e^{-\zeta'_{\Delta_\alpha}(0)}. \tag{85}$$

11.6 Variations of LogDet and curvature form

A holomorphic line bundle equipped with a Hermitian inner product has a canonical connection compatible with the two structures. This is also known as the Chern connection. The curvature form of this connection is given by $\bar\partial\partial \log \|\sigma\|^2$, where σ is any non-zero local holomorphic section.

In the case at hand, the second variation $\bar\partial\partial \log \|\sigma\|^2$ on the open set of invertible Cauchy-Riemann operators must be computed. Let us consider a holomorphic family of invertible Cauchy-Riemann operators $D_w = \bar\partial + \alpha_w$, where α_w depends holomorphically on the complex variable w. The second variation of logdet, that is $\delta_{\bar w}\delta_w\zeta'_\Delta(0)$, is computed in [25] as we recall now.

Lemma 11.1 *For the holomorphic family of Cauchy-Riemann operators D_w, the second variation of $\zeta'(0)$ is given by*

$$\delta_{\bar w}\delta_w\zeta'(0) = \frac{1}{2}\varphi_0\left(\delta_w D \delta_{\bar w}\mathrm{res}(\log\Delta\, D^{-1})\right).$$

□

The final step is to compute $\delta_{\bar w}\mathrm{res}(\log\Delta\, D^{-1})$. This combined with the above lemma will show that the curvature form of the determinant line bundle equals the Kähler form on the space of connections. We refer the reader to [25] for the proof which is long and technical. We emphasize that the original Quillen proof, based on Green function calculations, cannot be extended to the noncommutative case.

Lemma 11.2 *With the above definitions and notations, we have*

$$\sigma_{-2,0}(\log\Delta\, D^{-1}) = \frac{(\alpha+\alpha^*)\xi_1 + (\bar\tau\alpha + \tau\alpha^*)\xi_2}{(\xi_1^2 + 2\Re(\tau)\xi_1\xi_2 + |\tau|^2\xi_2^2)(\xi_1 + \tau\xi_2)} - \log\left(\frac{\xi_1^2 + 2\Re(\tau)\xi_1\xi_2 + |\tau|^2\xi_2^2}{|\xi|^2}\right)\frac{\alpha}{\xi_1 + \tau\xi_2},$$

and

$$\delta_{\bar{w}} \mathrm{res}(\log(\Delta) D^{-1}) = \frac{1}{2\pi \Im(\tau)} (\delta_w D)^*.$$

Now we can state the main result of [25] which computes the curvature of the determinant line bundle in terms of the natural Kähler form on the space of connections.

Theorem 11.1 *The curvature of the determinant line bundle for the noncommutative two torus is given by*

$$\delta_{\bar{w}} \delta_w \zeta'(0) = \frac{1}{4\pi \Im(\tau)} \varphi_0 \left(\delta_w D (\delta_w D)^* \right). \tag{86}$$

In order to recover the classical result of Quillen in the classical limit of $\theta = 0$, one has to notice that the volume form has changed due to a change of the metric. This means we just need to multiply the above result by $\Im(\tau)$.

12 Open problems

In this final section we formulate some of the open problems that we think are worthy of study for further understanding of local invariants of noncommutative manifolds.

1. Beyond dimension two and beyond conformally flat. The class of conformally flat metrics in dimensions bigger than two cover only a small part of all possible metrics. It would be very important to formulate large classes of metrics that are not conformally flat, but at the same time lend themselves to spectral analysis and to heat asymptotics techniques. It is also very important to have curvature formulas that work uniformly in all dimensions. The largest such class so far is the class of the so-called functional metrics introduced in [35] and surveyed in Section 9 of this paper. It is an interesting problem to further enlarge this class.
2. To extend the definition of curvature invariants to noncommutative spaces with non-integral dimension, including zero dimensional spaces. This would require rethinking the heat trace asymptotic expansion, and the nature of its leading and sub-leading terms. In particular since quantum spheres are zero dimensional, its spectrum is of exponential growth and does not satisfy the usual Weyl's asymptotic law. A first step would be to see how to formulate a Gauss-Bonnet type theorem for quantum spheres.
3. Weyl tensor and full curvature tensor. It is not clear that the classical differential geometry would, or should, give us a blueprint in the noncommutative case. One should be prepared for new phenomena. Having that in mind, one should still look for analogues of Weyl and full Riemann curvature tensors. The problem

is that the components of these tensors are quite entangled in the heat trace expansion, and separating and identifying their different components seem to be a hard task, if not impossible. One needs new ideas to make progress here.

4. Gauss-Bonnet terms in higher dimensions. The Gauss-Bonnet density in two dimension is particularly simple and is in fact equal to the scalar curvature multiplied by the volume form. In dimensions four and above this term is classically more complicated, being the Pfaffian of the curvature tensor. In dimension four it is a linear combination of norms of the Riemann tensor, the Ricci tensor, and the Ricci scalar. It is not clear how this can be expressed in terms of the heat kernel coefficients.
5. Higher genus noncommutative Riemann surfaces. It is highly desirable to define noncommutative Riemann surfaces of higher genus equipped with a spectral triple and check the Gauss-Bonnet theorem for them. This would greatly extend our understanding of local invariants of noncommutative spaces.
6. Noncommutative uniformization theorem. The study of curved noncommutative 2-tori suggests a natural problem in noncommutative geometry. At least for the class of noncommutative 2-tori it is desirable to know to what extent the uniformization theorem holds, or what form and shape it would take.
7. Analytic versus algebraic curvature. In classical differential geometry, as we saw in this paper, there are algebraic as well as analytic techniques (based on the heat equation) to define the scalar and Ricci curvature. The two approaches give the same results. This is not so in the noncommutative case. For noncommutative tori, when the deformation parameter satisfies some diophantine condition, Rosenberg in [54] proved a Levi-Civita type theorem and hence gets an algebraic definition of curvature. The resulting formula is very different from the formula of Connes-Moscovici-Fathizadeh-Khalkhali [21, 32] surveyed in this paper. It is important to see if there is any relation at all between these formulas and what this means for the study of curved noncommutative tori.

Acknowledgements F.F. acknowledges support from the Marie Curie/ SER Cymru II Cofund Research Fellowship 663830-SU-008. M.K. would like to thank Azadeh Erfanian for providing the original picture used in page four, Saman Khalkhali for his support and care, and Asghar Ghorbanpour for many discussions on questions related to this article.

References

1. M. Adler. On a trace functional for formal pseudo differential operators and the symplectic structure of the Korteweg-de Vries type equations. *Invent. Math.*, 50(3):219–248, 1978/79.
2. M. Atiyah, R. Bott, and V. K. Patodi. On the heat equation and the index theorem. *Invent. Math.*, 19:279–330, 1973.
3. T. A. Bhuyain and M. Marcolli. The Ricci flow on noncommutative two-tori. *Lett. Math. Phys.*, 101(2):173–194, 2012.
4. T. P. Branson and B. Ø rsted. Conformal indices of Riemannian manifolds. *Compositio Math.*, 60(3):261–293, 1986.

5. A. Buium. *Foundations of arithmetic differential geometry*, volume 222 of *Mathematical Surveys and Monographs*. American Mathematical Society, Providence, RI, 2017.
6. A. H. Chamseddine and A. Connes. The spectral action principle. *Comm. Math. Phys.*, 186(3):731–750, 1997.
7. A. H. Chamseddine and A. Connes. Scale invariance in the spectral action. *J. Math. Phys.*, 47(6):063504, 19, 2006.
8. P. B. Cohen and A. Connes. Conformal geometry of the irrational rotation algebra. *Preprint MPI*, 1992.
9. A. Connes. C^* algèbres et géométrie différentielle. *C. R. Acad. Sci. Paris Sér. A-B*, 290(13):A599–A604, 1980.
10. A. Connes. Spectral sequence and homology of currents for operator algebras. *Oberwolfach Tagungsber, 41/81, Funktionalanalysis und C^*-Algebren, 27-9/3-10*, 1981.
11. A. Connes. Cohomologie cyclique et foncteurs Ext^n. *C. R. Acad. Sci. Paris Sér. I Math.*, 296(23):953–958, 1983.
12. A. Connes. Noncommutative differential geometry. *Inst. Hautes Études Sci. Publ. Math.*, (62):257–360, 1985.
13. A. Connes. Cyclic cohomology and the transverse fundamental class of a foliation. In *Geometric methods in operator algebras (Kyoto, 1983)*, volume 123 of *Pitman Res. Notes Math. Ser.*, pages 52–144. Longman Sci. Tech., Harlow, 1986.
14. A. Connes. *Noncommutative geometry*. Academic Press, Inc., San Diego, CA, 1994.
15. A. Connes. On the spectral characterization of manifolds. *J. Noncommut. Geom.*, 7(1):1–82, 2013.
16. A. Connes and C. Consani. The Riemann-Roch strategy, Complex lift of the Scaling Site. *arXiv e-prints*, page arXiv:1805.10501, May 2018.
17. A. Connes and F. Fathizadeh. The term a_4 in the heat kernel expansion of noncommutative tori. *ArXiv e-prints*, Nov. 2016.
18. A. Connes and M. Marcolli. *Noncommutative geometry, quantum fields and motives*, volume 55 of *American Mathematical Society Colloquium Publications*. American Mathematical Society, Providence, RI; Hindustan Book Agency, New Delhi, 2008.
19. A. Connes and H. Moscovici. The local index formula in noncommutative geometry. *Geom. Funct. Anal.*, 5(2):174–243, 1995.
20. A. Connes and H. Moscovici. Type III and spectral triples. In *Traces in number theory, geometry and quantum fields*, Aspects Math., E38, pages 57–71. Friedr. Vieweg, Wiesbaden, 2008.
21. A. Connes and H. Moscovici. Modular curvature for noncommutative two-tori. *J. Amer. Math. Soc.*, 27(3):639–684, 2014.
22. A. Connes and P. Tretkoff. The Gauss-Bonnet theorem for the noncommutative two torus. In *Noncommutative geometry, arithmetic, and related topics*, pages 141–158. Johns Hopkins Univ. Press, Baltimore, MD, 2011.
23. L. Dabrowski and A. Sitarz. An asymmetric noncommutative torus. *SIGMA Symmetry Integrability Geom. Methods Appl.*, 11:Paper 075, 11, 2015.
24. R. Dong, A. Ghorbanpour, and M. Khalkhali. The Ricci Curvature for Noncommutative Three Tori. *arXiv e-prints*, page arXiv:1808.02977, Aug. 2018.
25. A. Fathi, A. Ghorbanpour, and M. Khalkhali. Curvature of the determinant line bundle for the noncommutative two torus. *Math. Phys. Anal. Geom.*, 20(2):Art. 4, 20, 2017.
26. A. Fathi and M. Khalkhali. On Certain Spectral Invariants of Dirac Operators on Noncommutative Tori. *arXiv e-prints*, page arXiv:1504.01174, Apr. 2015.
27. F. Fathi Zadeh and M. Khalkhali. The algebra of formal twisted pseudodifferential symbols and a noncommutative residue. *Lett. Math. Phys.*, 94(1):41–61, 2010.
28. F. Fathizadeh. On the scalar curvature for the noncommutative four torus. *J. Math. Phys.*, 56(6):062303, 14, 2015.
29. F. Fathizadeh and O. Gabriel. On the Chern-Gauss-Bonnet theorem and conformally twisted spectral triples for C^*-dynamical systems. *SIGMA Symmetry Integrability Geom. Methods Appl.*, 12:Paper No. 016, 21, 2016.

30. F. Fathizadeh and M. Khalkhali. The Gauss-Bonnet theorem for noncommutative two tori with a general conformal structure. *J. Noncommut. Geom.*, 6(3):457–480, 2012.
31. F. Fathizadeh and M. Khalkhali. Scalar curvature for the noncommutative two torus. *J. Noncommut. Geom.*, 7(4):1145–1183, 2013.
32. F. Fathizadeh and M. Khalkhali. Scalar curvature for noncommutative four-tori. *J. Noncommut. Geom.*, 9(2):473–503, 2015.
33. F. Fathizadeh and M. W. Wong. Noncommutative residues for pseudo-differential operators on the noncommutative two-torus. *J. Pseudo-Differ. Oper. Appl.*, 2(3):289–302, 2011.
34. R. Floricel, A. Ghorbanpour, and M. Khalkhali. The Ricci Curvature in Noncommutative Geometry. *To appear in the Journal of Noncommutative Geometry, arXiv e-prints*, page arXiv:1612.06688, Dec. 2016.
35. A. Ghorbanpour and M. Khalkhali. Spectral geometry of functional metrics on noncommutative tori. *arXiv e-prints*, page arXiv:1811.04004, Nov. 2018.
36. P. B. Gilkey. *Invariance theory, the heat equation, and the Atiyah-Singer index theorem*. Studies in Advanced Mathematics. CRC Press, Boca Raton, FL, second edition, 1995.
37. R. Høegh-Krohn, M. B. Landstad, and E. Stø rmer. Compact ergodic groups of automorphisms. *Ann. of Math. (2)*, 114(1):75–86, 1981.
38. B. Iochum and T. Masson. Heat trace for Laplace type operators with non-scalar symbols. *J. Geom. Phys.*, 116:90–118, 2017.
39. M. Kac. Can one hear the shape of a drum? *Amer. Math. Monthly*, 73(4, part II):1–23, 1966.
40. M. Khalkhali. *Basic noncommutative geometry*. EMS Series of Lectures in Mathematics. European Mathematical Society (EMS), Zürich, second edition, 2013.
41. M. Khalkhali and A. Moatadelro. A Riemann-Roch theorem for the noncommutative two torus. *J. Geom. Phys.*, 86:19–30, 2014.
42. M. Khalkhali, A. Moatadelro, and S. Sadeghi. A Scalar Curvature Formula For the Noncommutative 3-Torus. *ArXiv e-prints*, Oct. 2016.
43. M. Khalkhali and A. Sitarz. Gauss-Bonnet for matrix conformally rescaled Dirac. *J. Math. Phys.*, 59(6):063505, 10, 2018.
44. A. W. Knapp. *Lie groups, Lie algebras, and cohomology*, volume 34 of *Mathematical Notes*. Princeton University Press, Princeton, NJ, 1988.
45. M. Kontsevich and S. Vishik. Geometry of determinants of elliptic operators. In *Functional analysis on the eve of the 21st century, Vol. 1 (New Brunswick, NJ, 1993)*, volume 131 of *Progr. Math.*, pages 173–197. Birkhäuser Boston, Boston, MA, 1995.
46. M. Lesch. Divided differences in noncommutative geometry: rearrangement lemma, functional calculus and expansional formula. *J. Noncommut. Geom.*, 11(1):193–223, 2017.
47. M. Lesch and H. Moscovici. Modular curvature and Morita equivalence. *Geom. Funct. Anal.*, 26(3):818–873, 2016.
48. Y. Liu. Modular curvature for toric noncommutative manifolds. *J. Noncommut. Geom.*, 12(2):511–575, 2018.
49. J. I. Manin. Algebraic aspects of nonlinear differential equations. pages 5–152. (errata insert), 1978.
50. H. P. McKean, Jr. and I. M. Singer. Curvature and the eigenvalues of the Laplacian. *J. Differential Geometry*, 1(1):43–69, 1967.
51. S. Minakshisundaram and A. . Pleijel. Some properties of the eigenfunctions of the Laplace-operator on Riemannian manifolds. *Canadian J. Math.*, 1:242–256, 1949.
52. S. Paycha and S. Scott. A Laurent expansion for regularized integrals of holomorphic symbols. *Geom. Funct. Anal.*, 17(2):491–536, 2007.
53. D. Quillen. Determinants of Cauchy-Riemann operators on Riemann surfaces. *Funktsional. Anal. i Prilozhen.*, 19(1):37–41, 96, 1985.
54. J. Rosenberg, Levi-Civita's theorem for noncommutative tori. *SIGMA Symmetry Integrability Geom. Methods Appl.*, 9, 2013.
55. M. Wodzicki. Noncommutative residue. I. Fundamentals. In *K-theory, arithmetic and geometry (Moscow, 1984–1986)*, volume 1289 of *Lecture Notes in Math.*, pages 320–399. Springer, Berlin, 1987.

Index theory and noncommutative geometry: a survey

Alexander Gorokhovsky and Erik Van Erp

Dedicated to Alain Connes, with admiration

Abstract This chapter is an introductory survey of selected topics in index theory in the context of noncommutative geometry, focusing in particular on Alain Connes' contributions. This survey has two parts. In the first part, we consider index theory in the setting of K-theory of C^* algebras. The second part focuses on the local index formula of A. Connes and H. Moscovici in the context of noncommutative geometry.

1 Introduction

The contributions of Alain Connes to index theory are fundamental, broad, and deep. This article is an introductory survey of index theory in the context of noncommutative geometry. The selection of topics is determined by the interests of the authors, and we make no claim to being exhaustive.

This survey has two parts. In the first half, we consider index theory in the setting of K-theory. If P is an elliptic linear differential operator on a closed smooth manifold X, then the highest order part of P determines a cohomological object that is most naturally understood as an element in K-theory,

$$[\sigma(P)] \in K^0(T^*X)$$

A. Gorokhovsky (✉)
University of Colorado Boulder, Boulder, CO, USA
e-mail: gorokhov@euclid.colorado.edu

E. Van Erp
Dartmouth College, Hanover, NH, USA

A. Chamseddine et al. (eds.), *Advances in Noncommutative Geometry*,
https://doi.org/10.1007/978-3-030-29597-4_7

The index of P is a function of this K-theory class $[\sigma(P)]$. Atiyah and Singer derived a topological formula for the index map

$$K^0(T^*X) \to \mathbb{Z} \qquad [\sigma(P)] \mapsto \text{Index } P$$

Developments in noncommutative geometry have shed considerable light on the nature of this map. Here we emphasize, on the one hand, the organizing role of Connes' tangent groupoid and its generalizations, and on the other hand his study of the wrong-way functoriality of the Gysin map in K-theory. An important fruit of these investigations of index theory in noncommutative geometry is the Baum–Connes conjecture, which concerns the K-theory of the reduced C^*-algebra of locally compact groups. This work has deep ties to geometry and algebraic topology.

A different and purely analytic approach to index theory is based on the McKean–Singer formula [55] which expresses the index of an elliptic operator P in terms of the trace of the heat kernel,

$$\text{Index } P = \text{Tr}\, e^{-tP^*P} - \text{Tr}\, e^{-tPP^*} \qquad t > 0$$

The heat kernel has an asymptotic expansion in powers of t (as $t \to 0^+$). This implies that its trace, and hence the index of P, can be computed as the integral of a density which is locally determined by the coefficients of the operator P. This density is called the *local index* of P.

For Dirac-type operators, the local index density can be explicitly computed, and is identified with differential forms that represent the characteristic classes that appear in the Atiyah–Singer formula. Thus, one obtains not just a formula for the index, which is a *global* invariant, but a representation of the index as the integral of a density that is canonically and *locally* determined by the operator. This makes it possible to extend index theory to manifolds with boundary, as in the Atiyah–Patodi–Singer theorem [5]. Bismut extended this approach to elliptic families [18], using the superconnection approach developed for this purpose by Quillen [62]. The local index formula in this case describes an explicit differential form representing the Chern character of an elliptic family.

The extension of local index theory to the noncommutative framework necessitates the construction of the Chern character for noncommutative algebras. In this case, a natural receptacle for the Chern character is cyclic homology theory. Cyclic theory has been closely connected to index theory from its very beginning [27].

The second half of the present paper focuses on a more recent development: the local index formula of Connes and Moscovici in the context of noncommutative geometry. Our goal is to give a brief yet detailed exposition of the proof of this theorem. We follow closely the arguments of the original paper [25]. A distinctive feature of the Connes–Moscovici paper is its detailed and illuminating explicit calculations, illustrating the cocycle property of the local index formula, and the renormalization procedure used to derive it. For an alternative approach to the Connes–Moscovici formula, we refer the reader to [45, 46].

2 Atiyah and Singer

In 1963, Michael Atiyah and Isadore Singer announced their formula for the index of an elliptic operator on a compact manifold [6]. On a manifold X with complex vector bundles E, F, a linear differential operator of order d

$$P : C^{\infty}(X, E) \to C^{\infty}(X, F)$$

is, in local coordinates on $U \subset X$ and with local trivializations of E, F, of the form

$$P = \sum_{\alpha_1+\cdots+\alpha_n=0}^{d} a_{\alpha} \frac{\partial^{\alpha_1}}{\partial x_1^{\alpha_1}} \cdots \frac{\partial^{\alpha_n}}{\partial x_n^{\alpha_n}} : C^{\infty}\left(U, \mathbb{C}^r\right) \to C^{\infty}\left(U, \mathbb{C}^r\right)$$

where the coefficients a_{α} are smooth functions on U with values in the algebra $M_r(\mathbb{C})$ of $r \times r$ matrices. The principal symbol of P is a matrix-valued polynomial in the formal variable $\xi = (\xi_1, \ldots, \xi_n) \in \mathbb{R}^n$,

$$\sigma(P)(x, \xi) = \sum_{\alpha_1+\cdots+\alpha_n=d} a_{\alpha}(x)\, \xi_1^{\alpha_1} \cdots \xi_n^{\alpha_n} \in M_k(\mathbb{C})$$

By definition, the operator P is elliptic if the matrix $\sigma(P)(x, \xi) \in M_k(\mathbb{C})$ is invertible for all $\xi \neq 0$, and at all points of X. If X is a compact manifold without boundary, an elliptic operator on X has finite dimensional kernel and cokernel. The index of P is

$$\operatorname{Index} P = \dim\ker P - \dim\operatorname{coker} P = \dim\ker P - \dim\ker P^*$$

Interpreting (x, ξ) as coordinates for the cotangent bundle T^*X, $\sigma(P)$ has an invariant interpretation as an endomorphism of vector bundles

$$\sigma(P) : \pi^*E \to \pi^*F \qquad \pi : T^*X \to X$$

where π^*E, π^*F are E, F pulled back to T^*X. Since $\sigma(P)$ is invertible outside the zero section of T^*X, which is a compact subset, the triple $(\pi^*E, \pi^*F, \sigma(P))$ determines an element in the Atiyah–Hirzebruch K-theory group of T^*X,

$$[\sigma(P)] \in K^0(T^*X)$$

The Chern character

$$\mathrm{ch} : K^j(T^*X) \otimes \mathbb{Q} \to \bigoplus_{k=0}^{n} H_c^{2k+j}(T^*X, \mathbb{Q}) \qquad j = 0, 1$$

is a natural ring isomorphism of (rational) K-theory and cohomology with compact supports.

Proposition 1 ([6]) *If P is an elliptic operator on a closed manifold X, then*

$$\text{Index } P = \{\text{ch}(\sigma(P)) \cup \text{Td}(\pi^* TX \otimes \mathbb{C})\}[T^*X]$$

where Td *is the Todd class,* $TX \otimes \mathbb{C}$ *the complexified tangent bundle of* X*, and* $[T^*X] \in H_{2n}(T^*X, \mathbb{Q})$ *the fundamental cycle for* T^*X*, canonically oriented as a symplectic manifold.*

The formula of Atiyah and Singer answered a question of Gel′fand, who proposed in [32] the investigation of the relation between homotopy invariants and analytical invariants of elliptic operators. Moreover, a number of classical theorems in geometry and topology are special cases of the Atiyah–Singer index formula. If we take

$$P = d + d^* : C^\infty(X, \Lambda^{\text{even}} T^*M) \to C^\infty(X, \Lambda^{\text{odd}} T^*M)$$

with d the de Rham differential, we recover the curvature formula of Gauss–Bonnet for the Euler number of a closed oriented surface, and its generalization to higher dimensional manifolds due to Hopf and Chern. With

$$P = \bar{\partial} + \bar{\partial}^* : C^\infty(X, \Lambda^{\text{even}} T^{1,0} M) \to C^\infty(X, \Lambda^{\text{odd}} T^{1,0} M)$$

with $\bar{\partial}$ the Dolbeault operator, we get the Riemann–Roch theorem for Riemann surfaces and Hirzebruch's generalization to projective algebraic varieties. Finally, if P is the signature operator, we obtain Hirzebruch's signature theorem.

3 The Gysin map

The original proof of Proposition 1 followed the strategy of Hirzebruch's proof of the signature theorem, relying on the calculation of cobordism groups by René Thom. The proof published in [7] is independent of cobordism theory, and makes use of Atiyah–Hirzebruch K-theory.

If X is compact Hausdorff, $K^0(X)$ is the abelian group generated by isomorphism classes of complex vector bundles on X. Vector bundles naturally pull back along continuous maps. K-theory is a cohomology theory for locally compact Hausdorff spaces, and a contravariant functor for proper maps $f : X \to Y$. If X, Y are smooth manifolds, then K-theory is also *covariant* for (not necessarily proper) smooth maps $f : X \to Y$ that are K-oriented. A K-orientation of a smooth map $f : X \to Y$ is a spin^c structure for the vector bundle $TX \oplus f^*TY$. A K-orientation of f induces a wrong-way functorial Gysin map

$$f_! : K^\bullet(X) \to K^\bullet(Y)$$

In three special cases, the construction of the Gysin map is straightforward. If $f : U \to V$ is the inclusion of U as an open submanifold of V, then $f_!$ is the map in analytic K-theory determined by the inclusion $C_0(U) \subset C_0(V)$. If E is the total space of a spin^c vector bundle on Z, and $f : Z \to E$ is the zero section, then $f_!$ is the K-theory Thom isomorphism. Similarly, if $f : E \to Z$ is the base point projection of a spin^c vector bundle, then $f_!$ is the inverse of the Thom isomorphism. In general, every K-oriented map can be factored as a composition of three maps of the above type.

The cotangent bundle T^*X of a closed smooth manifold X has a canonical symplectic form, and is therefore a spin^c manifold. Thus, the map $p : T^*X \to$ point is K-oriented, and we have a Gysin map

$$p_! : K^0(T^*X) \to K^0(\text{point}) = \mathbb{Z}$$

Proposition 2 ([7]) *If P is an elliptic operator on a closed manifold X, then*

$$\text{Index}\, P = p_!(\sigma(P))$$

A purely topological calculation, carried out in [8], shows that the characteristic class formula of Proposition 1 follows from Proposition 2.

4 The tangent groupoid

One of many significant contributions of Alain Connes to index theory is the introduction of the tangent groupoid as a fundamental tool (see [28, chapter II.5]).

A groupoid G is a small category in which all morphisms are isomorphisms. We refer to the morphisms $G^{(1)}$ of the category as the elements of the groupoid G, and composition of morphisms as multiplication. An object $x \in G^{(0)}$ in the category is identified with its identity morphism $\text{Id}_x \in G^{(1)}$ and thought of as a unit element $x \in G$. Each morphism $\gamma \in G$ has a source object $s(\gamma) \in G^{(0)}$ and range object $r(\gamma) \in G^{(0)}$. A pair $\gamma_1, \gamma_2 \in G$ can be multiplied if $s(\gamma_1) = r(\gamma_2)$.

A Lie groupoid is a groupoid such that G is a smooth manifold, $G^{(0)} \subset G$ is a smooth submanifold, the source and target maps are smooth submersions $G \to G^{(0)}$ making the set of composable elements $G^{(2)} = \{(\gamma_1, \gamma_2) \mid s(\gamma_1) = r(\gamma_2)\}$ into a smooth submanifold of $G \times G$, and multiplication $G^{(2)} \to G : (\gamma_1, \gamma_2) \mapsto \gamma_1\gamma_2$ and inversion $G \to G : \gamma \mapsto \gamma^{-1}$ are smooth maps.

If G is a Lie groupoid, and $f, g \in C_c^\infty(G)$ are two smooth functions with compact support, then the convolution product $f * g$ is defined as on a group,

$$(f * g)(\gamma) = \int_{G_x} f\left(\gamma\beta^{-1}\right) g(\beta) d\lambda_x(\beta) \qquad x = s(\gamma) = s(\beta)$$

For this formula to make sense we must choose a smooth density λ_x on the submanifold $G_x = s^{-1}(x) \subset G$ for each unit $x \in G^{(0)}$. The support of λ_x must be all of G_x. To obtain an associative product the family of densities $\{\lambda_x, x \in G^{(0)}\}$ must depend smoothly on x, and be invariant under right multiplication $G_x \to G_y$: $\gamma_1 \to \gamma_1\gamma_2$ with $r(\gamma_2) = x, s(\gamma_2) = y$.

Involution is $f^*(\gamma) = \overline{f(\gamma^{-1})}$, and the $*$-algebra $C_c^\infty(G)$ completes to a C^*-algebra $C^*(G)$. A different choice of densities λ_x results in an isomorphic C^*-algebra $C^*(G)$. Many, if not most, C^*-algebras in noncommutative geometry arise as groupoid C^*-algebras.

Example 3 Let X be a smooth manifold. The tangent bundle TX, conceived as a smooth family of abelian Lie groups $T_xX \cong \mathbb{R}^n$, is a Lie groupoid. The set of units is the zero section $X \subset TX$, and source and range maps are the projection $TX \to X$. Composition is addition of vectors in the same fiber T_xX. The convolution algebra $C^*(TX)$ is commutative, because $\gamma_1\gamma_2 = \gamma_2\gamma_1$. Fourier transform in each fiber T_xX determines a natural isomorphism of C^*-algebras,

$$C^*(TX) \cong C_0(T^*X)$$

Example 4 The Cartesian product $X \times X$ is a groupoid with multiplication

$$(x, y)(y, z) = (x, z) \qquad x, y, z \in X$$

The units are the elements (x, x) of the diagonal in $X \times X$. Thus, the space of units can be identified with X. Convolution of functions $f, g \in C_c^\infty(X \times X)$ is

$$(f * g)(x, y) = \int_X f(x, z)g(z, y)d\lambda(z)$$

with arbitrary choice of positive smooth density λ on $X = X \times \{y\} = s^{-1}(y)$. This is the rule for composition of Schwartz kernels. The convolution algebra $C_c^\infty(X \times X)$ is naturally represented as operators on $L^2(X, \lambda)$, and these operators are compact because their Schwartz kernels are square integrable. After completion in norm we have

$$C^*(X \times X) \cong \mathcal{K}(L^2(X))$$

Algebraically, the tangent groupoid $\mathbb{T}X$ is the disjoint union of the abelian groups $T_xX \cong \mathbb{R}^n$ parametrized by $x \in X$, and a family of pair groupoids $X \times X$ parametrized by $t \in (0, 1]$,

$$\mathbb{T}X = TX \times \{0\} \sqcup X \times X \times (0, 1]$$

The nontrivial aspect of the construction of $\mathbb{T}X$ is the way the tangent bundle TX is glued, as a smooth boundary at $t = 0$, to the family of pair groupoids for $t \in (0, 1]$, by "blowing up" the diagonal in $X \times X$.

Evaluation at $t = 0$ gives a $*$-homomorphism

$$\mathrm{ev}_0 : C^*(\mathbb{T}X) \to C^*(TX) \cong C_0(T^*X)$$

The kernel of this map is a contractible C^*-algebra $C_0((0, 1], \mathcal{K})$, and so ev_0 induces an isomorphism in K-theory,

$$(\mathrm{ev}_0)_* : K_0(C^*(\mathbb{T}X)) \cong K^0(T^*X)$$

Evaluation at $t = 1$ determines a homomorphism

$$(\mathrm{ev}_1)_* : K_0(C^*(\mathbb{T}X)) \to K_0(C^*(X \times X)) \cong K_0(\mathcal{K}(L^2(X)) \cong \mathbb{Z}$$

Proposition 5 ([28, II.5]) *The homomorphism*

$$(\mathrm{ev}_1)_* \circ (\mathrm{ev}_0)_*^{-1} : K^0(T^*X) \to \mathbb{Z}$$

is the analytic index, i.e., it maps the principal symbol of an elliptic operator P to the index of P,

$$\mathrm{Index}\, P = (\mathrm{ev}_1)_* \circ (\mathrm{ev}_0)_*^{-1}[\sigma(P)]$$

Proof Let P be a linear differential operator on X. For simplicity, we assume that E, F are trivial line bundles. P determines a differential operator $\mathbb{P}$ on $\mathbb{T}X$ as follows. The space of units of $\mathbb{T}X$ is $X \times [0, 1]$. $\mathbb{P}$ differentiates only in the direction of the source fibers $s^{-1}(x, t)$ with $(x, t) \in X \times [0, 1]$. If $t > 0$, then $\mathbb{P}$ restricted to $s^{-1}(x, t) = X \times \{x\} \times \{t\}$ is $t^d P$. If $t = 0$, then on each fiber $s^{-1}(x, 0) = T_x X$ of the tangent bundle, $\mathbb{P}$ restricts to the constant coefficient operator P_x homogeneous of degree d, obtained by freezing the coefficients of P at x, and retaining only the highest order terms,

$$P_x = \sum_{\alpha_1 + \cdots + \alpha_n = d} a_\alpha(x) \frac{\partial^{\alpha_1}}{\partial x_1^{\alpha_1}} \cdots \frac{\partial^{\alpha_n}}{\partial x_n^{\alpha_n}}$$

(where $a_\alpha(x)$ are constant matrices). Note that the principal symbol $\sigma(P)(x, \xi)$ at $x \in X$ is the Fourier transform of P_x.

The smooth structure of $\mathbb{T}X$ is constructed in precisely such a way that the coefficients of $\mathbb{P}$ are smooth functions on $\mathbb{T}X$. In fact, $\mathbb{P}$ is a right invariant differential operator on the groupoid $\mathbb{T}X$, and is a multiplier of the convolution algebra $C_c^\infty(\mathbb{T}X)$. This construction generalizes to pseudodifferential operators. If

P is elliptic, and if Q is a parametrix of P, then $\mathbb{Q}$ is an inverse of $\mathbb{P}$ modulo $C_c^\infty(\mathbb{T}X)$. Therefore, $\mathbb{P}$ determines an element in algebraic K-theory

$$[\mathbb{P}] \in K_0^{\mathrm{alg}}(C_c^\infty(\mathbb{T}X))$$

The inclusion $C_c^\infty(\mathbb{T}X) \subset C^*(\mathbb{T}X)$ induces a map in K-theory, and we obtain

$$[\mathbb{P}] \in K_0(C^*(\mathbb{T}X))$$

Evaluation at $t = 0$ maps the K-theory class $[\mathbb{P}]$ to the class $[\sigma(P)]$ defined by Atiyah and Singer,

$$(\mathrm{ev}_0)_*([\mathbb{P}]) = [\sigma(P)] \in K^0(T^*X)$$

Evaluation at $t = 1$, on the other hand, sends $[\mathbb{P}]$ to an element in $K_0(\mathcal{K}) \cong \mathbb{Z}$ that corresponds to the index of P,

$$(\mathrm{ev}_1)_*([\mathbb{P}]) = \mathrm{Index}\, P \in \mathbb{Z}$$

□

In [28, II.5] Connes gives a short and conceptually elegant proof that the map $(\mathrm{ev}_1)_* \circ (\mathrm{ev}_0)_*^{-1} : K^0(T^*X) \to \mathbb{Z}$ determined by the tangent groupoid agrees with the Gysin map $p_! : K^0(T^*X) \to \mathbb{Z}$. This proves Proposition 2. Connes' proof makes use of the Baum–Connes isomorphism for the tangent groupoid $G = \mathbb{T}X$,

$$K^0(BG) \cong K_0(C^*(G))$$

(see Section 9). Here BG is a classifying space for free and proper actions of G. This isomorphism replaces the groupoid $G = \mathbb{T}X$ by the topological space BG, which turns the index map into one that can be computed topologically.

In subsequent work by many authors, the tangent groupoid has proven to be a very useful tool in index theory. Analogs of the tangent groupoid have been constructed to deal with index problems in various contexts. As a random (neither exhaustive nor fully representative) sample of publications ranging from 1987 to 2018, see [48, 56, 61, 1, 63, 65, 47, 30].

Remark 6 The tangent groupoid formalism is intimately connected to deformation theory. The C^*-algebra $C^*(\mathbb{T}X)$ is the algebra of sections of a continuous field of C^*-algebras over $[0, 1]$ with fiber $C_0(T^*X)$ at $t = 0$ and $\mathcal{K}(L^2(X))$ at all $t > 0$. This field encodes a strong deformation quantization of the algebra of classical observables (functions on the symplectic phase space T^*X) to quantum observables (operators on $L^2(X)$).

5 K-homology

The wrong-way functoriality of the Gysin map suggests that the topological index is perhaps more naturally regarded as a map in K-homology. Atiyah–Hirzebruch K-theory is associated with the Bott spectrum,

$$K^j(X) = \left[X, K_j\right] \qquad K_{2j} = BU \times \mathbb{Z},\ K_{2j+1} = U$$

Abstractly, K-homology $K_j(X)$ is then defined as the stable homotopy group of the smash product of X with the Bott spectrum,

$$K_j^{top}(X) = \pi_j(X \wedge \underline{K}) = \varinjlim \pi_{j+k}(X \wedge K_k)$$

K-homology is naturally covariant. The Gysin map $p_! : K^0(T^*X) \to \mathbb{Z}$ in K-theory corresponds in K-homology to the map determined by $\epsilon : X \to$ point,

$$\epsilon_* : K_0^{top}(X) \to K_0^{top}(\text{point}) \cong \mathbb{Z}$$

composed with the Poincaré duality isomorphism

$$K^0(T^*X) \cong K_0^{top}(X)$$

5.1 *Analytic K-homology*

The homotopy theoretic definition of K-homology lacks a concrete model for cycles in $K_j(X)$ that is useful in calculations. In [3] Atiyah shows that elliptic operators can represent K-cycles. If P is an elliptic operator on a closed manifold X, and V is a complex vector bundle on X, we can twist P by V to obtain a new elliptic operator

$$P_V : C^\infty(M, E \otimes V) \to C^\infty(M, F \otimes V)$$

Thus, P determines a homomorphism

$$K^0(X) \to \mathbb{Z} \qquad [V] \mapsto \text{Index } P_V$$

In [3] Atiyah axiomatized the notion of elliptic operator such that it still makes sense if X is not a smooth manifold, but only a compact Hausdorff space. Following Atiyah, an element in the even K-homology group $K_0(X)$ is represented by a bounded Fredholm operator $F : H^0 \to H^1$, where the Hilbert spaces H^j are equipped with a $*$-representation $\phi_j : C(X) \to \mathcal{B}(H^j)$, and F is *"an operator on X"* in the sense that

$$F\phi_0(f) - \phi_1(f)F : H^0 \to H^1 \text{ is a compact operator for all } f \in C(X) \tag{5.1}$$

This axiom guarantees that the index of the operator F twisted by a vector bundle V on X is a well-defined integer. In fact, twisting implements the cap product, i.e., the structure of K-homology as a module over the K-theory ring,

$$\cap : K^0(X) \times K_0(X) \to K_0(X) \qquad [V] \cap [P] = [P_V]$$

For an elliptic operator $P : C^\infty(E) \to C^\infty(F)$, the Hilbert spaces H^0, H^1 are appropriate Sobolev spaces of sections in E, F, such that $P : H^0 \to H^1$ is a bounded Fredholm operator, and Atiyah's axiom is satisfied because of the Rellich Lemma. The correct equivalence relation for Atiyah's group of abstract elliptic operators was identified by Kasparov [51]. The Atiyah–Kasparov analytic K-homology of X is the KK-group $KK(C(X), \mathbb{C})$. For finite CW-complexes X there is an isomorphism

$$KK(C(X), \mathbb{C}) \cong K_0^{top}(X)$$

Kasparov developed the theory far beyond Atiyah's ansatz. For two separable (possibly noncommutative) C^*-algebras A, B, elements in the abelian group $KK(A, B)$ are represented by generalized Fredholm operators $F : H^0 \to H^1$. Now H^0, H^1 are Hilbert C^*-modules over B, and $\phi_j : A \to \mathcal{L}(H^j)$ are $*$-representations of A by adjointable operators on H^j. F is invertible modulo "compact operators" in the sense of the theory of operators on Hilbert-modules. The power of KK-theory rests on the existence of an associative product

$$\# : KK(A, B) \times KK(B, C) \to KK(A, C)$$

For example, K-theory is

$$K_0(A) \cong KK(\mathbb{C}, A)$$

and an element in the bivariant group $\xi \in KK(A, B)$ induces, by right composition, a map of K-theory groups

$$K_0(A) \to K_0(B) : \alpha \mapsto \alpha \# \xi$$

Kasparov showed that Proposition 1 of Atiyah–Singer can be understood in K-homology as follows. The Dirac operator D_{T^*X} of the (symplectic) spin^c manifold T^*X determines a class in analytic K-homology,

$$[D_{T^*X}] \in KK(C_0(T^*X), \mathbb{C})$$

and hence determines a map $K^0(T^*X) \to \mathbb{Z}$. This map is equal to the topological index of Atiyah–Singer.[1] But the principal symbol $\sigma(P)$ of an elliptic operator P determines not only an element in K-theory,

$$[\sigma(P)] \in KK(\mathbb{C}, C_0(T^*X))$$

but in fact gives the slightly better

$$[\sigma(P)] \in KK(C(X), C_0(T^*X))$$

Proposition 7 ([19, section 24.5]) *The element in analytic K-homology determined by an elliptic operator P on a closed manifold X is the product*

$$[P] = [\sigma(P)]\#[D_{T^*X}] \in KK(C(X), \mathbb{C})$$

The importance of KK-theory to index theory is hard to overstate. For an excellent first introduction to KK-theory see [44].

5.2 Geometric K-homology

A geometric model for cycles in K-homology was developed by Baum and Douglas [12]. In their theory, a K-cycle is represented by a triple (M, V, φ), where M is a closed spin^c manifold, V is a complex vector bundle on M, and $\varphi : M \to X$ is a continuous map. The nontrivial aspect of the theory is the equivalence relation on such triples (see [12] for details). The groups $K_0^{\text{geo}}(X)$ and $K_1^{\text{geo}}(X)$ consist of equivalence classes of triples (M, V, φ) with M even or odd dimensional, respectively. Remarkably, Baum–Douglas geometric K-homology is equivalent to topological K-homology for *all* topological spaces X [50],

$$K_j^{\text{geo}}(X) \cong \pi_j(X \wedge \underline{K})$$

A proof of the special case of the Atiyah–Singer index theorem for Dirac operators using geometric K-homology is as follows. Propositions 7 and 8 together imply Proposition 1.

Proposition 8 *If M is a spin^c manifold with Dirac operator D, and V is a smooth complex vector bundle on M, then*

$$\text{Index}\, D_V = \{\text{ch}(V) \cup \text{Td}(M)\}[M]$$

[1] This is a restatement of Bott's version of the Atiyah–Singer formula in his review of [6] for the AMS mathematical reviews.

where Td(M) *is the Todd class of the spin*^{c} *vector bundle* TM.

Proof The geometric K-homology of a point $K_0^{\text{geo}}(\text{point})$ consists of pairs (M, V), with M a spin^c manifold with complex vector bundle V. There is an isomorphism of abelian groups,

$$K_0^{\text{geo}}(\text{point}) \cong \mathbb{Z} \qquad (M, V) \mapsto \{\text{ch}(V) \cup \text{Td}(M)\}[M]$$

The proof that this is an isomorphism is purely topological, and Bott periodicity plays a key role. There is a second homomorphism, namely the analytic index

$$K_0^{\text{geo}}(\text{point}) \to \mathbb{Z} \qquad (M, V) \mapsto \text{Index } D_V$$

The proof that this map is well-defined relies on bordism invariance of the index of Dirac operators. To prove that the two homomorphisms of abelian groups $K_0^{\text{geo}}(\text{point}) \to \mathbb{Z}$ are equal, it suffices to verify this by direct calculation in one example where D_V has nonzero index (for example, the Dolbeault operator of $\mathbb{C}P^1$). (See [15] for details.) □

The point of view proposed in [12] is that index theory, in general, is based on the equivalence between analytic and geometric K-homology. For a compact manifold X (or, more generally, a finite CW-complex), there is an isomorphism [13]

$$\mu : K_0^{\text{geo}}(X) \cong KK(C(X), \mathbb{C})$$

If (M, V, φ) is a geometric K-cycle for X, the spin^c manifold M has a Dirac operator D; D twisted by V determines an element in analytic K-homology $[D_V] \in KK(C(M), \mathbb{C})$; finally $\phi : M \to X$ maps this element to

$$\mu(M, V, \varphi) = \varphi_*(D_V) \in KK(C(X), \mathbb{C})$$

In [12], Baum and Douglas conceptualize index theory as follows:

> Given an element in analytic K-homology, $\xi \in KK_j(C(X), \mathbb{C})$, explicitly compute the unique $\tilde{\xi}$ in geometric K-homology, $\tilde{\xi} \in K_j^{\text{geo}}(X)$, corresponding to ξ.

In concrete terms, this proposes that index problems are solved by reduction to the index of a Dirac operator, possibly on a different manifold—as, for example, in Proposition 7. This perspective on index theory was used, more recently, to solve the index problem for hypoelliptic operators in the Heisenberg calculus on contact manifolds [14]. This is a class of (pseudo)differential operators that are Fredholm, but not elliptic. Yet these operators satisfy Atiyah's axiom (5.1), and therefore determine an analytic K-cycle. The index problem for this class of hypoelliptic operators was solved by computing an equivalent geometric K-cycle.

6 Elliptic families

The suggestion in [12] that index problems are solved by reduction to Dirac operators is intimately related to the point of view that the topological index is a Gysin map in K-theory. To see the connection between the two perspectives, we consider the families index theorem of Atiyah and Singer [9].

Suppose P_b is a smooth family of elliptic operators on a closed manifold X, parametrized by points $b \in B$ in another compact manifold B. Alternatively, we may think of such an elliptic family $\{P_b, b \in B\}$ as a single differential operator P on $X \times B$ that differentiates along the fibers of $X \times B \to B$. The kernel $V_b = \ker P_b$ of each individual operator P_b is a finite dimensional vector space. If the dimension of V_b is independent of $b \in B$, then the collection $\{V_b, b \in B\}$ defines a vector bundle $V = \ker P$ on B. The same holds for the cokernels of P_b, and the index of the elliptic family P is a formal difference of two vector bundles on B,

$$\text{Index } P = [\ker P] - [\text{coker } P] \in K^0(B)$$

The construction can be modified so that it makes sense even if the dimension of V_b is not constant. See [9], where the product $X \times B$ is generalized to fiber bundles $Z \to B$, and B is allowed to be a compact Hausdorff space.

The analog of Proposition 7 holds for families. As a single operator, $P : H^0 \to H^1$ is a generalized Fredholm operator, where H^0, H^1 are Hilbert modules over the C^*-algebra $C(B)$. As such, P determines a class in $KK(\mathbb{C}, C(B))$. The families index of P is precisely the isomorphism

$$\text{Index} : KK(\mathbb{C}, C(B)) \cong K^0(B)$$

If $V \to Z$ is a complex vector bundle on Z, then each elliptic operator P_b can be twisted by the restriction $V|Z_b$. We obtain the twisted family P_V, whose families index is an element in $K^0(B)$. Thus, P determines a group homomorphism

$$[V] \to \text{Index} P_V : K^0(Z) \to K^0(B)$$

This homomorphism is the product, in KK-theory, with a KK-element determined by the elliptic family P,

$$\text{Index} P_V = [V]\#[P] \qquad [P] \in KK(C(Z), C(B))$$

As with a single operator, the principal symbol of an elliptic family on $Z \to B$ is an element in K-theory,

$$[\sigma(P)] \in K^0\left(T^*_{\text{vert}} Z\right)$$

where $T_{\mathrm{vert}}Z$ is the bundle of vertical tangent vectors, i.e., the kernel of the map $d\pi : TZ \to TB$. We have in fact

$$[\sigma(P)] \in KK\left(C(Z), C_0\left(T^*_{\mathrm{vert}}Z\right)\right)$$

The cotangent bundles T^*Z_b of the fibers Z_b are symplectic and hence spin^c manifolds. The family of Dirac operators $D_{T^*Z_b}$ is an elliptic family parametrized by $b \in B$, and determines a KK-element

$$\left[D_{T^*_{\mathrm{vert}}Z}\right] \in KK\left(C_0\left(T^*_{\mathrm{vert}}Z\right), C(B)\right)$$

Then, as in Kasparov's Proposition 7, a version of the families index theorem is

$$[P] = [\sigma(P)]\#\left[D_{T^*_{\mathrm{vert}}Z}\right] \in KK(C(Z), C(B))$$

To derive the families index theorem of [9], we need a topological characterization of the KK-element $[D_{T^*_{\mathrm{vert}}Z}]$. This topological characterization is provided by the Gysin map.

Proposition 9 *Let $\pi : W \to B$ be a submersion of manifolds. Assume that $\ker d\pi \subset TW$ is a spin^c vector bundle, so that each fiber W_b is a spin^c manifold, and π is K-oriented. If D is the elliptic family of Dirac operators D_b of W_b, then the KK-element*

$$[D] \in KK(C_0(W), C(B))$$

has the property that the corresponding map in K-theory

$$K^0(W) \to K^0(B) : [V] \mapsto [V]\#[D] = \mathrm{Index}\, D_V$$

is the Gysin map

$$\pi_! : K^0(W) \to K^0(B)$$

Proposition 9 is a special case of the families index theorem in [9]. Together with the equality $[P] = [\sigma(P)]\#[D_{T^*_{\mathrm{vert}}Z}]$, it gives the index theorem for general elliptic families as,

$$\mathrm{Index}\, P = \pi_!(\sigma(P)) \in K^0(B) \qquad \pi : T^*_{\mathrm{vert}}Z \to B$$

7 The adiabatic groupoid

The tangent groupoid formalism can be adapted to the families index problem. The relevant generalization of Connes' tangent groupoid is developed by Nistor, Weinstein, and Xu in [61]. For an arbitrary Lie groupoid G, let T^sG be the vector bundle on G of vectors tangent to the source fibers, i.e., vectors in the kernel of $ds : TG \to TG^{(0)}$, where $s : G \to G^{(0)}$ is the source map of the groupoid G. Let AG be the restriction of T^sG to the space of objects $G^{(0)} \subset G$. The vector bundle AG has the structure of a Lie algebroid, but this is not relevant for our purposes here.

A "blow-up" construction that closely follows [28, II.5] produces a groupoid

$$G_{\mathrm{ad}} = AG \times \{0\} \sqcup G \times (0, 1]$$

called the *adiabatic groupoid* of G. Algebraically, G_{ad} is the disjoint union of a family of copies of G parametrized by $t \in (0, 1]$ with the vector bundle AG as the boundary at $t = 0$. Connes' tangent groupoid is the special case with $G = X \times X$ and $AG = TX$.

As in Section 4 above, the adiabatic groupoid gives rise to a map in K-theory

$$K^0(A^*G) \to K_0(C^*(G))$$

from the K-theory of the vector bundle A^*G to the K-theory of the Lie groupoid G. This map can be interpreted as an analytic index. If P is a (pseudo)differential operator on G that differentiates only in the direction of the fibers of the source map $s : G \to G^{(0)}$, and if P is right invariant, then if the principal symbol $\sigma(P)(x, \xi)$ is invertible for all groupoid units $x \in G^{(0)}$ and all $\xi \in A^*G$ with $\xi \neq 0$, then P has an index,

$$\mathrm{Index}\, P \in K_0(C^*(G))$$

The principal symbol of P determines a class in topological K-theory

$$[\sigma(P)] \in K^0(A^*G)$$

The analytic index map determined by the adiabatic groupoid maps $[\sigma(P)]$ to Index P. For differential operators, the proof of these facts is essentially the same as the proof of Proposition 5.

Example 10 Given a smooth fiber bundle $\pi : Z \to B$, consider the Lie groupoid

$$G = Z \times_B Z = \{(z, z') \in Z \times Z \mid \pi(z) = \pi(z')\}$$

with multiplication $(z, z')(z', z'') = (z, z'')$. The Lie algebroid of G is $AG = T_{\mathrm{vert}}Z$. The groupoid $Z \times_B Z$ is Morita equivalent to the manifold B (as a Lie groupoid in which all elements are units), and the adiabatic groupoid of G determines an index map

$$K^0(A^*G) = K^0\left(T^*_{\text{vert}}Z\right) \to K^0(B) \cong K_0(C^*(G))$$

This map is the analytic index map for families of elliptic operators on the fibers of $Z \to B$.

8 Foliations

A foliation of a smooth manifold Z is a subbundle $F \subseteq TZ$ such that $[F, F] \subseteq F$. Equivalently, Z is a disjoint union of immersed submanifolds $L \subset Z$, called the leaves of the foliation, such that at every point $x \in L$ for any leaf $L \subset Z$ we have $T_xL = F_x$. The leaf space is the set of leaves with the quotient topology inherited from Z. If $\pi : Z \to B$ is a fibration, then Z is foliated with $F = T_{\text{vert}}Z = \ker d\pi$. The leafs are the fibers $Z_b = \pi^{-1}(b)$, and the leaf space is B. In general, the leaf space of a foliation may have pathological topology. For example, if Z contains at least one leaf L that is dense in Z (e.g., the Kronecker foliation of a torus), then the leaf space has only two open sets.

The holonomy groupoid $G_{Z/F}$ of a foliated manifold (Z, F) is a Lie groupoid whose elements are equivalence classes of paths $\gamma : [0, 1] \to L$ connecting points $x = s(\gamma), y = r(\gamma)$ in the same leaf L. Two paths are equivalent if they have the same holonomy (see [23]). Multiplication in $G_{Z/F}$ is composition of paths. In simple cases where there is no holonomy (e.g., the Kronecker foliation), the groupoid $G_{Z/F}$ is algebraically a disjoint union of the pair groupiods $L \times L$ of the leaves. If the leaf space (with its quotient topology) is Hausdorff, then it is Morita equivalent to the groupoid $G_{Z/F}$. But in general the holonomy groupoid $G_{Z/F}$ (as a "smooth stack") is a better representation of the space of leaves than the leaf space with its quotient space.

By definition, the K-theory of the leaf space of a foliation is the K-theory of the reduced C^*-algebra of the holonomy groupoid,

$$K^0(Z/F) := K_0\left(C^*_r\left(G_{Z/F}\right)\right)$$

The Lie algebroid of $G_{Z/F}$ is $AG = F$, as a vector bundle over $G^{(0)} = Z$. Thus, by the general procedure of [61], we have an analytic index

$$K^0(F^*) \to K^0(Z/F)$$

This map generalizes the analytic index of elliptic families. It is defined, for example, for differential operators on Z that differentiate in the leaf direction, and are elliptic along the leafs. Such operators are called *longitudinally elliptic*.

Assume that the leaves are even dimensional spin^c manifolds. This means that F is a spin^c vector bundle of even rank, and so in particular we have the Thom isomorphism,

$$K^0(F^*) \cong K^0(Z)$$

The spin^c Dirac operators of the leaves form a longitudinally elliptic family D on Z. Given a vector bundle V on Z, we can twist the Dirac operator of each leaf L by the restriction $V|L$, and we obtain a new elliptic family D_V. The longitudinal index problem consists in identifying a "topological index" that equals the analytic index map

$$K^0(Z) \to K^0(Z/F) \qquad V \mapsto \mathrm{Index} D_V$$

The solution was suggested in [23] and proven in [26] by Connes and Skandalis. A smooth maps between foliated manifolds

$$f : Z_1/F_1 \to Z_2/F_2$$

is defined as a smooth groupoid homomorphism $G'_{Z_1/F_1} \to G_{Z_2/F_2}$, where G'_{Z_1/F_1} is a groupoid that is Morita equivalent to G_{Z_1/F_1}. (This is a morphism of stacks.) The notion of K-orientation naturally extends to such maps. The difficulty is showing that such a K-oriented map determines a Gysin map in K-theory that is (wrong-way) functorial,

$$f_! : K^\bullet(V_1/F_1) \to K^\bullet(V_2/F_2)$$

If F is a spin^c vector bundle, then the identity map $l : Z \to Z$ is K-oriented when considered as a morphism from the manifold Z to the foliation Z/F. In the case of a fiber bundle $\pi : Z \to B$, where $Z/F = B$, this is just the map π. The topological index for longitudinally elliptic operators is the Gysin map

$$l_! : K^0(Z) \to K^0(Z/F)$$

Thus, the index theorem of Connes–Skandalis for longitudinally elliptic operators generalizes Proposition 9.

9 The Baum–Connes conjecture

Among the most significant developments of index theory in the context of noncommutative geometry is the Baum–Connes conjecture. To fit it in the framework discussed so far, consider a manifold X with fundamental cover $p : \tilde{X} \to X$. The fundamental groupoid of X is the Lie groupoid

$$\tilde{X} \times_X \tilde{X} = \{(a, b) \in \tilde{X} \times \tilde{X} \mid p(a) = p(b)\}$$

with multiplication $(a, b)(b, c) = (a, c)$. Equivalently, elements of the fundamental groupoid are homotopy classes of paths $\gamma : [0, 1] \to X$ (with no base point), and multiplication is composition of paths. The fundamental groupoid is Morita equivalent to the fundamental group $\Gamma = \pi_1(X)$, and so

$$K_0\left(C^*\left(\tilde{X} \times_X \tilde{X}\right)\right) \cong K_0(C^*(\Gamma))$$

The Lie algebroid of $\tilde{X} \times_X \tilde{X}$ is TX, i.e., it is the same as the Lie algebroid of the pair groupoid $X \times X$. The adiabatic groupoid of $\tilde{X} \times_X \tilde{X}$ thus gives rise to an index map in K-theory

$$K^0(T^*X) \to K_0(C^*(\Gamma))$$

If P is an elliptic operator on X, then P lifts to a Γ-equivariant elliptic operator $\tilde{P}$ on $\tilde{X}$. The analytic index maps the symbol $[\sigma(P)] \in K^0(T^*X)$ to the Γ-index of $\tilde{P}$. By Poincaré duality, we may conceive of this Γ-index as a map

$$K_0(X) \to K_0(C^*(\Gamma))$$

(where $K_0(X)$ is K-homology, $K_0(C^*(\Gamma))$ is K-theory).

Now let Γ be an arbitrary countable discrete group. The classifying space $B\Gamma$ for principal Γ-bundles is not generally a smooth manifold, but one can still define a generalized index map

$$\mu : K_0(B\Gamma) \to K_0(C^*(\Gamma))$$

called the assembly map. By composition with the natural map $C^*(\Gamma) \to C_r^*(\Gamma)$, we may replace the full C^*-algebra $C^*(\Gamma)$ on the right-hand side by the reduced C^*-algebra $C_r^*(\Gamma)$,

$$\mu_r : K_0(B\Gamma) \to K_0\left(C_r^*(\Gamma)\right)$$

The conjecture asserts that μ_r is an isomorphism if Γ is a countable discrete group without torsion elements [52]. If Γ has torsion, the left-hand side of the conjecture is replaced by the Γ-equivariant K-homology (with Γ-compact supports) of the classifying space $\underline{E}\Gamma$ for *proper* (instead of principal) Γ-actions. If Γ is torsion free, then $K_0^\Gamma(\underline{E}\Gamma) \cong K_0(B\Gamma)$. In general, the assembly map

$$\mu : K_j^\Gamma(\underline{E}\Gamma) \to K_j\left(C_r^*(\Gamma)\right) \qquad j = 0, 1$$

is conjectured to be an isomorphism for *any* second countable locally compact group Γ [11]. Early versions of the Baum–Connes conjecture concerned the K-theory of groupoids (the holonomy groupoid of a foliation [23] and crossed product groupoids [10]), but we now have counterexamples for the conjecture for groupoids. For groups, to this day no counterexample has been found, and the conjecture has been verified for large classes of groups. Since injectivity of the assembly map for a discrete group Γ implies the Novikov higher signature conjecture, the Baum–Connes conjecture is a development within noncommutative geometry with significant implications in algebraic topology.

10 Cohomological index formulas and local index theory

One can obtain index formulas for the index of a single operator or an elliptic family by applying the Chern character to K-theoretical index formulas. A different analytic approach has been proposed by Atiyah and Bott [4]. It is based on the following fundamental observation. Let P, as before, be an elliptic operator on a manifold X acting between sections of bundles E and F. We can then form two positive operators P^*P and PP^* and consider two ζ-functions $\mathrm{Tr}(1+P^*P)^{-s}$ and $\mathrm{Tr}(1+P^*P)^{-s}$. Both of these functions are defined and holomorphic for $\Re s$ sufficiently large and admit meromorphic continuation to the entire complex plane. Moreover, they are holomorphic at $s=0$ and the value of each ζ-function at $s=0$ can be computed by an explicit integral of a well-defined local density, see [64]. The operators P^*P and PP^* have the same sets of nonzero eigenvalues with the same multiplicities. The multiplicities of 0 as an eigenvalue on the other hand differ for two operators, and the difference of multiplicities is precisely the index of operator P. It follows that for any s with $\Re s$ sufficiently large

$$\mathrm{Tr}(1+P^*P)^{-s}-\mathrm{Tr}(1+PP^*)^{-s}=\mathrm{index}\,P$$

Writing $\zeta(s):=\mathrm{Tr}(1+P^*P)^{-s}-\mathrm{Tr}(1+PP^*)^{-s}$ we thus obtain by analytic continuation

$$\mathrm{index}\,P=\zeta(0).$$

It is useful to rewrite these considerations in the following $\mathbb{Z}_2$-graded notations. Let

$$\mathcal{H}:=L^2(X,E)\oplus L^2(X,F) \tag{10.1}$$

be a Hilbert space. On $\mathcal{H}$ one has a grading operator γ given by the matrix $\begin{bmatrix}1&0\\0&-1\end{bmatrix}$ with respect to the decomposition (10.1) and an operator $D=\begin{bmatrix}0&P^*\\P&0\end{bmatrix}$. In these terms we can write $\zeta(s)=\mathrm{Tr}\,\gamma(1+D^2)^{-s}$. McKean and Singer [55] proposed using a closely related formula

$$\text{index}\, P = \operatorname{Tr} \gamma e^{-tD^2},$$

which holds for any $t > 0$, for index computation. Since the trace of the heat kernel e^{-tD^2} admits an asymptotic expansion as $t \to 0^+$ with the coefficients being canonically defined local densities X this formula allows in principle to compute the index as an integral over X. Explicitly, the index formula for Dirac-type operators has been obtained by computing the local densities using invariant theory in [37, 2]. Later significantly simpler methods allowing direct calculations of the local index density were found in [33, 34, 17], cf. [16]. It is important to note that this approach provides not just cohomology classes appearing in a cohomological index formula, but rather canonically defined differential forms representing these classes, thus justifying the name local index formulas. This locality has been exploited to extend index theory, e.g., to manifolds with boundary [5]. The extension of local index techniques to the elliptic families case has been carried out in [18] based on the superconnection formalism proposed in [62]. Further extension of these results to the case of foliations presents an additional difficulty. Since the algebras involved are noncommutative, de Rham theory is no longer a proper receptacle for the Chern character in K-theory. As we will see, this role is played by cyclic theory. Finally, we mention that a different approach to local index theory based on the study of cyclic theory of deformation quantization algebras has been developed by R. Nest and B. Tsygan. For their approach as well as some applications we refer to the original papers [57, 58, 59, 20].

11 Cyclic complexes

In this section, we give a very brief overview of periodic cyclic homological and cohomological complexes, mostly to fix the notations. The standard reference for this material is [54].

For a complex unital algebra $\mathcal{A}$ set $C_l(\mathcal{A}) := \mathcal{A} \otimes (\mathcal{A}/(\mathbb{C} \cdot 1))^{\otimes l}$, $l \geq 0$. For a topological algebra, e.g., a normed algebra, one needs to take an appropriately completed tensor product. One defines differentials $b\colon C_l(\mathcal{A}) \to C_{l-1}(\mathcal{A})$ and $B\colon C_l(\mathcal{A}) \to C_{l+1}(\mathcal{A})$ by

$$b\,(a_0 \otimes a_1 \otimes \ldots a_l) := \sum_{i=0}^{l-1} (-1)^i a_0 \otimes \ldots a_i a_{i+1} \otimes \ldots a_l + (-1)^l a_l a_0 \otimes a_1 \otimes \ldots a_{l-1}$$

$$B(a_0 \otimes a_1 \otimes \ldots a_l) := \sum_{i=0}^{l} (-1)^{li}\, 1 \otimes a_i \otimes a_{i+1} \otimes \ldots a_{i-1} \text{ (with } a_{-1} := a_l).$$

One verifies directly that b, B are well defined and satisfy $b^2 = 0$, $B^2 = 0$, $Bb + bB = 0$. We will be primarily interested in the periodic cyclic complex, which is a totalization of a bicomplex defined as follows. Set $C_{kl}(\mathcal{A}) := C_{l-k}(\mathcal{A})$, $k, l \in \mathbb{Z}$ ($C_{kl}(\mathcal{A}) = 0$ if $k > l$) and note that b, B define maps $b\colon C_{kl}(\mathcal{A}) \to C_{k,(l-1)}(\mathcal{A})$ and $B\colon C_{kl}(\mathcal{A}) \to C_{(k-1),l}(\mathcal{A})$. $\big(C_{\bullet,\bullet}(\mathcal{A}), b, B\big)$ is thus a bicomplex. To obtain the periodic cyclic complex it is totalized as follows:

$$CC_i^{per}(\mathcal{A}) := \prod_{k+l=i} C_{kl}(\mathcal{A})$$

with the differential $b + B\colon CC_i^{per}(\mathcal{A}) \to CC_{i-1}^{per}(\mathcal{A})$. Note $CC_i^{per}(\mathcal{A}) = CC_{i+2m}^{per}(\mathcal{A})$ for every $m \in \mathbb{Z}$, and therefore the complex is (indeed) 2-periodic. Its homology is denoted by $HC_0^{per}(\mathcal{A})(\cong HC_{2m}^{per}(\mathcal{A}))$ and $HC_1^{per}(\mathcal{A})(\cong HC_{1+2m}^{per}(\mathcal{A}))$, $m \in \mathbb{Z}$. We will write a chain in $CC_i^{per}(\mathcal{A})$, $i = 0$ or 1 as $\alpha = \sum_{m=0}^{\infty} \alpha_{i+2m}$, where $\alpha_k \in C_k(\mathcal{A})$.

Two important examples of classes in cyclic homology are the following. Let $e \in M_n(\mathcal{A}) = \mathcal{A} \otimes M_n(\mathbb{C})$ be a idempotent. Then

$$\mathrm{Ch}(e) := \mathrm{tr}\, e + \sum_{l=1}^{\infty} (-1)^l \frac{(2l)!}{l!} \,\mathrm{tr}\left(e - \frac{1}{2}\right) \otimes e^{\otimes 2l} \in \prod_{l \geq 0} C_{-l,l}(\mathcal{A}) = CC_0^{per}(\mathcal{A})$$

where $\mathrm{tr}\colon (\mathcal{A} \otimes M_n(\mathbb{C}))^{\otimes k} \to \mathcal{A}^{\otimes k}$ is the map given by

$$\mathrm{tr}(a_0 \otimes m_0) \otimes (a_1 \otimes m_1) \otimes \ldots (a_k \otimes m_k) = (\mathrm{tr}\, m_0 m_1 \ldots m_k) a_0 \otimes a_1 \otimes \ldots a_k.$$

The class of $\mathrm{Ch}(e)$ in $HC_0^{per}(\mathcal{A})$ depends only on the class of e in $K_0(\mathcal{A})$ and thus defines the Chern character morphism

$$\mathrm{Ch}\colon K_0(\mathcal{A}) \to HC_0^{per}(\mathcal{A})$$

Similarly, if $u \in M_n(\mathcal{A})$ is invertible on can define

$$\mathrm{Ch}(u) := \frac{1}{\sqrt{2\pi i}} \sum_{l=0}^{\infty} (-1)^l l! \,\mathrm{tr}\left(u^{-1} \otimes u\right)^{\otimes(l+1)} \in CC_1^{per}(\mathcal{A})$$

This defines a homomorphism

$$\mathrm{Ch}\colon K_1(\mathcal{A}) \to HC_1^{per}(\mathcal{A}).$$

from topological K-theory to periodic cyclic cohomology in the odd case. Note that one can define Chern characters for K_0 and K_1 by the same formulas in the case of algebraic K-theory as well, but in the case of K_1 the target of the Chern character should be a different flavor of cyclic theory, the negative cyclic homology.

Example 11 Let $\mathcal{A} = C^\infty(X)$ where X is a compact manifold. We have a map $\lambda\colon C_k(\mathcal{A}) \to \Omega^k(X)$ given by

$$\lambda(a_0 \otimes a_1 \otimes \ldots a_k) = \frac{1}{k!} a_0 da_1 \ldots da_k$$

which intertwines the differential B with de Rham differential d and b with 0. It follows that if we consider a bicomplex $\mathcal{D}_{kl} := \Omega^{l-k}(X)$ with the differentials given by d and 0, the map λ will induce a morphism of bicomplexes $C_{kl}(\mathcal{A})$ and $\mathcal{D}_{kl}$. It can be shown that this map is a quasi-isomorphism [27] and hence $HC_i^{per}(C^\infty(X)) \cong \oplus_{m\in\mathbb{Z}} H^{i+2m}(X)$. The map $K^i(X) \cong K_i(C^\infty(X)) \xrightarrow{\mathrm{Ch}} HC_i^{per}(C^\infty(X)) \xrightarrow{\lambda} \oplus_{m\in\mathbb{Z}} H^{i+2m}(X)$ recovers the ordinary Chern character in K-theory (up to normalization).

Dually, one considers cyclic cohomology. Set $C^k(\mathcal{A}) := (C_k(\mathcal{A}))'$—the space of continuous multilinear functionals $\phi(a_0, a_1, \ldots, a_k)$ on $\mathcal{A}$ with the property that if for some $i \geq 1$ $a_i = 1$, then $\phi(a_0, a_1, \ldots, a_k) = 0$. The transpose of the differentials b, B induce maps, also denoted b, B: $b\colon C^k(\mathcal{A}) \to C^{k+1}(\mathcal{A})$, $B\colon C^k(\mathcal{A}) \to C^{k-1}(\mathcal{A})$. One then forms a bicomplex $C^{kl}(\mathcal{A}) := C^{l-k}(\mathcal{A}), k, l \in \mathbb{Z}$ with the differentials $b\colon C^{kl}(\mathcal{A}) \to C^{k,(l+1)}(\mathcal{A})$ and $B\colon C^{kl}(\mathcal{A}) \to C^{(k+1),l}(\mathcal{A})$. Again dually to the homological case it is totalized using the direct sums rather than products:

$$CC^i_{per}(\mathcal{A}) := \bigoplus_{k+l=i} C^{kl}(\mathcal{A}).$$

Here again the complex we consider is 2-periodic. An even (resp. odd) cyclic cochain is thus given by a collection of multilinear functionals $\phi_k = \phi_k(a_0, \ldots, a_k) \in C^k(\mathcal{A})$, $k = 0, 2, \ldots$ (resp. $k = 1, 3, \ldots$), only finitely many of which are nonzero. A cochain is a cocycle if it satisfies $b\phi_k + B\phi_{k+2} = 0$. We denote the cohomology of the periodic cyclic complex $CC^\bullet_{per}(\mathcal{A})$ by $HC^\bullet_{per}(\mathcal{A})$; these again take two distinct values $HC^0_{per}(\mathcal{A})$ and $HC^1_{per}(\mathcal{A})$.

Note that it is important that the cyclic cohomological bicomplex is totalized by using direct sums and not products: if we remove the requirement that only finitely many of the ϕ_k are nonzero we obtain a complex with vanishing cohomology. One can however obtain a nontrivial theory with infinite cochains as follows. Assume that $\mathcal{A}$ is a Banach (or normed) algebra. Denote by $CC^i_{entire}(\mathcal{A})$ the space of cochains $\phi_k \in C^k(\mathcal{A})$, k same parity as i, satisfying

$$\sum \sqrt{k!}\|\phi_k\| r^k < \infty \text{ for every } r \geq 0.$$

The space of such cochains is preserved by the differential $b + B$ and we thus obtain a complex $CC^\bullet_{entire}(\mathcal{A})$ with the cohomology denoted by $HC^\bullet_{entire}(\mathcal{A})$.

$CC^{\bullet}_{per}(\mathcal{A})$ is clearly a subcomplex of $CC^{\bullet}_{entire}(\mathcal{A})$, and we therefore have a morphism $HC^{\bullet}_{per}(\mathcal{A}) \to HC^{\bullet}_{entire}(\mathcal{A})$ induced by the inclusion map.

12 The longitudinal index formula in cyclic theory

Let (Z, F) be a foliated manifold, and D a longitudinally elliptic operator on Z. Choose a (possibly disconnected) transversal T to the foliation. By restricting the foliation groupoid to T, i.e., considering only paths which start and end in T we obtain an etale groupoid G_T, Morita equivalent to the holonomy groupoid $G_{Z/F}$. One can consider the convolution algebra $C_c^\infty(G_T)$ of smooth compactly supported functions on G_T and define the index of a leafwise family D as an element

$$\text{index}\, D \in K_0\left(C_c^\infty(G_T) \otimes \mathcal{R}\right)$$

where $\mathcal{R}$ is the algebra of rapidly decaying infinite matrices. We can apply a Chern character to obtain a class in cyclic homology $\text{Ch}(\text{index}\, D) \in HC_0^{per}(C_c^\infty(G_T) \otimes \mathcal{R}) \cong HC_0^{per}(C_c^\infty(G_T))$. To obtain numerical information one has to pair this class with cyclic cohomology. The cyclic cohomology of etale groupoid convolution algebras has been completely described in [21, 29], based on earlier computations for group algebras and cross-products [22, 31, 60, 35]. Earlier Connes constructed a canonical imbedding $\Phi\colon H^\bullet(BG_T, \tau) \to HC_{per}^{\bullet - \dim T}(C_c^\infty(G_T))$. Here BG_T is the classifying space of the groupoid G_T and τ is the local system on BG_T induced by the orientation bundle of T. We can now state Connes' index formula for a longitudinally elliptic family. Assume for simplicity that D_V is a family of longitudinal Dirac operators with coefficients in a auxiliary bundle V. Then we have [28]

$$\langle \Phi(c), \text{Ch}(\text{index}\, D_V)\rangle = \int_Z \widehat{A}(F)\,\text{Ch}(V)\nu^*(c)$$

Here $\nu\colon Z \to BG_T$ is the classifying map of G_T and $\widehat{A}(F)$ is the $\widehat{A}$ genus of the bundle F.

We note that unlike the case of, for example, elliptic families, this result cannot in general be deduced from the K-theoretical Connes–Skandalis index theorem. This is due to the fact that while index D_V can be defined in $K_0(C_c^\infty(G_T) \otimes \mathcal{R})$, only its image in the K-theory of the C^*-completion of $C_c^\infty(G_T)$ is computed by the Connes–Skandalis index theorem. Cyclic homology and the Chern character on the other hand are nontrivial only for the smaller algebra $C_c^\infty(G_T) \otimes \mathcal{R}$.

The local index theory approach to the cohomological index formula for foliations and etale groupoids has been developed in [40, 41, 39].

13 Finitely summable Fredholm modules

Definition 12 An odd Fredholm module $(\mathcal{H}, F)$ over an algebra $\mathcal{A}$ is given by the following data:

A Hilbert space $\mathcal{H}$ and a representation on it of an algebra $\mathcal{A}$, i.e., a homomorphism $\pi : \mathcal{A} \to \mathrm{End}(\mathcal{H})$.

An operator F, such that

$$\pi(a)(1 - F^2) \in \mathcal{K} \text{ for every } a \in \mathcal{A} \tag{13.1}$$

$$\pi(a)[F, \pi(b)] \in \mathcal{K} \text{ for every } a, b \in \mathcal{A} \tag{13.2}$$

The set of (equivalence classes) of odd Fredholm modules with the appropriate equivalence relation (cf. [19]) and an operation of direct sum becomes the K-homology group $K^1(\mathcal{A})$. For a unital algebra $\mathcal{A}$ (but not necessarily unital representation) one can replace the Fredholm module by an equivalent one satisfying

$$1 - F^2 \in \mathcal{K} \text{ for every } a \in \mathcal{A} \tag{13.3}$$

$$[F, \pi(a)] \in \mathcal{K} \text{ for every } a \in \mathcal{A} \tag{13.4}$$

Definition 13 An even Fredholm module $(\mathcal{H}, F, \gamma)$ over an algebra $\mathcal{A}$ is given by the following data:

An odd Fredholm module $(\mathcal{H}, F)$ over an algebra $\mathcal{A}$

A $\mathbb{Z}_2$ grading on the Hilbert space $\mathcal{H}$, i.e., an operator γ satisfying $\gamma^2 = 1$. This data has to satisfy the following conditions: the operator F is odd with respect to this grading, i.e.,

$$F\gamma + \gamma F = 0$$

and elements of the algebra $\mathcal{A}$ are even, i.e.,

$$\pi(a)\gamma = \gamma\pi(a) \text{ for every } a \in \mathcal{A}.$$

As in the odd case, equivalence classes of even Fredholm modules form the group $K^1(\mathcal{A})$. Let $p \geq 1$.

Definition 14 An (odd or even) or Fredholm module is p-summable if the following stronger conditions hold:

$$\pi(a)(1 - F^2) \in \mathcal{L}^p \text{ for every } a \in \mathcal{A} \tag{13.5}$$

$$\pi(a)[F, \pi(b)] \in \mathcal{L}^p \text{ for every } a, b \in \mathcal{A} \tag{13.6}$$

where $\mathcal{L}^p$ is the Schatten-ideal of operators T with $\mathrm{Tr}(|T|^p) < \infty$. In [27] Connes shows that with every Fredholm module (called pre-Fredholm module in [27]) one can canonically, by changing the Hilbert space and representation, associate a different Fredholm module, satisfying

$$F^2 = 1 \tag{13.7}$$

representing the same K-homology class. If the original Fredholm module was p-summable, the new one will also be p-summable.

With the Fredholm module satisfying (13.7) Connes associates a cyclic cocycle called the character of the Fredholm module. Choose $n > p-1$ and the same parity as the Fredholm module. Then, viewed in the periodic cyclic bicomplex, Connes' character has only one component of degree n, τ_n. defined by the equations

for an even Fredholm module

$$\tau_n(F)(a_0, a_1, \ldots, a_n) = \frac{(n/2)!}{n!}\operatorname{Tr}'\left(\gamma\pi(a_0)[F,\pi(a_1)]\ldots[F,\pi(a_n)]\right) \tag{13.8}$$

and for an odd Fredholm module

$$\tau_n(F)(a_0, a_1, \ldots, a_n) = \sqrt{2i}\frac{\Gamma(n/2+1)}{n!}\operatorname{Tr}'\left(\pi(a_0)[F,\pi(a_1)]\ldots[F,\pi(a_n)]\right) \tag{13.9}$$

Here we use the notation $\operatorname{Tr}'(T) = 1/2\operatorname{Tr}(F(FT+TF))$. The cohomology class of the cocycle τ_n in periodic cyclic cohomology does not depend on the choice of n (of appropriate parity).

A fundamental property of the character of a Fredholm module is the following theorem of Connes [27]. Let e be an idempotent in $M_N(\mathcal{A})$, and $(\mathcal{H}, F, \gamma)$ be an even Fredholm module over $\mathcal{A}$. Construct a Fredholm operator

$$F_e = \pi(e)(F \otimes 1)\pi(e) \colon \mathcal{H}^+ \otimes \mathbb{C}^N \to \mathcal{H}^- \otimes \mathbb{C}^N$$

(where $\mathcal{H}^+$ and $\mathcal{H}^-$ denote the ± 1 eigenspaces of γ). Then

Theorem 15 *Let $(\mathcal{H}, F, \gamma)$ be an even p-summable Fredholm module satisfying $F^2 = 1$, $n > p-1$. Then*

$$\operatorname{index}(F_e) = \langle \tau_n(F), \operatorname{Ch}(e)\rangle \tag{13.10}$$

Here $\operatorname{Ch}(e)$ is the Chern character in cyclic homology.

Similarly in the odd case pairing of the character of a Fredholm module with a class in odd K-theory computes the spectral flow. More precisely, let $(\mathcal{H}, F)$ be an odd Fredholm module and u an invertible element of $M_N(\mathcal{A})$. Equivalently, if P is the spectral projection on the positive spectrum of F (i.e., $P = 1/2(1+F)$ if $F^2 = 1$) consider the Fredholm operator

$$T_u := (P \otimes 1)\pi(u)(P \otimes 1) \in \operatorname{End}\left(P\mathcal{H} \otimes \mathbb{C}^N\right).$$

Then

Theorem 16 *Let $(\mathcal{H}, F)$ be an odd p-summable Fredholm module satisfying $F^2 = 1$, $n > p - 1$. Then*

$$\operatorname{index} T_u = \langle \tau_n(F), \operatorname{Ch}(u) \rangle$$

When π is unital one can show that index T_u coincides with the spectral flow between two self-adjoint operators $F \otimes 1$ and $\pi(u)((F \otimes 1)\pi(u)^{-1}$ acting on the space $\mathcal{H} \otimes \mathbb{C}^N$.

In what follows we fix the representation of $\mathcal{A}$ on $\mathcal{H}$ and will therefore drop it from the notation. Moreover, we will assume that the algebra $\mathcal{A}$ is unital and its representation is unital as well.

Assume now that we have an even or odd Fredholm module satisfying

$$(1 - F^2) \in \mathcal{L}^{\frac{p}{2}} \tag{13.11}$$

$$[F, a] \in \mathcal{L}^p \text{ for } a \in \mathcal{A}. \tag{13.12}$$

We remark that for any p summable Fredholm module one can achieve these summability conditions by perturbing the operator F and keeping all the other data intact. Then one can directly construct cyclic cocycles on $\mathcal{A}$ representing the class of Connes' character of a Fredholm module [38].

In the even case: choose even $n > p$. Define a periodic cyclic cocycle with components $\operatorname{Ch}_n^k(F)$, $k = 0, 2, \ldots n$ by

$$\begin{aligned} &\operatorname{Ch}_n^k(F)(a_0, a_1, \ldots a_k) \\ &= \frac{(\frac{n}{2})!}{(\frac{n+k}{2})!} \sum_{i_0+i_1+\cdots+i_k=\frac{n-k}{2}} \operatorname{Tr} \gamma a_0 (1 - F^2)^{i_0} [F, a_1] \\ &\times (1 - F^2)^{i_1} \ldots [F, a_k](1 - F^2)^{i_k} \end{aligned} \tag{13.13}$$

Similarly in the odd case choose odd $n > p$. Define a periodic cyclic cocycle with components $\operatorname{Ch}_n^k(F)$, $k = 1, 3, \ldots n$ by

$$\begin{aligned} &\operatorname{Ch}_n^k(F)(a_0, a_1, \ldots, a_k) \\ &= \frac{\Gamma\left(\frac{n}{2} + 1\right)\sqrt{2i}}{\left(\frac{n+k}{2}\right)!} \sum_{i_0+i_1+\cdots+i_k=\frac{n-k}{2}} \operatorname{Tr} a_0 (1 - F^2)^{i_0} [F, a_1] \\ &\times (1 - F^2)^{i_1} \ldots [F, a_k](1 - F^2)^{i_k} \end{aligned} \tag{13.14}$$

One can slightly improve the summability assumptions and require only $n > p - 1$ (rather than $n > p$) by replacing Tr with Tr$'$ where now $\operatorname{Tr}'(T) = 1/2 \operatorname{Tr}(F(FT + TF)) + \operatorname{Tr}(1 - F^2)T$.

These cocycles represent the class of the character of a Fredholm module in the following sense. With every even p-summable Fredholm module $(\mathcal{H}, F, \gamma)$ one can associate by Connes' construction a Fredholm module $(\mathcal{H}', F', \gamma')$ satisfying $(F')^2 = 1$.

Theorem 17 *Let $(\mathcal{H}, F, \gamma)$ be a p-summable Fredholm module satisfying condition* (13.11). *Then*

$$[\mathrm{Ch}_n(F)] = \left[\tau_n(F')\right], \; n > p - 1.$$

An analogous result holds in the odd case.

14 Unbounded picture

Definition 18 A p-summable *spectral triple* (or unbounded Fredholm module) $(\mathcal{A}, \mathcal{H}, D)$ consists of a unital algebra $\mathcal{A}$ represented on a Hilbert space $\mathcal{H}$ and a self-adjoint operator D such that

$$\text{For any } a \in \mathcal{A}, a(\mathrm{Dom}\, D) \subset \mathrm{Dom}\, D \text{ and } [D, a] \text{ is bounded, and} \tag{14.1}$$

$$(D^2 + 1)^{-1} \in \mathcal{L}^{p/2} \tag{14.2}$$

An *even* p-summable spectral triple is a spectral triple $(\mathcal{A}, \mathcal{H}, D)$ with the following additional data: $\mathcal{H}$ is equipped with a $\mathbb{Z}_2$ grading given by an operator γ satisfying $\gamma^2 = 1$. We denote by $\mathcal{H}^{\pm}$ the ± 1 eigenspaces of γ. The operator D is odd with respect to this grading, i.e., $\gamma(\mathrm{Dom}\, D) \subset \mathrm{Dom}\, D$ and $D\gamma + \gamma D = 0$. We will assume for simplicity that $\mathcal{A}$ is not graded and represented by even operators.

Otherwise a spectral triple is *odd*.

Example 19 Let M be a compact $spin^c$ manifold and E the associated spinor bundle. With it one can associate a spectral triple as follows: $\mathcal{H} = L^2(E)$ is the Hilbert space of L^2-sections of E. $\mathcal{A} = C^\infty(M)$ acts on $\mathcal{H}$ by multiplication. Finally D is the Dirac operator on E. This spectral triple will be p-summable for every $p > dim M$. If the dimension of M is even, E has a natural grading which anticommutes with D. Thus we obtain an even finitely summable spectral triple for an even-dimensional M and odd for an odd-dimensional one.

If $(\mathcal{A}, \mathcal{H}, D)$ is a p-summable spectral triple, one can form an associated Fredholm module $(\mathcal{A}, \mathcal{H}, F)$ where $F = D(D^2 + 1)^{-1/2}$. It is easy to see that $(\mathcal{A}, \mathcal{H}, F)$ is p-summable and, moreover, satisfies condition (13.11). One can therefore define its character using (13.13) and (13.14). An alternative approach to the character in this case is given by the JLO formula [49, 36]. It is applicable in a more general context of θ-summable Fredholm modules, i.e., triples $(\mathcal{A}, \mathcal{H}, D)$ satisfying Equation (14.1) and with finite summability condition (14.2) replaced by

$$e^{-\theta D^2} \in \mathcal{L}^1 \text{ for any } \theta > 0. \tag{14.3}$$

Note that condition (14.2) implies (14.3) since $e^{-\theta D^2} = (1 + D^2)^{-p/2} \left((1+D^2)^{p/2} e^{-\theta D^2}\right)$ and the operator $(1 + D^2)^{p/2} e^{-\theta D^2}$ is bounded (by the spectral theorem).

The JLO (Jaffe–Lesniewski–Osterwalder, [49]) formula associates with every even θ-summable Fredholm module an infinite cochain in the (b, B) bicomplex of $\mathcal{A}$ given for $k = 0, 2, \ldots$ by

$$\begin{aligned} &\mathrm{Ch}^k(D)(a_0, a_1, \ldots a_k) \\ &\quad = \int_{\Delta^k} \mathrm{Tr}\, \gamma a_0 e^{-t_0 D^2} [D, a_1] e^{-t_1 D^2} \ldots [D, a_k] e^{-t_k D^2} dt_1 dt_2 \ldots dt_k \end{aligned} \tag{14.4}$$

where $t_i \geq 0, i = 0, \ldots, k, \sum t_i = 1$ are the barycentric coordinates on the simplex Δ^k.

The formula in the odd case is similar:

$$\begin{aligned} &\mathrm{Ch}^k(D)(a_0, a_1, \ldots a_k) \\ &\quad = \sqrt{2i} \int_{\Delta^k} \mathrm{Tr}\, a_0 e^{-t_0 D^2} [D, a_1] e^{-t_1 D^2} \ldots [D, a_k] e^{-t_k D^2} dt_1 dt_2 \ldots dt_k \end{aligned} \tag{14.5}$$

$k = 1, 3, \ldots$.

Theorem 20 ([49])

(1) *The cochain* $\mathrm{Ch}_\bullet(D)$ *is an entire cyclic cocycle.*
(2) *The cocycles* $\mathrm{Ch}_\bullet(D)$ *and* $\mathrm{Ch}_\bullet(\epsilon D)$ *are canonically cohomologous.*

The second statement follows from the following transgression formula:

$$-\frac{d}{d\epsilon} \mathrm{Ch}^k(D) = b\, \mathrm{Ch}^{k-1}(\epsilon D, D) + B\, \mathrm{Ch}^{k+1}(\epsilon D, D)$$

where, in the even case $\mathrm{Ch}(\epsilon D, D)$ is an odd cochain given by (k-odd)

$$\begin{aligned} \mathrm{Ch}^k(\epsilon D, D)(a_0, a_1, \ldots a_k) = \sum_{l=0}^{k} (-1)^l \int_{\Delta^{k+1}} \mathrm{Tr}\, \gamma a_0 e^{-t_0 D^2} [D, a_1] \\ e^{-t_1 D^2} \ldots e^{-t_l D^2} D e^{-t_{l+1} D^2} \ldots [D, a_k] e^{-t_{k+1} D^2} dt_1 dt_2 \ldots dt_{k+1}. \end{aligned} \tag{14.6}$$

In the odd case $\mathrm{Ch}(\epsilon D, D)$ is an even cochain given by (k-even)

$$\mathrm{Ch}^k(\epsilon D, D)(a_0, a_1, \ldots a_k) = \sqrt{2i}\sum_{l=0}^{k}(-1)^l \int_{\Delta^{k+1}} \mathrm{Tr}\, a_0 e^{-t_0 D^2}[D, a_1]$$

$$e^{-t_1 D^2} \ldots e^{-t_l D^2} D e^{-t_{l+1} D^2} \ldots [D, a_k] e^{-t_{k+1} D^2} dt_1 dt_2 \ldots dt_{k+1}. \qquad (14.7)$$

Finally, the following result of [24] shows that the JLO cocycle indeed computes Connes' character in the following sense:

Theorem 21 ([24]) *If* $(\mathcal{A}, \mathcal{H}, D)$ *is p-summable and* $F = D(1 + D^2)^{-1/2}$ *then the image of* $[\mathrm{Ch}_\bullet(F)]$ *in* $HC^\bullet_{entire}(\mathcal{A})$ *is* $[\mathrm{Ch}_\bullet(D)]$.

15 Locality and spectral invariants

It is well known that indices of elliptic operators, e.g., the Dirac operator from Example 19, can be computed as integrals of well-defined local densities. All of the formulas previously described above apply in the case of Example 19 and in conjunction with Equation (13.10) provide formulas for the index of the Dirac operator twisted with a vector bundle. A common feature of these formulas however is that they express the index in terms of traces of operators and thus are not local. One can however obtain a local expression from these formulas, e.g., by replacing D by ϵD in the JLO formula, $\epsilon > 0$ and considering the limit when $\epsilon \to 0^+$. One therefore is naturally led [28] to the question of obtaining a local formula for Connes' Chern character for spectral triples. Invariants of noncommutative geometry naturally have a spectral nature. A prototypical example of a local spectral invariant appearing in geometry is the noncommutative residue introduced in [66], cf. also [43]. We recall the definition.

Let X be a compact manifold. Denote by $\Psi(X)$ the algebra of classical pseudodifferential operators of integral orders, $\Psi^k(X)$ denotes the space of operator of order k. Choose a positive pseudodifferential operator R of order 1. For $A \in \Psi(X)$ consider the function

$$\zeta_A(s) := \mathrm{Tr}\, A R^{-s}.$$

$\zeta(s)$ is defined for $\Re(s) > \dim X + \operatorname{ord} A$ and admits a meromorphic extension to the entire complex plane [64]. It has at most a simple pole at $s = 0$.

Introduce

$$\mathrm{Res}\, A := \mathrm{Res}_{s=0}\, \zeta(s)$$

It has the following properties:

(1) $\mathrm{Res}\, A$ does not depend on the choice of $R \in \Psi^1(X)$,
(2) Res is a trace on the algebra $\Psi(X)$:

$$\operatorname{Res} AB = \operatorname{Res} BA \text{ for } A, B \in \Psi(X),$$

(3) $\operatorname{Res} A = \operatorname{Res} B$ whenever $A - B \in \Psi^{-\dim X - 1}(X)$.

The last property is a manifestation of the locality of Res: it depends only on a (part of) the complete symbol of an operator.

Explicitly in local coordinates noncommutative residue can be described as follows. Let $U \subset X$ be a coordinate neighborhood, and (x, ξ)—the standard coordinates on T^*U. Let

$$a(x,\xi) \sim \sum_{k=-\infty}^{m} a_k(x,\xi)$$

be the asymptotic expansion of the complete symbol $a(x,\xi)$ of the operator $A \in \Psi^m(M)$, $a_k(x,\xi)$ homogeneous of order k on in ξ. Then the expression

$$\left(\int_{|\xi|=1} a_{-n}(x,\xi) \operatorname{vol}_S\right) |dx|,$$

where vol_S denotes the normalized volume form on the unit sphere $|\xi| = 1$, defines a density on U independent of the choice of local coordinates.

$$\operatorname{Res}(A) = \int_X \left(\int_{|\xi|=1} a_{-n}(x,\xi) \operatorname{vol}_S\right) |dx|.$$

16 Pseudodifferential calculus for spectral triples

Motivated by connections between the noncommutative residue and the Dixmier trace, Connes and Moscovici extended it to more general spectral triples. As a first step they construct an analog of the pseudodifferential calculus for spectral triples.

Let $P_{\operatorname{Ker}(D^2)}$ be orthogonal projection onto $\operatorname{Ker}(D^2)$. Define

$$|D| := \sqrt{D^2} + P_{\operatorname{Ker}(D^2)}. \tag{16.1}$$

If D is invertible, then $|D|$ has the usual meaning, but for us $|D|$ is always a strictly positive operator.

For nonnegative s, put $\mathcal{H}^s = \operatorname{Dom}(|D|^s)$, with the inner product

$$\langle v_1, v_2\rangle_{\mathcal{H}^s} = \langle |D|^s v_1, |D|^s v_2\rangle_{\mathcal{H}}. \tag{16.2}$$

For $s < 0$, put $\mathcal{H}^s = \left(\mathcal{H}^{-s}\right)^*$. Put $\mathcal{H}^\infty = \bigcap_{s\geq 0} \mathcal{H}^s$, a dense subspace of $\mathcal{H}$.

Following [25, Appendix B], we can consider operators acting in a controlled way in this scale.

Let op^k be the set of closed operators F such that

(1) $\mathcal{H}^\infty \subset \mathrm{Dom}(F)$,
(2) $F(\mathcal{H}^\infty) \subset \mathcal{H}^\infty$, and
(3) For all s, the operator $F\colon \mathcal{H}^\infty \to \mathcal{H}^\infty$ extends to a bounded operator from $\mathcal{H}^s$ to $\mathcal{H}^{s-k}$.

Introduce now a derivation δ on $\mathcal{B}(\mathcal{H})$ by

$$\delta(T) := [|D|, T]$$

with the domain $\mathrm{Dom}\,\delta$ consisting of $T \in \mathcal{B}(\mathcal{H})$ such that $T(\mathrm{Dom}\,D) \subset \mathrm{Dom}\,D$ and $[|D|, T]$ extends to a bounded operator. Define $OP^0 := \cap\{\mathrm{Dom}\,\delta^n \mid n \in \mathbb{N}\}$. It is shown in [25, Appendix B] that

$$OP^0 \subset op^0.$$

Define now $OP^\alpha, \alpha \in \mathbb{R}$ as the set of closed operators P for which $|D|^{-\alpha} P \in OP^0$. Note that $OP^\alpha \subset op^\alpha$. Connes and Moscovici show that with

$$\nabla(T) := [D^2, T]$$

$\nabla(OP^\alpha) \subset OP^{\alpha+1}$. On the other hand $\delta(OP^\alpha) \subset OP^\alpha$.

Carefully estimating the remainder one can prove that for $T \in OP^\alpha$ there is an asymptotic expansion

$$|D|^{2z} \cdot T \cdot |D|^{-2z} \simeq \sum_{k=0}^{\infty} \frac{z(z-1)\dots(z-k)}{k!} \nabla^k(T)|D|^{-2k} \tag{16.3}$$

The precise meaning of the asymptotic expansion is that for every $N \in \mathbb{N}$

$$|D|^{2z} \cdot T \cdot |D|^{-2z} - \sum_{k=0}^{N} \frac{z(z-1)\dots(z-k+1)}{k!} \nabla^k(T)|D|^{-2k} \in OP^{\alpha-N-1}$$

(note that $\nabla^k(T)|D|^{-2k} \in OP^{\alpha-k}$). For an invertible D we have $\nabla(T)|D|^{-2} = 2\delta(T)|D|^{-1} + \delta^2(T)|D|^{-2}$; for not necessarily invertible D $\nabla(T)|D|^{-2} - 2\delta(T)|D|^{-1} - \delta^2(T)|D|^{-2}$ is a finite rank operator and hence is in $OP^{-\infty}$. Using this we can rewrite the asymptotic expansion as

$$|D|^{z} \cdot T \cdot |D|^{-z} \simeq \sum_{k=0}^{\infty} \frac{z(z-1)\dots(z-k)}{k!} \delta^k(T)|D|^{-k} \tag{16.4}$$

It will also be convenient to consider a derivation

$$L(T) := [\log |D|^2, T] = \frac{d}{dz}|_{z=0}|D|^{2z} \cdot T \cdot |D|^{-2z}$$

It follows from (16.3), (16.4) that for $T \in OP^\alpha$ $L(T) \in OP^{\alpha-1}$ and there are asymptotic expansions (in the above sense)

$$L(T) \simeq \sum_{k=1}^{\infty} \frac{(-1)^{k-1}}{k} \nabla^k(T)|D|^{-2k}, \; L(T) \simeq 2\sum_{k=1}^{\infty} \frac{(-1)^{k-1}}{k} \delta^k(T)|D|^{-k}.$$

Finally, for $T \in OP^\alpha$ we have the following equality:

$$|D|^{2z} \cdot T \cdot |D|^{-2z} = \sum_{k=0}^{\infty} \frac{z^k}{k!} L^k(T)$$

where convergence of the series on the right is in each of the norms $T \mapsto \|\delta^n(|D|^{-\alpha}T)\|, n = 0, 1, 2 \ldots$.

Impose from now on the following smoothness assumption on the spectral triple: for every $a \in \mathcal{A}$ we have $a \in OP^0$, $[D, a] \in OP^0$. Let $\mathcal{B}$ be the algebra generated by $\delta^n(a)$, $\delta^n([D, a])$, $a \in \mathcal{A}$, $n \geq 0$. Clearly $\mathcal{B} \subset OP^0$. One defines the pseudodifferential operators of order α by

$$\Psi^\alpha(\mathcal{A}) = \left\{ P \in OP^{\Re\alpha} \mid P \simeq \sum_{k=0}^{\infty} b_k |D|^{\alpha-k}, \; b_k \in \mathcal{B} \right\}.$$

It follows from the asymptotic expansion (16.4) that $\Psi^\alpha \cdot \Psi^\beta \subset \Psi^{\alpha+\beta}$. In particular we can consider the algebra

$$\Psi(\mathcal{A}) := \bigcup_{k \in \mathbb{Z}} \Psi^k(\mathcal{A}).$$

17 Dimension Spectrum

From now on we assume that the spectral triple $(\mathcal{A}, \mathcal{H}, D)$ is p-summable. Then $OP^\alpha \subset \mathcal{L}^1(\mathcal{H})$ f for $\alpha \leq -p$. It follows that for every $b \in \mathcal{B}$ the function

$$\zeta_b(s) := \operatorname{Tr} b|D|^{-s}$$

is defined and holomorphic for $\Re s > p$.

Definition 22 A spectral triple $(\mathcal{A}, \mathcal{H}, D)$ has a discrete dimensional spectrum $Sd \subset \mathbb{C}$ if Sd is a discrete set and $\zeta_b(s)$ extends holomorphically to $\mathbb{C} \setminus Sd$ for every $b \in \mathcal{B}$.

We will assume at the beginning that each $\zeta_b(s)$ actually extends meromophically to $\mathbb{C}$ with poles only in Sd and the order of every pole is at most $q \in \mathbb{N}$, q independent of the pole and b. It is immediate from the definition that for every $P \in \Psi^\alpha(\mathcal{A})$ the function $\zeta_P(s) := \operatorname{Tr} P|D|^{-s}$ is holomorphic for $\Re s$ sufficiently large and extends meromophically to $\mathbb{C}$ with possible poles in the set $\bigcup \{Sd + \alpha - k\}$, where k runs through nonnegative integers.

We now define a collection of linear functionals τ_i on $\Psi^\alpha(\mathcal{A})$ for every α, $-1 \le i \le q-1$ by the following equality:

$$\zeta_P(2z) = \tau_{q-1}(P)z^{-q} + \tau_{q-2}(P)z^{-q+1} + \ldots \tau_0(P)z^{-1} + \tau_{-1}(P) + O(|z|) \text{ near } z = 0$$

Note that for $\Re z$ sufficiently large

$$\begin{aligned} \operatorname{Tr}\left(P_1 P_2 |D|^{-2z}\right) &= \operatorname{Tr}\left(P_2\left(|D|^{-2z} P_1 |D|^{2z}\right)|D|^{-2z}\right) \\ &= \sum_{k=0}^{\infty} \frac{(-z)^k}{k!} \operatorname{Tr}\left(P_2 L^k(P_1)|D|^{-2z}\right) \end{aligned}$$

By meromorphic continuation this equality holds for every $z \in \mathbb{C}$, except for a discrete set. Comparing the Laurent series at $z = 0$ of the right- and left-hand sides we conclude that

$$\tau_i(P_1 P_2) = \tau_i(P_2 P_1) + \sum_{k=1}^{q-1-k} \frac{(-1)^k}{k!} \tau_{i+k}$$

In particular, τ_{q-1} is a trace on $\bigcup_\alpha \Psi^\alpha(\mathcal{A})$. It is important to note that τ_i, $i \ge 0$ are *local* in the following sense:

$$\tau_i(P) = 0 \text{ if } P \in \mathcal{L}^1(\mathcal{H}), i \ge 0.$$

The family τ_i thus generalizes the noncommutative residue to the spectral triple framework. τ_{-1} on the other hand does not have this property:

$$\tau_{-1}(P) = \operatorname{Tr} P \text{ if } P \in \mathcal{L}^1(\mathcal{H}).$$

If $\alpha \notin \{k - Sd\}$, $k \ge 0$, then τ_{-1} defines a trace on Ψ^α—a generalization of the Kontsevich–Vishik trace.

18 The local index formula in the odd case

We are now ready to sketch the derivation of the Connes–Moscovici local index formula in noncommutative geometry. We start with the odd case, as it is in this case that one can obtain a formula that is fully local. As in the geometric situation of the spectral triple defined by a Dirac operator, we obtain the formula by studying the behavior of the JLO formula under rescaling $D \to \epsilon D$ as $\epsilon \to 0^+$.

The starting point is the following result of [24]. First we recall the notion of the finite part of a function. Assume that a function $g(\epsilon)$, $\epsilon \in (0, T]$, can be written as

$$g(\epsilon) = \sum_{i,j\geq 0} \alpha_{i,j}(\log \epsilon)^j \epsilon^{-\lambda_i} + \sum_{k\geq 1} \beta_k (\log \epsilon)^k + \psi(\epsilon) \tag{18.1}$$

where the sum is finite, $\Re\lambda_i \leq 0$, $\lambda_i \neq 0$ and $\psi \in C[0, T]$. Then the finite part of g at 0 is defined by

$$\mathrm{PF}(g) := \psi(0).$$

Theorem 23 *Let $(\mathcal{A}, \mathcal{H}, D)$ be a p-summable spectral triple. Assume that the components of* Ch(ϵD) *and* Ch$(\epsilon D, D)$ *(see Equations* (14.5) *and* (14.7)*) have asymptotic behavior as in* (18.1). *Define a cochain ψ_k by*

$$\psi_k(a_0, a_1, \ldots, a_k) := \mathrm{PF}(\mathrm{Ch}(\epsilon D)(a_0, a_1, \ldots, a_k))$$

(here k takes odd values for odd spectral triples and even values for even spectral triples). Then $\psi_k = 0$ for $k > p$ and ψ_k is a periodic cyclic cocycle whose cohomology class coincides with $[\mathrm{Ch}(F)]$, $F = D(D^2+1)^{-1/2}$.

Therefore, we study the asymptotic behavior of the expression

$$\mathrm{Ch}^k(\epsilon D) = \int_{\Delta^k} \mathrm{Tr}\, a_0 e^{-t_0\epsilon^2 D^2}[\epsilon D, a_1]e^{-t_1\epsilon^2 D^2} \ldots [\epsilon D, a_k]e^{-t_k\epsilon^2 D^2} dt_1 dt_2 \ldots dt_k.$$

We start with the following identity (here $P \in OP^\alpha$):

$$e^{-sD^2} P = \sum_{m=0}^{N} \frac{(-s)^m}{m!} \nabla^m(P) e^{-sD^2}$$
$$+ \frac{(-s)^{N+1}}{N!} \int_0^1 (1-t)^N e^{-tsD^2} \nabla^{N+1}(P) e^{-(1-t)sD^2} dt.$$

Applying it repeatedly to Ch(ϵD) to move all the exponentials in

$$a_0 e^{-t_0\epsilon^2 D^2}[D, a_1]e^{-t_1\epsilon^2 D^2} \ldots [D, a_k]e^{-t_k\epsilon^2 D^2}$$

to the end of the expression we obtain

$$\mathrm{Ch}^k(\epsilon D) = \sqrt{2i} \sum_{0 \leq m_i \leq N} C_{\mathbf{m}} \epsilon^{k+2m} \operatorname{Tr} a_0 \nabla^{m_1}([D, a_1]) \dots \nabla^{m_k}([D, a_k]) e^{-\epsilon^2 D^2} + R_N$$

where $m = \sum_{i=1}^k m_i$ and

$$C_{\mathbf{m}} = (-1)^m \int_{\Delta^k} \frac{t_0^{m_1}(t_0 + t_1)^{m_2} \dots (t_0 + t_1 + \dots t_{k-1})^{m_k}}{m_1! m_2! \dots m_k!} dt_1 dt_2 \dots dt_k$$
$$= (-1)^m \frac{1}{(m_1 + 1)(m_1 + m_2 + 2) \dots (m_1 + m_2 + \dots m_k + k) m_1! m_2! \dots m_k!}.$$

The remainder R_N can be bounded by a constant times $\epsilon^{(k+1)N-p}$ (cf. [42]) and thus does not contribute to $\mathrm{PF}(\mathrm{Ch}(\epsilon D))$ if sufficiently large N is chosen. Hence it is sufficient to determine the finite part of

$$\epsilon^{k+2m} \operatorname{Tr} A e^{-\epsilon^2 D^2}, \quad A = a_0 \nabla^{m_1}([D, a_1]) \dots \nabla^{m_k}([D, a_k]) \in OP^m$$

Note that

$$e^{-\epsilon^2 D^2} = e^{-\epsilon^2 |D|^2} + (1 - e^{-\epsilon^2}) P_{\mathrm{Ker}(D^2)}.$$

Hence

$$\operatorname{Tr} A e^{-\epsilon^2 D^2} = \operatorname{Tr} A e^{-\epsilon^2 |D|^2} + O(\epsilon^2)$$

and the finite parts of $\epsilon^{k+2m} \operatorname{Tr} A e^{-\epsilon^2 D^2}$ and $\epsilon^{k+2m} \operatorname{Tr} A e^{-\epsilon^2 |D|^2}$ coincide for every k, m. The expression $\operatorname{Tr} A e^{-\epsilon^2 D^2}$ is related to the ζ-function $\zeta_A(s) = \operatorname{Tr} A |D|^{-s}$ by the Mellin transform: the equation $\Gamma(s)|D|^{-2s} = \int_0^\infty e^{-t|D|^2} t^{s-1} dt$ implies that

$$\Gamma(s) \zeta_A(2s) = \Gamma(s) \operatorname{Tr} A |D|^{-2s} = \int_0^\infty \operatorname{Tr} A e^{-t|D|^2} t^{s-1} dt$$

To deduce the asymptotic expansion of $\operatorname{Tr} A e^{-t|D|^2}$ at $t = 0$ from the information on the poles of $\Gamma(s)\zeta_A(2s)$ we need a technical assumption on the decay of $\Gamma(s)\zeta_A(2s)$ on vertical lines in the complex plane. Under this assumption we obtain that

$$\epsilon^{k+2m} \operatorname{Tr} A e^{-\epsilon^2 |D|^2} = \sum_j \sum_{k=0}^{q_j - 1} \alpha_{j,k} \epsilon^{-2a_j} \log^k \epsilon + o(1) \text{ as } \epsilon \to 0^+$$

where a_j are the poles of $\Gamma(s)\zeta_A(2s)$ in the half plane $\Re s \geq k/2 + m$ and q_j—the order of the pole at a_j. Moreover the constant term in this expansion is equal to

$\mathrm{Res}_{s=k/2+m}\,\Gamma(s)\zeta_A(2s)$. A virtually identical argument shows that $\mathrm{Ch}(\epsilon D, D)$ has the desired asymptotic behavior as $\epsilon \to 0^+$.

Summarizing, we obtain the following result of Connes and Moscovici:

Theorem 24 (Connes–Moscovici) *The formulas*

$$\begin{aligned}
&\psi_k(a_0, a_1, \dots a_k) \\
&\quad := \sqrt{2i} \sum_{m_i \geq 0} C_{\mathbf{m}} \,\mathrm{Res}_{s=k/2+m}\, \Gamma(s) \,\mathrm{Tr}\, a_0 \nabla^{m_1}([D, a_1]) \dots \nabla^{m_k}([D, a_k])|D|^{-2s} \\
&\quad = \sqrt{2i} \sum_{m_i \geq 0} C_{\mathbf{m}} \,\mathrm{Res}_{s=0}\, \Gamma(s + k/2 + m) \\
&\qquad \times \mathrm{Tr}\, a_0 \nabla^{m_1}([D, a_1]) \dots \nabla^{m_k}([D, a_k])|D|^{-2s-k-2m}
\end{aligned} \tag{18.2}$$

$k = 1, 3, \dots$ *define an odd periodic cyclic cocycle cohomologous to* $\mathrm{Ch}(F)$, $F = D(D^2+1)^{-1/2}$.

One can rewrite this result to obtain the formula in terms of generalized residues. Since for $h(s)$ holomorphic at $s = 0$

$$\mathrm{Res}_{s=0}\, h(s)\zeta_A(2s) = \sum_{l \geq 0} \frac{h^{(l)}(0)}{l!} \tau_l(A)$$

We therefore obtain the following version of the previous theorem.

Theorem 25 (Connes–Moscovici) *The formulas*

$$\begin{aligned}
\psi_k(a_0, a_1, \dots a_k) = \sqrt{2i} \sum_{m_i \geq 0, l \geq 0} &\frac{\Gamma^{(l)}(k/2+m)}{l!} \\
&\times C_{\mathbf{m}} \tau_l \left(a_0 \nabla^{m_1}([D, a_1]) \dots \nabla^{m_k}([D, a_k])|D|^{-k-2m} \right)
\end{aligned} \tag{18.3}$$

$k = 1, 3, \dots$ *define an odd periodic cyclic cocycle cohomologous to* $\mathrm{Ch}(F)$, $F = D(D^2+1)^{-1/2}$.

19 Renormalization

Here we outline the Connes–Moscovici process of renormalization which allows one to replace the local index cocycle ψ_k by a cohomologous one of the form

$$\psi'_k(a_0, a_1, \dots a_k) :=$$

$$\sum_{m_i \geq 0, l \geq 0} C'(l, m_1, m_2, \dots m_k) \tau_l \left(a_0 \nabla^{m_1}([D, a_1]) \dots \nabla^{m_k}([D, a_k]) |D|^{-k-2m} \right)$$

where the constants $C'(l, m_1, m_2, \dots m_k)$ have the following properties: they are rational multiples of $\sqrt{2\pi i}$ and the summation over l is finite even if the ζ-functions have isolated essential singularities and/or there is no bound on the order of poles that can occur in the ζ-functions. The starting point is the observation that the cohomology class of ψ_k does not change under rescaling $D \to e^{-\mu/2} D$, $\mu \in \mathbb{R}$. The cocycle ψ_k is then replaced by a cohomologous cocycle

$$\begin{aligned}\psi_k^{\mu}(a_0, a_1, \dots a_k) &:= \sqrt{2i} \sum_{m_i \geq 0} C_{\mathbf{m}} \operatorname{Res}_{s=0} e^{\mu s} \Gamma\left(s + \frac{k}{2} + m\right) \\ &\quad \times \operatorname{Tr} a_0 \nabla^{m_1}([D, a_1]) \dots \nabla^{m_k}([D, a_k]) |D|^{-2s-k-2m} \\ &= \psi_k(a_0, a_1, \dots a_k) + \sum_{\mu \geq 1} \frac{\mu^q}{q!} \phi_k^q(a_0, a_1, \dots a_k). \qquad (19.1)\end{aligned}$$

where

$$\begin{aligned}\phi_k^q(a_0, a_1, \dots a_k) &:= \sqrt{2i} \sum_{m_i \geq 0} C_{\mathbf{m}} \operatorname{Res}_{s=0} s^q \Gamma\left(s + \frac{k}{2} + m\right) \\ &\quad \times \operatorname{Tr} a_0 \nabla^{m_1}([D, a_1]) \dots \nabla^{m_k}([D, a_k]) |D|^{-2s-k-2m}\end{aligned}$$

Here we used the identity

$$\left| e^{-\mu/2} D \right|^{-2s} = e^{\mu s} |D|^{-2s} + (1 - e^{\mu s}) P_{\ker D^2}$$

to replace $\left| e^{-\mu/2} D \right|^{-2s}$ by $e^{\mu s} |D|^{-2s}$ without changing the residues. Hence for each $q = 1, 2, \dots$, ϕ_k^q is an odd cyclic cocycle cohomologous to 0. It follows that for every function $g(s)$ holomorphic at 0 with $g(0) = 1$ the formula

$$\begin{aligned}\psi'_k(a_0, a_1, \dots a_k) &:= \sqrt{2i} \sum_{m_i \geq 0} C_{\mathbf{m}} \operatorname{Res}_{s=0} g(s) \Gamma\left(s + \frac{k}{2} + m\right) \\ &\quad \times \operatorname{Tr} a_0 \nabla^{m_1}([D, a_1]) \dots \nabla^{m_k}([D, a_k]) |D|^{-2s-k-2m} \qquad (19.2)\end{aligned}$$

defines a periodic cyclic cocycle cohomologous to ψ_k. In Connes–Moscovici renormalization one chooses $g(s) = \frac{\Gamma(1/2)}{\Gamma(1/2+s)}$. One can then express the resulting formula in terms of generalized residues using the identity

$$\frac{\Gamma\left(\frac{1}{2}\right)}{\Gamma\left(\frac{1}{2}+s\right)}\Gamma\left(s+\frac{k}{2}+m\right)=\Gamma\left(\frac{1}{2}\right)\left(\frac{1}{2}+s\right)\left(\frac{3}{2}+s\right)\ldots\left(\frac{k+2m-2}{2}+s\right)$$

Denote by $\sigma_j(n)$ the coefficients in the expansion

$$\left(\frac{1}{2}+s\right)\left(\frac{3}{2}+s\right)\ldots\left(\frac{2n-1}{2}+s\right)=\sum_{l=0}^{\infty}\sigma_l(n)s^l$$

Theorem 26 (Connes–Moscovici) *The formulas*

$$\psi_k'(a_0, a_1, \ldots a_k) := \sqrt{2i}\sum_{m_i\geq 0, l\geq 0}\sigma_l\left(\frac{k-1}{2}+m\right)$$
$$\times\, C_{\mathbf{m}}\tau_l\left(a_0\nabla^{m_1}([D, a_1])\ldots\nabla^{m_k}([D, a_k])|D|^{-k-2m}\right)$$

$k = 1, 3, \ldots$ *define an odd periodic cyclic cocycle cohomologous to* $\mathrm{Ch}(F)$, $F = D(D^2+1)^{-1/2}$,

Note that $\sigma_l\left(\frac{k-1}{2}+m\right) = 0$ for $l > \frac{k-1}{2}+m$ and

$$a_0\nabla^{m_1}([D, a_1])\ldots\nabla^{m_k}([D, a_k])|D|^{-k-2m} \in OP^{-k-m} \subset \mathcal{L}^1$$

when $k+m \geq p$ (recall that p denotes the summability degree of the spectral triple). Since each τ_l vanishes on trace class operators, only τ_l with $l < p$ may appear in the renormalized formula.

Example 27 Let $(\mathcal{A}, \mathcal{H}, D)$ be a spectral triple with discrete dimension spectrum. Let $\widetilde{\mathcal{A}}$ be the algebra generated by $\mathcal{A}$ and operators of the form $|D|^{-k}(\log|D|)^l$, where k, l are integers, $k \geq 1, l \geq 0$. Since $\mathrm{Tr}\,(A\log|D|)\,|D|^{-s} = -\frac{d}{ds}\mathrm{Tr}\,A|D|^{-s}$, $(\widetilde{\mathcal{A}}, \mathcal{H}, D)$ again will be a spectral triple with discrete dimension spectrum. Moreover, if (at least for some $A \in \Psi(\mathcal{A})$) ζ-function $\mathrm{Tr}\,A|D|^{-s}$ is not entire, the ζ-functions with $A \in \Psi(\widetilde{\mathcal{A}})$ will have poles of arbitrarily high orders.

In the situation of Example 19 the corresponding algebra of pseudodifferential operators is contained in the algebra of pseudodifferential operators with log-polyhomogeneous symbols constructed and studied in detail in [53].

20 The even case

Most of the discussion above extends to the even case verbatim, so here we just state the relevant results and indicate the point where a difference with the odd case arises.

One first obtains the following result: the formulas:

$$\psi_k(a_0, a_1, \dots a_k) := \sum_{m_i \geq 0} C_{\mathbf{m}} \operatorname{Res}_{s=0} \Gamma(s + k/2 + m)$$
$$\times \operatorname{Tr} \gamma a_0 \nabla^{m_1}([D, a_1]) \dots \nabla^{m_k}([D, a_k])|D|^{-2s-k-2m} \tag{20.1}$$

$k = 0, 2, \dots$ define an even periodic cyclic cocycle cohomologous to $\operatorname{Ch}(F)$, $F = D(D^2 + 1)^{-1/2}$. Here, as in the odd case,

$$C_{\mathbf{m}} = \frac{(-1)^m}{(m_1 + 1)(m_1 + m_2 + 2) \dots (m_1 + m_2 + \dots m_k + k) m_1! m_2! \dots m_k!},$$

$m = m_1 + m_2 + \dots + m_k$

The renormalization process allows us to replace the cocycle ψ_k by a cohomologous one

$$\psi_k'(a_0, a_1, \dots a_k) := \sum_{m_i \geq 0} C_{\mathbf{m}} \operatorname{Res}_{s=0} \frac{\Gamma\left(s + \frac{k}{2} + m\right)}{\Gamma(s+1)}$$
$$\times \operatorname{Tr} \gamma a_0 \nabla^{m_1}([D, a_1]) \dots \nabla^{m_k}([D, a_k])|D|^{-2s-k-2m} \tag{20.2}$$

Denote by $\sigma_j(n)$ coefficients in the expansion

$$\frac{\Gamma(s+n)}{\Gamma(s+1)} = (1+s)(2+s) \dots (n-1+s) = \sum_{l=0}^{\infty} \sigma_l(n) s^l$$

Then we can write an expression for the cocycle ψ_k' in terms of linear functionals τ_l:

Theorem 28 (Connes–Moscovici) *The formulas*

$$\psi_0'(a_0) := \tau_{-1}(\gamma a_0)$$

$$\psi_k'(a_0, a_1, \dots a_k) := \sum_{m_i \geq 0, l \geq 0} \sigma_l \left(\frac{k}{2} + m\right) C_{\mathbf{m}} \tau_l \left(a_0 \nabla^{m_1}\right.$$
$$\left. \times ([D, a_1]) \dots \nabla^{m_k}([D, a_k])|D|^{-k-2m}\right) \textit{ for } k = 2, 4, \dots$$

define an even periodic cyclic cocycle cohomologous to $\operatorname{Ch}(F)$, $F = D(D^2 + 1)^{-1/2}$.

Note the important difference with the odd case: in addition to generalized residues this formula contains a nonlocal term involving τ_{-1}. Some nonlocality is unavoidable in the even case as can be seen by considering the case when $\mathcal{H}$ is finite dimensional. All the linear functionals τ_i, $i \geq 0$ vanish, as every operator is of trace class. Nevertheless one can have operators with nonzero index between finite dimensional vector spaces and thus the Connes character of a finite dimensional spectral triple can be nonzero.

References

1. J. Aastrup, R. Nest, and E. Schrohe, *A continuous field of C^*-algebras and the tangent groupoid for manifolds with boundary*, J. Funct. Anal. **237**(2006), no. 2, 482–506. MR2230347
2. M. Atiyah, R. Bott, and V. K. Patodi, *On the heat equation and the index theorem*, Invent. Math. **19** (1973), 279–330. MR0650828
3. M. F. Atiyah, *Global theory of elliptic operators*, Proc. Internat. Conf. on Functional Analysis and Related Topics (Tokyo, 1969), 1970, pp. 21–30.
4. M. F. Atiyah and R. Bott, *A Lefschetz fixed point formula for elliptic complexes. I*, Ann. of Math. (2) **86** (1967), 374–407. MR0212836
5. M. F. Atiyah, V. K. Patodi, and I. M. Singer, *Spectral asymmetry and Riemannian geometry. I*, Math. Proc. Cambridge Philos. Soc. **77** (1975), 43–69. MR0397797
6. M. F. Atiyah and I. M. Singer *The index of elliptic operators on compact manifolds*, Bull. Amer. Math. Soc. **69** (1963), 422–433. MR0157392
7. ——, *The index of elliptic operators. I*, Ann. of Math. (2) **87** (1968), 484–530.
8. ——, *The index of elliptic operators. III*, Ann. of Math. (2) **87** (1968), 546–604.
9. ——, *The index of elliptic operators. IV*, Ann. of Math. (2) **93** (1971), 119–138. MR0279833
10. P. Baum and A. Connes, *Chern character for discrete groups*, A fête of topology 1988, pp. 163–232. MR928402
11. P. Baum, A. Connes, and N. Higson, *Classifying space for proper actions and K-theory of group C^*-algebras*, C^*-algebras: 1943–1993 (San Antonio, TX, 1993), 1994, pp. 240–291. MR1292018
12. P. Baum and R. G. Douglas, *K homology and index theory*, Operator algebras and applications, Part I (Kingston, Ont., 1980), 1982, pp. 117–173.
13. P. Baum, N. Higson, and T. Schick, *On the equivalence of geometric and analytic K-homology*, Pure Appl. Math. Q. **3** (2007), no. 1, part 3, 1–24.
14. P. F. Baum and E. van Erp, *K-homology and index theory on contact manifolds*, Acta Math. **213** (2014), no. 1, 1–48. MR3261009
15. ——, *K-homology and Fredholm operators I: Dirac operators*, J. Geom. Phys. **134** (2018), 101–118. MR3886929
16. N. Berline, E. Getzler, and M. Vergne, *Heat kernels and Dirac operators*, Grundlehren Text Editions, Springer-Verlag, Berlin, 2004. Corrected reprint of the 1992 original. MR2273508
17. N. Berline and M. Vergne, *A computation of the equivariant index of the Dirac operator*, Bull. Soc. Math. France **113** (1985), no. 3, 305–345. MR834043
18. J.-M. Bismut, *The Atiyah-Singer index theorem for families of Dirac operators: two heat equation proofs*, Invent. Math. **83** (1986), no. 1, 91–151. MR813584
19. B. Blackadar, *K-theory for operator algebras* Second, Mathematical Sciences Research Institute Publications, vol. 5, Cambridge University Press, Cambridge, 1998.
20. P. Bressler, R. Nest, and B. Tsygan, *Riemann-Roch theorems via deformation quantization. I, II*, Adv. Math. **167** (2002), no. 1, 1–25, 26–73. MR1901245

21. J.-L. Brylinski and V. Nistor, *Cyclic cohomology of étale groupoids*, K-Theory **8** (1994), no. 4, 341–365. MR1300545
22. D. Burghelea, *The cyclic homology of the group rings*, Comment. Math. Helv. **60** (1985), no. 3, 354–365. MR814144
23. A. Connes, *A survey of foliations and operator algebras*, Operator algebras and applications, Part I (Kingston, Ont., 1980), 1982, pp. 521–628. MR679730
24. A. Connes and H. Moscovici, *Transgression and the Chern character of finite-dimensional K-cycles*, Comm. Math. Phys. **155** (1993), no. 1, 103–122. MR1228528
25. ——, *The local index formula in noncommutative geometry*, Geom. Funct. Anal. **5** (1995), no. 2, 174–243.
26. A. Connes and G. Skandalis, *The longitudinal index theorem for foliations*, Publ. Res. Inst. Math. Sci. **20** (1984), no. 6, 1139–1183. MR775126
27. A. Connes, *Noncommutative differential geometry*, Inst. Hautes Études Sci. Publ. Math. **62** (1985), 257–360. MR823176
28. ——, *Noncommutative geometry*, Academic Press Inc., San Diego, CA, 1994.
29. M. Crainic, *Cyclic cohomology of étale groupoids: the general case*, K-Theory **17** (1999), no. 4, 319–362. MR1706117
30. C. Debord and G. Skandalis, *Lie groupoids, exact sequences, Connes-Thom elements, connecting maps and index maps*, J. Geom. Phys. **129** (2018), 255–268. MR3789251
31. B. L. Feĭgin and B. L. Tsygan, *Cyclic homology of algebras with quadratic relations, universal enveloping algebras and group algebras*, K-theory arithmetic and geometry (Moscow, 1984–1986), 1987, pp. 210–239. MR923137
32. I. M. Gel′ fand, *On elliptic equations*, Russian Math. Surveys **15** (1960), no. 3, 113–123. MR0123085
33. E. Getzler, *Pseudodifferential operators on supermanifolds and the Atiyah-Singer index theorem*, Comm. Math. Phys. **92** (1983), no. 2, 163–178. MR728863
34. ——, *A short proof of the local Atiyah-Singer index theorem*, Topology **25** (1986), no. 1, 111–117. MR836727
35. E. Getzler and J. D. S. Jones, *The cyclic homology of crossed product algebras*, J. Reine Angew. Math. **445** (1993), 161–174. MR1244971
36. E. Getzler and A. Szenes, *On the Chern character of a theta-summable Fredholm module*, J. Funct. Anal. **84** (1989), no. 2, 343–357. MR1001465
37. P. B. Gilkey, *Curvature and the eigenvalues of the Laplacian for elliptic complexes*, Advances in Math. **10** (1973), 344–382. MR0324731
38. A. Gorokhovsky, *Characters of cycles, equivariant characteristic classes and Fredholm modules*, Comm. Math. Phys. **208** (1999), no. 1, 1–23. MR1729875
39. ——, *Bivariant Chern character and longitudinal index*, J. Funct. Anal. **237** (2006), no. 1, 105–134. MR2239260
40. A. Gorokhovsky and J. Lott, *Local index theory over étale groupoids*, J. Reine Angew. Math. **560** (2003), 151–198. MR1992804
41. ——, *Local index theory over foliation groupoids*, Adv. Math. **204** (2006), no. 2, 413–447. MR2249619
42. ——, *A Hilbert bundle description of differential K-theory*, Adv. Math. **328** (2018), 661–712. MR3771139
43. V. Guillemin, *A new proof of Weyl's formula on the asymptotic distribution of eigenvalues*, Adv. in Math. **55** (1985), no. 2, 131–160. MR772612
44. N. Higson, *A primer on KK-theory*, Operator theory: operator algebras and applications, Part 1 (Durham, NH, 1988), 1990, pp. 239–283. MR1077390
45. ——, *The local index formula in noncommutative geometry*, Contemporary developments in algebraic K-theory, 2004, pp. 443–536. MR2175637
46. ——, *The residue index theorem of Connes and Moscovici*, Surveys in noncommutative geometry, 2006, pp. 71–126. MR2277669
47. ——, *The tangent groupoid and the index theorem*, Quanta of maths, 2010, pp. 241–256. MR2732053

48. M. Hilsum and G. Skandalis, *Morphismes K-orientés d'espaces de feuilles et fonctorialité en théorie de Kasparov (d'après une conjecture d'A. Connes)*, Ann. Sci. École Norm. Sup. (4) **20** (1987), no. 3, 325–390.
49. A. Jaffe, A. Lesniewski, and K. Osterwalder, *Quantum K-theory. I The Chern character*, Comm. Math. Phys. **118** (1988), no. 1, 1–14. MR954672
50. M. Jakob, *A bordism-type description of homology*, Manuscripta Math. **96** (1998), no. 1, 67–80. MR1624352
51. G. G. Kasparov, *Topological invariants of elliptic operators. I. K-homology*, Izv. Akad. Nauk SSSR Ser. Mat. **39** (1975), no. 4, 796–838. MR0488027
52. ——. *Operator K-theory and its applications: elliptic operators, group representations, higher signatures, C^*-extensions*, Proceedings of the International Congress of Mathematicians, Vol. 1, 2 (Warsaw, 1983), 1984, pp. 987–1000. MR804752
53. M. Lesch, *On the noncommutative residue for pseudodifferential operators with log-polyhomogeneous symbols*, Ann. Global Anal. Geom. **17** (1999), no. 2, 151–187. MR1675408
54. J.-L. Loday *Cyclic homology*, Second, Grundlehren der Mathematischen Wissenschaften [Fundamental Principles of Mathematical Sciences], vol. 301, Springer-Verlag, Berlin, 1998. Appendix E by María O. Ronco, Chapter 13 by the author in collaboration with Teimuraz Pirashvili. MR1600246
55. H. P McKean Jr. and I. M. Singer, *Curvature and the eigenvalues of the Laplacian*, J. Differential Geometry **1** (1967), no. 1, 43–69. MR0217739
56. B. Monthubert and F. Pierrot, *Indice analytique et groupoïdes de Lie*, C. R. Acad. Sci. Paris Sér. I Math. **325** (1997), no. 2, 193–198. MR1467076
57. R. Nest and B. Tsygan, *Algebraic index theorem*, Comm. Math. Phys. **172** (1995), no. 2, 223–262. MR1350407
58. ——, *Algebraic index theorem for families*, Adv. Math. **113** (1995), no. 2, 151–205. MR1337107
59. ——, *Formal versus analytic index theorems*, Internat. Math. Res. Notices **11** (1996), 557–564. MR1405974
60. V. Nistor, *Group cohomology and the cyclic cohomology of crossed products*, Invent. Math. **99** (1990), no. 2, 411–424. MR1031908
61. V. Nistor, A. Weinstein, and P. Xu, *Pseudodifferential operators on differential groupoids*, Pacific J. Math. **189** (1999), no. 1, 117–152. MR1687747
62. D. Quillen, *Superconnections and the Chern character*, Topology **24** (1985), no. 1, 89–95. MR790678
63. P. C. Rouse, *Compactly supported analytic indices for Lie groupoids*, J. K-Theory **4** (2009), no. 2, 223–262. MR2551909
64. R. T. Seeley, *Complex powers of an elliptic operator*, Singular Integrals (Proc. Sympos. Pure Math., Chicago, Ill., 1966), 1967, pp. 288–307. MR0237943
65. E. van Erp, *The Atiyah-Singer formula for subelliptic operators on a contact manifold, Part I*, Ann. of Math. **171** (2010), 1647–1681.
66. M. Wodzicki, *Noncommutative residue. I. Fundamentals*, K-theory arithmetic and geometry (Moscow, 1984–1986), 1987, pp. 320–399. MR923140

Modular Gaussian curvature

Matthias Lesch and Henri Moscovici

Dedicated to Alain Connes with admiration and much appreciation

Abstract This is a brief survey of the main developments that led to the emergence of the quantized analogue of Gaussian curvature for the noncommutative torus and to its current understanding. It highlights the role of Connes' pseudodifferential calculus as the crucial technical tool for the explicit computation of the modular Gaussian curvature, the effectiveness of the variational methods, and it sheds more light on the intrinsic geometric meaning of the Morita equivalence in this context.

1 Introduction

The genesis of natural but noncommuting coordinates can be traced back to Heisenberg's uncertainty principle in quantum mechanics, which limits the accuracy of the simultaneous determination of the position and momentum (q, p) of a subatomic particle. As Heisenberg argued [HEI27] (and Kennard rigorously derived [KEN27]), the inherent imprecision of such a measurement "is a straightforward mathematical consequence of the quantum mechanical commutation rule for the position and the corresponding momentum operators $qp - pq = i\hbar$," where $\hbar = \frac{h}{2\pi}$ is the reduced Planck constant. Such an identity cannot be satisfied by matrices (over $\mathbb{C}$), which is obvious, but not even by bounded operators in Hilbert space. Assuming q and p self-adjoint, this can be seen by passing to the Weyl integrated form [WEY28, §45],

M. Lesch
Mathematisches Institut, Universität Bonn, Bonn, Germany
www.matthiaslesch.de; www.math.uni-bonn.de/people/lesch
e-mail: ml@matthiaslesch.de; lesch@math.uni-bonn.de

H. Moscovici (✉)
Department of Mathematics, The Ohio State University, Columbus, OH, USA
e-mail: henri@math.ohio-state.edu

A. Chamseddine et al. (eds.), *Advances in Noncommutative Geometry*,
https://doi.org/10.1007/978-3-030-29597-4_8

$$V_s U_t = e^{2\pi i \hbar t s} U_t V_s, \qquad t, s \in \mathbb{R}. \tag{1.1}$$

Moreover, the latter relation determines a unitary representation $\pi_\hbar$ of the (implicitly defined) group $H_3(\mathbb{R})$, called the Heisenberg group. By a celebrated theorem of Stone and von Neumann all such irreducible representations are unitarily equivalent. The restriction to the lattice $H_3(\mathbb{Z}) \subset H_3(\mathbb{R})$ of an irreducible unitary representation π_θ, $\theta \in \mathbb{R}$, generates the C^*-algebra A_θ nowadays known as the *noncommutative torus* of slope θ. When $\theta \in [0, 1] \setminus \mathbb{Q}$, as shall be assumed throughout this paper, the C^*-algebra A_θ is also known as the *irrational rotation algebra* and is the (unique up to isomorphism) C^*-algebra generated by a pair of unitary operators U_1, U_2 satisfying

$$U_2 U_1 = e^{2\pi i \theta} U_1 U_2. \tag{1.2}$$

Moreover A_θ is a simple C^*-algebra, thus typifying the coordinates of a purely noncommutative space. For this reason on the one hand, and due to its accessibility on the other, A_θ has received much attention during the last several decades and has been a favorite testing ground for quite a number of fruitful mathematical investigations.

Although the habitual geometric intuition is rendered utterly inoperative in a "space without points" such as the one represented by A_θ, the curvature as a "measure of deviation from flatness" (in Riemann's own words) could still make some sense. It is the goal of this brief survey to review the recent developments that led to the emergence of a quantized version of Gaussian curvature for the noncommutative torus. Many of the essential ideas presented below have their origin in Alain Connes' 1980 C. R. Acad. Sc. Paris Note [Con80], which effectively constitutes the birth certificate of noncommutative differential geometry. That foundational article not only established the most basic geometric concepts and constructions, such as the geometric realization of the finitely generated projective modules over A_θ, the explicit construction of constant curvature connections for them and the definition and calculation of their Chern classes, but also provided the crucial computational tool, in the form of a pseudodifferential calculus adapted to C^*-dynamical systems.

The specific line of research whose highlights we are about to summarize was sparked by a paper by Connes and Paula Cohen (*Conformal geometry of the irrational rotation algebra*, Preprint MPI Bonn, 1992–93) which showed how the passage from the (unique) trace of A_θ to a non-tracial conformal weight associated with a Weyl factor (or "dilaton") gives rise to a non-flat geometry on the noncommutative torus, which can be investigated with the help of the adapted pseudodifferential calculus of [Con80]. In a later elaboration [CoTr11] of that paper, the passage from flatness to conformal flatness was placed in the setting of spectral triples (see Sections 2.1 and 2.2 below), which in the intervening years has emerged as the proper framework for the metric aspect in noncommutative geometry (cf. [Con94, Ch. 6], [Con13], [CoMo08]). Completing the calculations begun in the 1992 preprint they proved in [CoTr11] an analogue of the Gauss–Bonnet formula for the conformally twisted (called "modular") spectral triples. The full calculation of the modular Gaussian curvature was first done by A. Connes in 2009, with the aid of Wolfram's Mathematica, and is included in [CoMo14]. Fathizadeh and Khalkhali [FaKh13] independently performed the same calculation with the help of a different computing software.

Apart from computing the expression of the modular curvature (see Section 2.3 below), Connes and Moscovici showed in [CoMo14] that one can make effective use of variational methods even in the abstract operator-theoretic context of the spectral triple encoding the geometry of the noncommutative torus. After giving a variational proof of the modular Gauss–Bonnet formula which requires no computations (see Section 2.4), they related the modular Gaussian curvature to the gradient of the Ray–Singer log-determinant of the Laplacian viewed as a functional on the space of Weyl factors. As a consequence, they obtained an a priori proof of an internal consistency relation for the constituents of the modular curvature. In addition they showed by purely operator-theoretic arguments that, as in the case of Riemann surfaces (cf. [OPS88]), the normalized log-determinant functional attains its extreme value only at the trivial Weyl factor, in other words for the flat "metric" (see Section 2.5 below).

For reasons which will soon become transparent (see Section 3.1), the natural equivalence relation between noncommutative spaces is that of Morita equivalence between their respective algebras of coordinates. For noncommutative tori the Morita equivalence is implemented by the Heisenberg bimodules described by Connes [Con80] and Rieffel [Rie81]. Lesch and Moscovici extended in [LeMo16] the results of [CoMo14] to spectral triples on noncommutative tori associated with Heisenberg equivalence bimodules (see Sections 3.2 and 3.3). Moreover, in doing so they managed to dispose of any computer-aided calculations (see Section 5.1). Most notably they showed (see Section 3.4 below) that whenever A_θ is realized as the endomorphism algebra of a Heisenberg $A_{\theta'}$-module endowed with the $A_{\theta'}$-valued Hermitian structure obtained by twisting the canonical one by a positive invertible element in A_θ, the curvature of A_θ with respect to the corresponding spectral triple over $A_{\theta'}$ is equal to the modular curvature associated with the same element of A_θ viewed as conformal factor. In a certain sense this is reminiscent of Gauss's Theorema Egregium "If a curved surface is developed upon any other surface whatever, the measure of curvature in each point remains unchanged."

The fundamentals of Connes' pseudodifferential calculus as well as its extension to twisted C^*-dynamical systems, which provide the essential device for proving all the above results, are explained in Section 4. Finally, Section 5 clarifies how to use the affiliated symbol calculus in order to compute the resolvent trace expansion, or equivalently the heat trace expansion, for the relevant Laplace-type operators.

2 Curvature of modular spectral triples

2.1 Flat spectral triples

In noncommutative geometry a metric structure on a space with C^*-algebra of coordinates A is represented by a triad of data $(A, \mathcal{H}, D)$ called *spectral triple*, modeled on the Dirac operator on a manifold: A is realized as a norm-closed subalgebra of bounded operators on a Hilbert space $\mathcal{H}$, D is an unbounded self-adjoint operator whose resolvent belongs to any p-Schatten ideal with $p > d$ where $d > 0$ signifies the dimension, and D interacts with the coordinates by having bounded commutators (or more generally bounded twisted commutators) $[D, a]$ for any a in a dense subalgebra of A. The Dirac operator was chosen as model since it represents the fundamental class in K-homology and at the same time plays the role of a quantized inverse line element (see [CON13]). In the case of A_θ one can obtain such a triad by simply reproducing the construction of the $\bar\partial + \bar\partial^*$ operator on the ordinary torus $\mathbb{T}^2 = (\mathbb{R}/2\pi\mathbb{Z})^2$.

To fix the notation we briefly review some basic properties of the C^*-algebra A_θ with $\theta \in \mathbb{R} \setminus \mathbb{Q}$.

First of all, the torus $\mathbb{T}^2$ acts on A_θ via the representation by automorphisms defined on the basis elements by

$$\alpha_r(U_1^n U_2^m) = e^{i(r_1 n + r_2 m)} U_1^n U_2^m, \quad r = (r_1, r_2) \in \mathbb{R}^2. \tag{2.1}$$

By analogy with the action of $\mathbb{T}^2$ on $A_0 \equiv C(\mathbb{T}^2)$, we call these automorphisms *translations*.

The smooth vectors of this representation of $\mathbb{T}^2$ are precisely the elements of the form $a = \sum_{m,n\in\mathbb{Z}} a_{m,n} U_1^m U_2^n$ with rapidly decaying coefficients, i.e., such that $(1 + |m|)^k (1 + |n|)^\ell |a_{m,n}| \le C_{k,\ell}$, $k, \ell > 0$. These elements form a subalgebra $\mathcal{A}_\theta$ which is the analogue of $C^\infty(\mathbb{T}^2)$ viewed in Fourier transform. The assignment

$$\mathcal{A}_\theta \ni a = \sum_{m,n\in\mathbb{Z}} a_{m,n} U_1^m U_2^n \quad \mapsto \quad \varphi_0(a) = a_{0,0},$$

determines the unique normalized trace φ_0 of the C^*-algebra A_θ.

The image of the differential of the above representation on $\mathcal{A}_\theta$ is the Lie algebra generated by the outer derivations δ_1 and δ_2, uniquely determined by the relations $\delta_i(U_j) = \delta_i^j U_j$, $i, j \in \{1, 2\}$.

By analogy with the ordinary torus, one defines on A_θ a translation invariant complex structure with modular parameter $\tau \in \mathbb{C}$, $\Im\tau > 0$, by means of the pair of derivations

$$\delta_\tau = \delta_1 + \bar\tau\delta_2, \quad \delta_\tau^* = \delta_1 + \tau\delta_2; \tag{2.2}$$

these are the counterparts of the operators $\frac{1}{i}\,(\partial/\partial x + \bar{\tau}\partial/\partial y)$, and $\frac{1}{i}\,(\partial/\partial x + \tau\partial/\partial y)$ acting on $C^\infty(\mathbb{T}^2)$.

To obtain the analogue of the corresponding flat metric on $\mathbb{T}^2$, we let $\mathcal{H}_0 \equiv L^2(A_\theta, \varphi_0)$ denote the Hilbert space completion of A_θ with respect to the scalar product

$$\langle a, b\rangle = \varphi_0(a^* b), \quad a, b \in \mathcal{A}_\theta.$$

The space $\Omega^1\mathcal{A}_\theta$ of formal 1-forms $\sum a\, db$, $a, b \in \mathcal{A}_\theta$, is also endowed with a semi-definite inner product defined by

$$\langle adb, a'db'\rangle = \varphi_0(a^*\,(a')\delta_\tau(b')\,\delta_\tau b^*)\,, \quad a, a', b, b' \in \mathcal{A}_\theta.$$

On completing its quotient modulo the subspace of those elements $\omega \in \Omega^1\mathcal{A}_\theta$ such that $\langle\omega, \omega\rangle = 0$, one obtains a Hilbert space denoted $\mathcal{H}^{(1,0)}$. $\mathcal{H}^{(1,0)}$ is also an A_θ-bimodule under the natural left and right action of A_θ. Both these actions are unitary. Moreover, the linear map from $\Omega^1\mathcal{A}_\theta$ to $\mathcal{A}_\theta$ defined by sending the class of $\sum adb$ in $\mathcal{H}^{(1,0)}$ to $\sum a\delta_\tau(b) \in \mathcal{H}_0$ induces an A_θ-bimodule isomorphism between $\mathcal{H}^{(1,0)}$ and $\mathcal{H}_0$.

Denoting by ∂_τ the closure of the operator $\delta_\tau : \mathcal{A}_\theta \to \mathcal{H}_0$ viewed as unbounded operator from $\mathcal{H}_0$ to $\mathcal{H}^{(1,0)}$ one obtains a spectral triple $(A_\theta, \widetilde{\mathcal{H}}, D_\tau)$ by taking $\widetilde{\mathcal{H}} = \mathcal{H}_0 \oplus \mathcal{H}^{(1,0)}$ and as unbounded self-adjoint operator $D_\tau = \begin{pmatrix} 0 & \partial_\tau^* \\ \partial_\tau & 0 \end{pmatrix}$. Concurrently, the triad $(A_\theta^{\mathrm{op}}, \widetilde{\mathcal{H}}, D_\tau)$ is a spectral triple with respect to the right action of A_θ. One can turn it into a spectral triple for the left action of A_θ by passing to its transposed $(A_\theta, \bar{\mathcal{H}}, \bar{D}_\tau)$ (see [CoMo14, §1.2] for the general definition), where $\bar{\mathcal{H}}$ is the complex conjugate of $\widetilde{\mathcal{H}}$ and $\bar{D}_\tau = \begin{pmatrix} 0 & \bar{\partial}_\tau^* \\ \bar{\partial}_\tau & 0 \end{pmatrix}$.

2.2 *Modular spectral triples*

To implement the analogue of a conformal change of metric structure, we choose a self-adjoint element $h = h^* \in \mathcal{A}_\theta$ and use it to replace the trace φ_0 by the positive linear functional $\varphi \equiv \varphi_h$ defined by

$$\varphi(a) \equiv \varphi_h(a) = \varphi_0(ae^{-h})\,, \quad a \in A_\theta. \tag{2.3}$$

Then φ determines an inner product $\langle\ ,\ \rangle_\varphi$ on A_θ,

$$\langle a, b\rangle_\varphi = \varphi(a^* b)\,, \quad a, b \in A_\theta,$$

which by completion gives rise to a Hilbert space $\mathcal{H}_\varphi$. The latter is again an A_θ-bimodule but, since φ is no longer tracial, the right action is no longer unitary.

The non-unimodularity of φ is expressed by Tomita's modular operator Δ, which in this case is

$$\Delta(x) = e^{-h} x e^{h}, \quad x \in A_\theta,$$

and gives rise to the 1-parameter group of inner automorphisms

$$\sigma_t(x) = \Delta^{-it} = e^{ith} x e^{-ith}, \quad x \in A_\theta. \tag{2.4}$$

Instead of the tracial property φ satisfies the KMS condition

$$\varphi(ab) = \varphi(b\sigma_i(a)) = \varphi(be^{-h}ae^{h}), \quad a, b \in A_\theta.$$

To restore the unitarity of the right action one redefines it by setting

$$a^{\mathrm{op}} := J_\varphi a^* J_\varphi \in \mathcal{L}(\mathcal{H}_\varphi), \quad a \in A_\theta,$$

where $J_\varphi(a) = \Delta^{1/2}(a^*) = k^{-1}a^*k$, $a \in A_\theta$, and $k = e^{h/2}$.

While keeping $\mathcal{H}^{(1,0)}$ unchanged, we now view δ_τ as a densely defined operator from $\mathcal{H}_\varphi$ to $\mathcal{H}^{(1,0)}$. Its closure ∂_φ is then used to define $D_\varphi = \begin{pmatrix} 0 & \partial_\varphi^* \\ \partial_\varphi & 0 \end{pmatrix}$ giving rise to the triad $(A_\theta^{\mathrm{op}}, \widetilde{\mathcal{H}}_\varphi, D_\varphi)$, where $\widetilde{\mathcal{H}}_\varphi = \mathcal{H}_\varphi \oplus \mathcal{H}^{(1,0)}$. This is a *twisted* spectral triple (see [CoMo08] for the general definition) over A^{op}, with the twisted commutators $D_\varphi\, a^{\mathrm{op}} - (k^{-1}ak)^{\mathrm{op}} D_\varphi$, $a \in \mathcal{A}_\theta$ bounded. Its *transposed*, formed as in the flat case, yields the *modular spectral triple* over $\mathcal{A}_\theta$, with operator $\bar{D}_\varphi = \begin{pmatrix} 0 & k\partial_\varphi \\ \partial_\varphi^* k & 0 \end{pmatrix}$, where the conformal factor k acts by left multiplication, and with underlying Hilbert space $\overline{\widetilde{\mathcal{H}}}_\varphi$.

By a series of identifications, it is shown in [CoMo14, §1.3] that the modular spectral triple associated with φ, or equivalently to the conformal factor $k = e^{h/2}$, is canonically isomorphic to the twisted spectral triple $(A_\theta, \widetilde{\mathcal{H}}_0, D_k)$ with $\widetilde{\mathcal{H}}_0 := \mathcal{H}_0 \oplus \mathcal{H}_0$ and $D_k := \begin{pmatrix} 0 & k\delta_\tau \\ \delta_\tau^* k & 0 \end{pmatrix}$.

We finally note that $D_k^2 = \Delta_k \oplus \Delta_k^{(0,1)}$, where

$$\Delta_k := k\Delta_\tau k \equiv k\delta_\tau\delta_\tau^* k \quad \text{and} \quad \Delta_k^{(0,1)} = \delta_\tau^* k^2 \delta_\tau, \tag{2.5}$$

are the counterparts of the Laplacian on functions, respectively, the Laplacian on (0, 1)-forms.

2.3 Modular curvature

The meaning of locality in noncommutative geometry is guided by the analogy with the Fourier transform, which interrelates the local behavior of functions with the decay at infinity of their coefficients. In a similar way, in the noncommutative formalism the local invariants of a spectral triple $(A, \mathcal{H}, D)$ are encoded in the high frequency behavior of the spectrum of the "inverse line element" D coupled with the action of the algebra of coordinates. For example, the local index formula in noncommutative geometry [CoMo95, Part II] expresses the Connes–Chern character of a spectral triple with finite dimension spectrum in terms of multilinear functionals given by residues of zeta functions defined by

$$z \mapsto \mathrm{Tr}\left(a_0[D, a_1]^{(k_1)} \dots [D, a_p]^{(k_1)} |D|^{-z}\right), \qquad \Re(z) >> 0,$$

where $a_0, \dots, a_p \in \mathcal{A}$ and $[D, a]^{(k)} = [D^2, \dots, [D^2, [D, a]] \cdots]$ with D^2 repeated k-times; the existence of the meromorphic continuation of such zeta functions is built in the definition of *finite dimension spectrum* for a spectral triple. Clearly, perturbing D by a trace class operator will not affect these residue functionals, whence the local nature of the index formula described in their terms.

In the specific case of the noncommutative torus the concept of locality can be pushed much closer to the customary one. Namely, if $(A_\theta, \widetilde{\mathcal{H}}_0, D_k)$ is a modular spectral triple as in Section 2.2, for its Laplacian "on functions" there is an asymptotic expansion

$$\mathrm{Tr}\left(a\, e^{-t\Delta_k}\right) \sim_{t \searrow 0} \sum_{q=0}^{\infty} \mathrm{a}_{2q}(a, \Delta_k)\, t^{q-1}, \qquad a \in \mathcal{A}_\theta, \tag{2.6}$$

whose functional coefficients a_{2q} are not only local in the above sense, but they are also absolutely continuous with respect to the unique trace, i.e., of the form

$$\mathcal{A}_\theta \ni a \longmapsto \mathrm{a}_{2q}(a, \Delta_k) = \varphi_0\left(a\, \mathcal{K}_k^{(q)}\right), \qquad \mathcal{K}_k^{(q)} \in \mathcal{A}_\theta,$$

with "Radon–Nikodym derivatives" $\mathcal{K}_k^{(q)} \in \mathcal{A}_\theta$ computable by means of symbolic calculus. The technical apparatus which justifies the heat expansion Equation (2.6) as well as the explicit computation of $\mathcal{K}_k^{(0)}$ will be discussed in Sections 4 and 5.

In particular, the Radon–Nikodym derivative of the term a_2, which classically delivers the scalar curvature, was fully computed in [CoMo14, FaKh13] and represents the modular scalar curvature. Abbreviating its notation to $\mathcal{K}_k$ instead of $\mathcal{K}_k^{(0)}$, it has the following expression:

$$\mathcal{K}_k = -\frac{\pi}{2\Im\tau}\left(K_0(\nabla)(\Delta(h)) + \frac{1}{2} H_0\left(\nabla^{(1)}, \nabla^{(2)}\right)(\square_{\Re}(h))\right), \tag{2.7}$$

where $\nabla = \log \Delta$ is the inner derivation implemented by $-h$,

$$\Delta(h) = \delta_\tau \delta_\tau^* = \delta_1^2(h) + 2\Re\tau\, \delta_1\delta_2(h) + |\tau|^2 \delta_2^2(h),$$

$\Box_\Re$ is the Dirichlet quadratic form

$$\Box_\Re(\ell) := (\delta_1(\ell))^2 + \Re\tau\, (\delta_1(\ell)\delta_2(\ell) + \delta_2(\ell)\delta_1(\ell)) + |\tau|^2 (\delta_2(\ell))^2,$$

and $\nabla^{(i)}$, $i = 1, 2$, signifies that ∇ is acting on the ith factor. The functions $K_0(s)$ and $H_0(s,t)$, whose expressions resulted from the symbolic computations, are given by

$$K_0(s) = \frac{-2 + s \coth\left(\frac{s}{2}\right)}{s \sinh\left(\frac{s}{2}\right)} \qquad \text{and} \qquad H_0(s,t) =$$

$$\frac{t(s+t)\cosh(s) - s(s+t)\cosh(t) + (s-t)(s+t+\sinh(s)+\sinh(t)-\sinh(s+t))}{st(s+t)\sinh\left(\frac{s}{2}\right)\sinh\left(\frac{t}{2}\right)\sinh\left(\frac{s+t}{2}\right)^2}.$$

The second function is related to the first by the functional identity

$$\begin{aligned} -\frac{1}{2}\widetilde{H}_0(s_1, s_2) = {} & \frac{\widetilde{K}_0(s_2) - \widetilde{K}_0(s_1)}{s_1 + s_2} \\ & + \frac{\widetilde{K}_0(s_1+s_2) - \widetilde{K}_0(s_2)}{s_1} - \frac{\widetilde{K}_0(s_1+s_2) - \widetilde{K}_0(s_1)}{s_2}, \end{aligned} \tag{2.8}$$

where

$$\begin{aligned} \widetilde{K}_0(s) &= 4\frac{\sinh(s/2)}{s} K_0(s) \qquad \text{and} \\ \widetilde{H}_0(s,t) &= 4\frac{\sinh((s+t)/2)}{s+t} H_0(s,t). \end{aligned} \tag{2.9}$$

A noteworthy feature of the main curvature-defining function is that, up to a constant factor, $\widetilde{K}_0$ is a generating function for the Bernoulli numbers; precisely,

$$\widetilde{K}_0(t) = 8 \sum_{n=1}^{\infty} \frac{B_{2n}}{(2n)!} t^{2n-2}. \tag{2.10}$$

2.4 *Modular Gauss–Bonnet formula*

Since the K-groups of the noncommutative torus are the same as of the ordinary torus, its Euler characteristic vanishes. Thus, the analogue of the Gauss–Bonnet theorem for the modular spectral triple is the identity

$$\varphi_0\left(\mathcal{K}_k^{(0)}\right) = 0.$$

This can be directly checked by making use of the fact that the group of modular automorphisms σ_t (cf. Equation (2.4)) preserves the trace φ_0 and fixes the dilaton h, in conjunction with the "integration by parts" rule

$$\varphi_0(a\delta_j(b)) = -\varphi_0(\delta_j(a)b), \qquad a, b \in \mathcal{A}_\theta.$$

(see [CoMo14, Lemma 4.2] for the precise identity to be used).

An alternative variational argument, given in [CoMo14, §1.4]), runs as follows. Consider the family of Laplacians

$$\triangle_s := k^s \triangle k^s = e^{\frac{sh}{2}} \triangle_\tau e^{\frac{sh}{2}}, \qquad s \in \mathbb{R}. \tag{2.11}$$

One has $\frac{d}{ds}\triangle_s = \frac{1}{2}(h\triangle_s + \triangle_s h)$. By Duhamel's formula one can interchange the derivative with the trace, hence

$$\frac{d}{ds} \operatorname{Tr}\left(e^{-t\triangle_s}\right) = -t \operatorname{Tr}\left(h \triangle_s e^{-t\triangle_s}\right) = t \frac{d}{dt} \operatorname{Tr}\left(h e^{-t\triangle_s}\right).$$

Differentiating term-by-term the asymptotic expansion Equation (2.6) (with $a = 1$ omitted in notation) yields

$$\frac{d}{ds}\mathrm{a}_j(\triangle_s) = \frac{1}{2}(j-2)\,\mathrm{a}_j(h, \triangle_s), \quad j \in \mathbb{Z}^+.$$

In particular, $\mathrm{a}_2(\triangle_s) = \mathrm{a}_2(\triangle_\tau)$. The latter vanishes because $\triangle_\tau$ is isospectral to the Laplacian of the ordinary torus with the same complex structure and, as is well-known, if $\triangle_M$ is the Laplacian on a Riemann surface then $\mathrm{a}_2(\triangle_M) = \frac{\chi(M)}{6}$, where $\chi(M)$ is the Euler characteristic of M.

2.5 Variation of determinant and modular Gaussian curvature

The zeta function $\zeta_{\triangle_k}(a, z) = \operatorname{Tr}\left(a\,\triangle_k^{-z}(1 - P_k)\right)$, $\Re z > 2$ where P_k stands for the orthogonal projection onto $\operatorname{Ker}\triangle_k$, is related to the corresponding theta function by the Mellin transform

$$\zeta_{\triangle_k}(a, z) = \frac{1}{\Gamma(z)} \int_0^\infty t^{z-1} \operatorname{Tr}\left(a\left(e^{-t\triangle_k} - P_k\right)\right) dt.$$

The asymptotic expansion Equation (2.6) ensures that it has meromorphic continuation and its value at 0 is

$$\zeta_{\Delta_k}(a,0) = \mathrm{a}_2(a,\Delta_\varphi) - \mathrm{Tr}(P_k\, a\, P_k) = \mathrm{a}_2(a,\Delta_\varphi) - \frac{\varphi_0(ak^{-2})}{\varphi_0(k^{-2})}. \tag{2.12}$$

In particular for $a = 1$ (suppressed in notation), one has

$$\zeta_{\Delta_k}(0) = -1, \tag{2.13}$$

and also the Ray–Singer log-determinant is well-defined:

$$\log \mathrm{Det}\, \Delta_k := -\zeta'_{\Delta_k}(0).$$

Differentiating the 1-parameter family of zeta functions corresponding to (2.11) one obtains the identity

$$\frac{d}{ds}\zeta_{\Delta_{sh}}(z) = -z\,\zeta_{\Delta_{sh}}(h,z), \quad \forall z \in \mathbb{C},$$

which in turn yields the variation formula

$$-\frac{d}{ds}\zeta'_{\Delta_{sh}}(0) = \zeta_{\Delta_{sh}}(h,0).$$

From Equations (2.12) and (2.7) applied to the weights φ_s with dilaton sh one obtains

$$\begin{aligned}\log \mathrm{Det}\, \Delta_k = \log \mathrm{Det}\, \Delta + \log\varphi(1) - \frac{\pi}{\Im\tau}\int_0^1 \varphi_0\Big(h\big(sK_0(s\nabla)(\Delta(\log k))\\ + s^2 H_0(s\nabla^{(1)}, s\nabla^{(2)})(\square_{\Re}(\log k))\big)\Big)ds\end{aligned}$$

The first term is the same as for the corresponding elliptic curve and by the Kronecker limit formula has the expression (cf. [RaSi73, Theorem 4.1])

$$\log \mathrm{Det}\, \Delta = -\frac{d}{ds}\Big|_{s=0} \sum_{(n,m)\neq(0,0)} |n+m\tau|^{-2s} = \log\left(4\pi^2\,|\eta(\tau)|^4\right),$$

where η is the Dedekind eta function $\eta(\tau) = e^{\frac{\pi i}{12}\tau}\prod_{n>0}\left(1-e^{2\pi i n\tau}\right)$. After a series of technical manipulations of the last term (see [CoMo14, §4.1]), one obtains the *modular analogue of Polyakov's anomaly formula*:

$$\log \mathrm{Det}\, \Delta_k = \log\left(4\pi^2\,|\eta(\tau)|^4\right) + \log\varphi(1) - \frac{\pi}{4\Im\tau}\varphi_0\left(K_+(\nabla^{(1)})(\square_{\Re}(h))\right), \tag{2.14}$$

where $K_+(s) := \frac{4}{s^2} - \frac{2\coth(\frac{s}{2})}{v} \geq 0$, $s \in \mathbb{R}$. Furthermore, it is shown in [CoMo14, Proof of Theorem 4.6] that the positivity of the function K_+ can be upgraded to operator positivity, implying the inequality

$$\varphi_0\left(K_+(\nabla^{(1)})(\square_{\Re}(\log k))\right) \geq 0, \tag{2.15}$$

with equality only for $k = 1$.

The (negative of) log-determinant can be turned into a scale invariant functional by adding the area term:

$$F(\log k) := \zeta'_{\Delta_k}(0) + \log\varphi(1) \;=\; -\log\mathrm{Det}(\Delta_k) + \log\varphi(1). \tag{2.16}$$

Due to the equality Equation (2.12), the corrected functional F remains unchanged when the Weyl factor k is multiplied by a scalar. In the new notation the identity Equation (2.14) reads as follows:

$$F(h) = -\log\left(4\pi^2\,|\eta(\tau)|^4\right) + \frac{\pi}{4\Im\tau}\varphi_0\left(K_+(\nabla^{(1)})(\square_{\Re}(h))\right). \tag{2.17}$$

In view of the inequality Equation (2.15) one concludes that, as in the case of the ordinary torus (cf. [OPS88]), *the scale invariant functional F attains its extremal value only for the trivial Weyl factor*, in other words at the flat metric.

The gradient of F is defined by means of the inner product of $L^2(A_\theta, \varphi_0)$ via the pairing

$$\langle \mathrm{grad}_h\, F, a\rangle \equiv \varphi_0(a\ \mathrm{grad}_h\, F) = \frac{d}{d\varepsilon}\Big|_{\varepsilon=0} F(h + \varepsilon a), \qquad a = a^* \in \mathcal{A}_\theta.$$

A direct computation of the gradient, using the definition Equation (2.16) combined with the identities Equations (2.12) and (2.7), yields the following explicit expression (cf. [CoMo14, Theorem 4.8]):

$$\mathrm{grad}_h\, F = \frac{\pi}{4\Im(\tau)}\left(\widetilde{K}(\nabla)(\Delta(h)) + \widetilde{H}(\nabla^{(1)}, \nabla^{(2)})(\square_{\Re}(h))\right). \tag{2.18}$$

In the case of the ordinary torus the gradient of the corresponding functional (cf. [OPS88, (3.8)]) gives precisely the Gaussian curvature. This makes it compelling to take the above formula as definition of the *modular Gaussian curvature*.

Finally, computing the gradient of F out of its explicit formula Equation (2.17), and then comparing with the expression Equation (2.18), produces the functional identity Equation (2.8) relating $\widetilde{H}$ and $\widetilde{K}$.

3 Morita invariance of the modular curvature

3.1 Foliation algebras and Heisenberg bimodules

The most suggestive depiction of the noncommutative torus was given by Connes in [CON82], where he described it as the "space of leaves" for the Kronecker foliation $\mathcal{F}_\theta$ of the ordinary torus $\mathbb{T}^2 = \mathbb{R}^2/\mathbb{Z}^2$, given by the differential equation $dy - \theta dx = 0$ with $\theta \in \mathbb{R} \setminus \mathbb{Q}$. The holonomy groupoid $\mathcal{G}_\theta$ of this foliation identifies with the smooth groupoid determined by the flow of the above equation. Its convolution C^*-algebra $C^*(\mathcal{G}_\theta)$, which represents the (coordinates of the) space of leaves, coincides with the crossed product $C(\mathbb{T}^2) \rtimes_\theta \mathbb{R}$, where the action of $\mathbb{R}$ on $\mathbb{T}^2$ is given by the flow Equation (2.1). $C^*(\mathcal{G}_\theta)$ is isomorphic to $A_\theta \otimes \mathcal{K}$, where $\mathcal{K}$ denotes the C^*-algebra of compact operators, and thus strongly Morita equivalent to A_θ.

Finer geometric representations of the space of leaves are obtained by passing to reduced C^*-algebras associated with complete transversals. Any pair of relatively prime integers $(d, c) \in \mathbb{Z}^2$ determines a family of lines of slope $\frac{d}{c}$, which project onto simple closed geodesics in the same free homotopy class, and the free homotopy classes of closed geodesics on $\mathbb{T}^2$ are parametrized by the rational projective line $P^1(\mathbb{Q}) \equiv \mathbb{Q} \cup \{\frac{1}{0}\}$. Letting $N_{c,d}$ denote the primitive closed geodesic of slope $\frac{d}{c}$ passing through the base point of $\mathbb{T}^2$, one obtains a complete transversal for $\mathcal{F}_\theta$. The convolution algebra of the corresponding étale holonomy groupoid identifies with the crossed product algebra $C(\mathbb{R}/\mathbb{Z}) \rtimes_{\theta'} \mathbb{Z}$, where $1 \in \mathbb{Z}$ acts by the rotation of angle $\theta' = \frac{a\theta+b}{c\theta+d}$ with $a, b \in \mathbb{Z}$ chosen such that $ad - bc = 1$. This C^*-algebra is none other than $A_{\theta'}$. In particular, $A_\theta = C(\mathbb{R}/\mathbb{Z}) \rtimes_\theta \mathbb{Z}$ is the reduced C^*-algebra associated with $N_{0,1}$. By construction all algebras $A_{\theta'}$ with $\theta' = g \cdot \theta$, $g \in \mathrm{SL}(2, \mathbb{Z})$, are Morita equivalent, and they actually exhaust (cf. [RIE81]) all the noncommutative tori Morita equivalent to A_θ.

In the same framework Connes [CON82, §13] gave a geometric description of the $(A_{\theta'}, A_\theta)$-bimodules $E(g, \theta)$ implementing the Morita equivalence of A_θ with $A_{\theta'}$. $E(g, \theta)$ is a completion of the $(\mathcal{A}_{\theta'}, \mathcal{A}_\theta)$-bimodule $\mathcal{E}(g, \theta) := \mathcal{S}(\mathbb{R})^{|c|} \equiv \mathcal{S}(\mathbb{R} \times \mathbb{Z}_c)$, $\mathbb{Z}_c := \mathbb{Z}/c\mathbb{Z}$, with the actions defined as follows:

$$(fU_1)(t, \alpha) := e^{2\pi i(t - \frac{\alpha d}{c})} f(t, \alpha), \; (fU_2)(t, \alpha) := f\left(t - \frac{c\theta + d}{c}, \alpha - 1\right);$$

$$(V_1 f)(t, \alpha) := e^{2\pi i\left(\frac{t}{c\theta+d} - \frac{\alpha}{c}\right)} f(t, \alpha), \; (V_2 f)(t, \alpha) := f\left(t - \frac{1}{c}, \alpha - a\right).$$

If $c = 0$, then $E(g, \theta) = A_\theta^{\mathrm{op}}$ is the trivial $(A_\theta^{\mathrm{op}}, A_\theta)$-bimodule. By analogy with the vector bundles over elliptic curves, one defines the rank, degree, and slope of $\mathcal{E}(g, \theta)$ by $\mathrm{rk}(g, \theta) = c\theta + d$, $\deg(g, \theta) = c$, resp. $\mu(g, \theta) := \frac{\deg(g,\theta)}{\mathrm{rk}(g,\theta)}$.

The L^2-scalar product on $E(g, \theta)$

$$< f_1, f_2 >:= \int_{\mathbb{R}\times\mathbb{Z}_c} \overline{f_1(t,\alpha)} f_2(t,\alpha) dt d\alpha$$

where the integration is with respect to the Lebesgue measure on $\mathbb{R}$ and the counting measure on $\mathbb{Z}_c$, determines uniquely A_θ-valued and $A_{\theta'}$-valued inner products satisfying the double equality

$$|\mathrm{rk}(g,\theta)|\, \varphi_0'({}_{A_{\theta'}}< f_2, f_1 >) = < f_1, f_2 > = \varphi_0(< f_1, f_2 >_{A_\theta}), \tag{3.1}$$

where φ_0' stands for the trace of $A_{\theta'}$. The completion $E(g,\theta)$ with respect to $| < \cdot,\cdot >_{A_\theta} |^{1/2}$ is a full right C^*-module over A_θ, and $\mathrm{End}_{A_\theta} = A_{\theta'}$. In addition, $\mathcal{H}_0(g,\theta) := E(g,\theta) \otimes_{A_\theta} L^2(\mathcal{A}_\theta, \varphi_0)$ is the Hilbert space $L^2(\mathbb{R}\times\mathbb{Z}_c)$.

Instead of the $\mathbb{R}^2$-action Equation (2.1), the nontrivial bimodules $\mathcal{E}(g,\theta)$ are acted upon by the Heisenberg group $H_3(\mathbb{R})$. Equivalently, $\mathbb{R}^2$ acts projectively on $\mathcal{E}(g,\theta)$, and this action is compatible with the natural $\mathbb{R}^2$-actions on $\mathcal{A}_\theta$ and $\mathcal{A}_{\theta'}$. At the Lie algebra level, this action gives rise to the standard connection $\nabla^{\mathcal{E}}$ on $\mathcal{E}(g,\theta)$, given by the derivatives $(\nabla_1 f)(t,\alpha) = \frac{\partial}{\partial t} f(t,\alpha)$, $(\nabla_2 f)(t,\alpha) = 2\pi i\, t\, \mu(g,\theta)\, f(t,\alpha)$ with constant curvature: $[\nabla_1, \nabla_2] = 2\pi i \mu(g,\theta)$ Id. Furthermore, this connection is bi-Hermitian, in the sense that it preserves both the A_θ-valued and the $A_{\theta'}$-valued inner product.

3.2 *Modular Heisenberg spectral triples*

Each bimodule $\mathcal{E} = \mathcal{E}(g,\theta)$ gives rise to a double spectral triple, by coupling it with the flat Dirac D_τ by means of its standard connection. Specifically, $\nabla^{\mathcal{E}}$ splits into holomorphic and anti-holomorphic components, $\nabla^{\mathcal{E}} = \partial_{\mathcal{E}} \oplus \partial_{\mathcal{E}}^*$, where $\partial_{\mathcal{E}} := \nabla_1 + \bar{\tau}\nabla_2$. One then forms the operator $D_{\mathcal{E}} = \begin{pmatrix} 0 & \partial_{\mathcal{E}}^* \\ \partial_{\mathcal{E}} & 0 \end{pmatrix}$ acting on the Hilbert space $\widetilde{\mathcal{H}}(g,\theta) = \mathcal{H}_0(g,\theta) \oplus \mathcal{H}^{(1,0)}(g,\theta)$, where $\mathcal{H}^{(1,0)}(g,\theta) := E(g,\theta) \otimes_{\mathcal{A}_\theta} \mathcal{H}^{(1,0)}(\mathcal{A}_\theta)$. Together with the natural right action of A_θ, these data define a spectral triple of constant curvature $\left(A_\theta^{\mathrm{op}}, \widetilde{\mathcal{H}}(g,\theta), D_{\mathcal{E}}\right)$. We note that from the spectral point of view the operator $D_{\mathcal{E}}$ resembles the Hodge–de Rham operator of an elliptic curve with coefficients in a line bundle. In particular, its Laplacian $\Delta_{\mathcal{E}} = \partial_{\mathcal{E}}^* \partial_{\mathcal{E}}$ is a direct sum of $|\deg(\mathcal{E})|$ copies of the harmonic oscillator

$$H := -\frac{d^2}{dt^2} + 4\pi^2 \mu(\mathcal{E})^2 |\tau|^2 t^2 - 4\pi i \mu(\mathcal{E}) \Re(\tau)\, t\, \frac{d}{dt} - 2\pi i \mu(\mathcal{E}) \bar{\tau}\, \mathrm{Id}\,.$$

Now turning on the conformal change Equation (2.3) from φ_0 to φ_h, one replaces $D_{\mathcal{E}}$ by $D_{\mathcal{E},\varphi}$ in the same way as in Section 2.2. The resulting spectral triple over the algebra A_θ^{op} is again a twisted one. After correcting for the lack of unitarity of the action of A_θ^{op} again as in Section 2.2, the operator $D_{\mathcal{E},\varphi}$ is being canonically identified with $D_{\mathcal{E},k} := \begin{pmatrix} 0 & R_k \partial_{\mathcal{E}}^* \\ \partial_{\mathcal{E}} R_k & 0 \end{pmatrix}$ acting on $\widetilde{\mathcal{H}}_0(g,\theta) = \mathcal{H}_0(g,\theta) \oplus \mathcal{H}_0(g,\theta)$.

The appropriate *transposed* in this setting is constructed using the canonical anti-isomorphism from $\mathcal{E} = \mathcal{E}(g,\theta)$ to $\mathcal{E}' := \mathcal{E}(g^{-1},\theta')$,

$$J_{g,\theta}(f)(x,\alpha) = \overline{f((c\theta+d)x,\, -d^{-1}\alpha)}, \quad f \in \mathcal{E}(g,\theta),$$

which switches the $(\mathcal{A}_{\theta'},\mathcal{A}_\theta)$-action on the first into the $(\mathcal{A}_\theta,\mathcal{A}_{\theta'})$-action on the second. We thus arrive at the modular Heisenberg spectral triple $\big(A_\theta,\, \widetilde{\mathcal{H}}_0(g^{-1},\theta'),\, \overline{D}_{\mathcal{E}',k}\big)$ with $\overline{D}_{\mathcal{E}',k} = -\operatorname{rk}(\mathcal{E}') \begin{pmatrix} 0 & k\partial_{\mathcal{E}'} \\ \partial_{\mathcal{E}'}^* k & 0 \end{pmatrix}$. Its Laplacian on sections is $\triangle_{\mathcal{E}',k} = \operatorname{rk}(\mathcal{E}')^2\, k\partial_{\mathcal{E}'}\partial_{\mathcal{E}'}^* k$.

A moment of reflection shows that replacing φ_0 by φ is equivalent to changing the Hermitian structure on $\mathcal{E}'$ by $k^{-2} \in \mathcal{A}_\theta \equiv \operatorname{End}_{\mathcal{A}_{\theta'}}(\mathcal{E}')$. Indeed, in view of Equation (3.1) applied to $\mathcal{E}'$, the passage to the $\mathcal{A}_{\theta'}$-valued Hermitian inner product

$$(f_1', f_2')_{\mathcal{A}_{\theta'},k} = |\operatorname{rk}(\mathcal{E}')|^{-1} < k^{-2} f_1', f_2' >_{\mathcal{A}_{\theta'}}, \quad f_1', f_2' \in \mathcal{E}', \tag{3.2}$$

has the same effect on the L^2-inner product, since

$$\begin{aligned}\varphi_0'\left((f_1', f_2')_{\mathcal{A}_{\theta'},k}\right) &= |\operatorname{rk}(\mathcal{E}')|^{-1}\varphi_0'(< k^{-2} f_1', f_2' >_{\mathcal{A}_{\theta'}}) = \varphi_0({}_{\mathcal{A}_\theta}\!< k^{-2} f_1', f_2' >) \\ &= \varphi_0({}_{\mathcal{A}_\theta}\!< f_1', f_2' > k^{-2}) = \varphi({}_{\mathcal{A}_\theta}\!< f_1', f_2' >).\end{aligned}$$

In conclusion, the passage from the "constant curvature metric" on A_θ represented by the Heisenberg spectral triple $\big(A_\theta^{\text{op}}, \widetilde{\mathcal{H}}(g,\theta), D_{\mathcal{E}}\big)$ to the "curved metric" represented by the modular Heisenberg spectral triple $\big(A_\theta, \widetilde{\mathcal{H}}_0(g^{-1},\theta'), \overline{D}_{\mathcal{E}',k}\big)$ can be interpreted as being *effected by changing the Hermitian structure of* $\mathcal{E}'$ *according to* Equation (3.2). Note that this interpretation remains valid even when $c = 0$, i.e., for $\mathcal{E} = \mathcal{A}_\theta$.

The extended version of Connes' pseudodifferential calculus (see Section 4) allows to establish the heat asymptotic expansion

$$\operatorname{Tr}\left(a\, e^{-t\triangle_{\mathcal{E}',k}}\right) \sim_{t\searrow 0} \sum_{q=0}^{\infty} \mathrm{a}_{2q}(a, \triangle_{\mathcal{E}',k})\, t^{q-1}, \quad a \in \mathcal{A}_\theta, \tag{3.3}$$

and express its functional coefficients in local form. In particular, the curvature functional is of the form

$$\mathrm{a}_2(a, \triangle_{\mathcal{E}',k}) = \frac{1}{4\pi\Im\tau}\varphi_{\mathcal{E}'}(a\,\mathcal{K}_{\mathcal{E}',k}) = \frac{1}{4\pi\Im\tau}\operatorname{rk}(\mathcal{E}')\varphi_0(a\,\mathcal{K}_{\mathcal{E}',k}), \quad a \in \mathcal{A}_\theta,$$

where $\varphi_{\mathcal{E}'} := \mathrm{rk}(\mathcal{E}')\varphi_0$ is the natural trace on $\mathcal{A}_\theta = \mathrm{End}_{\mathcal{A}_{\theta'}}(\mathcal{E}')$, and the curvature density has the expression (cf. [LEMO16, Theorem 2.12])

$$\mathcal{K}_{\mathcal{E}',k} = K(\nabla)(\triangle(h)) + H(\nabla^1, \nabla^2)\left(\square^{\Re}(h)\right) + \mu(\mathcal{E}')1. \tag{3.4}$$

3.3 Ray–Singer determinant vs. Yang–Mills functional

To obtain the variation formula of the Ray–Singer log-determinant functional

$$\mathcal{A}_\theta^{\mathrm{sa}} \ni h^* = h \longmapsto \log \mathrm{Det}(\triangle_{\mathcal{E}',k}) := -\zeta'_{\triangle_{\mathcal{E}',k}}(0),$$

one proceeds as in Section 2.5, starting with the insertion of the curvature expression (3.4) in the derivative $-\frac{d}{ds}\zeta'_{\triangle_{\mathcal{E}',k^s}}(0) = \zeta_{\triangle_{\mathcal{E}',k^s}}(h, 0)$. After integrating the resulting expression one arrives at the following exact formula for the Ray–Singer determinant (cf. [LEMO16, Theorem 2.15])

$$\begin{aligned}\log \mathrm{Det}(\triangle_{\mathcal{E}',k}) = \frac{1}{2}|\deg(\mathcal{E}')| \log\left(2|\mu(\mathcal{E}')|\Im(\bar{\tau})\right) - \frac{1}{2}|\deg(\mathcal{E}')|\, \varphi_0(h) \\ - \frac{|\mathrm{rk}(\mathcal{E}')|}{16\pi\Im\tau}\left(\frac{1}{3}\varphi_0(h\triangle h) + \varphi_0\left(K_2(\nabla_h^1)(\square^{\Re}(h))\right)\right).\end{aligned}$$

The scale invariant form of the functional is

$$F_{\mathcal{E}'}(h) = -\log \mathrm{Det}(\triangle_{\mathcal{E}',k}) - \frac{1}{2}|\deg(\mathcal{E}')|\varphi_0(h).$$

Using the preceding formula, its exact expression is seen to be

$$\begin{aligned}F_{\mathcal{E}'}(h) = &-\log \mathrm{Det}(\triangle_{\mathcal{E}'}) \\ &+ \frac{|\mathrm{rk}(\mathcal{E}')|}{16\pi\Im\tau}\left(\frac{1}{3}\varphi_0(h\triangle h) + \varphi_0\left(K_2(\nabla_h^1)(\square^{\Re}(h))\right)\right).\end{aligned} \tag{3.5}$$

When viewed as a functional on the (positive cone of) metrics on the Heisenberg left $\mathcal{A}_\theta$-module $\mathcal{E}'$, $F_{\mathcal{E}'}$ *attains its minimum only at the metric whose corresponding connection compatible with the holomorphic structure has constant curvature* (cf. [LEMO16, Theorem 2.16]).

Thus the Ray–Singer functional behaves in the same manner as the Yang–Mills functional of Connes and Rieffel (cf. [CORI87]), which however is defined on the space of connections on the noncommutative torus.

3.4 Invariance of the Gaussian curvature

The gradient of the functional $F_{\mathcal{E}'}$ now defined via the equation

$$\langle \mathrm{grad}_h F, a \rangle_{\mathcal{E}'} \equiv \frac{1}{4\pi \Im \tau} \varphi_{\mathcal{E}'}(a \cdot \mathrm{grad}_h F_{\mathcal{E}'}) := \frac{d}{d\epsilon}\Big|_{\epsilon=0} F(h + \epsilon a),$$

Its explicit expression can be computed as in [CoMo14, §4.2] and the answer turns out to be exactly the same as in the case of trivial coefficients, cf. Equation (2.18):

$$\mathrm{grad}_h F_{\mathcal{E}'} = \frac{\pi}{4\Im(\tau)} \Big(\widetilde{K}(\nabla)(\Delta(h)) + \widetilde{H}(\nabla^{(1)}, \nabla^{(2)})(\square_{\Re}(h)) \Big) = \mathrm{grad}_h F.$$

This result can be interpreted as expressing the invariance of the modular Gaussian curvature under Morita equivalence in two different ways. First it shows that *the Gaussian curvature associated with a change of Hermitian metric on a Heisenberg equivalence bimodule* $\mathcal{E}'$ *by a fixed positive invertible* $k \in \mathcal{A}_\theta$, *viewed as an element of* $\mathrm{End}_{\mathcal{A}_{\theta'}}(\mathcal{E}')$, *is independent of* $\mathcal{E}'$. Second, regarding the Heisenberg spectral triples with inverse line-element $D_{\mathcal{E}',k}$ as *right spectral triples, conferring metrics to* $\mathcal{A}_{\theta'}$, it proves that *the entire collection of Morita equivalent algebras* $\{\mathcal{A}_{g\cdot\theta};\ \theta \in \mathrm{SL}(2,\mathbb{Z})\}$ *inherits the same modular curvature as the intrinsic one of* $\mathcal{A}_\theta$.

4 Pseudodifferential multipliers and symbol calculus

The main technical device which was used for proving the above results is a pseudodifferential calculus adapted to twisted C^*-dynamical systems, extending the well-known calculi due to Connes [Con80].

Originally, pseudodifferential operators (ΨDO) were invented (see Kohn and Nirenberg [KoNi65] or for a textbook Shubin [Shu01]) to study elliptic partial differential operators. ΨDO form an algebra which contains differential operators *and* the parametrices to elliptic differential operators. They come with a symbolic calculus: while the (complete) symbols of differential operators are *polynomials* in the covariables, ΨDO are obtained by allowing more general types of symbol functions, e.g., of Hörmander type.

4.1 *Ordinary ΨDO in $\mathbb{R}^n$ from the point of view of C^*-dynamical systems*

4.1.1 Standard representation on the L^2-space (GNS space)

Connes' pseudodifferential calculus on a C^*-dynamical system $(A, \mathbb{R}^n, \alpha)$ should be viewed as a pseudodifferential calculus on $\mathbb{R}^n$. To motivate the defining formulas and to connect to the standard pseudodifferential calculus, we briefly recast the latter in the language of C^*-dynamical systems such that the link becomes apparent. Recall that for a suitably nice symbol function $\sigma(\xi, s)$, $s, \xi \in \mathbb{R}^n$[1] one defines the *pseudodifferential operator with complete symbol* σ as

$$\big(\mathrm{Op}(\sigma)u\big)(s) := \int_{\mathbb{R}^n} e^{i\langle s,\xi\rangle}\sigma(\xi, s)\hat{u}(\xi)đ\xi = \int_{\mathbb{R}^n}\int_{\mathbb{R}^n} e^{i\langle s-y,\xi\rangle}\sigma(\xi, s)u(y)dyđ\xi. \tag{4.1}$$

Now let us abuse this formula a little. Let $\mathcal{A}^\infty := \mathcal{S}(\mathbb{R}^n) \subset C_0(\mathbb{R}^n)$ be the Schwartz space viewed as a $*$-subalgebra of $C_0(\mathbb{R}^n)$. It acts on itself by left multiplication. Furthermore, there is a one parameter group of $*$-automorphisms $\alpha_x(f) := f(\cdot - x)$ and a one parameter family of operators $\pi_x(f) := f(\cdot - x)$ satisfying $\pi_y a \pi_{-y} = \alpha_{-y}(a)$, $a \in \mathcal{A}^\infty$. This gives rise to a covariant representation of the dynamical system $(\mathcal{A}^\infty, \mathbb{R}^n, \alpha)$ on the Hilbert space $L^2(\mathbb{R}^n)$ which is the GNS space of the α-invariant tracial weight $\varphi_0(f) = \int_{\mathbb{R}^n} f$, i.e., the completion of $\mathcal{A}^\infty$ with respect to the inner product $\langle f, g\rangle_{\varphi_0} = \varphi_0(f^*g) = \int_{\mathbb{R}^n} \overline{f} g$.

Now given $u \in \mathcal{A}^\infty$ and a symbol $\sigma \in \mathcal{S}(\mathbb{R}^n, \mathcal{A}^\infty) = \mathcal{S}(\mathbb{R}^n, \mathcal{S}(\mathbb{R}^n)) = \mathcal{S}(\mathbb{R}^n \times \mathbb{R}^n)$ we continue from Equation (4.1) and compute

$$\begin{aligned}\big(\mathrm{Op}(\sigma)u\big)(s) &= \int_{\mathbb{R}^n}\Big(\int_{\mathbb{R}^n} e^{i\langle y,\xi\rangle}\sigma(\xi, s)đ\xi\Big)u(s-y)dy \\ &= \int_{\mathbb{R}^n}\big(\mathcal{F}^{-1}_{\xi\to y}\sigma(y)\big)(s)u(s-y)dy \\ &= \int_{\mathbb{R}^n}\sigma^\vee_{\xi\to y}(y)\pi_y u dy(s),\end{aligned} \tag{4.2}$$

with $\sigma^\vee_{\xi\to y} := \mathcal{F}^{-1}_{\xi\to y}\sigma$.

Thus symbols in the Schwartz space $\mathcal{S}(\mathbb{R}^n, \mathcal{A}^\infty)$ act, after a Fourier transform in the first variable, covariantly with respect to the natural representation of the covariance algebra $\mathcal{S}(\mathbb{R}^n, \mathcal{A}^\infty) \rtimes_\alpha \mathbb{R}$ on the GNS space of the weight φ_0.

---

[1]For consistency with the later exposition we deliberately use a somewhat unusual order and naming convention for the variables ξ, s. ξ plays the role of the covariable and the spacial variable s is normally called x in ΨDO textbooks.

We want to view the function $\sigma(\xi, \cdot)$ as an algebra valued function on $\mathbb{R}^n_\xi$.

We note furthermore that for $\sigma \in \mathcal{S}(\mathbb{R}^n, \mathcal{A}^\infty) = \mathcal{S}(\mathbb{R}^n \times \mathbb{R}^n)$ the operator $\mathrm{Op}(\sigma)$ is trace class and from the calculation Equation (4.2) we see that the Schwartz kernel of $\mathrm{Op}(\sigma)$ on the diagonal is given by $\int_{\mathbb{R}^n} \sigma(\xi, \cdot) đ\xi$, hence

$$\mathrm{Tr}\big(\mathrm{Op}(\sigma)\big) = \int_{\mathbb{R}^n} \int_{\mathbb{R}^n} \sigma(\xi, s) ds đ\xi = \varphi_0\big((\sigma^\vee_{\xi \to y}(0)\big) = \int_{\mathbb{R}^n} \varphi_0\big(\sigma(\xi, \cdot) đ\xi. \tag{4.3}$$

We now take the Schwartz functions $\sigma^\vee_{\xi \to y}$ as basic objects. Identifying $f \in \mathcal{S}(\mathbb{R}^n, \mathcal{S}(\mathbb{R}^n))$ with $\pi(f) = \int_{\mathbb{R}^n} f(x) \pi_x dx$ the space $\mathcal{S}(\mathbb{R}^n, \mathcal{S}(\mathbb{R}^n))$ becomes a $*$-algebra with $*$-representation $f \mapsto \pi(f)$ on $L^2(\mathbb{R}^n)$. Explicitly, $\pi(f) \circ \pi(g) = \pi(f * g)$ and $\pi(f^*) = \pi(f^*)$, where

$$f^*(x) = \alpha_x\big(f(-x)^*\big), \quad (f * g)(x) = \int_{\mathbb{R}^n} f(y) \alpha_y\big(g(x - y)\big) dy, \tag{4.4}$$

resp. with the second variables spelled out, $\mathcal{S}(\mathbb{R}^n \times \mathbb{R}^n)$ becomes a $*$-algebra with involution and product given by

$$f^*(x, s) = \overline{f(-x, s - x)}, (f * g)(x, s) = \int_{\mathbb{R}^n} f(y, s) g(x - y, s - y) dy. \tag{4.5}$$

4.1.2 Pseudodifferential multipliers

We now lift the previous $*$-representation to a "universal" multiplier representation as follows:

$\mathcal{S}(\mathbb{R}^n, \mathcal{S}(\mathbb{R}^n))$ is a pre-C^*-module with inner product $\langle f, g \rangle = \int_{\mathbb{R}^n} f(x)^* g(x) dx$. Put

$$(a\, f)(x) = \alpha_{-x}(a) f(x), (\boldsymbol{U}_y f)(x) = f(x - y), \qquad a \in \mathcal{S}(\mathbb{R}^n). \tag{4.6}$$

Since $\boldsymbol{U}_x a \boldsymbol{U}_{-x} = \alpha_x(a)$ this gives rise to a covariant representation of the $*$-algebra $\mathcal{S}(\mathbb{R}^n, \mathcal{S}(\mathbb{R}^n))$ by associating to $f \in \mathcal{S}(\mathbb{R}^n, \mathcal{S}(\mathbb{R}^n))$ the multiplier $M_f = \int_{\mathbb{R}^n} f(x) \boldsymbol{U}_x dx$.

If φ is a α-invariant trace on $\mathcal{S}(\mathbb{R}^n)$, then the *dual trace* $\widehat{\psi}$ on $\mathcal{S}(\mathbb{R}^n, \mathcal{S}(\mathbb{R}^n))$ is given by

$$\widehat{\psi}(f) = \psi\big(f(0)\big) = \int_{\mathbb{R}^n} \psi\big(\hat{f}(\xi)\big) đ\xi. \tag{4.7}$$

Note that $đ\xi$ is the Plancherel measure of the dual group $(\mathbb{R}^n)^\wedge$ w.r.t. the duality pairing $(x, \xi) \mapsto e^{i\langle x, \xi \rangle}$.

In case of the trace $\varphi_0 = \int_{\mathbb{R}^n}$ from the previous section the dual trace equals the trace Equation (4.3) on the Hilbert space representation $L^2(\mathbb{R}^n)$. This equality should be viewed as a coincidence. In general, the dual trace does not coincide with the Hilbert space trace on a representation, resp. this depends on the representation, see Section 5.2.

By associating to $f \in \mathcal{S}(\mathbb{R}^n, \mathcal{A}^\infty)$ the multiplier $M_f = \int_{\mathbb{R}^n} f(x) \boldsymbol{U}_x dx$ the space $\mathcal{S}(\mathbb{R}^n, \mathcal{A}^\infty)$ becomes a $*$-algebra. Putting $P_f := M_{f^\vee}$ and allowing f to be a symbol of Hörmander class $\mathrm{S}^m(\mathbb{R}^n, \mathcal{A}^\infty)$ we obtain an algebra of multipliers which, via the representation π from above, is isomorphic to an algebra of pseudodifferential operators in $\mathbb{R}^n$. We deliberately say "an" and not "the" here as in $\mathbb{R}^n$ there are various versions of such algebras which differ only by the behavior of symbols as the spacial variable $s \to \infty$, cf. [SHU01, Chap. IV].

4.2 *Pseudodifferential multipliers on twisted crossed products*

The action of the Heisenberg group on $\mathcal{E}(g, \theta)$ induces a C^*-dynamical system $(\mathcal{A}, \mathbb{R}^{n=2}, \alpha)$ ($\mathcal{A} = \mathcal{A}_\theta$ or $\mathcal{A} = \mathcal{A}_{\theta'}$). Equivalently, $\mathbb{R}^n$ acts by a *projective* representation with cocycle $e(x, y) := e^{i\langle Bx, y\rangle}$, with a *skew-* symmetric matrix $B = (b_{kl})_{k,l=1}^n$. In order to construct the resolvent of elliptic differential operators (i.e., Laplacians) on Heisenberg modules one therefore extends the previous considerations to twisted C^*-dynamical systems. In the previous two sections we have formulated the standard pseudodifferential operator conventions in such a way that they carry over almost ad verbatim to the twisted case.

Consider a C^*-dynamical system $(A, \mathbb{R}^n, \alpha)$ with, now for simplicity, unital A. Furthermore, let

$$e(x, y) := e^{i\sigma(x,y)} = e^{i\langle Bx, y\rangle}, \quad \sigma(x, y) := \langle Bx, y\rangle \tag{4.8}$$

with a skew-symmetric real $n \times n$-matrix $B = (b_{kl})_{k,l=1}^n$. By $\mathcal{A}^\infty$ we denote the smooth subalgebra, i.e., those $a \in \mathcal{A}$ for which $t \mapsto \alpha_t(a)$ is smooth.

As before the Schwartz space $\mathcal{S}(\mathbb{R}^n, \mathcal{A}^\infty)$ is a pre-C^*-module with inner product $\langle f, g\rangle = \int_{\mathbb{R}^n} f(x)^* g(x) dx$. Putting $(\boldsymbol{U}_y f)(x) = e(x, -y) f(x - y)$ we obtain a projective family of unitaries $\boldsymbol{U}_x^* = \boldsymbol{U}_{-x}$, $\boldsymbol{U}_x \boldsymbol{U}_y = e(x, y) \boldsymbol{U}_{x+y}$, $x, y \in \mathbb{R}^n$, $\boldsymbol{U}_x a \boldsymbol{U}_{-x} = \alpha_x(a)$, $a \in \mathcal{A}^\infty$. Together with $(a f)(x) = \alpha_{-x}(a) f(x)$, $a \in \mathcal{A}^\infty$ and associating to $f \in \mathcal{S}(\mathbb{R}^n, \mathcal{A}^\infty)$ the multiplier $M_f = \int_{\mathbb{R}^n} f(x) \boldsymbol{U}_x dx$ the space $\mathcal{S}(\mathbb{R}^n, \mathcal{A}^\infty)$ becomes a $*$-algebra. Explicitly, cf. Equation (4.4)

$$f^*(x) = \alpha_x\big(f(-x)^*\big), \quad (f * g)(x) = \int_{\mathbb{R}^n} f(y) \alpha_y\big(g(x - y)\big) e(y, x) dy. \tag{4.9}$$

Note that the formula for f^* is the same as in the untwisted case.

As in the untwisted case, a α-invariant trace ψ on $\mathcal{A}$ induces a dual trace $\widehat{\psi}$ on $\mathcal{S}(\mathbb{R}^n, \mathcal{A}^\infty)$ which is given by the same formula as Equation (4.7).

To define the pseudodifferential operator convention we now read Equation (4.2) backwards. Namely, given Schwartz functions $f, u \in \mathcal{S}(\mathbb{R}^n, \mathcal{A}^\infty)$, and abbreviating $f^\vee := \mathcal{F}_{\xi \to y}^{-1} f$ the inverse Fourier transform of f, we find

$$
\begin{aligned}
(M_{f^\vee}u)(x) &:= \Big(\int_{\mathbb{R}^n} f^\vee(y)\boldsymbol{U}_y u\,dy\Big)(x)\\
&= \int_{\mathbb{R}^n} \alpha_{-x}(f^\vee(y))u(x-y)e(x,-y)\,dy\\
&= \int_{\mathbb{R}^n} \alpha_{-x}(f^\vee(x-y))u(y)e(x,y)\,dy \qquad (4.10)\\
&= \int_{\mathbb{R}^n}\int_{\mathbb{R}^n} e^{i\langle x-y,\xi-Bx\rangle}\alpha_{-x}(f(\xi))u(y)dy đ\xi \qquad (4.11)
\end{aligned}
$$

$$
\begin{aligned}
&= \int_{\mathbb{R}^n} e^{i\langle x,\xi\rangle}\alpha_{-x}(f(\xi))\hat{u}(\xi-Bx)\,đ\xi\\
&= \int_{\mathbb{R}^n} e^{i\langle x,\xi\rangle}\alpha_{-x}(f(\xi+Bx))\hat{u}(\xi)\,đ\xi \qquad (4.12)\\
&=: (P_f u)(x) \qquad (4.13)
\end{aligned}
$$

and call the so defined multiplier P_f a *(twisted) pseudodifferential multiplier with symbol* f. This should be compared to Equation (4.1).

Strictly speaking, so far we have only dealt with smoothing operators as all symbols were Schwartz functions. One now has to extend P_f to a larger class of functions f. The purpose of the somewhat lengthy exposition so far was to show that, at least in $\mathbb{R}^n$ but there in a rather broad sense, smoothing operators are nothing but convolution operators and their symbols are obtained by applying a partial Fourier transform. General ΨDO are therefore nothing but *singular* convolution operators. This is not surprising as ΨDO may, via the Schwartz Kernel Theorem, also be viewed as singular integral operators.

The extension to general symbol functions now follows the standard route. Putting $P_f := M_{f^\vee}$ and allowing f to be a symbol of Hörmander class $S^m(\mathbb{R}^n, \mathcal{A}^\infty)$ we obtain a class of multipliers extending the pseudodifferential multipliers à la Connes [CON80] and Baaj's [BAA88A, BAA88B]. Later we will also need the so-called classical (1-step polyhomogeneous) symbols $f \in CS^m(\mathbb{R}^n, \mathcal{A}^\infty)$ which have an asymptotic expansion

$$f \sim \sum_{j=0}^{\infty} f_{m-j}$$

with $f_{m-j}(\lambda\xi) = \lambda^{m-j} \cdot f_{m-j}(\xi)$, $|\xi| \geq 1, \lambda \geq 1$.

Thus for $f \in S^m(\mathbb{R}^n, \mathcal{A}^\infty)$ we obtain a well-defined multiplier P_f acting on the pre-C^*-module $\mathcal{S}(\mathbb{R}^n, \mathcal{A}^\infty)$ with *complete symbol* f. The usual stationary phase arguments (e.g., [SHU01, § I.3]) then allow to prove that the space $L^\bullet_\sigma(\mathbb{R}^n, \mathcal{A}^\infty) = \bigcup_{n\in\mathbb{Z}} L^m_\sigma(\mathbb{R}^n, \mathcal{A}^\infty)$ of twisted pseudodifferential multipliers (as well as its classical counterpart where the symbols f are 1-step polyhomogeneous) is a $*$-algebra.

For symbols $f \in S^m(\mathbb{R}^n, \mathcal{A}^\infty)$, $g \in S^{m'}(\mathbb{R}^n, \mathcal{A}^\infty)$ the composition $P_f \circ P_g$ is a pseudodifferential multiplier with symbol $h \in S^{m+m'}(\mathbb{R}^n, \mathcal{A}^\infty)$ and h has the asymptotic expansion

$$h(t) \sim \sum_\gamma \frac{i^{-|\gamma|}}{\gamma!} (\partial^\gamma f)(t) \partial_y^\gamma \big|_{y=0} \Big(\alpha_{-y}\big(g(t+By)\big)\Big). \tag{4.14}$$

Furthermore, P_f^* is a pseudodifferential multiplier with symbol

$$\sigma(P_f^*) \sim \sum_\gamma \frac{1}{\gamma!} \partial_t^\gamma \delta^\gamma f(t)^*. \tag{4.15}$$

Here δ^γ denotes the basic derivative on $\mathcal{A}$ induced by the flow α: For $a \in \mathcal{A}^\infty$ and a multiindex $\gamma \in \mathbb{Z}_+^n$ it is defined by

$$\delta^\gamma a := i^{|\gamma|} \partial_x^\gamma \big|_{x=0} \alpha_x(a) = i^{-|\gamma|} \partial_x^\gamma \big|_{x=0} \alpha_{-x}(a). \tag{4.16}$$

δ^γ plays the role of the partial derivative $i^{-|\gamma|} \frac{\partial^\gamma}{\partial x^\gamma}$.

4.3 *Differential multipliers*

In the standard calculus differential operators are characterized as those pseudodifferential operators whose complete symbols are polynomial in the covariables ξ. Adopting this as a definition for *differential multipliers* it turns out that the (multiplier counterparts) of the natural first and second order differential operators discussed in Section 2 are differential multipliers in this sense.

Somewhat more formally we call $P_f \in L_\sigma^\bullet(\mathbb{R}^n, \mathcal{A}^\infty)$ a *differential multiplier* of order m if

$$f(\xi) = \sum_{|\gamma| \le m} a_\gamma \xi^\gamma; \quad a_\gamma \in \mathcal{A}^\infty, \tag{4.17}$$

$f \in \mathcal{A}^\infty[\xi_1, \ldots, \xi_n]$ is a polynomial of degree at most m. Here the sum runs over all multiindices $\gamma \in \mathbb{Z}_+^n$ with $|\gamma| \le m$. Clearly, polynomials in ξ are 1-step polyhomogeneous and hence differential multipliers are *classical* pseudodifferential multipliers.

Recall that in the ordinary pseudodifferential calculus the symbol of the basic derivatives $i^{-|\gamma|} \partial_x^\gamma$ is given by ξ^γ. Therefore, for a multiindex γ we put $\boldsymbol{\partial}^\gamma := P_{\xi^\gamma}$. Explicitly, we find from Equation (4.12) for $u \in \mathcal{S}(\mathbb{R}^n, \mathcal{A}^\infty)$

$$\begin{aligned}(\partial^\gamma u)(x) = (P_{\xi^\gamma} u)(x) &= \int_{\mathbb{R}^n} e^{i\langle x,y\rangle} (\xi + Bx)^\gamma \hat{u}(\xi) đ\xi \\ &= i^{-|\gamma|} \partial_y^\gamma \big|_{y=0} \int_{\mathbb{R}^n} e^{i\langle x+y, \xi+Bx\rangle} \hat{u}(\xi) đ\xi \\ &= i^{-|\gamma|} \partial_y^\gamma \big|_{y=0} \big(e(x,y) u(x+y) \big) = i^{|\gamma|} \partial_y^\gamma \big|_{y=0} U_y u(x).\end{aligned} \tag{4.18}$$

It is important to note that due to the twisting in general $\partial^\gamma \partial^{\gamma'} \neq \partial^{\gamma+\gamma'}$, as can be seen either directly or by the just proved product formula.

As in the ordinary pseudodifferential calculus it is in general not true that $P_f^* = P_{f^*}$. However, $\sigma(P_f)^* = \sigma(P_f^*) \mod \mathrm{S}^{m-1}(\mathbb{R}^n, \mathcal{A}^\infty)$.

Furthermore, ∂^γ is formally self-adjoint and thus for any *differential* multiplier we have indeed $P_f^* = P_{f^*}$.

4.4 Differential multipliers of order 1 and 2

We look more closely at the most relevant case of differential multipliers of order 1 and 2. Let e_j, $j = 1, \ldots, n$ be the canonical basis vectors of $\mathbb{R}^n$. We abbreviate $\partial_j := \partial^{e_j}$ and recall that b_{jk} denotes the entries of the skew-symmetric structure matrix of the twisting Equation (4.8). Then by Equation (4.18)

$$\partial_j u(x) = i^{-1} \partial_{y_j} \big|_{y=0} e^{i\langle Bx, y\rangle} u(x+y) = \Big(\frac{1}{i} \partial_{x_j} + b_{jl} x_l \Big) u(x), \tag{4.19}$$

where summing over repeated indices is understood. Thus

$$\partial_j \partial_k = -\partial_{x_j} \partial_{x_k} - i b_{js} x_s \partial_{x_k} - i b_{ks} x_s \partial_{x_j} - i b_{kj} + b_{js} b_{kr} x_s x_r. \tag{4.20}$$

In particular we have the “curvature identity”

$$[\partial_j, \partial_k] = 2i\, b_{jk}. \tag{4.21}$$

The twisting and the noncommutativity has an interesting effect on the symbol calculus. The symbol of $\partial_j \cdot \partial_k$ is not $\xi_j \cdot \xi_k$ but rather it is a consequence of the formula Equation (4.14) that

$$\sigma\big(\partial_j \cdot \partial_k\big) = \xi_j \cdot \xi_k + i b_{jk} = \sigma\big(\partial^{e_j+e_k}\big) + i b_{jk}, \tag{4.22}$$

hence $\partial^{e_j+e_k} = \partial_j \cdot \partial_k - i b_{jk}$. From this the curvature identity Equation (4.21) also follows.

4.4.1 Differential multipliers in dimension $n = 2$

Specializing further to dimension $n = 2$ it is most convenient to make use of the complex Wirtinger derivatives. Furthermore, the structure matrix b_{jk} has only one interesting entry b_{12}. Fixing $\tau \in \mathbb{C}$ with $\Im\tau > 0$ (a complex structure!) we have the following basic differential multipliers:

$$\begin{aligned} \partial_\tau &:= \partial_1 + \overline{\tau}\partial_2, \quad \partial_\tau^* = \partial_1 + \tau\partial_2, \quad \partial_1 := \partial^{1,0}, \partial_2 := \partial^{0,1} \\ [\partial_\tau, \partial_\tau^*] &= -4 \cdot \Im\tau \cdot b_{12} =: c_\tau, \\ \Delta_\tau &:= \frac{1}{2}(\partial_\tau^*\partial_\tau + \partial_\tau\partial_\tau^*) = \partial_1^2 + |\tau|^2\partial_2^2 + \Re\tau(\partial_1\partial_2 + \partial_2\partial_1). \end{aligned}$$

We will first analyze these operators acting as multipliers on the Hilbert module completion of $\mathcal{S}(\mathbb{R}^n, \mathcal{A}^\infty)$. Later on we will have to pass to their concrete counterparts acting on the Heisenberg modules.

5 The resolvent expansion and trace formula

The resolvent trace, or equivalently the heat trace, expansion for second order Laplace type operators goes back at least to Minakshisundaram and Pleijel [MiPl49]. Via Karamata's tauberian theorem there is a connection to the eigenvalue counting function whose asymptotic analysis is quite subtle. The best remainder term for the counting function of general elliptic operators led Hörmander to develop his beautiful theory of Fourier integral operators [Hör71]. Later the resolvent trace (aka heat equation) method led to the development of local index theory [ABP73] with an enormous flow of publications.

In our opinion, by now the most streamlined approach to the resolvent expansion of elliptic *differential* operators is the calculus of parameter dependent pseudodifferential operators which essentially goes back to Seeley's seminal complex powers paper [See67] and which is presented very nicely in Shubin's book [Shu01, § II.9].[2] We will come back to this soon. Our goal here is to show that this calculus carries over to twisted pseudodifferential multipliers and that the second coefficient in the expansion can be calculated quite easily without any computer aid.

We consider the differential multiplier $P = P_{\varepsilon_1,\varepsilon_2} := k^2\Delta_\tau + \varepsilon_1(\partial_\tau k^2)\partial_\tau^* + \varepsilon_2(\partial_\tau^* k^2)\partial_\tau + a_0$, where $a_0 \in \mathcal{A}$ and $\varepsilon_1, \varepsilon_2$ are real parameters. This multiplier contains all conformal Laplace type multipliers, which occur on Heisenberg mod-

[2] In Seeley's paper a subtle oversight caused a certain confusion which, at least among non-experts, seems to exist to this day. The resolvents of elliptic pseudodifferential operators in general only belong to a "weakly parametric" calculus. This difference between the resolvent calculi for differential resp. true pseudodifferential operators was clarified almost 30 years after Seeley's original paper [GrSe95].

ules over noncommutative tori, as special cases. The symbol of P takes the form $\sigma_P(\xi) := a_2(\xi) + a_1(\xi) + a_0$, where $a_0 \in \mathcal{A}^\infty$ is the same as above and

$$\begin{aligned} a_2(\xi) &= k^2|\xi_1 + \overline{\tau}\xi_2|^2 =: k^2|\eta|^2, \\ a_1(\xi) &= \varepsilon_1(\partial_\tau k^2)\overline{\eta} + \varepsilon_2(\partial_\tau^* k^2)\eta, & \eta &:= \xi_1 + \overline{\tau}\xi_2, \\ &=: \varrho_1\overline{\eta} + \varrho_2\eta, & \varrho_1 &:= \varepsilon_1\partial_\tau k^2, \varrho_2 := \varepsilon_2\partial_\tau^* k^2. \end{aligned}$$

The resolvent $(P - \lambda)^{-1}$ belongs to the parameter dependent pseudodifferential calculus and therefore its symbol has a polyhomogeneous expansion $\sigma_{(P-\lambda)^{-1}} \sim b_{-2} + b_{-3} + b_{-4} + \ldots$, where $b_{-k}(\xi, \lambda) \in \mathcal{A}^\infty$ depends smoothly on (ξ, λ) and is *homogeneous* of degree $-k$: $b_{-k}(r\xi, r^2\lambda) = r^{-k}b(\xi, \lambda)$. As a consequence we obtain for the $a \in \mathcal{A}^\infty$ with respect the *dual trace* Equation (4.7) $\hat{\varphi}_0$ (φ_0 is the invariant trace on $\mathcal{A}^\infty$) an asymptotic expansion

$$\varphi_0\Big(e^{-tP}\Big) \sim_{t\searrow 0} \sum_{j=0}^{\infty} a_{2j}(P, a)t^{j-1}, \tag{5.1}$$

where it follows from the homogeneity[3]

$$\begin{aligned} a_{2j}(P, a) &= \int_{R^2}\int_C e^{-t\lambda}\varphi_0\big((b_{-2j}(\xi, gl)\big)đ\lambda đ\xi \\ &= \int_{\mathbb{R}^2}\varphi_0\Big(b_{-2j-2}(\xi, -1)\Big)đ\xi\int_C e^{-t\lambda}(-\lambda)^{-j}đ\lambda. \end{aligned} \tag{5.2}$$

Here C is a contour in the complex plane encircling the positive semiaxis clockwise such that $\int_C e^{-t\lambda}(r - \lambda)^{-1}\, đ\lambda = e^{-tr}$. The second line is a consequence of the homogeneity of the b_{-k} (see [CoMo95, §6]). For the second nontrivial heat coefficient one therefore obtains up to a sign (see loc. cit.)

$$a_2(P, a) = \int_{\mathbb{R}^2}\varphi_0(b_{-4}(\xi, -1))đ\xi.$$

Due to this formula, it will be convenient to compute $b_{-4}(\xi, -1)$ modulo functions of total ξ-integral 0. Up to a function of total ξ-integral 0 we have the following closed formulas for the first three terms in the symbol expansion of $(P - \lambda)^{-1}$:

$$b_{-2} = b = (k^2|\eta|^2 - \lambda)^{-1}, \quad b_{-3} = -bk^2\big(\eta\partial_\tau^* + \overline{\eta}\partial_\tau\big)b - ba_1b,$$

[3] Note that heat/resolvent invariants are enumerated from 0. We are after a_2 which is the second nontrivial heat invariant, as a_1 is always 0 for differential operators, but in the counting of the recursion system it is the third term.

$$b_{-4} = \big(2bk^2|\eta|^2 - 1 - \varepsilon_1 - \varepsilon_2\big)bk^2\Delta_\tau b + \lambda bk^2\big((\partial_\tau^* b)(\partial_\tau b) + (\partial_\tau b)(\partial_\tau^* b)\big)$$
$$+ \varepsilon_1 \cdot \lambda b(\partial_\tau k^2)b\partial_\tau^* b + \varepsilon_2 \cdot \lambda b(\partial_\tau^* k^2)b\partial_\tau b$$
$$+ \varepsilon_1\varepsilon_2 \cdot |\eta|^2 b \cdot \big((\partial_\tau k^2)b(\partial_\tau^* k^2) + (\partial_\tau^* k^2)b\partial_\tau k^2\big) \cdot b - ba_0 b.$$

The proof is straightforward, completely computer free, and fits on two pages, cf. [LeMo16, § 3.3].

5.1 Second heat coefficient

Integrating b_{-4} over ξ is still a little involved and it requires the Rearrangement Lemma [CoMo95, § 6.2]. This was recast and generalized in [Les17]. The calculus of divided differences allows to compute the many explicit integrals in a systematic way. As a result there exist entire functions $K(s)$, $H^{\Re}(s,t)$, $H^{\Im}(s,t)$, such that with $h := \log k^2$ the second heat coefficient of P (w.r.t. the natural dual trace on the twisted crossed product) takes the form

$$a_2(P,a) = \frac{1}{4\pi|\Im\tau|}\varphi_0\Big[a\Big(K(\nabla)(\Delta_\tau h) - k^{-2}a_0$$
$$+ H^{\Re}(\nabla^{(1)},\nabla^{(2)})(\square^{\Re}(h)) + H^{\Im}(\nabla^{(1)},\nabla^{(2)})(\square^{\Im}(h))\Big)\Big].$$

Here, $\square^{\Re/\Im}(h) := \frac{1}{2}\big(\partial_\tau h \cdot \partial_\tau^* \pm \partial_\tau^* h \cdot \partial_\tau h\big)$, $\nabla = -\operatorname{ad}(h)$, and $\nabla^{(i)}$ signifies that it acts on the i-th factor (*cf.* [CoMo14], [Les17]).

The functions K, $H^{\Re}$, $H^{\Im}$ depend only on P but not on τ. They can naturally be expressed in terms of simple divided divided differences of log.

5.2 Effective pseudodifferential operators and trace formulas

We consider the noncommutative torus $\mathcal{A}_\theta$ with generators U_1, U_2 and normalized trace φ_0. Let $f : \mathbb{R}^2 \to \mathcal{A}^\infty$ be a symbol function (or Schwartz function) of sufficiently low order. Recall the trace Equation (4.7) of the *multiplier* P_f: $\hat{\varphi}(P_f) = \int_{\mathbb{R}^2}\varphi_0(f(x))dx$. However, the multiplier P_f is canonically represented as an operator on the GNS space $L^2(A_\theta,\varphi_0)$ by $\mathrm{Op}(f) = \int_{\mathbb{R}^2} f^\vee(x)\pi_x dx$, where $\pi_x(U_1^{n_1}U_2^{n_2}) = e^{i\langle x,n\rangle}U_1^{n_1}U_2^{n_2}$ is the unitary which implements the natural $\mathbb{R}^2$-action on A_θ, cf. Equation (2.1). $\mathrm{Op}(f)$ acts as a trace class operator on $L^2(A_\theta,\varphi_0)$. More concretely, one computes $\mathrm{Op}(f)U_1^{n_1}U_{n_2} = f(-n_1,-n_2)U_1^{n_1}U_2^{n_2}$. Since $(U_1^{n_1}U_2^{n_2})_{n\in\mathbb{Z}^2}$ is an orthonormal basis of $L^2(\mathcal{A}^\infty,\varphi_0)$ we obtain the trace formula

$$\begin{aligned}\operatorname{Tr}(\operatorname{Op}(f)) &= \sum_{n\in\mathbb{Z}^2} \langle U_1^{n_1}U_2^{n_2}, f(-n_1,-n_2)U_1^{n_1}U_2^{n_2}\rangle \\ &= \sum_{n\in\mathbb{Z}^2} \varphi_0\Big(\big(U_1^{n_1}U_2^{n_2}\big)^* f(-n_1,-n_2)U_1^{n_1}U_2^{n_2}\Big) \\ &= \sum_{n\in\mathbb{Z}^2} \varphi_0\big(f(n)\big).\end{aligned} \tag{5.3}$$

This looks a little different from the formula for the dual trace. However, for a parameter dependent symbol $f(x,\lambda)$ we can take advantage of the Poisson summation formula. Then we find

$$\begin{aligned}\sum_{n\in\mathbb{Z}^2} f(k,\lambda) = \sum_{k\in\mathbb{Z}} \hat{f}(2\pi k,\lambda) &\\ &= \hat{f}(0,\lambda) + \sum_{k\in\mathbb{Z}^2\setminus\{0\}} \hat{f}(2\pi k,\lambda) \\ &= \int_{\mathbb{R}^2} f(\xi,\lambda)d\xi + O(\lambda^{-N})\end{aligned} \tag{5.4}$$

for any N. The latter follows from integration by parts in the Fourier transform and the symbol estimates.

Thus the upshot is that for trace class symbols in the parameter dependent calculus the multiplier trace and the trace in the Hilbert space representation coincide only *asymptotically*. However, for computing heat and resolvent trace asymptotics this is good enough.

Furthermore, this observation has a far reaching generalization. Namely, the effective implementation of the pseudodifferential calculus amounts to passing from its realization on multipliers to a direct action on projective representation spaces (Heisenberg modules) or on $L^2(\mathcal{A},\varphi_0)$ itself. More concretely, let $\boldsymbol{\pi}: G \to \mathcal{L}(\mathcal{H})$ be a projective unitary representation of $G = \mathbb{R}^n \times (\mathbb{R}^n)^\wedge$. For a symbol $f \in S^m(\mathbb{R}^n, \mathcal{A}^\infty)$ the assignment $S^m(\mathbb{R}^n,\mathcal{A}^\infty) \ni f \mapsto \operatorname{Op}(f) := \int_G f^\vee(y)\boldsymbol{\pi}(y)dy$ represents pseudodifferential multipliers as concrete operators in $\mathcal{H}$.

By exploiting the representation theory of the Heisenberg group we are able to relate the Hilbert space trace of parameter dependent pseudodifferential operators to the trace of the corresponding multiplier acting on $\mathcal{S}(\mathbb{R}^n, \mathcal{A}^\infty)$. For details see [LeMo16, § 5 and Appendix A].

Acknowledgements The work of M.L. was partially supported by the Hausdorff Center for Mathematics, Bonn, and the work of H. M. was partially supported by the National Science Foundation award DMS-1600541.

References

[ABP73] M. Atiyah, R. Bott, and V. K. Patodi, *On the heat equation and the index theorem*, Invent. Math. **19** (1973), 279–330. MR 0650828

[Baa88a] S. Baaj, *Calcul pseudo-différentiel et produits croisés de C^*-algèbres. I*, C. R. Acad. Sci. Paris Sér. I Math. **307** (1988), no. 11, 581–586. MR 967366 (90a:46171)

[Baa88b] ———, *Calcul pseudo-différentiel et produits croisés de C^*-algèbres. II*, C. R. Acad. Sci. Paris Sér. I Math. **307** (1988), no. 12, 663–666. MR 967808 (90a:46172)

[CoMo95] A. Connes and H. Moscovici, *The local index formula in noncommutative geometry*, Geom. Funct. Anal. **5** (1995), no. 2, 174–243. MR 1334867 (96e:58149)

[CoMo08] A. Connes and H. Moscovici, *Type III and spectral triples*, Traces in number theory, geometry and quantum fields, Aspects Math., E38, Friedr. Vieweg, Wiesbaden, 2008, pp. 57–71. MR 2427588

[CoMo14] ———, *Modular curvature for noncommutative two-tori*, J. Amer. Math. Soc. **27** (2014), no. 3, 639–684. arXiv:1110.3500 [math.QA], MR 3194491

[Con80] A. Connes, *C^* algèbres et géométrie différentielle*, C. R. Acad. Sci. Paris Sér. A-B **290** (1980), no. 13, A599–A604. MR 572645 (81c:46053)

[Con82] A. Connes, *A survey of foliations and operator algebras*, Operator algebras and applications, Part I (Kingston, Ont., 1980), Proc. Sympos. Pure Math., vol. 38, Amer. Math. Soc., Providence, R.I., 1982, pp. 521–628. MR 679730 (84m:58140)

[Con94] A. Connes, *Noncommutative geometry*, Academic Press Inc., San Diego, CA, 1994. MR 1303779 (95j:46063)

[Con13] ———, *On the spectral characterization of manifolds*, J. Noncommut. Geom. **7** (2013), no. 1, 1–82. MR 3032810

[CoRi87] A. Connes and M. A. Rieffel, *Yang-Mills for noncommutative two-tori*, Operator algebras and mathematical physics (Iowa City, Iowa, 1985), Contemp. Math., vol. 62, Amer. Math. Soc., Providence, RI, 1987, pp. 237–266. MR 878383 (88b:58033)

[CoTr11] A. Connes and P. Tretkoff, *The Gauss-Bonnet theorem for the noncommutative two torus*, Noncommutative geometry, arithmetic, and related topics, Johns Hopkins Univ. Press, Baltimore, MD, 2011, pp. 141–158. arXiv:0910.0188 [math.QA], MR 2907006

[FaKh13] F. Fathizadeh and M. Khalkhali, *Scalar curvature for the noncommutative two torus*, J. Noncommut. Geom. **7** (2013), no. 4, 1145–1183. arXiv:1110.3511 [math.QA], MR 3148618

[GrSe95] G. Grubb and R. T. Seeley, *Weakly parametric pseudodifferential operators and Atiyah-Patodi-Singer boundary problems*, Invent. Math. **121** (1995), no. 3, 481–529. MR 1353307 (96k:58216)

[Hei27] W. Heisenberg, *Über den anschaulichen Inhalt der quantentheoretischen Kinematik und Mechanik*, Z. Phys. **43** (1927), 172–198.

[Hör71] L. Hörmander, *Fourier integral operators. I*, Acta Math. **127** (1971), no. 1-2, 79–183. MR 0388463 (52 #9299)

[Ken27] E. H. Kennard, *Zur Quantenmechanik einfacher Bewegungstypen*, Z. Phys. **44** (1927), 326–352.

[KoNi65] J. J. Kohn and L. Nirenberg, *An algebra of pseudo-differential operators*, Comm. Pure Appl. Math. **18** (1965), 269–305. MR 0176362

[LeMo16] M. Lesch and H. Moscovici, *Modular curvature and Morita equivalence*, Geom. Funct. Anal. **26** (2016), no. 3, 818–873. arXiv:1505.00964 [math.QA]

[Les17] M. Lesch, *Divided differences in noncommutative geometry: rearrangement lemma, functional calculus and expansional formula*, J. Noncommut. Geom. **11** (2017), no. 1, 193–223.

[MiPl49] S. Minakshisundaram and Å. Pleijel, *Some properties of the eigenfunctions of the Laplace-operator on Riemannian manifolds*, Canadian J. Math. **1** (1949), 242–256. MR 0031145 (11,108b)

[OPS88] B. OSGOOD, R. PHILLIPS, and P. SARNAK, *Extremals of determinants of Laplacians*, J. Funct. Anal. **80** (1988), no. 1, 148–211. MR 960228 (90d:58159)

[RASI73] D. B. RAY and I. M. SINGER, *Analytic torsion for complex manifolds*, Ann. of Math. (2) **98** (1973), 154–177. MR 0383463

[RIE81] M. A. RIEFFEL, *C^*-algebras associated with irrational rotations*, Pacific J. Math. **93** (1981), no. 2, 415–429. MR 623572 (83b:46087)

[SEE67] R. T. SEELEY, *Complex powers of an elliptic operator*, Singular Integrals (Proc. Sympos. Pure Math., Chicago, Ill., 1966), Amer. Math. Soc., Providence, R.I., 1967, pp. 288–307. MR 0237943 (38 #6220)

[SHU01] M. A. SHUBIN, *Pseudodifferential operators and spectral theory*, second ed., Springer-Verlag, Berlin, 2001, Translated from the 1978 Russian original by Stig I. Andersson. MR 1852334 (2002d:47073)

[WEY28] H. WEYL, *Gruppentheorie und Quantenmechanik*, Hirzel, Leipzig, 1928.

Advances in Dixmier traces and applications

Steven Lord, Fedor A. Sukochev, and Dmitriy Zanin

Abstract Jacques Dixmier constructed a trace in the 1960s on an ideal larger than the trace class. In 1988 Alain Connes developed Dixmier's trace and used it centrally in noncommutative geometry, extending classical Yang-Mills actions, the noncommutative residue of Adler, Manin, Wodzicki and Guillemin, and integration of differential forms.

Independent of Dixmier's construction and Connes development, Albrecht Pietsch identified a bijective correspondence between traces on two-sided ideals and shift invariant functionals in the 1980s. At the same time Kalton and Figiel identified the commutator subspace of trace class operators, showing that there exist traces different from 'the trace' on the trace class ideal. The commutator approach was subsequently developed in the 1990s for arbitrary ideals by Dykema, Figiel, Weiss and Wodzicki.

We survey recent advances in singular traces, of which Dixmier's trace is an example, based on the approaches of Dixmier, Connes, Pietsch, Kalton, Figiel and the approach of Dykema, Figiel, Weiss and Wodzicki. The results include the bijective association of positive traces with Banach limits, the characterisation of Dixmier traces within this bijection, Lidskii and Fredholm formulations of singular traces as the summation of divergent sums of eigenvalues and expectation values, and their calculation using zeta function residues, heat semigroup asymptotics and symbols of integral operators.

There are basic implications of these advances for users in noncommutative geometry such as the redundancy of the requirement for invariance properties of the extended limit used in Dixmier's trace, the capacity to calculate traces for resolvents of non-smooth partial differential operators and the characterisation of independence from which singular trace is used in terms of the rate of log divergence of the series of energy expectation values—a more physically suitable criteria to impose, or to test the satisfaction of, than series of generally intractable singular values of products of operators. We also survey recent applications in

---

S. Lord · F. A. Sukochev (✉) · D. Zanin
School of Mathematics and Statistics, University of New South Wales, Sydney, NSW, Australia
e-mail: steven.lord@unsw.edu.au; f.sukochev@unsw.edu.au; d.zanin@unsw.edu.au

A. Chamseddine et al. (eds.), *Advances in Noncommutative Geometry*,
https://doi.org/10.1007/978-3-030-29597-4_9

noncommutative geometry such as calculation of traces using noncommutative symbols, that Connes' Hochschild Character formula holds for any trace, and extensions of Connes' results for quantum differentiability for Euclidean space and the noncommutative torus.

Keywords Singular trace · Dixmier trace · Noncommutative geometry

2000 Mathematics Subject Classification Primary 46L51

1 Introduction

The work of famous mathematicians, Fredholm, Hilbert, Weyl, von Neumann, led to the definition and study of the trace on a separable Hilbert space and the ideal of trace class operators. Early in the 1950s Halmos proved that every bounded operator of the Hilbert space to itself is the sum of two commutators involving bounded operators [98, 99]. Brown, Pearcy and Topping extended this in the 1970s to show that every compact operator is a sum of four commutators involving compact operators [21, 100, 165, 72]. Therefore, because a trace vanishes on commutators, the ideals of bounded and compact operators cannot have a non-zero trace. A fundamental question in functional analysis became whether the trace on the trace class operators was the only (non-zero) trace on a two-sided ideal of bounded operators.

Jacques Dixmier constructed such a trace in the 1960s on the ideal of compact operators whose series of singular values diverge as $\log(n)$ [61]. From 1988 Alain Connes developed the theory of Dixmier's trace [39, 38], with also Henri Moscovici [45]. Alain Connes coined the name "Dixmier trace" in the 1988 publication [39] (and Dixmier's trace was the topic of Connes' course at College de France the year before). An extensive body of work followed with other authors generalising the construction and applications initiated by Connes, e.g. [3, 4, 162, 65, 96, 95, 163, 147, 164].

Connes used the Dixmier trace centrally in noncommutative geometry, demonstrating its role in a unique and remarkable theory of noncommutative integration based on differential geometry. In [39] and [38] Connes applied the Dixmier trace to extensions of classical Yang-Mills and Polyakov actions, extended the noncommutative residue of Adler, Manin, Wodzicki and Guillemin, showed an equivalent expression for the Hochschild class of the Chern Character in terms of the Dixmier trace, and introduced the fundamental relation between the noncommutative integral defined by the Dixmier trace and Voiculescu's obstruction to the Berg-Weyl-von Neumann discretisation theorem for tuples.

After Dixmier's trace became central in Connes' noncommutative geometry [47], interest in the topic of traces exploded. Most users of noncommutative geometry still refer to the original presentation of Dixmier's trace by Connes from 1988 [39, 38] (or the monograph "Noncommutative Geometry" from 1994 [40]). With the condition of translation and dilation invariance of the extended limit removed [199] [150, Ch. 9], and the highly non-trivial fact that eigenvalues can be used instead of singular values [121, 70, 125, 200, 217], Dixmier's functional is

$$\omega\left(\frac{1}{\log(n)}\sum_{j=0}^{n}\lambda(j, A)\right) \tag{1.1}$$

where $\lambda(j, A)$, $j \geq 0$ are the eigenvalues of a compact operator A in any order so that $|\lambda(j, A)|$, $j \geq 0$ is non-increasing and $\omega \in \ell_\infty^*$ is an extended limit. That is, ω is a state on the algebra of bounded complex-valued sequences ℓ_∞ vanishing on the ideal c_0 of sequences converging to 0. These conditions provide that $\omega(x) = c$ whenever $\lim_{n\to\infty} x_n = c$, hence the term extended limit. The dependence on ω cannot be removed in general, so there are in fact many Dixmier traces. Formula (1.1) is finite if, ordered as before,

$$n\lambda(n, |A|) = O(1), \quad n \geq 0$$

(though not only if, see the discussion in Section 3.2.1) and defines a trace on the ideal $\mathcal{L}_{1,\infty}$ of compact operators satisfying this condition. The formula is the realisation of the formal idea of log divergence of the trace series

$$\lim_{n\to\infty}\frac{1}{\log(n)}\sum_{j=0}^{n}\lambda(j, A). \tag{1.2}$$

Despite Dixmier's trace playing an outstanding and central role in noncommutative geometry, its role is not unique. Some results have been shown to hold for larger classes of traces. Connes' trace theorem recovering the noncommutative residue and integration of forms [38], recovery of the Lebesgue integral [14, 71, 146], and the Hochschild character formula [39, 45, 108, 54, 24, 23] have been proven for all traces on $\mathcal{L}_{1,\infty}$ [125, 30, 149]. This relates to the rate of log divergence of eigenvalues in known applications, which we explain in the text. Though it is not straightforward to show, what we do know about the unique features of Dixmier's trace, and will outline below, is that the formulation of Dixmier's trace in (1.1) describes every positive trace on $\mathcal{L}_{1,\infty}$ monotone with respect to Hardy-Littlewood submajorisation [122, 216], it is in bijective correspondence with the set of factorisable Banach limits [203], and it coincides with the set of so-called heat functionals based on extracting the first coefficient of an asymptotic expansion of the heat semigroup partition function [27, 215]. Dixmier's trace is an example of a singular trace. A singular trace is a trace that vanishes on all finite rank operators.

Other approaches to singular traces developed independently. During the 1980s Albrecht Pietsch, working in the more general area of continuous operators between Banach spaces [167], identified a bijective correspondence between traces on two-sided ideals of continuous operators between Hilbert spaces and translation invariant functionals [171]. Nigel Kalton answered Pietsch's question from 1981 [170] concerning continuous traces different from 'the trace' on proper quasi-Banach ideals of the trace class ideal $\mathcal{L}_1$ in 1987 by constructing singular traces [119, Theorem 6] on certain quasi-Banach ideals. Kalton also identified which quasi-Banach ideals within $\mathcal{L}_1$ possessed no singular traces [119, Theorem 6].

Kalton in 1989 identified the commutator subspace of trace class operators studied by Gary Weiss [236, 120] and Anderson and Vaserstein [8, 7]. In doing so Kalton proved the existence of traces different from 'the trace' on $\mathcal{L}_1$ [120]. An alternative unpublished proof of the same fact using shift invariant functionals on ℓ_1 is due to Tadeusz Figiel [120, p. 73]. Figiel learnt of Pietsch's approach in 1981 (Srni, January 1981) [170, p. 89] and at a meeting in Georgenthal (East-Germany, April 1986). The commutator approach was subsequently developed in the 1990s for arbitrary ideals by Dykema et al. [66]. The approach can be used to define traces and prove their existence [68, 115, 241, 66]. Kalton in 1998 [121] developed the spectral version of the commutator characterisation to investigate which traces are determined completely by eigenvalues [121, 67]. Spectrality of traces had been raised earlier in the Banach space setting by Pietsch [170]. The 1989 paper of Varga [229] contained an approach to the existence of traces on ideals generated by a single compact operator. Another direction involves symmetric functionals [65, 76, 62, 64, 124, 123], where the correspondence of J. W. Calkin is used to transfer the construction and existence of symmetric functionals on sequence or function spaces to the study of traces on operator ideals.

The authors published a monograph on the general study of singular traces in 2012 [150] including some of the above results on Dixmier traces, on the heat semigroup and zeta function formulas associated with them [27, 28, 198, 215] [150, Ch. 8], and when the limit in (1.2) exists and the dependence on the extended limit in (1.1) is removed [147, 199, 223] [150, Ch. 9]. Prior to this there was a survey in 2006 [31]. Since then there have been several major advances. Pietsch's bijective identification of traces has come to the fore [168, 177, 203, 178], which is more constructive than the bijective identification between traces and symmetric functionals on the Calkin sequence space [150, Ch. 4]. The question of when traces on ideals of compact operators are determined solely by eigenvalues has been completely solved [217] (there exist traces which are not determined by eigenvalues). The integral product formula for tensor products of spectral triples is now known to be false without analytic conditions on the heat semigroup [218].

On the application side, integration formulas for forms in Connes' quantised calculus for $\mathbb{R}^d$ and the noncommutative torus have been proven for every trace on $\mathcal{L}_{1,\infty}$ [149], and Connes' Hochschild character formula has been shown for every trace on $\mathcal{L}_{1,\infty}$ in the compact [30] and non-compact case [220]. Singular traces have been applied outside of noncommutative geometry; in perturbation formulas [183], and in Banach geometry arising from the identification of the trace of the Haagerup L_1-space of a type III von Neumann algebra with a singular trace on the $L_{1,\infty}$-space of the type II invariant von Neumann algebra [182].

To this end, we survey these recent advances to Dixmier traces gained from using all the approaches of Dixmier, Connes, Pietsch, Kalton, and subsequently Dykema, Figiel, Weiss and Wodzicki, together with new and resurrected devices such as uniform and logarithmic submajorisation. There are basic implications for users such as the redundancy of translation and dilation invariance in Dixmier's original formulation (Section 3.2), and the capacity to use eigenvalues and expectation values in the formula (Section 3.3). This leads to a characterisation of independence from which trace is used in terms of the rate of log divergence of energy expectation values—a more physically suitable criteria to impose, or to test the satisfaction of, than the same statement for the generally intractable singular values of products of operators.

The main interest for the user are calculation formulas. The construction formulas using singular or eigenvalues sequences seldom provide a format by which the trace of the operators used in noncommutative geometry can be calculated; the almost sole exception being that the logarithmic divergence in (1.2) can be calculated for the Dirichlet Laplacian on a bounded domain in $\mathbb{R}^d$ using Weyl's asymptotics formula for the eigenvalues [237, 112, 97] [188, XIII.15]. Extending Connes trace theorem for classical pseudodifferential operators [38, Theorem 1], we discuss the class of integral operators with L_2-symbols on $L_2(\mathbb{R}^d)$ that have finite log divergence in (free) energy and the calculation of their Dixmier trace using an extended noncommutative residue (Section 5.1.1). This permits the calculation of the noncommutative integral of a bounded linear operator in $\mathcal{L}(H)$, and certain unbounded ones, using symbols (Section 5.1.2). Such symbols support an extended notion of principal symbol (Theorem 5.2), which we discuss and use to calculate the trace of quantised differential d-forms in $\mathbb{R}^d$ (Theorem 5.7). Following Connes identification of the Dixmier trace with the first residue of the zeta function and the first term in the asymptotic expansion of the heat partition function [40, 45, IV] the user should also note new conditions linking the heat semigroup and zeta function formulas to measurability and the calculation of a Dixmier trace as a genuine residue (Section 4.2.3).

After preliminaries on ideals and the singular value function in Section 2, Section 3 discusses existence and construction of singular traces. Here we summarise the situation for traces on $\mathcal{L}_{1,\infty}$ and the position of Dixmier traces. To not distract from the central case we concentrate mostly on the ideal $\mathcal{L}_{1,\infty}$. Section 4 concerns calculation of traces and which operators in the ideal have the same value when a non-zero trace is applied to that operator independent of which non-zero trace is applied—also called measurability [40, p. 308]. We illustrate the calculation formulas and measurability by discussing Connes' trace theorem from [38] for general classes of non-smooth integral operators (Theorems 5.1–5.4), and universal measurability in the Hochschild character formula (Theorem 5.8).

Throughout we highlight some open questions concerning singular traces and noncommutative geometry, and potential areas for new applications.

2 Preliminaries

Denote by $\mathcal{L}(H)$ the algebra of bounded linear operators on a separable Hilbert space H equipped with the uniform norm $\|\cdot\|$. Two operators A and B in $\mathcal{L}(H)$ are unitarily equivalent if

$$A = UBU^*, \quad \text{for some } U^*U = UU^* = 1, U \in \mathcal{L}(H).$$

Fix an orthonormal basis e_n, $n \geq 0$ of H. We identify the algebra l_∞ of bounded sequences with the subalgebra of all diagonal operators with respect to this basis. Define

$$\operatorname{diag}(x) = \sum_{n=0}^{\infty} x_n P_n, \quad x = (x_0, x_1, x_2, \dots) \in \ell_\infty$$

where P_n, $n \geq 0$, are the one-dimensional projections $P_n h = \langle e_n, h \rangle e_n$, $h \in H$. The choice of basis will be inessential, the diagonal operator defined with another basis is unitarily equivalent.

2.1 *Ideals and singular values*

A subspace $\mathcal{E}$ of $\mathcal{L}(H)$ for which

$$BAC \in \mathcal{E} \text{ for all } B, C \in \mathcal{L}(H), A \in \mathcal{E}$$

is an *ideal*. All ideals considered will be two-sided. Proper ideals of $\mathcal{L}(H)$ contain only compact operators. A quasi-norm $\|\cdot\|_{\mathcal{E}}$ on an ideal $\mathcal{E}$ is called *symmetric* if

$$\|BAC\|_{\mathcal{E}} \leq \|B\|\|A\|_{\mathcal{E}}\|C\|, \quad B, C \in \mathcal{L}(H), A \in \mathcal{E}.$$

If $\mathcal{E}$ is complete in a symmetric (quasi-)norm $\|\cdot\|_{\mathcal{E}}$ then $\mathcal{E}$ is called a (quasi-)Banach ideal.

The non-zero eigenvalues of a compact operator A form either a sequence converging to 0 or a finite set. In the former case we define an *eigenvalue sequence* $\lambda(A)$ of A as the sequence of eigenvalues $\lambda(n, A)$, $n \geq 0$ each repeated according to algebraic multiplicity, and arranged in an order such that $|\lambda(n, A)|$ is non-increasing [207, p. 7]. In the latter case we construct a finite sequence $\lambda(n, A)$, $0 \leq n \leq N$ of the non-zero eigenvalues and set $\lambda(n, A) = 0$ for $n > N$. The ordering may not be unique but all eigenvalues sequences are unitarily equivalent (using the implicit embedding of ℓ_∞ in $\mathcal{L}(H)$ given by diag). Since UAU^* has the same eigenvalues with multiplicity as A for any unitary $U \in \mathcal{L}(H)$, then $\lambda(n, A)$ and $\lambda(n, UAU^*)$ are unitarily equivalent. Denoted by $\mu(A)$ the *sequence of singular values* of A,

$\mu(n, A)$, $n \geq 0$, that is, an eigenvalue sequence of $|A|$ [207, Ch. 1] [88, II] [150, Chapter 2].

For $A \in \mathcal{L}(H)$ we also denote by $\mu(A)$ the non-increasing right continuous singular value function [150, Chapter 2]

$$\mu(t, A) = \inf\{\|A(1-P)\| : P = P^* = P^2 \in \mathcal{L}(H), \ \mathrm{Tr}(P) \leq t\}, \quad t > 0.$$

It is easy to see from the formula for the singular value function that

$$\mu(t, UAU^*) = \mu(t, A), \quad t > 0, UU^* = U^*U = 1, U \in \mathcal{L}(H).$$

and that $\mu(t, A)$, $t > 0$ is a step function with

$$\mu(t, A) = \mu(n, A), \quad t \in [n, n+1), n \geq 0.$$

The singular value function generalises the singular values of a compact operator.

For $A, B \in \mathcal{L}(H)$ we say A is Hardy-Littlewood *submajorised* by B, and write $A \prec\prec B$, when

$$\sum_{j=0}^{n} \mu(j, A) \leq \sum_{j=0}^{n} \mu(j, B), \quad \forall n \geq 0.$$

For $x \in \ell_\infty$, the sequence $\mu(n, \mathrm{diag}(x))$, $n \geq 0$ is the decreasing rearrangement of x. The decreasing rearrangement is often denoted x^*; as a special case of the singular value function however, we denote the decreasing rearrangement by $\mu(x)$ using the implicit embedding of ℓ_∞ in $\mathcal{L}(H)$ given by diag. A linear subspace E of ℓ_∞ is called a *symmetric sequence space* if $\mu(x_1) \leq \mu(x_2)$ for $x_2 \in E$ and $x_1 \in \ell_\infty$ implies that $x_1 \in E$. A quasi-norm $\|\cdot\|_E$ on E is symmetric if $\mu(x_1) \leq \mu(x_2)$ implies that $\|x_1\|_E \leq \|x_2\|_E$ for any $x_1, x_2 \in E$.

2.2 *Trace class and the trace*

Let Tr denote the standard trace on $\mathcal{L}(H)$

$$\mathrm{Tr}(A) = \sum_{n=0}^{\infty} \langle e_n, Ae_n \rangle, \quad A \in \mathcal{L}(H).$$

The trace is unitarily invariant

$$\mathrm{Tr}(UAU^*) = \mathrm{Tr}(A), \quad U^*U = UU^* = 1, U \in \mathcal{L}(H)$$

and independent of the choice of basis.

The space of all trace class operators

$$\mathcal{L}_1 = \{A \in \mathcal{L}(H) : \mathrm{Tr}(|A|) < \infty\}$$

forms an ideal of $\mathcal{L}(H)$ [189, p. 207]. The norm

$$\|A\|_1 = \mathrm{Tr}(|A|), \quad A \in \mathcal{L}_1$$

is a symmetric norm on $\mathcal{L}_1$, and $\mathcal{L}_1$ is complete in $\|\cdot\|_1$.

From the existence of a complete basis of eigenvectors for a normal compact operator, it follows that

$$\mathrm{Tr}(A) = \sum_{n=0}^{\infty} \lambda(n, A), \quad AA^* = A^*A, A \in \mathcal{L}_1$$

It is a non-trivial result of Lidskii [142, 184] [207, Sect. 3] that for arbitrary $A \in \mathcal{L}_1$

$$\mathrm{Tr}(A) = \sum_{n=0}^{\infty} \lambda(n, A), \quad A \in \mathcal{L}_1.$$

2.3 Weak trace class and Dixmier's trace

The notion of dimension in noncommutative geometry is described in the compact case by a self-adjoint operator $D : \mathrm{Dom}(D) \to H$ with compact resolvent such that

$$|D|^{-p} \in \mathcal{L}_{1,\infty}$$

for some $p > 0$ [40, p. 546], where

$$\mathcal{L}_{1,\infty} = \{A \in \mathcal{L}(H) : \mu(n, A) \leq c \cdot n^{-1}, c \text{ is a const.}\}.$$

Here $|D|^{-p}$ acts trivially on the finite dimensional kernel of D. Specifying

$$(1 + D^2)^{-p/2} \in \mathcal{L}_{1,\infty}$$

is an equivalent statement [27, Sect. 6], which we often choose to do.

2.3.1 Weak trace class

The assignment

$$\|A\|_{\mathcal{L}_{1,\infty}} = \sup_{n \geq 0} (n+1)\mu(n, A), \quad A \in \mathcal{L}_{1,\infty}$$

is a symmetric quasi-norm on $\mathcal{L}_{1,\infty}$ in whose topology $\mathcal{L}_{1,\infty}$ is closed.

The ideals $\mathcal{L}_{p,\infty}$, $p > 1$, defined as follows, are solid under submajorisation,

$$\begin{aligned}\mathcal{L}_{p,\infty} &= \{A \in \mathcal{L}(H) : \mu(n, A) \leq c \cdot n^{-1/p}, c \text{ is a const.}\}\\ &= \{A \in \mathcal{L}(H) : \mu(n, A) \prec\prec c \cdot n^{-1/p}, c \text{ is a const.}\}\end{aligned}$$

and are Banach ideals. The statement that $|D|^{-p} \in \mathcal{L}_{1,\infty}$ for some $p \geq 1$ is equivalent to $|D|^{-1} \in \mathcal{L}_{p,\infty}$. The ideal

$$\mathcal{M}_{1,\infty} = \{A \in \mathcal{L}(H) : \mu(n, A) \prec\prec c \cdot n^{-1}, c \text{ is a const.}\}$$

is the submajorisation closure of $\mathcal{L}_{1,\infty}$

$$\mathcal{M}_{1,\infty} = \{A \in \mathcal{L}(H) : \mu(A) \prec\prec \mu(B), B \in \mathcal{L}_{1,\infty}\}.$$

It is a Banach ideal under the norm

$$\|A\|_{\mathcal{M}_{1,\infty}} = \inf\{c \geq 0 : \mu(n, A) \prec\prec c \cdot n^{-1}\} = \sup_{n\geq 0} \frac{1}{\log(n+2)} \sum_{j=0}^{n} \mu(j, A).$$

The ideals $\mathcal{L}_{1,\infty}$ and $\mathcal{M}_{1,\infty}$ are not the same [196, 124]. $\mathcal{L}_{1,\infty}$ is not solid under submajorisation and is not a Banach ideal. $\mathcal{M}_{1,\infty}$ is not the Banach envelope of $\mathcal{L}_{1,\infty}$ [172]. The ideal $\mathcal{M}_{1,\infty}$ is often referred to as the dual of the Macaev ideal.

2.3.2 Extended limits

A state on a unital C^*-algebra is a bounded linear functional such that $\omega(1) = 1$. An *extended limit* $\omega \in \ell_\infty^*$ is a state on the algebra of bounded complex-valued sequences ℓ_∞ vanishing on the ideal c_0 of sequences converging to 0. These conditions provide that $\omega(x) = c$ whenever $\lim_{n\to\infty} x_n = c$, c const., hence the term extended limit.

Banach [11, p. 34] and Mazur [154, p. 103] proved the existence of extended limits. Not all the properties of the classical limit can be preserved by extended limits. Mazur [155] noted that an extended limit that preserves the shift invariance of the classical limit, that is

$$\lim_{n\to\infty} x_{n+1} = \lim_{n\to\infty} x_n,$$

cannot preserve the product formula

$$\lim_{n\to\infty} x_n y_n = \lim_{n\to\infty} x_n \cdot \lim_{n\to\infty} y_n.$$

That the product of limits being equal to the limit of the product is equivalent to ω being a character, or pure state, of ℓ_∞ that vanishes on c_0. We leave the discussion of pure states, that is, ultrafilters, to another occasion, and concentrate on extended limits for shift and additional invariances.

Many of the invariances are easier to phrase on $\mathbb{R}_+$ rather than $\mathbb{N} \cup \{0\}$. There is no loss considering states on the C^*-algebra $L_\infty(\mathbb{R}_+)$ and extended limits in the sense that $\omega \in L_\infty(\mathbb{R}_+)^*$ is a state that vanishes on the ideal $C_0(\mathbb{R}_+)$ of functions vanishing at infinity. Banach and Mazur proved the existence of extended limits on $L_\infty(\mathbb{R}_+)$ that extend taking the limit at infinity.

Define the shift operator

$$S_a : L_\infty(\mathbb{R}_+) \to L_\infty(\mathbb{R}_+)\,, \; S_a f(s) = f(s+a), \quad a \geq 0$$

and the dilation operator

$$D_b : L_\infty(\mathbb{R}_+) \to L_\infty(\mathbb{R}_+)\,, \; D_b f(s) = f(b^{-1}s), \quad b > 0.$$

An extended limit $\omega \in L_\infty(\mathbb{R}_+)^*$ is *shift invariant* if

$$\omega(f) = \omega(S_a f), \qquad \text{for all } a \geq 0.$$

Banach [11, p. 34] and Mazur [154, p. 103] proved the existence of shift invariant extended limits, originally termed Banach-Mazur limits but now referred to as Banach limits.

There are other operations on ℓ_∞ or $L_\infty(\mathbb{R}_+)$ required to understand the definition and calculation of Dixmier traces in noncommutative geometry, the most important being the Cesaro mean (or arithmetic mean) operator and the logarithmic mean operator. Discussion of these additional operators we refer to [27, 62, 63, 202, 222, 199, 224, 6]. We define them in the text as they are needed. The existence of extended limits invariant under the operations we consider is provided by Carey et al. [27, Theorem 1.5].

2.3.3 Dixmier trace

For any extended limit $\omega \in \ell_\infty^*$ the functional

$$\mathrm{Tr}_\omega(A) = \omega\left(\frac{1}{\log(n)} \sum_{j=0}^{n} \mu(n, A)\right), \qquad 0 \leq A \in \mathcal{L}_{1,\infty}$$

is positive, unitarily invariant, and normalised in the sense that

$$\mathrm{Tr}_\omega(\mathrm{diag}(n^{-1})) = 1.$$

It is also additive on the positive cone [61] [150, Sect. 9.7]

$$\mathrm{Tr}_\omega(A+B) = \mathrm{Tr}_\omega(A) + \mathrm{Tr}_\omega(B), \quad 0 \le A, B \in \mathcal{L}_{1,\infty}.$$

Therefore it has a unique linear extension $\mathrm{Tr}_\omega : \mathcal{L}_{1,\infty} \to \mathbb{C}$ such that

$$\mathrm{Tr}_\omega(UAU^*) = \mathrm{Tr}_\omega(A), \quad A \in \mathcal{L}_{1,\infty}, UU^* = U^*U = 1, U \in \mathcal{L}(H).$$

It is immediate from linear extension and the definition that

$$|\mathrm{Tr}_\omega(A)| \le \mathrm{Tr}_\omega(|A|) \le \|A\|_{\mathcal{L}_{1,\infty}},$$

and

$$\mathrm{Tr}_\omega(A) = 0, \quad A \in \mathcal{L}_1.$$

Hence Tr_ω is continuous on $\mathcal{L}_{1,\infty}$ and vanishes on the ideal of trace class operators $\mathcal{L}_1$ contained in $\mathcal{L}_{1,\infty}$.

3 Existence and construction of traces

Weyl's formula describes the asymptotic spectral behaviour of a general Laplacian Δ on a closed manifold (a compact manifold without boundary) Ω of dimension d [97] [188, XIII.15] [16, 138] [204, p. 117] (supressing some absolute constants),

$$\lambda(n, -\Delta) \sim \mathrm{Vol}(\Omega)^{-2/d} \cdot n^{2/d}, \quad n \ge 0$$

where λ describes the discrete eigenvalues of the Laplacian ordered, with multiplicity, as an increasing sequence. An inverse power of the negative of such a Laplacian (technically, any pseudodifferential parametrix of $(1-\Delta)^{d/2}$ of order $-d$) has harmonic asymptotic behaviour proportional to the volume

$$\lambda(n, (1-\Delta)^{-d/2}) \sim \mathrm{Vol}(\Omega) \cdot n^{-1}, \quad n \ge 0.$$

From a practical perspective, the log divergence of partial sums in Dixmier's formula in (1.1) (or Section 2.3.3 above) is naturally associated with volumes and with the inherent philosophy of the pseudodifferential calculus that operators of order $-d$ represent infinitesimals on Ω. Connes showed that this philosophy is borne out factually, in that it is indeed true that applying a Dixmier trace yields [38, 14, 148]

$$\mathrm{Tr}_\omega(A(1-\Delta)^{-d/2}) \sim \int_{S^*\Omega} a(v)dv \tag{3.1}$$

for any zero order *classical* pseudodifferential operator $A : C^\infty(\Omega) \to C^\infty(\Omega)$ with principal symbol a, and dv is the Liouville measure on the sphere bundle $S^*\Omega$ [35, VII]. Heuristically, $A(1-\Delta)^{-d/2}$ is the quantisation of $a(v)dv$ and the Dixmier trace acts as integration [93, 41]. The lhs of (3.1) is a genuine noncommutative extension of integration; it is well defined and, in fact, norm continuous for any bounded linear operator $A : L_2(\Omega) \to L_2(\Omega)$ paired with the bounded operator defined by the functional calculus $(1-\Delta)^{-d/2} : L_2(\Omega) \to L_2(\Omega)$.

From pseudodifferential theory, e.g. [204], it may be asked why the Dixmier trace is necessary. The operator $(1-\Delta)^{-\alpha/2}$ is trace class for $\alpha > d$; is

$$\mathrm{Tr}(A(1-\Delta)^{-\alpha/2}) \tag{3.2}$$

not sufficient to recover the integral? Without curvature, e.g. on the flat torus, the answer is yes. Generally, the Laplace-Beltrami operator, for example, contains first and zero order differential terms relating to the metric on Ω. In the expansion of the pseudodifferential operator $(1-\Delta)^{-\alpha/2}$ these become lower order trace class terms that contain additional geometric information [86, 16]. These additional terms do not generally vanish under the trace. However, when $\alpha = d$, the lower order trace class terms do vanish under the Dixmier trace, leaving the leading or asymptotic term relating to volume. Most readers will recognise the zeta function in (3.2) defined on $\alpha > d$ [158, 187, 109]. Poles of its analytic extension relate to the noncommutative residue [239, 240], and have many connections with mathematical physics included zeta-function regularisation [103], Seeley-deWitt coefficients [233] and the Einstein-Hilbert action [127, 1]. From a substitution in the Tauberian theorem of Hardy and Littlewood [102, p. 155], as noted by Connes [40, p. 306], the residue at the first pole $\alpha = d$ indeed becomes equivalent to formula (3.1). However, this only applies to certain Dixmier traces, or is limited to certain linear operators A admitting asymptotics [40, p. 545] [125]. Recent work has developed simplified and natural criteria for operators A such that (3.2) has a meromorphic extension to $\mathrm{Re}(z) > \alpha - 1$, and when the residue of a simple pole at $z = d$ is calculated by a trace which is singular, meaning that it vanishes on finite rank operators (see Section 4.2.3). Extension criteria for $\mathrm{Re}(z) \leq \alpha - 1$, and associations between residues at higher poles and singular traces is an open topic [211].

The use of Dixmier's trace in (3.1), and as the residue at the first pole of (3.2), is not unique [125, 228]. Other singular traces can provide the same result. The structure of singular traces therefore becomes a necessary component of a well-formulated theory of integration in Alain Connes' generalisation of pseudodifferential calculus and algebraic geometry. This structure, which recently has been identified, see below, for positive traces at least, with the infinite-dimensional lattice of Banach limits on ℓ_∞, collapses to the same functional on the operators of classical geometry. What are sufficient, if not necessary, conditions on operators for this same collapse of a lattice of traces to a unique functional as observed in classical geometry? What significance, if any, is there to the existence of the lattice? Operators that separate points in the lattice are not that exotic.

It is not hard to construct a *non-classical* zero order pseudodifferential operator $A : C^\infty(\Omega) \to C^\infty(\Omega)$ such that the lhs of (3.1) depends explicitly on the state $\omega \in \ell_\infty^*$ [125, 203].

From a theoretical point of view, Dixmier's construction is neither injective, in the sense that distinct extended limits can provide the same trace on $\mathcal{L}_{1,\infty}$, nor surjective in the sense that it does not describe all positive traces on $\mathcal{L}_{1,\infty}$ up to a scalar multiple. Instead, the spectral characterisation of sums of commutators developed during the late 1990s [76, 121, 67, 66] and Pietsch's dyadic averaging construction [167, 171, 229] developed during the late 1980s have been the basis for a rapid advance in the study of singular traces—especially for the ideal $\mathcal{L}_{1,\infty}$ of compact operators of weak trace class which feature prominently in Connes' noncommutative geometry. Over the last 5 years a series of papers [125, 217, 175, 177, 203] have solved the problem of bijective identification of positive traces and their spectrality. We highlight these existence and construction results as they have formed the basis for new results for integrals and residues in applications.

3.1 Existence of traces and the commutator subspace

J. W. Calkin first noted the bijective correspondence [22] [206, Ch. 2] [150] between two-sided ideals of compact operators and proper symmetric sequence spaces.[1] The correspondence extends to traces on ideals and rearrangement invariant functionals on symmetric sequence spaces.

3.1.1 Calkin correspondence

A two-sided ideal is completely determined by its diagonal; if $\mathcal{E}$ is a two-sided ideal of compact operators then

$$E = \mathrm{diag}(\mathcal{E}) := \{a \in \ell_\infty : \mathrm{diag}(a) \in \mathcal{E}\}$$

is a symmetric sequence space, conversely, if E is a proper symmetric sequence space then

$$\mathcal{E} := \{A \in \mathcal{L}(H) : \mu(A) \in E\}$$

[1] Symmetric sequence spaces are also called rearrangement invariant sequence spaces. The reader of the literature should be warned that some texts refer to rearrangement invariant spaces solid under Hardy-Littlewood submajorisation as symmetric spaces. The same object in other texts is referred to as fully symmetric spaces.

determines a two-sided ideal of compact operators. Every two-sided ideal of compact operators and every proper symmetric sequence space arise in this way,

$$\mathcal{E} \underset{\mu}{\overset{\text{diag}}{\rightleftarrows}} E.$$

John von Neumann in the 1930s first posed the question whether every symmetric norm on the symmetric sequence space E defines a symmetric norm on $\mathcal{E}$ by the assignment

$$\|A\|_{\mathcal{E}} = \|\mu(A)\|_{E}, \quad A \in \mathcal{E}.$$

This was solved in 2008 [124]. If E is complete, then $\mathcal{E}$ is complete. Hence all Banach ideals are in bijective correspondence with symmetric Banach sequence spaces. The question was resolved for symmetric quasi-norms in 2013 [221] [140].

3.1.2 Commutator subspace

Using the Calkin correspondence, traces on an ideal $\mathcal{E}$ associate with functionals on the sequence space E. A trace on $\mathcal{E}$ is a linear functional $\phi : \mathcal{E} \to \mathbb{C}$ such that

$$\phi(UAU^*) = \phi(A), \quad A \in \mathcal{E}$$

for all unitaries $U \in \mathcal{L}(H)$. That is, it vanishes on the subspace of $\mathcal{E}$

$$\text{linear span}\{A - UAU^* : A \in \mathcal{E}, UU^* = U^*U = 1, U \in \mathcal{L}(H)\}$$

which is equivalent to the subspace of finite sums of commutators

$$\text{Com}(\mathcal{E}) = \text{linear span}\{[A, B] : A \in \mathcal{E}, B \in \mathcal{L}(H)\}.$$

Under the Calkin correspondence $\text{Com}(\mathcal{E})$ corresponds to the subspace

$$Z(E) = \text{linear span}\{x - y : 0 \leq x, y \in E, \mu(x) = \mu(y)\}$$

of E called the centre. This is a consequence of a result due to Dykema et al. [66], see also the Figiel-Kalton theorem [76] and [121, Theorem 3.1]. If $0 \leq A, B \in \mathcal{E}$ then $A - B \in \text{Com}(\mathcal{E})$ if and only if (Theorem 3.2 below, based on [125, Theorem 3.2] which is a refinement of the main result of [66])

$$C\,(\mu(A) - \mu(B)) \in E$$

where C is the Cesaro mean

$$C : \ell_\infty \to \ell_\infty\,, \ (Cx)_n = \frac{1}{n+1}\sum_{j=0}^{n} x_j \quad x = (x_0, x_1, \ldots) \in \ell_\infty.$$

The Figiel-Kalton theorem [76] essentially states that $C\,(\mu(A) - \mu(B)) \in E$ if and only if the difference $\mu(A) - \mu(B)$ belongs to $Z(E)$.

By diagonalisation, a positive compact operator A is unitarily equivalent to $\operatorname{diag}(\mu(A))$, hence

$$A - \operatorname{diag}(\mu(A)) \in \operatorname{Com}(\mathcal{E}), \quad 0 \le A \in \mathcal{E}.$$

The difference between a positive operator A and its Calkin projection onto the diagonal is in the commutator subspace. For a trace ϕ on $\mathcal{E}$ this implies

$$\phi(A) = \phi \circ \operatorname{diag}(\mu(A)), \quad 0 \le A \in \mathcal{E}.$$

Hence, on the positive cone, the trace ϕ on $\mathcal{E}$ can be replaced by the functional $\phi \circ \operatorname{diag}$ on E. A linear functional $f : E \to \mathbb{C}$ is called symmetric if f vanishes on $Z(E)$, and it turns out that traces on the ideal $\mathcal{E}$ are bijective with symmetric functionals on E via the assignment $\phi \circ \operatorname{diag}$ [123, Theorem 5.2] (see also [180, Theorem 2.2]).

To sketch how this result is shown, consider f symmetric on E. Define the functional $f \circ \mu$ on $\mathcal{E}$ by

$$f \circ \mu(A) = f(\mu(A)), \quad 0 \le A \in \mathcal{E}.$$

Unitary invariance and scalar homogeneity on the positive cone are obvious. The challenge is additivity, and it needs to be shown that

$$\mu(A) + \mu(B) - \mu(A + B) \in Z(E) \quad 0 \le A, B \in \mathcal{E}.$$

Inequalities of Hersch [105, 106], or [150, Theorem 3.3, Theorem 3.4] show that

$$0 \le \sum_{j=0}^{n} \mu(j, A) + \mu(j, B) - \mu(j, A+B) \le \sum_{j=n+1}^{2n+1} \mu(j, A+B) \le n\mu(n+1, A+B)$$

which implies

$$C(\mu(A) + \mu(B) - \mu(A + B)) \in E.$$

By the Figiel-Kalton Theorem,

$$\mu(A) + \mu(B) - \mu(A + B) \in Z(E).$$

Similarly, if ϕ is a trace on $\mathcal{E}$, define the functional $\phi \circ \mathrm{diag}$ on E by

$$\phi \circ \mathrm{diag}(x) = \phi(\mathrm{diag}(x)), \quad 0 \leq x \in E.$$

Scalar homogeneity and additivity are obvious. The challenge is showing that $\phi \circ \mathrm{diag}$ vanishes on $Z(E)$. We have

$$\mathrm{diag}(x) - \mathrm{diag}(\mu(x)) = \mathrm{diag}(x) - \mu(\mathrm{diag}(x)) \in \mathrm{Com}(\mathcal{E}).$$

If $\mu(x) = \mu(y)$, $0 \leq x, y \in E$, then

$$\phi(\mathrm{diag}(x)) = \phi(\mathrm{diag}(\mu(x))) = \phi(\mathrm{diag}(\mu(y))) = \phi(\mathrm{diag}(y)).$$

Hence $\phi \circ \mathrm{diag}$ is a symmetric functional on E.

So, the Calkin correspondence also extends to a bijection between traces and symmetric functionals

$$\phi \underset{\circ\mu}{\overset{\circ\mathrm{diag}}{\rightleftarrows}} f.$$

If $\mathcal{E}$ is a quasi-Banach ideal, then the bijection preserves continuity. If $\phi \in \mathcal{E}^*$ is a continuous trace, then $f = \phi \circ \mathrm{diag}$ is continuous and a symmetric functional [124, 221]. Similarly, if $f \in E^*$ is symmetric, then $\phi = f \circ \mu$ is a continuous trace.

3.1.3 Existence of traces

An application of this bijection is the existence of non-trivial traces on ideals. If $Z(E) = E$, equivalently $C(E) \subset E$, then the zero functional is the only symmetric functional. If $Z(E) \neq E$, then for any $x \in E \setminus Z(E)$ there exists a symmetric functional f such that $f(x) \neq 0$.[2] The invariance of the diagonal E under the Cesaro mean can be tested for many ideals. The quest that led Dixmier to his trace was whether the algebraic properties of the type I factor $\mathcal{L}(H)$ implied Tr was the unique semifinite trace. The answer is no. In the following, existence of a trace should be read as existence of a trace different from the zero functional.

Remark 3.1

1. Traces do not require continuity. There are many traces on $\mathcal{L}_1$ [119]. This follows as there exist sequences $x \in \ell_1$ such that $Cx \notin \ell_1$ yet $\sum_{n=0}^{\infty} x_n = 0$. Then

[2] $Z(E)$ is a linear subspace of E and the algebraic dual of $E/Z(E)$ admits a functional $\tilde{f}$ such that $\tilde{f}([x]) \neq 0$. Let f be the extension of $\tilde{f}$ to E vanishing on $Z(E)$.

diag$(x) \in \ker \mathrm{Tr}$ but diag$(x) \notin \mathrm{Com}(\mathcal{L}_1)$. The commutator subspace $\mathrm{Com}(\mathcal{L}_1)$ is a strict subspace of ker Tr. For an example of such a sequence x, set

$$x_0 = -\sum_{n=1}^{\infty} \frac{1}{(n+1)\log^2(n+1)}, \quad x_n = \frac{1}{(n+1)\log^2(n+1)}, \quad n \geq 1.$$

Tr is the unique continuous trace on $\mathcal{L}_1$.

2. The ideal $\mathcal{L}_{1,\infty}^0$ is the maximal ideal (for the partial order of set inclusion) supporting an extension of the trace Tr on finite rank operators [68]. Recall $\mathcal{L}_{1,\infty}^0$ denotes the closure of the finite rank operators in the quasi-norm of $\mathcal{L}_{1,\infty}$.
3. The ideal $\mathcal{L}_{1,\infty}$ is the minimal ideal such that all traces are singular. Recall a trace is singular if it vanishes on all finite rank operators. There is no distinction between a trace and a singular trace on $\mathcal{L}_{1,\infty}$. In fact, every trace on $\mathcal{L}_{1,\infty}$ vanishes on $\mathcal{L}_1$ [150, Theorem 5.7.8]. Every ideal larger than $\mathcal{L}_{1,\infty}$ admits only singular traces that vanish on $\mathcal{L}_1$.
3. There are many unbounded traces and many quasi-norm continuous traces on $\mathcal{L}_{1,\infty}$. Let the subspace K denote the common kernel of all quasi-norm continuous trace on $\mathcal{L}_{1,\infty}$. Since $\mathcal{L}_{1,\infty}^0 \subset K$, the positive cone of $\mathcal{L}_{1,\infty}^0$ is contained in the positive cone of K. Since $C(\mu(x)) \in \ell_{1,\infty}$ for $0 \leq x \in \ell_\infty$ implies that $x \in \ell_1$, the positive cone of the commutator subspace $\mathrm{Com}(\mathcal{L}_{1,\infty})$ is the positive cone of $\mathcal{L}_1$,

$$\mathrm{Com}(\mathcal{L}_{1,\infty})_+ \subsetneq (\mathcal{L}_{1,\infty}^0)_+ \subset K_+$$

Hence $\mathrm{Com}(\mathcal{L}_{1,\infty})$ is not identical to $\mathcal{L}_{1,\infty}^0$ and strictly contained in K. There are unbounded traces on $\mathcal{L}_{1,\infty}$ that do not vanish on $\mathcal{L}_{1,\infty}^0$.
4. Every quasi-norm continuous trace on $\mathcal{L}_{1,\infty}$ is a linear combination of four positive traces on $\mathcal{L}_{1,\infty}$ [44, Corollary 2.2]. Subsequent results that mention only positive traces—construction in Section 3.2, measurability in Section 4 and applications in Section 5, apply therefore to all quasi-norm continuous traces.
5. There exist unbounded traces ϕ on $\mathcal{L}_{1,\infty}$ such that the trace $\mathrm{Tr}+\phi$ is a non-trivial trace on $\mathcal{L}_{1,\infty}^0$ identical to Tr on $\mathcal{L}_1$. Hence the extension of Tr on $\mathcal{L}_1$ to $\mathcal{L}_{1,\infty}^0$ is not unique.
6. Exact conditions for the existence of continuous traces on Banach ideals [150, Theorem 4.1.3] and quasi-Banach ideals [179, Theorem 2.1] are known. There are Banach ideals that admit traces, but no continuous traces at all [150, Example 5.6.9]. Are there Banach ideals where all traces are continuous? A. Pietsch has shown that all traces being continuous is equivalent to a finite number of linearly independent singular traces on a quasi-Banach ideal [176, Theorem 8.14]. Banach ideals admit either an infinite number of linearly independent singular traces, or none [176, Proposition 8.17]. Therefore, Banach ideals that admit a singular trace must also admit a discontinuous trace. This dichotomy, that generally an ideal admits either an infinite number of linearly independent singular traces or none, is an open conjecture [115] [176, Problem 4.6].

3.2 Construction of traces and dyadic averages

The Calkin correspondence provides an identification between traces on an ideal of compact operators and symmetric functionals on the diagonal of the ideal (as a sequence space when given a fixed orthonormal basis of the separable Hilbert space H). The correspondence yields no information about the form of the symmetric functional.

Dixmier's construction is the most approachable of the trace constructions on $\mathcal{L}_{1,\infty}$. We emphasise again the distinction made in the preliminaries; we concentrate on the quasi-Banach ideal of compact operators whose singular value sequence is order $O(n^{-1})$, $n \geq 1$. This is the classical ideal of weak trace class operators whose diagonal corresponds to weak ℓ_1. The term "Dixmier ideal" is often used in noncommutative geometry, with the same notation, to refer to the dual of the Macaev ideal [152]; compact operators whose series of singular values is at most logarithmically divergent. The two ideals are not the same [172], and the properties of traces on them are different.

3.2.1 Dixmier trace

Dixmier's trace was originally defined by him [61] as

$$\mathrm{Tr}_\omega(A) = \omega\left(\frac{1}{\log(n)}\sum_{j=0}^{n}\mu(j,A)\right), \quad A \geq 0 \tag{3.3}$$

and extended from the positive cone to all of $\mathcal{L}_{1,\infty}$ by linearity. According to Dixmier's letter to the 2012 Luminy workshop on singular traces, reproduced in the notes of [150, Ch. 6], Dixmier first considered Banach limits applied to the bounded sequence

$$\frac{\mu(n,A)}{\frac{1}{n+1}} = (n+1)\mu(n,A), \quad n \geq 0. \tag{3.4}$$

Dixmier was unable to prove additivity. N. Aronszajn suggested to Dixmier to use partial sums instead,

$$\frac{\sum_{j=0}^{n}\mu(j,A)}{\sum_{j=0}^{n}\frac{1}{j+1}}, \quad n \geq 0.$$

Under the assumption that the extended limit ω be shift and dilation invariant, Dixmier was able to prove additivity of (3.3). Recalling the Hardy-Littlewood submajorisation preorder, $A \prec\prec B$, $A, B \in \mathcal{L}(H)$

$$\sum_{j=0}^{n} \mu(j, A) \leq \sum_{j=0}^{n} \mu(j, B), \quad n \geq 0,$$

this latter ratio of partial sums being bounded for the compact operator A defines the dual of the Macaev ideal

$$\mathcal{M}_{1,\infty} = \{A \in \mathcal{L}(H) : A \prec\prec \operatorname{diag}(n^{-1})\}$$

and the formula (3.3) provides a trace on the dual of the Macaev ideal. As yet, however, no applications in noncommutative geometry have used operators from $\mathcal{M}_{1,\infty}$ that do not already belong to $\mathcal{L}_{1,\infty}$.

The condition of shift and dilation invariance of the extended limit in (3.3) can be removed completely [199] [150, Ch. 9]. This fact applies to $\mathcal{L}_{1,\infty}$ only. If ω is any state on ℓ_∞ that vanishes on c_0, there exists a dilation and translation invariant state ω_0 such that $\mathrm{Tr}_\omega = \mathrm{Tr}_{\omega_0}$ [222, Theorem 40] [199] [150, Ch. 9]. This displays the lack of injectivity in Dixmier's construction on $\mathcal{L}_{1,\infty}$, the set of Dixmier traces defined by any extended limit is the same as the set defined by translation and dilation invariant extended limits. On the dual of the Macaev ideal the construction still lacks injectivity [222, Theorem 40]; the set of dilation invariant states provides the same set as that of translation and invariant states.

Dixmier's trace is monotone for submajorisation

$$\mathrm{Tr}_\omega(A) \leq \mathrm{Tr}_\omega(B), \quad A \prec\prec B, 0 \leq A, B \in \mathcal{L}_{1,\infty}.$$

It is known that, up to scalar multiple, every positive trace on $\mathcal{L}_{1,\infty}$ that is monotone for submajorisation is a Dixmier trace [203, Corollary 5.7] [122]. This identification with submajorisation monotone traces is the first characterisation of Dixmier's trace. What it shows below is that Dixmier's construction is not surjective. There are positive traces on $\mathcal{L}_{1,\infty}$ that are not monotone for submajorisation.

3.2.2 Dyadic averages

It has turned out that an injective and surjective construction comes, not from the full partial sums as in Aronszajn's suggestion, but from dyadic partial sums. Considering instead the bounded sequence

$$\frac{\sum_{j=2^n-1}^{2^{n+1}-2} \mu(j, A)}{\sum_{j=2^n-1}^{2^{n+1}-2} \frac{1}{j+1}}, \quad n \geq 0, \tag{3.5}$$

define for any Banach limit $\theta \in \ell_\infty^*$ (recall Banach limits are the shift invariant states on ℓ_∞) the functional

$$\tau_\theta(A) = \theta\left(\frac{1}{\log 2}\sum_{j=2^n-1}^{2^{n+1}-2}\mu(j,A)\right), \quad A \geq 0. \tag{3.6}$$

This functional is additive on the positive cone of $\mathcal{L}_{1,\infty}$, and hence τ_θ extends by linearity to a positive trace on $\mathcal{L}_{1,\infty}$. Remarkably, this construction bijectively describes every positive trace on $\mathcal{L}_{1,\infty}$ [177, Theorem 4], see also [203, Corollary 3.8, Corollary 4.2].

Define the operators

$$D : \ell_\infty \to \ell_{1,\infty}\,, \ D' : \ell_{1,\infty} \to \ell_\infty$$

by

$$D(x) := \log 2 \cdot \left(\frac{x_0}{2^0}, \underbrace{\frac{x_1}{2^1}, \frac{x_1}{2^1}}_{2 \text{ times}}, \underbrace{\frac{x_2}{2^2}, \frac{x_2}{2^2}, \frac{x_2}{2^2}, \frac{x_2}{2^2}}_{4 \text{ times}}, \ldots, \underbrace{\frac{x_n}{2^n}, \ldots, \frac{x_n}{2^n}}_{2^n \text{ times}}, \ldots\right), \quad x \in \ell_\infty$$

and

$$D'(x) = \frac{1}{\log 2}\sum_{j=2^n-1}^{2^{n+1}-2}\mu(j,x), \quad x \in \ell_{1,\infty}.$$

The key element of the bijection is the result that [203, Lemma 3.6] (a variation of [175, Lemmas 5.3–5.5])

$$x - DD'(x) \in Z(\ell_{1,\infty}), \quad 0 \leq x \in \ell_{1,\infty}$$

where $Z(\ell_{1,\infty})$ is the centre of $\ell_{1,\infty}$. If f is a symmetric functional on $\ell_{1,\infty}$ and we assign $\theta = f \circ D$ then

$$f(x) = fDD'(x) = \theta(D'(x)) = \theta\left(\frac{1}{\log 2}\sum_{j=2^n-1}^{2^{n+1}-2}\mu(j,x)\right), \quad 0 \leq x \in \ell_{1,\infty}.$$

Hence $\theta = f \circ D$ provides the explicit isometry between symmetric functionals f on $\ell_{1,\infty}$ and shift invariant states θ on ℓ_∞. The Calkin correspondence between symmetric functionals on $\ell_{1,\infty}$ and traces on $\mathcal{L}_{1,\infty}$ then identifies (3.6) as a bijective construction between shift invariant functionals on ℓ_∞ and traces acting on the positive cone of $\mathcal{L}_{1,\infty}$. Those shift invariant functionals that are states (i.e. Banach limits) then correspond with positive traces that are normalised (i.e. $\tau_\theta(\mathrm{diag}(n^{-1})) = 1$).

This bijection is due to A. Pietsch who introduced the dyadic partial sums in the late 1980s on quasi-Banach ideals, summarised in the 1990 publication [171]. Pietsch later showed that the dyadic approach produces the bijection between traces on ideals and shift-monotone invariant linear functionals on shift-monotone invariant sequence spaces for *all* ideals [177, Theorem 4] [178, Theorem 4.6] (see [174] and historical comments in [177]). The bijection obtained by Pietsch is a consequence of his approach to the Calkin correspondence using dyadic decompositions [169, 171, 175], which have provided more results and more direct proofs than using the Schmidt decomposition of a compact operator. Shift-monotone invariant functionals and shift-monotone invariant sequence spaces introduced by Pietsch were largely unstudied compared to rearrangement invariance, symmetric functionals and symmetric sequence spaces, though their equivalence is established [175, Theorem 6.3]. There is, as yet, no neat classification for general shift-monotone invariant states such as their identification with Banach limits that occurs in the case of $\ell_{1,\infty}$. The 1995 doctoral thesis of Varga [230] also contains an approach based on dyadic partial sums. The bijection $\{a_n\}_{n=1}^{\infty} \to \{2^n a_n\}_{n=1}^{\infty}$ transfers between Pietsch's shift-monotone ideals and those introduced by Varga, and the bijection between traces on operator ideals and shift-monotone invariant functionals on associated unbounded shift-monotone invariant sequence space is proved as [230, Theorem 4] in Varga's thesis. Varga assumes that the operator ideals are full, in his terms, a condition that eventually is satisfied by all proper ideals of compact operators (private communication A. Pietsch). The paper [203] repeats many of Pietsch's existing results for the ideal $\mathcal{L}_{1,\infty}$. The format of the operators D and D' in [203] is chosen so that equivalence between the shift invariant states and the classically known set of Banach limits becomes overt in (3.6).

The set of Banach limits is known to have cardinality of beth two $2^{2^{\mathbb{N}}}$ [117, Theorem 3]. Therefore, due to the association with Banach limits, there are a lot of positive traces on $\mathcal{L}_{1,\infty}$; the set of positive traces on $\mathcal{L}_{1,\infty}$ has cardinality $2^{2^{\mathbb{N}}}$, see also [176, Theorem 9.6].

The bijective correspondence (3.6) between positive normalised traces on $\mathcal{L}_{1,\infty}$ and Banach limits provides a very neat classification for Dixmier's trace, and the smaller set of Dixmier traces used by Connes in the monograph "Noncommutative Geometry" in 1994 [40]; first noticed by Pietsch [173].

3.2.3 Characterisation of traces

A Banach limit θ is called a factorisable Banach limit if it is of the form

$$\theta = \gamma \circ C$$

where γ is an extended limit and $C : \ell_\infty \to \ell_\infty$ is the Cesaro operator. Raimi studied factorisable Banach limits in 1980, to determine whether every Banach limit extended Cesaro summability. He gave an example of a Banach limit that was not factorisable [186].

If θ is factorisable, then in (3.6) we have

$$\tau_\theta(A) = \gamma\left(\frac{1}{\log 2^n}\sum_{j=0}^{2^n}\mu(j,A)\right), \quad A \geq 0.$$

This is nearly the formula for a Dixmier trace and shows that τ_θ is submajorisation monotone (recall Dixmier's trace describes all submajorisation monotone traces on $\mathcal{L}_{1,\infty}$). The converse requires showing that if f is a submajorisation monotone symmetric functional on $l_{1,\infty}$, then $f \circ D : \ell_\infty \to \ell_\infty$ is factorisable [203, Theorem 3.13].

Hence the second characterisation of Dixmier's trace is that it is in bijective correspondence with the set of factorisable Banach limits.

On page 305 of "Noncommutative Geometry" [40], the Dixmier trace

$$\mathrm{Tr}_{\omega\circ M}(A) = \omega \circ M\left(\frac{1}{\log(n)}\sum_{j=0}^{n}\mu(j,A)\right), \quad A \geq 0$$

is used where M is the logarithmic form of the Cesaro operator

$$M(x) = \frac{1}{\log(n+2)}\sum_{j=1}^{n}\frac{x_j}{j}, \quad x = (x_1, x_2, \dots) \in \ell_\infty. \tag{3.7}$$

Connes' construction is not equivalent to Dixmier's [147]. This is revealed clearly in the particularly neat classification that in the bijection correspondence of positive normalised traces on $\mathcal{L}_{1,\infty}$ and Banach limits on ℓ_∞, Connes' trace is bijective with twice factorisable Banach limits

$$\theta = \gamma \circ C^2$$

where γ is an extended limit and C is the Cesaro operator [203, Theorem 5.13].

Further details on descendingly smaller sets of traces with ascendingly stronger invariance in the extended limit can be found in[3] [202, 222, 203, 223].

Raimi's construction of a Banach limit that is not factorisable has the consequence that there exist positive normalised traces on $\mathcal{L}_{1,\infty}$ that are not Dixmier traces. Dixmier's construction is not surjective. A similar argument shows that Connes' construction is not equivalent to Dixmier's. The cardinality of sets of once and twice factorisable Banach limits is an open question, as is the cardinality of the set differences of Banach limits and factorisable Banach limits. Hence the

[3] One of the strongest invariances generally considered are traces Tr_ω where $\omega = \omega \circ M$ is an extended limit invariant under M. This set of traces are characterised by what might be considered 'infinitely' factorisable Banach limits where $\theta = \theta \circ C$ is a Banach limit invariant under C [203].

cardinality of the sets of Dixmier traces discussed and the size of the difference between the sets of positive normalised traces and Dixmier traces is not precisely known [173].

3.3 Spectrality and expectation values

Eigenvalues can be used instead of singular values and linear extension in the construction of every positive trace on $\mathcal{L}_{1,\infty}$ since $\mathcal{L}_{1,\infty}$ is a solid set under logarithmic submajorisation [200, 121, 125, 217].

3.3.1 Spectral formulation and limits of expectation values at infinity

Explicitly, for any positive trace on $\mathcal{L}_{1,\infty}$ [203, Theorem 6.2]

$$\tau_\theta(A) = \theta\left(\frac{1}{\log 2}\sum_{j=2^n-1}^{2^{n+1}-2} \lambda(j,A)\right), \quad A \in \mathcal{L}_{1,\infty} \tag{3.8}$$

for any Banach limit θ. We reiterate that $\lambda(j, A)$ is an eigenvalue sequence; the eigenvalues of the compact operator A in any order so that $|\lambda(j, A)|$, $j \geq 0$ is non-increasing.

By the identification with factorisable Banach limits eigenvalues can be used instead of singular values and linear extension in Dixmier's construction. Dixmier's trace on all of $\mathcal{L}_{1,\infty}$, not just the positive cone of the ideal, is [125, Lemma 6.31]

$$\mathrm{Tr}_\omega(A) = \omega\left(\frac{1}{\log(n)}\sum_{j=0}^{n} \lambda(j,A)\right), \quad A \in \mathcal{L}_{1,\infty}. \tag{3.9}$$

The state ω is again any extended limit. Fack in 2004 noted that Dixmier's trace, specifically, was spectral [70]. Spectrality of all traces was proven earlier [67]. Almost no texts in noncommutative geometry mention it in the construction of the Dixmier trace, despite the strong interaction between spectral theory and pseudodifferential operators [204]. We show its utility below by pairing the log divergence of the spectrum of a pseudodifferential operator with integration of its symbol over increasingly larger cylinders in phase space in Section 5.

If we return to the noncommutative integration formula (3.1), it is also true for every bounded operator $A \in \mathcal{L}(L_2(\Omega))$ that [125, Theorem 7.6]

$$\mathrm{Tr}_\omega(A(1-\Delta)^{-d/2}) = \omega\left(\frac{1}{\log(n)}\sum_{j=0}^{n} \langle e_n, Ae_n\rangle (1-\lambda(j,-\Delta))^{-d/2}\right), \quad A \in \mathcal{L}(H) \tag{3.10}$$

where $\lambda(n, -\Delta)$ are the non-increasing eigenvalues of the Laplace-Beltrami operator Δ on the closed manifold Ω, and e_n are the eigenvectors ordered so that $-\Delta e_n = \lambda(n, -\Delta)e_n$, $n \geq 0$. The eigenvectors must be ordered; the order is not uniquely determined but there exist orthonormal bases such that the equality is false [150, Section 7.5]. Using Weyl's law (up to the constant $\mathrm{Vol}(\mathbb{S}^{d-1} \times \Omega)(d(2\pi)^d)^{-1}$ which we suppress)

$$(1 - \lambda(n, -\Delta))^{-d/2} \sim (n+1)^{-1}, \quad n \to \infty,$$

this simplifies even further [150, Theorem 12.1.2]

$$\mathrm{Tr}_\omega(A(1-\Delta)^{-d/2}) = (\omega \circ M)\left(\langle e_n, Ae_n\rangle\right), \quad A \in \mathcal{L}(H)$$

where M is the logarithmic mean operator in (3.7). Hence in noncommutative geometry the notion of a factorisable dilation invariant limit also becomes important. A feature of the identification (3.1), which we will cover next section, is that the equality is independent of the extended limit ω. If $x \in \ell_\infty$ and $\omega(x) = \alpha$ for fixed α for every extended limit ω, then x is convergent such that $\lim_{n\to\infty} x = \alpha$. So, in fact, for every zero order pseudodifferential operator on $A : C^\infty(\Omega) \to C^\infty(\Omega)$

$$\mathrm{Tr}_\omega(A(1-\Delta)^{-d/2}) = \lim \circ M\left(\langle e_n, Ae_n\rangle\right)$$

and the Liouville integral on the sphere bundle of the closed manifold Ω is the logarithmic mean limit of the free energy expectation values of A.

These features come from the identification of differences in the commutator subspace. The Calkin correspondence yields the bijection

$$\phi \underset{\circ\mu}{\overset{\circ\mathrm{diag}}{\rightleftarrows}} f.$$

between a trace ϕ on an ideal of compact operators $\mathcal{E}$ and a symmetric functional f on the ideal's diagonal E, such that

$$\phi(A) = f(\mu(A)), \quad 0 \leq A \in \mathcal{E}$$

on the positive cone of $\mathcal{E}$. If $\lambda(A)$ is any eigenvalue sequence of a compact operator $A \in \mathcal{E}$, under what conditions is it true that the extension of ϕ from the positive cone to the whole ideal is given by

$$\phi(A) = f(\lambda(A)), \quad A \in \mathcal{E}?$$

The question is not trivial; for an operator $A \in \mathcal{E}$ written as the sum of four positive operator $A = A_1 - A_2 + iA_3 - iA_4$ then it is not automatic that the difference

$$(\mu(A_1) - \mu(A_2) + i\mu(A_3) - i\mu(A_4)) - \lambda(A)$$

should belong to the centre $Z(E)$ of E. Or, more succinctly written,

$$A - \text{diag}(\lambda(A)) \in \text{Com}(\mathcal{E}). \tag{3.11}$$

The problem is even more basic; there is no guarantee that $\lambda(A)$ belongs to E.

3.3.2 Log submajorisation

Kalton, using the identification of the commutator subspace for normal operators due to [66], identified in 1998 the necessary and sufficient conditions for (3.11) for a countably generated ideal [121]. With Dykema it was extended to the semifinite case in [67, 70]. The problem was fully solved for any ideal $\mathcal{E}$ of compact operators in 2014 [217].

An arbitrary ideal of compact operators $\mathcal{E}$ is called *logarithmic submajorisation closed* if $A \in \mathcal{E}$ and B is a compact operator such that

$$\prod_{k=0}^{n} \mu(k, B) \leq \prod_{k=0}^{n} \mu(k, A), \quad n \geq 0$$

implies that $B \in \mathcal{E}$.

The following refinement of [66] from [125], updated with the result of [217], identifies when the difference of two operators belong to the commutator subspace in terms of the difference in their eigenvalue sequences.

Theorem 3.2 *Suppose $\mathcal{E}$ is an ideal of compact operators with diagonal E and either: (a) $A, B \in \mathcal{E}$ are normal; or (b) $\mathcal{E}$ is logarithmic submajorisation closed and $A, B \in \mathcal{E}$ are arbitrary. Then the following statements are equivalent:*

(1) *$A - B \in \text{Com}(\mathcal{E})$;*
(2) *for any eigenvalue sequences $\{\lambda(j, A)\}_{j=0}^{\infty}$ of A and $\{\lambda(j, B)\}_{j=0}^{\infty}$ of B,*

$$\left\{ \frac{1}{n+1} \left(\sum_{j=1}^{n} \lambda(j, A) - \sum_{j=0}^{n} \lambda(j, B) \right) \right\}_{n=1}^{\infty} \in E$$

(3) *for any eigenvalue sequences $\{\lambda(j, A)\}_{j=0}^{\infty}$ of A and $\{\lambda(j, B)\}_{j=0}^{\infty}$ of B,*

$$\left| \sum_{j=0}^{n} \lambda(j, A) - \sum_{j=0}^{n} \lambda(j, B) \right| \leq (n+1)\mu_n$$

for a positive decreasing sequence $\mu = \{\mu_n\}_{n=1}^{\infty} \in E$.

When $A, B \geq 0$ the statement that $A - B \in \mathrm{Com}(\mathcal{E})$ if and only if $C(\mu(A) - \mu(B)) \in E$ is the result used earlier. If $B = \mathrm{diag}(\lambda(A)) \in \mathcal{E}$, the statement of the theorem provides (3.11). From [217, Theorem 8]

Corollary 3.3 (Lidskii) *If $\mathcal{E}$ is a logarithmic submajorisation closed ideal of $\mathcal{L}(H)$, then $\lambda(A) \in E$ for every operator $A \in \mathcal{E}$ and*

$$\phi(A) = \phi \circ \mathrm{diag}(\lambda(A))$$

for every trace ϕ on $\mathcal{E}$.

If $\mathcal{E}$ is not logarithmic submajorisation closed, then there is an operator $A \in \mathcal{E}$ such that $\lambda(A) \notin E$ [217, Theorem 8]. This ends any hope of (3.11) for every operator in an ideal that is not logarithmically submajorisation closed. Logarithmic submajorisation was introduced by Weyl [238] [88, Lemma 3.3], who showed that the singular value sequence logarithmic submajorises an eigenvalue sequence. A formal definition was given by Ando and Hiai in [9], and expounded in the survey [107].

All Banach and quasi-Banach ideals, including therefore $\mathcal{L}_{1,\infty}$, are logarithmic submajorisation closed [121] [217, Lemma 35]. An arbitrary ideal $\mathcal{E}$ has a logarithmic submajorisation closure,

$$LE(\mathcal{E}) := \left\{ B \in \mathcal{L}(H) : \prod_{j=0}^{n} \mu(j, B) \leq \prod_{j=0}^{n} \mu(j, A), \forall n \geq 0, \text{ for some } A \in \mathcal{E} \right\}.$$

Then $LE(\mathcal{E})$ is the smallest logarithmic submajorisation closed ideal containing $\mathcal{E}$ [217, Lemma 33]. Consequently $LE(\mathcal{E})$ is the smallest ideal whose diagonal contains the eigenvalue sequences of $\mathcal{E}$, i.e. $A \in \mathcal{E}$ implies $\mathrm{diag}(\lambda(A)) \in LE(\mathcal{E})$. While a native spectral formulation is denied to traces on an arbitrary ideal $\mathcal{E}$, it is natural to ask which traces on $\mathcal{E}$ extend to traces on $LE(\mathcal{E})$ and thereby obtain a spectral formulation in $LE(\mathcal{E})$. This is still an open problem.

If $A \in \mathcal{L}(H)$ and $B = B^* \in \mathcal{E}$ for an arbitrary two-sided ideal $\mathcal{E}$, another corollary of Theorem 3.2 is [125, Corollary 4.6]

$$AB - \mathrm{diag}(\{\langle e_n, ABe_n\rangle\}_{n=1}^{\infty}) \in \mathrm{Com}(\mathcal{E}) \tag{3.12}$$

where e_n, $n \geq 0$ is an orthonormal basis of H such that $Be_n = \lambda(n, B)e_n$, $n \geq 0$. Formula (3.10) for the noncommutative integral follows setting $B = (1 - \Delta)^{-d/2} \in \mathcal{L}_{1,\infty}$.

4 Calculation and independence from the singular trace

The calculation of a Dixmier trace is rarely done using the explicit construction. Most cases in noncommutative geometry use the zeta function approach or the heat kernel asymptotics approach to calculate the trace. These approaches originated early in noncommutative geometry and are well studied. We quickly restate the known facts from [150, Ch. 8] in Theorem 4.1. We then concentrate on developments in the last 5 years.

For brevity in this section, we pass to continuous extended limits. A functional $\omega \in L_\infty(\mathbb{R}_+)^*$ is called an extended limit if it vanishes on $C_0(\mathbb{R}_+)$. Using the singular value function $\mu(t, A)$, $t \geq 0$ of a compact operator defined in the preliminaries, which is a step function with values $\mu(n, A)$, $n \geq 0$ given by the singular value sequence, define Dixmier's trace as

$$\mathrm{Tr}_\omega(A) = \omega\left(\frac{1}{\log(1+t)}\int_0^t \mu(s, A)ds\right), \quad 0 \leq A \in \mathcal{L}_{1,\infty}$$

for any extended limit $\omega \in L_\infty(\mathbb{R}_+)^*$. This produces the same set of traces [147, Section 2, Theorem 6.2] with correspondence given simply by[4]

$$\omega\left(\frac{1}{\log(1+t)}\int_0^t \mu(s, A)ds\right) = \omega \circ p\left(\frac{1}{\log(1+n)}\sum_{j=0}^{n} \mu(j, A)\right), 0 \leq A \in \mathcal{L}_{1,\infty}$$

where $p : \ell_\infty \to L_\infty(\mathbb{R}_+)$ is the piecewise linear extension

$$p(x)(t) = \sum_{n=0}^{\infty} (x_n + (x_{n+1} - x_n)(t-n))\, \chi_{[n,n+1)}(t), \quad t \geq 0.$$

4.1 Zeta functions and heat kernels

Connes noted [40, p. 306] that a substitution in the Tauberian theorem of Hardy and Littlewood [102, Theorem 95, p. 156] shows, for $0 \leq A \in \mathcal{L}_{1,\infty}$,

$$\frac{1}{\log(1+t)}\int_0^t \mu(s, A)ds \to c, \quad t \to \infty \tag{4.1}$$

[4]That this is a bijective correspondence follows from using the identification of Dixmier traces with fully symmetric functionals in [122].

for a value $c \geq 0$, if and only if

$$(s-1)\mathrm{Tr}(A^s) \to c, \quad s \to 1^+.$$

Hence, for certain $0 \leq A \in \mathcal{L}_{1,\infty}$ at least, the behaviour of the zeta function determines the Dixmier trace of A.

4.1.1 Zeta function and residues

From Minakshisundaram and Pleijel in 1949 [158], and more generally Seeley in 1967 [201], it was known that the zeta function of a $-d/m$ power of an order m elliptic differential operator on a closed Riemannian manifold Ω of dimension d has a meromorphic extension from $\mathrm{Re}(s) > 1$ with a simple pole at $s = 1$ [204, p. 112] [109]. For such elliptic differential operators, the residue of the zeta function at its first pole offers a way to both calculate the Dixmier trace, and through existence of the actual limit in (4.1), shows that the result does not depend on which Dixmier trace is chosen.

The residue at the first pole is extended to any classical pseudodifferential operator of order $-d$ on Ω by the noncommutative residue Res developed by Wodzicki in 1983 and Guillemin 1985 [239] [240, p. 384] [97] as an extension of Adler and Manin's work in one dimension [2, 153]. Hearing of the existence of the singular traces from Dixmier in 1985 [150, p. 218], Connes published the following trace theorem in 1988 [38, Theorem 1]

$$\mathrm{Tr}_\omega(Q) = \frac{1}{d}\mathrm{Res}(Q) \tag{4.2}$$

where Q is a classical scalar-valued pseudodifferential operator of order $-d$ on Ω that, as a compact operator acting on $L_2(\Omega)$, belongs to $\mathcal{L}_{1,\infty}$. Wodzicki's noncommutative residue $\mathrm{Res}(Q)$ can be calculated as follows or by integrating the principal symbol of Q over the sphere bundle (examined further below).

A scalar-valued pseudodifferential operator Q of order $-d$ on Ω can be written as

$$Q = Q_0(1-\Delta)^{-d/2}$$

where Q_0 is pseudodifferential and scalar-valued of order 0. The noncommutative residue $\mathrm{Res}(Q)$ of Q is equivalent to [239, p. 387]

$$d \cdot \mathrm{res}_{s=1}\mathrm{Tr}(Q_0(1-\Delta)^{-sd/2}),$$

so that

$$\mathrm{Tr}_\omega(Q) = \mathrm{res}_{s=1}\mathrm{Tr}(Q_0(1-\Delta)^{-sd/2}) = \lim_{s\to 1^+}(s-1)\mathrm{Tr}(Q_0(1-\Delta)^{-sd/2}).$$

Can this approach to calculating a Dixmier trace be applied to other operators on other Hilbert spaces? The natural question examined first in [27, 14, 28] concerns the behaviour of the function

$$(s-1)\mathrm{Tr}(BA^s), \quad s \to 1^+, B \in \mathcal{L}(H), 0 \le A \in \mathcal{L}_{1,\infty}.$$

The function $s \mapsto \mathrm{Tr}(BA^s)$, $\mathrm{Re}(s) > 1$ does not have a meromorphic analytic continuation generally. Counterexamples are provided by Carey and Sukochev [29, Lemma 17] or Kalton et al. [125, Corollary 6.34]. The function $(s-1)\mathrm{Tr}(BA^s)$ is bounded for $s \ge 1$ if $0 \le A \in \mathcal{L}_{1,\infty}$[5] [27, 28]. Taking an "extended limit as $s \to 1^+$" we can define the functional

$$\zeta_\omega(B, A) = \omega\left(\frac{1}{t}\mathrm{Tr}(BA^{1+1/t})\right), \quad B \in \mathcal{L}(H), 0 \le A \in \mathcal{L}_{1,\infty} \tag{4.3}$$

for an extended limit $\omega \in L_\infty(\mathbb{R}_+)$. Whether the functional $\mathrm{Tr}_\omega(BA)$ is equal to the functional $\zeta_\omega(B, A)$ for the same extended limit ω has been examined in [27, 25, 14, 28, 29, 215, 222, 228]. What is known is described in Theorem 4.1 below.

4.1.2 Heat kernel

Minakshisundaram published a shorter proof in 1953 [157] of the meromorphicity of the zeta function of the Laplace-Beltrami operator Δ on a closed manifold Ω using the Mellin transform to relate the asymptotics of the diagonal of the heat kernel and the poles of the (analytic continuation of the) zeta function. For any $s > 0$, the operator

$$e^{s\Delta} : L_2(\Omega) \to C^\infty(\Omega)$$

is infinitely smoothing [201]. It is compact as an operator $L_2(\Omega) \to L_2(\Omega)$ and trace class [204, p. 203]. For each $s > 0$ it has smooth real kernel $K(s, \cdot, \cdot) \in C^\infty(\Omega \times \Omega)$ of the form

$$K(s, x, y) = \sum_{n=0}^{\infty} e^{-s\lambda(n,-\Delta)} e_n(x) e_n(y), \quad s > 0, x, y \in \Omega$$

where $\lambda(n, -\Delta)$ are the eigenvalues of the Laplace-Beltrami operator in non-decreasing order with multiplicity and $e_n \in C^\infty(\Omega)$, $n \ge 0$ is an eigenbasis such that $-\Delta e_n = \lambda(n, -\Delta) e_n$, $n \ge 0$ [204, Theorem 8.3] [85, Lemma 1.6.3].

[5] The function $(s-1)\mathrm{Tr}(BA^s)$ is bounded for $s \ge 1$ if and only if $0 \le A \in \mathcal{M}_{1,\infty}$ [28, Theorem 4.5].

Minakshisundaram identified an asymptotic expansion in s of the kernel on the diagonal of $\Omega \times \Omega$ [85, Lemma 1.7.4] [204, p. 119],

$$K(s, x, x) \sim s^{-d/2} \sum_{k=0}^{\infty} a_k(x) s^k, \quad s \to 0^+, x \in \Omega$$

where $a_k(x) \in C^\infty$, $k \geq 0$ can be expressed in terms of combinatorial expressions of the derivatives of the symbol of Δ. They are local invariants of the differential operator Δ but difficult to calculate. Investigation of the Seeley-deWitt coefficient functions $a_k(x)$ has a long history in spectral geometry [84] [85, Section 4.9] [15, IV]. The asymptotic expansion of the kernel becomes an asymptotic expansion of the partition function

$$\mathrm{Tr}(e^{s\Delta}) = \int_\Omega K(s, x, x) dx \sim s^{-d/2} \sum_{n=0}^{\infty} a_k s^k, \quad s \to 0^+ \tag{4.4}$$

where the sequence of Seeley-deWitt numbers

$$a_k = \int_\Omega a_k(x) dx, \quad k \geq 0$$

are the integrals of the smooth coefficient functions. Minakshisundaram could calculate the first two coefficients for the Laplace-Beltrami operator [85, Theorem 4.8.18]

$$\mathrm{Tr}(e^{s\Delta}) \sim (4\pi)^{-d/2} s^{-d/2} \times \left(\mathrm{Vol}(\Omega) + \frac{1}{6} \int_\Omega R_g(x) dx \cdot s + O(s^2) + \cdots \right), \quad s \to 0^+ \tag{4.5}$$

where $R_g(x)$ denotes the scalar curvature of the metric g at the point $x \in \Omega$. So the invariants of dimension, volume and total scalar curvature of a closed Riemannian manifold are contained in the expansion. Further coefficients depend on combinatorial terms and derivatives of the curvature tensor.

Seeley showed that Minakshisundaram's approach works for a general positive elliptic differential operator P of order m on the closed manifold Ω, and under the Mellin transform the scalar coefficients $a_k(P)$ in the partition function asymptotic expansion correspond to the residues of the zeta function of P [85, Lemma 1.10.1]

$$a_k(P) = \mathrm{res}_{s=\frac{d-2k}{m}} \Gamma(s) \mathrm{Tr}(P^{-s}), \quad k \geq 0.$$

For $0 \leq k < d/2$ the leading poles are in the positive half-plane where the Gamma function Γ is regular. In terms of the noncommutative residue [240, p. 384]

$$a_k(P) = d^{-1}\Gamma\left(\frac{d-2k}{m}\right)\mathrm{Res}\left(P^{-\frac{d-2k}{m}}\right), \quad 0 \le k < d/2.$$

If $Q_0 : C^\infty(\Omega) \to C^\infty(\Omega)$ is a zero order differential operator, then

$$\mathrm{Tr}(Q_0 e^{-sP})$$

also has an asymptotic expansion as $t \to 0^+$ with coefficients $a_k(Q_0, P)$, $k \ge 0$ [85, 94, Lemma 1.7.7] such that

$$a_k(Q_0, P) = \mathrm{res}_{s=\frac{d-2k}{m}}\Gamma(s)\mathrm{Tr}(Q_0 P^{-s}), \quad k \ge 0.$$

If an order $-d$ pseudodifferential operator is of the form $Q = Q_0(1-\Delta)^{-d/2}$ where Q_0 is differential, with $P = (1-\Delta)^{d/2}$ we obtain

$$\begin{aligned} a_0(Q_0, (1-\Delta)^{d/2}) &= \lim_{s\to 0^+} s^{-d/2}\mathrm{Tr}(Q_0 e^{-s(1-\Delta)^{d/2}}) \\ &= \lim_{s\to 1^+}(s-1)\mathrm{Tr}(Q_0(1-\Delta)^{-sd/2}) = \mathrm{Tr}_\omega(Q). \end{aligned} \tag{4.6}$$

The first co-efficient in the heat kernel asymptotic expansion calculates a Dixmier trace of an order $-d$ pseudodifferential operator.

The same question as before was asked; can this approach to calculating a Dixmier trace be applied to other operators on other Hilbert spaces? When $0 \le A \in \mathcal{L}_{1,\infty}$ we can define the operator

$$e^{-sA^{-1}} \in \mathcal{L}_1, \quad s > 0$$

by the functional calculus, understanding that $e^{-sA^{-1}}\eta = 0$ for every $\eta \in \ker A$. The behaviour of the function

$$s \mapsto s\mathrm{Tr}(Be^{-sA^{-1}}), \quad s \to 0^+, B \in \mathcal{L}(H), 0 \le A \in \mathcal{L}_{1,\infty}$$

is bounded if and only if $A \in \mathcal{L}_{1,\infty}$ [198, 23] but not convergent in general.[6] Taking an "extended limit as $s \to 0^+$" we can define the functional [27, 14, 28, 198, 215]

$$\xi_\omega(B, A) = \omega\left(t^{-1}\mathrm{Tr}(Be^{-(tA)^{-1}})\right), \quad B \in \mathcal{L}(H), 0 \le A \in \mathcal{L}_{1,\infty} \tag{4.7}$$

for an extended limit $\omega \in L_\infty(\mathbb{R}_+)^*$.

[6]Note that this function is not bounded when $0 \le A \in \mathcal{M}_{1,\infty}$ belongs to the dual of the Macaev ideal. The logarithmic mean is applied to obtain a bounded function in this case [28, 198].

4.1.3 Calculation of leading terms

The extent to which the functional $\varsigma_\omega(B, A)$ defined in (4.3), and $\xi_\omega(B, A)$ defined in (4.7), determine the Dixmier trace of BA, and, as in (4.6) above, define the same value of the Dixmier trace, is answered in Theorem 4.1. Define the (continuous) logarithmic mean operator $M : L_\infty(\mathbb{R}_+) \to L_\infty(\mathbb{R}_+)$ by

$$Mf(t) = \frac{1}{\log(1+t)} \int_0^t \frac{f(s)}{s} ds, \quad t \geq 0, f \in L_\infty(\mathbb{R}_+).$$

Define $P^a : L_\infty(\mathbb{R}_+) \to L_\infty(\mathbb{R}_+)$ for any $a > 0$ by

$$P^a(f)(t) = f(t^a), \quad t \geq 0, f \in L_\infty(\mathbb{R}_+).$$

Define $\log : L_\infty(\mathbb{R}) \to L_\infty(\mathbb{R}_+)$ by

$$\log(f)(t) = f(\log(t)), \quad t > 0, f \in L_\infty(\mathbb{R}).$$

Note that $\log : L_\infty(\mathbb{R}) \to L_\infty(\mathbb{R}_+)$ is an isomorphism that intertwines the exponentiation operator P^a, $a > 0$ on $L_\infty(\mathbb{R}_+)$ and the dilation operator D_a on $L_\infty(\mathbb{R})$ [27, Prop. 1.3].

Theorem 4.1 *Suppose $B \in \mathcal{L}(H)$ and $0 \leq A \in \mathcal{L}_{1,\infty}$. Let $\varsigma_\omega(B, A)$ be as defined in (4.3), and $\xi_\omega(B, A)$ be as defined in (4.7). Then:*

(1) *the set of Dixmier traces is identical to the set of all heat kernel asymptotic functionals $\xi_\omega(1, \cdot)$, $\omega \in L_\infty(\mathbb{R}_+)^*$ an extended limit, acting on the positive cone of $\mathcal{L}_{1,\infty}$. Further*

(a) *if $\omega \in L_\infty(\mathbb{R}_+)^*$ is an extended limit then*

$$\mathrm{Tr}_\omega(BA) = \omega \circ M\left(\frac{1}{t}\mathrm{Tr}(Be^{-(tA)^{-1}})\right), \quad \forall B \in \mathcal{L}(H), 0 \leq A \in \mathcal{L}_{1,\infty}$$

(b) *however the logarithmic mean cannot be removed, there exists an extended limit $\omega \in L_\infty(\mathbb{R}_+)^*$ such that*

$$\mathrm{Tr}_\omega(A) \neq \omega\left(\frac{1}{t}\mathrm{Tr}(e^{-(tA)^{-1}})\right), \quad 0 \leq A \in \mathcal{L}_{1,\infty}.$$

(2) *the set of Dixmier traces is* strictly larger *than the set of all zeta residue functionals $\varsigma_\omega(1, \cdot)$, $\omega \in L_\infty(\mathbb{R}_+)^*$ an extended limit, acting on the positive cone of $\mathcal{L}_{1,\infty}$. Further*

(a) *if $\omega \in L_\infty(\mathbb{R}_+)^*$ is a P^a-invariant limit, $a > 0$, that is $\omega \circ P^a = \omega$ then*

$$\mathrm{Tr}_\omega(BA) = \omega \circ \log\left(\frac{1}{t}\mathrm{Tr}(BA^{1+1/t})\right), \qquad \forall B \in \mathcal{L}(H), 0 \le A \in \mathcal{L}_{1,\infty}$$

(b) *there exists an extended limit* $\omega \in L_\infty(\mathbb{R}_+)^*$ *such that*

$$\mathrm{Tr}_\omega(A) > \limsup_{t\to\infty} \frac{1}{t}\mathrm{Tr}(A^{1+1/t}), \quad 0 \le A \in \mathcal{L}_{1,\infty}.$$

(3) *if* $\omega \in L_\infty(\mathbb{R}_+)^*$ *is an* M*-invariant and* P^a*-invariant limit,* $a > 0$*, that is* $\omega \circ M = \omega \circ P^a = \omega$*, then*

$$\mathrm{Tr}_\omega(BA) = \omega\left(\frac{1}{t}\mathrm{Tr}(Be^{-(tA)^{-1}})\right) = \omega \circ \log\left(\frac{1}{t}\mathrm{Tr}(BA^{1+1/t})\right)$$

for all $B \in \mathcal{L}(H)$ *and* $0 \le A \in \mathcal{L}_{1,\infty}$.

The theorem combines result from [150, Section 8] and [198, Theorem 5] [224, Theorem 16] [222, Theorem 33] [203, p. 599] [214, Theorem 3.10], from a history of previous results in [91, 27, 25, 14, 28, 198, 215, 29]. It is not known if the statements of 1(a), 2(a) and 3 are optimal.

Theorem 4.1 provides the third characterisation of Dixmier's trace, they correspond to the set of heat kernel asymptotic functionals. Meaning they generally provide a "leading term of the asymptotic expansion" of the partition function

$$\mathrm{Tr}(e^{-sA^{-1}}), \quad s \to 0^+, 0 \le A \in \mathcal{L}_{1,\infty}$$

when there are no well-defined asymptotics. Statement 3 generalises the association between the noncommutative residue, the leading term of the heat kernel expansion of a positive elliptic differential operator on a closed Riemannian manifold, and the residue at the first pole of the zeta function of that operator. Unlike the case of differential operators, proper asymptotics do not exist and calculation is complicated by the presence of various extended limits; this is the best that can be done for arbitrary $0 \le A \in \mathcal{L}_{1,\infty}$. The complimentary question is to consider smaller sets of operators instead of extending the limits. For what smaller set of operators in the positive cone of $\mathcal{L}_{1,\infty}$ do the limits actually exist in Theorem 4.1, and the extended limits can be dispensed with?

We note some history of the proof of Theorem 4.1. In [102] Hardy proves the Hardy-Littlewood Tauberian theorem [102, Theorem 95, p. 155] following Karamata [126] [102, Theorem 98, p. 156]. Karamata's Tauberian theorem provides the equivalence in (4.1) between the logarithmically diverging series and zeta functions. The Mellin transform relates zeta functions to heat kernel asymptotics. This method underlies the proofs from [27] to [198], where a weak*-Karamata theorem was proved for extended limits and applied to zeta functions [27, Theorem 2.2] using substitution or used directly for heat kernels [198, Theorem 2]. In 2010 [215], by which time it was realised Dixmier traces provide all Hardy-Littlewood submajorisation monotone traces, the method transferred to identification of heat and zeta functionals with submajorisation monotone traces. Different computations yield the association of heat kernel functionals with Dixmier traces and the Mellin transform is not used. The bijective association with factorisable Banach limits was used recently to identify the set of zeta residue functions [214, Lemma 3.5].

4.1.4 The second term

For the Laplace-Beltrami operator Δ on a closed Riemannian manifold Ω the second term in (4.5) is

$$\frac{1}{6}\int_\Omega R_g(x)dx,$$

which is proportional to the Einstein-Hilbert action whose variation describes the vacuum field equations of general relativity. Replacing $-\Delta$ acting on $C^\infty(\Omega)$ by a generalised Laplacian D^2 where D is a general Dirac operator on the smooth sections $C^\infty(\Omega, E)$ of a Clifford bundle E over the closed manifold Ω [138, Sect. 5] only changes the constant of proportionality and introduces a cosmological constant to the action; noted by Kalau and Walze [118, Sect. 5] and seen in [85, Theorem 4.8.18]. Therefore the second term in the expansion of $\mathrm{Tr}(e^{-tD^2})$ for any standard Laplacian in differential geometry is equivalent to the vacuum gravity action. Following Connes [42], and Kastler's computation in [127], Kalau and Walze [118, Sect. 5] further noted that the partition function for the square of the Dirac operator of the Connes-Lott model of a four-dimensional manifold tensored by a matrix algebra representing the standard model [40, VI] has second term proportional to the vacuum Einstein-Hilbert action. In 1996 Chamseddine-Connes introduced the spectral action, where the first, second and third terms of an asymptotic expansion of the partition function for the Connes-Lott model were associated with a (bosonic) action incorporating the standard model, gravity and additional terms [34].

Obtaining the second and higher terms of the asymptotics of the partition function for more general geometries than those associated with elliptic differential operators is an open problem. Given the partition function $\mathrm{Tr}(e^{-sH})$, $s > 0$ of an arbitrary positive operator H with finite dimensional kernel and $H^{-d/2} \in \mathcal{L}_{1,\infty}$ for some $d \geq 4$, Dixmier's trace extracts the "leading term of the asymptotic expansion". Presently there are no analogue formulas involving singular traces that extract the second or higher terms. It was noted, in [240, p. 389] [118] and [34] and reiterated in [1] after [127], that in the case of Dirac operators on closed Riemannian manifolds, as above, the noncommutative residue

$$\mathrm{Res}((D^2)^{-d/2+1})$$

calculates the second term and hence Einstien-Hilbert action. However, for general positive operators such that $H^{-d/2} \in \mathcal{L}_{1,\infty}$ and $d \geq 4$, there is no analogue formula for the noncommutative residue of $H^{-d/2+1}$ in the same manner as the Dixmier trace is the noncommutative residue of $H^{-d/2}$.

4.2 *Measurability of operators*

Connes' trace theorem (4.2) shows that all Dixmier traces applied to a classical pseudodifferential operator of order $-d$ on a d-dimensional closed Riemannian manifold Ω have the same value, namely the residue of that pseudodifferential operator. Having a large and central class of operators in differential geometry where the calculation of the Dixmier trace does not depend on which Dixmier trace is chosen, warrants the introduction of the notion of a *measurable* operator [40, p. 308]. An operator $A \in \mathcal{L}_{1,\infty}$ is *measurable* if

$$\mathrm{Tr}_\omega(A) = \omega\left(\frac{1}{\log(n)}\sum_{j=0}^{n}\lambda(n, A)\right) = c, \qquad c \text{ const.}$$

for every extended limit $\omega \in L_\infty(\mathbb{R}_+)^*$.

Given the properties of pseudodifferential operators seen last section some natural questions arise. Does measurability imply the existence of an asymptotic expansion of order 1 of the associated partition function $\mathrm{Tr}(e^{-sA^{-1}})$, $s > 0$? Does measurability imply the existence of an analytic continuation and simple pole at $s = 1$ of the zeta function $\mathrm{Tr}(A^s)$, $\mathrm{Re}\, s > 1$. Are non-classical pseudodifferential operators of order $-d$ measurable? Are there measurable integral operators not associated with smooth symbols? Does measurability imply the existence of the limit in (1.2)?

4.2.1 Measurability

By virtue of the fact that the set of Dixmier traces can be associated surjectively to every extended limit, and that for every α such that

$$\liminf_{t\to\infty} f(t) < \alpha < \limsup_{t\to\infty} f(t), \qquad f = f^* \in L_\infty(\mathbb{R}_+)$$

there exists some extended limit $\omega \in L_\infty(\mathbb{R}_+)^*$ such that $\alpha = \omega(f(t))$ [27] [150, Lemma 9.3.6], then we can note immediately that $0 \leq A \in \mathcal{L}_{1,\infty}$ is measurable if and only if

$$\lim_{t\to\infty} \frac{1}{\log(1+t)} \int_0^t \mu(s, A)ds = c.$$

The existence of this limit is equivalent to the existence of all the limits in Theorem 4.1 in the case $B = 1$ [150, Theorem 9.3.1] [198, Theorem 6]. The operator $0 \leq A \in \mathcal{L}_{1,\infty}$ is measurable if and only if

$$\lim_{s\to\infty} M\left(\frac{1}{t}\mathrm{Tr}(e^{-(tA)^{-1}})\right)(s) = c,$$

and if and only if

$$\lim_{s\to 1^+} (s-1)\mathrm{Tr}(A^s) = c.$$

The logarithmic mean M cannot be removed from these equivalent conditions [150, Example 9.3.3]. Measurability does not imply the existence of a leading term in the partition function associated with $0 \leq A \in \mathcal{L}_{1,\infty}$ nor an analytic continuation of the zeta function with simple pole at $s = 1$.

It follows from an analysis of the Mellin transform that the existence of the limit

$$\lim_{s\to 0^+} s\mathrm{Tr}(e^{-sA^{-1}}) = c, \quad c \text{ const.}, 0 \leq A \in \mathcal{L}_{1,\infty},$$

or equivalently

$$\mathrm{Tr}(e^{-sA^{-1}}) = \frac{c}{s} + o\left(\frac{1}{s}\right), \quad c \text{ const.}, 0 \leq A \in \mathcal{L}_{1,\infty}, \tag{4.8}$$

is stronger than measurability [198, Theorem 6]. It implies

$$\mathrm{Tr}(A^s) = \frac{c}{s-1} + o\left(\frac{1}{s-1}\right), \quad c \text{ const.}$$

but is not strong enough to establish the nature of the singularity of the zeta function at $s = 1$. We discuss this further below. The existence of the limit (4.8) is equivalent however to spectral asymptotics of the operator $0 \leq A \in \mathcal{L}_{1,\infty}$, i.e. a Weyl formula [198, Theorem 6]

$$\mu(n, A) \sim \frac{c}{n}, \quad n \to \infty.$$

4.2.2 Universal measurability

Written another way, $A \in \mathcal{L}_{1,\infty}$ is measurable if and only if

$$\sum_{j=0}^{n} \lambda(j, A) = c \cdot \log(n) + o(\log(n)), \quad n \geq 0, c \text{ const.}$$

We recall from Section 3 that Dixmier's trace does not describe every trace on $\mathcal{L}_{1,\infty}$. An operator $A \in \mathcal{L}_{1,\infty}$ is *universally measurable* if

$$\phi(A) = c, \quad c \text{ const.}$$

for every trace ϕ on $\mathcal{L}_{1,\infty}$ such that $\phi(\mathrm{diag}(n^{-1})) = 1$, i.e. normalised trace. This form of measurability is equivalent to

$$A - c \cdot \mathrm{diag}(n^{-1}) \in \mathrm{Com}(\mathcal{L}_{1,\infty}).$$

By Theorem 3.2, $A \in \mathcal{L}_{1,\infty}$ is universally measurable if and only if

$$\sum_{j=0}^{n} \lambda(j, A) = c \cdot \log(n) + O(1), \quad n \geq 0, c \text{ const.}$$

The difference between universal measurability and measurability is therefore in the remainder of the logarithmic divergence of the partial sums of an eigenvalue sequence. All classical pseudodifferential operators P of order $-d$ on a closed manifold Ω of dimension d are universally measurable. In fact, it was shown that Connes' trace theorem (4.2) reads as [125, Corollary 7.22]

$$\phi(P) = \frac{1}{d}\mathrm{Res}(P) \tag{4.9}$$

for every normalised trace ϕ on $\mathcal{L}_{1,\infty}$. This result is not surprising, given the uniqueness of the noncommutative residue as a trace on classical pseudodifferential operators. The new feature behind this result is the asymptotic behaviour of the sums of eigenvalues of pseudodifferential operators P of order $-d$ in Theorem 5.1 below. There are non-classical pseudodifferential operators of order $-d$ that are measurable, but not universally measurable [203, Theorem 8.13], and non-classical pseudodifferential operators of order $-d$ that are not measurable [125, Corollary 7.23].

4.2.3 Products

Measurability of a product of operators is challenging. That $0 \leq A \in \mathcal{L}_{1,\infty}$ is measurable does not imply that BA is measurable for a given $B \in \mathcal{L}(H)$. There are counterexamples. On a d-dimensional closed manifold Ω the positive operator $A = (1 - \Delta)^{-d/2} \in \mathcal{L}_{1,\infty}$ is measurable. There exists a non-classical zero order pseudodifferential operator $B = Q_0 : C^\infty(\Omega) \to C^\infty(\Omega)$ such that the operator $Q_0(1 - \Delta)^{-d/2} \in \mathcal{L}_{1,\infty}$ is not measurable [147, Corollary 7.23].

This is problematic, since in noncommutative geometry the integral is a continuous linear functional on $\mathcal{L}(H)$ of the form

$$B \mapsto \mathrm{Tr}_\omega(BA), \quad B \in \mathcal{L}(H)$$

for some fixed measurable $0 \le A \in \mathcal{L}_{1,\infty}$ with $\mathrm{Tr}_\omega(A) = 1$. The equivalences for measurability of an operator $0 \le A \in \mathcal{L}_{1,\infty}$ listed above (4.8) become unclear [214, Sect. 1] if B is naively inserted into the formulas. For all $0 \le A \in \mathcal{L}_{1,\infty}$, $B \in \mathcal{L}(H)$, we have (3.12)

$$BA - \mathrm{diag}(\langle e_n, Be_n\rangle)\mathrm{diag}(\mu(n, A)) \in \mathrm{Com}(\mathcal{L}_{1,\infty}) \tag{4.10}$$

where e_n, $n \ge 0$ is an orthonormal basis of eigenvector ordered so that $Ae_n = \mu(n, A)e_n$, $n \ge 0$. Under the condition (4.8)—the Weyl formula for $0 \le A \in \mathcal{L}_{1,\infty}$, we have [150, Theorem 12.1.2]

$$\mathrm{Tr}_\omega(BA) = (\omega \circ M)(\langle e_n, Be_n\rangle)$$

for an extended limit $\omega \in \ell_\infty^*$. Hence the measurability of BA for a given operator $B \in \mathcal{L}(H)$ when A satisfies (4.8) is equivalent to logarithmic mean convergence

$$\lim_{j\to\infty} M(\langle e_n, Be_n\rangle)(j) = c, \quad c \text{ const.}$$

of the expectation values of $B \in \mathcal{L}(H)$.

Recent advances on measurability of products use the bijection of traces with Banach limits (3.6) and the fact that, due to (4.10), any product can be replaced with the product of a commuting normal and positive diagonal operator.

Theorem 4.2 *Suppose $B \in \mathcal{L}(H)$ and $0 \le A \in \mathcal{L}_{1,\infty}$. Let e_n, $n \ge 0$ denote the eigenbasis such that $Ae_n = \mu(n, A)e_n$. Let $\lambda(n, V)$, $n \ge 0$ denote an eigenvalue sequence of a compact operator $V \in \mathcal{L}_{1,\infty}$.*

(1) *The following statements are equivalent*

(a) *BA is measurable and*

$$\mathrm{Tr}_\omega(BA) = c, \quad c \text{ const.}$$

(b)

$$\lim_{n\to\infty} \frac{1}{\log(n+1)} \sum_{k=0}^{n} \lambda(k, BA) = c$$

(c)

$$\lim_{n\to\infty} \frac{1}{\log(n+1)} \sum_{k=0}^{n} \langle e_k, Be_k\rangle \lambda(k, A) = c$$

(d)

$$\lim_{s\to\infty} M(\frac{1}{t}\mathrm{Tr}(Be^{-(tA)^{-1}}))(s) = c$$

(e)

$$\lim_{s\to 1+} (s-1)\mathrm{Tr}(BA^s) = c.$$

(2) *The following statements are equivalent*

(a) *BA is universally measurable and for any normalised trace ϕ on $\mathcal{L}_{1,\infty}$*

$$\phi(BA) = c, \quad c \text{ const.}$$

(b)

$$\mathrm{Tr}(BAe^{-(tA)^{-2}}) = c \cdot \log t + O(1), \quad t \to \infty.$$

(3) (a) *If BA is universally measurable and for any normalised trace ϕ on $\mathcal{L}_{1,\infty}$*

$$\phi(BA) = c, \quad c \text{ const.},$$

then the following asymptotics hold

$$\mathrm{Tr}(BA^s) = \frac{c}{s-1} + O(1),\ s \to 1^+.$$

(b) *If the function*

$$z \mapsto \mathrm{Tr}(BA^z) - \frac{c}{z-1} \text{ is regular at } z = 1, \quad c \text{ const.}$$

then BA is universally measurable and

$$\phi(BA) = c$$

for any normalised trace ϕ on $\mathcal{L}_{1,\infty}$.

The equivalences in the statement of theorem can be false for the dual of the Macaev ideal $\mathcal{M}_{1,\infty}$.[7] The theorem combines result from [30, 214, 215] [125, Sect.

---

[7]For example, the statement 1. (a) ⇔ (b) in Theorem 4.2, that

$$\mathrm{Tr}_\omega(A) = c, \quad c \text{ const.}, A \in \mathcal{L}_{1,\infty}$$

is equivalent to

3] [220, Theorem 1.2.7] (actually the proof of [220, Theorem 1.2.7], which can be extended to satisfy the condition in Theorem 4.2(b)), from a history of results cited after Theorem 4.1.

Theorem 4.2 indicates that the property of the zeta function $z \mapsto \mathrm{Tr}(BA^z)$ having analytic extension to a neighbourhood of $z = 1$ and a genuine pole at that point

$$\mathrm{Tr}(BA^z) - \frac{c}{z-1} \text{ is regular at } z = 1, c \text{ const.} \tag{4.11}$$

is stronger than universal measurability. The condition of a first order expansion in s

$$\mathrm{Tr}(Be^{-(sA)^{-1}}) = \frac{c}{s} + O(s^{-\epsilon}), \quad s \to 0^+, \epsilon < 1, c \text{ const.} \tag{4.12}$$

is stronger than both universal measurability and condition (4.11) [29, Sect. 4].

Universal measurability and equivalent spectral conditions for (4.11) and (4.12) are current topics of research. There are other measurability conditions developed over the last 10 years, some weaker than measurability, some between measurability and universal measurability.

4.2.4 Dyadic averaging and remainders

We note from (3.6) that an operator $A \in \mathcal{L}_{1,\infty}$ with eigenvalue sequence $\lambda(j, A)$ has the same value on every positive trace on $\mathcal{L}_{1,\infty}$ if and only if

$$\theta\left(\frac{1}{\log 2} \sum_{j=2^n-1}^{2^{n+1}-2} \lambda(j, A)\right) = c, \quad c \text{ const.}$$

for every Banach limit θ. Lorentz [151] introduced the notion of almost convergence when a sequence has the same value for every Banach limit. Sucheston [213] gave concrete criteria for almost convergence. Directly from (3.6) it follows that $A \in \mathcal{L}_{1,\infty}$ with eigenvalue sequence $\lambda(j, A)$ has the same value c for every positive trace on $\mathcal{L}_{1,\infty}$ if and only if

---

$$\lim_{n\to\infty} \frac{1}{\log(n+1)} \sum_{k=0}^{n} \lambda(k, A) = c$$

for an eigenvalue sequence $\lambda(n, A)$, $n \geq 0$, is false on $\mathcal{M}_{1,\infty}$. It is true when $0 \leq A \in \mathcal{M}_{1,\infty}$ [147], but false for arbitrary operators in the ideal [199, Corollary 11]. The maximal ideal on which the statement 1. (a) $\Leftrightarrow$ (b) remains true has been identified [199, p. 3058]—it is not $\mathcal{L}_{1,\infty}$.

$$\sum_{j=2^n-1}^{2^{n+1}-2} \lambda(j, A) \text{ almost converges to } c \cdot \log 2, \quad n \to \infty$$

This criteria provides a notion weaker than universal measurability but stronger than measurability [203, Theorem 7.4]. The bijection between Dixmier traces on $\mathcal{L}_{1,\infty}$ and factorisable Banach limits in Section 3.2 provides an alternative criteria for measurability and could be added to the equivalent statements in Theorem 4.1(1)

$$\sum_{j=2^n-1}^{2^{n+1}-2} \lambda(j, A) \text{ is Cesaro convergent to } c \cdot \log 2, \quad n \to \infty.$$

Connes' version of Dixmier's trace corresponds to a twice factorisable Banach limit $\theta = \omega \circ C^2$ for some extended limit $\omega \in \ell_\infty^*$. Despite being a strictly smaller set of traces on $\mathcal{L}_{1,\infty}$ Connes' Dixmier trace does not provide a weaker notion of measurability. A Tauberian theorem on Cesaro summability due to Hardy and Landau [102, Theorem 63, Theorem 64], first used for the present purpose in 2004 [147, Theorem 3.16], indicates that the sequence $a_n = \sum_{j=2^n-1}^{2^{n+1}-2} \lambda(j, A)$, $n \geq 1$ is $(C, 1)$-convergent if and only if it is $(C, 2)$-convergent [203, Theorem 7.7]. This applies to further factorisation, so there is no weaker notion of measurability associated with the increasingly smaller sets of n-factorisable Banach limits of the form $\theta = \omega \circ C^n$, $n \geq 3$. Traces associated with Cesaro invariant Banach limits, $\theta = \theta \circ C$, which may be consider "infinitely factorisable" do, however, provide a weaker notion of measurability [203, Theorem 7.9].

It is perhaps too much to hope that the variety of conditions above and in prior sections can be captured in the remainder term of the log divergence;

$$\sum_{j=0}^{n} \lambda(n, A) = c \cdot \log(n) + r(n), \quad n \to \infty, A \in \mathcal{L}_{1,\infty}$$

where various orders of $r(n)$ provide a meaningful partial order on sets of traces or properties of the leading divergence of the zeta function of $|A|$ or asymptotic expansion of the partition function of $|A|$. Measurability and universal measurability fit into this scale. The condition of universal measurability shows the requirement to go beyond Tauberian theorems in the book of Hardy, which provide no resolution of the remainder beyond $o(\log(n))$.

Tauberian theorems with remainder [212] were utilised in [214] to prove Statement 3(b) in Theorem 4.2. An extension of Hardy and Littlewood's Tauberian theorem [212, Theorem 2.3.1] shows that the condition

$$\mathrm{Tr}(A^s) = \frac{c}{s-1} + O(1), \quad s \to 1^+, 0 \leq A \in \mathcal{L}_{1,\infty}$$

implies

$$\sum_{k=0}^{n} \lambda(k, A) = c \cdot \log(n) + O\left(\frac{\log(n)}{\log(\log(n))}\right), \quad n \geq 0.$$

Further, the remainder term is optimal. This precludes an if and only if statement in Theorem 4.2 (3)(a). Condition (4.11) and a Fatou theorem for Dirichlet series [212, Theorem 2.3.2] provides an $O(1)$ remainder, and hence universal measurability in Statement 3(b).

4.2.5 Examples from fractals

The asymptotic behaviour of Laplacians on fractals provide examples that distinguish between measurability and universal measurability [40, Sect. IV.3.ϵ] [214, Sect.5]. Fractal examples also distinguish between conditions (4.11) and (4.12) as noted in [214, Sect. 5] and [29, Sect. 4].

Let L be the Laplacian associated with a self-similar Dirichlet form on the Hilbert space $L_2(K, \mu)$ of a p.c.f. self-similar set $K \subset \mathbb{R}^d$, $d \geq 2$ with Bernoulli measure μ [132]. Recall that a self-similar set is a compact subset K of $\mathbb{R}^d$ such that

$$K = \cup_{i=1}^{N} F_i(K)$$

where $F_i : \mathbb{R}^d \to \mathbb{R}^d$, $i = 1, \ldots, N$ are contractions and $N \geq 2$. Recall the Bernoulli measure is defined by weights μ_i, $i = 1, \ldots, n$ such that

$$\sum_{i=1}^{N} \mu_i = 1$$

and, for all $m \geq 0$,

$$\mu(F_{w_1} \circ \ldots \circ F_{w_m}(K)) = \mu_{w_1} \cdots \mu_{w_m}, \quad w = (w_1, \ldots, w_m) \in \times_{j=1}^{m}\{1, \ldots N\}.$$

Denote by

$$\pi : \times_{j=1}^{\infty}\{1, \ldots N\} \to \mathcal{P}(K)$$

the map

$$\pi(w) = \cap_{n\geq 1} F_{w_1} \circ \ldots \circ F_{w_n}(K).$$

Denote by S the shift on $\times_{j=1}^{\infty}\{1, \ldots N\}$ given by

$$S(w)_n = w_{n+1}, \quad w \in \times_{j=1}^{\infty}\{1, \ldots N\}.$$

The self-similar set K is p.c.f. if [131]

$$P = \cup_{n \geq 1} S^n \left(\pi^{-1}(\cup_{i \neq j} F_i(K) \cap F_j(K))\right)$$

is a finite set where π^{-1} denotes the preimage of the surjection π. Set

$$V_0 = \pi(P).$$

The fractal K has an increasing sequence of approximating sets

$$V_m = \cup_{w \in \times_{j=1}^{m}\{1,\ldots N\}} F_{w_1} \circ \cdots \circ F_{w_m}(V_0), \quad m \geq 1.$$

K is the closure of $V_* = \cup_{m \geq 0} V_m$. A sequence of Dirichlet forms $\mathcal{D}_m$ [132, Definition 4.1, Lemma 6.1] [79] is self-similar if there exist $\lambda > 0$ and sequence $r = (r_1, \ldots, r_N)$ of positive numbers such that

$$\mathcal{D}_{m+1}(u, v) = \lambda \sum_{k=1}^{N} r_i^{-1} \mathcal{D}_m(u \circ F_i, v \circ F_i), \quad u, v : V_* \to \mathbb{R}$$

The sequence defines a Dirichlet form $\mathcal{D}$ with dense domain $\mathrm{Dom}(L^{1/2}) \subset L_2(K, \mu)$ by Kigami and Lapidus [132, Prop. 1.7] by setting

$$\mathrm{Dom}(L^{1/2}) = \{u : V_* \to \mathbb{R} : \lim_{m \to \infty} \mathcal{D}_m(u|_{V_m}, u|_{V_m}) \text{ exists}, u|_{V_0} = 0\}$$

and

$$\mathcal{D}(u, v) = \lim_{m \to \infty} \mathcal{D}_m(u|_{V_m}, v|_{V_m}), \quad u, v \in \mathrm{Dom}(L^{1/2}).$$

If a self-similar sequence of Dirichlet forms exists, the Dirichlet Laplacian L is the Friedrich's operator associated with the quadratic form $\mathcal{D}$. It has compact resolvent [132, Sect. 4] [80, Theorem 4.2]. Denote the increasing sequence of eigenvalues of L repeated with multiplicity by $\lambda(n, L)$, $n \geq 0$, and denote the spectral counting function

$$N_L(s) = \sup\{n \geq 0 : \lambda(n, L) \leq s\}, \quad s \geq 0.$$

The sequence of self-similar Dirichlet forms, the scale λ, and the constants $(r_1, \ldots r_N)$ are properties of the fractal K. The Laplacian L depends on the weights $(\mu_1, \ldots \mu_N)$, but we suppress the dependence. It is unknown if every p.c.f. self-similar set admits a self-similar Dirichlet form. The Sierpinski Gasket and nested fractals are known to have self-similar Dirichlet forms, amongst other examples in [132, Sect. 3]. Associate with the Laplacian L on K the numbers

$$\gamma_i = \left(\frac{r_i \mu_i}{\lambda}\right)^{1/2}, \quad i = 1, \dots, N.$$

Following Fukishima and Shima's results for the Sierpinski Gasket [80], and Kigami's construction of p.c.f. self-similar sets, Kigami and Lapidus in 1993 [132, Theorem 2.4] proved the following analogue on fractals of Weyl's asymptotic formula. The two cases come from the use of the renewal theorem from probability theory [75, Theorem 2.1].

Theorem 4.3 (Kigami and Lapidus [132]) *Let K be a p.c.f. self-similar set with Bernoulli measure μ admitting a self-similar Dirichlet form as above. Let L be the associated Dirichlet Laplacian on the Hilbert space $L^2(K, \mu)$ and let d_S be the unique number such that*

$$\sum_{i=1}^{N} \gamma_i^{d_S} = 1.$$

Then

$$0 < \liminf_{s \to \infty} s^{-d_S/2} N_L(s) \leq \limsup_{s \to \infty} s^{-d_S/2} N_L(s) < \infty \tag{4.13}$$

and

(1) *[non-lattice] if $\sum_{i=1}^{N} \log(\gamma_i)\mathbb{Z}$ is not a discrete subgroup of $\mathbb{R}$, then*

$$N_L(s) = (c + o(1))\, s^{d_S/2}, \quad s \to \infty$$

where c is a constant.

(2) *[lattice] if $\sum_{i=1}^{N} \log(\gamma_i)\mathbb{Z}$ is a discrete subgroup of $\mathbb{R}$ with generator τ (the smallest number τ such that the discrete subgroup is $\tau\mathbb{Z}$)*

$$N_L(s) = (G(\log s) + o(1))\, s^{d_S/2}, \quad s \to \infty$$

where G is a 2τ-periodic function.

Kigami and Lapidus conjectured that G is a non-constant function in the lattice case (equivalently, that the inequality in (4.13) is strict) when d_S is not an integer [132, p. 105]. The conjecture is still open. In [132, Sect. 3] examples of self-similar fractals and self-similar Dirichlet forms are given for all cases. There exist self-similar fractals in the interval [0, 1] which satisfy both the lattice and non-lattice case. The Modified Sierpinski Gasket in $\mathbb{C}$ associated with the parameter $1/3 < \alpha < 1/2$ and with normalised Hausdorff measure satisfies the non-lattice case for almost all values of α. The Koch snowflake in $[0, 1]$ with Hausdorff measure has $d_S = 1$, satisfies the lattice case, but the asymptotics for N_L exist. In 1991 Fukushima and Shima [80] defined Dirichlet forms associated with Kigami's construction of the standard Laplacian L on the Sierpinski Gasket in $\mathbb{R}^{N-1}$ [131]. For the standard Laplacian on the Sierpinski Gasket associated with Hausdorff measure then $\lambda = \frac{N+2}{N}$, and $r_i = 1$, $\mu_i = N^{-1}$, $i = 1, \ldots, d$ and $d_S = 2\log(N)/\log(N+2)$. Fukishima and Shima showed explicitly [80, Theorem 4.2] that the inequality in (4.13) is strict. Hence the $\log(N+2)$-periodic function G in Theorem 4.3(2) is not constant and no asymptotics exist for the standard Laplacian on the Sierpinski Gasket. Example 3.5 in [132] describes the Laplacian associated with nested fractals introduced by Lindstrøm [144], which all satisfy the lattice case. For Laplacians that admit a localised eigenfunction, including nested fractals, it has subsequently been shown that the inequality in (4.13) is strict [12]. Hence the periodic function G for a Laplacian on a nested fractal is also non-constant.

Note that Theorem 4.3 implies, in all cases, that $(1+L)^{-d_S/2} \in \mathcal{L}_{1,\infty}$. In the non-lattice case, or when G is constant, then the positive operator $(1+L)^{-d_S/2}$ satisfies the Weyl asymptotics (4.8). Hence $(1+L)^{-d_S/2}$ is measurable. In fact, for many known cases, the operator $(1+L)^{-d_S/2}$ satisfies (4.11) with $A = (1+L)^{-d_S/2}$ and $B = 1$, and hence is universally measurable. However, these cases fail to satisfy (4.12) for $A = (1+L)^{-d_S/2}$ and $B = 1$. We briefly explain.

From Theorem 4.3, in all cases, the partition function of L has the asymptotics

$$\mathrm{Tr}(e^{-tL}) = (G_1(-\log t) + o(1))t^{-d_S/2}, \quad t \to 0^+$$

for a possibly non-constant periodic positive bounded function G_1. Similar asymptotics for the partition function are known for non p.c.f. self-similar fractals such as generalised Sierpinski carpets [101, Theorem 1.1] [116]. From these asymptotics Steinhurst and Teplyaev [210, Theorem 2] proved, under assumptions that $r_i = 1$, $\mu_i = 1/N$, $i = 1, \ldots, N$ and that the self-similar fractal is intersection finite, that the zeta function

$$\mathrm{Tr}((1+L)^{-s/2})$$

has a meromorphic extension to $\mathrm{Re}(s) > d_S - \epsilon$ with simple poles on the line $\mathrm{Re}(s) = d_S$ at the points

$$d_S + i\frac{4\pi k}{\log(\lambda N)}, \quad k \in \mathbb{Z}.$$

Laplacians on the Sierpinski Gasket and nested fractals satisfy the assumptions of [210, Theorem 2]. They therefore satisfy (4.11), and hence the Laplacian is universally measurable with dimension, according to Connes, d_S. The value, for any trace ϕ on $\mathcal{L}_{1,\infty}$ is given by Theorem 4.2(3b)

$$\phi((1+L)^{-d_S/2}) = \mathrm{res}_{s=d_S}\mathrm{Tr}((1+L)^{-s/2}) = \mathrm{res}_{s=1}\mathrm{Tr}((1+L)^{-sd_S/2}).$$

Laplacians on the Sierpinski Gasket and nested fractals do not satisfy (4.12), or even (4.8), however.

Generally, condition (4.11) states the existence of the analytic continuation of the zeta function $\mathrm{Tr}(A^z)$, $0 \leq A \in \mathcal{L}_{1,\infty}$, in some neighbourhood of the line $\mathrm{Re}\, z = 1$ and the presence of a simple pole at $z = 1$. Condition (4.12) implies that the simple pole at $z = 1$ can be the only pole on the line $\mathrm{Re}\, z = 1$[8] [28, Sect. 5.2]. The fractal examples show that, in general, there can be Laplacians whose zeta functions have meromorphic extensions with poles on the line $\mathrm{Re}\, z = 1$. Lapidus and Frank prescribe fractal geometry as the presence of poles with imaginary components in a meromorphic continuation of the spectral zeta function [136].

A fractal string is an open bounded subset $\Omega \subset \mathbb{R}$ which is written as an disjoint countable union of open intervals of not necessarily distinct lengths l_j , $j \geq 0$, ordered in non-increasing order [136, Chap. 1]. Note that $\sum_{j=0}^{\infty} l_j$ is finite and equals the Lebesgue measure of Ω. Denote by D the Minkowski dimension of the boundary $\partial\Omega$. We have

$$0 \leq H \leq D \leq 1$$

where H is the Hausdorff dimension of $\partial\Omega$. We assume that $0 < D < 1$.

Let Δ_Ω denote the Dirichlet Laplacian on Ω, that is the Freidrich's extension associated with the Dirichlet form

$$(u, u) = \|u\|_{1,2}^2 = \int_\Omega |x|^2 |\hat{u}(x)|^2 dx,$$

with dense domain

$$\mathcal{F} = \left\{ u \in C(\overline{\Omega}) : (u, u) < \infty, u|_{\partial\Omega} = 0 \right\}.$$

The zeta function $\zeta_\Omega(s) = \mathrm{Tr}(\Delta_\Omega^{-s/2})$, $s > 1$ has a meromorphic extension on $\mathrm{Re}(s) > D$ and a simple pole at $s = 1$ [136, Theorem 1.21]. The residue at the simple pole calculates the volume of Ω [136, Theorem 1.22]. Hence $\Delta_\Omega^{-1/2} \in \mathcal{L}_{1,\infty}$ is universally measurable and every trace calculates the volume of Ω. The so-called geometric zeta function of Ω is more interesting.

We define a Laplacian whose zeta function corresponds to the geometric zeta function of Ω as defined in [136, Chap. 1]. The eigenvalues of $\Delta_\Omega^{-1/2}$ correspond to the positive double sequence [136, p. 24]

[8]A spectral triple $(\mathcal{A}, D, H)$ consists of a $*$-algebra $\mathcal{A} \subset \mathcal{L}(H)$ and a self-adjoint operator $D : \mathrm{Dom}(D) \to H$ such that $[D, a] \in \mathcal{L}(H)$. Connes and Moscovici defined the dimension spectrum [55, II.1]. Let $\mathrm{Sd} = \cup_B \mathrm{Sd}(B)$ where B belongs to the algebra generated by a, $[D, a]$, $a \in \mathcal{A}$ and $\mathrm{Sd}(B) \subset \mathbb{C}$ is the set such that $\mathrm{Tr}(B|D|^z)$ is analytic on $\mathbb{C} \setminus \mathrm{Sd}(B)$. It is usually assumed that Sd is a discrete set, but not that poles with imaginary components should be excluded.

$$a = \{k, j \geq 1 : a_{k,j} = \pi^{-1} k^{-1} l_j\}.$$

Let e_n, $n \geq 0$ be an orthonormal basis of eigenfunctions of the Laplacian Δ_Ω ordered so that $\Delta_\Omega^{-1/2} e_n = \mu(n, a) e_n$. Define the geometric Laplacian L on $L_2(\Omega)$ as the Friedrich's extension associated with the Dirichlet form

$$(u, u)_L = \sum_{n=0}^{\infty} l_n^{-2} \langle u, e_n \rangle \langle e_n, u \rangle$$

with dense domain

$$\mathcal{F}_L = \left\{ u \in C(\overline{\Omega}) : (u, u)_L < \infty, u|_{\partial\Omega} = 0 \right\}.$$

L has trivial kernel and compact resolvent. By construction, $L^{-1/2}$ is compact with eigenvalue sequence $\lambda(j, L^{-1/2}) = l_j$, $j \geq 0$, and the geometric zeta function of [136, p. 17] is

$$\zeta_L(s) := \mathrm{Tr}(L^{-s/2}), \quad s > D.$$

The relation between the spectral and the geometric zeta function is [136, Theorem 1.21]

$$\zeta(s) = \frac{\zeta_\Omega(s)}{\zeta_L(s)}, \quad s > D$$

where ζ is the Riemann zeta function. The geometric zeta function ζ_L has a singularity at $s = D$ [136, Theorem 1.10]. The nature of the singularity is at most simple. The boundary $\partial\Omega$ of a fractal string Ω has Minkowski dimension D if and only if $l_j^D = O(j^{-1})$, $j \to \infty$ [135, Theorem 2.4] [214, Sect. 5]. Hence

$$\partial\Omega \text{ has Minkowski dimension } D \text{ if and only if } L^{-D/2} \in \mathcal{L}_{1,\infty}.$$

For a detailed discussion of Minkowski measurability, Minkowski dimension and Minkowski content, we refer to [136, Chap. 1]. We also have [135, Theorem 2.2]

$$\partial\Omega \text{ is Minkowski measurable if and only if } \lim_{j\to\infty} j^{-1/D} l_j = c \tag{4.14}$$

for some positive constant c. The Minkowski content of $\partial\Omega$ then equals

$$\mathcal{M}(\partial\Omega) = \frac{2^{1-D}}{1-D} c^D.$$

The condition in (4.14) is the Weyl condition (4.8) discussed previously. We obtain the result noted by Connes in 1994 [40, p. 327] [214, Corollary 5.7]; if $\partial\Omega$ is Minkowski measurable, then the operator $L^{-D/2} \in \mathcal{L}_{1,\infty}$ is measurable and

$$\mathrm{Tr}_\omega(L^{-D/2}) = \frac{1-D}{2^{1-D}}\mathcal{M}(\partial\Omega)$$

for every Dixmier trace Tr_ω on $\mathcal{L}_{1,\infty}$ where $\mathcal{M}(\partial\Omega)$ is the Minkowski content of $\partial\Omega$.

The converse is not true [40, p. 328]. Measurability of $L^{-D/2}$ does not imply $\partial\Omega$ is Minkowski measurable since (4.8) is stronger than measurability. Choose $0 < D < 1$ and let Ω be the fractal string associated with the non-increasing numbers l_j with

$$l_j = \begin{cases} m^{-1/D} & j \in [2^{2m}, 2^{2m+1}] \\ 2^{1/D} m^{-1/D} & j \in [2^{2m+1}, 2^{2m+2}] \end{cases}, \quad j \geq 0.$$

Then

$$\sum_{j=0}^{n} \lambda(j, L^{-D/2}) = \frac{3}{2} \cdot \log(n) + O(1), \quad n \geq 0$$

and $L^{-D/2}$ is measurable, in fact, universally measurable, but $j^{-1/D} l_j$ oscillates between 1 and $2^{1/D}$ as $j \to \infty$.

Sequences l_j, $j \geq 0$ can be constructed so that $\partial\Omega$ is Minkowski measurable but $L^{-D/2}$ is not universally measurable. Take the fractal string Ω corresponding to the sequence

$$l_j = c^{1/D} \cdot \left((j+1)^{-1} + ((j+2)\log(j+2))^{-1}\right)^{1/D}, \quad j \geq 0$$

for a constant $c > 0$. Then $j^{-1/D} l_j \to c^{1/D}$ as $j \to \infty$, but

$$\sum_{j=0}^{n} \lambda(j, L^{-D/2}) = c \cdot \log(n) + O(\log(\log(n))), \quad n \geq 0$$

and $L^{-D/2}$ is not universally measurable. Geometric Laplacians on fractal strings with Minkowski measurable boundary provide examples of operators between measurable and universally measurable operators.

4.3 Fubini theorem

Tensor products are a central way to construct geometries. The Connes-Lott semi-classical approach is a product of four-dimensional space-time continuum and the standard model [40, p. 562]. It might be expected from the tensor product behaviour of the heat kernel asymptotic expansion [85, Lemma 1.7.5], and the association we have found in previous sections between Dixmier traces and the leading term of the asymptotic expansion, that the integral of the tensor product of two noncommutative geometries is equal to the product of the integrals. A counterexample published in 2017 [218] found that this is false without caveats.

Extending the formula for the tensor product of leading terms in the partition function asymptotic expansion [85, Lemma 1.7.5], in 1994 Connes on p. 563 of "Noncommutative Geometry" [40] proposed the statement

Statement 1 Suppose $D_1 : \mathrm{Dom}(D_1) \to H$ is a self-adjoint operator with compact resolvent and trivial kernel such that $|D_1|^{-p_1} \in \mathcal{L}_{1,\infty}$, $p_1 > 0$. Suppose D_2 satisfies the same condition for a number $p_2 > 0$ and $B_1, B_2 \in \mathcal{L}(H)$. If one of the terms $B_1|D_1|^{-p_1}$ or $B_2|D_2|^{-p_2}$ is "convergent", then

$$\Gamma\left(1+\frac{p_1+p_2}{2}\right)\mathrm{Tr}_\omega\left((B_1\otimes B_2)(|D_1|\otimes 1+1\otimes|D_2|)^{-(p_1+p_2)}\right)$$
$$=\Gamma\left(1+\frac{p_1}{2}\right)\mathrm{Tr}_\omega(B_1|D_1|^{-p_1})\cdot\Gamma\left(1+\frac{p_2}{2}\right)\mathrm{Tr}_\omega(B_2|D_2|^{-p_2})$$

for a Dixmier trace Tr_ω on $\mathcal{L}_{1,\infty}$ (as an ideal of $\mathcal{L}(H\otimes H)$ on the left side of the formula, and $\mathcal{L}(H)$ on the right).

The assumption of trivial kernel is irrelevant to the following discussion. As noted by Sukochev and Zanin [218], the term "convergent" used in [40] is unclear. It was shown in [218, Theorem 1.14] under the conditions that one, or both, of the terms $B_i|D_i|^{-p_i}$, $i = 1, 2$ are measurable (Section 4.2.1 above), or one, or both, of the terms are universally measurable (Section 4.2.2), then the statement is false. The counterexample given in [218] is for $B_1 = B_2 = 1$. Volume in noncommutative geometry, as defined using the Dixmier trace, can be badly behaved under products.

The convergence implied by Connes [218, p. 1231] is equivalent to the heat kernel asymptotic condition (4.8).

Statement 2 Suppose the conditions of Statement 1 and that

$$\mathrm{Tr}(T_1 e^{-sD_1^2}) = c\cdot s^{-\frac{p_1}{2}} + o\left(s^{-\frac{p_1}{2}}\right), \quad s\to 0^+, c \text{ const.}$$

Then

$$\Gamma\left(1+\frac{p_1+p_2}{2}\right)\mathrm{Tr}_\omega\left((B_1\otimes B_2)(|D_1|\otimes 1+1\otimes|D_2|)^{-(p_1+p_2)}\right)$$
$$=\Gamma\left(1+\frac{p_1}{2}\right)\mathrm{Tr}_\omega(B_1|D_1|^{-p_1})\cdot\Gamma\left(1+\frac{p_2}{2}\right)\mathrm{Tr}_\omega(B_2|D_2|^{-p_2}).$$

Statement 2 is still false [218, Theorem 1.15] without some restriction on the Dixmier trace [218, Theorem 1.9]. Recall the (continuous) logarithmic mean operator $M : L_\infty(\mathbb{R}_+) \to L_\infty(\mathbb{R}_+)$,

$$Mf(t) = \frac{1}{\log(1+t)} \int_0^t \frac{f(s)}{s} ds, \quad t \geq 0, f \in L_\infty(\mathbb{R}_+).$$

Define $P^a : L_\infty(\mathbb{R}_+) \to L_\infty(\mathbb{R}_+)$ for any $a > 0$ by

$$P^a(f)(t) = f(t^a), \quad t \geq 0, f \in L_\infty(\mathbb{R}_+).$$

Suppose ω is a state on $L_\infty(\mathbb{R}_+)$ that is M-invariant and P^a-invariant for all $a > 0$. Then Statement 2 is true [218, Theorem 1.10].

We note that only one of the terms needs to satisfy the convergence condition in Statement 2. We also note that the set of traces associated with extended limits satisfying M-invariance and P^a-invariance is strictly smaller than the subsets of Dixmier's trace considered so far, i.e. Connes' version of Dixmier's trace.

Condition (4.12) is stronger than (4.8) and universal measurability.

Statement 3 Suppose the conditions of Statement 1 and

$$\mathrm{Tr}(T_i e^{-sD_i^2}) = c_i \cdot s^{-\frac{p_i}{2}} + O\left(s^{-\frac{p_i}{2}+\epsilon}\right), \quad s \to 0^+, \epsilon > 0, c_i \text{ const.}$$

for *both* $i = 1, 2$. Then

$$\Gamma\left(1 + \frac{p_1 + p_2}{2}\right) \phi\left((B_1 \otimes B_2)(|D_1| \otimes 1 + 1 \otimes |D_2|)^{-(p_1+p_2)}\right)$$
$$= \Gamma\left(1 + \frac{p_1}{2}\right) \phi(B_1 |D_1|^{-p_1}) \cdot \Gamma\left(1 + \frac{p_2}{2}\right) \phi(B_2 |D_2|^{-p_2})$$

for any normalised trace ϕ on $\mathcal{L}_{1,\infty}$.

Statement 3 is true [218, Theorem 1.10]. Statement 3 is close to the leading term calculation in [85, Lemma 1.7.5]. If only one of the terms satisfies the convergence condition in Statement 3, then it is false [218, Theorem 1.15]. Statement 2 is therefore the nearest statement to a Fubini theorem for noncommutative geometry. Presently it is not known the degree to which the sufficient condition of M-invariance and P^a-invariance can be relaxed in Statement 2.

From the heat kernel asymptotic expansion of Minakshisundaram and Seeley, every generalised Laplacian $D^2 : C^\infty(\Omega, E) \to C^\infty(\Omega, E)$ and order zero pseudodifferential operator $B : C^\infty(\Omega, E) \to C^\infty(\Omega, E)$ over a vector bundle E on a closed Riemannian manifold Ω satisfies the condition of Statement 3 on the Hilbert space $L_2(\Omega, E)$ of square integrable sections. Noncommutative geometries that are essentially isospectral to their commutative counterparts such as noncommutative tori and $SU_q(2)$ satisfy the conditions of Statement 3 [218, Sect. 4].

Partition function estimates for Laplacians are known for some finitely and infinitely ramified self-similar fractals (Section 4.2.5). From those examples there are universally measurable Laplacians belonging to $\mathcal{L}_{1,\infty}$ that satisfy the conditions of Statement 2 but not Statement 3, and some that do not satisfy the conditions of Statement 2.

5 Recent applications

Dixmier's trace is not singled out by applications. Recent results [125, 203, 149, 30, 220] using the advances in traces above show that Connes' trace theorems [38] concerning the noncommutative residue and integration of forms hold for all traces on $\mathcal{L}_{1,\infty}$. The Hochschild character formula [45, 108, 54, 24, 23] has also been proven for any trace on $\mathcal{L}_{1,\infty}$ and under conditions weaker than the original.

5.1 *Integral operators and symbols*

The utility of the noncommutative residue is that it can be calculated locally by integrating the $-d$ symbol in the asymptotic expansion of a classical symbol over an atlas of the sphere bundle of a closed d-dimensional Riemannian manifold Ω [239]. By virtue of the residue vanishing on smoothing operators, integrating the $-d$ symbol is independent of which local charts the symbol is calculated in [91, Theorem 7.5]. When $A : C^\infty(\Omega) \to C^\infty(\Omega)$ is a classical pseudodifferential operator of order $-d$, its principal symbol a_{-d} is the order $-d$ symbol, and the noncommutative residue is obtained by integrating the principal symbol over the sphere bundle. Connes' original version of the trace theorem [38, Theorem 1] in (4.2) says that A belongs to the ideal $\mathcal{L}_{1,\infty}$ and the Dixmier trace of A is calculated from the principal symbol

$$\mathrm{Tr}_\omega(A) = \frac{1}{d} \int_{S^*\Omega} a_{-d}(v)dv.$$

Can traces of other weak-trace class integral operators be obtained from symbols?

5.1.1 Noncommutative residue

Trace theorems have advanced through the association between partial sums of eigenvalues of Hilbert-Schmidt operators and integration of their associated L_2-symbols over cylinders in the tangent bundle of a manifold Ω [125]. For illustration on $\mathbb{R}^d$; if $A \in \mathcal{L}_2(L_2(\mathbb{R}^d))$ is Hilbert-Schmidt, then it admits an L_2-symbol $p_A \in L_2(\mathbb{R}^d \times \mathbb{R}^d)$ such that

$$(Ah)(x) = \int_{\mathbb{R}^d} e^{ix\cdot\xi} p_A(x,\xi)\hat{h}(\xi)d\xi, \quad h \in L_2(\mathbb{R}^d)$$

where $\hat{h}$ denotes the Fourier transform. A non-trivial proof originating with Kalton shows that [125, Prop. 6.3] [125, Theorem 6.23]

Theorem 5.1 *If $A \in \mathcal{L}_2(L_2(\mathbb{R}^d))$ is Hilbert-Schmidt with symbol $p_A \in L_2(\mathbb{R}^d \times \mathbb{R}^d)$ of decay*

$$\int_{\mathbb{R}^d} \int_{\|\xi\|>n^{1/d}} |p_A(x,\xi)|^2 d\xi dx = O(n^{-1}), \quad n \geq 0, \tag{5.1}$$

and $AM_\psi = A$ for some $\psi \in C_c^\infty(\mathbb{R}^d)$, then $A \in \mathcal{L}_{1,\infty}$ and

$$\sum_{j=0}^{n} \lambda(j, A) = \int_{\mathbb{R}^d} \int_{\|\xi\| \leq n^{1/d}} p_A(x,\xi)\psi(x) d\xi dx + O(1), \quad n \geq 0 \tag{5.2}$$

where $\lambda(j, A)$, $j \geq 0$ is an eigenvalue sequence of A.

Here

$$(M_\psi h)(x) = \psi(x)h(x), \quad \psi \in C_c^\infty(\mathbb{R}^d), h \in L_2(\mathbb{R}^d)$$

is the operator giv en by pointwise multiplication and $C_c^\infty(\mathbb{R}^d)$ denotes smooth functions of compact support.

If a Hilbert-Schmidt operator A and its adjoint A^* both satisfy (5.1), then the condition that $M_\psi A = A$ for some $\psi \in C_c^\infty(\mathbb{R}^d)$ (instead of $AM_\psi = A$) is sufficient for $A \in \mathcal{L}_{1,\infty}$ and (5.2). That $M_\psi A = A$ for some $\psi \in C^\infty(\mathbb{R}^d)$ is equivalent to the symbol $p_A(x,\xi) = \psi(x)p_A(x,\xi)$ having compact support almost everywhere in the first variable. Integration of the L_2-symbol over cylinders in phase space provides estimates of the log divergence and the remainder term of the partial sums of eigenvalues. From Section 4, the divergence calculates a trace and the remainder term specifies degree of measurability. Specifically, under the conditions mentioned, $A \in \mathcal{L}_{1,\infty}$ and [125, Theorem 6.32]

$$\mathrm{Tr}_\omega(A) = \omega\left(\frac{1}{\log(n)} \int_{\mathbb{R}^d} \int_{\|\xi\| \leq n^{1/d}} p_A(x,\xi)\psi(x) d\xi dx\right)$$

for any extended limit $\omega \in \ell_\infty^*$. For any positive trace on $\mathcal{L}_{1,\infty}$ there is a similar formula involving the integral over increasing dyadic annuli in phase space [203, Theorem 8.10]; recalling the bijection (3.6) in Section 3 then

$$\tau_\theta(A) = \theta\left(\frac{1}{\log 2} \int_{\mathbb{R}^d} \int_{2^{\frac{n}{d}} \leq \|\xi\| \leq 2^{\frac{n+1}{d}}} p_A(x,\xi)\psi(x) d\xi dx\right)$$

for a Banach limit θ.

This approach can be viewed as a generalisation of the noncommutative residue. When $A : C^\infty(\Omega) \to C^\infty(\Omega)$ is a *classical* uniform pseudodifferential operator of order $-d$, with leading term $a_{-d}(x,\xi)$ homogenous order $-d$ in ξ outside a neighbourhood of the origin and compactly supported in x, the log divergence of the integral over the cylinders coincides with the integral over the sphere bundle [125, Prop. 6.16],

$$\sum_{j=0}^{n} \lambda(j, A) = \frac{1}{d} \int_{\mathbb{R}^d \times \mathbb{S}^{d-1}} a_{-d}(x, s) dx ds \cdot \log(n) + O(1), \quad n \geq 0$$

and it follows from the discussion in Section 4 that A is universally measurable with

$$\phi(A) = \frac{1}{d} \int_{\mathbb{R}^d \times \mathbb{S}^{d-1}} a_{-d}(x, s) dx ds$$

for every normalised trace ϕ on $\mathcal{L}_{1,\infty}$. Symbols with oscillatory leading term behaviour provide examples of operators that are not measurable or not universally measurable. Let $Q : C^\infty(\mathbb{R}^d) \to C^\infty(\mathbb{R}^d)$ be an order $-d$ *nonclassical* pseudodifferential operator with symbol $q(x, \xi) = q_1(x) q_2(\xi)$ where $q_1, q_2 \in C^\infty(\mathbb{R}^d)$, q_1 has compact support with

$$\int_{\mathbb{R}^d} q_1(x) dx = 1$$

and

$$q_2(\xi) = |\xi|^{-d} \left(\sin(\log\log|\xi|) + \cos(\log\log|\xi|)\right), \quad |\xi| \geq 4.$$

Then $Q \in \mathcal{L}_{1,\infty}$,

$$\mathrm{Tr}_\omega(Q) = \omega\left(\sin(\log\log(n^{1/d}))\right)$$

and Q is not measurable since $\sin(\log\log(n^{1/d}))$, $n \geq 0$ is not convergent [125, p. 29]. Another example is given in [203, Theorem 8.13], where Q is as described above except with the replacement

$$q_2(x) = |\xi|^{-d} \sin\left(\frac{\log|\xi|}{\log\log|\xi|}\right), \quad |\xi| \geq 4.$$

The pseudodifferential operator Q is measurable, in fact $\mathrm{Tr}_\omega(Q) = 0$ for all Dixmier traces, but $\tau_\theta(Q) \neq 0$ for some positive trace τ_θ on $\mathcal{L}_{1,\infty}$ [203, Lemma 8.12].

The log divergent term depends only on the principal symbol of the total symbol p_A of the Hilbert-Schmidt operator A, where the principal symbol is defined as an equivalence class providing the same log divergent behaviour [125, p. 48]. When A is a classical pseudodifferential operator of order $-d$, its principal symbol in the classical sense determines this equivalence class. Though this approach seems restricted to integral operators on $\mathbb{R}^d$, these estimates can be described solely in terms of the operator A and the Laplacian without reference to the underlying space $\mathbb{R}^d$ [150, Chap. 11].

The Laplacian can be replaced by generalisations such as the Laplace-Beltrami operator. By using charts, the $\mathbb{R}^d$ theory above can be transferred to a closed manifold Ω and trace formulas expressed using local expressions of symbols integrated over cylinder bundles within the tangent bundle [125, Theorem 7.13] [150, Sect. 11.6]. One obtains the earlier mentioned formula (4.9); for any classical pseudodifferential operator $A : C^\infty(\Omega) \to C^\infty(\Omega)$ of order $-d$

$$\phi(A) = \frac{1}{d}\int_{S^*\Omega} a_{-d}(v)dv = \frac{1}{d}\mathrm{Res}(A)$$

for every normalised trace ϕ on $\mathcal{L}_{1,\infty}$. There are alternative expressions using an ordered eigenvectors of the Laplace-Beltrami operator e_n, $n \geq 0$ such that $-\Delta e_n = \lambda(n, -\Delta)e_n$; if a Hilbert-Schmidt operator $A \in \mathcal{L}_2(L_2(\Omega))$ satisfies the decay [125, Prop. 5.13]

$$\sum_{j=n}^{\infty} \|Ae_j\|^2 = O(n^{-1}), \quad n \geq 0 \tag{5.3}$$

then $A \in \mathcal{L}_{1,\infty}$ [125, Theorem 5.2]

$$\sum_{j=0}^{n} \lambda(j, A) = \sum_{j=0}^{n} \langle e_n, Ae_n\rangle + O(1), \quad n \geq 0 \tag{5.4}$$

and the Dixmier trace calculates the logarithmic divergence of expectation values [125, Theorem 7.6]

$$\mathrm{Tr}_\omega(A) = \omega\left(\frac{1}{\log(n)}\sum_{j=0}^{n}\langle e_n, Ae_n\rangle\right).$$

The decay in (5.1) and integration over cylinders in phase space and the logarithmic divergence of eigenvalue series in (5.2) are related to the "free energy" generalised eigenfunctions $e^{-ix\cdot\xi}$, $x, \xi \in \mathbb{R}^d$. Conditions (5.1) and (5.3) are both equivalent to [125, Sect. 5]

$$\|A(1 + nH^{-p/2})^{-1}\|_{\mathcal{L}_2}^2 = O(n^{-1}), \quad n \geq 0, \text{ some } p > 0$$

where H is one minus the Laplacian in the case of (5.1) and one minus the Laplace-Beltrami operator in the case of (5.3). The fundamental theory works when H is any positive operator with trivial kernel [125, Sect. 5] [150, Sect. 12.1]. As yet there has been no exploration of the theory for Schrödinger operators, e.g. the quantum harmonic oscillator $H = -\Delta + M_{x^2}$ acting on $L_2(\mathbb{R})$ and $p = 2$. The theory on

$\mathbb{R}^d$ and closed manifolds was developed in the setting of Kohn-Nirenberg, that is, in terms of the left symbol p_A above. The Weyl symbol was examined in [83].

5.1.2 Noncommutative integral and symbols

Suppose $B : C^\infty(\mathbb{R}^d) \to C^\infty(\mathbb{R}^d)$ is a uniform order 0 classical pseudodifferential operator on $\mathbb{R}^d$ with principal symbol $\sigma_0(B)$; a bounded smooth function with bounded derivatives on the sphere bundle $\mathbb{R}^d \times \mathbb{S}^{d-1}$. Denote by $\Psi^0_{\rm cl}$ the algebra of uniform classical pseudodifferential operators of order 0 and the principal symbol map [194, Chap. 2] [113]

$$\sigma_0 : \Psi^0_{\rm cl} \to C_b^\infty(\mathbb{R}^d \times \mathbb{S}^{d-1}).$$

Assume that $BM_\psi = B$ for a compactly supported function $\psi \in C^\infty(\mathbb{R}^d)$.[9] Then $\sigma_0(B)\psi = \sigma_0(B)$ has compact support. Last section (Section 5.1.1) indicates that

$$\phi(B(1-\Delta)^{-d/2}) = \frac{1}{d(2\pi)^d}\int_{\mathbb{R}^d\times\mathbb{S}^{d-1}} \sigma_0(B)(x,s)dxds \tag{5.5}$$

for any normalised trace ϕ on $\mathcal{L}_{1,\infty}$. Calculation of the noncommutative integral using symbols can be extended to more general bounded operators.

If $B \in \mathcal{L}(L_2(\mathbb{R}^d))$ is any bounded operator satisfying $BM_\psi = B$ for $\psi \in C^\infty(\mathbb{R}^d)$ of compact support, then, as the classical pseudodifferential operator $M_\psi(1-\Delta)^{-d/2}$ belongs to $\mathcal{L}_{1,\infty}$, we have

$$B(1-\Delta)^{-d/2} \in \mathcal{L}_{1,\infty}.$$

From the pseudodifferential calculus $[M_\psi, (1-\Delta)^{-d/2}] \in \mathcal{L}_1$ [150, Sect. 10.2], and we have

$$B(1-\Delta)^{-d/2} - M_\psi BM_\psi(1-\Delta)^{-d/2}M_\psi \in \mathrm{Com}(\mathcal{L}_{1,\infty}).$$

The operator $M_\psi BM_\psi(1-\Delta)^{-d/2}M_\psi$ satisfies the conditions of Theorem 5.1 and has the L_2-symbol [150, p. 349]

$$p(x,\xi) = \psi(x)e_{-\xi}(x)(Be_\xi)(x)(1+\|2\pi\xi\|^2)^{-d/2} + f(x,\xi), \quad x,\xi \in \mathbb{R}^d$$

where

[9]For pseudodifferential operators this can be weakened to $M_\psi B = B$; this implies $BM_\psi - B \in \mathcal{L}_1$ [150, Example 10.2.23], which is sufficient. Throughout this section the condition $BM_\psi = B$ on $B \in \mathcal{L}(L_2(\mathbb{R}^d))$ can be replaced by $BM_\psi - B \in \mathcal{L}_1$.

$$e_\xi(x) = e^{-ix\cdot\xi}, \quad x, \xi \in \mathbb{R}^d$$

and

$$f(x, \xi) \in L_1(\mathbb{R}^d \times \mathbb{R}^d).$$

Näively rewriting the result (5.2) using this symbol, the asymptotic behaviour as $n \to \infty$ of the sequence of functions in $L_1(\mathbb{R}^d \times \mathbb{S}^{d-1})$

$$p_n(B)(x,s) = \frac{1}{\log(n^{1/d})} \int_1^{n^{1/d}} e_{-rs}(x)(Be_{rs})(x)\frac{dr}{r}, \quad n \geq 0, x \in \mathbb{R}^d, s \in \mathbb{S}^{d-1}$$

generalises the role of the principal symbol. For example

$$\mathrm{Tr}_\omega(B(1-\Delta)^{-d/2}) = \frac{1}{d(2\pi)^d} \cdot \omega\left(\int_{\mathbb{R}^d \times \mathbb{S}^{d-1}} p_n(B)(x,s)\psi(x)dxds\right)$$

for any extended limit $\omega \in \ell_\infty^*$. Similarly, defining

$$m_n(B)(x,s) = \frac{1}{\log(2^{1/d})} \int_{2^{n/d}}^{2^{(n+1)/d}} e_{-rs}(x)(Be_{rs})(x)\frac{dr}{r}, \; n \geq 0, x \in \mathbb{R}^d, s \in \mathbb{S}^{d-1}$$

then

$$\tau_\theta(B(1-\Delta)^{-d/2}) = \frac{1}{d(2\pi)^d} \cdot \theta\left(\int_{\mathbb{R}^d \times \mathbb{S}^{d-1}} m_n(B)(x,s)\psi(x)dxds\right)$$

for a Banach limit θ.

If B is a zero order classical uniform pseudodifferential operator satisfying $BM_\psi = B$, then $p_n(B), n \geq 0$ and $m_n(B), n \geq 0$ are the same constant sequence of functions $(\sigma_0(B), \sigma_0(B), \ldots)$. Here $\sigma_0(B)$ is the principal symbol of B. For general $B \in \mathcal{L}(H)$ these "principal symbol as a sequence" maps lose many properties of the classical principal symbol map.

A recent paper [225] considered calculation of traces using an extension of the principal symbol map that is unique and an algebra homomorphism. Abstractions of the principal symbol are usually based on quotient maps of compact operators and commutator ideals [59, 57, 185]—an idea of Gohberg originating in 1960 [87, 69], or based on invariant actions by unitaries [160, 104]. Denote by π_0 the algebra homomorphism from the bounded operators to the Calkin algebra of the bounded operators quotient compact operators; a setting for the principal symbol established by Cordes [57, Chap. IV]

$$\pi_0 : \mathcal{L}(L_2(\mathbb{R}^d)) \to \mathcal{Q}(L_2(\mathbb{R}^d)) = \mathcal{L}(L_2(\mathbb{R}^d))/\mathcal{K}(L_2(\mathbb{R}^d)).$$

If Ψ is a C*-algebra of bounded operators in $\mathcal{L}(H)$ whose commutators $[B_1, B_2]$, $B_1, B_2 \in \Psi$ are compact operators in $\mathcal{K}(H)$, then $\pi_0(\Psi)$ is a commutative C*-algebra in $\mathcal{Q}(L_2(\mathbb{R}^d))$ [59]. Using the Gelfand representation [161, Theorem. 2.1.10], $\pi_0(\Psi)$ is isomorphic with the C*-algebra of functions $C_0(X)$, for X some locally compact Hausdorff space. Suppressing the Gelfand isomorphism, the combination

$$\pi_0 : \Psi \to C_0(X)$$

is a homomorphism of C*-algebras. Cordes established [57, p. 172, Theorem 9.1] [57, p. 177, Theorem 10.1] [58] that π_0 implements the principal symbol map;

$$\pi_0 : \{B \in \Psi^0_{\text{cl}} : B = M_\varphi B M_\psi\} \to C_c^\infty(\mathbb{R}^d \times \mathbb{S}^{d-1}),$$

that is, if $B \in \Psi^0_{\text{cl}}$ and $B = M_\varphi B M_\psi$ for some compactly supported $\varphi, \psi \in C^\infty(\mathbb{R}^d)$ then

$$\pi_0 : B \to \sigma_0(B).$$

The principal symbol map σ_0 on Ψ^0_{cl} without restriction can be obtained by

$$\sigma_0(B)(x, \xi) = \lim_{n\to\infty} \pi_0(M_{\psi_n} B M_{\psi_n})(x, \xi)$$

where $\psi_n \in C^\infty(\mathbb{R}^d)$ is an increasing sequence of smooth compactly supported nonnegative functions such that $\psi_n(x) \to 1^-$ as $n \to \infty$.[10] The principal symbol map investigated by Cordes, Herman, Power and others in the 1960s and 1970s in the C*-algebraic setting [104, 197, 160, 58, 185] began with natural extensions of the Kohn-Nirenberg pseudodifferential symbol map [113]. Define the representation

$$\pi_1 : L_\infty(\mathbb{R}^d) \to \mathcal{L}(L_2(\mathbb{R}^d))$$

by

$$\pi_1(f)h = fh, \quad f \in L_\infty(\mathbb{R}^d), h \in L_2(\mathbb{R}^d).$$

[10] Even though Ψ^0_{cl} is a *-subalgebra of bounded operators on $L_2(\mathbb{R}^d)$, the principal symbol map on Ψ^0_{cl} cannot be defined directly with π_0. There are operators in Ψ^0_{cl} whose commutators are not compact operators on $L_2(\mathbb{R}^d)$ [57, Lemma 10.5]. Cordes considers several subalgebras of Ψ^0_{cl} with compact commutators, the maximal one being the *-subalgebra of operators B whose total symbol $\sigma(B)(x, \xi) \in C_b^\infty(\mathbb{R}^d \times \mathbb{R}^d)$ has all derivatives in $C_0^\infty(\mathbb{R}^d \times \mathbb{R}^d)$ [57, p. 133], and the *-subalgebra of operators B whose total symbol satisfy [58] [204, Chap. IV]

$$\partial_x^\alpha \partial_\xi^\beta \sigma(B)(x, \xi) = O((1 + |x|^2)^{-|\alpha|}) O((1 + |\xi|^2)^{-|\beta|}), \quad x, \xi \in \mathbb{R}^d.$$

Let

$$s : \mathbb{R}^d \setminus \{0\} \to \mathbb{S}^{d-1}\ , \ s(x) = \|x\|^{-1} \cdot x, \quad x \in \mathbb{R}^d.$$

For any $g \in L_\infty(\mathbb{S}^{d-1})$ define the bounded operator

$$\pi_2(g) = F^{-1}\pi_1(g \circ s)F : L_2(\mathbb{R}^d) \to L_2(\mathbb{R}^d)$$

where $F : L_2(\mathbb{R}^d) \to L_2(\mathbb{R}^d)$ is the unitary defined by Fourier transform. This is a representation

$$\pi_2 : L_\infty(\mathbb{S}^{d-1}) \to \mathcal{L}(L_2(\mathbb{R}^d)).$$

Denote by

$$\Pi(\mathcal{A}_1, \mathcal{A}_2) \subset \mathcal{L}(L_2(\mathbb{R}^d))$$

the C*-algebra generated by a C*-subalgebra $\mathcal{A}_1$ of $L_\infty(\mathbb{R}^d)$ represented in $\mathcal{L}(L_2(\mathbb{R}^d))$ by π_1, and a C*-subalgebra $\mathcal{A}_2$ of $L_\infty(\mathbb{S}^{d-1})$ represented in $\mathcal{L}(L_2(\mathbb{R}^d))$ by π_2. Then [57, Theorem 1.6, p. 135]

$$\pi_0 : \Pi(C_0(\mathbb{R}^d), C(\mathbb{S}^{d-1})) \to (\mathbb{C} + C_0(\mathbb{R}^d)) \otimes C(\mathbb{S}^{d-1})$$

such that

$$\pi_0(\pi_1(f)) = f \otimes 1$$

and

$$\pi_0(\pi_2(g)) = 1 \otimes g.$$

Using mollifiers as before provides a unique algebraic homomorphism

$$\hat{\sigma}_0 : \Pi(C_b(\mathbb{R}^d), C(\mathbb{S}^{d-1})) \to C_b(\mathbb{R}^d \otimes \mathbb{S}^{d-1})$$

such that

$$\hat{\sigma}_0(\pi_1(f)) = f \otimes 1$$

and

$$\hat{\sigma}_0(\pi_2(g)) = 1 \otimes g$$

which extends the principal symbol map σ_0 of classical pseudodifferential operators. Other presentations of the principal symbol as a quotient map usually restrict to continuous functions, [197, Remark 1.8] [5, p. 134] [129, p. 197]. The extension, which we also denote by $\hat{\sigma}_0$,

$$\hat{\sigma}_0 : \Pi(L_\infty(\mathbb{R}^d), C(\mathbb{S}^{d-1})) \to L_\infty(\mathbb{R}^d) \otimes C(\mathbb{S}^{d-1})$$

such that

$$\hat{\sigma}_0(\pi_1(f)) = f \otimes 1$$

and

$$\hat{\sigma}_0(\pi_2(g)) = 1 \otimes g$$

was known in 1971 [160, p. 616] [185, Theorem 6.2].

The main result of [225] extends $\hat{\sigma}_0$ further using an invariance argument similar to [160].

Theorem 5.2 *There is a unique algebraic homomorphism of C*-algebras extending the principal symbol map σ_0 of zero order uniform pseudodifferential operators on $\mathbb{R}^d$,*

$$\hat{\sigma}_0 : \Pi(L_\infty(\mathbb{R}^d), L_\infty(\mathbb{S}^{d-1})) \to L_\infty(\mathbb{R}^d \times \mathbb{S}^{d-1})$$

such that

$$\hat{\sigma}_0(\pi_1(f)) = f \otimes 1, \quad f \in L_\infty(\mathbb{R}^d)$$

and

$$\hat{\sigma}_0(\pi_2(g)) = 1 \otimes g, \quad g \in L_\infty(\mathbb{S}^{d-1}).$$

If $B \in \Pi(L_\infty(\mathbb{R}^d), L_\infty(\mathbb{S}^{d-1}))$ with $B = BM_\psi$ for some compactly supported $\psi \in C^\infty(\mathbb{R}^d)$, then $B(1-\Delta)^{-d/2} \in \mathcal{L}_{1,\infty}$ and

$$\phi(B(1-\Delta)^{-d/2}) = \frac{1}{d(2\pi)^d} \int_{\mathbb{R}^d \times \mathbb{S}^{d-1}} \hat{\sigma}_0(B)(x,s)dxds \tag{5.6}$$

for any positive normalised trace ϕ on $\mathcal{L}_{1,\infty}$.

The result combines [225, Theorem 1.2] [225, Theorem 1.5] and [225, Lemma 8.3]; it provides a trace formula equivalent to (5.5). In (5.5) the trace was arbitrary and followed from the spectral estimate (5.2). The proof of [225, Theorem 1.5] is different and uses the Riesz-Markov Theorem [189, p. 111]. Essentially, the algebra $\Pi(L_\infty(\mathbb{R}^d), L_\infty(\mathbb{S}^{d-1}))$ has a unique noncommutative integral. There is no extension of $\hat{\sigma}_0$ to the weak closure with the same properties. The weak operator closure of the C*-algebra $\Pi(L_\infty(\mathbb{R}^d), L_\infty(\mathbb{S}^{d-1}))$ is

$\mathcal{L}(L_2(\mathbb{R}^d))$, and, because every bounded operator is a finite sum of commutators [98, 99], there is no non-zero homomorphism of $\mathcal{L}(L_2(\mathbb{R}^d))$ into a commutative C*-algebra [225, Lemma 3.5].

If $B \in \Pi(L_\infty(\mathbb{R}^d), L_\infty(\mathbb{S}^{d-1}))$, then Cordes and Beals [58, 13, 227] noted the condition that [57, Theorem 10.1]

$$\partial^{\alpha,\beta} B \in \mathcal{L}(L_2(\mathbb{R}^d)), \qquad \text{all multi-indices } \alpha, \beta,$$

identifies a zero order classical uniform pseudodifferential operator of Kohn-Nirenberg within $\Pi(L_\infty(\mathbb{R}^d), L_\infty(\mathbb{S}^{d-1}))$. Here

$$\partial^{\alpha,\beta} B = [-i\partial_{\alpha_1}, \ldots, [-i\partial_{\alpha_N}, [M_{x_{\beta_1}}, [\ldots, [M_{x_{\beta_M}}, B]]]]] \tag{5.7}$$

where α and β denote multi-indices and $-i\partial_j$, $j = 1, \ldots, d$ and M_{x_j}, $j = 1, \ldots, d$ are the unbounded self-adjoint operators densely defined on $L_2(\mathbb{R}^d)$ by partial derivatives and multiplication by coordinates, respectively.

5.1.3 Noncommutative symbols

Noncommutative versions of the principal symbol have been developed to prove trace theorems for noncommutative geometries; McDonald et al. [156] extends the compact commutator setting of Cordes and Herman [59] and Power [185]. Denote the quotient map

$$\pi_0 : \mathcal{L}(H) \to \mathcal{Q}(H)$$

for a separable Hilbert space H. Let $\mathcal{A}_1$ and $\mathcal{A}_2$ be unital C*-algebras represented in $\mathcal{L}(H)$ by π_1 and π_2 respectively, where $\mathcal{A}_2$ is commutative. Denote by

$$\Pi(\mathcal{A}_1, \mathcal{A}_2) \subset \mathcal{L}(H)$$

the C*-algebra generated by $\pi_1(\mathcal{A}_1)$ and $\pi_1(\mathcal{A}_2)$. Let $\mathcal{A}_1 \otimes \mathcal{A}_2$ denote the C*-algebra obtained by closing the algebraic tensor product $\mathcal{A}_1 \otimes_{\text{alg}} \mathcal{A}_2$ in the spatial norm[11] [161, Sect. 6]. Suppose that

1. $[\pi_1(a), \pi_2(b)] \in \mathcal{K}(H)$ for all $a \in \mathcal{A}_1$ and $b \in \mathcal{A}_2$, and
2. the map

$$\sum_{k=1}^n a_k \otimes b_k \mapsto \pi_0\left(\sum_{k=1}^n \pi_1(a_k)\pi_2(b_k)\right)$$

is injective for all $a_k \in \mathcal{A}_1, b_k \in \mathcal{A}_2$, $1 \le k \le n$.

[11] As one of the algebras is commutative, it is in particular nuclear and all C^*-norms on $\mathcal{A}_1 \otimes_{\text{alg}} \mathcal{A}_2$ coincide.

Then $\pi_0 : \mathcal{L}(H) \to \mathcal{Q}(H)$ implements a unique continuous *-homomorphism [161, Theorem 6.3.7] [156, Lemma 3.2]

$$\pi_0 : \Pi(\mathcal{A}_1, \mathcal{A}_2) \to \mathcal{A}_1 \otimes \mathcal{A}_2.$$

Here we suppressed the *-isomorphism between $\mathcal{A}_1 \otimes \mathcal{A}_2$ and $\pi_0(\Pi(\mathcal{A}_1, \mathcal{A}_2))$.

The purpose of using the symbol to calculate the noncommutative integral is in the following observation; if ρ is a continuous linear functional on $\Pi(\mathcal{A}_1, \mathcal{A}_2)$ which vanishes on commutators, and

$$\rho(\pi_1(a)\pi_2(b)) = \psi_1(a)\psi_2(b), \quad a \in \mathcal{A}_1, b \in \mathcal{A}_2$$

for some $\psi_1 \in \mathcal{A}_1^*$ and $\psi_2 \in \mathcal{A}_2^*$ then

$$\rho(B) = (\psi_1 \otimes \psi_2)(\pi_0(B)), \quad B \in \Pi(\mathcal{A}_1, \mathcal{A}_2).$$

Here $\psi_1 \otimes \psi_2$ is the unique continuous extension of $(\psi_1 \otimes \psi_2)(a \otimes b) = \psi_1(a)\psi_2(b)$, $a \in \mathcal{A}_1, b \in \mathcal{A}_2$ acting on the algebraic tensor product [161, Corollary 6.4.3].

If $A \in \mathcal{L}_{1,\infty}$ is fixed, then

$$\rho(B) = \phi(BA), \quad B \in \mathcal{L}(H)$$

is a continuous linear functional on $\mathcal{L}(H)$ for any positive trace ϕ on $\mathcal{L}_{1,\infty}$. The trace ϕ vanishes on finite rank operators; if B is finite rank, then BA is finite rank and $\rho(B) = 0$. Hence ρ vanishes on $\mathcal{K}(H)$. Therefore, to identify a noncommutative integral on a C*-algebra of the form $\Pi(\mathcal{A}_1, \mathcal{A}_2)$ generalising the *-algebra of zero-order pseudodifferential operators, it suffices to identify the individual states

$$\phi(\pi_1(a)A)\ , \ \phi(\pi_2(b)A), \quad a \in \mathcal{A}_1, b \in \mathcal{A}_2.$$

We illustrate using the noncommutative torus and the noncommutative plane.

5.1.4 Noncommutative torus and noncommutative plane

Let θ be a real antisymmetric $d \times d$ matrix with $\det \theta \neq 0$. In particular, d is even.

The noncommutative d-torus $C(\mathbb{T}_\theta^d)$ is the universal C*-algebra generated by a family of unitaries $\{u_n\}_{n \in \mathbb{Z}^d}$ satisfying the relation [191],

$$u_m u_n = e^{2in\cdot\theta m} u_n u_m, \quad n, m \in \mathbb{Z}^d.$$

It has a representation π_1 on the Hilbert space $\ell_2(\mathbb{Z}^d)$ using the Fourier dual of the discrete Moyal product; the realisations are

$$\pi_1(u_m)h(n) = e^{-im\cdot\theta n} h(n-m), \quad n, m \in \mathbb{Z}^d, h \in \ell_2(\mathbb{Z}^d).$$

The geometry of the noncommutative torus, including connections, curvature, a pseudodifferential calculus, and index pairings, was developed originally by Connes [36] in 1980, see also presentations in [42] [91, Sect. 12.3] [40, p. 758] [46]; the Moyal product itself goes back to Groenewold and Moyal in the 1940s [92, 159]. The derivatives on the noncommutative torus are realised in the Fourier dual presentation using multiplication operators

$$(M_j h)(n) = n_j h(n), \quad n = (n_1, n_2, \ldots, n_d) \in \mathbb{Z}^d, j = 1, \ldots, d$$

and the Laplacian $M^2 : w_{2,2}(\mathbb{Z}^d) \to \ell_2(\mathbb{Z}^d)$ defined by

$$M^2 = \sum_{j=1}^{d} M_j^2$$

acting on the domain

$$w_{2,2}(\mathbb{Z}^d) = \{h \in \ell_2(\mathbb{Z}^2) : \sum_{n \in \mathbb{Z}^d} \|n\|^4 |h(n)|^2 < \infty\}.$$

Define the representation π_2 of the C*-algebra $C(\mathbb{S}^{d-1})$ on $\ell_2(\mathbb{Z}^d)$ as the multiplication operator

$$(\pi_2(g)h)(n) = (g \circ s)(n)h(n), \quad n \in \mathbb{Z}^d, h \in \ell_2(\mathbb{Z}^d).$$

where $s(n) = \frac{n}{\|n\|}$. The trace τ on $C(\mathbb{T}_\theta^d)$ is defined by the state

$$\tau(a) = \langle e_0, \pi_1(a) e_0 \rangle = \langle e_n, \pi_1(a) e_n \rangle, \quad a \in C(\mathbb{T}_\theta^d), n \in \mathbb{Z}^d$$

where e_n, $n \geq 0$ is the standard basis of $\ell_2(\mathbb{Z}^d)$. The last equality follows from translation invariance of the Haar measure on $\mathbb{Z}^d$. The conditions (1) and (2) in Section 5.1.3 above can be checked directly [156].

Suppose that ϕ is a positive trace on $\mathcal{L}_{1,\infty}$. We may calculate the following noncommutative integrals:

$$\psi_1(a) = \phi(\pi_1(a)(1 - M^2)^{-d/2}), \quad a \in C(\mathbb{T}_\theta^d)$$

and

$$\psi_2(g) = \phi(\pi_2(g)(1 - M^2)^{-d/2}), \quad g \in C(\mathbb{S}^{d-1}).$$

From $(1 - M^2)^{-d/2} \in \mathcal{L}_{1,\infty}$ and the formula (5.4)

$$\begin{aligned}
\sum_{j=1}^{n} \lambda(j, \pi_1(a)(1-M^2)^{-d/2}) &= \sum_{\|m\| \le n^{1/d}} \langle e_m, \pi_1(a)(1-M^2)^{-d/2} e_m \rangle + O(1) \\
&= \tau(a) \sum_{\|m\| \le n^{1/d}} (1+\|m\|^2)^{-d/2} + O(1) \\
&= \tau(a) \cdot \frac{\mathrm{Vol}\,\mathbb{S}^{d-1}}{d} \log(n) + O(1), \quad a \in C(\mathbb{T}^d_\theta).
\end{aligned}$$

It follows that

$$\psi_1(a) = \phi(\pi_1(a)(1-M^2)^{-d/2}) = \frac{\mathrm{Vol}\,\mathbb{S}^{d-1}}{d} \tau(a), \quad a \in C(\mathbb{T}^d_\theta).$$

Since ϕ is assumed to be continuous, it follows that ψ_2 is a continuous linear functional on $C(\mathbb{S}^{d-1})$. By the Riesz-Markov Theorem [189, p. 111] and direct confirmation of the rotation invariance of ψ_2 [156, Lemma 6.3] yields:

$$\psi_2(g) = \phi(\pi_2(g)(1-M^2)^{-d/2}) = \frac{1}{d} \int_{\mathbb{S}^{d-1}} g(s) ds, \quad g \in C(\mathbb{S}^{d-1}).$$

We summarise with the following theorem [156, Lemma 6.4]. See also Fathizadeh and Khalkhali for Connes' trace theorem on the noncommutative torus [73, Theorem 5.3] using the notion of pseudodifferential operators on the noncommutative torus due to Connes [36] , and the general result that the noncommutative residue is an invariant of isospectral deformation [150, Theorem 12.4.1] (see also [74]).

Theorem 5.3 *On the C^*-algebra of operators of "order 0" for the noncommutative torus defined above*

$$\Pi(C(\mathbb{T}^d_\theta), C(\mathbb{S}^{d-1})) \subset \mathcal{L}(L_2(\mathbb{T}^d)),$$

we have the principal symbol homomorphism

$$\pi_0 : \Pi(C(\mathbb{T}^d_\theta), C(\mathbb{S}^{d-1})) \to C(\mathbb{S}^{d-1}, C(\mathbb{T}^d_\theta)),$$

all operators in $\Pi(C(\mathbb{T}^d_\theta), C(\mathbb{S}^{d-1}))$ have unique noncommutative integral, and we have the trace theorem for the noncommutative integral

$$\phi(B(1-M^2)^{-d/2}) = \frac{1}{d} \int_{\mathbb{S}^{d-1}} \tau(\pi_0(B)(s)) ds, \quad B \in \Pi(C(\mathbb{T}^d_\theta), C(\mathbb{S}^{d-1}))$$

for any positive trace ϕ on $\mathcal{L}_{1,\infty}$.

The noncommutative plane $C_0(\mathbb{R}^d_\theta)$ has equivalent presentations as a Moyal product algebra [90, 231, 232, 81, 82] and deformation quantisation of $\mathbb{R}^d$ [192], quantum Euclidean space [89], or a canonical commutation relation (CCR) algebra [20, Section 5.2.2.2].

We use the Fourier dual of the Moyal representation on the Hilbert space $L_2(\mathbb{R}^d)$ [192]; define the unitaries

$$u_t h(x) = e^{-it\cdot\theta x} h(x-t), \quad x, t \in \mathbb{R}^d, h \in L_2(\mathbb{R}^d).$$

We have suppressed the representation π_1. Denote by $L_\infty(\mathbb{R}^d_\theta)$ the von Neumann algebra in $\mathcal{L}(L_2(\mathbb{R}^d))$ generated by the unitaries $\{u_t\}_{t\in\mathbb{R}^d}$. Then $L_\infty(\mathbb{R}^d_\theta)$ can be identified with $\mathcal{L}(L_2(\mathbb{R}^{d/2}))$ through a spatial *-isomorphism i_θ [90, Theorem 2.2] [82, Proposition 2.13] [141, Theorem 6.5].

We define a faithful semifinite normal trace τ_θ on $L_\infty(\mathbb{R}^d_\theta)$ and noncommutative L_p-spaces $L_p(\mathbb{R}^d_\theta)$, $1 \le p < \infty$, using the spatial *-isomorphism. Define

$$\tau_\theta(a) = (2\pi)^{-d/4}\mathrm{Tr}(i_\theta(a)), \quad a \in L_\infty(\mathbb{R}^d_\theta)$$

where Tr is the trace on $\mathcal{L}(L_2(\mathbb{R}^{d/2}))$, and then

$$L_p(\mathbb{R}^d_\theta) = i_\theta^{-1}(\mathcal{L}_p),$$

where $\mathcal{L}_p$ are the Schatten ideals of $\mathcal{L}(L_2(\mathbb{R}^{d/2}))$. The C*-algebra $C_0(\mathbb{R}^d_\theta)$ is identified with the compact operators on $L_2(\mathbb{R}^{d/2})$;

$$C_0(\mathbb{R}^d_\theta) = i_\theta^{-1}(\mathcal{K}(L_2(\mathbb{R}^{d/2}))).$$

The noncommutative geometry of the noncommutative plane is realised in the Fourier dual presentation using multiplication operators

$$(M_j h)(x) = x_j h(x), \quad x = (x_1, x_2, \ldots, x_d) \in \mathbb{R}^d, j = 1, \ldots, d$$

and the Laplacian $M^2 : W_{2,2}(\mathbb{R}^d) \to L_2(\mathbb{R}^d)$ defined by

$$M^2 = \sum_{j=1}^d M_j^2$$

acting on the Sobolev space

$$W_{2,2}(\mathbb{R}^d) = \{h \in L_2(\mathbb{R}^d) : \int_{\mathbb{R}^d} \|x\|^4 |h(x)|^2 dx < \infty\}.$$

For any $g \in L_\infty(\mathbb{S}^{d-1})$ define the bounded operator

$$\pi_2(g) = M_{g\circ s}\,, \ (M_{g\circ s} h)(x) = (g\circ s)(x)h(x), \quad h \in L_2(\mathbb{R}^d).$$

We describe the trace theorem involving a principal symbol map. The conditions in Section 5.1.3 can be checked for the C*-algebras $\mathbb{C} + C_0(\mathbb{R}^d_\theta)$ (adjoined with unit) and $C(\mathbb{S}^{d-1})$ represented by π_1 and π_2 in $\mathcal{L}(L_2(\mathbb{R}^d))$ [156, Lemma 4.3, Theorem 5.3]. To obtain a state on $\mathbb{C} + C_0(\mathbb{R}^d_\theta)$ we need to weight the noncommutative integral. We choose to weight it by a Schwartz function and require the abstract Cwikel estimates described in Section 5.2.2 below.

If $x \in L_\infty(\mathbb{R}^d_\theta)$, then $x \in L_2(\mathbb{R}^d_\theta)$ if and only if

$$x = \int_{\mathbb{R}^d} f(t) u_t dt.$$

for some unique $f \in L_2(\mathbb{R}^d)$ [81]. Here the integral is the Bochner integral in the norm of $L_2(\mathbb{R}^d_\theta)$. When $z \in \mathcal{S}(\mathbb{R}^d)$ is a Schwartz function we denote the operator in $L_\infty(\mathbb{R}^d_\theta)$ corresponding to z by

$$L_\theta(z) = \int_{\mathbb{R}^d} z(t) u_t dt.$$

Denote $\mathcal{S}(\mathbb{R}^d_\theta) = L_\theta(\mathcal{S}(\mathbb{R}^d))$. Then $\mathcal{S}(\mathbb{R}^d_\theta)$ is dense in $L_1(\mathbb{R}^d_\theta)$ in the L_1-norm and dense in $C_0(\mathbb{R}^d_\theta)$ in the uniform norm [231, 219, 141]. Further

$$\tau_\theta(L_\theta(z)) = z(0).$$

Section 5.2.2 shows that $L_\theta(z)(1 - M^2)^{-d/2} \in \mathcal{L}_{1,\infty}$. Fix $z \in \mathcal{S}(\mathbb{R}^d)$. For a positive trace ϕ on $\mathcal{L}_{1,\infty}$ the weighted noncommutative integral ρ_z defined by

$$\rho_z(B) = \phi(B L_\theta(z)(1 - M^2)^{-d/2}), \quad B \in \mathcal{L}(L_2(\mathbb{R}^d))$$

satisfies

$$|\rho_z(B)| \leq \|B\| \|L_\theta(z)(1 - M^2)^{-d/2}\|_{1,\infty}$$

and is a continuous linear functional on $\mathcal{L}(L_2(\mathbb{R}^d))$ that vanishes on $\mathcal{K}(L_2(\mathbb{R}^d))$. We use Section 5.1.3 to identify ρ_z on the C*-algebra $\Pi(\mathbb{C} + C_0(\mathbb{R}^d_\theta), C(\mathbb{S}^{d-1}))$.

The functional

$$\phi(x L_\theta(z)(1 - M^2)^{-d/2}), \quad x \in \mathbb{C} + C_0(\mathbb{R}^d_\theta)$$

is identified in [219, Theorem 1.1] using unitary invariances of M^2, see Section 5.2.2 below. Specifically

$$\phi(x L_\theta(z)(1 - M^2)^{-d/2}) = \frac{\text{Vol}\,\mathbb{S}^{d-1}}{d} \tau_\theta(x L_\theta(z)).$$

The functional

$$\phi(\pi_2(g)L_\theta(z)(1-M^2)^{-d/2}), \quad g \in C(\mathbb{S}^{d-1}), z \in \mathcal{S}(\mathbb{R}^d)$$

for fixed z can be identified, by considering rotation invariance, with the Lebesgue integral on the sphere [156, Lemma 6.13]

$$\phi(\pi_2(g)L_\theta(z)(1-M^2)^{-d/2}) = \frac{1}{d}\tau_\theta(L_\theta(z))\int_{\mathbb{S}^{d-1}} g(s)ds.$$

To summarise [156, Theorem 6.15]

Theorem 5.4 *On the C^*-algebra of operators of "order 0" for the noncommutative plane defined above*

$$\Pi(C_0(\mathbb{R}^d_\theta), C(\mathbb{S}^{d-1})) \subset \mathcal{L}(L_2(\mathbb{R}^d)),$$

we have the principal symbol homomorphism

$$\pi_0 : \Pi(C_0(\mathbb{R}^d_\theta), C(\mathbb{S}^{d-1})) \to C(\mathbb{S}^{d-1}, \mathbb{C} + C_0(\mathbb{R}^d_\theta)).$$

For a Schwartz function $z \in \mathcal{S}(\mathbb{R}^d)$; all operators in $\Pi(C_0(\mathbb{R}^d_\theta), C(\mathbb{S}^{d-1}))L_\theta(z)$ have unique noncommutative integral, and we have the trace theorem for the noncommutative integral

$$\phi(BL_\theta(z)(1-M^2)^{-d/2}) = \frac{1}{d}\int_{\mathbb{S}^{d-1}} \tau_\theta(\pi_0(B)(s)L_\theta(z))ds,$$

for $B \in \Pi(C_0(\mathbb{R}^d_\theta), C(\mathbb{S}^{d-1}))$ and any positive trace ϕ on $\mathcal{L}_{1,\infty}$.

5.2 Integration of functions and Cwikel estimates

Integration of smooth compactly supported functions was known for Dixmier traces directly from Connes' trace theorem in 1988 [38] and highlighted by Benameur and Fack [14, p. 34] [91, Corollary 7.21]. Oversights in [71] and [91, Corollary 7.22] incorrectly extended the result to compactly supported $f \in L_1(\mathbb{R}^d)$ before [146] and [125]. The statements below in Section 5.2.1 are corollaries of [125, Theorems 6.32, 7.6] following Theorem 5.1.

5.2.1 Integration of functions

If $f \in L_2(\mathbb{R}^d)$ has compact support, then the Hilbert-Schmidt operator [125, p. 25] [207, Chap. 4]

$$A_f = M_f(1-\Delta)^{-d/2} \in \mathcal{L}_2$$

satisfies the conditions of Theorem 5.1 up to addition of a trace class operator; here

$$(M_f h)(x) = f(x)h(x), \quad f \in L_2(\mathbb{R}^d), h \in L_\infty(\mathbb{R}^d)$$

and the product is understood as the composition

$$(1-\Delta)^{-d/2} : L_2(\mathbb{R}^d) \to L_\infty(\mathbb{R}^d)\,,\ M_f : L_\infty(\mathbb{R}^d) \to L_2(\mathbb{R}^d).$$

We revert to the multiplier notation for the action of functions by pointwise product in this section as two different representations of Schwartz functions are discussed further below.

The square-integrable symbol of A_f is the function

$$p_{A_f}(x,\xi) = f(x)(1+\|2\pi\xi\|^2)^{-d/2}, \quad x,\xi \in \mathbb{R}^d.$$

It follows from Theorem 3.2 and Theorem 5.1 [125, p. 43] that $M_f(1-\Delta)^{-d/2} \in \mathcal{L}_{1,\infty}$ and

$$\phi(M_f(1-\Delta)^{-d/2}) = \frac{\operatorname{Vol}\mathbb{S}^{d-1}}{d(2\pi)^d} \cdot \int_{\mathbb{R}^d} f(x)dx$$

for any normalised trace ϕ on $\mathcal{L}_{1,\infty}$.

The same result has been shown on a closed manifold Ω (the condition for compact support of f can of course be removed), where Δ is the Laplace-Beltrami operator [125] [150, Sect. 11.7] [146]. In this case the condition $f \in L_2(\Omega)$ is necessary and sufficient for $M_f(1-\Delta)^{-d/2} \in \mathcal{L}_{1,\infty}$ [150, p. 359].

For the noncompact manifold $\mathbb{R}^d$, Cwikel-type estimates of Birman and Solomyak [17] indicate that $M_f(1-\Delta)^{-d/2} \in \mathcal{L}_{1,\infty}$ when f belongs to the function space

$$\ell_1(L_2)(\mathbb{R}^d) = \{f : \sum_{m\in\mathbb{Z}^d} \|f\chi_{Q_m}\|_2 < \infty,\ Q_m = \text{unit cube in } \mathbb{R}^d \text{ translated by } m\}.$$

Denoting the norm on $\ell_1(L_2)(\mathbb{R}^d)$

$$\|f\|_{\ell_1(L_2)} = \sum_{m\in\mathbb{Z}^d} \|f\chi_{Q_m}\|_2, \quad f \in \ell_1(L_2)(\mathbb{R}^d),$$

then

$$\|M_f(1-\Delta)^{-d/2}\|_{\mathcal{L}_{1,\infty}} \le c\|f\|_{\ell_1(L_2)}, \quad f \in \ell_1(L_2)(\mathbb{R}^d), c \text{ const.}$$

Compactly supported functions in $L_2(\mathbb{R}^d)$ are dense in $\ell_1(L_2)(\mathbb{R}^d)$. Using continuity properties of a positive trace ϕ, and Birman and Solomyak's estimate, the condition of compact support can be removed [225, Prop. 4.1]. To summarise:

Theorem 5.5 *Let $f \in \ell_1(L_2)(\mathbb{R}^d)$ be as above and Δ be the Laplacian on $\mathbb{R}^d$. Then $M_f(1-\Delta)^{-d/2} \in \mathcal{L}_{1,\infty}$ and*

$$\phi(M_f(1-\Delta)^{-d/2}) = \frac{\operatorname{Vol}\mathbb{S}^{d-1}}{d(2\pi)^d} \cdot \int_{\mathbb{R}^d} f(x)dx, \quad f \in \ell_1(L_2)(\mathbb{R}^d)$$

for every positive normalised trace ϕ on $\mathcal{L}_{1,\infty}$.

Let Δ be the Laplace-Beltrami operator on a d-dimensional closed manifold Ω. Then $M_f(1-\Delta)^{-d/2} \in \mathcal{L}_{1,\infty}$ if and only if $f \in L_2(\Omega)$, and

$$\phi(M_f(1-\Delta)^{-d/2}) = \frac{\operatorname{Vol}\mathbb{S}^{d-1}}{d(2\pi)^d} \cdot \int_{\Omega} f(x)dx, \quad f \in L_2(\Omega)$$

for every normalised trace ϕ on $\mathcal{L}_{1,\infty}$.

Cwikel-type estimates have been abstracted to noncommutative algebras as discussed in the next section (Section 5.2.2). They are currently the main device for extending from compact support. The first statement in Theorem 5.5 requires positivity of the trace so that the Cwikel estimate can be employed, while the second statement does not require any condition on the trace because it follows from Theorem 5.1. Weakening the condition that $AM_\psi = A$ in Theorem 5.1 to obtain the first statement in Theorem 5.5 directly from the spectral estimate (5.2) is still an open problem.

5.2.2 Cwikel estimates

Section 5.2.1 used estimates of Birman and Solomyak to extend the noncommutative integral for the Euclidean plane beyond functions of compact support. Section 5.1.4 used a similar estimate for the noncommutative plane $\mathbb{R}^d_\theta$ as explained below.

Cwikel's celebrated estimate is [60, 205]

$$\|M_f g(-i\nabla)\|_{\mathcal{L}_{p,\infty}} \le c_p\|f\|_p\|g\|_{p,\infty}, \quad f \in L_p(\mathbb{R}^d), g \in L_{p,\infty}(\mathbb{R}^d), p > 2$$

for a constant $c_p > 0$. Here

$$g(-i\nabla)h = F^{-1}M_g Fh, \quad g \in L_{p,\infty}(\mathbb{R}^d), h \in \mathcal{S}(\mathbb{R}^d)$$

where $F : L_2(\mathbb{R}^d) \to L_2(\mathbb{R}^d)$ is the unitary defined by Fourier transform. The estimate appears in mathematical physics, where it is used to obtain the Cwikel-Lieb-Rosenblum estimates on the number negative eigenvalues of Schroedinger operators in $\mathbb{R}^d$, $d \geq 3$ [207, Ch. 7] [137].

The estimate has been generalised in multiple directions with various approaches [18, 19, 234, 235, 78, 10, 111]. The most relevant is the one mentioned due to Birman and Solomyak [18, Theorem 11.1] where

$$\|M_f g(-i\nabla)\|_{\mathcal{L}_{1,\infty}} \leq c_1 \|f\|_{\ell_1(L_2)} \|g\|_{1,\infty}, \quad f \in \ell_1(L_2)(\mathbb{R}^d), g \in L_{1,\infty}(\mathbb{R}^d)$$

for a constant $c_1 > 0$. The operator $(1-\Delta)^{-d/2} = g(-i\nabla)$ where

$$g(t) = (1 + \|t\|^2)^{-d/2} \in L_{1,\infty}(\mathbb{R}^d).$$

This provides the estimate

$$\|M_f(1-\Delta)^{-d/2}\|_{\mathcal{L}_{1,\infty}} \leq c_1 \|f\|_{\ell_1(L_2)}$$

used in Section 5.2.1.

Cwikel's estimate for the noncommutative plane will require extending some notions from Section 2. Let $\mathcal{A} \subset \mathcal{L}(H)$ be a semifinite von Neumann algebra with a faithful normal semifinite trace τ. Denote the τ-measurable operators associated with $\mathcal{A}$ by $\mathcal{S}(\mathcal{A}, \tau)$ [150, Chapter 2]. The *singular value function* $\mu(A)$ of an operator $a \in \mathcal{S}(\mathcal{A}, \tau)$ is a non-increasing right continuous function defined by setting

$$\mu_{\mathcal{A}}(t, a) = \inf\{\|a(1-p)\| : p = p^* = p^2, \ \tau(p) \leq t\}, \quad t > 0.$$

When $\mathcal{A} = L_\infty(\mathbb{R}^d)$ and τ is the Lebesgue integral, the generalised singular value function $\mu(f)$ is the decreasing rearrangement of a function f [133, 143]. We drop the subscript on the singular value function in this case. Denote by S the space of τ-measurable functions.[12] A linear subspace E of τ-measurable functions on $\mathbb{R}^d$ is called *symmetric* if $\mu(f_1) \leq \mu(f_2)$ for $f_2 \in E$ and $f_1 \in S$ implies that $f_1 \in E$. A quasi-norm $\|\cdot\|_E$ on E is called symmetric if $\mu(f_1) \leq \mu(f_2)$ implies $\|f_1\|_E \leq \|f_2\|_E$.

[12] S is the subspace of measurable functions $f \in L_0(\mathbb{R}^d)$ with a distribution function

$$n_f(\lambda) = m(\{s \in \mathbb{R}^d : |f(s)| > \lambda\}), \quad \lambda > 0, m \text{ Lebesgue measure},$$

that is finite as $\lambda \to \infty$. Equivalently, the function $\mu(s, f), s > 0$ is finite valued.

When $\mathcal{A} = \ell_\infty(\mathbb{Z}^d)$ and τ is counting, the generalised singular value function $\mu(x)$ is the decreasing rearrangement of the sequence x. If E is a symmetric function space with symmetric quasi-norm $\|\cdot\|_E$ and F a symmetric sequence space with symmetric quasi-norm $\|\cdot\|_F$, then we define a function space

$$E(F)(\mathbb{R}^d) = \{f \in S : \|f\|_{E(F)} < \infty\}$$

where

$$\|f\|_{E(F)} = \|\mu(\{\|f\chi_{Q_m}\|_F\}_{m\in\mathbb{Z}^d})\|_E$$

and

$$Q_m = \text{unit cube in } \mathbb{R}^d \text{ translated by } m \in \mathbb{Z}^d.$$

Weidl [235, Theorem 4.1] noted that Cwikel's theorem could follow from a more refined estimate on the singular value function [141, Corollary 3.6]

$$\mu_{\mathcal{L}(L_2(\mathbb{R}^d))}(M_f g(-i\nabla))^2 \prec\prec \mu(f \otimes g)^2, \quad f \in L_p(\mathbb{R}^d),\, g \in L_{p,\infty}(\mathbb{R}^d),\, p > 2.$$

Here $\mu(f\otimes g)$ is the decreasing rearrangement of the function $f\otimes g$ in $L_{p,\infty}(\mathbb{R}^d \times \mathbb{R}^d)$. The paper [141] obtained the above estimate; and generalised Cwikel's and Birman and Solomyak's estimates to noncommutative versions of the operators M_f and $g(-i\nabla)$.

Theorem 5.6 *Let $\mathcal{A}_1$ and $\mathcal{A}_2$ be semifinite von Neumann algebras represented in $\mathcal{L}(H)$ by π_1 and π_2 respectively. Denote by $\mathcal{A}_1\bar{\otimes}\mathcal{A}_2$ their spatial tensor product. Let τ_1 and τ_2 be faithful normal semifinite traces on $\mathcal{A}_1$ and $\mathcal{A}_2$. Suppose that*

$$\|\pi_1(x)\pi_2(y)\|_{\mathcal{L}_2} \leq \|x\|_{L_2(A_1,\tau_1)}\|y\|_{L_2(A_2,\tau_2)},$$

for all $x \in L_2(A_1, \tau_1)$, $y \in L_2(\mathcal{A}_2, \tau_2)$.

(1) *There is a constant $c > 0$ such that*

$$\mu_{\mathcal{L}(L_2(H))}(\pi_1(x)\pi_2(y))^2 \prec\prec c\,\mu_{\mathcal{A}_1\bar{\otimes}\mathcal{A}_2}(x\otimes y)^2.$$

(2) *Suppose E is an (L_2, L_∞) interpolation space. The operator $\pi_1(x)\pi_2(y)$ belongs to the ideal $\mathcal{E}$ of compact operators corresponding to the symmetric function space E if $x \otimes y$ belongs to $E(\mathcal{A}_1\bar{\otimes}\mathcal{A}_2, \tau_1 \otimes \tau_2)$, and we have the estimate*

$$\|\pi_1(x)\pi_2(y)\|_{\mathcal{E}} \leq c_E\|x\otimes y\|_{E(\mathcal{A}_1\bar{\otimes}\mathcal{A}_2,\tau_1\otimes\tau_2)}$$

for some constant $c_E > 0$.

Suppose $1 \leq p < 2$.

(3) *Then*

$$\|M_f g(-i\nabla)\|_{\mathcal{L}_p} \le c_p \|f \otimes g\|_{\ell_p(L_2)(\mathbb{R}^d \times \mathbb{R}^d)}$$

and

$$\|M_f g(-i\nabla)\|_{\mathcal{L}_{p,\infty}} \le c_p \|f \otimes g\|_{\ell_{p,\infty}(L_2)(\mathbb{R}^d \times \mathbb{R}^d)}$$

for some constant $c_p > 0$.

(4) *Suppose* $x \in L_p(R^d_\theta) = L_p(L_\infty(R^d_\theta), \tau_\theta)$ *where* $L_\infty(R^d_\theta)$ *and* τ_θ *denote the noncommutative plane represented on* $L_2(\mathbb{R}^d)$ *and faithful normal semifinite trace from Section 5.1.4. Define*

$$W_{m,p}(R^d_\theta) = \{x \in L_p(\mathbb{R}^d_\theta) : \partial^{(0,\alpha)}(x) \in L_p(\mathbb{R}^d_\theta), |\alpha| \le m\}$$

and

$$\|x\|_{W_{m,p}} = \sum_{|\alpha| \le m} \|\partial^{(0,\alpha)}(x)\|_{L_p(\mathbb{R}^d_\theta)}$$

where $\partial^{(0,\alpha)}$ *are the derivations in* (5.7) *and* α *is a multi-index. Then*

$$\|xM_g\|_{\mathcal{L}_p} \le c_p \|x\|_{W_{d,p}} \|g\|_{\ell_p(L_\infty)(\mathbb{R}^d)}$$

and

$$\|xM_g\|_{\mathcal{L}_{p,\infty}} \le c_p \|x\|_{W_{d,p}} \|g\|_{\ell_{p,\infty}(L_\infty)(\mathbb{R}^d)}$$

for some constant $c_p > 0$. *Note that* $M_g = Fg(-i\nabla)F^{-1}$ *as the noncommutative plane in Section 5.1.4 is presented for the Fourier dual.*

The statement combines the results [141, Theorem 3.4] [141, Corollary 3.5] [141, Corollary 4.6] and [141, Theorems 7.6–7.7]. No estimate similar to Theorem 5.6(3) can hold for $p = 2$ [141, Theorem 5.1]. Properties of the noncommutative Sobolev spaces $W_{m,p}(\mathbb{R}^d_\theta)$ are listed in [141, Prop. 6.15]; note the construction there is Fourier dual as indicated in Theorem 5.6(4). If $f \in \mathcal{S}(\mathbb{R}^d)$, then $L_\theta(f) \in W_{d,p}(\mathbb{R}^d_\theta)$ for all $d, p \ge 1$.

Birman and Solomyak's estimate was used to extend the noncommutative integral to functions of noncompact support in Section 5.2.1. The equivalent estimate in Theorem 5.6(4) can extend the identification of the noncommutative integral on the noncommutative plane, as used in Section 5.1.4, beyond Schwartz functions [219, Theorem 1.1]; if $x \in W_{d,1}(\mathbb{R}^d_\theta)$ then $x(1 - M^2)^{-d/2} \in \mathcal{L}_{1,\infty}$ and

$$\phi(x(1 - M^2)^{-d/2}) = \frac{\operatorname{Vol} \mathbb{S}^{d-1}}{d} \tau_\theta(x)$$

for any positive trace ϕ on $\mathcal{L}_{1,\infty}$. For $f \in \mathcal{S}(\mathbb{R}^d)$,

$$\phi(L_\theta(f)(1 - M^2)^{-d/2}) = \frac{\mathrm{Vol}\,\mathbb{S}^{d-1}}{d} f(0).$$

That the noncommutative integral is an invariant of isospectral deformation was seen originally in [82, Theorem 6.1] for $f \in C_c^\infty(\mathbb{R}^d)$ and ϕ a Dixmier trace. The approach in [82] and [23] used approximation by trace class operators. The above extension from [219, Theorem 1.1] incorporates a wider class of operators and traces.

5.3 *Integration of forms*

Suppose Δ is the Laplacian on $\mathbb{R}^d$. So far we have considered the bounded operator $(1 - \Delta)^{-d/2} \in \mathcal{L}(L_2(\mathbb{R}^d))$ as a kind of volume element dv [40, Sect. VI] [41, 51, 52, 53, 43]. Taking a trace of the operator $B(1 - \Delta)^{-d/2}$, when it can be shown to belong to the ideal $\mathcal{L}_{1,\infty}$, indicates the transformed volumes can be summed, and the noncommutative integral is defined by

$$B \mapsto \phi(B(1 - \Delta)^{-d/2}), \quad B \in \mathcal{L}(L_2(\mathbb{R}^d))$$

where ϕ is a trace on $\mathcal{L}_{1,\infty}$. The text so far has indicated what is known concerning the dependence of this functional on the trace ϕ, and examples of operators and conditions which provide $B(1 - \Delta)^{-d/2} \in \mathcal{L}_{1,\infty}$.

When Δ is the Laplace-Beltrami operator for a closed d-dimensional Riemannian manifold Ω, then the situation is simpler in that $(1 - \Delta)^{-d/2} \in \mathcal{L}_{1,\infty}$. The volume is finite. The noncommutative integral is defined for any $B \in \mathcal{L}(L_2(\Omega))$. Consideration still needs to be given for dependence of the trace. From Sects. 5.1 and 5.2, however, a large class of operators of interest in differential geometry and mathematical physics possess a noncommutative integral that is independent of, at least, the positive normalised trace ϕ on $\mathcal{L}_{1,\infty}$ chosen.

Differential geometry has another notion of integration associated with differential forms on the tangent bundle [208, 195, 35, 139]. On a p-dimensional Riemannian manifold Ω with boundary $\partial\Omega$, the exterior derivative

$$d : C_c^\infty(\Omega) \to C_c^\infty(\Omega, T^*\Omega),$$

in local co-ordinates

$$df(x) = \sum_{i=1}^{p} (\partial_i f)(x) dx_i, \quad x \in \Omega, f \in C_c^\infty(\Omega)$$

allows consideration of the exterior algebra $C_c^\infty(\Omega, \Lambda^*\Omega)$ of pointwise anti-symmetric products of the one-forms df. The exterior derivative is extended

$$d : C_c^\infty(\Omega, \Lambda^n\Omega) \to C_c^\infty(\Omega, \Lambda^{n+1}\Omega)$$

by

$$d(f dx_{i_1} \wedge \ldots \wedge dx_{i_n}) \mapsto \sum_{i=1}^{p} (\partial_i f)(x) dx_i \wedge dx_{i_1} \wedge \ldots \wedge dx_{i_n},$$

defined locally and extended linearly. From anti-symmetry,

$$d^2 = 0,$$

the space $C_c^\infty(\Omega, \Lambda^{p+1}\Omega)$ is trivial, and every $w \in C_c^\infty(\Omega, \Lambda^p\Omega)$ can be written locally in the form

$$w(x) = f(x)\sqrt{\det(g(x))} dx_1 \wedge \ldots \wedge dx_m, \quad f \in C_c^\infty(\Omega).$$

The linear functional ρ_Ω on $w \in C_c^\infty(\Omega, \Lambda^p\Omega)$, defined in a chart $\varphi : U \to \varphi(U) \subset \mathbb{R}^p$ by

$$\rho_\Omega(w) = \int_{\varphi(U)} f \circ \varphi^{-1}(s)\sqrt{\det(g(\varphi^{-1}(s)))} ds$$

can be associated with a measure μ on Ω. The form $v_\Omega = \sqrt{\det(g(x))} dx_1 \wedge \ldots \wedge dx_p$ is called the volume form on Ω, and when Ω is orientable

$$\rho_\Omega(f v_\Omega) = \int_\Omega f(x) d\mu(x), \quad f \in C_c^\infty(\Omega).$$

Note if $w \in C_c^\infty(\Omega, \Lambda^{p-1}\Omega)$, by Stokes' Theorem [32, 128]

$$\rho_\Omega(dw) = \rho_{\partial\Omega}(w|_{\partial\Omega})$$

where $\partial\Omega$ is the smooth boundary of Ω. This identifies the integral ρ_Ω as a p-current on the space of differential p-forms [190]. The homology class of ρ_Ω in the de Rham homology $H_p(\Omega)$ is called the fundamental class of Ω.

Hochschild homology and cohomology provide abstract versions of differential forms and currents on manifolds [47], [110]. Let $\mathcal{A}$ be an Frechet algebra [130, 77]; that is, an algebra and Frechet space [189, Sect. V.2] for which multiplication is jointly continuous. The tensor powers of $\mathcal{A}$ are completed in the projective tensor product topology [130, 77].

The Hochschild boundary $b : \mathcal{A}^{\otimes(n+1)} \to \mathcal{A}^{\otimes n}$ is defined by setting

$$b(a_0 \otimes \cdots \otimes a_n) = a_0 a_1 \otimes a_2 \otimes \cdots \otimes a_n + \sum_{k=1}^{n-1} (-1)^k a_0 \otimes a_1 \otimes \cdots a_{k-1} \otimes a_k a_{k+1} \otimes a_{k+1} \otimes \cdots \otimes a_n + (-1)^n a_n a_0 \otimes a_1 \otimes a_2 \otimes \cdots \otimes a_{n-1}.$$

If $c \in \mathcal{A}^{\otimes(n+1)}$ is such that $bc = 0$, then c is called a Hochschild cycle. Two elements $d, e \in \mathcal{A}^{\otimes(n+1)}$ are equivalent if $b(d - e) = 0$. Let $H_n(\mathcal{A})$ denote the space of equivalence classes.

If $\theta : \mathcal{A}^{\otimes n} \to \mathbb{C}$ is a continuous multilinear functional, then the multilinear functional $b\theta : \mathcal{A}^{\otimes(n+1)} \to \mathbb{C}$ is defined by

$$(b\theta)(a_0 \otimes \cdots \otimes a_n) = \theta(a_0 a_1 \otimes a_2 \otimes \cdots \otimes a_n) + \sum_{k=1}^{n-1} (-1)^k \theta \times (a_0 \otimes a_1 \otimes \cdots a_{k-1} \otimes a_k a_{k+1} \otimes a_{k+1} \otimes \cdots \otimes a_n) + (-1)^n \theta(a_n a_0 \otimes a_1 \otimes a_2 \otimes \cdots \otimes a_{n-1}).$$

If θ vanishes on Hochschild cycles, then θ is called a Hochschild cocycle. Let $H^n(\mathcal{A})$ denote the space of equivalence classes under b. The relation

$$\theta(bc) = (b\theta)(c)$$

is an abstraction of the relationship in differential geometry between currents and the boundary map ∂, and differential forms and the exterior derivative d [33]. If Ω is a closed manifold, then Connes showed that the Hochschild cohomology $H^*(C^\infty(\Omega))$ is isomorphic to the de Rham homology $H_*(\Omega)$ of currents on differential forms [47, Lemma 45, p. 344] [91, p. 363]. The Hochschild boundary map b is not equivalent to ∂ under this isomorphism though; an exact generalisation is possible with the introduction of cyclic homology and cohomology [40, p. 207]. The fundamental class $[\rho_\Omega] \in H_p(\Omega) = H^p(C^\infty(\Omega))$, or integral, of Ω is identified in the case of a closed manifold as the sole generator of $H^p(C^\infty(\Omega))$.

For a general noncommutative Frechet algebra $\mathcal{A}$ the Hochschild cohomology $H^p(\mathcal{A})$ need not be one-dimensional. There is another identification. The fundamental class of a manifold is the Hochschild cohomology class of the Chern character [24, p. 123] acting on p-forms. This characterisation is used in noncommutative geometry.

Noncommutative differential geometry was formulated in terms of forms and characters [37, 47, 226]; if $\mathcal{A}$ is a *-algebra in $\mathcal{L}(H)$ with H a separable Hilbert space H suppose the existence of a bounded operator $F = F^* \in \mathcal{L}(H)$ such that

$$F^2 = 1$$

and

$$[F, a] \in \mathcal{L}_{p+1}$$

for an integer $p > 0$. The triple $(\mathcal{A}, H, F)$ is a Fredholm module. Form the graded algebra $\Omega^* = \oplus_{k=0}^{\infty}\Omega^k$ where Ω^k contains the linear span of forms

$$w = a_0[F, a_1]\dots[F, a_k], \quad a_0, \dots, a_k \in \mathcal{A}.$$

Then $\Omega^0 = \mathcal{A}$ and $\Omega^k \subset \mathcal{L}_{(p+1)/k}$, $k \geq 1$. The map

$$d : \Omega^k \to \Omega^{k+1}$$

defined by

$$dw = Fw + (-1)^k wF, \quad w \in \Omega^{k+1}$$

satisfies [40, p. 292]

$$d^2 = 0$$

such that

$$d\,(a_0[F, a_1]\dots[F, a_k]) = [F, a_0][F, a_1]\dots[F, a_k], \quad a_0, \dots, a_k \in \mathcal{A}.$$

In this setting, called the quantum calculus, d has the role of the exterior derivative and the compact operators $[F, a]$, $a \in \mathcal{A}$ are differential one forms. When p is even, assume the existence of a unitary $\Gamma = \Gamma^*$ such that $[\Gamma, a] = 0$ and $\{\Gamma, F\} = 0$. When p is odd, set $\Gamma = 1$. Then

$$\Gamma F \prod_{k=0}^{p}[F, a_k] \in \mathcal{L}_1, \quad a_0, \dots, a_p \in \mathcal{A}.$$

Connes defined the character [40, p. 293]

$$\tau_p(a_0, \dots, a_p) = \frac{1}{2}\mathrm{Tr}\left(\Gamma F \prod_{k=0}^{p}[F, a_k]\right),$$

as the leading term of an abstraction of the Chern character. The Hochschild class of this leading term is the abstraction of the fundamental class of a manifold. When Ω is an oriented closed spin^c Riemannian manifold [138] of dimension $d > 1$ with spinor bundle S, $F = \mathrm{sgn}(D)$ is the sign of the Dirac operator on smooth sections of S, and $\mathcal{A} = C^\infty(\Omega)$ is represented by pointwise multiplication on square integrable sections of S, then [40, p. 551]

$$\tau_p(f_0, \ldots, f_p) = \int_\Omega f_0 df_1 \wedge \ldots \wedge df_p, \quad f_0, \ldots, f_p \in C^\infty(\Omega).$$

Langmann showed the same result directly for F the sign of the Dirac operator on $\mathbb{R}^d$ and $\mathcal{A} = C_c^\infty(\mathbb{R}^d)$ [134, p. 3826]. It follows from Theorem 5.8 below that the identification holds for any complete Riemannian manifold Ω where f_0 has compact support [220, Theorem 3.4.1].

We discuss recent identifications of which non-smooth functions $f \in L_\infty(\mathbb{R}^d)$, $d \geq 2$ are quantum differentiable in the sense of Connes [149], and that Connes' Hochschild character theorem [40, p. 308], which identifies the Hochschild class of the leading term of the Chern character, holds for any normalised trace ϕ on $\mathcal{L}_{1,\infty}$ [30] in both unital and non-unital cases.

5.3.1 Quantum differentiability

Let $d \geq 1$ and $N = 2^{\lfloor d/2 \rfloor}$. Let $D : \mathbb{C}^N \otimes \mathcal{S}(\mathbb{R}^d) \to \mathbb{C}^N \otimes \mathcal{S}(\mathbb{R}^d)$ be the Dirac operator

$$D = \sum_{j=1}^{d} \gamma_j \otimes -i\partial_j,$$

where $\gamma_1, \ldots, \gamma_d$ are $N \times N$ self-adjoint complex matrices satisfying the anticommutation relation [138],

$$\gamma_j \gamma_k + \gamma_k \gamma_j = 2\delta_{j,k}. \quad 1 \leq j, k \leq d,$$

and ∂_j, $j = 1, \ldots, d$ denote the partial derivatives in $\mathbb{R}^d$. The Dirac operator is essentially self-adjoint [138, p. 117]. Denote the closure

$$D : \mathbb{C}^N \otimes W_{1,2}(\mathbb{R}^d) \to \mathbb{C}^N \otimes L_2(\mathbb{R}^d)$$

where the Sobolev spaces on $\mathbb{R}^d$ are

$$W_{p,2}(\mathbb{R}^d) = \{f \in L_2(\mathbb{R}^d) : \int_{\mathbb{R}^d} |\hat{f}(\xi)|^2 (1 + \|\xi\|^2)^p dx < \infty\}, \quad p \geq 0.$$

Using the signum function and the Borel functional calculus [189, Theorem VII.2], define the bounded operator

$$F = \operatorname{sgn}(D) : \mathbb{C}^N \otimes L_2(\mathbb{R}^d) \to \mathbb{C}^N \otimes L_2(\mathbb{R}^d).$$

When $d = 1$, F is the Hilbert transform [40, p. 314] [91, p. 330]. When $d > 1$, F is Clifford multiplication of Riesz transforms [91, p. 333]

$$F = \sum_{j=1}^{d} \gamma_j \otimes R_j, \quad R_j = -i(-\Delta)^{-1/2}\partial_j, 1 \leq j \leq d.$$

Define the representation

$$\pi_1 : L_\infty(\mathbb{R}^d) \to \mathcal{L}(L_2(\mathbb{R}^d) \otimes \mathbb{C}^N), \ (\pi_1(f)h)(x) = f(x)h(x).$$

The functions $f \in L_\infty(\mathbb{R}^d)$ such that

$$[F, \pi_1(f)] \in \mathcal{L}_p, \quad p > d$$

are quantum differentials associated with the Dirac operator. Results of Peller [166, Chap. 6] and Janson and Wolff [114] give necessary and sufficient conditions; $[F, \pi_1(f)] \in \mathcal{L}_p, d < p < 2d$ if and only $f \in L_\infty$ belongs to the Besov function space $B^{d/p}_{pp}$

$$B^{d/p}_{pp}(\mathbb{R}^d) = \{f \in L_{p,\mathrm{loc}}(\mathbb{R}^d) : \int_{\mathbb{R}^d \times \mathbb{R}^d} \frac{|f(x) - f(y)|^p}{\|x-y\|^{2d}} dxdy < \infty\}.$$

Janson and Wolfe also showed that if $p \leq d$, then $[F, \pi_1(f)] \in \mathcal{L}_p$ if and only if f is a constant. The dimension d is a lower bound for non-trivial behaviour; noted by Connes for the one-dimensional case in 1994 [40, p. 314] and for $d > 1$ in the monograph [91, p. 333].

The functions $f \in L_\infty(\mathbb{R}^d)$ satisfying

$$[F, \pi_1(f)] \in \mathcal{L}_{d,\infty}$$

are a more refined class of quantum differentials. When $d = 1$ a necessary and sufficient condition on $f \in L_\infty(\mathbb{R})$ can be derived from [166, p. 256]. The appendix of Connes, Sullivan and Teleman's paper [56, p. 679] provided necessary and sufficient conditions on $f \in L_\infty(\mathbb{R}^d)$ for $[F, \pi_1(f)] \in \mathcal{L}_{d,\infty}$ when $d > 1$. The sketched proof used results of Rochberg and Semmes [193, Corollary 2.8, Theorem 3.4] on behaviour of the mean oscillation of the function f.

A recent paper [149] used double operator integrals to confirm Connes, Sullivan and Teleman's necessary and sufficient conditions for $f \in L_\infty(\mathbb{R}^d)$ without the intermediary of mean oscillation. It also identified the seminorm [149, Theorem 1]

$$\|[F, \pi_1(f)]\|_{\mathcal{L}_{d,\infty}}$$

as being equivalent to the Sobolev $(1, d)$-seminorm. Following Connes' result for closed manifolds in [38, Theorem 3(3)], it proved Connes' trace theorem for forms on $\mathbb{R}^d$ [149, Theorem 2].

Theorem 5.7 *Let $d > 1$.*

(1) *If $f \in L_\infty(\mathbb{R}^d)$, then $[F, \pi_1(f)] \in \mathcal{L}_{d,\infty}$ if and only if $\nabla f \in L_d(\mathbb{R}^d, \mathbb{C}^d)$ with equivalent seminorms*

$$\|[F, \pi_1(f)]\|_{\mathcal{L}_{d,\infty}} \cong \|\nabla f\|_{L_d(\mathbb{R}^d,\mathbb{C}^d)}.$$

(2) *Let $f = f^* \in L_\infty(\mathbb{R}^d)$ be real valued with $\nabla f \in L_d(\mathbb{R}^d, \mathbb{C}^d)$. Then*

$$|[F, \pi_1(f)]|^d \in \mathcal{L}_{1,\infty}$$

and there is a constant $c_d > 0$ such that

$$\phi(|[F, \pi_1(f)]|^d) = c_d \int_{\mathbb{R}^d} \|\nabla f(x)\|_2^d \, dx. \tag{5.8}$$

for any positive normalised trace ϕ on $\mathcal{L}_{1,\infty}$.

The value of c_d can be found below. In the statement of Theorem 5.7,

$$\nabla f = (\partial_1 f, \ldots, \partial_d f)$$

where $\partial_j f$, $j = 1, \ldots, d$ denotes the distributional derivative when $f \in L_\infty(\mathbb{R}^d)$ is not smooth [209, Chap. V]. Writing

$$\nabla f \in L_d(\mathbb{R}^d, \mathbb{C}^d)$$

assumes that $f \in L_\infty(\mathbb{R}^d)$ has weak partial derivatives and that the Bochner norm of ∇f in $L_d(\mathbb{R}^d, \mathbb{C}^d)$,

$$\|\nabla f\|_{L_d(\mathbb{R}^d,\mathbb{C}^d)} = \left(\int_{\mathbb{R}^d} \|(\nabla f)(x)\|_d^d \, dx\right)^{1/d} = \left(\int_{\mathbb{R}^d} \sum_{j=1}^d |\partial_j f(x)|^d dx\right)^{1/d},$$

is finite. The Sobolev space $W_{1,d}(\mathbb{R}^d)$ has the equivalent norm $\|f\|_{1,d} = \|f\|_d + \|\nabla f\|_{L_d(\mathbb{R}^d,\mathbb{C}^d)}$.

Formula (5.8) is analogous to the following trace formula of Connes on forms [38, Theorem 3(3)]. Suppose Ω is an oriented closed spin^c Riemannian manifold [138] of dimension $d > 1$ with spinor bundle S and $F = \operatorname{sgn}(D)$ is the sign of the Dirac operator on smooth sections of S. Given $f = f^* \in C^\infty(\Omega)$, let $\pi_1(f)$ be the operator of pointwise multiplication by f on sections of S. Theorem 3(3) of [38] states that

$$\mathrm{Tr}_\omega(|[F,\pi_1(f)]|^d) = \lambda_d \int_\Omega \|df(x)\|^d dx, \quad f = f^* \in C^\infty(\Omega) \tag{5.9}$$

where Tr_ω is a Dixmier trace, and $\lambda_d > 0$ is a constant depending only on d.

To compare the results, formula (5.8) holds for non-smooth functions and any positive trace ϕ on $\mathcal{L}_{1,\infty}$ for the noncompact space $\mathbb{R}^d$. We note a proof of (5.8) using the principal symbol map $\hat{\sigma}_0$ of Theorem 5.2 in Section 5.1.2. Tensoring by matrices [225, Lemma 9.1]; if

$$A \in \Pi(L_\infty(\mathbb{R}^d), L_\infty(\mathbb{S}^{d-1})) \otimes M_N(\mathbb{C})$$

then

$$\phi(A\pi_1(\psi)(1+D^2)^{-d/2}) = \frac{1}{d}\int_{\mathbb{R}^d\times\mathbb{S}^{d-1}} \mathrm{Tr}(\hat{\sigma}_0 \otimes 1_N)(A)(x,s)\psi(x)dxds$$

for any positive normalised trace ϕ on $\mathcal{L}_{1,\infty}$ and compactly supported function $\psi \in C_c^\infty(\mathbb{R}^d)$. For $f \in C_c^\infty(\mathbb{R}^d)$ a short argument shows that there is an operator $A_f \in \Pi(L_\infty(\mathbb{R}^d), L_\infty(\mathbb{S}^{d-1})) \otimes M_N(\mathbb{C})$ such that

$$|[F,\pi_1(f)]|^d = |A_f|^d(1+D^2)^{-d/2} + \mathcal{L}_1$$

where

$$(\hat{\sigma}_0 \otimes 1_N)(|A_f|^d)(x,s) = \|(\nabla f)(x) - ((s\cdot\nabla)f)(x)s\|^d \otimes 1_N, \quad (x,s) \in \mathbb{R}^d \times \mathbb{S}^{d-1}.$$

Rotation invariance of the integral on $\mathbb{S}^{d-1}$ implies the result

$$\int_{\mathbb{S}^{d-1}} \mathrm{Tr}\left(\|(\nabla f)(x) - ((s\cdot\nabla)f)(x)s\|^d \otimes 1_N\right) ds = c_d\|\nabla(f)(x)\|^d$$

with constant

$$c_d = 2^{\lfloor d/2 \rfloor}\int_{\mathbb{S}^{d-1}} \|e_1 - (s\cdot e_1)s\|^d ds$$

where $e_1 = (1,0,\ldots,0) \in \mathbb{R}^d$. Approximation of $f \in L_\infty(\mathbb{R}^d)$ in the seminorm of Theorem 5.7(1) by smooth functions completes the proof. This argument mimics the original proof of [38, Theorem 3(3)].

5.3.2 Hochschild character formula

Suppose $(\mathcal{A}, H, F)$ is a Fredholm module with $[F,a] \in \mathcal{L}_{p+1}$. Here p is a positive integer. When p is even, assume the existence of a unitary $\Gamma = \Gamma^*$ such that $[\Gamma, a] = 0$ and $\{\Gamma, F\} = 0$. When p is odd, set $\Gamma = 1$. Then

$$\Gamma F \prod_{k=0}^{n} [F, a_k] \in \mathcal{L}_1, \quad a_0, \ldots, a_n \in \mathcal{A}, n \geq p.$$

Connes defined the characters $n \geq 0$ [40, p. 293]

$$\tau_{p+2n}(a_0, \ldots, a_{p+2n}) = \frac{1}{2}\mathrm{Tr}\left(\Gamma F \prod_{k=0}^{p+2n} [F, a_k]\right), \tag{5.10}$$

which are cyclic cocycles of order $p + 2n$ whose cyclic cohomology class in HC^{p+2n} [48] [40, Chap. III] [145] is obtained from the leading character τ_p by n applications of the periodicity operator $S : HC^m \to HC^{m+2}$, $m \geq 0$ [40, p. 294]

$$[\tau_{m+2}] = -\frac{2}{m+2}[S\tau_m], \quad m = p + 2n, n \geq 0.$$

The periodic cyclic cohomology class of

$$(-1)^{p(p-1)/2} c_p \Gamma(1 + p/2)\tau_p, \quad c_p = 1, p \text{ even}, c_p = \sqrt{2i}, p \text{ odd}$$

is called the Chern character [40, p. 295].

Suppose the existence of a spectral triple $(\mathcal{A}, H, D)$ [40, p. 310] [49, 50]; a *-algebra $\mathcal{A}$ of $\mathcal{L}(H)$ where H is an infinite- dimensional separable Hilbert space and a bounded operator $D : \mathrm{Dom}(D) \in \mathcal{L}(H)$ such that

$$[D, a] \text{ extends to a bounded operator in } \mathcal{L}(H)$$

and

$$a(1 + D^2)^{-p/2} \in \mathcal{L}_{1,\infty}.$$

When $B \in \mathcal{L}(H)$ and $B : \mathrm{Dom}(D) \to \mathrm{Dom}(D)$ define the derivations

$$\partial(B) = [D, B]\,,\ \delta(B) = [|D|, B].$$

Define

$$\mathrm{dom}(\delta^k) = \{B \in \mathcal{L}(H) : \delta^k(B) \in \mathcal{L}(H)\}, \quad k \geq 0,$$

and the increasing family of seminorms

$$\rho_k(B) = \sum_{j=0}^{k} \|\delta^j(B)\| + \|\delta^j(\partial(B))\|, \quad B, \partial(B) \in \mathrm{dom}(\delta^k).$$

Assume that $\mathcal{A}$ is a Frechet *-algebra for ρ_k, $k \geq 0$. When D has trivial kernel $F = \text{sgn}(D)$ generates a Fredholm module [40, p. 310] [26] [91, p. 327]. This assumption can be removed when D has compact resolvent [40, p. 310] and forms a pre-Fredholm module [91, Sect. 8.2]. For Fredholm modules associated with spectral triples, there is the refinement in compact behaviour [181]

$$[F, a] \in \mathcal{L}_{p,\infty}.$$

Set

$$\Theta(a_0, \ldots, a_p) = \Gamma a_0 \prod_{k=1}^{p} [D, a_k] \in \mathcal{L}(H), \quad a_0 \otimes \ldots \otimes a_p \in \mathcal{A}^{p+1},$$

which is a bounded operator. For any trace $\phi \in \mathcal{L}_{1,\infty}$ define the noncommutative integral

$$\phi(\Theta(a_0, \ldots, a_p)(1 + D^2)^{-p/2}) = \phi\left(\Gamma a_0 \prod_{k=1}^{p} [D, a_k](1 + D^2)^{-p/2}\right). \tag{5.11}$$

The Hochschild character theorem due to Connes [40, p. 308] states that

$$\phi(\Theta(c)(1 + D^2)^{-p/2}) = \tau_p(c)$$

for every Hochschild cycle c and any Dixmier trace ϕ on $\mathcal{L}_{1,\infty}$ associated with factorisable Banach limit in the bijection of Section 3.2.3 (Connes' variant of Dixmier's trace). Up to coboundaries, the integral on forms defined by the quantum calculus and the generating element of the Chern character agrees with forms defined by commutators with the operator D and integration in the sense of the noncommutative integral defined by $(1 + D^2)^{-p/2}$.

In 2016 the character theorem was proved for any normalised trace ϕ on $\mathcal{L}_{1,\infty}$ [30]. Additional conditions were added recently to prove the Hochschild character theorem for non-unital spectral triples as well [220].

Theorem 5.8 *Let $(\mathcal{A}, H, D)$ be a graded spectral triple with* $\ker(D) = \{0\}$. *Suppose $p \geq 1$ is an integer.*

(1) *Suppose $\mathcal{A}$ contains the identity 1 of $\mathcal{L}(H)$ and $|D|^{-p} \in \mathcal{L}_{1,\infty}$. Then*

$$\phi(\Theta(c)|D|^{-p}) = \tau_p(c)$$

for every normalised trace ϕ on $\mathcal{L}_{1,\infty}$ and every Hochschild cycle c.

(2) *Suppose the following conditions:*

(a) *for every* $a \in \mathcal{A}$,

$$a(D+i)^{-p} \in \mathcal{L}_{1,\infty},$$
$$\partial(a)(D+i)^{-p} \in \mathcal{L}_{1,\infty};$$

(b) *for every* $a \in \mathcal{A}$ *and for all* $k \geq 0$, *we have*

$$\left\| \delta^k(a)(D+i\lambda)^{-p-1} \right\|_{\mathcal{L}_1} = O(\lambda^{-1}), \quad \lambda \to \infty,$$
$$\left\| \delta^k(\partial(a))(D+i\lambda)^{-p-1} \right\|_{\mathcal{L}_1} = O(\lambda^{-1}), \quad \lambda \to \infty.$$

Then

$$\phi(\Theta(c)|D|^{-p}) = \tau_p(c)$$

for every normalised trace ϕ *on* $\mathcal{L}_{1,\infty}$ *and every Hochschild cycle* c *such that*

$$c = \sum_{j=1}^{m} a_0^j \otimes \cdots \otimes a_p^j \in \mathcal{A}^{\otimes(p+1)}$$

satisfies $\psi a_0^j = a_0^j$, $1 \leq j \leq m$ *for a positive element* $\psi \in \mathcal{A}$.

This version of the Hochschild character theorem combines [30, Theorem 16], with errata mentioned in [220, p. 3], and [220, Theorem 1.2.5]. Theorem 1.2.3 in [220] shows that the zeta function

$$\mathrm{Tr}(\Theta(c)|D|^{-z}), \quad \mathrm{Re}\, z > p$$

satisfies (4.11) under the hypothesis in Theorem 5.8(2); it has an analytic continuation to the punctured half-plane $\mathrm{Re}\, z > p-1$ with simple pole at $z = p$. Theorem 5.8(2) also considers only algebraic Hochschild cycles, i.e. the algebraic tensor product, and not the continuous Hochschild homology and cohomology using the projective tensor product. Theorem 5.8 follows a line of results [91, p. 479] [24, 108, 14] concerning Connes' original Hochschild character theorem [40, p. 308]. The character theorem implies a cohomological obstruction to triviality of the noncommutative integral [40, p. 309]. It also implies that the operators $\Theta(c)|D|^{-p}$ are universally measurable (Section 4.2.2) for every Hochschild cycle c.

Acknowledgements The authors gratefully thank Alain Connes, Albrecht Pietsch and Raphael Ponge for historical comments. The authors also thank Albrecht Pietsch and Aleksandr Usachev for discussion on the PhD thesis of Jozsef Varga.

This research was supported by the Australian Research Council.

References

1. Thomas Ackermann, *A note on the Wodzicki residue*, Journal of Geometry and Physics **20** (1996), no. 4, 404–406.
2. M. Adler, *On a trace functional for formal pseudo differential operators and the symplectic structure of the Korteweg-de Vries type equations*, Invent. Math. **50** (1979), no. 3, 219–248. MR 520927 (80i:58026)
3. P. M. Alberti and R. Matthes, *Noncommutative geometry and the standard model of elementary particle physics*, Lecture Notes in Phys., vol. 596, ch. Connes' trace formula and Dirac realisation of Maxwell and Yang-Mills action, pp. 40–74, Springer, Berlin; New-York, 2002.
4. S. Albeverio, D. Guido, A. Ponosov, and S. Scarlatti, *Singular traces and compact operators*, J. Funct. Anal. **137** (1996), 281–302.
5. Clara L. Aldana, Maxim Braverman, Bruno Iochum, and Carolina Neira Jiménez (eds.), *Analysis, geometry, and quantum field theory: international conference in honor of Steve Rosenberg's 60th birthday, September 26–30, 2011, University of Potsdam, Potsdam, Germany*, Contemporary Mathematics, vol. 584, American Mathematical Society, 2012.
6. Egor Alekhno, Evgenii Semenov, Fedor Sukochev, and Alexandr Usachev, *Invariant Banach limits and their extreme points*, Studia Math. **242** (2018), no. 1, 79–107. MR 3766571
7. Joel Anderson, *Commutators in ideals of trace class operators ii*, Indiana University Mathematics Journal **35** (1986), no. 2, 373–378 (eng).
8. Joel Anderson and L. N. Vaserstein, *Commutators in ideals of trace class operators*, Indiana University Mathematics Journal **35** (1986), no. 2, 345–372 (eng).
9. Tsuyoshi Ando and Fumio Hiai, *Log majorization and complementary Golden-Thompson type inequalities*, Linear Algebra and its Applications **197–198** (1994), 113–131.
10. G. Arsu, *On Schatten-von Neumann class properties of pseudodifferential operators. The Cordes-Kato method*, J. Operator Theory **59** (2008), no. 1, 81–114. MR 2404466 (2009e:35303)
11. S. Banach, *Théorie des opérations linéaires*, Éditions Jacques Gabay, Sceaux, 1993, Reprint of the 1932 original. MR 1357166 (97d:01035)
12. Martin T. Barlow and Jun Kigami, *Localized eigenfunctions of the Laplacian on p.c.f. self-similar sets*, J. London Math. Soc. **56** (1997), no. 2, 320–332.
13. Richard Beals, *Characterization of pseudodifferential operators and applications*, Duke Math. J. **44** (1977), no. 1, 45–57. MR 0435933
14. M. Benameur and T. Fack, *Type II non-commutative geometry. I. Dixmier trace in von Neumann algebras*, Adv. Math. **199** (2006), no. 1, 29–87. MR 2186918 (2006m:58035)
15. Marcel Berger, *Riemannian geometry during the second half of the twentieth century*, University Lecture Series, vol. 17, American Mathematical Society, Providence, R.I., 2000.
16. Nicole Berline, Ezra Getzler, and Michèle Vergne, *Heat kernels and Dirac operators*, Grundlehren der Mathematischen Wissenschaften [Fundamental Principles of Mathematical Sciences], vol. 298, Springer-Verlag, Berlin, 1992. MR 1215720 (94e:58130)
17. M. Birman, G. Karadzhov, and M. Solomyak, *Boundedness conditions and spectrum estimates for the operators $b(X)a(D)$ and their analogs*, Estimates and asymptotics for discrete spectra of integral and differential equations (Leningrad, 1989–90), Adv. Soviet Math., vol. 7, Amer. Math. Soc., Providence, RI, 1991, pp. 85–106. MR 1306510 (95g:47075)
18. M. Birman and M. Solomyak, *Estimates of singular numbers of integral operators*, Russ. Math. Surv. **32** (1977), no. 1, 15–89 (English).
19. ______, *Estimates for the difference of fractional powers of selfadjoint operators under unbounded perturbations*, Zap. Nauchn. Sem. Leningrad. Otdel. Mat. Inst. Steklov. (LOMI) **178** (1989), no. Issled. Linein. Oper. Teorii Funktsii. 18, 120–145, 185. MR 1037767 (91d:47006)
20. Ola Bratteli and Derek W. Robinson, *Operator algebras and quantum-statistical mechanics. II*, Springer-Verlag, New York, 1981. MR 611508 (82k:82013)

21. Arlen Brown, Paul R. Halmos, and Carl Pearcy, *Commutators of operators on Hilbert space*, Canad. J. Math. **17** (1965), 695–708. MR 0203460 (34 #3311)
22. J. Calkin, *Two-sided ideals and congruences in the ring of bounded operators in Hilbert space*, Ann. of Math. (2) **42** (1941), 839–873. MR 0005790 (3,208c)
23. A. Carey, V. Gayral, A. Rennie, and F. Sukochev, *Integration on locally compact noncommutative spaces*, J. Funct. Anal. **263** (2012), no. 2, 383–414. MR 2923417
24. A. Carey, J. Phillips, A. Rennie, and F. Sukochev, *The Hochschild class of the Chern character for semifinite spectral triples*, J. Funct. Anal. **213** (2004), no. 1, 111–153. MR 2069783 (2005d:58050)
25. ———, *The local index formula in semifinite von Neumann algebras. I. Spectral flow*, Adv. Math. **202** (2006), no. 2, 451–516. MR 2222358 (2007a:19003)
26. A. Carey, J. Phillips, and F. Sukochev, *On unbounded p-summable Fredholm modules*, Adv. Math. **151** (2000), no. 2, 140–163. MR 1758245 (2001h:46107)
27. ———, *Spectral flow and Dixmier traces*, Adv. Math. **173** (2003), no. 1, 68–113. MR 1954456 (2004e:58049)
28. A. Carey, A. Rennie, A. Sedaev, and F. Sukochev, *The Dixmier trace and asymptotics of zeta functions*, J. Funct. Anal. **249** (2007), no. 2, 253–283. MR 2345333 (2010i:46095)
29. A. Carey and F. Sukochev, *Measurable operators and the asymptotics of heat kernels and zeta functions*, J. Funct. Anal. **262** (2012), no. 10, 4582–4599. MR 2900479
30. Alan Carey, Adam Rennie, Fedor Sukochev, and Dmitriy Zanin, *Universal measurability and the Hochschild class of the Chern character*, J. Spectr. Theory **6** (2016), no. 1, 1–41. MR 3484376
31. Alan Carey and Fedor Sukochev, *Dixmier traces and some applications in non-commutative geometry*, Russ. Math. Surv. **61** (2006), 1039–1099.
32. E. Cartan, *Les systèmes differentiels extérieurs et leur applications géométriques*, Actualités Scientifiques et Industrielles, vol. 994, Hermann, Paris, 1971.
33. Henri Cartan and Samuel Eilenberg, *Homological algebra*, Princeton Landmarks in Mathematics, Princeton University Press, Princeton, NJ, 1999, With an appendix by David A. Buchsbaum, Reprint of the 1956 original. MR 1731415 (2000h:18022)
34. A. Chamseddine and A. Connes, *The spectral action principle*, Comm. Math. Phys. **186** (1997), no. 3, 731–750. MR 1463819 (99c:58010)
35. I. Chavel, *Riemannian geometry: A modern introduction*, Cambridge Studies in Advanced Mathematics, Cambridge University Press, Cambridge, UK, 2006.
36. A. Connes, *C^*- algèbres et géométrie différentielle*, C. R. Acad. Sci. Paris Sér. A-B **290** (1980), no. 13, A599–A604. MR 572645 (81c:46053)
37. ———, *Introduction to noncommutative differential geometry*, Workshop Bonn 1984 (Bonn, 1984), Lecture Notes in Math., vol. 1111, Springer, Berlin, 1985, pp. 3–16. MR 797413 (87e:58193)
38. ———, *The action functional in noncommutative geometry*, Comm. Math. Phys. **117** (1988), no. 4, 673–683. MR 953826 (91b:58246)
39. A Connes, *Trace de Dixmier, modules de Fredholm et geometrie Riemannienne*, Nuclear Physics B - Proceedings Supplements **5** (1988), no. 2, 65 – 70.
40. A. Connes, *Noncommutative geometry*, Academic Press Inc., San Diego, CA, 1994. MR 1303779 (95j:46063)
41. ———, *Geometry from the spectral point of view*, Lett. Math. Phys. **34** (1995), no. 3, 203–238. MR 1345552 (96j:46074)
42. ———, *Gravity coupled with matter and the foundation of non-commutative geometry*, Comm. Math. Phys. **182** (1996), no. 1, 155–176. MR 1441908 (98f:58024)
43. ———, *On the foundations of noncommutative geometry*, The unity of mathematics, Progr. Math., vol. 244, Birkhäuser Boston, Boston, MA, 2006, pp. 173–204. MR 2181806
44. A. Connes, E. McDonald, F. Sukochev, and D. Zanin, *Conformal trace theorem for Julia sets of quadratic polynomials*, Ergodic Theory and Dynamical Systems (2017), 1–26.
45. A. Connes and H. Moscovici, *The local index formula in noncommutative geometry*, Geom. Funct. Anal. **5** (1995), no. 2, 174–243. MR 1334867 (96e:58149)

46. A. Connes and M. Rieffel, *Yang-Mills for noncommutative two-tori*, Operator algebras and mathematical physics (Iowa City, Iowa, 1985), Contemp. Math., vol. 62, Amer. Math. Soc., Providence, RI, 1987, pp. 237–266. MR 878383 (88b:58033)
47. Alain Connes, *Noncommutative differential geometry*, Inst. Hautes Études Sci. Publ. Math. (1985), no. 62, 257–360. MR 823176 (87i:58162)
48. ______, *Cyclic cohomology and noncommutative differential geometry*, Géométrie différentielle (Paris, 1986), Travaux en Cours, vol. 33, Hermann, Paris, 1988, pp. 33–50. MR 955850 (89i:58138)
49. ______, *Noncommutative geometry and reality*, J. Math. Phys. **36** (1995), no. 11, 6194–6231. MR 1355905 (96g:58014)
50. ______, *Noncommutative differential geometry and the structure of space-time*, The geometric universe (Oxford, 1996), Oxford Univ. Press, Oxford, 1998, pp. 49–80. MR 1634504 (99h:58013c)
51. ______, *Noncommutative geometry: the spectral aspect*, Symétries quantiques (Les Houches, 1995), North-Holland, Amsterdam, 1998, pp. 643–686. MR 1616407 (2000d:58042)
52. ______, *Noncommutative geometry—year 2000*, Geom. Funct. Anal. (2000), no. Special Volume, Part II, 481–559, GAFA 2000 (Tel Aviv, 1999). MR 1826266 (2003g:58010)
53. ______, *A short survey of noncommutative geometry*, J. Math. Phys. **41** (2000), no. 6, 3832–3866. MR 1768641 (2001m:58016)
54. Alain Connes and Henri Moscovici, *Transgression du caractère de Chern et cohomologie cyclique*, C. R. Acad. Sci. Paris Sér. I Math. **303** (1986), no. 18, 913–918. MR 873393 (88c:58006)
55. Alain Connes and Henri Moscovici, *The Local Index Formula in Noncommutative Geometry*, Geometric and Functional Analysis **5** (1995), 174–243.
56. Alain Connes, Dennis Sullivan, and Nicolas Teleman, *Quasiconformal mappings, operators on Hilbert space, and local formulae for characteristic classes*, Topology **33** (1994), no. 4, 663–681. MR MR1293305 (95g:58232)
57. H. O. Cordes, *Elliptic pseudodifferential operators—an abstract theory*, Lecture Notes in Mathematics, vol. 756, Springer, Berlin, 1979. MR 551619
58. ______, *On pseudo-differential operators and smoothness of special lie-group representations*, manuscripta mathematica **28** (1979), no. 1, 51–69.
59. H. O. Cordes and E. A. Herman, *Gel'fand theory of pseudo differential operators*, American Journal of Mathematics **90** (1968), no. 3, 681–717.
60. M. Cwikel, *Weak type estimates for singular values and the number of bound states of Schrödinger operators*, Ann. of Math. (2) **106** (1977), no. 1, 93–100. MR 0473576 (57 #13242)
61. J. Dixmier, *Existence de traces non normales*, C. R. Acad. Sci. Paris Sér. A-B **262** (1966), A1107–A1108. MR 0196508 (33 #4695)
62. P. Dodds, B. de Pagter, A. Sedaev, E. Semenov, and F. Sukochev, *Singular symmetric functionals*, Zap. Nauchn. Sem. S.-Peterburg. Otdel. Mat. Inst. Steklov. (POMI) **290** (2002), no. Issled. po Linein. Oper. i Teor. Funkts. 30, 42–71, 178 (Russian, with English and Russian summaries). MR 1942537 (2004a:46031)
63. ______, *Singular symmetric functionals and Banach limits with additional invariance properties*, Izv. Ross. Akad. Nauk Ser. Mat. **67** (2003), no. 6, 111–136. MR 2032092 (2005a:46061)
64. P. Dodds, B. de Pagter, A. Sedaev, E. Semenov, and F. Sukochev, *Singular symmetric functionals*, J. Math. Sci. **2** (2004), 4867–4885.
65. P. Dodds, B. de Pagter, E. Semenov, and F. Sukochev, *Symmetric functionals and singular traces*, Positivity **2** (1998), no. 1, 47–75. MR 1655756 (99m:46072)
66. K. Dykema, T. Figiel, G. Weiss, and M. Wodzicki, *Commutator structure of operator ideals*, Adv. Math. **185** (2004), no. 1, 1–79. MR 2058779 (2005f:47149)
67. K. Dykema and N. Kalton, *Spectral characterization of sums of commutators. II*, J. Reine Angew. Math. **504** (1998), 127–137. MR 1656763 (99g:47079)

68. K. Dykema, G. Weiss, and M. Wodzicki, *Unitarily invariant trace extensions beyond the trace class*, Complex analysis and related topics (Cuernavaca, 1996), Oper. Theory Adv. Appl., vol. 114, Birkhäuser, Basel, 2000, pp. 59–65. MR 1748001 (2001c:47025)
69. Alexander Dynin, *Review: H. O. Cordes, Spectral theory of linear differential operators and comparison algebras*, Bull. Amer. Math. Soc. (N.S.) **21** (1989), no. 1, 105–108.
70. Thierry Fack, *Sums of commutators in non-commutative Banach function spaces*, J. Funct. Anal. **207** (2004), no. 2, 358–398. MR 2032994 (2005e:46123)
71. Thierry Fack, *Singular Traces in N.C.G.*, Educational Week on Noncommutative Integration, Thomas Stieltjes Institute for Mathematics, http://www.math.leidenuniv.nl/~mdejeu/NoncomIntWeek_2008_Fack_singular_traces_in_NCG.pdf, June 2008.
72. Peng Fan and Che Kao Fong, *Which operators are the self-commutators of compact operators?*, Proceedings of the American Mathematical Society **80** (1980), no. 1, 58–60 (eng).
73. Farzad Fathizadeh and Masoud Khalkhali, *Weyl's law and connes' trace theorem for noncommutative two tori*, Letters in Mathematical Physics **103** (2013), no. 1, 1–18.
74. Farzad Fathizadeh and M. W. Wong, *Noncommutative residues for pseudo-differential operators on the noncommutative two-torus*, Journal of Pseudo-Differential Operators and Applications **2** (2011), no. 3, 289–302.
75. W. Feller, *An introduction to probability theory and its applications. Vol. II.* , Second edition, John Wiley & Sons Inc., New York, 1971.
76. T. Figiel and N. Kalton, *Symmetric linear functionals on function spaces*, Function spaces, interpolation theory and related topics (Lund, 2000), de Gruyter, Berlin, 2002, pp. 311–332. MR 1943290 (2003i:46028)
77. Maria Fragoulopoulou, *Topological algebras with involution*, North-Holland Mathematics Studies, vol. 200, North-Holland, 2005.
78. Rupert L. Frank, *Cwikel's theorem and the CLR inequality*, J. Spectr. Theory **4** (2014), no. 1, 1–21. MR 3181383
79. M. Fukushima, *Dirichlet forms, diffusion processes and spectral dimensions for nested fractals*, Ideas and methods in mathematical analysis, stochastics, and applications (Oslo, 1988), Cambridge Univ. Press, Cambridge, 1992, pp. 151–161. MR 1190496
80. M. Fukushima and T. Shima, *On a spectral analysis for the Sierpinski gasket*, Potential Analysis **1** (1992), no. 1, 1–35.
81. V. Gayral, J. Gracia-Bondía, B. Iochum, T. Schücker, and J. Várilly, *Moyal planes are spectral triples*, Comm. Math. Phys. **246** (2004), no. 3, 569–623. MR 2053945 (2005f:58045)
82. V. Gayral, B. Iochum, and J. Várilly, *Dixmier traces on noncompact isospectral deformations*, J. Funct. Anal. **237** (2006), no. 2, 507–539. MR 2230348 (2007g:58029)
83. Victor Gayral and Fedor Sukochev, *Dixmier traces and extrapolation description of noncommutative Lorentz spaces*, Journal of Functional Analysis **266** (2014), no. 10, 6256 – 6317.
84. Peter B. Gilkey, *The spectral geometry of a Riemannian manifold*, J. Differential Geom. **10** (1975), no. 4, 601–618 (eng).
85. ______, *Invariance theory, the heat equation, and the Atiyah-Singer index theorem*, Mathematics Lecture Series, vol. 11, Publish or Perish Inc., Wilmington, DE, 1984. MR MR783634 (86j:58144)
86. ______, *Invariance theory, the heat equation, and the Atiyah-Singer index theorem*, second ed., Studies in Advanced Mathematics, CRC Press, Boca Raton, FL, 1995. MR 1396308 (98b:58156)
87. I. C. Gohberg, *On the theory of multidimensional singular integral equations*, Soviet Math. Dokl. **1** (1960), 960–963. MR 0124704
88. I. C. Gohberg and M. G. Kreĭn, *Introduction to the theory of linear nonselfadjoint operators*, Translated from the Russian by A. Feinstein. Translations of Mathematical Monographs, Vol. 18, American Mathematical Society, Providence, R.I., 1969. MR 0246142
89. Adrian M. Gonzalez-Perez, Marius Junge, and Javier Parcet, *Singular integrals in quantum Euclidean spaces*, Mem. Amer. Math. Soc., to appear.
90. J.. Gracia-Bondía and J. Várilly, *Algebras of distributions suitable for phase-space quantum mechanics. I*, J. Math. Phys. **29** (1988), no. 4, 869–879. MR 940351

91. J. Gracia-Bondía, J. Várilly, and H. Figueroa, *Elements of noncommutative geometry*, Birkhäuser Advanced Texts: Basler Lehrbücher. [Birkhäuser Advanced Texts: Basel Textbooks], Birkhäuser Boston Inc., Boston, MA, 2001. MR 1789831 (2001h:58038)
92. H. Groenewold, *On the principles of elementary quantum mechanics*, Physica **12** (1946), 405–460. MR 0018562 (8,301a)
93. A. Grossmann, G. Loupias, and E. M. Stein, *An algebra of pseudo-differential operators and quantum mechanics in phase space*, Annales de l'institut Fourier **18** (1968), no. 2, 343–368 (eng).
94. Grubb, Gerd Functional calculus of pseudodifferential boundary problems. Second edition. Progress in Mathematics, 65. Birkhäuser Boston, Inc., Boston, MA, 1996. x+522 pp.
95. D. Guido and T. Isola, *Dimensions and singular traces for spectral triples, with applications to fractals*, J. Funct. Anal. **203** (2003), no. 2, 362–400. MR 2003353 (2005b:58038)
96. Daniele Guido and Tommaso Isola, *On the domain of singular traces*, Internat. J. Math. **13** (2002), no. 6, 667–674. MR 1915524 (2003h:46104)
97. Victor Guillemin, *A new proof of Weyl's formula on the asymptotic distribution of eigenvalues*, Adv. Math. **55** (1985), no. 2, 131–160.
98. Paul R. Halmos, *Commutators of operators*, Amer. J. Math. **74** (1952), 237–240.
99. Paul R. Halmos, *Commutators of operators. II*, Amer. J. Math. **76** (1954), 191–198. MR 0059484 (15,538d)
100. ———, *Commutators of operators*, pp. 126–134, Springer US, New York, NY, 1974.
101. B. M. Hambly, *Asymptotics for functions associated with heat flow on the Sierpinski carpet*, Canadian Journal of Mathematics **63** (2011), no. 1, 153–180 (eng).
102. G. H. Hardy, *Divergent Series*, Oxford University Press, Oxford, UK, 1949. MR 0030620 (11,25a)
103. S. W. Hawking, *Zeta function regularization of path integrals in curved spacetime*, Comm. Math. Phys. **55** (1977), no. 2, 133–148.
104. E. Herman, *The symbol of the algebra of singular integral operators*, Journal of Mathematics and Mechanics **15** (1966), no. 1, 147–155.
105. Joseph Hersch, *Caractérisation variationnelle d'une somme de valeurs propres consécutives; généralisation d'inégalités de Pólya-Schiffer et de Weyl*, C. R. Acad. Sci. Paris **252** (1961), 1714–1716. MR 0126065
106. ———, *Inégalités pour des valeurs propres consécutives de systèmes vibrants inhomogènes allant "en sens inverse" de celles de Pólya-Schiffer et de Weyl*, C. R. Acad. Sci. Paris **252** (1961), 2496–2498. MR 0123908
107. Fumio Hiai, *Log-majorizations and norm inequalities for exponential operators*, Banach Center Publications **38** (1997), no. 1, 119–181 (eng).
108. Nigel Higson, *The local index formula in noncommutative geometry*, Contemporary developments in algebraic K-theory, ICTP Lect. Notes, XV, Abdus Salam Int. Cent. Theoret. Phys., Trieste, 2004, pp. 443–536 (electronic). MR 2175637 (2006m:58039)
109. Nigel Higson, *Meromorphic continuation of zeta functions associated to elliptic operators*, Contemp. Math. **365** (2004), 129–142.
110. Reinhold Hubl, *Traces of differential forms and Hochschild homology*, Lecture Notes in Mathematics, vol. 1368, Springer-Verlag, Berlin, 1989 (eng).
111. Dirk Hundertmark, *On the number of bound states for Schrödinger operators with operator-valued potentials*, Ark. Mat. **40** (2002), no. 1, 73–87. MR 1948887
112. Victor Ivrii, *100 years of Weyl's law*, Bulletin of Mathematical Sciences **6** (2016), no. 3, 379–452.
113. Kohn J. J. and Nirenberg L., *An algebra of pseudo-differential operators*, Comm. Pure Appl. Math. **18** (1965), 269–305.
114. Svante Janson and Thomas H. Wolff, *Schatten classes and commutators of singular integral operators*, Ark. Mat. **20** (1982), no. 2, 301–310.
115. V. Kaftal and G. Weiss, *Traces on operator ideals and arithmetic means*, J. Operator Theory **63** (2010), no. 1, 3–46. MR 2606881 (2011j:47223)

116. Naotaka Kajino, *Log-periodic asymptotic expansion of the spectral partition function for self-similar sets*, Comm. Math. Phys. **328** (2014), no. 3, 1341–1370.
117. S. Kakutani and M. Nakamura, *Banach limits and the Čech compactification of a countable discrete set*, Proc. Imp. Acad. Tokyo **19** (1943), 224–229. MR 0014587 (7,306g)
118. W. Kalau and M. Walze, *Gravity, non-commutative geometry and the Wodzicki residue*, Journal of Geometry and Physics **16** (1995), no. 4, 327 – 344.
119. N. Kalton, *Unusual traces on operator ideals*, Math. Nachr. **134** (1987), 119–130. MR 918672 (89a:47070)
120. ______, *Trace-class operators and commutators*, J. Funct. Anal. **86** (1989), no. 1, 41–74. MR 1013933 (91d:47022)
121. ______, *Spectral characterization of sums of commutators. I*, J. Reine Angew. Math. **504** (1998), 115–125. MR 1656767 (99g:47078)
122. N. Kalton, A. Sedaev, and F. Sukochev, *Fully symmetric functionals on a Marcinkiewicz space are Dixmier traces*, Adv. Math. **226** (2011), no. 4, 3540–3549. MR 2764897 (2012c:47057)
123. N. Kalton and F. Sukochev, *Rearrangement-invariant functionals with applications to traces on symmetrically normed ideals*, Canad. Math. Bull. **51** (2008), no. 1, 67–80. MR 2384740 (2009h:46123)
124. ______, *Symmetric norms and spaces of operators*, J. Reine Angew. Math. **621** (2008), 81–121. MR 2431251 (2009i:46118)
125. Nigel Kalton, Steven Lord, Denis Potapov, and Fedor Sukochev, *Traces of compact operators and the noncommutative residue*, Adv. Math. **235** (2013), 1–55. MR 3010049
126. J. Karamata, *Uber die hardy-littlewoodschen umkehrungen des abelschen stetigkeitssatzes*, Mathematische Zeitschrift **32** (1930), no. 1, 319–320 (ger).
127. D. Kastler, *The Dirac operator and gravitation*, Comm. Math. Phys. **166** (1995), no. 3, 633–643. MR 1312438 (95j:58181)
128. Victor J. Katz, *The history of Stokes' theorem*, Mathematics Magazine **52** (1979), no. 3, 146–156.
129. Masoud Khalkhali and Matilde Marcolli, *An invitation to noncommutative geometry*, World Scientific Publishing, Singapore, 2008.
130. A. I. Khelemskii, *Banach and locally convex algebras*, Oxford Science Publications, Oxford University Press, Oxford, UK, 1993 (eng).
131. Jun Kigami, *Analysis on fractals*, Cambridge Tracts in Mathematics, Cambridge University Press, Cambridge, UK, 2001.
132. Jun Kigami and Michel L. Lapidus, *Weyl's problem for the spectral distribution of Laplacians on P.C.F. self-similar fractals*, Comm. Math. Phys. **158** (1993), no. 1, 93–125.
133. S. Kreĭn, Yu. Petunin, and E. Semenov, *Interpolation of linear operators*, Translations of Mathematical Monographs, vol. 54, American Mathematical Society, Providence, R.I., 1982, Translated from the Russian by J. Szűcs. MR 649411 (84j:46103)
134. Edwin Langmann, *Noncommutative integration calculus*, Journal of Mathematical Physics **36** (1995), no. 7, 3822–3835.
135. Michel L. Lapidus and Carl Pomerance, *The Riemann Zeta-Function and the One-Dimensional Weyl-Berry Conjecture for Fractal Drums*, Proceedings of the London Mathematical Society **s3-66** (1993), no. 1, 41–69.
136. Michel L. Lapidus and Machiel van Frankenhuijsen, *Fractal geometry, complex dimensions and zeta functions*, second ed., Springer Monographs in Mathematics, Springer, New York, 2013. MR 2977849
137. Ari Laptev and Timo Weidl, *Recent results on Lieb-Thirring inequalities*, Journées "Équations aux Dérivées Partielles" (La Chapelle sur Erdre, 2000), Univ. Nantes, Nantes, 2000, pp. Exp. No. XX, 14. MR 1775696
138. Jr. Lawson, H. Blaine and Marie-Louise Michelsohn, *Spin geometry*, Princeton Mathematical Series, vol. 38, Princeton University Press, Princeton, NJ, 1989.
139. J. Lee, *Introduction to smooth manifolds*, Graduate Texts in Mathematics, vol. 218, Springer-Verlag, New York, 2003. MR 1930091 (2003k:58001)

140. G. Levitina, A. Pietsch, F. Sukochev, and D. Zanin, *Completeness of quasi-normed operator ideals generated by s-numbers*, Indagationes Mathematicae **25** (2014), no. 1, 49–58.
141. G. Levitina, F. Sukochev, and D. Zanin, *Cwikel estimates revisited*, ArXiv e-prints (2017).
142. V. Lidskii, *Conditions for completeness of a system of root subspaces for non-selfadjoint operators with discrete spectrum*, Trudy Moskov. Mat. Obšč. **8** (1959), 83–120. MR 0107817 (21 #6539)
143. J. Lindenstrauss and L. Tzafriri, *Classical Banach spaces. II*, Ergebnisse der Mathematik und ihrer Grenzgebiete [Results in Mathematics and Related Areas], vol. 97, Springer-Verlag, Berlin, 1979, Function spaces. MR 540367 (81c:46001)
144. Tom Lindstrom, *Brownian motion on nested fractals*, Mem. Amer. Math. Soc., vol. 420, American Mathematical Society, Providence, R.I., 1990 (eng).
145. Jean-Louis Loday, *Cyclic homology*, second ed., Grundlehren der Mathematischen Wissenschaften [Fundamental Principles of Mathematical Sciences], vol. 301, Springer-Verlag, Berlin, 1998, Appendix E by María O. Ronco, Chapter 13 by the author in collaboration with Teimuraz Pirashvili. MR MR1600246 (98h:16014)
146. S. Lord, D. Potapov, and F. Sukochev, *Measures from Dixmier traces and zeta functions*, J. Funct. Anal. **259** (2010), no. 8, 1915–1949. MR 2671116 (2011j:58048)
147. S. Lord, A. Sedaev, and F. Sukochev, *Dixmier traces as singular symmetric functionals and applications to measurable operators*, J. Funct. Anal. **224** (2005), no. 1, 72–106. MR 2139105 (2006e:46065)
148. S. Lord and F. Sukochev, *Noncommutative residues and a characterisation of the noncommutative integral*, Proc. Amer. Math. Soc. **139** (2011), no. 1, 243–257. MR 2729087 (2011j:46104)
149. Steven Lord, Edward McDonald, Fedor Sukochev, and Dmitry Zanin, *Quantum differentiability of essentially bounded functions on Euclidean space*, J. Funct. Anal. **273** (2017), no. 7, 2353–2387. MR 3677828
150. Steven Lord, Fedor Sukochev, and Dmitry Zanin, *Singular traces: theory and applications*, De Gruyter Studies in Mathematics, vol. 46, Walter de Gruyter, 2012, Theory and applications. MR 3099777
151. G. Lorentz, *A contribution to the theory of divergent sequences*, Acta Math. **80** (1948), 167–190. MR 0027868 (10,367e)
152. V. I. Macaev, *A class of completely continuous operators*, Dokl. Akad. Nauk SSSR **139** (1961), 548–551, (Russian).
153. Yu. I. Manin, *Algebraic aspects of non-linear differential equations*, J. Sov. Math. **11** (1979), 1–122.
154. S. Mazur, *O metodach sumomalności*, Ksi e ga Pami a tkowa I Polskiego Zjazdu Matematycznego. Suplément auz Annales de la Société Polonaise Mathématique, 1929.
155. ______, *On the generalized limit of bounded sequences*, Colloquium Mathematicae **2** (1951), no. 3-4, 173–175.
156. Edward McDonald, Fedor Sukochev, and Dmitry Zanin, *A C^*-algebraic approach to the principal symbol. II*, ArXiv e-prints (2018).
157. S. Minakshisundaram, *Eigenfunctions on Riemannian manifolds*, J. Indian Math. Soc. **17** (1953), 158–165.
158. S. Minakshisundaram and A. Pleijel, *Some properties of the eigenfunctions of the Laplace-operator on Riemannian manifolds*, Canad. J. Math. **1** (1949), 242–256.
159. J. Moyal, *Quantum mechanics as a statistical theory*, Proc. Cambridge Philos. Soc. **45** (1949), 99–124. MR 0029330 (10,582d)
160. R. D. Moyer, *Computation of symbols on C^*-algebras of singular integral operators*, Bull. Amer. Math. Soc. **77** (1971), no. 4, 615–620.
161. Gerard J Murphy, *C^*-algebras and operator theory*, Academic Press, Boston, 1990.
162. R. Nest and E. Schrohe, *Dixmier's trace for boundary value problems*, Manuscripta Math. **96** (1998), no. 2, 203–218. MR 1624529 (99f:58201)
163. Fabio Nicola, *Trace functionals for a class of pseudo-differential operators in $\mathbb{R}^n$*, Mathematical Physics, Analysis and Geometry **6** (2003), no. 1, 89–105.

164. Fabio Nicola and Luigi Rodino, *C^*-algebras and elliptic theory II*, ch. Dixmier traceability for general pseudo-differential operators, pp. 227–237, Birkhäuser Verlag, 2008.
165. C. Pearcy and D. Topping, *On commutators in ideals of compact operators*, Michigan Math. J. **18** (1971), 247–252. MR 0284853 (44 #2077)
166. Vladimir V. Peller, *Hankel operators and their applications*, Springer Monographs in Mathematics, Springer-Verlag, New York, 2003. MR MR1949210 (2004e:47040)
167. A. Pietsch, *Operator ideals with many traces*, Forschungsergebnisse FSU Jena, Nr. N/86/16, 1986.
168. ———, *Dixmier traces of operators on Banach and Hilbert spaces*, Math. Nachr. **285** (2012), no. 16, 1999–2028. MR 2995429
169. Albrecht Pietsch, *Einige neue Klassen von kompakten linearen Operatoren*, Rev. Math. Pures Appl. (Roumaine) **8** (1963), 427–447.
170. ———, *Operator ideals with a trace*, Mathematische Nachrichten **100** (1981), no. 1, 61–91 (eng).
171. ———, *Traces and shift invariant functionals*, Math. Nachr. **145** (1990), 7–43. MR 1069016
172. ———, *About the Banach envelope of $l_{1,\infty}$*, Rev. Mat. Complut. **22** (2009), no. 1, 209–226. MR 2499333 (2010g:46003)
173. ———, *Connes–Dixmier versus Dixmier Traces*, Integral Equations and Operator Theory **77** (2013), no. 2, 243–259.
174. ———, *Traces of operators and their history*, Acta et Commentationes Universitatis Tartuensis de Mathematica **18** (2014), no. 1, 51 – 64.
175. ———, *Traces on operator ideals and related linear forms on sequence ideals (part I)*, Indagationes Mathematicae **25** (2014), no. 2, 341–365, Zaanen Centennial Special Issue.
176. ———, *Traces on operator ideals and related linear forms on sequence ideals (part II)*, Integral Equations and Operator Theory **79** (2014), no. 2, 255–299.
177. ———, *Traces on operator ideals and related linear forms on sequence ideals (part III)*, Journal of Mathematical Analysis and Applications **421** (2015), no. 2, 971–981.
178. ———, *A new approach to operator ideals on Hilbert space and their traces*, Integral Equations and Operator Theory **89** (2017), no. 4, 595–606.
179. ———, *The spectrum of shift operators and the existence of traces*, Integral Equations and Operator Theory **90** (2018), no. 2, 17.
180. ———, *Traces and symmetric linear forms*, Archiv der Mathematik (2018).
181. Denis Potapov and Fedor Sukochev, *Unbounded Fredholm modules and double operator integrals*, J. Reine Angew. Math. **626** (2009), 159–185. MR MR2492993
182. Denis Potapov, Fedor Sukochev, Anna Tomskova, and Dmitriy Zanin, *Fréchet differentiability of the norm of L_p-spaces associated with arbitrary von Neumann algebras*, C. R. Math. Acad. Sci. Paris **352** (2014), no. 11, 923–927. MR 3268764
183. Denis Potapov, Fedor Sukochev, Alexandr Usachev, and Dmitriy Zanin, *Singular traces and perturbation formulae of higher order*, J. Funct. Anal. **269** (2015), no. 5, 1441–1481. MR 3369943
184. S. Power, *Another proof of Lidskii's theorem on the trace*, Bull. London Math. Soc. **15** (1983), no. 2, 146–148. MR 689247 (85g:47029)
185. S. C. Power, *Commutator ideals and pseudodifferential C^*-algebras*, The Quarterly Journal of Mathematics **31** (1980), no. 4, 467–489.
186. R. A. Raimi, *Factorization of summability-preserving generalized limits*, J. London Math. Soc. **22** (1980), 398–402.
187. Burton Randol, *On the Analytic Continuation of the Minakshisundaram-Pleijel Zeta Function for Compact Riemann Surfaces*, Tran. Amer. Math. Soc. **201** (1975), 241–246.
188. Michael Reed and Barry Simon, *Methods of modern mathematical physics. IV. Analysis of operators*, Academic Press [Harcourt Brace Jovanovich, Publishers], New York-London, 1978. MR 0493421
189. ———, *Methods of modern mathematical physics. I*, second ed., vol. I: Functional Analysis, revised and enlarged edition, Academic Press Inc. [Harcourt Brace Jovanovich Publishers], New York, 1980, Functional analysis. MR 751959 (85e:46002)

190. Georges de Rham, *Differentiable manifolds: forms, currents, harmonic forms*, Grundlehren der mathematischen Wissenschaften ; 266, Springer-Verlag, Berlin-New York, 1984.
191. M. Rieffel, *C^*-algebras associated with irrational rotations*, Pacific J. Math. **93** (1981), no. 2, 415–429. MR 623572 (83b:46087)
192. ______, *Deformation quantization for actions of $\mathbf{R}^d$*, Mem. Amer. Math. Soc. **106** (1993), no. 506, x+93. MR 1184061 (94d:46072)
193. Richard Rochberg and Stephen Semmes, *Nearly weakly orthonormal sequences, singular value estimates, and Calderon-Zygmund operators*, J. Funct. Anal. **86** (1989), no. 2, 237–306.
194. M. Ruzhansky and V. Turunen, *Pseudo-differential operators and symmetries*, Pseudo-Differential Operators. Theory and Applications, vol. 2, Birkhäuser Verlag, Basel, 2010, Background analysis and advanced topics. MR 2567604 (2011b:35003)
195. T. Sakai, *Riemannian geometry*, Translations of Mathematical Monographs, vol. 149, American Mathematical Society, Providence, RI, 1996, Translated from the 1992 Japanese original by the author. MR 1390760 (97f:53001)
196. W. Sargent, *Some sequence spaces related to the l^p spaces*, J. London Math. Soc. **35** (1960), 161–171. MR 0116206 (22 #7001)
197. Elmar Schrohe, *The symbols of an algebra of pseudodifferential operators.*, Pacific J. Math. **125** (1986), no. 1, 211–224 (eng).
198. A. Sedaev, *Generalized limits and related asymptotic formulas*, Mat. Zametki **86** (2009), no. 4, 612–627. MR 2591352 (2011e:58048)
199. A. Sedaev and F. Sukochev, *Dixmier measurability in Marcinkiewicz spaces and applications*, J. Funct. Anal. **265** (2013), no. 12, 3053–3066. MR 3110494
200. A. Sedaev, F. Sukochev, and D. Zanin, *Lidskii-type formulae for Dixmier traces*, Integral Equations Operator Theory **68** (2010), no. 4, 551–572. MR 2745479 (2011j:46108)
201. R. Seeley, *Complex powers of an elliptic operator*, Singular Integrals (Proc. Sympos. Pure Math., Chicago, Ill., 1966), Amer. Math. Soc., Providence, R.I., 1967, pp. 288–307. MR 0237943 (38 #6220)
202. Evgenii Semenov and Fedor Sukochev, *Invariant Banach limits and applications*, J. Funct. Anal. **259** (2010), 1517–1541.
203. Evgenii Semenov, Fedor Sukochev, Alexandr Usachev, and Dmitriy Zanin, *Banach limits and traces on $\mathcal{L}_{1,\infty}$*, Adv. Math. **285** (2015), 568–628. MR 3406510
204. M. A. Shubin, *Pseudodifferential operators and spectral theory*, second ed., Springer Series in Soviet Mathematics, Springer-Verlag, Berlin, 2001, Translated from the 1978 Russian original by Stig I. Andersson. MR 1852334 (2002d:47073)
205. Barry Simon, *Analysis with weak trace ideals and the number of bound states of Schrödinger operators*, Trans. Amer. Math. Soc. **224** (1976), no. 2, 367–380. MR 0423128
206. ______, *Trace ideals and their applications*, London Mathematical Society Lecture Note Series, vol. 35, Cambridge University Press, Cambridge-New York, Cambridge, 1979. MR MR541149 (80k:47048)
207. ______, *Trace ideals and their applications*, second ed., Mathematical Surveys and Monographs, vol. 120, American Mathematical Society, Providence, RI, 2005. MR 2154153 (2006f:47086)
208. Michael Spivak, *A comprehensive introduction to differential geometry. Vol. III*, second ed., Publish or Perish Inc., Wilmington, Del., 1979. MR 532832 (82g:53003c)
209. Elias M. Stein, *Singular integrals and differentiability properties of functions*, Princeton Mathematical Series, No. 30, Princeton University Press, Princeton, N.J., 1970. MR 0290095 (44 #7280)
210. Benjamin A. Steinhurst and Alexander Teplyaev, *Existence of a meromorphic extension of spectral zeta functions on fractals*, Letters in Mathematical Physics **103** (2013), no. 12, 1377–1388.
211. A. B. Stern, *Finite-rank approximations of spectral zeta residues*, ArXiv e-prints (2017).
212. M. A. Subkhankulov, *Tauberovy teoremy s ostatkom*, Izdat. "Nauka", Moscow, 1976. MR 0612819

213. L. Sucheston, *Banach limits*, Amer. Math. Monthly **74** (1967), 308–311. MR 0225145 (37 #740)
214. F Sukochev, A Usachev, and D Zanin, *Singular traces and residues of the zeta-function*, Indiana University Mathematics Journal **66** (2017), no. 4, 1107–1144 (English).
215. F. Sukochev and D. Zanin, *ζ-function and heat kernel formulae*, J. Funct. Anal. **260** (2011), no. 8, 2451–2482. MR 2772378 (2012c:58048)
216. ———, *Dixmier traces are weak* dense in the set of all fully symmetric traces*, J. Funct. Anal. **266** (2014), no. 10, 6158–6173. MR 3188711
217. ———, *Which traces are spectral?*, Adv. Math. **252** (2014), 406–428. MR 3144235
218. ———, *Fubini theorem in noncommutative geometry*, J. Funct. Anal. **272** (2017), no. 3, 1230–1264. MR 3579138
219. ———, *Connes integration formula for the noncommutative plane*, Comm. Math. Phys. **359** (2018), no. 2, 449–466.
220. F. Sukochev and D. Zanin, *The Connes character formula for locally compact spectral triples*, ArXiv e-prints (2018).
221. Fedor Sukochev, *Completeness of quasi-normed symmetric operator spaces*, Indag. Math. (N.S.) **25** (2014), no. 2, 376–388. MR 3151823
222. Fedor Sukochev, Alexandr Usachev, and Dmitriy Zanin, *Generalized limits with additional invariance properties and their applications to noncommutative geometry*, Adv. in Math. **239** (2013), 164 – 189.
223. Fedor Sukochev, Alexandr Usachev, and Dmitriy Zanin, *On the distinction between the classes of Dixmier and Connes-Dixmier traces*, Proc. Amer. Math. Soc. **141** (2013), no. 6, 2169–2179. MR 3034443
224. ———, *Dixmier traces generated by exponentiation invariant generalised limits*, Journal of Noncommutative Geometry **8** (2014), no. 2, 321–336 (eng).
225. Fedor. Sukochev and Dmitriy. Zanin, *A C^*-algebraic approach to the principal symbol*, Journal of Operator Theory (Accepted manuscript. 2018).
226. Hiroshi Takai and Toshikazu Natsume, *A. Connes' noncommutative differential geometry*, Sūgaku **35** (1983), no. 2, 97–112. MR 732432 (86e:58002)
227. Michael E. Taylor, *Beals-Cordes-type characterizations of pseudodifferential operators*, Proc. Amer. Math. Soc. **125** (1997), no. 6, 1711–1716. MR 1371144 (97g:35186)
228. Alexandr Usachev, Fedor Sukochev, and Dmitriy Zanin, *Singular traces and residues of the zeta-function*, Indiana Univ. Math. J. **66** (2017), 1107–1144.
229. J. Varga, *Traces on irregular ideals*, Proc. Amer. Math. Soc. **107** (1989), no. 3, 715–723. MR 984818 (91e:47046)
230. ———, *Traces and commutators of ideals of compact operators*, Ph.D. Dissertation. Ohio State University, Mathematics, 1995.
231. Joseph C. Várilly and José M. Gracia-Bondía, *Algebras of distributions suitable for phase-space quantum mechanics. II. Topologies on the Moyal algebra*, J. Math. Phys. **29** (1988), no. 4, 880–887. MR 940352
232. Joseph C. Várilly, José M. Gracia-Bondía, and Walter Schempp, *The Moyal representation of quantum mechanics and special function theory*, Acta Applicandae Mathematica **18** (1990), no. 3, 225–250.
233. D. V. Vassilevich, *Heat kernel expansion: user's manual*, Physics Reports **388** (2003), no. 5, 279–360.
234. Timo Weidl, *Cwikel type estimates in non-power ideals*, Math. Nachr. **176** (1995), 315–334. MR 1361143
235. ———, *Another look at Cwikel's inequality*, Differential operators and spectral theory, Amer. Math. Soc. Transl. Ser. 2, vol. 189, Amer. Math. Soc., Providence, RI, 1999, pp. 247–254. 1730517 (2000i:47095)
236. G. L. Weiss, *Commutators and operator ideals*, Ph.D. Dissertation. University of Michigan, 1975, p. 209. MR 2625442

237. Hermann Weyl, *Ueber die asymptotische verteilung der eigenwerte*, Nachrichten von der Gesellschaft der Wissenschaften zu Göttingen, Mathematisch-Physikalische Klasse **1911** (1911), 110–117.
238. ______, *Inequalities between the two kinds of eigenvalues of a linear transformation*, Proc. Nat. Acad. Sci. U. S. A. **35** (1949), 408–411. MR 0030693 (11,37d)
239. M. Wodzicki, *Local invariants of spectral asymmetry*, Invent. Math. **75** (1984), no. 1, 143–177. MR 728144 (85g:58089)
240. ______, *Noncommutative residue. I. Fundamentals*, K-theory, arithmetic and geometry (Moscow, 1984–1986), Lecture Notes in Math., vol. 1289, Springer, Berlin, 1987, pp. 320–399. MR 923140 (90a:58175)
241. ______, *Vestigia investiganda*, Mosc. Math. J. **2** (2002), no. 4, 769–798, 806, Dedicated to Yuri I. Manin on the occasion of his 65th birthday. MR 1986090 (2005j:47075)

Commutants mod normed ideals

Dan-Virgil Voiculescu

Dedicated to Alain Connes on the occasion of his 70th birthday.

Abstract To Alain Connes' non-commutative geometry the normed ideals of compact operators are purveyors of infinitesimals. A numerical invariant, the modulus of quasicentral approximation, plays a key role in perturbations from these ideals. New structure is provided by commutants mod normed ideals of n-tuples of operators and by their Calkin algebras. I review the modulus of quasicentral approximation, the relation to invariance of absolutely continuous spectra, to dynamical entropy and the hybrid generalization. I then discuss commutants mod normed ideals, their Banach space duality properties, K-theory aspects, the case of the Macaev ideal. Sample open problems are included.

1 Introduction

The full name of the normed ideals to which we refer in the title, is "symmetrically normed ideals of compact operators," among which the Schatten–von Neumann p-classes are the most familiar. The commutants are commutants modulo a normed ideal of n-tuples of self-adjoint operators, or equivalently of the algebras generated by these operators. When the normed ideal is $\mathcal{K}$, that is the ideal of compact operators, then up to factoring by $\mathcal{K}$, this is roughly how one arrives at the Paschke dual of a finitely generated C^*-algebra ([35], see also [25]), that is a basic duality construction in the K-theory of C^*-algebras, that in essence hails from the abstract elliptic operators of Atiyah [2]. What happens if $\mathcal{K}$ is replaced by a smaller normed ideal $\mathcal{J}$? As we will see, commutants mod $\mathcal{J}$ are not simply "smooth versions" of

D.-V. Voiculescu (✉)
Department of Mathematics, University of California at Berkeley, Berkeley, CA, USA
e-mail: dvv@math.berkeley.edu

A. Chamseddine et al. (eds.), *Advances in Noncommutative Geometry*,
https://doi.org/10.1007/978-3-030-29597-4_10

those mod $\mathcal{K}$, and they are often closer to C^*-algebras than one would expect, while they connect with many questions in perturbation theory.

From an operator theory perspective, following the Brown–Douglas–Fillmore work [7, 8] and the development of the K-theory of C^*-algebras all the way to Kasparov's bivariant theory ([26], see also [5, 23]) many things about compact perturbations can now be understood from this point of view.

When other normed ideals than $\mathcal{K}$ are considered, there are different aspects to be taken into account. For instance, Alain Connes' cyclic cohomology is the algebraic framework for trace-formulas like the Helton–Howe formula for almost normal operators [22, 15]. On the other hand, the invariance of Lebesgue absolutely continuous spectra under trace-class perturbations fits with our modulus of quasicentral approximation invariant which is one focus of the present article.

Starting with an adaptation [43] of the Voiculescu non-commutative Weyl–von Neumann type theorem [42] to normed ideals other than $\mathcal{K}$, we found that $k_{\mathcal{J}}(\tau)$, the modulus of quasicentral approximation for the n-tuple τ relative to the normed ideal $\mathcal{J}$, underlies many questions concerning perturbation of operators [43, 44, 49]. This quantity has also turned out to be connected to the Kolmogorov–Sinai dynamical entropy and to the supramenability of groups [47, 54]. We should also mention that, as we found in [43, 44, 45], in many of these questions Lorentz $(p, 1)$ ideals instead of Schatten–von Neumann p-classes give sharp results when $p > 1$.

To Alain Connes' non-commutative geometry, the normed ideals are purveyors of infinitesimals [15]. The machinery around $k_{\mathcal{J}}(\tau)$ has found technical uses in the spectral characterization of compact manifolds [16] (i.e., their characterization as non-commutative manifolds) and in results about unbounded Fredholm modules [13].

Since the numerical invariant $k_{\mathcal{J}}(\tau)$ turned out to have good properties and to unify several perturbation problems, we recently looked whether there is not actually more structure around this number. The commutant $\mathcal{E}(\tau; \mathcal{J})$ of an n-tuple of operators τ mod a normed ideal $\mathcal{J}$ provides such structure. The first step in the study of $\mathcal{E}(\tau; \mathcal{J})$ is to introduce $\mathcal{K}(\tau; \mathcal{J})$ its ideal of compact operators and the corresponding Calkin algebra $\mathcal{E}/\mathcal{K}(\tau; \mathcal{J})$. Under suitable assumptions on $k_{\mathcal{J}}(\tau)$, there are many functional analysis similarities between $(\mathcal{K}(\tau; \mathcal{J}), \mathcal{E}(\tau; \mathcal{J}), \mathcal{E}/\mathcal{K}(\tau; \mathcal{J}))$ and the usual $(\mathcal{K}, \mathcal{B}, \mathcal{B}/\mathcal{K})$ where $\mathcal{B}$ denotes the bounded operators. With this analogy as a guide we also took the first steps in computing some K-groups of such algebras. The K-theory can be quite rich and this demonstrates why the $\mathcal{E}/\mathcal{K}(\tau; \mathcal{J})$ should not be thought as being "smooth versions" of $\mathcal{E}/\mathcal{K}(\tau; \mathcal{K})$, i.e., essentially of Paschke duals.

As a general comment about algebras associated with perturbations from a normed ideal $\mathcal{J}$, it appears that $\mathcal{E}(\tau; \mathcal{J})$, which is closer to the abstract elliptic operators point of view of Atiyah, being a Banach algebra may have from a functional analysis point of view some advantages over dealing with homomorphisms of algebras into $B/\mathcal{J}$ in the Brown–Douglas–Fillmore style. The algebra $B/\mathcal{J}$ is not a Banach algebra, though smooth functional calculus of various kinds can still be performed in $B/\mathcal{J}$. On the other hand $\mathcal{E}/\mathcal{K}(\tau; \mathcal{J})$ if $k_{\mathcal{J}}(\tau) < \infty$ is actually isomorphic to a C^*-algebra.

Very recently we found that $k_{\mathfrak{J}}(\tau)$ remains an effective tool also when generalized to handling of hybrid perturbations. That is instead of a normed ideal $\mathfrak{J}$, we will have an n-tuple of normed ideals $(\mathfrak{J}_1, \dots, \mathfrak{J}_n)$ and the perturbation of τ to τ'_1 is such that $T_j - T'_j \in \mathfrak{J}_j$, $1 \le j \le n$. We will also very briefly mention a few results in this direction [57, 58], which are quite sharp.

Here is in brief how the survey proceeds. After this introduction we provide some background on the normed ideals we will use. We then as a motivating example recall the main facts about normed ideal perturbations of one self-adjoint operator. Then we introduce the invariant $k_{\mathfrak{J}}(\tau)$ and its basic properties. Next we give a version of the Voiculescu theorem adapted to normed ideals. After this we pass to the applications of this machinery to normed ideal perturbations of n-tuples of commuting Hermitian operators. Next we explain the endpoint properties of $k^-_\infty(\tau)$ which is the case when $\mathfrak{J}$ is the Macaev ideal. The way the Kolmogorov–Sinai entropy is related to $k^-_\infty(\tau)$ is explained after this. Then we discuss results for finitely generated groups and the result and open problem about k^-_∞ and supramenable groups. After this, we go over to commutants mod normed ideals. We explain that $k_{\mathfrak{J}}(\tau)$ is related to approximate units of the compact ideal $\mathcal{K}(\tau; \mathfrak{J})$ of $\mathcal{E}(\tau; \mathfrak{J})$. The Banach space properties of $\mathcal{E}(\tau; \mathfrak{J})$ are then discussed as well as properties of $\mathcal{E}/\mathcal{K}(\tau; \mathfrak{J})$ where $k_{\mathfrak{J}}(\tau) < \infty$ or $k_{\mathfrak{J}}(\tau) = 0$. We then look at the results about $K_0(\mathcal{E}(\tau; \mathfrak{J}))$ for n-tuples of commuting Hermitian operators, which, to simplify matters we choose here to be multiplication operators by the coordinate functions in the L^2-space with respect to Lebesgue measure on a hypercube. We then discuss a few results in the hybrid setting. Quite briefly finally some applications to unbounded Fredholm modules are then pointed out.

The study of commutants mod normed ideals and of the invariant $k_{\mathfrak{J}}(\tau)$ and of the hybrid generalization are still at an early stage and from a reading of this survey one realizes the multitude of open problems of varying degrees of difficulty. Still at the end we briefly mention some sample open problems which had not appeared with the presentation of results.

At various points in this exposition, in order to avoid technical detail, we did not aim at the most general or technically strongest version of the results. We hope this kind of simplification will make it easier for the reader to focus on the big picture.

2 Background on normed ideals

Throughout we shall denote by $\mathcal{H}$ an infinite dimensional separable complex Hilbert space and by $\mathcal{B}(\mathcal{H})$, $\mathcal{K}(\mathcal{H})$, $\mathcal{R}(\mathcal{H})$ or simply $\mathcal{B}, \mathcal{K}, \mathcal{R}$ the bounded operators, the compact operators and the finite rank operators on $\mathcal{H}$. The Calkin algebra is then $\mathcal{B}/\mathcal{K}$ and $p : \mathcal{B} \to \mathcal{B}/\mathcal{K}$ will be the canonical homomorphism. A normed ideal $(\mathfrak{J}, |\ |_{\mathfrak{J}})$ is an ideal of $\mathcal{B}$ so that $\mathcal{R} \subset \mathfrak{J} \subset \mathcal{K}$ and $|\ |_{\mathfrak{J}}$ is a Banach space norm on $\mathfrak{J}$ satisfying a set of axioms which can be found in [20, 40]. In particular if $A, B \in \mathcal{B}$ and $X \in \mathfrak{J}$ we have $|AXB|_{\mathfrak{J}} \le \|A\|\,|X|_{\mathfrak{J}}\,\|B\|$. If $\mathfrak{J}$ is a normed ideal, its dual as a Banach space is also a normed ideal $\mathfrak{J}^d$, where now we have to also allow the possibility that $\mathcal{J}^d = \mathcal{B}$, and the duality is given by $(X, Y) \to \mathrm{Tr}\, XY$. If $\mathcal{J}$ is a normed ideal, then the closure of $\mathcal{R}$ in $\mathcal{J}$ is also a normed ideal which we shall denote by $\mathcal{J}^{(0)}$ and which may be strictly smaller than $\mathcal{J}$.

If $1 \le p < \infty$, the Schatten–von Neumann p-class $\mathcal{C}_p$ is the normed ideal of operators $X \in \mathcal{K}$ so that $|X|_p = (\mathrm{Tr}((X^*X)^{p/2}))^{1/p} < \infty$ endowed with the norm $|\ |_p$. In particular, $\mathcal{C}_1$ is the trace-class and $\mathcal{C}_2$ is the ideal of Hilbert–Schmidt operator. Another scale of normed-ideals which we shall use here is $(\mathcal{C}_p^-, |\ |_p^-)$, $1 \le p \le \infty$ which are the Lorentz $(p, 1)$-ideals. If $s_1 \ge s_2 \ge \dots$ are the eigenvalues of the compact operator $(X^*X)^{1/2}$, then $|X|_p^- = \sum_{j \in \mathbb{N}} s_j j^{-1+1/p}$. Of particular interest is the case where $p = \infty$ and $|X|_\infty^- = \sum_{j\in\mathbb{N}} s_j j^{-1}$. If $p = 1$, $\mathcal{C}_1^- = \mathcal{C}_1$, but otherwise $\mathcal{C}_p^- \subset \mathcal{C}_p$ but $\mathcal{C}_p^- \ne \mathcal{C}_p$, $1 < p \le \infty$, where $\mathcal{C}_\infty = \mathcal{K}$. The ideal $\mathcal{C}_p^-$ can also be described as the smallest normed ideal for which the norm on projections is equivalent to the p-norm, when $1 \le p < \infty$ while $|P|_\infty^- \sim \log \mathrm{Tr}\, P$ when $p = \infty$ for a projection P. The ideal $\mathcal{C}_\infty^-$ is also called the Macaev ideal.

The normed ideals can also be viewed as the non-commutative analogue of classical Banach sequence spaces [29].

3 The theorems of Weyl–von Neumann–Kuroda and of Kato–Rosenblum

Let A and B be Hermitian operators on $\mathcal{H}$ and let $(\mathcal{J}, |\ |_\mathcal{J})$ be a normed ideal. Recall that p denotes the homomorphism onto the Calkin algebra. Thus the essential spectrum $\sigma(p(A))$ of the Hermitian operator A is obtained from its spectrum $\sigma(A)$ by removing the isolated points $\lambda \in \sigma(A)$ which corresponds to eigenvalues of finite multiplicity. By results of Weyl, von Neumann and Kuroda, see [27], we have:

Assume $\mathcal{J} \ne \mathcal{C}_1$, $\varepsilon > 0$, and $\sigma(A) = \sigma(B) = \sigma(p(A)) = \sigma(p(B))$. Then there is a unitary operator U so that

$$|UBU^* - A|_\mathcal{J} < \varepsilon.$$

Note that, given A, we may choose B to be diagonal in an orthonormal basis and thus we get $|X - A|_\mathcal{J} < \varepsilon$ where $X = UBU^*$ is an operator which can be diagonalized in an orthonormal basis.

The trace-class $\mathcal{C}_1$ is actually the smallest normed ideal. If $\mathcal{J} = \mathcal{C}_1$, the previous result fails because of the Lebesgue absolutely continuous spectrum, which is a conserved quantity under trace-class perturbations. This is a consequence of the Kato–Rosenblum theorem of abstract scattering theory [27, 37]. If $X = X^*$, we say that X has Lebesgue absolutely continuous spectrum, if its spectral measure $E(X; \cdot)$ is absolutely continuous wrtLebesgue measure (equivalently this is that the scalar measures $\langle E(X; \cdot)\xi, \eta\rangle$ for all $\xi, \eta \in \mathcal{H}$ are Lebesgue absolutely continuous). Given X, the Hilbert space $\mathcal{H}$ splits in a unique way $\mathcal{H} = \mathcal{H}_{ac} \oplus \mathcal{H}_{sing}$ into X-invariant subspaces so that $X \mid \mathcal{H}_{ac}$ has Lebesgue absolutely continuous spectrum, while

$X \mid \mathcal{H}_{sing}$ has singular spectrum, that is the spectral measure of $X \mid \mathcal{H}_{sing}$ is carried by a Borel set of Lebesgue measure zero. It is a corollary of the Kato–Rosenblum theorem that:

if $X - A \in \mathcal{C}_1$, then $X \mid \mathcal{H}_{ac}(X)$ and $A \mid \mathcal{H}_{ac}(A)$ are unitarily equivalent.

The Kato–Rosenblum theorem actually provides two intertwiners, to achieve the unitary equivalence, the generalized wave operators $W_{\pm}$:

under the assumption $X - A \in \mathcal{C}_1$ the strong limits

$$W_{\pm} = s - \lim_{t \to \pm\infty} e^{itA} e^{-itX} \mid \mathcal{H}_{ac}(X)$$

exist. Moreover we have $W_{\pm}\mathcal{H}_{ac}(X) = \mathcal{H}_{ac}(A)$ and

$$W_{\pm}(X \mid \mathcal{H}_{ac}(X)) = (A \mid \mathcal{H}_{ac}(A))W_{\pm}.$$

Thus the Lebesgue absolutely continuous part $A \mid \mathcal{H}_{ac}(A)$ is conserved up to unitary equivalence under perturbations in $\mathcal{C}_1$ and cannot be diagonalized.

On the other hand for the singular spectrum, the ideal $\mathcal{C}_1$ is not different from other ideals. For instance, the Weyl–von Neumann–Kuroda theorem holds also for $\mathcal{C}_1$ if the spectrum is singular that is if $\mathcal{H}_{ac}(A) = \mathcal{H}_{ac}(B) = 0$.

Concerning what goes into the proofs of these theorems, the Weyl–von Neumann–Kuroda results rely essentially on partitioning

$$I = E(A; \omega_1) + \cdots + E(A; \omega_n)$$

where $\omega_1, \ldots, \omega_n$ is a partition of $\sigma(A)$ into Borel sets of small diameter, while the Kato–Rosenblum theorem uses some Fourier analysis, which can be viewed as related to the L^2-boundedness of the Hilbert transform.

To conclude this discussion we should also say that the role of assumptions about spectra and essential spectra $\sigma(A) = \sigma(p(A))$ will become clearer in the next section when we consider C^*-algebras.

4 The theorem of Voiculescu

After the work of Brown–Douglas–Fillmore which solved the unitary conjugacy question for normal elements of the Calkin algebra, it became clear that it is preferable in this kind of question to view operators or n-tuples of operators as representations of the C^*-algebra which they generate. For instance, given an n-tuple of commuting Hermitian operators $A_1, \ldots, A_n$ this is just the representation of the commutative C^*-algebra $C(K)$ of continuous functions on their joint spectrum $K = \sigma(A_1, \ldots, A_n)$ which arises from functional calculus $\rho(f) = f(A_1, \ldots, A_n)$.

One version of the Voiculescu theorem ([42], see also [1]) is a non-commutative generalization of the perturbation results for one Hermitian operator in case $\mathcal{J} = \mathcal{K}$, the ideal of compact operators:

> If $\mathcal{A}$ is a unital separable C^*-algebra and $\rho_1, \rho_2 : \mathcal{A} \to \mathcal{B}(\mathcal{H})$ are unital $*$-homomorphisms such that $\ker p \circ \rho_1 = \ker p \circ \rho_2 = 0$, then there is a unitary operator U so that $\rho_1(a) - U\rho_2(a)U^*(a) \in \mathcal{K}$ for all $a \in \mathcal{A}$.

For instance, to recover the Weyl–von Neumann result for two Hermitian operators A, B with $\sigma(A) = \sigma(B) = \sigma(p(A)) = \sigma(p(B)) = K$ one takes $\mathcal{A} = C(K)$, $\rho_1(f) = f(A)$, $\rho_2(f) = f(B)$ and the conclusion is applied to the particular choice of f being the identical function.

This very general result for compact perturbations obviously leads to the question: What happens when the compact operators are replaced by a smaller normed ideal $\mathcal{J}$? We found that there is a key quantity which needs to be taken into account and which will be discussed in the next section.

5 The invariant $k_{\mathcal{J}}(\tau)$

Let $\tau = (T_j)_{1 \le j \le n}$ be an n-tuple of bounded operators on $\mathcal{H}$ and $(\mathcal{J}, |\ |_{\mathcal{J}})$ a normed ideal. Then the *modulus of quasicentral approximation* is defined as follows [43, 45]:

> $k_{\mathcal{J}}(\tau)$ is the least $C \in [0, \infty]$, such that there exist finite rank operators $0 \le A_m \le I$ so that $A_m \uparrow I$ and we have
>
> $$\lim_{m \to \infty} \max_{1 \le j \le n} |[A_m, T_j]|_{\mathcal{J}} = C.$$

If $\mathcal{J} = \mathcal{C}_p$ or $\mathcal{J} = \mathcal{C}_p^-$ we denote $k_{\mathcal{J}}(\tau)$ by $k_p(\tau)$ or respectively by $k_p^-(\tau)$. In particular in the case of the Macaev ideal $\mathcal{J} = \mathcal{C}_\infty^-$ we get $k_\infty^-(\tau)$. Here are some of the first properties of this invariant (see [45]).

1° $[1, \infty] \ni p \to k_p^-(\tau) \in [0, \infty]$ is a decreasing function of p.
2° for a given τ there is $p_0 \in [1, \infty]$ so that if $p \in [1, p_0)$ then $k_p^-(\tau) = \infty$, while if $p \in (p_0, \infty]$ then $k_p^-(\tau) = 0$. (Note that if $k_p^-(\tau) \in (0, \infty)$ then we must have $p_0 = p$.)
3° assuming $\mathcal{R}$ is dense in $\mathcal{J}$, then if τ, τ' are n-tuples so that $T_j - T_j' \in \mathcal{J}$, $1 \le j \le n$, then $k_{\mathcal{J}}(\tau) = k_{\mathcal{J}}(\tau')$.
4° assuming $\tau = \tau^*$ and that $\mathcal{R}$ is dense in $\mathcal{J}$, we have: $k_{\mathcal{J}}(\tau) > 0$ iff there exist $Y_j = Y_j^* \in \mathcal{J}^{\mathrm{dual}}$, $1 \le j \le n$ so that $i\sum_j [T_j, Y_j] \in \mathcal{C}_1 + \mathcal{B}(\mathcal{H})_+$ and $\operatorname{Tr} i\sum_j [T_j, Y_j] > 0$.

To prove that $k_{\mathcal{J}}(\tau) = 0$ or that $k_{\mathcal{J}}(\tau) < \infty$ one can use the definition of $k_{\mathcal{J}}(\tau)$ and find a suitable sequence of operators A_n. To prove that $k_{\mathcal{J}}(\tau) > 0$ is usually more difficult as one has to find suitable Y_j $(1 \le j \le n)$ which satisfy property 4°. For instance, in the case of $\mathcal{J} = \mathcal{C}_1$ and $n = 1$, this boils down to the boundedness of the Hilbert transform in L^2. Indeed, remark that if T is the multiplication operator in $L^2([0, 1], d\lambda)$ by the coordinate function and H is the compression of the Hilbert transform to $L^2([0, 1], d\lambda)$ then $[T, H] = iP$ where P is the rank one projection onto the constant functions.

The reader may have wondered why we did not pay more attention to $k_p(\tau)$ instead of focusing on $k_p^-(\tau)$. The reason is that if $p = 1$ then $\mathcal{C}_1^- = \mathcal{C}_1$ and we have $k_1(\tau) = k_1^-(\tau)$, while otherwise, if $p > 1$ we have that $k_p(\tau) \in \{0, \infty\}$ [45].

6 Some uses of $k_{\mathcal{J}}(\tau)$

A general result that uses $k_{\mathcal{J}}(\tau)$ is the adaptation of the Voiculescu theorem [43], to deal with other normed ideals than $\mathcal{K}$. Here is a version of such an adaptation:

> Let $\mathcal{A}$ be a C^*-algebra generated by X_k, $1 \le k \le n$ and let ρ_1, ρ_2 be unital $*$-representations on $\mathcal{H}$ of $\mathcal{A}$, so that $\ker p \circ \rho_j = 0$, $j = 1, 2$. Assume moreover that $k_{\mathcal{J}}(\rho_j(X_k)_{1 \le k \le n}) = 0$, $j = 1, 2$. Then there is a unitary operator U so that
>
> $$U\rho_1(X_k)U^* - \rho_2(X_k) \in \mathcal{J}, \ 1 \le k \le n.$$

Actually, this version which we chose for simplicity is a corollary of a more general absorption version (see [43]). Note also that one can also have that

$$|U\rho_1(X_k)U^* - \rho_2(X_k)|_{\mathcal{J}} < \varepsilon, \ 1 \le k \le n$$

for a given $\varepsilon > 0$.

Remark that the previous result when applied to $\mathcal{A} = C(K)$ where $K \subset \mathbb{R}$ is a compact set and X is the identical function $X(t) = t$ reduces the Weyl–von Neumann–Kuroda theorem to prove the following fact:

> if $T = T^*$ and $\mathcal{J} \neq \mathcal{C}_1$, then $k_{\mathcal{J}}(T) = 0$.

There is also the following general construction for a given n-tuple τ and a normed ideal $\mathcal{J}$.

If $(P_i)_{i \in I}$ are projections in the commutant $(\tau)'$ so that $k_{\mathcal{J}}(\tau \mid P_i\mathcal{H}) = 0$, then also $P = \bigvee_{i \in I} P_i$ is so that $k_{\mathcal{J}}(\tau \mid P\mathcal{H}) = 0$. In particular, there is a largest reducing subspace $\mathcal{H}_s$ of τ, so that $k_{\mathcal{J}}(\tau \mid \mathcal{H}_s) = 0$. The subspace $\mathcal{H}_s$ is called the $\mathcal{J}$-singular subspace of τ, while $\mathcal{H}_s = \mathcal{H} \ominus \mathcal{H}_s$ is called the $\mathcal{J}$-absolutely continuous subspace of τ.

The names given to these reducing subspaces of τ are motivated by the case of commuting Hermitian operators. In particular, in the simplest case of one Hermitian operator $T = T^*$ and $\mathcal{J} = \mathcal{C}_1$, the $\mathcal{C}_1$-singular and $\mathcal{C}_1$-absolutely continuous subspaces of T are precisely the Lebesgue singular and Lebesgue absolutely continuous subspaces of T. The use of k_1 reduces this to the following fact:

> $k_1(T) = 0 \Leftrightarrow$ the spectral measure of T is singular w.r.t. Lebesgue measure.

Note that $\Rightarrow$ relies essentially on the L^2-boundedness of the Hilbert transform.

Much of our initial motivation for developing a machinery based on the invariant $k_{\mathcal{J}}(\tau)$ for studying perturbations of n-tuples of operators was to extend perturbation results for one Hermitian operator to commuting n-tuples of Hermitian operators. For instance, for $n = 2$, which is equivalent to dealing with one normal operator N, there was the problem attributed to P. R. Halmos whether $N = \mathcal{D} + K$ where $\mathcal{D}$ was diagonalizable and $K \in \mathcal{C}_2$. In essence, this problem was whether $\mathcal{C}_2$ plays the same role for pairs of commuting Hermitian operators that $\mathcal{C}_1$ plays for singletons. It turned out that normal operators can be diagonalized $\mod \mathcal{C}_2$ [43], but it is better to go over to n-tuples of commuting Hermitian operators and describe the general results obtained with the modulus of quasicentral approximation.

7 Perturbations of commuting n-tuples of Hermitian operators

It turned out that for n-tuples of commuting Hermitian operators there is a threshold ideal $\mathcal{C}_n^-$. This also explains the $\mod \mathcal{C}_2$ diagonalization of normal operators. Here are some of the main results for such n-tuples [3, 7, 43, 44].

1° If $\mathcal{J}$ is a normed ideal and $\mathcal{J} \supset \mathcal{C}_n^-$, $\mathcal{J} \neq \mathcal{C}_n^-$ and τ is an n-tuple of commuting Hermitian operators then $k_{\mathcal{J}}(\tau) = 0$. In particular, there is a diagonalizable n-tuple δ so that $\tau \equiv \delta \mod \mathcal{J}$.

2° If τ and τ' are n-tuples of commuting Hermitian operators and $\tau \equiv \tau' \mod \mathcal{C}_n^-$, then their Lebesgue absolutely continuous parts τ_{ac} and τ'_{ac} are unitarily equivalent.

3° There is a universal constant $0 < \gamma_n < \infty$ so that if τ is an n-tuple of commuting Hermitian operators, then

$$\left(k_n^-(\tau)\right)^n = \gamma_n \int_{\mathbb{R}^n} m(s)d\lambda(s)$$

where λ is Lebesgue measure and $m(s)$ is the multiplicity function of the Lebesgue absolutely continuous part of τ. If $n = 1$, then $\gamma_1 = 1/\pi$.

The Kato–Rosenblum theorem was also generalized using $k_n^-(\tau)$, but the results (see [44]) are perhaps not complete. For $n \geq 3$ we have a very general result showing that some very general generalized wave operators exist and are actually all equal. For $n = 2$ we get the existence of enough non-trivial intertwiners, but the convergence is not strong. Note that we do not have a proof of the usual

Kato–Rosenblum theorem using $k_1(\tau)$. Its corollary about the unitary equivalence of absolutely continuous parts can be however recovered in case the multiplicity function of one of the operators is integrable by using the formula for k_1 applied to $f(A)$ for C^∞-functions f.

It is an open problem whether the very strong Kato–Rosenblum type results we proved for $n \geq 3$ in [44] also hold for $n = 2$.

The machinery based on $k_{\mathcal{J}}(\tau)$ for dealing with perturbations of commuting n-tuples of Hermitian operators applies as soon as we know the decomposition $\mathcal{H} = \mathcal{H}_a(\tau) \oplus \mathcal{H}_s(\tau)$ for a given normed ideal $\mathcal{J}$. This essentially means to study the n-tuple of multiplication operators by the coordinate functions in $L^2(\mathbb{R}^n, \mu)$ where μ is a more general measure, for instance, more like a Hausdorff p-dimensional measure $1 \leq p < n$, p not necessarily an integer. The essential difficulty is in showing that certain singular integrals give operators in certain normed ideals in order to show that $k_p^-(\tau) > 0$. Rather general results of this kind were obtained in our joint work with Guy David [17]. Here is the key result from [17]:

Let μ be a Radon probability measure with compact support on $\mathbb{R}^n$ so that the Ahlfors condition

$$\mu(B(x,r)) \leq Cr^p, \ \forall x \in \mathbb{R}^n, \ r \leq 1$$

holds for a certain $p > 1$. Let further τ_μ be the n-tuple of multiplication operators by the coordinate functions in $L^2(\mathbb{R}^n, d\mu)$. Then we have:

$$k_p^-(\tau_\mu) > 0.$$

8 $k_p^-(\tau)$ at the endpoint $p = \infty$ and dynamical entropy

In case $\mathcal{J} = \mathcal{C}_\infty^-$, the Macaev ideal, the invariant $k_\infty^-(\tau)$ has remarkable properties [45].

Let τ be an n-tuple of bounded operators. We have:

(1) $k_\infty^-(\tau) < \infty$, *more precisely* $k_\infty^-(\tau) \leq 2\|\tau\| \log(2n+1)$
(2) $k_\infty^-(\tau) = k_\infty^-(\tau \otimes I_{\mathcal{H}})$
(3) *if $\mathcal{J}$ is a normed ideal so that $\mathcal{J} \supset \mathcal{C}_\infty^-$ and $\mathcal{J} \neq \mathcal{C}_\infty^-$ then*

$$k_{\mathcal{J}}(\tau) = 0$$

(4) *if $S_1, \ldots, S_n$ are isometries with orthogonal ranges and $n \geq 2$ then*

$$k_\infty^-(S_1, \ldots, S_n) > 0.$$

We saw that k_p^- for finite integer p is related to p-dimensional Lebesgue measure and somewhat more loosely when p is not an integer to corresponding quantities of Hausdorff dimension p. When $p = \infty$ we have found that instead of a p-dimensional measure there are connections to dynamical entropy. Using k_∞^- a quantity "approximately" equivalent to the Kolmogorov–Sinai dynamical entropy can be obtained [47]. Here is how this *dynamical perturbation entropy* [47] is constructed.

Let θ be a measure-preserving automorphism of a probability measure space (Ω, Σ, μ), $\mu(\Omega) = 1$. Let further U_θ be the unitary operator in $L^2(\Omega, \Sigma, \mu)$ induced by θ and Φ the set of multiplication operators in $L^2(\Omega, \Sigma, \mu)$ by measurable numerical functions which take finitely many values. The dynamical perturbation entropy is defined by the formula

$$\mathcal{H}_p(\theta) = \sup_{\substack{\varphi \subset \Phi \\ \varphi \text{ finite}}} k_\infty^-(\varphi \cup \{U_\theta\}).$$

This is the definition from [49]. It is easy to show that it is equal to the quantity denoted by $\tilde{\mathcal{H}}_P(\theta)$ in [47, 48].

Comparing $\mathcal{H}_P(\theta)$ to the Kolmogorov–Sinai entropy $h(\theta)$ we have the following results [48].

(i) There are universal constants $0 < C_1 < C_2 < \infty$ so that

$$C_1 h(\theta) \le \mathcal{H}_P(\theta) \le C_2 h(\theta).$$

(ii) If θ is a Bernoulli shift, then

$$\mathcal{H}_p(\theta) = \gamma h(\theta)$$

where $0 < \gamma < \infty$ is a universal constant.

It is not known whether (ii) does not actually hold for all θ.

The definition of $\mathcal{H}_P(\theta)$ easily extends to more general non-singular transformations θ for which there may be no equivalent invariant probability measure. It is not known whether $\mathcal{H}_P(\theta)$ is a non-trivial invariant for transformations which are not equivalent to transformations with an invariant probability measure. As pointed out by Lewis Bowen to us, the results of [24] may be relevant to this question.

Weaker than the connection to the Kolmogorov–Sinai entropy, there is also a connection to the Avez entropy of random walks on groups.

Let $\mathcal{G}$ be a group with a finite generator $g_1, \dots, g_n$ and let μ be a probability measure with finite support on $\mathcal{G}$ and let $h(\mathcal{G}, \mu)$ be the Avez entropy of the random walk on $\mathcal{G}$ defined by μ. We have the following result [50]:

If $h(\mathcal{G}, \mu) > 0$ then $k_\infty^-(\lambda(g_1), \dots, \lambda(g_n)) > 0$ where λ is the regular representation of $\mathcal{G}$ on $\ell^2(\mathcal{G})$.

We should also point out to the reader that further results on k_∞^- for Gromov hyperbolic groups $\mathcal{G}$ and for the entropy of subshifts can be found in the papers [32, 33, 34] of Rui Okayasu.

The result about Avez entropy dealt with $k_\infty^-(\lambda(g_1), \dots, \lambda(g_n))$. We shall return to this quantity in the next section where we discuss $k_\mathcal{J}(\lambda(g_1), \dots, \lambda(g_n))$ more generally.

9 Finitely generated groups and supramenability

If $\mathcal{G}$ is a finitely generated group with generator $\gamma = \{g_1, \dots, g_n\}$ and $\mathcal{J}$ is a normed ideal, then, which of the following three possibilities takes place

$$k_\mathcal{J}(\lambda(\gamma)) = 0,$$
$$0 < k_\mathcal{J}(\lambda(\gamma)) < \infty,$$
$$k_\mathcal{J}(\lambda(\gamma)) = \infty$$

does not depend on the choice of the generator γ and is thus an invariant of the group $\mathcal{G}$. In particular if $\mathcal{J} = \mathcal{C}_p^-$, the number $p_0 \in [1, \infty]$ so that $p \in [1, p_0) \Rightarrow k_p^-(\lambda(\gamma)) = 0$ and $p \in (p_0, \infty] \Rightarrow k_p^-(\lambda(\gamma)) = \infty$ is an invariant of $\mathcal{G}$, a kind of dimension.

Here are three examples.

1° [43] If $\mathcal{G} = \mathbb{Z}^n$, then $k_n^-(\lambda(\gamma)) \in (0, \infty)$.
2° [4] If $\mathcal{G}$ is the discrete Heisenberg group of 3×3 upper triangular, unipotent matrices with integer entries, then $k_4^-(\lambda(\gamma)) \in (0, \infty)$.
3° [45] If $\mathcal{G}$ is a free group on $n \geq 2$ generators, then $k_\infty^-(\lambda(\gamma)) \in (0, \infty)$.

In view of the special features of k_∞^- and $\mathcal{C}_\infty^-$ it is natural to wonder for which finitely generated groups is $k_\infty^-(\lambda(\gamma)) = 0$?

Here is what we know [54]:

(i) *if $\mathcal{G}$ has subexponential growth the $k_\infty^-(\lambda(\gamma)) = 0$.*
(ii) *if $k_\infty^-(\lambda(\gamma)) = 0$, then $\mathcal{G}$ is supramenable.*

The fact that subexponential growth insures the vanishing of $k_\infty^-(\lambda(\gamma))$ is easy. The second assertion uses a recent result of Kellerhals–Monod–Rørdam [28] which is not easy. For the reader's convenience we include here a few things about the notion of supramenability introduced by Joseph Rosenblatt [38] (we will stay with finitely generated groups). The group $\mathcal{G}$ is supramenable if for every subset $\emptyset \neq A \subset \mathcal{G}$ there is a left invariant, finitely additive measure on the subsets of $\mathcal{G}$, taking values in $[0, \infty]$, so that $\mu(A) = 1$. In particular supramenable groups are amenable and groups with subexponential growth are supramenable. On the other hand there are amenable groups which are not supramenable. The Kellerhals–Monod–Rørdam theorem establishes the fact that supramenability of $\mathcal{G}$ is equivalent to the fact that there is no Lipschitz embedding of a free group on two generators F_2 into $\mathcal{G}$ with respect to the Cayley graph metric. It is also not known whether supramenability and subexponential growth are not actually equivalent properties.

Concerning the class of finitely generated groups for which $k_\infty^-(\lambda(\gamma))$ vanishes, it is natural to ask whether it coincides with the class of supramenable groups or with the class of groups with subexponential growth, with the possibility that actually all three classes coincide. Note also a fourth condition introduced by Monod [31] quite recently and which could be equivalent to some of the preceding.

Amusingly, there is a certain similarity in what we do not know about supramenability and about vanishing of $k_\infty^-(\lambda(\gamma))$. We do not know whether the supramenability of $\mathcal{G}_1$ and $\mathcal{G}_2$ implies that of $\mathcal{G}_1 \times \mathcal{G}_2$. Similarly, we do not know whether vanishing of k_∞^- for $\mathcal{G}_1$ and $\mathcal{G}_2$ implies this property for $\mathcal{G}_1 \times \mathcal{G}_2$. We should also point out that this is a question specifically for generators of groups, since there are n-tuples τ and τ' so that $k_\infty^-(\tau \otimes I, I \otimes \tau') > 0$, while $k_\infty^-(\tau) = k_\infty^-(\tau') = 0$ (actually $k_p^-(\tau) = k_p^-(\tau') = 0$ for some given $p > 1$) [54].

Finally, we should remark that the questions about $k_{\mathcal{J}}(\lambda(\gamma)) > 0$ are actually questions about functions on $\mathcal{G}$. If $\ell_{\mathcal{J}}(\mathcal{G})$ denotes the symmetrically normed Banach space on $\mathcal{G}$, which identifies with the diagonal operators in $\mathcal{J}$, then $k_{\mathcal{J}}(\lambda(\gamma)) > 0$ is equivalent to

$$0 < \inf\left\{ \max_{1\le j\le n} |f(\cdot) - f(g_j\cdot)|_{\mathcal{J}} \mid f : \mathcal{G} \to \mathbb{R},\ \text{supp}\, f \text{ finite},\ f(e) = 1 \right\}.$$

The quantity appearing above can be viewed as a generalization of Yamasaki hyperbolicity, which is the special case when $\mathcal{J} = \mathcal{C}_p$ [59].

From this point of view it is easy to see that $k_{\mathcal{J}}(\lambda(\gamma)) = 0$ is a property of the Cayley graph of $\mathcal{G}$. Actually, even more:

> *if $\mathcal{G}_1, \mathcal{G}_2$ are groups with finite generators γ_1, γ_2 and $\psi : \mathcal{G}_1 \to \mathcal{G}_2$ is an injection which is Lipschitz with respect to the Cayley graph metrics, then*
>
> $$k_{\mathcal{J}}(\lambda(\gamma_2)) = 0 \Rightarrow k_{\mathcal{J}}(\lambda(\gamma_1)) = 0$$
>
> *(this was used for $\mathcal{J} = \mathcal{C}_\infty^-$ in* [54]*).*

10 The commutant mod a normed ideal $\mathcal{E}(\tau; \mathcal{J})$ and its compact ideal $\mathcal{K}(\tau; \mathcal{J})$

To put more structure around the invariant $k_{\mathcal{J}}(\tau)$, we shall now introduce the commutant mod $\mathcal{J}$ of τ. We shall assume $\tau = \tau^*$.

The commutant mod $\mathcal{J}$ *of the n-tuple of Hermitian operators* τ *is the subalgebra of* $\mathcal{B}(\mathcal{H})$

$$\mathcal{E}(\tau;\mathcal{J}) = \{X \in \mathcal{B}(\mathcal{H}) \mid [X,T_j] \in \mathcal{J},\ 1 \le j \le n\}$$

which is a Banach algebra with isometric involution when endowed with the norm

$$|||X||| = \|X\| + \max_{1\le j\le n} |[T_j,X]|_{\mathcal{J}}.$$

The compact ideal of $\mathcal{E}(\tau;\mathcal{J})$ *is*

$$\mathcal{K}(\tau;\mathcal{J}) = \mathcal{E}(\tau;\mathcal{J}) \cap \mathcal{K}$$

and the Calkin algebra of $\mathcal{E}(\tau;\mathcal{J})$ *is the quotient Banach algebra with involution*

$$\mathcal{E}/\mathcal{K}(\tau;\mathcal{J}) = \mathcal{E}(\tau;\mathcal{J})/\mathcal{K}(\tau;\mathcal{J}).$$

Whether $k_{\mathcal{J}}(\tau)$ vanishes, is finite but > 0 or $= \infty$, often appears among the assumptions when studying properties of $\mathcal{E}(\tau;\mathcal{J})$. Actually these 3 situations can be expressed also in terms of approximate units for the compact ideal $\mathcal{K}(\tau;\mathcal{J})$ [55, 57]. *Assume* $\mathcal{R}$ *is dense in* $\mathcal{J}$.

(a) *The following are equivalent:*

  (i) $k_{\mathcal{J}}(\tau) = 0$
  (ii) *there are* $A_m \in \mathcal{K}(\tau;\mathcal{J})$, $m \in \mathbb{N}$ *so that*

$$\lim_{m\to\infty} |||A_mK - K||| = \lim_{m\to\infty} |||KA_m - K||| = 0$$

  for all $K \in \mathcal{K}(\tau;\mathcal{J})$ *and* $|||A_m||| \le 1$, $m \in \mathbb{N}$.
  (iii) *condition (ii) is satisfied and moreover the* A_m *are finite rank,* $0 \le A_m \le I$ *and* $A_m \uparrow I$ *as* $m \to \infty$.

(b) *The following are equivalent*

  (i) $k_{\mathcal{J}}(\tau) < \infty$
  (ii) *there are* $A_m \in \mathcal{K}(\tau;\mathcal{J})$, $m \in \mathbb{N}$ *so that*

$$\lim_{m\to\infty} |||A_mK - K||| = \lim_{m\to\infty} |||KA_m - K||| = 0$$

  for all $K \in \mathcal{K}(\tau;\mathcal{J})$ *and* $\sup_{m\in\mathbb{N}} |||A_m||| < \infty$.
  (iii) *condition* (ii) *is satisfied and moreover the* A_m *are finite rank* $0 \le A_m \le I$ *and* $A_m \uparrow I$ *as* $m \to \infty$.

In general $\mathcal{E}/\mathcal{K}(\tau;\mathcal{J})$ is only a Banach-algebra and for purely algebraic reasons it identifies with the $*$-subalgebra $p(\mathcal{E}(\tau;\mathcal{J}))$ of the usual Calkin algebra $\mathcal{B}/\mathcal{K}$. However, the connection between $k_{\mathcal{J}}(\tau)$ and approximate unit for $\mathcal{K}(\tau;\mathcal{J})$ has the following somewhat unexpected consequence for $\mathcal{E}/\mathcal{K}(\tau;\mathcal{J})$.

Assuming $\mathcal{R}$ is dense in $\mathcal{J}$, we have

(a) *if $k_{\mathcal{J}}(\tau) = 0$ then $p(\mathcal{E}(\tau;\mathcal{J}))$ is a C^*-subalgebra of $\mathcal{B}/\mathcal{K}$ which is isometrically isomorphic to $\mathcal{E}/\mathcal{K}(\tau;\mathcal{J})$.*
(b) *if $k_{\mathcal{J}}(\tau) < \infty$ then $p(\mathcal{E}(\tau;\mathcal{J}))$ is a C^*-subalgebra of $\mathcal{B}/\mathcal{K}$, which is isomorphic as a Banach algebra to $\mathcal{E}/\mathcal{K}(\tau;\mathcal{J})$ (the norms are equivalent).*
(c) *in particular if $\mathcal{J} = \mathcal{C}_\infty^-$, $p(\mathcal{E}(\tau;\mathcal{C}_\infty^-))$ is always a C^*-algebra canonically isomorphic to the Banach algebra $\mathcal{E}/\mathcal{K}(\tau;\mathcal{J})$ and if $k_\infty^-(\tau) = 0$, the isomorphism is isometric.*

Concerning the analogy with the usual Calkin algebra, we should also mention the result in [6] about the center of certain $\mathcal{E}/\mathcal{K}(\tau;\mathcal{J})$.

11 Banach space dualities

When $k_{\mathcal{J}}(\tau) = 0$ and certain additional conditions are satisfied there are many similarities between $(\mathcal{K}(\tau;\mathcal{J}), \mathcal{E}(\tau;\mathcal{J}))$, and $(\mathcal{K}, \mathcal{B})$ (which can be viewed as the case of $\tau = 0$). For instance, we have the following Banach space duality properties [52, 56]

1° *assuming $\mathcal{R}$ is dense in $\mathcal{J}$ and in $\mathcal{J}^d$ and $k_{\mathcal{J}}(\tau) = 0$, we have that $\mathcal{E}(\tau;\mathcal{J})$ identifies with the bidual of $\mathcal{K}(\tau;\mathcal{J})$*
2° *assuming $\mathcal{J}$ is reflexive and $k_{\mathcal{J}}(\tau) = 0$, the Banach space $\mathcal{E}(\tau;\mathcal{J})$ has unique predual.*

The second property, the uniqueness of predual result, is the analogue of the Grothendieck–Sakai uniqueness [21, 39]. It uses a decomposition into singular and ultraweakly continuous part of the functionals on $\mathcal{E}(\tau;\mathcal{J})$ which is analogous to a theorem of Takesaki [41] for von Neumann algebras. This result then can be used in conjunction with a general result of Pfitzner [36] on unique preduals.

The dual of $\mathcal{K}(\tau;\mathcal{J})$ which is implicit in the above results, under the assumption that $\mathcal{R}$ is dense in $\mathcal{J}$ and that $k_{\mathcal{J}}(\tau) < \infty$ [52, 55] can be identified with $(\mathcal{C}_1 \times (\mathcal{J}^d)^n)/\mathcal{N}$, where $\mathcal{N}$ is the subspace of elements $(x, (y_j)_{1\le j\le n})$ so that $x = \sum_{1\le j\le n}[T_j, y_j]$. The norm on $(\mathcal{C}_1 \times (\mathcal{J}^d)^n)$ is

$$\|(x, (y_j)_{1\le j\le n}\| = \max\left(|x|_1, \sum_{1\le j\le n} |y_j|_{\mathcal{J}^d}\right).$$

The duality pairings arise by mapping $\mathcal{E}(\tau; \mathcal{J})$ (and hence also $\mathcal{K}(\tau; \mathcal{J})$) into $\mathcal{B} \times \mathcal{J}^n$ by

$$X \to (X, ([X, T_j])_{1 \leq j \leq n}).$$

12 Multipliers

If $\mathcal{J}$ is a normal ideal, we recall that $\mathcal{J}^{(0)}$ denotes the closure of $\mathcal{R}$ in $\mathcal{J}$. Let also

$$\tilde{\mathcal{J}} = \left\{ X \in \mathcal{K} \mid \sup_{P \in \mathcal{P} \cap \mathcal{R}} |PX|_{\mathcal{J}} < \infty \right\}$$

where $\mathcal{P}$ is the set of Hermitian projections in $\mathcal{B}$. The sup in the definition of $\tilde{\mathcal{J}}$ is the definition of a norm on $\tilde{\mathcal{J}}$ which is also a normed ideal. The fact that $\mathcal{B}$ is the multiplier algebra (double centralizer) of $\mathcal{K}$ has the following analogue in our setting [52, 55]:

If $\mathcal{J}$ is a normed ideal and τ is an n-tuple of Hermitian operators, so that $k_{\mathcal{J}}(\tau) < \infty$, then $\mathcal{K}(\tau; \mathcal{J}^{(0)})$ is a closed ideal in $\mathcal{E}(\tau; \tilde{\mathcal{J}})$ and $\mathcal{E}(\tau; \tilde{\mathcal{J}})$ identifies with the multiplier algebra of $\mathcal{K}(\tau; \mathcal{J}^{(0)})$.

13 Countable degree −1-saturation

The countable degree −1-saturation property of C^*-algebras of Farah and Hart [18] is a model-theory property which is satisfied by the Calkin algebra and by many other corona algebras. Very roughly, if certain linear relations are satisfied approximately, they can also be satisfied exactly. We have shown in [52] the following:

If $\mathcal{R}$ is dense in $\mathcal{J}$ and $k_{\mathcal{J}}(\tau) = 0$, then $\mathcal{E}/\mathcal{K}(\tau; \mathcal{J})$ has the countable degree −1-saturation property of Farah–Hart.

Based on a result of [10], here is an example of the consequences of degree −1-saturation.

Assume $\mathcal{R}$ is dense in $\mathcal{J}$ and $k_{\mathcal{J}}(\tau) = 0$. If Γ is a countable amenable group and ρ is a bounded homomorphism

$$\rho : \Gamma \to GL(\mathcal{E}/\mathcal{K}(\tau; \mathcal{J}))$$

then ρ is unitarizable, that is, there is $s \in GL(\mathcal{E}/\mathcal{K}(\tau; \mathcal{J}))$ so that

$$s\rho(\Gamma)s^{-1} \subset \mathcal{U}(\mathcal{E}/\mathcal{K}(\tau; \mathcal{J})).$$

14 K-theory aspects

In case $\mathcal{J} = \mathcal{K}$ and $C^*(\tau) \cap \mathcal{K} = \{0\}$, the algebra $\mathcal{E}/\mathcal{K}(\tau; \mathcal{J})$ which is the commutant of $p(C^*(\tau))$ in the Calkin algebra $\mathcal{B}/\mathcal{K}$ is precisely the Paschke dual of $C^*(\tau)$. The first impulse when encountering the algebra $\mathcal{E}/\mathcal{K}(\tau; \mathcal{J})$ is to think that they are some kind of smooth subalgebras of the Paschke dual. Already the fact that when $k_{\mathcal{J}}(\tau) < \infty$, $\mathcal{E}/\mathcal{K}(\tau; \mathcal{J})$ is a C^*-algebra suggests we are dealing with a quite different situation. Results about the K-theory, which show the K-theory can be much richer than that of the Paschke dual make this difference quite clear.

To avoid technicalities, we will deal here with generic examples instead of the more general results in the original papers. We shall consider τ_n the n-tuple of multiplication operators by the coordinate functions in $L^2([0, 1]^n, d\lambda)$, where λ is Lebesgue measure. The C^*-algebra of τ_n is the C^*-algebra of continuous functions on $[0, 1]^n$. Since $[0, 1]^n$ is contractable we infer because of the properties of the Paschke dual construction that the K_0-group of $\mathcal{E}/\mathcal{K}(\tau_n; \mathcal{K})$ is the same as $K_0(\mathcal{B}/\mathcal{K}) = 0$.

We shall denote by $\mathcal{F}_n$ the ordered group of Lebesgue measurable functions $f : [0, 1]^n \to \mathbb{Z}$ up to almost everywhere equality and which are in $L^\infty([0, 1]^n, d\lambda)$, i.e., bounded. Note that $\mathcal{F}_n$ coincides with the ordered group $K_0((\tau_n)')$ where $(\tau_n)'$ is the von Neumann algebra which is the commutant of τ_n.

Example 1 ([55]) There is an order preserving isomorphism

$$K_0\left(\mathcal{E}\left(\tau; \mathcal{C}_1\right)\right) \to \mathcal{F}_1$$

where for each projection P in $\mathcal{M}_n(\mathcal{E}(\tau; \mathcal{C}_1))$ its K_0-class $[P]_0$ is mapped to the multiplicity function of the Lebesgue absolutely continuous part of $P(T_1 \otimes I_n)P$ where $\tau_1 = (T_1)$. Since $k_1(\tau_1) < \infty$ we have that the K-group of $\mathcal{K}(\tau_1; \mathcal{C}_1)$ and $\mathcal{K}$ are equal and since $\mathcal{E}(\tau_1; \mathcal{C}_1)$ contains a Fredholm operator of index 1, we have that

$$K_0(\mathcal{E}\left(\tau_1, \mathcal{C}_1\right)) \simeq K_0(\mathcal{E}/\mathcal{K}(\tau_1, \mathcal{C}_1)).$$

Example 2 ([55]) Assume $\mathcal{J} \neq \mathcal{C}_1$ and $\mathcal{R}$ is dense in $\mathcal{J}$, then we have

$$K_0(\mathcal{E}\left(\tau_1; \mathcal{J}\right)) = 0$$

and this also implies

$$K_0(\mathcal{E}/\mathcal{K}(\tau_1; \mathcal{J})) = 0$$

since $\mathcal{R}$ is dense in $\mathcal{K}(\tau_1, \mathcal{J})$ and $K_1(\mathcal{K}(\tau_1, \mathcal{J})) = 0$.

Example 3 ([55]) If $n \geq 3$ and $\mathcal{J} = \mathcal{C}_n^-$, then we have

$$K_0\left(\mathcal{E}\left(\tau_n, \mathcal{C}_n^-\right)\right) = \mathcal{F}_n \oplus \mathcal{X}_n$$

where the direct summand $\mathcal{X}_n$ is not known. This can also be stated saying that the map

$$K_0\left((\tau_n)'\right) \to K_0\left(\mathcal{E}\left(\tau_n; \mathcal{C}_n^-\right)\right)$$

is an injection and its range is complemented. This uses the results on $\mathcal{C}_n^-$ perturbations of commuting n-tuples of Hermitian operators. Using $k_n^-(\tau_n) < \infty$ we also have that $K_0(\mathcal{E}(\tau_n, \mathcal{C}_n^-) \simeq K_0(\mathcal{E}/\mathcal{K}(\tau_n; \mathcal{C}_n^-))$.

The next example will show that if $n > 1$, there is no analogue of the situation we had when $n = 1$ in Example 2, that is that K_0 be trivial if $\mathcal{J} \supset \mathcal{C}_n^-, \mathcal{J} \neq \mathcal{C}_n^-$.

Example 4 ([51]) Let $n = 2$ and $\mathcal{J} = \mathcal{C}_2$. Then, if P is a projection in $\mathcal{M}_n(\mathcal{E}(\tau_2, \mathcal{C}_2))$, the operator $P((T_1 + iT_2) \otimes I_n)P$ is an operator with trace-class self-commutator and associated with such an operator there is its Pincus principal function $g_{P(T\otimes I_n)P}$ which is in $L^1([0,1]^2, d\lambda)$. The map

$$[P]_0 \to g_{P(T\otimes I_n)P} \in L^1\left([0,1]^2, d\lambda\right)$$

turns out to be well defined and gives a homomorphism

$$K_0(\mathcal{E}(\tau_2, \mathcal{C}_2)) \to L^1\left([0,1]^2, d\lambda\right).$$

One can also show that the range is an uncountable subgroup of $L^1([0,1]^2, d\lambda)$. We refer the reader to [51, 46, 53] for a discussion about how this homomorphism relates $K_0(\mathcal{E}(\tau_2; \mathcal{C}_2))$ to problems on almost normal operators. Note also that the Pincus principal function [9, 30] is related to cyclic cohomology and thus at least some part of $K_0(\mathcal{E}(\tau_2; \mathcal{C}_2))$ is related to cyclic cohomology [14]. We should also point out that the algebras $\mathcal{E}(\tau_2; \mathcal{C}_2)$ are also related to non-commutative potential theory objects [11, 12, 51].

15 The hybrid generalization

In the recent papers [57, 58] we have shown that the machinery we developed for normed ideal perturbations extends to hybrid perturbations that is n-tuples of Hermitian operators τ and τ' such that $T_j - T_j' \in \mathcal{J}_j$ where $(\mathcal{J}_1, \ldots, \mathcal{J}_n)$ is an n-tuple of normed ideals. The surprising feature has been that the extension continues to produce sharp results. We shall illustrate this with a few examples of results for commuting n-tuples of Hermitian operators.

If $\tau = (T_j)_{1\le j\le n}$ is an n-tuple of Hermitian operators and $(\mathcal{J}_1, \dots, \mathcal{J}_n)$ is an n-tuple of normed ideals, then $k_{(\mathcal{J}_1,\dots,\mathcal{J}_n)}(\tau)$ is defined as the smallest $C \in [0, \infty]$ for which there are $A_m \uparrow I$, $0 \le A_m \le I$ finite rank operators so that

$$\lim_{m\to\infty} \max_{1\le j\le n} \|[A_m, T_j]\|_{\mathcal{J}_j} = C.$$

If the n-tuple of ideals is $(\mathcal{C}_{p_1}, \dots, \mathcal{C}_{p_n})$ or $(\mathcal{C}^-_{p_1}, \dots, \mathcal{C}^-_{p_n})$ we also use the notation $k_{p_1,\dots,p_n}(\tau)$ or $k^-_{p_1,\dots,p_n}(\tau)$, respectively.

1° [58] *Let τ and τ' be n-tuples of commuting Hermitian operators and $p_j \ge 1$, $1 \le j \le n$ so that $p_1^{-1} + \dots + p_n^{-1} = 1$ and $T_j - T'_j \in \mathcal{C}^-_{p_j}$ $1 \le j \le n$. Then the absolutely continuous parts τ_{ac} and τ'_{ac} are unitarily equivalent.*

2° [57] *Let $p_j \ge 1$, $1 \le j \le n$ be so that $p_n^{-1} + \dots + p_1^{-1} = 1$. Then there is a universal constant $0 < \gamma_{p_1,\dots,p_n} < \infty$ so that if τ is an n-tuple of commuting Hermitian operators and $m(x)$, $x \in \mathbb{R}^n$ is the multiplicity function of its Lebesgue absolutely continuous part, we have*

$$\left(k^-_{p_1,\dots,p_n}(\tau)\right)^n = \gamma_{p_1,\dots,p_n} \int_{\mathbb{R}^n} m(s)d\lambda(s).$$

16 Unbounded Fredholm modules

Alain Connes ([13], see also [15, 19]) has provided an upper bound for k^-_n based on unbounded Fredholm modules arising in his work on non-commutative geometry.

The unbounded Fredholm module with which one deals here is given by a $*$-algebra of bounded operators on a Hilbert space $\mathcal{H}$ and an unbounded densely defined self-adjoint operator D so that:

$[D, a]$ when $a \in \mathcal{A}$, is densely defined and bounded, and $|D|^{-1} \in \mathcal{J}$, where $\mathcal{J}$ is a normed ideal.

We refer to $(\mathcal{A}, D)$ as an unbounded $\mathcal{J}$-Fredholm-module on $\mathcal{H}$. (From the early papers [13, 14] the terminology has been fluid and other related names like spectral triple, K-cycle have also been used). Here is the Connes estimate:

Let τ be an n-tuple of operators in $\mathcal{A}$, where $(\mathcal{A}, D)$ is a $(\mathcal{C}^-_q)^{dual}$-unbounded Fredholm module, with $q = p(p-1)^{-1}$, then

$$k^-_p(\tau) \le \beta_p \|[D, \tau]\| \left(\mathrm{Tr}_\omega\left(|D|^{-p}\right)\right)^{1/p}$$

where β_p is a universal constant and Tr_ω is the Dixmier trace.

Note that the ideal $(\mathcal{C}^-_q)^{dual}$ is larger than $\mathcal{C}^-_p$, it is actually $\mathcal{C}^+_p$ the (p, ∞) ideal on the Lorenz scale. The estimate fits situations involving pseudodifferential operators D.

Unless some of the unbounded Fredholm module requirements are relaxed one should not expect a perfect fit between existence of Fredholm modules and k_p^-. On the other hand, unbounded Fredholm modules behave well with respect to tensor products, which is not the case for k_p^-. For other results around unbounded Fredholm modules and $k_{\mathcal{J}}$ see also the last part of [45].

For more on the Connes estimate and non-commutative geometry see [13, 15, 19].

17 Sample open problems

Besides the open questions which have come up in our exposition, there are certainly many more. Here are a few we would like to point out.

Problem 1 Find upper and lower bounds for the universal constants γ_n, $n \geq 2$ in the formula for $(k_n^-(\tau))^n$ where τ is an n-tuple of commuting Hermitian operators. More generally the same question for the universal constants $\gamma_{p_1,\ldots,p_n}$ in the hybrid setting is also open. Clearly, it would be of interest to have lower and upper bounds as close to each other.

Problem 2 Does the Farah–Hart degree -1-saturation property still hold for $p(\mathcal{E}(\tau;\mathcal{J}))$ when the assumption $k_{\mathcal{J}}(\tau) = 0$ is replaced by $0 < k_{\mathcal{J}}(\tau) < \infty$? If the answer to the preceding is negative, is there some weaker form of degree -1-saturation of $p(\mathcal{E}(\tau;\mathcal{J}))$ when $0 < k_{\mathcal{J}}(\tau) < \infty$? In particular, it would be of interest to know the answer to these questions in the case when $\mathcal{J} = \mathcal{C}_1$ and τ is a singleton, a Hermitian operator with Lebesgue absolutely continuous spectrum of multiplicity one.

Problem 3 The perturbation entropy of a measure-preserving transformation $\mathcal{H}_P(\theta)$ has a natural generalization [47] to an invariant $\mathcal{H}_P(\theta_1,\ldots,\theta_n)$ of an n-tuple of such transformations

$$\sup_{\substack{\varphi \subset \Phi \\ \varphi \text{ finite}}} k_\infty^-\left(\varphi \cup \{\mathcal{U}_{\theta_1},\ldots,\mathcal{U}_{\theta_n}\}\right).$$

What is the corresponding generalization of the Kolmogorov–Sinai entropy $h(\theta)$ so that

$$\mathcal{H}_P(\theta_1,\ldots,\theta_n) \underset{\frown}{\smile} h(\theta_1,\ldots,\theta_n)?$$

One possible candidate for $h(\theta_1,\ldots,\theta_n)$ could be the supremum over finite partitions $\mathcal{P}$ of the probability measure space of

$$\liminf_{m\to\infty} m^{-1} H(\mathcal{P}_m)$$

where $\mathcal{P}_n$ is defined recursively by $\mathcal{P}_1 = \mathcal{P}$ and $\mathcal{P}_{m+1} = \mathcal{P}_m \vee \theta_1\mathcal{P}_m \vee \cdots \vee \theta_n\mathcal{P}_m$.

Acknowledgements This research was supported in part by NSF Grant DMS-1665534. Part of this paper was written while visiting IPAM in Spring 2018 for the Quantitative Linear Algebra program, which was supported by a NSF grant.

References

1. W. B. Arveson, *Notes on extensions of C^*-algebras*, Duke Math. J. **44** (1977), 329–355.
2. M. F. Atiyah, *Global theory of elliptic operators*, Proc. Internat. Conf. on Functional Analysis and Related Topics (Tokyo 1969), pp. 21–30, Univ. of Tokyo Press, Tokyo, 1970.
3. H. Bercovici and D. V. Voiculescu, *The analogue of Kuroda's theorem for n-tuples*, in: Operator Theory: Advances and Applications, Vol. 41, Birkhäuser, Basel, 1989.
4. D. Bernier, *Quasicentral approximate units for the discrete Heisenberg group*, J. Operator Theory **29** (1993), No. 2, 225–236.
5. B. Blackadar, *K-theory for operator algebras*, Math. Sci. Res. Inst. Publ. 5, Springer, New York, 1986.
6. J. Bourgain and D. V. Voiculescu, *The essential centre of the* mod–*a diagonalization ideal commutant of an n-tuple of commuting Hermitian operators*, in: Noncommutative Analysis, Operator Theory and Applications, 77–80, Oper. Theory Adv. Appl. **252**, Linear Oper. Linear Syst., Birkhäuser/Springer, 2016.
7. L. G. Brown, R. G. Douglas, and P. A. Fillmore, *Unitary equivalence modulo the compact operators and extensions of C^*-algebras*, in: Lecture Notes in Math., Vol. 345, Springer Verlag, 1984, pp. 58–127.
8. L. G. Brown, R. G. Douglas, and P. A. Fillmore, *Extensions of C^*-algebras and K-homology*, Ann. of Math. (2) **105** (1977), 265–324.
9. R. W. Carey and J. D. Pincus, *Commutators, symbols and determining functions*, J. Funct. Anal. **19** (1975), 50–80.
10. Y. Choi, I. Farah and N. Ozawa, *A non-separable amenable operator algebra which is not isomorphic to a C^*-algebra*, Forum Math. Sigma 2 (2014), 12 pp.
11. F. Cipriani, *Dirichlet forms of non-commutative spaces*, in: Lecture Notes in Math., Vol. 1954, Springer Verlag, 2008, pp. 161–276.
12. F. Cipriani and J.–L. Sauvageot, *Derivations and square roots of Dirichlet forms*, J. Funct. Anal. **13**(3), (2003), 521–545.
13. A. Connes, *Trace de Dixmier, modules de Fredholm et géometrie riemannierre*, Nucl. Phys. B, Proc. Suppl. 5B (1988), 65–70.
14. A. Connes, *Non-commutative differential geometry*, IHES Publ. Math., Vol. 62, 1985, pp. 257–360.
15. A. Connes, "Non-commutative geometry", Academic Press, Inc., San Diego, CA, 1994.
16. A. Connes, *On the spectral characterization of manifolds*, J. Noncommut. Geom. **7** (2013), No. 1, 1–82.
17. G. David and D. Voiculescu, *s-numbers of singular integrals for the invariance of absolutely continuous spectra in fractional dimension*, J. Funct. Anal. **94** (1990), 14–26.
18. I. Farah and B. Hart, *Countable saturation of corona algebras*, C. R. Math. Rep. Acad. Sci., Canada **35**(2) (2013), 35–56.
19. J. M. Garcia–Bondia, J. C. Varilly and H. Figueroa, "Elements of Non-commutative Geometry", Birkhäuser Advanced Texts: Baseler Lehrbucher, Birkhäuser, 2001.

20. I. C. Gohberg and M. G. Krein, *Introduction to the theory of non-self-adjoint operators*, Translations of Mathematical Monographs, Vol. 18, AMS, Providence, RI, 2005.
21. A. Grothendieck, *Une characterization vectorielle des espace* L^1, Canadian J. of Math. **7**(1955), 552–561.
22. J. W. Helton and R. Howe, *Integral operators, commutator traces, index and homology*, in: Proceedings of a Conference on Operator Theory, in: Lecture Notes in Math., Vol. 345, Springer Verlag, 1973, pp. 141–209.
23. N. Higson and J. Roe, "Analytic K-Homology", Oxford University Press, 2004.
24. M. Hochman, *Every Borel automorphism without finite invariant measures admits a two-set generator*, arXiv: 1508.02335.
25. J. Kaminker and C. L. Schochet, *Spanier–Whitehead* K*-duality for* C^**-algebras*, arXiv: 1609.00409.
26. G. G. Kasparov, *The operator* K*-functor and extensions of* C^**-algebras*, Math. USSR Izvestiya (English translation) **16** (1981), 513–572.
27. T. Kato, *Perturbation theory for linear operator*, Classics of Mathematics, Springer Verlag, Berlin, 1995.
28. J. Kellerhals, N. Monod and M. Rørdam, *Non-supramenable groups acting on locally compact spaces*, Doc. Math. **18** (2013), 1597–1626.
29. J. Lindenstrauss and L. Tzafriri, "Classical Banach Spaces I and II", Springer, 2014.
30. M. Martin and M. Putinar, *Lectures on hyponormal operators*, in: Operator Theory: Advances and Applications, Vol. 39, Birkhäuser, 1989.
31. N. Monod, *Fixed points in convex cones*, arXiv: 1701.05537.
32. R. Okayasu, *Entropy of subshifts and the Macaev norm*, J. Math. Soc. Japan, Vol. 56, No. 1 (2004), 177–191.
33. R. Okayasu, *Gromov hyperbolic groups and the Macaev norm*, Pacific J. Math., Vol. 223, No. 1 (2006), 141–157.
34. R. Okayasu, *Perturbation theoretic entropy of the boundary actions of free groups*, Proc. Amer. Math. Soc. **134** (2006), 1771–1776.
35. W. L. Paschke, K*-theory for commutants in the Calkin algebra*, Pacific J. Math. **95** (1981), 427–434.
36. H. Pfitzner, *Separable L-embedded Banach spaces are unique preduals*, Bull. London Math. Soc. **39** (2007).
37. M. Reed and B. Simon, *Methods of modern mathematical physics*, Vol. III: Scattering theory, Academic Press, 1979.
38. J. M. Rosenblatt, *Invariant measures and growth conditions*, Trans. Amer. Math. Soc. **193** (1974), 33–53.
39. S. Sakai, *A characterization of* W^**-algebras*, Pacific J. of Mathematics **6** (1956), 763–773.
40. B. Simon, "Trace Ideals and Their Applications", 2nd Ed., Mathematical Surveys and Monographs **120**, AMS, Providence, RI, 2005.
41. M. Takesaki, *On the conjugate space of an operator algebra*, Tohoku Math. J. **10** (1958), 194–203.
42. D. V. Voiculescu, *A non-commutative Weyl–von Neumann theorem*, Rev. Roumaine Math. Pures Appl. **21** (1976), 97–113.
43. D. V. Voiculescu, *Some results on norm-ideal perturbations of Hilbert space operators* I, J. Operator Theory **2** (1979), 3–37.
44. D. V. Voiculescu, *Some results on norm-ideal perturbations of Hilbert space operators* II, J. Operator Theory **5** (1981), 77–100.
45. D. V. Voiculescu, *On the existence of quasicentral approximate units relative to normed ideals* I, J. Funct. Anal. **91**(1) (1990), 1–36.
46. D. V. Voiculescu, *Almost normal operators* mod γ_p, in: V. P. Havin, S. V. Hruscev, N. K. Nikoski (Eds.), Linear and Complex Analysis Problem Book, Lecture Notes in Math., Vol. 1043, Springer Verlag, 1984, pp. 227–230.
47. D. V. Voiculescu, *Entropy of dynamical systems and perturbations of operators* I, Ergodic Theory and Dynamical Systems **11** (1991), 779–786.

48. D. V. Voiculescu, *Entropy of dynamical systems and perturbations of operators* II, Houston Math. J. **17** (1991), 651–661.
49. D. V. Voiculescu, *Perturbations of operators, connections with singular integrals, hyperbolicity and entropy*, in: Harmonic Analysis and Discrete Potential Theory (Frascati, 1991), 181–191, Plenum Press, New York, 1992.
50. D. V. Voiculescu, *Entropy of random walks on groups and the Macaev norm*, Proc. AMS **119** (1993), 971–977.
51. D. V. Voiculescu, *Almost normal operators* mod *Hilbert–Schmidt and the K-theory of the algebras $E\Lambda(\Omega)$*, J. Noncommutative Geom. **8**(4) (2014), 1123–1145.
52. D. V. Voiculescu, *Countable degree -1 saturation of certain C^*-algebras which are coronas of Banach algebras*, Groups Geom. Dyn. **8** (2014), 985–1006.
53. D. V. Voiculescu, *Some C^*-algebras which are coronas of non-C^*-Banach algebras*, J. Geom. and Phys. **105** (2016), 123–129.
54. D. V. Voiculescu, *A remark about supramenability and the Macaev norm*, arXiv: 1605.02135.
55. D. V. Voiculescu, *K-theory and perturbations of absolutely continuous spectra*, arXiv: 1606.00520.
56. D. V. Voiculescu, *Lebesgue decomposition of functionals and unique preduals for commutants modulo normed ideals*, Houston J. Math. **43** (2017), No. 4, 1251–1262.
57. D. V. Voiculescu, *Hybrid normed ideal perturbations of n-tuples of operators* I, J. Geom. Phys. **128** (2018), 169–184.
58. D. V. Voiculescu, *Hybrid normed ideal perturbations of n-tuples of operators* II *weak wave operators*, arXiv: 1801.00490.
59. M. Yamasaki, *Parabolic and hyperbolic infinite networks*, Hiroshima Math. J. (1977), 135–146.

Quantum field theory on noncommutative spaces

Raimar Wulkenhaar

Abstract This survey tries to give a rigorous definition of Euclidean quantum field theory on a fairly large class of noncommutative geometries, namely nuclear AF Fréchet algebras. After a review of historical developments and current trends we describe in detail the construction of the Φ^3-model and explain its relation to the Kontsevich model. We review the current status of the construction of the Φ^4-model and present in an outlook a possible definition of Schwinger functions for which the Osterwalder–Schrader axioms can be formulated.

1 Introduction

1.1 Quantum field theory and gravity

History of sciences culminates in the discovery that the enormous complexity of structures observed on earth and in its nearby part[1] of the universe derives, through a hierarchy of models, from a tiny set of rules that we call the standard model of particle physics coupled to Einstein gravity. This standard model is described elsewhere in this collection of surveys. Here we stress that it has to be built in two stages: The first stage is classical field theory, which has an elegant mathematical formulation in terms of (traditional or noncommutative) differential geometry. The dynamics is governed by field equations which can be derived from an action functional. The second stage is quantisation, i.e. the implementation of field equations between operators on Hilbert space which satisfy natural axioms. *This is not yet achieved.* It is true that remarkable approximations have been

[1]The large-scale dynamics of the universe seems to require 'dark matter' which is not at all understood.

R. Wulkenhaar (✉)
Mathematisches Institut der Westfälischen Wilhelms-Universität, Münster, Germany
e-mail: raimar@math.uni-muenster.de

A. Chamseddine et al. (eds.), *Advances in Noncommutative Geometry*,
https://doi.org/10.1007/978-3-030-29597-4_11

established, for instance, the agreement to 11 significant digits between quantum field theoretical prediction and measurement of the magnetic moment of the electron (classical field theory agrees to one digit with measurement).

The problem is twofold. There is first the challenge to construct an interacting quantum field theory on 4-dimensional Euclidean or Minkowskian space, a difficult mathematical problem. Several approaches seem possible; we give more details in Section 2.

However, even if one of these programmes succeeds, there remains a profound physical problem: gravity has to be taken into account. We should distinguish at least four levels:

1. Gravity is ignored, the universe is flat Minkowski space. See above.
2. Quantum field theory in an external classical gravitational background field, i.e. on a curved Lorentz manifold, but without any back-reaction of the quantum field theoretical objects to the manifold. The local quantum field theory approach which goes back to Haag–Kastler can cope with this generality. The free quantum field on a general Lorentz manifold is under control.
3. Gravity is still described by classical Einstein theory in which quantum field theoretical objects influence the metric via the stress-energy tensor. Discussed below.
4. True quantum gravity. A driving force in contemporary mathematical physics, with many ideas on the market. Work on these programmes has produced spectacular mathematical results and will continue to do so. But a solution is not in sight. We will not discuss it in this survey.

This survey is about a conceptional problem which arises already in level 3. Quantum fields are operator-valued distributions, smeared over the support of a test function (see Section 2). How 'small' can we make the support? According to Heisenberg's principle, the extension Δx of support in position space and the extension Δp of support in momentum space are correlated by $\Delta x \Delta p \geq \hbar/2$ (where $\hbar$ is Planck's constant). We can sharply localise Δx at expense of undetermined momentum. In a certain sense, it is this momentum uncertainty which manifests itself in the experimentally confirmed quantum corrections. However, all this breaks down in dynamical gravity. A momentum uncertainty Δp comes with an uncertainty in the stress-energy tensor which, by Einstein's field equation, creates an uncertainty in the metric tensor. For a rough estimate we translate $\Delta p = c\Delta m$ in a mass uncertainty (c is the speed of light) which induces an uncertainty $\Delta x_s = \frac{2G\Delta m}{c^2}$ of the Schwarzschild radius (where G is Newton's constant). Its influence on the original geometry in which we localised the support of our quantum field to Δx can only be ignored as long as

$$\Delta x \gg \Delta x_s = \frac{2G}{c^3}\Delta p > \frac{G\hbar}{c^3}\frac{1}{\Delta x}.$$

In other words, we cannot localise the support of quantum fields better than the Planck length $\ell_P = \sqrt{\frac{G\hbar}{c^3}}$ if (classical!) gravity is taken into account.

These restrictions on the localisability of quantum fields must be incorporated into quantum field theory itself. This is what quantum field theories on noncommutative geometries try to do.

1.2 Noncommutativity

We know from quantum mechanics that any measurement uncertainty (enforced by principles of Nature and not due to lack of experimental skills) goes hand in hand with noncommutativity. To the best of my knowledge, the possibility that geometry loses its meaning in quantum physics was first[2] considered by Schrödinger [Sch34]. On the other hand, Heisenberg suggested to use coordinate uncertainty relations to ameliorate the short-distance singularities in the quantum theory of fields. His idea (which appeared later [Hei38]) inspired Peierls in the treatment of electrons in a strong external magnetic field [Pei33]. Via Pauli and Oppenheimer the idea came to Snyder who was the first to write down uncertainty relations between coordinates [Sny47]. The mutual interaction of quantum-mechanical and gravitational disturbances was first discussed by Wheeler [Whe55] in his model of 'geons'.

The uncertainty relations for coordinates were revived by Doplicher et al. [DFR95] as a means to avoid gravitational collapse when localising events with extreme precision. According to [DFR95], the coordinate uncertainties Δx^μ have to satisfy $\Delta x^0(\Delta x^1 + \Delta x^2 + \Delta x^3) \geq \ell_P^2$ and $\Delta x^1 \Delta x^2 + \Delta x^2 \Delta x^3 + \Delta x^3 \Delta x^1 \geq \ell_P^2$. These uncertainty relations are induced by noncommutative coordinate operators $\hat{x}^\mu = (\hat{x}^\mu)^*$ satisfying $[\hat{x}^\mu, \hat{x}^\nu] = \mathrm{i}\hat{\Theta}^{\mu\nu}$, where $\hat{\Theta}^{\mu\nu}$ are the components of a 2-form $\hat{\Theta}$ which is central and normalised to $\langle \hat{\Theta}, \hat{\Theta} \rangle = 0$ and $\langle \hat{\Theta}, *\hat{\Theta} \rangle = 8\ell_p^4$. Moreover, in [DFR95] first steps are taken towards a perturbative quantum field theory on the resulting (Minkowskian) quantum space-time.

The previous discussion suggests that space itself, and not only the phase space of quantum mechanics, should be noncommutative. The corresponding mathematical framework of *noncommutative geometry* [Con94] has been developed. It is today an integral part of mathematics. Noncommutative geometry is the reformulation of geometry and topology in an algebraic and functional-analytic language, thereby permitting an enormous generalisation. Many of its facets are presented in this collection of surveys.

1.3 Structure of the survey

The main structures and techniques in noncommutative geometry are already presented in other surveys so that we will not repeat them. We therefore start

[2] Actually, Riemann himself speculated in his famous Habilitationsvortrag [Rie92] about the possibility that the hypotheses of geometry lose their validity in the infinitesimal small.

in Section 2 with an informal introduction into basic concepts of quantum field theory (QFT) in its traditional sense, i.e. on ordinary Minkowskian or Euclidean space. After a sketch of the Wightman axioms we give a few details of the Euclidean formulation of QFT. In this framework we describe the crucial concept of renormalisation. For completeness we also sketch Feynman graphs, but our philosophy is to avoid them.

Section 3 generalises the Euclidean formulation to a class of noncommutative geometries. We argue that—to implement renormalisation—this should be the class of nuclear AF Fréchet algebras. They are the analogue of AF C^*-algebras, but with closure in a locally-convex topology rather than in norm topology. Our construction heavily uses two classical theorems: The Bochner–Minlos theorem 2.3 provides us with a measure on the space of Euclidean quantum fields. The Kōmura-Kōmura theorem 3.3 achieves the coexistence of QFT on a discrete noncommutative algebra with an apparently continuous universe.

We describe in Section 4 a couple of examples of such noncommutative geometries and also mention a few other popular geometries which are not of this class. One should retain from these discussions that we work with sequences of matrix algebras. As such, QFT on noncommutative geometry is closely related to matrix models. After Sections 3 and 4 the reader may directly jump to Section 7 where we introduce the main techniques and a prominent example in a class of matrix models we have in mind. In between we try to give in Section 5 a review of more physically oriented work on noncommutative quantum field theories. In Section 6 we give more details on a particular direction which branched on one hand into the axiomatic formulation of Section 3 and on the other hand into a novel approach to quantum gravity. Both Sections 5 and 6 are nearly independent from the others and could be skipped or exclusively read.

After a preparation on matrix models in Section 7, we present in Sections 8 and 9 two Euclidean QFT models on noncommutative geometries for which our programme succeeded completely (Section 8) or at least partially (Section 9). In a more speculative Section 10 we outline how a QFT on noncommutative geometry could possibly be projected onto a true QFT on Minkowski space. That section should be read with caution; the sketched ideas might go into a void direction.

1.4 Disclaimer

This survey has severe limitations. They are partly unavoidable because several lines of research were developed in parallel, whereas only a sequential presentation can be given. More severely, since an enormous amount of publications has been produced, we can only review a tiny fraction. Our apologies go to everyone who is not adequately acknowledged.

2 Quantum field theory

2.1 Axiomatic and algebraic quantum field theory

It is fair to say that noncommutative geometry, operator algebras and quantum field theory have a common root in von Neumann's axiomatic characterisation of quantum mechanics [vN32]. Quantum field theory (QFT) is an extension of quantum mechanics to infinitely many degrees of freedom which at the same time takes care of interactions between particles mediated by quantised fields. In fact the distinction between particles and fields is abandoned in favour of a unifying framework. The first spectacular confirmation of the new concept was Bethe's explanation [Bet47] of the Lamb shift.

In the early 1950s, Gårding and Wightman gave a rigorous mathematical foundation to quantum field theory by casting the unquestionable physical principles (locality, covariance, stability, unitarity) into a set of axioms. These ideas were published years later [Wig56, WG64, SW64]. There are several possibilities to group the data, but essentially one wants for a single scalar field:

Definition 2.1 A *scalar quantum field* φ on D-dimensional Minkowski space $\mathbb{R}^{1,D-1}$ is an unbounded operator-valued distribution, i.e. $\varphi(f), \varphi^*(f) : \mathcal{D} \to \mathcal{D}$ are linear on a dense subspace $\mathcal{D} \subset \mathcal{H}$ of a Hilbert space, for any test function $f \in \mathcal{S}(\mathbb{R}^{1,D-1})$. Moreover:

1. *Covariance.* There is a representation of the Poincaré group $\mathcal{P}_+^\uparrow \ni (t, R)$ by unitaries $U(t, R)$ in $\mathcal{H}$, which preserves $\mathcal{D}$ and satisfies $U(t, R)\varphi(f)U(t, R)^{-1} = \varphi(f_{t,R})$ with $f_{t,R}(x) := f(R^{-1}(x - t))$.
2. *Spectrum condition.* The joint spectrum of the generators of the translation subgroup of $\mathcal{P}_+^\uparrow$ lies in the forward lightcone $V_+ = \{(p_0, \mathbf{p}) \in \mathbb{R}^{1,D-1} : p_0 \geq 0,\ p_0^2 \geq \|\mathbf{p}\|^2\}$.
3. *Locality.* For f, g causally independent, $[\varphi(f), \varphi(g)] = 0$.

It is convenient to require that the subspace of $\mathcal{P}_+^\uparrow$-invariant vectors in $\mathcal{D}$ is 1-dimensional, and that for a $\mathcal{P}_+^\uparrow$-invariant unit vector Ω (the vacuum), the generated subspace span(polynomials($\varphi(f_i), \varphi(f_j)^*)\Omega$) is dense in $\mathcal{H}$.

From these data one builds the *Wightman functions*, i.e. vacuum expectation values of field operators

$$(\mathcal{S}(\mathbb{R}^{1,D-1}))^N \ni (f_1, \ldots, f_N) \mapsto W(f_1, \ldots, f_N) := \langle \Omega, \Phi(f_1) \cdots \Phi(f_N)\Omega \rangle. \tag{2.1}$$

They are also called N-point functions or correlation functions. The Wightman axioms induce covariance, locality, positivity, spectrum and cluster properties for the Wightman functions. Conversely, Wightman's reconstruction theorem allows to reconstruct the data of Definition 2.1 from Wightman functions with these properties. Some fundamental theorems such as PCT theorem and spin-statistics theorem can be proved in this framework. See [SW64]. The Wightman theory is the basis for a rigorous theory of scattering processes (Haag-Ruelle theory [Haa58, Rue62])—and this is exactly what the large accelerator facilities detect.

Unfortunately, it turned out very difficult to provide examples richer than the free field which satisfy these (very natural) Wightman axioms. This difficulty motivated the development of equivalent or more general frameworks. In particular, the Wightman axioms are not made for gauge fields so that generalisation is indeed necessary. One powerful generalisation is Algebraic QFT (or better *Local QFT*) which shifts the focus from the field operators to the Haag–Kastler net of operator algebras assigned to open regions in space-time [HK64]. Fields merely provide coordinates on the algebra. This has the advantage to work with (C^*, von Neumann) algebras of bounded operators where powerful mathematical tools are available. Over the years this point of view turned out to be very fruitful [Haa96]. It is, in particular, possible to describe quantum field theory on curved space-time [BFV03].

2.2 *Euclidean QFT*

As consequence of the spectrum condition 2 in Definition 2.1, Wightman functions (2.1) admit an analytic continuation in time. At purely imaginary time they become the Schwinger functions [Sch59] of a Euclidean quantum field theory. Symanzik emphasised the powerful Euclidean-covariant functional integral representation [Sym64]. In this way the Schwinger functions become the moments of a statistical physics probability distribution. This tight connection between Euclidean quantum field theory and statistical physics led to a fruitful exchange of concepts and methods, most importantly that of the renormalisation group [WK74].

It is sometimes possible to rigorously prove the existence of a Euclidean quantum field theory or of a statistical physics model without knowing or using that this model derives from a true relativistic quantum field theory. Sufficient conditions on the Euclidean model which guarantee the Wightman axioms were first given by Nelson [Nel73b, Nel73a]. These conditions based on Markoff fields turned out to be too strong or inconvenient. Shortly later, Osterwalder and Schrader established a set of axioms [OS73, OS75] by which the Euclidean quantum field theory is (up to a regularity subtlety) equivalent to a Wightman theory. In simplified terms, the following data are necessary:

Definition 2.2 Let $\mathcal{S}_{N0} \subset \mathcal{S}(\mathbb{R}^{ND})$ be the subspace of test functions which vanish, with all derivatives, on any diagonal $x_i = x_j$, for $1 \leq i < j \leq N$. For $x_i =: (x_i^0, \mathbf{x}_i) \in \mathbb{R}^D$, let $\mathcal{S}_{N0+} \subset \mathcal{S}_{N0}$ be the subspace of test functions with support in the cone $\{x_i^0 \geq 0\}$ for all $i = 1, \dots, N$. Moreover, we let $f^\sigma(x_1, \dots, x_N) := f(x_{\sigma(1)}, \dots, x_{\sigma(N)})$ be a permutation and $f^r(x_1, \dots, x_N) := f((-x_1^0, \mathbf{x}_1), \dots, (-x_N^0, \mathbf{x}_N))$ be the reflection of all first components.

A *Euclidean quantum field theory* consists of a family $\{S_N\}$ of *Schwinger N-point distributions*, where S_N is a linear functional on $\mathcal{S}_{N0}$ which satisfies

1. *Euclidean invariance.* $S_N(f) = S_N(f_{t,R})$ for any $f \in \mathcal{S}_{N0}$ and $(t, R) \in \mathbb{R}^D \rtimes \mathrm{SO}(D)$, where $f_{t,R}(x_1, \dots, x_N) = f(R^{-1}(x_1 - t), \dots, R^{-1}(x_N - t))$.
2. *Reflection positivity.* For any tuple $(f_0, f_1, \dots, f_K)$ of $f_k \in \mathcal{S}_{k0+}$, one has $\sum_{k,l=1}^{K} S_{k+l}(\overline{f_k^r} \times f_l) \geq 0$.
   [here, $(\overline{f_k^r} \times f_l)(x_1, \dots, x_{k+l}) = \overline{f_k^r}(x_1, \dots, x_k) f_l(x_{k+1}, \dots, x_{k+l})$]
3. *Symmetry.* $S_N(f) = S_N(f^\sigma)$ for any $f \in \mathcal{S}_{N0}$ and any permutation σ.

The Osterwalder–Schrader theorem asserts that Schwinger functions according to Definition 2.2 of factorial growth (i.e. $|S_N(f)| \leq c^N (N!)^L \|f\|_{\mathcal{S}_{N0}}$ for some seminorm defining $\mathcal{S}_{N0}$) are Laplace–Fourier transforms of Wightman functions in a relativistic quantum field theory. The properties of the vacuum Ω follow if the Schwinger functions cluster, $\lim_{t\to\infty} S_{k+l}((f_k)_{t,1} \times f_l) = S_k(f_k) S_l(f_l)$.

2.3 The free Euclidean scalar field

We describe in some detail the free Euclidean scalar field because it serves as example for the construction in the noncommutative setting. We start from the Schwinger 2-point function and its corresponding Schwinger distribution

$$S_2(x, y) := \int_{\mathbb{R}^D} \frac{dq}{(2\pi)^D} \frac{e^{\mathrm{i}\langle x-y, q\rangle}}{\|q\|^2 + \mu^2}, \quad S_2(f \times g) = \int_{\mathbb{R}^{2D}} d(x, y)\, S_2(x, y) f(x) g(y). \tag{2.2}$$

It satisfies reflection positivity on tuples $(0, f, 0, \dots, 0)$, where $f \in \mathcal{S}_{10+}$:

$$\begin{aligned} &S_2(\overline{f^r} \times f) \qquad (2.3)\\ &= \int_{\mathbb{R}^{2D}} d(x, y) \int_{\mathbb{R}^D} \frac{dq}{(2\pi)^D} \frac{e^{\mathrm{i}(x^0-y^0)q^0+\mathrm{i}\langle \mathbf{x}-\mathbf{y}, \mathbf{q}\rangle}}{\|q\|^2 + \mu^2} \overline{f(-x^0, \mathbf{x})} f(y_0, \mathbf{y})\\ &= \int_{(\mathbb{R}_+)^2} d(x^0, y^0) \int_{-\infty}^{\infty} dq^0 \int_{\mathbb{R}^{3(D-1)}} \frac{d(\mathbf{x}, \mathbf{y}, \mathbf{q})}{(2\pi)^D} \frac{e^{-\mathrm{i}(x^0+y^0)q^0+\mathrm{i}\langle \mathbf{x}-\mathbf{y}, \mathbf{q}\rangle}}{q_0^2 + \|\mathbf{q}\|^2 + \mu^2} \overline{f(x^0, \mathbf{x})} f(y_0, \mathbf{y})\\ &= \int_{\mathbb{R}^{D-1}} \frac{d\mathbf{q}}{(2\pi)^{D-1} \cdot 2\omega_\mu(\mathbf{q})} \left| \int_0^\infty dx_0 \int_{\mathbb{R}^{D-1}} d\mathbf{x}\, e^{-x^0 \omega_\mu(\mathbf{q}) - \mathrm{i}\langle \mathbf{x}, \mathbf{q}\rangle} f(x_0, \mathbf{x}) \right|^2 \geq 0, \end{aligned}$$

where $\omega_\mu(\mathbf{q}) := \sqrt{\|\mathbf{q}\|^2 + \mu^2}$. Here, from the 2nd to 3rd line, after $x^0 \mapsto -x^0$, the support property of $f \in \mathcal{S}_{10+}$ has been used. This allows to evaluate the q_0-integral via the residue theorem, resulting in the last line.

Next we introduce one of our most important tools:

Theorem 2.3 (Bochner–Minlos) *Let X be a real nuclear vector space. Let a continuous map $\mathcal{F} : X \to \mathbb{R}$ with $\mathcal{F}(0) = 1$ be of positive type, i.e. for any*

$x_1, \dots, x_K \in X$ *and* $c_1, \dots, c_K \in \mathbb{C}$ *one has* $\sum_{i,j=1}^{K} c_i \bar{c}_j \mathcal{F}(x_i - x_j) \geq 0$. *Then there exists a unique Radon probability measure* $\mathrm{d}\mathcal{M}$ *on the dual space* X' *with*

$$\mathcal{F}(x) = \int_{X'} \mathrm{e}^{\mathrm{i}\phi(x)} \mathrm{d}\mathcal{M}(\phi). \tag{2.4}$$

The theorem was proved in this generality by Minlos [Min59]. A proof can be found, e.g. in [GJ87, §A6], starting from Bochner's theorem [Boc32] which covers the case $X = X' = \mathbb{R}$.

Recall that the Schwartz spaces $\mathcal{S}(\mathbb{R}^D)$ are nuclear. Consider for $f \in \mathcal{S}_{10+}$ the continuous functional $\mathcal{F}(f) = \exp(-\frac{1}{2}S_2(f \times f))$ defined by the Schwinger 2-point function (2.2). Then

$$\sum_{i,j=1}^{K} c_i \overline{c_j} \mathcal{F}(f_i - f_j) = \sum_{i,j=1}^{K} \sum_{n=0}^{\infty} C_i \overline{C_j} \frac{S_2(f_i \times f_j)^n}{n!} \tag{2.5}$$

with $C_i := c_i \mathrm{e}^{-\frac{1}{2}S_2(f_i \times f_i)}$. Since $\langle f_i, f_j \rangle = S_2(f_i \times f_i)$ has all properties of a scalar product, $(S_2(f_i \times f_j))_{ij}$ is a positive definite Gram matrix. By the Schur product theorem, $(S_2(f_i, f_j))^n$ is, as Hadamard product of positive matrices, again positive.

In this way we have constructed out of a Schwinger 2-point function (2.2) a measure $\mathrm{d}\mathcal{M}(\phi)$ on the space $(\mathcal{S}(\mathbb{R}^D))'$ of *Euclidean quantum fields*. It gives rise to Schwinger N-point distributions via

$$\begin{aligned} S_N(f_1 \times \cdots \times f_N) &:= \int_{(\mathcal{S}(\mathbb{R}^D))'} \phi(f_1) \cdots \phi(f_N)\, \mathrm{d}\mathcal{M}(\phi) \\ &= (-\mathrm{i})^N \frac{\partial^N}{\partial t_1 \cdots \partial t_N} \mathcal{F}(t_1 f_1 + \cdots + t_N f_N)\Big|_{t_i=0}. \end{aligned} \tag{2.6}$$

The family $\{S_N\}$ satisfies all Osterwalder–Schrader axioms.

2.4 The interacting scalar field

For a polynomial P bounded from below, we would like to define an interacting scalar QFT by a 'deformed measure' on $(S(\mathbb{R}^D))'$:

$$\mathrm{d}\mathcal{M}_{\mathrm{int}}(\phi) := \frac{\mathrm{d}\mathcal{M}(\phi)\ \exp\big(-\int_{\mathbb{R}^D} dx\ \lambda(x) P(\phi(x))\big)}{\int_{X'} \mathrm{d}\mathcal{M}(\phi)\ \exp\big(-\int_{\mathbb{R}^D} dx\ \lambda(x) P(\phi(x))\big)}, \tag{2.7}$$

where $\mathrm{d}\mathcal{M}(\phi)$ is the previous Bochner–Minlos measure and λ a test function which, to achieve Euclidean invariance, eventually is sent to a coupling constant. The 'definition' (2.7) has *very many* problems. It is motivated by the successful quantum-mechanical Feynman–Kac formula [Kac49] which constructs deformations of the Wiener measure on the space of Hölder-continuous paths.

The problems with (2.7) are related to the fact that the pointwise product of distributions is not necessarily a distribution. Therefore, the integral in (2.7), and integrals involving $d\mathcal{M}_{\text{int}}(\phi)$, are meaningless; we have to modify the rules. Many such modification procedures are known; they are (or should be) all equivalent. The following steps are typical:

Programme 2.4

1. *Infrared regularisation.* Restrict $\mathbb{R}^D$ to a compact subset $\mathcal{K} \subset \mathbb{R}^D$. If $\mathcal{K}$ is a cube, the Schwinger 2-point function (2.2) will contain a sum over discrete $\{q_n\}$ instead of an integral. After integrating over $\mathcal{K}$, now safe, the summation over $\{q_n\}$ might still diverge. Therefore it is necessary to introduce an ultraviolet regularisation.
2. *Ultraviolet regularisation.* Restrict to finitely many discrete momenta $\|q_n\| \leq \Lambda$. Both regularisations together give rise to a *finite-dimensional problem*, which is important to retain for the treatment of quantum field theories on noncommutative geometries in the next section.
3. *Identify parameters.* After regularisation one hopes to see how to modify parameters in order to achieve a well-defined limit $\Lambda \to \infty$ and $\mathcal{K} \to \mathbb{R}^D$. These parameters can be the scalar coefficients in the polynomial $P(\phi)$, the mass μ^2 in (2.2) and a global field redefinition $\phi \mapsto \sqrt{Z}\phi$. Let us collectively call them $\lambda_1(\mathcal{K}, \Lambda), \ldots, \lambda_r(\mathcal{K}, \Lambda)$.
4. *Renormalisation.* Identify r moments (here a lot of experience is necessary to decide which) of the regularised measure to be kept constant. We say they are normalised to values $M_1, \ldots, M_r$ which all depend on $\lambda_1(\mathcal{K}, \Lambda), \ldots, \lambda_r(\mathcal{K}, \Lambda)$. By the implicit function theorem, this dependence can generically be inverted to

$$\lambda_1(\mathcal{K}, \Lambda, M_1, \ldots, M_r), \ldots, \lambda_r(\mathcal{K}, \Lambda, M_1, \ldots, M_r).$$

   In this way *all* moments of the regularised measure, i.e. our regularised Schwinger functions, depend on $\mathcal{K}, \Lambda, M_1, \ldots, M_r$; we say they are *renormalised.*

Here comes the

Challenge 2.5 (of QFT)

1. Prove that after these preparations the limit $\Lambda \to \infty$ and $\mathcal{K} \to \mathbb{R}^D$ (understood as convergence of nets) of all moments exists.
2. Prove that the resulting Schwinger functions satisfy the Osterwalder–Schrader axioms.

This programme succeeded in a few cases, all in dimension $D < 4$. In $D = 2$ dimensions and for *arbitrary* polynomial $P(\phi)$ bounded from below, this was achieved in groundbreaking works by Simon [Sim74] and Glimm–Jaffe–Spencer

[GJS74]. See also [GJ87]. They proved that it is essentially enough to replace $P(\phi)$ by its *normal ordering* $:P(\phi):$, a well-defined procedure which for monomials reads $:\phi^k: = \sum_{k_2+\dots+k_s=k} c_{k_1,\dots,k_s} \phi^{k_1}(\int \phi^{k_2} \mathrm{d}\mathcal{M}(\phi)) \cdots (\int \phi^{k_s} \mathrm{d}\mathcal{M}(\phi))$ for certain integers $c_{k_1,\dots,k_s}$. The resulting polynomial is no longer bounded from below so that the very assumption of the Feynman–Kac formula is lost. That the Challenge 2.5 is solvable for this so-called $P(\phi)_2$-model (the subscript 2 refers to $D = 2$ dimensions) is a highly non-trivial result.

For the ϕ^4_3-model, which means $P(\phi) = \phi^4$ in $D = 3$ dimensions, a similar existence proof can be established, but the work is already harder. The strategy fails for the ϕ^4_4-model (i.e. in 4 dimensions). Here the only possibility is to expand $\mathrm{e}^{-\lambda \int_{\mathbb{R}^4} \mathrm{d}x\, (\phi(x))^4} = \sum_{n=0}^{\infty} \frac{(-\lambda)^n}{n!} (\int_{\mathbb{R}^4} \mathrm{d}x \;\; (\phi(x))^4)^n$ into a power series and to *formally* exchange sum $\sum_n$ and integration $\int_{X'} \mathrm{d}\mathcal{M}(\phi)$. By a procedure known as renormalised perturbation theory, briefly sketched below, one can give a meaning to the $\int_{X'} \mathrm{d}\mathcal{M}(\phi)$ order by order in λ^n. However, the resulting series necessarily has zero radius of convergence.[3] In principle there exist summation techniques for series where $\lambda = 0$ is a boundary point of the holomorphicity domain. But in case of ϕ^4_4 this is also expected to fail because of the so-called triviality conjecture [Aiz81, Frö82]. The problem was first discovered by Landau et al for QED [LAK54]; it almost killed renormalised quantum field theory (rescued by the discovery of asymptotic freedom in QCD [GW73, Pol73]). Later we come back to that point.

Here we only mention that Yang–Mills theory in 4 dimensions is conjecturally free of the triviality problem and should exist as a quantum field theory. The proof is one of the millennium prize problems [JW00], left for the far future.

2.5 Feynman graphs and Feynman integrals

Here we briefly discuss interesting structures which arise when exchanging sum and integral of the expanded interaction term $\mathrm{e}^{-\frac{\lambda}{4!} \int_{\mathbb{R}^D} dx\, (\phi(x))^4}$. The integral in footnote 3 serves as a warning that this is not what one ultimately wants. Step 1 of Challenge 2.5 is often in reach order by order in λ^n; for that it is helpful that the implicit function theorem has an easy Taylor approximation. Step 2 is meaningless as long as the convergence of the series remains obscure.

[3]It is instructive to look at the following integral, which is a sort of ϕ^4_0-model:

$$I(\lambda) = \int_{-\infty}^{\infty} \mathrm{d}\phi\; \mathrm{e}^{-\phi^2 - \lambda\phi^4} \text{``}=\text{''} \sum_{k=0}^{\infty} \int_{-\infty}^{\infty} \mathrm{d}\phi\; \frac{(-\lambda)^k \phi^{4k}}{k!} \mathrm{e}^{-\phi^2} = \sum_{k=0}^{\infty} (-\lambda)^k \frac{\Gamma(2k + \frac{1}{2})}{\Gamma(k+1)},$$

where '=' results when exchanging sum and integral. The series diverges for any $\lambda \neq 0$, but the lhs is perfectly defined for $\mathrm{Re}(\lambda) \geq 0$ and evaluates into $I(\lambda) = \frac{1}{2\sqrt{\lambda}} \exp(\frac{1}{8\lambda}) K_{\frac{1}{4}}(\frac{1}{8\lambda})$, where K_ν is a modified Bessel function.

The various contributions are conveniently organised into *Feynman graphs* [Fey49]. These arise because the (divergent) integral

$$\lambda^n \int_{\mathbb{R}^{nD}} d(y_1, \dots, y_N) f_1(y_1) \cdots f_N(y_N) \int_{\mathbb{R}^{nD}} d(x_1, \dots, x_n) \\ \times \int_{X'} d\mathcal{M}(\phi)\, \phi(y_1) \cdots \phi(y_N)(\phi(x_1))^4 \cdots (\phi(x_n))^4$$

contributing to the perturbative order-n Schwinger N-point function factors into all possible pairings of $\phi(x_i)$ with $\phi(x_j)$ into $S_2(x_1, x_2)$. Up to combinatorial factors which we do not discuss, this is given by the sum over *Feynman graphs* Γ with

- n four-valent vertices located at $x_1, \dots, x_n \in \mathbb{R}^D$, each with a factor λ assigned;
- N one-valent vertices located at $y_1, \dots, y_N \in \mathbb{R}^D$;
- with factor $S_2(x_a, x_b)$ assigned to every edge between x_a, x_b (which can be x's and y's, also $x_a = x_b$ is allowed);
- integrated over $x_1, \dots, x_n$ (will diverge unless restricted to $\mathcal{K}$ and Λ);
- integrated against test functions $f_1(y_1), \dots, f_N(y_N)$ over $y_1, \dots, y_N$.

The resulting *Feynman integral* can be rearranged in several ways. One can keep the momentum variables q from (2.2) in every edge and move the Fourier kernels to the vertices, where the x-integrals give Dirac-δ distributions. The momentum rules are thus:

- assign oriented momentum q_{ab} to every (arbitrarily oriented) edge from vertex a to vertex b, assign $\frac{1}{\|q_{ab}\|^2+\mu^2}$ to that edge;
- assign $(2\pi)^D \lambda \delta(q_1 + \cdots + q_4)$ to every 4-valent vertex a, where $q_i := q_{bi}$ if the edge arrives from vertex b and $q_i := -q_{ib}$ if the edge goes to b.

Writing the weight factors as $\frac{1}{\|q\|^2+\mu^2} = \int_0^\infty d\alpha\ e^{-\alpha(\|q\|^2+\mu^2)}$ and returning to $(2\pi)^D \delta(q_1, \dots, q_4) = \int_{\mathbb{R}^D} dx\ e^{i\langle x, (q_1+\cdots+q_r)\rangle}$, all x-integrations and q-integrations for edges between 4-valent vertices are Gaußian and give rise to the *parametric representation* which can immediately be deduced from:

Theorem 2.6 *A connected graph Γ with L edges between 4-valent vertices contributes with weight*

$$\int_{\mathbb{R}^D} \frac{dp_1\ \hat{f}_1(p)}{|p_1\|^2 + \mu^2} \cdots \int_{\mathbb{R}^D} \frac{dp_1\ \hat{f}_N(p_N)}{|p_N\|^2 + \mu^2} \delta(p_1 + \ldots + p_N) \mathcal{A}_\Gamma(p_1, \dots, p_N), \quad \text{where}$$

$$\mathcal{A}_\Gamma(p_1, \dots, p_N) = \int_{(\mathbb{R}_+)^L} d(\alpha_1, \dots, \alpha_L)\ \frac{e^{-\mu^2(\alpha_1+\cdots+\alpha_L) - \frac{V_\Gamma(\alpha_\ell, p_v)}{U_\Gamma(\alpha_\ell)}}}{(U_\Gamma(\alpha_\ell))^{\frac{D}{2}}}, \tag{2.8}$$

$$U_\Gamma(\alpha_\ell) = \sum_{T_1 \in \Gamma} \prod_{\ell \notin T_1} \alpha_\ell, \qquad V_\Gamma(\alpha_\ell) = \sum_{T_2 \in \Gamma} \Big(\prod_{\ell \notin T_2} \alpha_\ell \Big) \Big\| \sum_{v \in T_{21}} p_v \Big\|^2.$$

Here the sum in U_Γ runs over all spanning trees T_1 of Γ (trees which meet every 4-valent vertex); the sums in V_Γ run over all forests T_2 of exactly two trees T_{21} and T_{22} which together contain all 4-valent vertices, and each of them at least one vertex with some incoming momentum p_ν.

The U_Γ, V_Γ are referred to as the Kirchhoff–Symanzik polynomials of Γ. At vanishing external momenta p_ν, the amplitude diverges for $\alpha \to 0$. This divergence is best controlled by a decomposition into Hepp sectors $\alpha_{\pi(1)} < \cdots < \alpha_{\pi(L)}$, which give rise to iterated integrals. Such iterated integrals are fascinating objects [Bro13b]. They form a Hopf algebra [Kre00], which relates them to noncommutative geometry [CK98], and they evaluate (for $\mu = 0$ and $p_\nu = 0$) to special values of analytic number theory, typically multiple zeta values [Bro13a]. The Symanzik polynomials provide connections between algebraic varieties and Feynman integrals [Blo15]. According to the Goncharov–Manin conjecture [GM04], there is a relation between periods in mixed Tate motives and residues of Feynman integrals.

The perturbative regularisation and renormalisation methods for the above Feynman integrals produce two problems which make a resummation of the perturbation series impossible. The same will apply to the perturbative treatment of QFTs on noncommutative geometries. One can do better, both in traditional and noncommutative QFT. First, the usual renormalisation at a single scale $p_\nu = 0$ produces amplitudes which grow as $\log \|p_\nu\|$. Inserting n such renormalised graphs γ as subgraphs into a bigger, convergent graph Γ with n vertices leads to amplitudes $\mathcal{A}_\Gamma = O(n!)$. This is the *renormalon problem*; it arises because one repairs too much. Constructive renormalisation theory [Riv91] avoids the renormalon problem by slicing the α-integrals and external momenta p_ν into multiple scales and only repairs if the α-scale is higher than the p-scale. The second problem is that the number of Feynman graphs with n vertices grows too fast with n. This problem is addressed by a reorganisation of the perturbation series into trees instead of graphs [GRM09]. The idea goes as follows: A Schwinger function is a sum of amplitudes indexed by graphs, $S = \sum_\Gamma S_\Gamma$. Let $T \subset \Gamma$ be the spanning trees, and assume there is a weight function with $\sum_{T\subset\Gamma} w(\Gamma, T) = 1$. Then formally $S = \sum_\Gamma \sum_{T\subset\Gamma} w(\Gamma, T) S_\Gamma = \sum_T S_T$ with $S_T = \sum_{\Gamma\supset T} w(\Gamma, T) S_\Gamma$.

3 Euclidean quantum fields on noncommutative geometries

3.1 *Nuclear AF Fréchet algebras*

A noncommutative geometry is for us a spectral triple $(\mathcal{A}, \mathcal{H}, \mathcal{D})$ [Con95] consisting of an associative $*$-algebra represented on Hilbert space $\mathcal{H}$, together with a self-adjoint unbounded operator $\mathcal{D}$ such that $[\mathcal{D}, a]$ extends to a bounded operator for all $a \in \mathcal{A}$. Often compactness of $a(D + \mathrm{i})^{-1}$ is required, and various topologies on and closures of $\mathcal{A}$ are considered.

In this survey we restrict ourselves to *Euclidean quantum field theory* which we intend to construct by analogy with Sections 2.3 and 2.4. The construction of a (Euclidean) QFT on a noncommutative geometry $(\mathcal{A}, \mathcal{H}, \mathcal{D})$ is a compromise between two contradictory requirements. Steps 1+2 of Programme 2.4 require a reduction to a finite-dimensional problem where everything is well-defined. This allows to adjust parameters so that a limit can be studied. Finite-dimensional algebras are matrix algebras, and the limiting procedure seems at first sight to be what is known as AF C^*-algebras [Bra72]. An important (non-unital) example is the algebra of compact operators which is the norm closure of finite-rank operators. On the other hand, for the initial definition of a free Euclidean scalar field via the Bochner–Minlos theorem 2.3 we need a scalar product on a *nuclear* vector space. But the vector space of compact operators is not nuclear, and consequently the norm closure is the wrong concept for our purpose.

What we rather need is a class of algebras $\mathcal{A}$ which we would like to call *nuclear AF Fréchet algebras*:

Definition 3.1 A *Fréchet space* is a locally convex vector space X topologised by a countable increasing family (p_n) of seminorms, which make X metrisable and complete. The Fréchet space is *nuclear* if

1. the topology is defined by a countable family (p_n) of Hilbert seminorms, i.e. for every n there is an inner product $\langle\ ,\ \rangle_n$ on X with $(p_n(x))^2 = \langle x, x\rangle_n$.
2. If X_n denotes the closure of X with respect to $\langle\ ,\ \rangle_n$, then for any p_n there is a larger p_m such that the natural map from X_m to X_n is trace-class.

Definition 3.2 A *(nuclear) Fréchet algebra* is an algebra $\mathcal{A}$ that, as a vector space, is a (nuclear) Fréchet space and in which the multiplication $\cdot : \mathcal{A} \times \mathcal{A} \to \mathcal{A}$ is continuous.

Nuclear Fréchet algebras can always be understood as a certain space of *smooth functions with deformed product*. Namely,

Theorem 3.3 *For a Fréchet space X are equivalent:*

1. *X is nuclear.*
2. *X is isomorphic to a closed subspace of* $C^\infty(U)$*, for any open* $U \subset \mathbb{R}^D$.

The equivalence is essentially due to T. Kōmura and Y. Kōmura [KK66] (for $D = 1$; the general case can be found in the literature, see e.g. [Vog00]). Given now a Fréchet algebra $\mathcal{A}$, then an isomorphism ι_U of vector spaces between $\mathcal{A}$ and a closed subspace of $C^\infty(U)$ induces a deformed product $\star_U$ on $\iota_U(\mathcal{A})$ by

$$\iota_U(a) \star_{\iota_U} \iota_U(b) := \iota_U(ab). \tag{3.1}$$

We will mainly be interested in $U = \mathbb{R}^D$. In case that $\iota_{\mathbb{R}^D}(\mathcal{A})$ is invariant under translations by $\mathbb{R}^D$, or even under the Euclidean group $\mathbb{R}^D \rtimes SO(D)$, or under a subgroup of them, we can define a corresponding group action on $\mathcal{A}$ by

$$\alpha_{t,R}(a) := \iota_{\mathbb{R}^D}^{-1}((\iota_{\mathbb{R}^D} a)_{t,R}), \qquad \text{where } f_{t,R}(x) = f(R^{-1}(x-t)). \tag{3.2}$$

Such a group action is very important for us because it is needed to formulate an analogue of the Osterwalder–Schrader axioms, see Section 10. There are clearly examples for Fréchet algebras carrying an action of the Euclidean group, but we do not know how generic they are. So let us formulate:

Question 3.4 Which assumptions on a Fréchet algebra guarantee the existence of an isomorphism to a closed subspace of $C^\infty(\mathbb{R}^D)$ that is invariant under (a subgroup of) the Euclidean group?

It remains to define our approximation property needed for renormalisation:

Definition 3.5 A nuclear Fréchet algebra $\mathcal{A}$ is called *AF* (for approximately finite-dimensional) if there is an increasing sequence $\mathcal{A}^0 \subset \mathcal{A}^1 \subset \mathcal{A}^2 \subset \dots$ of finite-dimensional subalgebras, embedded into each other by $*$-homomorphisms $\iota_N : \mathcal{A}^N \to \mathcal{A}^{N+1}$, such that $\bigcup_{N\in\mathbb{N}} \mathcal{A}^N$ is dense in $\mathcal{A}$ in the locally-convex topology induced by the Hilbert seminorms $\{p_n\}$ of $\mathcal{A}$.

This definition is inspired by the corresponding definition AF-C^*-algebras [Bra72], but we do not require $\mathcal{A}$ to be unital and we close in the locally-convex topology. The class of unital AF-C^*-algebras is very rich and classified by K-theoretic data [Ell76]. We have no idea about the corresponding landscape in the locally-convex setup:

Question 3.6

1. How rich is the class of nuclear AF Fréchet algebras?
2. How much do they depend on the choice of Hilbert seminorms?
3. Is there any chance to classify them, possibly under extra conditions?
4. Is there any relation to limits of compact quantum metric spaces [Rie04]?

Remark 3.7 The Kōmura-Kōmura theorem 3.3 is a typical example for the coexistence of the discrete and the continuum in noncommutative geometry [Con95]. It is undeniable that our universe is very close to a continuous space. But we cannot conclude that our universe is a manifold; it just means that smooth functions on a manifold are the universal model for a nuclear Fréchet space which could very well be inherently discrete.

3.2 *The free Euclidean scalar field on a noncommutative geometry*

Every nuclear Fréchet algebra $\mathcal{A}$ admits a free Euclidean scalar field. The vector space $\mathcal{A}_*$ of self-adjoint elements of $\mathcal{A}$ is a real nuclear vector space. Consider a continuous symmetric positive-semidefinite bilinear form C on $\mathcal{A}_*$, called the *covariance*. Such a C always exists; take, for instance, the inner product defining any of the Hilbert seminorms p_n on $\mathcal{A}$. In the same way as in Section 2.3, the continuous linear map $\mathcal{F}_C : \mathcal{A}_* \to \mathbb{R}$ defined by

$$\mathcal{F}_C(a) := \exp\Big(-\frac{1}{2}C(a,a)\Big), \qquad a = a^* \in \mathcal{A}_*, \tag{3.3}$$

satisfies the assumptions of the Bochner–Minlos theorem 2.3. Consequently, there exists a unique Radon probability measure $d\mathcal{M}_C$ on the dual space $\mathcal{A}'_*$ of *Euclidean scalar fields* with

$$\mathcal{F}_C(a) = \int_{\mathcal{A}'_*} \mathrm{d}\mathcal{M}_C(\Phi)\, \mathrm{e}^{\mathrm{i}\Phi(a)}. \tag{3.4}$$

It is straightforward to generalise this construction to finitely generated projective modules over $\mathcal{A}$, but for the sake of clarity we will not spell it out. The moments of $\mathrm{d}\mathcal{M}_C$ are, as before, given by

$$\begin{aligned}
&\int_{\mathcal{A}'_*} \mathrm{d}\mathcal{M}_C(\Phi)\, \Phi(a_1)\cdots\Phi(a_N) \\
&= (-\mathrm{i})^N \frac{\partial^N}{\partial t_1\cdots\partial t_N}\mathcal{F}(t_1a_1+\cdots+t_Na_N)\Big|_{t_i=0} \\
&= \begin{cases} \displaystyle\sum_{\text{pairings of } [N]} C(a_{i_1},a_{j_1})\cdots C(a_{i_{N/2}},a_{j_{N/2}}) & \text{for } N \text{ even.} \\ 0 & \text{for } N \text{ odd.}\end{cases}
\end{aligned} \tag{3.5}$$

A pairing is a partition of $\{1,2,\dots,N\}$ into $\frac{N}{2}$ subsets $(i_1,j_1),\dots,(i_{N/2},j_{N/2})$ with $i_k < j_k$. We interpret $S_N(a_1\otimes\cdots\otimes a_N) := \int_{\mathcal{A}'_*} \mathrm{d}\mathcal{M}_C(\Phi)\, \Phi(a_1)\cdots\Phi(a_N)$ as the *Schwinger N-point function of the free scalar field* of covariance C on the noncommutative algebra $\mathcal{A}$.

We postpone a discussion of Osterwalder–Schrader axioms for the free field to Section 10.

3.3 Towards an interacting scalar field on noncommutative geometry

Switching on interaction is, as in ordinary quantum field theory, a very hard problem. We proceed as in Section 2.4 and formally define correlation functions as moments of a Feynman–Kac perturbation of $d\mathcal{M}_C$. Given a functional S_{int} on $\mathcal{A}'_*$, bounded from below, we define

$$\begin{aligned}\langle a_1 \otimes \cdots \otimes a_N \rangle &:= \frac{\displaystyle\int_{\mathcal{A}'_*} d\mathcal{M}_C(\Phi)\, \Phi(a_1)\cdots\Phi(a_N)\exp(-S_{\text{int}}(\Phi))}{\displaystyle\int_{\mathcal{A}'_*} d\mathcal{M}_C(\Phi)\, \exp(-S_{\text{int}}(\Phi))} \\ &= \frac{(-\mathrm{i})^N}{\mathcal{Z}_C(0)} \frac{\partial^N \mathcal{Z}_C(t_1 a_1 + \cdots + t_N a_N)}{\partial t_1 \cdots \partial t_N}\bigg|_{t_i=0},\end{aligned} \tag{3.6}$$

where

$$\mathcal{Z}_C(J) = \int_{\mathcal{A}'_*} d\mathcal{M}_C(\Phi)\, \mathrm{e}^{\mathrm{i}\Phi(J) - S_{\text{int}}(\Phi)} \tag{3.7}$$

is the *partition function*. For the free theory $S_{\text{int}}(\Phi) \equiv 0$, it coincides with the characteristic function $\mathcal{F}_C(J)$ defined in (3.3), with $J \in \mathcal{A}_*$. For $S_{\text{int}}(\Phi) \neq 0$, however, these naïve correlation or partition functions do not make any sense. One meets the usual divergences whose treatment requires regularisation and renormalisation. As stressed in Section 2.4, we have to restrict in a first step to finite-dimensional subspaces, at least for a rigorous (non-perturbative) treatment. In perturbation theory one may hope to do less, but this depends on the situation.

Inspired by renormalisation in usual QFT, sketched in Section 2.4, we proceed as follows:

Programme 3.8 For a given nuclear AF Fréchet algebra $\mathcal{A}$, consider covariances $C^{(\mathcal{N})}$ and interaction functionals $S^{(\mathcal{N})}_{\text{int}}$ on the finite-dimensional subspace $\mathcal{A}^{\mathcal{N}}_*$ of $\mathcal{A}_*$. Parametrise them by real numbers $\lambda_1(\mathcal{N}), \ldots, \lambda_r(\mathcal{N})$, define correlation functions (3.6) by integrals with measure $d\mathcal{M}_{C^{(\mathcal{N})}}$ over $(\mathcal{A}^{\mathcal{N}}_*)'$. Identify r of these moments $M_1, \ldots, M_r$, all functions of $\lambda_1(\mathcal{N}), \ldots, \lambda_r(\mathcal{N})$, but considered as fixed. The implicit function theorem generically allows to invert to $\lambda_1(\mathcal{N}, M_1, \ldots, M_r), \ldots, \lambda_r(\mathcal{N}, M_1, \ldots, M_r)$. Consequently, *all* level-$\mathcal{N}$ correlation functions depend on $\mathcal{N}$ and $M_1, \ldots, M_r$.

Now we face part 1 of the fundamental Challenge 2.5: Prove that with these preparations, under the embeddings $\iota_{\mathcal{N}}$ and corresponding embeddings of $(C^{(\mathcal{N})})$ and $(S^{(\mathcal{N})}_{\text{int}})$, the limit $\mathcal{N} \to \infty$ of all moments (3.6) exists, thereby defining Schwinger functions of an interacting scalar QFT on $\mathcal{A}$.

Discussion of Step 2, the Osterwalder–Schrader axioms, has to be postponed.

We describe the parametrisation by $\lambda_1(\mathcal{N}), \ldots, \lambda_r(\mathcal{N})$ for the most relevant case of tracial interaction functionals. As with C^*-algebras, every $\mathcal{A}^{\mathcal{N}}$ is a direct sum of finitely many matrix algebras, $\mathcal{A}^{\mathcal{N}} = \bigoplus_i M_{n_{N,i}}(\mathbb{C})$. Let $(e^{(i)}_{kl})$ be the standard matrix basis of $M_{n_{N,i}}(\mathbb{C})$. We extend the real linear functionals Φ to $\mathcal{A}$ via $\Phi(a+ib) := \Phi(a) + \mathrm{i}\Phi(b)$. Then the restriction of $\Phi \in \mathcal{A}'_*$ to $\mathcal{A}^{\mathcal{N}}$ is uniquely specified by the complex numbers $\Phi^{(i)}_{kl} := \Phi(e^{(i)}_{kl}) \equiv \overline{\Phi^{(i)}_{lk}}$, and the following defines a functional on $(\mathcal{A}^{\mathcal{N}}_*)'$:

$$S^{(\mathcal{N})}_{\text{int}}(\Phi) := \sum_i \sum_{p_i} \frac{\lambda_{i,p_i}(\mathcal{N})}{p_i} \sum_{k^i_1,\ldots,k^i_{p_i}=1}^{n_{N,i}} \Phi^{(i)}_{k^i_1 k^i_2} \Phi^{(i)}_{k^i_2 k^i_3} \cdots \Phi^{(i)}_{k^i_{p_i-1} k^i_{p_i}} \Phi^{(i)}_{k^i_{p_i} k^i_1}. \tag{3.8}$$

Additional $\lambda_c(\mathcal{N})$ will parametrise the covariances and possible field redefinitions $\Phi^{(i)}_{kl} \mapsto \sqrt{Z_i(\mathcal{N})}\Phi^{(i)}_{kl}$.

An investigation in this generality has not yet been performed. We describe in Section 4.1 below the simplest case given by a single summand i (hence omitted) and an embedding $\iota_{\mathcal{N}}$ analogous to compact operators.

4 Some noncommutative geometries for QFT

4.1 *Simplest example: Moyal algebra*

Let $\mathcal{A}_\theta = \{a = (a_{kl}) \;:\; k,l = 1,2,3,\ldots\}$ be the vector space of double-indexed sequences, with involution $(a^*)_{kl} = \overline{a_{lk}}$, completed in the Fréchet topology induced by the family of inner products

$$\langle a, b\rangle_m := \sum_{k,l=1}^{\infty} \theta^{2m}(k+\tfrac{1}{2})^m (l+\tfrac{1}{2})^m \overline{a_{kl}} b_{kl}. \tag{4.1}$$

Let e_{rs} be the terminating sequence in $\mathcal{A}_\theta$ defined by $(e_{rs})_{kl} = \delta_{rk}\delta_{sl}$, then $(b^{(m)}_{rs})$ with $b^{(m)}_{rs} := \theta^{-m}(r+\frac{1}{2})^{-m/2}(s+\frac{1}{2})^{-m/2} e_{rs}$ form an orthonormal basis with respect to $\langle\ ,\ \rangle_m$. Let $\mathcal{A}_{\theta,m}$ be the closure of $\mathcal{A}_\theta$ with respect to $\langle\ ,\ \rangle_m$. Every $a \in \mathcal{A}_{\theta,m+3}$ has a representation

$$a = \sum_{r,s=1}^{\infty} \langle b^{(m+3)}_{rs}, a\rangle_{m+3} b^{(m+3)}_{rs} = \sum_{r,s=1}^{\infty} \frac{\langle b^{(m+3)}_{rs}, a\rangle_{m+3}}{\theta^3 (r+\frac{1}{2})^{\frac{3}{2}} (s+\frac{1}{2})^{\frac{3}{2}}} b^{(m)}_{rs},$$

which shows that, for any m, the natural map $\mathcal{A}_{\theta,m+3} \ni a \mapsto a \in \mathcal{A}_{\theta,m}$ is trace-class. Hence, $\mathcal{A}_\theta$ is a nuclear Fréchet space.

For $a = (a_{kl}), b = (b_{kl}) \in \mathcal{A}_\theta$ we consider $[ab]_{kl} := \sum_{n=1}^{\infty} a_{kn} b_{nk}$. Using Cauchy–Schwarz inequality several times one has

$$\begin{aligned}(p_m(ab))^2 &\leq \Big(\sum_{k,n=1}^{\infty} \theta^m (k+\tfrac{1}{2})^m |a_{kn}|^2\Big)\Big(\sum_{l,n'=1}^{\infty} \theta^m (l+\tfrac{1}{2})^m |b_{n'l}|^2\Big) \\ &\leq \frac{1}{\theta^{2m}} (p_m(a))^2 (p_m(b))^2. \end{aligned} \tag{4.2}$$

This shows that the sequence $ab := ([ab]_{kl})$ belongs to $\mathcal{A}_\theta$ and that the corresponding multiplication $\mathcal{A}_\theta \times \mathcal{A}_\theta \ni (a,b) \mapsto ab \in \mathcal{A}_\theta$ is continuous. Hence, $\mathcal{A}_\theta$ is a nuclear Fréchet algebra. The terminating sequences (e_{rs}) introduced above satisfy $e_{rs} e_{tu} = \delta_{st} e_{ru}$. Thus they play the rôle of matrix bases, and we will often expand $a = \sum_{k,l=1}^{\infty} a_{kl} e_{kl}$ where it is understood that $e_{kl} \in \mathcal{A}_\theta$ and $a_{kl} \in \mathbb{C}$.

By Theorem 3.3 there exists an isomorphism $\iota_\theta := \iota_{\mathbb{R}^2}$ of vector spaces between $\mathcal{A}_\theta$ and a closed subspace of $C^\infty(\mathbb{R}^2)$. A particular realisation is given by $\iota_\theta(e_{kl}) := f_{k-1,l-1}$, where

$$f_{kl}^{(\theta)}(x_1,x_2) := 2(-1)^k \sqrt{\frac{k!}{l!}} \left(\sqrt{\frac{2}{\theta}}(x_1 + \mathrm{i}x_2)\right)^{l-k} L_k^{l-k}\left(\frac{2\|x\|^2}{\theta}\right) \mathrm{e}^{-\frac{\|x\|^2}{\theta}}, \tag{4.3}$$

for $x = (x_1, x_2)$. The $L_m^\alpha(t)$ are associated Laguerre polynomials of degree m in t. One has $\int_{\mathbb{R}^2} \mathrm{d}x\, f_{mn}^{(\theta)}(x) = 2\pi\theta\delta_{mn}$. By linear extension we obtain an isomorphism between $\mathcal{A}_\theta$ and the nuclear vector space $\mathcal{S}(\mathbb{R}^2)$ of Schwartz functions. According to (3.1), this isomorphism induces an associative product $\star_\theta$ on $\mathcal{S}(\mathbb{R}^2)$, the *Moyal product*. One can verify [GBV88]

$$(\phi \star_\theta \psi)(x) = \int_{\mathbb{R}^2 \times \mathbb{R}^2} \frac{\mathrm{d}k\, \mathrm{d}y}{(2\pi)^2}\, \phi\left(x + \frac{1}{2}\Theta y\right) \psi(x+y) \mathrm{e}^{\mathrm{i}\langle y,k\rangle}, \tag{4.4}$$

where $\Theta = \begin{pmatrix} 0 & \theta \\ -\theta & 0 \end{pmatrix}$. This makes the Moyal product an example of a strict deformation quantisation by action of $\mathbb{R}^2$ [Rie93]. In particular, the action of the Euclidean group $\mathbb{R}^2 \rtimes SO(2)$ on $\mathcal{S}(\mathbb{R}^2)$ induces via (3.2) a corresponding group action $\alpha_{t,R}$ on $\mathcal{A}_\theta$ which has the important property to commute with the multiplication:

$$\alpha_{t,R}(ab) = (\alpha_{t,R}a)(\alpha_{t,R}b). \tag{4.5}$$

Remains to describe the AF structure of $\mathcal{A}_\theta$. We let $\mathcal{A}_\theta^N := \mathrm{span}(e_{kl} : 1 \leq k, l \leq N)$, then every $\mathcal{A}_\theta^N$ is a subalgebra for the multiplication in $\mathcal{A}^N$. The natural identification $\mathcal{A}_\theta^N \equiv M_N(\mathbb{C})$ defines via

$$\iota_{\mathcal{N}} : \mathcal{A}_\theta^{\mathcal{N}} = M_{\mathcal{N}}(\mathbb{C}) \ni a \mapsto \begin{pmatrix} a & 0 \\ 0 & 0 \end{pmatrix} \in M_{\mathcal{N}+1}(\mathbb{C}) = \mathcal{A}_\theta^{\mathcal{N}+1}$$

the connecting $*$-homomorphism $\iota_{\mathcal{N}}$. Given a finite family $(p_{n_1}, \ldots, p_{n_K})$ of Hilbert seminorms and $\epsilon > 0$, for every $a = (a_{kl}) \in \mathcal{A}_\theta$ the absolute convergence $p_{n_i}(a) < \infty$ guarantees the existence of an $N_{\epsilon;n_1,\ldots,n_K} \in \mathbb{N}$ with

$$p_{n_i}\Bigg(a - \sum_{k,l=1}^{N_{\epsilon;n_1,\ldots,n_K}} a_{kl} e_{kl}\Bigg) < \epsilon \qquad \text{for all } i = 1, \ldots, K.$$

Hence, $\bigcup_{\mathcal{N}=1}^{\infty} \mathcal{A}_\theta^{\mathcal{N}}$ is dense in $\mathcal{A}_\theta$ for the Fréchet topology, and $\mathcal{A}_\theta$ is indeed a nuclear AF Fréchet algebra.

4.2 *Quantum fields on the Moyal algebra*

We describe possible choices of covariances and interaction functionals on $\mathcal{A}_\theta$. The interaction functionals (3.8) specify to

$$S_{\text{int}}^{(\mathcal{N})}(\Phi) = \sum_p \frac{\lambda_p(\mathcal{N})}{p} \sum_{k_1,\ldots,k_p=1}^{\mathcal{N}} \Phi_{k_1k_2}\Phi_{k_2k_3}\cdots\Phi_{k_{p-1}k_{k_p}}\Phi_{k_pk_1}, \tag{4.6}$$

where $\Phi_{kl} := \Phi(e_{kl})$ and the sum over p is finite. The particularly relevant cases $p = 3$ and $p = 4$ are discussed in Sections 8 and 9. Passing via ι_θ to Schwartz functions and taking the Fourier transform, the functional (4.6) becomes in the limit $\mathcal{N} \to \infty$

$$\begin{aligned} &S_{\text{int}}(\Phi) &(4.7)\\ &= \sum_p \frac{\lambda_p}{p} \int_{\mathbb{R}^{2p}} \mathrm{d}(q_1, \ldots, q_p)\, \delta(q_1 + \ldots + q_p) \mathrm{e}^{\frac{\mathrm{i}}{2}\sum_{1\le k<l\le p}\langle q_k, \Theta q_l\rangle} \hat{\Phi}(q_1) \cdots \hat{\Phi}(q_p), \end{aligned}$$

assuming that $\lambda_p(\mathcal{N})$ has a limit (which is rarely the case in practice). This form has often been used in a perturbative treatment of QFTs on Moyal space (see Section 5). It is not suitable for a rigorous construction.

The covariances (3.3) could be chosen arbitrarily, but for a convenient interpretation we assume that they arise from a family $\{d_\nu\}_{\nu=1,\ldots,s}$ of continuous linear maps $d_\nu : \mathcal{A}_\theta \to \mathcal{A}_\theta$ of geometrical significance, for instance, induced by the Dirac operator $\mathcal{D}$ of a spectral triple $(\mathcal{A}, \mathcal{H}, \mathcal{D})$. For $a = \sum_{k,l=1}^{\mathcal{N}} a_{kl} e_{kl} \in \mathcal{A}_\theta^{\mathcal{N}} \equiv M_{\mathcal{N}}(\mathbb{C})$, consider

$$\sum_{\nu=1}^{s} \mathrm{Tr}((d_\nu a)^* d_\nu a) =: \sum_{k,l,m,n=1}^{\mathcal{N}} D_{kl;mn} a_{kl} a_{mn}. \tag{4.8}$$

The covariance is the inverse of that matrix, $\sum_{m',n'=1}^{\mathcal{N}} C^{(\mathcal{N})}(e_{kl}, e_{m'n'}) D_{n'm';mn} = \delta_{kn}\delta_{lm} = \sum_{m',n'=1}^{\mathcal{N}} D_{kl;m'n'} C^{(\mathcal{N})}(e_{n'm'}, e_{mn})$.

The following choices are of particular relevance:

- For an arbitrary sequence (E_k) of positive real numbers, define $d_E(e_{kl}) = \sqrt{E_k + E_l} e_{kl}$. It follows $C_E(e_{kl}, e_{mn}) = \frac{\delta_{kn}\delta_{lm}}{E_k+E_l}$. This covariance together with $p = 3$ in (4.6) defines the Kontsevich model [Kon92] which is of paramount importance in algebraic geometry. We discuss it in Sections 7.2 and 8. The same covariance but with quartic interaction $p = 4$ has also received considerable attention and will be discussed in Section 9.
- The action $\alpha_{t,1}$ of translations by $t \in \mathbb{R}^2$ defined in (3.2) can be shown to be generated by

$$(\partial_1 - \mathrm{i}\partial_2)(e_{kl}) = \sqrt{\frac{2}{\theta}}(\sqrt{l-1}e_{k,l-1} - \sqrt{k}e_{k+1,l}),$$

$$(\partial_1 + \mathrm{i}\partial_2)(e_{kl}) = \sqrt{\frac{2}{\theta}}(\sqrt{k-1}e_{k-1,l} - \sqrt{l}e_{k,l+1}). \tag{4.9}$$

  In this way the covariance of the Laplacian $\langle a, -\Delta a\rangle = \sum_{\nu=1}^{2} \mathrm{Tr}((\partial_\nu a)^* \partial_\nu a)$ can be defined. The calculation is lengthy; one has to diagonalise the resulting matrix $\Delta_{kl;mn}$ via Meixner polynomials. We briefly describe these steps in Section 6.1.
- Pointwise multiplication $(M_1\phi)(x) = x_1\phi(x)$ and $(M_2\phi)(x) = x_2\phi(x)$ of Schwartz functions defines continuous linear maps which translate via ι_θ^{-1} into the following action on the matrix bases:

$$(M_1 + \mathrm{i}M_2)(e_{kl}) = \sqrt{\frac{\theta}{2}}(\sqrt{l-1}e_{k,l-1} + \sqrt{k}e_{k+1,l}),$$

$$(M_1 - \mathrm{i}M_2)(e_{kl}) = \sqrt{\frac{\theta}{2}}(\sqrt{k-1}e_{k-1,l} + \sqrt{l}e_{k,l+1}). \tag{4.10}$$

  Instead of the Laplacian one can consider the slightly more general covariance of the operator $\langle a, H^\Omega a\rangle = \sum_{\nu=1}^{2} \mathrm{Tr}((\partial_\nu a)^* \partial_\nu a + \frac{4\Omega^2}{\theta^2}(M_\nu a)^* M_\nu a)$. See Section 6.1.

Remark 4.1 The Moyal product (4.4) has its origin in quantum mechanics, in particular in Weyl's operator calculus [Wey28]. Wigner introduced the useful concept of the phase space distribution function [Wig32]. Then, Groenewold [Gro46] and Moyal [Moy49] showed that quantum mechanics can be formulated on classical phase space using the *twisted product* concept. In particular, Moyal proposed the 'sine-Poisson bracket' (nowadays called Moyal bracket), which is the analogue of the quantum mechanical commutation relation. The twisted product was extended from Schwartz class functions to (appropriate) tempered distributions by Gracia-Bondía and Várilly [GBV88, VGB88]. The programme of Groenewold and Moyal culminated in the axiomatic approach of *deformation quantisation* [BFF$^+$78a, BFF$^+$78b]. The problem to lift a given Poisson structure to an associative $\star$-product was solved by Kontsevich [Kon03]. Cattaneo and Felder [CF00] found a physical derivation of Kontsevich's formula in terms of a path integral quantisation of a Poisson sigma model [SS94]. The Moyal product is a strict deformation by $\mathbb{R}^D$-action [Rie93], not only a formal deformation. The Moyal plane is a spectral triple [GGBI$^+$04], and the spectral action has been computed [Vas04, GI05].

4.3 4-Dimensional Moyal space

The generalisation of the Moyal algebra introduced in Section 4.1 to 4 dimensions is achieved by double-double-indexed sequences $\mathcal{A}_\Theta = \{a = (a_{\underline{k}\underline{l}}) \;:\; \underline{k}, \underline{l} \in \mathbb{N}^2_{\geq 1}\}$ with $(a^*)_{\underline{k}\underline{l}} = \overline{a_{\underline{l}\underline{k}}}$ and completed in the Fréchet topology induced by

$$\langle a, b\rangle_m := \sum_{\underline{k},\underline{l}\in\mathbb{N}^2_{\geq 1}} \Big(\theta_1^{2m}(k_1+\tfrac{1}{2})^m(l_1+\tfrac{1}{2})^m + \theta_2^{2m}(k_2+\tfrac{1}{2})^m(l_2+\tfrac{1}{2})^m\Big)\overline{a_{\underline{k}\underline{l}}}b_{\underline{k}\underline{l}}. \tag{4.11}$$

We introduce already here the block-diagonal matrix $\Theta = \mathrm{diag}\big(\Theta_1, \Theta_2\big)$ with $\Theta_i = \big(\begin{smallmatrix} 0 & \theta_i \\ -\theta_i & 0\end{smallmatrix}\big)$. A multiplication on the nuclear Fréchet space $\mathcal{A}_\Theta$ is again introduced via multi-indexed matrix bases $(e_{\substack{r_1\, s_1\\ r_2\, s_2}})_{\substack{k_1\, l_1\\ k_2\, l_2}} = \delta_{r_1k_1}\delta_{r_2k_2}\delta_{s_1l_1}\delta_{s_2l_2}$ and $e_{\substack{k_1\, l_1\\ k_2\, l_2}} e_{\substack{m_1\, n_1\\ m_2\, n_2}} := \delta_{l_1m_1}\delta_{l_2m_2} e_{\substack{k_1\, n_1\\ k_2\, n_2}}$. For the isomorphism $\iota_\Theta := \iota_{\mathbb{R}^4}$ of Theorem 3.3 we arrange $\iota_\Theta : \mathcal{A}_\Theta \to \mathcal{S}(\mathbb{R}^4)$ by defining $\iota_\Theta(e_{\substack{k_1\, l_1\\ k_2\, l_2}}) := f^{(\theta_1)}_{k_1-1,l_1-1} \times f^{(\theta_2)}_{k_2-1,l_2-1}$ and linear extension, where $(f^{(\theta_1)}_{kl} \times f^{(\theta_2)}_{mn})(x_1, x_2, x_3, x_4) := f^{(\theta_1)}_{kl}(x_1, x_2) f^{(\theta_1)}_{mn}(x_3, x_4)$. Then the resulting $\star$-product (3.1) on $\mathcal{S}(\mathbb{R}^4)$ takes the form

$$(\phi \star_\Theta \psi)(x) = \int_{\mathbb{R}^4\times\mathbb{R}^4} \frac{dk\, dy}{(2\pi)^4}\, \phi(x + \tfrac{1}{2}\Theta y)\psi(x+y)e^{\mathrm{i}\langle y,k\rangle},$$

generalising (4.4). The AF-structure is obtained via the Cantor polynomial which implements the bijection between $\mathbb{N}^2$ and $\mathbb{N}$. After a shift, the Cantor bijection reads $P\big(\begin{smallmatrix}k_1\\ k_2\end{smallmatrix}\big) := \frac{1}{2}((k_1+k_2)^2 - 3k_1 - k_2 + 2)$. Accordingly, we identify $e_{\substack{k_1\, l_1\\ k_2\, l_2}}$ with the standard matrix basis $e_{\frac{1}{2}((k_1+k_2)^2-3k_1-k_2+2),\frac{1}{2}((l_1+l_2)^2-3l_1-l_2+2)}$. Symmetry between both components selects $\mathcal{A}^N_\Theta \equiv M_{N(N+1)/2}(\mathbb{C})$; the embedding $\iota_N : \mathcal{A}^N_\Theta \to \mathcal{A}^{N+1}_\Theta$ is given by filling up with zeros.

Question 4.2 Is it true that $\lim_{\mathcal{N}\to\infty}\mathcal{A}_\Theta^{\mathcal{N}}$ described here is, as nuclear AF Fréchet algebra, different from $\lim_{\mathcal{N}\to\infty}\mathcal{A}_\theta^{\mathcal{N}}$ introduced in Section 4.1?

Under the isomorphism ι_Θ we obtain finite-dimensional subalgebras of the 4-dimensional Moyal algebras $\iota_\Theta(\mathcal{A}_\Theta^{\mathcal{N}}) = \mathrm{span}(f^{(\theta_1)}_{k_1l_1} \times f^{(\theta_2)}_{k_2l_2} \;:\; k_1+l_1 \leq \mathcal{N}-1,\; k_2+l_2 \leq \mathcal{N}-1)$. Interestingly, this dependence on the length $|{}^{k_1}_{k_1}| := k_1 + k_2$ of double indices will be respected by the covariances chosen in Sections 6, 8 and 9.

4.4 Gauge models

Gauge models arise very naturally in noncommutative geometry [CR87]. With spectral triples either the older formulation [Con94, Con95] via the Dixmier trace or the spectral action [Con96, CC97] is available. Everything works for the Moyal space [Gay03, Vas04, GI05]. But this defines only the classical action which does *not* give rise to a covariance for the free gauge field. Gauge fixing [FP67] is required and can be implemented in noncommutative geometry [Wul00, Per07].

In D dimensions one needs D gauge fields $A_1, \ldots, A_D \in \mathcal{A}$ which have to be extended by a Faddeev–Popov ghost c (a Maurer–Cartan form for the BRST-differential s) and two auxiliary objects $\bar{c}, B$. Then a covariance for $(A_1, \ldots, A_D, B)$ and for $(c, \bar{c})$ exists, and quantum gauge theory can *formally* be defined along the same lines as before. Partial results for perturbative renormalisation have been achieved, also numerical results have been obtained, but nothing rigorous. Some of these investigations will be reviewed in Section 5.

4.5 Fuzzy spaces

The fuzzy sphere [Mad92] is one of the simplest noncommutative spaces. It is obtained by truncating representations of $su(2)$. The algebra $S^2_{\mathcal{N}}$ is identified with mappings from the representation space $\frac{\mathcal{N}}{2}$ of $su(2)$ to itself, thus with the algebra $M_{\mathcal{N}+1}(\mathbb{C})$. The fuzzy sphere $S^2_{\mathcal{N}}$ is generated by $\hat{X}_\nu$, $\nu = 1, 2, 3$, which form an $su(2)$-Lie algebra with suitable rescaling, identified by the requirement that the Casimir operator still fulfils the defining relation of the two-sphere as an operator:

$$[\hat{X}_\mu, \hat{X}_\nu] = \sum_{\kappa=1}^{3} \mathrm{i}\lambda\epsilon_{\mu\nu\kappa}\hat{X}_\kappa, \qquad \sum_{\nu=1}^{3}\hat{X}_\nu\hat{X}_\nu = R^2, \qquad \frac{R}{\lambda} = \sqrt{\tfrac{\mathcal{N}}{2}(\tfrac{\mathcal{N}}{2}+1)}. \tag{4.12}$$

The philosophy about the limit $\mathcal{N} \to \infty$ is quite different than before, namely the ordinary commutative sphere should arise in the limit. The necessary framework was worked out by Rieffel [Rie04]. Its main steps are the realisation of the $S^2_{\mathcal{N}}$

as compact quantum metric spaces, where Lipschitz seminorms are relevant. The embeddings of $S^2_{\mathcal{N}}$ into bigger structures are achieved via "bridges", and in the end it is shown that the sequence of $S^2_{\mathcal{N}}$ converges to S^2 in the Gromov–Hausdorff topology.

The Lie algebra $su(2)$ generated by J_ν, $\nu = 1, 2, 3$, acts on $a \in S^2_{\mathcal{N}}$ by the adjoint action $J_\nu a = \frac{1}{\lambda}[\hat{X}_\nu, a]$. Thus, an element $a \in S^2_{\mathcal{N}}$ can be represented by $a = \sum_{l=0}^{\mathcal{N}} \sum_{m=-l}^{l} a_{lm} \Psi_{lm}$, where

$$\sum_{\nu=1}^{3} J_\nu^2 \Psi_{lm} = l(l+1)\Psi_{lm}, \quad J_3 \Psi_{lm} = m\Psi_{lm}, \quad \frac{4\pi}{\mathcal{N}+1}\mathrm{Tr}\big(\overline{\Psi_{lm}}\Psi_{l'm'}\big) = \delta_{ll'}\delta_{mm'}. \tag{4.13}$$

Other fuzzy spaces include the fuzzy $\mathbb{C}P^2$ [GS99, ABIY02] and the q-deformed fuzzy sphere [GMS01, GMS02].

Remark 4.3 Any quantum field theory shows divergences in some way. The first step to treat them is regularisation. Typically a regularisation destroys the symmetries of the theory so that the limit $\mathcal{N} \to \infty$ is considered. Fuzzy noncommutative spaces [Mad92, Mad95] achieve a regularisation of quantum field theory models without losing symmetry [GM92, GKP96b, GKP96a, GS99]. Of course, the usual divergences of a quantum field theory on S^2 will reappear in the limit $\mathcal{N} \to \infty$. This limit was investigated in [CMS01]. For the one-loop self-energy in the ϕ^4-model, a finite but non-local difference between the $\mathcal{N} \to \infty$ limit of the fuzzy sphere and the ordinary sphere was found. See [Haw99] for similar calculations. Gauge models on the fuzzy sphere have been studied, e.g. in [IKTW01, Ste04]. Another approach to finite quantum field theories on noncommutative spaces is point-splitting via tensor products [CHMS00, BDFP03].

4.6 A non-example: the noncommutative torus

The noncommutative torus A_θ is the universal C^*-algebra generated by two unitaries U, V satisfying $UV = \mathrm{e}^{2\mathrm{i}\pi\theta} VU$, for some $\theta \in \mathbb{R} \setminus \mathbb{Q}$. Several equivalent presentations are known, for instance, as irrational rotation algebra. It is probably the best-studied noncommutative space [Rie90]. We are more interested in a Fréchet subalgebra of A_θ which consists of elements of the form

$$a = \sum_{q_1, q_2 \in \mathbb{Z}} a_{q_1 q_2} U^{q_1} V^{q_2}, \quad \langle a, b\rangle_n := \sum_{q_1, q_2 \in \mathbb{Z}} (1 + |q_1| + |q_2|)^n \overline{a_{q_1 q_2}} b_{q_1 q_2} < \infty. \tag{4.14}$$

The noncommutative torus is not an AF algebra (all AF-algebras have trivial K_1-group, whereas $K_1(A_\theta) = \mathbb{Z}$). However, there is an AF algebra into which A_θ embeds. The construction relies on the approximative continued fraction expansion of θ and is explained in [LLS01]. It is also shown there that objects of a quantum field theory on the noncommutative torus can conveniently be constructed as limits of finite-dimensional problems. This formulation is employed in [LLS04] to construct matrix models which approximate field theories on the noncommutative torus.

The computation of the spectral action [EILS08] and renormalisation of scalar fields [DPV16] on the noncommutative torus are quite involved. QFTs on projective modules over the noncommutative torus were treated in [GJKW07].

4.7 Other (non-) examples

Many other noncommutative spaces have been studied. A formal definition of free Euclidean scalar fields is mostly possible. An overview goes beyond the scope of this survey, but a few examples can be flashed: κ-deformation [IMSS11], quantum groups [ILS09].

5 QFT on NCG: the first years[4]

5.1 Very short overview about QFT on deformed Minkowski space

The initial work by Doplicher–Fredenhagen–Roberts [DFR95] mentioned in Section 1.2 also introduced free relativistic quantum fields on quantum space-time and prepared for a perturbative treatment of interactions. Later the Euclidean approach (see Section 5.2), formally obtained by a Wick rotation, became much more popular.

It was pointed out in [BDFP02] that a simple Wick rotation does *not* give a meaningful theory on Minkowskian space-time, first of all because formal (i.e. wrong) Wick rotation destroys unitarity [GM00, AGBZ01]. To obtain a consistent Minkowskian quantum field theory, it was proposed in [BDFP02] to iteratively solve the field equations à la Yang-Feldman. See also [Bah03]. Another possibility is time-ordered perturbation theory [LS02c, LS02b]. See also [BFG$^+$03, DS03]. Unfortunately, the resulting Feynman rules become so complicated that apart from tadpole-like diagrams [BFG$^+$03] it seems impossible to perform perturbative calculations in time-ordered perturbation theory. Moreover, it seems impossible to preserve Ward identities [ORZ04], and dispersion relations are severely distorted [Zah06].

[4] This is a slight adaptation from a previous review [Wul06].

A fascinating re-import of quantum fields on deformed Minkowski space back into usual Minkowski space was initiated by Grosse and Lechner [GL07]. They considered a family of free quantum fields indexed by the noncommutativity parameter and related by a Lorentz transform. They showed that the family can be considered as wedge-local on ordinary Minkowski space, satisfying the axioms, and possessing a non-trivial two-particle S-matrix. Their construction was generalised to the Haag–Kastler setting in [BS08] and termed 'warped convolution'. After further investigation in the Wightman setting [GL08], it was shown in [BLS11] that warped convolution is an isometric representation of Rieffel's strict deformation quantisation [Rie93] of C^*-dynamical systems.

5.2 Perturbative QFT on deformed Euclidean space

The Euclidean approach started with Filk [Fil96] who showed that the planar graphs of a field theory on the Moyal plane are identical to the commutative theory (and thus have the same divergences). Another achievement in [Fil96] was the definition of the *intersection matrix* of a graph which is read off from its reduction to a rosette. Later in [VGB99] the persistence of divergences was rephrased in the framework of noncommutative geometry, based on the general definition of a dimension and the noncommutative formulation of external field quantisation. At about the same time, Connes et al. [CDS98] investigated the possibility that M-theory is compactified on the noncommutative torus instead of on an ordinary torus. M-theory lives in higher dimensions so that some of them must be compactified to give a realistic model. Compactifying on a noncommutative instead of a commutative torus amounts to turn on a constant background 3-form C. An alternative interpretation based on D-branes on tori in presence of a Neveu-Schwarz B-field was given by Douglas and Hull [DH98].

Knowing that divergences persist in quantum field theories on the Moyal plane, the question arises whether these models are renormalisable. Martín and Sánchez-Ruiz [MSR99] investigated U(1) Yang–Mills theory on the noncommutative $\mathbb{R}^4$ (the same as the Moyal space) at the one-loop level. They found that all one-loop pole terms of this model in dimensional regularisation[5] can be removed by multiplicative renormalisation (minimal subtraction) in a way preserving the BRST symmetry. This is completely analogous to the situation on the noncommutative 4-torus [KW00] where ζ-function techniques and cocycle identities are used to extract pole parts of Feynman graphs, thereby proving multiplicative renormalisation of the initial action and verifying the Ward identities. Around the same time there appeared also an investigation of $(2+1)$-dimensional super-Yang–Mills theory with the 2-dimensional space being the noncommutative torus [SJ99].

[5] There is of course a problem extending Θ to complex dimensions, this is however discussed in [MSR99].

Inspired by [CDS98] and its companion [DH98], Schomerus [Sch99] observed that in string theory with D-branes, and a magnetic field on the branes, the field theory limit of string theory produces Kontsevich's formal $\star$-product [Kon03] of deformation quantisation. There are also other noncommutative spaces which arise as limiting cases of string theory [ARS99].

Shortly later, the appearance of noncommutative field theory in the zero-slope limit of type-II string theory was thoroughly investigated by Seiberg and Witten [SW99]. They noticed that passing to the zero-slope limit in two different regularisation schemes (point-splitting and Pauli–Villars) gives rise to a Yang–Mills theory either on noncommutative or on commutative $\mathbb{R}^D$. Since the regularisation scheme cannot matter, Seiberg and Witten argued that both theories must be gauge-equivalent. More general, under an infinitesimal transformation of θ one has to require that gauge-invariant quantities remain gauge-invariant. This requirement leads to the Seiberg–Witten differential equation

$$\frac{dA_\mu}{d\theta_{\rho\sigma}} = -\frac{1}{8}\{A_\rho, \partial_\sigma A_\mu + F_{\sigma\mu}\}_\star + \frac{1}{8}\{A_\sigma, \partial_\rho A_\mu + F_{\rho\mu}\}_\star, \tag{5.1}$$

where $\{a, b\}_\star = a \star b + b \star a$.

The Seiberg–Witten paper [SW99] made the connection between string theory and noncommutative geometry extremely popular. Several lines of research appeared. One important question concerns the extension of the one-loop renormalisation proof of quantum field theories on noncommutative $\mathbb{R}^D$ to any loop order. The main contributions to this programme are due to Chepelev and Roiban [CR00]. Their work uses ribbon graphs in an essential manner. Ribbon graphs were invented by t'Hooft [tH74] for strong interactions and were first employed for noncommutative field theories in [Haw99]. Such ribbon graphs can be drawn on an (oriented) Riemann surface with boundary to which the external legs of the graph are attached. See Section 7.1. Chepelev and Roiban derived the parametric representation of a ribbon graph Γ and found the analogues of the Symanzik polynomials U_Γ, V_Γ of Theorem 2.6, which now contain θ in a manner that depends on the topology of Γ. This makes the identification of divergent Hepp sectors more involved. The first conclusion in [CR00] was that a noncommutative field theory is renormalisable iff its commutative counterpart is renormalisable. However, by computing the non-planar one-loop graphs explicitly, Minwalla, Van Raamsdonk and Seiberg pointed out a serious problem in the renormalisation of ϕ^4-theory on noncommutative $\mathbb{R}^4$ and ϕ^3-theory on noncommutative $\mathbb{R}^6$ [MVRS00]. It turned out that this problem was simply overlooked in the first version of [CR00], with the power-counting analysis being correct. A refined proof of the power-counting theorem was given in [CR01]. Roughly, the problem discovered in [MVRS00] is the following: Non-planar graphs are regulated by the phase factors in the $\star$-product (4.7), but only if the external momenta of the graph are non-exceptional. Inserting non-planar graphs (declared as regular) as subgraphs into bigger graphs, external momenta of the subgraph are internal momenta for the total graph. As such, exceptional external momenta for the subgraph are realised in the loop integration, resulting in an divergent integral for the total graph. This is the so-called *UV/IR-mixing* problem [MVRS00].

The UV/IR mixing problem received considerable attention. In the following months an enormous number of articles doing (mostly) one-loop computations of all kind of models appeared of which only a few key results should be mentioned in this survey: the two-loop calculation of ϕ^4-theory [ABK00b]; the renormalisation of complex $\phi \star \phi^* \star \phi \star \phi^*$ theory [ABK00a], later explained by a topological analysis [CR01]; computations in noncommutative QED [Hay00]; the calculation of noncommutative $U(1)$ Yang–Mills theory [MST00], with an outlook to super-Yang–Mills theory; the one-loop analysis of noncommutative $U(N)$ Yang–Mills theory [BS01]. Several reviews of these activities appeared, for instance, by Konechny and Schwarz with focus on compactifications of M-theory on noncommutative tori as well as on instantons and solitons on noncommutative $\mathbb{R}^D$ [KS02], by Douglas and Nekrasov [DN01] as well as by Szabo [Sza03], both with focus on field theory on noncommutative spaces in relation to string theory. For reviews which include results discussed in Section 6, see [Wul06, Riv07b].

It was also investigated whether for gauge theories on noncommutative $\mathbb{R}^D$ the Seiberg–Witten θ-expansion defined in (5.1) can be helpful. In that approach one solves (5.1) as a formal power series in θ, in fact often truncated to finite order. The result is a local quantum field theory which has no relation to the original problem. Anyway, the Seiberg–Witten approach was made popular in [JSSW00] where it was argued that this is the only way to obtain a finite number of degrees of freedom in non-Abelian noncommutative Yang–Mills theory. The solution of (5.1) to all orders in θ and lowest order in $A^{(0)}$ was given in [Gar00]. A generating functional for the complete solution was derived in [JSW01]. The quantum field theoretical treatment of θ-expanded field theories was initiated in [BGP$^+$02]. In [BGG$^+$01] it was shown that the superficial divergences in the photon self-energy are field redefinitions to all orders in θ and any loop order [BGG$^+$01]. However, this fails for more complicated sectors [Wul02]. In fact, in the class of formal power series in θ, quantum field theoretical quantities are (up to field redefinition) the same with or without the Seiberg–Witten map [GW02]. Thus, the Seiberg–Witten map is merely an unphysical (but convenient) change of variables.

In [Ste07] an alternative interpretation of the UV/IR-mixing of $U(1)$-gauge fields was proposed: It does not describe a noncommutative photon but a sort of graviton. See [GSW08] and, for a review, [Ste10]. Phenomenological investigations of θ-expanded field theories have also been performed [CJS$^+$02, BDD$^+$03].

5.3 *Numerical simulations*

There is another approach which goes back to older work on the large-N limit of 2-dimensional $SU(N)$ lattice gauge theory. Here the number of degrees of freedom is reduced and corresponds to a zero-dimensional model [EK82], under the condition that no spontaneous breakdown of the $[U(1)]^4$-symmetry appears.

As shown in [GAO83], a spontaneous symmetry breakdown does not appear when twisted boundary conditions are used. This construction was adapted in [AII+00] to type-IIB matrix models. It was shown in [AMNS99] that, imposing a natural constraint for the (finite) matrices, the twisted Eguchi–Kawai construction [GAO83] can be generalised to noncommutative Yang–Mills theory on a toroidal lattice. The appearing gauge-invariant operators are the analogues of Wilson loops [Wil74]. This formulation enabled numerical simulations [BHN02, BHN03] of the various limiting procedures which confirmed conjectures [GS01] about striped and disordered patterns in the phase diagram of spontaneously broken noncommutative ϕ^4-theory.

6 Renormalisation of noncommutative ϕ^4-theory to all orders

With Harald Grosse we started in summer 2002 an investigation of the UV/IR-mixing problem in the matrix basis (4.3) [GBV88] of the Moyal space. In combination with Polchinski's implementation [Pol84] of Wilson's renormalisation group equations [WK74], we hoped to disentangle various limit procedures which occur in the renormalisation of Feynman graphs. Our programme succeeded, but not for the anticipated reason.

6.1 QFT with harmonic oscillator covariance on Moyal space

The Laplace kernel defined in (4.8) with (4.9) takes for the $\mathcal{A}_\Theta^N$-approximation of 4-dimensional Moyal space (see Section 4.3) the form [GW05b]

$$\sum_{\nu=4}^{2} \mathrm{Tr}((\partial_\nu a)^* \partial_\nu a) + \mu^2 \mathrm{Tr}(X^* X) =: \sum_{k_i,l_i,m_i,n_i=1}^{N} \Delta_{\substack{k_1 \, l_1 \\ k_2 \, l_2};\substack{m_1 \, n_1 \\ m_2 \, n_2}} a_{\substack{k_1 \, l_1 \\ k_2 \, l_2}} a_{\substack{m_1 \, n_1 \\ m_2 \, n_2}},$$

$$\begin{aligned}
\Delta_{\substack{k^1 \, l^1 \\ k^2 \, l^2};\substack{m^1 \, n^1 \\ m^2 \, n^2}} &= \big(\mu^2 + \tfrac{2}{\theta}(m^1+n^1+m^2+n^2+2)\big)\delta_{n^1k^1}\delta_{m^1l^1}\delta_{n^2k^2}\delta_{m^2l^2} \qquad (6.1)\\
&\quad - \tfrac{2}{\theta}\big(\sqrt{k^1l^1}\,\delta_{n^1+1,k^1}\delta_{m^1+1,l^1} + \sqrt{m^1n^1}\,\delta_{n^1-1,k^1}\delta_{m^1-1,l^1}\big)\delta_{n^2k^2}\delta_{m^2l^2}\\
&\quad - \tfrac{2}{\theta}\big(\sqrt{k^2l^2}\,\delta_{n^2+1,k^2}\delta_{m^2+1,l^2} + \sqrt{m^2n^2}\,\delta_{n^2-1,k^2}\delta_{m^2-1,l^2}\big)\delta_{n^1k^1}\delta_{m^1l^1}.
\end{aligned}$$

We call the line (6.1) the local interaction, the last two lines the nearest-neighbour interaction. When deriving Feynman rules for ribbon graphs on assigns to the edges the covariance, which is the inverse of $\Delta_{\substack{k^1 \, l^1 \\ k^2 \, l^2};\substack{m^1 \, n^1 \\ m^2 \, n^2}}$. As shown in [GW05a], renormalisability requires a sufficiently fast decay of the covariance $C(e_{\substack{k^1 \, l^1 \\ k^2 \, l^2}}, e_{\substack{m^1 \, n^1 \\ m^2 \, n^2}})$ with $\max(k_i, l_i, m_i, n_i)$ and a bound on partial sums $\sum_{k^1,k^2} \max_{l_i,m_i} C(e_{\substack{k^1 \, l^1 \\ k^2 \, l^2}}, e_{\substack{m^1 \, n^1 \\ m^2 \, n^2}})$.

It turned out that this would be the case if only the local interaction was present, but the nearest-neighbour interaction spoils it. We therefore decided to scale down, completely ad hoc, the nearest-neighbour terms. The resulting kernel

$$\begin{aligned} H^{\Omega}_{\substack{m^1\ n^1\\ m^2\ n^2};\substack{k^1\ l^1\\ k^2\ l^2}} &= \big(\mu^2+\tfrac{2+2\Omega^2}{\theta}(m^1+n^1+m^2+n^2+2)\big)\delta_{n^1k^1}\delta_{m^1l^1}\delta_{n^2k^2}\delta_{m^2l^2} \\ &\quad - \tfrac{2-2\Omega^2}{\theta}\big(\sqrt{k^1l^1}\,\delta_{n^1+1,k^1}\delta_{m^1+1,l^1}+\sqrt{m^1n^1}\,\delta_{n^1-1,k^1}\delta_{m^1-1,l^1}\big)\delta_{n^2k^2}\delta_{m^2l^2} \\ &\quad - \tfrac{2-2\Omega^2}{\theta}\big(\sqrt{k^2l^2}\,\delta_{n^2+1,k^2}\delta_{m^2+1,l^2}+\sqrt{m^2n^2}\,\delta_{n^2-1,k^2}\delta_{m^2-1,l^2}\big)\delta_{n^1k^1}\delta_{m^1l^1} \end{aligned} \tag{6.2}$$

turned out to describe the harmonic oscillator Schrödinger operator

$$\begin{aligned} &\sum_{\nu=1}^{4}\Big(\mathrm{Tr}((\partial_\nu a)^*\partial_\nu a)+\frac{4\Omega}{\theta^2}\mathrm{Tr}((M_\nu a)^*M_\nu a)\Big)+\mu^2\mathrm{Tr}(a^*a) \\ &=: \sum_{k_i,l_i,m_i,n_i=1}^{\mathcal{N}} H^{\Omega}_{\substack{k_1\ l_1\\ k_2\ l_2};\substack{m_1\ n_1\\ m_2\ n_2}}\, a_{\substack{k_1\ l_1\\ k_2\ l_2}}\, a_{\substack{m_1\ n_1\\ m_2\ n_2}}\,, \end{aligned} \tag{6.3}$$

where M_ν is the pointwise multiplication introduced in (4.10). The introduction of Ω was completely ad hoc. It has, however, one appealing property. The interaction $\mathrm{Tr}(\Phi^n)$, for n even, is invariant under a duality transformation of the Moyal product discovered by Langmann and Szabo [LS02a]. This transformation transforms $\sum_{\nu=1}^4 \mathrm{Tr}((\partial_\nu\Phi)^*\partial_\nu\Phi)$ into $\sum_{\nu=1}^4 \mathrm{Tr}((M_\nu\Phi)^*M_\nu\Phi)$ and vice versa, thus achieving duality-covariance of the model with Ω-term.

For renormalisation a fine control of the covariance is necessary. To invert (6.3) one first diagonalises (6.2) by noticing that the corresponding 3-term relation defines the Meixner polynomials [KS96]. Then the inverse is computed to

$$\begin{aligned} C(e_{\substack{m^1\ n^1\\ m^2\ n^2}}, e_{\substack{k^1\ l^1\\ k^2\ l^2}}) &= \frac{\theta}{2(1+\Omega)^2}\delta_{m^1+k^1,n^1+l^1}\delta_{m^2+k^2,n^2+l^2} \\ &\times \sum_{v^1=\frac{|m^1-l^1|}{2}}^{\frac{m^1+l^1}{2}} \sum_{v^2=\frac{|m^2-l^2|}{2}}^{\frac{m^2+l^2}{2}} B\big(1+\tfrac{\mu^2\theta}{8\Omega}+\tfrac{1}{2}(m^1+k^1+m^2+k^2)-v^1-v^2,\ 1+2v^1+2v^2\big) \\ &\times {}_2F_1\left(\begin{matrix} 1+2v^1+2v^2,\ \frac{\mu^2\theta}{8\Omega}-\frac{1}{2}(m^1+k^1+m^2+k^2)+v^1+v^2 \\ 2+\frac{\mu^2\theta}{8\Omega}+\frac{1}{2}(m^1+k^1+m^2+k^2)+v^1+v^2 \end{matrix}\,\middle|\, \frac{(1-\Omega)^2}{(1+\Omega)^2}\right) \\ &\times \prod_{i=1}^{2}\Big(\frac{1-\Omega}{1+\Omega}\Big)^{2v^i}\sqrt{\binom{n^i}{v^i+\frac{n^i-k^i}{2}}\binom{k^i}{v^i+\frac{k^i-n^i}{2}}\binom{m^i}{v^i+\frac{m^i-l^i}{2}}\binom{l^i}{v^i+\frac{l^i-m^i}{2}}}\,. \end{aligned} \tag{6.4}$$

The free theory is now under control. In the limit $\Omega \to 0$ to the covariance of the Laplacian a confluent hypergeometric function arises.

Formally we define an interacting QFT via the perturbation (3.6) of the Bochner–Minlos measure $\mathrm{d}\mathcal{M}_C$ associated with the covariance (6.4):

$$G_{\substack{k^1 l^1\\ m^1 n^1};\dots;\substack{k^N l^N\\ m^N n^N}} := \frac{\displaystyle\int \mathrm{d}\mathcal{M}_C(\Phi)\, \Phi_{\substack{k^1 l^1\\ m^1 n^1}} \cdots \Phi_{\substack{k^N l^N\\ m^N n^N}} \exp\Big(-\frac{\lambda}{4}\mathrm{Tr}(\Phi^4)\Big)}{\displaystyle\int \mathrm{d}\mathcal{M}_C(\Phi)\, \exp\Big(-\frac{\lambda}{4}\mathrm{Tr}(\Phi^4)\Big)}, \qquad (6.5)$$

where $\Phi_{\substack{k^1 l^1\\ m^1 n^1}} := \Phi(e_{\substack{k^1 l^1\\ m^1 n^1}})$.

At this stage we are interested in a perturbative expansion as formal power series $G_{\substack{k^1 l^1\\ m^1 n^1};\dots;\substack{k^N l^N\\ m^N n^N}} = \sum_{v=0}^{\infty}\lambda^v \sum_{g=0}^{\infty}\sum_{B=1}^{N} G^{(v,B,g)}_{\substack{k^1 l^1\\ m^1 n^1};\dots;\substack{k^N l^N\\ m^N n^N}}$ in which we collect contributions of ribbon graphs with v vertices, B boundary components and genus g. We postpone a discussion of such ribbon graphs to Section 7.1. As pointed out in Programme 3.8, for renormalisation we have to restrict to finite matrix size $\mathcal{N}$ and to take a conditional limit $\mathcal{N} \to \infty$ where certain correlation functions (6.5) are held fixed. In the original work [GW05b], instead of a sharp restriction $\mathcal{N}$ a smooth cut-off of matrix indices near $\theta\Lambda^2$ was chosen. This allows to derive first-order Polchinski differential equations [Pol84] which describe the flow of correlation functions when varying the scale Λ. The key strategy is to integrate these differential equation for mixed boundary conditions. Finitely many correlation functions which are termed relevant or marginal are integrated from $\Lambda = 0$ to Λ. These are $G^{(v,1,0)}_{\substack{00\\00};\substack{00\\00};\substack{00\\00};\substack{00\\00}}$, $G^{(v,1,0)}_{\substack{00\\00};\substack{00\\00}}$ (which has a relevant and a marginal contribution) as well as $G^{(v,1,0)}_{\substack{10\\00};\substack{01\\00}} = G^{(v,1,0)}_{\substack{00\\10};\substack{00\\01}} = G^{(v,1,0)}_{\substack{01\\00};\substack{10\\00}} = G^{(v,1,0)}_{\substack{00\\01};\substack{00\\10}}$. The remaining infinitely many irrelevant correlation functions are integrated from $\Lambda = \infty$ down to Λ. Here a subtlety has be taken into account: A local planar four-point function (which is not the generic case; adjacent indices are the same) must be split as

$$\begin{aligned} &G^{(v,1,0)}_{\substack{k^1 k^2\\ m^1 m^2};\substack{k^2 k^3\\ m^2 m^3};\substack{k^3 k^4\\ m^3 m^4};\substack{k^4 k^1\\ m^4 m^1}} \\ &\equiv \Big(G^{(v,1,0)}_{\substack{k^1 k^2\\ m^1 m^2};\substack{k^2 k^3\\ m^2 m^3};\substack{k^3 k^4\\ m^3 m^4};\substack{k^4 k^1\\ m^4 m^1}} - G^{(v,1,0)}_{\substack{00\\00};\substack{00\\00};\substack{00\\00};\substack{00\\00}}\Big) + G^{(v,1,0)}_{\substack{00\\00};\substack{00\\00};\substack{00\\00};\substack{00\\00}}. \end{aligned}$$

The final term is marginal and integrated from 0 to Λ, whereas the difference of the first two terms must be proved to be irrelevant, integrated from ∞ down to Λ. Similar mixed integrations are necessary for the local and nearest-neighbour planar 2-point function. After all one achieves, order by order in the coupling constant, bounds which allow to take the limit $\Lambda \to \infty$ of any Λ-dependent correlation function. Here bounds on the covariance enter, which in [GW05b] were only numerically achieved. In [RVTW06] rigorous analytic bounds for Ω close to 1 were proved.

Renormalisability of the 2-dimensional case has been proved in [GW03]. In this case the oscillator frequency required in intermediate steps can be switched of at the end.

6.2 The β-function

In total we have four marginal and relevant correlation functions integrated from initial values at $\Lambda_R = 0$ to Λ. They can be interpreted as produced by Λ-dependent parameters in the $S_{\text{int}}(\Phi)$-perturbed measure. These are

1. a scale-dependent mass $\mu(\Lambda)$,
2. a scale-dependent oscillator frequency $\Omega(\Lambda)$ in (6.2) or (6.4),
3. a wave-function renormalisation $\Phi \mapsto \sqrt{Z(\Lambda)}\Phi$ which induces a global prefactor $\frac{1}{Z(\Lambda)}$ in front of (6.4) and
4. a combined factor $\lambda \mapsto \lambda(\lambda)Z(\Lambda)^2$ in $S_{\text{int}}(\Phi)$.

The logarithmic derivatives $\beta_\Omega = \Lambda\frac{\partial}{\partial\Lambda}\Omega(\mu_R, \Omega_R, \lambda_R, \Lambda)$ and $\beta_\lambda = \Lambda\frac{\partial}{\partial\Lambda}\lambda(\mu_R, \Omega_R, \lambda_R, \Lambda)$ are referred to as β-functions (of oscillator frequency and coupling constant). Here $\mu_R, \Omega_R, \lambda_R$ are the initial values of mass, oscillator frequency and coupling constant corresponding to the moments held fixed at $\Lambda_R = 0$. At one-loop order one finds [GW04]

$$\beta_\Omega = \frac{\lambda_R\Omega_R}{96\pi^2}\frac{(1-\Omega_R^2)}{(1+\Omega_R^2)^3} \qquad \beta_\lambda = \frac{\lambda_R^2}{48\pi^2}\frac{(1-\Omega_R^2)}{(1+\Omega_R^2)^3}. \tag{6.6}$$

These relations have far-reaching consequences. Namely, $\frac{\lambda(\Lambda)}{\Omega^2(\Lambda)}$ is *constant* under the renormalisation group flow (first noticed by David Broadhurst). Solving the coupled system of differential equations one finds $\lim_{\Lambda\to\infty}\Omega(\Lambda) = 1$ and consequently $\lim_{\Lambda\to\infty}\lambda(\Lambda) = \frac{\lambda_R}{\Omega_R^2}$. The finiteness of $\lambda(\infty)$ is in sharp contrast with the usual (commutative) ϕ_4^4-model which is believed to suffer from the triviality problem. Strictly speaking, triviality is only proved in $4+\epsilon$ dimensions [Aiz81, Frö82], but the perturbative renormalisation group flow indicates triviality also in 4 dimensions. Triviality means that the running coupling constant $\lambda(\Lambda)$ diverges already at finite Λ_0, referred to as the Landau pole [LAK54]. The only possibility to extend the model to $\Lambda \to \infty$ is to let the initial coupling $\lambda_R \to 0$, resulting in a free (or trivial) field theory.

The one-loop absence of the triviality problem had considerable impact on the further development of the subject. It seemed that implementing the constructive (as opposed to perturbative) approach [Riv91] to quantum field theory, the Φ^4-model on 4-dimensional Moyal space could possibly become the first constructed interacting quantum field theory model in 4 dimensions. The first step, the multiscale-slicing of the covariance, was introduced in [RVTW06]. We describe in the next subsection the further development into this direction. Here we focus on the progress which the enlarged community achieved for the β-function.

At the fixed point $\Omega = 1$ of the renormalisation group flow, the covariance simplifies enormously: The matrix Schrödinger operator (6.2) becomes $H^{\Omega=1}_{\substack{m^1 n^1 \\ m^2 n^2};\substack{k^1 l^1 \\ k^2 l^2}} = \big(\mu^2+\frac{4\Omega^2}{\theta}(m^1+n^1+m^2+n^2+2)\big)\delta_{n^1k^1}\delta_{m^1l^1}\delta_{n^2k^2}\delta_{m^2l^2}$ with inverse

$$C^{\Omega=1}\left(e_{\substack{m^1 n^1 \\ m^2 n^2}}, e_{\substack{k^1 l^1 \\ k^2 l^2}}\right) = \frac{\delta_{n^1k^1}\delta_{m^1l^1}\delta_{n^2k^2}\delta_{m^2l^2}}{\mu^2+\frac{4\Omega^2}{\theta}(m^1+n^1+m^2+n^2+2)}. \tag{6.7}$$

This simplification has early been exploited by Langmann et al. [LSZ03, LSZ04] to make contact with the theory of matrix models [DFGZJ95]. We return to this point in Section 7. Also the perturbative calculation of the β-function simplifies enormously. Disertori and Rivasseau proved in [DR07] that at $\Omega = 1$ the β-function remains zero up to three-loop order.

This result clearly suggested the existence of a symmetry transformation which implies $\beta_\lambda = 0$ to all orders in perturbation theory. The transformation was soon identified by Disertori et al. in [DGMR07], inspired by 1-dimensional Fermi liquid [BM04]. We will derive these Ward–Takahashi identities in slightly generalised form in Section 7.3. In [DGMR07] a special case was considered and thereby proved that the divergent part of the 4-point function is, graph by graph, completely determined by the divergent part of the 2-point function. Therefore, it is enough to renormalise the 2-point function; no infinite renormalisation of the coupling constant λ is necessary. This means that the β-function at $\Omega = 1$ vanishes to all orders in perturbation theory.

Research bifurcated at this point. With H. Grosse we developed a solution strategy for models in the Kontsevich class (to be reviewed in Sections 8 and 9). The authors of [DGMR07] tailored constructive renormalisation theory to the noncommutative situation. We briefly review some achievements in the next subsection.

6.3 Constructive renormalisation

Several aspects of constructive renormalisation are best understood in position space. The bosonic covariance is the Mehler kernel [GRVT06],

$$C(x,y) = \frac{\Omega^2}{\pi^2\theta^2}\int_0^\infty \frac{dt}{\sin^2\frac{4\Omega t}{\theta}}\, e^{\left(-\frac{\Omega}{2\theta}\|x-y\|^2\coth(\frac{2\Omega}{\theta})-\frac{\Omega}{2\theta}\|x+y\|^2\tanh(\frac{2\Omega}{\theta})-\mu^2 t\right)}. \tag{6.8}$$

In [GRVT06] also the fermionic covariance was evaluated, which was used in [VT07] to prove renormalisability to all orders of the orientable noncommutative Gross–Neveu model. In [GR07] the parametric representation was derived, which in particular identified the analogues of the Symanzik polynomials (see Theorem 2.6). Besides linking QFT to algebraic geometry, these Symanzik polynomials are also particular multivariate versions of the Tutte polynomial in graph theory [KRTW10]. The graph-theoretical interpretation of the noncommutative analogue of the Symanzik polynomial (Bollobás-Riordan polynomials) was clarified in [KRVT11].

Traditional bosonic constructive renormalisation employs two technical tools: the cluster expansion and the Mayer expansion [GJS74, BK87]. They are designed for usual Euclidean space which is divided into cubes to test the localisation of vertices. Since vertices of a QFT on Moyal space are not localised, these traditional tools cannot be applied. In [Riv07a], Rivasseau developed the *loop vertex expansion* which serves as substitute for cluster and Mayer expansion. It was made for constructive matrix theory, but it is also a conceptional simplification for traditional constructive renormalisation [MR08]. The loop vertex expansion combines the Hubbard–Stratonovich transform, the BKAR forest formula [BK87, AR95] and the replica trick. For $\mathcal{N} \times \mathcal{N}$-matrices Φ, the Hubbard–Stratonovich transform is based on the following identity:

$$\mathrm{e}^{-\lambda/4\,\mathrm{Tr}(\Phi^4)} \int d\sigma\; \mathrm{e}^{-\frac{1}{2}\mathrm{Tr}(\sigma^2)} = \int d\sigma\; \mathrm{e}^{-\frac{1}{2}\mathrm{Tr}(\sigma^2)-\mathrm{i}\sqrt{\lambda/2}\,\mathrm{Tr}(\sigma\Phi^2)},$$

where $d\sigma$ is the translation-invariant Lebesgue measure on $\mathbb{R}^{\mathcal{N}^2}$. Then the Euclidean scalar field Φ is integrated with measure $d\mathcal{M}_C(\Phi)$, resulting in an effective potential $\mathrm{e}^{-V(\sigma)}$, called the loop vertex, for the intermediate field σ. In the limit $\mathcal{N} \to \infty$ divergences reappear and must be treated by multiscale slicing [GR15]. The factorisation over slices is automated by Grassmann integrals over fermionic variables. Then, in an expansion of the exponential $\mathrm{e}^{-W} = \sum_{n=0}^{\infty} \frac{1}{n!}(-W)^n = \sum_{n=0}^{\infty} \frac{1}{n!} \prod_{a=1}^{n} (-W_a)\Big|_{W_a=W}$, one artificially distinguishes the factors (replica trick). The replica measure is degenerate, consisting of an $n \times n$-matrix with all entries 1. The BKAR forest formula [BK87, AR95] allows to write such matrices as sum of positive matrices indexed by forests. Here a two-level forest formula, bosonic and fermionic, is necessary. Taking the logarithm amounts to restricting the forest to a sum of trees. In this way the organisation of the perturbative series into trees, briefly outlined at the end of Section 2.5, is achieved. For an overview about this and other new methods in constructive QFT, see [GRS14]. In [RW12, RW15] this new constructive renormalisation method was successfully applied to the ϕ_2^4-model. The considerably harder problem, the constructive renormalisation of the Φ^4-model on 2-dimensional Moyal space with harmonic propagation of critical frequency $\Omega = 1$, was achieved by Wang [Wan18]. He proved that the logarithm of the partition function is the Borel sum of its perturbation series, analytic in a cardioid domain $|\lambda| < \rho\cos^2(\frac{1}{2}\arg(\lambda))$, excluding the negative reals.

6.4 Other developments

In [dGWW07] (using Mehler kernels in position space) and in [GW07] (using the matrix basis) an induced gauge theory with Ω-term was derived by coupling quantum scalar fields to classical gauge fields A_μ. The induced class of actions can be formulated using covariant coordinates [MSSW00] $X_\mu(x) = (\Theta^{-1})_{\mu\nu}x^\nu + A_\mu$ as (with Einstein's sum convention)

$$S = \int_{\mathbb{R}^4} \mathrm{d}x \,\Big(c_1 F_{\mu\nu} \star F^{\mu\nu} + c_2\{X_\mu, X_\nu\}_\star \star \{X^\mu, X^\nu\}_\star + c_3 X_\mu \star X^\mu\Big)(x), \tag{6.9}$$

where $F_{\mu\nu} = (\Theta^{-1})_{\mu\nu} - \mathrm{i}[X_\mu, X_\nu]_\star = \partial_\mu A_\nu - \partial_\nu A_\mu - \mathrm{i}[A_\mu, A_\nu]_\star$ is the curvature and $[a,b]_\star := a \star b - b \star a$, $\{a,b\}_\star := a \star b + b \star a$. These gauge models were abandoned because of their complicated vacuum structure.

In [GMRT09] another possibility to cure the UV/IR-mixing problem on Moyal space was suggested. It relies on a covariance which in position coordinates reads $C(x,y) = \int \mathrm{d}q \, \frac{e^{\mathrm{i}\langle q, x-y\rangle}}{\|q\|^2+\mu^2+\frac{\alpha}{\theta^2\|q\|^2}}$. Renormalisability to all orders was also proved in [GMRT09]. It was generalised to gauge theories in [BGK+08], where however the renormalisation is much more involved [BKR+10].

The extension of the harmonic oscillator potential to Minkowski space was discussed in [FS09]. Corresponding field theory models are problematic [Zah11].

A spectral noncommutative geometry which leads to (6.3) was analysed in [GW12, GW13a]. It lives in a Clifford algebra of doubled dimension which unites the standard Dirac operator with the 'Feynman slash', central in a new proposal for quanta of geometry [CCM15, CCM14].

6.5 Tensor models

Work on quantum field theories on noncommutative geometries inspired a new research topic: coloured random tensor models. Tensor models were introduced in [ADJ91] to extend the success of matrix models in describing 2-dimensional quantum gravity [DFGZJ95] (see also Section 7.2) to higher dimension. However, they were essentially useless because no analogue of the $1/N$-expansion [tH74] was found. In 2009, Gurau [Gur11a] introduced the *colouring* of tensor models. The colouring allowed Gurau [Gur11b, Gur12] and with Rivasseau in [GR11] shortly after to show that the tensor models have an analytically controlled $1/N$-expansion indexed by a positive integer called the *degree*. Then, Ben Geloun and Rivasseau proved in [BGR13] that a certain rank-4 tensor model is renormalisable to all orders in perturbation theory. The proof uses multiscale analysis and relies on experience with topological aspects in QFTs on noncommutative geometries. Soon many more renormalisable tensor models have been found; and they generically show asymptotic freedom [BG14]. First analytical results were established in [BGRR11]: tensor models undergo a phase transition to a theory of continuous random spaces when tuning to criticality. Also the loop vertex expansion [Riv07a] was generalised to tensor models [Gur14].

With these initial achievements, the subject of coloured tensor field theory took a spectacular development. For more information we refer to several reviews already available. There is an early review [GR12] by Gurau and Ryan, a sequence of 'Tensor track' lectures by Rivasseau (e.g. [Riv14]), Gurau's book [Gur17]. Most recently connections to the Sachdev-Ye-Kitaev model [SY93, Kit15a, Kit15b, MS16] have been established [Wit16]. The enormous activity goes beyond this survey; we only refer to recent lectures [KPT18].

7 Structures and techniques in matrix models

7.1 Riemann surfaces and ribbon graphs

A *Riemann surface* is a complex-analytic manifold of complex dimension 1 (hence real dimension 2). We are only interested in compact (and connected) Riemann surfaces on which we distinguish a (possibly empty) set $x_1, \dots, x_s$ of marked points. Two such Riemann surfaces are isomorphic if there exists a biholomorphic map between them which sends marked points into marked points. Two (marked) Riemann surfaces are homeomorphic if and only if they have the same number s of marked points and the same genus $g \in \mathbb{N}$. Their common Euler characteristics is $\chi_{g,s} = 2 - 2g - s$. The isomorphism classes of Riemannian manifolds of genus g with s marked points form for $\chi_{g,s} < 0$ a complex orbifold $\mathcal{M}_{g,s}$ of complex dimension $d_{g,s} = 3g - 3 + s$, called the *moduli space of complex curves*.

As conjectured by Witten [Wit91] and proved by Kontsevich [Kon92] the topology of the moduli spaces $\mathcal{M}_{g,s}$ is deeply related to matrix models, and therefore, as argued in the previous sections, to QFT on noncommutative geometries. For the Witten–Kontsevich relation we have to introduce two further structures: a compactification $\overline{\mathcal{M}}_{g,s}$ of the moduli spaces (which we sketch in Section 7.2) and ribbon graphs drawn on Riemann surfaces.

A *ribbon graph* is a simplicial 2-complex Γ made of $|\mathcal{V}_\Gamma|$ vertices, $|\mathcal{E}_\Gamma|$ edges and $|\mathcal{F}_\Gamma|$ faces. An edge connects two vertices (possible the same) and separates two faces (possibly the same). Ribbon graphs arise in several variants, depending on presence of marked faces or boundaries. First consider absent boundaries and a ribbon graph Γ with a total number $|\mathcal{F}_\Gamma|$ of faces, s of them marked. This ribbon graph can be drawn on a compact genus-g Riemann surface $\Sigma_{g,s}$ with s marked points. The genus is determined by $\chi = 2-2g-s = |\mathcal{V}_\Gamma| - |\mathcal{E}_\Gamma| + (|\mathcal{F}_\Gamma| - s)$. The drawing partitions $\Sigma_{g,s}$ into $|\mathcal{F}_\Gamma|$ closed subsurfaces, each topologically a disk, and with s of these disks containing precisely one marked point. Conversely, a Riemann

Fig. 1 Left: a ribbon graph with $|\mathcal{V}| = 2$ vertices (both tri-valent), $|\mathcal{E}| = 3$ edges and $|\mathcal{F}| = 3$ faces of which $s = 2$ are marked, drawn on a sphere (a compact Riemann surface of genus $g = 0$). One marked point is the north pole, the other one near the equator. The Euler characteristics is $\chi = |\mathcal{V}| - |\mathcal{E}| + (|\mathcal{F}| - s) = 2-3+(3-2) = 0 = 2-2g-s$. Centre: Removing small open disks around the marked points produces a surface with boundary, here of $B = 2$ components (topologically a cylinder). The marked faces become an annulus. Right: 4 half-edges, each connecting a vertex to one of the boundary components, are added to the central picture. In total there are $|\mathcal{V}| = 2$ vertices (one 4-valent and one 6-valent), $|\mathcal{E}| = 7$ (half-)edges and $|\mathcal{F}| = 5$ faces (4 of them external; the remaining internal face coincides with the unmarked face on the left). The Euler characteristics is $\chi = |\mathcal{V}| - |\mathcal{E}| + |\mathcal{F}| = 2 - 7 + 5 = 0$. We later say that this ribbon graph describes a contribution to the planar (1+3)-point function in a QFT model with both Φ^4- and Φ^6-interactions

surface is the gluing of topological disks into the faces of a ribbon graph. See the left picture in Figure 1 for an example.

Closely related are Riemann surfaces and ribbon graphs with boundaries. They arise by removing from the Riemann surface a small open disk (inside the disks glued into the ribbon graph) around a marked point. The previously marked face thus becomes an annulus (see central picture in Figure 1).

We extend the previous ribbon graphs by admitting half-edges in the annulus. Half-edges connect with its true end to a vertex on the previous marked face and with the other virtual end to the boundary. Crossings of half-edges with other (half-) edges are forbidden. See the right picture in Figure 1 for an example. We have two equivalent interpretations of the Euler characteristics. Either we ignore the half-edges (consider them as amputated), or we count them as ordinary edges but also include the additional external faces between half-edges and parts of the boundary.

Ribbon graphs with half-edges ending at boundary components can be contracted by subsequently gluing a pair of half-edges to form a true edge. Two cases must be distinguished:

I. half-edges ending at different boundary components (of the same surface or of disconnected surfaces) are glued; see Figure 2;
II. half-edges ending at the same boundary component are glued; see Figure 3.

We see that the subcase where all boundary components carry exactly one half-edge corresponds to the usual framework of bordisms. This framework is relevant for the Atiyah–Segal formulation [Ati88, Seg01] of *topological quantum field theory (TQFT)* [Wit88].

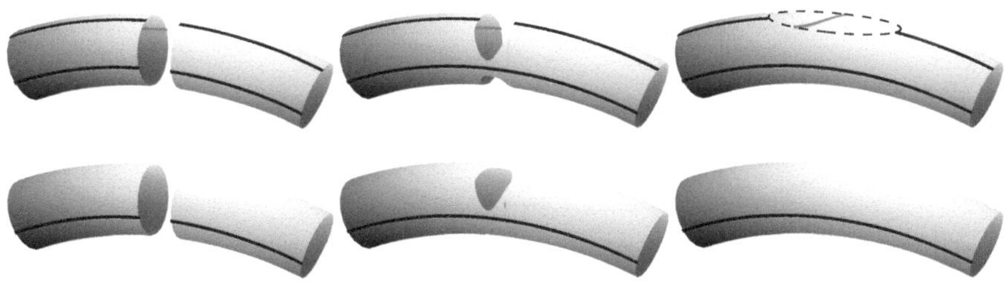

Fig. 2 Gluing of half-edges from different boundary components. (Top) [at least one of the previous boundary components carries ≥ 2 half-edges]: The neighbourhoods of the boundary components where we want to glue half-edges can be deformed to a half-cylinder (left). We glue two half-edges to an edge and also the faces bordering the previous half-edges (centre). The result is deformed to a single common boundary component (right). The total number of (half-)edges is reduced by 1 ($\Delta|\mathcal{E}| = -1$), the number of faces is reduced by 2, ($\Delta|\mathcal{F}| = -2$), the number of boundary components is reduced by 1 ($\Delta B = -1$). Vertices are unchanged. If boundary components of a connected surface are glued this way, its Euler characteristics change by $\Delta|\mathcal{V}| - \Delta|\mathcal{E}| + \Delta|\mathcal{F}| = 0 - (-1) + (-2) = -2(\Delta g) - \Delta B$, i.e. the genus is increased by $\Delta g = +1$. If two disconnected surfaces of topology (g_1, B_1) and (g_2, B_2) are glued, the resulting connected surface has Euler characteristics $(|\mathcal{V}_1|+|\mathcal{V}_2|) - (|\mathcal{E}_1|+|\mathcal{E}_2|-1) + (|\mathcal{F}_1|+|\mathcal{F}_2|-2) = (2-2g_1-B_1)+(2-2g_2-B_2)-1 = 2-2(g_1+g_2)-(B_1+B_2-1)$. Hence, the genus is additive. (Bottom) [both previous boundary components carry a single half-edge]: After deforming the neighbourhood to a half-cylinder (left), we glue both half-edges to an edge and also the faces bordering the previous half-edges (centre). The resulting boundary component no longer carries half-edges and by convention is shrunk to the empty set (right). We have $\Delta|\mathcal{E}| = -1$, $\Delta|\mathcal{F}| = -1$, $\Delta B = -2$. In case of the same surface, the Euler characteristics change by $\Delta|\mathcal{V}| - \Delta|\mathcal{E}| + \Delta|\mathcal{F}| = 0 - (-1) + (-1) = -2(\Delta g) - \Delta B$, i.e. the genus is increased by $\Delta g = +1$. In case of different surfaces, $(|\mathcal{V}_1|+|\mathcal{V}_2|) - (|\mathcal{E}_1|+|\mathcal{E}_2|-1) + (|\mathcal{F}_1|+|\mathcal{F}_2|-1) = (2-2g_1-B_1)+(2-2g_2-B_2) = 2-2(g_1+g_2)-(B_1+B_2-2)$. Hence, the genus is additive

We consider it worthwhile to extend this axiomatisation to the richer case where several half-edges end at the boundary. Namely, individual ribbon graphs correspond to a single contribution to the perturbative expansion of correlation functions (3.6) in a QFT on noncommutative geometries (see Section 3). For a non-perturbative formulation we are interested in the sum over all contributions encoded in ribbon graphs with the same boundary structure, or better we do not want to perturbatively expand at all. This means we encode a non-perturbative amplitude of a QFT on noncommutative geometries in a Riemann surface with boundary and *defects* on the boundary components. Such surfaces can be glued *along the defects*, not along the boundary as a whole. The corresponding rules can be read off from Figures 2 and 3 by reduction to the end points of half edges. To such a surface $\Sigma^g_{N_1,\dots,N_B}$ of genus g with B boundaries of $N_1, \dots, N_B$ defects, all $N_\beta \geq 1$, we associate an amplitude

$$G^{(g)}_{N_1,\dots,N_B} : \bigotimes_{\beta=1}^{B} \underbrace{\mathcal{A}_* \otimes_c \cdots \otimes_c \mathcal{A}_*}_{N_\beta} \to \mathbb{C}, \tag{7.1}$$

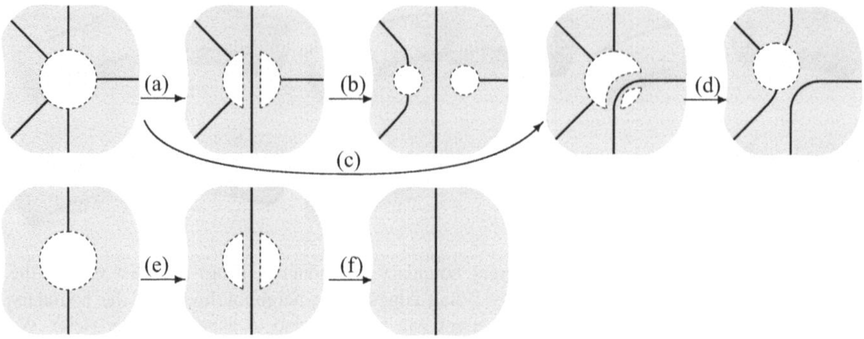

Fig. 3 Gluing of a pair of half-edges at the same boundary component which carries N half-edges. (Top) (a) We glue for $N \geq 4$ two non-neighboured half-edges to an edge and also join the faces bordering the previous half-edges. The boundary component splits into two. (b) The result is deformed into two disjoint boundary components with at least one and in total $N - 2$ half-edges. (c) We glue for $N \geq 3$ two neighboured half-edges to an edge and also join the faces bordering the previous half-edges. The boundary component splits into two, but one of them no longer carries any half-edge. (d) The boundary component without half-edge is shrunk to the empty set, the other one deformed into a boundary component with $N - 2$ half-edges. (Bottom) (e) We glue for $N = 2$ both half-edges to an edge and also join the faces bordering the previous half-edges. The boundary component splits into two, but none of them contains any half-edge. (f) Both boundary component without half-edges are shrunk to the empty set

where $\otimes_c$ is a cyclic tensor product and the leading $\bigotimes$ a symmetric tensor product. As such we have the first ingredient of a hypothetical functor from the category of Riemann surface with boundary and defects to the category of vector spaces. The gluing of such surfaces along defects is mapped to tensor products with contraction of vector spaces. Such an axiomatic setting analogous to TQFT could be called *noncommutative quantum field theory (NCQFT)* because it exactly captures the non-perturbative formulation of Section 3.

Question 7.1 Can these ideas be turned into a consistent axiomatisation? Is it useful in other areas?

In practice we have more structures on the vector space side:

- The vector spaces we are interested in have trace functionals $T_n(a_1, \dots, a_n) = \lambda_n \mathrm{Tr}(a_1 \cdots a_N)$. An n-valent vertex in a ribbon graph is mapped to T_n. These vertices alone do not describe any surface, but they encode another building block: an elementary n-disk, i.e. a sphere ($g = 0$) with one boundary component ($B = 1$) and n defects on it. As part of the rules one has to implement the removal of an elementary n-disk with at least one of its defects located on a boundary component. This removal translates to the Dyson–Schwinger equations in quantum field theory. See Sections 7.3, 8 and 9.

- The Ward–Takahashi identities present on the vector space side (see Corollary 7.6 later) should also be transferred to the category of surfaces.

We will see later that Dyson–Schwinger equations and Ward–Takahashi identities completely determine the NCQFT models, at least for Φ^4- and Φ^3-interaction.

7.2 The Kontsevich model

This section gives a short introduction into the Kontsevich model [Kon92]. It became a classical topic which is reviewed and discussed in nearly every book and review on matrix models and 2-dimensional quantum gravity. More details than given here can be found, e.g. in the books by Lando and Zvonkin [LZ04], by Eynard [Eyn16] as well as in the review [DFGZJ95] by Di Francesco, Ginsparg and Zinn-Justin. The Kontsevich model comes close to a quantum field theory; it ignores however renormalisation and is understood as a formal power series only. In Section 8 we show how to non-perturbatively construct renormalised correlation functions out of a quantum field theory closely related to the Kontsevich model. This construction heavily uses prior work on the original Kontsevich model, most importantly an exact solution [MS91] of a non-linear integral equation and the topological recursion [EO07, Eyn14, Eyn16].

Euclidean quantum gravity is an attempt to give a meaning to the partition function

$$\sum_{\text{topologies}} \int_{\text{metrics}} dg \, \exp(-S_{EH}(g)),$$

where $S_{EH}(g)$ is the Einstein–Hilbert action with cosmological constant. In $D = 2$ dimensions, where the Einstein–Hilbert action reduces (by the Gauß-Bonnet theorem) to the Euler characteristics and (from the cosmological constant) the area of the surface, it was argued at the end of [LPW88] and further elaborated in [Wit90, MP90, Wit91] that topological gravity in 2 dimensions reduces to topological data of the moduli spaces $\{\mathcal{M}_{g,s}\}$ (more precisely their compactifications). Particularly significant are the intersection numbers which we briefly introduce below. More details can be found in [Wit91, LZ04, Eyn16].

For the *Deligne–Mumford compactification* on adds to $\mathcal{M}_{g,s}$ degenerate surfaces, so-called *nodal curves*. They arise from gluing (smaller) Riemann surfaces $\Sigma_1 \cup \cdots \cup \Sigma_\ell$ of Euler characteristics $\chi_i = 2 - 2g_i - s_i \leq 0$ along each two of their marked points. The resulting nodal curve contributes to $\overline{\mathcal{M}}_{g,s}$ if $2 - 2g - s = \sum_{i=1}^{\ell} \chi_i$ and s is the number of marked points of $\Sigma_1, \ldots, \Sigma_\ell$ which are not glued. For example, a sphere with three marked points glued along two of them gives rise to a pinched torus of genus $g = 1$ and one remaining marked point: $\overline{\mathcal{M}}_{1,1} = \mathcal{M}_{1,1} \cup \mathcal{M}_{0,3}$. In general, $\overline{\mathcal{M}}_{g,s}$ has subsets of smaller dimension than $d_{g,s}$; it is called a stack.

On $\overline{\mathcal{M}}_{g,s}$ there is a natural family $\{\mathcal{L}_i\}_{i=1,\dots,s}$ of complex line bundles obtained by taking as fibre of $\mathcal{L}_i$ at $x \in \overline{\mathcal{M}}_{g,s}$ the cotangent space $T^*_{z_i}C$ at the marked point z_i of the curve $x \equiv C$. Complex line bundles are classified by their first Chern class $c_1(\mathcal{L}_i) \in H^2(\overline{\mathcal{M}}_{g,s}, \mathbb{Q})$. The (commutative) wedge product of $\dim(\overline{\mathcal{M}}_{g,s}) = 3g-3+s$ of these 2-forms $c_1(\mathcal{L}_i)$ is of top degree $2(3g-3+s)$, equal to the real dimension of $\overline{\mathcal{M}}_{g,s}$. So the following integral is meaningful:

$$\langle \tau_{d_1} \cdots \tau_{d_s} \rangle := \int_{\overline{\mathcal{M}}_{g,s}} \prod_{j=1}^{s} \big(c_1(\mathcal{L}_j)\big)^{d_j}, \tag{7.2}$$

which is non-zero only if $d_1 + \cdots + d_s = 3g-3+s$. These rational numbers are called *intersection numbers* and provide topological invariants of $\overline{\mathcal{M}}_{g,s}$.

Since the order of marked points does not matter, the intersection numbers can be collected to $\langle \tau_0^{k_0} \tau_1^{k_1} \cdots \rangle$. Their generating function is defined by

$$F(t_0, t_1, \dots) = \sum_{k_0,k_1,\dots=0}^{\infty} \langle \tau_0^{k_0} \tau_1^{k_1} \dots \rangle \prod_{i=0}^{\infty} \frac{t_i^{k_i}}{k_i!}. \tag{7.3}$$

The simplest cases and an analogy to the *Hermitean one-matrix model* [BK90, DS90, GM90] led Witten to the following conjecture:

Conjecture 7.2 ([Wit91])

1. F obeys the string equation

$$\frac{\partial F}{\partial t_0} = \frac{t_0^2}{2} + \sum_{i=0}^{\infty} t_{i+1} \frac{\partial F}{\partial t_i}. \tag{7.4}$$

2. $U(\{t\}) := \frac{\partial^2}{\partial t_0^2} F(\{t\})$ satisfies the Korteweg-de Vries equations

$$\frac{\partial U}{\partial t_n} = \frac{\partial}{\partial t_0} R_{n+1}(U, \partial_{t_0} U, \partial_{t_0}^2 U, \dots), \tag{7.5}$$

where the R_n are polynomials in U and their t_0-derivatives which are recursively defined by $R_1(U) = U$ and

$$\frac{\partial}{\partial t_0} R_{n+1} = \frac{1}{2n+1} \Big(R_n \frac{\partial U}{\partial t_0} + 2U \frac{\partial R_n}{\partial t_0} + \frac{1}{4} \frac{\partial^3 R_n}{\partial t_0^3} \Big).$$

Kontsevich [Kon92] achieved a proof of the Witten's conjecture 7.2 by relating F to the partition function of a new type of matrix model, nowadays called the Kontsevich model. Starting point is a theorem by Strebel [Str67] which provides a stratification of decorated moduli spaces by ribbon graphs:

Theorem 7.3 ([Str84]) *On any Riemann surface $C \in \mathcal{M}_{g,s}$ (where $s > 0$ and $\chi_{g,s} < 0$) with marked points $z_1, \dots, z_s$ there is, for any given perimeters $L_1, \dots, L_s \in \mathbb{R}_+$, a unique quadratic differential $\Omega(z) = f(z)(\mathrm{d}z)^2$ such that*

- *f is meromorphic on C, with poles of order 2 at z_i, and no other poles;*
- *horizontal trajectories of Ω, defined by $\mathrm{Im}(\int^z \sqrt{\Omega}) = const$, are either circles about the marked points or critical trajectories which form a ribbon graph with s faces drawn on C.*

The jth face has perimeter L_j when measured with the metric $\frac{1}{2\pi}|\sqrt{\Omega}|$.

A k-fold zero of the quadratic differential gives rise to a $(k+2)$-valent vertex of the ribbon graph of critical trajectories. The ribbon graph of a generic surface $C \in \mathcal{M}_{g,s}$ has only 3-valent vertices (corresponding to simple zeros) and $2(3g-3+s)+s$ edges (combine $2-2g = v-e+s$ with $3v = 2e$) whose lengths $\ell_1, \dots, \ell_{2(3g-3+s)+s} > 0$ are measured by $\frac{1}{2\pi}|\sqrt{\Omega}|$. An edge between different vertices of valencies k_1, k_2 may have degenerate length 0; this collapses the vertices to a single $(k_1 + k_2 - 2)$-valent vertex and corresponds to a (k_1+k_2-4)-fold zero of the quadratic differential. The topology (g, s) is unchanged by such contractions. It turns out that this assignment of ribbon graphs with 3-valent vertices and (possibly degenerate) edge lengths ℓ_e to a Riemann surface with face perimeters L_i defines (for $s \geq 1$) an isomorphism of orbifolds (see [LZ04, Eyn16])

$$\mathcal{M}_{g,s} \times (\mathbb{R}_+^\times)^s \sim \bigcup_{\mathcal{RG}^3_{g,s}} (\mathbb{R}_+)^{s+2(3g-3+s)}, \tag{7.6}$$

where we denote by $\mathcal{RG}^3_{g,s}$ the set of (connected) genus-g ribbon graphs with s faces and only 3-valent vertices. In particular, the top degree differential forms must be proportional to each other. Kontsevich proved in [Kon92] that

$$\frac{2^{3-3g-s}}{(3g-3+s)!}\left(\sum_{i=1}^{s} L_i^2 c_1(\mathcal{L}_i)\right)^{3g-3+s} \wedge \mathrm{d}L_1 \wedge \cdots \wedge \mathrm{d}L_s = 2^{2g-2+s} \bigwedge_{e=1}^{s+2(3g-3+s)} \mathrm{d}\ell_e, \tag{7.7}$$

independently of the ribbon graph. Since $L_i = \sum_{e \in \text{edges around face } i} \ell_e$ and every edge e separates two faces $i(e), i'(e)$ (possibly the same), one has $\prod_{i=1}^{s} \mathrm{e}^{-E_i L_i} = \prod_{e \in \mathcal{E}_\Gamma} \mathrm{e}^{-\ell_e(E_{i(e)}+E_{i'(e)})}$. Inserted into the cell decomposition (7.6) gives after integration with volume forms (7.7) the following:

Theorem 7.4 ([Kon92]) *The intersection numbers of line bundles on $\overline{\mathcal{M}}_{g,s}$ are generated by*

$$\sum_{d_1+\cdots+d_s=3g-3+s} \langle \tau_{d_1} \cdots \tau_{d_s} \rangle \prod_{i=1}^{s} \frac{(2d_i-1)!!}{E_i^{2d_i+1}} = \sum_{\Gamma \in \mathcal{RG}^3_{g,s}} \frac{2^{2g+s-2}}{\#\mathrm{Aut}(\Gamma)} \prod_{e \in \mathcal{E}_\Gamma} \frac{1}{E_{i(e)} + E_{i'(e)}}, \tag{7.8}$$

where $\mathcal{E}_\Gamma$ denotes the set of edges of Γ and $\#\mathrm{Aut}(\Gamma)$ is the order of the automorphism group of Γ. The faces are labelled by positive real numbers $E_1, \dots, E_s$, and $E_{i(e)}, E_{i'(e)}$ are the labels of the two faces $i(e), i'(e)$ separated by the edge e.

The sum over ribbon graphs with weight $\frac{1}{E_i+E_{i'}}$ for an edge separating faces i, i' is easily interpreted as the perturbative expansion of a partition function. For a diagonal $\mathcal{N} \times \mathcal{N}$-matrix $E = (E_i\delta_{ij})$ and $\mathrm{d}\Phi$ the usual Lebesgue measure on the vector space $M_\mathcal{N}(\mathbb{C})_* \simeq \mathbb{R}^{\mathcal{N}^2}$ of self-adjoint matrices, the Gauß measure

$$\mathrm{d}\mathcal{M}_{C_E}(\Phi) := \frac{\mathrm{d}\Phi\ \exp\Big(-\mathcal{N}\mathrm{Tr}\big(E\Phi^2 + \frac{1}{3}\lambda\Phi^3\big)\Big)}{\int_{(M_\mathcal{N}(\mathbb{C})_*)'} \mathrm{d}\Phi\ \exp\Big(-\mathcal{N}\mathrm{Tr}\big(E\Phi^2\big)\Big)} \tag{7.9}$$

is precisely the Borel measure on $(M_\mathcal{N}(\mathbb{C})_*)'$ for a covariance $C_E(e_{kl}, e_{mn}) = \frac{\delta_{lm}\delta_{kn}}{\mathcal{N}(E_k+E_l)}$ introduced in Section 3.2. In particular, its moments are given by (3.6). It is then a combinatorial exercise to establish

$$\log\Big(\int_{(M_\mathcal{N}(\mathbb{C})_*)'} \mathrm{d}\mathcal{M}_{C_E}(\Phi)\ \mathrm{e}^{-\frac{\lambda}{3}\mathcal{N}\mathrm{Tr}(\Phi^3)}\Big)$$
$$= \sum_{g=0}^{\infty}\sum_{s=1}^{\infty}\frac{1}{s!}\Big(\frac{\lambda^2}{\mathcal{N}}\Big)^{2g-2+s} \sum_{i_1,\dots,i_s=1}^{\mathcal{N}} \Big[\sum_{\Gamma\in\mathcal{RG}_{g,s}} \frac{1}{\#\mathrm{Aut}(\Gamma)} \prod_{e\in\mathcal{E}_\Gamma} \frac{1}{E_{i(e)}+E_{i'(e)}}\Big], \tag{7.10}$$

where the innermost sum is over labelled ribbon graphs Γ of genus g with s faces labelled $E_{i_1}, \dots, E_{i_s}$. These face labels are subsequently summed over its indices from 1 to $\mathcal{N}$.

Inserting (7.8) for [ ] on the rhs of (7.10) shows that the intersection numbers $\langle\tau_{d_1}\cdots\tau_{d_s}\rangle$ are generated by the cubic matrix model (7.9). Strictly speaking, independence of the formal variables t_i in (7.3) is only achieved in the limit $\mathcal{N} \to \infty$. On the other hand, convergence of the sums over $i_1, \dots, i_s$ on the rhs and a meaningful integral on the lhs of (7.10) are not guaranteed for $\mathcal{N} \to \infty$. For these reasons the Kontsevich model is not yet a quantum field theory, but as shown in Section 8, it can be turned into one. We remark that (7.10) describes only the vacuum contributions. True correlation functions do arise in the proof [Wit92, DFIZ93] of the string equation (7.4) and the KdV equation (7.5). Some of these correlation functions have a topological interpretation as κ-classes [AC96].

7.3 The Ward–Takahashi identity in matrix models

The Ward–Takahashi identities to be derived in this section play a key rôle in the exact solutions of QFT models in Sections 8 and 9.

Lemma 7.5 *Let $\mathcal{A}$ be a nuclear AF Fréchet algebra (see Section 3) with matrix basis (e_{kl}). Let $\mathcal{F}_{C_E}(J)$ be the Bochner–Minlos characteristic function (3.3) for a covariance $C_E(e_{kl}, e_{mn}) = \frac{\delta_{kn}\delta_{lm}}{E_k+E_l}$, where (E_k) is a sequence of positive real numbers and $J \in \mathcal{A}_*$. Then*

$$(E_k - E_l)\frac{\partial^2 \mathcal{F}_{C_E}(J)}{\partial J_{kn}\partial J_{nl}} = J_{ln}\frac{\partial \mathcal{F}_{C_E}(J)}{\partial J_{kn}} - J_{nk}\frac{\partial \mathcal{F}_{C_E}(J)}{\partial J_{nl}}. \tag{7.11}$$

Proof Expanding $J = \sum_{p,q} J_{pq} e_{pq}$, the characteristic function reads

$$\mathcal{F}_{C_E}(J) = \int_{\mathcal{A}'_*} \mathrm{d}\mathcal{M}_{C_E}(\Phi)\, \mathrm{e}^{\mathrm{i}\sum_{p,q} J_{pq}\Phi(e_{pq})},$$

where $\Phi(e_{pq}) := \frac{1}{2}\Phi(e_{pq} + e_{qp}) - \frac{\mathrm{i}}{2}\Phi(\mathrm{i}e_{pq} - \mathrm{i}e_{qp})$. Then

$$\begin{aligned}(E_k - E_l)\frac{\partial^2 \mathcal{F}_{C_E}(J)}{\partial J_{kn}\partial J_{nl}} &= \frac{\partial}{\partial J_{nl}}(E_k + E_n)\frac{\partial \mathcal{F}_{C_E}(J)}{\partial J_{kn}} - \frac{\partial}{\partial J_{kn}}(E_l + E_n)\frac{\partial \mathcal{F}_{C_E}(J)}{\partial J_{nl}} \\ &= \frac{\partial}{\partial J_{nl}}\int_{\mathcal{A}'_*} \mathrm{d}\mathcal{M}_{C_E}(\Phi)\; \mathrm{i}(E_k + E_n)\Phi(e_{kn})\mathrm{e}^{\mathrm{i}\sum_{p,q} J_{pq}\Phi(e_{pq})} \\ &\quad - \frac{\partial}{\partial J_{kn}}\int_{\mathcal{A}'_*} \mathrm{d}\mathcal{M}_{C_E}(\Phi)\; \mathrm{i}(E_l + E_n)\Phi(e_{nl})\mathrm{e}^{\mathrm{i}\sum_{p,q} J_{pq}\Phi(e_{pq})}.\end{aligned}$$

Now observe that, expanding the exponential and evaluating the pairings (3.5), we have

$$\int_{\mathcal{A}'_*} \mathrm{d}\mathcal{M}_{C_E}(\Phi)\; \Phi(e_{kn})\mathrm{e}^{\mathrm{i}\sum_{p,q} J_{pq}\Phi(e_{pq})} = \frac{\mathrm{i}J_{nk}}{E_k + E_n}\int_{\mathcal{A}'_*} \mathrm{d}\mathcal{M}_{C_E}(\Phi)\; \mathrm{e}^{\mathrm{i}\sum_{p,q} J_{pq}\Phi(e_{pq})}$$

and similarly for the other term. The $(E_k + E_n)$ and $(E_l + E_n)$ terms cancel, and derivative and multiplication with J commute up to a term which also cancels. We end up in (7.11). □

It is now remarkable that, at least formally, Lemma 7.5 extends to interacting QFT models. Namely, the partition function (3.7) can be realised as a derivative operator applied to the characteristic function:

$$\mathcal{Z}_E(J) = \int_{\mathcal{A}'_*} \mathrm{d}\mathcal{M}_C(\Phi)\; \mathrm{e}^{\mathrm{i}\Phi(J) - S_{\mathrm{int}}(\{\Phi(e_{pq})\})} = \exp\Big(-S_{\mathrm{int}}\Big(\Big\{\frac{\partial}{\mathrm{i}\partial J_{pq}}\Big\}\Big)\Big)\mathcal{F}_{C_E}(J). \tag{7.12}$$

Since the derivative operator commutes with the lhs of (7.11), we conclude:

Corollary 7.6 *Under the conditions of Lemma 7.5, the partition function (7.12) satisfies*

$$(E_k - E_l)\frac{\partial^2 \mathcal{Z}_E(J)}{\partial J_{kn}\partial J_{nl}} = J_{ln}\frac{\partial \mathcal{Z}_E(J)}{\partial J_{kn}} - J_{nk}\frac{\partial \mathcal{Z}_E(J)}{\partial J_{nl}} \tag{7.13}$$
$$+\Big[J_{ln}, S_{\text{int}}\Big(\Big\{\frac{\partial}{\mathrm{i}\partial J_{pq}}\Big\}\Big)\Big]\frac{\partial \mathcal{Z}_E(J)}{\partial J_{kn}} - \Big[J_{nk}, S_{\text{int}}\Big(\Big\{\frac{\partial}{\mathrm{i}\partial J_{pq}}\Big\}\Big)\Big]\frac{\partial \mathcal{Z}_E(J)}{\partial J_{nl}}.$$

If $S_{\text{int}} = S_{\text{int}}^{(\mathcal{N})}$ *is a linear combination of traces (4.6), then*

$$\sum_{n=1}^{\mathcal{N}}(E_k - E_l)\frac{\partial^2 \mathcal{Z}_E(J)}{\partial J_{kn}\partial J_{nl}} = \sum_{n=1}^{\mathcal{N}}\Big(J_{ln}\frac{\partial \mathcal{Z}_E(J)}{\partial J_{kn}} - J_{nk}\frac{\partial \mathcal{Z}_E(J)}{\partial J_{nl}}\Big). \tag{7.14}$$

Proof Only (7.14) is to show. For $S_{\text{int}}^{(\mathcal{N})}$ given by (4.6) one has

$$S_{\text{int}}\Big(\Big\{\frac{\partial}{\mathrm{i}\partial J_{pq}}\Big\}\Big) = \sum_p \frac{\lambda_p(\mathcal{N})}{\mathrm{i}^p p}\sum_{k_1,\dots,k_p=1}^{\mathcal{N}}\frac{\partial^p}{\partial J_{k_1k_2}\cdots\partial J_{k_{p-1}k_p}\partial J_{k_pk_1}}.$$

Now both terms in the 2nd line of (7.13) give, when summed over n, the same term

$$\sum_p \frac{\lambda_p(\mathcal{N})}{\mathrm{i}^p}\sum_{k_2,\dots,k_p=1}^{\mathcal{N}}\frac{\partial^p}{\partial J_{kk_2}\partial J_{k_2k_3}\cdots\partial J_{k_{p-1}k_p}\partial J_{k_pl}}\mathcal{Z}_E(J).$$

(Without the sum, we have $k_2 \mapsto n$ in the first term, not summed, and $k_p \mapsto n$ in the second term, not summed. Then the difference does not cancel.) □

The *Ward–Takahashi identity* (7.14) was originally proved in [DGMR07] starting from invariance of the matrix Lebesgue measure under unitary transformations. That such transformations are not necessary was only recently observed in [HW18].

Recall from (3.6) that correlation functions are obtained by directional derivatives of the partition function (3.7) for which we have the representation (7.12). A series expansion of exp in (7.12) gives rise to expressions encoded by ribbon graphs on Riemann surfaces with possibly several boundary components and half-edges ending at defects on the boundaries. See Section 7.1. As discussed there we recollect the contributions with the same defect structure, which amounts to the same contractions of test functions $J \in \mathcal{A}_*$:

$$\log\frac{\mathcal{Z}_E^{(\mathcal{N})}(J)}{\mathcal{Z}_E^{(\mathcal{N})}(0)} \tag{7.15}$$
$$= \sum_{B=1}^{\infty}\sum_{g=0}^{\infty}\sum_{N_1\leq\dots\leq N_B}\frac{\mathcal{N}^{2-B-2g}}{S_{N_1\dots N_B}}\sum_{k_1^1,\dots,k_{N_B}^B}^{(\mathcal{N})} G^{(g)}_{|k_1^1\dots k_{N_1}^1|\dots|k_1^B\dots k_{N_B}^B|}\prod_{\beta=1}^{B}\frac{\mathbb{J}_{k_1^\beta\dots k_{N_\beta}^\beta}}{N_\beta},$$

where $\mathbb{J}_{k_1^\beta \dots k_{N_\beta}^\beta} = \mathrm{i}^{N_\beta} \prod_{i=1}^{N_\beta} J_{k_i^\beta k_{i+1}^\beta}$ with cyclic identification $k_{N_\beta+1}^\beta \equiv k_1^\beta$. The sums over k_i^β are over a finite set determined by $\mathcal{N}$, and $S_{N_1 \dots N_B} = \nu_1! \cdots \nu_s!$ if each ν_j of the N_β coincide. Conversely, the correlation functions $G^{(g)}_{\dots}$ are for pairwise different k_i^β recovered via

$$\sum_{g=0}^{\infty} \mathcal{N}^{-2g} G^{(g)}_{|k_1^1 \dots k_{N_1}^1| \dots |k_1^B \dots k_{N_B}^B|} = \frac{1}{\mathcal{N}^{2-B}} \frac{\partial^{N_B}}{\partial \mathbb{J}_{k_{N_B}^B \dots k_1^B}} \cdots \frac{\partial^{N_1}}{\partial \mathbb{J}_{k_{N_1}^1 \dots k_1^1}} \log \mathcal{Z}_E^{(\mathcal{N})}(J)\Big|_{J=0}, \tag{7.16}$$

where $\frac{\partial^{N_\beta}}{\partial \mathbb{J}_{k_{N_\beta}^\beta \dots k_1^\beta}} = (-\mathrm{i})^{N_\beta} \frac{\partial^{N_\beta}}{\partial J_{k_{N_\beta}^\beta k_{N_\beta-1}^\beta} \cdots J_{k_1^\beta k_{N_\beta}^\beta}}$.

Dyson–Schwinger equations result from the interplay between the J-derivatives in (7.16) with the internal J-derivatives in $S_{\mathrm{int}}\big(\{\frac{\partial}{\mathrm{i}\partial J_{pq}}\}\big)$ according to the representation (7.12) of $\mathcal{Z}_E(J)$. Following an observation in [GW14a], the following programme arises:

Programme 7.7 For QFT models with covariance $C_E(e_{kl}, e_{mn}) = \frac{\delta_{kn}\delta_{lm}}{E_k+E_l}$, the interplay of J-derivatives gives rise to expressions known from the Ward–Takahashi identity (7.14). In particular cases, which include the Φ^3 and Φ^4 interactions, the tower of Dyson–Schwinger equations decouples into a closed non-linear equation for the simplest function $G_{\dots}$ and a hierarchy of affine equations for all other functions. The whole model can then (at least in principle) be recursively solved starting from the solution of a single non-linear equation.

This programme succeeded completely for the Φ^3-model (reviewed in Section 8) and partially for the Φ^4-model (reviewed in Section 9).

8 Exact solution of the Φ^3-model

8.1 Preliminary remarks

It was first stressed in [GS06b] that results about the Kontsevich model can be used to define a quantum field theory on noncommutative Moyal space with Φ^3-interaction and harmonic oscillator covariance (see Section 6.1) at critical frequency $\Omega = 1$. By including a linear term proportional to $\mathrm{Tr}(\Phi)$ with carefully adjusted singular coefficient, Grosse and Steinacker were able to renormalise the divergence in the 1-point function. They derived exact formulae for the low-genus one-point function from the intersection numbers computed in [IZ92]. Via quantum equations of motions, higher correlation functions were related to the 1-point function. Shortly later the renormalisation in dimensions $D = 4$ [GS06a] and $D = 6$ [GS08] was also understood. Below we review an extension [GSW17, GSW18, GHW19] of these techniques based on the Ward–Takahashi identity proved in Corollary 7.6.

Recall the harmonic oscillator Hamiltonian (6.2) which at critical frequency $\Omega = 1$ and in dimension $D \in \{2,4,6\}$ reads $H^1(e_{\underline{k}\underline{l}}) = (E_{\underline{k}}+E_{\underline{l}})e_{\underline{k}\underline{l}}$ with $E_{\underline{k}} = \frac{\mu^2}{2}+\frac{D}{\theta}+\frac{4}{\theta}|\underline{k}|$. By $\underline{k} = (k_1,\dots,k_{D/2})$ we understand a $(D/2)$-tuple of natural numbers, of length $|\underline{k}| := k_1+\dots+k_{D/2}$, which parametrises the matrix bases $(e_{\underline{k}\underline{l}})_{\underline{k},\underline{l}\in\mathbb{N}^{D/2}}$ of the D-dimensional Moyal space (see Section 4.3 for $D = 4$). The $E_{\underline{k}}$ will be identified with the labels E_i in the Kontsevich formula (7.8). In fact we can construct QFT models for a more general label function (than resulting from H^1)

$$E = (\tilde{E}_{\underline{k}}\delta_{\underline{k},\underline{l}}), \qquad \tilde{E}_{\underline{k}} := \frac{\tilde{\mu}^2}{2} + \mu^2 e\Big(\frac{|\underline{k}|}{\mu^2 V^{\frac{2}{D}}}\Big), \qquad e(0) \equiv 0, \tag{8.1}$$

where $e : \mathbb{R}_+ \to \mathbb{R}_+$ is a monotonously increasing differentiable function. For the covariance of the harmonic oscillator on Moyal space we have $e(x) = x$ independent of D and $V^{\frac{2}{D}} = \frac{\theta}{4}$. The parameter $\mu > 0$ will become the renormalised mass, whereas the bare mass $\tilde{\mu} \equiv \tilde{\mu}(\mathcal{N})$ is a function of $(V, \mathcal{N}, \lambda, \mu)$ identified later. For such label functions $\tilde{E}_{\underline{k}}$ we consider a quantum field theory on a noncommutative geometry, understood as nuclear AF Fréchet algebra $\mathcal{A}$ (see Section 3), with generalised matrix basis $(e_{\underline{k}\underline{l}})_{\underline{k},\underline{l}\in\mathbb{N}^{D/2}}$. It is defined by a covariance C_E and an interaction functional which according to Programme 3.8 is parametrised by sequences $\tilde{\mu}, \tilde{Z}, \tilde{\kappa}, \tilde{\nu}, \tilde{\zeta}, \tilde{\lambda}$ in $\mathcal{N}$ which implement the embeddings $\iota_{\mathcal{N}} : \mathcal{A}^{\mathcal{N}} \to \mathcal{A}^{\mathcal{N}+1}$. We have $\mathcal{A}^{\mathcal{N}} = \mathrm{span}(e_{\underline{k}\underline{l}} \;:\; \underline{k},\underline{l} \in \mathbb{N}^{D/2}_{\mathcal{N}})$, where $\mathbb{N}^{D/2}_{\mathcal{N}}$ consists of the $\frac{D}{2}$-tuples $\underline{k}$ with $|\underline{k}| \leq \mathcal{N}$. For the Φ^3-model we choose the covariance

$$C_E^{(\mathcal{N})}(e_{\underline{k}\underline{l}}, e_{\underline{m}\underline{n}}) = \frac{\delta_{\underline{k},\underline{n}}\delta_{\underline{l},\underline{m}}}{V\tilde{Z}(\tilde{E}_{\underline{k}} + \tilde{E}_{\underline{l}})} \tag{8.2}$$

(in which the $\tilde{E}_{\underline{k}}$ also depend on $\mathcal{N}$ via $\tilde{\mu}$) and the interaction functional on $\mathcal{A}'_*$

$$S^{(\mathcal{N})}_{\text{int}}(\Phi) := V\Big(\sum_{\underline{n}\in\mathbb{N}^{D/2}_{\mathcal{N}}} (\tilde{\kappa}+\tilde{\nu}\tilde{E}_{\underline{n}}+\tilde{\zeta}\tilde{E}^2_{\underline{n}})\Phi_{\underline{n}\underline{n}} + \frac{\tilde{\lambda}\tilde{Z}^{\frac{3}{2}}}{3} \sum_{\underline{n},\underline{m},\underline{l}\in\mathbb{N}^{D/2}_{\mathcal{N}}} \Phi_{\underline{n}\underline{m}}\Phi_{\underline{m}\underline{l}}\Phi_{\underline{l}\underline{n}}\Big), \tag{8.3}$$

where $\Phi_{\underline{k}\underline{l}} := \Phi(e_{\underline{k}\underline{l}})$.

8.2 Solution of the planar sector

We return to Equation (7.16), adapted to multi-indices $\underline{k}$. Evaluation of the rightmost derivative gives with (7.12) and $\mathcal{F}_{C_E}(J) = \exp\big(-\frac{V}{2}\sum_{\underline{k},\underline{l}\in\mathbb{N}^{D/2}_{\mathcal{N}}} \frac{J_{\underline{k}\underline{l}}J_{\underline{l}\underline{k}}}{\tilde{Z}(\tilde{E}_{\underline{k}}+\tilde{E}_{\underline{l}})}\big)$

$$(-\mathrm{i})\frac{\partial}{\partial J_{\underline{k}_1^1\underline{k}_{N_1}^1}}\log \mathcal{Z}_E(J) = \frac{\mathrm{i}VJ_{\underline{k}_{N_1}^1\underline{k}_1^1}}{\tilde{Z}(\tilde{E}_{\underline{k}_1^1}+\tilde{E}_{\underline{k}_{N_1}^1})} - \frac{V\big(\tilde{\kappa}+\tilde{\nu}\tilde{E}_{\underline{k}_1^1}+\tilde{\zeta}\tilde{E}^2_{\underline{k}_1^1}\big)}{2\tilde{Z}\tilde{E}_{\underline{k}_1^1}}\delta_{\underline{k}_{N_1}^1,\underline{k}_1^1}$$
$$+\frac{\tilde{\lambda}\tilde{Z}^{\frac{1}{2}}}{V\mathcal{Z}_E(J)(\tilde{E}_{\underline{k}_1^1}+\tilde{E}_{\underline{k}_{N_1}^1})}\sum_{\underline{n}\in\mathbb{N}_N^{D/2}}\frac{\partial^2}{\partial J_{\underline{k}_{N_1}^1\underline{n}}\partial J_{\underline{n}\underline{k}_1^1}}\mathcal{Z}_E(J). \tag{8.4}$$

The first line only contributes to $B=1$ and $N_1 \le 2$ in (7.16). Inserting (7.15) into the second line and evaluating the remaining derivatives in (7.16) gives exact (non-perturbative) *Dyson–Schwinger equations* between the correlation functions $G^{(g)}_{\dots}$. However, since the last line of (8.4) has one more derivative than the lhs, these equations relate an N-point function to the not yet known $(N+1)$-point function. This would make the Dyson–Schwinger equations rather useless. We are rescued by the Ward–Takahashi identity of Corollary 7.6. There we have to replace $E_k \mapsto V\tilde{Z}\tilde{E}_{\underline{k}}$ to account for the different conventions in (8.2) and Lemma 7.5. We also rescale $J_{kl} \mapsto VJ_{\underline{k}\underline{l}}$. The second line of (8.3) is a trace and does not contribute to the second line of (7.13) when summed over $\underline{n}$. For $|\underline{k}| \neq |\underline{l}|$ we can divide by $\tilde{E}_{\underline{k}}-\tilde{E}_{\underline{l}}\neq 0$ (by the assumptions on $e(x)$ in (8.1)):

$$\sum_{\underline{n}\in\mathbb{N}_N^{D/2}}\frac{\partial^2\mathcal{Z}_E(J)}{\partial J_{\underline{k}\underline{n}}\partial J_{\underline{n}\underline{l}}} = \frac{V}{\tilde{Z}(\tilde{E}_{\underline{k}}-\tilde{E}_{\underline{l}})}\sum_{\underline{n}\in\mathbb{N}_N^{D/2}}\Big(J_{\underline{l}\underline{n}}\frac{\partial\mathcal{Z}_E(J)}{\partial J_{\underline{k}\underline{n}}} - J_{\underline{n}\underline{k}}\frac{\partial\mathcal{Z}_E(J)}{\partial J_{\underline{n}\underline{l}}}\Big) \tag{8.5}$$
$$+\frac{V}{\mathrm{i}\tilde{Z}}(\tilde{\nu}+\tilde{\zeta}(\tilde{E}_{\underline{k}}+\tilde{E}_{\underline{l}}))\frac{\partial\mathcal{Z}_E(J)}{\partial J_{\underline{k}\underline{l}}},\qquad \text{for } |\underline{k}|\neq|\underline{l}|.$$

The lhs is (assuming $N_1>1$; the case $N_1=1$ implies $|\underline{k}|=|\underline{l}|$) precisely of the form needed in the second line of (8.4). Starting with the 1-point function $G_{|\underline{k}|}$ which needs a special treatment, one obtains a hierarchy of equations which only depend on data known by induction. Hence, if $G_{|\underline{k}|}$ can be determined, the exact solution of the Φ^3-matricial QFT model is possible.

For $B=1$, with N_1-1 further derivatives applied to (8.4), one obtains after insertion of (8.5) and suppression of the upper index $\underline{k}_i^1 \equiv \underline{k}_i$

$$\tilde{Z}(1-\tilde{\lambda}\tilde{Z}^{-\frac{1}{2}}\tilde{\zeta})\Big((\tilde{E}_{\underline{k}_1}+\tilde{E}_{\underline{k}_{N_1}}) - \frac{\tilde{\lambda}\tilde{Z}^{-\frac{1}{2}}\tilde{\nu}}{(1-\tilde{\lambda}\tilde{Z}^{-\frac{1}{2}}\tilde{\zeta})}\Big)G^{(g)}_{|\underline{k}_1\dots\underline{k}_{N_1}|}$$
$$=\delta_{g,0}\delta_{N_1,2}+\tilde{\lambda}\tilde{Z}^{\frac{1}{2}}\frac{G^{(g)}_{|\underline{k}_1\dots\underline{k}_{N_1-1}|}-G^{(g)}_{|\underline{k}_2\dots\underline{k}_{N_1}|}}{\tilde{E}_{\underline{k}_1}-\tilde{E}_{\underline{k}_{N_1}}}. \tag{8.6}$$

This equation fixes $\tilde{\mu}(\mathcal{N}), \tilde{\lambda}(\mathcal{N}), \tilde{\zeta}(\mathcal{N})$ in terms of $\mu, \lambda, \tilde{Z}(\mathcal{N}), \tilde{\nu}(\mathcal{N})$ to

$$\tilde{Z}(1-\tilde{\lambda}\tilde{Z}^{-\frac{1}{2}}\tilde{\zeta})=1, \qquad \tilde{\lambda}\tilde{Z}^{\frac{1}{2}}=\lambda, \qquad \tilde{\mu}^2=\mu^2+\frac{\tilde{\lambda}\tilde{Z}^{-\frac{1}{2}}\tilde{\nu}}{(1-\tilde{\lambda}\tilde{Z}^{-\frac{1}{2}}\tilde{\zeta})}.$$

The recursion can be solved explicitly [GSW17]:

$$G^{(g)}_{|\underline{k}_1\dots\underline{k}_{N_1}|}=\sum_{i=1}^{N_1}\frac{W^{(g)}_{\underline{k}_i}}{2\lambda}\prod_{j=1,j\neq i}^{N_1}\frac{\lambda}{E^2_{\underline{k}_i}-E^2_{\underline{k}_j}}, \qquad \frac{W^{(g)}_{\underline{k}}}{2\lambda}:=G^{(g)}_{|\underline{k}|}+\frac{\delta_{g,0}E_{\underline{k}}}{\lambda}, \tag{8.7}$$

where $E_{\underline{k}}:=\tilde{E}_{\underline{k}}-\frac{1}{2}\tilde{\lambda}\tilde{Z}^{\frac{1}{2}}\tilde{\nu}=\mu^2(\frac{1}{2}+e(\frac{|\underline{k}|}{\mu^2V^{2/D}}))$ is the renormalisation of (8.1) which replaces $\tilde{\mu}$ by μ.

The 1-point function is directly obtained from (8.4) at $\underline{k}^1_1=\underline{k}^1_{N_1}\equiv\underline{k}$ and $J\equiv 0$. After renormalisation and insertion of (8.7) one arrives at

$$\begin{aligned}&\sum_{h=0}^{g}W^{(h)}_{\underline{k}}W^{(g-h)}_{\underline{k}}+2\tilde{\nu}\lambda W^{(g)}_{\underline{k}}+\frac{2\lambda^2}{V}\sum_{\underline{n}\in\mathbb{N}^{D/2}_{\mathcal{N}}}\frac{W^{(g)}_{\underline{k}}-W^{(g)}_{\underline{n}}}{E^2_{\underline{k}}-E^2_{\underline{n}}}\\&=\Big(\frac{4E^2_{\underline{k}}}{\tilde{Z}}-\tilde{\nu}^2\lambda^2\Big(1+\frac{1}{\tilde{Z}}\Big)-\frac{4\tilde{\kappa}\lambda}{\tilde{Z}}\Big)\delta_{g,0}-4\lambda^2G^{(g-1)}_{|\underline{k}|\underline{k}|}.\end{aligned} \tag{8.8}$$

The following observation is crucial. It was already employed in [MS91] and brought to perfection in topological recursion [Eyn16]:

Observation 8.1 For a real parameter c soon to be determined, replacing $4E^2_{\underline{k}}+c\mapsto z^2$ by a complex variable, the Equations (8.8) have a continuation $W^{(g)}_{|\underline{k}|}\mapsto W^{(g)}(z)$ which are holomorphic outside the support of $\{(4E^2_{\underline{n}}+c)^{\frac{1}{2}}\}$. All other Dyson–Schwinger equations extend similarly to several complex variables and define holomorphic functions $G^{(g)}(z^1_1,\dots,z^1_{N_1}|\dots|z^B_1,\dots,z^B_{N_B})$ of $z^\beta_i\in\mathbb{C}\setminus\{0\}$, possibly with the exception of diagonals $z^\beta_i=\pm z^\beta_j$. The original matricial correlation functions are recovered from $W^{(g)}_{\underline{k}}=W^{(g)}((4E^2_{\underline{k}}+c)^{\frac{1}{2}})$ and

$$\begin{aligned}&G^{(g)}_{|\underline{k}^1_1\dots\underline{k}^1_{N_1}|\dots|\underline{k}^B_1\dots\underline{k}^B_{N_B}|}\\&=G^{(g)}\Big((4E^2_{\underline{k}^1_1}+c)^{\frac{1}{2}},\dots,(4E^2_{\underline{k}^1_{N_1}}+c)^{\frac{1}{2}}\Big|\dots\Big|(4E^2_{\underline{k}^B_1}+c)^{\frac{1}{2}},\dots,(4E^2_{\underline{k}^B_{N_B}}+c)^{\frac{1}{2}}\Big).\end{aligned}$$

We pass to mass-dimensionless quantities via multiplication by specified powers of μ [GSW18]. This amounts to choose the mass scale as $\mu=1$. Also $V=(\frac{\theta}{4})^{D/2}$ is dimensionless from now on. It is convenient to introduce a measure

$$d\varrho(y) = \frac{8\lambda^2}{V} \sum_{\underline{n}\in\mathbb{N}_{\mathcal{N}}^{D/2}} \delta(y^2 - (4E_{\underline{n}}^2 + c))dy^2 = \frac{8\lambda^2}{V} \sum_{\underline{n}\in\mathbb{N}_{\mathcal{N}}^{D/2}} \delta\big(y - (4E_{\underline{n}}^2 + c)^{\frac{1}{2}}\big)dy. \tag{8.9}$$

The measure has support on $[\sqrt{1+c}, \sqrt{\Lambda_{\mathcal{N}}^2 + c}]$, where $\Lambda_{\mathcal{N}}^2 = \max(4E_{\underline{n}}^2 : |\underline{n}| = \mathcal{N})$. From now on we drop $\mathcal{N}$ in favour of a dependence of correlation functions on a scale Λ that in the end has to be sent to ∞ by the same renormalisation procedure of Programme 3.8.

After these reparametrisations, Equation (8.8) takes the form

$$\sum_{h=0}^{g} W^{(h)}(z)W^{(g-h)}(z) + 2\tilde{\nu}\lambda W^{(g)}(z) + \int_{\sqrt{1+c}}^{\sqrt{\Lambda^2+c}} d\varrho(y)\, \frac{W^{(g)}(z) - W^{(g)}(y)}{z^2 - y^2}$$
$$= \Big(\frac{z^2 - c}{\tilde{Z}} - \tilde{\nu}^2\lambda^2\Big(1 + \frac{1}{\tilde{Z}}\Big) - \frac{4\tilde{\kappa}\lambda}{\tilde{Z}}\Big)\delta_{g,0} - 4\lambda^2 G^{(g-1)}(z|z). \tag{8.10}$$

For $g = 0$ one obtains in this way a closed non-linear equation for a sectionally holomorphic function $W^{(0)}(z)$ which (at $\tilde{\kappa} = \tilde{\nu} = 1 - \tilde{Z} = 0$) was solved by Makeenko and Semenoff [MS91] using techniques for boundary values of sectionally holomorphic functions. Alternatively, it can be solved by residue techniques for meromorphic functions [Eyn16]. These methods are easily extended to include $\tilde{\nu}, \tilde{\kappa}, \tilde{Z}$ and give

$$W^{(0)}(z) = \frac{z}{\sqrt{\tilde{Z}}} - \lambda\tilde{\nu} + \frac{1}{2}\int_{\sqrt{1+c}}^{\sqrt{\Lambda^2+c}} \frac{d\varrho(y)}{y(z+y)}, \tag{8.11}$$

$$\text{where} \quad \frac{c}{\tilde{Z}} + \frac{1}{\sqrt{\tilde{Z}}}\int_{\sqrt{1+c}}^{\sqrt{\Lambda^2+c}} \frac{d\varrho(y)}{y} = -\frac{4\lambda\tilde{\kappa}}{\tilde{Z}} - \frac{\lambda^2\tilde{\nu}^2}{\tilde{Z}}. \tag{8.12}$$

We have eventually reached the point where we can describe the renormalisation procedure. It depends on a spectral dimension which characterises the growth rate of $|\underline{n}| \mapsto E_{\underline{n}}$:

Definition 8.2

- *dimension 0*: $\sum_{\underline{n}\in\mathbb{N}^{D/2}} \frac{1}{E_{\underline{n}}}$ converges. No renormalisation is necessary, $\tilde{\kappa} = \tilde{\nu} = \tilde{Z} - 1 = 0$. The finite number c is determined from the consistency equation $c + \int_{\sqrt{1+c}}^{\infty} \frac{d\varrho(y)}{y} = 0$. This is the case considered in [MS91] and [Eyn16] for the usual Kontsevich model. It is not realised on Moyal space.
- *dimension 2*: $\sum_{\underline{n}\in\mathbb{N}^{D/2}} \frac{1}{E_{\underline{n}}}$ diverges but $\sum_{\underline{n}\in\mathbb{N}^{D/2}} \frac{1}{E_{\underline{n}}^2}$ converges. We can set $\tilde{\nu} = \tilde{Z} - 1 = 0$ and determine $\tilde{\kappa}(c, \Lambda)$ as the solution of (8.12). The finite parameter c translates into a normalisation condition. A natural choice is $G^{(0)}_{|\underline{0}|} = 0$, which

by (8.7) translates into $W^{(0)}(\sqrt{1+c}) = 1$. The equation for c is then the limit $\Lambda \to \infty$ of (8.11) at $z = \sqrt{1+c}$.

- *dimension 4*: $\sum_{\underline{n}\in\mathbb{N}^{D/2}} \frac{1}{E_{\underline{n}}^2}$ diverges but $\sum_{\underline{n}\in\mathbb{N}^{D/2}} \frac{1}{E_{\underline{n}}^3}$ converges. We can set $\tilde{Z} = 1$ and determine $\tilde{\nu}(c,\Lambda)$ by the condition that the rhs of (8.11) at $z = \sqrt{1+c}$ equals $1 = W^{(0)}(\sqrt{1+c})$ for any Λ. Then determine $\tilde{\kappa}(c,\Lambda)$ as the solution of (8.12). The finite parameter c is typically obtained from a condition $\frac{\mathrm{d}}{\mathrm{d}E_{\underline{n}}} G^{(0)}_{|\underline{n}|}\big|_{\underline{n}=0} = 0$ which by (8.7) translates into $\frac{1}{z}\frac{\mathrm{d}}{\mathrm{d}z} W^{(0)}(z)\big|_{z=\sqrt{1+c}} = 1$.
- *dimension 6*: $\sum_{\underline{n}\in\mathbb{N}^{D/2}} \frac{1}{E_{\underline{n}}^3}$ diverges but $\sum_{\underline{n}\in\mathbb{N}^{D/2}} \frac{1}{E_{\underline{n}}^4}$ converges. We determine $\tilde{\nu}(c,\Lambda,\tilde{Z})$ and $\tilde{\kappa}(c,\Lambda,\tilde{Z})$ as for dimension 4 and now fix $\tilde{Z}(c,\Lambda)$ from $\frac{1}{z}\frac{\mathrm{d}}{\mathrm{d}z} W^{(0)}(z)\big|_{z=\sqrt{1+c}} = 1$. The finite parameter c is typically obtained from another condition $\frac{\mathrm{d}^2}{\mathrm{d}E_{\underline{n}}^2} G^{(0)}_{|\underline{n}|}\big|_{\underline{n}=0} = 0$ with by (8.7) translates into $(\frac{c}{z}\frac{\mathrm{d}}{\mathrm{d}z} + (z^2 - c)\frac{d^2}{dz^2}) W^{(0)}(z)\big|_{z=\sqrt{1+c}} = 0$.
- *dimension* ≥ 8: $\sum_{\underline{n}\in\mathbb{N}^{D/2}} \frac{1}{E_{\underline{n}}^4}$ diverges. No quantum field theory can be achieved in this case.

The normalisation conditions can for $D \in \{0,2,4,6\}$ be summarised to the following equation for the crucial parameter c:

$$(-c)\Big(\frac{2}{1+\sqrt{1+c}}\Big)^{\delta_{D,2}+\delta_{D,4}} = \int_{\sqrt{1+c}}^{\infty} \frac{\mathrm{d}\varrho(y)}{y(\sqrt{1+c}+y)^{D/2}}. \tag{8.13}$$

Recalling the prefactor λ^2 in (8.9), the implicit function theorem guarantees a smooth solution $c(\lambda, \{E_{\underline{n}}\})$ of (8.13) inside a disk of radius λ_c.

Remark 8.3 It is remarkable that in this QFT model on noncommutative geometry $\mathcal{A} = \bigcup_N \mathcal{A}^N$ such a non-perturbative renormalisation procedure can be established. Usually one can only renormalise individual (ribbon) graphs with recursive diving into subgraphs [BP57]. This recursive prescription in encoded in a Hopf algebra [Kre98, CK98] and relates to other occurrences of Hopf algebras in noncommutative geometry [CM98]. Here, this diving into subgraphs is completely avoided. One can show [GSW18] that, breaking down these exact formulae into ribbon graphs, there is perfect agreement with the usual BPHZ renormalisation [BP57, Hep66, Zim69], including the handling of overlapping divergences by Zimmermann's forest formula [Zim69].

Remark 8.4 Furthermore, it turns out that in $D = 6$ dimensions and for $\lambda \in \mathbb{R}$ the β-function of the coupling constant λ is *positive* [GSW18]. Nonetheless, there is no triviality problem and the model can rigorously be constructed. We see this as indication that also for realistic quantum field theories (such as QED and the Higgs sector of the standard model) with positive β-function a construction is not completely impossible.

The Dyson–Schwinger equations for higher correlation functions have a simple solution in terms of $1 + \cdots + 1$-point functions [GSW17]:

$$
\begin{aligned}
&G^{(g)}\big(z_1^1,\dots,z_{N_1}^1\big|\dots\big|z_1^B,\dots,z_{N_B}^B\big) \\
&= \sum_{i_1=1}^{N_1}\cdots\sum_{i_B=1}^{N_B} G^{(g)}\big(z_{i_1}^1\big|\dots\big|z_{i_B}^B\big) \\
&\times\Big(\prod_{j_1=1,j_1\neq i_1}^{N_1}\frac{4\lambda}{(z_{i_1}^1)^2-(z_{j_1}^1)^2}\Big)\cdots\Big(\prod_{j_B=1,j_B\neq i_B}^{N_B}\frac{4\lambda}{(z_{i_B}^B)^2-(z_{j_B}^B)^2}\Big).
\end{aligned}
\tag{8.14}
$$

For $B = 1$ on has to replace $G(z_i)$ by $\frac{1}{2\lambda}W^{(g)}(z_i)$, see (8.7). The remaining Dyson–Schwinger equations for the $1 + \cdots + 1$-point functions all involve the integral operator $\hat{K}_z$ defined by

$$
\hat{K}_z f(z) := (W^{(0)}(z) + \lambda\tilde{\nu}) f(z) + \frac{1}{2}\int_{\sqrt{1+c}}^{\sqrt{\Lambda^2+c}} d\varrho(y)\,\frac{f(z)-f(y)}{z^2-y^2}. \tag{8.15}
$$

For instance, the Dyson–Schwinger equation for the planar (i.e. $g = 0$) $(1+1)$-point function becomes $\hat{K}_{z_1} G^{(0)}(z_1|z_2) = -\lambda G^{(0)}(z_1, z_2, z_2)$ and has the solution

$$
G^{(0)}(z_1|z_2) = \frac{4\lambda^2}{z_1 z_2 (z_1+z_2)^2}. \tag{8.16}
$$

Formula (8.16) is the same as the cylinder amplitude [Eyn16, Thm. 6.4.3] in the usual Kontsevich model! This is a clear indication that all topological sectors other than the disk of the Φ^3-QFT model on noncommutative geometries are governed by a universal structure called *topological recursion*. This indication was fully confirmed in [GHW19]. We give more details in the next subsection. The sector $(g = 0, B \geq 3)$ is also accessible by combinatorial techniques [GSW17] which give the following result:

$$
\begin{aligned}
G^{(0)}(z^1|\dots|z^B) &= \frac{d^{B-3}}{dt^{B-3}}\bigg|_{t=0}\bigg(\frac{(-2\lambda)^{3B-4}}{(R(t))^{B-2}\prod_{\beta=1}^{B}((z^\beta)^2-2t)^{\frac{3}{2}}}\bigg),
\end{aligned}
\tag{8.17}
$$

$$
R(t) := \lim_{\Lambda\to\infty}\bigg(\frac{1}{\sqrt{\tilde{Z}(\Lambda)}} - \int_{\sqrt{1+c}}^{\sqrt{\Lambda^2+c}} \frac{d\varrho(y)}{y\big(y+\sqrt{y^2-2t}\big)\sqrt{y^2-2t}}\bigg).
$$

The t-differentiation produces a polynomial in $\frac{1}{z_\beta}$, of odd degree in each variable, with coefficients in rational functions of the moments

$$
\varrho_l := \lim_{\Lambda\to\infty}\bigg(\frac{\delta_{l,0}}{\sqrt{\tilde{Z}(\Lambda)}} - \frac{1}{2}\int_{\sqrt{1+c}}^{\sqrt{\Lambda^2+c}}\frac{d\varrho(y)}{y^{3+2l}}\bigg). \tag{8.18}
$$

These moments play a key rôle in the solution of the non-planar sector.

8.3 The non-planar sector

The main tool is a differential operator identified in [GHW19],

$$\hat{\mathrm{A}}^{\dagger g}_{z_1,\dots,z_B} := \sum_{l=0}^{3g+B-4} \Big(-\frac{(3+2l)\varrho_{l+1}}{\varrho_0 z_B^3} + \frac{3+2l}{z_B^{5+2l}}\Big)\frac{\partial}{\partial \varrho_l} + \sum_{i=1}^{B-1} \frac{1}{\varrho_0 z_B^3 z_i}\frac{\partial}{\partial z_i}. \tag{8.19}$$

It is understood to act on Laurent polynomials in $z_2, \dots, z_B$, bounded at ∞, with coefficients in rational functions of the moments ϱ_l defined in (8.18). These differential operators play the rôle of 'boundary creation operators':

Theorem 8.5 ([GHW19]) *The* $1+\ldots+1$*-point function at genus* $g \geq 1$ *is given by*

$$G^{(g)}(z_1|\ldots|z_B) = (2\lambda)^{3B-4}\hat{\mathrm{A}}^{\dagger g}_{z_1,\dots,z_B}\big(\hat{\mathrm{A}}^{\dagger g}_{z_1,\dots,z_{B-1}}\big(\cdots\hat{\mathrm{A}}^{\dagger g}_{z_1,z_2}W^{(g)}(z_1)\ldots\big)\big), \tag{8.20}$$

for $z_i \neq 0$.

The proof is lengthy. It consists in checking that taking (8.20) as an ansatz, all Dyson–Schwinger equations for functions with $B \geq 2$ boundary components are identically fulfilled, provided that $W^{(g)}(z_1)$ is an odd Laurent polynomial of z_1, bounded at ∞, which depends only on $\varrho_0, \dots, \varrho_{3g-2}$. The assumptions are later confirmed via solution of (8.21).

Equation (8.10) takes with $\hat{K}_z$ defined in (8.15) for $g \geq 1$ the form

$$\hat{K}_z W^{(g)}(z) = -\frac{1}{2}\sum_{h=1}^{g-1} W^{(h)}(z)W^{(g-h)}(z) - 2\lambda^2 G^{(g-1)}(z|z). \tag{8.21}$$

We recall $G^{(0)}(z|z) = \frac{\lambda^2}{z^4}$ and $G^{(g-1)}(z|z) = (2\lambda)^2\hat{\mathrm{A}}^{\dagger g-1}_{z,z}W^{(g-1)}(z)$ for $g \geq 2$. Thus, all $W^{(g)}(z)$ can recursively evaluated if $\hat{K}_z$ has a tractable inverse. This is the case:

Proposition 8.6 *Let* $f(z) = \sum_{k=0}^{\infty}\frac{a_{2k}}{z^{2k}}$ *be an even Laurent series about* $z = 0$ *bounded at* ∞*. Then the inverse of the integral operator* $\hat{K}_z$ *is given by the residue formula*

$$\Big[z^2\hat{K}_z\frac{1}{z}\Big]^{-1} f(z) = -\operatorname*{Res}_{z'\to 0}\big[K(z,z')f(z')dz'\big], \tag{8.22}$$

$$\textit{where}\quad K(z,z') := \frac{2}{(W^{(0)}(z') - W^{(0)}(-z'))(z'^2 - z^2)}.$$

The proof can be directly achieved from the series expansion of $K(z, z')$ [GHW19]. Inspiration for a residue formula (8.22) comes from *topological recursion*:

Remark 8.7 A $(1 + 1 + \ldots + 1)$-point function of genus g with B boundary components fulfils a universal structure when expressed in terms of $\omega_{g,B}$ defined by

$$\omega_{g,B}(z_1, \ldots, z_B) := \Big(\prod_{i=1}^{B} z_i\Big)\Big(G^{(g)}(z_1|\ldots|z_B) + 16\lambda^2 \frac{\delta_{g,0}\delta_{2,B}}{(z_1^2 - z_2^2)^2}\Big), \qquad B > 1$$

$$\omega_{g,1}(z) := \frac{zW^{(g)}(z)}{2\lambda}.$$

Furthermore, let $y(x)$ be the *spectral curve* defined by $x(z) = z^2$ and

$$y(z) := \frac{W^{(0)}(z)}{2\lambda} = \frac{z}{2\lambda\sqrt{Z}} - \frac{\tilde{\nu}}{2} + \frac{1}{4\lambda}\int_{\sqrt{1+c}}^{\sqrt{1+\Lambda^2}} \frac{d\varrho(t)}{t(t+z)}.$$

It can be checked that with these definitions, up to trivial redefinitions by powers of 2λ, the theorems proved in topological recursion [Eyn16] apply. These determine all $\omega_{g,B}$ with $2 - 2g - B < 0$ out of the initial data $y(z)$ and $\omega_{0,2}$:

Theorem 8.8 ([Eyn16, Thm. 6.4.4]) *For a subset $I = \{i_1, \ldots, i_{|I|}\} \subset \{1, \ldots, B\}$ let $z_I := (z_{i_1}, \ldots, z_{i_{|I|}})$. Then for $2 - 2g - (1 + B) < 0$ the function $\omega_{g,B+1}(z_0, \ldots, z_B)$ is given by the topological recursion*

$$\omega_{g,B+1}(z_0, \ldots, z_B) = \operatorname*{Res}_{z\to 0}\Big[K(z_0, z)\,dz\Big(\omega_{g-1,B+2}(z, -z, z_1, \ldots, z_B) + \sideset{}{'}\sum_{\substack{h+h'=g\\ I\uplus I'=\{1,\ldots,B\}}} \omega_{h,|I|+1}(z, z_I)\omega_{h',|I'|+1}(-z, z_{I'})\Big)\Big],$$

where $K(z_0, z) = \frac{1}{(z^2 - z_0^2)(y(z) - y(-z))}$ and the sum $\sum'$ excludes $(h, I) = (0, \emptyset)$ and $(h, I) = (g, \{1, \ldots, B\})$.

Similar topological recursions have been established in various topics, for instance, in the one-matrix model [Eyn04], the two-matrix model [CEO06], in the theory of Gromov–Witten invariants [BKMP09] and for hyperbolic volumes of moduli spaces [Mir07].

Proposition 8.6 applied to (8.21) provide with Theorem 8.5 and (8.14) the recursive solution of the planar sector. One can achieve more:

Proposition 8.9 ([GHW19]) *There is a unique function F_g of $\{\varrho_l\}$ satisfying $W^{(g)}(z) = (2\lambda)^4 \hat{A}_z^{\dagger g} F_g(\varrho)$,*

$$F_1(\varrho) = -\frac{1}{24}\log \varrho_0, \qquad F_g(\varrho) = \frac{1}{(2-2g)(2\lambda)^4}\sum_{l=0}^{\infty} \operatorname*{Res}_{z\to 0}\Big[\frac{z^{4+2l}\varrho_l}{3+2l} W^{(g)}(z)\mathrm{d}z\Big].$$

Here we close the circle because the $F_g(\varrho)$ are, after a change of variables, nothing but the restriction to genus g of the generating functions (7.3) of intersection numbers. The change of variables turns out to be $\varrho_0 = 1 - t_0$ and $-(2l+1)!!\varrho_l = t_{l+1}$. This follows essentially from comparison between (8.18) and (7.8) at infinitesimally small λ, i.e. $c = 0$, or from a similar relation in topological recursion. Therefore, given the usual generating function of intersection numbers (see [IZ92]),

$$F_g(t_0, t_2, t_3, \dots, t_{3g-2})$$
$$:= \sum_{(k)} \frac{\langle \tau_2^{k_2}\tau_3^{k_2}\dots\tau_{3g-2}^{k_{3g-2}}\rangle}{(1-t_0)^{2g-2+\sum_i k_i}}\prod_{i=2}^{3g-2}\frac{t_i^{k_i}}{k_i!}, \qquad \sum_{i\geq 2}(i-1)k_i = 3g-3, \tag{8.23}$$

we have $F_g(\varrho) := (2\lambda)^{4g-4} F_g(t)\big|_{1-t_0=\varrho_0,\, t_l=-(2l-1)!!\varrho_{l-1}}$, and from there we build $W^{(g)} = (2\lambda)^4 \hat{A}_z^{\dagger g} F_g(\varrho)$ and higher functions via Theorem 8.5.

On the other hand, the solution of Equation (8.21) via Proposition 8.6 also permits to derive a formula for F_g. For that it is convenient to collect all genera to $\hat{A}_z^\dagger := \sum_{g=1}^\infty \hat{A}_z^{\dagger,g}$ and $\mathcal{Z}_V^{np} := \exp\big(\sum_{g=1}^\infty V^{2-2g} F_g(\varrho)\big)$. Then (8.21) is equivalent to

$$0 = \Big(\frac{2V^2}{(2\lambda)^4}\hat{K}_z\hat{A}_z^\dagger + \Big(\hat{A}_z^\dagger + \frac{1}{\varrho_0 z^4}\frac{\partial}{\partial z}\Big)\hat{A}_z^\dagger + \frac{V^2}{4(2\lambda)^4 z^4}\Big)\mathcal{Z}_V^{np}. \tag{8.24}$$

Inverting $\hat{K}_z\hat{A}_z^\dagger$ via Propositions 8.6 and 8.9, and separating the case $g = 1$, the following result can be established:

Theorem 8.10 ([GHW19]) *The generating function (8.23) of intersection numbers on the moduli spaces $\overline{\mathcal{M}}_{g,n}$ of complex curves of genus g [Wit91, Kon92] is obtained from*

$$\boxed{\exp\Big(\sum_{g=2}^{\infty} N^{2-2g} F_g(t)\Big) = \exp\Big(-\frac{1}{N^2}\Delta_t + \frac{F_2(t)}{N^2}\Big)1} \tag{8.25}$$

where $F_2(t) = \frac{7}{240}\cdot\frac{t_2^3}{3!T_0^5} + \frac{29}{5760}\frac{t_2t_3}{T_0^4} + \frac{1}{1152}\frac{t_4}{T_0^3}$ *with* $T_0 := (1-t_0)$ *generates the intersection numbers of genus 2 and* $\Delta_t = -\sum_{i,j}\hat{g}^{ij}\partial_i\partial_j - \sum_i \hat{\Gamma}^i\partial_i$ *is a Laplacian on the formal parameters* $t_0, t_2, t_3, \dots$ *given by*

$$
\begin{aligned}
\Delta_t := &-\Big(\frac{2t_2^3}{45T_0^3}+\frac{37t_2t_3}{1050T_0^2}+\frac{t_4}{210T_0}\Big)\frac{\partial^2}{\partial t_0^2}-\Big(\frac{2t_2^3}{27T_0^4}+\frac{1097t_2t_3}{12600T_0^3}+\frac{41t_4}{2520T_0^2}\Big)\frac{\partial}{\partial t_0}\\
&-\sum_{k=2}^{\infty}\Big(\Big(\frac{2t_2^2}{45T_0^3}+\frac{2t_3}{105T_0^2}\Big)t_{k+1}+\frac{t_2\mathcal{R}_{k+1}(t)}{2T_0}+\frac{3\mathcal{R}_{k+2}(t)}{2(3+2k)}\Big)\frac{\partial^2}{\partial t_k\partial t_0}\\
&-\sum_{k,l=2}^{\infty}\Big(\frac{t_2t_{k+1}t_{l+1}}{90T_0^2}+\frac{t_{k+1}\mathcal{R}_{l+1}(t)}{4T_0}+\frac{t_{l+1}\mathcal{R}_{k+1}(t)}{4T_0}\\
&\qquad\qquad+\frac{(1+2k)!!(1+2l)!!\mathcal{R}_{k+l+1}(t)}{4(1+2k+2l)!!}\Big)\frac{\partial^2}{\partial t_k\partial t_l}\\
&-\sum_{k=2}^{\infty}\Big(\Big(\frac{19t_2^2}{540T_0^4}+\frac{5t_3}{252T_0^3}\Big)t_{k+1}+\frac{t_2\mathcal{R}_{k+1}(t)}{48T_0^2}+\frac{\mathcal{R}_{k+2}(t)}{16(3+2k)T_0}+\frac{t_2t_{k+2}}{90T_0^3}\\
&\qquad\qquad+\frac{\mathcal{R}_{k+2}(t)}{2T_0}\Big)\frac{\partial}{\partial t_k}
\end{aligned}
$$

$$
\textit{with } \mathcal{R}_m(t) := \frac{2}{3}\sum_{k=1}^{m}\frac{(2m-1)!!\,kt_{k+1}}{(2k+3)!!T_0}\sum_{l=0}^{m-k}\frac{l!}{(m-k)!}B_{m-k,l}\Big(\Big\{\frac{j!t_{j+1}}{(2j+1)!!T_0}\Big\}_{j=1}^{m-l+1}\Big).
$$

The $F_g(t)$ are recursively extracted from $\mathcal{Z}_g(t) := \frac{1}{(g-1)!}(-\Delta_t + F_2(t))^{g-1}1$ and

$$
F_g(t) = \mathcal{Z}_g(t) - \frac{1}{(g-1)!}\sum_{k=2}^{g-1} B_{g-1,k}\Big(\big\{h!F_{h+1}(t)\big\}_{h=1}^{g-k}\Big). \tag{8.26}
$$

Here and in Theorem 8.10, $B_{m,k}(\{x\})$ are the Bell polynomials.

Theorem 8.10 seems to be closely related with $\exp(\sum_{g\geq 0} F_g) = \exp(\hat{W})1$ proved by Alexandrov [Ale11], where $\hat{W} := \frac{2}{3}\sum_{k=1}^{\infty}(k+\frac{1}{2})t_k\hat{L}_{k-1}$ involves the generators $\hat{L}_n$ of the Virasoro algebra. Including N and moving $\exp(N^2F_0+F_1)$ to the other side, Δ_t is in principle obtained via Baker–Campbell–Hausdorff formula from Alexandrov's equation, but evaluating the necessary commutators is not viable.

Theorem 8.10 and Equation (8.26) are easily implemented in any computer algebra system and quickly allow to compute intersection numbers to moderately large g. The result is confirmed by other implementations such as [DSvZ18].

8.4 Summary

The construction of the renormalised Φ_D^3-QFT model on noncommutative geometries of dimension $D \leq 6$ is complete. Given the mass-renormalised sequence $(E_{\underline{n}})$ for the covariance and renormalised coupling constant λ, the planar 1-point function

$G^{(g)}(z) = \frac{W^{(0)}(z)-\sqrt{z^2-c}}{2\lambda}$ is described by (8.11) with parameters chosen according to Definition 8.2 and Equation (8.13). It gives rise to planar functions with several boundary components by (8.16), (8.17) and (8.14). The non-planar sector is obtained by the following steps:

1. Compute the free energy $F_g(t)$ via Theorem 8.10 and the note thereafter. Take $F_1 = -\frac{1}{24}\log T_0$ for $g = 1$. Alternatively, start from intersection numbers obtained by other methods (e.g. [DSvZ18]).
2. Change variables to $\varrho_0 = 1 - t_0$ and $\varrho_l = -\frac{t_l+1}{(2l+1)!!}$, where ϱ_l are given by (8.18) for the measure (8.9) and with c implicitly defined by (8.13).
3. Apply to the resulting $F_g(\varrho)$ according to Proposition 8.9 and Theorem 8.5 the boundary creation operators $\hat{\mathrm{A}}^{\dagger g}_{z_1,\dots,z_B} \circ \dots \hat{\mathrm{A}}^{\dagger g}_{z_1,z_2} \circ \hat{\mathrm{A}}^{\dagger g}_{z_1}$ defined in (8.19). Multiply by $(2\lambda)^{4g+3B-4+\delta_{B,1}}$ to obtain $G^{(g)}(z_1|\cdots|z_B)$.
4. Pass to $G^{(g)}(z^1_1 \dots z^1_{N_1}|\dots|z^B_1 \dots z^B_{N_B})$ via difference quotients (8.14).

Finally, evaluate at $z^\beta_{k_\beta} \mapsto (4E^2_{\underline{p}^\beta_{k_\beta}} + c)^{1/2}$ to obtain $G^{(g)}_{|\underline{p}^1_1 \dots \underline{p}^1_{N_1}|\dots|\underline{p}^B_1 \dots \underline{p}^B_{N_B}|}$, where $E_{\underline{p}}$ arises by mass-renormalisation from the $\tilde{E}_{\underline{p}}$ in the initial action (8.3) of the model.

There remains a final problem. We have achieved exact formulae for any correlation function at any fixed genus, which corresponds to a convergent sum over amplitudes encoded in infinitely many ribbon graphs. It remains to understand the sum $\sum_{g=0}^\infty V^{-2g} G^{(g)}_{\dots}$ over genera in (7.16), which by the Steps 1–4 is derived from the sum $\sum_{g=2}^\infty N^{2-2g} F_g(t)$ in (8.25). The generating function F_g contains $p(3g-3) \sim \frac{1}{12\sqrt{3}(g-1)} \exp(\pi\sqrt{2g-2})$ terms, which are too many for ordinary convergence:

Question 8.11 Is it possible to Borel-sum the series $\sum_{g=1}^\infty V^{2-2g} F_g(\varrho)$ for $\varrho_l > 0$? Note that this corresponds to $t_l < 0$ for $l \geq 2$ and $\lambda \in \mathrm{i}\mathbb{R}$ (see (8.9)). One should use asymptotic estimates [MZ15] of intersection numbers or the heat flow of Δ_t given in Theorem 8.10, or a recent estimate by Eynard [Eyn19].

One could also ask whether the metric $\hat{g}$ in $\Delta_t = -\sum_{i,j} \hat{g}^{ij}\partial_i\partial_j - \sum_i \hat{\Gamma}^i \partial_i$ has any significance:

Question 8.12 Is $\hat{\Gamma}^i$ a Levi-Civita connection for $\hat{g}^{ij}$? Does $\hat{g}^{ij}$ admit a reasonable definition of a volume and a curvature? Is there any relation to the Weil–Petersson volumina which are deeply connected with intersection numbers [AC96, Mir07]?

9 Exact solution of the Φ^4-model

9.1 The planar sector

It would have far-reaching consequences if the Φ^4-model admitted a similar construction as the Φ^3-model. After a decade of work and many failed attempts, such a construction is now in reach. Combining Dyson–Schwinger equations and Ward–Takahashi identity, we derived in [GW09] a closed equation for the planar two-point function of the Φ^4-model. This equation is complicated. A considerable simplification to an angle functions of essentially only one variable was achieved in [GW14a]. In [PW18] the exact solution was found for the important special case of a scaling limit of two-dimensional Moyal space. In [GHW19b] it was understood how to generalise this construction to any covariance of dimension ≤ 4. For finite matrices a representation by rational functions arises. This rationality is a strong support for the conjecture that the Φ^4-model is integrable. Below we give a few details. It remains to be seen whether correlation functions at higher (g, B) satisfy any sort of topological recursion [EO07], and to identify the integrable structure.

The Φ^4-model is defined by the identical covariance $C_E^{(\mathcal{N})}$ as given before in (8.2) with (8.1) but instead of (8.3) by a *quartic* interaction functional

$$S_{\text{int}}^{(\mathcal{N})}(\Phi) := V\frac{\lambda\tilde{Z}^2}{4}\sum_{\underline{k},\underline{l},\underline{m},\underline{n}\in\mathbb{N}_{\mathcal{N}}^{D/2}} \Phi_{\underline{k}\underline{l}}\Phi_{\underline{l}\underline{m}}\Phi_{\underline{m}\underline{n}}\Phi_{\underline{n}\underline{k}}\,, \qquad \Phi_{\underline{k}\underline{l}} := \Phi(e_{\underline{k}\underline{l}})\,. \tag{9.1}$$

By the same techniques as described in Section 7.3—Dyson–Schwinger equations combined with Ward–Takahashi identity—exact non-perturbative equations for correlation functions are obtained. Below we give these equations for planar functions with a single boundary component ($g = 0$, $B = 1$):

Proposition 9.1 ([GW14a])

$$G^{(0)}_{|\underline{k}\underline{l}|} = \frac{1}{\tilde{Z}(\tilde{E}_{\underline{k}}+\tilde{E}_{\underline{l}})} - \frac{\tilde{Z}\lambda}{\tilde{E}_{\underline{k}}+\tilde{E}_{\underline{l}}}\frac{1}{V}\sum_{\underline{n}\in\mathbb{N}_{\mathcal{N}}^{D/2}}\Big(G^{(0)}_{|\underline{k}\underline{l}|}G^{(0)}_{|\underline{k}\underline{n}|} - \frac{G^{(0)}_{|\underline{n}\underline{l}|}-G^{(0)}_{|\underline{k}\underline{l}|}}{\tilde{Z}(\tilde{E}_{\underline{n}}-\tilde{E}_{\underline{k}})}\Big)\,, \tag{9.2}$$

$$G^{(0)}_{|\underline{k}_0\underline{k}_1\dots\underline{k}_{N-1}|} \tag{9.3}$$

$$= (-\lambda)\sum_{l=1}^{\frac{N-2}{2}}\frac{G^{(0)}_{|\underline{k}_0\underline{k}_1\dots\underline{k}_{2l-1}|}G^{(0)}_{|\underline{k}_{2l}\underline{k}_{2l+1}\dots\underline{k}_{N-1}|} - G^{(0)}_{|\underline{k}_{2l}\underline{k}_1\dots\underline{k}_{2l-1}|}G^{(0)}_{|\underline{k}_0\underline{k}_{2l+1}\dots\underline{k}_{N-1}|}}{(E_{\underline{k}_0}-E_{\underline{k}_{2l}})(E_{\underline{k}_1}-E_{\underline{k}_{N-1}})}\,.$$

The general case including higher-genus contributions can be found in [GW14a]. A manifestly symmetric variant of (9.2) was derived in [PW18].

The first Equation (9.2) requires renormalisation (see below), whereas (9.3) is automatically expressed in terms of the renormalised 2-point function $G^{(0)}_{|\underline{k}\underline{l}|}$ and the mass-renormalised sequence $E_{\underline{k}} := \tilde{E}_{\underline{k}} + \frac{\mu^2-\tilde{\mu}^2}{2}$ with $E_{\underline{n}} - E_{\underline{k}} \equiv \tilde{E}_{\underline{n}} - \tilde{E}_{\underline{k}}$. Thus, no renormalisation of the coupling constant λ is necessary, which means that the β-function vanishes identically (provided that the 2-point function can be renormalised). This is the non-perturbative proof of [DGMR07].

Equation (9.3) is the analogue of (8.6). Its explicit solution is a sum of monomials in $G^{(0)}_{|\underline{k}_{2i}\underline{k}_{2j+1}|}$, $\frac{1}{E_{\underline{k}_{2i}}-E_{\underline{k}_{2j}}}$ and $\frac{1}{E_{\underline{k}_{2i+1}}-E_{\underline{k}_{2j+1}}}$. As proved in [dJHW19], these monomials are in one-to-one correspondence with *Catalan tables* of length $\frac{N}{2}$, which are iterations of Catalan tuples.

Definition 9.2 A *Catalan tuple* of length k is a $(k+1)$-tuple $\tilde{t} = (t_0, \ldots, t_k)$ with $\sum_{i=0}^{k} t_i = k$ and $\sum_{i=0}^{l} t_i > l$ for any $l < k$. We let $|\tilde{t}|$ be the length of $\tilde{t}$.

A *Catalan table* of length k is a $k+1$-tuple $T = \langle \tilde{t}_0, \ldots, \tilde{t}_k \rangle$ of Catalan tuples $\tilde{t}_i$ such that $(|\tilde{t}_0|+1, |\tilde{t}_1|, \ldots, |\tilde{t}_1|)$ is itself a Catalan tuple.

There are $C_k = \frac{1}{k+1}\binom{2k}{k}$ different Catalan tuples of length k (whence its name) and $\frac{1}{k+1}\binom{3k+1}{k}$ different Catalan tables T of length k. A Catalan table of length $\frac{N}{2}$ simultaneously encodes a pocket tree for the monomials in $G^{(0)}_{\underline{k}_{2i}\underline{k}_{2j+1}}$, a rooted tree for the monomials in $\frac{1}{E_{\underline{k}_{2i}}-E_{\underline{k}_{2j}}}$ and an opposite tree for the monomials in $\frac{1}{E_{\underline{k}_{2i+1}}-E_{\underline{k}_{2j+1}}}$.

By the same reasoning as before, any solution of (9.2) extends to a sectionally holomorphic function $G^{(0)}(\zeta_1, \zeta_2)$ with $G^{(0)}_{|\underline{k}\underline{l}|} = G^{(0)}(\zeta_1, \zeta_2)\big|_{\zeta_1=\tilde{E}_{\underline{k}}-\tilde{\mu}^2/2, \zeta_2=\tilde{E}_{\underline{l}}-\tilde{\mu}^2/2}$ which satisfies

$$(\tilde{\mu}^2+\zeta_1+\zeta_2)\tilde{Z}G^{(0)}(\zeta_1,\zeta_2) \tag{9.4}$$
$$= 1 - \lambda \int_0^{\Lambda^2} dt\, \varrho(t) \left(\tilde{Z}G^{(0)}(\zeta_1,\zeta_2)\, \tilde{Z}G^{(0)}(\zeta_1,t) - \frac{\tilde{Z}G^{(0)}(t,\zeta_2) - \tilde{Z}G^{(0)}(\zeta_1,\zeta_2)}{(t-\zeta_1)} \right),$$

where $\varrho(t) := \frac{1}{V}\sum_{\underline{n}\in\mathbb{N}_N^{D/2}} \delta(t - (\tilde{E}_{\underline{n}} - \frac{1}{2}\tilde{\mu}^2))$ and $\tilde{E}_{\underline{n}} \in [\frac{1}{2}\tilde{\mu}^2, \Lambda^2 + \frac{1}{2}\tilde{\mu}^2]$ for all $\underline{n} \in \mathbb{N}_N^{D/2}$. Next we temporarily assume that ϱ can be approximated by a Hölder-continuous function. The strategy developed in [GW14a] consists in an ansatz

$$ZG^{(0)}(a,b) = \frac{e^{\mathcal{H}_a^{\Lambda}[\tau_b(\bullet)]}\sin\tau_b(a)}{\lambda\pi\varrho(a)} = \frac{e^{\mathcal{H}_b^{\Lambda}[\tau_a(\bullet)]}\sin\tau_a(b)}{\lambda\pi\varrho(b)}, \tag{9.5}$$

where the angle function $\tau_a : (0, \Lambda^2) \to [0, \pi]$ for $\lambda > 0$ and $\tau_a : (0, \Lambda^2) \to [-\pi, 0]$ for $\lambda < 0$ remains to be determined. Here,

$$\mathcal{H}^{\Lambda}_{a}[f(\bullet)] := \frac{1}{\pi}\lim_{\epsilon\to 0}\int_{[0,\Lambda^2]\setminus[a-\epsilon,a+\epsilon]}\frac{dt\; f(t)}{t-a} = \lim_{\epsilon\to 0}\mathrm{Re}\Big(\frac{1}{\pi}\int_0^{\Lambda^2}\frac{dt\; f(t)}{t-(a+\mathrm{i}\epsilon)}\Big) \tag{9.6}$$

denotes the finite Hilbert transform. We go with the ansatz (9.5) into (9.4) at $\zeta_1 = a + \mathrm{i}\epsilon$ and $\zeta_2 = b$:

$$\begin{aligned}&\Big(\tilde{\mu}^2 + a + b + \lambda\pi\mathcal{H}^{\Lambda}_{a}[\varrho(\bullet)] + \frac{1}{\pi}\int_0^{\Lambda^2} dt\; e^{\mathcal{H}^{\Lambda}_{t}[\tau_a(\bullet)]}\sin\tau_a(t)\Big) ZG^{(0)}(a,b)\\ &= 1 + \mathcal{H}^{\Lambda}_{a}\big[e^{\mathcal{H}^{\Lambda}_{\bullet}[\tau_b]}\sin\tau_b(\bullet)\big] .\end{aligned} \tag{9.7}$$

A Hölder-continuous function $\tau : (0,\Lambda^2) \to [0,\pi]$ or $\tau : (0,\Lambda^2) \to [-\pi,0]$ satisfies

$$\begin{aligned}\mathcal{H}^{\Lambda}_{a}\big[e^{\mathcal{H}^{\Lambda}_{\bullet}[\tau]}\sin\tau(\bullet)\big] &= e^{\mathcal{H}^{\Lambda}_{a}[\tau]}\cos\tau(a) - 1 ,\\ \int_0^{\Lambda^2} dt\; e^{\pm\mathcal{H}^{\Lambda}_{t}[\tau(\bullet)]}\sin\tau(t) &= \int_0^{\Lambda^2} dt\; \tau(t) .\end{aligned} \tag{9.8}$$

The first identity appeared in [Tri57], the second one was proved in [PW18]. Inserting both identities into (9.7) gives with (9.5) a consistency relation for the angle function:

$$\tau_a(p) = \arctan\left(\frac{\lambda\pi\varrho(p)}{\tilde{\mu}^2 + a + p + \lambda\pi\mathcal{H}^{\Lambda}_{p}[\varrho(\bullet)] + \frac{1}{\pi}\int_0^{\Lambda^2} dt\; \tau_p(t)}\right) , \tag{9.9}$$

where the arctan-branch in $[0,\pi]$ is selected for $\lambda > 0$ and the branch in $[-\pi,0]$ for $\lambda < 0$.

The dependence on a in (9.9) is relatively simple so that the first attempts focused on the resulting equation for $\tau_0(p)$. This allowed to prove, for the case $\varrho(x) = x$ of 4-dimensional Moyal space with harmonic propagation, existence of a solution [GW16]. Also an interesting phase structure was detected [GW14b], but a solution was missed.

9.2 *Exact solution of the planar 2-point function*

A breakthrough was achieved in [PW18], where the special case $\varrho(x) = 1$ was solved that describes a scaling limit of the 2-dimensional Moyal space with harmonic propagation:

Theorem 9.3 ([PW18]) *For $\varrho(x) = 1$ and with $\tilde{\mu}^2 = 1 - 2\lambda \log(1 + \Lambda^2)$, the consistency equation* (9.9) *has in the limit $\Lambda \to \infty$ for $\lambda > 0$ the solution*

$$\tau_a(p) = \arctan\left(\frac{\lambda\pi}{a + \lambda\, W_0\big(\frac{1}{\lambda}e^{\frac{1+p}{\lambda}}\big) - \lambda \log\big(\lambda W_0\big(\frac{1}{\lambda}e^{\frac{1+p}{\lambda}}\big) - 1\big)}\right), \tag{9.10}$$

where W_0 denotes the principal branch of the Lambert function [Lam58, CGH[+]96].

The HyperInt package [Pan15] was used to push a perturbative solution of (9.10) far enough to guess the whole perturbation series. The series is resumed by Lagrange-Bürmann formula [Lag70, Bür99] to Lambert-W. The result is confirmed by the residue theorem. The 2-point function $G^{(0)}(a, b)$ is then evaluated from (9.5) via deformation of complex contour integrals (the result will be given below for any ϱ).

Building on [PW18], in [GHW19b] the exact solution of the non-linear equation (9.4) was achieved for *any* density ϱ (which encodes a sequence (E_n) of dimension $D \leq 4$ according to Definition 8.2). Starting point was the observation that the denominator of (9.10) can be written (up to a shift $1 + a$) as $-\mathrm{Re}(-f(p) + \lambda \log(-f(p)))$, where $f(p) = \lambda W_0\big(\frac{1}{\lambda}e^{\frac{1+p}{\lambda}}\big) - 1$ solves $1 + p = 1 + f(p) + \lambda \log(1 + f(p)))$. The logarithm is the renormalised Stieltjes transform of the measure $\varrho(t) = 1$. This suggested to try the same combination of reflection $z \leftrightarrow -\mu^2 - z$ with the Stieltjes transform of the given density ϱ. But this was not enough: The general case requires a *deformation of ϱ to an implicitly defined measure function ϱ_λ*:

Definition 9.4 Given a real λ in some open neighbourhood of 0, a scale $\mu^2 > 0$ and a Hölder-continuous function $\varrho : [0, \Lambda^2] \to \mathbb{R}_+$ of dimension $D \in \{0, 2, 4\}$. Then a function ϱ_λ on $[\nu_\lambda, \Lambda_\lambda^2]$ is implicitly defined by

$$\varrho(t) =: \varrho_\lambda(R_\lambda^{-1}(t)), \qquad \Lambda_\lambda^2 := R_\lambda^{-1}(\Lambda^2), \qquad \nu_\lambda := R_\lambda^{-1}(0), \tag{9.11}$$

where $R_\lambda : \mathbb{C} \setminus [-\mu^2 - \Lambda_\lambda^2, -\mu^2 - \nu_\lambda] \to \mathbb{C}$ is defined via the same function ϱ_λ by

$$R_\lambda(z) := z - \lambda(-z)^{\frac{D}{2}} \int_{\nu_\lambda}^{\Lambda_\lambda^2} \frac{dt\, \varrho_\lambda(t)}{(t + \mu^2)^{\frac{D}{2}}(t + \mu^2 + z)}. \tag{9.12}$$

The definition is consistent because for $|\lambda|$ small enough, R_λ is a biholomorphic map from the half-plane $\mathrm{Re}(z) > -\frac{\mu^2}{2}$ onto a domain which contains $[0, \Lambda^2]$.

Using the same complex analysis techniques as in [PW18], including Lagrange inversion theorem and Bürmann formula, the following generalisation of Theorem 9.3 can be achieved:

Theorem 9.5 ([GHW19b]) *Let $\varrho : [0, \Lambda^2] \to \mathbb{R}_+$ be a Hölder-continuous measure of dimension $D \in \{0, 2, 4\}$ and ϱ_λ its deformation according to Definition 9.4* *for a real coupling constant λ with $|\lambda| < \big(\int_{\nu_\lambda}^{\Lambda_\lambda^2} \mathrm{d}t \frac{\varrho_\lambda(t)}{(t+\mu^2/2)^2} + \delta_{D,4} \int_{\nu_\lambda}^{\Lambda_\lambda^2} \mathrm{d}t \frac{\varrho_\lambda(t)}{(t+\mu^2)^2}\big)^{-1}$. Then the consistency equation (9.9) for the angle function is solved by*

$$\tau_a(p) = \lim_{\epsilon \to 0} \mathrm{Im}\big(\log(a - R_\lambda(-\mu^2 - R_\lambda^{-1}(p + \mathrm{i}\epsilon)))\big) , \tag{9.13}$$

where R_λ is defined by (9.12) and $\tilde{\mu}$ is renormalised according to

$$\tilde{\mu}^2 = \mu^2\Big(1 - \lambda\delta_{D,4} \int_{\nu_\lambda}^{\Lambda_\lambda^2} \frac{\mathrm{d}t\ \varrho_\lambda(t)}{(t+\mu^2)^2}\Big) - 2\lambda(\delta_{D,2} + \delta_{D,4}) \int_{\nu_\lambda}^{\Lambda_\lambda^2} \frac{\mathrm{d}t\ \varrho_\lambda(t)}{t+\mu^2} . \tag{9.14}$$

A constant measure such as $\varrho(t) = 1$ remains undeformed to $\varrho_\lambda(x) = 1$, and (9.13) reduces for $D = 2$ and $\mu = 1$ to (9.10).

Evaluation of the Hilbert transform of (9.13) yields for (9.5):

Proposition 9.6 ([GHW19b]) *The renormalised planar two-point function of the D-dimensional Φ^4-model is given by*

$$G^{(0)}(a, b) := \frac{(\mu^2)^{\delta_{D,4}}(\mu^2 + a + b)\exp(N_\lambda(a, b))}{(\mu^2 + b + R_\lambda^{-1}(a))(\mu^2 + a + R_\lambda^{-1}(b))} , \tag{9.15}$$

where R_λ is built via (9.12) with the deformed measure ϱ_λ defined in (9.11) and

$$\begin{aligned} N_\lambda(a, b) := \frac{1}{2\pi\mathrm{i}} \int_{-\infty}^{\infty} \mathrm{d}t \Bigg\{ &\log\left(\frac{a - R_\lambda(-\frac{\mu^2}{2} - \mathrm{i}t)}{a - (-\frac{\mu^2}{2} - \mathrm{i}t)}\right) \frac{\mathrm{d}}{\mathrm{d}t} \log\left(\frac{b - R_\lambda(-\frac{\mu^2}{2} + \mathrm{i}t)}{b - (-\frac{\mu^2}{2} + \mathrm{i}t)}\right) \\ &- \delta_{D,4} \log\left(\frac{R_\lambda(-\frac{\mu^2}{2} - \mathrm{i}t)}{(-\frac{\mu^2}{2} - \mathrm{i}t)}\right) \frac{\mathrm{d}}{\mathrm{d}t} \log\left(\frac{R_\lambda(-\frac{\mu^2}{2} + \mathrm{i}t)}{(-\frac{\mu^2}{2} + \mathrm{i}t)}\right) \Bigg\} . \end{aligned} \tag{9.16}$$

In $D = 4$ dimensions, $G^{(0)}(a, b)$ is only determined up to a multiplicative constant (the finite part of $\tilde{Z}$) which here is normalised to $G^{(0)}(0, 0) = 1$ independently of μ.

Moyal space in dimension $D = 2$ corresponds to $\varrho(x) = 1$ and accordingly $R_\lambda(x) = x + \lambda\log(1 + x)$ (when setting $\mu = 1$). The perturbative expansion of $N_\lambda(a, b)$ involves Nielsen's generalised polylogarithms [Nie09] and Riemann zeta values.

Moyal space in dimension $D = 4$ corresponds to $\varrho(t) = t$, which by (9.11) and (9.12) results (for $\Lambda \to \infty$) in

$$\varrho_\lambda(x) = R_\lambda(x) = x - \lambda x^2 \int_0^\infty \frac{\mathrm{d}t\ R_\lambda(t)}{(t + \mu^2)^2(t + \mu^2 + x)} . \tag{9.17}$$

Proposition 9.7 ([GHW19c]) *The Fredholm integral equation (9.17) has the solution*

$$R_\lambda(x) = x\,{}_2F_1\Big(\begin{matrix}\alpha_\lambda,\ 1-\alpha_\lambda\\ 2\end{matrix}\Big|-\frac{x}{\mu^2}\Big), \tag{9.18}$$

$$\textit{where}\quad \alpha_\lambda := \begin{cases} \frac{\arcsin(\lambda\pi)}{\pi} & \textit{for } |\lambda| \le \frac{1}{\pi},\\ \frac{1}{2} + \mathrm{i}\frac{\operatorname{arcosh}(\lambda\pi)}{\pi} & \textit{for } \lambda \ge \frac{1}{\pi}. \end{cases}$$

Inserting (9.18) into (9.15) provides an integral representation[6] for the planar two-point function, which is *exact* in $\lambda \ge -\frac{1}{\pi}$. Its perturbative expansion involves a particular class of hyperlogarithms in alternating letters $0, -1$ [GHW19c].

Remark 9.8 It is very important that $R_\lambda : \mathbb{R}_+ \to \mathbb{R}_+$ given by (9.18) is bijective. This was not expected in the beginning: If ϱ_λ in (9.12) had the same asymptotic behaviour $\varrho_\lambda(x) \propto x$ as $\varrho(x) = x$ for $D = 4$, then R_λ would reach a maximum on $\mathbb{R}_+$ and could not be inverted globally, unless $\lambda = 0$ is trivial. The Φ^4-model on 4-dimensional Moyal space avoids the triviality [Aiz81, Frö82] of the commutative ϕ^4_4-model by a significant modification of the spectral dimension. Defining $D_{\text{spec}}(\rho) = \inf\{p : \int_0^\infty dt\, \frac{\rho(t)}{(1+t)^{p/2}} < \infty\}$, then $D_{\text{spec}}(\varrho_\lambda) = 4 - 2\frac{\arcsin(\lambda\pi)}{\pi}$ for $|\lambda| \le \frac{1}{\pi}$ but $D_{\text{spec}}(\varrho) = 4$.

The final result (9.15) and (9.16) does not require anymore that ϱ is Hölder-continuous. It also holds for ϱ a finite sum of Dirac measures, and in this case one can even evaluate the remaining integral (9.16):

Theorem 9.9 ([GHW19b]) *Consider the Φ^4-model for $N \times N$-matrices in which the covariance is defined by a d-tupel $(E_1, \dots, E_d)$ of positive real numbers, where E_k arises with multiplicity r_k, and $\sum_{k=1}^d r_k = N$. These data encode a rational function*

$$R(z) := z - \frac{\lambda}{N}\sum_{k=1}^{d} \frac{\varrho_k}{\varepsilon_k + z}, \tag{9.19}$$

where $\{\varepsilon_k, \varrho_k\}_{k=1,\dots,d}$ are the unique solutions in a neighbourhood of $\lambda = 0$ of

$$E_k = R(\varepsilon_k)\,, \quad r_k = \varrho_k R'(\varepsilon_k) \qquad \textit{with} \quad \lim_{\lambda\to 0}\varepsilon_k = E_k\,, \quad \lim_{\lambda\to 0}\varrho_k = r_k. \tag{9.20}$$

[6]The inverse function $R_\lambda^{-1}(x)$ in (9.15) can be combined with $\mathcal{N}_\lambda$ to another integral representation [GHW19b].

Then the planar two-point function has in an open neighbourhood of $\lambda = 0$ *the explicit solution* $G^{(0)}_{ab} = \mathcal{G}^{(0)}(\varepsilon_a, \varepsilon_b)$, *where* $\mathcal{G}^{(0)} : \bar{\mathbb{C}} \times \bar{\mathbb{C}} \to \bar{\mathbb{C}}$ *is the rational function*

$$\mathcal{G}^{(0)}(z,w) = \frac{1 - \dfrac{\lambda}{\mathcal{N}} \displaystyle\sum_{k=1}^{d} \frac{r_k}{(R(z) - R(\varepsilon_k))(R(\varepsilon_k) - R(-w))} \prod_{j=1}^{d} \frac{R(w) - R(-\widehat{\varepsilon_k}^j)}{R(w) - (\varepsilon_j)}}{R(w) - R(-z)} \tag{9.21}$$

in which $z \in \{u, \hat{u}^1, \ldots, \hat{u}^d\}$ *is the list of roots of* $R(z) = R(u)$. *The 2-point function is symmetric,* $\mathcal{G}^{(0)}(z,w) = \mathcal{G}^{(0)}(w,z)$, *and defined outside poles located at* $z + w = 0$, *at* $z = \widehat{\varepsilon_k}^m$ *and at* $w = \widehat{\varepsilon_l}^n$, *for* $k, l, m, n = 1, \ldots, d$.

Theorem 9.9 undeniably establishes that the Φ^4-model is exactly solvable in surprisingly close analogy with the Φ^3-model (i.e. the Kontsevich model). The rationality achieved in (9.21) is overwhelming support for the conjecture that the Φ^4-model is integrable, too, which means it descends from a τ-function satisfying a Hirora equation [Miw82].

The simplest case $E = \frac{\mu^2}{2} = \text{const}$ of a single $r_1 = \mathcal{N}$-fold degenerate eigenvalue $E_1 = \frac{\mu^2}{2}$ is already interesting. One finds [GHW19b]

$$G^{(0)}_{11} = \frac{4}{3} \frac{\mu^2 + 2\sqrt{\mu^4 + 12\lambda}}{\left(\mu^2 + \sqrt{\mu^4 + 12\lambda}\right)^2}. \tag{9.22}$$

It agrees with corresponding formulae in the literature [BIPZ78].

9.3 Outlook

It remains to recursively solve the Dyson–Schwinger equations for $B > 1$. In [GW14a] this problem was already reduced to affine equations for $N_1 + \cdots + N_B$-point functions with all $N_\beta \leq 2$, where $N_\beta = 1$ is much simpler than $N_\beta = 2$. One should start with finite matrices where the initial data of the recursion is known from (9.21). The same change of variables (9.19) brings the equation for the $1+1$-point function into the form

$$\big(R(z) - R(-z)\big)\mathcal{G}^{(0)}(z|w) - \frac{\lambda}{\mathcal{N}} \sum_{k=1}^{d} \frac{r_k \mathcal{G}^{(0)}(\varepsilon_k|w)}{R(\varepsilon_k) - R(z)} = \lambda \frac{\mathcal{G}^{(0)}(z,w) - \mathcal{G}^{(0)}(w,w)}{R(z) - R(w)}. \tag{9.23}$$

The lhs agrees exactly with the corresponding operator in topological recursion [EO07, Eyn16], provided that one chooses the *classical spectral curve* $\mathcal{E}(x(z), y(z)) = 0$ as

$$x(z) = R(z) , \qquad y(z) = -R(-z). \tag{9.24}$$

But the rhs of (9.23) is completely different. Its poles at $z, w \in \{\widehat{\varepsilon_k}^l\}$ have no counterpart in topological recursion. These poles are expected to proliferate into all functions of higher topology. Moreover, topological recursion assumes that the branched covering $x : \overline{\mathbb{C}} \to \overline{\mathbb{C}}$ is invariant under the local Galois involution, here $z \mapsto -z$. This invariance does not hold in (9.24). Finally, higher N-point functions arising as inhomogeneity of the recursion are (via the generalisation of (9.3)) non-linear in the basic $N_1 + \cdots + N_B$-point functions with all $N_\beta \leq 2$, whereas in topological recursion the analogous dependence is linear.

Thus, in spite of striking similarities with *the* topological recursion, the recursion of the Φ^4-model differs significantly when looking closer. Solving the Φ^4-recursion from scratch will be a fascinating programme for the next years. The remarkable fact that also the Φ^4-model is exactly solvable raises several questions:

Question 9.10

- Why are the exact solutions of the Φ^4- and Φ^3-models so similar, whereas their perturbative treatment falls completely apart?
- Do all contributions to correlation functions with topology (g, B) have a significance in topological recursion, or only those with an odd number of defects per boundary component?
- What is the integrable structure of the Φ^4-model? Is the logarithm of the partition function of the Φ^4-model a τ-function for a Hirota equation?
- Is the logarithm of the partition function a series in certain t_i with rational coefficients? If so, do these rational numbers describe some intersection numbers of some characteristic classes on some moduli space?
- Can one identify a Virasoro algebra, or some generalisation, in the Φ^4-model?
- Can the standard model of particle physics learn something from the integrable Φ^4-model? Does the enormous complexity concerning polylogarithms and other transcendental functions in the standard model possibly arise through the perturbative solution of implicitly defined problems similar to Definition 9.4?

10 Osterwalder–Schrader axioms

Strictly speaking, the programme outlined in Section 3, completely solved for the Φ^3-model in Section 8 and partially solved for the Φ^3-model in Section 9, does not yet produce any quantum field theory. It gives consistent continuum limits of statistical physics models, but not a QFT. For a true quantum field theory a time evolution is necessary. On standard Euclidean space $\mathbb{R}^D$, time evolution is a deep consequence of the Osterwalder–Schrader axioms [OS73, OS75] (see Definition 2.2). The most decisive axiom is *reflection positivity*, a variant of the Hausdorff–Bernstein–Widder theorem:

Theorem 10.1 (Hausdorff–Bernstein–Widder) *Let S be a convex cone in a real vector space X, containing 0. Then for a continuous function $F : S \to \mathbb{R}$ the following are equivalent:*

1. *F is decreasing and positive definite, i.e. $\sum_{i,j=1}^{K} \overline{c_i} c_j F(t_i + t_j) \geq 0$ for all $c_i, c_j \in \mathbb{C}$ and $t_i, t_j \in S$.*
2. *F is the Laplace transform $F(t) = \int_{X'} d\mu(\lambda)\, \mathrm{e}^{-\lambda(t)}$ of a positive measure on X'.*
3. *F is completely monotonic, i.e. $\prod_{i=1}^{K}(\mathrm{id} - T_{\delta_i})F \geq 0$ for all $\delta_i \in S$, where $(T_{\delta_i}F)(t) = F(t + \delta_i)$.*

Consequently, F is smooth, and 3. can be replaced by $(-1)^{|n|} f^{(n)}(t) \geq 0$ for any multi-index n.

The original publications [Hau23, Ber29, Wid31] prove the theorem for $S = \mathbb{R}_+$. The higher-dimensional version $S = (\mathbb{R}_+)^N$ is given in Bochner's book [Boc55]. It was extended to fairly general Abelian semigroups by Nussbaum [Nus55] and to operators in Hilbert space by several authors, for instance [KL81]. The main feature of the Laplace integral is that it provides a holomorphic extension of F into the tube $S + \mathrm{i}X$. It is this purely imaginary $\mathrm{i}X$ what we refer to as 'time'. Reflection positivity is a challenging topic in mathematics and physics. We refer to [NO18] and the Oberwolfach reports [JNOS17] for more details and for an overview about current research activities.

To develop an Osterwalder–Schrader setup for a (noncommutative) nuclear AF Fréchet algebra $\mathcal{A} = \bigcup_N \mathcal{A}^{\mathcal{A}}$ on which we constructed QFT models we need a continuous linear map $q_* : \mathcal{A}_* \to C^\infty(U)$ into a vector space of smooth functions. It is tempting to identify q_* with the isomorphism ι_U provided by the Kōmura-Kōmura theorem 3.3. In this case we would need that the image $\iota_U(\mathcal{A}_*)$ is invariant under translations. This is the case for the Moyal algebras in Sections 4.1 and 4.3, but we do not know it in general (see Question 3.4). We briefly review in Section 10.1 what is known for the choice $q_* = \iota_{\mathbb{R}^D}$. The lesson will be to proceed differently. We develop first ideas in Section 10.2.

10.1 Previous approaches to reflection positivity

Reflection positivity in QFTs on Moyal space has been studied in several contexts. We admit, however, that a satisfactory picture was not yet given. If one had a reflection-positive QFT on ordinary $\mathbb{R}^3$ or $\mathbb{R}^4$, then one can choose to Moyal-deform only a 2-dimensional subspace orthogonal to the time direction. In this case analytic continuation ('Wick rotation') and deformation commute up to an isomorphism of the Minkowskian Moyal algebra [GLLV13]. The restriction to degenerate Moyal space is an essential requirement. Without it the continuation of Moyal-deformed Wightman functions leads to Euclidean functions with twists in mass-shell momenta [Bah10].

Another approach proposed in [GW13b] consists in using the isomorphism $\iota_\Theta = \iota_{\mathbb{R}^D}$ to map matrix correlation functions $\langle e_{\underline{k}_1\underline{l}_1} \otimes \cdots \otimes e_{\underline{k}_N\underline{l}_N}\rangle$ defined in (3.6) into $\mathbb{R}^D$-labelled candidate Schwinger functions

$$S_c(x_1,\dots,x_n) \tag{10.1}$$

$$:= \sum_{\underline{k}_1,\dots,\underline{l}_N\in\mathbb{N}^{D/2}} f_{\underline{k}_1\underline{l}_1}(x_1)\cdots f_{\underline{k}_N\underline{l}_N}(x_N)\frac{(-\mathrm{i})^N}{\mathcal{Z}_C(0)}\frac{\partial^N\mathcal{Z}_C(t_1e_{\underline{k}_1\underline{l}_1}+\dots+t_Ne_{\underline{k}_N\underline{l}_N})}{\partial t_1\cdots\partial t_N}\Big|_{t_i=0},$$

where the $f_{\underline{kl}}$ are extension of $f_{kl}^{(\theta)}$ defined in (4.3) to $D/2$ components. The function $\mathcal{Z}_C(J)$ is expanded by (7.15). This involves the covariance C defined in (6.7) which essentially relies on the harmonic oscillator Schrödinger operator $H^{\Omega=1} = -\Delta + \frac{4}{\theta^2}x^2$. The explicit dependence on the position x in this operator makes the candidate Schwinger functions (10.1) not even translation-invariant. However, the dangerous term vanishes in the limit $V = (\frac{\theta}{4})^{D/2} \to \infty$. Note that the limit $V \to \infty$ is highly singular for the matrix basis functions (4.3). It was proved in [GW13b] that the limit $V \to \infty$ of (10.1) is well-defined and gives with a convention $\frac{\delta J_{\underline{mn}}}{\delta J(x)} := \mu^D f_{\underline{mn}}(x)$ the following formula for connected Schwinger functions:

$$S_c(x_1,\dots,x_N)$$

$$= \frac{1}{(8\pi)^{\frac{D}{2}}}\sum_{\substack{N_1+\dots+N_B=N\\ N_\beta \text{ even}}}\sum_{\sigma\in\mathcal{S}_N}\Big(\prod_{\beta=1}^{B}\frac{4^{N_\beta}}{N_\beta}\int_{\mathbb{R}^4}\frac{dp_\beta}{4\pi^2}\,e^{\mathrm{i}\langle p_\beta,\sum_{i=1}^{N_\beta}(-1)^{i-1}x_{\sigma(s_\beta+i)}\rangle}\Big)$$

$$\times G_{\big|\underbrace{\frac{\|p_1\|^2}{2},\dots,\frac{\|p_1\|^2}{2}}_{N_1}\big|\dots\big|\underbrace{\frac{\|p_B\|^2}{2},\dots,\frac{\|p_B\|^2}{2}}_{N_B}\big|}. \tag{10.2}$$

Euclidean invariance is manifest. The most interesting sector is $N_\beta = 2$ in every boundary component. This $(2+\ldots+2)$-sector describes the propagation and interaction of B Euclidean 'particles' without any momentum exchange. Absence of momentum transfer is characteristic to integrable models [Mos75, Kul76], but in 4 dimensions a sign of *triviality* [Aks65]. However, not all assumptions of this triviality proof are satisfied in the models under consideration.

For the Φ^3-model constructed in Section 8, the explicit formulae (8.17) and (8.14) admit a direct verification of complete monotonicity (property 3. of Theorem 10.1). For the 2-point function this amounts to prove that $a \mapsto G_{|aa|} \equiv \int_0^\infty \frac{d\rho(m^2)}{a+m^2}$ is a Stieltjes function [GW13b], i.e. the Stieltjes transform of a positive measure $d\rho(m^2)$. Surprisingly, the 2-point function of the Φ^3_D-model is Stieltjes for $D = 4$ and $D = 6$ (and λ real where the partition function is meaningless), *but not for* $D = 2$ or λ purely imaginary [GSW17]. Numerical evidence was given [GW14b] that the same is true for the Φ^4_4-model: The 2-point function is definitely not reflection positive in the stable case $\lambda > 0$, whereas for $\lambda < 0$ positivity seems to hold.

Reflection positivity *cannot be expected to hold* for the whole set of Schwinger functions (10.2) for the Φ^3-model. The reason is the fast decay in $E_{\underline{k}}$ established in (8.17) which contradicts complete monotonicity in Theorem 10.1.

10.2 A proposal

In a sort of outlook we sketch ideas about another approach to the Osterwalder–Schrader axioms in QFTs on noncommutative algebras $\mathcal{A}$. The failure of (10.1) to produce reflection-positive Schwinger functions suggests that the dequantisation $q_* : \mathcal{A}_* \to C^\infty(\mathbb{R}^D)$ *should be different from the Kōmura-Kōmura isomorphism* ι_U.

Instead we propose to build the dequantisation as a *positive map* $(q^*(a^*a))(x) \geq 0$ for all $a \in \mathcal{A}$ and $x \in \mathbb{R}^D$. Adjusting the norm we choose it in the class of $\mathbb{R}^D$-labelled states $\{\omega_v \;:\; v \in \mathbb{R}^D\}$. There are good reasons for this ansatz. On the Moyal algebra $(\mathcal{S}(\mathbb{R}^2), \star_\theta)$ one can check that $\tilde{\omega}_\gamma(f) := \frac{1}{\pi\gamma\theta}\int_{\mathbb{R}^2} dx\; e^{-\frac{|x|^2}{\gamma\theta}} f(x)$ is a state if and only if $\gamma \geq 1$. Pointwise evaluation, a state for the commutative product, would be recovered by $\lim_{\gamma\to 0}\tilde{\omega}_\gamma(f) = f(0)$, but it is not positive for the Moyal product. We argued in the very beginning (Section 1.1) that sharp localisation in a QFT is incompatible with gravity. This observation was precisely the reason to introduce QFTs on noncommutative geometries (Section 1.2). The smearing in an area $|x|^2 \geq \theta$ via the state $\tilde{\omega}_\gamma$ implements the localisability restrictions, with $\theta = \ell_p^2$ being the Planck area. Moreover, states provide the correct framework to pass from the noncommutative topology encoded in an algebra $\mathcal{A}$ to the metric aspects of noncommutative geometry [Con94]. Given a spectral triple $(\mathcal{A}, \mathcal{H}, \mathcal{D})$, a metric structure is defined on (an appropriate subspace of) states on $\mathcal{A}$ via Connes' distance formula [Con94]

$$\mathrm{dist}(\omega_1, \omega_2) = \sup\{|\omega_1(a) - \omega_2(a)| \;:\; \|[\mathcal{D}, a]\| \leq 1\}. \tag{10.3}$$

We therefore propose:

Definition 10.2 Let $\mathcal{A}$ be nuclear AF Fréchet algebra generated by orthonormal matrix bases $\{e_{\underline{k}\underline{l}}\}$. Then renormalised correlation functions $\langle e_{\underline{k}_1\underline{l}_1} \otimes \cdots \otimes e_{\underline{k}_N\underline{l}_N}\rangle$ defined on $\mathcal{A}$ via (3.6) and renormalisation gives rise to Schwinger functions by

$$S(\upsilon_N) := \sum_{\underline{k}_1,\dots,\underline{l}_N \in \mathbb{N}^{D/2}} \omega_{\upsilon_N}(e_{\underline{k}_1\underline{l}_1} \otimes \cdots \otimes e_{\underline{k}_N\underline{l}_N})\langle e_{\underline{k}_1\underline{l}_1} \otimes \cdots \otimes e_{\underline{k}_N\underline{l}_N}\rangle, \tag{10.4}$$

where ω_{υ_N} is a state on $\mathcal{A}^{\otimes N}$.

We assume that the Kōmura-Kōmura isomorphism endows $\mathcal{A}$ by an action $\alpha : \mathbb{R}^D \times \mathcal{A} \to \mathcal{A}$ of translations. It is probably not necessary that the $\mathbb{R}^D$-action commutes with multiplication in $\mathcal{A}$. We obtain an $\mathbb{R}^D$-action on the states ω_{υ_N} by duality,

$$\big(\alpha_{t_1,\dots,t_N}\omega_{\upsilon_N}\big)\big(a_1 \otimes \cdots \otimes a_N\big) := \omega_{\upsilon_N}\big(\alpha_{t_1}a_1 \otimes \cdots \otimes \alpha_{t_N}a_N\big), \tag{10.5}$$

for $t_i \in \mathbb{R}^D$. Hence, it is enough to specify a single reference state $\omega_{\hat{\upsilon}_N}$ which via $\omega_{t_1,\dots,t_N} := \alpha_{t_1,\dots,t_N}\omega_{\hat{\upsilon}_N}$ induces an $\mathbb{R}^{ND}$-indexed family of states. This would make the $\mathbb{R}^D$ a universal model of noncommutative geometries defined via the distance formula (10.3). It was shown in [MT13] that for $\mathcal{A}$ the 2D-Moyal algebra one has $\mathrm{dist}(\omega_t, \omega_{t'}) = \|t - t'\|$ for the noncommutative distance (10.3) between any such translates $\omega_t, \omega_{t'}$ of a reference state on the Moyal algebra. In short, everything is consistent.

One of the Osterwalder–Schrader axioms requires translation invariance of the Schwinger functions (10.4), i.e. $S(t_1, \dots, t_N) = S(t_1 + t_0, \dots, t_N + t_0)$ for any $t_0 \in \mathbb{R}^D$. This is not automatic for our proposal, but can be achieved for the Moyal algebra following an observation in [BDFP03]. Namely, the tensor product of Moyal algebras factorises into $\mathcal{A}^{\otimes N} = \mathcal{A} \otimes \mathcal{A}^{\otimes N-1}$, where the first tensor factor describes the centre-of-motion coordinate and the second one depends only on coordinate differences. Let $\iota_c : \mathcal{A}^{\otimes N} \to \mathcal{A} \otimes \mathcal{A}^{\otimes N-1}$ be this factorisation isomorphism, then translation-invariant Schwinger functions can be defined as

$$\begin{aligned} &S(t_1, \dots, t_N) \\ &:= \sum_{\underline{k}_1,\dots,\underline{l}_N \in \mathbb{N}^{D/2}} \big(\hat{\omega}\otimes\omega_{t_1-t_2,\dots,t_{N-1}-t_N}\big)\big(\iota_c(e_{\underline{k}_1\underline{l}_1}\otimes\ldots\otimes e_{\underline{k}_N\underline{l}_N})\big)\langle e_{\underline{k}_1\underline{l}_1}\otimes\ldots\otimes e_{\underline{k}_N\underline{l}_N}\rangle, \end{aligned} \tag{10.6}$$

where $\hat{\omega}$ averages over the centre-of-motion and $\omega_{t_1-t_2,\dots,t_{N-1}-t_N}$ is a $\mathbb{R}^{D(N-1)}$-translate of a reference state on $\mathcal{A}^{\otimes N-1}$. The Schwinger functions (10.6) are translation-invariant by construction. It thus remains:

Question 10.3 Is it possible to find examples of reference states, or even to classify them, for which the free theory with covariance C_E is reflection positive? Does it extend to reflection positivity of Schwinger 2-point functions (10.6) for the moments $\langle e_{\underline{k}_1 \underline{l}_1} \otimes e_{\underline{k}_2 \underline{l}_2} \rangle$ of the Φ^3 and Φ^4-models?

The dream would be to prove reflection positivity of all Schwinger functions.

Acknowledgements I thank Alain Connes for the initiative to write such a survey and for constant support and encouragement through the years. I am most grateful to my long-term collaborator Harald Grosse with whom nearly all of the presented results have been achieved. After the two years 2000/2001 together in Vienna, our collaboration has been supported by the Erwin-Schrödinger-Institute, by the Max-Planck-Institute for Mathematics in the Sciences and by the Deutsche Forschungsgemeinschaft (DFG) via the coordinated programmes SFB 478 and SFB 878. Sections 8 and 9 are based on recent results obtained with Harald Grosse, Akifumi Sako, Erik Panzer and Alexander Hock. The survey was finally assembled within the Cluster of Excellence[7] "Mathematics Münster".

References

ABIY02. G. Alexanian, A. P. Balachandran, G. Immirzi, and B. Ydri. Fuzzy CP^2. *J. Geom. Phys.*, 42:28–53, 2002, http://arxiv.org/abs/hep-th/0103023. doi:10.1016/S0393-0440(01)00070-5.

ABK00a. I. Ya. Aref'eva, D. M. Belov, and A. S. Koshelev. A note on UV / IR for noncommutative complex scalar field. 2000, http://arxiv.org/abs/hep-th/0001215.

ABK00b. I. Ya. Aref'eva, D. M. Belov, and A. S. Koshelev. Two loop diagrams in noncommutative ϕ_4^4 theory. *Phys. Lett.*, B476:431–436, 2000, http://arxiv.org/abs/hep-th/9912075. doi:10.1016/S0370-2693(00)00169-6.

AC96. E. Arbarello and M. Cornalba. Combinatorial and algebro-geometric cohomology classes on the moduli spaces of curves. *J. Algebraic Geom.*, 5(4):705–749, 1996.

ADJ91. J. Ambjørn, B. Durhuus, and T. Jónsson. Three-dimensional simplicial quantum gravity and generalized matrix models. *Mod. Phys. Lett.*, A6:1133–1146, 1991. doi:10.1142/S0217732391001184.

AGBZ01. L. Alvarez-Gaume, J. L. F. Barbon, and R. Zwicky. Remarks on time space noncommutative field theories. *JHEP*, 05:057, 2001, http://arxiv.org/abs/hep-th/0103069. doi:10.1088/1126-6708/2001/05/057.

AII$^+$00. H. Aoki, N. Ishibashi, S. Iso, H. Kawai, Y. Kitazawa, and T. Tada. Noncommutative Yang-Mills in IIB matrix model. *Nucl. Phys.*, B565:176–192, 2000, http://arxiv.org/abs/hep-th/9908141. doi:10.1016/S0550-3213(99)00633-1.

Aiz81. M. Aizenman. Proof of the triviality of ϕ_d^4 field theory and some mean field features of Ising models for $d > 4$. *Phys. Rev. Lett.*, 47:1–4, 1981. doi:10.1103/PhysRevLett.47.1.

[7]"Gefördert durch die Deutsche Forschungsgemeinschaft (DFG) im Rahmen der Exzellenzstrategie des Bundes und der Länder EXC 2044–390685587, Mathematik Münster: Dynamik–Geometrie–Struktur".

Aks65. S. Ø. Aks. Proof that scattering implies production in quantum field theory. *J. Math. Phys.*, 6:516–532, 1965. doi:10.1063/1.1704305.

Ale11. A. Alexandrov. Cut-and-join operator representation for Kontsevich-Witten tau-function. *Mod. Phys. Lett.*, A26:2193–2199, 2011, http://arxiv.org/abs/1009.4887. doi:10.1142/S0217732311036607.

AMNS99. J. Ambjørn, Y. M. Makeenko, J. Nishimura, and R. J. Szabo. Finite N matrix models of noncommutative gauge theory. *JHEP*, 11:029, 1999, http://arxiv.org/abs/hep-th/9911041. doi:10.1088/1126-6708/1999/11/029.

AR95. A. Abdesselam and V. Rivasseau. Trees, forests and jungles: A botanical garden for cluster expansions. In *Constructive physics: Results in field theory, statistical mechanics and condensed matter physics*, volume 446 of *Lect. Notes Phys.*, pages 7–36. 1995, http://arxiv.org/abs/hep-th/9409094. doi:10.1007/3-540-59190-7_20.

ARS99. A. Yu. Alekseev, A. Recknagel, and V. Schomerus. Noncommutative world volume geometries: Branes on SU(2) and fuzzy spheres. *JHEP*, 09:023, 1999, http://arxiv.org/abs/hep-th/9908040. doi:10.1088/1126-6708/1999/09/023.

Ati88. M. Atiyah. Topological quantum field theories. *Inst. Hautes Études Sci. Publ. Math.*, 68:175–186, 1988. URL http://www.numdam.org/item?id=PMIHES_1988__68__175_0.

Bah03. D. Bahns. Unitary quantum field theory on the noncommutative Minkowski space. *Fortsch. Phys.*, 51:658–663, 2003, http://arxiv.org/abs/hep-th/0212266. doi:10.1002/prop.200310079.

Bah10. D. Bahns. Schwinger functions in noncommutative quantum field theory. *Annales Henri Poincaré*, 11:1273–1283, 2010, http://arxiv.org/abs/0908.4537. doi:10.1007/s00023-010-0061-4.

BDD$^+$03. W. Behr, N. G. Deshpande, G. Duplancić, P. Schupp, J. Trampetić, and J. Wess. The $Z \to \gamma\gamma, gg$ decays in the noncommutative standard model. *Eur. Phys. J.*, C29:441–446, 2003, http://arxiv.org/abs/hep-ph/0202121. doi:10.1140/epjc/s2003-01207-4.

BDFP02. D. Bahns, S. Doplicher, K. Fredenhagen, and G. Piacitelli. On the unitarity problem in space-time noncommutative theories. *Phys. Lett.*, B533:178–181, 2002, http://arxiv.org/abs/hep-th/0201222. doi:10.1016/S0370-2693(02)01563-0.

BDFP03. D. Bahns, S. Doplicher, K. Fredenhagen, and G. Piacitelli. Ultraviolet finite quantum field theory on quantum space-time. *Commun. Math. Phys.*, 237:221–241, 2003, http://arxiv.org/abs/hep-th/0301100. doi:10.1007/s00220-003-0857-x.

Ber29. S. Bernstein. Sur les fonctions absolument monotones. *Acta Math.*, 52(1):1–66, 1929. doi:10.1007/BF02547400.

Bet47. H. A. Bethe. The electromagnetic shift of energy levels. *Phys. Rev.*, 72:339–341, 1947. doi:10.1103/PhysRev.72.339.

BFF$^+$78a. F. Bayen, M. Flato, C. Fronsdal, A. Lichnerowicz, and D. Sternheimer. Deformation theory and quantization. 1. Deformations of symplectic structures. *Annals Phys.*, 111:61, 1978. doi:10.1016/0003-4916(78)90224-5.

BFF$^+$78b. F. Bayen, M. Flato, C. Fronsdal, A. Lichnerowicz, and D. Sternheimer. Deformation theory and quantization. 2. Physical applications. *Annals Phys.*, 111:111, 1978. doi:10.1016/0003-4916(78)90225-7.

BFG$^+$03. H. Bozkaya, P. Fischer, H. Grosse, M. Pitschmann, V. Putz, M. Schweda, and R. Wulkenhaar. Space-time noncommutative field theories and causality. *Eur. Phys. J.*, C29:133–141, 2003, http://arxiv.org/abs/hep-th/0209253. doi:10.1140/epjc/s2003-01210-9.

BFV03. R. Brunetti, K. Fredenhagen, and R. Verch. The generally covariant locality principle: A new paradigm for local quantum field theory. *Commun. Math. Phys.*, 237:31–68, 2003, http://arxiv.org/abs/math-ph/0112041. doi:10.1007/s00220-003-0815-7.

BG14. J. Ben Geloun. Renormalizable models in rank $d \geq 2$ tensorial group field theory. *Commun. Math. Phys.*, 332:117–188, 2014, http://arxiv.org/abs/1306.1201. doi:10.1007/s00220-014-2142-6.

BGG+01. A. Bichl, J. Grimstrup, H. Grosse, L. Popp, M. Schweda, and R. Wulkenhaar. Renormalization of the noncommutative photon self-energy to all orders via Seiberg-Witten map. *JHEP*, 06:013, 2001, http://arxiv.org/abs/hep-th/0104097. doi:10.1088/1126-6708/2001/06/013.

BGK+08. D. N. Blaschke, F. Gieres, E. Kronberger, M. Schweda, and M. Wohlgenannt. Translation-invariant models for non-commutative gauge fields. *J. Phys.*, A41:252002, 2008, http://arxiv.org/abs/0804.1914. doi:10.1088/1751-8113/41/25/252002.

BGP+02. A. A. Bichl, J. M. Grimstrup, L. Popp, M. Schweda, and R. Wulkenhaar. Perturbative analysis of the Seiberg-Witten map. *Int. J. Mod. Phys.*, A17:2219–2232, 2002, http://arxiv.org/abs/hep-th/0102044. doi:10.1142/S0217751X02010649.

BGR13. J. Ben Geloun and V. Rivasseau. A renormalizable 4-dimensional tensor field theory. *Commun. Math. Phys.*, 318:69–109, 2013, http://arxiv.org/abs/1111.4997. doi:10.1007/s00220-012-1549-1.

BGRR11. V. Bonzom, R. Gurau, A. Riello, and V. Rivasseau. Critical behavior of colored tensor models in the large N limit. *Nucl. Phys.*, B853:174–195, 2011, http://arxiv.org/abs/1105.3122. doi:10.1016/j.nuclphysb.2011.07.022.

BHN02. W. Bietenholz, F. Hofheinz, and J. Nishimura. The renormalizability of 2-D Yang-Mills theory on a noncommutative geometry. *JHEP*, 09:009, 2002, http://arxiv.org/abs/hep-th/0203151. doi:10.1088/1126-6708/2002/09/009.

BHN03. W. Bietenholz, F. Hofheinz, and J. Nishimura. Noncommutative field theories beyond perturbation theory. *Fortsch. Phys.*, 51:745–752, 2003, http://arxiv.org/abs/hep-th/0212258. doi:10.1002/prop.200310092.

BIPZ78. E. Brézin, C. Itzykson, G. Parisi, and J. B. Zuber. Planar diagrams. *Commun. Math. Phys.*, 59:35–51, 1978. doi:10.1007/BF01614153.

BK87. D. C. Brydges and T. Kennedy. Mayer expansions and the Hamilton-Jacobi equation. *J. Statist. Phys.*, 48(1-2):19–49, 1987. doi:10.1007/BF01010398.

BK90. E. Brezin and V. A. Kazakov. Exactly Solvable Field Theories of Closed Strings. *Phys. Lett.*, B236:144–150, 1990. doi:10.1016/0370-2693(90)90818-Q.

BKMP09. V. Bouchard, A. Klemm, M. Mariño, and S. Pasquetti. Remodeling the B-model. *Commun. Math. Phys.*, 287:117–178, 2009, http://arxiv.org/abs/0709.1453. doi:10.1007/s00220-008-0620-4.

BKR+10. D. N. Blaschke, E. Kronberger, A. Rofner, M. Schweda, R. I. P. Sedmik, and M. Wohlgenannt. On the problem of renormalizability in non-commutative gauge field models: A critical review. *Fortsch. Phys.*, 58:364–372, 2010, http://arxiv.org/abs/0908.0467. doi:10.1002/prop.200900102.

Blo15. S. Bloch. Feynman amplitudes in mathematics and physics. 2015, http://arxiv.org/abs/arXiv:1509.00361.

BLS11. D. Buchholz, G. Lechner, and S. J. Summers. Warped convolutions, Rieffel deformations and the construction of quantum field theories. *Commun. Math. Phys.*, 304:95–123, 2011, http://arxiv.org/abs/1005.2656. doi:10.1007/s00220-010-1137-1.

BM04. G. Benfatto and V. Mastropietro. Ward identities and vanishing of the beta function for $d = 1$ interacting Fermi systems. *J. Statist. Phys.*, 115(1–2):143–184, 2004. doi:10.1023/B:JOSS.0000019825.37968.23.

Boc32. S. Bochner. *Vorlesungen über Fouriersche Integrale*, volume 12 of *Mathematik und ihre Anwendungen in Monographien und Lehrbüchern*. Akademische Verlagsgesellschaft m.b.h., 1932.

Boc55. S. Bochner. *Harmonic analysis and the theory of probability*. University of California Press, Berkeley and Los Angeles, 1955.

BP57. N. N. Bogoliubow and O. S. Parasiuk. Über die Multiplikation der Kausalfunktionen in der Quantentheorie der Felder. *Acta Math.*, 97:227–266, 1957. doi:10.1007/BF02392399.

Bra72. O. Bratteli. Inductive limits of finite dimensional C^*-algebras. *Trans. Amer. Math. Soc.*, 171:195–234, 1972. doi:10.2307/1996380.

Bro13a. D. Broadhurst. Multiple zeta values and modular forms in quantum field theory. In *Computer algebra in quantum field theory*, Texts Monogr. Symbol. Comput., pages 33–73. Springer, Vienna, 2013.

Bro13b. F. Brown. Iterated integrals in quantum field theory. In *Geometric and topological methods for quantum field theory*, pages 188–240. Cambridge Univ. Press, Cambridge, 2013. URL http://www.ihes.fr/~brown/ColombiaNotes7.pdf.

BS01. L. Bonora and M. Salizzoni. Renormalization of noncommutative U(N) gauge theories. *Phys. Lett.*, B504:80–88, 2001, http://arxiv.org/abs/hep-th/0011088. doi:10.1016/S0370-2693(01)00281-7.

BS08. D. Buchholz and S. J. Summers. Warped convolutions: A novel tool in the construction of quantum field theories. In *Quantum field theory and beyond: Essays in honor of Wolfhart Zimmermann*, pages 107–121, 2008, http://arxiv.org/abs/0806.0349. doi:10.1142/9789812833556_0007.

Bür99. H. Bürmann. Essai de calcul fonctionnaire aux constantes ad-libitum. *Mem. Inst. Nat. Sci Arts. Sci. Math. Phys.*, 2:13–17, 1799.

Car22. T. Carleman. Sur la résolution de certaines équations intégrales. *Arkiv for Mat., Astron. och Fysik*, 16:19pp, 1922.

CC97. A. H. Chamseddine and A. Connes. The Spectral action principle. *Commun. Math. Phys.*, 186:731–750, 1997, http://arxiv.org/abs/hep-th/9606001. doi:10.1007/s002200050126.

CCM14. A. H. Chamseddine, A. Connes, and V. Mukhanov. Geometry and the Quantum: Basics. *JHEP*, 12:098, 2014, http://arxiv.org/abs/1411.0977. doi:10.1007/JHEP12(2014)098.

CCM15. A. H. Chamseddine, A. Connes, and V. Mukhanov. Quanta of Geometry: Noncommutative Aspects. *Phys. Rev. Lett.*, 114(9):091302, 2015, http://arxiv.org/abs/1409.2471. doi:10.1103/PhysRevLett.114.091302.

CDS98. A. Connes, M. R. Douglas, and A. S. Schwarz. Noncommutative geometry and matrix theory: Compactification on tori. *JHEP*, 02:003, 1998, http://arxiv.org/abs/hep-th/9711162. doi:10.1088/1126-6708/1998/02/003.

CEO06. L. Chekhov, B. Eynard, and N. Orantin. Free energy topological expansion for the 2-matrix model. *JHEP*, 12:053, 2006, http://arxiv.org/abs/math-ph/0603003. doi:10.1088/1126-6708/2006/12/053.

CF00. A. S. Cattaneo and G. Felder. A path integral approach to the Kontsevich quantization formula. *Commun. Math. Phys.*, 212:591–611, 2000, http://arxiv.org/abs/math/9902090. doi:10.1007/s002200000229.

CGH+96. R. M. Corless, G. H. Gonnet, D. E. G. Hare, D. J. Jeffrey, and D. E. Knuth. On the Lambert W function. *Adv. Comput. Math.*, 5:329–359, 1996. doi:10.1007/BF02124750.

CHMS00. S. Cho, R. Hinterding, J. Madore, and H. Steinacker. Finite field theory on noncommutative geometries. *Int. J. Mod. Phys.*, D9:161–199, 2000, http://arxiv.org/abs/hep-th/9903239. doi:10.1142/S0218271800000153.

CJS+02. X. Calmet, B. Jurčo, P. Schupp, J. Wess, and M. Wohlgenannt. The standard model on noncommutative space-time. *Eur. Phys. J.*, C23:363–376, 2002, http://arxiv.org/abs/hep-ph/0111115. doi:10.1007/s100520100873.

CK98. A. Connes and D. Kreimer. Hopf algebras, renormalization and noncommutative geometry. *Commun. Math. Phys.*, 199:203–242, 1998, http://arxiv.org/abs/hep-th/9808042. doi:10.1007/s002200050499.

CM98. A. Connes and H. Moscovici. Hopf algebras, cyclic cohomology and the transverse index theorem. *Comm. Math. Phys.*, 198(1):199–246, 1998, http://arxiv.org/abs/math/9806109. doi:10.1007/s002200050477.

CMS01. C.-S. Chu, J. Madore, and H. Steinacker. Scaling limits of the fuzzy sphere at one loop. *JHEP*, 08:038, 2001, http://arxiv.org/abs/hep-th/0106205. doi:10.1088/1126-6708/2001/08/038.

Con94. A. Connes. *Noncommutative geometry*. Academic Press, Inc., San Diego, CA, 1994.

Con95. A. Connes. Noncommutative geometry and reality. *J. Math. Phys.*, 36:6194–6231, 1995. doi:10.1063/1.531241.

Con96. A. Connes. Gravity coupled with matter and foundation of noncommutative geometry. *Commun. Math. Phys.*, 182:155–176, 1996, http://arxiv.org/abs/hep-th/9603053. doi:10.1007/BF02506388.

CR87. A. Connes and M. A. Rieffel. Yang-Mills for noncommutative two-tori. In *Operator algebras and mathematical physics (Iowa City, Iowa, 1985)*, volume 62 of *Contemp. Math.*, pages 237–266. Amer. Math. Soc., Providence, RI, 1987. doi:10.1090/conm/062/878383.

CR00. I. Chepelev and R. Roiban. Renormalization of quantum field theories on noncommutative $\mathbb{R}^d$. 1. Scalars. *JHEP*, 05:037, 2000, http://arxiv.org/abs/hep-th/9911098. doi:10.1088/1126-6708/2000/05/037.

CR01. I. Chepelev and R. Roiban. Convergence theorem for noncommutative Feynman graphs and renormalization. *JHEP*, 03:001, 2001, http://arxiv.org/abs/hep-th/0008090. doi:10.1088/1126-6708/2001/03/001.

DFGZJ95. P. Di Francesco, P. H. Ginsparg, and J. Zinn-Justin. 2-D gravity and random matrices. *Phys. Rept.*, 254:1–133, 1995, http://arxiv.org/abs/hep-th/9306153. doi:10.1016/0370-1573(94)00084-G.

DFIZ93. P. Di Francesco, C. Itzykson, and J. B. Zuber. Polynomial averages in the Kontsevich model. *Commun. Math. Phys.*, 151:193–219, 1993, http://arxiv.org/abs/hep-th/9206090. doi:10.1007/BF02096753.

DFR95. S. Doplicher, K. Fredenhagen, and J. E. Roberts. The quantum structure of space-time at the Planck scale and quantum fields. *Commun. Math. Phys.*, 172:187–220, 1995, http://arxiv.org/abs/hep-th/0303037. doi:10.1007/BF02104515.

DGMR07. M. Disertori, R. Gurau, J. Magnen, and V. Rivasseau. Vanishing of beta function of non-commutative ϕ_4^4 theory to all orders. *Phys. Lett.*, B649:95–102, 2007, http://arxiv.org/abs/hep-th/0612251. doi:10.1016/j.physletb.2007.04.007.

dGWW07. A. de Goursac, J.-C. Wallet, and R. Wulkenhaar. Noncommutative induced gauge theory. *Eur. Phys. J.*, C51:977–987, 2007, http://arxiv.org/abs/hep-th/0703075. doi:10.1140/epjc/s10052-007-0335-2.

DH98. M. R. Douglas and C. M. Hull. D-branes and the noncommutative torus. *JHEP*, 02:008, 1998, http://arxiv.org/abs/hep-th/9711165. doi:10.1088/1126-6708/1998/02/008.

dJHW19. J. de Jong, A. Hock, and R. Wulkenhaar. work in progress. 2019.

DN01. M. R. Douglas and N. A. Nekrasov. Noncommutative field theory. *Rev. Mod. Phys.*, 73:977–1029, 2001, http://arxiv.org/abs/hep-th/0106048. doi:10.1103/RevModPhys.73.977.

DPV16. D. D'Ascanio, P. Pisani, and D. V. Vassilevich. Renormalization on noncommutative torus. *Eur. Phys. J.*, C76(4):180, 2016, http://arxiv.org/abs/1602.01479. doi:10.1140/epjc/s10052-016-4022-z.

DR07. M. Disertori and V. Rivasseau. Two- and three-loops beta function of non-commutative ϕ_4^4 theory. *Eur. Phys. J.*, C50:661–671, 2007, http://arxiv.org/abs/hep-th/0610224. doi:10.1140/epjc/s10052-007-0211-0.

DS90. M. R. Douglas and S. H. Shenker. Strings in less than one dimension. *Nucl. Phys.*, B335:635–654, 1990. doi:10.1016/0550-3213(90)90522-F.

DS03. S. Denk and M. Schweda. Time ordered perturbation theory for nonlocal interactions: Applications to NCQFT. *JHEP*, 09:032, 2003, http://arxiv.org/abs/hep-th/0306101. doi:10.1088/1126-6708/2003/09/032.

DSvZ18. V. Delecroix, J. Schmitt, and J. van Zelm. `admcycles` – a Sage program for calculations in the tautological ring of $\overline{\mathcal{M}}_{g,n}$ and computation of cycles of admissible covers, 2018. URL https://people.math.ethz.ch/~schmittj/admcycles.html.

EILS08. D. Essouabri, B. Iochum, C. Levy, and A. Sitarz. Spectral action on noncommutative torus. *J. Noncommut. Geom.*, 2(1):53–123, 2008, http://arxiv.org/abs/0704.0564. doi:10.4171/JNCG/16.

EK82. T. Eguchi and H. Kawai. Reduction of dynamical degrees of freedom in the large N gauge theory. *Phys. Rev. Lett.*, 48:1063, 1982. doi:10.1103/PhysRevLett.48.1063.

Ell76. G. A. Elliott. On the classification of inductive limits of sequences of semisimple finite-dimensional algebras. *J. Algebra*, 38(1):29–44, 1976. doi:10.1016/0021-8693(76)90242-8.

EO07. B. Eynard and N. Orantin. Invariants of algebraic curves and topological expansion. *Commun. Num. Theor. Phys.*, 1:347–452, 2007, http://arxiv.org/abs/math-ph/0702045. doi:10.4310/CNTP.2007.v1.n2.a4.

Eyn04. B. Eynard. Topological expansion for the 1-Hermitian matrix model correlation functions. *JHEP*, 11:031, 2004, http://arxiv.org/abs/hep-th/0407261. doi:10.1088/1126-6708/2004/11/031.

Eyn14. B. Eynard. Invariants of spectral curves and intersection theory of moduli spaces of complex curves. *Commun. Num. Theor. Phys.*, 8:541–588, 2014, http://arxiv.org/abs/1110.2949. doi:10.4310/CNTP.2014.v8.n3.a4.

Eyn16. B. Eynard. *Counting surfaces*, volume 70 of *Progress in Mathematical Physics*. Birkhäuser/Springer, 2016. doi:10.1007/978-3-7643-8797-6. CRM Aisenstadt chair lectures.

Eyn19. B. Eynard. Large genus behavior of topological recursion. 2019, http://arxiv.org/abs/1905.11270.

Fey49. R. P. Feynman. Space-time approach to quantum electrodynamics. *Phys. Rev.*, 76:769–789, 1949. doi:10.1103/PhysRev.76.769.

Fil96. T. Filk. Divergencies in a field theory on quantum space. *Phys. Lett.*, B376:53–58, 1996. doi:10.1016/0370-2693(96)00024-X.

FP67. L. D. Faddeev and V. N. Popov. Feynman diagrams for the Yang-Mills field. *Phys. Lett.*, B25:29–30, 1967. doi:10.1016/0370-2693(67)90067-6.

Frö82. J. Fröhlich. On the triviality of $\lambda\phi_d^4$ theories and the approach to the critical point in $d \geq 4$ dimensions. *Nucl. Phys.*, B200:281–296, 1982. doi:10.1016/0550-3213(82)90088-8.

FS09. A. Fischer and R. J. Szabo. Duality covariant quantum field theory on noncommutative Minkowski space. *JHEP*, 02:031, 2009, http://arxiv.org/abs/0810.1195. doi:10.1088/1126-6708/2009/02/031.

GAO83. A. Gonzalez-Arroyo and M. Okawa. The twisted Eguchi-Kawai model: A reduced model for large N lattice gauge theory. *Phys. Rev.*, D27:2397, 1983. doi:10.1103/PhysRevD.27.2397.

Gar00. M. R. Garousi. Noncommutative world volume interactions on D-branes and Dirac-Born-Infeld action. *Nucl. Phys.*, B579:209–228, 2000, http://arxiv.org/abs/hep-th/9909214. doi:10.1016/S0550-3213(99)00826-3.

Gay03. V. Gayral. The action functional for Moyal planes. *Lett. Math. Phys.*, 65:147–157, 2003, http://arxiv.org/abs/hep-th/0307220. doi:10.1023/B:MATH.0000004380.57824.94.

GBV88. J. M. Gracia-Bondía and J. C. Várilly. Algebras of distributions suitable for phase space quantum mechanics. 1. *J. Math. Phys.*, 29:869–879, 1988. doi:10.1063/1.528200.

GGBI+04. V. Gayral, J. M. Gracia-Bondía, B. Iochum, T. Schücker, and J. C. Várilly. Moyal planes are spectral triples. *Commun. Math. Phys.*, 246:569–623, 2004, http://arxiv.org/abs/hep-th/0307241. doi:10.1007/s00220-004-1057-z.

GHW19. H. Grosse, A. Hock, and R. Wulkenhaar. A Laplacian to compute intersection numbers on $\overline{\mathcal{M}}_{g,n}$ and correlation functions in NCQFT. 2019, http://arxiv.org/abs/1903.12526.

GHW19b. H. Grosse, A. Hock, and R. Wulkenhaar. Solution of all quartic matrix models. 2019, http://arxiv.org/abs/1906.04600.

GHW19c. H. Grosse, A. Hock, and R. Wulkenhaar. Solution of the self-dual Φ^4 QFT-model on four-dimensional Moyal space. 2019, http://arxiv.org/abs/1908.04543.

GI05. V. Gayral and B. Iochum. The spectral action for Moyal planes. *J. Math. Phys.*, 46:043503, 2005, http://arxiv.org/abs/hep-th/0402147. doi:10.1063/1.1855401.

GJ87. J. Glimm and A. Jaffe. *Quantum physics. A functional integral point of view*. Springer-Verlag, New York, second edition, 1987. doi:10.1007/978-1-4612-4728-9.

GJKW07. V. Gayral, J.-H. Jureit, T. Krajewski, and R. Wulkenhaar. Quantum field theory on projective modules. *J. Noncommut. Geom.*, 1(4):431–496, 2007, http://arxiv.org/abs/hep-th/0612048. doi:10.4171/JNCG/13.

GJS74. J. Glimm, A. Jaffe, and T. Spencer. The Wightman axioms and particle structure in the $\mathcal{P}(\phi)_2$ quantum field model. *Ann. of Math. (2)*, 100:585–632, 1974. doi:10.2307/1970959.

GKP96a. H. Grosse, C. Klimčík, and P. Prešnajder. On finite 4-D quantum field theory in noncommutative geometry. *Commun. Math. Phys.*, 180:429–438, 1996, http://arxiv.org/abs/hep-th/9602115. doi:10.1007/BF02099720.

GKP96b. H. Grosse, C. Klimčík, and P. Prešnajder. Towards finite quantum field theory in noncommutative geometry. *Int. J. Theor. Phys.*, 35:231–244, 1996, http://arxiv.org/abs/hep-th/9505175. doi:10.1007/BF02083810.

GL07. H. Grosse and G. Lechner. Wedge-local quantum fields and noncommutative Minkowski space. *JHEP*, 11:012, 2007, http://arxiv.org/abs/0706.3992. doi:10.1088/1126-6708/2007/11/012.

GL08. H. Grosse and G. Lechner. Noncommutative deformations of Wightman quantum field theories. *JHEP*, 09:131, 2008, http://arxiv.org/abs/0808.3459. doi:10.1088/1126-6708/2008/09/131.

GLLV13. H. Grosse, G. Lechner, T. Ludwig, and R. Verch. Wick rotation for quantum field theories on degenerate Moyal space(-time). *J. Math. Phys.*, 54:022307, 2013, http://arxiv.org/abs/1111.6856. doi:10.1063/1.4790886.

GM90. D. J. Gross and A. A. Migdal. Nonperturbative two-dimensional quantum gravity. *Phys. Rev. Lett.*, 64:127–130, 1990. doi:10.1103/PhysRevLett.64.127.

GM92. H. Grosse and J. Madore. A noncommutative version of the Schwinger model. *Phys. Lett.*, B283:218–222, 1992. doi:10.1016/0370-2693(92)90011-R.

GM00. J. Gomis and T. Mehen. Space-time noncommutative field theories and unitarity. *Nucl. Phys.*, B591:265–276, 2000, http://arxiv.org/abs/hep-th/0005129. doi:10.1016/S0550-3213(00)00525-3.

GM04. A. B. Goncharov and Y. I. Manin. Multiple ζ-motives and moduli spaces $\overline{\mathfrak{M}}_{0,n}$. *Compos. Math.*, 140(1):1–14, 2004. doi:10.1112/S0010437X03000125.

GMRT09. R. Gurau, J. Magnen, V. Rivasseau, and A. Tanasa. A translation-invariant renormalizable non-commutative scalar model. *Commun. Math. Phys.*, 287:275–290, 2009, http://arxiv.org/abs/0802.0791. doi:10.1007/s00220-008-0658-3.

GMS01. H. Grosse, J. Madore, and H. Steinacker. Field theory on the q deformed fuzzy sphere. 1. *J. Geom. Phys.*, 38:308–342, 2001, http://arxiv.org/abs/hep-th/0005273. doi:10.1016/S0393-0440(00)00068-1.

GMS02. H. Grosse, J. Madore, and H. Steinacker. Field theory on the q deformed fuzzy sphere. 2. Quantization. *J. Geom. Phys.*, 43:205–240, 2002, http://arxiv.org/abs/hep-th/0103164. doi:10.1016/S0393-0440(02)00023-2.

GR07. R. Gurau and V. Rivasseau. Parametric representation of noncommutative field theory. *Commun. Math. Phys.*, 272:811–835, 2007, http://arxiv.org/abs/math-ph/0606030. doi:10.1007/s00220-007-0215-5.

GR11. R. Gurau and V. Rivasseau. The 1/N expansion of colored tensor models in arbitrary dimension. *Europhys. Lett.*, 95(5):50004, 2011, http://arxiv.org/abs/1101.4182. doi:10.1209/0295-5075/95/50004.

GR12. R. Gurau and J. P. Ryan. Colored tensor models - A review. *SIGMA*, 8:020, 2012, http://arxiv.org/abs/1109.4812. doi:10.3842/SIGMA.2012.020.

GR15. R. Gurau and V. Rivasseau. The multiscale loop vertex expansion. *Annales Henri Poincaré*, 16(8):1869–1897, 2015, http://arxiv.org/abs/1312.7226. doi:10.1007/s00023-014-0370-0.

GRM09. R. Gurau, V. Rivasseau, and J. Magnen. Tree quantum field theory. *Annales Henri Poincaré*, 10:867–891, 2009, http://arxiv.org/abs/0807.4122. doi:10.1007/s00023-009-0002-2.

Gro46. H. J. Groenewold. On the principles of elementary quantum mechanics. *Physica*, 12:405–460, 1946. doi:10.1016/S0031-8914(46)80059-4.

GRS14. R. Gurau, V. Rivasseau, and A. Sfondrini. Renormalization: an advanced overview. 2014, http://arxiv.org/abs/1401.5003.

GRVT06. R. Gurau, V. Rivasseau, and F. Vignes-Tourneret. Propagators for noncommutative field theories. *Annales Henri Poincaré*, 7:1601–1628, 2006, http://arxiv.org/abs/hep-th/0512071. doi:10.1007/s00023-006-0296-2.

GS99. H. Grosse and A. Strohmaier. Towards a nonperturbative covariant regularization in 4-D quantum field theory. *Lett. Math. Phys.*, 48:163–179, 1999, http://arxiv.org/abs/hep-th/9902138. doi:10.1023/A:1007518622795.

GS01. S. S. Gubser and S. L. Sondhi. Phase structure of noncommutative scalar field theories. *Nucl. Phys.*, B605:395–424, 2001, http://arxiv.org/abs/hep-th/0006119. doi:10.1016/S0550-3213(01)00108-0.

GS06a. H. Grosse and H. Steinacker. A nontrivial solvable noncommutative ϕ^3 model in 4 dimensions. *JHEP*, 08:008, 2006, http://arxiv.org/abs/hep-th/0603052. doi:10.1088/1126-6708/2006/08/008.

GS06b. H. Grosse and H. Steinacker. Renormalization of the noncommutative ϕ^3 model through the Kontsevich model. *Nucl. Phys.*, B746:202–226, 2006, http://arxiv.org/abs/hep-th/0512203. doi:10.1016/j.nuclphysb.2006.04.007.

GS08. H. Grosse and H. Steinacker. Exact renormalization of a noncommutative ϕ^3 model in 6 dimensions. *Adv. Theor. Math. Phys.*, 12(3):605–639, 2008, http://arxiv.org/abs/hep-th/0607235. doi:10.4310/ATMP.2008.v12.n3.a4.

GSW08. H. Grosse, H. Steinacker, and M. Wohlgenannt. Emergent gravity, matrix models and UV/IR mixing. *JHEP*, 04:023, 2008, http://arxiv.org/abs/0802.0973. doi:10.1088/1126-6708/2008/04/023.

GSW17. H. Grosse, A. Sako, and R. Wulkenhaar. Exact solution of matricial Φ^3_2 quantum field theory. *Nucl. Phys.*, B925:319–347, 2017, http://arxiv.org/abs/1610.00526. doi:10.1016/j.nuclphysb.2017.10.010.

GSW18. H. Grosse, A. Sako, and R. Wulkenhaar. The Φ^3_4 and Φ^3_6 matricial QFT models have reflection positive two-point function. *Nucl. Phys.*, B926:20–48, 2018, http://arxiv.org/abs/1612.07584. doi:10.1016/j.nuclphysb.2017.10.022.

Gur11a. R. Gurau. Colored group field theory. *Commun. Math. Phys.*, 304:69–93, 2011, http://arxiv.org/abs/0907.2582. doi:10.1007/s00220-011-1226-9.

Gur11b. R. Gurau. The 1/N expansion of colored tensor models. *Annales Henri Poincaré*, 12:829–847, 2011, http://arxiv.org/abs/1011.2726. doi:10.1007/s00023-011-0101-8.

Gur12. R. Gurau. The complete 1/N expansion of colored tensor models in arbitrary dimension. *Annales Henri Poincaré*, 13:399–423, 2012, http://arxiv.org/abs/1102.5759. doi:10.1007/s00023-011-0118-z.

Gur14. R. Gurau. The $1/N$ expansion of tensor models beyond perturbation theory. *Commun. Math. Phys.*, 330:973–1019, 2014, http://arxiv.org/abs/1304.2666. doi:10.1007/s00220-014-1907-2.

Gur17. R. Gurau. *Random Tensors*. Oxford University Press, Oxford, 2017.

GW73. D. J. Gross and F. Wilczek. Ultraviolet behavior of non-Abelian gauge theories. *Phys. Rev. Lett.*, 30:1343–1346, 1973. doi:10.1103/PhysRevLett.30.1343.

GW02. J. M. Grimstrup and R. Wulkenhaar. Quantisation of θ-expanded noncommutative QED. *Eur. Phys. J.*, C26:139–151, 2002, http://arxiv.org/abs/hep-th/0205153. doi:10.1140/epjc/s2002-01038-9.

GW03. H. Grosse and R. Wulkenhaar. Renormalisation of ϕ^4-theory on noncommutative $\mathbb{R}^2$ in the matrix base. *JHEP*, 12:019, 2003, http://arxiv.org/abs/hep-th/0307017. doi:10.1088/1126-6708/2003/12/019.

GW04. H. Grosse and R. Wulkenhaar. The β-function in duality covariant noncommutative ϕ^4-theory. *Eur. Phys. J.*, C35:277–282, 2004, http://arxiv.org/abs/hep-th/0402093. doi:10.1140/epjc/s2004-01853-x.

GW05a. H. Grosse and R. Wulkenhaar. Power-counting theorem for non-local matrix models and renormalisation. *Commun. Math. Phys.*, 254:91–127, 2005, http://arxiv.org/abs/hep-th/0305066. doi:10.1007/s00220-004-1238-9.

GW05b. H. Grosse and R. Wulkenhaar. Renormalisation of ϕ^4-theory on noncommutative $\mathbb{R}^4$ in the matrix base. *Commun. Math. Phys.*, 256:305–374, 2005, http://arxiv.org/abs/hep-th/0401128. doi:10.1007/s00220-004-1285-2.

GW07. H. Grosse and M. Wohlgenannt. Induced gauge theory on a noncommutative space. *Eur. Phys. J.*, C52:435–450, 2007, http://arxiv.org/abs/hep-th/0703169. doi:10.1140/epjc/s10052-007-0369-5.

GW09. H. Grosse and R. Wulkenhaar. Progress in solving a noncommutative quantum field theory in four dimensions. 2009, http://arxiv.org/abs/0909.1389.

GW12. H. Grosse and R. Wulkenhaar. 8D-spectral triple on 4D-Moyal space and the vacuum of noncommutative gauge theory. *J. Geom. Phys.*, 62:1583–1599, 2012, http://arxiv.org/abs/0709.0095. doi:10.1016/j.geomphys.2012.03.005.

GW13a. V. Gayral and R. Wulkenhaar. Spectral geometry of the Moyal plane with harmonic propagation. *J. Noncommut. Geom.*, 7:939–979, 2013, http://arxiv.org/abs/1108.2184. doi:10.4171/JNCG/140.

GW13b. H. Grosse and R. Wulkenhaar. Solvable limits of a 4D noncommutative QFT. 2013, http://arxiv.org/abs/1306.2816.

GW14a. H. Grosse and R. Wulkenhaar. Self-dual noncommutative ϕ^4-theory in four dimensions is a non-perturbatively solvable and non-trivial quantum field theory. *Commun. Math. Phys.*, 329:1069–1130, 2014, http://arxiv.org/abs/1205.0465. doi:10.1007/s00220-014-1906-3.

GW14b. H. Grosse and R. Wulkenhaar. Solvable 4D noncommutative QFT: phase transitions and quest for reflection positivity. 2014, http://arxiv.org/abs/1406.7755.

GW16. H. Grosse and R. Wulkenhaar. On the fixed point equation of a solvable 4D QFT model. *Vietnam J. Math.*, 44(1):153–180, 2016, http://arxiv.org/abs/1505.05161. doi:10.1007/s10013-015-0174-7.

Haa58. R. Haag. Quantum field theories with composite particles and asymptotic conditions. *Phys. Rev.*, 112:669–673, 1958. doi:10.1103/PhysRev.112.669.

Haa96. R. Haag. *Local quantum physics. Fields, particles, algebras*. Texts and Monographs in Physics. Springer-Verlag, Berlin, second edition, 1996. doi:10.1007/978-3-642-61458-3.

Hau23. F. Hausdorff. Momentprobleme für ein endliches Intervall. *Math. Z.*, 16(1):220–248, 1923. doi:10.1007/BF01175684.

Haw99. E. Hawkins. Noncommutative regularization for the practical man. 1999, http://arxiv.org/abs/hep-th/9908052.

Hay00. M. Hayakawa. Perturbative analysis on infrared aspects of noncommutative QED on $\mathbb{R}^4$. *Phys. Lett.*, B478:394–400, 2000, http://arxiv.org/abs/hep-th/9912094. doi:10.1016/S0370-2693(00)00242-2.

Hei38. W. Heisenberg. Die Grenzen der Anwendbarkeit der bisherigen Quantentheorie. *Z. Phys.*, 110:251–266, 1938. doi:10.1007/BF01342872.

Hep66. K. Hepp. Proof of the Bogolyubov-Parasiuk theorem on renormalization. *Commun. Math. Phys.*, 2:301–326, 1966. doi:10.1007/BF01773358.

HK64. R. Haag and D. Kastler. An algebraic approach to quantum field theory. *J. Math. Phys.*, 5:848–861, 1964. doi:10.1063/1.1704187.

HW18. A. Hock and R. Wulkenhaar. Noncommutative 3-colour scalar quantum field theory model in 2D. *Eur. Phys. J.*, C78(7):580, 2018, http://arxiv.org/abs/1804.06075. doi:10.1140/epjc/s10052-018-6042-3.

IKTW01. S. Iso, Y. Kimura, K. Tanaka, and K. Wakatsuki. Noncommutative gauge theory on fuzzy sphere from matrix model. *Nucl. Phys.*, B604:121–147, 2001, http://arxiv.org/abs/hep-th/0101102. doi:10.1016/S0550-3213(01)00173-0.

ILS09. B. Iochum, C. Levy, and A. Sitarz. Spectral action on $SU_q(2)$. *Commun. Math. Phys.*, 289:107–155, 2009, http://arxiv.org/abs/0803.1058. doi:10.1007/s00220-009-0810-8.

IMSS11. B. Iochum, T. Masson, T. Schucker, and A. Sitarz. Compact κ-deformation and spectral triples. *Rept. Math. Phys.*, 68:37–64, 2011, http://arxiv.org/abs/1004.4190. doi:10.1016/S0034-4877(11)60026-8.

IZ92. C. Itzykson and J. B. Zuber. Combinatorics of the modular group. 2. The Kontsevich integrals. *Int. J. Mod. Phys.*, A7:5661–5705, 1992, http://arxiv.org/abs/hep-th/9201001. doi:10.1142/S0217751X92002581.

JNOS17. A. Jaffe, K.-H. Neeb, G. Olafsson, and B. Schlein, editors. *Reflection positivity*, volume 14 of *Oberwolfach Rep.*, 2017. doi:10.4171/OWR/2017/55.

JSSW00. B. Jurčo, S. Schraml, P. Schupp, and J. Wess. Enveloping algebra valued gauge transformations for non-Abelian gauge groups on noncommutative spaces. *Eur. Phys. J.*, C17:521–526, 2000, http://arxiv.org/abs/hep-th/0006246. doi:10.1007/s100520000487.

JSW01. B. Jurčo, P. Schupp, and J. Wess. Non-Abelian noncommutative gauge theory via noncommutative extra dimensions. *Nucl. Phys.*, B604:148–180, 2001, http://arxiv.org/abs/hep-th/0102129. doi:10.1016/S0550-3213(01)00191-2.

JW00. A. M. Jaffe and E. Witten. Quantum Yang-Mills theory. 2000. URL https://www.claymath.org/sites/default/files/yangmills.pdf.

Kac49. M. Kac. On distributions of certain Wiener functionals. *Trans. Amer. Math. Soc.*, 65:1–13, 1949. doi:10.2307/1990512.

Kit15a. A. Kitaev. A simple model of quantum holography. 1. Talk at KITP, 2015. URL http://online.kitp.ucsb.edu/online/entangled15/kitaev/.

Kit15b. A. Kitaev. A simple model of quantum holography. 2. Talk at KITP, 2015. URL http://online.kitp.ucsb.edu/online/entangled15/kitaev2/.

KK66. T. Kōmura and Y. Kōmura. Über die Einbettung der nuklearen Räume in $(s)^A$. *Math. Ann.*, 162:284–288, 1966. doi:10.1007/BF01360917.

KL81. A. Klein and L. J. Landau. Construction of a unique self-adjoint generator for a symmetric local semigroup. *J. Funct. Anal.*, 44(2):121–137, 1981. doi:10.1016/0022-1236(81)90007-0.

Kon92. M. Kontsevich. Intersection theory on the moduli space of curves and the matrix Airy function. *Commun. Math. Phys.*, 147:1–23, 1992. doi:10.1007/BF02099526.

Kon03. M. Kontsevich. Deformation quantization of Poisson manifolds. 1. *Lett. Math. Phys.*, 66:157–216, 2003, http://arxiv.org/abs/q-alg/9709040. doi:10.1023/B:MATH.0000027508.00421.bf.

KPT18. I. R. Klebanov, F. Popov, and G. Tarnopolsky. TASI lectures on large N tensor models. *PoS*, TASI2017:004, 2018, http://arxiv.org/abs/1808.09434. doi:10.22323/1.305.0004.

Kre98. D. Kreimer. On the Hopf algebra structure of perturbative quantum field theories. *Adv. Theor. Math. Phys.*, 2:303–334, 1998, http://arxiv.org/abs/q-alg/9707029. doi:10.4310/ATMP.1998.v2.n2.a4.

Kre00. D. Kreimer. Chen's iterated integral represents the operator product expansion. *Adv. Theor. Math. Phys.*, 3:627–670, 2000, http://arxiv.org/abs/hep-th/9901099. doi:10.4310/ATMP.1999.v3.n3.a7.

KRTW10. T. Krajewski, V. Rivasseau, A. Tanasă, and Z. Wang. Topological graph polynomials and quantum field theory. I. Heat kernel theories. *J. Noncommut. Geom.*, 4(1):29–82, 2010, http://arxiv.org/abs/0811.0186. doi:10.4171/JNCG/49.

KRVT11. T. Krajewski, V. Rivasseau, and F. Vignes-Tourneret. Topological graph polynomials and quantum field theory, part II. Mehler kernel theories. *Annales Henri Poincaré*, 12:483–545, 2011, http://arxiv.org/abs/0912.5438. doi:10.1007/s00023-011-0087-2.

KS96. R. Koekoek and R. F. Swarttouw. The Askey-scheme of hypergeometric orthogonal polynomials and its q-analogue. 1996, http://arxiv.org/abs/math/9602214.

KS02. A. Konechny and A. S. Schwarz. Introduction to M(atrix) theory and noncommutative geometry. *Phys. Rept.*, 360:353–465, 2002, http://arxiv.org/abs/hep-th/0012145. doi:10.1016/S0370-1573(01)00096-5.

Kul76. P. P. Kulish. Factorization of the classical and quantum S-matrix and conservation laws. *Theor. Math. Phys.*, 26:132, 1976. doi:10.1007/BF01079418. [also in Teor. Mat. Fiz. 26:198, 1976].

KW00. T. Krajewski and R. Wulkenhaar. Perturbative quantum gauge fields on the noncommutative torus. *Int. J. Mod. Phys.*, A15:1011–1030, 2000, http://arxiv.org/abs/hep-th/9903187. doi:10.1142/S0217751X00000495.

Lag70. J. L. Lagrange. Nouvelle méthode pour résoudre des équations littérales par le moyen de séries. *Mém. Acad. Roy. des Sci. et Belles-Lettres de Berlin*, 24, 1770.

LAK54. L. D. Landau, A. A. Abrikosov, and I. M. Khalatnikov. On the removal of infinities in quantum electrodynamics (in russ.). *Dokl. Akad. Nauk SSSR*, 95:497–500, 1954.

Lam58. J. H. Lambert. Observationes variae in mathesin puram. *Acta Helvetica, physico-mathematico-anatomico-botanico-medica*, 3:128–168, 1758.

LLS01. G. Landi, F. Lizzi, and R. J. Szabo. From large N matrices to the noncommutative torus. *Commun. Math. Phys.*, 217:181–201, 2001, http://arxiv.org/abs/hep-th/9912130. doi:10.1007/s002200000356.

LLS04. G. Landi, F. Lizzi, and R. J. Szabo. A new matrix model for noncommutative field theory. *Phys. Lett.*, B578:449–458, 2004, http://arxiv.org/abs/hep-th/0309031. doi:10.1016/j.physletb.2003.10.059.

LPW88. J. M. F. Labastida, M. Pernici, and E. Witten. Topological gravity in two dimensions. *Nucl. Phys.*, B310:611–624, 1988. doi:10.1016/0550-3213(88)90094-6.

LS02a. E. Langmann and R. J. Szabo. Duality in scalar field theory on noncommutative phase spaces. *Phys. Lett.*, B533:168–177, 2002, http://arxiv.org/abs/hep-th/0202039. doi:10.1016/S0370-2693(02)01650-7.

LS02b. Y. Liao and K. Sibold. Time ordered perturbation theory on noncommutative spacetime. 2. Unitarity. *Eur. Phys. J.*, C25:479–486, 2002, http://arxiv.org/abs/hep-th/0206011. doi:10.1007/s10052-002-1018-7.

LS02c. Y. Liao and K. Sibold. Time ordered perturbation theory on noncommutative spacetime: Basic rules. *Eur. Phys. J.*, C25:469–477, 2002, http://arxiv.org/abs/hep-th/0205269. doi:10.1007/s10052-002-1017-8.

LSZ03. E. Langmann, R. J. Szabo, and K. Zarembo. Exact solution of noncommutative field theory in background magnetic fields. *Phys. Lett.*, B569:95–101, 2003, http://arxiv.org/abs/hep-th/0303082. doi:10.1016/j.physletb.2003.07.020.

LSZ04. E. Langmann, R. J. Szabo, and K. Zarembo. Exact solution of quantum field theory on noncommutative phase spaces. *JHEP*, 01:017, 2004, http://arxiv.org/abs/hep-th/0308043. doi:10.1088/1126-6708/2004/01/017.

LZ04. S. K. Lando and A. K. Zvonkin. *Graphs on surfaces and their applications*, volume 141 of *Encyclopaedia of Mathematical Sciences*. Springer-Verlag, Berlin, 2004. doi:10.1007/978-3-540-38361-1. With an appendix by Don B. Zagier, Low-Dimensional Topology, II.

Mad92. J. Madore. The fuzzy sphere. *Class. Quant. Grav.*, 9:69–88, 1992. doi:10.1088/0264-9381/9/1/008.

Mad95. J. Madore. *An introduction to noncommutative differential geometry and its physical applications*, volume 206 of *London Mathematical Society Lecture Note Series*. Cambridge University Press, Cambridge, 1995.

Min59. R. A. Minlos. Generalized random processes and their extension in measure. *Trudy Moskov. Mat. Obšč.*, 8:497–518, 1959. URL http://www.mathnet.ru/php/getFT.phtml?jrnid=mmo&paperid=94&what=fullt&option_lang=eng.

Mir07. M. Mirzakhani. Simple geodesics and Weil-Petersson volumes of moduli spaces of bordered Riemann surfaces. *Invent. Math.*, 167(1):179–222, 2007. doi:10.1007/s00222-006-0013-2.

Miw82. T. Miwa. On Hirota's difference equations. *Proc. Japan Acad. Ser. A Math. Sci.*, 58(1):9–12, 1982. http://projecteuclid.org/euclid.pja/1195516178.

Mos75. J. Moser. Three integrable Hamiltonian systems connected with isospectral deformations. *Adv. Math.*, 16:197–220, 1975. doi:10.1016/0001-8708(75)90151-6.

Moy49. J. E. Moyal. Quantum mechanics as a statistical theory. *Proc. Cambridge Phil. Soc.*, 45:99–124, 1949. doi:10.1017/S0305004100000487.

MP90. R. C. Myers and V. Periwal. Topological gravity and moduli space. *Nucl. Phys.*, B333:536–550, 1990. doi:10.1016/0550-3213(90)90050-N.

MR08. J. Magnen and V. Rivasseau. Constructive ϕ^4 field theory without tears. *Annales Henri Poincaré*, 9:403–424, 2008, http://arxiv.org/abs/0706.2457. doi:10.1007/s00023-008-0360-1.

MS91. Yu. Makeenko and G. W. Semenoff. Properties of Hermitean matrix models in an external field. *Mod. Phys. Lett.*, A6:3455–3466, 1991. doi:10.1142/S0217732391003985.

MS16. J. Maldacena and D. Stanford. Remarks on the Sachdev-Ye-Kitaev model. *Phys. Rev.*, D94(10):106002, 2016, http://arxiv.org/abs/1604.07818. doi:10.1103/PhysRevD.94.106002.

MSR99. C. P. Martín and D. Sánchez-Ruiz. The one loop UV divergent structure of U(1) Yang-Mills theory on noncommutative $\mathbb{R}^4$. *Phys. Rev. Lett.*, 83:476–479, 1999, http://arxiv.org/abs/hep-th/9903077. doi:10.1103/PhysRevLett.83.476.

MSSW00. J. Madore, S. Schraml, P. Schupp, and J. Wess. Gauge theory on noncommutative spaces. *Eur. Phys. J.*, C16:161–167, 2000, http://arxiv.org/abs/hep-th/0001203. doi:10.1007/s100520050012.

MST00. A. Matusis, L. Susskind, and N. Toumbas. The IR / UV connection in the noncommutative gauge theories. *JHEP*, 12:002, 2000, http://arxiv.org/abs/hep-th/0002075. doi:10.1088/1126-6708/2000/12/002.

MT13. P. Martinetti and L. Tomassini. Noncommutative geometry of the Moyal plane: Translation isometries, Connes' distance on coherent states, Pythagoras equality. *Commun. Math. Phys.*, 323:107–141, 2013, http://arxiv.org/abs/1110.6164. doi:10.1007/s00220-013-1760-8.

MVRS00. S. Minwalla, M. Van Raamsdonk, and N. Seiberg. Noncommutative perturbative dynamics. *JHEP*, 02:020, 2000, http://arxiv.org/abs/hep-th/9912072. doi:10.1088/1126-6708/2000/02/020.

MZ15. M. Mirzakhani and P. Zograf. Towards large genus asymptotics of intersection numbers on moduli spaces of curves. *Geom. Funct. Anal.*, 25(4):1258–1289, 2015, http://arxiv.org/abs/1112.1151. doi:10.1007/s00039-015-0336-5.

Nel73a. E. Nelson. Construction of quantum fields from Markoff fields. *J. Functional Analysis*, 12:97–112, 1973. doi:10.1016/0022-1236(73)90091-8.

Nel73b. E. Nelson. Quantum fields and Markoff fields. In *Partial differential equations (Proc. Sympos. Pure Math., Vol. XXIII, Univ. California, Berkeley, Calif., 1971)*, pages 413–420. Amer. Math. Soc., Providence, R.I., 1973.

Nie09. N. Nielsen. Der Eulersche Dilogarithmus und seine Verallgemeinerungen. *Nova acta - Kaiserlich Leopoldinisch-Carolinische Deutsche Akademie der Naturforscher*, 90(3):121–212, 1909.

NO18. K.-H. Neeb and G. Ólafsson. *Reflection positivity*, volume 32 of *SpringerBriefs in Mathematical Physics*. Springer, 2018. doi:10.1007/978-3-319-94755-6.

Nus55. A. E. Nussbaum. The Hausdorff-Bernstein-Widder theorem for semi-groups in locally compact Abelian groups. *Duke Math. J.*, 22:573–582, 1955. URL http://projecteuclid.org/euclid.dmj/1077466543.

ORZ04. T. Ohl, R. Rückl, and J. Zeiner. Unitarity of time - like noncommutative gauge theories: The violation of Ward identities in time ordered perturbation theory. *Nucl. Phys.*, B676:229–242, 2004, http://arxiv.org/abs/hep-th/0309021. doi:10.1016/j.nuclphysb.2003.10.022.

OS73. K. Osterwalder and R. Schrader. Axioms for Euclidean Green's functions. *Commun. Math. Phys.*, 31:83–112, 1973. doi:10.1007/BF01645738.

OS75. K. Osterwalder and R. Schrader. Axioms for Euclidean Green's functions. 2. *Commun. Math. Phys.*, 42:281–305, 1975. doi:10.1007/BF01608978.

Pan15. E. Panzer. Algorithms for the symbolic integration of hyperlogarithms with applications to Feynman integrals. *Computer Physics Communications*, 188:148–166, 2015, http://arxiv.org/abs/1403.3385. doi:10.1016/j.cpc.2014.10.019. maintained and available at https://bitbucket.org/PanzerErik/hyperint.

Pei33. R. Peierls. Zur Theorie des Diamagnetismus von Leitungselektronen. *Z. Phys.*, 80:763, 1933.

Per07. D. Perrot. Anomalies and noncommutative index theory. In *Geometric and topological methods for quantum field theory*, volume 434 of *Contemp. Math.*, pages 125–160. Amer. Math. Soc., Providence, RI, 2007, http://arxiv.org/abs/hep-th/0603209. doi:10.1090/conm/434/08344.

Pol73. H. D. Politzer. Reliable perturbative results for strong interactions? *Phys. Rev. Lett.*, 30:1346–1349, 1973. doi:10.1103/PhysRevLett.30.1346.

Pol84. J. Polchinski. Renormalization and effective Lagrangians. *Nucl. Phys.*, B231:269–295, 1984. doi:10.1016/0550-3213(84)90287-6.

PW18. E. Panzer and R. Wulkenhaar. Lambert-W solves the noncommutative Φ^4-model. *Commun. Math. Phys.* (in press), 2019/20, http://arxiv.org/abs/1807.02945, doi:10.1007/s00220-019-03592-4.

Rie92. B. Riemann. Ueber die Hypothesen, welche der Geometrie zugrunde liegen. In *Bernhard Riemann's gesammelte mathematische Werke und wissenschaftlicher Nachlaß*, Leipzig, 1892. B. G. Teubner. URL https://www.emis.de/classics/Riemann/Geom.pdf.

Rie90. M. A. Rieffel. Noncommutative tori—a case study of noncommutative differentiable manifolds. In *Geometric and topological invariants of elliptic operators (Brunswick, ME, 1988)*, volume 105 of *Contemp. Math.*, pages 191–211. Amer. Math. Soc., Providence, RI, 1990. doi:10.1090/conm/105/1047281.

Rie93. M. A. Rieffel. Deformation quantization for actions of $\mathbf{R}^d$. *Mem. Amer. Math. Soc.*, 106(506):x+93, 1993. doi:10.1090/memo/0506.

Rie04. M. A. Rieffel. Gromov-Hausdorff distance for quantum metric spaces. *Mem. Amer. Math. Soc.*, 168(796):viii+91, 2004, http://arxiv.org/abs/math/0011063. doi:10.1090/memo/0796.

Riv91. V. Rivasseau. *From Perturbative to Constructive Renormalization.* Princeton University Press, New Jersey, 1991.

Riv07a. V. Rivasseau. Constructive matrix theory. *JHEP*, 09:008, 2007, http://arxiv.org/abs/0706.1224. doi:10.1088/1126-6708/2007/09/008.

Riv07b. V. Rivasseau. Non-commutative renormalization. In *Quantum Spaces*, volume 53 of *Progress in Mathematical Physics*, pages 19–107. Birkhäuser, Basel, 2007, http://arxiv.org/abs/0705.0705. doi:10.1007/978-3-7643-8522-4_2.

Riv14. V. Rivasseau. The tensor track, III. *Fortsch. Phys.*, 62:81–107, 2014, http://arxiv.org/abs/1311.1461. doi:10.1002/prop.201300032.

Rue62. D. Ruelle. On the asymptotic condition in quantum field theory. *Helv. Phys. Acta*, 35:147–163, 1962. doi:10.5169/seals-113272.

RVTW06. V. Rivasseau, F. Vignes-Tourneret, and R. Wulkenhaar. Renormalization of noncommutative ϕ^4-theory by multi-scale analysis. *Commun. Math. Phys.*, 262:565–594, 2006, http://arxiv.org/abs/hep-th/0501036. doi:10.1007/s00220-005-1440-4.

RW12. V. Rivasseau and Z. Wang. Constructive renormalization for Φ_2^4 theory with loop vertex expansion. *J. Math. Phys.*, 53:042302, 2012, http://arxiv.org/abs/1104.3443. doi:10.1063/1.4705689.

RW15. V. Rivasseau and Z. Wang. Corrected loop vertex expansion for Φ_2^4 theory. *J. Math. Phys.*, 56(6):062301, 2015, http://arxiv.org/abs/1406.7428. doi:10.1063/1.4922116.

Sch34. E. Schrödinger. Über die Unanwendbarkeit der Geometrie im Kleinen. *Naturwiss.*, 31:518–520, 1934. doi:10.1007/BF01494946.

Sch59. J. Schwinger. Euclidean quantum electrodynamics. *Phys. Rev.*, 115:721–731, 1959. doi:10.1103/PhysRev.115.721.

Sch99. V. Schomerus. D-branes and deformation quantization. *JHEP*, 06:030, 1999, http://arxiv.org/abs/hep-th/9903205. doi:10.1088/1126-6708/1999/06/030.

Seg01. G. Segal. Topological structures in string theory. *R. Soc. Lond. Philos. Trans. Ser. A Math. Phys. Eng. Sci.*, 359(1784):1389–1398, 2001. doi:10.1098/rsta.2001.0841. Topological methods in the physical sciences (London, 2000).

Sim74. B. Simon. *The $P(\phi)_2$ Euclidean (quantum) field theory*. Princeton University Press, Princeton, N.J., 1974. Princeton Series in Physics.

SJ99. M. M. Sheikh-Jabbari. Renormalizability of the supersymmetric Yang-Mills theories on the noncommutative torus. *JHEP*, 06:015, 1999, http://arxiv.org/abs/hep-th/9903107. doi:10.1088/1126-6708/1999/06/015.

Sny47. H. S. Snyder. Quantized space-time. *Phys. Rev.*, 71:38–41, 1947. doi:10.1103/PhysRev.71.38.

SS94. P. Schaller and T. Strobl. Poisson structure induced (topological) field theories. *Mod. Phys. Lett.*, A9:3129–3136, 1994, http://arxiv.org/abs/hep-th/9405110. doi:10.1142/S0217732394002951.

Ste04. H. Steinacker. Quantized gauge theory on the fuzzy sphere as random matrix model. *Nucl. Phys.*, B679:66–98, 2004, http://arxiv.org/abs/hep-th/0307075. doi:10.1016/j.nuclphysb.2003.12.005.

Ste07. H. Steinacker. Emergent gravity from noncommutative gauge theory. *JHEP*, 12:049, 2007, http://arxiv.org/abs/0708.2426. doi:10.1088/1126-6708/2007/12/049.

Ste10. H. Steinacker. Emergent geometry and gravity from matrix models: an introduction. *Class. Quant. Grav.*, 27:133001, 2010, http://arxiv.org/abs/1003.4134. doi:10.1088/0264-9381/27/13/133001.

Str67. K. Strebel. On quadratic differentials with closed trajectories and second order poles. *J. Analyse Math.*, 19:373–382, 1967. doi:10.1007/BF02788726.

Str84. K. Strebel. *Quadratic differentials*, volume 5 of *Ergebnisse der Mathematik und ihrer Grenzgebiete*. Springer-Verlag, Berlin, 1984. doi:10.1007/978-3-662-02414-0.

SW64. R. Streater and A. Wightman. *PCT, Spin and Statistics, and All that*. W. A. Benjamin Inc., 1964.

SW99. N. Seiberg and E. Witten. String theory and noncommutative geometry. *JHEP*, 09:032, 1999, http://arxiv.org/abs/hep-th/9908142. doi:10.1088/1126-6708/1999/09/032.

SY93. S. Sachdev and J. Ye. Gapless spin fluid ground state in a random, quantum Heisenberg magnet. *Phys. Rev. Lett.*, 70:3339, 1993, http://arxiv.org/abs/cond-mat/9212030. doi:10.1103/PhysRevLett.70.3339.

Sym64. K. Symanzik. Application of functional integrals to Euclidean quantum field theory. In *Proc. Conf. on Theory and Appl. of Analysis in Function Space (Dedham, Mass. 1963)*, pages 197–206. The M.I.T. Press, Cambridge, Mass., 1964.

Sza03. R. J. Szabo. Quantum field theory on noncommutative spaces. *Phys. Rept.*, 378:207–299, 2003, http://arxiv.org/abs/hep-th/0109162. doi:10.1016/S0370-1573(03)00059-0.

tH74. G. 't Hooft. A planar diagram theory for strong interactions. *Nucl. Phys.*, B72:461–473, 1974. doi:10.1016/0550-3213(74)90154-0.

Tri57. F. G. Tricomi. *Integral equations*. Interscience, New York, 1957.

Vas04. D. V. Vassilevich. Noncommutative heat kernel. *Lett. Math. Phys.*, 67:185–194, 2004, http://arxiv.org/abs/hep-th/0310144. doi:10.1023/B:MATH.0000035037.50663.b1.

VGB88. J. C. Várilly and J. M. Gracia-Bondía. Algebras of distributions suitable for phase-space quantum mechanics. 2. Topologies on the Moyal algebra. *J. Math. Phys.*, 29:880–887, 1988. doi:10.1063/1.527984.

VGB99. J. C. Várilly and J. M. Gracia-Bondía. On the ultraviolet behavior of quantum fields over noncommutative manifolds. *Int. J. Mod. Phys.*, A14:1305, 1999, http://arxiv.org/abs/hep-th/9804001. doi:10.1142/S0217751X99000671.

vN32. J. von Neumann. *Mathematische Grundlagen der Quantenmechanik.* Springer-Verlag, Berlin, 1932. URL http://resolver.sub.uni-goettingen.de/purl?PPN379400774.

Vog00. D. Vogt. Lectures on Fréchet spaces, 2000. URL http://www2.math.uni-wuppertal.de/~vogt/vorlesungen/fs.pdf.

VT07. F. Vignes-Tourneret. Renormalization of the orientable non-commutative Gross-Neveu model. *Annales Henri Poincaré*, 8:427–474, 2007, http://arxiv.org/abs/math-ph/0606069. doi:10.1007/s00023-006-0312-6.

Wan18. Z. Wang. Constructive renormalization of the 2-dimensional Grosse-Wulkenhaar model. *Annales Henri Poincaré*, 19(8):2435–2490, 2018, http://arxiv.org/abs/1805.06365. doi:10.1007/s00023-018-0688-0.

Wey28. H. Weyl. *Gruppentheorie und Quantenmechanik.* S. Hirzel, Leipzig, 1928.

WG64. A. Wightman and L. Gårding. *Fields as operator-valued distributions in relativistic quantum theory.* Arkiv för fysik. Almqvist & Wiksell, 1964.

Whe55. J. A. Wheeler. Geons. *Phys. Rev.*, 97:511–536, 1955. doi:10.1103/PhysRev.97.511.

Wid31. D. V. Widder. Necessary and sufficient conditions for the representation of a function as a Laplace integral. *Trans. Amer. Math. Soc.*, 33(4):851–892, 1931. doi:10.2307/1989513.

Wig32. E. P. Wigner. On the quantum correction for thermodynamic equilibrium. *Phys. Rev.*, 40:749–760, 1932. doi:10.1103/PhysRev.40.749.

Wig56. A. S. Wightman. Quantum field theory in terms of vacuum expectation values. *Phys. Rev.*, 101:860–866, 1956. doi:10.1103/PhysRev.101.860.

Wil74. K. G. Wilson. Confinement of quarks. *Phys. Rev.*, D10:2445–2459, 1974. doi:10.1103/PhysRevD.10.2445.

Wit88. E. Witten. Topological quantum field theory. *Commun. Math. Phys.*, 117:353, 1988. doi:10.1007/BF01223371.

Wit90. E. Witten. On the structure of the topological phase of two-dimensional gravity. *Nucl. Phys.*, B340:281–332, 1990. doi:10.1016/0550-3213(90)90449-N.

Wit91. E. Witten. Two-dimensional gravity and intersection theory on moduli space. In *Surveys in differential geometry (Cambridge, MA, 1990)*, pages 243–310. Lehigh Univ., Bethlehem, PA, 1991. doi:10.4310/SDG.1990.v1.n1.a5.

Wit92. E. Witten. On the Kontsevich model and other models of two-dimensional gravity. In *Differential geometric methods in theoretical physics. Proceedings, 20th International Conference, New York, 1991*, pages 176–216. World Sci. Publ., River Edge, NJ, 1992.

Wit16. E. Witten. An SYK-like model without disorder. 2016, http://arxiv.org/abs/1610.09758.

WK74. K. G. Wilson and J. B. Kogut. The renormalization group and the ϵ expansion. *Phys. Rept.*, 12:75–199, 1974. doi:10.1016/0370-1573(74)90023-4.

Wul00. R. Wulkenhaar. Slavnov-Taylor identity in noncommutative geometry. *Int. J. Mod. Phys.*, B14:2503–2509, 2000. doi:10.1142/S0217979200002053.

Wul02. R. Wulkenhaar. Non-renormalizability of θ-expanded noncommutative QED. *JHEP*, 03:024, 2002, http://arxiv.org/abs/hep-th/0112248. doi:10.1088/1126-6708/2002/03/024.

Wul06. R. Wulkenhaar. Field theories on deformed spaces. *J. Geom. Phys.*, 56:108–141, 2006. doi:10.1016/j.geomphys.2005.04.019.

Zah06. J. W. Zahn. *Dispersion relations in quantum electrodynamics on the noncommutative Minkowski space*. PhD thesis, Hamburg U., 2006, http://arxiv.org/abs/0707.2149. URL http://www-library.desy.de/cgi-bin/showprep.pl?thesis06-037.

Zah11. J. Zahn. Divergences in quantum field theory on the noncommutative two-dimensional Minkowski space with Grosse-Wulkenhaar potential. *Annales Henri Poincaré*, 12:777–804, 2011, http://arxiv.org/abs/1005.0541. doi:10.1007/s00023-011-0089-0.

Zim69. W. Zimmermann. Convergence of Bogolyubov's method of renormalization in momentum space. *Commun. Math. Phys.*, 15:208–234, 1969. doi:10.1007/BF01645676.

Higher invariants in noncommutative geometry

Zhizhang Xie and Guoliang Yu

Dedicated to Alain Connes with great admiration

Abstract We give a survey on higher invariants in noncommutative geometry and their applications to differential geometry and topology.

1 Introduction

Geometry and topology of a smooth manifold is often governed by natural differential operators on the manifold. When a smooth manifold M is closed (compact without boundary), a basic invariant of these differential operators is their Fredholm index. Roughly speaking the Fredholm index measures the size of the solution space for an infinite dimensional linear system associated to the operator D. More precisely, the Fredholm index of D by the formula: $\operatorname{index}(D) = \dim(\operatorname{kernel}(D)) - \dim(\operatorname{kernel}(D^*))$. The beauty of the Fredholm index is its invariance under small perturbations and homotopy equivalence. The Fredholm index is an obstruction to invertibility of the operator. The Fredholm index of such an operator D is computed by the well-known Atiyah–Singer index theorem [AS]. The Atiyah–Singer index theorem has important applications to geometry, topology, and mathematical physics.

Alain Connes' powerful noncommutative geometry provides the framework for a much more refined index theory, called higher index theory [BC, BCH, C, CM, K]. Higher index theory is a far-reaching generalization of classic Fredholm index

Z. Xie
Department of Mathematics, Texas A&M University, College Station, TX, USA
e-mail: xie@math.tamu.edu

G. Yu (✉)
Department of Mathematics, Texas A&M University, College Station, TX, USA
Shanghai Center for Mathematical Sciences, Shanghai, China
e-mail: guoliangyu@math.tamu.edu

A. Chamseddine et al. (eds.), *Advances in Noncommutative Geometry*,
https://doi.org/10.1007/978-3-030-29597-4_12

theory by taking into consideration of the symmetries given by the fundamental group. Let D be an elliptic differential operator on a closed manifold M of dimension n. If $\widetilde{M}$ is the universal cover of M, and $\widetilde{D}$ is the lift of D onto $\widetilde{M}$, then we can define a higher index of $\widetilde{D}$ in $K_n(C_r^*(\pi_1 M))$, where $\pi_1 M$ is the fundamental group of M and $K_n(C_r^*(\pi_1 M))$ is the K-theory of the reduced group C^*-algebra $C_r^*(\pi_1 M)$. This higher index is an obstruction to the invertibility of $\widetilde{D}$ and is invariant under homotopy. Higher index theory plays a fundamental role in the studies of problems in geometry and topology such as the Novikov conjecture on homotopy invariance of higher signatures and the Gromov–Lawson conjecture on nonexistence of Riemmanian metrics with positive scalar curvature. Higher indices are often referred to as primary invariants due to its homotopy invariance property. The Baum–Connes conjecture provides an algorithm for computing the higher index [BC, BCH] while the strong Novikov conjecture predicts when the higher index vanishes [K]. When a closed manifold M carries a Riemannian metric with positive scalar curvature, by the Lichnerowicz formula, the Dirac operator $\widetilde{D}$ on $\widetilde{M}$ is invertible and hence its higher index vanishes. If M is aspherical, i.e., its universal cover is contractible, then the strong Novikov conjecture predicts that the higher index of the Dirac operator is nonzero and hence implies the Gromov–Lawson conjecture stating that any closed aspherical manifold cannot carry a Riemannian metric with positive scalar curvature [R]. Another important application of higher index theory is the Novikov conjecture [N], a central problem in topology. Roughly speaking, the Novikov conjecture claims that compact smooth manifolds are rigid at an infinitesimal level. More precisely, the Novikov conjecture states that the higher signatures of compact oriented smooth manifolds are invariant under orientation preserving homotopy equivalences. Recall that a compact manifold is called aspherical if its universal cover is contractible. In the case of aspherical manifolds, the Novikov conjecture is an infinitesimal version of the Borel conjecture, which states that all compact aspherical manifolds are topologically rigid, i.e., if another compact manifold N is homotopy equivalent to the given compact aspherical manifold M, then N is homeomorphic to M. A theorem of Novikov states that the rational Pontryagin classes are invariant under orientation preserving homeomorphisms [N1]. Thus the Novikov conjecture for compact aspherical manifolds follows from the Borel conjecture and Novikov's theorem, since for aspherical manfolds, the information about higher signatures is equivalent to that of rational Pontryagin classes. In general, the Novikov conjecture follows from the strong Novikov conjecture when applied to the signature operator. With the help of noncommutative geometry, spectacular progress has been made on the Novikov conjecture.

When the higher index of an elliptic operator is trivial with a given trivialization, a secondary index theoretic invariant naturally arises [HR2, Roe1]. This secondary invariant is called the higher rho invariant. It serves as an obstruction to locality of the inverse of an invertible elliptic operator. For example, when a closed manifold M carries a Riemannian metric with positive scalar curvature, the Dirac operator $\widetilde{D}$ on its universal cover $\widetilde{M}$ is invertible, hence its higher index is trivial. In this case, the positive scalar curvature metric gives a specific trivialization of the higher index, thus naturally defines a higher rho invariant. Such a secondary index theoretic invariant is of fundamental importance for studying the space of positive scalar curvature metrics of a given closed spin manifold. For instance, this secondary invariant is an essential ingredient for measuring the size of the moduli space (under diffeomorphism group) of positive scalar curvature metrics on a given closed spin manifold [XY1]. The following is another typical situation where higher rho invariants naturally arise. Given an orientation preserving homotopy equivalence between two oriented closed manifolds, the higher index of the signature operator on the disjoint union of the two manifolds (one of them equipped with the opposite orientation) is trivial with a trivialization given by the homotopy equivalence. Hence such a homotopy equivalence naturally defines a higher rho invariant for the signature operator [HR2, Roe1]. More generally, the notion of higher rho invariants can be defined for homotopy equivalences between topological manifolds [Z], and these invariants serve as a powerful tool for detecting whether a homotopy equivalence can be "deformed" into a homeomorphism. Furthermore, the authors proved in [WXY] that the higher rho invariant defines a group of homomorphism on the structure group of a *topological* manifold. As an application, one can use the higher rho invariant to measure the degree of non-rigidity of a topological manifold.

Connes' cyclic cohomology theory provides a powerful method to compute higher rho invariants. It turns out that the pairing of cyclic cohomology with higher rho invariants can be computed in terms of John Lott's higher eta invariants. This relation can be used to give an elegant approach to the higher Atiyah-Patodi-Singer index theory for manifolds with boundary and provide a potential way to construct counter examples to the Baum–Connes conjecture.

The purpose of this article is to give a friendly survey on these recent developments of higher invariants in noncommutative geometry and their applications to geometry and topology.

2 Geometric C^*-algebras

In this section, we give an overview of several C^*-algebras naturally arising from geometry and topology. The K-theory groups of these C^*-algebras serve as receptacles of our higher invariants.

Let X be a proper metric space. That is, every closed ball in X is compact. An X-module is a separable Hilbert space equipped with a $*$-representation of $C_0(X)$, the algebra of all continuous functions on X which vanish at infinity. An X-module is called nondegenerate if the $*$-representation of $C_0(X)$ is nondegenerate. An X-module is said to be standard if no nonzero function in $C_0(X)$ acts as a compact operator.

We shall first recall the concepts of propagation, local compactness, and pseudo-locality.

Definition 2.1 Let H_X be an X-module and T a bounded linear operator acting on H_X.

(i) The propagation of T is defined to be $sup\{d(x, y) \mid (x, y) \in supp(T)\}$, where $supp(T)$ is the complement (in $X \times X$) of the set of points $(x, y) \in X \times X$ for which there exist $f, g \in C_0(X)$ such that $gTf = 0$ and $f(x) \neq 0$, $g(y) \neq 0$;
(ii) T is said to be locally compact if fT and Tf are compact for all $f \in C_0(X)$;
(iii) T is said to be pseudo-local if $[T, f]$ is compact for all $f \in C_0(X)$.

Pseudo-locality is the essential property for the concept of an abstract "differential operator" in K-homology theory [A, K].

The following concept was introduced by Roe in his work on higher index theory for noncompact spaces [Roe].

Definition 2.2 Let H_X be a standard nondegenerate X-module and $\mathcal{B}(H_X)$ the set of all bounded linear operators on H_X. The Roe algebra of X, denoted by $C^*(X)$, is the C^*-algebra generated by all locally compact operators with finite propagations in $\mathcal{B}(H_X)$.

The following localization algebra was introduced by [Y].

Definition 2.3 The localization algebra $C_L^*(X)$ is the C^*-algebra generated by all bounded and uniformly norm-continuous functions $f : [0, \infty) \to C^*(X)$ such that

$$\text{propagation of } f(t) \to 0, \text{ as } t \to \infty.$$

We define $C_{L,0}^*(X)$ to be the kernel of the evaluation map

$$e\colon C_L^*(X) \to C^*(X), \quad e(f) = f(0).$$

In particular, $C_{L,0}^*(X)$ is an ideal of $C_L^*(X)$.

The localization algebra was motivated by local index theory.

Now we take symmetries into consideration. Let us assume that a discrete group Γ acts properly on X by isometries. Let H_X be an X-module equipped with a covariant unitary representation of Γ. If we denote the representation of $C_0(X)$ by φ and the representation of Γ by π, this means

$$\pi(\gamma)(\varphi(f)v) = \varphi(f^\gamma)(\pi(\gamma)v),$$

where $f \in C_0(X)$, $\gamma \in \Gamma$, $v \in H_X$, and $f^\gamma(x) = f(\gamma^{-1}x)$. In this case, we call (H_X, Γ, φ) a covariant system.

Definition 2.4 ([Y3]) A covariant system (H_X, Γ, φ) is called admissible if

(1) the Γ-action on X is proper and cocompact;
(2) H_X is a nondegenerate standard X-module;
(3) for each $x \in X$, the stabilizer group Γ_x acts on H_X regularly in the sense that the action is isomorphic to the action of Γ_x on $l^2(\Gamma_x) \otimes H$ for some infinite

dimensional Hilbert space H. Here Γ_x acts on $l^2(\Gamma_x)$ by translations and acts on H trivially.

We remark that for each locally compact metric space X with a proper and cocompact isometric action of Γ, there exists an admissible covariant system (H_X, Γ, φ). Also, we point out that the condition (3) above is automatically satisfied if Γ acts freely on X. If no confusion arises, we will denote an admissible covariant system (H_X, Γ, φ) by H_X and call it an admissible (X, Γ)-module.

Definition 2.5 Let X be a locally compact metric space X with a proper and cocompact isometric action of Γ. If H_X is an admissible (X, Γ)-module, we denote by $\mathbb{C}[X]^\Gamma$ the $*$-algebra of all Γ-invariant locally compact operators with finite propagations in $\mathcal{B}(H_X)$. We define the equivariant Roe algebra $C^*(X)^\Gamma$ to be the completion of $\mathbb{C}[X]^\Gamma$ in $\mathcal{B}(H_X)$.

We remark that if the Γ-action on X is cocompact, then the equivariant Roe algebra $C^*(X)^\Gamma$ is $*$-isomorphic to $C_r^*(\Gamma) \otimes K$, where $C_r^*(\Gamma)$ is the reduced group C^*-algebra of Γ and K is the C^*-algebra of all compact operators. We also point out that, up to isomorphism, $C^*(X) = C^*(X, H_X)$ does not depend on the choice of the standard nondegenerate X-module H_X. The same statement holds for $C_L^*(X)$, $C_{L,0}^*(X)$, and their Γ-equivariant versions.

We can also define the maximal versions of the geometric C^*-algebras in this section by taking the norm completions over all $*$-representations of their algebraic counterparts.

3 Higher index theory and localization

In this section, we construct the higher index of an elliptic operator. We also introduce a local index map from the K-homology group to the K-group of the localization algebra and explain that this local index map is an isomorphism.

3.1 K-homology

We first discuss the K-homology theory of Kasparov. Let X be a locally compact metric space with a proper and cocompact isometric action of Γ. The K-homology groups $K_j^\Gamma(X)$, $j = 0, 1$, are generated by the following cycles modulo certain equivalence relations (cf. [K]):

1. an even cycle for $K_0^\Gamma(X)$ is a pair (H_X, F), where H_X is an admissible (X, Γ)-module and $F \in \mathcal{B}(H_X)$ such that F is Γ-equivariant, $F^*F - I$ and $FF^* - I$ are locally compact and $[F, f] = Ff - fF$ is compact for all $f \in C_0(X)$.

2. an odd cycle for $K_1^{\Gamma}(X)$ is a pair (H_X, F), where H_X is an admissible (X, Γ)-module and F is a Γ-equivariant self-adjoint operator in $\mathcal{B}(H_X)$ such that $F^2 - I$ is locally compact and $[F, f]$ is compact for all $f \in C_0(X)$.

Roughly speaking, the K-homology group of X is generated by abstract elliptic operators over X [A, K].

In the general case where the action of Γ on X is not necessarily cocompact, we define

$$K_i^{\Gamma}(X) = \varinjlim_{Y \subseteq X} K_i^{\Gamma}(Y),$$

where Y runs through all closed Γ-invariant subsets of X such that Y/Γ is compact.

3.2 K-theory and boundary maps

In this subsection, we recall the standard construction of the index maps in K-theory of C^*-algebras. For a short exact sequence of C^*-algebras $0 \to \mathcal{J} \to \mathcal{A} \to \mathcal{A}/\mathcal{J} \to 0$, we have a six-term exact sequence in K-theory:

$$\begin{array}{ccccc} K_0(\mathcal{J}) & \longrightarrow & K_0(\mathcal{A}) & \longrightarrow & K_0(\mathcal{A}/\mathcal{J}) \\ {\scriptstyle \partial_0} \uparrow & & & & \downarrow {\scriptstyle \partial_1} \\ K_1(\mathcal{A}/\mathcal{J}) & \longleftarrow & K_1(\mathcal{A}) & \longleftarrow & K_1(\mathcal{J}) \end{array}$$

Let us recall the definition of the boundary maps ∂_i.

1. Even case. Let u be an invertible element in $\mathcal{A}/\mathcal{J}$. Let v be the inverse of u in $\mathcal{A}/\mathcal{J}$. Now suppose $U, V \in \mathcal{A}$ are lifts of u and v. We define

$$W = \begin{pmatrix} 1 & U \\ 0 & 1 \end{pmatrix} \begin{pmatrix} 1 & 0 \\ -V & 1 \end{pmatrix} \begin{pmatrix} 1 & U \\ 0 & 1 \end{pmatrix} \begin{pmatrix} 0 & -1 \\ 1 & 0 \end{pmatrix}.$$

Notice that W is invertible and a direct computation shows that

$$W - \begin{pmatrix} U & 0 \\ 0 & V \end{pmatrix} \in \mathcal{J}.$$

Consider the idempotent

$$P = W \begin{pmatrix} 1 & 0 \\ 0 & 0 \end{pmatrix} W^{-1} = \begin{pmatrix} UV + UV(1-UV) & (2-UV)(1-UV)U \\ V(1-UV) & (1-VU)^2 \end{pmatrix}. \tag{3.1}$$

We have

$$P - \begin{pmatrix} 1 & 0 \\ 0 & 0 \end{pmatrix} \in \mathcal{J}.$$

By definition,

$$\partial([u]) := [P] - \left[\begin{pmatrix} 1 & 0 \\ 0 & 0 \end{pmatrix}\right] \in K_0(\mathcal{J}).$$

2. Odd case. Let q be an idempotent in $\mathcal{A}/\mathcal{J}$ and Q a lift of q in $\mathcal{A}$. Then

$$\partial([q]) := [e^{2\pi i Q}] \in K_1(\mathcal{J}).$$

3.3 *Higher index map and local index map*

In this subsection, we describe the constructions of the higher index map [BC, BCH, K, FM] and the local index map [Y, Y1].

Let (H_X, F) be an even cycle for $K_0^{\Gamma}(X)$. Choose a Γ-invariant locally finite open cover $\{U_i\}$ of X with diameter $(U_i) < c$ for some fixed $c > 0$. Let $\{\phi_i\}$ be a Γ-invariant continuous partition of unity subordinate to $\{U_i\}$. We define

$$\mathcal{F} = \sum_i \phi_i^{1/2} F \phi_i^{1/2},$$

where the sum converges in strong operator norm topology. It is not difficult to see that $(H_X, \mathcal{F})$ is equivalent to (H_X, F) in $K_0^{\Gamma}(X)$. By using the fact that $\mathcal{F}$ has finite propagation, we see that $\mathcal{F}$ is a multiplier of $C^*(X)^{\Gamma}$ and, is a unitary modulo $C^*(X)^{\Gamma}$. Consider the short exact sequence of C^*-algebras

$$0 \to C^*(X)^{\Gamma} \to \mathcal{M}(C^*(X)^{\Gamma}) \to \mathcal{M}(C^*(X)^{\Gamma})/C^*(X)^{\Gamma} \to 0,$$

where $\mathcal{M}(C^*(X)^{\Gamma})$ is the multiplier algebra of $C^*(X)^{\Gamma}$. By the construction in Section 3.2 above, $\mathcal{F}$ produces a class $\partial([\mathcal{F}]) \in K_0(C^*(X)^{\Gamma})$. We define the higher index of (H_X, F) to be $\partial([\mathcal{F}])$. From now on, we denote $[\mathcal{F}]$ by $\mathrm{Ind}(H_X, F)$ or simply $\mathrm{Ind}(F)$, if no confusion arises.

To define the local index of (H_X, F), we need to use a family of partitions of unity. More precisely, for each $n \in \mathbb{N}$, let $\{U_{n,j}\}$ be a Γ-invariant locally finite open cover of X with diameter $(U_{n,j}) < 1/n$ and $\{\phi_{n,j}\}$ be a Γ-invariant continuous partition of unity subordinate to $\{U_{n,j}\}$. We define

$$\mathcal{F}(t) = \sum_j (1-(t-n))\phi_{n,j}^{1/2}\mathcal{F}\phi_{n,j}^{1/2} + (t-n)\phi_{n+1,j}^{1/2}\mathcal{F}\phi_{n+1,j}^{1/2} \tag{3.2}$$

for $t \in [n, n+1]$.

Then $\mathcal{F}(t), 0 \leq t < \infty$, is a multiplier of $C_L^*(X)^\Gamma$ and a unitary modulo $C_L^*(X)^\Gamma$. By the construction in Section 3.2 above, we define $\partial([\mathcal{F}(t)]) \in K_0(C_L^*(X)^\Gamma)$ to be the local index of (H_X, F). If no confusion arises, we denote this local index class by $\mathrm{Ind}_L(H_X, F)$ or simply $\mathrm{Ind}_L(F)$.

Now let (H_X, F) be an odd cycle in $K_1^\Gamma(X)$. With the same notation from above, we set $q = \frac{\mathcal{F}+1}{2}$. Then the index class of (H_X, F) is defined to be $[e^{2\pi i q}] \in K_1(C^*(X)^\Gamma)$. For the local index class of (H_X, F), we use $q(t) = \frac{\mathcal{F}(t)+1}{2}$ in place of q.

We have the following commutative diagram:

$$\begin{array}{ccc} & K_*^\Gamma(X) & \\ {\scriptstyle Ind_L}\swarrow & & \searrow{\scriptstyle Ind} \\ K_*(C_L^*(X)^\Gamma) & \xrightarrow{\quad e_* \quad} & K_*(C^*(X)^\Gamma), \end{array}$$

where e_* is the homomorphism induced by the evaluation map e at 0.

The following result was proved in the case of simplicial complexes in [Y] and the general case in [QR].

Theorem 3.1 *If a discrete group Γ acts properly on a locally compact space X, then the local index map is an isomorphism from the K-homology group $K_*^\Gamma(X)$ to the K-group of the localization algebra $K_*(C_L^*(X)^\Gamma)$.*

4 The Baum–Connes assembly and a local-global principle

In this section, we formulate the Baum–Connes conjecture as a local-global principle and discuss its connection to the Novikov conjecture.

We first recall the concept of Rips complexes.

Definition 4.1 Let Γ be a discrete group, let $F \subseteq \Gamma$ be a finite symmetric subset containing the identity (symmetric in the sense if $g \in F$, then $g^{-1} \in F$). The Rips complex $P_F(\Gamma)$ is a simplicial complex such that

(i) the set of vertices is Γ;
(ii) a finite subset $\{\gamma_0, \cdots, \gamma_n\}$ span a simplex if and only if $\gamma_i^{-1}\gamma_j \in F$ for all $0 \leq i, j \leq n$.

We endow the Rips complex with the simplicial metric, i.e., the maximal metric whose restriction to a maximal simplex is the standard Euclidean metric on the simplex.

The Baum–Connes conjecture [BC, BCH] can be stated as follows.

Conjecture 4.2 (Baum–Connes Conjecture) The evaluation map e induces an isomorphism e_* from the K-group of the equivariant localization algebra $\varinjlim_F K_*(C_L^*(P_F(\Gamma))^\Gamma)$ to the K-group of the equivariant Roe algebra $\varinjlim_F K_*(C^*(P_F(\Gamma))^\Gamma)$, where the limit is taken over the directed set of all finite symmetric subset F of Γ containing the identity.

Note that $\varinjlim_F K_*(C^*(P_F(\Gamma))^\Gamma)$ is isomorphic to K-group of $C_r^*(\Gamma)$, the reduced group C^*-algebra of Γ since the Γ action on the Rips complex is cocompact.

While the K-theory of the equivariant Roe algebra is global and hard to compute, the K-theory of the localization algebra is local and completely computable. Thus the Baum–Connes conjecture is a local-global principle. If true, the conjecture provides an algorithm for computing K-groups of equivariant Roe algebras and higher indices of elliptic operators. In particular, in this case, we see that every element in the K-group of the equivariant Roe algebra can be localized.

More generally, if A is a C^*-algebra with an action of Γ, then we can define the equivariant Roe algebra with coefficients in A, denoted by $C^*(P_F(\Gamma), A)^\Gamma$. The equivariant Roe algebra with coefficients in A is $*$-isomorphic to $(A \rtimes \Gamma) \otimes \mathcal{K}$, where $\mathcal{K}$ is the algebra of compact operators on a Hilbert space. We can similarly introduce an equivariant localization algebra with coefficients to formulate the Baum–Connes conjecture with coefficients.

Higson and Kasparov developed an index theory of certain differential operators on an infinite-dimensional Hilbert space and proved the following spectacular result [HK].

Theorem 4.3 *If a discrete group Γ acts on Hilbert space properly and isomentrically, then the Baum–Connes conjecture with coefficients holds for Γ.*

Recall that an isometric action α of a group Γ on a Hilbert space H is said to be proper if $\|\alpha(\gamma)h\| \to \infty$ when $\gamma \to \infty$ for any $h \in H$, i.e., for any $h \in H$ and any positive number $R > 0$, there exists a finite subset F of Γ such that $\|\alpha(\gamma)h\| > R$ if $\gamma \in \Gamma - F$. A theorem of Bekka–Cherix–Valette states that an amenable group acts properly and isometrically on a Hilbert space [BCV]. Roughly speaking, a group is amenable if there exist large finite subsets of the group with small boundary. The concept of amenability is a large scale geometric property and was introduced by von Neumann. We refer the readers to the book [NY] as a general reference for geometric group theory related to the Novikov conjecture.

The following deep theorem is due to Lafforgue [L1].

Theorem 4.4 *The Baum–Connes conjecture with coefficients holds for hyperbolic groups.*

Earlier Lafforgue developed a Banach KK-theory to attack the Baum–Connes conjecture [L]. This approach yielded the Baum–Connes conjecture for hyperbolic groups [L, MY].

The Baum–Connes conjecture with coefficients actually fail for general groups. Higson–Lafforgue–Skandalis gave a counterexample in [HLS]. On the other hand, the Baum–Connes conjecture (without coefficients) is still open.

5 The Novikov conjecture

A central problem in topology is the Novikov conjecture. Roughly speaking, the Novikov conjecture claims that compact smooth manifolds are rigid at an infinitesimal level. More precisely, the Novikov conjecture states that the higher signatures of compact oriented smooth manifolds are invariant under orientation preserving homotopy equivalences. Recall that a compact manifold is called aspherical if its universal cover is contractible. In the case of aspherical manifolds, the Novikov conjecture is an infinitesimal version of the Borel conjecture, which states that all compact aspherical manifolds are topologically rigid, i.e., if another compact manifold N is homotopy equivalent to the given compact aspherical manifold M, then N is homeomorphic to M. A theorem of Novikov says that the rational Pontryagin classes are invariant under orientation preserving homeomorphisms [N1]. Thus the Novikov conjecture for compact aspherical manifolds follows from the Borel conjecture and Novikov's theorem, since for aspherical manfolds, the information about higher signatures is equivalent to that of rational Pontryagin classes. In general, the Novikov conjecture follows from the (rational) strong Novikov conjecture.

The (rational) strong Novikov conjecture can be stated as follows.

Conjecture 5.1 (Strong Novikov Conjecture) The evaluation map e induces an injection e_* from the K-group of the equivariant localization algebra $\varinjlim_F K_*(C_L^*(P_F(\Gamma))^\Gamma)$ to the K-group of the equivariant Roe algebra $\varinjlim_F K_*(C^*(P_F(\Gamma))^\Gamma)$, where the limit is taken over the directed set of all finite symmetric subset F of Γ containing the identity. The rational strong Novikov conjecture states that e_* is an injection after tensoring with $\mathbb{Q}$.

The strong Novikov conjecture predicts when the higher index of an elliptic operator is nonzero. The strong Novikov conjecture implies the following analytic Novikov conjecture.

Conjecture 5.2 (Analytic Novikov Conjecture) The evaluation map e induces an injection e_* from the K-group of the equivariant localization algebra $K_*(C_L^*(E\Gamma)^\Gamma)$ to the K-group of the equivariant Roe algebra $K_*(C^*(E\Gamma)^\Gamma)$, where $K_*(C_L^*(E\Gamma)^\Gamma)$ is defined to be $\varinjlim_X K_*(C_L^*(X)^\Gamma)$ with the limit to be taken over the directed set of locally compact, Γ-equivariant, Γ-cocompact subset X of the universal space $E\Gamma$ for free Γ-action, and similarly $K_*(C^*(E\Gamma)^\Gamma)$ is defined to be the limit $\varinjlim_X K_*(C^*(X)^\Gamma)$. The rational analytic Novikov conjecture states that e_* is an injection after tensoring with $\mathbb{Q}$, that is,

$$e_*\colon K_*(C_L^*(E\Gamma)^\Gamma) \otimes \mathbb{Q} \to K_*(C^*(E\Gamma)^\Gamma) \otimes \mathbb{Q}$$

is an injection.

The classical Novikov conjecture follows from the rational analytic Novikov conjecture. With the help of noncommutative geometry, spectacular progress has been made on the Novikov conjecture. It has been proven that the Novikov conjecture holds when the fundamental group of the manifold lies in one of the following classes of groups:

1. groups acting properly and isometrically on simply connected and non-positively curved manifolds [K],
2. hyperbolic groups [CM],
3. groups acting properly and isometrically on Hilbert spaces [HK],
4. groups acting properly and isometrically on bolic spaces [KS],
5. groups with finite asymptotic dimension [Y1],
6. groups coarsely embeddable into Hilbert spaces [Y2, H, STY],
7. groups coarsely embeddable into Banach spaces with property (H) [KY],
8. all linear groups and subgroups of all almost connected Lie groups [GHW],
9. subgroups of the mapping class groups [Ha, Ki],
10. subgroups of $\mathrm{Out}(F_n)$, the outer automorphism groups of the free groups [BGH],
11. groups acting properly and isometrically on (possibly infinite dimensional) admissible Hilbert–Hadamard spaces, in particular geometrically discrete subgroups of the group of volume preserving diffeomorphisms of any smooth compact manifold [GWY].

In the first three cases, an isometric action of a discrete group Γ on a metric space X is said to be *proper* if for some $x \in X$, $d(x, gx) \to \infty$ as $g \to \infty$, i.e., for any $x \in X$ and any positive number $R > 0$, there exists a finite subset F of Γ such that $d(x, gx) > R$ if $g \in \Gamma - F$.

In a tour de force, Connes proved a striking theorem that the Novikov conjecture holds for higher signatures associated to Gelfand-Fuchs classes [C1]. Connes, Gromov, and Moscovici proved the Novikov conjecture for higher signatures associated to Lipschitz group cohomology classes [CGM]. Hanke–Schick and Mathai proved the Novikov conjecture for higher signatures associated to group cohomology classes with degrees one and two [HS, Ma].

J. Rosenberg discovered an important application of the (rational) strong Novikov conjecture to the existence problem of Riemannian metrics with positive scalar curvature [R]. We refer to Rosenberg's survey [R1] for recent developments on this topic.

5.1 Non-positively curved groups and hyperbolic groups

In this subsection, we give a survey on the work of A. Mishchenko, G. Kasparov, A. Connes and H. Moscovici, G. Kasparov and G. Skandalis on the Novikov conjecture for non-positively curved groups and Gromov's hyperbolic groups.

In [M], A. Mishchenko introduced a theory of infinite dimensional Fredholm representations of discrete groups to prove the following theorem.

Theorem 5.3 *The Novikov conjecture holds if the fundamental group of a manifold acts properly, isometrically and cocompactly on a simply connected manifold with non-positive sectional curvature.*

In [K], G. Kasparov developed a bivariant K-theory, called KK-theory, to prove the following theorem.

Theorem 5.4 *The Novikov conjecture holds if the fundamental group of a manifold acts properly and isometrically on a simply connected manifold with non-positive sectional curvature.*

As a consequence, G. Kasparov proved the following striking theorem.

Theorem 5.5 *The Novikov conjecture holds if the fundamental group of a manifold is a discrete subgroup of a Lie group with finitely many connected components.*

The theory of hyperbolic groups was developed by Gromov [G]. Gromov's hyperbolic groups are generic among all finitely presented groups. A. Connes and H. Moscovici proved the following spectacular theorem using powerful techniques from noncommutative geometry [CM].

Theorem 5.6 *The Novikov conjecture holds if the fundamental group of a manifold is a hyperbolic group in the sense of Gromov.*

The proof of Theorem 5.6 uses Connes' theory of cyclic cohomology in a crucial way. Cyclic homology theory plays the role of de Rham theory in noncommutative geometry, and is the natural receptacle for the Connes–Chern character [C].

The following theorem of G. Kasparov and G. Skandalis unified the above results [KS].

Theorem 5.7 *The Novikov conjecture holds if the fundamental group of a manifold is bolic.*

Bolicity is a notion of non-positive curvature. Examples of bolic groups include groups acting properly and isometrically on simply connected manifolds with non-positive sectional curvature and Gromov's hyperbolic groups.

5.2 *Amenable groups, groups with finite asymptotic dimension and coarsely embeddable groups*

In this subsection, we give a survey on the work of Higson–Kasparov on the Novikov conjecture for amenable groups, the work of G. Yu on the Novikov conjecture for groups with finite asymptotic dimension, and the work of G. Yu, N. Higson, Skandalus-Tu-Yu on the Novikov conjecture for groups coarsely embeddable into Hilbert spaces. Finally we discuss the work of Kasparov–Yu on the connection of the Novikov conjecture with Banach space geometry.

As mentioned above (Theorem 4.3), Higson and Kasparov proved that the Baum–Connes conjecture holds for groups that act properly and isometrically on a Hilbert space [HK]. As a consequence, the Novikov conjecture holds for these groups.

Theorem 5.8 *The Novikov conjecture holds if the fundamental group of a manifold acts properly and isometrically on a Hilbert space.*

Since amenable groups act properly and isometrically on a Hilbert space [BCV], the above theorem has the following immediate corollary.

Corollary *The Novikov conjecture holds if the fundamental group of a manifold is amenable.*

This corollary is quite striking since the geometry of amenable groups can be very complicated (for example, the Grigorchuk's groups [Gr]).

Next we recall a few basic concepts from geometric group theory. A non-negative function l on a countable group G is called a length function if (1) $l(g^{-1}) = l(g)$ for all $g \in G$; (2) $l(gh) \leq l(g) + l(h)$ for all g and h in G; (3) $l(g) = 0$ if and only if $g = e$, the identity element of G. We can associate a left-invariant length metric d_l to l: $d_l(g, h) = l(g^{-1}h)$ for all $g, h \in G$. A length metric is called proper if the length function is a proper map (i.e., the inverse image of every compact set is finite in this case). It is not difficult to show that every countable group G has a proper length metric. If l and l' are two proper length functions on G, then their associated length metrics are coarsely equivalent. If G is a finitely generated group and S is a finite symmetric generating set (symmetric in the sense that if an element is in S, then its inverse is also in S), then we can define the word length l_S on G by

$$l_S(g) = \min\{n : g = s_1 \cdots s_n, s_i \in S\}.$$

If S and S' are two finite symmetric generating sets of G, then their associated proper length metrics are quasi-isometric.

The following concept is due to Gromov [G1].

Definition 5.9 The asymptotic dimension of a proper metric space X is the smallest integer n such that for every $r > 0$, there exists a uniformly bounded cover $\{U_i\}$ for which the number of U_i intersecting each r ball $B(x, r)$ is at most $n + 1$.

For example, the asymptotic dimension of $\mathbb{Z}^n$ is n and the asymptotic dimension of the free group $\mathbb{F}_n$ with n generators is 1. The asymptotic dimension is invariant under coarse equivalence. The Lie group $GL(n, \mathbb{R})$ with a left-invariant Riemannian metric is quasi-isometric to $T(n, \mathbb{R})$, the subgroup of invertible upper triangular matrices. By permanence properties of asymptotic dimension [BD1], we know that the solvable group $T(n, \mathbb{R})$ has finite asymptotic dimension. As a consequence, every countable discrete subgroup of $GL(n, \mathbb{R})$ has finite asymptotic dimension (as a metric space with a proper length metric). More generally one can prove that every discrete subgroup of an almost connected Lie group has finite asymptotic dimension (a Lie group is said to be almost connected if the number of its connected components is finite). Gromov's hyperbolic groups have finite asymptotic dimension [Roe2]. Mapping class groups also have finite asymptotic dimension [BBF].

In [Y1], G. Yu developed a quantitative operator K-theory to prove the following theorem.

Theorem 5.10 *The Novikov conjecture holds if the fundamental group of a manifold has finite asymptotic dimension.*

The basic idea of the proof is that the finiteness of asymptotic dimension allows us to develop an algorithm to compute K-theory in a quantitative way. This strategy has found applications to topological rigidity of manifolds [GTY].

The following concept of Gromov makes precise of the idea of drawing a good picture of a metric space in a Hilbert space.

Definition 5.11 (Gromov): Let X be a metric space and H be a Hilbert space. A map $f : X \to H$ is said to be a coarse embedding if there exist non-decreasing functions ρ_1 and ρ_2 on $[0, \infty)$ such that

(1) $\rho_1(d(x, y)) \leq d_H(f(x), f(y)) \leq \rho_2(d(x, y))$ for all $x, y \in X$;
(2) $\lim_{r\to+\infty} \rho_1(r) = +\infty$.

Coarse embeddability of a countable group is independent of the choice of proper length metrics. Examples of groups coarsely embeddable into Hilbert space include groups acting properly and isometrically on a Hilbert space (in particular amenable groups [BCV]), groups with Property A [Y2], countable subgroups of connected Lie groups [GHW], hyperbolic groups [S], groups with finite asymptotic dimension, Coxeter groups [DJ], mapping class groups [Ki, Ha], and semi-direct products of groups of the above types.

The following theorem unifies the above theorems.

Theorem 5.12 *The Novikov conjecture holds if the fundamental group of a manifold is coarsely embeddable into Hilbert space.*

Roughly speaking, this theorem says if we can draw a good picture of the fundamental group in a Hilbert space, then we can recognize the manifold at an infinitesimal level. This theorem was proved by G. Yu when the classifying space of the fundamental group has the homotopy type of a finite CW complex [Y2] and this finiteness condition was removed by N. Higson [H], Skandalis-Tu-Yu [STY]. The original proof of the above result makes heavy use of infinite dimensional analysis. More recently, R. Willett and G. Yu found a relatively elementary proof within the framework of basic operator K-theory [WiY].

E. Guentner, N. Higson and S. Weinberger proved the beautiful theorem that linear groups are coarsely embeddable into Hilbert space [GHW]. Recall that a group is called linear if it is a subgroup of $GL(n, k)$ for some field k. The following theorem follows as a consequence [GHW].

Theorem 5.13 *The Novikov conjecture holds if the fundamental group of a manifold is a linear group.*

More recently, Bestvina–Guirardel–Horbez proved that $\mathrm{Out}(F_n)$, the outer automorphism groups of the free groups, is coarsely embeddable into Hilbert space. This implies the following theorem [BGH].

Theorem 5.14 *The Novikov conjecture holds if the fundamental group of a manifold is a subgroup of* $\mathrm{Out}(F_n)$.

We have the following open question.

Open Question 5.15 Is every countable subgroup of the diffeomorphism group of the circle coarsely embeddable into Hilbert space?

Let $\mathfrak{C}$ be the smallest class of groups which include all groups coarsely embeddable into Hilbert space and is closed under direct limit. Recall that if I is a directed set and $\{G_i\}_{i\in I}$ is a direct system of groups over I, then we can define the direct limit $\lim G_i$. We emphasize that here the homomorphism $\phi_{ij} : G_i \to G_j$ for $i \leq j$ is not necessarily injective.

The following result is a consequence of Theorem 5.12.

Theorem 5.16 *The Novikov conjecture holds if the fundamental group of a manifold is in the class* $\mathfrak{C}$.

The following open question is a challenge to geometric group theorists.

Open Question 5.17 Is there any countable group not in the class $\mathfrak{C}$?

We mention that the Gromov monster groups are in the class $\mathfrak{C}$ [G2, G3, AD, O].

Next we shall discuss the connection of the Novikov conjecture with geometry of Banach spaces.

Definition 5.18 A Banach space X is said to have Property (H) if there exist an increasing sequence of finite dimensional subspaces $\{V_n\}$ of X and an increasing sequence of finite dimensional subspaces $\{W_n\}$ of a Hilbert space such that

(1) $V = \cup_n V_n$ is dense in X,
(2) if $W = \cup_n W_n$, $S(V)$ and $S(W)$ are respectively the unit spheres of V and W, then there exists a uniformly continuous map $\psi : S(V) \to S(W)$ such that the restriction of ψ to $S(V_n)$ is a homeomorphism (or more generally a degree one map) onto $S(W_n)$ for each n.

As an example, let X be the Banach space $l^p(\mathbb{N})$ for some $p \geq 1$. Let V_n and W_n be respectively the subspaces of $l^p(\mathbb{N})$ and $l^2(\mathbb{N})$ consisting of all sequences whose coordinates are zero after the n-th terms. We define a map ψ from $S(V)$ to $S(W)$ by

$$\psi(c_1, \cdots, c_k, \cdots) = (c_1|c_1|^{p/2-1}, \cdots, c_k|c_k|^{p/2-1}, \cdots).$$

ψ is called the Mazur map. It is not difficult to verify that ψ satisfies the conditions in the definition of Property (H). For each $p \geq 1$, we can similarly prove that C_p, the Banach space of all Schatten p-class operators on a Hilbert space, has Property (H).

Kasparov and Yu proved the following.

Theorem 5.19 *The Novikov conjecture holds if the fundamental group of a manifold is coarsely embeddable into a Banach space with Property (H).*

Let c_0 be the Banach space consisting of all sequences of real numbers that are convergent to 0 with the supremum norm.

Open Question 5.20 Does the Banach space c_0 have Property (H)?

A positive answer to this question would imply the Novikov conjecture since every countable group admits a coarse embedding into c_0 [BG].

A less ambitious question is the following.

Open Question 5.21 Is every countable subgroup of the diffeomorphism group of a compact smooth manifold coarsely embeddable into C_p for some $p \geq 1$?

For each $p > q \geq 2$, it is also an open question to construct a bounded geometry space which is coarsely embeddable into $l^p(\mathbb{N})$ but not $l^q(\mathbb{N})$. Beautiful results in [JR] and [MN] indicate that such a construction should be possible. Once such a metric space is constructed, the next natural question is to construct countable groups which coarsely contain such a metric space. These groups would be from another universe and would be different from any group we currently know.

5.3 Gelfand-Fuchs classes, the group of volume preserving diffeomorphisms, Hilbert–Hadamard spaces

In this subsection, we give an overview on the work of A. Connes, Connes–Gromov–Moscovici on the Novikov conjecture for Gelfand-Fuchs classes and the recent work of Gong–Wu–Yu on the Novikov conjecture for groups acting properly and isometrically on a Hilbert–Hadamard spaces and for any geometrically discrete subgroup of the group of volume preserving diffeomorphisms of a compact smooth manifold.

A. Connes proved the following deep theorem on the Novikov conjecture [C1].

Theorem 5.22 *The Novikov conjecture holds for higher signatures associated to the Gelfand-Fuchs cohomology classes of a subgroup of the group of diffeomorphisms of a compact smooth manifold.*

The proof of this theorem uses the full power of noncommutative geometry [C].

Open Question 5.23 Does the Novikov conjecture hold for any subgroup of the group of diffeomorphisms of a compact smooth manifold?

Motivated in part by this open question, S. Gong, J. Wu and G. Yu proved the following theorem [GWY].

Theorem 5.24 *The Novikov conjecture holds for groups acting properly and isometrically on an admissible Hilbert–Hadamard space.*

Roughly speaking, Hilbert–Hadamard spaces are (possibly infinite dimensional) simply connected spaces with non-positive curvature. We will give a precise definition a little later. We say that a Hilbert–Hadamard space M is *admissible* if it has a sequence of subspaces M_n, whose union is dense in M, such that each M_n, seen with its inherited metric from M, is isometric to a finite-dimensional Riemannian manifold. Examples of admissible Hilbert–Hadamard spaces include all simply connected and non-positively curved Riemannian manifold, the Hilbert space, and certain infinite dimensional symmetric spaces. Theorem 5.24 can be viewed as a generalization of both Theorems 5.4 and 5.8.

Infinite dimensional symmetric spaces are often naturally admissible Hilbert–Hadamard spaces. One such an example is

$$L^2(N, \omega, \mathrm{SL}(n, \mathbb{R})/\mathrm{SO}(n)),$$

where N is a compact smooth manifold with a given volume form ω. This infinite-dimensional symmetric space is defined to be the completion of the space of all smooth maps from N to $X = \mathrm{SL}(n, \mathbb{R})/\mathrm{SO}(n)$ with respect to the following distance:

$$d(\xi, \eta) = \left(\int_N (d_X(\xi(y), \eta(y)))^2 \, d\omega(y) \right)^{\frac{1}{2}},$$

where d_X is the standard Riemannian metric on the symmetric space X and ξ and η are two smooth maps from N to X. This space can be considered as the space of L^2-metrics on N with the given volume form ω and is a Hilbert–Hadamard space. With the help of this infinite dimensional symmetric space, the above theorem can be applied to study the Novikov conjecture for geometrically discrete subgroups of the group of volume preserving diffeomorphisms on such a manifold.

The key ingredients of the proof for Theorem 5.24 include a construction of a C^*-algebra modeled after the Hilbert–Hadamard space, a deformation technique for the isometry group of the Hilbert–Hadamard space and its corresponding actions on

K-theory, and a KK-theory with real coefficient developed by Antonini, Azzali, and Skandalis [AAS].

Let $\mathrm{Diff}(N, \omega)$ denote the group of volume preserving diffeomorphisms on a compact orientable smooth manifold N with a given volume form ω. In order to define the concept of geometrically discrete subgroups of $\mathrm{Diff}(N, \omega)$, let us fix a Riemannian metric on N with the given volume ω and define a length function λ on $\mathrm{Diff}(N, \omega)$ by

$$\lambda_+(\varphi) = \left(\int_N (\log(\|D\varphi\|))^2 d\omega\right)^{1/2}$$

and

$$\lambda(\varphi) = \max\left\{\lambda_+(\varphi), \lambda_+(\varphi^{-1})\right\}$$

for all $\varphi \in \mathrm{Diff}(N, \omega)$, where $D\varphi$ is the Jacobian of φ, and the norm $\|\cdot\|$ denotes the operator norm, computed using the chosen Riemannian metric on N.

Definition 5.25 A subgroup Γ of $\mathrm{Diff}(N, \omega)$ is said to be a geometrically discrete subgroup if $\lambda(\gamma) \to \infty$ when $\gamma \to \infty$ in Γ, i.e., for any $R > 0$, there exists a finite subset $F \subset \Gamma$ such that $\lambda(\gamma) \geq R$ if $\gamma \in \Gamma \setminus F$.

Observe that although the length function λ depends on our choice of the Riemannian metric, the above notion of geometric discreteness does not. Also notice that if γ preserves the Riemannian metric we chose, then $\lambda(\gamma) = 0$. This suggests that the class of geometrically discrete subgroups of $\mathrm{Diff}(N, \omega)$ does not intersect with the class of groups of isometries. Of course, we already know the Novikov conjecture for any group of isometries on a compact Riemannian manifold. This, together with the following result, gives an optimistic perspective on the open question on the Novikov conjecture for groups of volume preserving diffeomorphisms.

Theorem 5.26 *Let N be a compact smooth manifold with a given volume form ω, and let* $\mathrm{Diff}(N, \omega)$ *be the group of all volume preserving diffeomorphisms of N. The Novikov conjecture holds for any geometrically discrete subgroup of* $\mathrm{Diff}(N, \omega)$.

Now let us give a precise definition of Hilbert–Hadamard space. We will first recall the concept of CAT(0) spaces. Let X be a geodesic metric space. Let Δ be a triangle in X with geodesic segments as its sides. Δ is said to satisfy the CAT(0) inequality if there is a comparison triangle Δ' in Euclidean space, with sides of the same length as the sides of Δ, such that distances between points on Δ are less than or equal to the distances between corresponding points on Δ'. The geodesic metric space X is said to be a CAT(0) space if every geodesic triangle satisfies the CAT(0) inequality.

Let X be a geodesic metric space. For three distinct points $x, y, z \in X$, we define the comparison angle $\widetilde{\angle} xyz$ to be

$$\tilde{\angle}xyz = \arccos\left(\frac{d(x,y)^2 + d(y,z)^2 - d(x,z)^2}{2d(x,y)d(y,z)}\right).$$

In other words, $\tilde{\angle}xyz$ can be thought of as the angle at y of the comparison triangle Δxyz in the Euclidean plane.

Given two nontrivial geodesic paths α and β emanating from a point p in X, meaning that $\alpha(0) = \beta(0) = p$, we define the angle between them, $\angle(\alpha, \beta)$, to be

$$\angle(\alpha, \beta) = \lim_{s,t\to 0} \tilde{\angle}(\alpha(s), p, \beta(t)),$$

provided that the limit exists. For CAT(0) spaces, since the $\tilde{\angle}(\alpha(s), p, \beta(t))$ decreases with s and t, the angle between any two geodesic paths emanating from a point is well-defined. These angles satisfy the triangle inequality.

For a point $p \in X$, let Σ'_p denote the metric space induced from the space of all geodesics emanating from p equipped with the pseudometric of angles, that is, for geodesics α and β, we define $d(\alpha, \beta) = \angle(\alpha, \beta)$. Note, in particular, from our definition of angles, that $d(\alpha, \beta) \leq \pi$ for any geodesics α and β.

We define Σ_p to be the completion of Σ'_p with respect to the distance d. The *tangent cone* K_p at a point p in X is then defined to be a metric space which is, as a topological space, the cone of Σ_p. That is, topologically

$$K_p \simeq \Sigma_p \times [0, \infty)/\Sigma_p \times \{0\}.$$

The metric on it is given as follows. For two points $p, q \in K_p$ we can express them as $p = [(x, t)]$ and $q = [(y, s)]$. Then the metric is given by

$$d(p, q) = \sqrt{t^2 + s^2 - 2st\cos(d(x, y))}.$$

The distance is what the distance would be if we went along geodesics in a Euclidean plane with the same angle between them as the angle between the corresponding directions in X.

The following definition is inspired by [FS].

Definition 5.27 A *Hilbert–Hadamard space* is a complete geodesic CAT(0) metric space (i.e., an Hadamard space) all of whose tangent cones are isometrically embedded in Hilbert spaces.

Every connected and simply connected *Riemannian–Hilbertian manifold with non-positive sectional curvature* is a separable Hilbert–Hadamard space. In fact, a Riemannian manifold without boundary is a Hilbert–Hadamard space if and only if it is complete, connected, and simply connected, and has non-positive sectional curvature. We remark that a CAT(0) space X is always uniquely geodesic.

Recall that a subset of a geodesic metric space is called *convex* if it is again a geodesic metric space when equipped with the restricted metric. We observe that a closed convex subset of a Hilbert–Hadamard space is itself a Hilbert–Hadamard space.

Definition 5.28 A separable Hilbert–Hadamard space M is called *admissible* if there is a sequence of convex subsets isometric to finite-dimensional Riemannian manifolds, whose union is dense in M.

The notion of Hilbert–Hadamard spaces is more general than simply connected Riemannian–Hilbertian space with non-positive sectional curvature. For example, the infinite dimensional symmetric space $L^2(N, \omega, \mathrm{SL}(n, \mathbb{R})/\mathrm{SO}(n))$ is a Hilbert–Hadamard space but not a Riemannian–Hilbertian space with non-positive sectional curvature.

6 Secondary invariants for Dirac operators and applications

We have been mainly concerned with the primary invariants, i.e., the higher index invariants, till now. Starting this section, we shall shift our focus to secondary invariants. We try to keep the discussion relatively self-contained, which hopefully will give a better sense of some of the more recent development on secondary invariants.

In this section, we introduce a secondary invariant for Dirac operators on manifolds with positive scalar curvature and apply the invariant to measure the size of the moduli space of Riemmanian metrics with positive scalar curvature on a given spin manifold.

We carry out the construction in the odd-dimensional case. The even dimensional case is similar. Suppose that X is an odd-dimensional complete spin manifold without boundary and we fix a spin structure on X. Assume that there is a discrete group Γ acting on X properly and cocompactly by isometries. In addition, we assume the action of Γ preserves the spin structure on X. A typical such example comes from a Galois cover $\widetilde{M}$ of a closed spin manifold M with Γ being the group of deck transformations.

Let S be the spinor bundle over M and D be the associated Dirac operator on X. Let $H_X = L^2(X, S)$ and

$$F = D(D^2 + 1)^{-1/2}.$$

(H_X, F) defines a class in $K_1^{\Gamma}(X)$. Note that F lies in the multiplier algebra of $C^*(X)^{\Gamma}$, since F can be approximated by elements of finite propagation in the multiplier algebra of $C^*(X)^{\Gamma}$. As a result, we can directly work with[1]

[1] In other words, there is no need to pass to the operator $\mathcal{F}$ or $\mathcal{F}(t)$ as in the general case.

$$F(t) = \sum_j ((1-(t-n))\phi_{n,j}^{1/2} F \phi_{n,j}^{1/2} + (t-n)\phi_{n+1,j}^{1/2} F \phi_{n+1,j}^{1/2}) \tag{6.1}$$

for $t \in [n, n+1]$. The same index construction as before defines the index class and the local index class of (H_X, F). We shall denote them by $\mathrm{Ind}(D) \in K_1(C^*(X)^\Gamma)$ and $\mathrm{Ind}_L(D) \in K_1(C_L^*(X)^\Gamma)$ respectively.

Now suppose in addition X is endowed with a complete Riemannian metric g whose scalar curvature κ is positive everywhere, then the associated Dirac operator in fact naturally defines a class in $K_1(C_{L,0}^*(X)^\Gamma)$. Indeed, recall that

$$D^2 = \nabla^*\nabla + \frac{\kappa}{4},$$

where $\nabla : C^\infty(X, S) \to C^\infty(X, T^*X \otimes S)$ is the associated connection and ∇^* is the adjoint of ∇. If $\kappa > 0$, then it follows immediately that D is invertible. So, instead of $D(D^2+1)^{-1/2}$, we can use

$$F := D|D|^{-1}.$$

Note that $\frac{F+1}{2}$ is a genuine projection. Define $F(t)$ as in formula (6.1), and define $q(t) := \frac{F(t)+1}{2}$. We form the path of unitaries $u(t) = e^{2\pi i q(t)}, 0 \le t < \infty$, which defines an element in $(C_L^*(X)^\Gamma)^+$. Notice that $u(0) = 1$. So this path $u(t), 0 \le t < \infty$, in fact lies in $(C_{L,0}^*(X)^\Gamma)^+$, therefore defines a class in $K_1(C_{L,0}^*(X)^\Gamma)$.

Let us now define the higher rho invariant. It was first introduced by Higson and Roe [Roe1, HR3]. Our formulation is slightly different from that of Higson and Roe. The equivalence of the two definitions was shown in [XY].

Definition 6.1 The higher rho invariant $\rho(D, g)$ of the pair (D, g) is defined to be the K-theory class $[u(t)] \in K_1(C_{L,0}^*(X)^\Gamma)$.

The definition of higher rho invariant in the even dimensional case is similar, where one needs to work with the natural $\mathbb{Z}/2\mathbb{Z}$-grading on the spinor bundle.

Next we shall apply the higher rho invariant to estimate the size of the moduli space of Riemannian metrics with positive scalar curvature on a given spin manifold. Let M be a closed smooth manifold. Suppose that M carries a metric of positive scalar curvature. It is well known that the space of all Rimennian metrics on M is contractible, hence topologically trivial. To the contrary, the space of all positive scalar curvature metrics on M, denoted by $\mathcal{R}^+(M)$, often has very nontrivial topology. In particular, $\mathcal{R}^+(M)$ is often not connected and in fact has infinitely many connected components [BoG, LP1, PS, RS]. For example, by using the Cheeger–Gromov L^2-rho invariant and Lott's delocalized eta invariant, Piazza and Schick showed that $\mathcal{R}^+(M)$ has infinitely many connected components, if M is a closed spin manifold with $\dim M = 4k+3 \ge 5$ and $\pi_1(M)$ contains torsion [PS].

Following Stolz [St], Weinberger and Yu introduced an abelian group $P(M)$ to measure the size of the space of positive scalar curvature metrics on a manifold M

[WY]. In addition, they used the finite part of K-theory of the maximal group C^*-algebra $C^*_{\max}(\pi_1(M))$ to give a lower bound of the rank of $P(M)$. A special case of their theorem states that the rank of $P(M)$ is ≥ 1, if M is a closed spin manifold with $\dim M = 2k+1 \geq 5$ and $\pi_1(M)$ contains torsion. In particular, this implies the above theorem of Piazza and Schick.

For convenience of the reader, we recall the definition of the abelian group $P(M)$. Let M be an oriented smooth closed manifold with $\dim M \geq 5$ and its fundamental group $\pi_1(M) = \Gamma$. Assume that M carries a metric of positive scalar curvature. We denote it by g_M. Let I be the closed interval $[0, 1]$. Consider the connected sum $(M \times I)\sharp(M \times I)$, where the connected sum is performed away from the boundary of $M \times I$. Note that $\pi_1\big((M \times I)\sharp(M \times I)\big) = \Gamma * \Gamma$ the free product of two copies of Γ.

Definition 6.2 We define the generalized connected sum $(M \times I)\natural(M \times I)$ to be the manifold obtained from $(M \times I)\sharp(M \times I)$ by removing the kernel of the homomorphism $\Gamma * \Gamma \to \Gamma$ through surgeries away from the boundary.

Note that $(M \times I)\natural(M \times I)$ has four boundary components, two of them being M and the other two being $-M$, where $-M$ is the manifold M with its reversed orientation. Now suppose g_1 and g_2 are two positive scalar curvature metrics on M. We endow one boundary component M with g_M, and endow the two $-M$ components with g_1 and g_2. Then by the Gromov–Lawson and Schoen–Yau surgery theorem for positive scalar curvature metrics [GL, SY], there exists a positive scalar curvature metric on $(M \times I)\natural(M \times I)$ which is a product metric near all boundary components. In particular, the restriction of this metric on the other boundary component M has positive scalar curvature. We denote this metric on M by g.

Definition 6.3 Two positive scalar curvature metrics g and h on M are concordant if there exists a positive scalar curvature metric on $M \times I$ which is a product metric near the boundary and restricts to g and h on the two boundary components respectively.

One can in fact show that if g and g' are two positive scalar curvature metrics on M obtained from the same pair of positive scalar curvature metrics g_1 and g_2 by the above procedure, then g and g' are concordant [WY].

Definition 6.4 Fix a positive scalar curvature metric g_M on M. Let $P^+(M)$ be the set of all concordance classes of positive scalar metrics on M. Given $[g_1]$ and $[g_2]$ in $P^+(M)$, we define the sum of $[g_1]$ and $[g_2]$ (with respect to $[g_M]$) to be $[g]$ constructed from the procedure above. Then it is not difficult to verify that $P^+(M)$ becomes an abelian semigroup under this addition. We define the abelian group $P(M)$ to be the Grothendieck group of $P^+(M)$.

Recall that the group of diffeomorphisms on M, denoted by Diff(M), acts on $\mathcal{R}^+(M)$ by pulling back the metrics. The moduli space of positive scalar curvature metrics is defined to be the quotient space $\mathcal{R}^+(M)/\text{Diff}(M)$. Similarly, Diff($M$) acts on the group $P(M)$ and we denote the coinvariant of the action by $\widetilde{P}(M)$. That is, $\widetilde{P}(M) = P(M)/P_0(M)$, where $P_0(M)$ is the subgroup of $P(M)$ generated by elements of the form $[x] - \psi^*[x]$ for all $[x] \in P(M)$ and all $\psi \in \mathrm{Diff}(M)$. We call $\widetilde{P}(M)$ the moduli group of positive scalar curvature metrics on M. It measures the size of the moduli space of positive scalar curvature metrics on M. The following conjecture gives a lower bound for the rank of the abelian group $\widetilde{P}(M)$.

Conjecture 6.5 Let M be a closed spin manifold with $\pi_1(M) = \Gamma$ and $\dim M = 4k - 1 \geq 5$, which carries a positive scalar curvature metric. Then the rank of the abelian group $\widetilde{P}(M)$ is $\geq N_{\mathrm{fin}}(\Gamma)$, where $N_{\mathrm{fin}}(\Gamma)$ is the cardinality of the following collection of positive integers:

$$\big\{d \in \mathbb{N}_+ \mid \exists \gamma \in \Gamma \text{ such that } \mathrm{order}(\gamma) = d \text{ and } \gamma \neq e\big\}.$$

In [XY1], we apply the higher rho invariants of the Dirac operator to prove the following result.

Theorem 6.6 *Let M be a closed spin manifold which carries a positive scalar curvature metric with* $\dim M = 4k - 1 \geq 5$. *If the fundamental group $\Gamma = \pi_1(M)$ of M is strongly finitely embeddable into Hilbert space, then the rank of the abelian group $\widetilde{P}(M)$ is $\geq N_{\mathrm{fin}}(\Gamma)$.*

To prove this theorem, we need index theoretic invariants that are insensitive to the action of the diffeomorphism group. The index theoretic techniques used in [WY], for example, do not produce such invariants. The key idea of the proof is that the higher rho invariant remains unchanged in a certain K-theory group under the action of the diffeomorphism group, allowing us to distinguish elements in $\widetilde{P}(M)$.

We now recall the concept of strongly finite embeddability into Hilbert space for groups [XY1]. This concept is a stronger version of the notion of finite embeddability into Hilbert space introduced in [WY], a concept more flexible than the notion of coarse embeddability.

Definition 6.7 A countable discrete group Γ is said to be finitely embeddable into Hilbert space H if for any finite subset $F \subseteq \Gamma$, there exist a group Γ' that is coarsely embeddable into H and a map $\phi : F \to \Gamma'$ such that

(1) if γ, β and $\gamma\beta$ are all in F, then $\phi(\gamma\beta) = \phi(\gamma)\phi(\beta)$;
(2) if γ is a finite order element in F, then $\mathrm{order}(\phi(\gamma)) = \mathrm{order}(\gamma)$.

As mentioned above, Weinberger and Yu proved that Conjecture 6.5 holds for all groups that are finitely embeddable into Hilbert space [WY].

If $g \in \Gamma$ has finite order d, then we can define an idempotent in the group algebra $\mathbb{Q}\Gamma$ by

$$p_g = \frac{1}{d}\Big(\sum_{k=1}^{d} g^k\Big).$$

For the rest of this survey, we denote the maximal group C^*-algebra of Γ by $C^*(\Gamma)$.

Definition 6.8 We define $K_0^{\text{fin}}(C^*(\Gamma))$, called the finite part of $K_0(C^*(\Gamma))$, to be the abelian subgroup of $K_0(C^*(G))$ generated by $[p_g]$ for all elements $g \neq e$ in G with finite order.

We remark that rationally all representations of a finite group are induced from its finite cyclic subgroups [Serre]. This explains that the finite part of K-theory, despite being constructed using only cyclic subgroups, rationally contains all K-theory elements which can be constructed using finite subgroups.

Definition 6.9 Let $\mathcal{J}_0^{\text{fin}}(C^*(\Gamma))$ be the abelian subgroup of $K_0^{\text{fin}}(C^*(\Gamma))$ generated by elements $[p_\gamma] - [p_\beta]$ with $\text{order}(\gamma) = \text{order}(\beta)$. We define the reduced finite part of $K^0(C^*(\Gamma))$ to be

$$\widetilde{K}_0^{\text{fin}}(C^*(\Gamma)) = K_0^{\text{fin}}(C^*(\Gamma))/\mathcal{J}_0^{\text{fin}}(C^*(\Gamma)).$$

An argument in [WY] can be used to prove the following result, which plays a crucial role in the proof of Theorem 6.6.

Proposition 6.10 *Let $\{\gamma_1, \cdots, \gamma_n\}$ be a collection of nontrivial elements (i.e., $\gamma_i \neq e$) with distinct finite order in Γ. We define $\mathcal{M}_{\gamma_1,\cdots,\gamma_n}$ to be the abelian subgroup of $K_0^{fin}(C^*(\Gamma))$ generated by $\{[p_{\gamma_1}], \cdots, [p_{\gamma_n}]\}$. Let $\widetilde{\mathcal{M}}_{\gamma_1,\cdots,\gamma_n}$ be the image of $\mathcal{M}_{\gamma_1,\cdots,\gamma_n}$ in $\widetilde{K}_0^{\text{fin}}(C^*(\Gamma))$. If Γ is finitely embeddable into Hilbert space, then*

(1) the abelian group $\widetilde{\mathcal{M}}_{\gamma_1,\cdots,\gamma_n}$ has rank n,
(2) any nonzero element in $K_0^{\text{fin}}(C^*(\Gamma))$ is not in the image of the assembly map

$$\mu : K_0^\Gamma(E\Gamma) \to K_0(C^*(\Gamma)),$$

where $E\Gamma$ is the universal space for proper and free Γ-action.

So one is led to the following conjecture.

Conjecture 6.11 Let Γ be a countable discrete group. Suppose $\{\gamma_1, \cdots, \gamma_n\}$ is a collection of elements in Γ with distinct finite orders and $\gamma_i \neq e$ for all $1 \leq i \leq n$. Then

(1) the abelian group $\widetilde{\mathcal{M}}_{\gamma_1,\cdots,\gamma_n}$ has rank n,
(2) any nonzero element in $K_0^{\text{fin}}(C^*(\Gamma))$ is not in the image of the assembly map

$$\mu : K_0^\Gamma(E\Gamma) \to K_0(C^*(\Gamma)),$$

where $E\Gamma$ is the universal space for proper and free Γ-action.

We are now ready to introduce the notion of strongly finitely embeddability for groups. Since we are interested in the fundamental groups of manifolds, all groups are assumed to be finitely generated in the following discussion.

Let Γ be a countable discrete group. Then any set of n automorphisms of Γ, say, $\psi_1, \cdots, \psi_n \in \mathrm{Aut}(\Gamma)$, induces a natural action of F_n the free group of n generators on Γ. More precisely, if we denote the set of generators of F_n by $\{s_1, \cdots, s_n\}$, then we have a homomorphism $F_n \to \mathrm{Aut}(\Gamma)$ by $s_i \mapsto \psi_i$. This homomorphism induces an action of F_n on Γ. We denote by $\Gamma \rtimes_{\{\psi_1,\cdots,\psi_n\}} F_n$ the semi-direct product of Γ and F_n with respect to this action. If no confusion arises, we shall write $\Gamma \rtimes F_n$ instead of $\Gamma \rtimes_{\{\psi_1,\cdots,\psi_n\}} F_n$.

Definition 6.12 A countable discrete group Γ is said to be strongly finitely embeddable into Hilbert space H, if $\Gamma \rtimes_{\{\psi_1,\cdots,\psi_n\}} F_n$ is finitely embeddable into Hilbert space H for all $n \in \mathbb{N}$ and all $\psi_1, \cdots, \psi_n \in \mathrm{Aut}(\Gamma)$.

We remark that all coarsely embeddable groups are strongly finitely embeddable. Indeed, if a group Γ is coarsely embeddable into Hilbert space, then $\Gamma \rtimes_{\{\psi_1,\cdots,\psi_n\}} F_n$ is coarsely embeddable (hence finitely embeddable) into Hilbert space for all $n \in \mathbb{N}$ and all $\psi_1, \cdots, \psi_n \in \mathrm{Aut}(\Gamma)$.

If a group Γ has a torsion free normal subgroup Γ' such that Γ/Γ' is residually finite, then Γ is strongly finitely embeddable into Hilbert space. Indeed, recall that any finitely generated group has only finitely many distinct subgroups of a given index. Let Γ_m be the intersection of all subgroups of Γ with index at most m. Then Γ/Γ_m is a finite group. Moreover, for given $\psi_1, \cdots, \psi_n \in \mathrm{Aut}(\Gamma)$, the induced action of F_n on Γ descends to an action of F_n on Γ/Γ_m. In other words, we have a natural homomorphism

$$\phi_m : \Gamma \rtimes F_n \to (\Gamma/\Gamma_m) \rtimes G_m,$$

where G_m is the image of F_n under the homomorphism $F_n \to \mathrm{Aut}(\Gamma/\Gamma_m)$. It follows that, for any finite set $F \subseteq \Gamma$, there exists a sufficiently large m such that the map

$$\phi_m : F \subset \Gamma \rtimes F_n \to (\Gamma/\Gamma_m) \rtimes G_m$$

satisfies

(1) if γ, β and $\gamma\beta$ are all in F, then $\phi(\gamma\beta) = \phi(\gamma)\phi(\beta)$;
(2) if γ is a finite order element[2] in F, then $\mathrm{order}(\phi(\gamma)) = \mathrm{order}(\gamma)$.

Notice that $(\Gamma/\Gamma_m) \rtimes G_m$ is a finite group, which is obviously coarsely embeddable into Hilbert space. This shows that Γ is strongly finitely embeddable into Hilbert space.

To summarize, we see that the class of strongly finitely embeddable groups includes all residually finite groups, virtually torsion free groups (e.g., $Out(F_n)$), and groups that coarsely embed into Hilbert space, where the latter contains all amenable groups and Gromov's hyperbolic groups.

[2]Note that in this case, all finite order elements in $\Gamma \rtimes_{\{\psi_1,\cdots,\psi_n\}} F_n$ come from Γ.

The notion of sofic groups is a generalization of amenable groups and residually finite groups. It is an open question whether sofic groups are (strongly) finitely embeddable into Hilbert space. Narutaka Ozawa, Denise Osin and Thomas Delzant have independently constructed examples of groups which are not finitely embeddable into Hilbert space. An affirmative answer to the above question would imply that there exist non-sofic groups.

By definition, strongly finite embeddability implies finite embeddability. It is an open question whether the converse holds:

Open Question 6.13 If a group is finitely embeddable into Hilbert space, then does it follow that the group is also strongly finitely embeddable into Hilbert space?

In fact, it was shown in [WY] that Gromov's monster groups and any group of analytic diffeomorphisms of an analytic connected manifold fixing a given point are finitely embeddable into Hilbert space. It is still an open question whether these groups are strongly finitely embeddable into Hilbert space.

Now let us proceed to prove Theorem 6.6. One of main ingredients of the proof is the following proposition,[3] which, combined with a surgery technique [GL, SY] and the relative higher index theorem [B, XY2], allows us to construct genuinely "new" positive scalar curvature metrics from old ones. For a finite group F, an F-manifold Y is called F-connected if the quotient Y/F is connected. Let $\mathbb{Z}_d$ be the cyclic group of order d.

Proposition 6.14 *Given positive integers d and k, there exist $\mathbb{Z}_d$-connected closed spin $\mathbb{Z}_d$-manifolds $\{Y_1, \cdots, Y_n\}$ with $\dim Y_i = 2k$ such that*

(*a*) *the $\mathbb{Z}_d$-equivariant indices of the Dirac operators on $\{Y_1, \cdots, Y_n\}$ rationally generate $KO(\mathbb{Z}_d) \otimes \mathbb{Q}$,*
(*b*) *$\mathbb{Z}_d$ acts on Y_i freely except for finitely many fixed points.*

Let M be a closed spin manifold with a positive scalar curvature metric g_M and $\dim M \geq 5$ as before. For each nontrivial finite order element $\gamma \in \Gamma$, one can construct a new positive scalar curvature metric h_γ on M such that the relative higher index $Ind_\Gamma(g_M, h_\gamma) = [p_\gamma] \in K_0(C^*(\Gamma))$, where $p_\gamma = \frac{1}{d}\sum_{k=1}^{d} \gamma^k$ with $d = order(\gamma)$. The detailed construction will be given in the next paragraphs. Here let us recall the definition of this relative higher index $Ind_\Gamma(g_M, h_\gamma)$. We endow $M \times \mathbb{R}$ with the metric $g_t + (dt)^2$, where g_t is a smooth path of Riemannian metrics on M such that

$$g_t = \begin{cases} g_M & \text{for } t \leq 0, \\ h_\gamma & \text{for } t \geq 1, \\ \text{any smooth path of metrics from } g_M \text{ to } h_\gamma & \text{for } 0 \leq t \leq 1. \end{cases}$$

[3] Proposition 6.14 first appeared in [WY]. The original statement in [WY] seems to contain a minor error when d is even, the version we state in this survey and its proof can be found in [XYZ].

Then $M \times \mathbb{R}$ becomes a complete Riemannian manifold with positive scalar curvature away from a compact subset. Denote by $D_{M\times\mathbb{R}}$ the corresponding Dirac operator on $M \times \mathbb{R}$ with respect to this metric. Then the higher index of $D_{M\times\mathbb{R}}$ is well-defined and is denoted by $Ind_{\Gamma}(g_M, h_\gamma)$ (cf. the discussion at the beginning of Section 7 below).

Next we shall describe a construction of a new positive scalar curvature metric h_γ on M associated to a nontrivial finite order element $\gamma \in \Gamma$. Let $\widetilde{M}$ be the universal cover of M. For each finite order element g in G with order d. By Proposition 6.14, there exist $\mathbb{Z}/d\mathbb{Z}$-connected compact smooth spin $\mathbb{Z}/d\mathbb{Z}$-manifolds $\{N_1, \cdots, N_n\}$ such that the dimension of each N_i is $4k$ and the sum of the $\mathbb{Z}/d\mathbb{Z}$-equivariant indices of the Dirac operators on $\{N_1, \cdots, N_n\}$ is a nonzero multiple of the trivial representation of $\mathbb{Z}/d\mathbb{Z}$.

Let $N_{g,l} = G \times_{\mathbb{Z}/d\mathbb{Z}} N_l$, where $\mathbb{Z}/d\mathbb{Z}$ acts on N_l as in Proposition 6.14 and $\mathbb{Z}/d\mathbb{Z}$ acts on G by $[m]h = hg^m$ for all $h \in G$ and $[m] \in \mathbb{Z}/d\mathbb{Z}$. Observe that $N_{g,l}$ is a G-manifold.

Let $\{g_1, \cdots, g_r\}$ be a collection of finite order elements such that $\{[p_{g_1}], \cdots, [p_{g_r}]\}$ generates an abelian subgroup of $K_0(C^*(G))$ with rank r. Let $N_{g_i} = \bigsqcup_{l=1}^{j_i} N_{g_i,l}$ be the disjoint union of all G-manifolds described as above. Let I be the unit interval $[0, 1]$. We first form a generalized G-equivariant connected sum $(\widetilde{M} \times I)\natural N_{g_i}$ along a free G-orbit of each $N_{g_i,l}$ and away from the boundary of $\widetilde{M} \times I$ as follows. We first obtain a G^{*j_i}-equivariant connected sum $(\widetilde{M} \times I)\sharp N_{g_i}$ along a free G-orbit of each $N_{g_i,l}$ and away from the boundary of $\widetilde{M} \times I$, where G^{*j_i} is the free product of j_i copies of G. More precisely, we inductively form the G^{*j_i}-equivariant connected sum $(\cdots((\widetilde{M} \times I)\sharp N_{g_i,1})\cdots)\sharp N_{g_i,j_i}$, where the equivariant connected sum is inductively taken along a free orbit and away from the boundary. We denote this space by $(\widetilde{M} \times I)\sharp N_{g_i}$. We then perform surgeries on $(\widetilde{M} \times I)\sharp N_{g_i}$ to obtain a G-equivariant cobordism between two copies of G-manifold $\widetilde{M}$.

For any positive scalar curvature metric h on M, by [RW, Theorem 2.2], the above cobordism gives us another positive scalar curvature metric h_i on M. Now the relative higher index theorem [B, XY2] implies that the relative higher index of the Dirac operator $M \times \mathbb{R}$ associated to the positive scalar curvature metrics of h_i and g_M is $[p_{g_i}]$ in $K_0(C^*(G))$. As a consequence, we know that $\{[h_1], \cdots, [h_r]\}$ generates an abelian subgroup of $P(M)$ with rank r.

To summarize, one can construct distinct elements in $P(M)$ by surgery theory and the relative higher index theorem. Moreover, these elements are distinguished by their relative higher indices (with respect to g_M). However, to prove Theorem 6.6, that is, to show that these concordance classes of positive scalar curvature metrics remain distinct even after modulo the action of diffeomorphisms, we will need to use higher rho invariants (instead of relative higher indices) in an essential way.

Proof of Theorem 6.6 Consider the following short exact sequence:

$$0 \to C^*_{L,0}(\widetilde{M})^{\Gamma} \to C^*_L(\widetilde{M})^{\Gamma} \to C^*(\widetilde{M})^{\Gamma} \to 0$$

where $\widetilde{M}$ is the universal cover of M. It induces the following six-term long exact sequence:

$$\begin{array}{ccccc} K_0(C^*_{L,0}(\widetilde{M})^\Gamma) & \longrightarrow & K_0(C^*_L(\widetilde{M})^\Gamma) & \xrightarrow{\mu_M} & K_0(C^*(\widetilde{M})^\Gamma) \\ \uparrow & & & & \downarrow \partial \\ K_1(C^*(\widetilde{M})^\Gamma) & \longleftarrow & K_1(C^*_L(\widetilde{M})^\Gamma) & \longleftarrow & K_1(C^*_{L,0}(\widetilde{M})^\Gamma) \end{array}$$

Recall that we have $K_0(C^*_L(\widetilde{M})^\Gamma) \cong K_0^\Gamma(\widetilde{M})$ and $K_0(C^*(\widetilde{M})^\Gamma) \cong K_0(C^*(\Gamma))$.

Fix a positive scalar curvature metric g_M on M. For each finite order element $\gamma \in \Gamma$, we can construct a new positive scalar curvature metric h_γ on M such that the relative higher index $Ind_\Gamma(g_M, h_\gamma) = [p_\gamma] \in K_0(C^*(\Gamma))$ as described as above. Let us still denote by h_γ (resp. g_M) the metric on $\widetilde{M}$ lifted from the metric h_γ (resp. g_M) on M. Let $\rho(D, h_\gamma)$ and $\rho(D, g_M)$ be the higher rho invariants for the pairs (D, h_γ) and (D, g_M), where D is the Dirac operator on $\widetilde{M}$. Then we have

$$\partial([p_\gamma]) = \partial\big(Ind_\Gamma(g_M, h_\gamma)\big) = \rho(D, h_\gamma) - \rho(D, g_M), \tag{6.2}$$

(cf. [PS1, XY]).

One of the key points of the proof is to construct a certain group homomorphism on $\widetilde{P}(M)$ which can be used to distinguish elements in $\widetilde{P}(M)$. First, we define a map $\varrho : P(M) \to K_1(C^*_{L,0}(\widetilde{M})^\Gamma)$ by

$$\varrho(h) := \rho(D, h) - \rho(D, g_M)$$

for all $h \in P(M)$. It follows from the definition of $P(M)$ and [XY, Theorem 4.1] that the map ϱ is a well-defined group homomorphism. Now recall that a diffeomorphism $\psi \in \mathrm{Diff}(M)$ induces a homomorphism

$$\psi_* : K_1(C^*_{L,0}(\widetilde{M})^\Gamma) \to K_1(C^*_{L,0}(\widetilde{M})^\Gamma).$$

Let $\mathcal{I}_1(C^*_{L,0}(\widetilde{M})^\Gamma)$ be the subgroup of $K_1(C^*_{L,0}(\widetilde{M})^\Gamma)$ generated by elements of the form $[x] - \psi_*[x]$ for all $[x] \in K_1(C^*_{L,0}(\widetilde{M})^\Gamma)$ and all $\psi \in \mathrm{Diff}(M)$. We see that ϱ descends to a group homomorphism

$$\widetilde{\varrho} : \widetilde{P}(M) \to K_1(C^*_{L,0}(\widetilde{M})^\Gamma)/\mathcal{I}_1(C^*_{L,0}(\widetilde{M})^\Gamma).$$

To see this, it suffices to verify that

$$\varrho(h) - \varrho(\psi^*(h)) \in \mathcal{I}_1(C^*_{L,0}(\widetilde{M})^\Gamma)$$

for all $[h] \in P(M)$ and $\psi \in \mathrm{Diff}(M)$. Indeed, we have

$$\varrho(h) - \varrho(\psi^*(h)) = \rho(D, h) - \rho(D, g_M) - \big(\rho(D, \psi^*(h)) - \rho(D, g_M)\big)$$

$$= \rho(D,h) - \rho(D,\psi^*(h))$$
$$= \rho(D,h) - \psi_*(\rho(D,h)) \in \mathcal{I}_1(C^*_{L,0}(\widetilde{M})^\Gamma).$$

We remark that it is crucial to use the higher rho invariant, instead of the relative higher index, to construct such a group homomorphism. Let us explain the subtlety here. Note that there is in fact a well-defined group homomorphism Ind_{rel} : $P(M) \to K_0(C^*(\Gamma))$ by $Ind_{rel}(h) = Ind_\Gamma(D; g_M, h)$. The well-definedness of Ind_{rel} follows from the definition of $P(M)$ and the relative higher index theorem [B, XY2]. However, in general, it is *not* clear at all whether Ind_{rel} descends to a group homomorphism $\widetilde{P}(M) \to K_0(C^*(\Gamma))/\mathcal{I}_0(C^*(\Gamma))$, where $\mathcal{I}_0(C^*(\Gamma))$ is the subgroup of $K_0(C^*(\Gamma))$ generated by elements of the form $[x] - \psi_*[x]$ for all $[x] \in K_0(C^*(\Gamma))$ and all $\psi \in \mathrm{Diff}(M)$.

Now for a collection of elements $\{\gamma_1, \cdots, \gamma_n\}$ with distinct finite orders, we consider the associated collection of positive scalar curvature metrics $\{h_{\gamma_1}, \cdots, h_{\gamma_n}\}$ as before. To prove the theorem, it suffices to show that for any collection of elements $\{\gamma_1, \cdots, \gamma_n\}$ with distinct finite orders, the elements

$$\widetilde{\varrho}(h_{\gamma_1}), \cdots, \widetilde{\varrho}(h_{\gamma_n})$$

are linearly independent in $K_1(C^*_{L,0}(\widetilde{M})^\Gamma)/\mathcal{I}_1(C^*_{L,0}(\widetilde{M})^\Gamma)$.

Let us assume the contrary, that is, there exist $[x_1], \cdots, [x_m] \in K_1(C^*_{L,0}(\widetilde{M})^\Gamma)$ and $\psi_1, \cdots, \psi_m \in \mathrm{Diff}(M)$ such that

$$\sum_{i=1}^n c_i \varrho(h_{\gamma_i}) = \sum_{j=1}^m \big([x_j] - (\psi_j)_*[x_j]\big), \tag{6.3}$$

where $c_1, \cdots, c_n \in \mathbb{Z}$ with at least one $c_i \neq 0$.

We denote by W the wedge sum of m circles. The fundamental group $\pi_1(W)$ is the free group F_m of m generators $\{s_1, \cdots, s_m\}$. We denote the universal cover of W by $\widetilde{W}$. Clearly, $\widetilde{W}$ is the Cayley graph of F_m with respect to the generating set $\{s_1, \cdots, s_m, s_1^{-1}, \cdots, s_m^{-1}\}$. Notice that F_m acts on M through the diffeomorphisms $\psi_1, \cdots, \psi_m$. In other words, we have a homomorphism $F_m \to \mathrm{Diff}(M)$ by $s_i \mapsto \psi_i$. We define

$$X = M \times_{F_m} \widetilde{W}.$$

Notice that $\pi_1(X) = \Gamma \rtimes_{\{\psi_1,\cdots,\psi_m\}} F_m$. Let us write $\Gamma \rtimes F_m$ for $\Gamma \rtimes_{\{\psi_1,\cdots,\psi_m\}} F_m$, if no confusion arises.

Let $\widetilde{X}$ be the universal cover of X. We have the following short exact sequence:

$$0 \to C^*_{L,0}(\widetilde{X})^{\Gamma\rtimes F_m} \to C^*_L(\widetilde{X})^{\Gamma\rtimes F_m} \to C^*(\widetilde{X})^{\Gamma\rtimes F_m} \to 0.$$

Recall that $K_0(C^*_L(\widetilde{X})^{\Gamma\rtimes F_m}) \cong K_0^{\Gamma\rtimes F_m}(\widetilde{X})$ and $K_0(C^*(\widetilde{X})^{\Gamma\rtimes F_m}) \cong K_0(C^*(\Gamma \rtimes F_m))$. So we have the following six-term long exact sequence:

$$\begin{array}{ccccc}
K_0(C^*_{L,0}(X)^{\Gamma\rtimes F_m}) & \longrightarrow & K_0^{\Gamma\rtimes F_m}(\widetilde{X}) & & K_0(C^*(\Gamma\rtimes F_m)) \\
 & & & & \big\downarrow \partial \\
K_1(C^*(\Gamma\rtimes F_m)) & \longleftarrow & K_1^{\Gamma\rtimes F_m}(\widetilde{X}) & \longleftarrow & K_1(C^*_{L,0}(\widetilde{X})^{\Gamma\rtimes F_m})
\end{array} \tag{6.4}$$

Now recall the following Pimsner-Voiculescu exact sequence [PV]:

$$\begin{array}{ccccc}
\bigoplus_{j=1}^m K_0(C^*(\Gamma)) & \xrightarrow{\sum_{j=1}^m 1-(\psi_j)_*} & K_0(C^*(\Gamma)) & \xrightarrow{i_*} & K_0(C^*(\Gamma\rtimes F_m)) \\
\big\uparrow & & & & \big\downarrow \\
K_1(C^*(\Gamma\rtimes F_m)) & \longleftarrow & K_1(C^*(\Gamma)) & \xleftarrow{\sum_{j=1}^m 1-(\psi_j)_*} & \bigoplus_{j=1}^m K_1(C^*(\Gamma))
\end{array}$$

where $(\psi_j)_*$ is induced by ψ_j and i_* is induced by the inclusion map of Γ into $\Gamma \rtimes F_m$. Similarly, we also have the following two Pimsner-Voiculescu type exact sequences for K-homology and the K-theory groups of $C^*_{L,0}$-algebras in the diagram (6.4) above.

$$\begin{array}{ccccc}
\bigoplus_i^m K_0^\Gamma(\widetilde{M}) & \xrightarrow{\sum_{j=1}^m 1-(\psi_j)_*} & K_0^\Gamma(\widetilde{M}) & \xrightarrow{i_*} & K_0^{\Gamma\rtimes F_m}(\widetilde{X}) \\
\big\uparrow & & & & \big\downarrow \\
K_1^{\Gamma\rtimes F_m}(\widetilde{X}) & \longleftarrow & K_1^\Gamma(\widetilde{M}) & \xleftarrow{\sum_{j=1}^m 1-(\psi_j)_*} & \bigoplus_{i=1}^m K_1^\Gamma(\widetilde{M})
\end{array}$$

$$\begin{array}{ccccc}
\bigoplus_{i=1}^m K_0(C^*_{L,0}(\widetilde{M})^\Gamma) & \xrightarrow{\sum_{j=1}^m 1-(\psi_j)_*} & K_0(C^*_{L,0}(\widetilde{M})^\Gamma) & \xrightarrow{i_*} & K_0(C^*_{L,0}(X)^{\Gamma\rtimes F_m}) \\
\big\uparrow & & & & \big\downarrow \\
K_1(C^*_{L,0}(\widetilde{X})^{\Gamma\rtimes F_m}) & \longleftarrow & K_1(C^*_{L,0}(\widetilde{M})^\Gamma) & \xleftarrow{\sum_{j=1}^m 1-(\psi_j)_*} & \bigoplus_{i=1}^m K_1(C^*_{L,0}(\widetilde{M})^\Gamma)
\end{array}$$

where again $(\psi_j)_*$ and i_* are defined in the obvious way.

Combining these Pimsner-Voiculescu exact sequences together, we have the following commutative diagram:

$$\begin{array}{ccccccccc}
 & & \vdots & & \vdots & & \vdots & & \\
 & & \big\downarrow & & \big\downarrow & & \big\downarrow & & \\
\longrightarrow & & \bigoplus_{j=1}^m K_0^\Gamma(\widetilde{M}) & \xrightarrow{\sigma} & K_0^\Gamma(\widetilde{M}) & \xrightarrow{i_*} & K_0^{\Gamma\rtimes F_m}(\widetilde{X}) & \longrightarrow & \\
 & & \big\downarrow & & \big\downarrow & & \big\downarrow \mu & & \\
\longrightarrow & & \bigoplus_{j=1}^m K_0(C^*(\Gamma)) & \xrightarrow{\sigma} & K_0(C^*(\Gamma)) & \xrightarrow{*} & K_0(C^*(\Gamma\rtimes F_m)) & \longrightarrow & \\
 & & \big\downarrow & & \big\downarrow & & \big\downarrow \partial_{\Gamma\rtimes F_m} & & \\
\longrightarrow & & \bigoplus_{j=1}^m K_1(C^*_{L,0}(\widetilde{M})^\Gamma) & \xrightarrow{\sigma} & K_1(C^*_{L,0}(\widetilde{M})^\Gamma) & \xrightarrow{i_*} & K_1(C^*_{L,0}(\widetilde{X})^{\Gamma\rtimes F_m}) & \longrightarrow & \\
 & & \big\downarrow & & \big\downarrow & & \big\downarrow & & \\
 & & \vdots & & \vdots & & \vdots & &
\end{array} \tag{6.5}$$

where $\sigma = \sum_{j=1}^{m} 1 - (\psi_j)_*$. Notice that all rows and columns are exact.

Now on one hand, if we pass Equation (6.3) to $K_1(C^*_{L,0}(\widetilde{X})^{\Gamma \rtimes F_m})$ under the map i_*, then it follows immediately that

$$\sum_{k=1}^{n} c_k \cdot i_*[\varrho(h_{\gamma_k})] = 0 \text{ in } K_1(C^*_{L,0}(\widetilde{X})^{\Gamma \rtimes F_m}),$$

where at least one $c_k \neq 0$. On the other hand, by assumption, Γ is strongly finitely embeddable into Hilbert space. Hence $\Gamma \rtimes F_m$ is finitely embeddable into Hilbert space. By Proposition 6.10, we have the following:

1. $\{[p_{\gamma_1}], \cdots, [p_{\gamma_n}]\}$ generates a rank n abelian subgroup of $K_0^{\text{fin}}(C^*(\Gamma \rtimes F_m))$, since $\gamma_1, \cdots, \gamma_n$ have distinct finite orders. In other words,

$$\sum_{k=1}^{n} c_k[p_{\gamma_k}] \neq 0 \in K_0^{\text{fin}}(C^*(\Gamma \rtimes F_m))$$

   if at least one $c_k \neq 0$.
2. Every nonzero element in $K_0^{\text{fin}}(C^*(\Gamma \rtimes F_m))$ is not in the image of the assembly map

$$\mu : K_0^{\Gamma \rtimes F_m}(E(\Gamma \rtimes F_m)) \to K_0(C^*(\Gamma \rtimes F_m)),$$

   where $E(\Gamma \rtimes F_m)$ is the universal space for proper and free $\Gamma \rtimes F_m$-action. In particular, every nonzero element in $K_0^{\text{fin}}(C^*(\Gamma \rtimes F_m))$ is not in the image of the map

$$\mu : K_0^{\Gamma \rtimes F_m}(\widetilde{X}) \to K_0(C^*(\Gamma \rtimes F_m))$$

   in diagram (6.5). It follows that the map

$$\partial_{\Gamma \rtimes F_m} : K_0^{\text{fin}}(C^*(\Gamma \rtimes F_m)) \to K_1(C^*_{L,0}(\widetilde{X})^{\Gamma \rtimes F_m})$$

   is injective. In other words, $\partial_{\Gamma \rtimes F_m}$ maps a nonzero element in $K_0^{\text{fin}}(C^*(\Gamma \rtimes F_m))$ to a nonzero element in $K_1(C^*_{L,0}(\widetilde{X})^{\Gamma \rtimes F_m})$.

To summarize, we have

(a) $\sum_{k=1}^{n} c_k[p_{\gamma_k}] \neq 0$ in $K_0^{\text{fin}}(C^*(\Gamma \rtimes F_m))$,
(b) $\sum_{k=1}^{n} c_k \cdot i_*[\varrho(h_{\gamma_k})] = 0$ in $K_1(C^*_{L,0}(\widetilde{X})^{\Gamma \rtimes F_m})$,
(c) the map $\partial_{\Gamma \rtimes F_m} : K_0^{\text{fin}}(C^*(\Gamma \rtimes F_m)) \to K_1(C^*_{L,0}(\widetilde{X})^{\Gamma \rtimes F_m})$ is injective,
(d) and by Equation (6.2), $\partial_{\Gamma \rtimes F_m}\Big(\sum_{k=1}^{n} c_k[p_{\gamma_k}]\Big) = \sum_{k=1}^{n} c_k \cdot i_*[\varrho(h_{\gamma_k})]$.

Therefore, we arrive at a contradiction. This finishes the proof. □

7 Higher index, higher rho and positive scalar curvature at infinity

In this section, we will first describe a construction of the higher index for the Dirac operator on a complete manifold with positive scalar curvature at infinity. This construction is due to Gromov–Lawson in the classic Fredholm case [GL1] and its generalization to higher index case is due to Bunke [B] (see also [Roe1, BW, Roe3]). We will then discuss a connection between the higher index for the Dirac operator on a manifold with boundary and the higher rho invariant of the Dirac operator on the boundary.

Let M be a complete Riemannian spin manifold with a proper and isometric action of a discrete group Γ. We assume that M has positive scalar curvature at infinity relative to the action of Γ, i.e., there exists a Γ-cocompact subset Z of M and a positive number a such that the scalar curvature k of M is greater than or equal to a outside Z. Let D be the Dirac operator M.

We need some preparations in order to define the higher index. The following useful lemma is due to Roe [Roe3].

Lemma 7.1 *With the notation as above, suppose that $f \in S(\mathbb{R})$ has its Fourier transform $\hat{f}$ supported in $(-r, r)$. Let $\phi \in C_0(M)$ have support disjoint from a $4r$-neighborhood of Z. We have*

$$||f(D)\phi|| \leq 2||\phi|| \sup\{|f(\lambda)| : |\lambda| \geq a\}.$$

Proof Let us first deal with the case where f is an even function. In the case, the Fourier transform formula gives us

$$f(D) = \int_{-r}^{r} \hat{f}(t)\cos(tD)dt.$$

Let us define

$$U = \{x \in M : d(x, Z) > r\} \text{ and } U' = \{x \in M : d(x, Z) > 2r\}.$$

Consider the unbounded symmetric operator D^2 with domain $C_c^\infty(U)$. This operator is bounded below by $a^2 I$ and has a Friedrichs extension on the Hilbert space $L^2(U, S)$, where S is the spinor bundle. We denote this extension by E. Clearly, E is bounded below by the same lower bound $a^2 I$.

A standard finite propagation speed argument shows that if s is smooth and compactly supported on U', then

$$\cos(tD)s = \cos(t\sqrt{E})s$$

for $-r \leq t \leq r$. Since the spectrum of $\sqrt{E}$ is bounded below by a, we have

$$||f(\sqrt{E})|| \leq \sup\{|f(\lambda)| : |\lambda| \geq a\}.$$

This implies the following inequality:

$$||f(D)\phi|| \leq ||\phi|| \sup\{|f(\lambda)| : |\lambda| \geq a\}.$$

If f is an odd function, we have

$$||f(D)\phi||^2 \leq ||\bar{\phi}|| \quad || f^2(D)\phi ||.$$

In this case, the function f^2 is even, belongs to $S(\mathbb{R})$, and has Fourier transform supported in $(-2r, 2r)$. Hence we have the following inequality:

$$||f^2(D)\phi|| \leq ||\phi|| \sup\{|f(\lambda)|^2 : |\lambda| \geq a\}.$$

It follows that

$$||f(D)\phi|| \leq ||\phi|| \sup\{|f(\lambda)| : |\lambda| \geq a\}.$$

The general case follows from the above two special cases by writing f as a sum of even and odd functions. □

With the help of the above lemma, we can prove the following result.

Lemma 7.2 *For any $f \in C_c(-a, a)$, we have $f(D) \in \lim_{R\to\infty} C^*(N_R(Z))^\Gamma$, where $N_R(Z)$ is the R-neighborhood*[4] *of Z and $\lim_{R\to\infty} C^*(N_R(Z))^\Gamma$ is the C^*-algebra limit of the equivariant Roe algebras.*

Proof For any $\epsilon > 0$, there exists a smooth function g such that its Fourier transform is compactly supported, and

$$\sup\{|g(\lambda) - f(\lambda)| : \lambda \in \mathbb{R}\} < \epsilon.$$

It follows that $|g(\lambda)| < \epsilon$ for $|\lambda| > a$. Let r be a positive number such that $\text{Supp}(\hat{g}) \subseteq (-r, r)$ and let $\psi : M \to [0, 1]$ be a continuous Γ-invariant function equal to 1 on a $4r$-neighborhood of Z and vanishing outside a $5r$-neighborhood of Z. We write

$$f(D) = \psi g(D)\psi + (1 - \psi)g(D)\psi + g(D)(1 - \psi) + (f(D) - g(D)).$$

Note that the first term is a Γ-equivariant and locally compact operator with finite propagation supported near Z, the second and third terms have norm bounded by 2ϵ by Lemma 7.1. This implies the desired result. □

[4]Without loss of generality, we can assume $N_R(Z)$ is Γ-invariant.

We remark $\lim_{R\to\infty} C^*(N_R(Z))^\Gamma$ is isomorphic to the reduced group C^*-algebra $C_r^*(\Gamma)$.

A normalizing function $\chi : \mathbb{R} \to [-1, 1]$ is, by definition, a continuous odd function that goes to ± 1 as $x \to \infty$. Now choose a normalizing function χ such that $\chi^2 - 1$ is supported in $(-a, a)$ and let

$$F = \chi(D).$$

By Lemma 7.2, the same construction from Section 3 defines a higher index $Ind(D) \in K_*(C_r^*(\Gamma))$.

The following question is wide open.

Open Question 7.3 Let M be a complete spin manifold with a proper and isometric action of a discrete group Γ. Let D be the Dirac operator on M. Assume that M has positive scalar curvature at infinity relative to the action of Γ. Is $Ind(D)$ an element in the image of the Baum–Connes assembly map?

Let N be a spin manifold with boundary, where the boundary ∂N is endowed with a positive scalar curvature metric. We will explain that the K-theoretic "boundary" of the higher index class of the Dirac operator on N is equal to the higher rho invariant of the Dirac operator on ∂N. More generally, let M be an m-dimensional complete spin manifold with boundary ∂M such that

(i) the metric on M has product structure near ∂M and its restriction on ∂M, denoted by h, has positive scalar curvature;
(ii) there is a proper and cocompact isometric action of a discrete group Γ on M;
(iii) the action of Γ preserves the spin structure of M.

We denote the associated Dirac operator on M by D_M and the associated Dirac operator on ∂M by $D_{\partial M}$. With the positive scalar curvature metric h on the boundary ∂M, we can define the higher index class $Ind(D_M)$ of D_M in $K_*(C_r^*(\Gamma))$ as follows. We can attach a cylinder $\partial M \times [0, \infty)$ to the boundary of M to form a complete Riemannian manifold (without boundary) $\bar{M}$, where the Riemannian metric on M is naturally extended to $\bar{M}$ such that Riemannian metric on the cylinder is a product. The action of Γ on M naturally extends to an action on $\bar{M}$. By construction, $\bar{M}$ has positive scalar curvature at infinity relative to the action of M. We can therefore define the higher index $Ind(D_M)$ of D_M to be the higher index of the Dirac operator on $\bar{M}$.

Notice that the short exact sequence of C^*-algebras

$$0 \to C_{L,0}^*(M)^\Gamma \to C_L^*(M)^\Gamma \to C^*(M)^\Gamma \to 0$$

induces the following long exact sequence in K-theory:

$$\cdots \to K_i(C_L^*(M)^\Gamma) \to K_i(C^*(M)^\Gamma) \xrightarrow{\partial_i} K_{i-1}(C_{L,0}^*(M)^\Gamma) \to K_{i-1}(C_L^*(M)^\Gamma) \to \cdots.$$

Also, by functoriality, we have a natural homomorphism

$$\iota_* : K_{m-1}(C^*_{L,0}(\partial M)^\Gamma) \to K_{m-1}(C^*_{L,0}(M)^\Gamma)$$

induced by the inclusion map $\iota : \partial M \hookrightarrow M$. With the above notation, one has the following theorem.

Theorem 7.4 $\partial_m(Ind(D_M)) = \iota_*(\rho(D_{\partial M}, h))$ *in* $K_{m-1}(C^*_{L,0}(M)^\Gamma)$.

This theorem is due to Piazza and Schick [PS1] when the dimension of M is even and to Xie and Yu [XY] in the general case. As an immediate application, one sees that nonvanishing of the higher rho invariant is an obstruction to extension of the positive scalar curvature metric from the boundary to the whole manifold. Moreover, the higher rho invariant can be used to distinguish whether or not two positive scalar curvature metrics are connected by a path of positive scalar curvature metrics.

8 Secondary invariants of the signature operators and topological non-rigidity

In this section, we introduce the higher rho invariants for a pair of closed manifolds which are homotopic equivalent to each other. Roughly speaking, we consider the relative signature operator associated to this pair of manifolds. This relative signature operator has trivial higher index with a natural trivilialization given by the homotopy equivalence. This trivialization allows us to define a higher rho invariant, which can be used to detect whether a homotopy equivalence can be deformed into a homeomorphism.

We shall focus on the case of smooth manifolds. General topological manifolds can be handled in a similar way with the help of Lipschitz structures [Su].

Let M and N be two closed oriented smooth manifolds of dimension n. We will only discuss the odd-dimensional case; the even dimensional case is completely similar. We denote the de Rham complex of differential forms on M by

$$\Omega^0(M) \xrightarrow{d} \Omega^1(M) \xrightarrow{d} \cdots \xrightarrow{d} \Omega^n(M),$$

whose L^2-completion is

$$\Omega^0_{L^2}(M) \xrightarrow{d} \Omega^1_{L^2}(M) \xrightarrow{d} \cdots \xrightarrow{d} \Omega^n_{L^2}(M).$$

We shall write d_M if we need to specify d is the differential associated to the de Rham complex of M. Similarly, we have

$$\Omega^0_{L^2}(N) \xrightarrow{d_N} \Omega^1_{L^2}(N) \xrightarrow{d_N} \cdots \xrightarrow{d_n} \Omega^n_{L^2}(N)$$

for the manifold N.

Let $T = *_M : \Omega^k_{L^2}(M) \to \Omega^{n-k}_{L^2}(M)$ be the Hodge star operator on M, which is defined by

$$\langle T\alpha, \beta\rangle = \int_M \alpha \wedge \overline{\beta},$$

where $\overline{\beta}$ is the complex conjugation of β. The Hodge star operator T satisfies the following properties:

(1) $T^*\alpha = (-1)^{k(n-k)}T\alpha$ for any $\alpha \in \Omega^k_{L^2}(M)$;
(2) $Td\alpha + (-1)^k d^*T\alpha = 0$ for any smooth $\beta \in \Omega^k(M)$;
(3) $T^2\alpha = (-1)^{nk+k}\alpha$ for any $\alpha \in \Omega^k_{L^2}(M)$;

where T^* is the adjoint of T, and d^* is the adjoint of d. More generally, a bounded operator T satisfying conditions (1) and (2) is said to be a *duality operator* of the chain complex $(\Omega^*_{L^2}(M), d)$ if in addition, it satisfies the condition

(3)′ T induces a chain homotopy equivalence from the dual complex of $(\Omega^*_{L^2}(M), d)$ to the complex $(\Omega^*_{L^2}(M), d)$, where the dual complex is defined to be

$$\Omega^n_{L^2}(M) \xrightarrow{d^*} \Omega^{n-1}_{L^2}(M) \xrightarrow{d^*} \cdots \xrightarrow{d^*} \Omega^0_{L^2}(M).$$

In this case, we call $(\Omega^*_{L^2}(M), d)$ together with the duality operator T a (unbounded) Hilbert–Poincaré complex.

Define $S = i^{k(k-1)+m}T$, where $m = (n-1)/2$. It follows from properties (1) and (3) above that S is a self-adjoint involution.

Definition 8.1 The signature operator D of M is defined to be $i(d + d^*)S$ acting on even degree differential forms.

All the above discussion generalizes to the universal covering $\widetilde{M}$ of M. We denote the corresponding $\pi_1(M)$-equivariant signature operator of $\widetilde{M}$ by $\widetilde{D}$.

In the standard K-theoretic construction of the index of $\widetilde{D}$ (cf. Section 3), let us choose the normalizing function $\chi(t) = \frac{2}{\pi}\arctan(t)$. In this case, we have

$$Ind(\widetilde{D}) = e^{2\pi i \frac{\chi(\widetilde{D})+1}{2}} = (\widetilde{D} - i)(\widetilde{D} + i)^{-1}.$$

Let $B = d + d^*$. The above formula implies the following index formula:

$$Ind(\widetilde{D}) = (B - S)(B + S)^{-1} \in K_1(C^*_r(\pi_1(M))). \tag{8.1}$$

The above index formula in fact holds for general Hilbert–Poincaré complexes, that is, chain complexes with general duality operators. We shall not get into the technical details regarding the notion of Hilbert–Poincaré complexes, but instead refer the reader to [HR] for details. A key feature of the notion of Hilbert–Poincaré complexes is that it allows us to use a much larger class of duality operators besides the Hodge star operators. In the case of general Hilbert–Poincaré complexes, the well-definedness of the above index formula is justified by the following lemma [HR, Lemma 3.5].

Lemma 8.2 *$B + S$ and $B - S$ are invertible.*

Proof Consider the mapping cone complex associated to the chain map

$$S : (\Omega^*_{L^2}(M), -d^*) \to (\Omega^*_{L^2}(M), d)$$

with the differential

$$b = \begin{pmatrix} d & 0 \\ S & d^* \end{pmatrix}.$$

Since S is an isomorphism on the homology, the mapping cone complex is acyclic. Therefore the operator $b + b^*$ is invertible. Recall that S is self-adjoint. Hence we have

$$b + b^* = \begin{pmatrix} d + d^* & S \\ S & d + d^* \end{pmatrix}.$$

Note that

$$\begin{pmatrix} d + d^* & S \\ S & d + d^* \end{pmatrix} \begin{pmatrix} v \\ v \end{pmatrix} = \begin{pmatrix} (d + d^* + S)v \\ (d + d^* + S)v \end{pmatrix}.$$

This implies that $B + S$ is invertible. We can similarly show that $B - S$ is invertible. □

Suppose $f: M \to N$ is an orientation preserving homotopy equivalence between M and N. It is known that $Ind(\widetilde{D}_M) = Ind(\widetilde{D}_N)$ in $K_1(C_r^*(\Gamma))$, where $\Gamma = \pi_1(M) = \pi_1(N)$, cf. [K1, KM]. Intuitively speaking, one can use the homotopy equivalence f together with the signature operators on M and N to produce an invertible operator D_f on $M \cup (-N)$ such that the index of D_f coincides with the index of the signature operator on $M \cup (-N)$, which is $Ind(\widetilde{D}_M) - Ind(\widetilde{D}_N)$, cf. [HiS]. Here $-N$ is the manifold N with the reversed orientation and $M \cup (-N)$ stands for the disjoint union of the two. In particular, the invertibility of D_f is the reason that $Ind(\widetilde{D}_M) = Ind(\widetilde{D}_N)$, by giving a specific trivialization of the index class of $D_{M \cup (-N)}$. Thus the homotopy equivalence f naturally defines a higher rho invariant. In the following, we shall take a different but simpler approach to construct the higher rho invariant of f. Although this process does not produce an invertible operator D_f, but it does provide a trivialization at the K-theory level. Our choice of such an approach is mainly its simplicity, which will hopefully convey the key ideas with more clarity.

We denote the induced pullback map on differential forms by $f^*\colon \Omega^*(N) \to \Omega^*(M)$. In general, f^* does not extend to a bounded linear map between the spaces of L^2 forms $\Omega^*_{L^2}(N)$ and $\Omega^*_{L^2}(M)$. In order to fix this issue, we need the following construction due to Hilsum and Skandalis [HiS]. First, suppose $\varphi : X \to Y$ is a submersion between two closed manifolds. It is easy to see that φ^* does extend to a bounded linear operator from $\Omega^*_{L^2}(Y)$ to $\Omega^*_{L^2}(X)$. Now let $\iota\colon N \to \mathbb{R}^k$ be an embedding. Suppose U is a tubular neighborhood of N in $\mathbb{R}^k$ and $\pi : U \to N$ is the associated projection. Without loss of generality, we assume $\iota(N) + \mathbb{B}^k \subset U$, where $\mathbb{B}^k$ is the unit ball of $\mathbb{R}^k$. Let $p\colon M \times \mathbb{B}^k \to N$ be the submersion defined by $p(x,t) = \pi(f(x)+t)$. Furthermore, let ω be a volume form on $\mathbb{B}^k$ whose integral is 1. Then the formula

$$\alpha \to \int_{\mathbb{B}^k} p^*(\alpha) \wedge \omega$$

defines a morphism of chain complexes $A\colon \Omega^*(N) \to \Omega^*(M)$, where $\int_{\mathbb{B}^k}$ denotes fiberwise integration along $\mathbb{B}^k$. It is easy to see that A extends to a bounded linear operator from $\Omega^*_{L^2}(N)$ to $\Omega^*_{L^2}(M)$. We shall still denote this extension by $A\colon \Omega^*_{L^2}(N) \to \Omega^*_{L^2}(M)$.

Now a routine calculation shows that A is a homotopy equivalence between the two complexes $(\Omega_{L^2}(M), d_M)$ and $(\Omega_{L^2}(N), d_N)$ such that ATA^* is chain homotopy equivalent to T', where T' is the Hodge star operator on N. It follows that the operator

$$S = \begin{pmatrix} 0 & AT \\ TA^* & 0 \end{pmatrix}$$

together with the chain complex $(\Omega^*_{L^2}(M) \oplus \Omega^*_{L^2}(N), d_M \oplus d_N)$ gives rise to an (unbounded) Hilbert–Poincaré complex.

We have the following lemma due to Higson and Roe [HR].

Lemma 8.3 *If we write*

$$B = \begin{pmatrix} B_M & 0 \\ 0 & B_N \end{pmatrix} = \begin{pmatrix} d_M + d_M^* & \\ 0 & d_N + d_N^* \end{pmatrix},$$

then the element

$$(B - S)(B + S)^{-1}$$

is equal to $Ind(\widetilde{D}_M) - Ind(\widetilde{D}_N)$ *in* $K_1(C_r^*(\Gamma))$.

Proof Note that T' and ATA^* induce the same map on homology. It follows that the path

$$\begin{pmatrix} T & 0 \\ 0 & (s-1)T' - sATA^* \end{pmatrix}$$

is an operator homotopy connecting the duality operator $T \oplus -T'$ to the duality operator $T \oplus -ATA^*$. The path

$$\begin{pmatrix} \cos(s)T & \sin(s)TA^* \\ \sin(s)AT & -\cos(s)ATA^* \end{pmatrix}$$

is an operator homotopy connecting the duality operator $T \oplus -ATA^*$ to the duality operator $\begin{pmatrix} 0 & AT \\ TA^* & 0 \end{pmatrix}$, where $s \in [0, \pi/2]$. Now the lemma follows from the explicit index formula in line (8.1). □

For each $t \in [0, \pi]$, the following operator

$$S_t = \begin{pmatrix} 0 & e^{it}AT \\ e^{-it}TA^* & 0 \end{pmatrix}$$

defines a duality operator for the chain complex $(\Omega^*_{L^2}(M) \oplus \Omega^*_{L^2}(N), d_M \oplus d_N)$. It is not difficult to verify that

$$(B - S_0)(B + S_t)^{-1}$$

defines a continuous path of invertible elements in $(C_r^*(\Gamma) \otimes \mathcal{K})^+$. Note that $S_\pi = -S_0$, thus $(B - S_0)(B + S_\pi)^{-1} = 1$. Therefore, the path $(B - S_0)(B + S_t)^{-1}$ gives a specific trivialization of the index class $Ind(\widetilde{D}_M) - Ind(\widetilde{D}_N)$. This trivialization in turn induces a higher rho invariant as follows. Let $\{v(t)\}_{1 \le t \le 2}$ be the path of invertible elements connecting $(B - S_0)(B + S_0)^{-1}$ to

$$\left(B - \left(\begin{smallmatrix} T & 0 \\ 0 & T' \end{smallmatrix}\right)\right)\left(B + \left(\begin{smallmatrix} T & 0 \\ 0 & T' \end{smallmatrix}\right)\right)^{-1}.$$

We define

$$\rho(t) = \begin{cases} (B - S_0)(B + S_{(1-t)\pi})^{-1}, & \text{for } 0 \le t \le 1, \\ v(t) & \text{for } 1 \le t \le 2, \\ \begin{pmatrix} e^{\pi i(\chi(\frac{1}{t}\widetilde{D}_M)+1)} & 0 \\ 0 & e^{\pi i(\chi(\frac{1}{t}\widetilde{D}_{-N})+1)} \end{pmatrix} & \text{for } 2 \le t < \infty. \end{cases}$$

Definition 8.4 We define the higher rho invariant of a given homotopy equivalence $f: M \to N$ to be the above element $[\rho]$ in $K_1(C^*_{L,0}(\widetilde{N})^\Gamma)$. Here we have used f to map elements in $C^*(\widetilde{M})^\Gamma$ to $C^*(\widetilde{N})^\Gamma$.

The fact that ρ is an element in the matrix algebra of $(C^*_{L,0}(\widetilde{N})^\Gamma)^+$ follows from a standard finite propagation speed argument. The even dimensional higher rho invariant can be defined in a similar way. Zenobi generalized the concept of higher rho invariant to homotopy equivalences between closed topological manifolds with the help of Lipschitz structures [Z].

Given a closed oriented manifold N, the higher rho invariant in fact defines a map from the structure set of N to $K_n(C^*_{L,0}(\widetilde{N})^\Gamma)$, where $n = \dim N$. On the other hand, when N is a *topological* manifold, the structure set of N carries a natural abelian group structure. It was long standing open problem whether the higher rho inviariant map is a group homomorphism from the structure group of N to $K_1(C^*_{L,0}(\widetilde{N})^\Gamma)$. This was answered in positive in complete generality by Weinberger–Xie–Yu [WXY]. In the following we shall briefly discuss some of the key ideas of their proof and also some applications to topology.

Let X be a closed oriented connected *topological* manifold of dimension n. The structure group $\mathcal{S}(X)$ is the abelian group of equivalence classes of all pairs (f, M) such that M is a closed oriented manifold and $f: M \to X$ is an orientation preserving homotopy equivalence. Recall that the abelian group structure on $\mathcal{S}(X)$ is originally described through the Siebenmann periodicity map, which is an injection from $\mathcal{S}(X)$ to $\mathcal{S}_\partial(X \times D^4)$, where D^4 is the 4-dimensional Euclidean unit ball and $\mathcal{S}_\partial(X \times D^4)$ is the rel ∂ version of structure set of $X \times D^4$. The set $\mathcal{S}_\partial(X \times D^4)$ carries a natural abelian group structure by stacking, and induces an abelian group structure on $\mathcal{S}(X)$ by Nicas' correction map to the Siebenmann periodicity map [Ni]. Both $\mathcal{S}(X)$ and $\mathcal{S}_\partial(X \times D^4)$ carry a higher rho invariant map. It is not difficult to verify that the higher rho invariant map on $\mathcal{S}_\partial(X \times D^4)$ is additive, i.e., a homomorphism between abelian groups. One possible approach to show the additivity of the higher rho invariant map on $\mathcal{S}(X)$ is to prove the compatibility of higher rho invariant maps on $\mathcal{S}(X)$ and $\mathcal{S}_\partial(X \times D^4)$. However, there are some essential analytical difficulties to *directly* prove such a compatibility, due to the subtleties of the Siebenmann periodicity map.[5] A main novelty of Weinberger–Xie–Yu's approach [WXY] is to give a new description of the topological structure group in terms of smooth manifolds with boundary. This new description uses more objects and an equivalence relation broader than h-cobordism, which allows us to replace topological manifolds in the usual definition of $\mathcal{S}(X)$ by smooth manifolds with boundary. Such a description leads to a transparent group structure, which is given by disjoint union. The main body of Weinberger–Xie–Yu's work [WXY] is devoted to proving that the new description coincides with the classical description of the topological structure group; and to developing the theory of higher rho invariants in

[5] A geometric construction of the Siebenmann periodicity map was given by Cappell and Weinberger [CW].

this new setting, in which higher rho invariants are easily seen to be additive. As a consequence, the higher rho invariant maps on $\mathcal{S}(X)$ and $\mathcal{S}_\partial(X \times D^4)$ are indeed compatible.

Theorem 8.5 ([WXY, Theorem 4.40]) *The higher rho invariant map is a group homomorphism from $\mathcal{S}(X)$ to $K_n(C^*_{L,0}(\widetilde{X})^\Gamma)$.*

As mentioned above, the above theorem solves the long standing open problem whether the higher rho inviariant map defines a group homomorphism on the topological structure group. As an application, Weinberger–Xie–Yu applied the above theorem to prove that the structure groups of certain manifolds are infinitely generated [WXY].

Theorem 8.6 *Let M be a closed oriented topological manifold of dimension $n \geq 5$, and Γ be its fundamental group. Suppose the rational strong Novikov conjecture holds for Γ. If $\oplus_{k\in\mathbb{Z}} H_{n+1+4k}(\Gamma, \mathbb{C})$ is infinitely generated, then the topological structure group of $S(M)$ is infinitely generated.*

We refer to the article [WXY] for examples of groups satisfying the conditions in the above theorem.

9 Non-rigidity of topological manifolds and reduced structure groups

The structure group measures the degree of non-rigidity and the reduced structure group estimates the size of non-rigidity modulo self-homotopy equivalences. In this section, we apply the higher rho invariants of signature operators to give a lower bound of the free rank of reduced structure groups of closed oriented topological manifolds. Our key tool is the additivity property of higher rho invariants from the previous section. There are in fact two different versions of reduced structure groups, $\widetilde{\mathcal{S}}_{alg}(X)$ and $\widetilde{\mathcal{S}}_{geom}(X)$, whose precise definitions will be given below. The group $\widetilde{\mathcal{S}}_{alg}(X)$ is functorial and fits well with the surgery long exact sequence. On the other hand, the group $\widetilde{\mathcal{S}}_{geom}(X)$ is more geometric in the sense that it measures the size of the collection of closed manifolds homotopic equivalent but not homeomorphic to X.

Since we will be using the maximal version of various C^*-algebras throughout this section, we will omit the subscript "max" for notational simplicity.

Let X be an n-dimensional oriented closed topological manifold. Denote the monoid of orientation preserving self-homotopy equivalences of X by $Aut_h(X)$. There are two different actions of $Aut_h(X)$ on $\mathcal{S}(X)$, which induce two different versions of reduced structure groups as follows.

On one hand, $Aut_h(X)$ acts naturally on $\mathcal{S}(X)$ by

$$\alpha_u(\theta) = u_*(\theta)$$

for all $u \in Aut_h(X)$ and all $\theta \in \mathcal{S}(X)$, where u_* is the group homomorphism from $\mathcal{S}(X)$ to $\mathcal{S}(X)$ induced by the map u [KiS]. This action α is compatible with the actions of $Aut_h(X)$ on other terms in the topological surgery exact sequence.

On the other hand, $Aut_h(X)$ also naturally acts on $\mathcal{S}(X)$ by compositions of homotopy equivalences, that is,

$$\beta_u(\theta) = (u \circ f, M)$$

for all $u \in Aut_h(X)$ and all $\theta = (f, M) \in \mathcal{S}(X)$. Note that

$$\beta_u : \mathcal{S}(X) \to \mathcal{S}(X)$$

only defines a bijection of sets, and is not a group homomorphism in general.

Definition 9.1 With the same notation as above, we define the following reduced structure groups.

(1) Define $\widetilde{\mathcal{S}}_{alg}(X)$ to be the quotient group of $\mathcal{S}(X)$ by the subgroup generated by elements of the form $\theta - \alpha_u(\theta)$ for all $\theta \in \mathcal{S}(X)$ and all $u \in Aut_h(X)$.
(2) we define $\widetilde{\mathcal{S}}_{geom}(X)$ to be the quotient group of $\mathcal{S}(X)$ by the subgroup generated by elements of the form $\theta - \beta_u(\theta)$ for all $\theta \in \mathcal{S}(X)$ and all $u \in Aut_h(X)$.

Next we recall a method of constructing elements in the structure group by the finite part of K-theory [WY, Theorem 3.4].

Let M be a $(4k-1)$-dimensional closed oriented connected topological manifold with $\pi_1 M = \Gamma$. Suppose $\{g_1, \cdots, g_m\}$ is a collection of elements in Γ with distinct finite orders such that $g_i \neq e$ for all $1 \leq i \leq m$. Recall the topological surgery exact sequence:

$$\cdots \to H_{4k}(M, \mathbb{L}_\bullet) \to L_{4k}(\Gamma) \xrightarrow{\mathscr{S}} \mathcal{S}(M) \to H_{4k-1}(M, \mathbb{L}_\bullet) \to \cdots .$$

For each finite subgroup H of Γ, we have the following commutative diagram:

$$\begin{array}{ccc} H^H_{4k}(\underline{E}H, \mathbb{L}_\bullet) & \xrightarrow{A} & L_{4k}(H) \\ \uparrow & & \downarrow \\ H^G_{4k}(\underline{E}\Gamma, \mathbb{L}_\bullet) & \xleftarrow{A} & L_{4k}(\Gamma), \end{array}$$

where the vertical maps are induced by the inclusion homomorphism from H to Γ. For each element g in H with finite order d, the idempotent $p_g = \frac{1}{d}(\sum_{k=1}^{d} g^k)$ produces a class in $L_0(\mathbb{Q}H)$, where $L_0(\mathbb{Q}H)$ is the algebraic definition of L-groups using quadratic forms and formations with coefficients in $\mathbb{Q}$. Let $[q_g]$ be the corresponding element in $L_{4k}(\mathbb{Q}H)$ given by periodicity. Recall that

$$L_{4k}(H) \otimes \mathbb{Q} \simeq L_{4k}(\mathbb{Q}H) \otimes \mathbb{Q}.$$

For each element g in H with finite order, we use the same notation $[q_g]$ to denote the element in $L_{4k}(H) \otimes \mathbb{Q}$ corresponding to $[q_g] \in L_{4k}(\mathbb{Q}H)$ under the above isomorphism.

We also have the following commutative diagram:

$$\begin{array}{ccc} H^{\Gamma}_{4k}(E\Gamma, \mathbb{L}_\bullet) \otimes \mathbb{Q} & \xrightarrow{A} & L_{4k}(\Gamma) \otimes \mathbb{Q} \\ \uparrow & & \downarrow \\ K^{\Gamma}_0(E\Gamma) \otimes \mathbb{Q} & \xleftarrow{\mu_*} & K_0(C^*(\Gamma)) \otimes \mathbb{Q}, \end{array}$$

where the left vertical map is induced by a map at the spectra level and the right vertical map is induced by the inclusion map:

$$L_{4k}(\Gamma) \to L_{4k}(C^*(\Gamma)) \cong K_0(C^*(\Gamma))$$

(see [R2] for the last identification).

Now if Γ is finitely embeddable into Hilbert, then the abelian subgroup of $K_0(C^*(\Gamma))$ generated by $\{[p_{g_1}], \cdots, [p_{g_m}]\}$ is not in the image of of the map

$$\mu_* : K^{\Gamma}_0(E\Gamma) \to K_0(C^*(\Gamma)).$$

It follows that

1. any nonzero element in the abelian subgroup of $L_{4k}(\Gamma) \otimes \mathbb{Q}$ generated by the elements $\{[q_{g_1}], \cdots, [q_{g_m}]\}$ is not in the image of the rational assembly map
$$A : H^{\Gamma}_{4k}(E\Gamma, \mathbb{L}_\bullet) \otimes \mathbb{Q} \to L_{4k}(\Gamma) \otimes \mathbb{Q};$$
2. the abelian subgroup of $L_{4k}(\Gamma) \otimes \mathbb{Q}$ generated by $\{[q_{g_1}], \cdots, [q_{g_m}]\}$ has rank m.

By exactness of the surgery sequence, we know that the map

$$\mathscr{S} : L_{4k}(\Gamma) \otimes \mathbb{Q} \to \mathcal{S}(M) \otimes \mathbb{Q}, \tag{9.1}$$

is injective on the abelian subgroup of $L_{4k}(\Gamma) \otimes \mathbb{Q}$ generated by $\{[q_{g_1}], \cdots, [q_{g_n}]\}$.

In order to prove the main result of this section, we need to apply the above argument not only to Γ, but also to certain semi-direct products of Γ with free groups of finitely many generators.

Recall that $N_{\text{fin}}(\Gamma)$ is the cardinality of the following collection of positive integers:

$$\{d \in \mathbb{N}_+ \mid \exists \gamma \in \Gamma \text{ such that } \gamma \neq e \text{ and } order(\gamma) = d\}.$$

We have the following result [WXY]. At the moment, we are only able to prove the theorem for $\widetilde{\mathcal{S}}_{alg}(M)$. We will give a brief discussion to indicate the difficulties in proving the version $\widetilde{\mathcal{S}}_{geom}(M)$ after the theorem.

Theorem 9.2 *Let M be a closed oriented topological manifold with dimension $n = 4k-1$ ($k > 1$) and $\pi_1 M = \Gamma$. If Γ is strongly finitely embeddable into Hilbert space (cf. Definition 6.12), then the free rank of $\widetilde{\mathcal{S}}_{alg}(M)$ is $\geq N_{fin}(\Gamma)$.*

Proof A key point of the argument below is to use a semi-direct product $\Gamma \rtimes F_m$ to turn certain outer automorphisms of Γ into inner automorphisms of $\Gamma \rtimes F_m$.

Consider the higher rho invariant homomorphism from Theorem 8.5:

$$\rho : \mathcal{S}(M) \to K_1(C^*_{L,0}(\widetilde{M})^\Gamma).$$

Note that every self-homotopy equivalence $\psi \in Aut_h(M)$ induces a homomorphism[6]

$$\widetilde{\psi}_* : K_1(C^*_{L,0}(\widetilde{M})^\Gamma) \to K_1(C^*_{L,0}(\widetilde{M})^\Gamma).$$

Let $\mathcal{I}_1(C^*_{L,0}(\widetilde{M})^\Gamma)$ be the subgroup of $K_1(C^*_{L,0}(\widetilde{M})^\Gamma)$ generated by elements of the form $[x] - \widetilde{\psi}_*[x]$ for all $[x] \in K_1(C^*_{L,0}(\widetilde{M})^\Gamma)$ and all $\psi \in Aut_h(M)$. Note that, by the definition of the higher rho invariant, we have

$$\rho(\alpha_\psi(\theta)) = \widetilde{\psi}_*(\rho(\theta)) \in K_1(C^*_{L,0}(\widetilde{M})^\Gamma)$$

for all $\theta \in \mathcal{S}(M)$ and $\psi \in Aut_h(M)$. It follows that ρ descends to a group homomorphism $\widetilde{\mathcal{S}}_{alg}(M) \to K_1(C^*_{L,0}(\widetilde{M})^\Gamma)/\mathcal{I}_1(C^*_{L,0}(\widetilde{M})^\Gamma)$.

Now for a collection of elements $\{\gamma_1, \cdots, \gamma_\ell\}$ with distinct finite orders, we consider the elements $\mathscr{S}(p_{\gamma_1}), \cdots, \mathscr{S}(p_{\gamma_\ell}) \in \mathcal{S}(M)$ as in line (9.1). To be precise, the elements $\mathscr{S}(p_{\gamma_1}), \cdots, \mathscr{S}(p_{\gamma_\ell})$ actually lie in $\mathcal{S}(M) \otimes \mathbb{Q}$. Consequently, all abelian groups in the following need to be tensored by the rationals $\mathbb{Q}$. For simplicity, we shall omit $\otimes\mathbb{Q}$ from our notation, with the understanding that the abelian groups below are to be viewed as tensored with $\mathbb{Q}$. Also, let us write

$$\rho(\gamma_i) := \rho(\mathscr{S}(p_{\gamma_i})) \in K_1(C^*_{L,0}(\widetilde{M})^\Gamma).$$

[6]Let us review how the homomorphism $\psi_* : K_1(C^*_{L,0}(\widetilde{M})^\Gamma) \to K_1(C^*_{L,0}(\widetilde{M})^\Gamma)$ is defined. The map $\psi : M \to M$ lifts to a map $\widetilde{\psi} : \widetilde{M} \to \widetilde{M}$. However, to view $\widetilde{\psi}$ as a Γ-equivariant map, we need to use two different actions of Γ on $\widetilde{M}$. Let τ be a right action of Γ on $\widetilde{M}$ through deck transformations. Then we define a new action τ' of Γ on $\widetilde{M}$ by $\tau'_g = \tau_{\psi_*(g)}$, where $\psi_* : \Gamma \to \Gamma$ is the automorphism induced by ψ. It is easy to see that $\widetilde{\psi} : \widetilde{M} \to \widetilde{M}$ is Γ-equivariant, when Γ acts on the first copy of $\widetilde{M}$ by τ and the second copy of $\widetilde{M}$ by τ'. Let us denote the corresponding C^*-algebras by $C^*_{L,0}(\widetilde{M})^\Gamma_\tau$ and $C^*_{L,0}(\widetilde{M})^\Gamma_{\tau'}$. Observe that, despite the two different actions of Γ on $\widetilde{M}$, the two C^*-algebras $C^*_{L,0}(\widetilde{M})^\Gamma_\tau$ and $C^*_{L,0}(\widetilde{M})^\Gamma_{\tau'}$ are canonically identical, since an operator is invariant under the action τ if and only if it is invariant under the action τ'.

To prove the theorem, it suffices to show that for any collection of elements $\{\gamma_1, \cdots, \gamma_\ell\}$ with distinct finite orders, the elements

$$\rho(\gamma_1), \cdots, \rho(\gamma_\ell)$$

are linearly independent in $K_1(C^*_{L,0}(\widetilde{M})^\Gamma)/\mathcal{I}_1(C^*_{L,0}(\widetilde{M})^\Gamma)$.

Let us assume the contrary, that is, there exist $[x_1], \cdots, , [x_m] \in K_1(C^*_{L,0}(\widetilde{M})^\Gamma)$ and $\psi_1, \cdots, \psi_m \in Aut_h(M)$ such that

$$\sum_{i=1}^{\ell} c_i \rho(\gamma_i) = \sum_{j=1}^{m} \big([x_j] - (\widetilde{\psi}_j)_*[x_j]\big), \tag{9.2}$$

where $c_1, \cdots, c_\ell \in \mathbb{Z}$ with at least one $c_i \neq 0$. In fact, we shall study Equation (9.2) in the group $K_1(C^*_{L,0}(E(\Gamma \rtimes F_m))^{\Gamma \rtimes F_m})$ and arrive at a contradiction, where $\Gamma \rtimes F_m$ is a certain semi-direct product of Γ with the free group of m generators F_m and $E(\Gamma \rtimes F_m)$ is the universal space for free and proper $\Gamma \rtimes F_m$-actions.

Let us fix a map $\sigma : M \to B\Gamma$ that induces an isomorphism of their fundamental groups, where $B\Gamma$ is the classifying space of Γ. Suppose $\varphi\colon M \to M$ is an orientation preserving self-homotopy equivalence of M. Then φ induces an automorphism[7] of Γ, also denoted by $\varphi \in \mathrm{Aut}(\Gamma)$. Now consider the semi-direct product $\Gamma \rtimes_\varphi \mathbb{Z}$ and its associated classifying space $B(\Gamma \rtimes_\varphi \mathbb{Z})$. Let $\hat{\varphi}$ be the element in $\Gamma \rtimes_\varphi \mathbb{Z}$ that corresponds to the generator $1 \in \mathbb{Z}$. We write

$$\Phi\colon B(\Gamma \rtimes_\varphi \mathbb{Z}) \to B(\Gamma \rtimes_\varphi \mathbb{Z})$$

for the map induced by the automorphism $\Gamma \rtimes_\varphi \mathbb{Z} \to \Gamma \rtimes_\varphi \mathbb{Z}$ defined by $a \to \hat{\varphi} a \hat{\varphi}^{-1}$. Suppose $\iota\colon B\Gamma \to B(\Gamma \rtimes_\varphi \mathbb{Z})$ is the map induced by the inclusion $\Gamma \hookrightarrow \Gamma \rtimes_\varphi \mathbb{Z}$. Then the map

$$\iota \circ \sigma \circ \varphi\colon M \xrightarrow{\varphi} M \xrightarrow{\sigma} B\Gamma \xrightarrow{\iota} B(\Gamma \rtimes_\varphi \mathbb{Z})$$

is homotopy equivalent to the map

$$\Phi \circ \iota \circ \sigma\colon M \xrightarrow{\sigma} B\Gamma \xrightarrow{\iota} B(\Gamma \rtimes_\varphi \mathbb{Z}) \xrightarrow{\Phi} B(\Gamma \rtimes_\varphi \mathbb{Z}),$$

since they induce the same map on fundamental groups. Let $\widetilde{\sigma} : \widetilde{M} \to E\Gamma$ be the lift of the map $\sigma : M \to B\Gamma$. Similarly, $\widetilde{\varphi}\colon \widetilde{M} \to \widetilde{M}$ is the lift of $\varphi\colon M \to M$, and $\widetilde{\Phi}\colon E(\Gamma \rtimes_\varphi \mathbb{Z}) \to E(\Gamma \rtimes_\varphi \mathbb{Z})$ is the lift of $\Phi\colon B(\Gamma \rtimes_\varphi \mathbb{Z}) \to B(\Gamma \rtimes_\varphi \mathbb{Z})$.

Since $\Phi\colon B(\Gamma \rtimes_\varphi \mathbb{Z}) \to B(\Gamma \rtimes_\varphi \mathbb{Z})$ is induced by an inner conjugation morphism on $\Gamma \rtimes_\varphi \mathbb{Z}$, the map[8] $\widetilde{\Phi}_*\colon K_1(C^*_{L,0}(E\Gamma)^\Gamma) \to K_1(C^*_{L,0}(E\Gamma)^\Gamma)$ is the identity map.

---

[7]Precisely speaking, φ only defines an outer automorphism of Γ, and one needs to make a specific choice of a representative in $\mathrm{Aut}(\Gamma)$. In any case, any such choice will work for the proof.

[8]The C^*-algebra $C^*_{L,0}(E\Gamma)^\Gamma$ is the inductive limit of $C^*_{L,0}(Y)^\Gamma$, where Y ranges over all Γ-cocompact subspaces of $E\Gamma$.

It follows that for each $[x] \in K_1(C^*_{L,0}(\widetilde{M})^\Gamma)$, we have

$$\widetilde{\iota}_*\widetilde{\sigma}_*(\widetilde{\varphi}_*[x]) = \widetilde{\Phi}_*\widetilde{\iota}_*\widetilde{\sigma}_*([x]) = \widetilde{\iota}_*\widetilde{\sigma}_*([x])$$

in $K_1(C^*_{L,0}(E(\Gamma \rtimes_\varphi \mathbb{Z}))^{\Gamma \rtimes_\varphi \mathbb{Z}})$, where $\widetilde{\iota}_*\widetilde{\sigma}_*$ is the composition

$$K_1(C^*_{L,0}(\widetilde{M})^\Gamma) \xrightarrow{\widetilde{\sigma}_*} K_1(C^*_{L,0}(E\Gamma)^\Gamma) \xrightarrow{\widetilde{\iota}_*} K_1(C^*_{L,0}(E(\Gamma \rtimes_\varphi \mathbb{Z}))^{\Gamma \rtimes_\varphi \mathbb{Z}}).$$

The same argument also works for an arbitrary finite number of orientation preserving self-homotopy equivalences $\psi_1, \cdots, \psi_m \in Aut_h(M)$ simultaneously, in which case we have

$$\widetilde{\iota}_*\widetilde{\sigma}_*((\widetilde{\psi}_i)_*[x]) = \widetilde{\iota}_*\widetilde{\sigma}_*([x]) \text{ in } K_1(C^*_{L,0}(E(\Gamma \rtimes_{\{\psi_1,\cdots,\psi_m\}} F_m))^{\Gamma \rtimes_{\{\psi_1,\cdots,\psi_m\}} F_m}),$$

for all $[x] \in K_1(C^*_{L,0}(\widetilde{M})^\Gamma)$. In other words, $(\widetilde{\psi}_i)_*[x]$ and $[x]$ have the same image in $K_1(C^*_{L,0}(E(\Gamma \rtimes_{\{\psi_1,\cdots,\psi_m\}} F_m))^{\Gamma \rtimes_{\{\psi_1,\cdots,\psi_m\}} F_m})$. For notational simplicity, let us write $\Gamma \rtimes F_m$ for $\Gamma \rtimes_{\{\psi_1,\cdots,\psi_m\}} F_m$. If no confusion is likely to arise, we shall still write $[x]$ for its image $\widetilde{\iota}_*\widetilde{\sigma}_*([x])$ in $K_1(C^*_{L,0}(E(\Gamma \rtimes F_m))^{\Gamma \rtimes F_m})$.

If we pass Equation (9.2) to $K_1(C^*_{L,0}(E(\Gamma \rtimes F_m))^{\Gamma \rtimes F_m})$ under the map

$$K_1(C^*_{L,0}(\widetilde{M})^\Gamma) \xrightarrow{\widetilde{\sigma}_*} K_1(C^*_{L,0}(E\Gamma)^\Gamma) \xrightarrow{\widetilde{\iota}_*} K_1(C^*_{L,0}(E(\Gamma \rtimes F_m))^{\Gamma \rtimes F_m}),$$

then it follows from the above discussion that

$$\sum_{k=1}^{\ell} c_k \rho(\gamma_k) = 0 \text{ in } K_1(C^*_{L,0}(E(\Gamma \rtimes F_m))^{\Gamma \rtimes F_m}),$$

where at least one $c_k \neq 0$. We have

$$\partial_{\Gamma \rtimes F_m}\Big(\sum_{k=1}^{\ell} c_k[p_{\gamma_k}]\Big) = 2 \cdot \big(\sum_{k=1}^{\ell} c_k \rho(\gamma_k)\big) = 0, \tag{9.3}$$

where $\partial_{\Gamma \rtimes F_m}$ is the connecting map in the following long exact sequence:

$$\begin{array}{ccccc} K_0(C^*_{L,0}(E(\Gamma \rtimes F_m))^{\Gamma \rtimes F_m}) & \longrightarrow & K_0^{\Gamma \rtimes F_m}(E(\Gamma \rtimes F_m)) & \xrightarrow{\mu} & K_0(C^*(\Gamma \rtimes F_m)) \\ \uparrow & & & & \downarrow \partial_{\Gamma \rtimes F_m} \\ K_1(C^*(\Gamma \rtimes F_m)) & \longleftarrow & K_1^{\Gamma \rtimes F_m}(E(\Gamma \rtimes F_m)) & \longleftarrow & K_1(C^*_{L,0}(E(\Gamma \rtimes F_m))^{\Gamma \rtimes F_m}) \end{array} \tag{9.4}$$

Now by assumption Γ is strongly finitely embeddable into Hilbert space. Hence $\Gamma \rtimes F_m$ is finitely embeddable into Hilbert space. By Proposition 6.10, we have the following.

1. $\{[p_{\gamma_1}], \cdots, [p_{\gamma_\ell}]\}$ generates a rank n abelian subgroup of $K_0^{\text{fin}}(C^*(\Gamma \rtimes F_m))$, since $\gamma_1, \cdots, \gamma_n$ have distinct finite orders. In other words,

$$\sum_{k=1}^{n} c_k[p_{\gamma_k}] \neq 0 \in K_0^{\text{fin}}(C^*(\Gamma \rtimes F_m))$$

if at least one $c_k \neq 0$.
2. Every nonzero element in $K_0^{\text{fin}}(C^*(\Gamma \rtimes F_m))$ is not in the image of the assembly map

$$\mu\colon K_0^{\Gamma \rtimes F_m}(E(\Gamma \rtimes F_m)) \to K_0(C^*(\Gamma \rtimes F_m)).$$

In particular, we see that $\partial_{\Gamma \rtimes F_m}\colon K_0^{\text{fin}}(C^*(\Gamma \rtimes F_m)) \to K_1(C^*_{L,0}(\widetilde{X})^{\Gamma \rtimes F_m})$ is injective.

It follows that $\partial_{\Gamma \rtimes F_m}\left(\sum_{k=1}^{\ell} c_k[p_{\gamma_k}]\right) \neq 0$, which contradicts Equation (9.3). This finishes the proof. □

It is tempting to use a similar argument to prove an analogue of Theorem 9.2 above for $\widetilde{\mathcal{S}}_{geom}(M)$. However, there are some subtleties. First, note that

$$\alpha_\varphi(\theta) + [\varphi] = \beta_\varphi(\theta)$$

for all $\theta = (f, N) \in \mathcal{S}(M)$ and all $\varphi \in Aut_h(M)$, where $[\varphi] = (\varphi, M)$ is the element given by $\varphi\colon M \to M$ in $\mathcal{S}(M)$. It follows that

$$\rho(\beta_\varphi(\theta)) = \rho(\alpha_\varphi(\theta)) + \rho([\varphi]) = \varphi_*(\rho(\theta)) + \rho([\varphi]).$$

In other words, in general, $\rho(\beta_\varphi(\theta)) \neq \varphi_*(\rho(\theta))$, and consequently the homomorphism

$$\rho\colon \mathcal{S}(M) \to K_1(C^*_{L,0}(\widetilde{M})^\Gamma)$$

does *not* descend to a homomorphism from $\widetilde{\mathcal{S}}_{geom}(M)$ to $K_1(C^*_{L,0}(\widetilde{M})^\Gamma)/\mathcal{I}_1(C^*_{L,0}(\widetilde{M})^\Gamma)$. New ideas are needed to take care of this issue. On the other hand, there is strong evidence which suggests an analogue of Theorem 9.2 for $\widetilde{\mathcal{S}}_{geom}(M)$. For example, this has been verified by Weinberger and Yu for residually finite groups [WY, Theorem 3.9]. Also, Chang and Weinberger showed that the free rank of $\widetilde{\mathcal{S}}_{geom}(M)$ is at least 1 when $\pi_1 X = \Gamma$ is not torsion free [ChW, Theorem 1].

The above discussion motivates the following conjecture.

Conjecture 9.3 Let M be a closed oriented topological manifold with dimension $n = 4k - 1$ ($k > 1$) and $\pi_1 M = \Gamma$. Then the free ranks of $\widetilde{\mathcal{S}}_{alg}(M)$ and $\widetilde{\mathcal{S}}_{geom}(M)$ are $\geq N_{\text{fin}}(\Gamma)$.

We conclude this section by proving the following theorem, which is an analogue of the theorem of Chang and Weinberger cited above [ChW, Theorem 1].

Theorem 9.4 *Let X be a closed oriented topological manifold with dimension $n = 4k-1$ $(k > 1)$ and $\pi_1 X = \Gamma$. If Γ is not torsion free, then the free rank of $\widetilde{S}_{alg}(X)$ is ≥ 1.*

Proof Recall that for any non-torsion-free countable discrete group G, if $\gamma \neq e$ is a finite order element of G, then $[p_\gamma]$ generates a subgroup of rank one in $K_0(C^*(G))$ and any nonzero multiple of $[p_\gamma]$ is not in the image of the assembly map $\mu\colon K_0^\Gamma(EG) \to K_0(C^*(G))$ [WY]. Using this fact, the statement follows from the same proof as in Theorem 9.2. □

10 Cyclic cohomology and higher rho invariants

Connes' cyclic cohomology theory provides a powerful method to compute higher rho invariants. In this section, we give a survey of recent work on the pairing between Connes' cyclic cohomology and C^*-algebraic secondary invariants. In the case of higher rho invariants given by invertible[9] operators on manifolds, this pairing can be computed in terms of Lott's higher eta invariants. We apply these results to the higher Atiyah-Patodi-Singer index theory and discuss a potential way to construct counter examples to the Baum–Connes conjecture.

We shall first discuss the zero dimensional cyclic cocycle case. Let M be a spin Riemannian manifold with positive scalar curvature and let D be the Dirac operator on M. Let $\widetilde{M}$ be the universal cover of M and $\widetilde{D}$ the lifting of D. Lott introduced the following delocalized eta invariant $\eta_{\langle h\rangle}(\widetilde{D})$ [Lo1]:

$$\eta_{\langle h\rangle}(\widetilde{D}) := \frac{2}{\sqrt{\pi}} \int_0^\infty \mathrm{tr}_{\langle h\rangle}(\widetilde{D}e^{-t^2\widetilde{D}^2})dt, \tag{10.1}$$

under the condition that the conjugacy class $\langle h\rangle$ of $h \in \Gamma = \pi_1 M$ has polynomial growth. Here $\Gamma = \pi_1 M$ is the fundamental group of M, and the trace map $\mathrm{tr}_{\langle h\rangle}$ is defined as follows:

$$\mathrm{tr}_{\langle h\rangle}(A) = \sum_{g\in\langle h\rangle} \int_{\mathcal{F}} A(x, gx)dx$$

on Γ-equivariant Schwartz kernels $A \in C^\infty(\widetilde{M} \times \widetilde{M})$, where $\mathcal{F}$ is a fundamental domain of $\widetilde{M}$ under the action of Γ.

We have the following theorem [XY3].

[9] Here "invertible" means being invertible on the universal cover of the manifold.

Theorem 10.1 *Let M be a closed odd-dimensional spin manifold equipped with a positive scalar curvature metric g. Suppose $\widetilde{M}$ is the universal cover of M, $\tilde{g}$ is the Riemannnian metric on $\widetilde{M}$ lifted from g, and $\widetilde{D}$ is the associated Dirac operator on $\widetilde{M}$. Suppose the conjugacy class $\langle h\rangle$ of a non-identity element $h \in \pi_1 M$ has polynomial growth, then we have*

$$\tau_h(\rho(\widetilde{D}, \widetilde{g})) = -\frac{1}{2}\eta_{\langle h\rangle}(\widetilde{D}),$$

where $\rho(\widetilde{D}, \widetilde{g})$ is the K-theoretic higher rho invariant of $\widetilde{D}$ with respect to the metric $\tilde{g}$, and τ_h is a canonical determinant map associated to $\langle h\rangle$.

As an application of Theorem 10.1 above, we have the following algebraicity result concerning the values of delocalized eta invariants [XY3].

Theorem 10.2 *With the same notation as above, if the rational Baum–Connes conjecture holds for Γ, and the conjugacy class $\langle h\rangle$ of a non-identity element $h \in \Gamma$ has polynomial growth, then the delocalized eta invariant $\eta_{\langle h\rangle}(\widetilde{D})$ is an algebraic number. Moreover, if in addition h has infinite order, then $\eta_{\langle h\rangle}(\widetilde{D})$ vanishes.*

This theorem follows from the construction of the determinant map τ_h and a L^2-Lefschetz fixed-point theorem of B.-L. Wang and H. Wang [WW, Theorem 5.10]. When Γ is torsion-free and satisfies the Baum–Connes conjecture, and the conjugacy class $\langle h\rangle$ of a non-identity element $h \in \Gamma$ has polynomial growth, Piazza and Schick have proved the vanishing of $\eta_{\langle h\rangle}(\widetilde{D})$ by a different method [PS, Theorem 13.7].

In light of this algebraicity result, we propose the following open question.

Open Question 10.3 If the conjugacy class $\langle h\rangle$ of a non-identity element $h \in \Gamma$ has polynomial growth, what values can the delocalized eta invariant $\eta_{\langle h\rangle}(\widetilde{D})$ take in general? Are they always algebraic numbers?

In particular, if a delocalized eta invariant is transcendental, then it will lead to a counterexample to the Baum–Connes conjecture [BC, BCH, C]. Note that the above question is a reminiscent of Atiyah's question concerning rationality of ℓ^2-Betti numbers [A1]. Atiyah's question was answered in negative by Austin, who showed that ℓ^2-Betti numbers can be transcendental [Au].

So far, we have been assuming the conjugacy class $\langle h\rangle$ has polynomial growth, which guarantees the convergence of the integral in (10.1). In general, the integral in (10.1) fails to converge. The following theorem of Chen–Wang–Xie–Yu [CWXY] gives a sufficient condition for when the integral in (10.1) converges.

Theorem 10.4 *Let M be a closed manifold and $\widetilde{M}$ the universal covering over M. Suppose D is a self-adjoint first-order elliptic differential operator over M and $\widetilde{D}$ the lift of D to $\widetilde{M}$. If $\langle h\rangle$ is a nontrivial conjugacy class of $\pi_1(M)$ and $\widetilde{D}$ has a sufficiently large spectral gap at zero, then the delocalized eta invariant $\eta_{\langle h\rangle}(\widetilde{D})$ defined in line* (10.1) *converges absolutely.*

We would like to emphasis that the theorem above works for all fundamental groups. In the special case where the conjugacy class $\langle h\rangle$ has sub-exponential growth, then any nonzero spectral gap is in fact sufficiently large, hence in this case $\eta_{\langle h\rangle}(\widetilde{D})$ is well-defined as long as $\widetilde{D}$ is invertible.

Let us make precise of what "sufficiently large spectral gap at zero" means. Fix a finite generating set S of Γ. Let ℓ be the corresponding word length function on Γ determined by S. Since S is finite, there exist C and $K_{\langle h\rangle} > 0$ such that

$$\#\{g \in \langle h\rangle : \ell(g) = n\} \leqslant Ce^{K_{\langle h\rangle}\cdot n}. \tag{10.2}$$

We define $\tau_{\langle h\rangle}$ to be

$$\tau_{\langle h\rangle} = \liminf_{\substack{g\in\langle h\rangle \\ \ell(g)\to\infty}} \Big(\inf_{x\in\widetilde{M}} \frac{\operatorname{dist}(x, gx)}{\ell(g)} \Big). \tag{10.3}$$

Since the action of Γ on $\widetilde{M}$ is free and cocompact, we have $\tau_{\langle h\rangle} > 0$.

We denote the principal symbol of D by $\sigma_D(x, v)$, for $x \in M$ and cotangent vector $v \in T_x^*M$. We define the propagation speed of D to be the positive number

$$c_D = \sup\{\|\sigma_D(x, v)\| : x \in M, v \in T_x^*M, \|v\| = 1\}.$$

Definition 10.5 With the above notation, let us define

$$\sigma_{\langle h\rangle} := \frac{2K_{\langle h\rangle} \cdot c_D}{\tau_{\langle h\rangle}}. \tag{10.4}$$

Recall that $\widetilde{D}$ is said to have a spectral gap at zero if there exists an open interval $(-\varepsilon, \varepsilon) \subset \mathbb{R}$ such that spectrum$(\widetilde{D}) \cap (-\varepsilon, \varepsilon)$ is either $\{0\}$ or empty. Moreover, $\widetilde{D}$ is said to have a *sufficiently large* spectral gap at zero if its spectral gap is larger than $\sigma_{\langle h\rangle}$.

Again it is natural to ask the following question.

Open Question 10.6 With $\widetilde{D}$ as in the above theorem, what values can the delocalized eta invariant $\eta_{\langle h\rangle}(\widetilde{D})$ take in general? Are they always algebraic numbers?

A special feature of traces is that they always have uniformly bounded representatives, when viewed as degree zero cyclic cocycles. In fact, our proof of Theorem 10.4 allows us to generalize Theorem 10.4 to cyclic cocycles of higher degrees, as long as they have at most exponential growth. Recall that the cyclic cohomology of a group algebra $\mathbb{C}\Gamma$ has a decomposition respect to the conjugacy classes of Γ ([Nis]):

$$HC^*(\mathbb{C}\Gamma) \cong \prod_{\langle h\rangle} HC^*(\mathbb{C}\Gamma, \langle h\rangle),$$

where $HC^*(\mathbb{C}\Gamma, \langle h\rangle)$ denotes the component that corresponds to the conjugacy class $\langle h\rangle$. If $\langle h\rangle$ is a nontrivial conjugacy class, then a cyclic cocycle in $HC^*(\mathbb{C}\Gamma, \langle h\rangle)$ will be called a delocalized cyclic cocycle at $\langle h\rangle$.

Theorem 10.7 *Assume the same notation as in Theorem 10.4. Let φ be a delocalized cyclic cocycle at a nontrivial conjugacy class $\langle h\rangle$. If φ has exponential growth and $\widetilde{D}$ has a sufficiently large spectral gap at zero, then a higher analogue $\eta_\varphi(\widetilde{D})$ (cf. [CWXY, Definition 3.17]) of the formula* (10.1) *converges absolutely.*

For higher degree cyclic cocycles, the precise meaning of "sufficiently large spectral gap at zero" is similar to but slightly different from that of the case of traces. We refer the reader to [CWXY, Section 3.2] for more details. For now, we simply point out that if both Γ and φ have sub-exponential growth, then any nonzero spectral gap is in fact sufficiently large, hence in this case $\eta_\varphi(\widetilde{D})$ is well-defined as long as $\widetilde{D}$ is invertible. The explicit formula for $\eta_\varphi(\widetilde{D})$ is described in terms of the transgression formula for the Connes–Chern character [C, C2]. It is essentially[10] a periodic version of the delocalized part of Lott's noncommutative-differential higher eta invariant. We shall call $\eta_\varphi(\widetilde{D})$ a *delocalized higher eta invariant* from now on.

Formally speaking, just as Lott's delocalized eta invariant $\eta_{\langle h\rangle}(\widetilde{D})$ can be interpreted as the pairing between the degree zero cyclic cocycle $\mathrm{tr}_{\langle h\rangle}$ and the higher rho invariant $\rho(\widetilde{D})$, so can the delocalized higher eta invariant $\eta_\varphi(\widetilde{D})$ be interpreted as the pairing between the cyclic cocycle φ and the higher rho invariant $\rho(\widetilde{D})$. A key analytic difficulty here is to verify when such a pairing is well-defined, or more ambitiously, to verify when one can extend this pairing to a pairing between the cyclic cohomology of $\mathbb{C}\Gamma$ and the K-theory group $K_*(C^*_{L,0}(\widetilde{M})^\Gamma)$. The group $K_*(C^*_{L,0}(\widetilde{M})^\Gamma)$ consists of C^*-algebraic secondary invariants; in particular, it contains all higher rho invariants from the discussion above. Such an extension of the pairing is important, often necessary, for many interesting applications to geometry and topology (cf. [PS1, XY1, WXY]).

In [CWXY], such an extension of the pairing, that is, a pairing between delocalized cyclic cocycles of *all degrees* and the K-theory group $K_*(C^*_{L,0}(\widetilde{M})^\Gamma)$ was established, in the case of Gromov's hyperbolic groups. More precisely, we have the following theorem [CWXY].

Theorem 10.8 *Let M be a closed manifold whose fundamental group Γ is hyperbolic. Suppose $\langle h\rangle$ is nontrivial conjugacy class of Γ. Then every element $[\alpha] \in HC^{2k+1-i}(\mathbb{C}\Gamma, \langle h\rangle)$ induces a natural map*

$$\tau_{[\alpha]}\colon K_i(C^*_{L,0}(\widetilde{M})^\Gamma) \to \mathbb{C}$$

[10] We refer the reader to [CWXY] for details on how to identify the formula for $\eta_\varphi(\widetilde{D})$ in Theorem 10.7 with the periodic version of Lott's noncommutative-differential higher eta invariant.

such that the following are satisfied:

(i) $\tau_{[S\alpha]} = \tau_{[\alpha]}$, *where S is Connes' periodicity map*

$$S\colon HC^*(\mathbb{C}\Gamma, \langle h\rangle) \to HC^{*+2}(\mathbb{C}\Gamma, \langle h\rangle);$$

(ii) *if D is an elliptic operator on M such that the lift $\widetilde{D}$ of D to the universal cover $\widetilde{M}$ of M is invertible, then we have*

$$\tau_{[\alpha]}(\rho(\widetilde{D})) = \eta_{[\alpha]}(\widetilde{D}),$$

where $\rho(\widetilde{D})$ is the higher rho invariant of $\widetilde{D}$ and $\eta_{[\alpha]}(\widetilde{D})$ is the delocalized higher eta invariant from Theorem 10.7. In particular, in the case of hyperbolic groups, the delocalized higher eta invariant $\eta_{[\alpha]}(\widetilde{D})$ converges absolutely, as long as $\widetilde{D}$ is invertible.

The construction of the map $\tau_{[\alpha]}$ in the above theorem uses Puschnigg's smooth dense subalgebra for hyperbolic groups [P1] in an essential way. In more conceptual terms, the above theorem provides an explicit formula to compute the delocalized Connes–Chern character of C^*-algebraic secondary invariants. More precisely, the same techniques developed in [CWXY] actually imply[11] that there is a well-defined delocalized Connes–Chern character $Ch_{deloc}\colon K_i(C^*_{L,0}(\widetilde{M})^\Gamma) \to \overline{HC}^{deloc}_*(\mathcal{B})$, where $\mathcal{B}$ is Puschnigg's smooth dense subalgebra of $C^*_r(\Gamma)$ and $\overline{HC}^{deloc}_*(\mathcal{B})$ is the delocalized part of the cyclic homology[12] of $\mathcal{B}$. Now for Gromov's hyperbolic groups, every cyclic cohomology class of $\mathbb{C}\Gamma$ continuously extends to cyclic cohomology class of $\mathcal{B}$ (cf. [P1] for the case of degree zero cyclic cocycles and [CWXY] for the case of higher degree cyclic cocycles). Thus the map $\tau_{[\alpha]}$ can be viewed as a pairing between cyclic cohomology and delocalized Connes–Chern characters of C^*-algebraic secondary invariants. As a consequence, this unifies Higson–Roe's higher rho invariant and Lott's higher eta invariant for invertible operators.

We point out that the proof of Theorem 10.8 does *not* rely on the Baum–Connes isomorphism for hyperbolic groups [L, MY], although the theorem is closely connected to the Baum–Connes conjecture and the Novikov conjecture. On the other hand, if one is willing to use the full power of the Baum–Connes isomorphism for hyperbolic groups, there is in fact a different, but more indirect, approach to the

[11] In fact, even more is true. One can use the same techniques developed in [CWXY] to show that if $\mathcal{A}$ is smooth dense subalgebra of $C^*_r(\Gamma)$ for any group Γ (not necessarily hyperbolic) and in addition $\mathcal{A}$ is a Fréchet locally m-convex algebra, then there is a well-defined delocalized Connes–Chern character $Ch_{deloc}\colon K_i(C^*_{L,0}(\widetilde{M})^\Gamma) \to \overline{HC}^{deloc}_*(\mathcal{A})$. Of course, in order to pair such a delocalized Connes–Chern character with a cyclic cocycle of $\mathbb{C}\Gamma$, the key remaining challenge is to continuously extend this cyclic cocycle of $\mathbb{C}\Gamma$ to a cyclic cocycle of $\mathcal{A}$.

[12] Here the definition of cyclic homology of $\mathcal{B}$ takes the topology of $\mathcal{B}$ into account, cf. [C2, Section II.5].

delocalized Connes–Chern character map. First, observe that the map $\tau_{[\alpha]}$ factors through a map

$$\tau_{[\alpha]}\colon K_i(C^*_{L,0}(\underline{E}\Gamma)^\Gamma)\otimes\mathbb{C}\to\mathbb{C}$$

where $\underline{E}\Gamma$ is the universal space for proper Γ-actions. Now the Baum–Connes isomorphism $\mu\colon K^G_*(\underline{E}\Gamma)\xrightarrow{\cong}K_*(C^*_r(\Gamma))$ for hyperbolic groups implies that one can identify $K_i(C^*_{L,0}(\underline{E}\Gamma)^\Gamma)\otimes\mathbb{C}$ with $\bigoplus_{\langle h\rangle\neq 1}HC_*(\mathbb{C}\Gamma,\langle h\rangle)$, where $HC_*(\mathbb{C}\Gamma,\langle h\rangle)$ is the delocalized cyclic homology at $\langle h\rangle$ and the direct sum is taken over all nontrivial conjugacy classes. In particular, after this identification, it follows that the map $\tau_{[\alpha]}$ becomes the usual pairing between cyclic cohomology and cyclic homology. However, for a specific element, e.g., the higher rho invariant $\rho(\widetilde{D})$, in $K_i(C^*_{L,0}(\underline{E}\Gamma)^\Gamma)$, its identification with an element in $\bigoplus_{\langle h\rangle\neq 1}HC_*(\mathbb{C}\Gamma,\langle h\rangle)$ is rather abstract and implicit. More precisely, the computation of the number $\tau_{[\alpha]}(\rho(\widetilde{D}))$ essentially amounts to the following process. Observe that if a closed spin manifold M is equipped with a positive scalar curvature metric, then stably it bounds (more precisely, the universal cover $\widetilde{M}$ of M becomes the boundary of another Γ-manifold, after finitely many steps of cobordisms and vector bundle modifications). In principle, the number $\tau_{[\alpha]}(\rho(\widetilde{D}))$ can be derived from a higher Atiyah-Patodi-Singer index theorem for this bounding manifold. Again, there is a serious drawback of such an indirect approach—the explicit formula for $\tau_{[\alpha]}(\rho(\widetilde{D}))$ is completely lost. In contrast, a key feature of the construction of the delocalized Connes–Chern character map in Theorem 10.8 is that the formula is explicit and intrinsic.

In [DG], Deeley and Goffeng also constructed an implicit delocalized Chern character map for C^*-algebraic secondary invariants. Their approach is in spirit similar to the indirect method just described above (making use of the Baum–Connes isomorphism for hyperbolic groups), although their actual technical implementation is different.

As an application, we use this delocalized Connes–Chern character map from Theorem 10.8 to derive a delocalized higher Atiyah-Patodi-Singer index theorem for manifolds with boundary. More precisely, let W be a compact n-dimensional spin manifold with boundary ∂W. Suppose W is equipped with a Riemannian metric g_W which has product structure near ∂W and in addition has positive scalar curvature on ∂W. Let $\widetilde{W}$ be the universal covering of W and $g_{\widetilde{W}}$ the Riemannian metric on $\widetilde{W}$ lifted from g_W. With respect to the metric $g_{\widetilde{W}}$, the associated Dirac operator $\widetilde{D}_W$ on $\widetilde{W}$ naturally defines a higher index $Ind_\Gamma(\widetilde{D}_W)$ (as in Section 7) in $K_n(C^*(\widetilde{W})^\Gamma)=K_n(C^*_r(\Gamma))$, where $\Gamma=\pi_1(W)$. Since the metric $g_{\widetilde{W}}$ has positive scalar curvature on $\partial\widetilde{W}$, it follows from the Lichnerowicz formula that the associated Dirac operator $\widetilde{D}_\partial$ on $\partial\widetilde{W}$ is invertible, hence naturally defines a higher rho invariant $\rho(\widetilde{D}_\partial)$ in $K_{n-1}(C^*_{L,0}(\widetilde{W})^\Gamma)$. We have the following delocalized higher Atiyah-Patodi-Singer index theorem.

Theorem 10.9 *With the notation as above, if* $\Gamma = \pi_1(W)$ *is hyperbolic and* $\langle h \rangle$ *is a nontrivial conjugacy class of* Γ*, then for any* $[\varphi] \in HC^*(\mathbb{C}\Gamma, \langle h \rangle)$*, we have*

$$Ch_{[\varphi]}(Ind_\Gamma(\widetilde{D}_W)) = -\frac{1}{2}\eta_{[\varphi]}(\widetilde{D}_\partial), \tag{10.5}$$

where $Ch_{[\varphi]}(Ind_\Gamma(\widetilde{D}_W))$ *is the Connes–Chern pairing between the cyclic cohomology class* $[\varphi]$ *and the higher index class* $Ind_\Gamma(\widetilde{D}_W)$.

Proof This follows from Theorems 10.8 and 7.4. □

By using Theorem 10.8, we have derived Theorem 10.9 as a consequence of a K-theoretic counterpart. This is possible only because we have realized $\eta_{[\varphi]}(\widetilde{D}_\partial)$ as the pairing between the cyclic cocycle φ and the C^*-algebraic secondary invariant $\rho(\widetilde{D}_\partial)$ in $K_1(C^*_{L,0}(\widetilde{W})^\Gamma)$.

Alternatively, one can also derive Theorem 10.9 from a version of higher Atiyah-Patodi-Singer index theorem due to Leichtnam and Piazza [LP3, Theorem 4.1] and Wahl [Wa, Theorem 9.4 & 11.1]. This version of higher Atiyah-Patodi-Singer index theorem is stated in terms of noncommutative differential forms on a smooth dense subalgebra of $C_r^*(\Gamma)$; or noncommutative differential forms on a certain class of smooth dense subalgebras (if exist) of general C^*-algebras (not just group C^*-algebras) in Wahl's version. In the case of Gromov's hyperbolic groups, one can choose such a smooth dense subalgebra to be Puschnigg's smooth dense subalgebra $\mathcal{B}$. As mentioned before, for Gromov's hyperbolic groups, every cyclic cohomology class of $\mathbb{C}\Gamma$ continuously extends to a cyclic cohomology class of $\mathcal{B}$ (cf. [P1] for the case of degree zero cyclic cocycles and [CWXY] for the case of higher degree cyclic cocycles). Now Theorem 10.9 follows by pairing the higher Atiyah-Patodi-Singer index formula of Leichtnam-Piazza and Wahl with the delocalized cyclic cocycles of $\mathbb{C}\Gamma$.

One can also try to pair the higher Atiyah-Patodi-Singer index formula of Leichtnam-Piazza and Wahl with group cocycles of Γ, or equivalently cyclic cocycles in $HC^*(\mathbb{C}\Gamma, \langle 1 \rangle)$, where $\langle 1 \rangle$ stands for the conjugacy class of the identity element of Γ. In this case, for fundamental groups with property RD, Gorokhovsky, Moriyoshi, and Piazza proved a higher Atiyah-Patodi-Singer index theorem for group cocycles with polynomial growth [Gr].

For other related interesting development, we refer the reader to the following papers [BL, BLH, BR, BD2, BW1, CG, CP, CFY, C3, CH, D, DFW, FH, FJ1, FJ2, FJ3, FJ4, FP, FW1, FW2, GMP, GWY, HR1, J, LP1, LP2, Lo, OY, Pi, P, W, W1, WiY1, WiY2, RTY].

Acknowledgements The authors wish to thank Alain Connes for numerous inspiring discussions. The author "Zhizhang Xie" is partially supported by NSF 1500823, NSF 1800737. The author "Guoliang Yu" is partially supported by NSF 1700021, NSF 1564398.

References

[Al] P. ALBIN. *Stratified surgery and K-theory invariants of the signature operator.* https://arxiv.org/abs/1710.00934.

[ALMP] P. ALBIN, É. LEICHTNAM, R. MAZZEO, P. PIAZZA. *The signature package on Witt spaces.* Ann. Sci. Éc. Norm. Supér. (4) 45 (2012), no. 2, 241–310.

[ALMP1] P. ALBIN, E. LEICHTNAM, R. MAZZEO, P. PIAZZA. *The Novikov conjecture on Cheeger spaces.* J. Noncommut. Geom. 11 (2017), no. 2, 451–506.

[AD] G. ARZHANTSEVA, T. DELZANT. *Examples of random groups.* Preprint, 2008.

[A] M. ATIYAH. *Global theory of elliptic operators.* 1970 Proc. Internat. Conf. on Functional Analysis and Related Topics (Tokyo, 1969) pp. 21–30 Univ. of Tokyo Press, Tokyo.

[A1] M. ATIYAH. *Elliptic operators, discrete groups and von Neumann algebras.* Colloque "Analyse et Topologie" en l'Honneur de Henri Cartan (Orsay, 1974), pp. 43–72. Astérisque, No. 32–33, Soc. Math. France, Paris, 1976.

[AS] M. ATIYAH, I. SINGER. *The index of elliptic operators.* I. Ann. of Math. (2) 87 1968 484–530.

[AAS] P. ANTONINI, S. AZZALI, G. SKANDALIS. *The Baum-Connes conjecture localised at the unit element of a discrete group.* arXiv:1807.05892, 2018.

[Au] T. AUSTIN *Rational group ring elements with kernels having irrational dimension. Proc. Lond. Math. Soc. (3)*, 107(6):1424–1448, 2013.

[BC] P. BAUM, A. CONNES. *K-theory for discrete groups* Operator algebras and applications, Vol. 1, volume 135 of London Mathematical Society, pages 1–20. Cambridge Univ. Press, Cambridge, 1988.

[BCH] P. BAUM, A. CONNES, N. HIGSON. *Classifying space for proper actions and K-theory of group C^*-algebras.* C^*-algebras: 1943–1993 (San Antonio, TX, 1993), 240–291, Contemp. Math., 167, Amer. Math. Soc., Providence, RI, 1994.

[BL] A. BARTELS, W. LÜCK. *The Borel Conjecture for hyperbolic and CAT(0)-groups.* Ann. of Math. (2) 175 (2012), no. 2, 631–689.

[BLH] A. BARTELS, W. LÜCK, R. HOLGER. *The K-theoretic Farrell-Jones conjecture for hyperbolic groups.* Invent. Math. 172 (2008), no. 1, 29–70.

[BR] A. BARTELS, D. ROSENTHAL. *On the K-theory of groups with finite asymptotic dimension.* J. Reine Angew. Math. 612 (2007), 35–57.

[BD1] G. BELL, A. DRANISHNIKOV. *On asymptotic dimension of groups.* Algebr. Geom. Topol. 1 (2001), 57–71 (electronic).

[BD2] G. BELL, A. DRANISHNIKOV. *A Hurewicz-type theorem for asymptotic dimension and applications to geometric group theory.* Transactions American Math. Soc. Volume 358, Number 11, 2006, 4749–4764.

[BCV] M. BEKKA, P. CHERIX, A. VALETTE. *Proper affine isometric actions of amenable groups.* Novikov conjectures, index theorems and rigidity, Vol. 2 (Oberwolfach, 1993), 1–4, London Math. Soc. Lecture Note Ser., 227, Cambridge Univ. Press, Cambridge, 1995.

[BBF] M. BESTVINA, K. BROMBERG, K. FUJIWARA. *Constructing group actions on quasi-trees and applications to mapping class groups.* Publ. Math. Inst. Hautes Études Sci. 122 (2015), 1–64.

[BGH] M. BESTVINA, V. GUIRARDEL, C. Horbez. *Boundary amenability of Out(F_N).* arXiv:1705.07017, 2017.

[BoG] B. BOTVINNIK , P. B. GILKEY. *The eta invariant and metrics of positive scalar curvature.* Math. Ann., 302(3):507–517, 1995.

[BG] N. BROWN, E. GUENTNER. *Uniform embeddings of bounded geometry spaces into reflexive Banach space.* Proc. Amer. Math. Soc. 133 (2005), no. 7, 2045–2050.

[B] U. BUNKE. *A K-theoretic relative index theorem and Callias-type Dirac operators.* Math. Ann., 303(2):241–279, 1995.

[BW] J. Block, S. Weinberger. *Arithmetic manifolds of positive scalar curvature.* J. Differential Geom. 52 (1999), no. 2, 375–406.

[BW1] J. Block, S. Weinberger. *Aperiodic tilings, positive scalar curvature and amenability of spaces.* J. Amer. Math. Soc. 5 (1992), no. 4, 907–918.

[CW] S. Cappell, S. Weinberger. *A geometric interpretation of Siebenmann's periodicity phenomenon.* In Geometry and topology (Athens, Ga., 1985), volume 105 of Lecture Notes in Pure and Appl. Math., pages 47–52. Dekker, New York, 1987.

[CG] G. Carlsson, B. Goldfarb. *The integral K-theoretic Novikov conjecture for groups with finite asymptotic dimension.* Invent. Math. 157 (2004), no. 2, 405–418.

[CP] G. Carlsson, E. Pedersen. *Controlled algebra and the Novikov conjectures for K- and L-theory.* Topology 34 (1995), no. 3, 731–758.

[CFY] S. Chang, S. Ferry, G. Yu. *Bounded rigidity of manifolds and asymptotic dimension growth.* J. K-Theory 1 (2008), no. 1, 129–144.

[ChW] S. Chang, S. Weinberger. *On invariants of Hirzebruch and Cheeger-Gromov.* Geom. Topol. 7 (2003), 311–319.

[Ch] W. Charlotte. *Higher-invariants and the surgery structure set.* J. Topol. 6.

[CWXY] X. Chen, J. Wang, Z. Xie, Yu. *Delocalized eta invariants, cyclic cohomology and higher rho invariants.* arXiv:1901.02378, 2019.

[C] A. Connes. *Noncommutative Geometry.* Academic Press, 1994.

[C1] A. Connes. *Cyclic cohomology and the transverse fundamental class of a foliation.* Geometric methods in operator algebras (Kyoto, 1983), pages 52–144, volume 123, Pitman Res. Notes Math. Ser., 1986.

[C2] A. Connes. *Noncommutative differential geometry.* Inst. Hautes Études Sci. Publ. Math., (62):257–360, 1985.

[C3] A. Connes. *On the Chern character of θ summable Fredholm modules.* Comm. Math. Phys. 139 (1991), no. 1, 171–181.

[CH] A. Connes, N. Higson. *Déformations, morphismes asymptotiques et K-théorie bivariante. (French) [Deformations, asymptotic morphisms and bivariant K-theory]* C. R. Acad. Sci. Paris Sér. I Math. 311 (1990), no. 2, 101–106.

[CM] A. Connes, H. Moscovici. *Cyclic cohomology, the Novikov conjecture and hyperbolic groups.* Topology 29 (1990), no. 3, 345–388.

[CGM] A. Connes, M. Gromov, H. Moscovici. *Group cohomology with Lipschitz control and higher signatures.* Geom. Funct. Anal. 3 (1993), no. 1, 1–78.

[D] M. Davis. *Groups generated by reflections and aspherical manifolds not covered by Euclidean space.* Ann. of Math. (2) 117 (1983), no. 2, 293–324.

[DG] R. J. Deeley, M. Goffeng. *Realizing the analytic surgery group of Higson and Roe geometrically part III: higher invariants.* Math. Ann., 366(3-4):1513–1559, 2016.

[DFW] A. Dranishnikov, S. Ferry, S. Weinberger. *An etale approach to the Novikov conjecture.* Comm. Pure Appl. Math. 61 (2008), no. 2, 139–155.

[DJ] A. Dranishnikov, T. Januszkiewicz. *Every Coxeter group acts amenably on a compact space.* Proceedings of the 1999 Topology and Dynamics Conference (Salt Lake City, UT). Topology Proc. 24 (1999), Spring, 135–141.

[FH] F. T. Farrell, W. C. Hsiang. *On Novikov's conjecture for nonpositively curved manifolds.* Ann. of Math. (2) 113 (1981), no. 1, 199–209.

[FJ1] T. Farrell, L. Jones. *A topological analogue of Mostow's rigidity theorem.* J. Amer. Math. Soc. 2 (1989), no. 2, 257–370.

[FJ2] T. Farrell, L. Jones. *Classical aspherical manifolds.* CBMS Regional Conference Series in Mathematics, 75. Published for the Conference Board of the Mathematical Sciences, Washington, DC; by the American Mathematical Society, Providence, RI, 1990.

[FJ3] T. Farrell, L. Jones. *Topological rigidity for compact non-positively curved manifolds.* Differential geometry: Riemannian geometry (Los Angeles, CA, 1990), 229–274, Proc. Sympos. Pure Math., 54, Part 3, Amer. Math. Soc., Providence, RI, 1993.

[FJ4] T. FARRELL, L. JONES. *Rigidity for aspherical manifolds with* $\pi_1 \subset \mathrm{GL}_m(R)$. Asian J. Math. 2 (1998), no. 2, 215–262.

[FM] A. T. FOMENKO, A. S. MISHCHENKO. *The index of elliptic operators over C^*-algebras*. (Russian) Izv. Akad. Nauk SSSR Ser. Mat. 43 (1979), no. 4, 831–859, 967.

[FP] S. FERRY, E. PEDERSEN. *Epsilon surgery theory*. Novikov conjectures, index theorems and rigidity, Vol. 2 (Oberwolfach, 1993), 167–226, London Math. Soc. Lecture Note Ser., 227, Cambridge Univ. Press, Cambridge, 1995.

[FS] D. FISHER, L. SILBERMAN. *Groups not acting on manifolds*. Int. Math. Res. Not. IMRN 2008, no. 16, Art. ID rnn060, 11 pp.

[FW1] S. FERRY, S. WEINBERGER. *Curvature, tangentiality, and controlled topology*. Invent. Math. 105 (1991), no. 2, 401–414.

[FW2] S. FERRY, S. WEINBERGER. *A coarse approach to the Novikov conjecture*. Novikov conjectures, index theorems and rigidity, Vol. 1 (Oberwolfach, 1993), 147–163, London Math. Soc. Lecture Note Ser., 226, Cambridge Univ. Press, Cambridge, 1995.

[GWY] S. GONG, J. WU, G. YU. *The Novikov conjecture, the group of volume preserving diffeomorphisms and Hilbert-Hadamard spaces*. arXiv:1811.02086, 2018.

[Gr] R. I. GRIGORCHUK. *Degrees of growth of finitely generated groups and the theory of invariant means*. (Russian) Izv. Akad. Nauk SSSR Ser. Mat. 48 (1984), no. 5, 939–985.

[G] M. GROMOV. *Hyperbolic groups*. In Gersten, Steve M. Essays in group theory. Mathematical Sciences Research Institute Publications. 8. New York: Springer. pp. 75–263.

[G1] M. GROMOV. *Asymptotic invariants of infinite groups*. Geometric group theory, Vol. 2 (Sussex, 1991), 1–295, London Math. Soc. Lecture Note Ser., 182, Cambridge Univ. Press, Cambridge, 1993.

[G2] M. GROMOV. *Spaces and questions*. GAFA 2000 (Tel Aviv, 1999). Geom. Funct. Anal. 2000, Special Volume, Part I, 118–161.

[G3] M. GROMOV. *Random walks in random groups*. Geom. Funct. Anal. 13 (2003), no. 1, 73–146.

[GL] M. GROMOV, B. LAWSON. *The classification of simply connected manifolds of positive scalar curvature*. Ann. of Math. (2), 111(3):423–434, 1980.

[GL1] M. GROMOV, B. LAWSON. *Positive scalar curvature and the Dirac operator on complete Riemannian manifolds*. Inst. Hautes Études Sci. Publ. Math. No. 58 (1983), 83–196 (1984).

[GMP] A. GOROKHOVSKY, H. MORIYOSHI , P. PIAZZA. *A note on the higher Atiyah-Patodi-Singer index theorem on Galois coverings*. J. Noncommut. Geom., 10(1):265–306, 2016.

[GHW] E. GUENTNER, N. HIGSON, S. WEINBERGER. *The Novikov Conjecture for Linear Groups*. Publ. Math. Inst. Hautes Études Sci. No. 101 (2005), 243–268.

[GTY] E. GUENTNER, R. TESSERA, G. YU. *A notion of geometric complexity and its application to topological rigidity*. Invent. Math. 189 (2012), no. 2, 315–357.

[GWY] E. GUENTNER, R. WILLETT, G. YU. *Dynamical complexity and controlled operator K-theory*. Groups, Geometry and Dynamics, Vol. 7, 2 (2013) 377–402.

[Ha] U. HAMENSTÄDT. *Geometry of the mapping class groups. I. Boundary amenability*. Invent. Math. 175 (2009), no. 3, 545–609.

[HS] S. HANKE, T. SCHICK. *The strong Novikov conjecture for low degree cohomology*. Geom. Dedicata 135 (2008), 119–127.

[H] N. HIGSON. *Bivariant K-theory and the Novikov conjecture*. Geom. Funct. Anal. 10 (2000), no. 3, 563–581.

[HK] N. HIGSON, G. KASPAROV. *E -theory and KK-theory for groups which act properly and isometrically on Hilbert space*. Invent. Math. 144 (2001), no. 1, 23–74.

[HR] N. HIGSON, J. ROE. *Mapping surgery to analysis. I. Analytic signatures*. *K-Theory*, 33(4):277–299, 2005.

[HR1] N. HIGSON, J. ROE. *Mapping surgery to analysis. II. Geometric signatures*. *K-Theory*, 33(4):301–324, 2005.

[HR2] N. HIGSON, J. ROE. *Mapping surgery to analysis. III. Exact sequences. K-Theory*, 33(4):325–346, 2005.

[HR3] N. HIGSON, J. ROE. *N. Higson and J. Roe. K-homology, assembly and rigidity theorems for relative eta invariants*. Pure Appl. Math. Q., 6 (2, Special Issue: In honor of Michael Atiyah and Isadore Singer):555–601, 2010.

[HLS] N. HIGSON, V. LAFFORGUE, G. SKANDALIS. *Counterexamples to the Baum-Connes conjecture*. Geom. Funct. Anal. 12 (2002), no. 2, 330–354.

[HiS] M. HILSUM, G. SKANDALIS. *Georges Invariance par homotopie de la signature à coefficients dans un fibré presque plat. (French) [Homotopy invariance of the signature with coefficients in an almost flat fiber bundle]*. J. Reine Angew. Math. 423 (1992), 73–99.

[J] L. JI. *The integral Novikov conjectures for linear groups containing torsion elements*. J. Topol. 1 (2008), no. 2, 306–316.

[JR] W. B. JOHNSON, N. L. RANDRIANARIVONY. l_p $(p > 2)$ *does not coarsely embed into a Hilbert space*. Proc. Amer. Math. Soc. 134 (2006), 1045–1050 (electronic).

[KM] J. KAMINKER, J. MILLER. Homotopy invariance of the analytic index of signature operators over C^*-algebras. *J. Operator Theory*, 14(1):113–127, 1985.

[K] G. KASPAROV. *Equivariant KK-theory and the Novikov conjecture*. Invent. Math. 91 (1988), no. 1, 147–201.

[K1] G. KASPAROV. K-theory, group C^*-algebras, and higher signatures (conspectus). In *Novikov conjectures, index theorems and rigidity, Vol. 1 (Oberwolfach, 1993)*, volume 226 of *London Math. Soc. Lecture Note Ser.*, pages 101–146. Cambridge Univ. Press, Cambridge, 1995.

[KS] G. KASPAROV, G. SKANDALIS. *Groups acting properly on "bolic" spaces and the Novikov conjecture*. Ann. of Math. (2) 158 (2003), no. 1, 165–206.

[KiS] R. C. KIRBY, L. C. SIEBENMANN. *Foundational essays on topological manifolds, smoothings, and triangulations*. With notes by John Milnor and Michael Atiyah. Annals of Mathematics Studies, No. 88. Princeton University Press, Princeton, N.J.; University of Tokyo Press, Tokyo, 1977. vii+355 pp.

[KY] G. KASPAROV, G. YU. *The Novikov conjecture and geometry of Banach spaces*. Geometry and Topology 16 (2012), no. 3, 1859–1880.

[Ki] Y. KIDA. *The mapping class group from the viewpoint of measure equivalence theory*. Mem. Amer. Math. Soc. 196 (2008), no. 916.

[L] V. LAFFORGUE. *K-théorie bivariante pour les algèbres de Banach et conjecture de Baum-Connes*. (French) [Bivariant K-theory for Banach algebras and the Baum-Connes conjecture] Invent. Math. 149 (2002), no. 1, 1–95.

[L1] V. LAFFORGUE. *La conjecture de Baum-Connes à coefficients pour les groupes hyperboliques*. (French) [The Baum-Connes conjecture with coefficients for hyperbolic groups] J. Noncommut. Geom. 6 (2012), no. 1, 1–197.

[LLP] E. LEICHTNAM, J. LOTT, P. PIAZZA. *On the homotopy invariance of higher signatures for manifolds with boundary*. J. Differential Geom. 54 (2000), no. 3, 561–633.

[LP] E. LEICHTNAM, P. PIAZZA. *Rho-classes, index theory and Stolz' positive scalar curvature sequence*. J. Topol. 7 (2014), no. 4, 965–1004.

[LP1] E. LEICHTNAM, P. PIAZZA. *On higher eta-invariants and metrics of positive scalar curvature*. K-Theory, 24(4):341–359, 2001.

[LP2] E. LEICHTNAM, P. PIAZZA. *Spectral sections and higher Atiyah-Patodi-Singer index theory on Galois coverings*. Geom. Funct. Anal. 8 (1998), no. 1, 17–58.

[LP3] E. LEICHTNAM, P. PIAZZA. *Homotopy invariance of twisted higher signatures on manifolds with boundary*. Bull. Soc. Math. France, 127(2):307–331, 1999.

[Lo] J. LOTT. *Higher eta-invariants*. K-Theory, 6(3):191–233, 1992.

[Lo1] J. LOTT. *Delocalized L^2-invariants*. J. Funct. Anal., 169(1):1–31, 1999.

[Ma] V. MATHAI. *The Novikov conjecture for low degree cohomology classes*. Geom. Dedicata 99 (2003), 1–15.

[MY] I. MINEYEV, G. YU. *The Baum-Connes conjecture for hyperbolic groups*. Invent. Math. 149 (2002), no. 1, 97–122.

[M] A. S. MISHCHENKO. *Infinite-dimensional representations of discrete groups, and higher signatures*. (Russian) Izv. Akad. Nauk SSSR Ser. Mat. 38 (1974), 81–106.

[MN] M. MANOR, A. NAOR. *Metric cotype*. Ann. of Math. (2) 168 (2008), no. 1, 247–298.

[Ni] A. NICA. *Induction theorems for groups of homotopy manifold structures*. Mem. Amer. Math. Soc., 39(267):vi+108, 1982.

[Nis] V. NISTOR. *Group cohomology and the cyclic cohomology of crossed products*. Invent. Math., 99(2):411–424, 1990.

[N] S. P. NOVIKOV. *Algebraic construction and properties of Hermitian analogs of k-theory over rings with involution from the point of view of Hamiltonian formalism. Some applications to differential topology and to the theory of characteristic classes*. Izv. Akad. Nauk SSSR, v. 34, 1970 I N2, pp. 253–288; II: N3, pp. 475–500.

[N1] S. P. NOVIKOV. *Topological invariance of rational classes of Pontrjagin*. (Russian) Dokl. Akad. Nauk SSSR 163 1965 298–300.

[NY] P. NOWAK, G. YU. *Large scale geometry*. European Mathematical Society Publishing House, 2012.

[O] D. OSAJDA. *Small cancellation labellings of some infinite graphs and applications*. arXiv:1406.5015, 2014.

[OY] H. OYONO-OYONO, G. YU. *On quantitative operator K-theory*. Ann. Inst. Fourier (Grenoble) 65 (2015), no. 2, 605–674.

[PS] P. PIAZZA, T. SCHICK. *Groups with torsion, bordism and rho invariants*. Pacific J. Math., 232(2):355–378, 2007.

[PS1] P. PIAZZA, T. SCHICK. *Rho-classes, index theory and Stolz' positive scalar curvature sequence*. J. Topol. 7 (2014), no. 4, 965–1004.

[PS2] P. PIAZZA, T. SCHICK. *The surgery exact sequence, K-theory and the signature operator*. Ann. K-Theory 1 (2016), no. 2.

[PZ] P. PIAZZA, V. F. ZENOBI. *Singular spaces, groupoids and metrics of positive scalar curvature*. J. Geom. Phys. 137.

[Pi] M. PIMSNER. *KK-groups of crossed products by groups acting on trees*. Invent. Math. 86 (1986), no. 3, 603–634.

[PV] M. PIMSNER, D. VOICULESCU. *K-groups of reduced crossed products by free groups*. J. Operator Theory 8 (1982), no. 1, 131–156.

[P] M. PUSCHNIGG. *The Kadison-Kaplansky conjecture for word-hyperbolic groups*. Invent. math. 149 (2002), 153–194.

[P1] M. PUSCHNIGG. *New holomorphically closed subalgebras of C^*-algebras of hyperbolic groups*. Geom. Funct. Anal. 20 (2010), no. 1, 243–259.

[QR] Y. QIAO, J. ROE. *On the localization algebra of Guoliang Yu*. Forum Math. 22 (2010), no. 4, 657–665.

[RTY] D. RAMRAS, R. TESSERA, G. YU. *Finite decomposition complexity and the integral Novikov conjecture for higher algebraic K-theory*. J. Reine Angew. Math. 694 (2014), 129–178.

[Roe] J. ROE. *Coarse cohomology and index theory on complete Riemannian manifolds*. Mem. Amer. Math. Soc. 104 (1993), no. 497.

[Roe1] J. ROE. *Index theory, coarse geometry, and topology of manifolds*. CBMS Regional Conference Series in Mathematics, 90. Published for the Conference Board of the Mathematical Sciences, Washington, DC; by the American Mathematical Society, Providence, RI, 1996.

[Roe2] J. ROE, *Hyperbolic groups have finite asymptotic dimension*. Proc. Amer. Math. Soc. 133 (2005), no. 9, 2489–2490.

[Roe3] J. ROE, *Positive curvature, partial vanishing theorems, and coarse indices*. Proc. Edinb. Math. Soc. (2) 59 (2016), no. 1, 223–233.

[R] J. ROSENBERG. *C^*-algebras, positive scalar curvature, and the Novikov conjecture*. Inst. Hautes Études Sci. Publ. Math. No. 58 (1983), 197–212 (1984).

[R1] J. ROSENBERG. *Manifolds of positive scalar curvature: a progress report*. Surveys in differential geometry. Vol. XI, 259–294, Surv. Differ. Geom., 11, Int. Press, Somerville, MA, 2007.

[R2] J. ROSENBERG. *Analytic Novikov for topologists*. Novikov conjectures, index theorems and rigidity, Vol. 1 (Oberwolfach, 1993), 338–372, London Math. Soc. Lecture Note Ser., 226, Cambridge Univ. Press, Cambridge, 1995.

[RS] J. ROSENBERG, S. STOLZ. *Metrics of positive scalar curvature and connections with surgery*. In Surveys on surgery theory, Vol. 2, volume 149 of Ann. of Math. Stud., pages 353–386. Princeton Univ. Press, Princeton, NJ, 2001.

[RW] J. ROSENBERG, S. WEINBERGER. *Higher G-indices and applications*. Ann. Sci. École Norm. Sup. (4) 21 (1988), no. 4, 479–495.

[S] Z. SELA. *Uniform embeddings of hyperbolic groups in Hilbert spaces*. Israel J. Math. 80 (1992), no. 1-2, 171–181.

[Serre] J. -P. SERRE. *Linear representations of finite groups*. Translated from the second French edition by Leonard L. Scott. Graduate Texts in Mathematics, Vol. 42. Springer-Verlag, New York-Heidelberg, 1977.

[SY] R. SCHOEN, S. T. YAU *On the structure of manifolds with positive scalar curvature*. Manuscripta Math., 28(1-3):159–183, 1979.

[STY] G. SKANDALIS, J. TU, G. YU. *The coarse Baum-Connes conjecture and groupoids*. Topology 41 (2002), no. 4, 807–834.

[St] S. STOLZ. *Concordance classes of positive scalar curvature metrics*. Preprint.

[Su] D. SULLIVAN. *Hyperbolic geometry and homeomorphisms*. Geometric topology (Proc. Georgia Topology Conf., Athens, Ga., 1977), pp. 543–555, Academic Press, New York-London, 1979.

[Wa] C. WAHL. *The Atiyah-Patodi-Singer index theorem for Dirac operators over C^*-algebras*. Asian J. Math., 17(2):265–319, 2013.

[WW] B.-L. WANG, H. WANG. Localized index and L^2-Lefschetz fixed-point formula for orbifolds. *J. Differential Geom.*, 102(2):285–349, 2016.

[W] S. WEINBERGER. *Variations on a theme of Borel*. Book draft, available at http://math.uchicago.edu/~shmuel/VTBdraft.pdf.

[W1] S. WEINBERGER. *Aspects of the Novikov conjecture*. Geometric and topological invariants of elliptic operators (Brunswick, ME, 1988), 281–297, Contemp. Math., 105, Amer. Math. Soc., Providence, RI, 1990.

[WXY] S. WEINBERGER, Z. XIE, G. YU. *Additivity of higher rho invariants and nonrigidity of topological manifolds*. To appear in Communications on Pure and Applied Mathematics.

[WY] S. WEINBERGER, G. YU. *Finite part of operator K-theory for groups finitely embeddable into Hilbert space and the degree of nonrigidity of manifolds*. Geom. Topol. 19 (2015), no. 5, 2767–2799.

[WiY] R. WILLETT, G. YU. *Higher index theory*. Book draft, available at https://math.hawaii.edu/~rufus/Skeleton.pdf, 2018.

[WiY1] R. WILLETT, G. YU. *Higher index theory for certain expanders and Gromov monster groups, I*. Adv. Math. 229 (2012), no. 3, 1380–1416.

[WiY2] R. WILLETT, G. YU. *Higher index theory for certain expanders and Gromov monster groups, II*. Adv. Math. 229 (2012), no. 3, 1762–1803.

[XY] Z. XIE, G. YU. *Positive scalar curvature, higher rho invariants and localization algebras*. Adv. Math. 262 (2014), 823–866.

[XY1] Z. XIE, G. YU. *Higher rho invariants and the moduli space of positive scalar curvature metrics*. Adv. Math. 307 (2017), 1046–1069.

[XY2] Z. XIE, G. YU. *A relative higher index theorem, diffeomorphisms and positive scalar curvature*. Adv. Math., 250: 35–73, 2014.

[XY3] Z. XIE, G. YU. *Delocalized eta invariants, algebraicity, and K-theory of group C^*-algebras*. arXiv:1805.07617.

[XYZ] Z. Xie, G. Yu, R. Zeidler. *On the range of the relative higher index and the higher rho invariant for positive scalar curvature*. arXiv:1712.03722.

[Y] G. Yu. *Localization algebras and the coarse Baum-Connes conjecture*. K-Theory 11 (1997), no. 4, 307–318.

[Y1] G. Yu. *The Novikov conjecture for groups with finite asymptotic dimension*. Ann. of Math. (2) 147 (1998), no. 2, 325–355.

[Y2] G. Yu. *The coarse Baum-Connes conjecture for spaces which admit a uniform embedding into Hilbert space*. Invent. Math. 139 (2000), no. 1, 201–240.

[Y3] G. Yu. *A characterization of the image of the Baum-Connes map*. Quanta of maths, 649–657, Clay Math. Proc., 11, Amer. Math. Soc., Providence, RI, 2010.

[Ze] R. Zeidler. *Positive scalar curvature and product formulas for secondary index invariants*. J. Topol. 2016.

[Z] V. Zenobi. *Mapping the surgery exact sequence for topological manifolds to analysis*. J. Topol. Anal. 9 (2017), no. 2, 329–361.

[Z1] V. F. Zenobi. *Adiabatic groupoid and secondary invariants in K-theory*. Adv. Math. 347.

The manufacturer's authorised representative in the EU is Springer Nature Customer Service Centre GmbH, Europaplatz 3, 69115 Heidelberg, Germany. If you have any concerns regarding our products, please contact ProductSafety@springernature.com

Printed and bound by CPI Group (UK) Ltd, Croydon, CR0 4YY
30/09/2026
02225508-0020